International Conference on New Energy and Future Energy System 2016 (NEES 2016)

IOP Conference Series: Earth and Environmental Science
Volume 40

Beijing, China
19 – 22 August 2016

Editor:

Grigorios L. Kyriakopoulos

ISBN: 978-1-5108-3050-9
ISSN: 1755-1307

Printed from e-media with permission by:

Curran Associates, Inc.
57 Morehouse Lane
Red Hook, NY 12571

Some format issues inherent in the e-media version may also appear in this print version.

Printed by Curran Associates, Inc. (2016)

For permission requests, please contact the Institute of Physics
at the address below.

Institute of Physics
Dirac House, Temple Back
Bristol BS1 6BE UK

Phone: 44 1 17 929 7481
Fax: 44 1 17 920 0979

techtracking@iop.org

Additional copies of this publication are available from:

Curran Associates, Inc.
57 Morehouse Lane
Red Hook, NY 12571 USA
Phone: 845-758-0400
Fax: 845-758-2633
Email: curran@proceedings.com
Web: www.proceedings.com

Table of contents

Volume 40

2016 International Conference on New Energy and Future Energy System (NEFES 2016)

19–22 August 2016, Beijing, China

Accepted papers received: 8 August 2016
Published online: 26 September 2016

Preface

Papers

Preface

The *2016 International Conference on New Energy and Future Energy System (NEFES 2016)* will be held from August 19 to 22, 2016 in Beijing, the capital of China.

The 2016 International Conference on New Energy and Future Energy System (NEFES 2016) is organized to provide a platform for researchers and practitioners in both industry and academia to develop ideas and thoughts, share their latest achievements and discuss the possiblities and challenges in terms of new energy and future energy systems. NEFES 2016 aims to cover various fields of new energy materials and resources, and their networking and computing issues. The key focus of the conference will be on the areas of "New Energy", "InterGrid" and "New Energy Material".

This proceedings of the NEFES 2016 conference contain 87 specially selected manuscripts submitted to the conference. We have organized the peer review and plagiarism check for each article to ensure the paper quality.

We would like to take this opportunity to thank all authors who contributed to this conference, and particularly those of all accepted papers for their high quality and successful contributions. Special thanks go to all reviewers for their careful and critical reading of the manuscripts and useful comments and suggestions. We do hope that these proceedings will be beneficial for readers in their future research endeavours and careers. We also gratefully acknowledge the efforts and dedication of many individuals, especially IOP colleagues, including Anete Ashton and all the Editorial Board members in IOP Publishing for their support in producing the proceedings of this event.

Editor

Dr. Grigorios L. Kyriakopoulos

National Technical University of Athens, Greece

Peer review statement

All papers published in this volume of *IOP Conference Series: Earth and Environmental Science* have been peer reviewed through processes administered by the proceedings Editors. Reviews were conducted by expert referees to the professional and scientific standards expected of a proceedings journal published by IOP Publishing.

An improved thermal network model of the IGBT module for wind power converters considering the effects of base plate solder fatigue

H Li[1,4], Y G Hu[1, 4], S Q Liu[2], Y Li[1], X L Liao[1] and Z X Liu[3]
[1]The State Key Laboratory of Power Transmission Equipment & System Security and New Technology (Chongqing University), Chongqing, China
[2]The State Grid Jiangxi Electric Power Corporation Ganzhou Power Supply Company, Ganzhou, China
[3]Chongqing KK-QIANWEI Wind Power Equipment Co., Ltd, Chongqing, China

E-mail: cqulh@163.com, huyaogang345@163.com

Abstract. This study presents an improved thermal network model of the IGBT module that considers the effects of base plate solder fatigue on the junction temperature of the said module used in wind power converters. First, the coupling thermal structure 3D finite element model of the IGBT module is established based on the structure and material parameters of the module used in the wind power converters of a doubly fed induction generator. The junction temperature of the module is also investigated at different thermal desquamating degrees of the base plate solder. Second, the thermal resistance parameters are determined at different desquamating degrees, and the improved thermal network model that considers the effects of base plate solder fatigue is established. Finally, the two results of the calculation of the junction temperature are compared in different fatigue stages through the improved thermal network model and the 3D finite element model, which testify to the effectiveness of the improved thermal network model.

1. Introduction

As an energy transfer interface, the wind power converter is the core component of large power wind turbines. Given the successful development of wind farms and particularly offshore wind turbines [1-2], some critical components of gearbox, generator, and power converter are subjected to serious operational environments. Especially power converters are also prone to drastic thermal stress shocks, such as the continuous random torque in wind turbines. The safe operation and reliability of wind power converters have received increase attention in recent years [3–5]. The reliability of IGBT modules, which highly depend on accurate information on the junction temperature, is vital for power converters and the whole wind turbine system [5]. A thermal network model, such as a foster network, is generally used to calculate the junction temperature of an IGBT module [4, 6]. However, the thermal stress between different materials results in the desquamation of the base plate solder, which may in turn change the thermal resistance of the module, particularly during the later period of its operational lifetime. The resisting normal thermal-electronic thermal model is based on the fixed thermal parameters supported by power module manufacturers [6] and may not be suitable for the

[4] Address for correspondence: H Li, Y G Hu, the State Key Laboratory of Power Transmission Equipment & System Security and New Technology (Chongqing University), Chongqing, China. E-mail: cqulh@163.com, huyaogang345@163.com

monitoring of the condition of the module in different lifetime periods. Thus, investigations of the thermal model of the base plate solder fatigue and thermal analysis have an important academic significance to the reliability evaluation and monitoring of the condition of wind power converters.

Some researchers have focused on the junction temperature and solder fatigue. The lumped parameter method is used to establish the thermal network model for transient thermal conduction [6]. In [7], the influences of the impulse load on the junction temperature of the IGBT module is compared under the current hysteretic control, SPWM, and SVPWM control methods. The multichip thermal effects on the power cycling ability of an IGBT module are considered, and the evaluation error among different thermal models is analysed in [8]. A method of monitoring the condition of wind turbines based on case temperature is proposed in [9–10]. However, these investigations cannot effectively calculate the junction temperature at different lifetimes because of the fixed thermal resistance assumption, and the relationship between the thermal parameters and solder fatigue remains unknown. Although [11] monitors the condition of a power module with variations in the junction temperature at different power cycling times, this power module still uses the fixed thermal resistance. In [12], the junction temperature is calculated by analysing the heat conduction path. However, extracting the aging thermal parameters from 2D analysis is insufficient and needs further improvement.

This study presents an improved thermal network model of IGBT modules that considers the effects of base plate solder fatigue in large wind turbine power converters. In Section II, the actual structure of an IGBT module and the coupling thermal structure 3D finite element model (FEM) are introduced. In Section III, the thermal resistance of the module is fully investigated, and its improved thermal network model is proposed. In Section IV, the FEM and improved thermal network model of the module are verified by comparing them with the actual case monitoring temperature and multiple FEM results, respectively. In Section V, the conclusion is presented.

2. Actual structure of the IGBT module and FEM

2.1. Actual structure of the IGBT module

The actual structure of the IGBT module (FZ1600R17HP4) of a large-power doubly fed induction generator (DFIG) wind power converter is introduced to investigate comprehensively the IGBT module of a wind power converter. The inner structure of the wind power converter and the electrical connection of its IGBT module are shown in Figure 1. Figure 1 shows that each IGBT module consists of multiple IGBT chips and freewheeling diode (FWD) chips and that the structure of the layer of the IGBT module includes seven layers. Both the copper and ceramic layers comprise the direct bonded copper layer, which is connected to the base plate by the solder. Each layer of the IGBT module has different coefficients of thermal expansion (CTE), which result in the occurrence of alternating thermal stress between the two layers with temperature changes [13–15]. This kind of CTE mismatch causes fatigue cracks, which affect the heat dissipation of the IGBT chips. The concentration of extra temperature and the severe mechanical deformations enlarge the area of fatigue cracks and cause the desquamating part of the base plate solder.

IOP Conf. Series: Earth and Environmental Science **40** (2016) 012001 doi:10.1088/1755-1315/40/1/012001

(a) 2 MW DFIG wind power converter

(b) Inner structure of the IGBT module

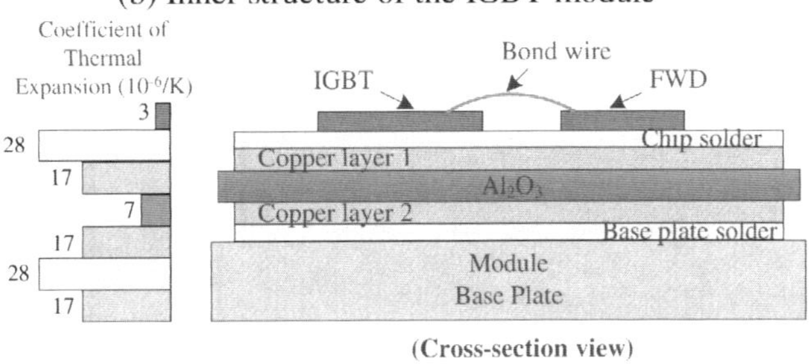

(c) Cross-section view of the IGBT module

Figure 2. Chip sizes and FEM of the IGBT module

2.2. FEM of the IGBT module

The 3D FEM of the IGBT module is established with ANSYS to analyze further the relationship between thermal resistance and solder desquamation. According to the material properties of the IGBT module in Table 1 [16], the geometric model of the IGBT module is established with the tool of the design model in the ANSYS/workbench.

Table 1. Material properties of the IGBT module

IGBT module	Material	Thermal Conductivity /W•(m•K)$^{-1}$	Poisson's ratio	Young's modulus /MPa
IGBT	Si	139	0.28	112000
FWD	Si	139	0.28	112000
Solder1	Sn-Ag-Cu	78	0.4	43000
DCB Copper1	Cu	386	0.34	110000
Ceramic	Al_2O_3	18	0.22	370000
DCB Copper2	Cu	386	0.34	110000
Solder2	Sn-Ag-Cu	78	0.4	43000
Base Plate	Cu	386	0.34	110000

Given the structural symmetry and independence of its location, the 1/4 units marked with T1–T4 and D1–D4 of the whole IGBT module structure are chosen and investigated here. The sizes and distances of these chips are shown in Figure 2. (a), and the 3D FEM of the IGBT module is established, as shown in Figure 2. (b).

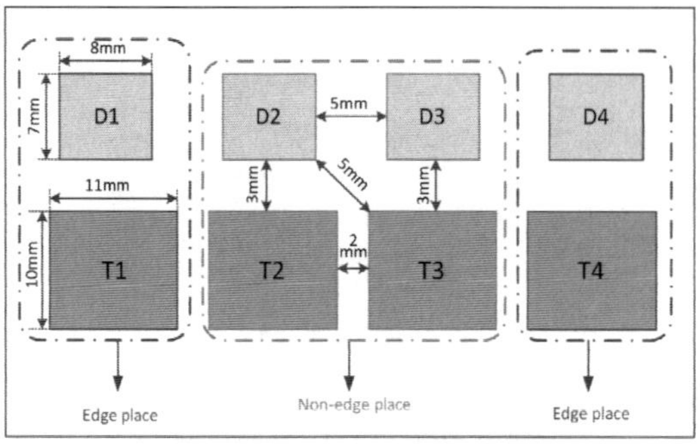

(a) Chip sizes of the IGBT module

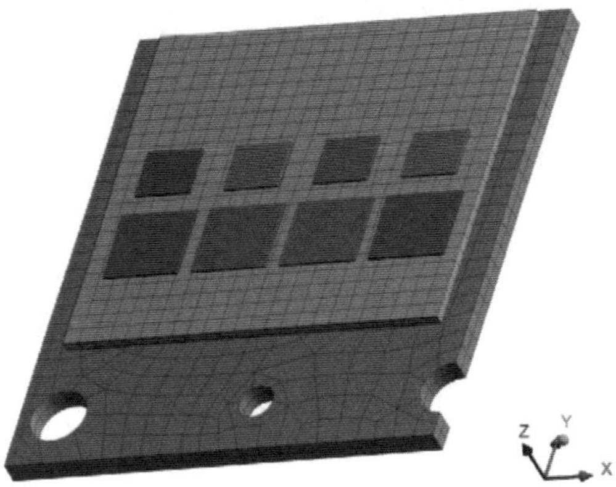

(b) FEM of the IGBT module

Figure 2. Chip sizes and FEM of the IGBT module

3. Thermal resistance increment and the improved thermal network model of the IGBT module

3.1. Thermal analysis of the IGBT module at different desquamating degrees

The junction temperature is analyzed to investigate the thermal resistance increment of the IGBT module in different solder fatigue periods by introducing the solder desquamating degree, which indicates the proportion of the desquamation area to the whole solder area, as shown in Figure 3.

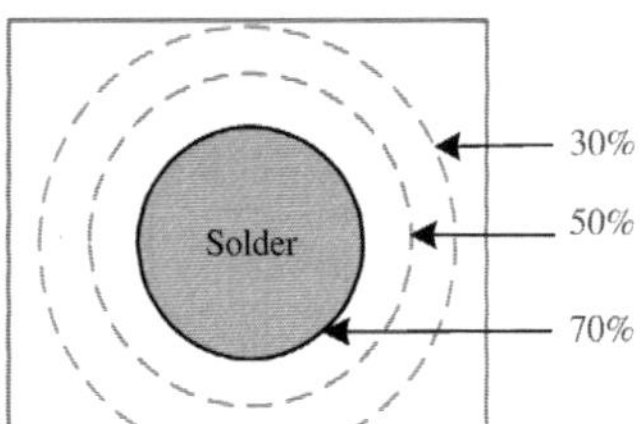

Figure 3. Solder desquamating degree

If the wind speed is 12 m/s, then the power losses of the IGBT modules can be calculated with the power loss calculation equations described in [17], which are 2.27 W/mm3 in IGBT and 1.5 W/mm3 in FWD for this DFIG wind power converter. Each layer is presumably connected perfectly with no relative movements. The heat spreading through the silica gel is ignored. All heat flows also presumably derive from the active junctions and flow to the cooling air of the heat plate because of the sealed package. The ambient temperature is 50 °C. The above simulation condition is used for FEM, and the expected emulation result is obtained at different desquamating degrees of 0%, 30%, 50%, and 70%, as shown in Figure 4.

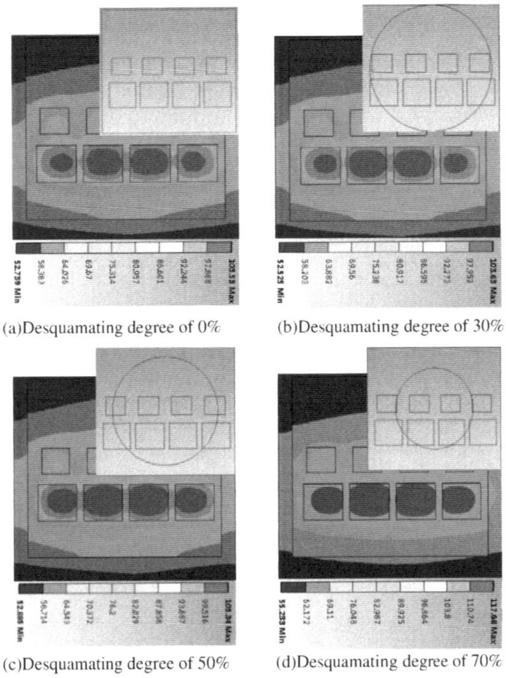

(a)Desquamating degree of 0% (b)Desquamating degree of 30%

(c)Desquamating degree of 50% (d)Desquamating degree of 70%

Figure 4. Temperature distribution of the IGBT module at different desquamating degrees of the base plate solder

Figure 4 shows that the junction temperature of IGBT changes slightly under the condition of 30% solder desquamating compared to that under the healthy condition of 0%. When the solder desquamation reaches 50%, the junction temperature of the IGBT is 105.3 °C, and the 2 °C increment is larger than that under the healthy condition. Although the junction temperature reaches 117.6 °C at the desquamating degree of 70%, the 14 °C increment is larger than that under the healthy condition. The junction temperature characteristics of IGBT caused by different solder desquamating conditions and wind speeds are also investigated, as shown in Figure 5. At the same solder desquamating degree, the junction temperature increases as the wind speed increases, as shown in Figure 5. At the same wind speed, the junction temperature of the IGBT also tends to increase as the solder desquamation increases, particularly when the solder desquamating degree reaches 50%.

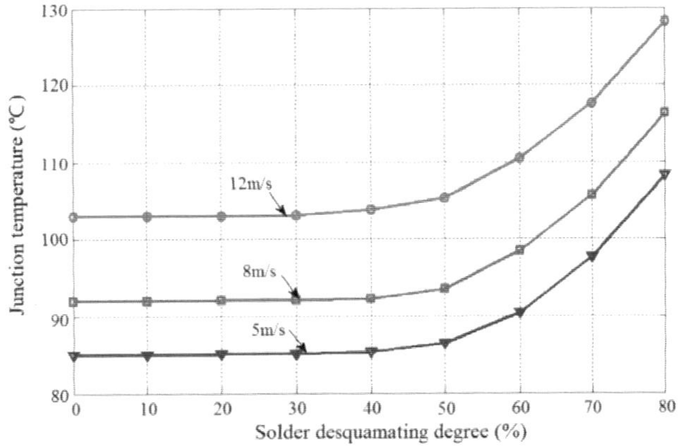

Figure 5. Junction temperature curves of the IGBT at different desquamating degrees of the base plate solder and wind speeds

3.2. Thermal resistance increment

To further investigate the relation of steady thermal resistance and solder fatigue, the junction to the case thermal resistance R_{th} can be described as follows:

$$R_{th} = \frac{T_j - T_c}{P} \quad (1)$$

The failure mechanism of the IGBT module results in a change in the fatigue thermal resistance $R_{th}(t)$ in different lifetime periods. The thermal resistance increment can be depicted as

$$\Delta R_{th}(t) = R_{th}(t) - R_{th0},\ t \geq 0 \quad (2)$$

where R_{th0} means the steady thermal resistance of the IGBT module under healthy condition.

According to the simulation results of FEM at different solder desquamating degrees, a thermal resistance increment in the wind speed of 12 m/s is calculated with (1) and (2) to further determine the influence of the solder fatigue on the thermal resistance of the IGBT module; the results are shown in Table 2.

Table 2. Junction temperature and thermal resistance increment

Desquamating degree	T_{j_IGBT} (°C)	ΔR_{th} (°C/W)	Proportion
0%	103.00	0	0%
10%	103.00	0	0%
20%	103.06	0.0002	0.10%
30%	103.09	0.0005	0.20%
40%	103.80	0.0038	1.57%
50%	105.30	0.0108	4.50%
60%	111.47	0.0348	14.50%
70%	117.63	0.0679	28.30%

The junction temperature and the thermal resistance increase as the solder desquamating degree increases. In general, the IGBT module is considered a failure when the thermal resistance increment reaches 20% [5, 9, 10]. The thermal resistance increment of the base plate solder slightly changes when the solder desquamating degree is within 30%. The reason is that the thermal conduction area is not influenced under that condition. However, the thermal resistance increment is 14.5% when the solder desquamating degree reaches 60%. It also tends to increase immediately. The thermal resistance increment curves at desquamating degrees of 5, 8, and 12 m/s are shown in Figure 6 to further study the influence of wind speed on thermal resistance.

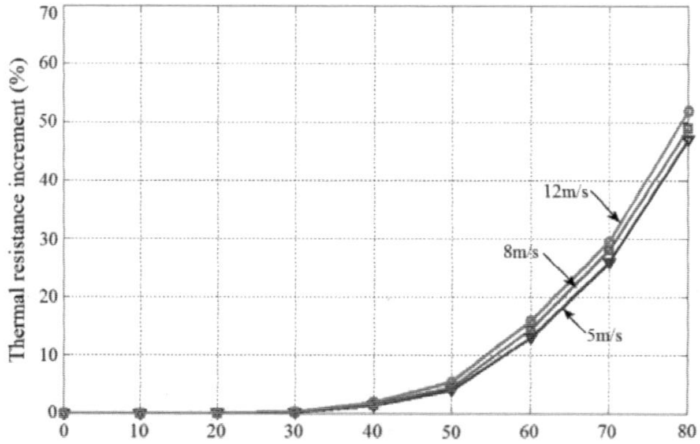

Figure 6. Thermal resistance increment of the IGBT at different solder desquamating degrees and wind speeds considering the change in thermal resistance.

Figure 6 illustrates that, at the same solder desquamating degree, wind speed apparently has no influence on thermal resistance increments, which are determined by different solder desquamating degrees.

3.3. Improved thermal network model

According to the structure of the layer of the IGBT module shown in Figure 1(c), the thermal impedance from the junction to case Z_{th_jc} is modeled as a four-layer Foster RC network, as shown in Figure 7.

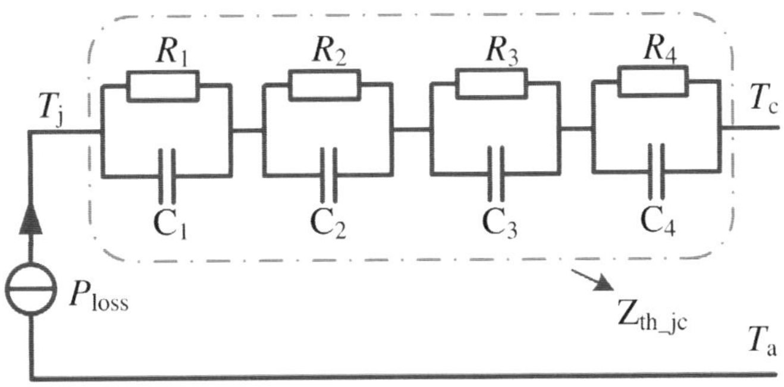

Figure 7. General foster thermal model of the IGBT module.

In Figure 7, the thermal resistance R_j determines the steady state mean value of the junction temperature, and the thermal capacitance C_j determines the dynamic change or fluctuation of the junction temperature. T_j and T_a denote the junction temperature and ambient temperature, respectively.

The thermal resistance increment must be considered during the process of thermal network modeling to accurately obtain the junction temperature of the IGBT module. According to the structure of the multiple IGBT chips and FWD chips at the base plate, the thermal resistance should be composed of inherent resistance increment $R_{th\text{-}sc}$ and increment ΔR_{th_tr} when the heat transfer process of the IGBT module stabilizes and the thermal capacitance can be ignored. Thus, the improved thermal network model of the IGBT module is as shown in Figure 8.

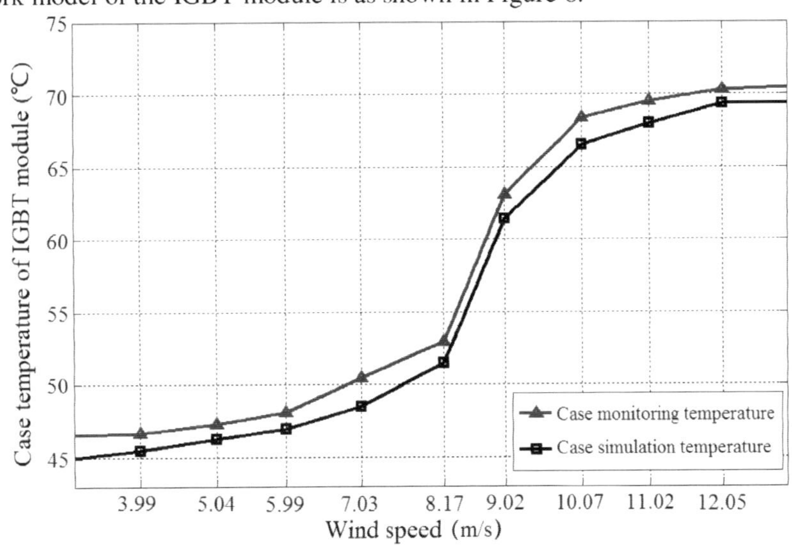

Figure 8. Improved thermal network model of the IGBT module considering the change in thermal resistance.

According to the improved thermal network model in Figure 8, R_{aged} is used as the junction case thermal resistance, which can be described as

$$R_{aged} = R_{th\_jc} + \Delta R_{th} \quad (3)$$

Thus, the junction temperature of the IGBT module can be calculated with (4):

$$T_{j\_aged} = P_{loss} \cdot (R_{th\_jc} + \Delta R_{th}) + T_c \quad (4)$$

where T_{j_aged} means the aged junction temperature, and P_{loss} denotes the power loss of the IGBT module.

4. Model validation

4.1. Validation of FEM

To further prove the effectiveness of the FEM and thermal analysis, the case temperature is calculated through FEM and compared with the actual case monitoring temperature of a 2 MW DFIG wind power converter without solder desquamation. The comparative results are depicted in Figure 9.

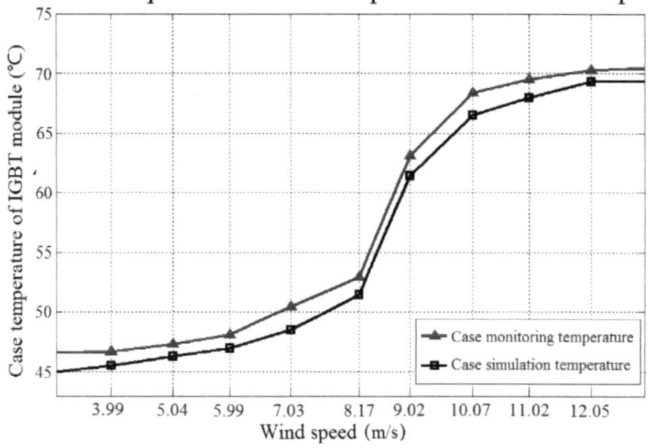

Figure 9. Monitoring data and simulation results of the IGBT module case temperature of a wind power converter.

The curve of the case monitoring temperature is obtained with the average case monitoring temperature at the same wind speed for 60 s, and the curve of the case simulation temperature is obtained through FEM. Figure 9 shows that the case temperature obtained through FEM matches the actual case monitoring temperature well under different operation conditions. This finding verifies the effectiveness of FEM and thermal analysis of the IGBT module.

4.2. Validation of the improved thermal network model

To verify the effectiveness of the thermal resistance increment and the proposed improved thermal network model, the improved thermal network model of the IGBT module is established in PLECS. The junction temperature of the IGBT module at different solder desquamating degrees is also calculated at the wind speeds of 6.8, 11.3, and 12 m/s. The simulation results with the same power losses are also calculated for comparison, as shown in Figure 10. The steady junction temperature of the IGBT with the improved thermal network model that considers the effects of solder fatigue matches the FEM thermal results well under the same operational condition. This finding further verifies the effectiveness of the improved thermal network model. Because the result from the thermal network is the mean value, whereas that from FEM is the maximum value, the two curves are slightly different.

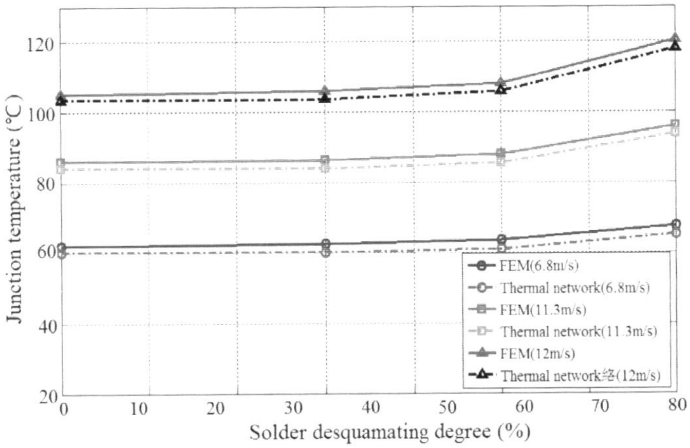

Figure 10. Comparison of the junction temperature of the IGBT at different fatigue stages.

5. Conclusion

This study proposes an improved thermal network model of the IGBT model for wind power converters, which considers the effects of base plate solder fatigue. The junction temperature and thermal resistance are analysed at different solder desquamating degrees through the proposed FEM. The actual case temperature and simulation results are used to verify the effectiveness of the improved thermal network model. The simulation results show that the junction temperature increases as the solder desquamating degree increases under the same operational condition. The thermal resistance increment also increases as the solder desquamation increases, and the junction temperature of the IGBT module can be calculated effectively with the proposed improved thermal network model. The proposed model is further verified with the FEM thermal results under multiple operational conditions. The future work will be to study the calculation of junction temperature considering effects of chip solder fatigue.

Acknowledgments

This study was supported by the Chongqing Graduate Student Research Innovation Project (CYB14014), the International Science and Technology Cooperation Program of China (2013DFG61520), the Fundamental Research Funds for the Central Universities (CDJXS10151152), the Fundamental Research Funds for the Central Universities (CDJZR12150074), and the Integration and Demonstration Program of Chongqing (CSTC2013JCSF70003). The authors are grateful for the supports.

References

[1] Oshaba A S and Ali E S 2013 *Research Journal of Applied Sciences Engineering & Technology* **18** 4594-606
[2] E. S. Ali 2015 *Energy* **89** 593-600
[3] Yang S Y, Angus Bryant, Philip Mawby, Xiang D W, Ran L and Tavner P 2011 *IEEE Transactions on Industry Applications* **47** 1441-51
[4] Blaabjerg F, Liserre M, and Ma K 2012 *IEEE Transactions on Industry Applications* **48** 708-19
[5] Yang S, Xiang D, Bryant A, Mawby P A, Ran L and Tavner P J. 2010 *IEEE Transactions on Power Electronics* **25** 2734-52
[6] Wei K X and Du M X 2011 *Transactions of China Electrotechnical Society* **26** 79-84
[7] Nejadpak A, Mirafzal B, Mohammed O and Lixiang W 2010 *IECON Annual Conference on IEEE Industrial Electronics Society* (Glendale, Arizona, USA) **36** 451-6
[8] Poller T, D'Arco S, Hernes M and Lutz J 2012 *International Conference on Integrated Power Electronics Systems* (Nuremberg, German) 1-6

[9] Xiang D, Ran L, Tavner P J, Bryant A, Yang S and Mawby P 2011 *IEEE Transactions on Industry Applications* **47** 2578-91
[10] Tian B, Wang Z and Qiao W 2014 *IEEE Applied Power Electronics Conference and Exposition–Apec* (TX, USA) 2564-68
[11] Tian B, Qiao W, Wang Z, Gachovska T and Hudgins J L 2014 *IEEE Applied Power Electronics Conference and Exposition–Apec* (TX, USA) 2550-55
[12] Wang Z, Qiao W, Tian B and Qu L Y 2014 *IEEE Applied Power Electronics Conference and Exposition–Apec* (TX, USA) 513-8
[13] Ciappa M. 2002 *Microelectronics Reliability* **42** 653-667
[14] Guo H, Watson S, Tavner P and Xiang J 2009 *Reliability Engineering & System Safety* **94** 1057-63
[15] Tang Y, Wang B,Chen M and Liu B 2014 *Transactions of China Electrotechnical Society* **29** 17-23
[16] Wintrich A, Nicolai U, Tursky W, and Reimann T 2011 *Application manual power semiconductors* (Nuremberg: ISLE Verlag)
[17] Mao P, Xie S J and Xu Z G 2010 *Pro of the CSEE* **30** 40-47

Experimental study on ignition characteristics of pulverized coal under high-temperature oxygen condition

G W Liu[1,3], Y H Liu[1] and P Dong[2]

[1]School of Energy and Power Engineering, Northeast Dianli University, Jilin 132000, China
[2]College of Energy Science and Engineering, Harbin Institute of Technology, Harbin 150001, China

E-mail: 2388403013@qq.com

Abstract. The high-temperature oxygen ignition technology of pulverized coal, which can replace the oil gun and achieve oil-free pulverized coal ignition by mixing the high-temperature oxygen and the pulverized coal stream directly, was proposed and a relevant ignition experimental system was built. The ignition characteristics of pulverized coal under high-temperature oxygen condition were investigated: the ignition process was described and analyzed, the influence of relevant parameters on the pulverized coal stream ignition were obtained and analyzed. The results showed: when the oxygen heating temperature is over 750 °C, the pulverized coal stream could be ignited successfully by high-temperature oxygen; increasing the pulverized coal concentration, primary air temperature and oxygen volume flow rate or decreasing the primary air velocity is helpful for the ignition and combustion of the pulverized coal stream.

1. Introduction

With the rising price of fuel oil, a consensus has been obtained on cutting down electricity generation costs by reducing fuel oil consumption in pulverized coal-fired utility boiler ignition. Various pulverized coal oil-free ignition technologies have become a focus of research. Sugimoto et al. [1] investigated the stabilization of pulverized coal combustion assisted by plasma, which proved that the plasma ignition technology is practical and feasible in the field of utility boilers ignition. Kanilo et al. [2] studied the ignition and combustion of pulverized coal using a microwave plasma assisted burner and gave some suggestions for the improvement of burner performance. In China, Zhang et al. [3] described the application of plasma ignition technologies in bituminous coal-fired boilers. Plasma and microwave assisted burners both have problems such as short lifetime and needing regular maintenance. Li et al. [4] proposed the induction-heating ignition of pulverized coal, which is also called high-temperature air oil-free pulverized coal ignition. In induction-heating ignition, the high-temperature air is the main and most reliable heat source for pulverized coal ignition.

Compared with conventional air combustion, oxygen-enriched combustion has a lower ignition temperature and higher combustion intensity and flame temperature. In addition, with the development of pressure swing adsorption technology [5-6], the cost of oxygen generation has been reduced greatly.

[3]Address for correspondence: School of Energy and Power Engineering, Northeast Dianli University, Jilin 132000, China. Email: 2388403013@qq.com.

Therefore, the application of oxygen-enriched combustion technology in the pulverized coal ignition of utility boilers is feasible and acceptable. The main principle of pressure swing adsorption technology [7-8] is that the adsorption capacity of adsorbents (zeolites) on nitrogen is higher than oxygen under certain pressure, so oxygen and nitrogen can be separated. The operation process of pressure swing adsorption includes air inflation, pressurization and adsorption and vacuum desorption. The entire process is carried out under conditions of normal temperature, so the cost of oxygen generation is low: the power consumption of 1 Nm^3 oxygen generation is approximately 0.35–0.45 kW·h. Preliminary calculations indicate that compared with oil ignition, the cost of oxygen ignition is cheaper at only 30–40% that of oil ignition[9-10].

In this paper, the high-temperature oxygen ignition technology of pulverized coal, which can replace the oil gun and achieve oil-free pulverized coal ignition by mixing the high-temperature oxygen and the pulverized coal stream directly, was proposed and a relevant ignition experimental system was built. At the same time, the ignition characteristics of pulverized coal under high-temperature oxygen condition were investigated.

2. Technical design philosophy of ignition device

The basic technical design philosophy of high-temperature oxygen pulverized coal ignition is summarized as follows: Pulverized coal is ignited directly in the primary air-coal mixture by mixing high-temperature oxygen as secondary air. This can improve the temperature and oxygen concentration of the flow. Thus for both high-temperature oxygen and air pulverized coal ignition, the main feature is to use the high-temperature combustion-supporting gas as the main heat source for pulverized coal ignition[11-12]. Compared with high-temperature air pulverized coal ignition, high-temperature oxygen pulverized coal ignition has more advantages. First, the heating gas volume is reduced substantially because of the same oxygen supply. The heating equipment thermal power and corresponding costs are also reduced. Second, with increase in oxygen concentration, the ignition temperature and corresponding ignition heat of the pulverized coal stream decrease, so difficulties in ignition are also reduced[13-14]. In addition, the combustion and flue-gas temperature of the pulverized coal stream increases sharply under oxygen-enriched conditions, so the radiation and convection heat transfer in the flow field is enhanced greatly. Meanwhile, the combustion efficiency of pulverized coal will also increase close to the efficiency of oil combustion, so the stabilization of pulverized coal ignition will be enhanced significantly.

To reduce oxygen consumption in ignition, a small special ignition device was designed that acts as a small oil gun for high-temperature oxygen pulverized coal ignition[15]. A schematic of the ignition device is shown in figure 1.

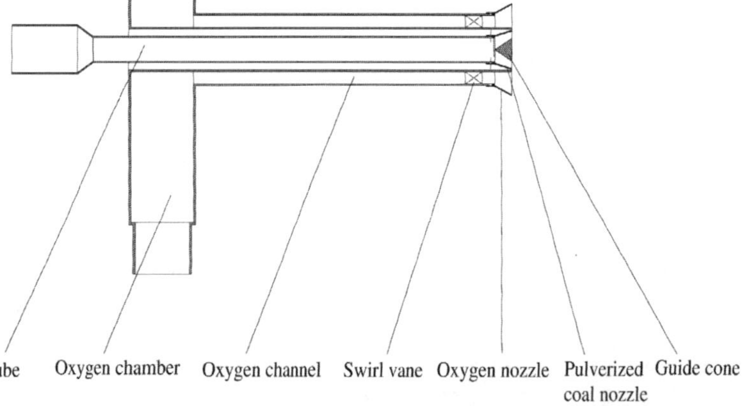

Figure 1. Schematic of high-temperature oxygen pulverized coal ignition device.

Its basic structure is as follows: a central tube for the primary air-coal mixture is located at the center position and a guide cone is arranged in the center tube nozzle. The swirl high-temperature oxygen is ejected through the outer annular channel and swirl vane. The main reason for using this structure is that the combustion intensity and pulverized coal flame temperature increase under oxygen-enriched conditions. The primary air-coal mixture and oxygen therefore need to be supplied separately. The precombustion or premix chamber cannot be used in the ignition device to prevent sparking. The combustion-supporting gas volume is reduced substantially in the high-temperature oxygen pulverized coal ignition, and the gas density decreases with increase in gas heating temperature, so the gas penetration capacity becomes worse. The guide cone is necessary to strengthen the mixing between the primary air-coal mixture and the high-temperature oxygen and then to ensure pulverized coal ignition.

In high-temperature oxygen pulverized coal ignition, the ignition device is ignited by the high-temperature oxygen first, and then the entire pulverized coal burner is ignited by the ignition device torch. After the entire pulverized coal burner is ignited, the high-temperature oxygen can be replaced by ordinary secondary air.

3. Experimental setup

An experimental system was built for high-temperature oxygen pulverized coal ignition. It consists of an ignition device, gas heating system, powder supply system and measurement equipment. The rated thermal power of the experimental ignition device is 0.1 MW.

Primary air and oxygen are heated by two electric heaters of 4.0kW and 8.0 kW maximum thermal power, respectively. The thermal power of the two electric heaters is controlled by a silicon-controlled rectifier cabinet. Commissioning tests of the electric heaters showed that the adjustable temperature scales of the primary air and oxygen are 20–400 °C and 20–950 °C, respectively, with relative errors less than 1.0 % and 3.0 %, respectively. A schematic of the electric heater is shown in figure 2.

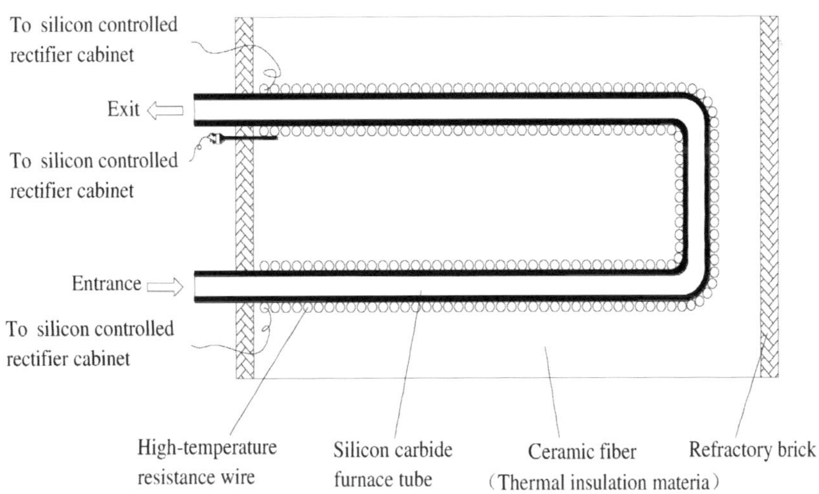

Figure 2. Schematic of electric heater.

Pulverized coal is supplied by a screw feeder in a uniform range of 110–270 g/min with a supply error of less than ± 8.0 g/min.

The flame temperatures of experimental ignition were measured using an IR-AHS portable digital radiation thermometer that is a narrowband radiation thermometer for high temperature. The measuring scale of the radiation thermometer is 600–3000°C with a relative error of less than 2.0 %.

The radiation thermometer position was fixed during the ignition process (one-sided 1.5 m to the ignition device) to ensure accuracy and repeatability of measured results.

Four pulverized coals, including lignite, bituminous coal, lean coal and mixed coal (lignite: bituminous coal = 3:1) were selected as experimental coal samples. Igniting anthracite coal is difficult in the above experimental ignition device and a special ignition device design was required. The proximate and ultimate analysis of coal samples is given in table 1[9].

Table 1. Proximate and ultimate analysis of coal samples.

Coal	Ultimate analysis (%)					Proximate analysis (%)			Calorific value (kJ/kg)
	C_{daf}	H_{daf}	O_{daf}	N_{daf}	S_{daf}	M_{ar}	A_d	V_{daf}	$Q_{net,ar}$
Lignite	73.00	4.85	20.30	0.89	0.96	35.4	15.2	43.0	14038
Bituminous coal	83.11	5.69	10.00	0.82	0.38	5.5	24.0	26.0	23865
Lean coal	87.62	3.44	2.16	1.36	5.42	6.5	26.0	13.3	23073
Mixed coal	75.53	5.06	17.73	0.87	0.81	27.9	17.4	38.8	16495

The four pulverized coals are all ground by an experimental rod mill. The pulverized coal fineness R90 and average particle sizes of the four pulverized coals are shown in Table 2[9]. R90 is the mass percentage of particles that cannot pass through a 90 μm aperture sieve, the average diameters of the coal particles were calculated using the method of reference [16].

Table 2. Coal particle size.

	R_{90}/(%)	Average diameter/(μm)
Lignite	21.7	79.1
Bituminous coal	17.9	72.4
Lean coal	14.4	69.7
Mixed coal	15.4	67.1

4. Ignition process analysis

The design operation condition for the ignition device is a pulverized coal concentration of primary air-coal mixture of 0.8 kg coal/kg air. The volume of oxygen satisfies the combustion stoichiometric ratio and the primary air temperature is 100 °C. The high-temperature oxygen pulverized coal ignition process was studied by increasing the oxygen temperature gradually. The appearance of a stable flame is defined as the point at which pulverized coal ignition is successful. The results show that with an increase in oxygen heating temperature, three different stages appear in the process of pulverized coal stream ignition; the ignition, transition and stable ignition stages.

No ignition occurs when the oxygen heating temperature is below 300 °C, and the pulverized coal stream cannot be ignited completely.

The transition stage occurs when the oxygen heating temperature is 300–450 °C, the entire pulverized coal stream is still not ignited and the pulverized coal stream temperature does not increase visibly, but some pulverized coal particles, especially the small particles, are ignited. A photograph of the particle ignition is shown in figure 3. When the oxygen heating temperature is higher than 450 °C, the number of burning particles increases and black smoke is generated in the pulverized coal stream. The essence of black smoke is a large number of fine coke particles. Due to the heating effect of the oxygen, the pulverized coal particles lose moisture and volatile, then become coke particles and the quality reduce greatly. At the same time, because of the thermal cracking property of the pulverized coal particles, the pulverized coal particles eventually become very small coke particles. These fine coke particles form the black smoke. Sometimes deflagration occurs after generation of the black smoke, and the oxygen heating temperature 450-650 °C is a dangerous range that is prone to deflagration. The main reason for this phenomenon is that with an increase in oxygen heating temperature, the average temperature of the entire flow field increases and the volatile matter of the pulverized coal is released rapidly in the pulverized coal stream. At the same time, because of turbulence instability and temperature nonuniformity in the flow field, the basic ignition conditions could be satisfied in some parts of the flow field, so the pulverized coal stream can be ignited. However, because the heat quantity is insufficient for igniting the entire pulverized coal stream stably, deflagration appears. Because the lack of heat quantity is the main reason for deflagration, methods such as increasing primary air and oxygen temperatures and preheating the ignition device before ignition are reasonable to avoid deflagration in the ignition.

The stable ignition stage occurs when the oxygen heating temperature is over 750 °C and the pulverized coal stream could be ignited successfully by high-temperature oxygen. The flame color is bright white and the flame temperature is above 2000 °C. A photograph of the flame torch in the stable ignition stage is shown in figure 4.

Figure 3. Photograph of particle ignition in the transition stage.

Figure 4. Photograph of flame torch in the stable ignition stage.

5. Influence of operational parameters

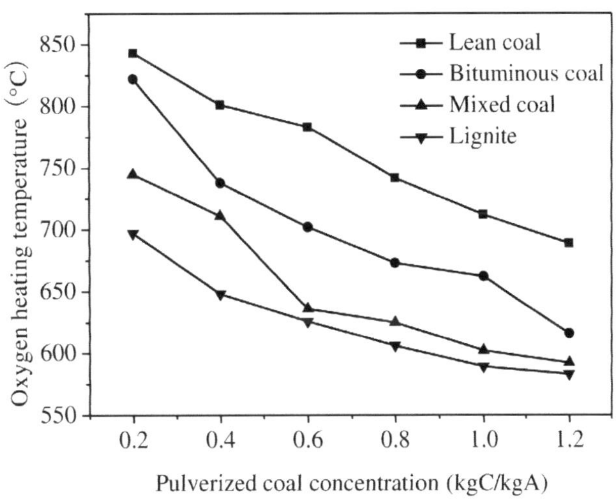

Figure 5. Influence of pulverized coal concentration on minimum oxygen heating temperature.

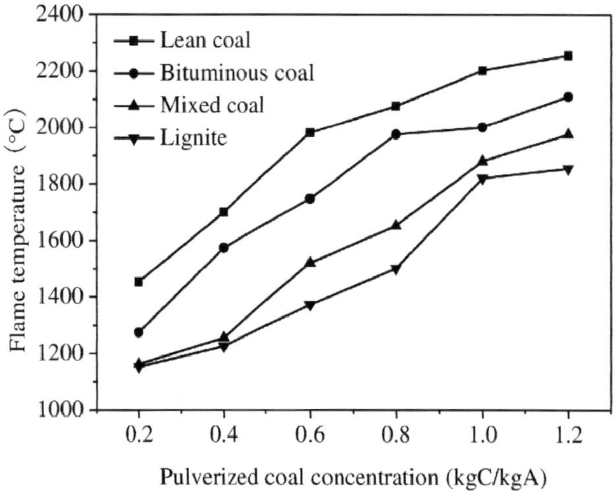

Figure 6. Influence of pulverized coal concentration on flame temperature.

Figure 5 and 6 show the influence of pulverized coal concentration on the experimental ignition. With an increase in pulverized coal concentration, the minimum oxygen heating temperature for igniting the entire pulverized coal stream successfully decreases and the corresponding flame temperature increases. The main reasons for this phenomenon include the following: first, the average spacing of pulverized coal particles decreases in the pulverized coal stream with an increase in pulverized coal concentration. The heat release becomes concentrated and is helpful for forming a high-temperature area to provide a heat source for pulverized coal ignition. Second, when the thermal power of the ignition device is constant, which means that the coal consumption and heat release remain unchanged in unit time, both the primary air volume and ignition heat for igniting the entire pulverized coal stream decrease with increase in pulverized coal concentration. The pulverized coal stream can therefore be ignited with a lower external heat supplement. In addition, the average oxygen concentration of the flow field increases because of the decrease in primary air volume. The corresponding ignition temperature of the pulverized coal stream therefore decreases and the flame temperature increases.

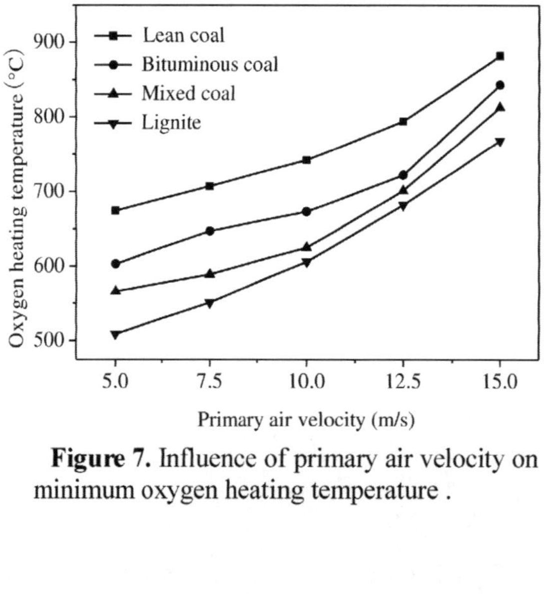

Figure 7. Influence of primary air velocity on minimum oxygen heating temperature .

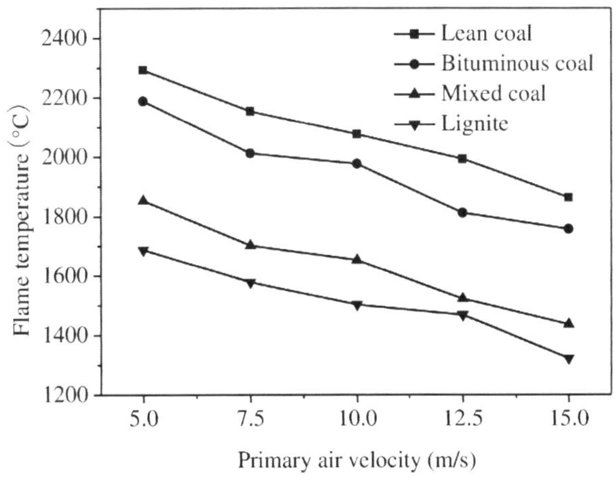

Figure 8. Influence of primary air velocity on flame Temperature.

Figure 7 and 8 show the influence of primary air velocity on the experimental ignition. With an increase in primary air velocity, the minimum oxygen heating temperature for igniting the entire pulverized coal stream increases. When the primary air velocity is greater than 15 m/s, the pulverized coal stream cannot be ignited steadily even if the oxygen heating temperature is up to 900 °C. The main reasons for this phenomenon are that when the pulverized coal concentration is constant, the heat supplement is insufficient for providing the ignition heat for igniting the entire pulverized coal stream, which increases substantially with increase in primary air velocity. Second, because the primary air temperature is only 100 °C, an increase of primary air velocity decreases both the average oxygen concentration and temperature in the entire flow field and the ignition condition of the pulverized coal stream worsens. With an increase in primary air velocity, the corresponding flame temperature decreases. This is because the residence time and corresponding heat release of pulverized coal particles both decrease in the flow field with increase in primary air velocity. At the same time, the low average oxygen concentration also limits the combustion intensity and flame temperature of the pulverized coal stream in the flow field.

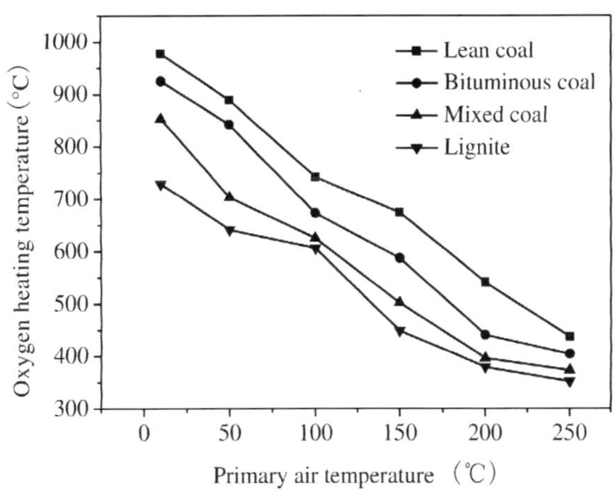

Figure 9. Influence of primary air temperature on minimum oxygen heating temperature.

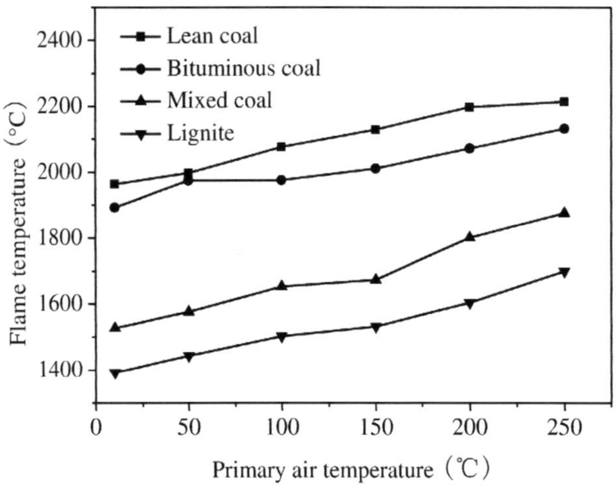

Figure 10. Influence of primary air temperature on flame temperature.

Figure 9 and 10 show the influence of primary air temperature on experimental ignition. With an increase in primary air temperature, the minimum oxygen heating temperature for igniting the entire pulverized coal stream decreases. Compared with other parameters, the influence of primary air temperature is more obvious. The main reason is that the heat for igniting the entire pulverized coal stream decreases significantly with an increase in primary air temperature. The corresponding minimum oxygen heating temperature therefore decreases. When the primary air temperature is above 200 °C, some pulverized coal particles, especially small particles, will start to burn in the center tube. When the burning pulverized coal particles are ejected into the flow field and mix with the high-temperature oxygen, both the combustion intensity and heat release rate are enhanced significantly, so the pulverized coal stream is easier to ignite by high-temperature oxygen. However, when the primary air temperature reaches 300 °C, too many pulverized coal particles start to burn in the center tube and deflagration tends to occur in the ignition process. In addition, with an increase in primary air temperature, the corresponding flame temperature increases slightly. This occurs because the average temperature of the combustion-supporting gas increases with increase in primary air temperature in the flow field and the flame and flue-gas temperatures increase.

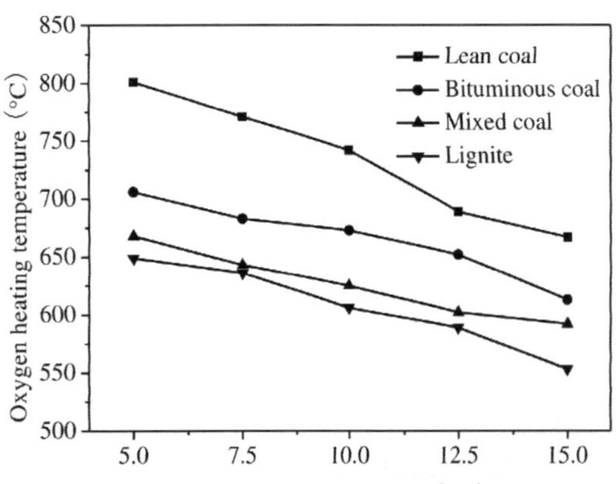

Figure 11. Influence of high-temperature oxygenheating temperature volume flow rate on minimum oxygen.

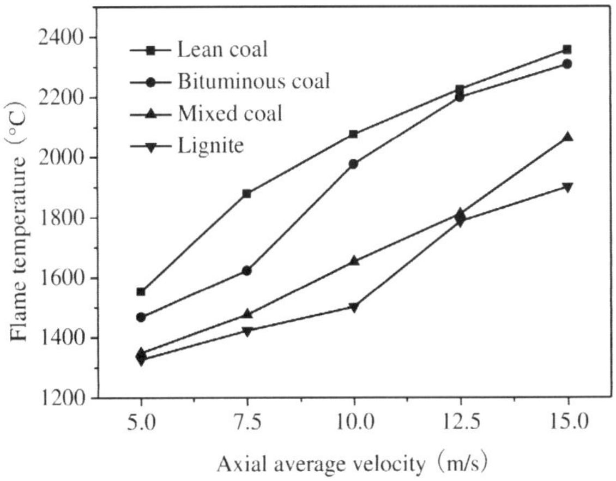

Figure 12. Influence of high-temperature oxygen volume flow rate on flame temperature.

Figure 11 and 12 show the influence of oxygen volume flow rate on experimental ignition. Because the high-temperature oxygen is a swirl flow in the ignition, the axial average velocity of high-temperature oxygen is used to indicate the high-temperature oxygen volume flow rate change. With an increase in oxygen volume flow rate, the minimum oxygen heating temperature for igniting the entire pulverized coal stream decreases, but the trend is insignificant. The main reasons for the phenomenon include the fact that when the heat and oxygen supplement are insufficient, mixing of the pulverized coal stream and high-temperature oxygen is the primary factor in controlling pulverized coal stream ignition. With an increase in high-temperature oxygen volume flow rate, the heat quantity, average oxygen concentration and average temperature of the flow field all increase, but mixing does not yield an obvious improvement. In addition, with an increase in high-temperature oxygen volume flow rate, the average spacing of pulverized coal particles increases. This leads to a reduced heat transfer between the pulverized coal particles. On the contrary, the influence of high-temperature oxygen volume flow rate on flame temperature is significant: the flame temperature increases with increase in oxygen volume flow rate. The main reason is that the average oxygen concentration of the entire flow field increases with increase in oxygen volume flow rate, so the combustion intensity and flame temperature of the pulverized coal stream increase.

The required oxygen heating temperature for igniting the lignite is lowest, followed by the mixed coal and bituminous coal, and the highest is the lean coal. The maximum temperature difference between the lean coal and lignite reaches nearly 300 °C. The corresponding flame temperatures show the same trend with the average temperature difference between the lean coal and the lignite being 400–500 °C.

6. Conclusions

In this paper, the high-temperature oxygen ignition technology of pulverized coal was proposed and a relevant ignition experimental system was built. Then the ignition characteristics of pulverized coal under high-temperature oxygen condition were investigated. The results are as follow:

- With an increase in oxygen heating temperature, three consecutive stages appear in the pulverized coal stream ignition: no ignition, transition and stable ignition stages.
- When the oxygen heating temperature is over 750 °C, the pulverized coal stream could be ignited successfully by high-temperature oxygen.
- Increasing the pulverized coal concentration, primary air temperature and oxygen volume flow rate or decreasing the primary air velocity is helpful for the ignition and combustion of the pulverized coal stream.

Nomenclature

Symbol:
C - mass fraction of carbon (%);
H - mass fraction of hydrogen (%);
O - mass fraction of oxygen (%);
N - mass fraction of nitrogen (%);
S - mass fraction of sulfur (%);
M - mass fraction of moisture (%);
A - mass fraction of ash (%);
V - mass fraction of volatiles (%);
Q - calorific value (kJ/kg).
Subscripts:
daf - dry ash-free basis;
ar - as received basis;
d - dry basis;
net - net value.

Acknowledgments

This paper is a part of the research that is sponsored by The Scientific Research Foundation for Doctors of Northeast Dianli University (Contract No. BSJXM-201432).

References

[1] Sugimoto M, Maruta K, Takeda K, Solonenko O P, Sakashita M and Nakamura M 2002 Stabilization of pulverized coal combustion by plasma assist *Thin Solid Films* **407** 186–91
[2] Kanilo P M, Kazanesev V I, Rasyuk N I, Schunemann K and Vavriv D M 2003 Microwave plasma combustion of coal *Fuel* **82**187–93
[3] Zhang X Y, Luo Z B, Zhang S K, Zou G W and Jiang B H 2003 Application testing and study of plasma combustion technology in coal fired boilers with double inlet and outlet tube mill and whirl burner *China Power* **36** 25–29
[4] Li W J, Cen K F, Zheng C G, Zhou J H and Cao X Y 2004 Induction-heating of pulverized coal stream *Fuel* **83**:2103–07
[5] Tondeur D and Wankat P C 1985 Gas purification by pressure swing adsorption *Sep Purif Methods* **14**(2):157-212
[6] White D H and Barkley P G 1989 The design of pressure swing adsorption systems *Chem Eng Progress* (1) 25-33
[7] Yoshida S, Hirano S and Harada A 2001 Nitrogen adsorption properties of cubic and orthorhombic Li-exchanged low silica X. *Microporous Mesoporous Mater* **46** 203-09
[8] Shen DM, Bulow M, Jale SR 2001 Therm odynamics of nitrogen and oxygen sorption on zeolites LiLSX and CaA *Microporous Mesoporous Mater* **48** 211-21
[9] Liu GW, Qu DZ, Dong P, Bie RS 2013 Experimental study on combustion characteristics of pulverized coal under enriched-oxygen condition by entrained flow reactor *Journal of Harbin Institute of Technology* **20**(1) 46-52
[10] Yan WP, Mi CL 2010 Economic analysis of a 300MW utility boiler with oxygen-enriched combustion Journal of *Chinese Society of Power Engineering* **30**(3) 184-81(in Chinese)
[11] Nie X, Zhou ZJ, Lv M, et al 2008 Experimental study on pulverized coal flow ignition and flameout in high temperature air. *Proceedings of the CSEE* **28**(14) 67-72(in Chinese)
[12] Yewen TN, Eric C, Mark A 2005 Combustion characteristics of coal in a mixture of oxygen and recycled flue gas *Fuel* **85** 507-12
[13] Maja BT, Jacob B, Peter AJ, et al 2010 Oxy-fuel combustion of solid fuels *Progress in Energy and Combustion Science* **3** 581-625
[14] Wu KK, Chang YC, Chen CH, Chen YD 2010 High-efficiency combustion of natural gas with

21-30% oxygen-enriched air *Fuel* **89** 2455-62
[15] Liu GW, Dong P, Bie RS 2013 Experiment study on high-temperature oxygen-enriched oil-free pulverized coal ignition Journal of *Combustion Science and Technology* **19** 399-406(in Chinese)
[16] Wang QG, Yan WP, Chen YY 2007 The method of calculating the diameter of coal dust based on the sole coal fineness's sizing data *Boiler Technology* **38** 43-45(in Chinese)

IOP Conf. Series: Earth and Environmental Science **40** (2016) 012003 doi:10.1088/1755-1315/40/1/012003

A Wolf Pack Algorithm for Active and Reactive Power Coordinated Optimization in Active Distribution Network

H M Zhuang[1] and X J Jiang[1]

School of Control Engineering, Chengdu University of Information Technology, Chengdu 610225, Sichuan Province, China

E-mail: zhmcyj@163.com, jxiujie@cuit.edu.cn

Abstract. This paper presents an active and reactive power dynamic optimization model for active distribution network (ADN), whose control variables include the output of distributed generations (DGs), charge or discharge power of energy storage system (ESS) and reactive power from capacitor banks. To solve the high-dimension nonlinear optimization model, a new heuristic swarm intelligent method, namely wolf pack algorithm (WPA) with better global convergence and computational robustness, is adapted so that the network loss minimization can be achieved. In this paper, the IEEE33-bus system is used to show the effectiveness of WPA technique compared with other techniques. Numerical tests on the modified IEEE 33-bus system show that WPA for active and reactive multi-period optimization of ADN is exact and effective.

1. Introduction

With increasing penetration of decentralized generation, distributed energy storage devices (ESD), control equipments and advanced communication networks, the distribution system is gradually becoming to the active one from traditionally passive network [1]. But high penetration of DGs access to the power grid may lead to voltage fluctuation or overvoltage, which seriously restrict the consumptive capacity of ADN for renewable energy generations (RESs). Therefore, it is necessary that the active and reactive power of dispatchable DGs need to be optimally dispatched by distribution network operators (DNOs). So the active and reactive power coordinated optimization (ARPCO) has become a research topic.

In the past years, many researchers have also engaged in optimal programming and operation. Numbers of programming methods or intelligent techniques were applied to optimize reactive power dispatch (OPRD). For example, particle swarm optimization (PSO) [2, 3], artificial bee colony (ABC) [4] and the hybrid differential evolution with ant system applied by C.-M. Huang[5] to solve OPRD. On the compelling problem of ADNs control, Reference [6] has proposed a new bidirectional converter as the interface of combined generating and electric storage systems with the grid. In reference [7], a broadcast-based unified control algorithm was brought forward to provide reactive power support for the grid by a seamless control of heterogeneous energy resources such as distributed storage systems and demand-responsive loads. In [8], a real-time distributed generation control

[1] Address for correspondence: H M Zhuang and X J Jiang, School of Control Engineering, Chengdu University of Information Technology, Chengdu 610225, Sichuan Province, China. Email: zhmcyj@163.com, jxiujie@cuit.edu.cn

scheme and a real-time decentralized load control scheme have been designed for power networks with tap-changing transformers.

Distribution networks are characterized by reduced line lengths with a non-negligible resistance over reactance (R/X) ratio, so the ARPCO of ADNs is more significant. In [9], a two-stage optimal schedule model for ADN has been proposed to dispatch active and reactive power of DGs. Marco Bronzini has presented a centralized control model for distributed resources in ADN and solved the problem using quasi-Newton method [10]. Levron et al. [11] presented a dynamic search algorithm based on a load flow for the OPF problem in microgrids with ESDs and RESs, but the method worked well only for systems with a small scale system. A solution for ARPCO in ADNs was explicated by Reference [12], but the charging and discharging periods of ESS were fixed so that the solutions were low-quality. In [13], a mixed-integer second-order cone programming (MISOCP) model has been proposed for the optimal operation problem in ADN. This method can convergence to the optimal value by using existing optimization software, but the solution is not accurate enough. Graditi G. [14] models the 24h behavior of appliances by means of a real-valued function and then extended the GSO algorithm to solve multi-objective energy management optimization problems for smart grids with direct control over shiftable loads. This method was verified comparable or even better than NSGA-II [15]. In [16], a comprehensive multi-objective optimal technique was proposed for optimal operation of ADN. The method was included the internal procedure and the external one. The former applied the Nondominated Sorting Genetic Algorithm II for the optimal management, and the latter was responsible for choosing the design features such as ratings and/or types.

This paper proposed the ARP coordinated dynamic multi-period optimization model, and applies a new swarm intelligent optimization algorithm- WPA to solve ARPCO problem. WPA has been proposed by [17] and is especially suitable for solving high-dimension optimization problems. This paper is organized as follows: Section 2 discusses the ARPCO model, and section 3 briefly describes WPA technique. Then, WPA implementation in solving ARPCO model is explicated in section 4. Section 5 presents the simulation results and discussion. At last, section 6 states the conclusion of this paper. The main contribution of this paper includes the realistic ARPCO model considering ESDs and application of the WPA to solve it, whose main advantages are: 1) a realistic and precise model; 2) computational robustness; and 3) good convergence performance.

2. ARPCO Model

The equipments such as dispatchable distributed energy resources (DER) and switchable shunt capacitor banks (SCB) are usually controlled for optimal operation of ADN. So the active power of DER and reactive power of SCB are chosen as control variables, ie, $\mathbf{u}=[\boldsymbol{P}_{\mathrm{DG}}{}^{t}, \boldsymbol{P}_{\mathrm{ESS}}{}^{t}, \boldsymbol{Q}_{\mathrm{CB}}{}^{t}]^{\mathrm{T}}$.

In general, the minimum energy loss in the scheduling period is selected as the objective function for the optimal operation of ADN, f expressed as follow:

$$f=\boldsymbol{P}_{\mathrm{LOS}}=\sum_{t=1}^{T}\sum_{i=1}^{N_l}\boldsymbol{G}_{i(m,n)}[(V_m^t)^2+(V_n^t)^2-2V_m^tV_n^t\cos\theta_{mn}^t] \tag{1}$$

Where T is the period number of a complete dispatching cycle, N_l is the number of transmission lines, V^t and θ^t , as state variables, are respectively the bus voltage amplitude and phase angel in the period-t. $G_{i(m,n)}$ is the conductance of the branch-i connecting node-m and node-n.

The equality constraint equations include load flow suggested in [8] and the operation equality constraints of distributed generator (DG) and ESS, as follows:

Load flow:

$$\begin{cases} P_{Gi}{}^{t}-P_{Li}{}^{t}=V_i^t\Sigma V_j^t(G_{ij}\cos\theta_{ij}{}^{t}+B_{ij}\sin\theta_{ij}{}^{t}) & (2.1)\\ Q_{Gi}{}^{t}-Q_{Li}{}^{t}=V_i^t\Sigma V_j^t(B_{ij}\cos\theta_{ij}{}^{t}-G_{ij}\sin\theta_{ij}{}^{t}) & (2.2)\end{cases}$$

Where P_{G} may be the active power from DG, ESS or substation. For the ESS, its charge power is positive and discharge power is negative.

Distributed generator (DG):

$$Q_{\mathrm{DG}i}{}^{t}=P_{\mathrm{DG}i}{}^{t}\tan\varphi \tag{3}$$

Where P_{DGi}, Q_{DGi}, φ are the active power , reactive power and power factor angle of DG-i respectively.

ESS:

$$\begin{cases} E_i^t = E_i^{t-1} - \Delta t P_{ESi}^{t-1}, \quad t=2,\ldots,T,\ i=2,\ldots,N_s & (4.1) \\ E_i^1 = E_i^0 & (4.2) \end{cases}$$

Where E_i^t is the remaining energy of the ESS-i in the period-t, N_s is the number of ESS.

In addition, the inequality constraints include operation constraints of ADN, as follow:

DG constraints:

$$P_{DGi}^{\min} \leq P_{Dgi}^{t} \leq P_{DGi}^{\max} \tag{5}$$

Where $P_{DGi}^{\min}$ and $P_{DGi}^{\max}$ are the upper and lower limits of active power generation respectively.

ESS constraints: Charge or discharge power and energy are restricted by their upper and lower limits, as follow:

$$\begin{cases} P_{ESj}^{\min} \leq P_{ESj}^{t} \leq P_{ESj}^{\max} & (6.1) \\ E_j^{\max} \times 20\% \leq E_j^t \leq E_j^{\max} \times 90\% & (6.2) \end{cases}$$

Where $E_j^{\max}$ is the max energy stored in ESS. For longer life of the ESS, the energy stored in ESS is generally from 20% to 90% of $E_j^{\max}$ in fact [13].

Constraints of power from substation: In order to restrain the influence of the AND power fluctuation on the transmission network, the constraint of power from substation is considered as follow:

$$\begin{cases} 0 \leq P_{EX}^{t} \leq P_{EX}^{\max} & (7.1) \\ 0 \leq Q_{EX}^{t} \leq Q_{EX}^{\max} & (7.2) \end{cases}$$

Shunt capacitors restricted by their limits:

$$Q_{Ci}^{\min} \leq Q_{Ci}^{t} \leq Q_{Ci}^{\max} \tag{8}$$

Operation safety constraint of AND:

$$\begin{cases} V_i^{\min} \leq V_i^t \leq V_i^{\max} & (9.1) \\ I_{ij}^{t} \leq I_{ij}^{\max} & (9.2) \end{cases}$$

Where V_i^t is the node-i voltage amplitude in the period-t; I_{ij}^t is the current of the buanch-ij in the period-t; $V_i^{\min}$ and $V_i^{\max}$ are the voltage limits, $I_{ij}^{\max}$ is the max branch current.

3. WPA

Based on the cooperative hunting characteristics of wolves, Wu Husheng, etc put forward a new intelligent algorithm-Wolf pack algorithm (WPA) [17]. It has been proven to has better global convergence and computational robustness, and especially suitable for solving high-dimension and multimodal function optimization problems with other classical intelligent algorithm such as PSO, fish swarm algorithm, genetic algorithm (GA) and so on.

In nature, the wolf belongs to candidate family and lives in a pack consisting of 5-12 wolves on average. In general, the common wolf pack can be divided into three parts: a lead wolf, some scout wolves and ferocious wolves. The lead wolf, as a leader, is always the smartest and most ferocious one. Scout wolves hunt around for prey. Ferocious ones are responsible for rounding up the prey.

WPA is different from previous swarm intelligence optimization algorithm. It has three different search abilities: hunting behavior, calling behavior, siege behavior, and two intelligent rules: a winner-takes-all productive rule for lead wolf and a randomly renewable mechanism named survival of the stronger for the wolf pack.

The parameters such as X_i, N, $k_{\max}$, $T_{\max}$, S , L_{near} , α, β and the objective function value Y have been defined explicitly and all the components have been discussed in detail in [8], so the main computation steps are only described below.

Step1: Initialization. For a D dimension of the search space, the follow parameters are initialized: X_i, N, $k_{\max}$, $T_{\max}$, S , L_{near}, β.

Step2: The wolf with the maxium Y value (ie, Y_{lead}) is selected as lead wolf, and the rest N_{sw} (an inter value between $N/(\alpha+1)$ and N/α) wolves with better Y value, as scount wolves, begin to scout in D dimension solution space according to equation (10) until $Y_i>Y_{lead}$ or $T_{\max}$ is reached, then go to step3;

$$x_{id}^{p} = x_{id} + \sin(2\pi \times p/h) \times step_{a}^{d},\quad p = \{1, 2, \ldots, h\} \tag{10}$$

Where x_{id}^{p} is the dth variable value of the wolf-i after moving towards the pth direction; $step_a$ is the scouting step length, h is randomly selected in $[h_{min}, h_{max}]$ and must be an integer.

Step3: The rest N_{fw} ($=N-N_{sw}-1$) wolves take calling behavior according to equation (11). If $Y_i \geq Y_{lead}$, go to step2; otherwise the wolf-i continue running until $L(i, l) \leq L_{near}$, then go to step 4;

$$x_{id}^{k+1} = x_{id}^{k} + setp_{b}^{d} \times (g_{d}^{k} - x_{id}^{k}) / \left| g_{d}^{k} - x_{id}^{k} \right| \tag{11}$$

Where x_{id}^{k+1} is the position of ferocious-wolf- i at the (k+1)th iteration; $step_b$ is the summoning step length, g_d^k is the position of lead wolf in the dth variable space at the kth iteration. $L(i, l)$ is the distance between the ith wolf and the lead wolf and expressed as Manhattan distance.

Step4: The position of siege wolves is updated according to equation (12);

$$x_{id}^{k+1} = x_{id}^{k} + \lambda \times setp_{c}^{d} \times \left| G_{d}^{k} - x_{id}^{k} \right| \tag{12}$$

Where λ is a random number uniformly distributed at the interval [-1, 1], $step_c$ is the siege step length.

The relationship between the step length $step_a$, $step_b$ and $step_c$ in dth variable space is as follows:

$$step_{a}^{d} = step_{b}^{d}/2 = 2step_{c}^{d} = \left| x_{d\max} - x_{d\min} \right| / S \tag{13}$$

Step5: Update the lead wolf to the one with maximum objective function value, and then delete R wolves with worst objective function value from the wolf pack and meanwhile produce R wolves according to equation (14) to replace them;

$$x_{id} = g_d \bullet \text{rand}(-0.1, 0.1),\ i = \{1, 2, \ldots, R\} \tag{14}$$

Where R is an integer and randomly selected at the interval $[n/(2\beta), n/\beta]$.

Step6: If the program reaches the precision requirement or the maximum number of iterations, the position and Y_{lead} of lead wolf, the problem optimal solution, will be outputted, otherwise go to step2 to continue iteration until termination condition is met.

In the above steps, step 2 is fine search, reflecting the local optimal solution search precision; step 3 is a rough search, reflecting the local optimal solution search efficiency; step 4 is gradually refined search, which reflects the accuracy that the local optimal solution is also global one; step 5 produces a new generation of wolves, which not only retains the excellence of the local optimal solution founded by original wolves, but also increases the probability of reaching the optimal solution to guarantee the global optimality of the algorithm.

4. WPA for solving ARPCO problem

In order to apply the WPA to find the optimal control variables of the ARPCO problem , the vector of polulation can be expressed as follows:

$$\mathbf{X} = \begin{bmatrix} x_{11} & \Lambda & x_{1D} \\ \mathrm{M} & \Lambda & \mathrm{M} \\ x_{1D} & \Lambda & x_{ND} \end{bmatrix} \tag{15}$$

The objective function Y is evaluated according to equation (16) as follow:

$$Y = -f = -P_{LOS} \tag{16}$$

The initial position of each artificial wolf can be formed as follow:

$$x_{id} = x_{d\min} + \text{rand}(0,1)(x_{d\max} - x_{d\min}) \tag{17}$$

Where rand (0, 1) is a random number uniformly distributed at the interval [0, 1], x_{dmin}, x_{dmax} is the lower and upper limit of the dth control variable respectively.

L_{near} is evaluated according to equation (18) as follow:

$$L_{near} = \frac{1}{D g \omega} \sum_{d=1}^{D} \left| x_{d\max} - x_{d\min} \right| \tag{18}$$

Where ω is distance determinant factor.

For the ARPCO problem, the iteration termination condistion is expresse as follow:

$$\left|Y_{\text{lead}}^{k} - Y_{\text{lead}}^{k-3}\right| \leq \varepsilon \quad 或 \quad k = k_{\max} \tag{19}$$

Where is the convergence control constant number.

In order to calculate the objective function, position variables from each wolf are firstly mapped into the power flow equations and the flow program is executed to obtain the network loss by MATPOWER. Y is evaluated according to equation (16). Then the simulation is processed according to the WPA solving procedure from step1 to setp6 described above. During the process, these variables out of bound are tagged at their boundaries. The simulation is repeated until meeting the termination condition. The implementation of WPA for ARPCO problem is depicted in Figure 1.

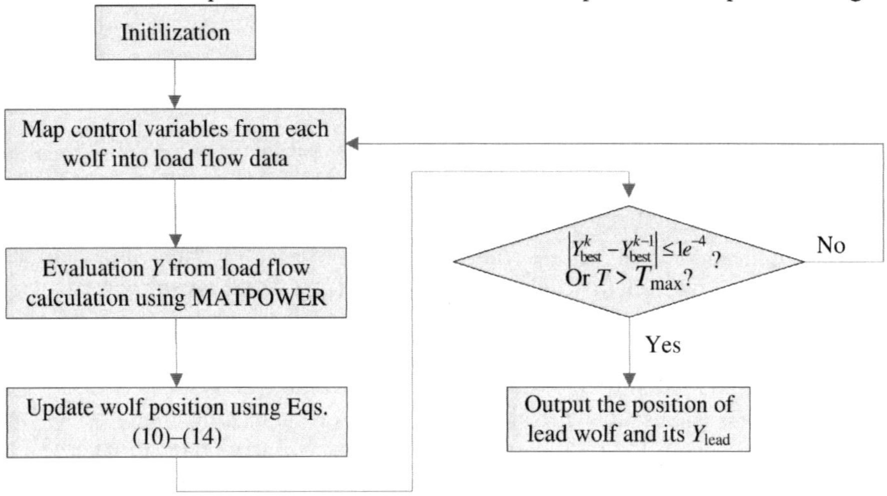

Figure 1. Flowchart of propsed WPA for solving ARPCO

5. Simulations

In order to apply the simulations for solving ARPCO problem by WPA are implemented using Matlab R2012b on a Windows 7 Professional Intel i5-3210M CPU 2.5GHz 8GB RAM.

In this paper, the IEEE 33 node radial distribution system which PV, SVC, ESS, CB are integrated into is selected as the example system, as shown in figure 2. The network with 37 branches is radially operated. The total load power: P_{load} = 3635kW, Q_{load} = 2265kvar; for each PV: $P_{\text{PV}}^{\max}$ = 300kW, cosφ = 0.95; for each shunt capacitor (SC): $Q_{\text{SC}}^{\min}$ = 25kvar, $Q_{\text{SC}}^{\max}$ = 100kvar; for each ESS: $P_{\text{ES}}^{\max}$ = 240 kW, $P_{\text{ES}}^{\min}$ = -200 kW, $E^{\max}$ = 1200kWh.

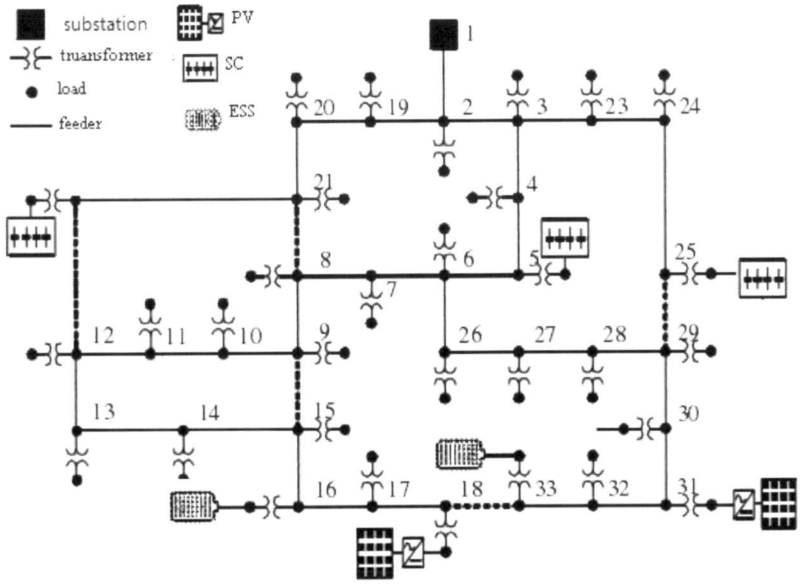

Figure 2. Modified IEEE 33-bus test system

The load curve and the sunshine intensity curve of the system at the moment of 24 are shown as figure 3. For simplifying the analysis, all of the PV use the same sunshine curve and load node use the same load curve.

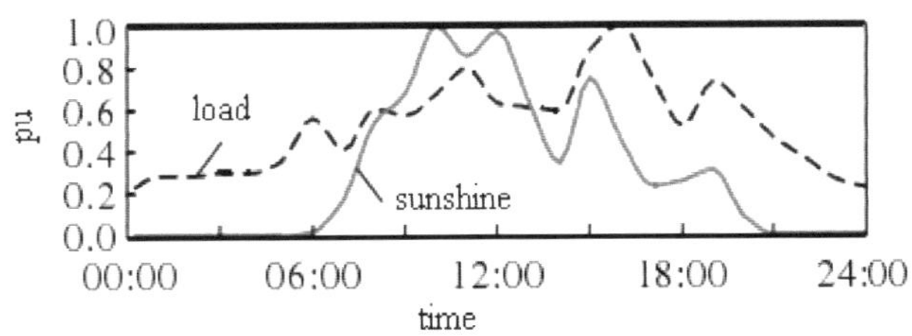

Figure 3. Load and sunshine profile

It is seen from figure 3 that the peak value of the load and the peak point of the light does not appear at the same time. If there are no control equipments such as energy storage, PV will abandon light due to the light load so that the load absorb power only from the substation during the heavy load period , and then the heavy load node voltage will drop and the voltage difference further rise, which causes a large number of network loss.

To test the robustness of this method under the large volume data condition, MATLAB cubic spline interpolation method is used to interpolate the above 24 hour data as 96 data points so that the interval acquiring the data for the day-ahead scheduling is 15 minutes.

WPA parameters were specified as according to reference [17]: $N = 50$, $K_{\max} = 1000$, $S = 200$, $T_{\max} = 15$, $\alpha = 4$, $\beta = 5$, $\varepsilon = 10^{-3}$. After 68 iteration steps (shown as fig. 4), the global optimal solution were found and the mean square error is about 0.0005, the total time is 20. 517s. The ARPCO results are shown as figure 5(a) ~ 5(e). PV is almost in max power state because of ESS.

Figure 5. (a) shows that ESS start to charge at load trough period and to discharge at load peak period so as to reduce the total loss in the scheduling period. Again, energy is stored when output of PV increases and injected into the system when load power increases, shown as figure 5. (b). Network loss is smaller because of smaller voltage difference at load trough period and is larger because of larger voltage difference at load peak period. As a result, the total network loss reduces compared with no optimization, shown as figure 5. (e) with less total power loss than the loss in [13]. Figure 5. (c) shows that SCB operation occurs when the load demand and PV generation increase, and their operation numbers are strictly restricted in the allowable range. Figure. 5(d) shows that the distribution system absorbs less active and reactive power from the substation after ARPCO process in order to improve the penetration of PV.

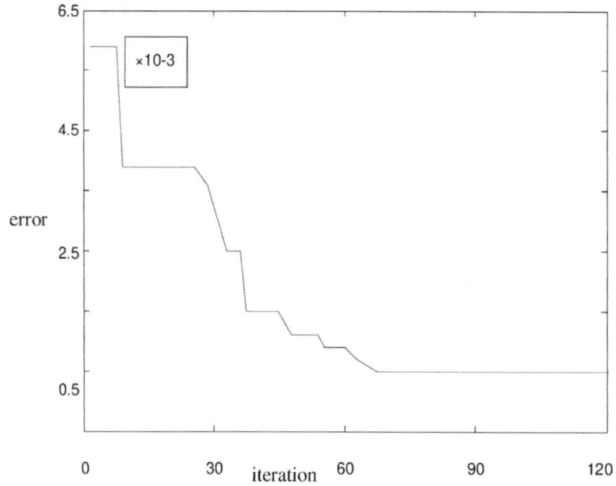

Figure 4. The WPA error curve

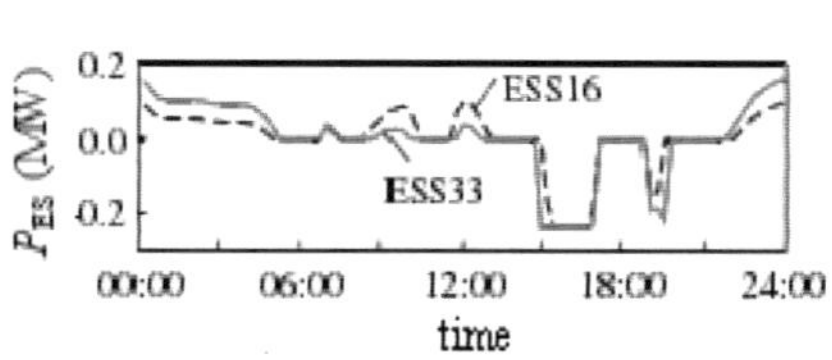

Figure 5. (a) charge or discharge power profile

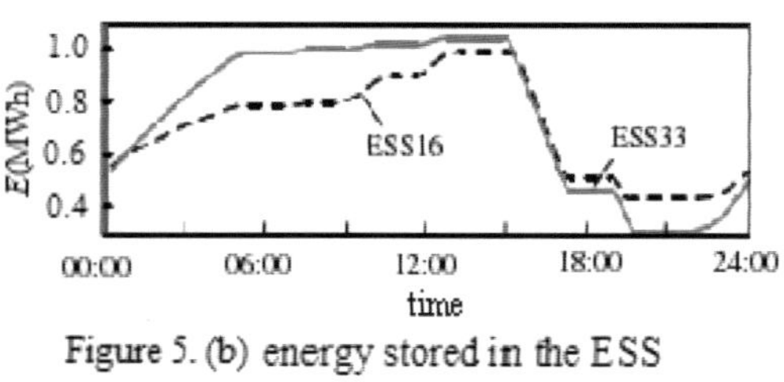

Figure 5. (b) energy stored in the ESS

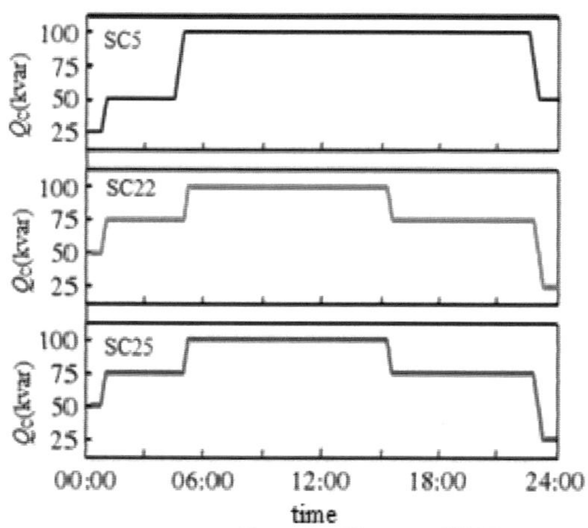

Figure 5. (c) Reactive Power of SCB

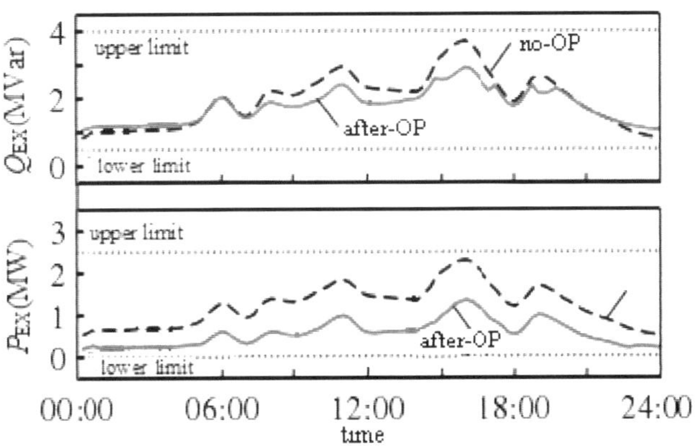

Figure 5. (d) power from substation

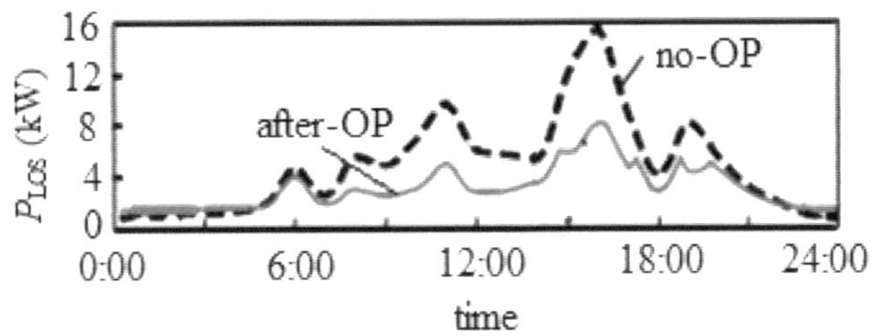

Figrue 5. (e) power loss profile

Figure 5. ARPCO results

From table 1, it can be seen that the energy loss obtained by WPA is the lowest and the WPA can converge to the optimal solution of the problem in a reasonable amount of time. Therefore the WPA for solving ARPCO problem is a competitive algorithm.

Table 1. Comparison of three algorithm

Algorithm	WPA	PSO	SOCP
Energy Loss(kWh)	109.642	118.169	125.705
Computational Time (s)	20. 517	23. 314	17.161

6. Conclusions

This paper proposed the ARPCO model considering the output of DG, charge or discharge power of ESS and reactive power from SBs as control variables. And minimizing the energy loss of the distribution network was taken as the optimal objective. The use of WPA technique guarantees convergence to optimality in a reasonable amount of time.

The modified IEEE 33-bus test system was used to demonstrate the accuracy of the ARPCO mathematical models and the efficiency of the WPA for solving it. When compared with PSO and SOCP technique used in [13], WPA shows more effective because of minimum total energy loss and reasonable convergence speed.

In future, the stochastic of renewable resources can be considered in solving ARPCO problem to demonstrate the robustness of WPA. In addition, an advanced WPA with better performance will be proposed in future.

Acknowledgments

This research was financially supported by the Scientific Research Foundation of CUIT.

References

[1] Hatziargyriou N. et al. 2011 CIGRE WG: Network of the future, electricity supply systems of the future *Electra* **256** 42–9

[2] Zhao B 2005 A multiagent-based particle swarm optimization approach foroptimal reactive power dispatch *IEEE Trans. Power Syst* **20** 1070–8

[3] Vlachogiannis J G and Lee K Y 2006 A comparative study on particle swarm optimiza-tion for optimal steady-state performance of power systems *IEEE Trans. PowerSyst.* **21** 1718–28

[4] Ayan K 2012 Artificial bee colony algorithm solution for optimal reactivepower flow *Appl. Soft Comput* **12** 1477–82

[5] Huang C M and Huang Y C 2012 Combined differential evolution algorithm andant system for optimal reactive power dispatch *Energy Procedia* **14** 1238–43

[6] Graditi G. *et al.* 2015 An Innovative Conversion Device to the Grid Interface of Combined RES-based generators and Electric Storage Systems *IEEE Transactions on Industrial Electronics* **62** 2540-50

[7] Christakou K 2016 A unified control strategy for active distribution networks via demand response and distributed energy storage systems *Sustainable Energy, Grids and Networks* **6** 1–6

[8] Zhang X 2016 Real-time active and reactive power regulation in power systems with tap-changing transformers and controllable loads *Sustainable Energy, Grids and Networks* **5** 27–38

[9] Zhuang H M and Xiao J 2015 Two-stage optimal schedule model for active distribution network *Journal of Southwest Jiaotong Univerity* **50** 928–34

[10] Bronzini M, La Scala M and Sbrizzai R 2011 Coordination of active and reactive distributed resouces in smart grid *IEEE Trans. Power Syst.* **3** 381–8

[11] Levron Y, Guerrero J M and Beck Y 2013 Optimal power flow in microgrids with energy storage *IEEE Trans. Power Syst* **28** 3226–34

[12] Gabash A and Li P 2015 Active-reactive optimal power flow in distribution networks with embedded generation and battery storage *IEEE Trans. Power Syst* **27** 2026–35

[13] Liu Y B. *et al.* 2014 A Mixed Integer Second-order Cone Programming Based Active and Reactive Power Coordinated Multi-period Optimization for Active Distribution Network *Proceedings of the CSEE* **34** 2575–83

[14] Graditi G, Di Silvestre M L and Gallea R 2015 Heuristic-based shiftable loads optimal management in smart micro-grids *IEEE Transactions on Industrial Informatics* **11** 271-80

[15] Deb K, Pratap A, Agarwal S and Meyarivan T 2002 A fast elitist multi-objective genetic algorithm: Nsga-II *IEEE Trans. Evol. Comput* **6 (2)** 182–97

[16] Di Silvestre M L, Graditi G and Sanseverino E R 2014 A generalized framework for optimal sizing of Distributed Energy Resources in micro-grids using an Indicator-Based Swarm Approach *IEEE Transactions on Industrial Informatics* **10** 152-62

[17] Wu H, Zhang F and Wu L 2013 New swarm intelligence algorithm–wolf pack algorithm *Systems Engineering and Electronics* **35** 2430–8

IOP Conf. Series: Earth and Environmental Science **40** (2016) 012004 doi:10.1088/1755-1315/40/1/012004

Field survey analysis of the public's cognition on the new energy industry in China

Q F Guo[1] and Y Song

School of business administration, China University of Petroleum (Beijing) No. 18 Fuxue Road Changping District, Beijing 102249, China

E-mail: guo18901106799@163.com

Abstract. The public enjoy an important role in the development of China's new energy industry. However, the role has not attracted sufficient attention. By the way of field investigation, the paper acquired the first hand data of the public cognition on the China's new energy industry. Survey data showed that the public enjoyed awareness of China's new energy industry to some extent. And the public had optimistic expectations on the future development of new energy industry. Moreover, there were obvious differences in the degree of public's familiarity with different new energy varieties. The education level and age of the individual public had a significant impact on his awareness of China's new energy industry. To raise public participation in China's new energy industry, it entailed highlighting the status of the public in China's new energy industry, increasing the publicity of the new energy industry with different measures for different types of public group.

1. Introduction

With the increasing depletion of traditional fossil fuels, unstable energy supply, serious pollution, countries all over the world began to attach importance to the development of new energy sources.① In order to ensure the energy security, promote the adjustment of economic development mode, China also vigorously developed new energy sources. A variety of subsidy policies continued to be introduced, so that all kinds of enterprises flooded into the new energy industry. In 2014, China's clean energy investment reached $767 million, increasing by 20%, and more than the United States to become the world's largest clean energy investment nation. The hydropower, wind power and nuclear power construction scale ranks the first in the world. Along with the rapid development of China's new energy industry,

[1] Address for correspondence: Q F Guo, School of business administration, China University of Petroleum (Beijing), No.18 Fuxue Road Changping District Beijing city, 102249, China. E-mail: guo18901106799@163.com.

problems were constantly emerging, such as irrational industrial structure, the imperfect industry chain, the lack of core technology, institutional obstacles and heavy market hindering.

New energy development cannot be separated from the behavior of economic agents. With respect to the governments (including the central government and local governments), enterprises and other economic entities, the status and role of the public had not caused full attention in the China's new energy industry development. In fact, the role of the public in the development of China's new energy industry is still very important. First of all, the public is an important part of the end product consumption in the new energy industry. Secondly, the public also effects on the supply side activities (investment, R&D and production, etc) in new energy industry by influencing related policies such as tax, price, environment and so on. Therefore, to realize the sound and rapid development of China's new energy industry, it is required to understand the relevant behaviors of the public in the new energy industry.

There were a few related literatures on the public's behavior in the new energy industry. Sovacool et al. listed thirteen myths that the American public misunderstood the energy fields including new energy industry. Schmid showed that the extensive participation of the private sector played a more and more important role in the development of the new energy industry through empirical studies. Kitzing et al. showed that the new energy policy in Europe was moving towards an integration trend of top-down. Pfeiffer indicated that economic development and effective management, higher level of education, stable democratic system had contributed to the development of non-hydro renewable energy power generation technology through the study of 108 developing countries from 1998 to 2010. Mallon took public participation as one of the ten major factors in the success of the new energy market Li Meng et al. gave the power mechanism of the development of new energy industry, and thought that the important role of social factors in the development of new energy industry had not been fully played out. Gehrke analyzed how to use renewable energy in the family. As a whole, the researches of new energy industry development of public behaviors are still preliminary, fragmented and lack of systematic. This paper attempts to make a systematic study on the public behaviors of the new energy industry in China. However, because the public are complex and diverse groups, it is difficult to understand the behaviors of the public to the new energy industry. On the other hand, as China's new energy industry was still in the primary stage, and there was a certain gap between most of the public and new energy industry. Knowledge precedes action, and through field investigation and data analysis on the public perception of China's new energy industry, the paper attempts to understand a clue of public behaviors to China's new energy industry. Finally, the paper hoped to contribute to the draw up of policies concerned with China's new energy industry, especially those bottom-up policies.

2. The acquisition of field survey data

Field research or field investigation is the practice and application of the direct observation method. And it is also a pre-step for the acquisition of the original data. The good design of the questionnaire is the basis of the ideal field survey. China's new energy industry was still in its initial stage of development, and enjoyed low public participation. On the contrary, the composition of the public was more complex and diversified. Therefore, The design of the questionnaire in the paper not only took into account the key contents of public awareness of

China's new energy industry, but also the requirements of the respondents' cooperation. Finally, the questionnaire was revised and determined by a small scope pre-survey.

In 2015, a total of 300 people were interviewed on the spot, and the geographical distribution of respondents included Beijing, Xinjiang, Henan, Heilongjiang, Shandong and other provinces. The identity of sample public was very rich: Ages ranged from 20 to 75, and occupations included all walks of life such as students and teachers. The survey data would basically reflect the situation of public cognitive on China's new energy industry. Based on the brief records of interviewing process, 236 valid questionnaires were selected, and the rate of the valid questionnaire was 78.7%. Although the contents of the questionnaire was simple and clear, but there were still more than 20% of the respondents refused to fill in the questionnaire with no understanding or no interest in the new energy industry as excuses, which showed the public were indifference to China's new energy industry in a certain extent. Therefore, it is very urgent to strengthen the public awareness of new energy industry in China.

3. The frequency statistics analysis of the public's cognition on the China's new energy industry

3.1 Methodology

Frequency statistical analysis can reflect the public's cognition on new energy industry from the angle of probability as a whole. The frequency statistics are calculated by the following formula:

$$f_i = \frac{\sum_{i=1}^{N} \alpha_i}{N} \tag{1}$$

In terms (1): f_i denotes the probability of occurrence of a situation (e.g. the public knows very well about new energy industry); α_i denotes whether the respondents selected the option of i ; if selected, α_i=1, otherwise, α_i=0.N denotes the scale of total samples, and the total valid sample scale in the paper was 236. $\sum_{i=1}^{N} f_i = 1$.

3.2 The public overall understanding of China's new energy industry

In order to understand the situation of the public awareness of China's new energy industry, 4 individual options were designed in the questionnaire including "very understanding", "understand some", "heard" and "not clear". The results of frequency statistical showed that: 8.9% of the respondents were very understanding of new energy; 61.9% of the respondents said they understand some of the new energy; 28% of respondents said they only heard of new energy; and 1.3% of respondents said they did not know the new energy. More than 70% of the respondents had an understanding of new energy, which showed that the public's cognition on the new energy industry was relatively high, which meant the long-term sound development of China's new energy industry would have a good public foundation. However, it would not be very optimistic. Only 8.9% of the respondents were very understanding of new energy industry; on the contrary, nearly 30% of the respondents only heard or were not clear of it. Considering that there were more than 20% of the invalid questionnaires, the gap between China's new energy industry and the public was still very large.

China's great efforts to develop new energy industry had been over 10 years, and a good many national level policies had also been introduced to strongly encourage and support the new energy industry. However, there were still a considerable parts of the public said they did not quite understand the new energy. Therefore, how to make the new energy into the tens of thousands of China's households, accepted by the broad masses of the people, the promotion of the public in China's new energy industry would still need a lot of work.

3.3 The comparison of the public cognition on different varieties of new energy sources

In order to examine the situation of the public familiar with each new energy varieties, for the respondents who choose "very well", "understanding" of the new energy industry, five single options were designed in questionnaire to judge which was the most familiar one of new energy varieties, including "solar energy", "wind power" and "biomass", "new energy vehicles", "don't know". Survey data showed that solar energy was the best known to the general public with accounting for 51.47%; wind power followed with accounting for about 30.15%; the third was the new energy vehicles with accounting for about 13.97%; biomass was at least with accounting for only 2.94%; and another 1.47% of the respondents said they did not understand any kind of the four given new energy sources. The difference of public awareness of new energy sources was basically consistent with the development of new energy products in China. Solar energy in China enjoyed rapid development in recent years, and solar water heaters and other thermal products had been more popular in particular. Wind power also acquired rapid development in recent years, and the wind power was more popular in the area of advantageous natural conditions particularly. Biomass energy is an "old" energy. However, the modernization of China's biomass energy was very slow because of the constraints such as transformation of technology.

3.4 Public optimism about the future of the new energy industry

In order to examine the situation of public optimism about the future of the new energy industry, for the respondents who choose "very well", "understanding" of the new energy industry, three single options were designed in questionnaire including "Promising", "no future" and "not clear". Survey data showed that "Promising" accounted for 85.63%; "no future" accounted for 6.59%; and "not clear" accounted for 7.78%. The price of fossil energy and China's increasingly severe environmental problems in recent years made the public keep overall optimistic outlook on China's new energy industry. However, there were still some doubts, which also reflected many problems and difficulties in the development of China's new energy industry in this stage. For example, photovoltaic and wind power was difficult to enter the existing power grid; it was difficult to collect biomass materials; and the new energy vehicles existed technical bottlenecks; etc.

3.5 The public's perception of how long China's new energy turning into conventional energy

In order to examine the situation of the public's perception of how long China's new energy turning into conventional energy, for the respondents who choose "very well", "understanding" of the new energy industry, four single options were designed in questionnaire including "five years", "ten years" , "twenty years" and "not clear". Survey data showed that "five years" accounted for12.57%; "ten years" accounted for 44.31%; "twenty years" accounted for125.15%; and "not clear" accounted for 17.96%. There was considerable disagreement in the public's perception of how long China's new energy turning into conventional energy, which also indicated that the conventional time of the new energy industry had a lot of uncertainty. However, most of the public thought that the conventional time of the new energy industry would be relatively long, and this perception may be objective and fair.

3.6 The most important measure to develop new energy resources

In order to examine the situation of the public's perception of the most important measure to develop new energy resources, for the respondents who choose "very well", "understanding" of the new energy industry, four single options were designed in questionnaire including "technical innovation", "supporting enterprises", "expanding the market", and "natural development". Statistics show that 52.14% of respondents believed that technological innovation was the most important measure of the development of new energy industry in china, which indicated that the public believes that technological innovation is the biggest driving force for the development of new energy sources. 24.29% of respondents believed that supporting enterprises was the most important measure of the development of new energy industry in china. Because enterprises are the most direct agents of new energy commercialization, the status of enterprises in China's new energy industry had also been highly valued by the public. However, Only 20% of respondents attached great importance of the market in the China's new energy industry, which might reflect the current embarrassing situation with low degree of commercialization in China's new energy industry. Only 3.57% of respondents believed that the new energy industry should be allowed to develop naturally, which showed that the public had a high degree of awareness of the need for policy support for the new energy industry.

4. The interactive analysis between the subject characteristics of respondents and their familiarities with new energy industry in China

4.1 Methodology

Respondents' age, educational background and other main characteristics will affect their cognitive status of the new energy industry in China. In order to further analyze the relationship between personality characteristics of respondents and their cognitive status of China's new energy industry, the paper introduced the interactive analysis between the subject characteristics of respondents and their familiarities with new energy industry in China. The aim of interactive analysis was to provide some kind of behavioral cues that taking different measures for different types of the public groups in China's new energy industry. The formula of interactive analysis is as follows:

$$p_{ji} = \frac{\sum \alpha_i \beta_j}{\sum \beta_j} \tag{2}$$

In terms (1): p_{ji} denotes the probability for the specific respondent with j subject characteristic to choose the i option. β_j denotes whether the specific respondent was in line with the j subject characteristic; if satisfied, β_j=1, otherwise, β_j=0. α_i denotes whether the respondents selected the option of i ; if selected, α_i=1, otherwise, α_i=0. $\sum \beta_j$ is in fact the sum of the samples of respondents who satisfy j subject characteristic.

Because the terms that relating to the subject characteristics of respondents and their perception options of new energy in China were very complicated, there would be great number of combinations between subject characteristics of respondents and their perception options of new energy in China. Due to space limitations, the paper just conducted the interactive analysis between the age, educational background of respondents and their familiarities with new energy industry in China. The age of respondents was divided into four types including "less than25", "25-35", "36-50" and "more than 50". The educational background of respondents was divided into four types including "junior high school and below", "high school", "junior college or bachelor degree", "master degree and above".

4.2 The interactive analysis between the age of respondents and their familiar situations of new energy industry in China

The results of interactive analysis between the age of respondents and their familiarities with new energy industry in China were shown in table 1.

Table 1. The relationship between the age of respondents and their familiarities with new energy industry in China

Types of ages	Number of samples	Familiarities with new energy industry in China			
		very understanding	understand some	heard	not clear
less than25	126	9.5%	67.5%	22.2%	0.8%
25-35	33	3.0%	63.6%	30.3%	3.0%
36-50	56	10.7%	51.8%	37.5%	0.0%
more than 50	14	0.0%	50.0%	42.9%	7.1%

Note: In the process of interactive analysis, the samples without the age information of respondents were excluded.

It would be seen from table 1 that the age of respondents enjoyed significant effect on their familiarities of new energy industry in China. As a whole, the lower the respondents' age, the higher degree of his familiarity with new energy industry in China they kept, and the age group with the highest degree of familiarity with new energy industry in China was 36-50. The relationship between the ages of respondents and their familiarities with new energy industry in China had double meanings: The public with their ages at 36-50 would be the backbone of China's new energy industry. On the other hand, the younger public enjoying good familiarities with new energy industry in China would help the future development of this industry have a robust public basis.

4.3 The interactive analysis between the educational backgrounds of respondents and their familiarities with new energy industry in China

The results of interactive analysis between the educational background of respondents and their familiarities with new energy industry in China are shown in table 2.

Table 2. The relationship between the educational backgrounds of respondents and their familiarities with new energy industry in China

Types of educational backgrounds	Number of samples	Familiarities with new energy industry in China			
		very understanding	understand some	heard	not clear
junior high school and below	13	0.0%	23.1%	76.9%	0.0%
high school	48	6.5%	60.4%	31.0%	2.1%
junior college or bachelor degree	143	7.7%	65.0%	25.9%	1.4%
master degree and above	15	13.3%	66.7%	20.0%	0.0%

Note: In the process of interactive analysis, the samples without the educational background information of respondents were excluded.

It would be seen from table 1 that none of respondents with the educational background of junior high school and below was "not clear" in the China's new energy industry, and there were somehow discrepancies between the results above mentioned and actual situation. However, the educational backgrounds of respondents enjoyed a positive relationship with their familiarities with new energy industry in China as a whole. The respondents with higher educational background were more likely to be familiar with the new energy industry in China.

5. Conclusions and policy recommendations

In addition to the government policy support, the perfect market mechanism and the enterprise's initiative, it also requires the public recognition and acceptance in the development of new energy industry. The paper examined the public awareness of the development of new energy in china. Those findings suggested that most of the public understood and were hopeful about the new energy industry in China, but it still needed to be enhanced. China's public awareness of solar energy, wind power was relatively large, but their understanding of new energy vehicles, biomass was relatively inadequate. The subject characteristics of respondents could significantly influence on their familiarities with new energy industry in China.

Compared to conventional fossil fuels, the new energy industry is more closely linked with the public, especially the distributed new energy sources. The development of new energy needs the public's positive reaction in China, and the public awareness and behavior of the new energy industry will affect the development of China's new energy industry. To raise the public rational awareness of the new energy industry and the level of participation will be conducive to the marketing and the implementation of relevant policies in China's new energy industry.

First of all, it should be attached great importance to the public's position in the development of new energy industry. Especially, the status and role of the public should be paid attention during the governments' introducing the relevant policies and the enterprises' promoting their market in the new energy industry. Meanwhile, it is indispensable to establish effective public contact mechanisms, such as feedback, public surveys, etc.

Secondly, it is urgent to strengthen propaganda to the public in China's new energy industry. The Governments and related enterprises should vigorously promote the new energy industry so that the new energy industry can establish a close relationship with the public. Further, it should be strived to make the new energy industry become a part of the public's daily life. In the aspects of publicity of new energy industry, it should not favor one more than another for different varieties of new energy. Each variety of new energy sources has its own characteristics and advantages. All varieties of new energy should acquire importance to better play their own value with enlarging growth space.

Last but not least, it is to take different measures for different public groups. For example, the young public's current position on the new energy industry is relatively low, but their optimism of the new energy industry is very high. Therefore, based on the characteristics of young public, the governments and enterprises should increase awareness and experience of the new technology and new consumption styles for the young public in China's new energy industry.

References:

[1] Sovacool B K and Brown M A 2007 *Energy and American society: thirteen myths* (New York: Springer Science + Business Media BV)

[2] Schmid G 2012 The development of renewable energy power in India: which policies have been effective *Energy Policy* **45** 317–26

[3] Kitzing L, Mitchell C and Morthorst P E 2012 Renewable energy policies in Europe: Converging or diverging *Energy Policy* **51** 192–201

[4] Pfeiffer B and Mulder P 2013 Explaining the diffusion of renewable energy technology in developing countries *Energy Economics* **40** 285–96

[5] Li M and Deng X D 2014 China's new energy industry development: factor analysis of the impacts *Innovation* **2** 17-20

[6] Mallon K 2014 *Renewable energy policy and politics: A handbook for decision-making* (Beijing: Economy & Management Publishing House)

[7] Gehrke R 2015 *Renewable energies for your home* (Beijing: China Machine Press)

Economic-energy-industrial-environmental optimization (EEIEO) model for identification of optimal strategies - a case study of Beijing-Tianjin-Hebei region, China

C Chen[1,2,3], H L Long[1, 4], J Wan[3], JL Jia[3], X Li[3] and CJ Chu[3]

[1]Institute of Geographic Sciences and Natural Resources Research, Chinese Academy of Sciences, Beijing, 100101, China
[2]University of Science and Technology Beijing, Donlinks School of Economics and Management, Beijing, 100083, China
[3]China Academy for Environmental Planning, Beijing, 100012, China

E-mail: longhl@igsnrr.ac.cn

Abstract. An economic-energy-industrial-environmental optimization (EEIEO) model is proposed for identification of optimal economic, industry, energy and environment strategies. The EEIEO model is applied to a real case of Beijing-Tianjin-Hebei (BTH) region, which is the important economic growth pole of northern China. The EEIEO model could fully consider the interaction between industrial, energy, urbanization and environment sector, and generate the optimized economic development, industrial restructuring, energy consumption and environment management schemes. This is first attempt to introduce economic, energy, industrial, urbanization and environmental sectors into an optimization framework, while sustainable energy and environment development pathways are explored through EEIEO model. The results suggest that: (i) the GDP of BTH region would increase about 73.80% over the planning horizon; (ii) the contribution of tertiary industry for BTH region's economic development would gradually increase from 54.00% in 2015 to 65.00% in 2030; (iii) the consumption of coal would decrease by 36%, and the natural gas would obviously increase by 97.70% over the planning horizon; and (iv) the SO_2, smoke and dust emissions and CO_2 would reduce by 30.20%, 35.30% and 4.50% from 2015 to 2030, respectively.

1. Introduction

Energy resources are highly relevant to the economic, industrial structure, environment and urbanization, which importance for national prosperity development, people's living standards and social stability [1]. Beijing-Tianjin-Hebei (BTH) region, as one of the important poles of economic growth in China, is the economy, political and culture center of northern China. In the recent years, it

[4] Address for correspondence: H L Long, Institute of Geographic Sciences and Natural Resources Research, Chinese Academy of Sciences, Beijing, 100101, China. E-mail: longhl@igsnrr.ac.cn.

has achieved great economic development. According to China Statistical Bureau, the gross domestic product (GDP) reached up to 6.27 × 10^{13} RMB¥ in 2013, which occupied 11% of China's GDP. The rapid economic development results in the celerity growth of energy demands. Until 2013, the total terminal energy consumption amount was 346.26 million ton of coal equivalent (tce) until 2013, with a growth rate of 19.30% in the past 3 years.

Rapidly increasing of energy consumption demands caused serious environmental pollution problems. Currently, Beijing-Tianjin-Hebei region becomes one of the most polluted places of China. By 2012, the SO_2, NO_X and CO_2 emissions have reached up to 1.59 million, 1.43 million and 1.46 million tons, respectively [2]. About 90% of SO_2 emissions, 66.00% of NO_X emissions and 90.00% of CO_2 emissions are from burning coal. On December 20th, 2014, Chinese government implemented "Action plan for energy development strategy, 2014-2020", which set a target to mitigate about 80.00 million tons of coal consumption amounts comparing to that of 2005 in BTH region. However, in view of the stage of current economic development and economic operation pattern, the conflict between coal mitigation, pollutant emission and economic growth is inevitable, which will affect economic growth rate of BTH region [3]. Therefore, it is both important and necessary for the government to coordinate economic development and mainly pollutants (i.e., SO_2, NOx, smoke and dust), CO_2 emissions [3].

A great number of research efforts were undertaken for above problems [4-11]. Wang and Yang [3] quantitatively analyzed the delinking indicators on industry growth and environmental pressures in Beijing-Tianjin-Hebei economic band from 1996-2010, which based on the tow-level logarithmic mean divisia index method. Tian et al. [6] conducted both the regional analysis throughout the China and case studies focused on nine typical regions to identify regional patterns of industrial structure change and CO_2 emissions, the results indicate that structural change in industry structure was highly correlated with the stage of economic development. Chen et al. [7] proposed a tri-objective linear programming program for generation expansion planning by maximizing the total power generation, minimizing the total system cost and the ratio of the total power generation to the total CO_2 emission. Chen et al. [8] employed a multi-regional input-output model at the provincial level to evaluate the environmental costs of coal burning in China in 2007, in terms of its damages from climate change externality. Zhang et al. [10] developed an integrated assessment model including provincial energy conservation supply curves, the greenhouse gas and air pollution interactions and synergies model (which can be used to calculate air pollutant emissions) to assess the potential of energy saving in terms of emission mitigation of CO_2 and air pollutants an multiple benefits of energy efficiency measures at the provincial level during the period 2011-2030. In terms of the contents of research, the studies focused on the relationship between economics and CO_2 emissions, but often failed overall considering the economic development industrial structure restructuring, urbanization, energy consumption, energy structure adjustment and environment into a framework, as well as the impacts of economic development industrial structure restructuring, urbanization, energy consumption on the atmosphere pollutants are not fully considered. In real world, the relationships between industrial structure, urbanization, energy consumption, and environment are often complicated by the interactions of themselves; any changes in one sector could lead to a series of consequences. For example, the changes of industrial structure or urbanization could result in variations in economic, energy consumption and environmental. In planning such energy systems, independent consideration of one or several sectors would be unable to completely reflect the overall system characteristics.

Therefore, this study aims to develop an economic-energy-industrial-environmental optimization (EEIEO) model with taking a fully consideration of the interaction between industrial, energy, urbanization and environment sectors. The EEIEO model could effectively coordinate of the conflict among economic development, energy resources consumption, industrial restructuring and pollutant mitigation. In detail, the development of EEIEO model could: (i) reflect the relationship between economic, industrial and urbanization's impact on the energy and environment systems; (ii) generate the optimized economic development, industrial restructuring and energy consumption schemes; and (iii) evaluate the pollutants emissions under the optimized economic development, industrial restructuring and energy consumption schemes.

2. Overview of the study system

Beijing-Tianjin-Hebei (BTH) contains two municipalities (Beijing, Tianjin) and one province (Hebei). BTH, as one of the most economically vibrant regions in China, covers 2.28% of the Chinese territory, while generated 11.00% of the total national GDP in 2012. The population of BTH region was 110.50 million in 2013, accounting for 8.12% of the Chinese population. The urbanization rate amounted to 61.00%, which was higher than the average of Chinese (53.40%). The industry structure (primary industry, second industry and tertiary industry) proportion of BTH region was 6.00: 43.00: 51.00 in 2013. Figure 1 shows the GDP and industry structure of BTH region from 2001-2013 [12-14].

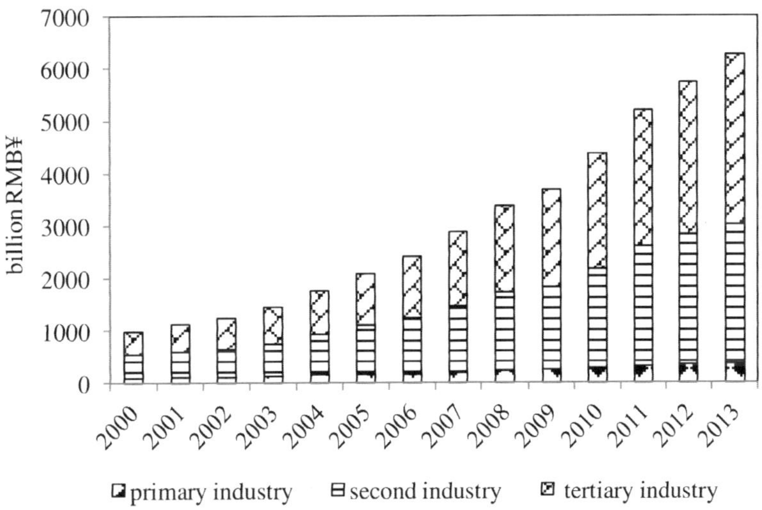

Figure 1. The GDP and industry structure of BTH region from 2001 to 2013.

With the increasing development of economy and society, the total amount of terminal energy demand has a rapidly growth. Figure 2 shows that the total terminal energy consumption amount was 346.26 million tons of coal equivalent until 2013, with a growth rate of 19.30% compared with 2010. The terminal consumption of coal was 82.77 million tons of coal equivalent until 2013, which accounted for 23.90% of the total energy consumption. The proportion of terminal coal consumption had decreased from 26.15% in 2010 to 23.90% in 2013. The terminal consumption amount of natural gas had shown an increase trend, which increase from 11.49 million tons of coal equivalent in 2010 to 15.72 million tons of coal equivalent in 2013. The terminal consumption amount of electricity had increased from 128.79 million tons of coal equivalent in 2010 to 160.05 million tons of coal equivalent in 2013, which proportion had increased from 44.37% in 2010 to 46.22% in 2013 [12-14]. Summarily, the city's terminal energy consumption structure has been improved in recent years.

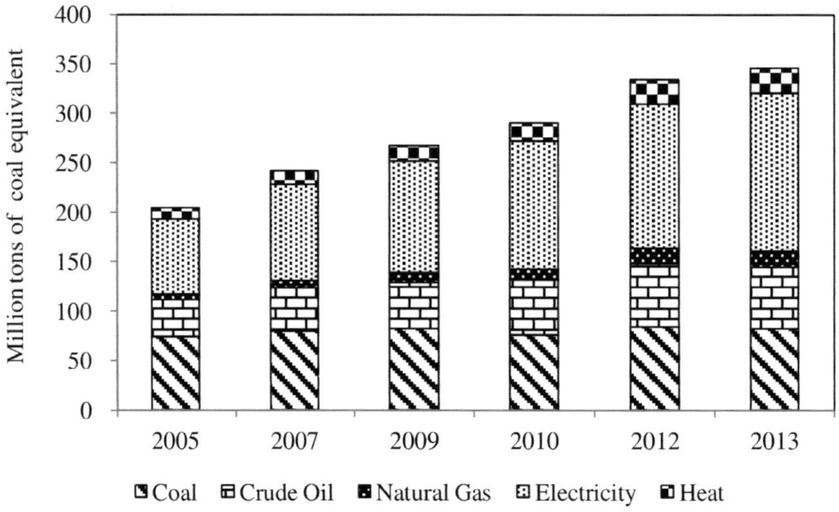

Figure 2. The total amount of terminal energy consumption amount from 2005 to 2013.

A large number of fossil fuels (such as coal, crude oil, and natural gas) lead to serious atmospheric environment issues of BTH region [15]. Currently, Beijing-Tianjin-Hebei region becomes one of the most polluted places of China. Until 2013, the SO_2 emissions reached to 1.59×10^6 tons; the NO_X emissions reached to 1.43×10^6 tons; the smoke and dust emissions reached to 1.46×10^6 tons. The mainly pollutants average concentration are obviously higher that the standard regulated, which have negative effects on the human health and hinder the sustainable development of BTH region [16]. Therefore, it is necessary to propose some low-carbon and environment friendly plans of economic, industrial as well as energy consumption to support the sustainable development and reduce pollutant emissions of BTH region.

3. EEIEO modelling formulation

The interactive relationship among economic, energy consumption, industry, urbanization and environmental is complex. Any change in one subsystem will cause the variations of others, and finally vary the economic, environmental, and social costs. For example, the economic and industry development policy can change a series of activities, such as the types of energy resources, energy consumption and pollutants emissions. To maximize the GDP with considering the requirements of air pollutants mitigation, an effective economic, energy and environment planning model is required to identify the optimal social development pattern.

The proposed EEIEO model could not only effectively coordinate of the conflict interactions among economic development, energy resources consumption, industrial restructuring and pollutant mitigation, but also generate optimized economic development, energy structure adjustment and industrial restructuring, urbanization, as well as environment management schemes. In the study systems, five terminal energy consumption resources (such as coal, oil, natural gas, electricity and heat) are considered to satisfy the energy demand for end-users. End users of energy resources can be categorized into agriculture, industry, construction, transportation, wholesale and retail trade and other industries. The objective of proposed EEIEO model is to maximize the GDP of BTH region, with considering the constraints of the energy, industry, population and environment aspects. In detail, the EEIEO model can be formulated as follows:

objective function:

Max $$GDP_t = XY_t + \sum_{i=1}^{n} XE_{it} + \sum_{j=1}^{m} XS_{jt} \tag{1}$$

constraints:

(1) Constraints for mas balance for energy resources:

$$XY_t \cdot EDY_{mt} + \sum_{i=1}^{n} XE_{it} \cdot EDE_{imt} + \sum_{j=1}^{m} XS_{jt} \cdot EDS_{jmt} + XU_t \cdot EDU_{mt} + XR_t \cdot EDR_{mt} \le ET_{mt} \tag{2}$$

$$XY_t \cdot EDY_{mt} + \sum_{i=1}^{n} XE_{it} \cdot EDE_{it} + \sum_{j=1}^{m} XS_{jt} \cdot EDS_{it} + XU_t \cdot EDU_{mt} + XR_t \cdot EDR_{mt} \ge ETX_{mt} \tag{3}$$

(2) Constraints for industry structure:

$$XY_t / (XY_t + \sum_{i=1}^{n} XE_{it} + \sum_{j=1}^{m} XS_{jt}) \le YCL_t \tag{4}$$

$$XY_t / (XY_t + \sum_{i=1}^{n} XE_{it} + \sum_{j=1}^{m} XS_{jt}) \ge YCS_t \tag{5}$$

$$\sum_{i=1}^{n} XE_{it} / (XY_t + \sum_{i=1}^{n} XE_{it} + \sum_{j=1}^{m} XS_{jt}) \le ECL_t \tag{6}$$

$$\sum_{i=1}^{n} XE_{it} / (XY_t + \sum_{i=1}^{n} XE_{it} + \sum_{j=1}^{m} XS_{jt}) \ge ECS_t \tag{7}$$

$$\sum_{j=1}^{m} XS_{jt} / (XY_t + \sum_{i=1}^{n} XE_{it} + \sum_{j=1}^{m} XS_{jt}) \le SCL_t \tag{8}$$

$$\sum_{j=1}^{m} XS_{jt} / (XY_t + \sum_{i=1}^{n} XE_{it} + \sum_{j=1}^{m} XS_{jt}) \ge SCS_t \tag{9}$$

(3) Constraints for population:

$$XR_t + XU_t \le RL_t \tag{10}$$

$$XR_t + XU_t \ge RLX_t \tag{11}$$

(4) Constraints for urbanization:

$$XR_t / (XR_t + XU_t) \le CZHL_t \tag{12}$$

$$XR_t / (XR_t + XU_t) \ge CZHS_t \tag{13}$$

(5) Constrains for environment

(a) Constraints for SO_2 emission amount:

$$(XY_t \cdot EDY_{mt} \cdot SPY_{mt} + \sum_{i=1}^{n} XE_{it} \cdot EDE_{imt} \cdot SPE_{imt} + \sum_{j=1}^{m} XS_{jt} \cdot EDS_{jmt} \cdot SPS_{imt} + XU_t \cdot EDU_{mt} \cdot SPU_{mt} + XR_t \cdot EDR_{mt} \cdot SPR_{mt}) \cdot (1 - \eta s_t) \le SPT_t^{\pm} \tag{14}$$

(b) Constraints for NO_X emission amount:

$$(XY_t \cdot EDY_{mt} \cdot NPY_{mt} + \sum_{i=1}^{n} XE_{it} \cdot EDE_{imt} \cdot NPE_{imt} + \sum_{j=1}^{m} XS_{jt} \cdot EDS_{jmt} \cdot NPS_{imt} + XU_t \cdot EDU_{mt} \cdot NPU_{mt} + XR_t \cdot EDR_{mt} \cdot NPR_{mt}) \cdot (1 - \eta n_t) \le NPT_t \tag{15}$$

(c) Constraints for smoke and dust emission amount:

$$(XY_t \cdot EDY_{mt} \cdot SDPY_{mt} + \sum_{i=1}^{n} XE_{it} \cdot EDE_{imt} \cdot SDPE_{imt} + \sum_{j=1}^{m} XS_{jt} \cdot EDS_{jmt} \cdot SDPS_{jmt} + XU_t \cdot EDU_{mt} \cdot SDPU_{mt} + XR_t \cdot EDR_{mt} \cdot SDPR_{mt}) \cdot (1 - \eta sd_t) \le SDPT_t \tag{16}$$

The study time considered four periods, t = 1 for 2015, t = 2 for 2020, t = 3 for 2025, t = 4 for 2030, which focused on the periods of China's 13th Five Year Plan (2016-2020), 14th Five Year Plan

(2021-2025) and 15th Five Year Plan (2026-2030). Table 1 shows the energy consumption intensity of terminal department (i.e., agriculture, industry, construction industry, transportation department, wholesale and retail trade and other industries) in planning periods. Table 2 shows the environment emission coefficient of terminal departments. Table 3 presents the upper bound of terminal energy consumption amount during planning periods, which obtained through Logistic prediction method based on data in past 18 years (1995-2013).

Table 1. Energy consumption intensity of terminal departments in 2015, 2020, 2025 and 2030.

Energy consumption intensity	Terminal department	2015	2020	2025	2030
Coal ($ton/10^6$ RMB¥)	Agriculture	5.14	4.05	3.73	3.62
	Industry	30.00	28.00	22.00	14.00
	Construction industry	1.96	1.54	1.42	1.42
	Transportation department	1.38	1.08	1.00	1.00
	Wholesale and retail trade	2.22	1.74	1.60	1.60
	Other industries	3.00	2.50	2.10	2.10
Oil ($ton/10^6$ RMB¥)	Agriculture	5.87	4.78	4.31	4.31
	Industry	6.08	6.20	6.20	6.20
	Construction industry	10.82	8.80	7.95	7.55
	Transportation department	22.96	18.67	16.85	16.85
	Wholesale and retail trade	1.32	1.08	0.97	0.92
	Other industries	1.25	1.01	0.92	0.87
Natural gas (cubic meter/10^6 RMB¥)	Agriculture	1.73	1.79	1.96	1.96
	Industry	3299.89	4000.00	4000.00	4000.00
	Construction industry	215.40	223.12	244.52	246.12
	Transportation department	1370.13	1419.27	1555.37	1756.40
	Wholesale and retail trade	1459.24	1511.57	1656.53	1870.63
	Other industries	1525.39	1580.10	1731.63	1955.43
Electricity (kWh/10^6 RMB¥)	Agriculture	3.02	2.76	2.52	2.52
	Industry	15.95	18.00	18.00	18.00
	Construction industry	2.10	1.92	1.75	1.67
	Transportation department	2.46	2.25	2.05	2.20
	Wholesale and retail trade	1.95	1.78	1.63	1.70
	Other industries	2.22	2.03	1.86	1.77
Heat (Million kJ/10^6 RMB¥)	Agriculture	0.00	0.00	0.00	0.00
	Industry	169.26	168.33	168.33	158.00
	Construction industry	9.43	9.38	10.15	9.00
	Transportation department	16.34	16.25	17.60	16.00
	Wholesale and retail trade	27.53	27.38	29.65	29.00
	Other industries	59.76	59.43	64.37	63.40

Table 2. The environment emission coefficients of terminal departments (kilogram/ton).

Terminal department	Energy type	CO_2	SO_2	NO_X	Smoke and dust
Agriculture	Coal	2715.18	27.97	2.93	112.05

	Oil	2126.27	5.83	2.93	2.00
	Natural gas	1634.81	0.00	1.47	0.00
Industry	Coal	2715.18	27.97	8.79	112.05
	Oil	2126.27	5.83	5.86	2.00
	Natural gas	1634.81	0.00	4.40	0.00
Wholesale and retail trade	Coal	2715.18	27.97	2.93	112.05
	Oil	2126.27	5.83	2.93	2.00
	Natural gas	1634.81	0.00	1.47	0.00
Transportation department	Coal	2147.53	5.41	23.45	2.00
	Oil	2009.32	0.58	17.58	2.00
	Natural gas	1634.81	0.00	17.58	0.00
Life	Coal	2715.18	27.97	2.93	2.00
	Oil	2126.27	5.83	2.93	2.00
	Natural gas	1634.81	0.00	1.47	0.00

Table 3. The upper bound of terminal energy consumption amount from 2015 to 2030.

	2015	2020	2025	2030
Coal (million tons)	10968.22	9824.22	8600.00	7500.00
Oil (million tons)	9000.00	10000.00	10000.00	8000.00
Natural gas (billion m^3)	180.00	250.00	300.00	380.00
Electricity (billion kWh)	6000.00	6500.00	7000.00	9000.00
Heat (million kJ)	83000.00	120000.00	140000.00	170000.00

4. Results analysis

The proposed EEIEO model could generate the optimized results for economic development, industry adjustment, energy consumption and air pollutants emissions in 2015, 2020, 2025 and 2030. In detail, it is as follows:

4.1. Economic development

Figure 3 shows the gross economic product (GDP) and economic structure of BTH region from 2015 to 2030. As results, the GDP of BTH region would obviously increase over the planning horizon, being 7080.00 billion RMB¥ in 2015, 8474.00 billion RMB¥ in 2020, 10505.00 billion RMB¥ in 2025, and 12307.00 billion RMB¥ in 2030. However, the economic growth rate would gradually slow down over the planning horizon, which would decrease from 20.00% (2015 to 2020) to 14.00% (2025 to 2030). From Figure 3, there would be a shift in BTH region's economy structure towards the tertiary industry. The proportion of tertiary industry would increase from 54.00% in 2015 to 65.00% in 2030. The proportion of secondary industry would decrease from 40.00% in 2015 to 31.00% in 2030.

IOP Conf. Series: Earth and Environmental Science **40** (2016) 012005 doi:10.1088/1755-1315/40/1/012005

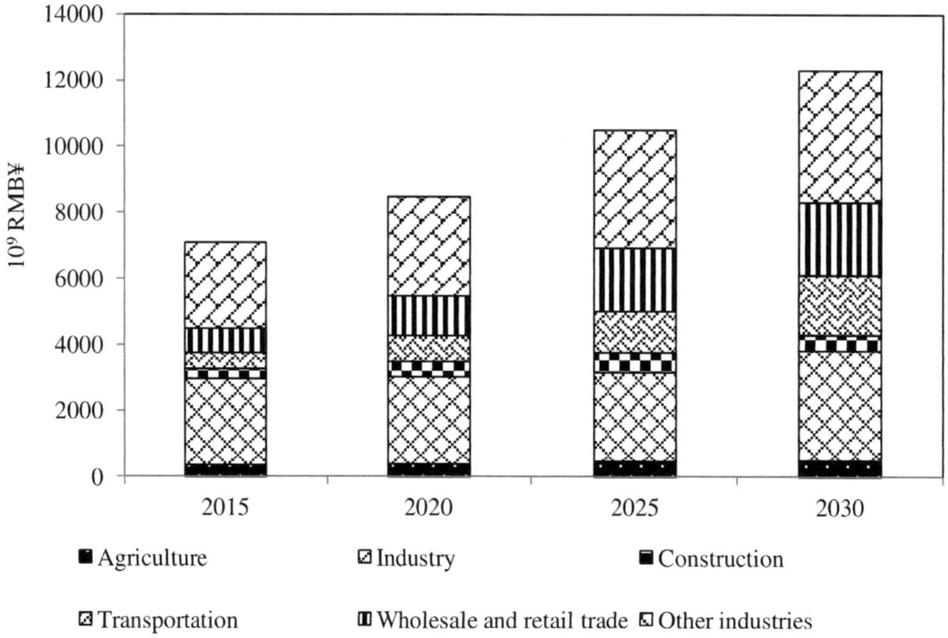

Figure 3. Solutions of the gross economic product (GDP) of BTH region from 2015 to 2030.

4.2. Population and urbanization

Figure 4 shows the population change of BTH region from 2015 to 2030. The population of BTH region would gradually increase from 111.00 million in 2015 from 125.00 million in 2030. The urban population would show an obviously increase trend. The urban population would increase from 69.60 million in 2015 to 88.15 million in 2030, with a growth rate of 32.00%. While, the rural population would appear a negative growth trend, being 44.40, 39.80, 38.40 and 36.80 million in 2015, 2020, 2025 and 2030, respectively. In the next few years, BTH region experienced rapidly urbanization process. According to results, the urbanization of BTH region would reach to 60.00% in 2015, 66.00% in 2020, 68.00% in 2025, and 71.00% in 2030.

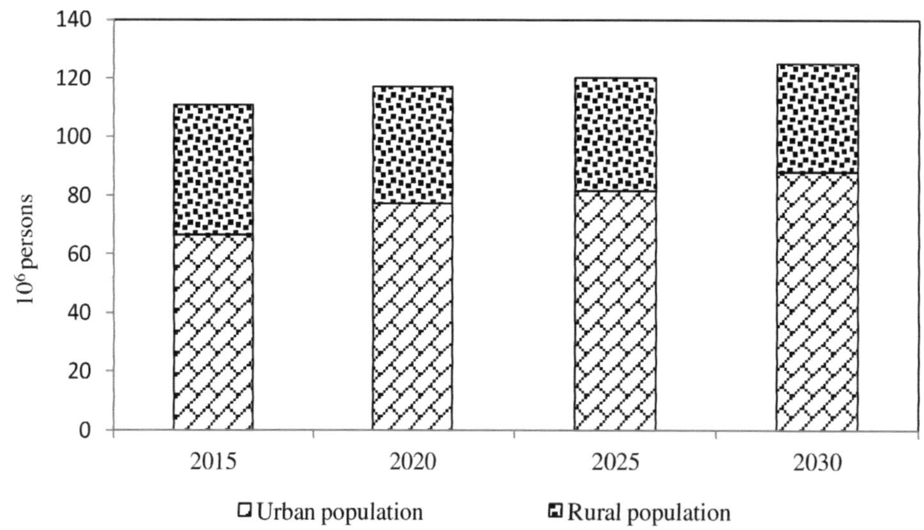

Figure 4. Solutions of the population of BTH region from 2015 to 2030.

4.3. Terminal energy consumption

The totally terminal energy consumption amount and terminal energy consumption structure are presented in Figure 5. Figure 5 shows that the terminal energy consumption amount would show an increased trend, being 368.87 million tons of coal equivalent (tce) in 2015, 409.55 million tce in 2020, 437.73 million tce in 2025, 447.19 million tce in 2030. The growth rate of energy consumption amount would show a sharply decrease trend, being 11.00% from 2015 to 2020, 7.00% from 2020 to 2025, 2.00% from 2025 to 2030, respectively. As results, the terminal energy consumption structure would obviously improve in planning periods. The terminal coal consumption amount would experience negative growth in planning periods, which would decrease from 74.88 million tce in 2015 to 47.60 million tce in 2030. And the proportion of terminal coal consumption amount would gradually decrease from 20.30% in 2015 to 10.60% in 2030. While, some relatively clean energy resources would be vigorously encouraged over the planning horizon. For example, the terminal consumption amount of natural gas would increase from 20.91 million tce to 41.35million tce, which a growth rate of 97.69%. The proportion of terminal natural gas consumption amount would increase from 5.67% in 2015 to 9.25% in 2030. The terminal electricity consumption amount would show an increased trend, being 183.52 million tce in 2015, 212.32 million tce in 2020, 227.16 million tce in 2025, 242.62 million tce in 2030, respectively. In conclusion, the terminal energy consumption structure would obviously improve because of the shift in BTH region's economy structure towards the tertiary industry.

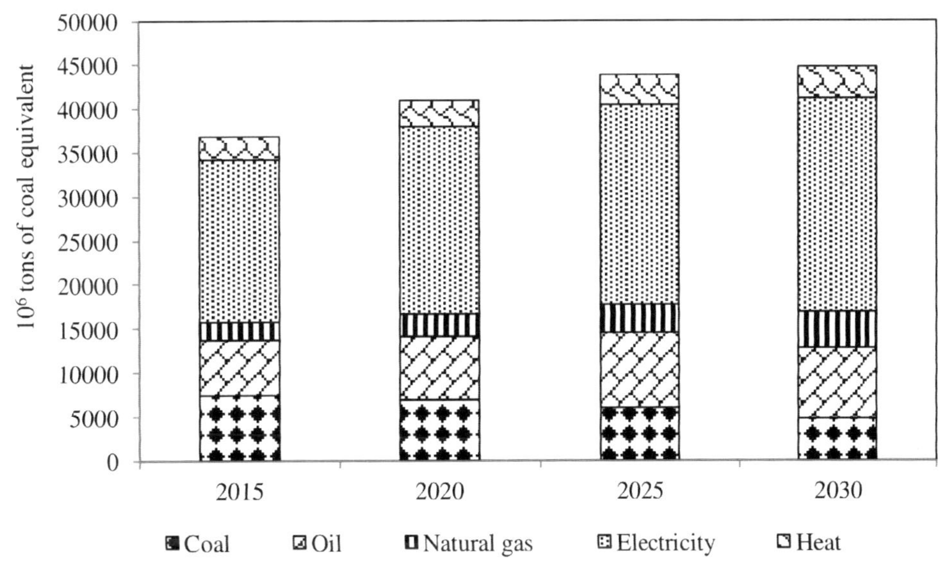

Figure 5. Solutions of the terminal energy consumption amount from 2015 to 2030.

4.4. Pollutant and carbon dioxide emissions

Figure 6 presents the results of pollutants (i.e., SO_2, NO_X, smoke and dust) and carbon dioxide (CO_2) emissions from terminal energy consumption. In Figure 6, the SO_2 emissions from terminal energy consumption would show a decrease trend during the planning horizon, being 840.12×10^3, 801.53×10^3, 724.82×10^3, and 586.19×10^3 tons in 2015, 2020, 2025 and 2030, with a decrease rate of 20.01% during the planning periods. The results indicate the coal would be the major contributor for SO_2 emissions. For example, coal consumption would generate 87.00% of total SO_2 emissions in

2015. While, the NO_X emission from terminal energy consumption would obviously increase from 458.20×10^3 tons in 2015 to 510.63×10^3 tons in 2030, which mainly because of the sharply increase of the terminal oil and natural gas consumption amount. The oil's contribution rate for NOx emissions would increase from 48.80% in 2015 to 67.70% in 2030. The smoke and dust emissions would decrease during planning periods, with a decrease rate of 35.30%. The results indicated that CO_2 emission would increase from 350.79 million tons in 2015 to 371.18 million tons in 2025. While, it would decrease about 9% from 2025 to 2030, which because the negative growth of oil consumption from 2025 to 2030. In the further 15 years, oil would become the major contributor for CO_2 emissions, which accounts for about 50% of total CO_2 emissions till 2030.

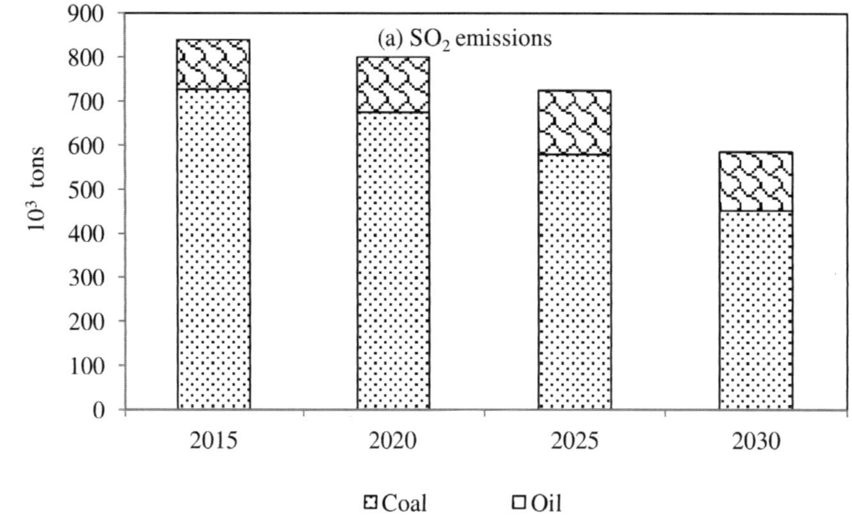

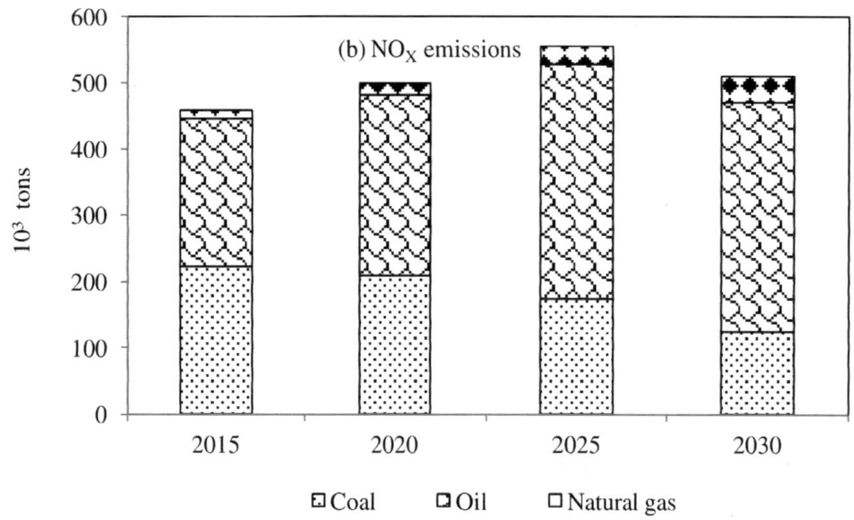

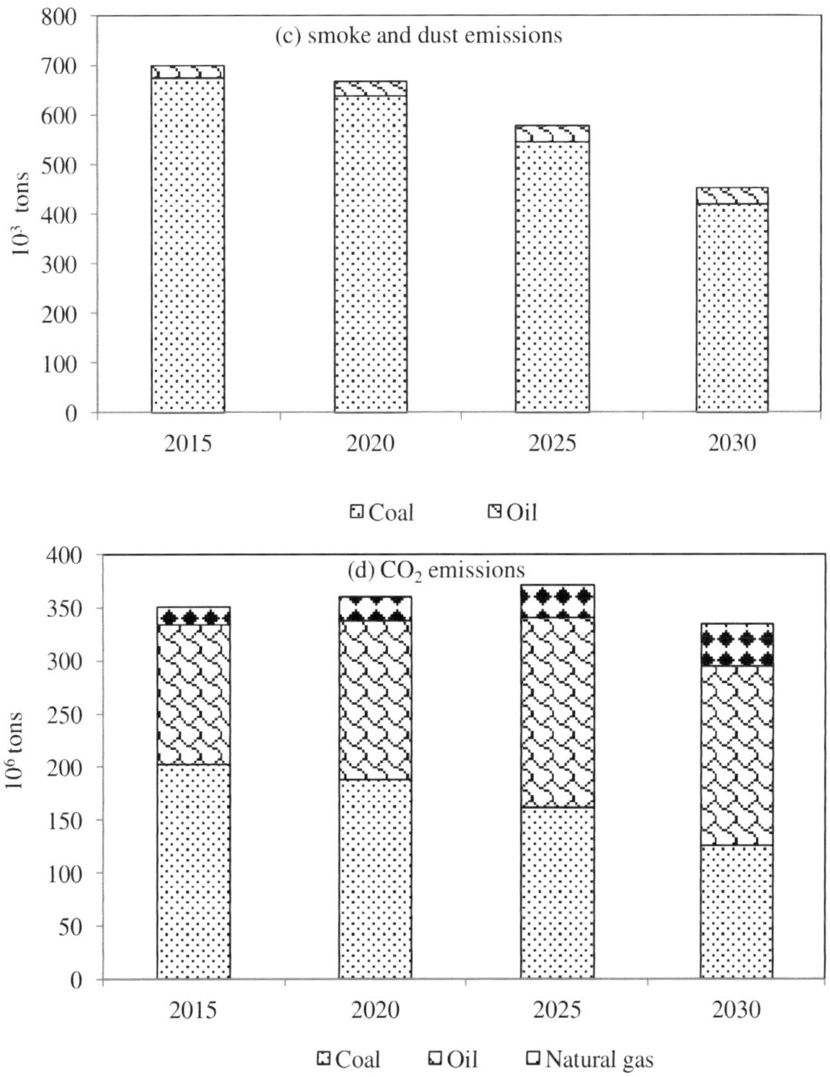

Figure 6. Solutions of the pollutants (i.e., SO_2, NO_X, smoke and dust) and carbon dioxide (CO_2) emissions from terminal energy consumption from 2015 to 2030.

5. Discussion

5.1. Verify the results with truth value

In this paper, in order to verify whether the results of proposed EIEEO model are reliable or not, the optimized schemes about economic (GDP, the added value of first industrial, the added value of secondary industry, the added value of tertiary industry), society (population, urban populations, rural population, urbanization) and energy (the energy consumption amount, coal consumption amount, oil consumption amount, natural gas consumption amount, electricity consumption amount, heat consumption amount) aspects would be compared with truth value in 2014 and 2015. Table 4 shows the optimized value from EIEEO model, truth value and error value. As table 4 shows, the error value is lower than 20.00%, except the natural gas consumption amount. This is mainly because EIEEO model would encourage faster speed for energy structure adjustment. The verification results show the proposed model is efficient for generating the optimized plans for economic development. In detail, it is as follows:

Table 4. The optimized value from EIEEO model, truth value and error value.

	Truth value	Optimized value	Error value
Society aspect			
Population (2015)	11142.37	11500.00	3.21%
Urban population (2014)	6748.00	7245.00	7.37%
Rural population (2014)	4304.00	4255.00	-1.14%
Urbanization (2014)	60.56%	63.00%	4.03%
Economic aspect			
GDP (in 2015, 10^9 RMB¥)	69312.89	70806.10	2.15%
The added value of first industry (10^9 RMB¥)	3790.11	3802.07	0.32%
The proportion of first industry added value	5.47%	5.37%	-1.80%
The added value of secondary industry (10^9 RMB¥)	26638.00	28768.73	8.00%
Industry (10^9 RMB¥)	23270.37	25766.66	10.73%
Construction (10^9 RMB¥)	3490.23	3002.07	-13.99%
The proportion of secondary industry added value	38.43%	40.63%	5.72%
The added value of tertiary industry (10^9 RMB¥)	38884.78	38235.29	-1.67%
The proportion of tertiary industry added value	56.10%	54.00%	-3.74%
Energy aspect (2014)			
Energy consumption amount (10^6 tons of coal equivalent)	34067.00	37093.04	8.88%
Coal consumption amount (10^6 tons of coal equivalent)	8090.80	7496.66	-7.34%
Oil consumption amount (10^6 tons of coal equivalent)	6321.44	6265.00	-0.89%
Natural gas consumption amount (10^6 tons of coal equivalent)	1698.93	2122.90	24.96%
Electricity consumption amount (10^6 tons of coal equivalent)	15560.69	18402.37	18.26%
Heat consumption amount (10^6 tons of coal equivalent)	2395.14	2806.11	17.16%

Note: the truth value for economic and society are acquired from [17-20]. The Energy Statistical Yearbook in 2016 is not publicly available, so the truth value for energy consumption in 2015 could not be obtained.

5.2. Compared the results with correlation research

In this paper, in order to further verify whether the results of proposed EIEEO model, some related study results were discussed in this paper. Several researchers have made efforts to predict the GDP, industry structure and population of BTH region with growth rate, regression analysis and grey predict methods. However, these studies could not fully consider the interaction between economic, industry, energy consumption and environment. For example, Li [21] used the growth rate and regression methods to forecast the proportion of firstly industry, secondary industry and tertiary industry. According to results, the proportion of firstly industry, secondary industry and tertiary industry would be reached to 3.45:55.70:40.80 in 2015; 3.00:54.30:42.70 in 2020, which has a deviation with the actual value (5.47%: 38.43%: 56.10% in 2015). Lin [22] predicted the urbanization rate of BTJ region with the scenario analysis method, which results indicated the urbanization rate would reach to 66.00-68.00% in 2015, 76.00-79.00% in 2020, 83.00-84.00% in 2025, and 86.00-90.00% in 2030, respectively. Yang [23] predicted the urbanization rate and proportion of tertiary industry would reach to 60.00% and 70.00% with fully consideration the "new normal" for Chinese economic. In conclusion, the results from EIEEO model are reasonable and reliable, which could propose the decision supports for local managers.

Table 5. The comparison between optimized value and research results from other researchers.

Author	Institution	Index	Model	Scenario	2015	2020	2025	2030
This paper	Institute of Geographic sciences and natural resources research	GDP	EIEEO model	—	7080.61	8474.58	10505.58	12307.70
		Industry structure		—	5.37:40:63:54.00	4.72:36.28:59.00	4.57:31.08:64.35	4.11:30.8:65.00
		Urbanization		—	60.00%	66.00%	68.00%	70.52%
		Population (10^6 persons)		—	111.00	117.16	120.00	125.00
Li M	Tianjin University	Population (10^6 persons)	Gray model	—	108.36	116.07	—	—
		Industry structure	Formulate growth rate; regression analysis	—	3.45:55.7:40.8	3.0:54.3:42.7	—	—
Lin W	China University of Geosciences	Urbanization	Scenario analysis	Low scenario	66.00%	76.00%	83.20%	86.00%
				Medium scenario	67.00%	77.50%	83.40%	87.00%
				Medium-low scenario	67.00%	78.00%	83.80%	88.00%
				High scenario	68.00%	78.50%	84.00%	90.00%
Yang KZ	Peking University	Tertiary industry	Formulate growth rate	—	—	60.00%	—	—
		Urbanization		—	—	70.00%	—	—

Acknowledgments
This research was supported by the "China Postdoctoral Science Foundation (2015M570140)", and "Key Projects of National Natural Science Foundation of China (41130748)". The authors are grateful to the editors and the anonymous reviewers for their insightful comments and suggestions.

A. 1 Nomenclatures for parameters and variables

$GDP_t^{\pm}$ Gross National Product in period t (10^9 RMB¥);

i Type of the secondary industry, $i = 1, 2$; $i = 1$ for industry, $i = 2$ for construction industry;

j Type of tertiary industry, j = 1, 2, 3; $j = 1$ for transportation department, $j = 2$ for wholesale and retail trade, $j = 3$ for other industries;

m Type of energy resource for terminal sector, $m = 1, 2, 3, 4, 5$; $m = 1$ for coal, $m = 2$ for oil, $m = 3$ for natural gas, $m = 4$ for electricity, $m = 5$ for heat;

t Time period, $t = 1, 2, 3, 4$; $t = 1$ for 2015, $t = 2$ for 2020; $t = 3$ for 2025; $t = 4$ for 2030;

A. 2 Model parameters

ET_{mt} The upper bound of energy resource m in period t (10^6 tons of coal equivalent);

ETX_{mt} The lower bound of energy resource m in period t (10^6 tons of coal equivalent);

EDY_{mt} Energy consumption intensity of energy type m for agriculture in period t (ton/10^6 RMB¥);

EDE_{imt} Energy consumption intensity of energy type m for second industry i in period t (ton/10^6 RMB¥);

$EDS_{jmt}^{\pm}$ Energy consumption intensity of energy type m for tertiary industry j in period t (ton/10^6 RMB¥);

EDU_{mt} Energy consumption intensity of energy type m for urban residents in period t (ton/person);

EDR_{mt} Energy consumption intensity of energy type m for rural residents in period t (ton/person);

YCL_t The lower bound proportion of agriculture in period t (%);

YCS_t The upper bound proportion of agriculture in period t (%);

ECL_t The lower bound proportion of second industry in period t (%);

ECS_t The upper bound proportion of second industry in period t (%);

SCL_t The lower bound proportion of tertiary industry in period t (%);

SCS_t The upper bound proportion of tertiary industry in period t (%);

RL_t The upper bound of population in period t;

RLX_t The lower bound of population in period t;

$CZHL_t$ The lower bound of urbanization in period t;

$CZHS_t$ The upper bound of urbanization in period t;

SPY_{mt} The SO_2 generation efficiency for energy type i of agriculture in period t (kilogram/ton);

SPE_{imt} The SO_2 generation efficiency for energy type i of second industry in period t (kilogram/ton);

SPS_{imt} The SO_2 generation efficiency for energy type i of tertiary industry in period t (kilogram/ton);

SPU_{mt} The SO_2 generation efficiency for energy type i of urban residents in period t (kilogram/ton);

SPR_{mt} The SO_2 generation efficiency for energy type i of rural residents in period t (kilogram/ton);

SPT_t Allowable emission level for SO_2 in period t (10^3 tons);

NPY_{mt} The NO_X generation efficiency for energy type i of agriculture in period t (kilogram/ton);

NPE_{imt} The NO_X generation efficiency for energy type i of second industry in period t (ton/10^3 tons);

NPS_{imt} The NO_X generation efficiency for energy type i of tertiary industry in period t (ton/10^3 tons);

NPU_{mt} The NO_X generation efficiency for energy type i of urban residents in period t (ton/person);

NPR_{mt} The NO_X generation efficiency for energy type i of rural residents in period t (ton/person);

NPT_t Allowable emission level for NO_X in period t (10^3 tons);

$SDPY_{mt}$ The smoke and dust generation efficiency for energy type i of agriculture in period t (kilogram/ton);

$SDPE_{imt}$ The smoke and dust generation efficiency for energy type i of second industry in period t (kilogram/ton);

$SDPS_{jmt}$ The smoke and dust generation efficiency for energy type i of second industry in period t (kilogram/ton);

$SDPU_{mt}$ The smoke and dust generation efficiency for energy type i of urban residents in period t (ton/person);

$SDPR_{mt}$ The smoke and dust generation efficiency for energy type i of rural residents in period t (ton/person);

$SDPT_t$ Allowable emission level for smoke and dust in period t (10^3 tons);

CPY_{mt} The CO_2 generation efficiency for energy type i of agriculture in period t (kilogram/ton);

CPE_{imt} The CO_2 generation efficiency for energy type i of second industry in period t (kilogram/ton);

CPS_{imt} The CO_2 generation efficiency for energy type i of second industry in period t (kilogram/ton);

CPU_{mt} The CO_2 generation efficiency for energy type i of urban residents in period t (kilogram/ton);

CPR_{mt} The CO_2 generation efficiency for energy type i of rural residents in period t (kilogram/ton);

A. 3 Decision variables

XY_t Total output value of the agriculture in period t (10^9 RMB¥);

XE_t Total output value of the second industry in period t (10^9 RMB¥);

XS_{jt} Total output value of the tertiary industry in period t (10^9 RMB¥);

XU_t The population of urban residents in period t (person);

XR_t The population of rural residents in period t (person);

$CZHR_t$ The urbanization rate in period t (%);

SEM_t The SO_2 emission amount in period t (10^3 tons);

NEM_t The NOX emission amount in period t (10^3 tons);

$SDEM_t$ The smoke and dust emission amount in period t (10^3 tons);

CEM_t The CO_2 emission amount in period t (10^3 tons);

References

[1] Zheng Y H and Luo D K 2013 Industrial structure and oil consumption growth path of China Empirical evidence *Energy* **57** 336-43

[2] China Statistical Bureau 2013 Chinese Statistical Yearbook. (Beijing: China Statistics Press)

[3] Wang Z H and Yang L 2015 Delinking indicators on regional industry development and carbon emissions Beijing-Tianjin-Hebei economic band case *Ecological Indicators* **48** 41-8

[4] Lang J L, Cheng S Y, Wei W, Zhou Y, Wei X and Chen D S 2012 A study on the trends of vehicular emissions in the Beijing-Tianjin-Hebei (BTH) region China *Atmospheric environment* **62** 605-14

[5] Sun L and Chen W 2013 The improved China CCS decision support system: A case study for Beijing–Tianjin–Hebei region of China *Applied Energy* **112** 793-9

[6] Tian X, Chang M, Shi F and Tanikawa H 2014 How does industrial structure change impact carbon dioxide emissions? A comparative analysis focusing on nine provincial regions in China *Environmental science & Policy* **37** 243-54

[7] Chen F, Huang G H and Fan Y R 2015 A linearization and parameterization approach to tri-objective linear programming problems for power generation expansion planning *Energy* **87** 240-50

[8] Chen Z M, Liu Y, Qin P, Zhang B, Lester L, Chen G H, Guo Y and Zheng X Y 2015 Environmenta externality of coal use in China: welfare effect and tax regulation *Applied Energy* **156** 16-31

[9] Wu D, Xu Y and Zhang S Q 2015 Will joint regional air pollution control be more cost-effective? An empirical study of China's Beijing-Tianjin-Hebei region *J. Environmental Management* **149** 27-36

[10] Zhang S H, Worrell E and Graus W C 2015 Mapping and modeling multiple benefits of energy efficiency and emission mitigation in China's cement industry at the provincial level *Applied Energy* **155** 35-8

[11] Zhang Y, Zheng H M, Yang Z F, Li Y X, Liu G Y, Su M R and Yin X 2016 Urban energy flow processes in the Beijing-Tianjin-Hebei (Jing-Jin-Ji)urban agglomeration: combining multi-reginal input-output tables with ecological network analysis *J. Cleaner Production* **114** 243-56

[12] Beijing Statistical Bureau 2014 Beijing Statistical Yearbook. (Beijing:China Statistics Press)

[13] Tianjin Statistical Bureau 2014 Tianjin Statistical Yearbook. (Tianjin: China Statistics Press)

[14] Hebei Statistical Bureau 2014 Hebei Statistical Yearbook. (Shijiazhuang: China Statistics Press)

[15] Chen C, Li Y P and Huang G H 2016 Interval-fuzzy municipal-scale energy model for identification of optimal strategies for energy management – A case study of Tianjin, China *Renewable Energy* **86** 1161-77

[16] Chinese Ministry of Environment Protection 2013 Chinese Environmental Quality Bulletin. (Beijing: Chinese Environmental Press)

[17] Beijing Statistical Bureau 2015 Beijing Statistical Bulletin for National Economic and Social Development. (Beijing: China Statistics Press)

[18] Tianjin Statistical Bureau 2015 Tianjin Statistical Bulletin for National Economic and Social Development. (Tianjin: China Statistics Press)

[19] Hebei Statistical Bureau 2015 Hebei Statistical Bulletin for National Economic and Social Development. (Shijiazhuang: China Statistics Press)

[20] China Statistical Bureau 2015 China Energy Statistical Yearbook. (China Statistics Press: Beijing)

[21] Li M 2015 Research on the development of regional econonical intergration of Beijing Tianjin and Hebei (Tianjin: Tianjin University)

[22] Lin W 2015 The influence of urbanization on the carying capacity of land resources of Beijing,

Tianjin and Hebei (Beijing: China University of Geosciences)
[23] Yang KZ 2014 http://news.xinhuanet.com/house/bj/2014-09-01/c_1112311301.htm

IOP Conf. Series: Earth and Environmental Science **40** (2016) 012006 doi:10.1088/1755-1315/40/1/012006

A mathematical procedure to predict optical performance of CPCs

Y M Yu, M J Yu and R S Tang [1]

Education Ministry Key Laboratory of Advanced Technology and Preparation for Renewable Energy Materials, Yunnan Normal University, Kunming 650500, PR China

Email: kingtang01@126.com

Abstract. To evaluate the optical performance of a CPC based concentrating photovoltaic system, it is essential to find the angular dependence of optical efficiency of compound parabolic concentrator (CPC-θ_e) where the incident angle of solar rays on solar cells is restricted within θ_e for the radiation over its acceptance angle. In this work, a mathematical procedure was developed to calculate the optical efficiency of CPC-θ_e for radiation incident at any angle based radiation transfer within CPC-θ_e. Calculations show that, given the acceptance half-angle (θ_a), the annual radiation of full CPC-θ_e increases with the increase of θ_e and the CPC without restriction of exit angle (CPC-90) annually collects the most radiation due to large geometry (C_t); whereas for truncated CPCs with identical θ_a and C_t, the annual radiation collected by CPC-θ_e is almost identical to that by CPC-90, even slightly higher. Calculations also indicate that the annual radiation on the absorber of CPC-θ_e at the angle larger than θ_e decrease with the increase of θ_e but always less than that of CPC-90, and this implies that the CPC-θ_e based PV system is more efficient than CPC-90 based PV system because the radiation on solar cells incident at large angle is poorly converted into electricity.

1. Introduction

[1] Address for correspondence: Education Ministry Key Laboratory of Advanced Technology and Preparation for Renewable Energy Materials, Yunnan Normal University, Kunming 650500, PR China. Email: kingtang01@126.com.

In recent years, solar photovoltaic (PV) have been widely used for electrical generation over the world due to emerging issues of conventional energy shortage [1]. However, the application of PV system is limited due to high cost of electricity generation as compared to the conventional electrical generation technologies [2]. Therefore, to lower the cost of electricity generating from PV systems is a unique solution to expanding applications of PV technology. Apart from seeking for new materials and production techniques of solar cells, the use of cheep optical concentrator is regarded as an effective way to cost reduction of electricity generation from PV systems.

Concentrating PV systems (CPV) are generally classified into high concentrating PV system (HCPV) and low concentrating systems (LCPV). CPVs seem simple in the mechanism but are difficult to implement, especially HCPVs, which require expensive and complex sun-tracking device, cooling technique of solar cells and specially designed solar cells [3, 4]. Thus LCPV is more attractive due to no need of sun-tracking system. In the past two decades, many theoretical and experimental studies were performed to employ compound parabolic concentrators (CPC) for concentrating solar radiation on solar cells [5-9]. Compared to CPCs, the V-trough concentrator shares the advantages of easy construction and more uniform solar irradiation on the base, but the increase in power output from V-trough based CPVs is limited [10-12]. To make the irradiation on solar cells of CPC based CPV uniform, Hatwaambo tested a CPV system where semi-diffuse reflective materials were used, and found that the increase of the fill factor of the photovoltaic system was insignificant [13, 14]. However, recent study by Yu and Tang found that the use of semi-diffuse reflectors in CPVs would lead the collectible radiation decrease greatly[15]. For a CPV system, the incident angle of solar rays reflecting from the lower part of reflectors is considerably large, thus can not be efficiently converted into electrical power due to poor solar absorption [16,17]. To increase the absorption of solar radiation by solar cells of CPC based PV system, CPCs with a restricted exit angle (CPC-θ_e) was first proposed [5,18], but it is in recent years that several researches on its performance are found in the literature [17,19,20,21]. For such CPC, all solar rays over its acceptance angle (θ_a) arrive on the absorber at the angle less than the desired value θ_e, whereas for solar radiation beyond its acceptance angle, a fraction of incident radiation will arrive on the receiver at the angle larger than θ_e, but it was not considered in previous works of the authors [20,21], leading its performance underestimated. In this work, a general mathematical procedure to determine the angular dependence of optical efficiency of CPC-θ_e with any θ_e was developed, and effects of θ_e on its performance was theoretical investigated in terms of annual collectible radiation.

2. Mathematical procedure to calculate the optical efficiency of CPC-θ_e

2.1. Equation of reflectors

As seen from figure 1, the reflectors of CPC-θ_e consist of upper parabola and lower plane mirror, and the plane mirror is tangent to the lower end of parabolic reflector. To be convenient analysis, the width of the absorber is set to be 1, thus, parabolic reflector can be expressed by:

$$\begin{cases} x = \dfrac{(\sin\theta_e + \sin\theta_a)\sin\theta}{1-\cos(\theta+\theta_a)} - 0.5 \\ y = \dfrac{(\sin\theta_e + \sin\theta_a)\cos\theta}{1-\cos(\theta+\theta_a)} \end{cases} \quad (\theta_l \le \theta \le \theta_e) \tag{1}$$

where θ is the polar angle at the absorber end; θ_e is the maximum exit angle for radiation within its acceptance angle; and θ_t is the edge-ray angle of CPC- θ_e . For full CPC- θ_e , $\theta_t = \theta_a$ and $C_t = \sin\theta_e / \sin\theta_a$; whereas for truncated CPCs with a given C_t, θ_a and θ_e, the θ_t can be calculated based on equation (1) [20]. Obviously, CPC-90, the CPC without restriction of exit angle, is a special case of CPC-θ_e for the case of $\theta_e = 90^\circ$.

It should be noted that the maximum θ_t should be less θ_e otherwise CPC-θ_e is reduced into a V-trough concentrator. The coordinate of lower end *D* of the parabola can be determined based on equation (1) by setting $\theta = \theta_e$ as follows:

$$\begin{cases} x_D = \dfrac{(\sin\theta_e + \sin\theta_a)\sin\theta_e}{1-\cos(\theta_e+\theta_a)} - 0.5 \\ y_D = \dfrac{(\sin\theta_e + \sin\theta_a)\cos\theta_e}{1-\cos(\theta_e+\theta_a)} \end{cases} \tag{2}$$

The plane mirror is expressed by:

$$y = c\tan\gamma_D(x-0.5) \qquad (0.5 \le x \le x_D) \tag{3}$$

The tilt angle of plane mirrors (γ_D) relative to y-axis is given by:

$$\gamma_D = 0.5(\theta_e - \theta_a) \tag{4}$$

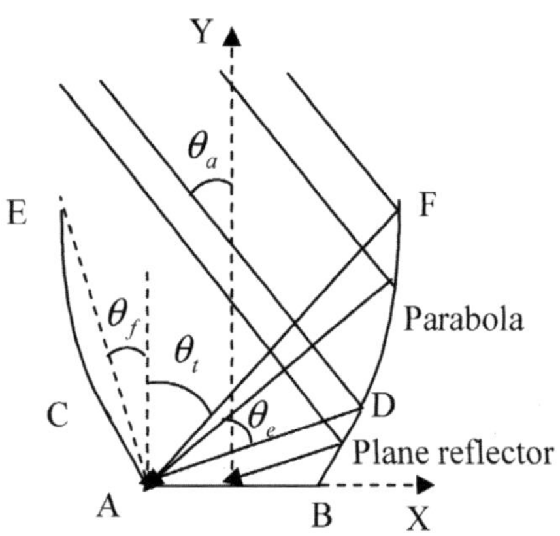

Figure1. Geometry of CPC-θ_e

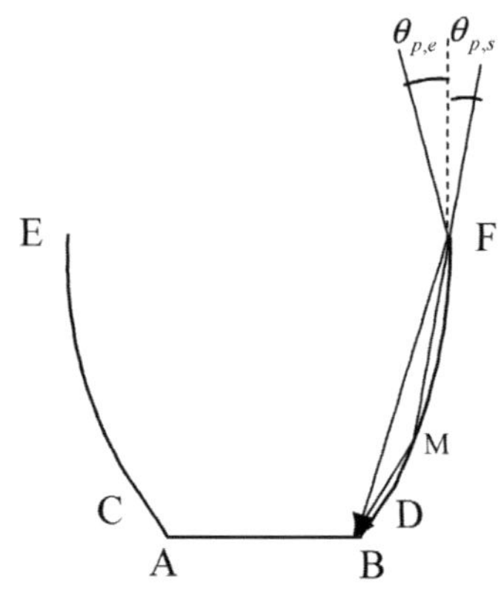

Figure 2. Angle range of solar rays that arrive on the absorber after more than two reflections.

2.2. Optical efficiency of CPC-θ_e

To calculate the collectible radiation of CPC-θ_e, it is essential to find the angular dependence of optical efficiency. For CPC-θ_e, all of radiation over its acceptance angle will arrive on the absorber at the incident angle (θ_{in}) less than θ_e as shown in figure 1; whereas for radiation incident at $\theta_p > \theta_a$, part of radiation incident on the plane reflector (see figure 2) and upper parabolic reflector (see figure 3) will arrive on the absorber at $\theta_{in} > \theta_e$. Thus the collectible radiation on the receiver includes the radiation incident at $\theta_{in} \le \theta_e$ (I_1), radiation reflecting from the plane reflector (I_2) at $\theta_{in} > \theta_e$ and that reflecting from the upper parabolic reflector to the opposite plane mirror first and then reflecting onto the absorber (I_3) at $\theta_{in} > \theta_e$. Therefore, the optical efficiency of CPC-θ_e is given by:

$$\eta(\theta_p) = \frac{I_1 + I_2 + I_3}{I_{ap}} = f_1 + f_2 + f_3 \tag{5}$$

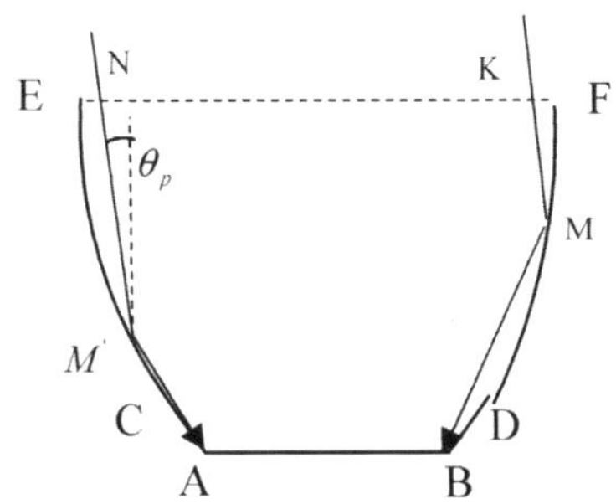

Figure 3. Fraction of radiation that arrive on the absorber after more than two reflections.

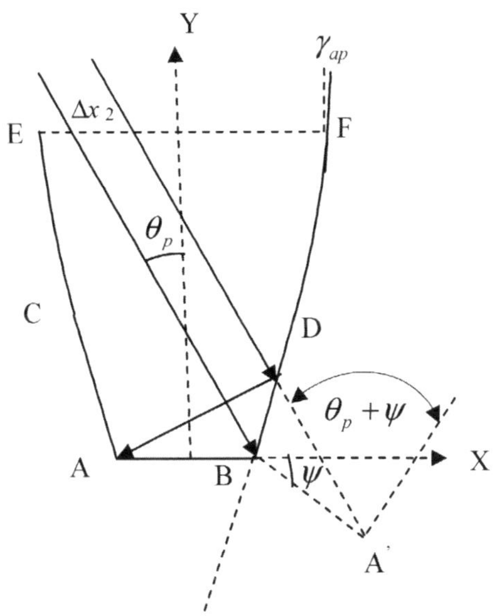

Figure 4. Transfer of radiation incident on the plane the mirror of CPC-θ_e at $\theta_a < \theta_p < \theta_t$

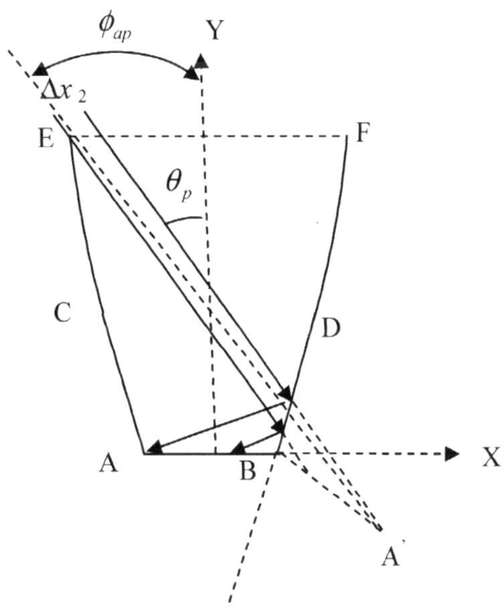

Figure 5. Transfer of radiation incident on mirror of CPC-θ_e at $\theta_t < \theta_p < \theta_{p,c1}$

In equation (5), the I_{ap} is the radiation incident on the aperture of CPC-θ_e, f_1 is the optical efficiency contributed by radiation arriving on the absorber at $\theta_{in} \le \theta_e$, f_2 is the efficiency contributed by radiation on the absorber after reflection from the plane reflectors at $\theta_{in} > \theta_e$, and f_3 is that contributed by radiation on the absorber after reflecting from the upper parabolic reflector first and then from its opposite plane reflector at $\theta_{in} > \theta_e$.

2.2.1. Calculation expressions of f_1. For radiation incident at small θ_p, partial radiation undergoes multi-reflection before arriving on the absorber [22]. To make calculations accurate, the two-reflection model, in which the radiation arriving on the absorber after more than two reflections is regarded as that arrives on the absorber after just two reflections, is employed, thus f_1 can be expressed by:

$$f_1 = f_0 + (f_{11} - f_{12})\rho + f_{12}\rho^2 \tag{6}$$

where f_0 is the fraction of radiation directly irradiating on the absorber; f_{11} and f_{12} stand for the fraction of radiation arriving on the absorber after more than one and two reflections, respectively. equation (6) also can be rewritten as:

$$f_1 = f_0 + f_{11}\rho - f_{12}\rho(1-\rho) = \eta_1 - f_{12}\rho(1-\rho) \tag{7}$$

where $\eta_1 = f_0 + f_{11}\rho = f_0 + (1-f_0)\rho$ is the f_1 of CPCs estimated based on the one-reflection model [20,22], and calculated by:

$$\eta_1 = \begin{cases} \rho + (1-\rho)/C_t & \theta_p \le \theta_f \\ \rho + 0.5(1-\rho)(1+C_t)(1-\tan\theta_p/\tan\theta_t)/C_t & \theta_f < \theta_p \le \theta_a \\ 0.5(1+C_t)(1-\tan\theta_p/\tan\theta_t)/C_t & \theta_a < \theta_p \le \theta_t \\ 0 & \theta_p > \theta_t \end{cases} \tag{8}$$

In the case of $\theta_f > \theta_a$, it is given by

$$\eta_1 = \begin{cases} \rho + (1-\rho)/C_t & \theta_p \le \theta_a \\ 1/C_t + 0.5\rho(1-1/C_t) - 0.5\rho(1+1/C_t)\tan\theta_p/\tan\theta_{t,2} & \theta_a < \theta_p \le \theta_f \\ 0.5(1+C_t)(1-\tan\theta_p/\tan\theta_t)/C_t & \theta_f < \theta_p \le \theta_t \\ 0 & \theta_p > \theta_t \end{cases} \tag{9}$$

where θ_f (see figure 1) is given by:

$$\tan\theta_f = \tan\theta_t(C_t - 1)/(C_t + 1) \tag{10}$$

As shown in figure 2, solar rays incident on the (right) reflector at $\theta_{p,e}$ and $-\theta_{p,s}$ just strike at the end (B) of the absorber after one reflection, thus solar rays incident at $-\theta_{p,s} < \theta_p < \theta_{p,e}$ will undergo more than two reflections before arriving on the absorber (see figure 3). Based on the reflection law of light, one has:

$$\theta_{p,e} = \theta_f - 2\gamma_{ap} \tag{11}$$

The γ_{ap} is the tilt-angle of the line tangent to the upper end of the parabolic reflector relative to y-axis (see figure 4), and can be found based on equation (1) as follow:

$$\tan\gamma_{ap} = \left.\frac{dx}{dy}\right|_{\theta=\theta_t} = \frac{\cos\theta_a - \cos\theta_t}{\sin\theta_a + \sin\theta_t} \tag{12}$$

The critical angle $\theta_{p,s}$ is subjected to following equation:

$$\gamma_M = 0.5(\theta_{p,s} + \gamma_{MB}) \tag{13}$$

where γ_M is the tilt-angle of the line tangent to M (see figure 3), γ_{MB} is the tilt-angle of line *MB*. For a given CPC, γ_M, γ_{MB} and $\theta_{p,s}$ as the function of θ_M can be respectively expressed by:

$$\tan\gamma_M = \frac{\cos\theta_a - \cos\theta_M}{\sin\theta_a + \sin\theta_M} \tag{14a}$$

$$\tan\gamma_{MB} = (x_m - 0.5)/y_M \tag{14b}$$

$$\tan\theta_{p,s} = (0.5C_t - x_M)/(h_t - y_M) \tag{14c}$$

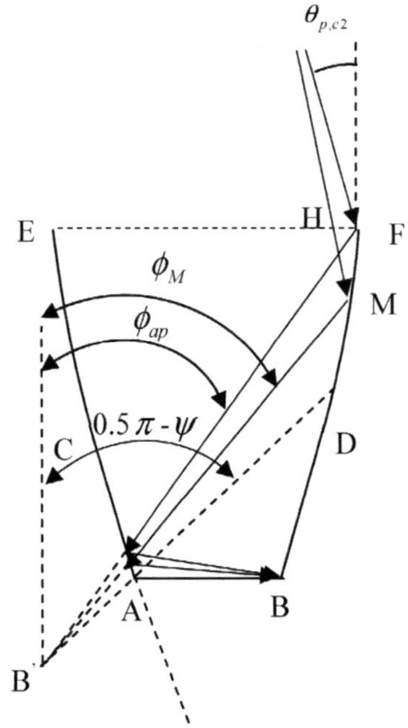

Figure 6. Transfer of radiation incident on the upper parabolic reflector of CPC-θ_e at $\theta_a < \theta_p < \theta_{p,c2}$

By substituting γ_M, γ_{MB} and $\theta_{p,s}$ obtained from equation (14) into equation (13) or by iterative calculations, one obtains θ_M, then $\theta_{p,s}$ is obtained from equation (14c). As shown in figure 3, the fraction of radiation arriving on the absorber after more than two reflections is given by $f_{12} = (KF + ED)/C_t$, and KF is given by:

$$KF = \begin{cases} 0.5C_t - x_M + (h_t - y_M)\tan\theta_p & (-\theta_{p,s} < \theta_p < \theta_{p,e}) \\ 0 & else \end{cases} \quad (15)$$

where x_M and y_M as the function of θ_M are x-, y-coordinates of critical point M, respectively, and θ_M can be calculated based on equation (13), (14a)-(14b) by setting $\theta_{p,s} = -\theta_P$. Similarly, ED can be calculated in the same way as finding KF by setting $\theta_p = -\theta_P$.

2.2.2. Calculation method of f_2. As shown in figure 4, BA' is the first image of the absorber formed by the plane mirror. According to the imaging principle of plane mirrors, rays pointing to the image of the absorber formed by the mirror must arrive on the absorber after reflecting from the mirror [12]. Therefore, when solar rays incident on the plane mirror at $\theta_p > \theta_a$, partial radiation will arrive on the absorber at $\theta_{in}=\theta_p+\psi>\theta_e$ (for radiation incident at $\theta_p=\theta_a$, the $\theta_{in}=\theta_e$, thus the opening angle of V-trough formed by two plane reflectors of the CPC-θ_e is $\psi=2\gamma_D=\theta_e-\theta_a$), the remaining will redirect to the opposite mirror and then escape from CPC-θ_e or arrive on the absorber at $\theta_{in}=(\theta_p+2\psi)>\theta_e$ [12]. Solar radiation arriving on the absorber after two reflections in between two plane mirrors must satisfy:

$$\begin{cases}\theta_p>\theta_a\\ \theta_p+2\psi\le 0.5\pi\end{cases} \tag{16}$$

And this leads:

$$\theta_a>2\theta_e-0.5\pi \tag{17}$$

This means that solar rays entering the V-trough are possible to arrive on the absorber after two reflections for CPC-θ_e with $\theta_a>2\theta_e-0.5\pi$. Thus for CPC-$\theta_e$ with $\theta_e=65^{\circ}$ and 60°, θ_a must be larger than 40° and 30°, respectively. In practical design of CPC-θ_e, θ_e is usually larger than 65° and $\theta_a<35^{\circ}$ [17], and such CPC-θ_e is subjected to $\theta_a<2\theta_e-0.5\pi$. In this work, the CPCs subjected to $\theta_a<2\theta_e-0.5\pi$ are considered for simplifying analysis, thus no radiation arrives on the absorber of CPC-θ_e after more than two reflections in between two mirrors.

To ensure solar rays reflecting from mirrors arrive on the absorber, the θ_p should be subjected to $\psi+\theta_p\le 0.5\pi$, i.e. $\theta_p\le 0.5\pi-\psi$; but in the other hand, θ_p should be less than ϕ_{ap} due to the shade of reflectors on the opposite mirrors (see figure5-6). Therefore, the critical incident angle $\theta_{p,c1}$, solar rays incident at the angle less than which will arrive on the absorber at $\theta_{in}>\theta_e$, should take the smaller one of ϕ_{ap} and $0.5\pi-\psi$, namely:

$$\theta_{p,c1}=Min(\phi,0.5\pi-\psi) \tag{18}$$

As shown in figure 5-6, ϕ_{ap} is the tilt-angle of the line linking the upper end *E* (*F*) of parabolic reflectors and the image of absorber end *A* (*B*), and is calculated by:

$$\tan\phi_{ap}=\frac{0.5+\cos\psi+0.5C_t}{h_t+\sin\psi} \tag{19}$$

Here $h_t=0.5(C_t+1)/\tan\theta_t$ is the height of CPCs. As shown in figure 4, the image (BA') of the absorber is fully irradiated by radiation incident at $\theta_p\le\theta_t$ and partially irradiated as $\theta_t<\theta_p<\theta_{p,c1}$

(see figure 5), therefore, the fraction of radiation on the absorber after reflection from the plane mirrors, Δx_2, is determined by:

$$\Delta x_2 / C_t = \begin{cases} (\cos\psi - \sin\psi \tan\theta_p)/C_t & (\theta_a < \theta_p \le \theta_t) \\ (h_t + \sin\psi)(\tan\varphi_{ap} - \tan\theta_p)/C_t & (\theta_t < \theta_p < \theta_{p,c1}) \\ 0 & \theta_p \ge \theta_{p,c1} or \theta_p \le \theta_a \end{cases} \tag{20}$$

$$f_2 = \rho \Delta x_2 / C_t \tag{21}$$

2.2.3. Calculation method of f_3. As shown in figure 6, solar rays incident on the upper end (F) of the parabolic reflector at $\theta_{p,c2}$ will just redirect to the image B' after reflection, thus, for solar rays entering CPC-θ_e at $\theta_a < \theta_p < \theta_{p,c2}$, the radiation incident on the upper parabola (FM) will redirect to the opposite mirror (AC) and then be reflected onto the absorber, and solar rays incident on lower part of parabola (MD) will finally escape from the cavity of the CPC-θ_e after multiple reflections within the CPC cavity. The polar angle θ_m of the critical point (M) on the parabola is subjected to following equation group:

$$\begin{cases} 2\gamma_m = \phi_m - \theta_p \\ \tan\phi_m = \dfrac{x_m + 0.5 + \cos\psi}{y_m + \sin\psi} \end{cases} \tag{22}$$

where x_m and y_m as the function of θ_m are *x* and *y* coordinates of point *M*, respectively; ϕ_m as the function of θ_m is the angle formed by line MB' and y-axis (see figure 6); γ_m, as the function of θ_m, is the slope of the line tangent to point *M* and determined by equation (12). Thus, given θ_p ($\theta_a < \theta_p < \theta_{p,c2}$), θ_m can be obtained by iterative calculations, then the fraction of solar radiation incident on the upper parabola (*FM*) that arrive on the absorber after reflection from the parabola first then from the opposite mirror, $\Delta x_3 / C_t$, can be calculated by:

$$\Delta x_3 / C_t \begin{cases} 0.5C_t - x_m + (h_t - y_m)\tan\theta_p & (\theta_a < \theta_p < \theta_{p,c2}) \\ 0 & else \end{cases} \tag{23}$$

$$f_3 = \rho^2 (\Delta x_3 / C_t) \tag{24}$$

According to the law of light reflection, the critical incident angle $\theta_{p,c2}$ (see figure 5) is given by:

$$\theta_{p,c2} = \phi_{ap} - 2\gamma_{ap} \tag{25}$$

It is noted that $\gamma_{ap} = 0$ and $\theta_{p,c2} = \theta_{p,c1}$ for full $CPC - \theta_e$. Analysis shows that, for truncated CPC-θ_e, $\theta_{p,c2}$ decreases with the increase of θ_t, and $\theta_{p,c1}$ is always larger than $\theta_{p,c2}$. It is also noted that no radiation will arrive on the absorber at $\theta_{in} > \theta_e$ after reflections from upper parabola first and then from the opposite mirror in the case of $\theta_t \geq 0.5\pi - \psi$ (see figure 6) because point M must be above the line $B'A$ to ensure solar rays reflecting from M redirect to the image B' of absorber end B.

3. Mathematical method to calculate daily collectible radiation

It is assumed that CPC-θ_e is oriented in the east-west direction with the aperture being tilted at β from the horizon, and radiation reflected from the ground is not considered. Thus, the collectible radiation on absorber at any moment for isotropic sky diffuse radiation is given by:

$$I = C_t I_b g(\theta_{ap}) \eta(\theta_p) \cos\theta_{ap} + C_t \int_{-(0.5\pi-\beta)}^{0.5\pi} iu(\theta_p) \cos(\theta_p) d\theta_p \tag{26}$$

where I_b is the intensity of beam radiation; $i = 0.5I_d$ is the directional intensity of sky diffuse radiation on the cross-section of CPC-troughs [12, 22]. Equation (26) can be rewritten as

$$I = C_t I_b g(\theta_{ap}) \eta(\theta_p) \cos\theta_{ap} + 0.5I_d (C_{d,1} + C_{d2}) \tag{27}$$

where I_d is sky diffuse radiation on the horizon, and $C_{d,1}$ and $C_{d,2}$ are calculated as follows:

$$C_{d,1} = C_t \int_0^{0.5\pi} \eta(\theta_p) \cos\theta_p d\theta_p \tag{28}$$

$$C_{d,2} = C_t \int_0^{0.5\pi-\beta} \eta(\theta_p) \cos\theta_p d\theta_p \tag{29}$$

For a given CPC-θ_e, $C_{d,1}$ is a constant but $C_{d,2}$ is dependent on β, and both can be obtained by numerical calculations. Similarly, radiation on the absorber of CPC-θ_e at $\theta_{in} \leq \theta_e$ is expressed by:

$$I(\theta_{in} \leq \theta_e) = C_t I_b g(\theta_{ap}) f_1 \cos\theta_{ap} + 0.5C_t I_d \int_{-(0.5\pi-\beta)}^{\theta_t} f_1 \cos\theta_p d\theta_p \tag{30}$$

The daily radiation on the absorber can be calculated by:

$$H_{day} = C_t \int_{-t_0}^{t_0} I_b g(\theta_{ap}) \eta(\theta_p) \cos\theta_p dt + 0.5H_d (C_{d,1} + C_{d2}) \tag{31}$$

$$H_{day}(\theta_{in} \leq \theta_e) = C_t \int_{-t_0}^{t_0} I_b g(\theta_{ap}) f_1(\theta_p) \cos\theta_p dt + 0.5C_t H_d \int_{-(0.5\pi-\beta)}^{\theta_t} f_1 \cos\theta_p d\theta_p \tag{32}$$

where H_d is the daily sky diffuse radiation on the horizon, $t_0 = \tau_{day}\omega_0 / 2\pi$ is the sunset time on the horizon, $\tau_{day}(24\times3600s)$ is the day length, ω_0 is the hour angle of sunset on the horizon [5]. Given the geometry of CPC-θ_e and tilt-angle β, the last term of right hand in equation (32) can be numerically calculated. The daily radiation on the absorber with $\theta_{in} > \theta_e$, $H_{day}(\theta_{in} > \theta_e)$, can be simply calculated by subtracting H_{day} $(\theta_{in} \leq \theta_e)$ from H_{day}. The incidence angle of solar rays on the aperture of CPC-θ_e (θ_{ap}) and projection incident angle (θ_p) at any moment of a day can be found based solar geometry [5, 22, 23]. The $g(\theta_{ap})$ in equation (26-27) is a control function, being 1 for $\cos\theta_{ap} > 0$ otherwise zero. On knowing θ_p and θ_{ap}, and f_1, f_2 and f_3 can be determined based mathematical procedure suggested in this work, then the H_{day} and $H_{day}(\theta_{in} > \theta_e)$ can be numerically calculated, finally summing H_{day} and $H_{day}(\theta_{in} > \theta_e)$ in all days of a year obtain the annual radiation on the absorber of CPC-θ_e ($S_{CPC-\theta_e}$) and annual radiation on the absorber with $\theta_{in} > \theta_e$ ($S(\theta_{in} > \theta_e)$). The annual radiation on the absorber of CPC-90 with $\theta_{in} > \theta_e$ can be obtained according to the method presented in the previous work of authors [20].

In this work, monthly horizontal radiation in Beijing ($\lambda = 39.95^\circ$), the capital of China, was used for the analysis [24], the daily sky diffuse radiation, H_d, and beam radiation (I_b) at any moment of a day are estimated based on the empirical correlations proposed by Collares-Pereira and Rabl [25]. The angle step for calculating $C_{d,1}$ and $C_{d,2}$ is take to be 0.005°; the time step to calculate daily radiation on the absorber is taken to be 1 min; and the θ_a of CPCs is set to 26° [26]. To investigate the optical performance of CPC-θ_e, two cases with the β being yearly fixed and yearly adjusted four times at three tilts are considered. For CPCs with β being yearly fixed (1T-CPC), the β is taken to be site latitude (λ) [26]; whereas for CPCs with β being yearly adjusted four times at three tilts (3T-CPC), the β is set to be λ during periods of 22 days around both equinoxes, and adjusted to $\lambda+23^\circ$ and $\lambda-23^\circ$ in winters and summers, respectively [12].

4. Results and discussions

4.1. Optical efficiency comparison between CPC-θ_e and CPC-90

As seen from figure 7-8, f_2 and f_3 is zero for radiation within the acceptance angle but not for radiation beyond the acceptance angle as expected. It is also seen that the f_3 is larger than f_2 for full CPC-65 and less than f_2 for truncated CPC-65. This means that for full CPC-θ_e, the contribution of upper parabola together with the opposite plane mirror to $S(\theta_{in} > \theta_e)$ is larger than that of plane mirror alone; whereas for truncated CPC-θ_e, the situation is reversed. Comparisons of optical efficiency between CPC-90 and CPC-65 are presented in figure 9 and 10. It is found that the optical efficiency of both CPCs is almost identical for radiation within the acceptance angle; whereas for radiation incident at $\theta_p > \theta_a$, the f of CPC-65 is always larger than that of CPC-90 for full CPCs, but for truncated CPCs with identical C_t and θ_a, the f of CPC-90 is larger than that of CPC-65 in the case of $\theta_p > 31^\circ$ due to the large edge-ray angle (θ_t) of CPC-90 and zero f_3 of CPC-65 (see figure 6)

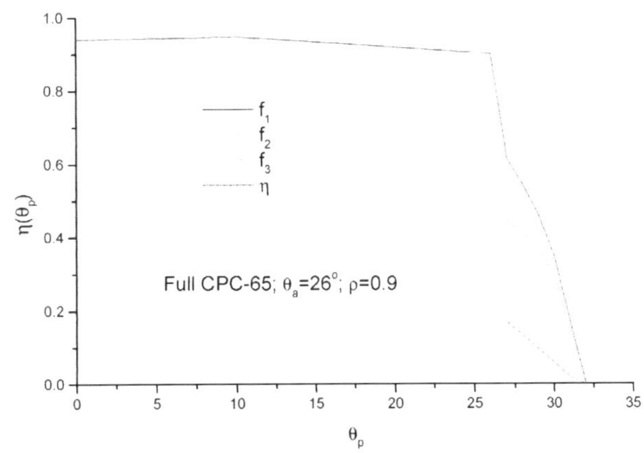

Figure 7. Angular variations of optical efficiency of full CPC-65

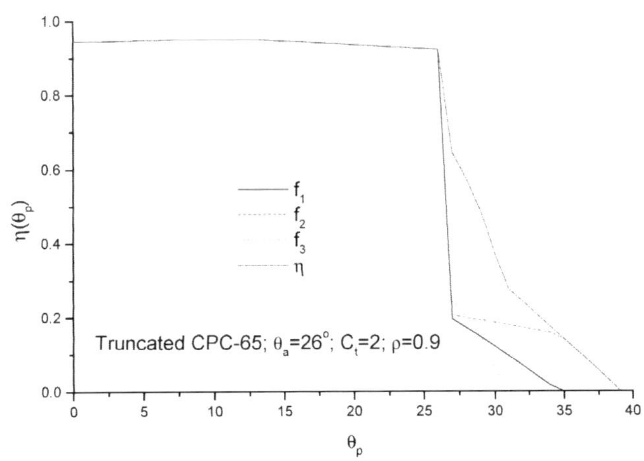

Figure 8. As in Fig.7 but for truncated CPC-65

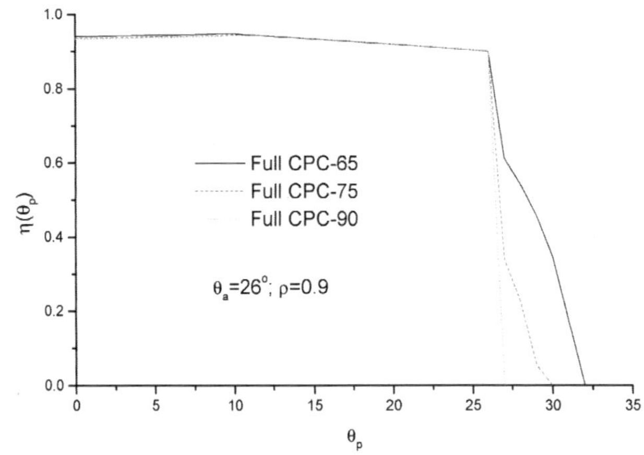

Figure 9. Angular variation of optical efficiency of full CPC- with different.

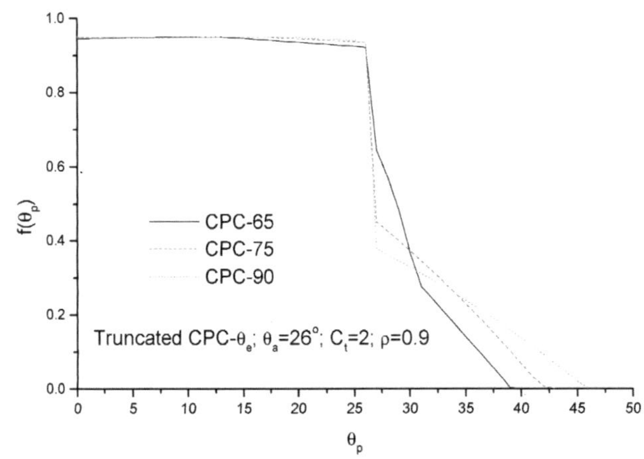

Figure 10. As in Fig.9 but truncated CPC-θ_e

4.2. Comparison of annual collectible radiation between CPC-θ_e and CPC-90

Figure 11 presents the ratio of annual radiation concentrated by CPC-θ_e to that by CPC-90 in the case of β being yearly fixed (1T-CPCs). It is seen that, for full CPCs with identical θ_a, the annual radiation collected by CPC-90 is always larger than that by CPC-θ_e due to large geometric concentration of CPC-90; whereas for truncated CPCs with identical θ_a and C_t, the annual radiation collected by CPC-θ_e are almost identical to that by CPC-90, and even slightly higher in the case of low C_t. The same results as in figure 11 are also found in figure12 in which β is yearly adjusted four times at three tilts (3T-CPCs).

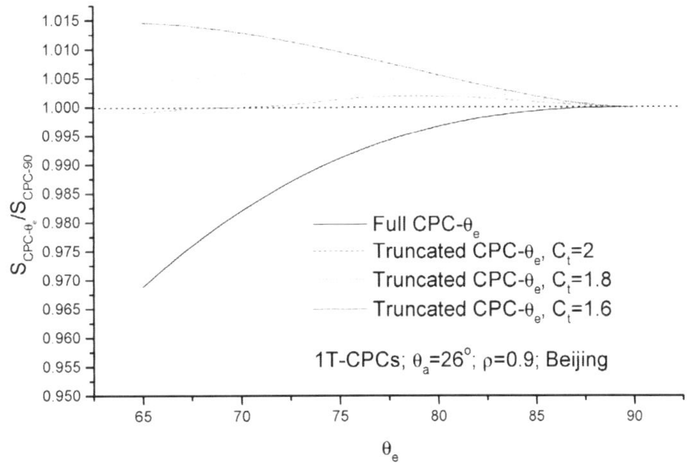

Figure 11. Comparison of annual collectibleradiation between 1T-CPC-θ_e and 1T-CPC-90.

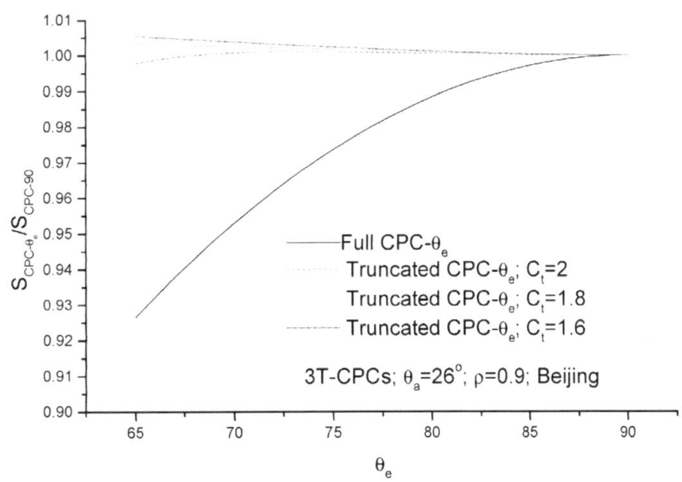

Figure12. Comparison of annual collectible radiation between 3T-CPC-θ_e and 3T-CPC-90.

4.3. Comparison of $S(\theta_{in} > \theta_e)$ *between CPC-*θ_e *and CPC-90*

The annual radiation on the absorber of CPCs with $\theta_{in} > \theta_e$ are presented in figure 13-14. It is seen that $S(\theta_{in} > \theta_e)$ of both CPC-θ_e and CPC-90 decreases with the increase of θ_e, and $S(\theta_{in} > \theta_e)$ collected by CPC-θ_e is much less than that by CPC-90 in the case of θ_e less than 70°, this implies that the CPC-θ_e based PV system should be more efficient than CPC-90 based PV system because the radiation on solar cells with $\theta_{in} > 65^\circ$ is poorly absorbed by solar cells.

It is noted that, given θ_e, the annual radiation on the absorber of CPC-90 is independent on its geometric concentration (C_t) and $S(\theta_{in} > \theta_e)$ keeps constant as a result of the fact that the $S(\theta_{in} > \theta_e)$ comes from the lower part of reflectors of CPC-90 [20], but $S(\theta_{in} > \theta_e)$ collected by CPC-θ_e is sensitive to C_t because critical angles $\theta_{p.c1}$ and $\theta_{p.c2}$ are sensitive to θ_t as shown in figure15. In fact, $S(\theta_{in} > \theta_e)$ collected by CPC-θ_e in a site depends on the angular dependence of ($f_2 + f_3$) and tilt-angle adjustment strategy, and in turn the angular dependence of ($f_2 + f_3$) is dependent on the edge-ray angle (θ_t) or C_t as seen from figure 16. As shown in figure 15, for 1T-CPC-65, $S(\theta_{in} > 65^o)$ increases with C_t due to high $f_2 + f_3$ for $\theta_p < 30^\circ$ (see figure15) and large C_t; whereas for 3T-CPC-65, $S(\theta_{in} > 65^o)$ decreases with C_t as a result of the fact that the sun is almost within the acceptance angle of CPCs (i.e. $\theta_p < \theta_a$) over the daytime in any day of a year, and $S(\theta_{in} > 65^o)$ mainly originates from the sky diffuse radiation, thus $S(\theta_{in} > 65^o)$ increases with θ_t or decreases with C_t. Figure14 also indicates that $S(\theta_{in} > 65^o)$ of 1T-CPC-65 is much larger than that of 3T-CPC-65, showing that the use of 3T-CPCs facilitates the improvement of photovoltaic performance CPC-θ_e based PV systems due to lower $S(\theta_{in} > \theta_e)$.

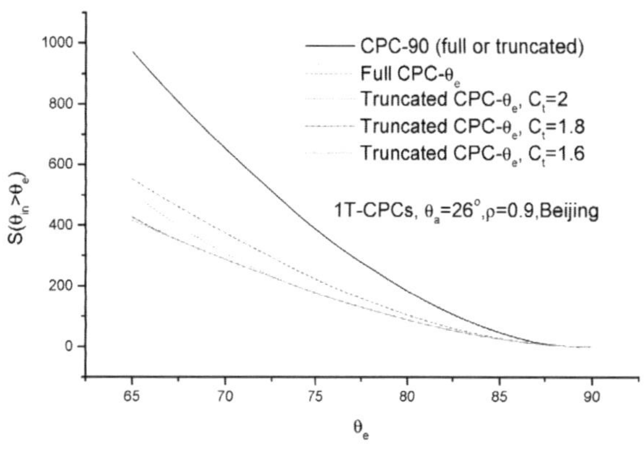

Figure 13. Annual radiation on the absorber of 1T-CPCs at $\theta_{in} > \theta_e$

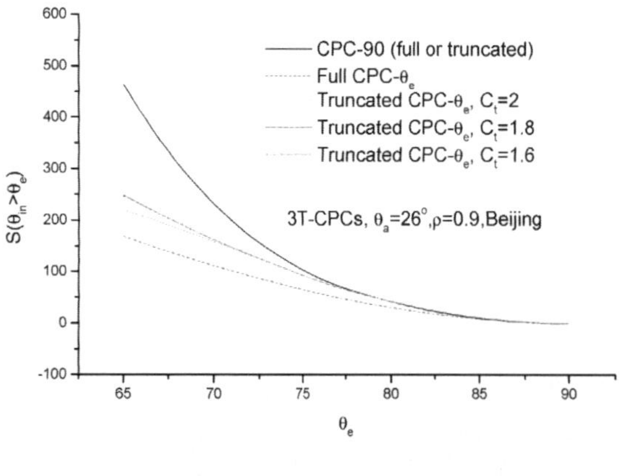

Figure 14. As in figure 13 but for 3T-CPCs

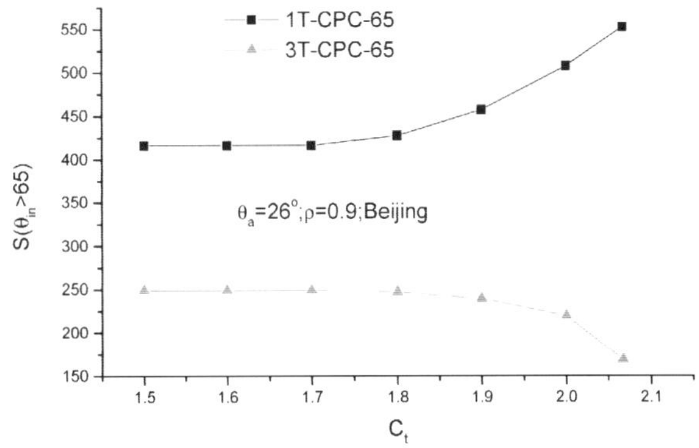

Figure 15. Effects of geometric concentration on S($\theta_{in} > 65$) of CPC-65

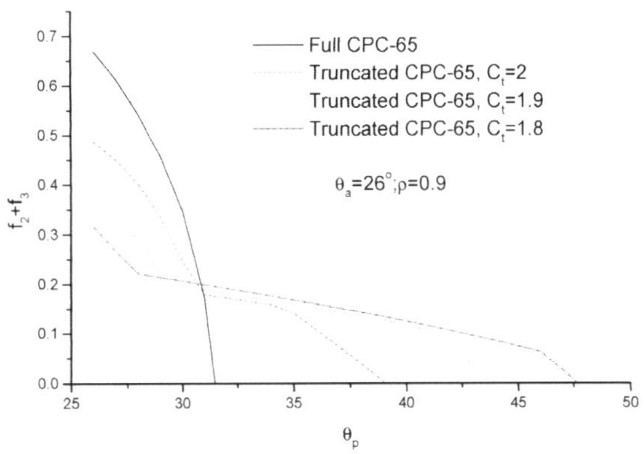

Figure 16. Optical efficiency for radiation arriving on the absorber at θ_{in}>65

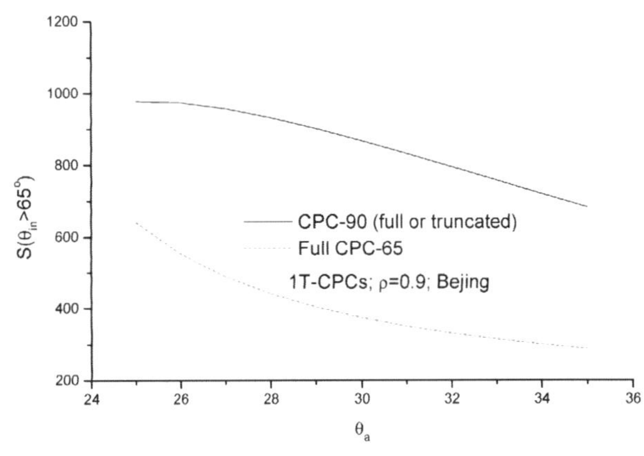

Figure 17. Effects of acceptance half-angle on S($\theta_{in} > 65$) of CPCs

Figure 17 shows the effect of acceptance angle (θ_a) on $S(\theta_{in} > \theta_e)$. It is seen that, given θ_e (such as 65°), $S(\theta_{in} > \theta_e)$ collected by CPC-90 and CPC- θ_e decreases with the increase of θ_a. This indicates that θ_a should be large as possible in the practical design of CPCs based on requirements of least daily operation hours, geometric concentration and strategy of tilt-angle adjustments [27].

5. Conclusions

The analysis in this work shows that part of radiation beyond the acceptance angle of CPC-θ_e would arrive on the absorber at $\theta_{in} > \theta_e$, the annual radiation on the absorber with $\theta_{in} > \theta_e$ depends on the acceptance angle, geometric concentration as well tilt-angle adjustment strategy of CPC-θ_e. Calculations indicate that, for full CPCs with identical θ_a, the annual radiation collected by CPC-θ_e is less than that by CPC-90 due to the large geometric concentration (C_t) of CPC-90; whereas for truncated CPCs with identical θ_a and C_t, the annual radiation collected by CPC-θ_e is almost identical to that by CPC-90, and even slightly higher. Calculations also show that the annual radiation on the absorber of CPC-θ_e with $\theta_{in} > \theta_e$ decrease with the increase of θ_e but always less than that of CPC-90, and this implies that the CPC-θ_e based PV system is more efficient than CPC-90 based PV system because the radiation on solar cells incident at large angle is poorly converted into electricity.

Acknowledgements

This work is partial fulfillment of funded research program 51466016, financially supported by Natural Science Foundation of China.

References

[1] Firdaus M S, Siti H A B, Roberto R I, Scott G B G S, Abu B M and Ruzairi A R 2013 Performance analysis of a mirror symmetrical dielectric totally internally reflecting concentrator for building integrated photovoltaic systems *Applied Energy* **111** 288-99

[2] Robert M C 2005 Concentrator photovoltaic technologies: Review and market prospects *Refocus* **6**(4) 35-39

[3] Chong K K, Lau S L, Yew T K and Tan P C L 2013 Design and development in optics of concentrator photovoltaic system *Renewable and Sustainable Energy Reviews* **19** 589-612

[4] Royne A, Dey C R and Mills D R. 2005 Cooling of photovoltaic cells under concentrated illumination: a critical review *Solar Energy Materials and Solar Cells* **86** 451-483

[5] Rabl A 1985 *Active solar collectors and their applications* Oxford: Oxford University Press

[6] Bahaidarah H M, Tanweer B, Gandhidasan P, Ibrahim N and Rehman S 2014 Experimental and numerical study on non-concentrating and symmetric unglazed compound parabolic photovoltaic concentration systems *Applied Energy* **136** 527-36

[7] Brogren M, Wennerberg J, Kapper R and Karlsson B 2003 Design of concentrating elements with CIS thin-film solar cells for façade integration *Solar Energy Materials and Solar Cells* **75** 567–75

[8] Mallick T K, Eames P C, Hyde T J and Norton B 2004 The design and experimental characterization of an asymmetric compound parabolic photovoltaic concentrator for building facade integration in the UK *Solar Energy* **77** 319–27

[9] Mallick T K, Eames P C and Norton B 2006 Non-concentrating and asymmetric compound parabolic concentrating building facade integrated photovoltaic: An experimental comparison *Solar Energy* **80** 834–49

[10] Hollands K G T 1971 A concentrator for thin-film solar cells *Solar Energy* **13** 149-63

[11] Fraidenraich N and Almeida G J 1991 Optical properties of V-trough concentrators *Solar Energy* **47** 147-55

[12] Tang R S, Liu X Y 2011 Optical performance and design optimization of V-trough concentrators for photovoltaic applications *Solar Energy* **85** 2154-66

[13] Hatwaambo S, Hakansson H, Nilsson J, Karlsson B 2008 Angular characterization of low concentrating PV-CPC using low-cost reflectors *Solar Energy Materials and Solar Cells* **92** 1347-51

[14] Hatwaambo S, Hakansson H, Roos A, Karlsson B 2009 Mitigating the non-uniform illumination in low concentrating CPCs using structured reflectors *Solar Energy Materials and Solar cells* **93(11)** 2020-24

[15] Yu Y M, Tang R S 2015 Diffuse reflections within CPCs and its effect on energy collection *Solar Energy* **120** 44-54

[16] Duffie J A, Beckman J A 1991 *Solar engineering of thermal processes*. 2nd edition (New York: Willey & Sons, Inc.)

[17] Hein M, Dimroth F, Siefer G, Bett AW 2003 Characterisation of a 300× photovoltaic concentrator system with one-axis tracking *Solar Energy Materials and Solar Cells* **75** 277-83

[18] Rabl A, Winston R 1976 Ideal concentration for finite sources and restricted exit angle *Applied Optics* **15** 2880-3

[19] Yu Y M, Tang R S, Xia C F 2012 Irradiation distribution on solar cells inside CPCs with a restricted exit angle *Advanced Materials Research* **588-589** 603-7

[20] Yu Y M, Liu N Y, Tang R S 2004 Optical performance of CPCs for concentrating solar radiation on flat receivers with a restricted incidence angle *Renewable Energy* **62** 679–88

[21] Yu Y M, Liu Ni Y, Li G H, Tang R S 2015 Performance comparison of CPCs with and without exit angle restriction for concentrating radiation on solar cells *Applied Energy* **155** 284-93

[22] Tang R S, Wang J F 2014 A note on multiple reflections of radiation within CPCs and its effect on calculations of energy collection *Renewable Energy* **57** 490–6.

[23] Tang R S, Yu Y M 2010 Feasibility and optical performance of one axis three positions sun-tracking polar-axis aligned CPCs for photovoltaic applications *Solar Energy* **84** 1666–75

[24] Chen Z Y 2001 *The Climatic Summarization of Yunnan* (Beijing: Weather Publishing House)

[25] Collares-Pereira M, Rabl A 1979 The average distribution of solar radiation: correlations between diffuse and hemispherical and betweenhourly and daily insolation values *Solar Energy* **22** 155–64

[26] Tang R S, Wu M G, Yu Y M, Li M 2010 Optical performance of fixed east–west aligned CPCs used in China *Renewable Energy* **35** 1837–41

Assessment of energy and economic performance of office building models: a case study

X Y Song[1,2], C T Ye[1,3], H S Li[1,2], X L Wang[1] and W B Ma[1]

[1]Key Laboratory of Renewable Energy and Gas Hydrate, Guangzhou Institute of Energy Conservation, Chinese Academy of Sciences, Guangzhou 510640, China
[2]University of Chinese Academy of Sciences, Beijing, 100049, China

E-mail address: yect@ms.giec.ac.cn

Abstract. Energy consumption of building accounts for more than 37.3% of total energy consumption while the proportion of energy-saving buildings is just 5% in China. In this paper, in order to save potential energy, an office building in Southern China was selected as a test example for energy consumption characteristics. The base building model was developed by TRNSYS software and validated against the recorded data from the field work in six days out of August-September in 2013. Sensitivity analysis was conducted for energy performance of building envelope retrofitting; five envelope parameters were analyzed for assessing the thermal responses. Results indicated that the key sensitivity factors were obtained for the heat-transfer coefficient of exterior walls (U-wall), infiltration rate and shading coefficient (SC), of which the sum sensitivity factor was about 89.32%. In addition, the results were evaluated in terms of energy and economic analysis. The analysis of sensitivity validated against some important results of previous studies. On the other hand, the cost-effective method improved the efficiency of investment management in building energy.

1. Introduction

According to the announcement of ASHRAE Handbook HVAC Applications, building sector is responsible for more than 40% of the global end-use energy consumption [1]. As the second largest building energy user around the world, China ascends first in residential energy consumption and ranks third in commercial energy consumption [2]. The final energy consumption in Chinese buildings grew by 37.3% in 2015; if this trend continues, the energy use could increase by 70% in 2050 [3-5].

Kelly [6] reported that 33% of all global carbon emission could be attributed to existing buildings.

[3] Address for correspondence: C T Ye, Key Laboratory of Renewable Energy and Gas Hydrate, Guangzhou Institute of Energy Conservation, Chinese Academy of Sciences, Guangzhou 510640, China. E-mail address: yect@ms.giec.ac.cn.

It is broadly recognized that conducting cost-effective retrofits on buildings exerts it crucially to achieve the goals of green buildings; this is due to the large ratio of existing buildings to new constructions [7, 8]. Additionally, functional reform or more stringent energy-efficiency standard embraces the retrofitting market of building energy conservation. For instance, the "Green Building Action Plan" released by the Chinese central government stated that the total floor area of retrofitting existing buildings will reach 570 million square meters by 2015 in China [9, 10]. Therefore, it is significant to enhance the existing buildings' energy efficiency.

Technology and management are the vital impact factors that influence the effectiveness of building energy retrofitting [11]. As to technology aspects, reference literatures in the related area were concerned about the thermic insulation materials, the windows replacement and thermal performances [12-15]. Different building environmental assessment tools by occupants or life cycle phases of buildings were also analyzed in literatures [16, 17]. Many of the studies were based on theoretical calculation. Additionally, the simulation-based integration of dynamic calculation and experimental studies guarantees the possibility of model-checking and the accuracy of further simulation, and TRNSYS is one of the appropriate tools [18]. As to the management aspects, the application of energy-saving measures (ESMs) usually needs to be supported by economic assessments. Gero et al. [19], for example, proposed a multi criteria model and explored the trade-offs between the building thermal performance and capital cost of the building. Goodacre et al. [20] applied a cost-effective method to assess the potential scale of the benefits from upgrading heating and hot-water energy efficiency in English buildings. A desired efficiency label at the minimum cost with the multi-objective optimization method was also obtained in literature [21]. Terés-Zubiaga et al. [22] employed different envelope retrofitting methods and evaluated the results under an economical NPV approach.

However, although the methods of envelope retrofitting have been extensively studied, there is little information available about the sensitivity of energy saving in office building envelopes in Southern China simulated by TRNSYS; and literature about the economic way of evaluating the retrofits in office building energy is still insufficient in Southern China. In this study, 27 combined methods have been applied with TRNSYS software to enhance the office building envelope. These measures are exterior wall insulation, infiltration repairs and windows replacement. Some important contributions of this paper have been found, including that the daily energy demand has been well recorded for the usage of model checking. On the other hand, unlike the traditional retrofits, the sensitivity factor of the envelope that signifies the capacity of energy saving in different parameters has been applied, playing a key role in combining diverse ESMs. Moreover, these combined scenarios' results are evaluated under an economic and energy analysis, which could be useful for the future building retrofitting practices and the policy making.

2. Methodology: case study selection

In this study, a multi-office building named Jicheng Mansion (JCM), which is located in Tianhe District of Guangzhou, China, was selected as the reference building. The building was constructed in 2003, a specific period in which a significant number of similar office buildings had been built up before legislative acts on energy efficiency was carried out.

The 10-storey building has a net floor area of 612 m², of which the overall floor area is 6120 m². The ceiling height of JCM is 33.6 m, and the window-wall ratio is 0.3. The construction is mainly consisted of reinforced concrete.

Figure 1. General view of JCM.

3. Monitoring

In sure of obtaining data to calibrate the building simulation model, the reference JCM building has been monitored. The four external walls of JCM oriented to the main four directions: South, West, North and East. Obviously, there are three zones in each floor, i.e., a storage room with two toilets, a workplace and a staircase with an elevator. The Workplace is the thermal zone, with a net floor area of 456 m². The building outline is depicted in Figure 2.

Figure 2. Layout of the studied building

Figure 2 shows the outline of each floor. As to Workplace, the south wall is an interior wall without solar heat gain. The office building is equipped with the central air-conditioning system. Southern China is located in a subtropical area, and the refrigeration season is yearly from May to November, thus cooling demand is solely concerned in this case. Due to the rated refrigerating capacity of the system, the chiller supplies cooling load for seven floors, and the 1st, 5th and 9th floors are not equipped with air conditioners.

The data of the inlet water temperature, outlet chilled water temperature and flow rate of chilled water has been well recorded from the field work in order to calculate the cooling load of the building. Detailed information on sensors' characteristics is presented in Table 1.

Table 1. Characteristics of the used sensors.

Parameter	Units	Sensor	Uncertainty
Temperature	K	SIN-RC-4HC/A	±0.50 K
Flow rate	m^3/h	LDG-MK	±0.04 m^3/h

The data has been integrated in a 10-minute frequency for the stable measurement of the JCM building energy consumption, and the accuracy of the obtained data is well guaranteed.

4. Model definition and validation

4.1 Building model definition

The building model has been developed by TRNSYS software. The first step is to set up a building model in TRNSYS with the actual JCM construction data; the second is to define the thermal parameters of the building and the operation schedule of the building; and the final procedure is to calculate the dynamic energy consumption with the TRNSYS simulation studio. The Workplace is defined as a thermal zone in which the temperature is set as 26 ℃. The internal heat gains are set as follows: a person occupancy density is 20 W/m^2, an equipment load is 16 W/m^2 and a lighting load is 11 W/m^2. The building system operated on a 5-day week basis of which the work schedule is between 9am and 5pm. Moreover, the detailed description of the related envelope data constructed in TRNSYS model is presented in Table 2, and the main characteristics of windows are described in Table 3.

Table 2. Detailed construction data.

Envelope	Thickness [m]	Conductance [kJ/(h·m·K)]	Capacitance [kJ/(kg·K)]	Density [kg/m³]
WALL (exterior walls)				
Fiberglass quilt	0.002	0.144	0.84	12
Gas_concrete	0.18	1.04	1	800
Fiberglass quilt	0.002	0.144	0.84	12
Exterior walls U-Value	1.120 W/(m²·K)			
Floor (roof and floors)				
Roof deck	0.06	0.504	0.9	530
Poly_urethane	0.01	0.11	1	40
Gas_concret block	0.08	0.97	1	700
Roof U-Value	0.831 W/(m²·K)			

Table 3. Windows constructed in TRNSYS model.

Frame (%)	U_{frame} [W/(m²·k)]	Glass	U_{glass} [W/(m²·k)]
15	3.06	Single	5.16

4.2 Building model validation

As mentioned above, the data of temperature and flow rate had been gathered from six days out of August and September, 2013. The selection of this period is according to its representativeness of the torrid weather conditions, during which the operation status of refrigerating unit is relatively typical. The basic setting of the simulated model is introduced in Table 4, in which the parameter of "vol" represents for the volume of Workplace. Meanwhile, the ventilation rate of the building is calculated by the function: 2*TIME, and the parameter of "TIME" is related to the work schedule of which the value is set as follow: from 8:00am to 9:00am is 0.5, from 9:00am to 12:00pm is 0.95, from 12:00pm to 14:00pm is 0.8, from 14:00pm to 17:00pm is 0.95.

Table 4. Parameters simulated in the model.

Parameters	Workplace
Infiltration (from outdoors) [vol/h]	2
Ventilation [vol/h]	2*TIME

To ensure the degree of accuracy, simulated results are validated by the measured electricity data which is demonstrated in Figure 3.

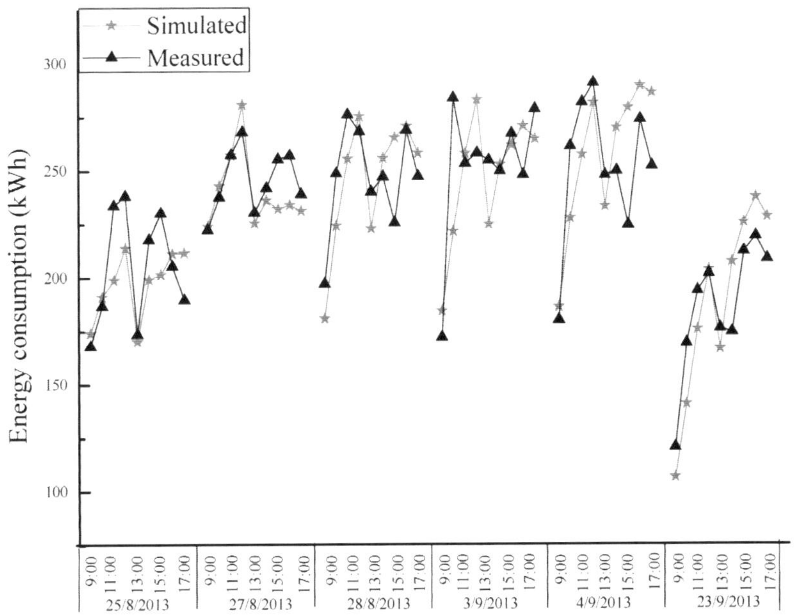

Figure 3. Simulated versus measured energy consumption.

The calculated values are likely to keep in pace with the measured energy consumption of the office building.

To quantify the differences, the percentage error and NMBE are calculated by equation (1):

$$\begin{cases} PE = \dfrac{x_i - y_i}{y_i} \\ NMBE = \sum_{i=1}^{n} \dfrac{x_i - y_i}{y_i} \end{cases} \quad (1)$$

where PE is the percentage of error between simulated and measured energy consumption, NMBE is the normalized mean bias error, x_i is the simulated electricity consumption (kWh), y_i is the measured electricity consumption (kWh) and n is the simulation hours. The result of percentage error is depicted in Figure 4.

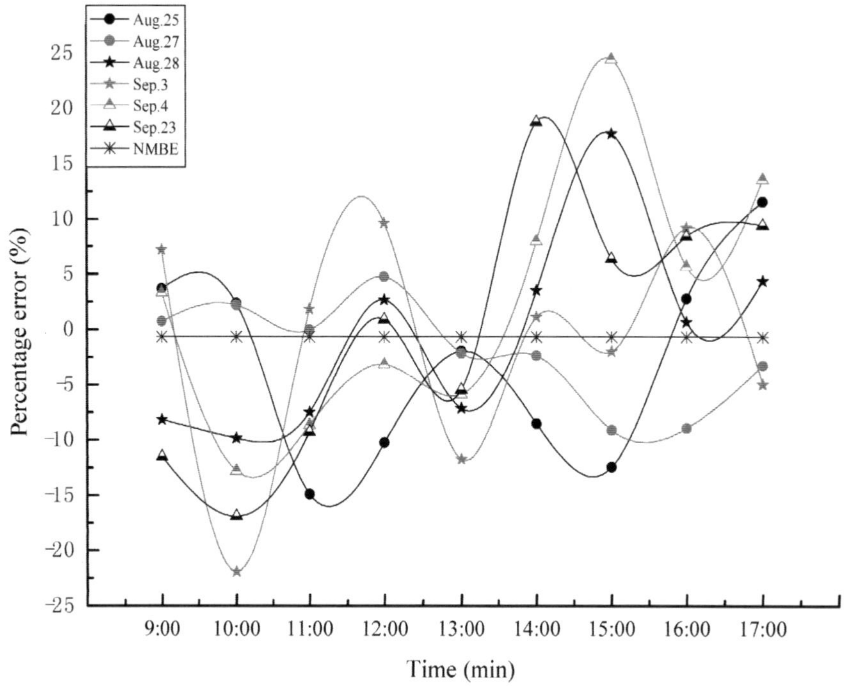

Figure 4. Simulated versus measured percentage error.

Refer to the simulated research on office buildings from J. Terés-Zubiaga and J. Yoon [22, 23], the usual practice should be to follow the criteria set by established professional standards [24, 25], and normalized mean bias error (NMBE) between simulated and measured energy consumption should not exceed ±5%.

As shown in Figure 4, the NMBE is -0.66%, which is within ±5% and thus the simulation model is reliable. The model accuracy is considered to be good in view of these results, and the simulation model turns out to be believable.

5. Sensitivity analysis of building envelope

After the validation of the model, the sensitivity of building envelope has been analyzed in order to indicate how sensitive the simulated output would be in response to changes made to the input parameter.

The sensitivity factor, denoted as IC, is essentially a ratio of the percentage change (with respect to the base case value) [26], as suggested in Eq. (2):

$$IC=\frac{OP-OP_{bc}}{OP_{bc}}\div\frac{IP-IP_{bc}}{IP_{bc}} \quad (2)$$

where OP and IP are values of corresponding output and input, OP_{bc} and IPbc are base case values of output and input respectively. IC is essentially a ratio of the percentage change (with respect to the base case value) in computed output (i.e., total annual energy consumption) to the percentage change in input design parameter.

Based on the aforementioned equation, the relative sensitivity has been worked out, as shown in Figure 5.

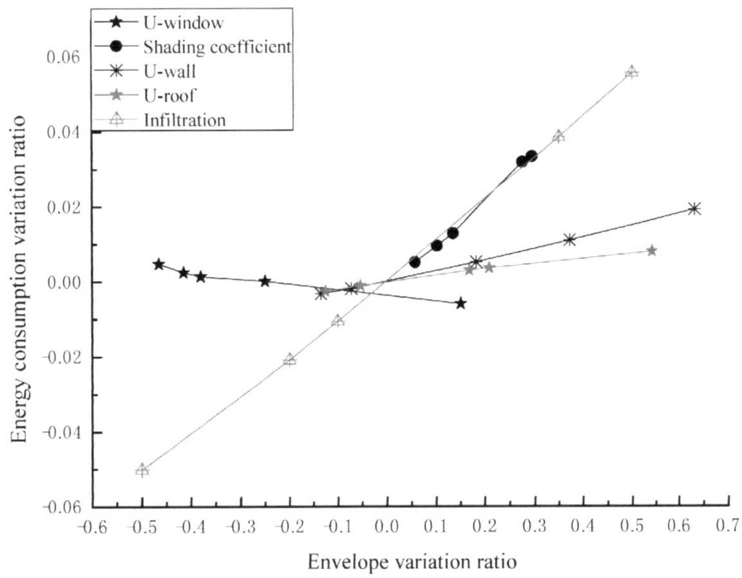

Figure. 5 Envelope sensitivity.

As Figure 5 illustrates, the slope of the linear regression signifies the sensitivity factor of the parameter [26].

The relative sensitivity, ω_i, is denoted as the quotient of the single sensitivity factor to the sum sensitivity of all parameters, as shown in Equation (3):

$$\omega_i = \frac{|IC_i|}{\sum_{i=1}^{n} |IC_i|} \tag{3}$$

where n is the amount of parameters.

The results are clearly summed up in Table 5.

Table 5. Envelope sensitivity factor and relative sensitivity.

Number	Parameter	Sensitivity factor	Relative sensitivity
1	Shading coefficient	0.1234	42.59%
2	Infiltration	0.1062	36.67%
3	U-wall	0.0291	10.06%
4	U-roof	0.0158	5.46%

5	U-window	-0.0151	5.22%

Table 5 indicates that the most effective impact factor of energy consumption is the shading coefficient. Southern China is located in a humid subtropical zone with durable insolation, thus the shading coefficient could pose significant impacts in energy consumption. The infiltration rate directly impacts building energy consumption, depending on the tightness of building enclosure. The most insensitive parameter lies in the U-window, because of the small ratio of window-wall and the particular location of building. Due to the low rate of the roof to wall, the heat-transfer coefficient of the roof exerts less influence than the wall in energy consumption. Indeed, different energy-saving measures can be fully analyzed with the comprehensive consideration of these aspects discussed above. The individual sensitivity could help building designers get some idea about the potential energy savings of prospective ESMs.

6. Energy saving measures of envelope

The ESMs for improving the building energy efficiency could be studied based on the results mentioned above. The sum relative sensitivity of U-window and U-roof are 10.68%, far less than the other factors. Therefore, the intensive ESM should focus on the enhancement of exterior walls, infiltration rate and shading coefficient. Additionally, the exterior walls are enhanced with the EPS, a common engineering insulation material with low thermal conductivity. The infiltration rate is repaired with the building glue. The shading coefficient is enhanced through the method of windows replacement; the windows are replaced with insulating glass windows of lower solar heat gain coefficients. Namely, economic results presented in this paper would be applicable.

According to the assumption given by Terés-Zubiaga et al. [22], three scenarios are proposed in this study: scenario 0, when no thermal improvement is carried out; scenario 1, when the building is slightly improved; scenario 2 when a highly improvement is conducted. The resulted models are named according to the combinations of the ESM adopted in each case. Thus, model (1 2 0) represents exterior wall scenario 1, infiltration rate scenario 2 and shading coefficient scenario 0.

It is assumed that the costs in each scenario are the material costs, other costs (e.g. scaffolding) are assumed to exist in the maintenance work of the building and are not taken into account of the calculations.

The detailed information about ESM is described in Table 6.

Table 6. Summary of data regarding ESM in exterior walls, infiltration rate and shading coefficient.

ESM in Exterior walls	Addition of EPS in walls		
	Currently	Slightly improved	High standard
Scenario	0	1	2
Thermal ins thickness [cm]	0	4	10
U[W/(m²·K)]	1.12	0.489	0.265
Total investment [Yuan]	0	10817.28	29797.72

ESM in Infiltration rate	Infiltration improvement		
	Currently	Slightly improved	High standard
Scenario	0	1	2
U[Vol/h)]	2	1.5	1.0
Total investment [Yuan]	0	26761.51	53723.42
ESM in Shading Coefficient	SC improvement		
	Currently	Slightly improved	High standard
Scenario	0	1	2
SC Value	0.78	0.55	0.23
Total investment [Yuan]	0	47754.72	96305.35

Table 6 reveals the detailed information of ESM. In the case of exterior walls, infiltration rate and shading coefficient, costs assumed in scenarios 1 and 2 are actually compared with scenario 0. Obviously, the investment is in positive correlation to the standard of retrofits.

7. Results

After the validation of the model, the simulation has been operated under 27 scenarios. The assumed operating conditions are based on the actual work schedule. Using a 1-hour time step, the simulation runs with TRNSYS software (Note: in the analysis, the input data are same as the typical values given in Tables 2, 3 and 4, except the parameter whose effect is discussed varies.). The cooling load is depicted in Figure 6.

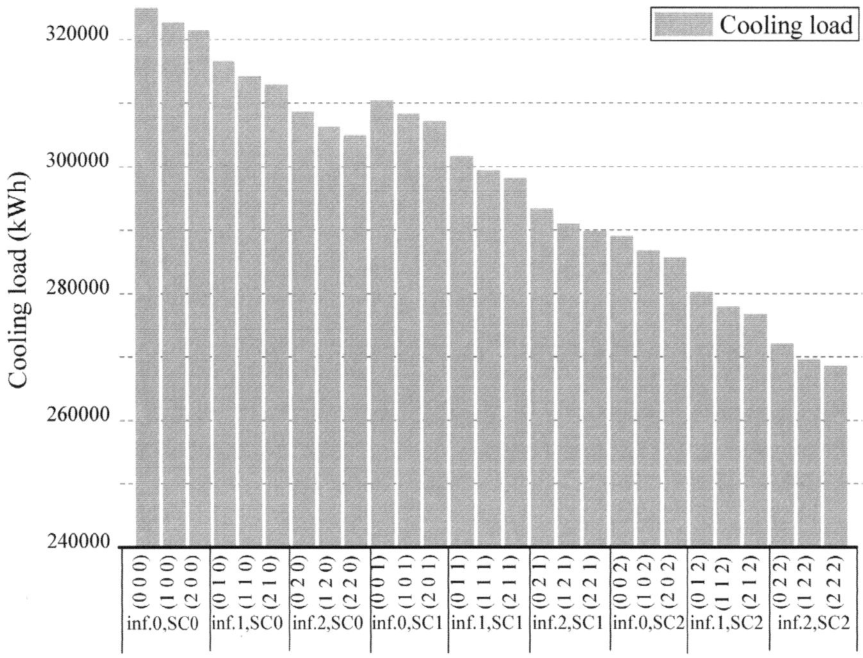

Figure 6. Simulated cooling load.

As Figure 6 illustrates, it is ascertainable that when enhancing the envelope gradually, the amount of cooling loads will decrease accordingly. The biggest amount of cooling load is the reference building (0 0 0) that climbs up to 324.8 MWh per year. The smallest amount of cooling demand is the all-high-standard enhanced building (2 2 2), which consumes 268.4 MWh.

The energy-saving ratio is the energy demand reduction compared to the reference scenario (0 0 0). As a result, the ratios range from 0.69% to 17.37%. The most energy-efficient method owns to the all-high-standard enhanced building (2 2 2), which saves about 17.37% of the cooling demand with respect to the reference case.

8. Discussion

The influence of exterior walls, infiltration rate and shading coefficient on the energy demand has been analyzed in detail. On the other hand, the application of ESM usually needs to be supported by the economic evaluation. Net present value (NPV), payback period and profit rate [23] are chosen as financial elements to analyze the economic feasibility of these measures. Payback period is denoted as the period of time that is required to repay the cost of the initial investment. The profit rate is the rate of NPV to initial investment. At the same time, the NPV sums the initial capital investment and the present net cash flows over the lifespan of the building. No maintenance costs are assumed in this study, the only cash flow existing there is the yearly-savings. Therefore, NPV is calculated as Equation (4):

$$\begin{cases} \text{NPV} = -I + \sum_{n=0}^{\text{LS}} \frac{S_n}{(1+r)^n} \\ S_n = S_0(1+p)^n \end{cases} \tag{4}$$

As the formula expresses, I is the initial investment, LS is the lifespan of the building and n is the time period, S_n is the savings of the year n and S_0 is the initial savings, r represents the cost of capital (%) and p is the increase rate of energy cost (i.e., the increase cost of natural gas or electricity). An investment should be undertaken only when the NPV is greater than 0, and the best solution comes to the highest NPV scenario in a fixed lifespan.

Different r values (4-8%) were assumed in [27]. In a brief premise, the cost of capital and the increase rates in this project are taken to be 4% in these scenarios, because each studied region could analyze the NPV according to the specific economic data of the actual p and r values. The lifespan of JCM building is about 30 years.

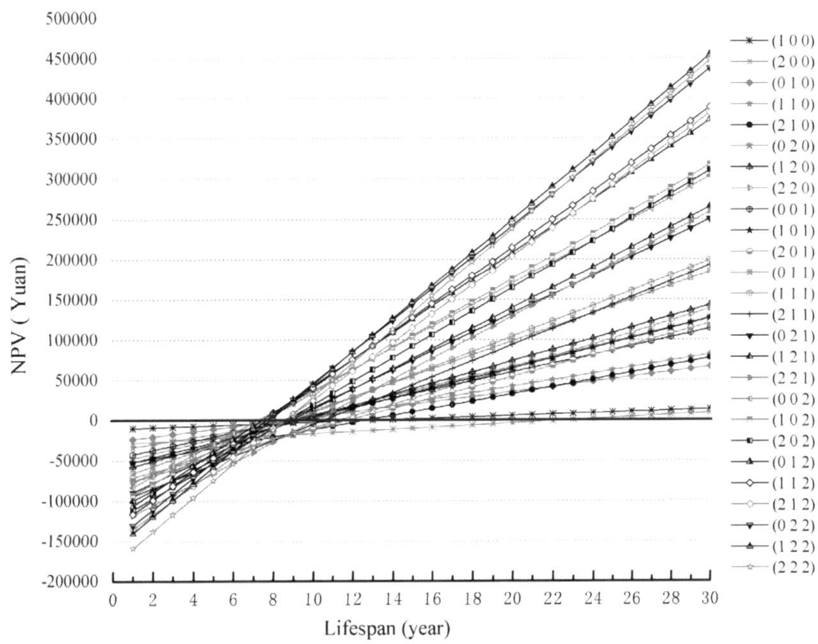

Figure 7. Cash flows of the 26 scenarios.

The cash flows of these 26 scenarios are illustrated in Figure 7 (scenario (0 0 0) is the reference case without cash flow). The intersection of the curve and Y axis (x=0) is the initial investment of the enhanced scenario and the node of the curve and the transverse line (y=0) is the payback period of the scenario. The performance of these scenarios is presented in Table 7.

Table 7. Solutions for optimal scenarios.

(Wall, Inf., SC)	Energy Saving (Yuan)	Initial investment (Yuan)	NPV (Yuan)	Payback period (year)	Profit rate
(1 0 0)	829.32	10817.28	14062.46	13.04	1.30
(2 0 0)	1311.10	29797.72	9535.27	22.73	0.32
(0 1 0)	3122.18	26761.51	66903.77	8.57	2.50
(1 1 0)	3990.02	37578.79	82121.91	9.42	2.19
(2 1 0)	4473.88	56559.22	77657.12	12.64	1.37
(0 2 0)	6034.93	53723.42	127324.50	8.90	2.37
(1 2 0)	6921.76	64540.69	143112.07	9.32	2.22
(2 2 0)	7400.80	83521.13	138502.77	11.29	1.66
(0 0 1)	5387.30	47754.72	113864.26	8.86	2.38
(1 0 1)	6177.39	58572.00	126749.72	9.48	2.16
(2 0 1)	6601.80	77552.44	120501.51	11.75	1.55
(0 1 1)	8645.49	74516.23	184848.60	8.62	2.48
(1 1 1)	9477.35	85333.51	198987.01	9.00	2.33
(2 1 1)	9904.34	104313.94	192816.13	10.53	1.85
(0 2 1)	11719.90	101478.14	250118.81	8.66	2.46
(1 2 1)	12595.01	112295.41	265554.86	8.92	2.36
(2 2 1)	13022.31	131275.85	259393.47	10.08	1.98
(0 0 2)	13311.62	96305.35	303043.15	7.23	3.15
(1 0 2)	14144.18	107122.63	317202.65	7.57	2.96
(2 0 2)	14542.59	126103.07	310174.59	8.67	2.46
(0 1 2)	16545.63	123066.86	373301.94	7.44	3.03
(1 1 2)	17424.40	133884.14	388847.82	7.68	2.90
(2 1 2)	17825.32	152864.57	381895.04	8.58	2.50
(0 2 2)	19576.83	150028.77	437276.02	7.66	2.91
(1 2 2)	20498.10	160846.04	454097.02	7.85	2.82
(2 2 2)	20899.84	179826.48	447168.65	8.60	2.49

Table 7 indicates that the cost of investment is in positive relation to the capacity of energy saving. The biggest energy-saving capacity comes to the scenario (2 2 2), which saves 20899.84 Yuan for each year; the amount of initial investments are also the maximum number, and the payback period is 8.60 years. The NPV of scenario (2 2 2), which wins a profit of more than double of its initial investment, ascends the 2nd among these scenarios. Thus the aforementioned method is cost-effective. As to the payback period, the shortest time is the scenario (0 0 2), which takes 7.23 years to pay for the initial investment. A point worth emphasizing is the comparison of scenario (2 2 1) and scenario (0 0 2), namely, the energy savings of these two plans are nearly the same, while the former scenario costs 1.4 times of the initial investment than the later one, which signifies the effectiveness of shading coefficient retrofitting.

When considering the rate of NPV to investment cost, scenario (0 0 2) is evidently the best profit rate, which obtains a profit rate of 3.15 times. The scenario profit rate of (0 2 0) is 2.37, and the least scenario (2 0 0) owns a profit rate of 0.32. Thus shading coefficient retrofitting is much cost-effective than the other two methods.

According to the results, in the lifespan of 30 years, the best solution comes to the scenario (1 2 2), which is the NPV winner with the profit of 454097.02 Yuan. The payback time of the case is 7.85 years, an economic time cost for shareholders or investors. Base on the TRNSYS simulation model, it is very helpful to find the best economical pathway for energy saving of building envelope retrofitting by changing the character of the building material and construction procedure in the model instead of reconstruction of the existing building. Generally, for homogeneous office buildings, this method is time saving and economical efficient; and for the other buildings, the process of researching an optimal solution is cost-effective. Therefore, the sufficient results are helpful for technology improvement and efficient management.

9. Conclusions

This paper presents the representative of the early 21st Century office building in southern China. The test model has been constructed and validated for implementing different retrofitting measures of the building envelope. The sensitivity of related parameters has been obtained in assistant of organizing different energy-saving measures. As a result, the placement of exterior wall enhancement, infiltrations repairing and windows replacement were three of the most significant parameters. Then the 27 ESM results were obtained to analyze the capacity and efficiency of the envelope retrofitting. Furthermore, these ESM results were appraised through an economic approach.

Results show that the saving ratios of cooling demand range from 0.69% to 17.37%, and the all-high-standard enhanced building scenario (2 2 2) could save the most energy. It appears that the retrofitting of shading coefficient is cost-effective; in a lifespan of 30 years, the scenario (1 2 2) with a 454097.02 Yuan NPV turns out to be the optimal scenario of envelope enhancement.

The results of this research are expected to be submitted to the Guangdong Provincial Science and Technology Department and Construction Department, and the government will promote a legislative program for office building energy-saving assessment. Finally, the optimum applicability of such energy and economic performance of office building models will be written in the engineering technical guide. The optimum applicability of such energy and economic performance is corresponding to the legislative protection and initiatives; the methodology is helpful for guidance of office building retrofit works.

Acknowledgements

This work was supported by the Science and Technology Planning Project of Guangdong Province, China (no.2013A011404007, no.2013B091500026, no.2013B010405013, no.2012A010800024).

References

[1] Ashrae 2011 *ASHRAE Handbook-HVAC Applications (SI)* (Atlanta, USA: ASHRAE Inc.)

[2] International Energy Agency 2007 *Energy Balances of OECD/non-OECD Countries* (Paris, France: International Energy Agency)

[3] Department of Energy statistics, National Bureau of Statistics, People's Republic of China 2015 *China Energy Statistical Yearbook* (Beijing, China :China Statistics Press)

[4] Tsinghua University 2015 *Building Energy Use in China* (Beijing, China: Tsinghua University) https://www.iea.org/publications/freepublications/publication/PARTNERCOUNTRYSERIESBuildingEnergy_WEB_FINAL.pdf
[5] International Energy Agency 2015 *Energy Efficiency Market Report* (Paris, France: International Energy Agency) https://www.iea.org/publications/freepublications/publication/energy-efficiency-market-report-2020-.html
[6] M J Kelly 2009 Retrofitting the existing UK building stock *Building Research and Information* **37** 196–200
[7] Huang Y, Niu J L and Chung T M 2011 Energy and carbon emission payback analysis for energy-efficient retrofitting in buildings-Overhang shading option *Energy and Buildings* **44** 94–103
[8] Koester R J, Eflin J and Vann J 2006 Greening of the campus: a whole-systems approach *J. Cleaner Production* **14** 769–79
[9] The Central People's Government of the People's Republic of China 2013 *Action Plan of Green Building* http://www.gov.cn/zwgk/2013-01/06/content_2305793.htm
[10] Zhou Z, Zhang S, Wang C, Zuo J and He Q 2015 Achieving energy efficient buildings via retrofitting of existing buildings *J. Cleaner Production* **112** 3605-15
[11] Papadopoulos A M 2005 State of the art in thermal insulation materials and aims for future developments *Energy and Buildings* **37** 77–86
[12] Al-Homoud M S 2005 Performance characteristics and practical applications of common building thermal insulation materials *Building and Environment* **40** 353–66
[13] Bojic M, Yik F and Sat P 2002 Energy performance of windows in high-riseresidential buildings in Hong Kong *Energy and Buildings* **34** 71–82
[14] Stegou-Sagia A, Antonopoulos K, Angelopoulou C and Kotsiovelos G 2007 The impact of glazing on energy consumption and comfort *Energy Conversion and Management* **48** 2844–52
[15] Happio A and Viitaniemi P 2008 A critical review of building environmental assessment tools *Environmental Impact Assessment Review* **28** 469-82
[16] Ng S T, Chen Y and Wong J M W 2013 Variability of building environmental assessment tools on evaluating carbon emissions *Environmental Impact Assessment Review* **38** 131–41
[17] C Buratti, E Moretti, E Belloni, F Cotana, Unsteady simulation of energy performance and thermal comfort in non-residential buildings, Building and Environment 59 (2013) 482–491
[18] Terziotti L T, Sweet M L and McLeskey J T 2012 Modeling seasonal solar thermal energy storage in a large urban residential building using TRNSYS 16 *Energy and Buildings* **45** 28–31
[19] Gero J, Neville D and Radford A 1983 Energy in context: a multi-criteria model for building design *Building and Environment* **18** 99–107
[20] Goodacre C, Sharples S and Smith P 2002 Integrating energy efficiency with the social agenda in sustainability *Energy and Buildings* **34** 53–61
[21] Asadi E, Silva M G D, Antunes C H and Dias L 2012 Multi-objective optimization for building retrofit strategies: a model and an application *Energy and Buildings* **44** 81–7

[22] Terés-Zubiaga J, Campos-Celador A, González-Pino I and Escudero-Revilla C 2015 Energy and economic assessment of the envelope retrofitting in residential buildings in northern Spain *Energy and Buildings* **86** 194–202

[23] Yoon J, Lee E J and Claridge D E 2003 Calibration procedure for energy performance simulation of a commercial building *J. Solar Energy Engineering* **125** 251-7

[24] Cowan J, Kromer S, Claridge D E, Schiller S 1984 International performance measurement & verification protocol: Concepts and options for determining energy and water savings *Journal of Futures Markets* **4** 531-557.

[25] Lam J C, Wan K K W, Yang L 2008 Sensitivity analysis and energy conservation measures implications *Energy Conversion and Management* **49** 3170-7

[26] Lam J C and Hui S C M 1996 Sensitivity analysis of energy performance of office buildings *Building and Environment* **31** 27-39

[27] Nikolaidis Y, Pilavachi P A and Chletsis A 2009 Economic evaluation of energy saving measures in a common type of Greek building *Applied Energy* **86** 2550–59

Study on the oxidative stability of poly α-olefin aviation lubricating base oil using PDSC method

N Wu[1,2], Y W Fei[2], H W Yang[2], Y M Wang[2] and Z M Zong[1,3]

[1] School of Chemical Engineering & Technology, China University of Mining and Technology, Jiangsu Xuzhou 221116, China
[2] Department of Aviation oil and material, Air Force Logistics Institute, Jiangsu Xuzhou 221000, China

Email: wunan_china@163.com

Abstract. The oxidation stability of the domestic and import PAO aviation lubricating base oil was studied by the method of pressurized differential scanning calorimetry testing the initial oxidation temperature. The effects of anti-oxidants were investigated, and the best ratio of antioxidants was determined.

1. Introduction

Poly-α-olefin (PAO) aviation lubricating oil has many high performances, such as good viscosity-temperature performance, low pour point, high flash point, low temperature fluidity, wide operating temperature range, high hydrolysis resistance, and good compatibility with petroleum refining base oil[1,2]. It can ensure the mechanical device playing excellent performance in harsh environment. During engine operation, the temperature of aviation lubricating oil in the tank will be up to 160℃, while the temperature will be higher between the friction disks. In high-temperature environment, the decay of the lubricating oil will lead to the viscosity-rising, the acid value-increasing, sediment and deposit in contact with air and under the catalytic action of copper, iron and other metals[3]. Eventually the lubricating oil will become invalid, which will affect the performance of aircraft engines, make the shaft-braking and other catastrophic failure [4,5]. Therefore, oxidation stability is an important index of lubricating oil quality, which directly affecting the performance of lubricating oil and determining the service life.

Methods to evaluate the oxidation stability of lube oil are various, such as rotary oxygen bomb test, openings in vitro oxidation test, etc. Most of these methods have the disadvantages of large amount of samples, operating complicated, measuring a long time, poor reproducibility and other shortcomings. Early differential scanning calorimetry (DSC) simulates the oxidation process under boundary lubrication conditions. But the method will cause great error analysis, due to the large number of lubricating oil evaporation and absorb heat under the condition of high temperature [7-10]. In recent years, high-pressure differential scanning calorimetry (PDSC) analysis develops on the basis of the

[3] Address for correspondence: Z M Zong, School of Chemical Engineering & Technology, China University of Mining and Technology, Jiangsu Xuzhou, 221116, China. Email: wunan_china@163.com

DSC, which is a very effective method to evaluate the oxidation stability of lubricating oil, overcoming the shortcoming of DSC with trace, rapid, accurate and other advantages [11].

In this paper, PDSC is used to investigate the oxidation stability of domestic and imported poly-α-olefin aviation lubricants, as well as examine both lubricating base oil sensitivity of three commonly used antioxidants.

2. Experimental

2.1. The main materials and equipment

The lubricating oil used in the experiment is from Shanghai FOX Chemical Technology Company and an oil research institute. The physical and chemical properties of the two base oils are shown in Table 1. The three antioxidants used in the experiment are shown in Table 2. The PDSC is PDSC8000 produced by Shanghai Perkin Elmer.

Table 1. Basic analysis of the domestic and import PAO aviation lubricating base oil

Test Item	Domestic PAO	Import PAO	Test Method
Kinematic Viscosity (40℃) ($mm^2\ s^{-1}$)	17.91	17.94	GB/T 265
Kinematic Viscosity (100℃) ($mm^2\ s^{-1}$)	4.817	4.018	GB/T 265
Viscosity Index	121	131	GB/T 1995
Flash Point(COC) (℃)	214	201	GB/T 3536
Pour Point (℃)	-62	-68	GB/T 3535
Acid Number (mg KOH g^{-1})	0.01	0.009	GB/T 7304

Table 2. The antioxidants used in the experiments

Item	Type	Formula
2,6-diisopropyl-4-methylphenol (T501)	Phenolic antioxidant	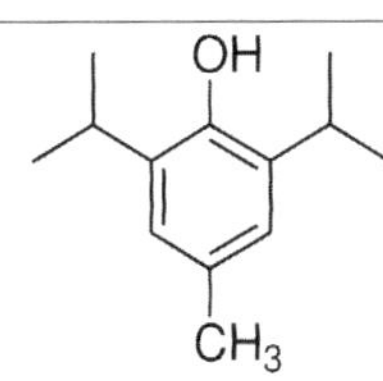
Bis (4-℃tylphenyl) amine (Tz516)	Amine antioxidant	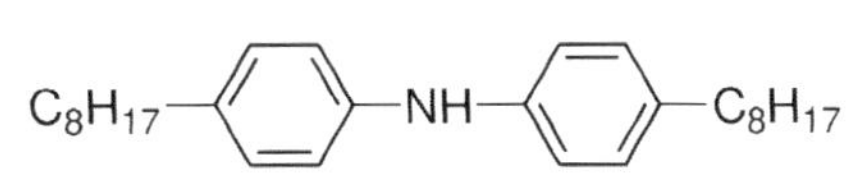
N-phenylnaphthalen-1-amine (T531)	Amine antioxidant	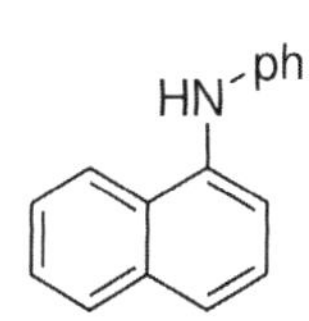

2.2. Experimental methods

Using the programmed temperature method, the oil is heated gently in a certain pressure of oxygen atmosphere. And the initial oxidation temperature (IOT) is measured to characterize oxidation stability of the sample. The higher the initial oxidation temperature is, the better the oxidation stability. Experimental measurement conditions: temperature rise rate is 10 ℃/min, oxygen pressure is 1.0-1.5MPa, oxygen flow rate is 100mL / min, 6mm opening aluminum pan, dosage of samples is 3 μg.

3. Results and discussion

3.1. Comparison of oxidation stability of two aviation lube base oil

The initial oxidation temperature of domestic and imported PAO aviation lubricating oil was shown in table 3.

Table 3. PDSC results of two PAO aviation lubricating base oil

Item	IOT (℃)	FOT (℃)	Enthalpy (J g^{-1})
domestic PAO4	202.79	218.80	-2572.83
imported PAO4	205.99	226.61	-3275.96

The oxidation of base oil is an exothermic reaction. When the oil is heated in the oxygen air, there will be an obvious peak in the PDSC curve as soon as the base oil begins to oxidize, and the rate of heat release and is corresponding to the speed of oxidation reaction. The initial oxidation temperature of base oil and the fastest oxidation temperature are lower, that the oil oxidation stability is poor. It can be seen from Table 3, the domestic PAO began to oxidation at 202 ℃, and reached maximum oxidation rate at 218 ℃; while the imported PAO began to oxidation at 205 ℃, and reached maximum oxidation rate at 226 ℃. As a result, the oxidation stability of the domestic PAO is similar to the imported with a little difference.

3.2. Study on the susceptibility of two base oil to oxidant T501

Add the oxidant T501 into domestic and imported PAO of 0.5%, 1.0%, 1.5% and 2.0% by volume, comparing the susceptibility of two lubricating oil receptivity. The results were shown in Table 4.

Table 4. The IOT of two PAO aviation lubricating base oil added T501

Antioxidant content (%)	IOT (℃)	
	Domestic PAO4	Imported PAO4
0	202.79	205.99
0.5	206.47	207.83
1.0	207.15	209.91
1.5	208.58	210.21
2.0	207.91	210.36

It can be seen from the table that the IOT of two kinds of PAO base oil with T501 increased slightly, and will gradually increase with the increase of the antioxidant content. But the upward trend is not obvious, the temperature changes all-around 5 ℃, when the antioxidant content is more than 1.5%, the rise in temperature tends to be gentle.

3.3. Study on the susceptibility of two base oil to oxidant T531

Add the oxidant T531 into domestic and imported PAO of 0.5%, 1.0%, 1.5% and 2.0% by volume, comparing the susceptibility of two lubricating oil receptivity. The results were shown in Table 5.

Table 5. The IOT of two PAO aviation lubricating base oil added T531

Antioxidant content (%)	IOT(℃)	
	Domestic PAO4	Imported PAO4
0	202.79	205.99
0.5	232.26	231.77
1.0	236.76	236.83
1.5	237.63	240.36
2.0	241.19	240.28

It can be seen from the table that the IOT of two kinds of PAO base oil with T531 increased sharply, while the domestic PAO aviation lubricating oil increased nearly 39℃, and the imported PAO aviation lubricating oil increased nearly 35℃. The IOT of domestic PAO will be higher, when the antioxidant content is more, as well as the imported. But when the antioxidant content exceeds 1.5%, the IOT of the imported PAO changed little.

3.4. Study on the susceptibility of two base oil to oxidant Tz516

Add the oxidant Tz516 into domestic and imported PAO of 0.5%, 1.0%, 1.5% and 2.0% by volume, comparing the susceptibility of two lubricating oil receptivity. The results were shown in Table 6.

Table 6. The IOT of two PAO aviation lubricating base oil added Tz516

Antioxidant content (%)	IOT (℃)	
	Domestic PAO4	Imported PAO4
0	202.79	205.99
0.5	228.97	230.87
1.0	235.71	236.67
1.5	247.53	246.05
2.0	248.86	250.92

It can be seen from the table that the IOT of two kinds of PAO base oil with Tz516 increased sharply, while the domestic PAO aviation lubricating oil increased nearly 46 ℃, and the imported PAO aviation lubricating oil increased nearly 45 ℃. And the IOT of both two kinds of PAO will be higher, when the antioxidant content is more. But when the antioxidant content exceeds 1.5%, the rise in temperature tends to be gentle.

3.5. Discussion

Oxidation of the lubricating oil is a chain reaction of free radicals, while some acidic materials such as aliphatic acids, aldehydes and ketones will be produced in the reaction process, and they will catalyst the oxidation reaction of lubricants[12]. In terms of molecular structure, T531 and Tz516 are amine antioxidants, T501 is phenolic antioxidants. The mechanism of phenolic antioxidants is to react with the free radical produced in the chain reaction, generating more stable chemical substances, thus interrupting the chain reaction, delaying the oxidation rate. While amine antioxidants will react with the acidic materials produced in the oxidation reactions, catalytic reduction of such compounds in the lubricating oil in the oxidation process, decreasing the oxidation reaction rate, improving the oxidation stability of the lubricating oil. The phenolic antioxidants are used in the relatively low temperature range, while the amine antioxidants are higher. The oxidation temperature of PAO usually reached to 200℃ or more. As a result the IOT increased greatly, antioxidant performance significantly, when added the amine antioxidants.

4. Conclusion

It can be find from the results of PDSC that the domestic PAO began to oxidize at about 202℃, accelerated oxidation at 218℃; while imported PAO began to oxidize at about 206℃, accelerated oxidation at 227℃. Thus, the oxidation stability of domestic PAO aviation lubricating oil is slightly lower than the imported ones.

When adding an antioxidant, oxidation stability of domestic and imported PAO aviation lubricants enhanced, both trends are consistent, but different sensitivity to different antioxidant. The IOT of two base oil will increase slightly adding the antioxidants T501 and with increase of the antioxidant content, base oil of incipient oxidation temperature is gradually increased, but the upward trend was not obvious; when adding antioxidants T531 and Tz516, the initial oxidation temperature of two base oils have significantly increased, and with the increase of the antioxidant content, the initial oxidation temperature of the base oil to improve significantly, but when the antioxidant content is more than 1.5%, a rise in temperature tends to be gentle. Thus, the antioxidant property of amine antioxidant is better than phenolic antioxidants for PAO aviation lubricants.

It has been found that the oxidation stability of both domestic PAO and imported PAO were similar and the sensitivity of antioxidants are basically the same. Considering the oxidation stability of the base oil will not change significantly as long as the antioxidant content increase, the best antioxidant dose is 1.5%. So as to maximize play the antioxidant properties of antioxidant and can guarantee the economy.

To sum up, characterized by less sample, simple operation, high sensitivity, high accuracy and low cost, PDSC is widely used in the study of lubricating oil. It can be used to determinate the antioxidant properties of base oil and antioxidants, analyse the correlation between antioxidant activity and antioxidant content of lubricating oil, to calculate the oxidation activation energy and so on.

But this method also has certain limitations. It is needed to combine with other analysis methods to better evaluate the performance of the lubricating oil, because of the difficulty to further analyze the trace oil sample being oxidated.

References

[1] Murphy W R, Blain D A and Galiano-Roth A S 2002 Synthetic basics: benefits of synthetic lubricants in industrial applications *J. Synthetic Lubrication* **18** 301-25
[2] Kioupis L I and Maginn E J 1999 Molecular simulation of poly-α-olefin synthetic lubricants: impact of molecular architecture on performance properties *J. Phys. Chemistry B* **103** 10781-90
[3] Li C X and Wei X 2009 Analysis of compositions and structures of PAO by infrared spectrum *Synthetic Lubricants* **36** 10-3
[4] Guo Q P, Fei Y W, Yao T and Cheng Z S. 2014 Application of pressurized differential scanning calorimetry on thermal-oxidative research of lubricants *Synthetic Lubricants* **41** 28-31
[5] Xiang H and Chen G X 2008 Study on the oxidative stability of coal-to-liquid base oil using pressure differential scanning calorimetry method *Lubrication Engineering* **33** 89-91
[6] Xue W G 2005 Oxidation property evaluation of base stock and antioxidant by pressure differential scanning calorimetry *Synthetic Lubricants* **32** 8-11
[7] Wei J J and Xue Q J 1991 Evaluation of thermal oxidation stability of base stocks and additives

by differential scanning calorimetry *Acta Petrolei Sinica* **7** 104-10
[8] Zhang M, Li J S and Li J Q 2006 Study on the oxidative stability of base oil for biodegradable lubricants using pressure differential scanning calorimetry method *Lubrication Engineering* **12** 152-4
[9] Yu X M, Tan G J and Wang Y J 1999 Studies on oxidation stability of lubricating oils and base oils by PDSC thermal analysis *Lubricants* **14** 48-51
[10] Yao J B 2000 Evaluation of High Temperature Antioxidation Performances For Lubricants by Differential Scanning Calorimetry (DSC) *Synthetic Lubricants* **2** 35-7
[11] Li J S and Zhang M 2008 Study on the efficiency of anti-oxidants in rapeseed oil using PDSC method *J. Analytical Science* **24** 189-92

IOP Conf. Series: Earth and Environmental Science **40** (2016) 012009 doi:10.1088/1755-1315/40/1/012009

Study on High Efficient Electric Vehicle Wireless Charging System

H X Chen[1, 3], Z Z Liu[1], H Zeng[1], X D Qu[1] and Y J Hou[2]

[1]School of Electrical Engineering, Shandong University, Jinan, 250061, China
[2]Shandong Academy of Sciences, LiXia District, Jinan, 250061, China

E-mail: chenhongxing1991@126.com

Abstract. Electric and unmanned is a new trend in the development of automobile, cable charging pile can not meet the demand of unmanned electric vehicle. Wireless charging system for electric vehicle has a high level of automation, which can be realized by unmanned operation, and the wireless charging technology has been paid more and more attention. This paper first analyses the differences in S-S (series-series) and S-P (series-parallel) type resonant wireless power supply system, combined with the load characteristics of electric vehicle, S-S type resonant structure was used in this system. This paper analyses the coupling coefficient of several common coil structure changes with the moving distance of Maxwell Ansys software, the performance of disc type coil structure is better. Then the simulation model is established by Simulink toolbox in Matlab, to analyse the power and efficiency characteristics of the whole system. Finally, the experiment platform is set up to verify the feasibility of the whole system and optimize the system. Based on the theoretical and simulation analysis, the higher charging efficiency is obtained by optimizing the magnetic coupling mechanism.

1. Introduction

At present, China's car ownership exceeded the 163million, ranking second in the world. In recent years, a wide range of concerns about the issue of haze have a greater relationship with car exhaust emissions, promotion of pure electric vehicles is an effective way to alleviate the shortage of energy and environmental pollution. Limited by the capacity of the battery, the current driving range of electric vehicles is relatively short, the construction of battery charging station has become the biggest bottleneck restricting the application and development of electric vehicles [1]. Wireless charging technology allows the moving car to achieve charging, can significantly reduce the requirements of EV battery energy density, and helps to promote the large-scale application of pure electric vehicles. Wireless charging technology has low charging efficiency [2-4], which restricts the development of wireless charging technology, it also brings a great challenge for the adaptation of the electric vehicle manufacturers.

2. Theoretical analysis

The wireless power supply technology is mainly divided into three types: RF or microwave wireless power supply technology, electromagnetic induction type wireless power supply technology and

[3] Address for correspondence: H X Chen, School of Electrical Engineering, Shandong University, Jinan, 250061, China. E-mail: chenhongxing1991@126.com.

IOP Conf. Series: Earth and Environmental Science **40** (2016) 012009 doi:10.1088/1755-1315/40/1/012009

electromagnetic resonant wireless power supply technology. Microwave power transmission is a kind of far field radiation energy transmission mode, because of its low transmission efficiency, and harmful to the human body, so it is not suitable for the electric vehicle wireless charging system [5]. Due to the short transmission distance of electromagnetic induction type wireless power supply technology, so its application in the electric vehicle wireless charging is still relatively small, but electromagnetic induction type wireless power supply technology has a high efficiency, in the field of wireless charging electric vehicles have good prospects [5,6]. Electromagnetic resonance wireless power supply technology has the advantages of long transmission distance and high coupling coefficient, at present, this technology is widely used in the field of electric vehicle wireless charging.

Electromagnetic resonance wireless power supply system has two basic coil structures, namely, two coil structure and four coil structure. Two coil structure is the most convenient way to realize electromagnetic resonance wireless power supply, the four coil structure is based on the 2 resonant coils, and the 2 induction coils are added, so that the power supply is isolated from the transmitting coil, and the load is isolated from the receiving coil [7-9]. This paper takes the two coil electromagnetic resonance wireless power supply system as an example to analyze.

2.1. Circuit structure selection

Electromagnetic resonant wireless power supply system has four common topology structure [11], namely: transmitting coil series resonant, receiving coil series resonant(S-S); transmitting coil series resonant, receiving coil parallel resonant(S-P); transmitting coil parallel resonant, receiving coil series resonant(P-S); transmitting coil parallel resonant, receiving coil parallel resonance(P-P) [12-13]. Each resonant structure is shown in figure 1.

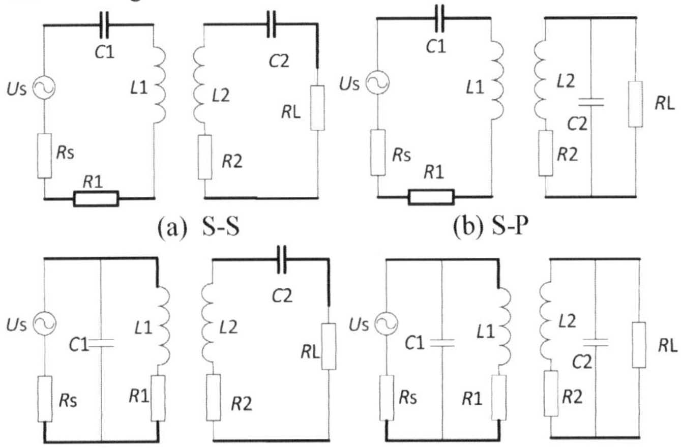

Figure 1. Four resonant structures

The advantages and disadvantages of four kinds of resonance structures are summarized in the literature [14], according to the load characteristics of electric vehicle, we are using more S-S and S-P type resonant structure. The formula of the power and efficiency of the S-S and S-P resonant structures is derived in the literature [15].

S-S type resonant structure:

$$P_{ss} = \frac{(\omega M)^2 R_L}{[r_1 R_L + r_1 r_2 + (\omega M)^2]^2} U^2 \tag{1}$$

$$\eta_{ss} = \frac{(\omega M)^2}{r_1(r_2 + R_L) + (\omega M)^2} \frac{R_L}{r_2 + R_L} \tag{2}$$

S-P type resonant structure:

$$P_{sp} = \frac{(\omega M)^2 R_L}{[(1+j\omega R_L C_2)((\omega M)^2 + r_1 r_2) + j\omega L_2 r_1]^2} U^2 \tag{3}$$

$$\eta_{sp} = \frac{(\omega M)^2 R_L}{(1+j\omega R_L C_2)((\omega M)^2 + r_1 r_2) + j\omega L_2 r_1} \times \frac{1}{(j\omega L_2 + r_2 + j\omega R_L r_2 C_2)} \tag{4}$$

The relationship between the coupling coefficient *K* and the mutual inductance value *M*:

$$M = k\sqrt{L_1 L_2} \tag{5}$$

In the above formula, M indicates the mutual inductance between the transmitting coil and the receiving coil, R_L indicates the load resistance, r_1 and r_2 indicates the resistance value of the transmitting and receiving coils, L_1 and L_2 indicates inductance value of transmitting coil and receiving coil, C_1 and C_2 indicates resonant capacitor at transmitter and receiver. According to the equation (2) and the equation (4), using Matlab to draw the curve of the S-S and the S-P type resonant structure system efficiency with the change of the equivalent load is shown in figure 2.

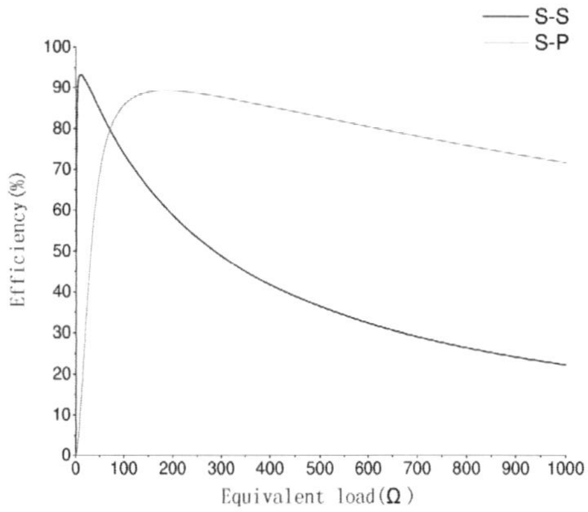

Figure 2. The relationship between system efficiency and equivalent load

From figure 2, when equivalent load resistance is small, the efficiency of S-S type resonant structures is higher than that of the S-P type resonant structures, when equivalent load resistance is large, the efficiency of S-S type resonant structures is lower than that of the S-P type resonant structures. The voltage of the electric vehicle in this system is 74V, if the charging power is 3Kw, the equivalent load is less than 1 ohm, S-S type resonant structure has a high system efficiency.

According to the power equation and efficiency equation of the S-S type resonant structure and the relationship between the coupling coefficient and the mutual inductance, the curve of the system efficiency and the coupling coefficient and the relationship between the transmission power and the coupling coefficient can be drawn. As shown in figure 3.

According to the curve in figure 3 shows: with the increase of coupling coefficient, the system efficiency first increased rapidly, and then gradually stabilized at around 93%, transmission power of the system is the first sharp increase, after a sharp decrease, the peak appears in the coupling coefficient of about 0.5. The system efficiency and transmission power should be considered when match the coupling coefficient between the receiving and transmitting coils. According to the curve shown in Figure 3, when the coupling coefficient is 0.15, the system efficiency is more than 91%, and the system transmission power is more than 7Kw. The system is suitable for high power electric vehicle wireless charging.

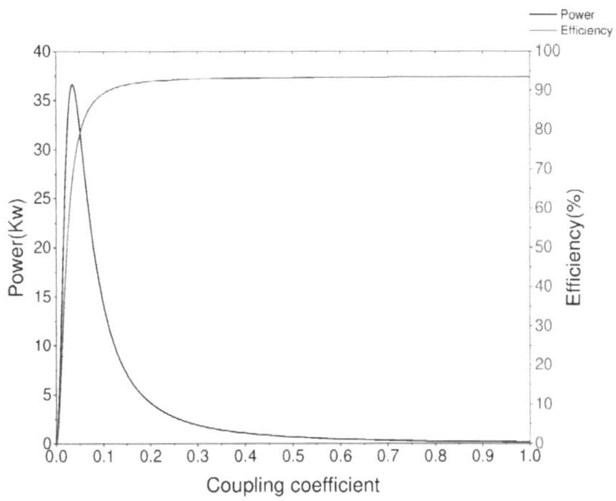

Figure 3. The relationship between system efficiency, transmission power and coupling coefficient

2.2. Optimization of coil structure

The design of coil structure is one of the key points of the design of wireless charging system for electric vehicle [16-17]. A good coil structure should have the characteristics of high level of freedom, moderate coupling coefficient, moderate distance between receiving and transmitting coils, less leakage, and no heating of the coil. At present, most of the research institutions in China and abroad on the coil part of the research are mostly the same as the literature [18], Tending to have a higher coupling coefficient structure in order to obtain a higher system efficiency. But in practical application, different coupling coefficients should be chosen according to different working conditions. Use Ansys Maxwell software to model the common coil structure as shown in figure 4.

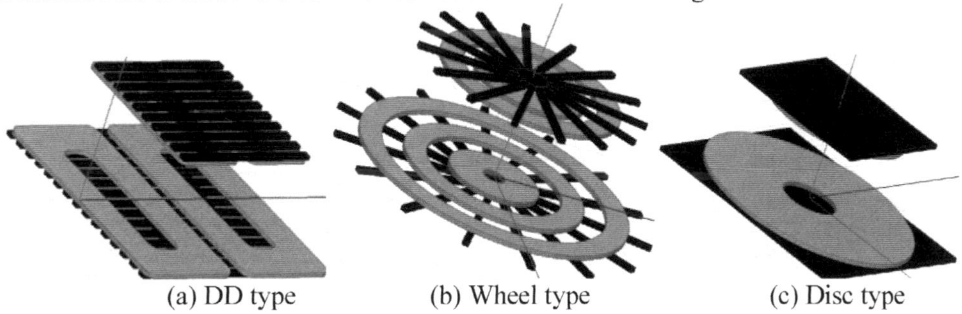

(a) DD type (b) Wheel type (c) Disc type

Figure 4. Coil structure

In figure 4, the black section represents the core of ferrite material, the yellow part represents the power coil. The diameter of the disc type transmitting coil is 600mm, and the diameter of the receiving coil is 400mm, the most outer coil diameter of the wheel transmission coil is 600mm, and the coil diameter of the receiving coil is 400mm, DD type transmitting part and the receiving part are two coils, the transmitting coil length was 600mm, the receiving coil length was 400mm. When the car parking, there will be a certain deviation, causing the transmission coil and the receiving coil has a certain distance. In this paper, under the premise of the vertical distance between the transmitter coil and the receiving coil is 200mm, horizontal migration experiment of three kinds of coils is carried out. Due to the asymmetry of the DD coil structure, the horizontal migration experiment of X and Y direction is required for three kinds of coils, the experimental results are shown in figure 5.

Combination of figure 5 and figure 3, in the X direction, the coupling coefficient of the DD type coil structure is maintained between 0.14 and 0.165, and the system efficiency is higher, in the Y direction, the coupling coefficient of the DD coil structure is maintained between 0.06 and 0.165, system efficiency decreased from 92% to about 70%. In the X (Y) direction, the coupling coefficient of the Disc type coil structure is maintained between 0.095 and 0.155, system efficiency is maintained at more than 85%. In the X (Y) direction, the coupling coefficient of the Wheel type coil structure is maintained between 0.04 and 0.14, system efficiency is gradually reduced to 65%. When the electric vehicle wireless charging, receiving coil and the transmitting coil will inevitably be some offset. Due to the symmetry of the disc structure, the efficiency in each direction of the shift will not be reduced sharply, the system has been in the state of high efficiency, so the disc type is more suitable for the actual electric vehicle wireless charging system.

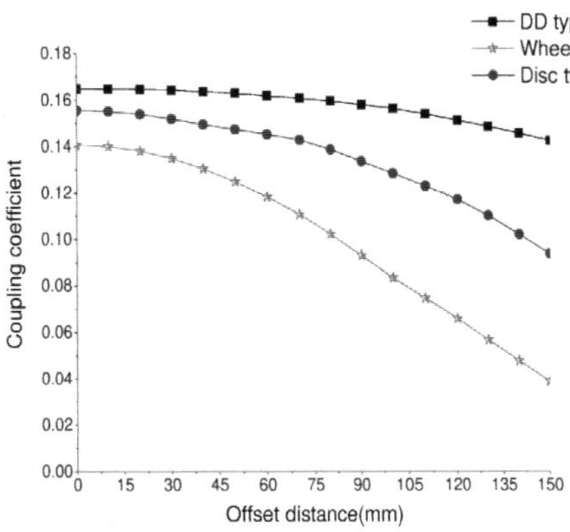

Figure 5. (a) X direction

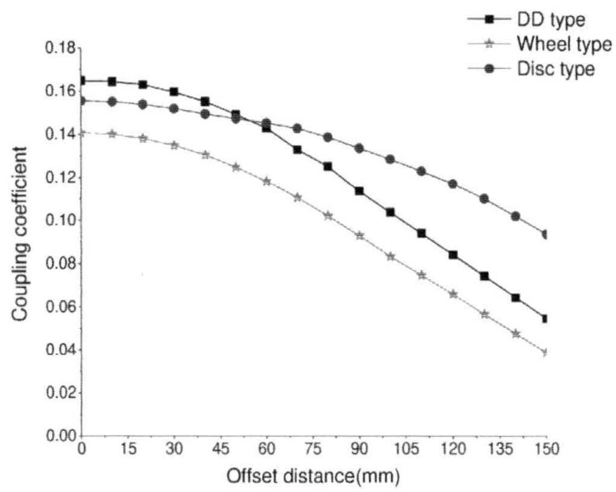

Figure 5. (b) Y direction

Figure 5. Relationship between offset distance and coupling coefficient

3. System simulation

According to the previous section, the simulation system with S-S resonator, coupling coil-coupling coefficient is set to 0.15, input transceiver coil inductance and resistance values, through the Simulink toolbox to build electric vehicle wireless charging simulation system as shown in figure 6.

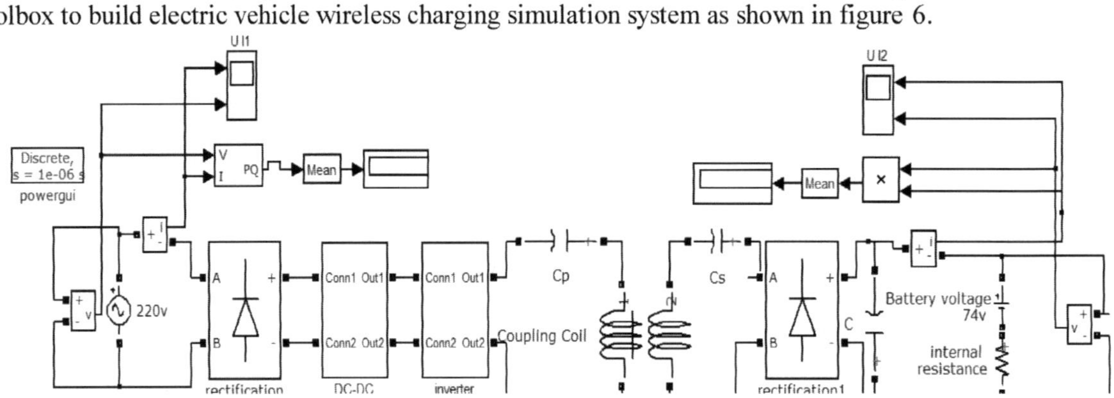

Figure 6. Wireless charging simulation system for electric vehicle

In order to ensure the system has a good applicability, this simulation system uses single-phase 220V AC power supply as the input power, after power frequency rectifier, DC chopper, and full bridge inverter and then input coupling coil part, the energy is transmitted from the transmitting coil to the receiving coil, after high frequency rectifier to charge the electric vehicle. In the simulation, due to the load characteristics of the battery are greatly changed, simulation using a resistor instead of a battery has a big difference between the actual situations. Through several experimental results found that the combination of DC source to represent the battery voltage and resistor represents the internal resistance of the battery closer to the battery load characteristics.

The output power of the system can be adjusted by adjusting the output voltage of the DC-DC, the output power and efficiency of the system under different input voltage can be calculated according to the power of the input side and the output side, as shown in figure 7.

According to the simulation curve of output power and efficiency of the system under different input voltage, it can be known that the output power of the electric vehicle wireless charging system can be more than 3Kw, the efficiency of the electric vehicle wireless charging system can be more than 90%, meet the demand of electric vehicle charging system for high power and high efficiency.

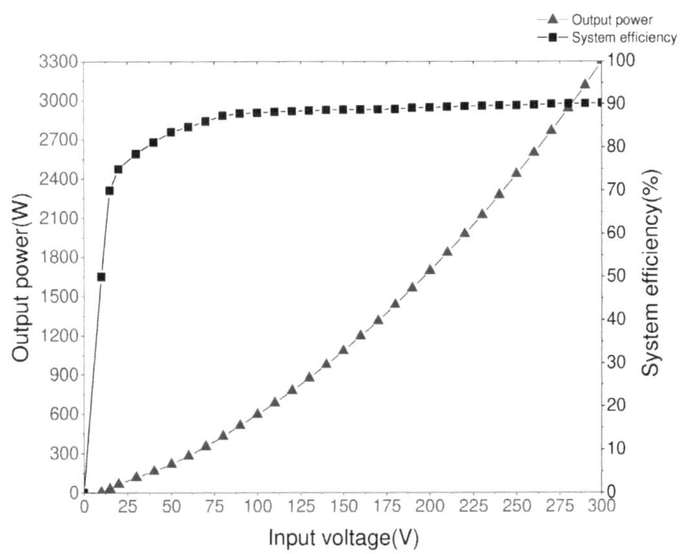

Figure 7. Simulation curve of output power and system efficiency

4. Experiment and conclusion

According to the analysis of the above chapters, this system chooses the S-S type resonant structure, and the disk type coil coupling structure to build the experimental system and carries on the loading experiment. The experimental platform is shown in figure 8.

Figure 8. (a) Battery pack experiment

Figure 8. (b) Installation experiments

Figure 8. Experiment platform

The battery voltage of the experimental vehicle is 74V, the vertical distance between the receiving and transmitting coils is 200mm, and the coupling coefficient K is 0.15. When the receiving coil horizontal offset 0 cm, 5 cm, 10 cm, the experimental results are shown in figure 9.

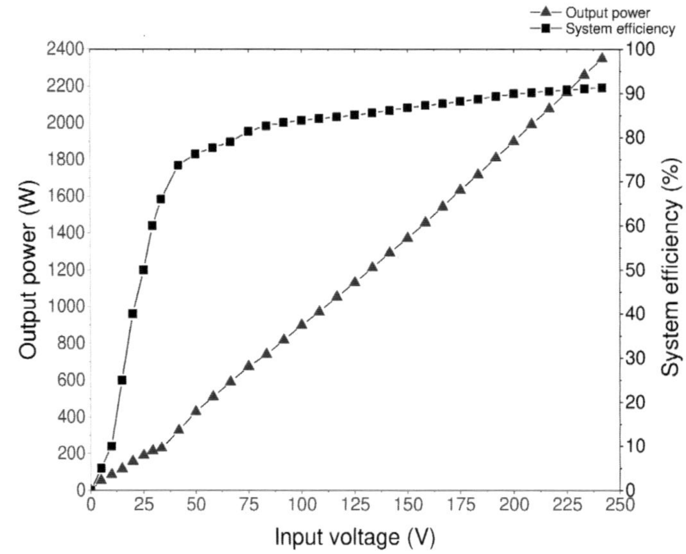

(a) Horizontal offset 0 cm

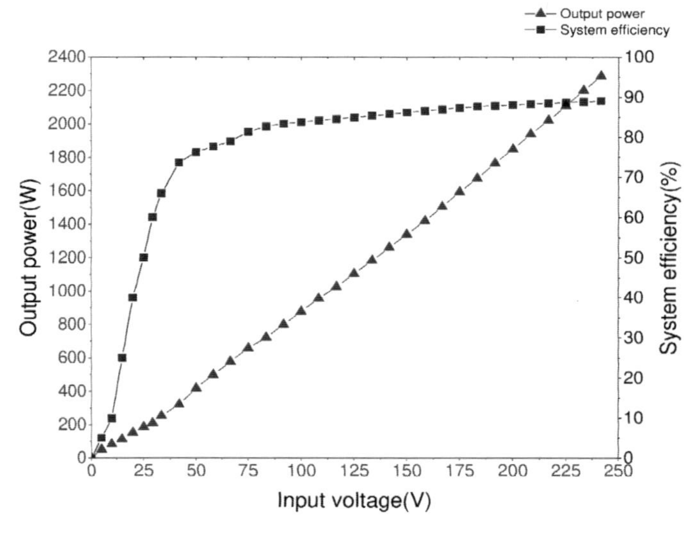

(b) Horizontal offset 5 cm

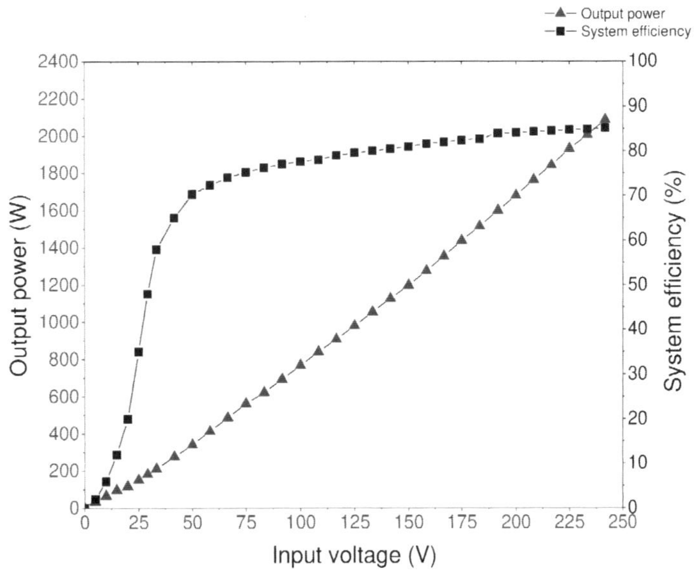

(c) Horizontal offset 10cm

Figure 9. Experiment result

Due to the limitation of the duty cycle of the DCDC circuit, the maximum input voltage of the experiment is 250V. It can be known from Figure 9 that when horizontal offset 0 cm, the input voltage is 250V, the system efficiency reaches 89%, and the output power exceeds 2200W; when horizontal offset 5 cm, the system efficiency and output power are increased slightly compared to the horizontal offset 0 cm; when horizontal offset 10 cm, the output power is reduced to about 2000W, and the system efficiency is reduced to about 85%. Thus, the electric vehicle wireless charging system designed in this paper can obtain higher system efficiency and output power in a large degree of freedom. The experimental results are very consistent with the simulation results; the system has reached the design requirements. But the input voltage is low and the battery voltage of the experimental car is too low, which results in the low output power of the system, in the following experiments, it is also needed to find a way to get more power and more efficient.

References

[1] Chen Z Z and Wu Z J 2008 Construction of EVcharging station *PopularUtilization of Electricity* **6** 34-6
[2] Wang J H *et al.* 2008 Research on detachable transformer for contactless energy transmission system *Transformer* **45** 26-9
[3] Fernandez C *et al.* 2008 Overview of different alternatives for the contact-less transmission of energy *Proc. the IEEE Annual Conference of the Industrial Electronics Society* 1318-23
[4] Wang S, Covic G A and Stielau O H 2004 Power transfer capability and bifurcation phoenomena of loosely coupled inductive power transfer systems *IEEE Transacitons on Inductirial Electronics* **51** 148-57
[5] Wang Z Y *et al.* 2014 Advances of wireless charging technology in electric vehicle *J. Power Supply* **3** 27-32
[6] Boys J T, Covic G A and Green A W 2000 Stability and control of inductively coupled power transfer systems *IEE ProcElectrical Power Application* **147** 37-43
[7] Zhao Z M, Zhang Y M and Chen K N 2013 New progress of magnetically-coupled resonant wireless power transfer technology *Proc. the Chinese Society for Electrical Engineering* **33**

1-13

[8] Lu W C *et al.* 2015 Comparison of Series Series Model and Series Parallel Model of Wireless Power Transmission Via Magnetic Resonance *Power Electronics* **49** 73-5

[9] Kurs A *et al.* 2017 Wireless power transfer via strongly coupled magnetic resonances *Science* **317** 83-6

[10] Moon S C *et al.* 2014 Analysis and design of a wireless power transfer system with an intermediate coil for high efficiency *IEEE Transactions on Industrial Electronics* **61** 5861-70

[11] Tan L *et al.* 2010 Study of wireless power transfer system through strongly coupled resonances *International Conference on Electrical and Control Engineering* 4275-78

[12] Li Y *et al.* 2012 Influence factors analysis on power and efficiency in wireless power transfer system *Advanced Technology of Electrical Engineering and Energy* **31** 31-4

[13] Chunbo Z *et al.* 2008 Simulation and experimental analysis on wireless energy transfer based on magnetic resonances *Proc. the Vehicle Power and Propulsion Conf.* 1-4

[14] Cheng L M and Cui Y L 2012 Magnetic Coupling Resonant Type Wireless Power Transmission Technology Study Progress *Electrotechnics Electric* **12** 1-5

[15] Ren X F 2014 *Desing and Performance Optimization of the Wireless Charging System in Electric Vehicles* (Harbin: Harbin Institute of Technology)

[16] Yu C L , Lu R G , Mao Y H, Zhu C 2009 Research on the model of magnetic resonance based wireless energy transfer system *Proc. the Vehicle Power and Propulsion Conf.* 414-8

[17] Zhang X *et al.* 2010 The Application of Non-contact Power Transmission Technology (NPT) in the Modern Transport System *Proc. the 2010 IEEE International Conf. on Mechatronics and Automation* 345-9

[18] Budhia M *et al.* 2011 Development and evaluation of single sided flux couplers for contactless electric vehicle charging *IEEE Energy Conversion Congress and Exposition* 614-21

Optimization of unipolar magnetic couplers for EV wireless power chargers

H Zeng[1, 2], Z Z Liu[1], H X Chen[1], B Zhou[1] and T Hei[1]

[1]School of Electrical Engineering, Shandong University, Jinan, 250061, China

E-mail: zenghaohust@163.com

Abstract. In order to improve the coupling coefficient of EV wireless power chargers, it's important to optimize the magnetic couplers. To improve the coupling coefficient, the relationship between coupling coefficient and efficiency is derived, and the expression of coupling coefficient based on magnetic circuit is deduced, which provide the basis for optimizing the couplers. By 3D FEM simulation, the optimal core structure and coils are designed for unipolar circular couplers. Experiments are designed to verify the correctness of the optimization results, and compared with previous coupler, the transmission efficiency is improved and weight is reduced.

1. Introduction

EV (Electric vehicle, EV) wireless charging (shown in figure 1) is an emerging technology in recent years. Compared to wired charging system, wireless charging technology can reduce aboveground space, apply to the harsh environment and have no risk of touch spark [1, 2]. Besides, it is expected to be used for mobile power supply for solving the problem that battery energy density is hard to be improved. Therefore, the research and application of wireless charging technology has been getting more and more attention. Generally, EV chargers are required to equip high transfer efficiency and high power output capability, but output power and efficiency is the key factor restricting the development of wireless charging technology.

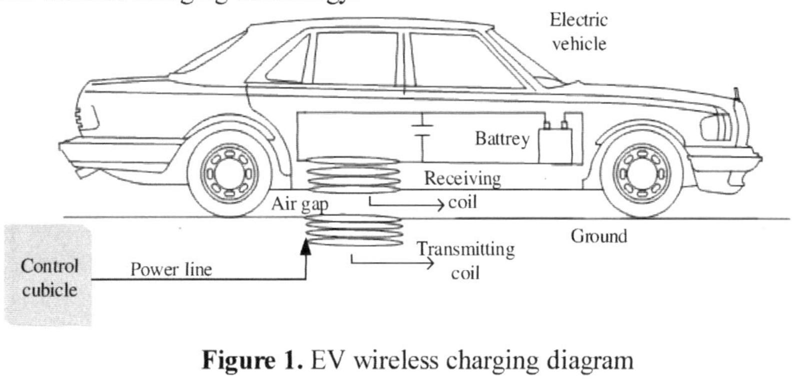

Figure 1. EV wireless charging diagram

[2] Address for correspondence: H Zeng, School of Electrical Engineering, Shandong University, Jinan, 250061, China. E-mail: zenghaohust@163.com.

To improve the efficiency and output power of wireless chargers, the design of magnetic couplers play a key role. In EV wireless charging systems, the air gap (shown in figure 1) between transmitting and receiving coils always changes from 150mm to 250mm. In this situation, leakage inductances of the coils are much larger than the mutual inductance, resulting that coupling coefficient is very small, which is always lower than 0.2. To improve the output power and efficiency, the key is to improve the coupling coefficient of loosely coupled coils.

At present, one of the common way to improve the coupling coefficient is changing the structure of coil, another is add more magnetic cores, and there are many papers have done some relevant research in designing the magnetic coupler. On the basis of assurance on the coupling coefficient, paper [3] optimized the core structure by finite element analysis and experiments, reducing the weight of the magnetic couplers. Paper [4] explored the impact of the number of cores on the coupling coefficient for DD (Double-D, DD) coils and unipolar coils, aiming at finding the maximum coupling coefficient. Paper [5] proposed a form of DLDD (Double Layer Double D-type, DLDD) coil structure, improving the transmission distance and the scope of the charge. But the above papers did not construct a theory to illustrate the number and the shape of magnetic cores how to influence coupling coefficient, so they always lack of direction of optimization. Besides, the above papers designed transmitting and receiving coils in same size, but in practice receiving coil is generally small.

Unipolar magnetic coupler is studied in this paper, which is one of the most popular couplers at present. Firstly, the paper establish relevant theoretical models of electric and magnetic circuits, providing the guidance for the design of the core structure of unipolar coils. Secondly, in order to achieve the desired maximum coupling coefficient with minimal use of materials, the core structure is designed and optimized by 3D finite element simulation. Finally, experiments are designed to verify the theoretical and simulation results. All the receiving coil in this paper is smaller than the transmitting coil.

2. Analysis of electric and magnetic circuits

Analysis of electric circuit is more common than magnetic circuits' analysis. Therefore, in this section a brief explanation is given about circuit analysis, but giving a more detailed description of the magnetic circuit analysis.

2.1. Circuit analysis

In EV wireless charging systems, depending on the connection type of capacitances and coils, resulting in four compensation ways, which are SS, SP, PS and PP. Figure 2 shows an equivalent circuit diagram of SS topology. Studies have shown that, circuit using SS topology may be equivalent to a voltage source to external circuit, and easy to achieve a large output power in a smaller coupling coefficient [6], which is more suitable for wireless charging of EV. This is why the paper selects SS topology as an object of study for EV wireless charging.

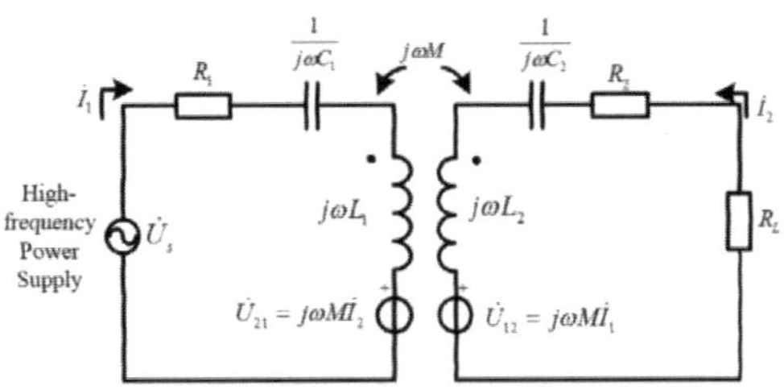

Figure 2. An equivalent circuit diagram of SS topology

From analysis of the circuit, the output power P_{out} can be derived as

$$P_{out} = I_2^2 R_L = \left(\frac{\omega M I_1}{R_2 + R_L} \right)^2 R_L \tag{1}$$

If coupling coefficient is represented by k, equation (1) can be written in another form:

$$P_{out} = \left(\frac{\omega k \sqrt{L_1 L_2} I_1}{R_2 + R_L} \right)^2 R_L \tag{2}$$

From the equation (1), we can get to a conclusion: in the condition that the primary current I_1 is constant, the output power is being large as the coupling coefficient increases.

The input power can be derived as

$$P_{in} = I_1^2 R_1 + I_2^2 (R_2 + R_L) \tag{3}$$

If the quality factors of transmitting and receiving coils are represented by Q_1 and Q_2, where

$$Q_1 = \frac{\omega L_1}{R_1}, Q_2 = \frac{\omega L_2}{R_2} \tag{4}$$

Then, the transmission efficiency η can be derived as

$$\eta = \frac{P_{out}}{P_{in}} = \frac{R_L}{\frac{(R_2 + R_L)^2}{k^2 Q_1 Q_2 R_2} + R_2 + R_L} \tag{5}$$

Equation (5) illustrates that in the case of the transmitting and receiving coils are determined, the transmission efficiency is only related with the coupling coefficient, and the bigger coupling coefficient, the higher the efficiency.

2.2. Magnetic circuit analysis

The magnetic circuit is path that flux go through. Because there is a big gap between the two coils, which are loosely coupled, so the flux path between the two coils can be basically divided into two parts: the first is the flux in magnetic cores near the coils, the second is that in the air gap. Analysis of magnetic circuits of the EV wireless charging systems can reduce flux path's reluctance in a clear direction, which plays a guiding role in optimization of coils and core.

Currently the unipolar magnetic couplers have many kind of shapes, like circular coils, rectangular coils and rectangular coils with rounded corners. But the magnetic field distribution of these couplers is approximately the same, in condition that in one coupler transmitting and receiving coils have only one coil respectively.

The paper below chooses a circular coil coupler with plate cores as an illustration, which is shown in figure 3, then establishes the model of equivalent magnetic circuit and derives the expression of coupling coefficient. Thus, according to the flux path and the expression of coupling coefficient, we can conclude the method that makes the coupling coefficient be larger.

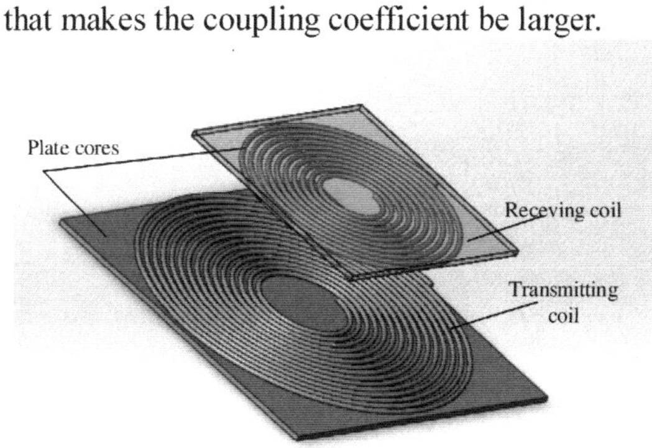

Figure 3. Structure diagram of circular coil coupler with plate cores

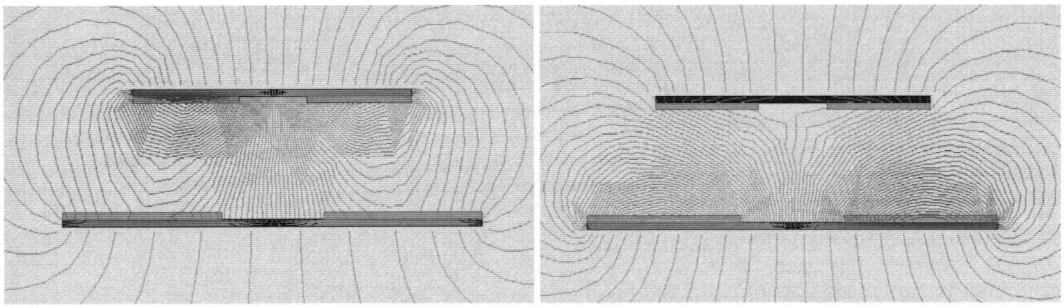

Figure 4. The magnetic field distribution at different times

Figure 4 shows the magnetic field of distribution of a circular coil coupler with plate cores at different times in a cross section. From the two pictures above, the magnetic field distribution of this type of couplers is very clear. Thus, a magnetic partition model is come up with and shown in figure 5.

From figure 5, the magnetic circuit can be deduced by expressing the mutual coupling areas' reluctance with R_m and expressing the self-coupling areas' reluctance with R_s, which is shown in figure 6.

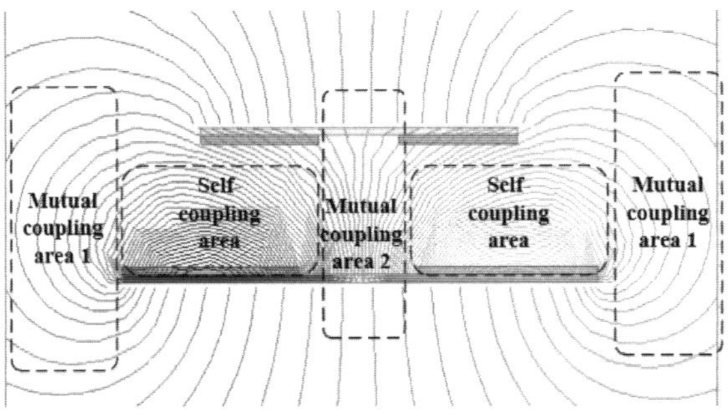

Figure 5. The approximate distribution of self-coupling areas and mutual coupling areas

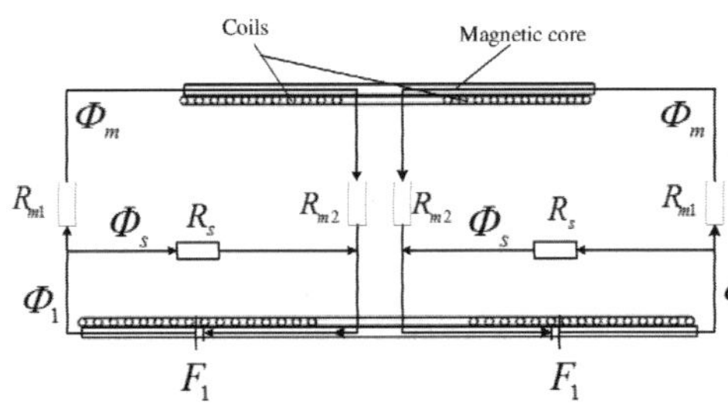

Figure 6. The equivalent magnetic circuit

From figure 3, 4 and 5, the circular coil coupler with plate cores is axial symmetry, so in a cross section its magnetic field distribution should be bilateral symmetry. Because of which that in figure 5 and 6, magnetic partition and equivalent magnetic circuit are modelled bilateral symmetry. In figure 5 and 6, R_{m1} represents the reluctance of mutual coupling area 1, and R_{m2} represents the reluctance of mutual coupling area 2. In figure 6, F_1 represents the MMF (Magnetomotive force, MMF), and Φ_s represents magnetic flux that is self-coupling, and Φ_m represents flux that is mutual coupling. Then, from the equivalent magnetic circuit, we can get

$$\Phi_1 = \Phi_s + \Phi_m \tag{6}$$

Φ_1 is the total flux of one side. Thus, the coupling coefficient can be express as

$$k = \frac{\Phi_m}{\Phi_1} = \frac{R_s // (R_{m1} + R_{m2})}{R_{m1} + R_{m2}} = \frac{1}{1 + \dfrac{R_{m1} + R_{m2}}{R_s}} \tag{7}$$

From equation (7), the method to increase k is to reduce the R_s and magnify ($R_{m1} + R_{m2}$). Then, according to figure 5 and 6, we can change parameters of coils and cores in the direction above-mentioned to increase k.

3. Optimization and simulation of circular coil coupler

3.1. Optimization of magnetic cores

At present the most widely used core structures for circular coil is the plate cores (shown in figure 3) or strip cores which are radial under the coils (shown in figure 7(a)). But the question is we don't know how many strip cores should be used, so firstly using plate cores as a reference, then designing the core structures of the circular coils.

Parameters of the transmitting and receiving coils used in this section are shown in table 1:

Table 1. Coil and core parameters of unipolar circular couplers

	Outer diameter(mm)	Inside diameter(mm)	The thickness of the magnetic cores(mm)	The thickness of the coils(mm)
Transmitting coils	600	200	10	6
Receiving coils	400	150	10	6

Besides, the gap between transmitting and receiving coils is 200mm, and the width of strip cores is 20mm.

3.1.1 Optimization of the number of cores.

At first, three different kind of cores with the same coils are simulated, which are coils with plate cores, coils with less strip cores (15/10, 15/10 means there are 15 strip cores under the transmitting coils and 10 strip cores on the receiving coils, the same below), and coils with more strip cores (20/15), and recording each coupling coefficient between the two coils when the receiving coil is misaligned from the transmitting coil by 10mm every time.

(a) Coils with strip cores (b) Coils with fan-shaped cores (c) Coils with long fan-shaped cores

Figure 7. Three shapes of cores

Then three different curves about coupling coefficient can be sketched, which are shown in figure 8(a).

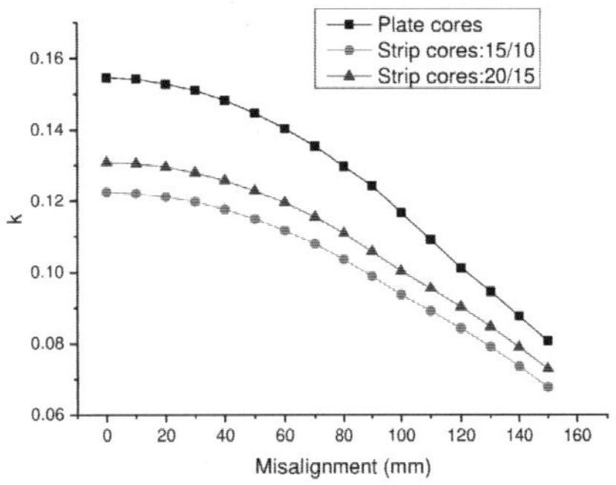

(a) Coupling coefficient with different number of strip cores

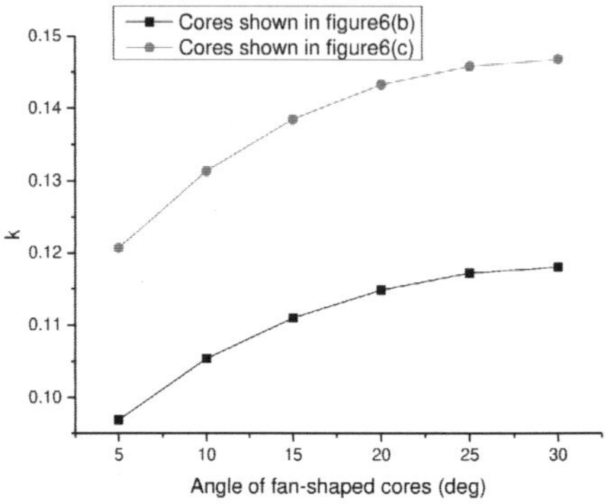

(b) Coupling coefficient with different angle of fan-shaped cores

Figure 8. The variation tendency of coupling coefficient with different cores

From figure 8(a), the number of strip cores has a great influence in coupling coefficient. Thus, models shown in figure 7(b) and figure 7(c) are used to study the impact. Model shown in figure 7(b) adopts 12 fan-shaped cores which are equivalent to strip cores, by the reason that it is convenience in simulation. In FEM simulation, coupling coefficient is calculated every 5 °extension of the fan-shaped cores, then the curves shown in figure 8(b) is obtained.

From figure 8(b), the slope of the curve is slowing as the angle of fan-shaped cores increases, almost unchanged from 25 degree to 30 degree. This means, it is unnecessary to cover the coils with a whole core, and the equivalent area of the core is suitable that designed to cover two-thirds of the area of the coil.

3.1.2 Optimization of the position of cores.

From figure 8(b), not only can we find the regulations of how many cores should be placed, but also we can notice that coupling coefficient of coils with long fan-shaped cores(shown in figure 7(b)) is much larger than coils with short fan-shaped cores(shown in figure 7(c)). It means that the extension of the core will significantly increase coupling coefficient. This is because, as shown in figure 5, extended core reduces the reluctance of mutual coupling area 1, and appropriately increase the reluctance of the self-coupling area, thereby increasing the coupling coefficient.

Thus, a new model is come up with abiding by the method found in 2.2 and shown in figure 9.

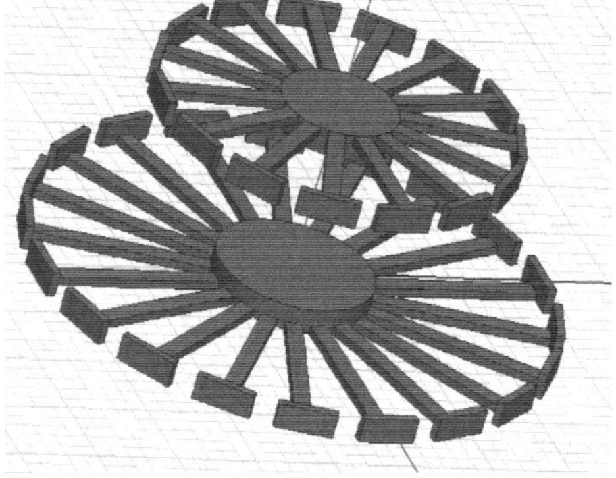

(a) A new model of the cores

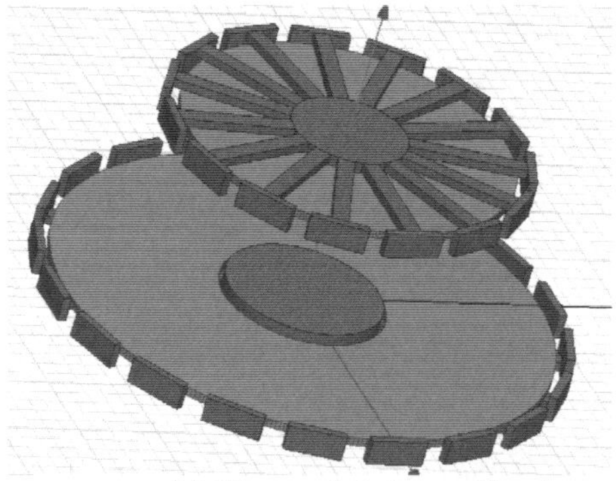

(b) The model in application

Figure 9. A kind of new model of the cores

From figure 9, many short but high strip cores are placed around the coils, and a cylindrical core is placed in the hole of the coil. From figure 5, the short strip cores and the cylindrical cores greatly reduce the reluctance of mutual coupling area, and the coupling coefficient is shown in figure 10.

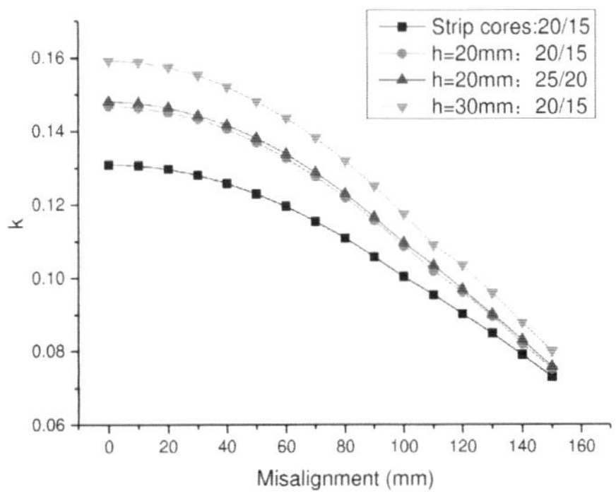

Figure 10. Comparison of coupling coefficient with different cores

In figure 10, "h=20mm" means the height of short strip cores and cylindrical cores is 20mm. Figure 10 compares the coupling coefficient in case of several cores, changing the number of long strip cores and thickness of short strip cores and cylindrical cores respectively, then the conclusion is that the factor that influence the coupling coefficient is mainly the height of the short strip cores and cylindrical cores, while the number of the long strip cores' influence is not obvious.

Thus, to improve the coupling coefficient, it is necessary to follow the rules that increasing reluctance of the self-coupling area and reducing the reluctance of the self-coupling area, and it is effective to make the strip cores' length is longer than the width of the coil, and to place short strip cores and cylindrical cores in the position which is shown in figure 9.

3.2. Optimization of coils

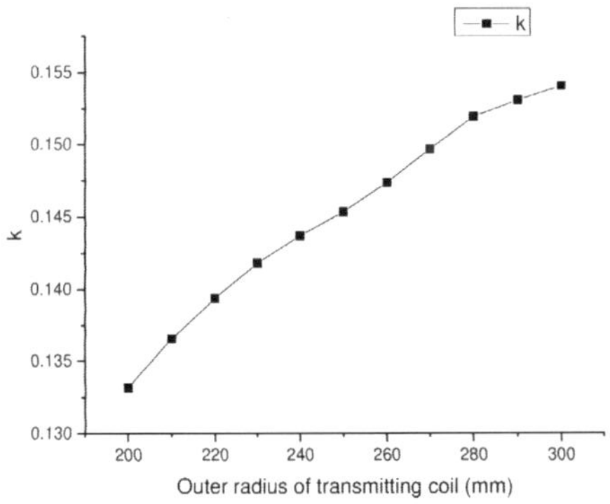

(a)The change of coupling coefficient by changing of transmitting coil's outer radius

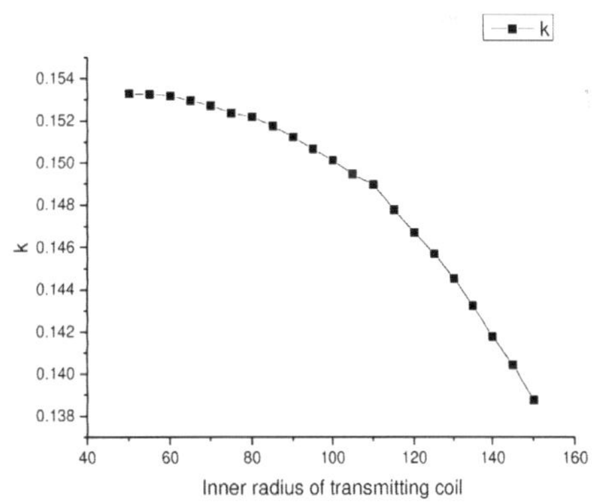

(b) The change of coupling coefficient by changing of transmitting coil's inner radius

Figure 11. The change of coupling coefficient in different situation

From figure 11, the conclusion is that if the transmitting coil inner diameter remains unchanged, the larger the outer diameter of the transmitting coil, the greater the coupling coefficient; and if the transmitting coil's outer diameter remains unchanged, the larger the inner diameter of the transmitter coil, the coupling coefficient is smaller. The conclusion can be also explained by the figure 5 and 6 and equation (7), which is that the larger the outer diameter of the transmitting coil, the bigger R_s is, so the coupling coefficient is larger; and the larger the inner diameter of the transmitter coil, the smaller R_s is, so the coupling coefficient is smaller. The conclusion is suitable for the receiving coil.

IOP Conf. Series: Earth and Environmental Science **40** (2016) 012010 doi:10.1088/1755-1315/40/1/012010

Thus, the bigger the better transmit coil if it is feasible; and the inner diameter of the transmitting coil should be appropriately reduced.

4. Experiment

Based on the above simulation, at first coils are wrapped and cores are placed to test the correctness of the simulation results.

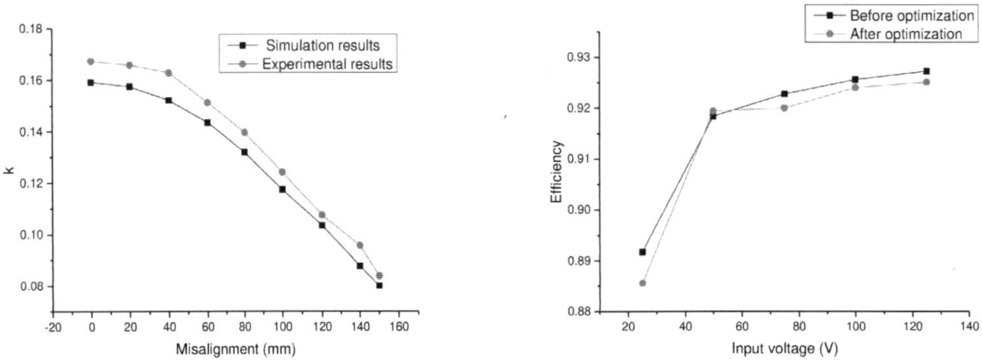

(a) Results comparison of simulation and experiment

(b) Efficiency comparison before and after optimization

Figure 12. Experimental data

From figure 12 (a), the experimental coupling coefficient of the structure shown in figure 9 fits the simulation results. Then, a compare experiment is designed and using the couplers before and after optimization respectively. Parameters of the couplers is shown in table 1, and core structures are plate cores and structure shown in figure 9. And before the experiment, the coupling coefficient of coupler that before optimization is 0.167, and the coupling coefficient of coupler that after optimization is 0.163.

The efficiency in figure 12 (b) refers to DC-in to DC-out. From figure 12(b), the efficiency curves of before and after optimization are basically the same, which proves the availability of structural optimization.

And the transmitting and receiving coil are packaged by aluminium case, which can be used for magnetic field shielding, and assembled on an EV whose model is Cloud 100 manufactured by ZOTYE AUTO. The optimized couplers performed as well as the primary coils, but the weight of the optimized couplers is much lighter than the primary couplers.

Receiving terminal

Vehicle chassis

Transmitting terminal

Garage floor

Figure 13. Transmitting and receiving coil assembled on an EV in experiments

5. Conclusion

In order to improve the efficiency and output power of EV wireless power chargers, the coupling coefficient needs to be large. In this paper, the unipolar magnetic couplers for EV wireless charging has been optimized. From the circuit of SS topology, the equations of output power and transmission efficiency is derived, illustrating the importance of the coupling coefficient; by means of 2D and 3D FEM simulation, the magnetic circuit of circular unipolar coupler is determined; then, the coupling coefficient's expression based on magnetic circuit is obtained. According to the theory, the parameters that influence the coupling coefficient has been researched and a new kind of core structure has been designed for circular unipolar couplers. Compared to the original couplers, the weight of the optimized couplers is reduced and the coupling coefficient is increased. Finally, the simulation results is verified by the experiment, and the practicability of the optimized coupler has been proved by the efficiency experiment.

References

[1] Mohrehkesh S and Nadeem T 2011 Toward a wireless charging for battery electric vehicles at traffic intersections *2011 14th International IEEE Conference on Intelligent Transportation Systems (Itsc)* 113-8
[2] Zhong W X, Xun L and Hui S Y R 2011 A novel single-layer winding array and receiver coil structure for contactless battery charging systems with free-positioning and localized charging features *IEEE Trans. Ind. Electron* **58** 4136–44
[3] Budhia M, Covic G A and Boys J T 2011 Design and optimization of circular magnetic structures for lumped inductive power transfer systems *IEEE Trans. Power Electron* **26** 3096–108
[4] Zhang W, White J C, Abraham A M and Mi C C 2015 Loosely coupled transformer structure and interoperability study for EV wireless charging systems *IEEE Trans. Power Electron* **30** 6356–8
[5] Wang Z H, Hu C, Sun Y and Dai X 2015 Design of magnetic coupler for inductive power transfer system based on output power and efficiency *Transactions of China Electrotechnical Society* **30** 29-30
[6] Li S Q and Mi C C 2015 Wireless power transfer for electric vehicle applications *IEEE J. Emerging and Selected Topics in Power Electronics* 3 10–1

Classifications of central solar domestic hot water systems

J Y Guo[1,4] , B Hao[2], C Peng[3] and S S Wang[3]

[1]School of Environment and Energy Engineering, Beijing University of Civil Engineering and Architecture, Beijing 100044, China
[2]Shen Zhen Institute of Building Research Co,LTD, Beijing 100080, China
[3]Center of Science and Technology of Construction, Ministry of Housing and Urban-Rural Development, Beijing 100835, China

E-mail: 15810564218@163.com

Abstract. Currently, there are many means by which to classify solar domestic hot water systems, which are often categorized according to their scope of supply, solar collector positions, and type of heat storage tank. However, the lack of systematic and scientific classification as well as the general disregard of the thermal performance of the auxiliary heat source is important to DHW systems. Thus, the primary focus of this paper is to determine a classification system for solar domestic hot water systems based on the positions of the solar collector and auxiliary heating device, both respectively and in combination. Field-testing data regarding many central solar DHW systems demonstrates that the position of the auxiliary heat source clearly reflects the operational energy consumption. The consumption of collective auxiliary heating hot water system is much higher than individual auxiliary heating hot water system. In addition, costs are significantly reduced by the separation of the heat storage tank and the auxiliary heating device.

1. Introduction

Solar energy resources are rich within China; more than two-thirds of the country receives at least 2,500 hours of sunshine each year, creating advantageous conditions for the development and utilization of solar energy. Currently solar energy heat utilization is used primarily for the supply of hot water for domestic purposes, which accounts for more than 90% of solar energy heat utilization in China [1]. Nevertheless, many challenges persist in the application of solar hot water systems in residential buildings. For example, the high price of the adaptive technology can be prohibitive [2]. The cost of hot water supply for 144 residents in Henan, Yunnan, Beijing, Guangdong, Shandong and other places was investigated. Results indicated average costs which exceed 20 yuan/ton, accounting for 18% of the survey samples. Unfortunately, energy management contracts require a price of 40-60 yuan/ton, which greatly exceeds the cost of electric water heaters. In addition, property management companies report great challenges, including high maintenance costs and difficult management of DHW systems [3]. Existing studies have primarily focused on solar collector efficiency, rather than system level analysis. Some researchers have reported the impact of energy consumption of the different systems. Suzhen Lu investigated numerous applications of central solar domestic hot water systems, and suggested that system form has great influence on energy consumption [4]. Rui Yu, et al.,

[4] Address for correspondence: J Y Guo, School of Environment and Energy Engineering, Beijing University of Civil Engineering and Architecture, 100044, China. E-mail: 15810564218@163.com.

simulated the energy consumption of various systems in order to investigate their varying suitability [5]. Some researchers have suggested that both heating methods and heating devices represent core components of hot water supply systems [6, 7]. Liuxiang Zhang, *et al.*, investigated the basic system characteristics in detail which, combined with appropriate technology, can be used in central solar domestic hot water systems [8]. These studies all indicate that the system form has a significant impact on practical applications.

The key to popularization and the reform of technical standards is to develop a system of classification for solar hot water systems. Foreign standards primarily refer to the international standard ISO 9459-2-1995, which is categorized according to the seven features of solar hot water systems: the relationship between solar energy and other energy sources in the system, whether the heat transfer medium in collectors can be applied directly to users, whether the heat transfer medium comes in contact with air, etc. Each of the seven characteristics is divided into 2-3 types, allowing the classification of a wide range of solar hot water systems.

Existing classifications in China are primarily derived from provisions within the international standard, including the following: <Technical code for solar water heating system> (GB50364-2005), <Solar-plus-supplementary water heating system> (GB/T29158-2012), <Guide of Engineering Testing of Construction Solar Water Heating System> (GB/T20095-2006), etc. Various standards employ different methods of classification. For example, <Technical code for solar water heating system > (GB50364-2005) suggests that classification is based on the scope of the hot-water supply, system operation mode, heat transfer medium in the collector and auxiliary heat source at the installation point, and start mode. There are many additional methods of classification. For example, according to the type of solar collectors employed, systems can be divided into numerous categories including flat solar, vacuum tube, or u-tube solar hot water systems; according to the volume of the storage tank, systems can be divided into solar water heaters with a tank capacity of less than $0.6m^3$, and those with a tank volume that is equal to or greater than $0.6m^3$.

The <Technical code for solar water heating system> (GB50364-2005) clearly indicates that the design of a solar hot water system should be primarily equipped according to the building water supply and available drainage, in accordance with the requirements of the <code for design of building water supply and drainage> (GB50015-2003). Experience will also simultaneously serve as the primary basis of expert design.

Above all, the emphasis of previous research on water characteristics and the solar collection system are not conducive to the design and operation of central solar domestic hot water systems. Therefore, the incorporation of the thermal performance of solar hot water systems is necessary to promote their development and practical application. Future research should also incorporate the household load when designing solar collection systems.

2. Thermal performance in solar hot water systems

Classification methods must take the nature of the solar hot water system into account. As previously discussed, solar hot water systems are comprehensive, and are comprised of three features: heat, energy and water. The thermal performance must also be taken into account, which reflects the thermal energy storage and heating capacity of the auxiliary heat source.

Due to thermal energy storage and regulations, the performance of the heat storage tank directly affects the operation of the entire system [12]. The heat storage tank stores heat collected during the day for use in the evening; the water temperature in the heat storage tank is approximately 50℃ or more. The relationship between the volume of the tank, the area of the solar panels and user water consumption should also be taken into account. Due to the instability of solar heating, an auxiliary heat source is typically included in the system. The primary function of the auxiliary heat source is to produce hot water when heat consumption is greater or the amount of available hot water has been reduced (i.e., as a result of bad weather). Without an auxiliary heat source, the system will not function properly. In terms of heat, solar hot water systems consist of collected solar energy and

IOP Conf. Series: Earth and Environmental Science **40** (2016) 012011 doi:10.1088/1755-1315/40/1/012011

heat provided by the auxiliary heat source, which are affected by user consumption and heat dissipation.

During system operation, water reflects heat flow as the medium flows through the system (Figure 1). Results indicated that solar energy cannot be efficiently distributed to the user in many applications. It can be assumed that heat provided by solar energy is 100 (GJ); however, due to the efficiency of solar collectors, only 40% of solar energy can actually be supplied. As the heat flows into the heat storage tank, approximately only 70-80% of collected heat is retained. Investigation and field-testing of central solar hot water system projects in Beijing, Inner Mongolia and Tianjin, indicated approximately 70% of collected heat is lost.

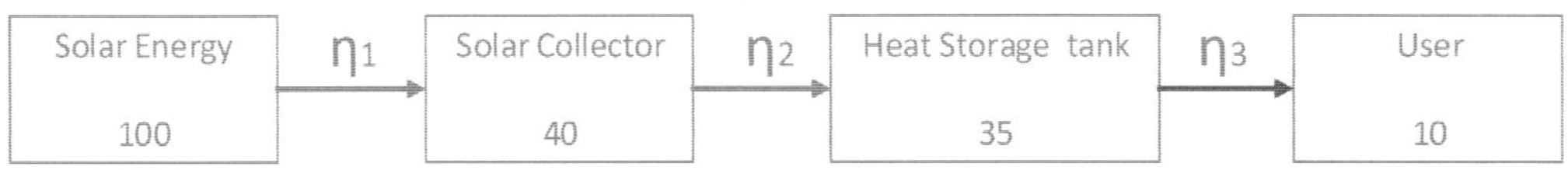

Figure 1. The energy flow schematic diagram of solar domestic hot water systems.

Therefore, it can be concluded that much solar energy is lost before it is distributed to the user, necessitating the inclusion of an auxiliary heat source. Thus, it is possible to evaluate the performance of a solar hot water system according to the consumption of the auxiliary heat source, which is an important system factor.

3. Analysis the problem of classification

At present, DHW systems are primarily classified according to the building water supply and drainage, while the energy supply is rarely taken into account. For example, according to the <Technical code for solar water heating system of civil buildings> (GB503642005), classification of the scope of hot water supply can be divided into three forms, as presented in Table 1.

Table 1. Categorization by scope of hot water supply.

No.	System	Solar collector	Heat storage tank
2.0.10	Collective hot water system	Central	Central
2.0.11	Collective-individual hot water system	Central	Individual
2.0.12	Individual hot water system	Individual	Individual

As shown in Table 1, this method of classification considers the features of solar energy collection and water use, but does not take into account the thermal energy storage and the auxiliary heat source, which are indispensable in the system. Specifically, solar energy provides the heat accumulated throughout the entire day; however, it is impossible to satisfy the hot water requirements of every day throughout the year using solar power alone. Therefore, the, auxiliary heat source is necessary to meet hot water demand, and which reflects the thermal performance of the heating action. Without heat storage, the collected heat would be lost, and which reflects the thermal performance of thermal energy storage.

3.1 Analysis the problem of classification

Ignoring the influence of the auxiliary heat source within the system may not reflect the thermal performance of the solar hot water system, complicating the existing classification methods. A solar fraction is typically used to evaluate the performance of solar hot water systems, which rely on the heat collected from solar energy to ensure that heat demands are met. Therefore, the influence of the auxiliary heating device in the system must be taken into account.

The primary concern of residents when evaluating choices of hot water systems is the cost of the hot water supply, which includes the price of water, management expenses, operation, maintenance, replacement parts, etc. Xiaohui Du reported that the high prices associated with some residential buildings discouraged some residents from use of their hot water systems. Meanwhile, this burden also exists for developers, property owners and property management companies.

Cost primarily reflects the consumption of auxiliary conventional energy [21]. As demonstrated by typical values, regardless of geography and with the exception of fixed costs (such as construction costs and the price of water), the greatest cost variable is the consumption of auxiliary conventional energy, i.e., water or gas.

According to the <Code for design of building water supply and drainage>, the solar fraction can be calculated by the following equation according to the heat collected and the total load:

$$f = \frac{Q_s}{Q} \tag{1}$$

Some researchers have reported that this indicator along cannot appropriately describe the system. In order to address this problem, the indicators of solar availability and auxiliary conventional energy substitution rate have been developed, as determined by the following equation [11, 13, 16]:

$$f_s = \frac{Q_s - Q_d}{Q_s} \tag{2}$$

$$f_a = \frac{Q_s - Q_d}{Q_u} \tag{3}$$

Field-testing of many projects with different system forms indicated a relationship between hot water cost and various indicts (Table 2).

Table 2. Relationship between hot water cost and evaluation indicts.

System	Project	Solar fraction	Price (yuan/ton)	Solar availability	Auxiliary conventional energy substitution rate
Central	Beijing	100%	25	6.2%	18%
	Chifeng	89%	20	21%	19%
Individual	Tianjin	79%	15	32%	25%

As shown in Table 2, the solar fraction of central hot water systems is less than that of individual hot water systems. In accordance with the original project evaluation, high solar fractions indicate not only high conventional energy substitution rates, but also indicate overall good system performance. However, the price of hot water remains high due to the greater use of auxiliary conventional energy and the lesser contribution of solar energy to the system, which is demonstrated by the auxiliary conventional energy substitution rate. As a result, a solar fraction based only on the evaluation index cannot accurately evaluation the overall system performance, indicating that the consumption of auxiliary heat and hot water price should also be taken into consideration.

Present methods of classification take into account only the solar energy, while neglecting the auxiliary heat source. Due to underestimation of the contribution by the auxiliary heat source in system specification and design, many projects have been developed which have high solar fractions but fail to conserve energy. Therefore, this paper proposes that consumption of the auxiliary heat source should be used as the primary basis for the evaluation of energy saving systems.

3.2 Influence of heat features

Classification methods often take into account the water features, but neglect the influence of heat, resulting in the current research trend which focuses primarily on water conservation in solar domestic hot water systems. Systems with heat features supply not only the water but also the heat, which must also take into account any heat lost in transfer. Some researchers have reported that heat loss accounts for more than 30% of the total heat included in the system. Ignoring the hot feature of philosophy would lead to the non-energy saving design. Therefore, the present classification methods cannot provide accurate and scientific classification of solar energy heat water systems.

4. Categorization of systems by auxiliary heat source position

Based on the discussion above, scientific classification of domestic solar hot water systems should take thermal performance into account. Thus, the position of the heat storage tank, auxiliary heating device, and the the centralized or individual nature of the system, should be considered in the classifications of the central solar hot water systems. This section introduces a classification principle based on the position of the auxiliary heat source rather than the tank capacity, which differs from the principle offered in <Technical code for solar water heating system of civil buildings> (GB503642005) and other standards.

4.1 Position and energy consumption

Guangwei Deng compared collective hot water systems and individual hot water systems, and determined that 24-hour supply introduced enormous heat transmission and distribution losses, and substantially increased costs in the system [9]. Rui Yu, et al., applied the model to analysis of collective hot water systems, and reported that consumption by individual auxiliary heat systems was far less that consumption by collective auxiliary heat systems [6]. By analyzing data from multiple solar hot water system projects, results indicate that the position of the auxiliary heat source has great influence on the energy consumption. In order to analyze and compare solar energy collection, Heat dissipation and Heat of auxiliary heat source, bench marking in user heat which sets 100 and the 4 parts were standardized. As we know, the size of the system would lead to the different hot water demands. So, by this way, we can see consumption of heat clear. The results are shown in Table 4.

Table 4. Various system parameters.

System	Position	Type	Supply time	Project	Solar energy collection	Heat of auxiliary heat source	Heat dissipation	Hot water cost （yuan/ton）
Collective hot water system	Collective	Electric	All day	Beijing	295	82	277	25
		Gas	Fixed time	Chi feng	100	79	79	20
Individual hot water system	Individual	Electric	All day	Tianjin	41	71	12	15

In terms of heat dissipation, an all-day hot water supply system in Beijing and Tianjin was compared to a collective auxiliary heating system with maximum heat loss. Based on the amount of heat dissipation, results indicate that solar energy is almost entirely lost, which prohibits residents from taking full advantage of their heating systems. The Tianjin project investigated the characteristics of an individual auxiliary heating system whose heat dissipation accounted for 12% of all collected heat, indicating that nearly all collected heat flowed to the residents. In general the energy consumed by all-day hot water supply systems is greater than consumed by fixed-time hot water supply systems. Due to the enormous heat dissipation experienced by all-day systems, the hot water supply system must employ the auxiliary heating device to guarantee that hot water demands can be met at any time, whereas fixed-time hot water supply systems can only provide hot water during the time allotted [11]. The heat loss of the Chi feng project is much greater than that of the Tianjin project, as related to supply time and auxiliary heat source position. For central solar hot water systems, the heat dissipation in the collective auxiliary heating system is greater than that of individual auxiliary heating systems. The location of the auxiliary heating device in the tank, heat loss in the pipe loop provided by the auxiliary heat source is avoided, resulting in a heat loss ratio less than that of collective auxiliary heating hot water systems.

In order to elaborate on the principle of classification, another system known as an instantaneous solar domestic hot water system is described below (Figure 2).

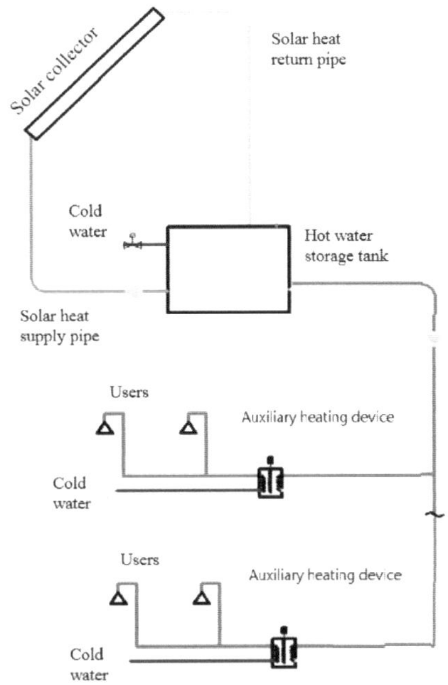

Figure 2. Collective heat collection and individual auxiliary heating solar hot water supply system.

According to the position of the auxiliary heating device, the above system is an individual auxiliary heating system. This system differs from the collective hot water system, which can be defined in relevant codes, due to the position of the tank.

Advantages of this model include cost reduction and increased ease of application. When the auxiliary heating device and heat storage tank are provided in combination, heat must be constantly provided by the auxiliary source to ensure the water temperature in the tank. This system uses heat from an auxiliary heat source to guarantee that needs are met rather than relying on solar energy collection, in order to reduce heat dissipation during delivery. Wang Guojian compared the various forms of solar domestic hot water systems, and determined that the above system (which is supplemented by gas) provides an annual coal consumption of approximately 37% as compared to electric auxiliary heating systems. Additionally, the above system can effectively reduce the initial investment cost and hot water cost to approximately 8.5 yuan/ton [15]. In regard to energy conservation and investment and user experience, this system has promising applications for the future development of central domestic solar hot water systems.

Based on analysis of heat loss and heat provided by the auxiliary heat source in the above system, it can be concluded that the position of the auxiliary heating device as collective or individual greatly influences system consumption.

4.2 Position and user demands

- Occupancy rate

  High occupancy residential buildings are suitable for collective-individual hot water systems [17]. When the heat consumption is small, effective utilization is realized. If the users bear all associated costs, the final price will be prohibitive. In order to ensure a sufficient number of users, the property company which operates the system at low cost will apply winter heating profits to compensate for the loss of hot water systems, inflicting heavy losses on both users and property owners.

- Operation and maintenance
  Hot water requirements cannot be guaranteed in the case of failure. As compared to individual auxiliary heating systems, collective auxiliary heating systems are typically maintained by property management, who guarantees the system operation. In order to make the system user-friendly and easy to troubleshoot, users must incur initial investment expenses. When the efficiency is low, the operational cost may even exceed that of individual auxiliary heating systems.
- User control
  The operation and maintenance of individual auxiliary heating systems are inconvenient; however, they offer the advantage of flexible control. According to actual heat consumed per household, expenses can be more accurately arranged. In the event of bad weather, heating and time needs change depened on the needs of individual households. In addition, property charges for water and electricity costs become unnecessary, thus simplifying the management mode.

5. Definitions

Through discussion of the system forms, energy consumption and user requirements, this essay recommends categorization of heating systems according to the positions of the auxiliary heating device and solar collector, which serve as criterion to distinguish whether the system is a collective or individual system. There are two types of collective heat collection systems: collective auxiliary heating, individual auxiliary heating. Thus, the following definitions are proposed:

Collective heat collection-collective auxiliary heating hot water system which refers to the integration of a solar collector, heat storage tank, auxiliary heating devices, and unified installation which is assigned to a water supply the solar hot water to each terminal. The schematic diagram of a collective heat collection-collective auxiliary heating hot water system is shown in Figure 3.

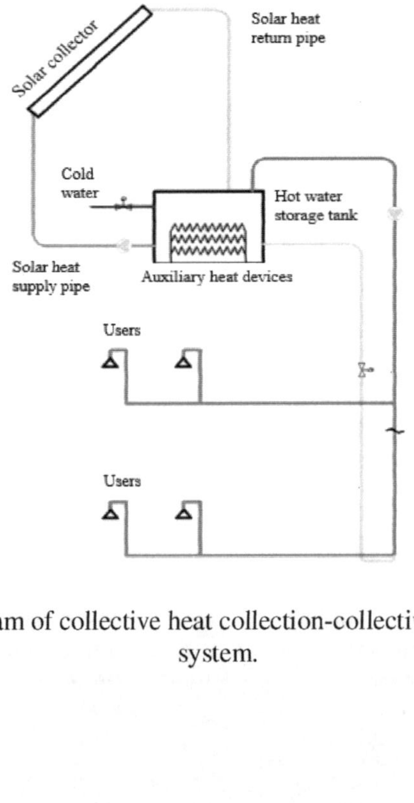

Figure 3. Schematic diagram of collective heat collection-collective auxiliary heating hot water system.

A collective heat collection-individual auxiliary heating hot water system refers to a system with a solar collector installed in the roof in addition to numerous other devices set independently for each user, which include a heat storage tank and auxiliary heating devices. The schematic diagram of a collective heat collection-individual auxiliary heating hot water system is shown in Figure 4.

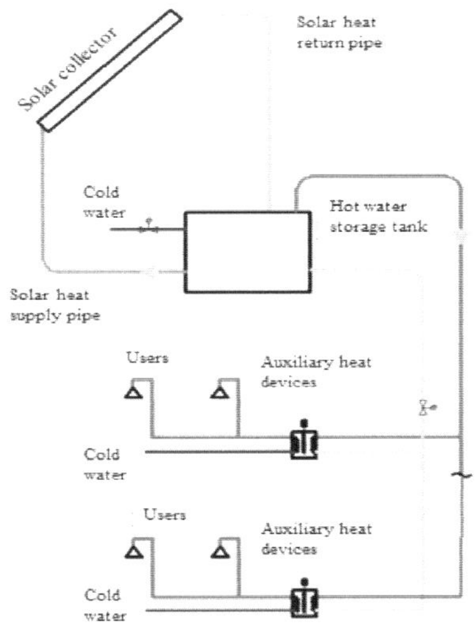

Figure 4. Collective heat collection-individual auxiliary heating solar hot water system.

6. Conclusions

Through the analysis of existing classification challenges, results indicate that the thermal performance of solar hot water systems have been largely ignored in previous research. Thus, to make the system of classification more scientific, the inclusion of thermal performance criteria in classification is proposed. For the classification of central solar domestic hot water systems, it is recommended that the position of the auxiliary heat source be located rather than in the storage tank, as recommended by <Technical code for solar water heating system of civil buildings> (GB503642005).

Alternatively, results indicate that different auxiliary heat source positions will greatly influence energy consumption and related costs. Therefore, for the development of central solar domestic hot water systems, collective heat collection-individual auxiliary heating solar hot water systems should be encouraged for wide application. Additionally, the separation of the heat storage tank from terminal auxiliary heating should also be promoted.

A. 1 Nomenclature

f	Solar fraction (%)
f_s	Solar availability (%)
f_a	Auxiliary conventional energy substitution rate (%)
Q_s	Solar energy collection (GJ)
Q	System load (GJ)
Q_d	Heat dissipation (GJ)
Q_u	Heat of auxiliary heat source (GJ)

References

[1] 2012 *Annual Report on China Building Energy Efficiency* (Beijing: China architecture and building press)

[2] Du X H 2010 *Design research of solar water heating system into high-rise residences* (Tianjin: Tianjin University)
[3] Li J, Zhu X and Li J S 2015 Study on central heating collection and individual heat supply solar water heating system *Water &Wastewater Engineering* **08** 80-8
[4] Lu S Z 2008 *Study on solar water heating system integrated with high-rise building for the large scale application* (Tianjin: Tianjin University)
[5] Wang X L 206 *Optimal design of high-rise building hot water supply systems* (Chongqing: Chongqing University)
[6] Yu R, Gao Y and Zhang Y ,Xiaohang F,Da Y 2015 Investigation of residential solar domestic hot water system and energy consumption model *Building Science* **10** 146-53
[7] Guojian W, Qingping L,Yiheng H,Wuqiang S 2015 Research and experimental analysis of solar domestic hot water systems *Housing Industry* **11** 61-4
[8] Zhang X L, Li Y H and Zhu D H 2012 Centralized integration of solar water heating system in a residential building in research **6**
[9] Deng G W 2013 *The effect of occupant behaviors on evaluating adaptability of centralized building energy saving technologies* (Beijing: Beijing University of technology)
[10] Wang S S, Hao B, Chen X L and Peng C 2015 Investigation on the way of living hot water demand and energy *Water &Wastewater Engineering* **11** 73-7
[11] Chen X L, Hao B, Liu S, Zhou W Y, Yao C N and Wang S S 2015 Study on demand-side design parameters of solar domestic hot water system in residential buildings *Energy Procedia* **70**
[12] Zhao S M and Gao F 2009 hot water storage tank for solar domestic hot water systems and auxiliary heat source design *Water &Wastewater Engineering* **04** 77-81
[13] Yao C N, Hao B, Liu S and Chen X L 2015 Analysis for common problems in Chinese solar domestic hot water system field-testing *Energy Procedia* **70** 402-8
[14] Zhang W J, Hao B, Li N P, Liu S and Yao CN 2015 Investigation on photovoltaic application in buildings in China *Energy Procedia* **70** 673-82
[15] Allen S R, Hammond G P, Harajli H A, McManusn M C and Winnett A B 2009 Integrated appraisal of a solar hot water system *Energy* 353
[16] Liu S, Hao B, Chen X L, Yao C N and Zhou W Y 2015 Analysis on limitation of using solar fraction as solar hot water system design index and evaluation *Energy Procedia* **70** 353-60
[17] Ji Jie, Han Chongwei, He Wei, Pei Gang Dynamic performance of parabolic trough solar collector. In: Proceedings of ISES Solar World Congress.Solar Energy and Human Settlement,2007(2):23~27
[18] Alfred H 2000 15 Years of R&D in central solar heating in Denmark *Solar Energy* **6**
[19] Papakostas K T, Papageorgiou N E and Sotiropoulos B A 1996 Residential hot water use patterns in Greece *Solar Energy* **54** 369-74
[20] Chow T T, Dong Z T, Chan L S, Fong K F and Bai Y 2011 Performance evaluation of evacuated tube solar domestic hot water systems in Hong Kong *Energy & Buildings* 4312
[21] Haillot D, Goetz V, Py X, Benabdelkarim M 2011 High performance storage composite for the enhancement of solar domestic hot water systems *Solar Energy* 855

The asymmetric effect of coal price on the China's macro economy using NARDL model

J C Hou[1] and M C Yang[1]

School of Economics and Management, Shanghai University of Electric Power, Shanghai, 200090, China

E-mail: yangmeichen0@163.com

Abstract. The present work endeavors to explore the asymmetric effect of coal price on the China's macro economy by applying nonlinear autoregressive distributed lag (NARDL) model for the period of January 2005 to June 2015. The obtained results indicate that the coal price has a strong asymmetric effect on China's macro economy in the long-run. Namely one percent increase in coal price leads to 0.6194 percent of the China's macro economy increase; and while the coal price is reduces by 1 percent, the China's macro economy will decrease by 0.008 percent. These data indicate that when coal price rises, the effect on China's macro economy is far greater than the price decline. In the short-run, coal price fluctuation has a positive effect on the China's macro economy.

1. Introduction

Due to the characteristics of China's energy structure—"rich coal, lean oil, less gas", the proportion of coal in primary energy production and consumption structure is more than 65% for a long time. Although the China is vigorously developing new energy and clean energy in recent years, it is difficult to improve, as the coal counts for a large proportion in energy production and consumption in the short term, lacking for other new energy resources system, immature new energy technology and imperfect new energy structure system. However, as the upstream industry of the national economy, the price fluctuations of coal is bound to price fluctuations effect of the market of the means of production, which directly leads to fluctuations in the cost of downstream industry and have an impact on the market operation and the general price level. Finally, it has an impact on the economic development of our country. Therefore, studying the influence of coal price on China's macro economy pay the role of a guide when improving coal enterprises and promote sustainable economic development.

In available literature, the CGE model, VAR model, impulse response function and other methods or models have been applied to study from different perspectives. For example, Gu Xiaohui used VAR model, Cointegration test and Granger causality test to analyze the impact of the coal, the international oil prices internationally on the price level (PPI,CPI), industrial added value (VAI) and monetary circulation [1]. Lin Boqiang, Mou Dunguo studied the effects of oil and coal prices on Chinese macro economy by CGE method [2]. Zhou Aiqian analyzed the reflect of the various sectors of the national economy on the coal price's changes by using input-output model and VAR's impulse response function [3]. Ding Zhiguo applied multiple

[1] Address for correspondence: M C Yang, School of Economics and Management, Shanghai University of Electric Power, Shanghai, 200090, China. E-mail: yangmeichen0@163.com.

econometric methods, Granger causality test, VAR model, ECM model and Impulse Response Function, etc. to make some general test to the conduction effect of coal price to the aggregates of entity economy, then used parameter-varying state spatial model to measure its time-varying efficiency [4]. Wang Lizhi and Gao Zhiyuan used CGE model to study the impact of coal price decline on China's macro economy [5]. Under the linear and nonlinear double perspective, Li Wenbo used Zou's test, Granger causality test and impulse response function, etc to quantitatively analyzed the influence of coal price fluctuation on China's economic growth[6]. However, domestic literature rarely recognizes that the impact of coal price on the macroeconomic isn't necessarilylinear, maybe it is nonlinear. In addition, because of the oil - based foreign energy characteristics, most of them are based on energy prices or oil prices as a research center. For example, Jammazi studied the relationship between international crude oil price and exchange rate based on the wavelet-based NARDL model [7].

How strong the impact of fluctuations in coal price bears on China's macro economy? Is the relationship between them linear or nonlinear? In order to solve these problems, this paper explores the asymmetric effect of coal price on the China's macroeconomic by applying NARDL model proposed by Shin from two levels of the long-run and short-run [8]. Then, we analyze the reasons behind it, and provide scientific evidence for improving coal enterprises and promoting the balanced development of China's economy.

2. Model and data

2.1 Nonlinear autoregressive distributed lag (NARDL) model

The NARDL model explores the long-run and short-run asymmetrical relationship between the non-stationary time series using a formula. Therefore, the NARDL model is suitable for the study and analysis the relationship between coal price and China's macro economy.

According to the theory of Granger and Johansen [9,10], we can construct the dynamic relationship between variables by using error correction model (ECM). Therefore, we can use linear ECM to describe the long-run and short-run effects of coal price on China's macro economy (GDP) under the assumption that there is no asymmetric cointegration.

$$\Delta Lgdp_t = \mu + \rho Lgdp_{t-1} + \delta Lp_{t-1} + \sum_{i=1}^{p-1} \alpha_i \Delta Lgdp_{t-i} + \sum_{i=0}^{q-1} \omega_i \Delta Lp_{t-i} + \varepsilon_t \tag{1}$$

Where, Lgdp=log(gdp), Lp=log(p), Δ=first difference, p represents maximum lag order number of Chinese macro Economy. And q represents maximum lag order number of Coal Price. We can see the long-run and short-run relationship between coal price and China's macroeconomic by Equation (1), but the formula is not suitable for the existence nonlinear relationship of two variables. In consequence, this paper applies NARDL model proposed by Shin which can be very precise description of the asymmetric effect of coal price on the China's macroeconomic from two levels of the long-run and short-run [9]. Taking the coal price as an exogenous variable and decomposing it into the two parts: the accumulation sum of positive impact and the accumulation sum of negative impact, representing increases and decreases.

$$P_t^+ = \sum_{j=1}^{t} \Delta P_j^+ = \sum_{j=1}^{t} \max(\Delta P_j, 0)$$
$$P_t^- = \sum_{j=1}^{t} \Delta P_j^- = \sum_{j=1}^{t} \min(\Delta P_j, 0) \tag{2}$$

According to the theory proposed by Shin, we can construct the asymmetric correction error model to explain the long-run and short-run asymmetric relations between the two variables through Equation (1) and Equation (2).

$$\Delta Lgdp_t = \mu + \rho Lgdp_{t-1} + \delta^+ Lp_{t-1}^+ + \delta^- Lp_{t-1}^- + \sum_{i=1}^{p-1} \alpha_i \Delta Lgdp_{t-i} + \sum_{i=0}^{q-1} \left(\omega_i^+ \Delta Lp_{t-i}^+ + \omega_i^- \Delta Lp_{t-i}^-\right) + \varepsilon_t \tag{3}$$

To validate a cointegrated and asymmetric relationship using the ARDL Bounds test approach, we used the Bound-coefficient test[12]. In the test, the significance of the associated variables was ascertained by jointly setting ρ=δ+=δ-=0, δ+=δ- . The long-run asymmetric coefficients in Equation (3) is tied by the following relationships: LP+=-δ+/ρ; LP-=-δ-/ρ.

2.2 The data

This paper selects the monthly data January from 2005 to June 2015 for analysis in China. For the coal price we make use of the coal mining and washing prices index with the data from the China Economic Information Network (http://cyk.cei.gov.cn/aspx Default.aspx). And we use Gross Domestic Product (GDP) on behalf of the Chinese macro economy whose the quarterly data is from National Database. To make use of quarterly data of GDP, we need to process it into the monthly data. Firstly, we adopt the monthly data of the industrial added value (Industrial added value of the monthly data from the national research network, which for the lack of the number of each year in January of the same period of the previous year, we use interpolation method to get) for processing quarterly data of GDP, and then calculate the weight of the industrial added value of each month. Ultimately, monthly GDP data can be calculated. To convert the data into 2005 as a fixed base data. In order to reduce the fluctuation, we take the natural logarithm for the GDP and the coal price variables respectively recorded as LP and LGDP.

3. **Empirical analysis**

3.1 Unit root test

The unit root test results using Augmented Dickey–Fuller test and Phillips–Perron test are shown in Table 1. It can be found that all the price series have unit root at their level values, but at their first difference they become stationary, namely LGDP ~ I(1), LP ~ I(1).

Table 1. Unit root test results of ADF test and PP test

Test sequence	ADF		PP	
	T- Statistics	Prob.	T- Statistics	Prob.
LGDP	-1.381887	0.8612	-0.907524	0.9511
ΔLGDP	-4.080820[a]	0.0015	-60.4224[a]	0.0001
LP	-1.739467	0.4088	-1.652080	0.4531
ΔLP	-4.773595[a]	0.0000	-10.07735[a]	0.0000

Note: a, b, c represent significance at 1%, 5% and 10% level respectively.

3.2 Granger Causality test

In order to determine the relationship between coal price and the Chinese economy, we use eviews9.0 software to carry out the Granger causality test. Because the variables lag length (m) is arbitrary in the regression model, so we respectively selected lag order about 2 or 4 in order to ensure that the test results are not affected by lag length (m).

Table 2. Results of Granger causality test for multiple lag lengths

null hypothesis	Lag Length	F- Statistics	Prob.	Judgment
LP ↛ LGDP	2	10.1080	0.0000[a]	Refuse
LGDP ↛ LP	2	3.73874	0.0266[b]	Refuse
LP ↛ LGDP	3	18.4825	0.0000[a]	Refuse
LGDP ↛ LP	3	3.25089	0.0244[b]	Refuse
LP ↛ LGDP	4	16.2745	0.0000[a]	Refuse
LGDP ↛ LP	4	5.25153	0.0006[a]	Refuse

Note: a, b, c represent significance at 1%, 5% and 10% level respectively.

From the analysis results in Table 2, in the 5% significant level, there is a two-way Granger causality between LP and LGDP, which shows that the economic growth of China is not only affected by the change of the coal price from 2005 to 2015; In contrast, the coal price also constrained by China's economic growth in a certain extent. The factors include that energy is a prerequisite for the development of the national economy and China depends on long-term energy consumption to promote the development of economy.

3.3 Empirical results using NARDL

Before using the NARDL model, it is necessary to ensure the lag order number of the time series used in Equation (3). Pesaran *et al.* recommended the use of SBC criteria or AIC criterion, so we use the AIC criterion which most of the literature are used to determine the lag order number of the variables in the model. At the same time, taking into account the length of the data, we found that the model is suitable for the lag order number (p,q)=(3, 2).

Table 3. Results of the NARDL model

Regressors	coefficient	Prob.
C	1.658252	0.0000
LGDP(-1)	-0.327212	0.0000
$LP^{+}(-1)$	0.202686	0.0000
$LP^{-}(-1)$	0.002629	0.9393
DLGDP(-1)	0.194328	0.0041
DLGDP(-2)	-0.018022	0.7790
DLGDP(-3)	0.148820	0.0217
DLP^{+}	1.995196	0.3718
DLP^{-}	1.285200	0.6779
$DLP^{+}(-1)$	-0.286816	0.9262
$DLP^{-}(-1)$	-1.366430	0.7503
$DLP^{+}(-2)$	-0.913357	0.6822
$DLP^{-}(-2)$	1.277002	0.6820
Test of Cointegration ($\rho=\delta^{+}=\delta^{-}=0$)	15.41418^{a}	
Test for Asymmetry ($\delta^{+}=\delta^{-}$)	6.340519^{a}	
R-Squared	0.199136	
Log likehood L	295.4183	

Wald test: a, b, c represent significance at 1%, 5% and 10% level respectively.

Table 3 has shown the estimated results for asymmetry effect and the Prob. of the Wald test for coal price and GDP. The coefficient of coal plus (δ +) is significant at 5% level, but the coefficient of coal minus (δ-) isn't significant at 10% level. While the coefficient of coal plus was greater than the coefficient of coal minus. This indicates that the China's macro economy reacts with a higher magnitude when coal price increases than when price decreases. Calculation of the asymmetric effect coefficients of the coal price increase and decrease, respectively, on the Chinese macro economy in the long-run: $LP+=-\delta+/\rho=-0.202686/-0.327212=0.6194$, $LP-=-\delta-/\rho=-0.002629/-0.327212=0.0080$. The significance of the difference of these two coefficients was again tested using the Wald test and the null hypothesis of equality of the co-efficient could not be rejected at the 1% level. Thus the impact of coal price changes on the increase and decrease of GDP was statistically significant.

The significant asymmetric effect of coal price on the China's macro economy is mainly caused by the following reasons. In the heavy chemical industry development stage, secondary industry remains the leading industry of our national economy, and is limited to the China's energy characteristics—"rich coal, lean oil, less gas". Thus, the China's economic growth to a large extent depends on the consumption of coal. However, as the upstream industry of the national economy, the rise in price may improve the price level of the means of production market and led directly to the growth of downstream industry costs. Thus, it promotes the rise of the price level of the whole national economy. However, the increase in coal price will bring negative influence to the related industries to consumption of coal, which can offset part of the positive impact of rising coal prices on the economy. Overall, the increase in coal price pays a certain stimulating role in our country economy. When the decline in coal prices, as the downstream industry of coal industry chain, the large changes in output of mining industry, construction etc. to a large extent contributed to the growth of these industrial economies. But industries related to the lesser extent of Coal Industry have small changes in output, such as agriculture, forestry, animal husbandry, fishery, services, etc.. Therefore, the positive impact of the consumption of coal based on industries greatly offset the negative impact of the decline in coal prices on the economy.

For the short-run effect, a 1% increase in coal price has respectively changed 1.995%, -0.287% and -0.913% of GDP current, first and second, the overall effectiveness: 0.795%. That is, the increase of coal price has a short-term positive effect on GDP. 1% decrease in coal price, GDP current, first and second respectively change 1.285%, -1.367% and 1.277%, the overall effectiveness: 1.195%. Thus the decrease of coal price has a short-term positive effect on GDP. Generally speaking, fluctuations of coal price have short - run positive effects on GDP in China.

4. Conclusions and suggestions

This paper uses the NARDL model to analyze the asymmetric effect of coal price on China's macro economy from January 2005 to June 2015. And get the following conclusions and suggestions:

There is a two-way Granger causality between LP and LGDP, which shows that the economic growth of China is not only affected by the change of the coal price from 2005 to 2015; in contrast, the coal price also constrained by China's economic growth to a certain extent. The economic growth is influenced by coal price, which is consistent with the irreplaceable of energy in the development of the productive factors, especially the energy characteristics of "rich coal, lean oil, less gas". For many years, China's economic growth has consumed a large amount of energy, which is contrary to the sustainable development strategy of China's economy. Therefore, our country should actively adjust the industrial structure and accelerate economic growth mode from extensive to intensive type.

There is a long-run asymmetric relationship between coal price and China's macroeconomic. The fluctuation of coal price has a positive effect on GDP, but the sharp rise in price leads directly to the cost growth of downstream industry. And the enterprise will maximize the pursued profits by raising price, which even causes cost-push inflation and recession. The continued decrease of coal price would bring about decline of prices and economy and have a strong impact on the economy. Therefore, we should properly control the coal price to avoid frequent fluctuations which can actively promote the growth of our economy.

In the short term, coal price fluctuation has a positive effect on China's macro economy. Although fluctuations in coal price have a positive effect on China's economic in the short term, we should take some method as far as possible to avoid frequent volatility from the long-term perspective, such as increasing energy control means, establishing and improving the coal storage system.

References

[1] Mehta P K and Pirtz D 1980 Magnesium oxide additive for producing self stress in mass concrete *J. Proceedings of the 7th International Congress on the Chemistry of Cement* **3** 6-9

[2] Gu X H and Qiao S J 2010 *The effects of international coal and crude oil price changing on Chinese economy* (Inner Mongolia: Inner Mongolia University)

[3] Lin B Q and Mou D G 2012 The Impact of Energy Price Increases on Macro-economy: An Analyses based on CGE Method *Foreign Economic Relations & Trade* **6** 38-40

[4] Zhou A Q and Zhang Q 2009 *Coal prices on the impact of economic growth in China* (Nanjing: Nanjing University of Aeronautics and Astronautics)

[5] Ding Z H and Zhou M H 2011 *Research on the influence effect of Coal Price Fluctuation on China entity Economy* (Xu zhou: China University of Mining Technology)

[6] Wang L J and Gao Z Y 2014 Research on the effect of coal price decline on China's

economy: based on CGE model *Scientific Decision-Making* **07** 48-62

[7] Ding Z H, Li W B, Zhou M H and He L Y 2015 Impact of coal price fluctuation on China's real economy *Resources Science* **37** 1973-82

[8] Jammazi R, Lahiani A and Nguyen D K 2015 A wavelet-based nonlinear ARDL model for assessing the exchange rate pass-through to crude oil prices *J. International Financial Markets, Institutions and Money* **34** 173–87

[9] Shin Y, Yu B and Greenwood-Nimmo M 2013 *Modelling asymmetric cointegration and dynamic multipliers in an ARDL framework* ed W C Horrace and Sickles R C (New York: Springer Science & Business Media) 281-314

[10] Engle R F and Granger C W J 1987 Co-integration and error correction: representation, estimation, and testing *Econometrica* **55** 251–76

[11] Johansen S 1988 Statistical analysis of cointegration vectors *J. Economic Dynamics and Control* **12** 231–54

[12] Pesaran M H, Shin Y and Smith R J 2001 Bounds testing approaches to the analysis of level relationships *J. Appl. Econom* **16** 289–326

The theoretical analysis and experimental research on the optimal condition of semiconductor refrigeration

Y F Pang[1]

School of Civil and Refrigeration Engineering, Harbin University of Commerce, Harbin 150028, China

E-mail: 76456974@qq.com

Abstract. The traditional limiting conditions have the biggest refrigeration quantity condition and the biggest refrigeration coefficient condition, there is a special operating mode during these conditions, enabling to both have the big refrigeration quantity and the small power loss, this operating mode is "Optimum condition". This article first carried on the theoretical analysis to the semiconductor's optimum condition, inferred optimum electric current's theoretical formula; Carried on the experiment again to a semiconductor refrigeration box by regulating current changing operating mode, which had analyzed performance parameter's change situation under 8 kinds of condition experiments, carried on the regression analysis to the experimental data, obtained the regression equation thus discovered optimum electric current corresponding optimum condition. Carried on working under this condition, and then obtained the big refrigeration quantity and small power, which enhanced the refrigeration performance of semiconductor refrigerator. The experimental result and the theoretical analysis result tallied basically.

1. Introduction

Semiconductor refrigeration has two conditions: the biggest refrigeration quantity condition and the biggest refrigeration coefficient condition [1~3]. In the biggest refrigeration quantity condition, although the refrigerating capacity is the largest, but consumed large power, get smaller refrigeration coefficient; In the biggest refrigeration coefficient condition, although the economy is good, consumed less power less, but the refrigerating capacity is small. Between the two conditions, it must have a special condition, which not only have larger refrigerating capacity, but also have small power consumption; this condition is the "optimal condition". Working in the optimal condition of semiconductor refrigeration, we can reduce the power and cost, so that the comprehensive benefit of semiconductor refrigeration achieve the optimal condition, improve the cooling performance of semiconductor refrigeration. So we need to find out the optimal condition of thermo-electric refrigeration, which has a realistic significance to save energy, reduce costs, and improve the economic benefit.

1. The theoretical analysis of semiconductor refrigeration's the optimal condition

The refrigerating capacity formula of the semiconductor refrigeration is known:

[1] Address for correspondence: Y F Pang, School of Civil and Refrigeration Engineering, Harbin University of Commerce, Harbin 150028, China. Email: 76456974@qq.com.

$$Q = \alpha T_c - \frac{1}{2} I^2 R - K\Delta T \tag{1}$$

Power is:

$$W = \alpha I \Delta T + I^2 R \tag{2}$$

To the type (1), let be $d(Q)/dI = 0$, abstained:

$$I = \alpha T_c / R \tag{3}$$

Put type (3) into type (1), the maximum refrigerating capacity can be obtained:

$$Q_m = \frac{\alpha^2}{R}[\frac{T_c^2}{2} - \frac{\Delta T}{Z}] \tag{4}$$

Maximum power consumption:

$$W_m = \frac{\alpha^2}{R} T_c T_h \tag{5}$$

Let be $\theta_Q = \frac{Q}{Q_m}$, θ_Q expressed the proximity of refrigerating capacity Q relative to maximum refrigerating capacity Q_m. $\theta_W = \frac{W}{W_m}$, θ_W expressed the proximity of The power consumption W Relative to the maximum power consumption W_m; $\theta = \theta_Q - \theta_W$, θ expressed comprehensive parameters of the capacity and power consumption. Obviously, the greater θ, shown that the greater the difference between θ_Q and θ_W, mean is that the value of θ_Q is big, the value of θ_W is small. Let be $\frac{d\theta}{dI} = 0$, we Can get the current value when θ is maximum, this is the optimal current, the corresponding condition is the optimal conditions [4~6]. There are:

$$\theta_Q = \frac{Q}{Q_m} = \frac{\alpha T_c - \frac{1}{2} I^2 R - K\Delta T}{\frac{\alpha^2}{R}[\frac{T_c^2}{2} - \frac{\Delta T}{Z}]} \tag{6}$$

and:

$$\theta_W = \frac{W}{W_m} = \frac{\alpha I \Delta T + I^2 R}{\frac{\alpha^2}{R} T_c T_h} \tag{7}$$

so:

$$\theta = \theta_Q - \theta_W = \frac{T_c T_h(\alpha T_c - \frac{1}{2} I^2 R - K\Delta T) - [\frac{T_c^2}{2} - \frac{\Delta T}{Z}](\alpha I \Delta T + I^2 R)}{\frac{\alpha^2}{R}[\frac{T_c^2}{2} - \frac{\Delta T}{Z}] T_c T_h} \tag{8}$$

Let be $d(\theta)/dI = 0$, abstained:

$$IR[T_c T_h + 2[\frac{T_c^2}{2} - \frac{\Delta T}{Z}]] = T_c^2 T_h \alpha - \alpha \Delta T[\frac{T_c^2}{2} - \frac{\Delta T}{Z}] \qquad (9)$$

$$I_S = I = \frac{\alpha}{R} \frac{[T_c^2 [T_h - \frac{\Delta T}{2}] + \frac{\Delta T^2}{Z}]}{[T_c (T_c + T_h) - 2\frac{\Delta T}{Z}]} \qquad (10)$$

Among them: T_c is the cold-side temperature of the semiconductor chip; T_h is the hot-side temperature; ΔT is the temperature difference of hot-side and cold-side; α is the thermo-electromotive force; R is the resistance.

This study selected thermopile model which is TEC1-12705 type, its performance parameters are: electric double logarithmic n=127; the thermo-electromotive force ratio $\alpha = 127 \times 2 \times 2.06 \times 10^{-4}$ V/K=0.0518 V/K; Thermopile resistance R = 2.25 Ω; thermal conductivity of Material K = 16.1×10^{-2} ($W / cm \cdot K$); The optimal value coefficient of material Z = 2.8×10^{-3} K^{-1}.

Take that the temperature difference of hot-side and cold-side ΔT = 20 k, the cold-side temperature T_c = 287 k, put above performance parameters into the formula can be concluded that the current I of optimal conditions I = 3.980753 A.

2. The construction of experimental device

The experiment object is a small refrigerating box of space 12 L, power supply is 0-12 V adjustable dc power supply which is produced of Shanghai Fudan, temperature measurement use thermocouple. Unearthed a rectangular openings on the edge of the box body arm side, the cold-side of refrigeration arm clingy with an aluminum block, and daub thermal conductive silicone, left-side close to the cooling box, and right-side close to the semiconductor refrigeration piece. The hot-side of refrigeration piece connected to the radiator, filling heat insulation material around the perimeter of the semiconductor.

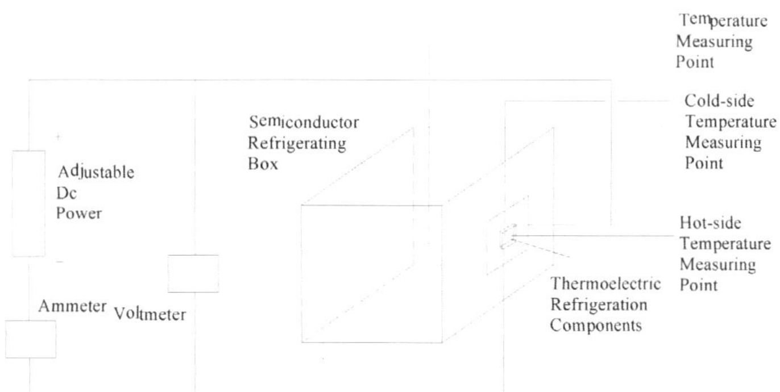

Figure 1. Experimental system structure.

In order to reduce the contact thermal resistance between radiator with semiconductor refrigeration piece, enhanced thermal conductivity, daub evenly a layer of thermal conductive silicone between them, and daub between the cold-side of refrigeration piece and heat dissipation aluminum as well. Bolt between heat conduction fin and heat insulation plate, daub thermal conductive silicone between the contact surfaces in order to reduce the contact thermal resistance. Figure 1 is experimental system structure.

3. The scheme determination of experiments

After experimental device is determined, do experimental research under different working conditions, change working condition of this system by adjusting the current. Conduct experiment in different electrical flow of 8 groups, measured cold-side and hot-side temperature of semiconductor refrigeration piece under different working conditions, under the corresponding conditions calculate performance parameters such as refrigerating capacity, refrigeration coefficient, power, etc. through comparison the parameters such as θ_Q 、 θ_W and θ, choose the optimal condition of semiconductor refrigeration.

The experiment use natural cooling heat at the cold-side and use air cooling at the hot-side, which fin add fan for heat dissipation.

4. The result of the experiment and analysis

Because the semiconductor refrigeration piece's working current of 2.2 A under biggest refrigeration efficiency condition, the condition of the working current of 5 A under maximum refrigerating capacity condition which this experiment selected. So conduct the experimental conditions working in 2.2, 2.6, 3, 3.4, 3.8, 4.2, 4.6 and 5 A. In eight experiments, the experimental data record every minute, after one hour temperature constant basically, get the results of the experiment is as follows:

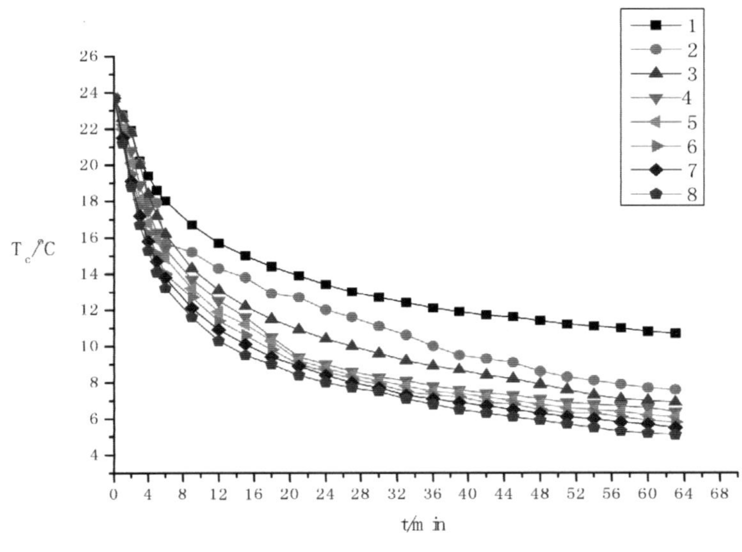

Figure 2. The cold-side temperature curve changing with time under 8 kinds of working condition.

As shown in Figure 2 is the cold-side temperature curve changing with time under 8 kinds of working condition, the Figure shows that the working current is 2.2 A, cold-side temperature of semiconductor refrigeration piece stability in around 11℃ after 1 hour, with the increase of working current I, when the current is 5 A, the cold-side temperature of semiconductor refrigeration piece stability in 5 ℃ after 1 hour, eventually the cold-side temperature is lower than which is in 2.2 A.

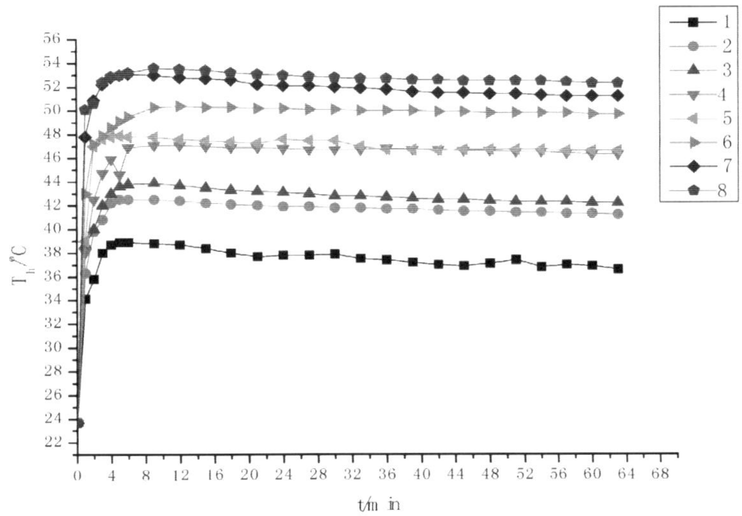

Figure 3. The hot-side temperature curve changing with time under eight kinds of working condition.

Figure 3 is the hot-side temperature curve changing with time under eight kinds of working condition, the Figure can be seen, when the working current is 2.2 A, hot-side temperature stability in 37.2 ℃ after 1 h, with the increase of working current, when the current is 5 A, the hot-side temperature stability in around 52.3 ℃ after 1 hour, is much higher than which is in the 2.2 A.

It is known from above analysis, with the increasing of working current I, The drop rate of semiconductor refrigeration piece's cold-side temperature increase, and the ultimately stability cold-side temperature gradually reduce. With the increasing of working current I, the rise rate of semiconductor refrigeration chip's the hot-side temperature also increase, finally stable hot-side temperature rise. Therefore, in order to obtain lower cold-side temperature, we can increase working current appropriately, therefore gain better refrigeration effect.

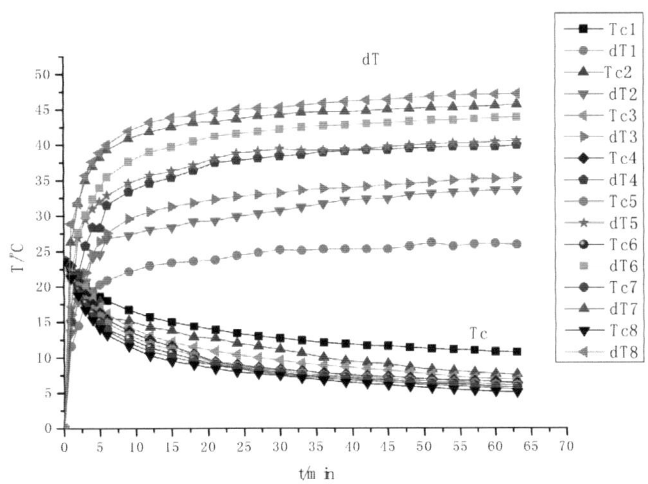

Figure 4. The curve of cold-side temperature and the temperature difference between hot-side and cold-side.

Figure 4 is the curve of cold-side temperature and the temperature difference between hot-side and cold-side, the Figure can be seen, when the working current is 2.2 A, the temperature difference between hot-side and cold-side finally stability in 26 ℃, along with the increasing of the working current, the temperature difference between hot-side and cold-side of semiconductor refrigeration piece increase, the cold-side temperature drop, when the current is increased to 5 A, the temperature difference between hot-side and cold-side of semiconductor refrigeration piece is 47.2 ℃. Thus, increase working current of the semiconductor refrigeration piece can increase the temperature difference between hot-side and cold-side, and decrease the cold-side temperature.

So it can be concluded, in order to improve the performance of semiconductor refrigeration, get lower cold-side temperature, we can improve the current, increase the temperature difference between hot-side and cold-side. But due to the increase of the working current, the hot-side temperature will gradually rise, the consumed power will also increase, so looking for the optimal current must be between 2.2 A and 5 A.

Take the temperature difference between hot-side and cold-side ΔT = 20k ,we can find cold-side temperature when the temperature difference between hot-side and cold-side is 20k under every working condition electric from diagram, calculate refrigerating capacity Q_m , power W_m and refrigeration coefficient ε , under the biggest refrigerating capacity condition (5A), then calculate the corresponding refrigerating capacity Q , power W and refrigeration coefficient ε under every working conditions, and according to $\theta_Q = \frac{Q}{Q_m}$ and $\theta_W = \frac{W}{W_m}$ calculate θ_Q and θ_W , work out $\theta = \theta_Q$ - θ_W eventually. Refrigerating capacity θ_Q , power θ_W , refrigeration coefficient θ under each working conditions are listed in the following table:

Table 1. The comparison table of experiment performance parameter under the eight kinds of working condition.

Working condition	current	Refrigerating capacity	Consumed power	ε	$\theta_Q = \frac{Q}{Q_m}$	$\theta_W = \frac{W}{W_m}$	θ
1	2.2	11.67693	13.76782	0.848132	0.343237	0.219582	0.123655
2	2.6	16.58728	20.37692	0.814023	0.487574	0.324991	0.162584
3	3	22.16013	28.64537	0.773603	0.651385	0.456864	0.194521
4	3.4	23.78448	30.45754	0.780906	0.699132	0.485766	0.213366
5	3.8	27.24993	35.96753	0.757626	0.800997	0.573645	0.22735
6	4.2	30.99592	43.57283	0.711359	0.911109	0.694941	0.214516
7	4.6	32.69867	53.62726	0.60974	0.96116	0.855299	0.105861
8	5	34.29128	62.7905	0.546122	1.007974	1.001443	0.006531

Because θ is greater, which represent that the difference between θ_Q and θ_W is greater, the larger θ_Q is, the smaller θ_W is, mean that the refrigerating capacity is larger, power is smaller, the condition which corresponding to the current is the optimal conditions. From table 1 it can be seen that with the increase of current, the refrigerating capacity has increased as well, but the consumed power also will increase, and the optimal condition require that refrigerating capacity is larger, and consumed power is smaller, that is θ maximum operating condition. So it can be seen from the table θ corresponding to 3.8 A is the maximum, we can deduce that the optimal current is about 3.8 A.

From the data of current I and θ, we can see that with the increase of current I, θ first increase then decrease, namely I and θ are parabola relationship, and there is a current value I which make θ is maximum θ_m. Let be the relationship equation between I and θ as follows: $\theta = a + b_1 I + b_2 I^2$.

It is a nonlinear equation, I and I^2 as two variables respectively, and then there is binary linear relationship between θ and I, I^2, so build forms as follows:

Use Excel software [return] function to do a regression analysis, the regression equation is established.

Table 2. Data statistical Figure.

θ	I	I^2
0.123655	2.2	4.84
0.162584	2.6	6.76
0.194521	3	9
0.213366	3.4	11.56
0.22735	3.8	14.44
0.214516	4.2	17.64
0.105861	4.6	21.16
0.006531	5	25

The regression equation is:

θ=-0.74858+0.572291I-0.08374 I^2

Optimal curve as follows:

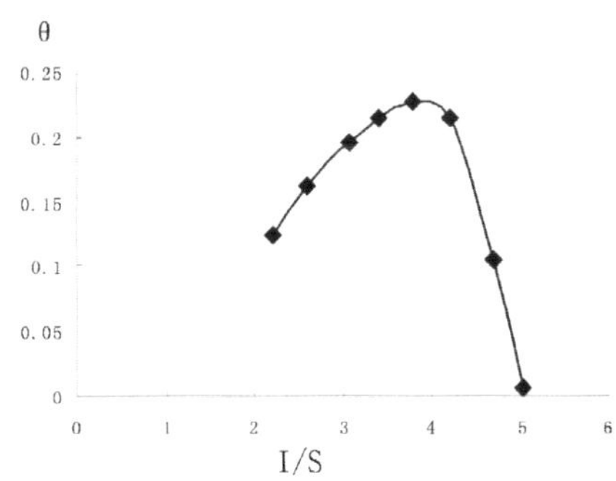

Figure 5. The optimal diagram.

Calculate the optimal current corresponding to θ reaches the maximum value by the statistics function of Excel software:

Table 3

θ=-0.74858+0.572291I-0.08374 I^2		
The objective function		
0.229199	I=	3.952173

When I = 3.952173, θ is maximum 0.229199, the experiment result I = 3.952173 basically tallies with the results of numerical calculation in front of I = 3.980753. So the condition corresponding electrical current is the optimal condition, the electric current is the optimal current value.

5. Conclusions

Due to the maximum refrigerating capacity condition and maximum refrigeration coefficient condition all belong to the theoretical limit conditions. In the maximum refrigerating capacity condition, although the refrigerating capacity is the largest, but it consume large power, get smaller refrigeration coefficient; in the maximum refrigeration coefficient condition, although the economy is good, consumption is less, but refrigerating capacity is small. Therefore, the optimal operating condition can make the semiconductor refrigeration have larger refrigerating capacity and low power consumption.

- In this paper, theoretically analysis on the optimal condition of semiconductor refrigeration, deduce the formula of optimal current, find out theoretical optimal condition.
- Experimental study was carried out on a small refrigeration space, by changing the current draw eight kinds of conditions, and conduct eight experiments, it is concluded that rule which the hot-side and cold-side temperature of semiconductor refrigeration piece changing with the time.
- According to the experimental data, using the regression analysis, it is concluded that the regression equation, find out the optimal current, and draw the optimal curve.
- It can be seen by the above data, the optimal current numerical experiment and the optimal current value of theoretical analysis are basically the same.

References

[1] Xu D S 1992 *Semiconductor Refrigeration and Application Technology* (Shanghai: Shanghai Jiaotong University Press)

[2] Wang Y, Pang Y F and Yin G 2009 Research on the Performance of Solar Semiconductor Refrigeration *The 4th Asian Conference on Refrigeration and Air-conditioning* 602-05

[3] Xuan X C, *et al.* 2001 The development of semiconductor refrigeration material *Refrigeration Technology* **2** 37-41+48

[4] Liu W H, *et al.* 2007 Theory analysis on Optimization of semiconductor refrigeration condition *Application of Energy Technology* **1** 5-7

[5] Qi C J, *et al. 2007* Experimental study on the optimization design status of semiconductor refrigerating machinery *Cryogenic Engineering* **155** 43-6

[6] Chen R B 1995 Comprehensive analysis on the best work state of semiconductor refrigerating machinery *Harbin Engineering College* **28** 83~9

Energy consumption quota management of Wanda commercial buildings in China

D B Sun[1], H Xiao[2, 4], X Wang[1], J J Liu[2], X Wang[2], X Q Jin[2], J Wang [3] and X K Xie[3]

[1] Wanda Commercial Management Co.,Ltd, 10/F Block, Wanda Plaza, 93 Jianguo Road, Chaoyang District, Beijing, 100022, P. R. China
[2] Persagy Technology Co.Ltd, 4-301, Guiguliangcheng, No.1 Nongda South Road, Haidian Distract, Beijing, 100084, P. R. China
[3] Liaoning Technical University, No. 47 Zhong Hua Rd, Fuxin City, Liaoning Province, 123000, P. R. China

E-mail: xiaohe@persagy.com

Abstract. There is limited research of commercial buildings' energy use data conducted based on practical analysis in China nowadays. Some energy consumption quota tools like Energy Star in U.S or VDI 3807 in Germany have limitation in China's building sector. This study introduces an innovative methodology of applying energy use quota model and empirical management to commercial buildings, which was in accordance of more than one hundred opened shopping centers of a real estate group in China. On the basis of statistical benchmarking, a new concept of "Modified coefficient", which considers weather, occupancy, business layout, operation schedule and HVAC efficiency, is originally introduced in this paper. Our study shows that the average energy use quota increases from north to south. The average energy use quota of sample buildings is 159 kWh/(m2.a) of severe cold climate zone, 179 kWh/(m2.a) of cold zone, 188 kWh/(m2.a) of hot summer and cold winter zone, and 200 kWh/(m2.a) of hot summer and warm winter zone. The energy use quota model has been validated in the property management for year 2016, providing a new method of commercial building energy management to the industry. As a key result, there is 180 million energy saving potential based on energy quota management in 2016, equals to 6.2% saving rate of actual energy use in 2015.

1. Introduction

[4] Address for correspondence: H Xiao, Persagy Technology Co.Ltd, 4-301, Guiguliangcheng, No.1 Nongda South Road, Haidian Distract, Beijing, 100084, P. R. China. E-mail: xiaohe@persagy.com.

1.1 China's policy status of building energy consumption quotas

Energy use of China's building sector has dramatically increased within this decade. It accounts for 16% of global consumption, which increased at an average rate of 2.4% per year from 1997 to 2008. The major direction of building energy saving during 12th Five Year Plan (FYP, 2011-2015) concentrates on improving building envelope of new constructions, energy saving technologies/products, retrofitting of existing buildings and renewable energy use[1,2]. MOHURD announced that a preliminary energy saving target of 116 Mtce for China's building sector could be achieved by the end of 12th FYP [3]. Meanwhile, some researchers pointed out that blind development of energy saving technologies ignoring local conditions cannot really save energy use. Tsinghua University studies actual energy use data of 103 LEED certificated buildings [4]. The result shows the average EUI of LEED buildings is 217 kWh/ (m^2.a), which is not lower than others.

Therefore, researcher and policy makers realize that energy consumption quotas system of new- and existing- buildings is a truly effective method to control actual energy consumption of China's building compared to the previous method. Shenzhen takes the lead in addressing commercial building energy quota standard in the beginning of 12th FYP. Shenzhen Construction Bureau (SCB) issues three energy quota standards of existing office building, shopping center and hotel in 2012 [5].Those standards are revised and combined as "Standard for Energy Quota of Existing Commercial Building in Shenzhen" in 2015 (SCB, 2015). For instance, the energy quota is 100~110 kWh/ (m^2.a) of commercial office building and 300~320 kWh/ (m^2.a) of shopping center in Shenzhen.

On the basis, MOHURD promulgated "Standard for Energy Consumption of Buildings (draft for comment)" in Jun. 2014 [3]. The standard has 6 major innovations: 1)it covers both commercial and residential buildings; 2) it sets two targets of each type of building ,the restrict value and guide target; 3) it separates the new- and existing- buildings' target; 4) it separately issues heating use quotas (kgce/(m2.a)) and electricity use quotas(kWh/(m^2.a)); 5) four climate zones have different energy consumption quotas; 6) revising method of commercial building is provided considering operation hours, occupant intensity, HVAC system design difference, etc. Also take commercial office building in the hot summer and warm winter area (HSWW) for example, the restrict EUI is 110 kWh/(m^2.a) for existing ones, and the highest guide target is 75 kWh/(m^2.a); the restrict EUI is 90 kWh/(m^2.a) for newly constructed ones, and the guide target EUI is 65 kWh/(m^2.a). The major energy consumption quotas of other types of buildings are compared in Table 1.

Table 1. Comparison of energy consumption quotas between two standards

Type	SCB,2015 (Shenzhen only)	MOHURD, 2014 (HSWW area)			
		Existing		Newly constructed	
		Restrict	Guide	Restrict	Guide
Commercial office	110/105/100	110	75	90	65
Government office	90/85/80	80	55	55	40
Shopping center	320/310/300	300	245	100	80
Supermarket	310/300/290	290	240	105	80
5-Star hotel	220/210/200	220	160	120	100
4-Star hotel	190/180/170	190	140	110	90

1.2 Literature review of building energy consumption quota

Energy consumption quota is used to give a reasonable suggestion to the building operator to control the total energy use, usually only concerning about the electricity in the beginning. The energy data is determined based on several basic informations, consisting of building area, climate zone, occupancy, operating hours, mechanical system information, etc.

Two typical analysis methods are used to evaluate the building energy use, model simulation and statistical analysis [6].

Model simulation method uses mathematical method to analyze, compare or predict building electricity consumption, to evaluate the operation status of buildings mechanical system [7]. Carriere *et al.* studied the energy use characteristics of large commercial buildings [8]. Several building basic information are collected including building function, climate, ME system information, running time and so on [9]. The EPA's energy efficiency evaluation software, Energy Star in the U.S, presents the method to calculate the energy consumption quota of various types of buildings [10]. By entering the energy consumption, building area, the number of staff, etc, energy star tool compares the energy use quota with similar buildings in the database of the software and get the scores in centesimal grade system of this building. Those studies point out that temperature and humidity, occupant density and air-conditioning style are three most important impact factors of commercial building energy use.

Statistical analysis model used mathematical method, based on the statistical analysis software, to determine a reasonable energy consumption rage for buildings. Sharp used 6 significant factors from CBECS database in U.S., applied linear regression model of historical building energy use data, to predict commercial building Energy Use Intensity (EUI), and finally use median value as benchmarking index [11,12]. Zhang determined energy consumption, which is lower than 60% equal value as the energy use data [13]. German VDI 3807 evaluation system gives a calculation method of the energy consumption quota according to the building type, equipment power capacity, operation hours, etc. by investigation [10]. Those researchers illustrate that median or statistical distribution range is accurate than average, which can be considered as energy quota.

1.3 Limitation of present studies

There are two major limitations of present studies about energy use quota [14]. First, there was a very limited scope survey of energy use in large-scaled commercial buildings in China, instead of small-scaled investigations, case studies or scenario analysis. Thus, there is no energy use quotas model of large-scaled commercial building in China yet. Second, energy consumption quota tools like Energy Star in the U.S and VDI3807 in Germany are based on a box model, whose physical significance of regression equations are not clear to explain the impact factor of actual building energy consumption.

The method of this paper is original works of author members. This is the first and precious chance collecting hourly actual energy use data of hundreds of commercial buildings in China. Thus, on the basis of the qualified data, this study firstly introduces an innovative methodology of applying energy use quota model and empirical management to commercial building. A new concept of "Modified coefficient", considering weather, occupancy, business layout, operation schedule and HVAC efficiency, is introduced to adjust energy use quota. Furthermore, simulation is adopted to predict the optimal energy use of a typical commercial building, so as to evaluate potential of energy savings. The energy use quota model has been validated in the property management for year 2016, providing a new method of commercial building energy management to the industry.

2. Methodology

2.1 Definition and data source

There are 135 large-scaled commercial buildings, which are divided by three dimensions.

- According to the definition of five climate zones, dividing into severe cold, cold, hot summer and cold winter, hot summer and warm winter, and temperate.
- According to the city classification, dividing into first-tier, second-tier and third-tier.
- According to business maturity in each city, further dividing into type-I, type-II, type-III and type-IV.

All sample buildings were installed with Persagy energy metering system. Each building installs more than 150 three-phase smart meters on low voltage power distribution branches. Hourly energy use data of each meter is transferred to the server. Based on those meter data, same electricity metering model is applied, as shown in Figure 1. First class breakdown is defined by management partition, including public area, anchor stores, small tenants and error. Second class breakdown is defined by mechanical system, including HVAC system, lighting, utility equipment, information center and others. Third class breakdown is defined by device, such as chiller plant, air-handling terminals, elevator, drainage pumps, indoor lighting, landscape illume, etc.

The Electricity Use Intensity (EUI) is calculated by annual electricity consumption divided by business area. The business area equals to total building area minus parking area and machine room space. All building energy use data is collected by electricity metering system. Each sample building contains at least one-year electricity use data.

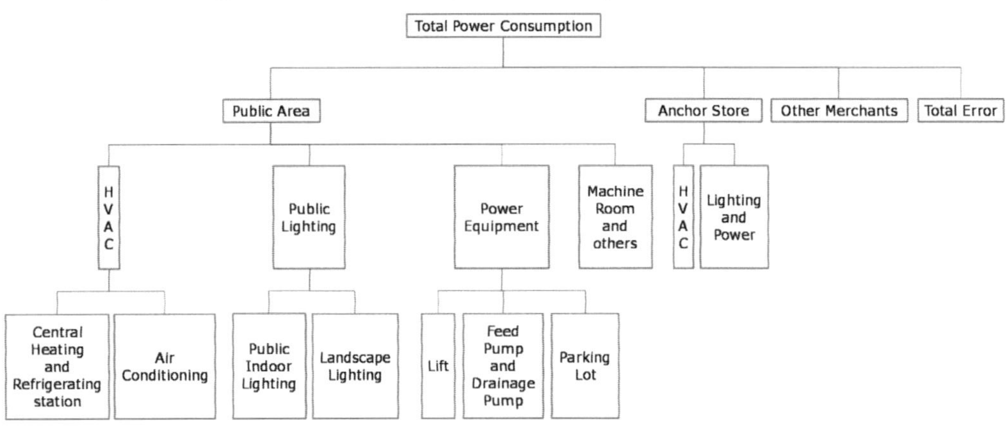

Figure 1. Electricity breakdown logical sketch map of sample buildings

2.2 Energy consumption quotas model based on historical data

135 sample buildings cover 21 provinces or cities in China, include five climate zones. The impact factors of building energy use can be separated into subjective and objective ones. The objective factor includes weather, architecture design, building envelope, mechanical system design, etc.; the subjective factor contains business hours and operation strategies. Hence, this

IOP Conf. Series: Earth and Environmental Science **40** (2016) 012014 doi:10.1088/1755-1315/40/1/012014

study conduct five "Modified coefficient", considering weather α_1, occupancy α_2, business layout α_3, operation schedule α_4 and HVAC system design efficiency α_5 ,so as to adjust energy use quota on the basis of historical energy data. EUI_{quota} is calculated according to the Equation (1) and Equation (2).

$$EUI_{quota} = EUI_0 \times max(\alpha_1, \alpha_2) \times \alpha_3 \times \alpha_4 \times \alpha_5 \quad (1)$$

$$EUI_0 = \frac{E_0}{A_0} \quad (2)$$

where

EUI_{quota}	annual electricity use quota per square meter, unit: kWh/(m^2.a)
EUI_0	annual actual electricity use per square meter, unit: kWh/(m^2.a)
E_0	annual actual electricity use, unit: kWh
A_0	Business area, equals to total building area minus parking area and machine room space
α_1	weather modifying coefficient
α_2	occupancy modifying coefficient
α_3	business layout modifying coefficient
α_4	operation schedule modifying coefficient
α_5	HVAC system design efficiency modifying coefficient

(1) Weather modifying coefficient: α_1

Ambient temperature, humidity and solar radiation are major climate impact factors of building cooling and heating load [15], especially to cooling load from ventilation [16]. Based on on-site investigation of commercial building in China, energy consumption of HVAC system takes 43% of public area energy use approximately. Hence, weather condition cannot be simplified in the energy quota model. The weather modifying model, firstly, calculating coefficient c_1. c_1 equals to the difference of outdoor and indoor enthalpy of sample building, divided by the average of this difference of all sample buildings which in same city type and climate zone, as shown in Equation 3. Secondly, this coefficient is only used to modify HVAC system energy use, as shown in Equation4. Thirdly, α_1 is calculated according to Equation (5).

$$c_1 = \frac{(i_{out} - i_{in})_{sample}}{(i_{0,out} - i_{0,in})} \quad (3)$$

$$E^{*}_{HVAC} = E_{HVAC} \times c_1 \tag{4}$$

$$\alpha_1 = \frac{\left(E^{*}_{HVAC} + E_{non\text{-}HVAC}\right)}{E_{Total}} \tag{5}$$

where

i_{out}	Outdoor enthalpy of sample building, unit: kJ/kg$_{dry\text{-}air}$. Data source is China Weather Reporting Website: http://www.cma.gov.cn/
i_{in}	Indoor enthalpy of sample building based on Wanda Plaza design standard, unit: kJ/kg$_{dry\text{-}air}$. In summer is DT=26℃, RH=60%, i_{in}=58.471 kJ/kg$_{dry\text{-}air}$. In winter is DT=22℃, RH=50%, i_{in}=43.122 kJ/kg$_{dry\text{-}air}$.
E_{HVAC}^{*}	adjusted HVAC energy consumption of sample building, unit: kWh
E_{HVAC}	actual HVAC energy use of sample building, unit: kWh
α_1	weather modifying coefficient
E_{Total}	total annual energy use of sample building, unit: kWh
$E_{non-HVAC}$	annual other energy use except HVAC of sample building, unit: kWh

(2) Occupancy modifying coefficient: α_2

Customers occupancy changes during business hours and varies from holidays and festivals in commercial building. By analyzing data from customer flow monitoring system, there is significant difference of occupancy index per square meter, due to the variation of city and district where sample building located. Customer occupancy is a key impact factor of building cooling and heating load which cannot be neglected as well. Based on daily customer occupancy data from monitoring system and chiller energy consumption data from electricity metering system, the nonlinear regression model is computed by Matlab, calculating regressed chiller energy consumption y^{*}, as shown in Equation (6). Secondly, c_2 equals to the y^{*} divided by actual chiller energy consumption by metering system, as shown in Equation 7. Thirdly, this coefficient is only used to modify HVAC system energy use, as shown in Equation7. Thirdly, α_2 is calculated according to Equation (8).

$$y^{*} = a \times e^{(bx_1)} \tag{6}$$

$$c_2 = \frac{y^{*}}{y_1} \tag{7}$$

$$E_{HVAC}^{*} = C_2 \times E_{HVAC} \tag{8}$$

$$\alpha_2 = \frac{E_{HVAC}^{*} + E_{Others}}{E_{HVAC}} \tag{9}$$

where

a、b	constant
x_1	daily cumulative passenger flow per unit area, unit: perple/m^2/day
y_1	daily actual electricity use of chiller per unit area, unit: Wh/m^2/day
y_1^*	adjusted daily actual electricity use of chiller per unit area, unit: Wh/m^2/day
α_2	occupancy modifying coefficient

(3) Business layout modifying coefficient: α_3

According to on-site investigation of large-scaled shopping mall opened during recent five years, catering tenants becomes more and more important, which takes 25%~40% of total tenants' area. Compared to retail shops, equipment heat and personal heat per square meter of catering area is higher and has obvious impact on building cooling load. Moreover, some catering tenants shut down make-up air fan, only turn on exhaust fan. It causes the cooling air of public area is extracted into catering area and exhausted, which increases building cooling load. Hence, α_3 modified HVAC system energy consumption according to the percentage of catering tenants' area.

Firstly, this study computes a simulation model of a typical sample building in DeST, simulates building cooling and heating load under different catering area scenarios. The regression of building load and catering area percentage is presented in Equation (10). The standard business layout scenario is catering tenants' area: other retail tenants' area: public area=40%: 20%: 40%. Based on simulation model, the building cooling load $y_{40\%} = 120W/m^2$.

Secondly, c_3 equals to the y divided by $y_{40\%}$, as shown in Equation 11. Thirdly, this coefficient c_3 is only used to modify HVAC system energy use, as shown in Equation (12). Usually, catering tenants' area percentage is lower than 40%, so if $y \le y_{40\%}$, then $c_3 \le 1, E^*_{HVAC} \le E_{HVAC}$; if $y > y_{40\%}$, then $c_3 > 1, E^*_{HVAC} > E_{HVAC}$. Finally, α_3 is calculated according to Equation (13).

$$y = 131.2x + 67.027 \tag{10}$$

$$c_3 = \frac{y_2}{y_{40\%}} \tag{11}$$

$$E_{HVAC}{}^* = \begin{cases} (1+c_3)\times E_{HVAC}, y_2 \le y_{40\%} \\ c_3 \times E_{HVAC}, y_2 > y_{40\%} \end{cases} \tag{12}$$

$$\alpha_3 = \frac{E_{HVAC}{}^* + E_{Others}}{E_{Total}} \tag{13}$$

Where

x_2	business layout of sample building
y_2	cooling load based on the simulation regression formula under the standard business layout of sample building, unit: W/m^2
$y_{40\%}$	cooling load based on the simulation regression formula under the standard business layout, unit: W/m^2
α_3	business layout modifying coefficient

(4) Operation schedule modifying coefficient: α_4

Operation schedule modifying coefficient is introduced to eliminate the differences of energy use caused by different business hours. Modified method is directly cited in session 5.3.4 of the national standard method "building energy consumption standard" (Draft for Comments, expected promulgation and implementation in 2016), as follows:

$$\alpha_4 = 0.3 + 0.7\frac{H}{H_0} \tag{14}$$

H	actual annual operation hours of sample building, unit: h/a
H_0	standard annual operation hours of building, commercial building of which is 4380 h/a
α_4	operation schedule modifying coefficient

(5) HVAC system design efficiency modifying coefficient: α_5

In order to eliminate the impact of system design differences, the model introduces HVAC system design efficiency modifying coefficient. The model has been simplified, only considering the design efficiency difference of chiller plant and HVAC terminals (AHU, PAU, FCU, etc.).

Firstly, calculating c_4. c_4 equals to the $\overline{EER_0}$ divided by EER, as shown in Equation (15). If the chiller plant's EER of sample building is higher than the average EER of the same climate zone, $c_4 < 1, E^*_{chillerplant} < E_{chillerplant}$, which means the chiller plant of sample building should be more energy saving than previous energy use, as shown in Equation (16). If $c_4 \geq 1, E^*_{chillerplant} = E_{chillerplant}$, which means although the chiller plant's efficiency of sample building is lower than the average, the modifying energy use of chiller plant remains the same.

Secondly, calculating c_5 and modifying HVAC terminals energy consumption, as shown in Equation (17) and Equation (18).

Thirdly, α_5 is calculated according to Equation (19).

$$c_4 = \frac{\overline{EER_0}}{EER} \tag{15}$$

$$E_{Chiller\ Plant}^{*} = \begin{cases} c_4 \times E_{Chiller\ Plant} & EER > \overline{EER_0} \\ E_{Chiller\ Plant} & EER \leq \overline{EER_0} \end{cases} \tag{16}$$

$$c_5 = \frac{\overline{ATF_0}}{ATF} \tag{17}$$

$$E_{Terminals}^{*} = \begin{cases} c_5 \times E_{Terminals} & ATF > \overline{ATF_0} \\ E_{Terminals} & ATF \leq \overline{ATF_0} \end{cases} \tag{18}$$

$$\alpha_5 = \frac{E_{Chiller\ Plant}^{*} + E_{Terminals}^{*} + E_{Others}}{E_{Total}} \tag{19}$$

$E_{Chiller\ Plant}^{*}$	adjusted chiller plant energy consumption of sample building, unit: kWh
$E_{Terminals}^{*}$	adjusted HVAC terminal device energy consumption of sample building, unit: kWh
$\overline{EER_0}$	average chiller plant energy efficiency of the same climate zone compared to sample building
EER	chiller plant energy efficiency of system of sample building
$\overline{ATF_0}$	average air transfer factor of the same climate zone compared to sample building
ATF	air transfer factor of sample building
α_5	HVAC system design efficiency modifying coefficient

2.3 Quota validation by simulation

2.3.1 DeST model introduction.

In order to simulate the reasonable energy use of a sample building, comparing it with the energy quota and find out the possible energy saving potential, this study select a commercial building in the hot summer and cold winter area as an example, establishing a DeST model to simulate cold and heat load of the sample building.

DeST can be used to simulate and analyze both building energy consumption and HVAC system. It is developed by Tsinghua University with the aims to benefit for practical and research use of building simulation related applications in China [17]. It is commonly used in China nowadays, the calculation efficiency and accuracy have evident advantages [18].

The total building area is about 148 thousand square meters. The parking lot area is 119 thousand square meters. The total air conditioning area is 193 thousand square meters. The 3-dimention DeST model of this building is shown in Figure 2.

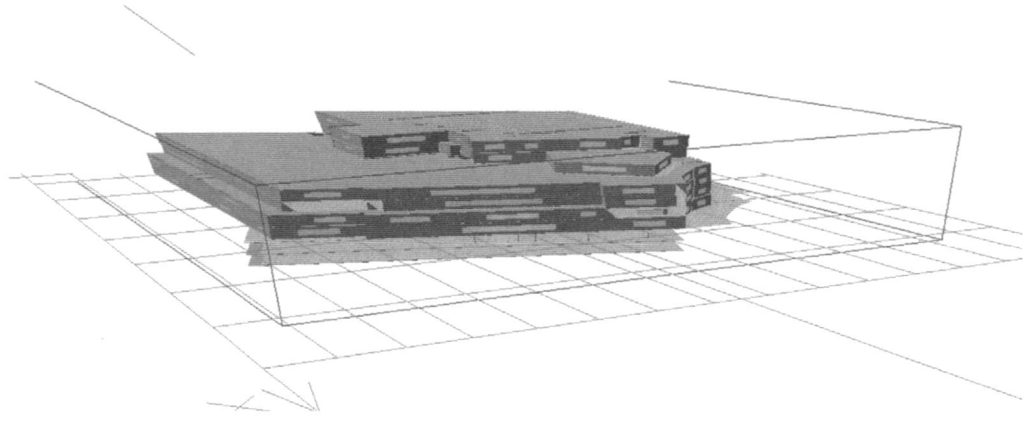

Figure 2. The simulation model of a commercial building in hot summer and cold winter area established by DeST

2.4 Simulation result of optimum energy quota

Based on the simulation result of hourly cold and heat load, and the on-site measurement of HVAC system efficiency, the optimized energy use of HVAC system is predicted, as shown in Table 2. Meanwhile, according to the on-site measurement and investigation, the optimum energy consumption of lighting, elevator, parking lots area, etc. is estimated, based on practical energy saving strategies.

The actual total energy use intensity of the sample building is 209 kWh/ (m^2.a), after optimizing, the optimum total energy consumption is 169 kWh/(m^2.a). The total energy saving rate is over 19%.

Table 2. Power consumption index comparison before and after optimization of the commercial building in hot summer and cold winter area

Type	Actual power consumption index	Optimize power consumption index	Area	Energy Saving	Partial energy saving rate
	(kWh/m^2.a)	(kWh/m^2.a)	(m^2)	(10^4kWh/a)	/
Indoor public lighting	25.6	15	26248	27.82	41.4%
Nightscape lighting	13	9.5	26248	9.19	26.9%
HVAC	46	38.5	67700	50.78	16.3%
Parking lot	11.2	7.6	47900	17.24	32.1%
Elevator	10	8.7	47900	6.23	13.0%
Tenants	72	61.2	45200	48.82	15.0%
Water supply and drainage pump	8	8	47900	0.00	0.0%
Machine room and others	23	20.5	47900	11.98	10.9%

Total (10^4kWh/a)	/	/	/	111.3	/
Total energy use intensity (EUI) (kWh/m^2.a)	208.8	169	/	/	19%

3. Results and discussion

3.1 Quotas for opened portfolio

Based on the historical energy data and modifying coefficient, energy use quota of sample buildings is calculated based on Equation 1. Energy use quota of 53 buildings with more than 1-year hourly energy data is shown in Figure 3. The average quota of each climate zone is increased from north to south.

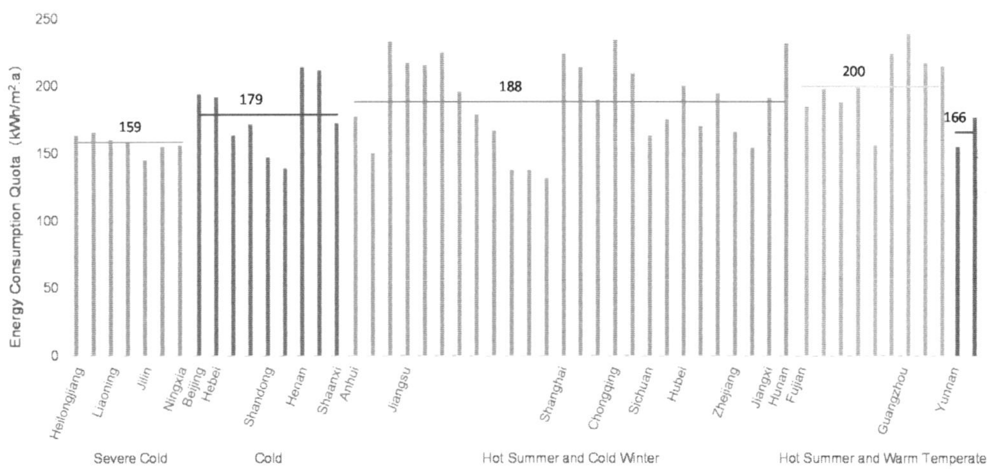

Figure 3. Energy consumption quota result of opened buildings (with hourly energy data more than one-year)

3.2 Quotas for newly-built portfolio

For buildings which opened in less than one year, since they are lack of whole year energy use data, the energy use quota is determined according to the quota of sample buildings in the same city district and climate zone, based on Figure 3. If there is no sample of the same matrix, the samples' average of the same province and climate zone is ensured as energy use quota approximately. The result of newly-built commercial buildings is shown in Figure 4. The average quota of each climate zone is increased from north to south.

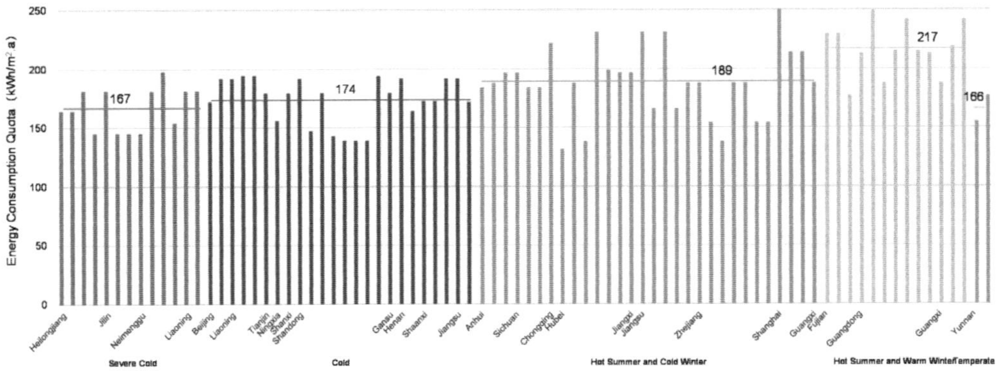

Figure 4. Energy consumption quota results of newly-built buildings (less than one-year)

3.3 Practical management in Wanda Plaza

3.3.1 Practical quotas management approach.

The energy use quota of buildings is used to guide their annual budget and provide practical suggestions of energy saving strategies. The deviation of energy metering data, energy bills and energy budget is compared to determine whether use energy quota calculated by session 3.1 or adjust quota due to special reasons.

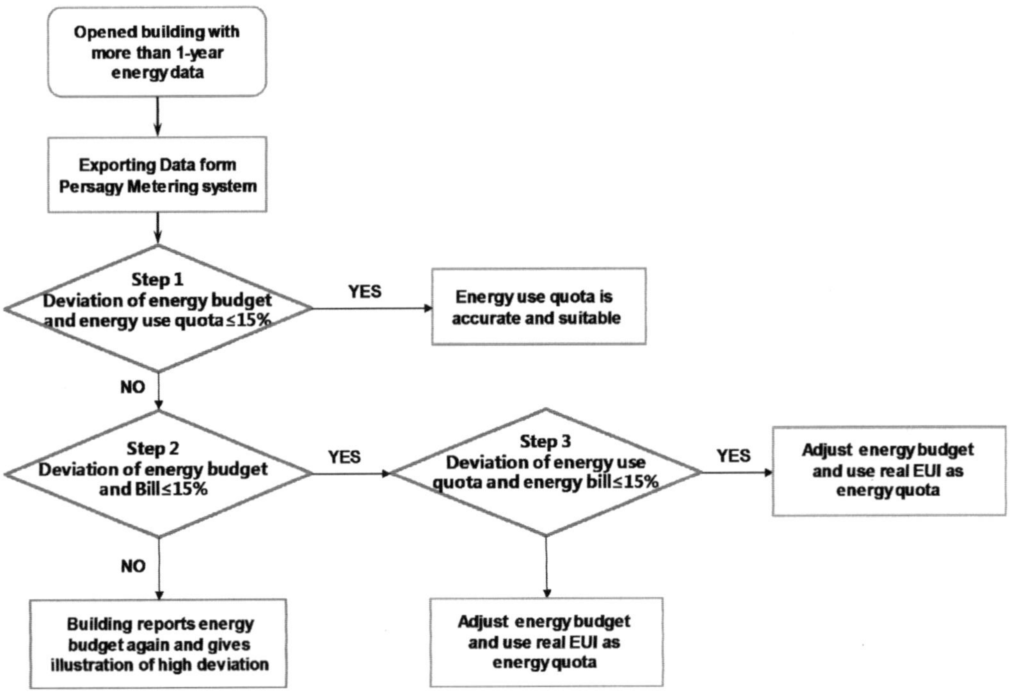

Figure 5. Logic diagram of energy quota determination for opened more than 1-year commercial building

Figure 5 illustrates logical process suitable for commercial buildings opened more than 1 year. If energy bills and energy budget can support that the energy quota is close to actual EUI and suitable for the building, the building manager should report further reason of their energy budget or change the budget to energy use quota instead. If energy bills and energy budget prove that energy quota is away from historical energy use, the energy use quota will be adjusted to the actual EUI.

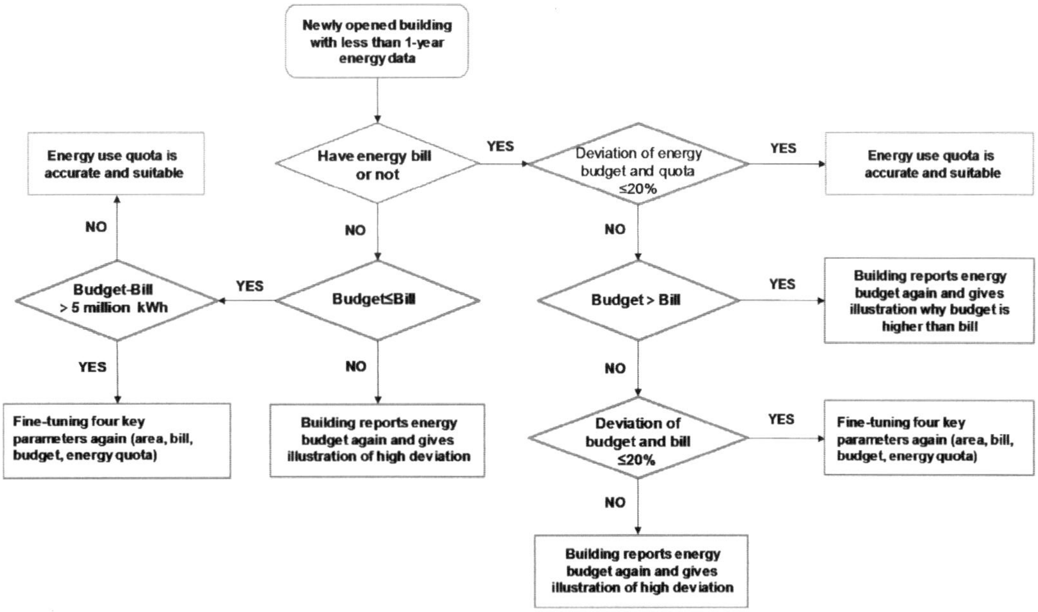

Figure 6. Logic diagram of energy quota determination for opened less than 1-year commercial building

Figure 6 reveals logical process suitable for commercial buildings opened within less than 1 year. If building has monthly energy bills and close to energy budget, then energy quota is suitable or need to be fine-tuned. If there is no energy bills or energy bill is away from budget and quota, then building needs to report further information for supportive or change the energy budget instead.

3.3.2. Empirical analysis of 2016 energy budget

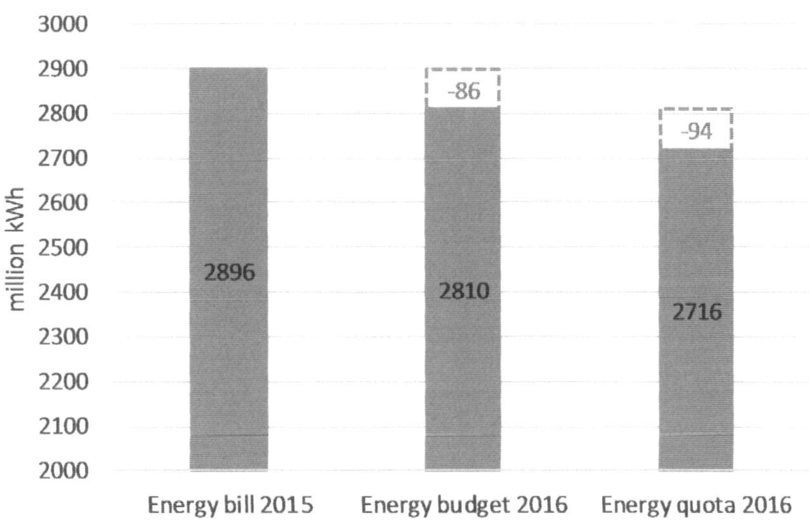

Figure 7. Comparison between energy bill 2015, energy budget 2016 and energy quota 2016

After applying practical quotas management approach, total energy budget of all sample commercial buildings in 2016 is 2810 million kWh, which is 3% energy saving compared to energy bill in 2015, as shown in Figure 7. If all building reaches energy quota 2716 million kWh in 2016, further 3.2% energy saving will be achieved. Hence, there is 180 million energy saving potential based on energy quota management in 2016, equals to 6.2% energy saving rate of actual energy use in 2015.

4. Conclusion

Energy consumption quota model, which considers a new concept of "Modified coefficient", considering weather, occupancy, business layout, operation schedule and HVAC efficiency, can be used for comparing and evaluating the performance of a large-scaled commercial building in China. The average energy use quota of shopping center is 159 kWh/(m^2.a) of severe cold climate zone, 179 kWh/(m^2.a) of cold zone, 188 kWh/(m^2.a) of hot summer and cold winter zone, and 200 kWh/(m^2.a) of hot summer and warm temperature zone. Phase- I of energy use quota management in this commercial building properties has successfully help building owners reducing 5%~12% annual budget approx.

However, because building energy use quota is analyzed based on historical data, if there is any wrong control logics or management problems caused energy waste, energy consumption quota will be inaccurate to some extent.

Besides, the phase-I study of energy use quota management doesn't consider the benchmarking with other properties EUI index in China, only compared within itself. The continued further study will collect a large number of commercial building samples with good energy metering data for benchmarking. In the upgraded Phase-II, historical energy data considering modified factor and benchmarking index compared with other properties will combined calculating energy use quota of each commercial building.

A. **1** Nomenclature Table

Symbol	Definition	Unit
EUI_{quota}	Annual electricity use quota per square meter	kWh/(m^2.a)
EUI_0	Annual actual electricity use per square meter	kWh/(m^2.a)
E_0	Annual actual electricity use	kWh
A_0	Business area, equals to total building area with subtraction of parking area and machine room space	m^2
α_1	weather modifying coefficient	-
α_2	occupancy modifying coefficient	-
α_3	business layout modifying coefficient	-
α_4	operation schedule modifying coefficient	-
α_5	HVAC system design efficiency modifying coefficient	-

i_{out}	outdoor enthalpy of sample building	kJ/kg$_{dry\text{-}air}$
i_{in}	indoor enthalpy of sample building	kJ/kg$_{dry\text{-}air}$
E_{HVAC}^{*}	adjusted HVAC energy consumption of sample building	kWh
E_{HVAC}	actual HVAC energy use of sample building	kWh
E_{Total}	total annual energy use of sample building	kWh
$E_{non-HVAC}$	annual other energy use except HVAC of sample building	kWh
x_1	daily cumulative passenger flow per unit area	people/m^2/day
y_1	daily actual electricity use of chiller per unit area	Wh/m^2/day
y_1^{*}	adjusted daily actual electricity use of chiller per unit area	Wh/m^2/day
x_2	business layout of sample building	-
y_2	cooling load based on the simulation regression formula under the standard business layout of sample building	W/m^2
$y_{40\%}$	cooling load based on the simulation regression formula under the standard business layout	W/m^2
H	actual annual operation hours of sample building	h/a
H_0	standard annual operation hours of buildings	h/a
$E_{Chiller\ Plant}^{*}$	adjusted chiller plant energy consumption of sample building	kWh
$E_{Terminals}^{*}$	adjusted HVAC terminal device energy consumption of sample building	kWh
$\overline{EER_0}$	Average Energy Efficiency Ratio: average chiller plant energy efficiency of the same climate zone compared to sample building	-
EER	Energy Efficiency Ratio: chiller plant energy efficiency of system of sample building	-
$\overline{ATF_0}$	average air transfer factor of the same climate zone compared to sample building	-
ATF	air transfer factor of sample building	-

References

[1] Yu Y, Li Y Y, Xiang D X and Jiang X Y 2014 Study on Energy Consumption Quota of Civil Buildings in Wuhan *Building Science* **30** 106-12

[2] Building Energy Conservation Research Center 2014 *2014 Annual Report on China*

Building Energy Efficiency (Beijing: China Architecture & Building Press)

[3] MOHURD 2014 *Standard for Energy Consumption of Buildings (draft for comment)* (http://www.ccsn.gov.cn)

[4] Lin B, Liu Y and Pei Z 2015 Field investigation of operation performance of green office buildings in China *Heating Ventilating & Air Conditioning* **45**

[5] Shenzhen Construction Bureau (SCB) 2015 *Standard for Energy Quota of Commercial Building in Shenzhen* (http://www.szjs.gov.cn/csml/bgs/xxgk/tzgg_1/201505/P020150508605786853667.pdf)

[6] Zhao J, Xin Y and Tong D 2012 Energy consumption quota of public buildings based on statistical analysis *Energy Policy* **43** 362-72

[7] Lee W 2008 Benchmarking the energy efficiency of government buildings with data envelopment analysis *Energy and Buildings* **40** 891-5

[8] Carriere M, Schoenau G and Besant R W 1999 Investigation of some large building energy conservation opportunities using the DOE-2 model *Energy Conservation and Management* **40** 861-72

[9] Zhang X L 2009 Establishing energy consumption quota and reducing energy consumption of public buildings to implement the effective supervision to the government office buildings and large-scale public buildings *Construction Science and Technology* **8** 22-3

[10] Cao Y, Liu Y M, Yu D, Wei Z, Liu H and Song Y H 2012 Application and Comparison of Energy Consumption Quota Determination of Air-conditioning System of Official Buildings in Beijing using Energy Consumption Benchmarking Methods of Germany and United States *Building Science* **28** 17-24

[11] Wen J W and Yang C Z 2008 Benchmarks and evaluation for energy performance of public buildings *Gas & Heat* **28** 12-5

[12] Sharp T and Oak Ridge National Laboratory 1996 *Energy benchmarking in commercial-office buildings* In: Proceedings from the 1996 ACEEE Summer Study on Energy-Efficiency in Buildings American Council for an Energy-Efficient Economy Washington D C **4** 321-9

[13] Zheng X W, Pan Y Q, Huang Z Z, Wu G and Lou Z F2006 Research and application of benchmarking building energy consumption based on database of energy consumption *Energy Conservation and Environmental Protection* **12** 10-2

[14] Zhu N, Zhu T L, Tong D D and Ding Y 2015 *D*etermination of building energy consumption baseline in China *Heating Ventilating & Air Conditioning* **45** 59-64

[15] Tian S Y and Li B Z 1992 The composition method of standard year of dynamic climatological data for heating and air conditioning energy consumption analysis *Heating Ventilating & Air Conditioning* **2** 19-22

[16] Li Q 2014 *Evaluation Method on Energy Consumption Quota in Public Buildings* (Chongqing: Chongqing University)

[17] Yan D, Xia J, Tang W Y, Song F T, Zhang X L and Jiang Y 2008 DeST-An integrated building simulation toolkit-Part I: Fundamentals *Building Simulation* **1** 95-110

[18] Jian Y W and Wang R F 2011 Comparison of thermal model between Energyplus and DeST, *Heating Ventilating & Air Conditioning* **41**

Catalytic performances of HZSM-5, NaY and MCM-41 in two-stage catalytic pyrolysis of pinewood

Y M Wang and J Wang[1]

Department of Chemical Engineering for Energy, East China University of Science and Technology, 130[#] Meilong Road, Shanghai 200237, PR China

E-mail: jwang2006@ecust.edu.cn

Abstract. Experiments were carried out in an atmospheric two-stage fixed bed reactor to investigate the catalytic cracking of pinewood pyrolysis vapour over three single catalysts, HZSM-5, NaY and MCM-41. The pinewood was pyrolyzed in the first stage reactor at a heating rate of 10 °C min^{-1} from room temperature to 700 °C, and the resultant vapour was cracked through the second reactor at a temperature of 500, 600 or 700 °C with and without catalyst. Both the gases and liquid compounds were thoroughly determined. It was found that all three catalysts had significant catalytic effects on the vapour cracking especially in the range of 500–600 °C. However, three catalysts showed dissimilarity to each other with respect to the distributions of products. Among three catalysts, HZSM-5 displayed the highest selectivity for the formation of olefins and light aromatics, with the least deposit of coke, though NaY showed the strongest capability of deoxygenation. The HZSM-5 cracking at 600 °C was preferred to balance the yield and quality of bio-products. MCM-41 behaved as a worse catalyst in the deoxygenation, and its resultant liquid product contained more heavy aromatics.

1. Introduction

With the non-sustainability of fossil energy, the conversion of renewable lignocellulosic biomass to a viable alternative for traditional fuels has drawn considerable attentions [1]. Lignocellulosic biomass can be utilized via various thermochemical processes. Pyrolysis is such an important process that can convert biomass to gaseous, liquid and solid products of more energy-intensive and more convenient form. However, the crude bio-oil derived from pyrolysis are chemically complex and unstable because the pyrolysis is generally terminated by a number of non-equilibrium state reactions [2]. The excessive presence of oxygen in pyrolysis bio-oil is a critical problem, which causes the bio-oil to have several drawbacks of low calorific value, acidity, viscosity and instability [3].

The deoxygenation of bio-oil is a key technical way to upgrading the pyrolysis bio-oil and converting it to more valuable chemicals. Several upgrading processes have been put forward by researchers, which mainly include pressurized treatment [4], hydroprocessing [5] and catalytic cracking [6]. Catalytic upgrading has some advantages over hydroprocessing of on need for expensive hydrogen and over pressurized treatment of operating at an atmospheric pressure [7]. It is widely proved that a large variety of oxygenate compounds derived from biomass could be converted into hydrocarbons over acidic zeolite catalysts at the expense of a low yield of bio-oil [7]. Williams *et al.*

[1] Address for correspondence: J Wang, Department of Chemical Engineering for Energy, East China University of Science and Technology, 130[#] Meilong Road, Shanghai 200237, PR China. E-mail: jwang2006@ecust.edu.cn.

observed that the yield of oil obtained from the pyrolysis of rice husks at the temperature of 550 °C reduced from 28.5 wt. % to 4.4 wt. % through the ZSM catalyst bed at the temperature of 500 °C. Adjaye and Bakhshi compared HZSM-5, HY and a silica alumina catalyst in production of hydrocarbons by upgrading of a fast pyrolysis bio-oil [8]. Iliopoulou et al invesitaged the effect of Al-MCM-41 on the biomass pyrolysis behaviour [9]. It appears that the yield and quality of bio-oil can be largely tailored using different catalysts.

In this work, we have examined the catalysis of three typical catalysts, HZSM-5, NaY and MCM-41 for cracking of pinewood pyrolysis vapour in a two-stage fixed bed reactor. We have tried to determine a wide spectrum of products including gases (CO, CO_2, CH_4, H_2, C2~C3) and main liquid products (water, small molecule oxygenates, light aromatic hydrocarbons, etc.). Therefore, a comprehensive evaluation can be performed with respect to their catalytic effects on bio-oil deoxygenation, aromatization and the selective formation of olefins.

2. Experimental

Table 1. The properties of pine wood.

Proximate analysis (wt. %, dry basis)			Ultimate analysis (wt. %, daf. basis)				
Ash	Volatile	Fixed carbon	C	H	O[a]	N	S
0.50	85.01	14.49	51.71	6.17	42.02	0.10	0.00

[a] By difference.

Table 2. The properties of zeolites.

Zeolites	Components (wt. %)			BET surface area ($m^2\ g^{-1}$)	Average pore width (nm)	Acidity (mmol $H^+\ g^{-1}$)	WA/SA ratio[a]
	SiO_2	Al_2O_3	Na_2O				
HZSM-5	83.69	7.05	0.08	224.67	1.756	1.81	2.84
NaY	66.04	17.96	6.29	383.87	1.905	2.83	5.41
MCM-41	99.09	0.61	0.00	1291.52	3.167	0.45	0.87

[a] WA/SA ratio, the ratio of weaken acidity (100~400 °C) to strong acidity (400~700 °C).

The properties of the pinewood sample and three catalysts used in this study are listed in Table 1 and Table 2, respectively. The dried pinewood sample was composed of 10.5% hemicellulose, 48.6% cellulose and 25.3% lignin, with 15.5% extractives. And the catalysts were provided by Yuanli Ltd, Co in China. The schematic diagram of the apparatus used for pyrolysis and vapour cracking was illustrated in our previous paper [10]. The main part of the apparatus was the two coupled stainless steel tubular reactors, which were heated independently with two electric furnaces. In each run, a 2.2 g sample of dried pinewood was wrapped in a piece of wire mesh and placed in a lower (first) reactor. Pinewood was pyrolyzed at a heating rate of 10 °C min^{-1} from room temperature to 700 °C. The resultant volatiles were purged with argon to the upper (second) reactor where the temperature was pre-set at 500, 600 or 700 °C. In the case of catalytic cracking, a 70 mm (about 6.4g of HZSM-5 and NaY, about 2.1g of MCM-41) high bed of catalyst wrapped in wire mesh was tightly inserted in the reactor. The space time of vapour through the hot catalyst bed was about 2.9~2.3s. The liquid product was captured in the cold traps (–12~–8 °C), the non-condensable gas was collected in thirteen gas bags, and the char product and catalyst were recovered at the end of experiment. Comparative experiments were also conducted without catalyst.

The four major gases (H_2, CO, CH_4 and CO_2) and the four light hydrocarbon gases (C_2H_4, C_2H_6, C_3H_6, C_3H_8) were quantitatively determined using a GC-TCD analyzer (Agilent 6820) and a GC-FID analyzer (Haixin, GC-950), respectively. The water content in the liquid product was determined by a Coulometry trace moisture analyzer (Mettler Toledo, C20). The quantitative determination of typical liquid compounds was carried out on a GC-FID analyzer (Haixin, GC-950), by an external calibration method. The distributions of PAHs in liquid product were measured by a fluorescence spectrophotometer (Lengguang, F97Pro). An elemental analyzer (Elementar Vario EL III) was used to characterize the elemental analysis of chars and pine wood. A Micromeritics ASAP 2020 instrument was used to measure the pore surface area of catalyst. The composition of catalyst was measured using a Sequential X-Ray Fluorescence Spectrometer (Shimadzu, XRF-1800). The acidities of three catalysts were measured by a NH_3-TPD method with the use of a Micromeritics ASAP 2920 instrument.

3. Results and discussions

3.1. Release rates of gaseous products

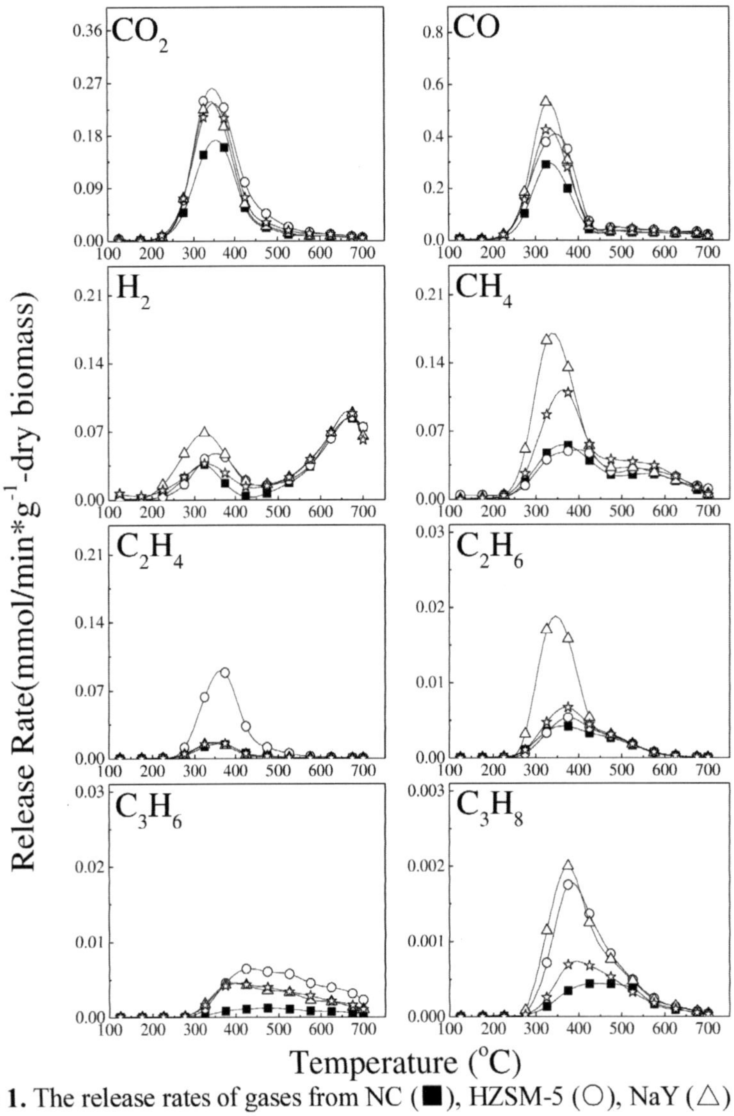

Figure 1. The release rates of gases from NC (■), HZSM-5 (○), NaY (△) and MCM-41 (☆) at varying pyrolysis temperatures. The cracking temperature, 600 °C.

Figure 1 shows the curve of release rate versus pyrolysis temperature for eight gases obtained at the cracking temperature of 600 °C with three types of catalyst and without catalyst. In the presence of each catalyst, the release rate of CO_2 and CO were increased significantly at the pyrolysis temperature range of 275-475 °C, where the pyrolysis volatiles were violently generated from pinewood [11]. The increase occurred mainly as the pyrolysis temperature increased from 325 °C to 375 °C, where the cellulose decomposed largely. It was interesting that the order of CO release rate at 325 °C was: NaY >>MCM-41 >HZSM-5, while the order at 375 °C became: HZSM-5 >NaY ≈MCM-41. It was also found that NaY was strongly inclined to produce H_2 at 225-525 °C, while HZSM-5 and MCM-41 showed an increase rate of H_2 at a later temperature range of 375-525 °C. This meant that NaY had a catalytic effect on dehydrogenation of volatiles derived during the entire pyrolysis process, whereas this effect of HZSM-5 and MCM-41 occurred only on the volatiles derived after 375 °C, where the pyrolysis reactions were composed primarily of the decomposition of lignin and non-ligninic residues. HZSM-5 showed distinct increase in the release rate of C_2H_4 at 275-475 °C and C_3H_6 at 325-700 °C, indicating its selectivity for production of olefins. In contrast, NaY and MCM-41 showed marked increases in the release rates of CH_4 and C_2H_6 at 275-425 °C. Moreover, the increase of CH_4 over MCM-41was lasted to 575 °C. It could be inferred that this large pore size catalyst had a catalytic effect on demethanation of lignin-derived volatile matter.

3.2. Overall distribution of products

The yields of four categories of products including gas, liquid, coke and char, which were produced under various cracking conditions, are shown in Figure 2. Due to the same condition of pyrolysis in the down reactor, the yields of char were close to each experiment. At the cracking temperatures of 500 °C and 600 °C, all three catalysts exhibited the catalytic effects on the deduction in the yield of liquid product, accompanied by the increases in the yield of gas and the yield of coke deposit. In sharp contrast, the cracking at 500 °C and 600 °C without catalyst only resulted in slight changes in the yields of products. Sun et al. [12] observed from a similar thermal cracking experiment that the elevation of temperature from 500 °C to 600 °C resulted in a significant degradation of some non-evaporated compounds to some evaporated compounds in the bio-oil. At the cracking temperature of 700 °C, the gaps in the yields of products with and without catalysts were no longer significant. It could be deduced that thermal effect became dominant at a higher cracking temperature. In addition, the yields of both coke deposit and liquid products decreased with increasing the cracking temperature. Among these three zeolites, the degree of coking was: NaY > MCM-41 >HZSM-5.

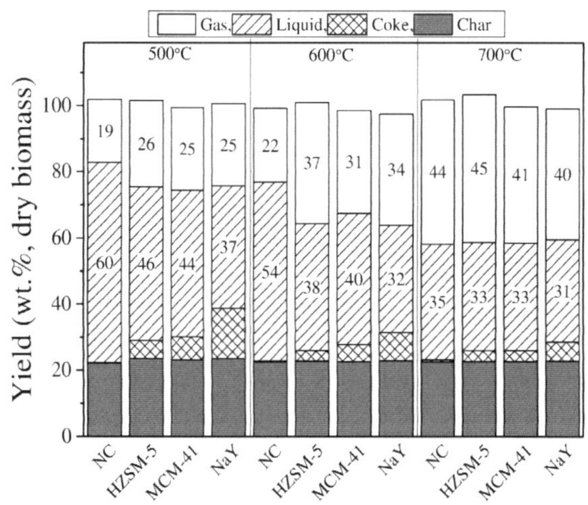

Figure 2. Yields of gas, tar, water, coke and char obtained at different cracking temperatures with and without zeolite catalysts

3.3. Yields of individual gas products

Table 3 shows the yields of eight gases obtained with three types of catalyst and without catalyst at different cracking temperatures. Under all cracking conditions, CO_2 and CO were the main gases. At the cracking temperature of 500 °C, through the catalyst beds, both the yields of CO_2 and CO increased to some extent, consistent with the increase in total gaseous product as shown in Figure 2. Increasing the cracking temperature from 500 °C to 700 °C gradually enhanced the yields of CO whether the catalyst was used or not, while the changes in the yields of CO_2 were different to each other. For instance, the marked increases of CO_2 with HZSM-5 and without catalyst were occurred at 500-600 °C and 600-700 °C, respectively; the yield of CO_2 gradually increased over MCM-41 from 500 °C to 700 °C, while that over NaY remained almost constant.

Among six hydrogen-containing gases, H_2 and CH_4 were the main gases at the cracking temperature of 500 °C without catalyst. Through the catalyst beds, the yield of all these gases slightly increased over MCM-41, while the yields of C_2H_4 and C_3H_6 over HZSM-5 and the yields of CH_4 and C_2H_6 over NaY increased significantly. At the cracking temperature of 600 °C, all these hydrogen-containing gases showed some increases in their yields for each catalyst. The yields of H_2 and CH_4 obtained with and without catalyst exhibited large gaps at this temperature. In addition, the yield of C_2H_4 over HZSM-5 and the yield of C_2H_6 over NaY increased significantly. At the cracking temperature of 700 °C, the gaps in the yields of H_2 and CH_4 obtained with and without became smaller due to the dominated thermal cracking effect at a higher cracking temperature. However, HZSM-5 still had a promoting effect on the formation of C_2H_4 and C_3H_6, in contrast to the inhibitory effect of NaY and MCM-41 on the formation of C_2H_4.

Table 3. Cumulative yields of gases at different cracking temperatures with and without zeolites. NC, no catalyst. (wt. %)

Cracking conditions		Gaseous yields (wt.%, dry biomass)							
		CO_2	CO	H_2	CH_4	C_2H_4	C_2H_6	C_3H_6	C_3H_8
500 °C	NC	9.57	7.28	0.29	1.51	0.21	0.17	0.05	0.03
	HZSM-5	10.66	11.00	0.26	1.49	1.27	0.17	0.75	0.14
	NaY	12.07	8.91	0.33	2.42	0.46	0.36	0.27	0.09
	MCM-41	11.76	10.58	0.31	1.72	0.40	0.20	0.11	0.04
600 °C	NC	9.25	9.97	0.35	1.90	0.51	0.23	0.14	0.04
	HZSM-5	14.61	14.67	0.41	2.15	2.76	0.28	0.63	0.13
	NaY	12.07	15.63	0.48	3.77	0.48	0.62	0.46	0.12
	MCM-41	12.41	13.95	0.40	3.03	0.54	0.30	0.49	0.05
700 °C	NC	12.31	22.99	0.62	4.19	2.53	0.42	0.18	0.04
	HZSM-5	14.71	20.49	0.72	4.71	3.08	0.41	0.58	0.04
	NaY	11.66	20.88	0.86	3.32	1.71	0.55	0.34	0.05
	MCM-41	13.21	20.77	0.62	4.91	1.08	0.43	0.17	0.04

3.4. Yields of main liquid products

Table 4. Yields of eight groups of compounds in bio-oil obtained with three types of catalysts at 600 °C (mg g^{-1}, dry biomass).

Groups	NC	HZSM-5	NaY	MCM-41
Acids	20.25	0.45	2.08	5.98
Ketones	5.08	1.86	0.34	3.06
Esters	4.75	0.29	2.49	1.66
Furans	8.07	0.05	0.24	2.33
Guaiacols	0.11	0.11	0.02	0.09
Phenols	6.83	3.71	5.03	8.23
MAHs	2.57	30.67	3.08	2.98
DAHs	3.54	6.89	1.02	3.14
TAHs	0.37	0.57	0.25	0.87

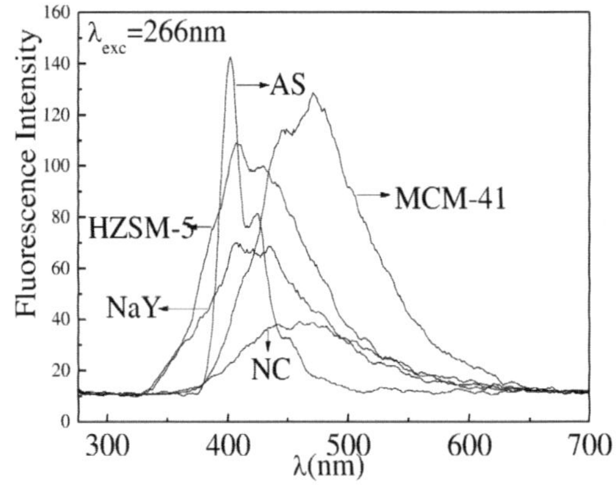

Figure 3. Fluorescence emission spectra of the liquid products produced with three types of catalyst (HZSM-5, NaY and MCM-41) and without catalyst (NC) at 600 °C by reference to the spectra of a mixture of PAH reagents. Exciting wavenumber, 266 nm.

The quantitative determination of nineteen compounds was carried on a GC-FID by using an external standard method. And other eighteen compounds were semi-quantitatively determined based on the effective carbon number method appropriate for FID signals. These compounds could be divided into nine groups, acids (methanoic acid, acetic acid and propanoic acid), ketones (hydroxyacetone and 2-cyclopenten-1-one), esters (methyl formate, 1,2-ethanediol,manoacetate and methyl acetate), furans (furfural, 2-furanmethanol and benzofuran), guaiacols (2-methoxyphenol, and 4-methyl-2-methoxyphenol), phenols (phenol, cresols, 1,2-benzenedi15ol and 2,4-xylenol), MAHs (benzene, toluene, m-xylenes, o-xylene, ethylbenzene and 1,2,3-trimethylbenzene), DAHs (indane, indene, naphthalene and 2-methylnaphthalene), and TAHs (fluorene, phenanthrene and anthracene). The yields of these groups of compounds obtained with three different catalysts at 600 °C are listed in Table 4. In the absence of catalyst with the cracking temperature of 600 °C, the main oxygen-containing compounds were still significantly present, such as acids, ketones, esters and furans. The acids were the main composition of the bio-oil, with the amount of 20.25 mg g^{-1}-dry biomass. Interestingly, using three zeolite catalysts reduced the main oxygen-containing compounds dramatically, but the extent of the catalytic effects by three catalysts was different. HZSM-5 showed the strongest ability on the degeneration of acids, esters and furans, while NaY showed the strongest ability on the degeneration of ketones and guaiacols. MCM-41 displayed a worst deoxygenation among three catalysts, with the highest content of acids, ketones and furans, due to its lowest acidity. However, MCM-41 promoted the generation of phenols. It could be deduced that these extra phenols were generated from the decomposition of the large molecule guaiacols in the large pore of MCM-41. In the presence of HZSM-5, interestingly, the yields of both MAHs and DAHs were significantly enhanced via the catalytic cracking. In the presence of NaY, the yields of both DAHs and TAHs were decreased, but the yield of MAHs was increased. In the presence of MCM-41, the yields of both MAHs and TAHs were increased, but the yield of DAHs was decreased. From these results, it could be inferred that HZSM-5 had a strong ability on the formation of monocyclic aromatic hydrocarbons and dicyclic aromatic hydrocarbons, and MCM-41 had a strong ability on the formation of tricyclic aromatic hydrocarbons, but NaY only had a slight ability on the formation of monocyclic aromatic hydrocarbons.

Figure 3 shows the UV-excited (λ=266 nm) fluorescence spectra of some typical liquid products obtained with three types of catalyst and without catalyst. we examined the fluorescence spectra of five mixtures dissolved in acetone, which contained a mixture of acetic acid, hydroxyacetone, furfural and 2-furanmethanol, a mixture of phenol, o-cresol, p-cresol, 2,4-xylenol and 2-methoxyphenol, a mixture of benzene, toluene and m-xylene, a mixture of indene, naphthalene and 2-methylnaphthalene, and a mixture of fluorene, phenanthrene and anthracene, respectively. It was observed that the former four solutions had no fluorescence at the excited wavelength. Only the last solution did the fluorescence occur, as shown in the figure (sample AS). Compared to the spectra of the liquid product obtained without catalyst, those of the liquid products obtained with HZSM-5 and NaY showed stronger intensity, particularly at a lower wavenumber which mainly corresponded to the fluorescence of tricyclic aromatics. In contrast, the spectrum of liquid product obtained with MCM-41 was peaked at a higher wavelength, reflective of the larger molecule aromatics. The spectrum of this liquid product was similar to that of liquid product obtained without catalyst, but with higher intensity. These results indicated the catalytic aromatization of all three catalysts. However, HZSM-5 and NaY were favourable for producing smaller molecule aromatics, unlike MCM-41.

4. Conclusions

HZSM-5 shows a better catalytic effect on the selective deoxygenation of bio-oil as well as the production of olefins and light aromatics. NaY also shows a strong ability on the deoxygenation of bio-oil, but with less light aromatics. MCM-41 displayed the worst ability on the deoxygenation and the formation of heavy aromatics among three catalysts. HZSM-5 showed less coking compared to NaY and MCM-41. Through the catalytic cracking with HZSM-5 at 600 °C, the yield of liquid product was 38.5 wt. %, with 24.0 wt. % water content (not shown) and 14.5 wt. % of bio-oil; the yield of monocyclic aromatic hydrocarbons was 3.0wt. % (all dry biomass basis).

Acknowledgments
This work is partially due to the National Natural Science Foundation of China (Grand No. 21376080).

Reference

[1] Brown T R 2015 A techno-economic review of thermochemical cellulosic biofuel pathways *Bioresource Technology* **178** 166-76
[2] Wang Z, Cao J and Wang J 2009 Pyrolytic characteristics of pine wood in a slowly heating and gas sweeping fixed-bed reactor. *J. Analytical and Applied Pyrolysis* **84** 179-84
[3] Mohan D, Pittman C U and Steele P H 2006 Pyrolysis of Wood/Biomass for Bio-oil: A Critical Review *Energy & Fuels* **20** 848-89
[4] Qian Y, Zhang J and Wang J 2014 Pressurized pyrolysis of rice husk in an inert gas sweeping fixed-bed reactor with a focus on bio-oil deoxygenation *Bioresource Technology* **174** 95-102
[5] Wang Y, He T, Liu K, Wu J and Fang Y 2012 From biomass to advanced bio-fuel by catalytic pyrolysis/hydro-processing: hydrodeoxygenation of bio-oil derived from biomass catalytic pyrolysis *Bioresource Technology* **108** 280-4
[6] Galadima A, Muraza O 2015 In situ fast pyrolysis of biomass with zeolite catalysts for bioaromatics/gasoline production: A review *Energy Conversion and Management* 105 338-54
[7] Rezaei P S, Shafaghat H and Daud W M A W 2014 Production of green aromatics and olefins by catalytic cracking of oxygenate compounds derived from biomass pyrolysis: A review *Applied Catalysis A: General* **469** 490-511
[8] Adjaye J D, Bakhshi N N 1995 Production of hydrocarbons by catalytic upgrading of a fast pyrolysis bio-oil. Part I: Conversion over various catalysts *Fuel Processing Technology* **45** 161-83
[9] Iliopoulou E F, Antonakou E V, Karakoulia S A, Vasalos I A, Lappas A A and Triantafyllidis K S 2007 Catalytic conversion of biomass pyrolysis products by mesoporous materials: Effect of steam stability and acidity of Al-MCM-41 catalysts *Chemical Engineering Journal* **134** 51-7
[10] Wang Y and Wang J 2016 Multifaceted effects of HZSM-5 (Proton-exchanged Zeolite Socony Mobil-5) on catalytic cracking of pinewood pyrolysis vapor in a two-stage fixed bed reactor. *Bioresource Technology* **214** 700-10
[11] Shi X and Wang J 2014 A comparative investigation into the formation behaviors of char, liquids and gases during pyrolysis of pinewood and lignocellulosic components *Bioresource Technology* **170** 262-9
[12] Sun Q, Yu S, Wang F and Wang J 2011 Decomposition and gasification of pyrolysis volatiles from pine wood through a bed of hot char *Fuel* **90** 1041-8

Effects of setting angle on performance of fish-bionic wind wheel

G S Li, Z X Yang[1], L Song, Q Chen, Y B Li and W Chen

Institute of Systems Science and Engineering, Henan Engineering Laboratory of Wind Power Systems Henan University of Science and Technology, No.48 Xiyuan Road, Luoyang 471003, China

Email: zxyang@haust.edu.cn

Abstract. With the energy crisis and the increasing environmental pollutionmore and more efforts have been made about wind power development. In this paper, a new type of vertical axis named the fish-bionic wind wheel was proposed, and the outline of wind wheel was constructed by curve of Fourier fitting and polynomial equations. This paper attempted to research the relationship between the setting angle and the wind turbine characteristics by computational fluid dynamics (CFD) simulation. The results showed that the setting angle of the fish-bionic wind wheel has some significant effects on the efficiency of the wind turbine. Within the range of wind speed from 13m/s to 15m/s, wind wheel achieves the maximum efficiency when the setting angle is at 37 degree. The conclusion will work as a guideline for the improvement of wind turbine design.

1. Introduction

Being clean and renewable, wind energy has great reserves in nature, and it has great potential as a new energy with good prospects for development [1-9]. In China, the demand for energy is increasing rapidly with the development of economy. The traditional oil and coal has caused serious environmental pollution, while renewable wind energy can provide complement to modern society without pollution. Therefore more and more countries have recognized the importance of wind energy[1, 3, 6, 8, 9]. Wind power is the most important way to use wind energy, and its wattage is increasing year by year.

The structure of wind turbines is very simple, but the aerodynamic characteristics of the work is extremely complex. At present, we have two main methods to research wind turbine flow field characteristics: numerical analysis and wind tunnel testing [10, 11].

There are many types of wind turbineswhich can be divided into horizontal axis wind turbine [12-15] and vertical axis wind turbine [16-23] depending on the wind direction and the axis direction.

Compared with the traditional horizontal axis wind turbine, the vertical axis wind turbine has the advantage of advanced design methods, high wind power utilization, low starting wind speed, low noise, and broad market

[1] Address for correspondence: Z X Yang, Institute of Systems Science and Engineering, Henan Engineering Laboratory of Wind Power Systems Henan University of Science and Technology, No.48 Xiyuan Road, Luoyang 471003, China. E-mail: zxyang@haust.edu.cn.

prospects [14, 24-26]. Vertical axis wind turbine includes Savonius type[24, 27], Darrieus type[28, 29], and H-type vertical axis wind turbine[30, 31].

This paper introduces a new type of fish-bionic wind wheel, and researches into the relationship between the setting angle and the wind turbine characteristics by CFD.

2. Structure and principle of fish- bionic wind wheel

The proposed wind wheel is come from the contour of fish, and the geometry is a combination of Fourier fitting, polynomial equations and peripheral curve. At both ends of the rotor blades, the rotor is provided with two threaded holes in the height direction respectively, and connected to the upper and lower cover with certain setting angle. Fish-bionic wind wheel structure is shown in Figure 1 and Figure 2.

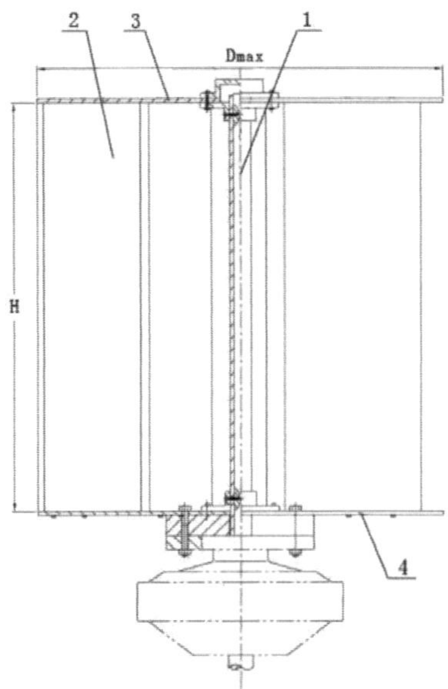

Figure 1. Front view of fish-bionic wind wheel.

In Figure 1, (1) is wind wheel center axis, (2) is wind turbine blade, (3) is wind wheel upper cover, (4) is wind wheel lower cover.

Upper cover and lower cover need to be installed in parallel, and perpendicular to the central axis of the wind turbine. Wind turbine blades are curved structures, where three blades must be the same bending direction, and it should install at least three blades in order to double the use of wind energy effectively. H is the distance between the upper cover and lower cover, and the ratio between the H and D_{max} is between 1 to 3.

In Figure 2, (5) is wind wheel front edge point, (6) is wind wheel rear edge point. D_{max} is diameter of circle which is composed by the front edge points of three blades. D_{min} is diameter of circle which is composed by the rear edge points of three blades. The length between the front edge point of blade and rear edge point of blade is chord. The ratio between the chord length and D_{max} is 0.5. 40 degree is the setting angle of the wind turbine, which is the angle between the connecting line of front edge and the rear edge point and the connecting line of front edge and the center axis, and setting angle will affect the performance of wind turbines.

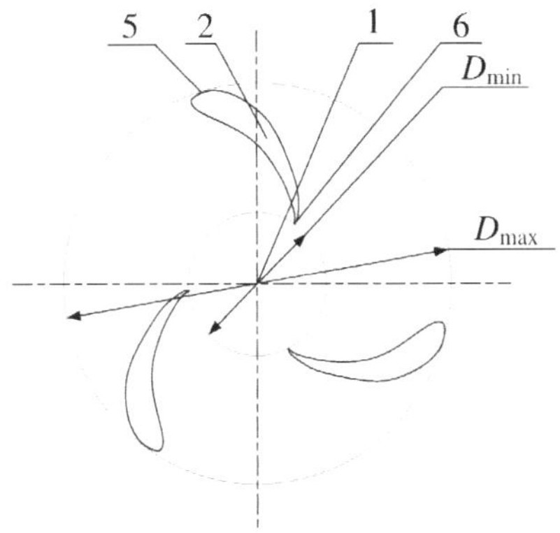

Figure 2. Plan view of fish-bionic wind wheel (without upper cover).

Figure 3. Fish swimming in the water.

As is shown in Figure 3, fish swimming in the water is similar to air flowing through something since fish tail swing generates driving force. According to the principle of bionics, the paper improves the utilization of wind energy by optimizing the end of the blade, changing the number of blades and the installation angle.

We get wind wheel rotor blades sample from bionics, and use an approximate equations to describe sample by curve fitting, then the polynomial equations of intersection line between upper cover and inner surface is:

$$f(x) = 13.22x^7 - 51.08x^6 + 79.58x^5 - 65.14x^4 + 30x^3 - 7.95x^2 + 1.135x + 0.007948 \quad (1)$$

The Fourier fitting of intersection line between upper cover and outer surface is:

$$f(x) = 48.51 - 54.53\cos(1.571x) - 56.82\sin(1.571x) - 1.339\cos(3.142x) + 40.82\sin(3.142x) + 8.975\cos(4.713x) - 8.37\sin(4.713x) - 1.619\cos(6.284x) - 0.05546\sin(6.284x) \quad (2)$$

Where the x is in the range [0-1].

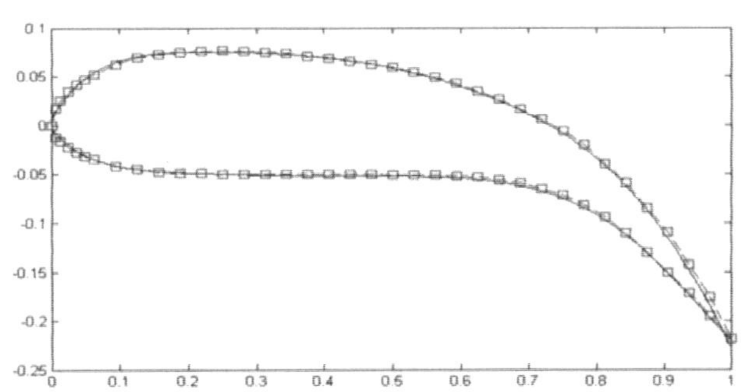

Figure 4. Sample points and approximate curve fitting of fish-bionic wind wheel.

According to the results of curve fitting, we have designed the rotor blades shown in Figure 4 and Figure 5.

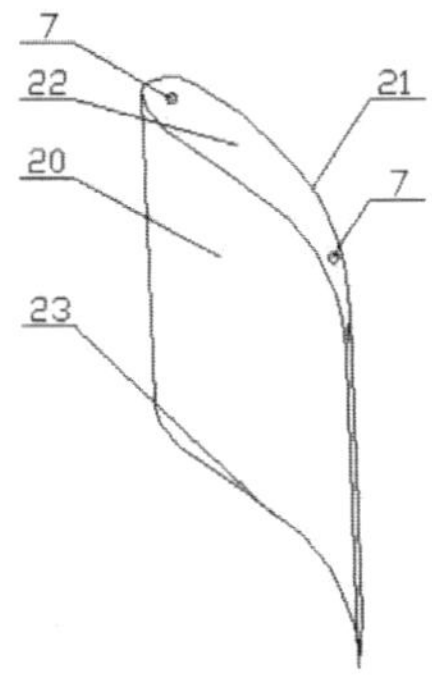

Figure 5. Wind turbine blade structure.

In Figure 5, (7) are threaded holes, the two threaded holes play a role in fixing the upper and lower cover. Both are located on the chord length. (20) is inner surface, (21) is outer surface, (22) is upper cover, (23) is lower cover, the upper cover and lower cover are parallel to each other and the projection of the covers are coincident, and inside surface and outside surface are perpendicular to the upper cover and lower cover, the thickness of the blade reduces gradually and intersects at a point from the front edge point to the rear edge point.

Wind wheel will work by the pressure which is different between the inside and outside surfaces of head. This will produce a pressure that will differentially push the rotating wind wheel, due to the asymmetrical circular surface of the head when the wind flows from the inlet into the wind wheel.

The front half of wind wheel is a lift-typed wind turbine because that has different crown of the shape at the two sides. The rear half of wind wheel is a resistance-typed wind turbine because the concave is located at the windward side of the wind. Traditionally, the lift-typed wind turbines have poor starting performance, and the resistance-typed wind turbines have lower speeds, but the special structure of fish-bionic wind wheel overcomes the above mentioned shortcomings.

3. CFD model

Computational fluid dynamics use discrete mathematical methods, and it can simulate the running state of model in theory, and uses powerful computing capability of computer to solve various issues of hydrodynamics calculations. CFD software simulations can greatly reduce the use of pre-investment projects, save time and improve work efficiency, thus it has become an important modern engineering software tool.

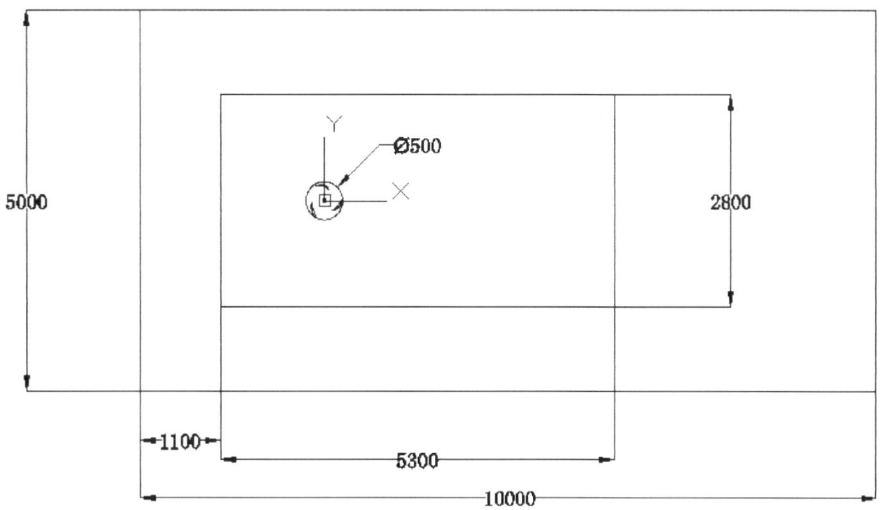

Figure 6. CAD drawing of wind wheel computational domain (unit: mm).

The fish-bionic wind wheel was constructed and computational domain was chosen according to equation (1) and equation (2), then transient and dynamic performance of wind wheel was analyzed by sliding mesh method of computational fluid dynamics.

Wind wheel diameter was 0.5 m, the distance bewteen the wind turbine and left boundary was five times of the diameter of the wind wheel, the distance bewteen the wind turbine and upper and lower bounds was five times of the diameter of the wind wheel, the distance bewteen the wind turbine and right boundary was fifteen times of the diameter of the wind wheel. Mesh quality is the main factor affecting the results of the simulation experiment. In this paper, the triangular unstructured grids are used because the outer contour of wind wheel is very complex. After the mesh independent verification, the number of mesh was 5.86E5.

Reynolds number was 2.50E5, turbulence kinetic energy was 0.24, turbulent dissipation rate was 1.34, and RNG k-e turbulence model was used in the simulation. And the inlet boundary condition was velocity inlet, wind speed set in the simulation, outlet boundary conditions was outflow, the upper and lower boundary conditions symmetry wall, the blade surface was moving wall, the central axis was stationary wall. The pressure-velocity coupling algorithm was SIMPLE, the transient formulation was discretized by the first order implicit scheme, and the turbulent kinetic energy and turbulent dissipation rate were discretized by the sencond order upwind scheme.

According to the simulation results, the last cycle of the wind wheel torque (final 360 torque values,)was selected, and the average as average torque of a wind wheel cycle was found, then we obtained the efficiency of wind turbines by following equation.

Wind power input

$$P_i = \frac{\rho H D \upsilon^3}{2 \times 10^3 \times 10^3} \tag{3}$$

Wind turbine power output

$$P_o = \frac{2\pi n T}{60} \tag{4}$$

Wind turbine efficiency

IOP Conf. Series: Earth and Environmental Science **40** (2016) 012016 doi:10.1088/1755-1315/40/1/012016

$$\eta = \frac{P_o}{P_i} \tag{5}$$

Where:

υ : Wind velocity (m/s).

ρ : Air density (1.225kg/m3).

H : Height of wind wheel (500mm).

D : Diameter of wind wheel (500mm).

n : Rotor speed (r/min).

T : Wind wheel output torque (N.m)

4. CFD simulation results

As it can be seen from Fig.7, there is large low-speed area at the rear end of wind wheel, which is caused by the rotation of the wind wheel.The low wind area is not only beneficial to the rapid flow of wind between the wind wheel blades, but also conducive to the wind wheel rotation. The rear end of wind wheel shows resistance-type characteristics when wind wheel movement, fishtail portion makes the air flow through the fish's belly impact on adjacent blades fish head, thereby increases the speed and output torque of the wind wheel.

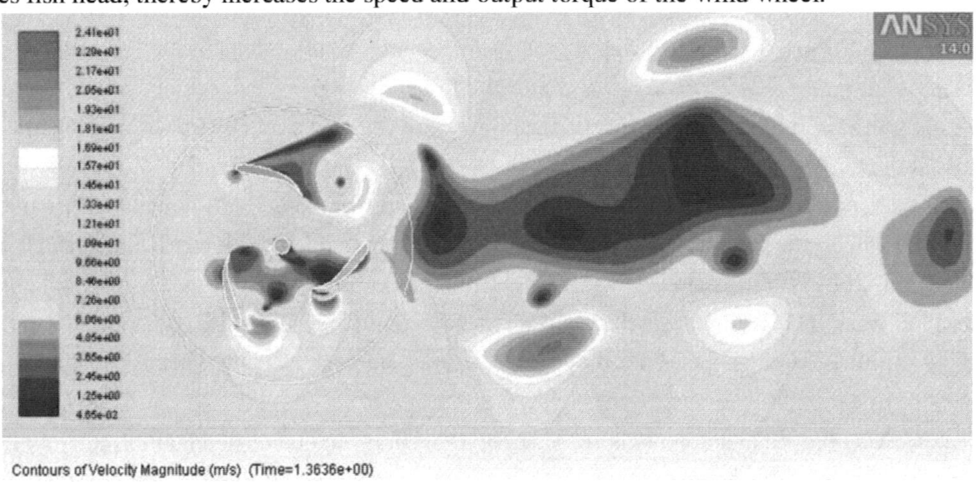

Figure 7. The velocity field (the setting angle is 38 degree)

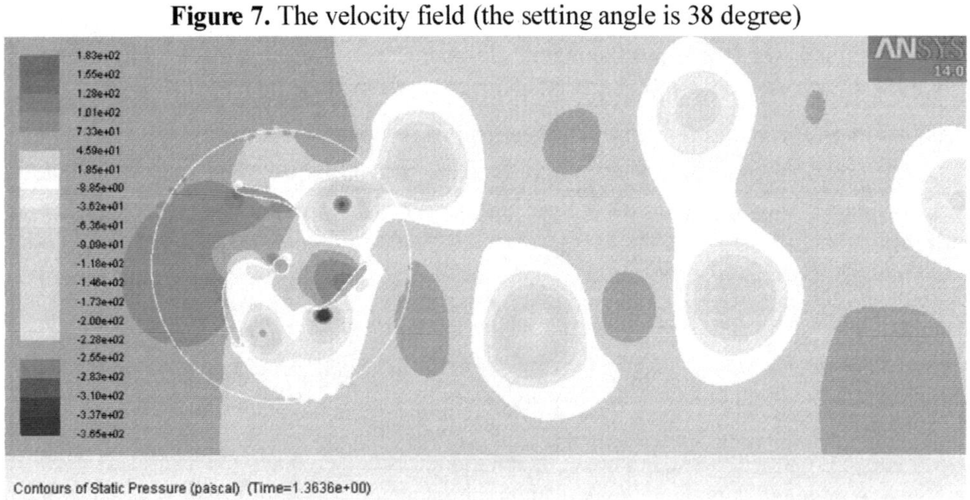

Figure 8. The pressure field (the setting angle is 38 degree)

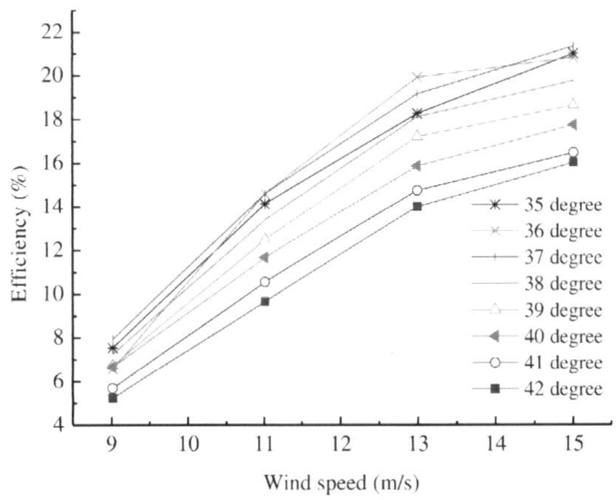

Figure 9. Relationship of wind speed and efficiency at different setting angles.

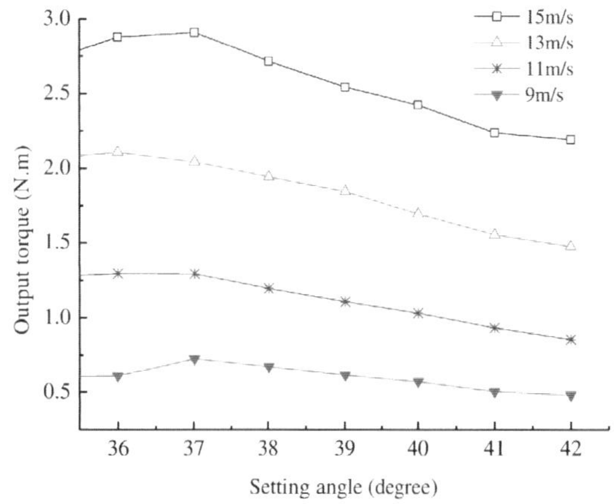

Figure 10. Relationship of setting angle and output torque.

We can see that the rotor surface pressure in the head is relatively larger in Fig.8, that is, the fish body portion and the surface inside and outside of head portion showing lift-typed wind turbine characteristics. The characteristics are related to the motion style of wind, and rotor rotation is relied on a pressure head inside and outside working surfaces; the pressure will push the rotating wind wheel, it improves the startup performance of lift-typed wind turbine because of the differential existence of pressure.

From the Figure 9, we can see that the setting angle has a greater impact on efficiency. and the setting angle is determined by the structure of the wind turbine, which affects secondary use of wind energy inside the wind wheel .

The efficiency increases obviously when wind speed goes from 9m/s to 13m/s, the efficiency increases gently when wind speed goes from 13m/s to 15m/s, and wind wheel achieves the maximum efficiency when setting angle is 37 degree.

As it can be seen from the Fig.10, the output torque of wind wheel increases at first and then decreases with the increase of setting angle.

5. Concluding remarks

The paper describes a kind of higher efficiency bionic structure wind wheel and gets wind turbine blade from bionic , then sliding mesh method of computational fluid dynamics was used to analyse the transient and dynamic performance of wind wheel.It was found that the setting angle has a direct impact on the efficiency of wind turbines and will achieve maximum efficiency in a particular installation angle.

Simulation results show that the proposed wind turbine structure has the advantage of doubled use of wind energy. The setting angle is an important factor which affects the efficiency of the wind turbine, and we can increase wind energy conversion rate by choosing the appropriate setting angle, and the angle is about 37 degrees.

Acknowledgements

We most gratefully acknowledge the financial support of the National Natural Science Foundation of China (NSFC) under grant No.71071078 & No.70671035, and the Doctoral Scientific Research Foundation of Henan University of Science and Technology.

References

[1] Brekken T K A, Yokochi A, Jouanne A V, Yen Z Z, Hapke H M, and Halamay D A 2011 Optimal Energy Storage

Sizing and Control for Wind Power Applications *IEEE Transactions on Sustainable Energy* **2** 69-77

[2] Dai T, Song B, and Shu S W 2011 Study on the wind energy resources assessment in wind power generation *Management Science and Electronic Commerce (AIMSEC), 2011 2nd International Conference on Artificial Intelligence* 6804-07

[3] Hu P, Karki R, and Billinton R 2009 Reliability evaluation of generating systems containing wind power and energy storage *IET Generation, Transmission & Distribution* **3** 783-91

[4] Hoffman D L and Molinski T S 2009 How new technology developments will lower wind energy costs *Integration of Wide-Scale Renewable Resources Into the Power Delivery System, 2009 CIGRE/IEEE PES Joint Symposium* 1-1

[5] Hui J, Bakhshai A, and Jain P 2014 Power management supervisory control algorithm for standalone wind energy systems *2014 IEEE 36th International Telecommunications Energy Conference (INTELEC)* 1-6

[6] Burke D J and Malley M J O 2011 Factors Influencing Wind Energy Curtailment *IEEE Transactions on Sustainable Energy* **2** 185-93

[7] Lerch E 2007 Storage of fluctuating wind energy *2007 European Conference on Power Electronics and Applications* 1-8

[8] Ni B and Sourkounis C 2009 Investigations on energy yield and power fluctuation from different control methods for wind energy converters at various wind conditions *2009. EPE '09. 13th European Conference on Power Electronics and Applications* 1-10

[9] Xu X, Chen Y, Zheng Y, Bai X, Wei M, and Liu Y A 2010 Observation and Simulation of Wind Energy Resource in Jiangsu Coastal Area *2010 Asia-Pacific Power and Energy Engineering Conference (APPEEC)* 1-4

[10] Karki R, Po H, and Billinton R 2006 A simplified wind power generation model for reliability evaluation *IEEE Transactions on Energy Conversion* **21** 533-40

[11] Lu M S, Chang C L, Lee W J, and Wang L 2009 Combining the Wind Power Generation System With Energy Storage Equipment *IEEE Transactions on Industry Applications* **45** 2109-15

[12] Guo Y, Hosseini S H, Tang C Y, Jiang J N, and Ramakumar R G 2013 An Approximate Wind Turbine Control System Model for Wind Farm Power Control *IEEE Transactions on Sustainable Energy* **4** 262-74

[13] Kusiak A, Zhang Z, and Li M 2010 Optimization of Wind Turbine Performance With Data-Driven Models *IEEE Transactions on Sustainable Energy* **1** 66-76

[14] Petersson A, Thiringer T, Harnefors L, and Petru T 2005 Modeling and experimental verification of grid interaction of a DFIG wind turbine *IEEE Transactions on Energy Conversion* **20** 878-86

[15] Petru T and Thiringer T 2002 Modeling of wind turbines for power system studies *IEEE Transactions on Power Systems* **17** 1132-39

[16] Dong L, Liao M, Li Y, Song X, and Xu K 2009 Study on Aerodynamic Design of Horizontal Axis Wind Turbine Generator System *2009. ICEET '09. International Conference on Energy and Environment Technology* 841-44

[17] Schreck S J and Robinson M C 2007 Horizontal Axis Wind Turbine Blade Aerodynamics in Experiments and Modeling *IEEE Transactions on Energy Conversion* **22** 61-70

[18] Khan S, Khan A, Irfan M, and Hussain S 2012 Aerodynamic analysis and dynamic modeling of small horizontal axis wind turbine *2012 International Conference on Robotics and Artificial Intelligence (ICRAI)* 117-24

[19] Mesemanolis A and Mademlis C 2014 Combined maximum power point and yaw control strategy for a horizontal axis wind turbine *2014 International Conference on Electrical Machines (ICEM)* 1704-10

[20] Murshed M, Arafat M Y, and Razzak M A 2016 Analysis of air foils and design of blades for a low-speed 250W horizontal axis wind turbine suitable for coastal areas of Bangladesh *2016 4th International Conference on the Development in the in Renewable Energy Technology (ICDRET)* 1-6

[21] Qiu M, Zhou Z, and Zhang J 2009 Three-dimensional flow simulation for horizontal axis wind turbine *2009 World Non-Grid-Connected Wind Power and Energy Conference* 1-5

[22] Yu H, Zhang L, Wang J, Gao Z, and You Z 2011 The simulation and comparison of bound vortex of horizontal axis wind turbine *2011 International Conference on Materials for Renewable Energy & Environment (ICMREE)* 575-79

[23] Galani P A, Teixeira J C, and Guerra S M G 2015 Influence of failure rate and downtime in annual energy produced on a horizontal axis wind turbine *IEEE Latin America Transactions* **13** 2949-54

[24] Shahariar G M H and Hasan M R 2014 Design and construction of a Vertical Axis Wind Turbine *2014 9th International Forum on Strategic Technology (IFOST)* 326-29

[25] Almohammadi K M, Ingham D B, Ma L, and Pourkashanian M 2015 2-D-CFD Analysis of the Effect of Trailing Edge Shape on the Performance of a Straight-Blade Vertical Axis Wind Turbine *IEEE Transactions on Sustainable Energy* **6** 228-35

[26] Preen R J and Bull L 2015 Toward the Coevolution of Novel Vertical-Axis Wind Turbines *IEEE Transactions on Evolutionary Computation* **19** 284-94

[27] Sharma M, Rieger B, Jou C, and Butka B 2012 Development of helical vertical axis wind turbine *2012 Proceedings of IEEE Southeastcon* 1-5

[28] Kumar A and Saini R P 2015 Investigation on performance of improved savonius rotor: An overview *2015 International Conference on Recent Developments in Control, Automation and Power Engineering (RDCAPE)* 151-56

[29] Lei S, Hong-Zhao L, and Zong-Xiao Y 2015 Performance comparison for savonius type wind turbines by numerical analysis approaches *2015 International Conference on Advanced Mechatronic Systems (ICAMechS)* 402-07

[30] Sunyoto A, Wenehenubun F, and Sutanto H 2013 The effect of number of blades on the performance of H-Darrieus type wind turbine *2013 International Conference on QiR (Quality in Research)* 192-96

[31] Dai H K, Yang Z X, and Song L 2014 Mathematical modeling for H-type vertical axis wind turbine *2014 IEEE 11th International Conference on Networking, Sensing and Control (ICNSC)* 695-700

Cutback sensitivity test for boron-free small modular PWR

J Choe[1], H C Shin[2], J E Jeong[2] and D Lee[1, 3]
[1]Department of Mechanical & Nuclear Engineering, Ulsan National Institute of Science and Technology, Ulsan, 44919, Republic of Korea
[2]Core and Fuel Analysis Group, Korea Hydro & Nuclear Power Central Research Institute, Daejeon, 34101, Republic of Korea

Email Address: deokjung@unist.ac.kr

Abstract. A soluble boron-free small modular pressurized water reactor (SMPWR) uses burnable absorbers (BA) instead of soluble boron to reduce excess reactivity. As a consequence, the fuel cycle length can be shortened by the residual penalty of BA. This paper performs cutback sensitivity tests to extend the cycle length. The influence of the height of the cutback, of the ^{235}U enrichment rate, and of the BA material on the power peaking factor (Fq), the axial offset (AO) and the fuel cycle length is analyzed with the reactor core design system, CASMO-4E/SIMULATE-3 code system.

1. Introduction

Soluble boron-free operation of small modular pressurized water reactor (SMPWR) has some advantages: reduction of the plant size owing to the removal of the chemical volume control system (CVCS), reduction of liquid radioactive waste, no corrosion issue due to boric acid [1]. However, large amounts of burnable absorber (BA) must be loaded in the core so as to reduce excess reactivity. As a consequence, the fuel cycle length can be shortened by the residual penalty of BA. This paper performs cutback sensitivity tests at the top and bottom regions of the core to extend the cycle length.

The influence of the height of the cutback, of the ^{235}U enrichment rate and of the BA material on the power peaking factor (Fq), the axial offset (AO) and the fuel cycle length has been analyzed. The Studsvik's reactor core design CASMO-4E/SIMULATE-3 code system is used for these simulations [2-4].

2. Reference SMPWR Design Parameters

Core design parameters of reference SMPWR are summarized in Table 1. The thermal power is 200 MW, and the power density is 58.4 kW/L. The base model of a fuel assembly (FA) is 17 pins by 17 pins, Westinghouse type. The fuel material is uranium oxide. The BA design uses new type of BA, R-BA, which is a thin Zr-^{167}Er layer coated inside the cladding, as shown in Figure 1 [5]. The target reactivity swing is below 1,000 pcm to secure enough shutdown margin and to secure stable control of excessive reactivity by control rods.

[3] Address for correspondence: Department of Mechanical & Nuclear Engineering, Ulsan National Institute of Science and Technology, Ulsan, 44919, Republic of Korea. Email: deokjung@unist.ac.kr.

IOP Conf. Series: Earth and Environmental Science **40** (2016) 012017 doi:10.1088/1755-1315/40/1/012017

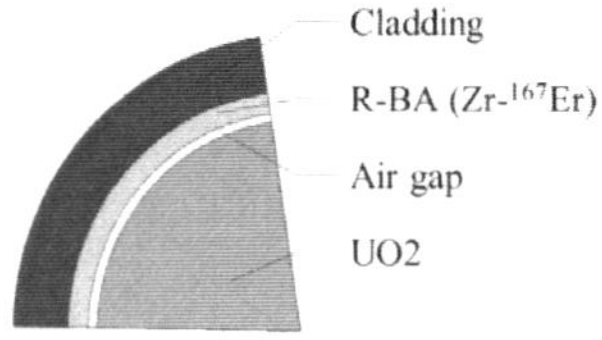

Figure 1. R-BA Cross Section View.

Table 1. Core Design Parameters of reference SMPWR.

Parameters	Target Value
Thermal power	200 MW
Power density	58.4 kW/L
Fuel cycle length	36 months
Fuel material	UO_2
BA material	R-BA with Zr-^{167}Er
Fuel enrichment	4.90 w/o ^{235}U
Number of FAs	37
Active core height	2.0 m
Reactivity($\Delta\rho$) Swing[a]	< 1,000 pcm
Boron concentration	0 ppm
3D pin peaking factor [b]	< 4.4
Axial Offset [b]	-0.4 < AO < +0.4

[a] ((max k_{eff} -1) / max k_{eff}) × 10^5 [pcm], [b] [6]

The two-batch loading pattern with 37 FAs is shown in Figure 2. This configuration uses 12 FAs of Type 01 and 25 FAs of Type 02. The specification of R-BA for each FA type is different: Type 01 FA uses 1 mm of ^{167}Er 5 w/o R-BA for 88 pins whereas Type 02 FA uses 2 mm of ^{167}Er 15 w/o R-BA for 216 pins, as shown in Figure 3. The reactivity swing of this core is 548 pcm and the cycle length reaches 798 effective full power days (EFPD, 26.6 months). The fuel cycle length of the reference core is shorter than the target value.

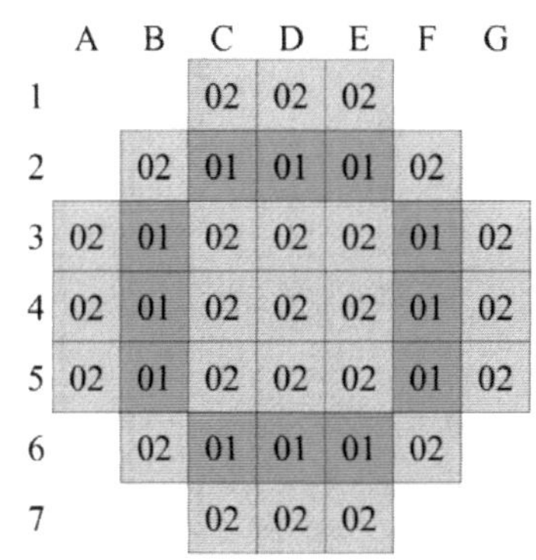

Figure 2. Two-batch Loading Pattern for SMPWR.

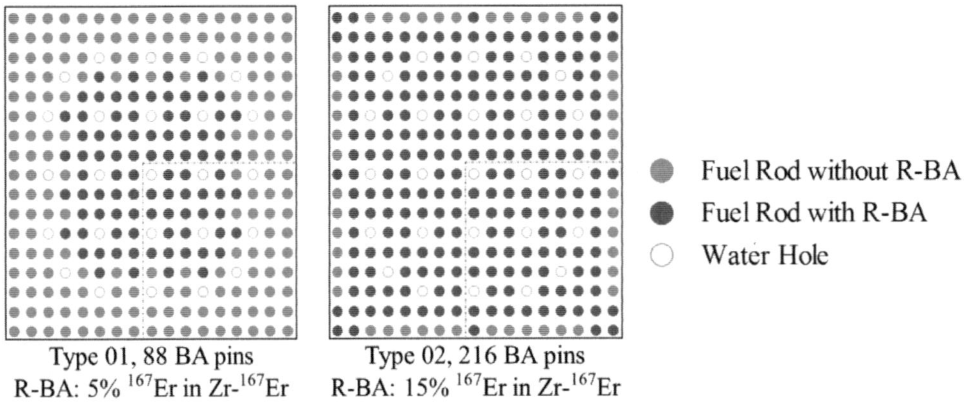

Figure 3. Fuel Assembly Pattern for Type 01 and Type 02.

3. Cutback Sensitivity Test

3.1 Description of Cutback Sensitivity Test

The R-BA is loaded from the top to the bottom in the reference core as shown in Figure 4. The cutback sensitivity test is performed at the top and bottom regions. The relative power is low at the top and bottom regions; thus those regions may not need to load large amount of BA.

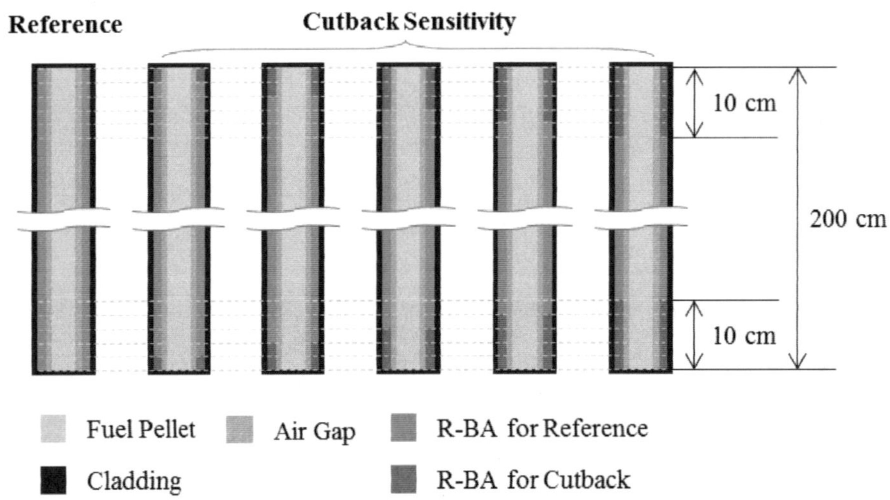

Figure 4. Axial View of BA Rods for Test Cases.

The test cases are summarized in Table 2. There is no cutback (0 cm cutback) in the reference core. The height of cutback was then changed from 2 cm to 10 cm by 2 cm step from both the top and the bottom, as shown in Figure 4. For each non-zero cutback, two sensitivity tests were carried out:

- First, no BA material was used in the cutback regions; the ^{235}U enrichment rate in the cutback regions was changed from 2.2 w/o to 4.9 w/o by 0.5 w/o steps.
- Second, different BA materials were tested in the cutback regions; the ^{235}U enrichment rate was fixed to 4.9 w/o ^{235}U.

In the 2nd sensitivity test, 2 different BA materials were tested, one after the other. The first BA material tested in the cutback regions is Zr-B instead of Zr-^{167}Er. The B content in Zr-B was changed from 5 w/o to 10 w/o by 5 w/o step. The second BA material tested is Zr-Gd instead of Zr-^{167}Er. The Gd content in Zr-Gd was changed from 5 w/o to 20 w/o by 5 w/o step. The content range of BA material in Zr-BA is determined by their phase diagrams [7-9].

Table 2. Cases description for Cutback Height and Material.

	Cases Description					
Height of Cutback [cm]	2.00	4.00	6.00	8.00	10.0	
^{235}U Enrichment rate (no BA) [w/o ^{235}U]	2.20	3.00	3.50	4.00	4.50	4.90
BA Content of Zr-B in BA (^{235}U enrich. rate = 4.9 w/o) [w/o]	5.00	10.0				
BA Content of Zr-Gd in BA (^{235}U enrich. rate = 4.9 w/o) [w/o]	5.00	10.0	15.0	20.0		

3.2 First Study: Fuel Enrichment variation without BA

The fuel cycle length and the reactivity swing as a function of cutback height and fuel enrichment in the cutback regions are presented in Figure 5. The higher the cutback height is, the longer the fuel cycle length is, due to the absence of BA in the cutback regions.

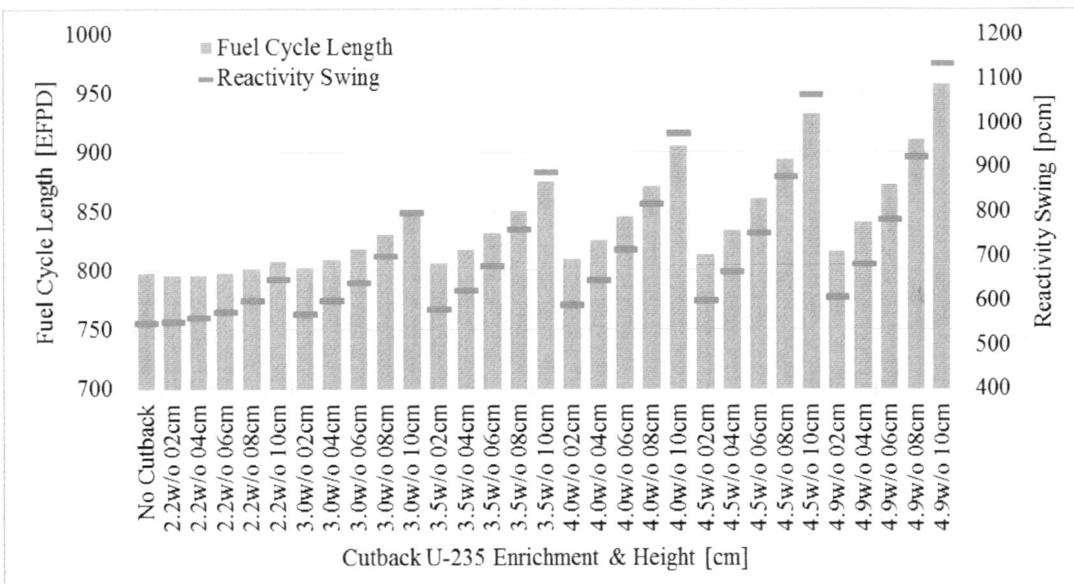

Figure 5. Fuel cycle length and reactivity swing as a function of fuel enrichment in the cutback regions.

3.3 Second Study: BA Material

The fuel cycle length and the reactivity swing variations, depending on the BA content in Zr-B or Zr-Gd cutback regions, are presented in Figure 6. Using Zr-Gd is better than using Zr-B. Indeed, the higher the B content in Zr-B is, the shorter the fuel cycle length is. On the other hand, the higher the height of Zr-Gd is, the longer the fuel cycle length is.

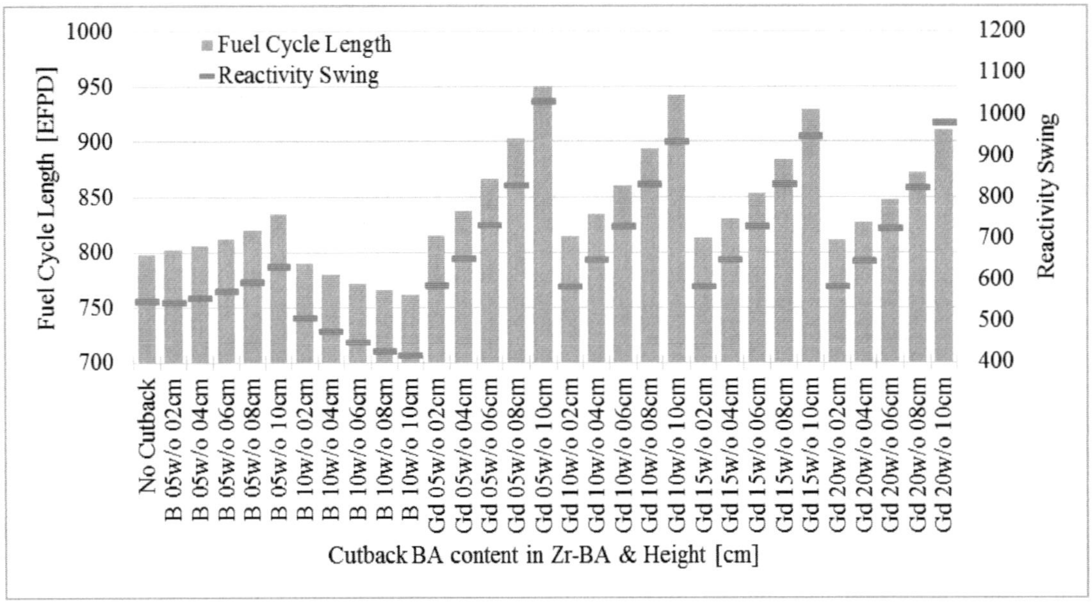

Figure 6. Fuel cycle length and reactivity swing change according to BA content.

Three cases of interest are presented in Table 3. Indeed, for those cases, the cycle length is increased by more than 140 EFPD compared to the reference cycle length. The cutback height for the three cases equals 10 cm and this makes the fuel cycle length longer. The cutback with 4.9 w/o ^{235}U without BA gives the longest cycle length, but the initial excess reactivity is higher than for the cases with Zr-Gd in cutback regions, as shown in Figure 7. The cutback with Zr-Gd makes the initial excess reactivity low, thus it is good for securing shutdown margin at the beginning of cycle (BOC). In the three cases, the axial offset and the power peaking factor are within target values ($-0.4<AO<+0.4$ and $Fq<4.4$ respectively), as shown in Figure 8 and 9. The cutback with Gd 10w/o in Zr-Gd is the best test case, because the reactivity swing of the two other cases is higher than the 1000-pcm reactivity swing target value.

Table 3. Cutback Sensitivity Test Results.

Test cases, Height	Cycle Length [EFPD]	Δρ Swing [pcm]
Reference	798	548
Gd 5w/o in Zr-Gd, 10cm	951	1029
Gd 10w/o in Zr-Gd, 10cm	942	934
4.9 w/o ^{235}U without BA, 10cm	959	1132

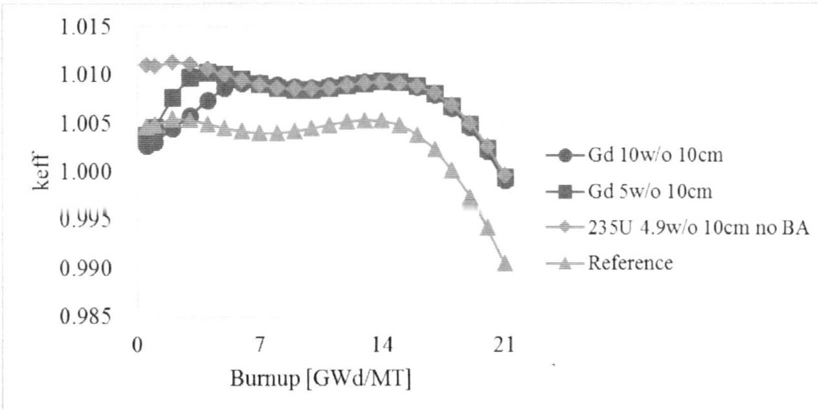

Figure 7. k_{eff} change behavior as burnup

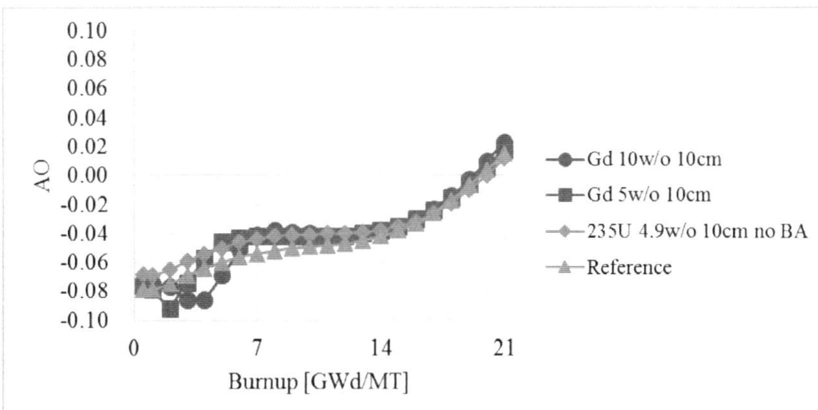

Figure 8. Axial offset change as burnup.

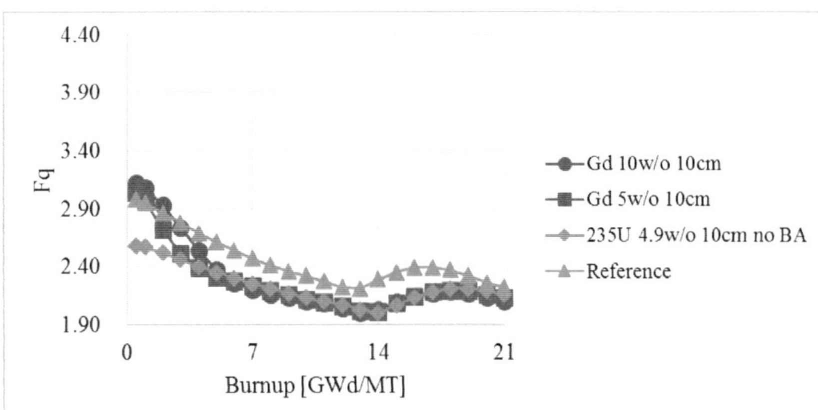

Figure 9. 3D pin peaking factor change as burnup.

4. Conclusions

This paper conducted cutback sensitivity tests in SMPWR to extend the cycle length. The cutback region may not need to load considerable amount of BA due to low relative power. At the bottom and top of the core pins, different heights of cutback were tested, along with different fuel enrichment rates and different BA materials (Zr-B or Zr-Gd). The reference core has no cutback region and uses only ^{167}Er-Zr as BA material, the cycle length of the reference core being 798 EFPD. An optimal cutback region of 10 cm height with natural Gd 10% in Zr-Gd was found, for which the cycle length increases to 942 EFPD. The core with this optimal cutback is satisfied with the design limits, which are the axial offset and the 3D pin peaking factor, over the operation time and the maximum excess reactivity is within 1000 pcm of target value. This is a 4.8-month increase of the reactor cycle length compared with the reference core.

References

[1] Choe J, Lee D and Shin H C 2016 *Preliminary Design of Boron-Free Small Modular Pressurized Water Reactor*, PHYSOR 2016 Sun Valley ID

[2] Rhodes J, Lee D and Smith K 2009 CASMO-4E: Extended capability CASMO-4 user's manual *Studsvik Scandpower*

[3] T. Bahadir 1999 CMS-LINK USer's Manual (Waltham: Studsvik Scandpower)

[4] BLJ Covington 2009 SIMULATE-3: Advanced Three-Dimensional Two-Group Reactor Analysis Code User's Manual (Waltham: Studsvik Scandpower)

[5] Choe J, Lee D and Shin H C 2016 New Burnable Absorber for Long-Cycle Low Boron Operation of PWRs *Annals of Nuclear Energy* **88** 272

[6] Mart J R 2013 *Feasibility Study on a Soluble Boron-Free Small Modular Reactor* (Oregon: Oregon State University)

[7] Franke P and Neuschütz D 2007 Thermodynamic Properties of Inorganic Materials *Springer Berlin Heidelberg*

[8] Mattern N *et al.* 2012 Experimental and thermodynamic assessment of the Gd–Zr system *Calphad: Computer Coupling of Phase Diagrams and Thermochemistry* **39** 27-32

[9] Brachet J C, Masclet C T, Hamon D, Guilbert T, Trego G, Jourdan J, Stern A, and Raepsaet C 2011 Oxygen, Hydrogen and Main Alloying Chemical Elements Partitioning Upon Alpha←→Beta Phase Transformation in Zirconium Alloys **172** 753

Hong Kong CIE sky classification and prediction by accessible weather data and trees-based methods

S Lou, D H W Li[1], J C Lam

Building Energy Research Group, Department of Architecture and Civil Engineering, City University of Hong Kong, Tat Chee Avenue, Kowloon, Hong Kong SAR, China

E-mail: bcdanny@cityu.edu.hk

Abstract. Solar irradiance and daylight illuminance are important for solar energy and daylighting designs. Recently, the International Commission of Illuminance (CIE) adopted a range of sky conditions to represent the possible sky distributions which are crucial to the estimation of solar irradiance and daylight illuminance on vertical building facades. The important issue would be whether the sky conditions are correctly identified by the accessible variables. Previously, a number of climatic parameters including sky luminance distributions, vertical solar irradiance and sky illuminance were proposed for the CIE sky classification. However, such data are not always available. This paper proposes an approach based on the readily accessible data that systematically recorded by the local meteorological station for many years. The performance was evaluated using measured vertical solar irradiance and illuminance. The results show that the proposed approach is reliable for sky classification.

1. Introduction

Daylight data are essential to the designs of daylight-linked artificial lighting controls in energy-efficient buildings [1] by reducing the lighting power consumption and interior heat gain. For daylighting applications, the skylight is always preferred since it has less potential to cause glare. The scattered daylight, however, can be a non-uniform light source whose intensity in terms of luminance varies over the whole sky hemisphere [2]. The anisotropy of skylight can cause significant differences of the daylight via fenestrations in different orientations. It is important to predict the sky luminance distributions. Since both sky luminance and solar radiance are of similar characteristics, the same mathematical approaches for determining daylight illuminance can also be applied to solar irradiance [3].

Sky distributions can be influenced by factors including solar position, turbidity and cloud amount. A usual approach to estimate the distribution is by the corresponding sky condition, since specified sky conditions induced from climatic factors at approaching levels are featured by similar luminance (and radiance) distributions. Generally, the sky conditions can be interpreted into overcast, partly cloudy and clear skies. In 2003, the International Commission on Illumination (CIE) adopted 15 standard skies proposed by Kittler and his colleagues [4]. Each of the CIE Standard Sky has its well defined luminance distribution pattern by mathematical expressions which have been reviewed by measurements in a number of world-wide regions [5-7]. Many studies reported that the 15 CIE

[1] Address for correspondence: D H W Li, Building Energy Research Group, Department of Architecture and Civil Engineering, City University of Hong Kong, Tat Chee Avenue, Kowloon, Hong Kong SAR, China. E-mail: bcdanny@cityu.edu.hk.

Standard Skies provide good and comprehensive frameworks of the actual skylight luminance distributions in the world.

The usual instrument for measuring sky luminance distribution is by mean of a sky scanner that records the luminance levels of 145 points of the whole sky dome. However, the sky-scanner is costly and maintenance intensive while the 145 points are not recorded simultaneously. Sky classifications based on different climatic indices were also proposed [3, 8, 9]. Sky luminance distributions particularly for the whole sky are seldom available for the sites under consideration or exist for very limited periods. It is expected that an approach based on the readily accessible horizontal solar radiations and available meteorological data would be more appropriate. This paper studies the CIE sky classification using measured data in our measuring station [10] and the Hong Kong Observatory (HKO) [11]. A sky classification approach was developed by the algorithm of Classification Tree (C_Tree). The performance of the proposed method was evaluated against measured data. The findings and design implications are discussed.

2. CIE Standard Skies

The set of 15 CIE Standard Skies [12] predicts the ratio between luminance at any point of the sky hemisphere ($L_{\alpha\phi}$) to that at the zenith point (L_Z). Since daylight is the visual part of solar radiation, tthe same mathematical models can also be used for the radiance distribution calculation. The standard formula of the ratio ($l_{\alpha\phi}$) is the product of a relative gradation expression $\phi(Z)/\phi(0)$ and a relative scattering indicatrix function $f(\chi)/f(Z_S)$ as expressed in Equations 1 to 3.

$$l_{\alpha\phi} = \frac{L_{\alpha\phi}}{L_Z} = \frac{f(\chi)\phi(Z)}{f(Z_S)\phi(0)} \tag{1}$$

$$\frac{\phi(Z)}{\phi(0)} = \frac{1 + a\exp(b/\cos Z)}{1 + a\exp b} \tag{2}$$

$$\frac{f(\chi)}{f(Z_S)} = \frac{1 + c\left[\exp(d\chi) - \exp(d\pi/2)\right] + e\cos^2\chi}{1 + c\left[\exp(dZ_S) - \exp(d\pi/2)\right] + e\cos^2 Z_S} \tag{3}$$

Where Z is the zenith angle of the sky point under consideration (rad), $Z = \pi/2 - \alpha$; Z_S is zenith angle of the sun; a, b, c, d and e are coefficients, the values of which can be adjusted to describe the 15 CIE Standard Skies from heavy overcast to clear skies; χ is the shortest angular distance between the sky element and the solar disc, i.e. scattering angle (rad), which can be calculated as:

$$\chi = \arccos\left(\cos Z_S \cdot \cos Z + \sin Z_S \cdot \sin Z \cdot \cos|\varphi - \varphi_S|\right) \tag{4}$$

Where φ_s and φ are the azimuth angle of the sun and the sky point under consideration, respectively. Detailed angles are given in Figure 1.

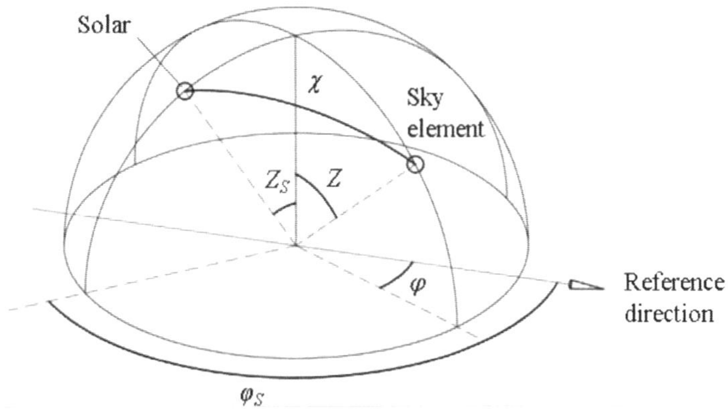

Figure 1. The angles of sun and sky elements

The gradation function (Equation 1) describes the sky luminance variation between horizon and zenith. In terms of overcast skies, greater luminance can be found at zenith. The trend is opposite for clear skies. The indicatrix function (Equation 2) peaks at the position of sun and drops rapidly when the scattering angle increases. By adjusting the value of coefficients from a to e, 15 groups of equations can be generated to denote the luminance distribution of a range of sky conditions, including 5 clear, 5 partly cloudy and 5 clear skies. The coefficients of the skies and their features are listed in Table A1 (Appendix A).

3. Sky classification by luminance

The most straightforward and comprehensive approach of sky classification is by comparing the luminance distribution pattern between the 15 CIE Standard Skies with measurement. The data of luminance distribution are usually acquired by a sky scanner which measures the luminance of 145 circular sky patches (L_i, cd m^{-2}, i = 1 to 145). As mentioned by Tregenza [13], using the measured zenith luminance (L_Z, cd m^{-2}) as the basis of a Standard Sky can cause uncertainty, because the measured L_Z may be biased due to the direct sunlight when the sun is close to zenith point. It is proposed to normalize L_i and the ratio of the CIE Standard Skies (l_i) by the corresponding horizontal sky-diffuse. The normalized distributions of the measurement ($L_{i,\ nor}$) and the CIE Standard Skies ($l_{i,\ nor}$) are illustrated in Equations 5 and 6.

$$L_{i,nor} = \frac{L_i}{V_h} \quad (5)$$

$$l_{i,nor} = \frac{l_i}{v_h} \quad (6)$$

Where V_h is the horizontal illuminance from all the 145 sky luminance by measurement, v_h is the horizontal illuminance from all the 145 sky luminance under individual CIE Standard Skies with L_Z set as one. Detailed calculations of the V_h and v_h by L_i and l_i were illustrated by Tregenza [13]. The Standard Sky is identified as the one with the minimum root mean square error between $L_{i,\ nor}$ and $l_{i,\ nor}$, as given in Equation 7.

$$rms = \sqrt{\frac{1}{145}\sum_{i=1}^{145}\left(L_{i,nor} - l_{i,nor}\right)^2} \quad (7)$$

Figure 2 shows the frequency of occurrence (FOC) of sky classified by the aforementioned approach. The sky luminance data were 10-minute readings recorded at the City University of Hong Kong (CityUHK) in 2004. Details of the station can be found in our previous work [10]. In total, 21,089 sets of data were employed, among which 37.4, 39.8 and 22.8% were identified as overcast, partly cloudy and clear skies, respectively. The most representative skies were Skies 1, 8 and 13 with the FOC of 20.1%, 31.9 and 15.7%, respectively. The results are consistent with a previous study by others [14].

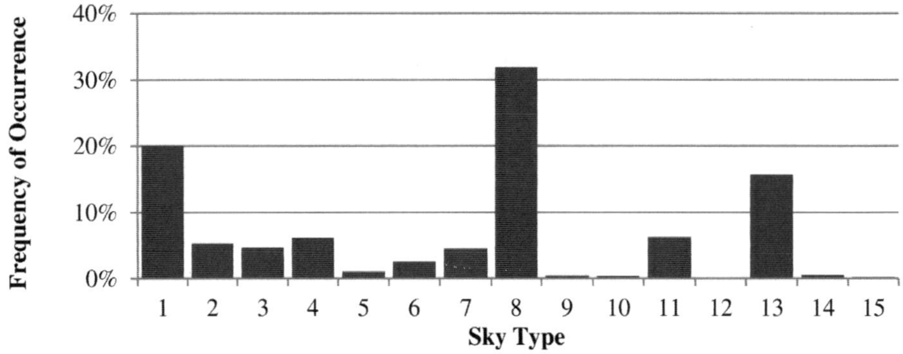

Figure 2. The frequency of occurrence of CIE Standard Skies in Hong Kong.

4. Sky classification by meteorological data

The classification by scanning the luminance distribution is perhaps the most comprehensive method but it has a number of limitations [3]. For instance, the long-term luminance measurements are rarely available while the meteorological features of a place are normally determined by the data recordings for 25 to 30 years [15]. To develop a long-term database of the CIE Standard Sky according to the accessible HKO measurements, the algorithm of Classification Tree was used to correlate the meteorological data sets with the sky classification results in previous section (Figure 1).

4.1. The Classification Tree

The algorithm of Classification Tree (C_Tree) was first developed by Breiman et al [16]. The algorithm categorizes data sets with a number of variables (Var_1, Var_2, ...) into various groups by a sequence of binary partitions (i.e. splits). Figure 3 illustrates a split that separates a single group with N_i data sets to two child groups according to the value of the selected *j*th variable (Var_j). The variable and its value for the split are selected to maximally reduce the size-weighted impurity of the CIE Skies in the child groups. By quantifying the impurity as the Gini index (G) defined in Equation 8, the reduction of size-weighted impurity can be quantified as Equation 9.

$$G = 1 - \sum_k \left[p(k) \right]^2 \tag{8}$$

$$\Delta G = G_i N_i - G_{i+1} N_{i+1} - G_{i+2} N_{i+2} \tag{9}$$

Where $p(k)$ is the fraction of datasets with sky type k that reaches a node. G is positive for a node that has more than a single type of the CIE Standard Sky. The classification by a single split repeats until the data in the node have purified a particular sky type, or the tree structure being established reaches the designated complexity.

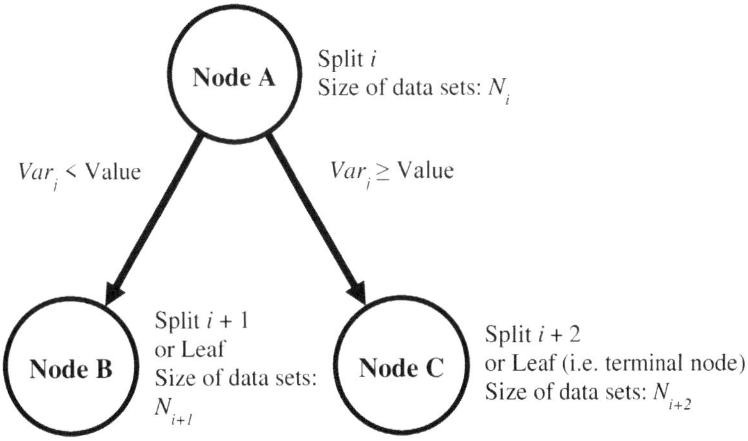

Figure 3. The topography of a single split in Classification Tree.

The complexity of the C_Tree can be controlled by the minimum leaf size (L_{MIN}) [17], which stops the splitting when the size of data sets in its child node is less than the setting. A lower L_{MIN} allows the algorithm to do more detailed classifications by a more complexed C_Tree which can improve the purity of sky type in a signal group. However, the model developed with too small L_{MIN} could be complicated to interpret and even over-fitted, which may give poor performance for new data sets. It is preferably to optimize the L_{MIN} in a case by case manner. In present study, the C_Trees were trained with L_{MIN} from 10 to 1,000.

4.2. The Development Classification Tree

The development of C_Tree that correlates the types of CIE Standard Sky with the meteorological data (variables) was implemented by the Machine Learning Toolbox of MATLAB 2015*b* [18]. The variables summarized in Table 1 were enrolled without a pre-selection since there was little knowledge on the importance of the variables and the algorithm itself can eliminate the unnecessary ones. The clearness index K_t and the diffuse fraction K are defined in Equations 10 and 11.

$$K = E_G / E_E \tag{10}$$

$$K_t = E_D / E_G \tag{11}$$

Where E_G is the horizontal global irradiance, E_E is the horizontal extraterritorial irradiance and E_D is the horizontal diffuse irradiance. The irradiance data were measured in our measuring station (CityUHK) every 10 minutes while the data of other variables were recorded by the HKO in 1-hour interval. The data from HKO were converted into 10-min by a shape-preserving piecewise cubic interpolation [19] based on the values of neighbouring points in MATLAB 2015*b* [20].

Table 1. Input variables for the development of Classification Tree (C_Tree)

Parameter	Notation	Unit	Parameter	Notation	Unit
Solar altitude angle	α_S	°	Relative humidity	*rh*	%
Clearness index	K_t		Visibility	*vis*	m
Diffuse fraction	K		Wind speed	*v*	m/s
Sunshine duration hour	H	hour	Mean sea level pressure	P	Pa
Air temperature	T_a	°C	Cloud amount	*cld*	%

To reduce the burden of computation, simplify the structure of C_Tree and improve reliability of the results, the framework given in Figure 4 were adopted to develop the approach for sky classification by C_Tree. As given in the figure, C_Tree 0 was developed to categorize the overcast (Skies 1 to 5), partly cloudy (Skies 6 to 10) and clear (Skies 11 to 15) before 3 C_Trees 1 to 3 were established to further specify the 15 Skies.

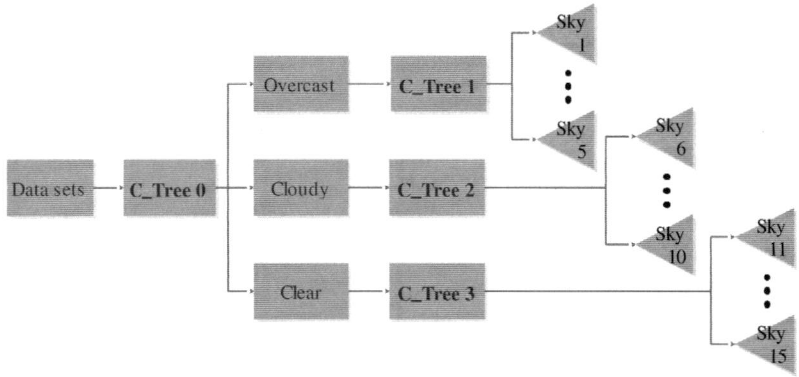

Figure 4. The framework of proposed sky classification approach by Classification Trees (C_Tree)

To evaluate the expected performance of every C_Tree with different L_{MIN}, we repeated a 5-fold cross validation [21] for 1,000 times which can reduce the uncertainty of random training data selection in sensible time and RAM consumption. The results include the median and range of percentage of skies being misclassified (%Mis) i, and the average number of splits with L_{MIN} being set from 10 to 500 as shown in Figures 5 to 8 for C_Trees 0 to 3.

For the development of all C_Trees, the %Mis drops initially then rises gently as L_{MIN} increases. The C_Trees are not powerful enough when L_{MIN} is too small and over-fitted when L_{MIN} is too large. The number of splits, however, reduces drastically when L_{MIN} increases indicating that less complexed trees are formed with the setting of lower L_{MIN}. Although the minimum %Mis can be identified for each C_Tree, the corresponding number of splits may be overwhelming. Setting L_{MIN} slightly greater can significantly lower down the number of splits and tree complexity, which makes the C_Trees more interpretable though the quality of prediction may be sacrificed slightly. By compromising the tree complexity against the performance, the L_{MIN} were set 130, 150, 425 and 100 for the C_Trees 0, 1, 2 and 3, respectively. The far larger %RMSE of C_Tree 1 compared with other C_Trees (i.e. C_Tree 0, C_Tree 2 and C_Tree 3) implies that it would be quite difficult to further identify the inconspicuous differences between the classified overcast skies (i.e. Skies 1 to 5) via meteorological data, especially based on the horizontal solar irradiance. The numbers of splits were controlled in a sensible level, which were around 27, 12, 3 and 7 respectively for C_Trees 0, 1, 2 and 3. The C_Trees developed by the designated settings are given in Figures B1 to B4 in Appendix B.

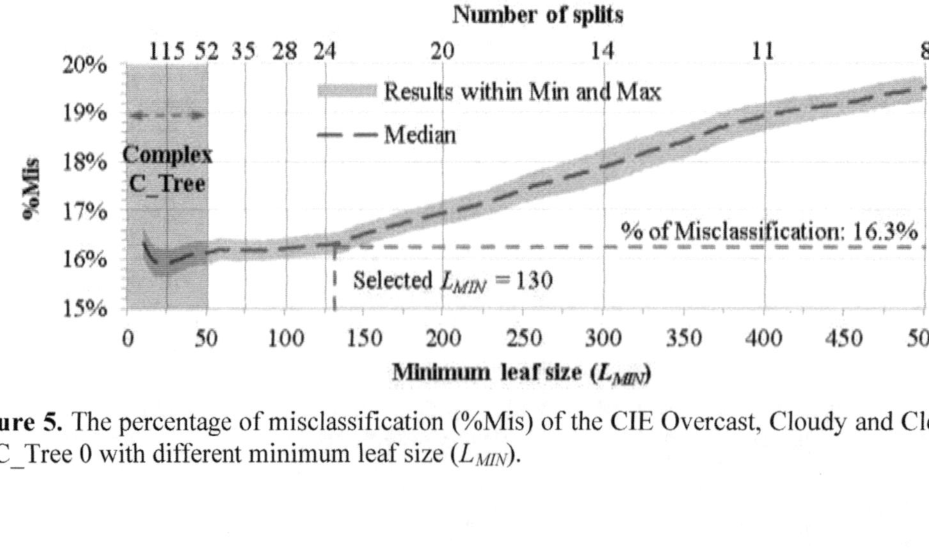

Figure 5. The percentage of misclassification (%Mis) of the CIE Overcast, Cloudy and Clear Skies by C_Tree 0 with different minimum leaf size (L_{MIN}).

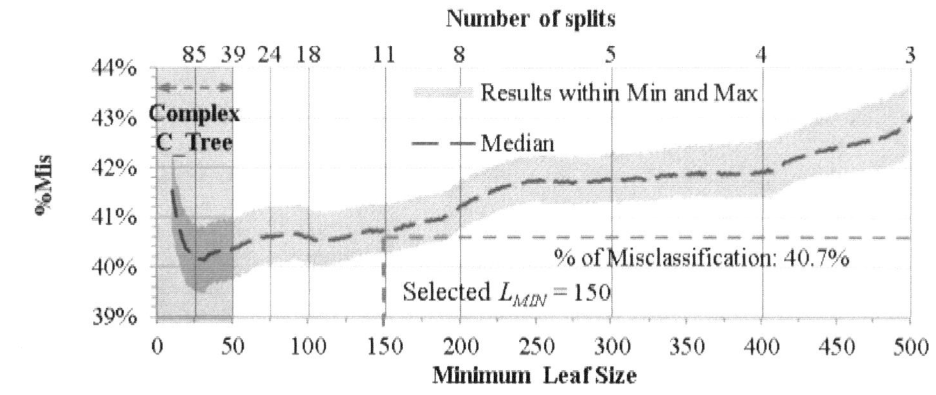

Figure 6. The percentage of misclassification (%Mis) of Sky 1 to 5 by C_Tree 1 with different minimum leaf size (L_{MIN})

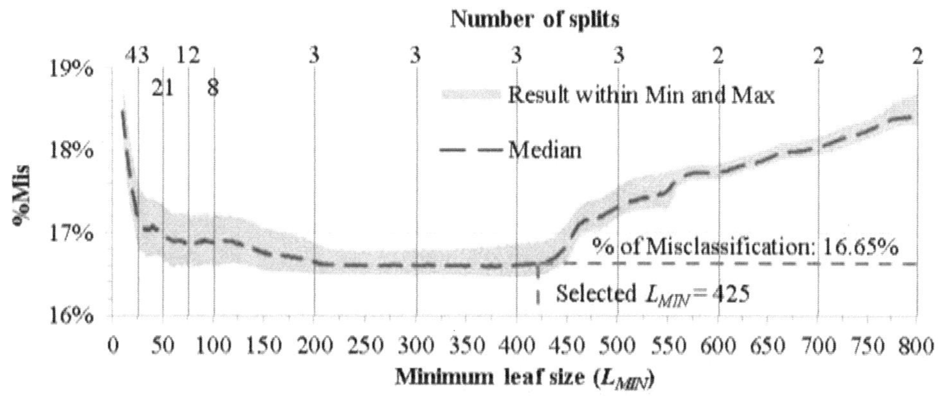

Figure 7. The percentage of misclassification (%Mis) of Sky 6 to 10 by C_Tree 2 with different minimum leaf size (L_{MIN})

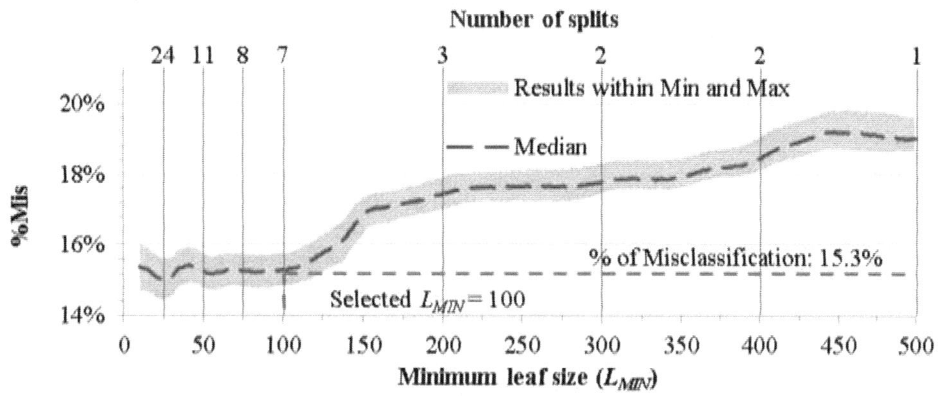

Figure 8. The percentage of misclassification (%Mis) of Sky 11 to 15 by C_Tree 3 with different minimum leaf size L_{MIN}

5. Performance evaluation

For the development of vertical building integrated photovoltaic panels and evaluation of solar heat gains and daylighting designs, accurate vertical global solar irradiance and illuminance data are required. To evaluate whether the skies are correctly identified, the global components for the vertical planes facing the 4 cardinal directions were modelled based on the sky types classified by the two approaches [3, 22]. The global solar irradiance and illuminance incident on a vertical surface (E_{VG}) is the sum of direct beam (E_{VB}), sky-diffuse (E_{VD}) and ground reflected components (E_{VR}), which can be expressed as Equation 12.

$$E_{VG} = E_{VB} + E_{VD} + E_{VR} \tag{12}$$

The calculation of E_{VB} is based on the solar altitude (α_S) angle, azimuth angle (φ_S) and the surface azimuth angle (φ_N). For the estimation of E_{VR}, it is assumed that the vertical plane receives half of the irradiance and illuminance being reflected isotropically from the ground. Denoting the ground reflectance ρ of 0.2 is a reasonable assumption in dense urban environment and adopted by a number of researchers [23-25], the 2 components can be calculated by Equations 13 and 14.

$$E_{VB} = \left[\left(E_G - E_D\right)/\sin\alpha_S\right]\cos\alpha_S\cos\left(\phi_S - \phi_N\right) \tag{13}$$

$$E_{VR} = \rho E_G/2 \tag{14}$$

The numerical method was used to compute E_{VD} [3]. The performance was assessed by the root mean square error with respect to the average of the simultaneously measured global solar irradiance and illuminance (%RMSE) and the results are summarized in Table 2. It is found the %RMSE of vertical irradiance and illuminance on vertical planes facing all directions are less than 20% and 25% respectively. The %RMSE of proposed approach is slightly greater than the luminance based approach and the difference is less than 2% for the four cardinal orientations. The findings show that the proposed approach can give reliable prediction of the CIE Standard Skies and the vertical illuminance and irradiance.

Table 2. %RMSE for vertical solar irradiance based on Skies classified by the 2 approaches

%RMSE	Vertical irradiance				Vertical illuminance			
	N	E	S	W	N	E	S	W
Luminance by sky scanner	15%	17%	13%	14%	23%	11%	12%	14%
Proposed	16%	19%	14%	15%	24%	13%	13%	16%

6. Conclusions

An approach that classifies the sky conditions of CIE Standard Skies was developed by the algorithm of Classification Tree, which correlates the readily accessible meteorological data with individual CIE Standard Skies. The CIE Skies being employed to develop the model were identified by the luminance data of a sky scanner. The solar radiation, daylight illuminance and sky luminance data were acquired in the City University of Hong Kong and the meteorological data were measured by the Hong Kong Observatory in 2004. The skies identified by the proposed approach give good estimations of irradiance and illuminance on vertical planes with the root mean square error of less than 25%. The findings show that the proposed approach is appropriate to classify the CIE Standard Sky in Hong Kong using readily accessible meteorological data. Future work will be conducted to reduce the complexity of C_Tree structure and sensitivity analysis will be implemented to introduce more important variables and remove those rarely used in the model.

Acknowledgements
The work described in this paper was fully supported by a grant from City University of Hong Kong (Project No. 7004543). Siwei LOU was supported by a City University of Hong Kong studentship.

Appendix A
The coefficients and features of the 15 CIE Standard Skies are given in Table A1

Table A1. The general features of the 15 CIE Standard Skies

Sky	Gradation		For indicatrix			Type of sky	
	a	b	c	d	e	General classification	Description of luminance distribution
1	4.0	-0.7	0	-1	0	Overcast	Steep gradation, azimuthal uniformity
2	4.0	-0.7	2	-1.5	0.15	Overcast	Steep gradation, slight brighten around sun
3	1.1	-0.8	0	-1	0	Overcast	Gentle gradation, azimuthal uniformity
4	1.1	-0.8	2	-1.5	0.15	Overcast	Gentle gradation, slight brighten around sun
5	0	-1	0	-1	0	Overcast	Uniform
6	0	-1	2	-1.5	0.15	Partly cloudy	No gradation, slight brighten around sun
7	0	-1	5	-2.5	0.3	Partly cloudy	No gradation, brighter circumsolar region
8	0	-1	10	-3	0.45	Partly cloudy	No gradation, distinct solar corona
9	-1	-0.55	2	-1.5	0.15	Partly cloudy	Sun position shaded/obscured
10	-1	-0.55	5	-2.5	0.3	Partly cloudy	Brighter circumsolar region
11	-1	-0.55	10	-3	0.45	Clear	White-blue sky, distinct solar corona
12	-1	-0.32	10	-3	0.45	Clear	Low turbidity
13	-1	-0.32	16	-3	0.3	Clear	Polluted atmosphere
14	-1	-0.15	16	-3	0.3	Clear	Cloudless turbid sky, broad sun corona
15	-1	-0.15	24	-2.8	0.15	Clear	White-blue turbid sky, wide sun corona

Appendix B

Structure of the Classification Trees (C_Trees 0 to 3) are given in Figures B1 to B4

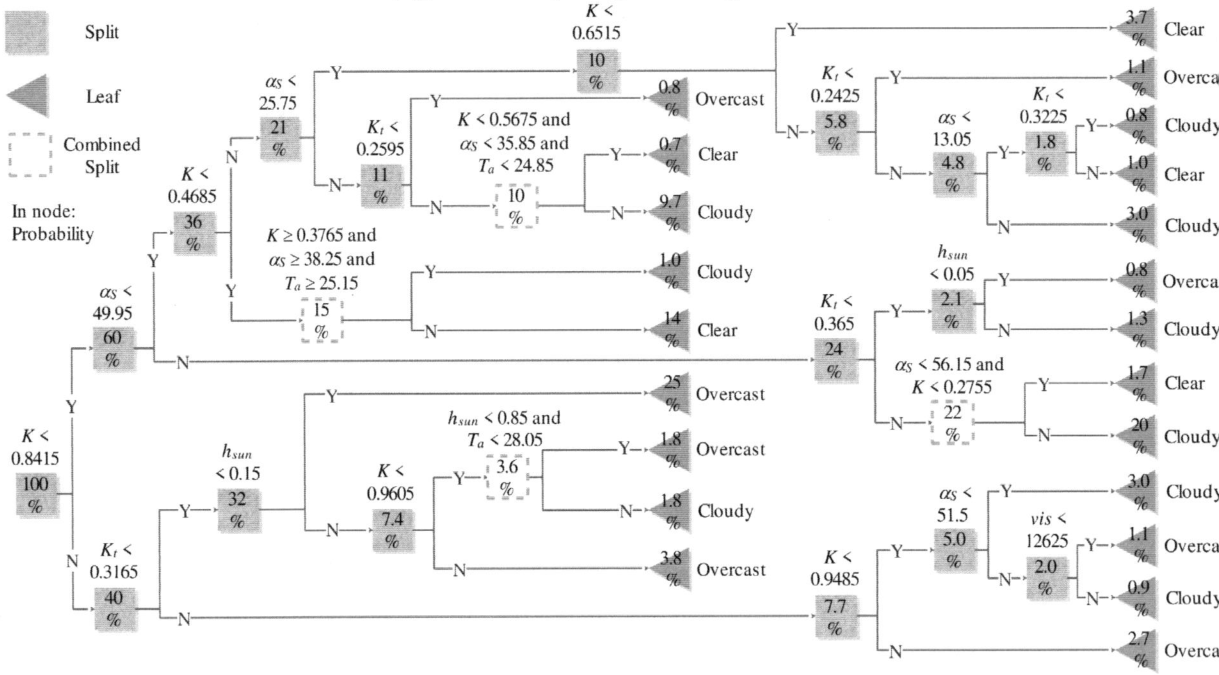

Figure B1. The C_Tree that determines the Overcast, Partly Cloudy and Clear Skies (i.e. C_Tree 0)

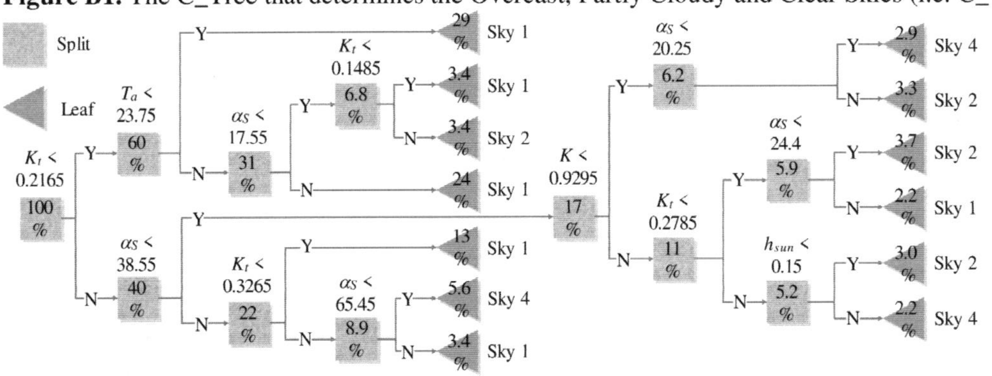

Figure B2. The C_Tree that determines the Sky 1 to 5 for the classified Overcast Skies (i.e. C_Tree 1)

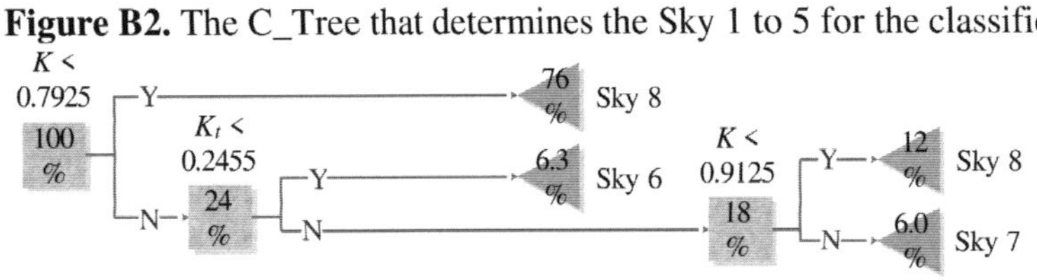

Figure B3. The C_Tree that determines the Sky 6 to 10 for the classified Cloudy Skies (i.e. C_Tree 2)

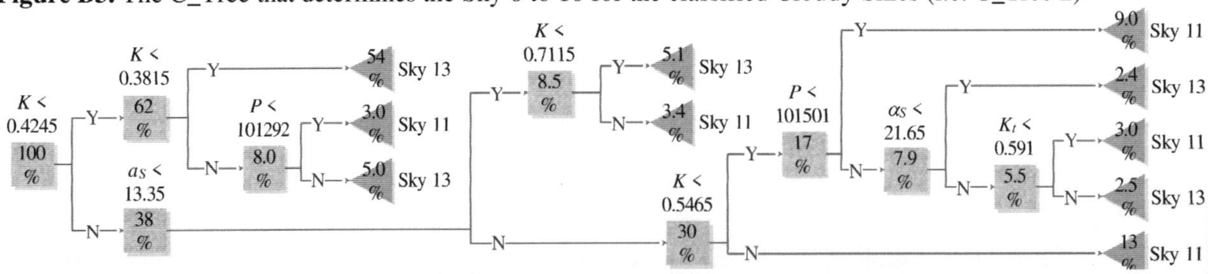

Figure B4. The C_Tree that determines the Sky 6 to 11 for the classified Clear Skies (i.e. C_Tree 3)

 doi:10.1088/1755-1315/40/1/012018

References

[1] Li D H W, Cheung K L, Wong S L and Lam T N T 2010 An analysis of energy-efficient light fittings and lighting controls *Appl. Energy* **87** 558–67

[2] Perez R, Seals R and Michalsky J 1993 All-weather model for sky luminance distribution–Preliminary configuration and validation *Sol. Energy* **50** 235–45

[3] Li D H W, Chau T C and Wan K K W 2014 A review of the CIE general sky classification approaches *Renew. Sust. Energ. Rev.* **31** 563–74

[4] Kittler R, Darula S and Pereze R 1998 *A set of standard skies* (Bratislava: Polygrafia SAC)

[5] Li D H W, Lau C C S and Lam J C 2003 A study of 15 sky luminance patterns against Hong Kong data *Architect. Sci. Rev.* **46** 61–8

[6] Tregenza PR 1999 Standard skies for maritime climates *Lighting Res. Technol.* **31**97–106

[7] Wittkopf S K and Soon L K 2007 Analysing sky luminance scans and predicting frequent sky patterns in Singapore. *Lighting Res. Technol.* **39** 31–51

[8] Kittler R and Darula S 2002 Parametric definition of the daylight climate *Renew. Energy* **26** 177–87

[9] Li D H W and Lau C C S and Lam J C 2004 Standard skies classification using common climatic parameters *J. Sol. Energy Eng.* **126** 957–64

[10] Li D H W and Tang H L 2008 Standard skies classification in Hong Kong *J. Atmos. Sol.-Terr. Phy.***70** 1222–30

[11] Li D H W, Lou S W and Lam J C 2015 An analysis of global, direct and diffuse solar radiation *Energy Procedia* **75** 388-93

[12] CIE 2003 *Spatial Distribution of daylight - CIE Standard General Sky* (Vienna: CIE Central Bureau)

[13] Tregenza P R 2004 Analysing sky luminance scans to obtain frequency distributions of CIE standard general skies *Lighting Res. Technol.* **36** 271–81

[14] Ng E, Cheng V, Gadi A, Mu J, Lee M and Gadi A 2007 Defining standard skies for Hong Kong. *Build. Environ.* **42** 866–76

[15] Chan A L S, Chow T T, Fong S K F and Lin J Z 2006 Generation of a typical meteorological year for Hong Kong *Energy Convers. Manag.* **47** 87–96

[16] Breiman L, Friedman J, Olshen R A and Stone CJ 1984 *Classification and regression trees* (Belmont, CA: CRC press)

[17] Khoshgoftaar T M, Allen E B, Deng J 2001 Controlling overfitting in software quality models: experiments with regression trees and classification *Proc. 7th Int. Software Metrics Symp. METRICS* (Los Alamitos, CA: IEEE Computer Society) p 190

[18] Mathworks 2015 *Machine learning toolbox user's guide* (Natick, MA: The MathWorks, Inc.)

[19] Fritsch F N and Carlson R E 1980 Monotone piecewise cubic interpolation *SIAM J. Numer. Anal.* **17** 238–46

[20] Mathworks 2015 *Curve fitting toolbox user's guide* (Natick, MA: The MathWorks, Inc.)

[21] Kohavi R 1995 A study of cross-validation and bootstrap for accuracy estimation and model selection *Int. joint conf. artificial intelligence (IJCAI)* **2** 1137-43

[22] Ho J, Ng E and Chan P 2015 Predicting the hourly Hong Kong representative sky from Typical Meteorological Year data for dynamic daylighting simulation *Lighting Res. Technol.* **47** 730–9

[23] Li D H W, Cheung G H W, Cheung K L and Lam J C 2009 Simple method for determining daylight illuminance in a heavily obstructed environment *Build. Environ.* **44** 1074-80

[24] Ng E 2001 A simplified daylighting design tool for high-density urban residential buildings *Lighting Res. Technol* **33** 259–72

[25] Tregenza P R 1995 Mean daylight illuminance in rooms facing sunlit streets *Build. Environ.* **30** 83-9

The compatibility evaluation of Cr^{3+} Gel system and polymer/surfactant system with alternating injection mode

J H Zhang[1], H K Li [1], Y N Wang [1,3], J Q Zhi[1] and Y Liu[2]

[1]Key Laboratory for Enhancing Oil/Gas Recovery of Ministry of Education in Northeast Petroleum University, Daqing, Heilongjiang, 163318,China
[2]Drilling & Production Technology Research Institute Petrochina Qinghai Oilfield Company, Dunhuang, Gansu, 736202, China

E-mail: xiaoying_2323@163.com

Abstract. Alternately injecting the slug of the gel and polymer/surfactant compound system is a new way to further enhance oil recovery after polymer flooding. The displacement system needs to produce an ultra low interfacial tension to oil and to enlarge swept volume significantly. Based on experimental analysis, the influence factors of Cr^{3+} gel system viscosity and the compatibility of gel with two types of surfactant compared with composite ion gel system has been studied. The experimental result shows that it has well stability, and the compatibility of gel with RMA-1 type surfactant is very well. It can produce an ultra low interfacial tension to oil so that enhanced oil recovery has been reached more than 10 percent by using the gel system to displace residual oil after polymer flooding in artificial large flat-panel model.

1. Introduction

Polymer Flooding has been applied in the Daqing Oilfield since 1996, its scale expands unceasingly and has become the leading technology for the development of Daqing Oilfield. Although the polymer flooding can enhance recovery more than 10% higher than that of water flooding, there are still some problems, such as polymer will be held up in channel and injection profile will reverse after injecting a certain amount of polymer solution, which lead to about 50% crude oil still can not be produced after polymer flooding [1-4]. Water breakthrough and water out degree is being aggravated by the more serious reservoir heterogeneity after polymer flooding. In the existing various kinds of deep profile control and flooding technology, it's economically feasible to use the crosslinked polymer as profile control and displacement agent. Although certain effects are obtained in pilot tests, its antiscour performance is weaker because of many influential factors in the reservoir after polymer flooding [5-6]. When gel is injected into the inhomogeneous formation, it will take the lead into the high permeable layers where the filtrational resistance is smaller and the permeability is higher [7]. The gel can plug the high permeability layers temporarily with its gel forming properties, which makes the subsequent injection of binary complex system turn and into the low permeability layers that are unswept [8-9].So

[3] Address for correspondence: Y N Wang, Key Laboratory for Enhancing Oil/Gas Recovery of Ministry of Education in Northeast Petroleum University, Daqing, Heilongjiang, 163318, China. E-mail: xiaoying_2323@163.com

the displacement technology, alternately injecting the slug of the gel and polymer/surfactant compound system, is presented to improve development effect [10].

Gel is the decentralized system, formed by the crosslinking reaction between crosslinking agent (such as chromium ion crosslinking agent) and polyacrylamide molecular functional groups. And the gelling process is the phenomenon that many individual growth clusters of substance form infinite network structure by the crosslinking. The concentration of crosslinker agent (chromium ion) will affect the relaxation time of the gel system [11]. Although the gel can reduce the oil flow rate well, the condition of high temperature can make the oil displacement performance reduced greatly [12].

The compatibility of gel with polymer/surfactant compound system is unclear; therefore the performance of Cr^{3+} gel system and its compatibility is studied, comparing with composite ion gel system.

2. Experimental condition

Laboratory chemicals & reagents includes polymer(molecular weight is 2.5×10^7), Cr^{3+} crosslinking agent, composite ion crosslinking agent, RMA-1 type surfactant, Heavy alkylbenzene sulfonate. The water used in experimental solutions is advanced treated sewage that degree of mineralization is 6495.15mg/L. Experiment equipment includes interfacial tensiometer, Brookfield viscometer, multi-start magnetic stirring apparatus, electronic balance, thermostat, etc. The experimental temperature is of 45℃.

3. Results and Discussion

3.1. Influence factors of viscosity of gel system

3.1.1. Mass concentration of polymer.

Using Brookfield viscometer, the viscosity of Cr^{3+} gel system (the concentration ratio of polymer to crosslinking agent is 10:1) at the initial time and 24 h is measured under different mass concentrations. The relation curves of viscosity with mass concentration are plotted in Figure.1. The regularity can be seen that the viscosity of gel system increases with the increase of mass concentration. This is due to the greater the mass concentration of polymer solution, the higher the carboxyl content of the solution, and the higher the probability of collision between Carboxyl and Cr^{3+} released from crosslinking agent, which accelerate the complexation between carboxyl and Cr^{3+}. Molecular chains of polymer solutions get entangled with each other and form an irregular mesh structure. With the increase of solution concentration, the entanglement among molecular chains becomes more closely, the relative intermolecular distance becomes smaller, and the arrangement becomes more closely, which results in the increase of gel strength. When the mass concentration of polymer is 1800 mg/L, the viscosity of the system is more than 16000mPa.s after preparing 24h. And then with the increase of mass concentration the rising range of the viscosity decreases. So it's best to prepare the gel system with polymer of which mass concentration is 1800mg/L from the perspective of economy.

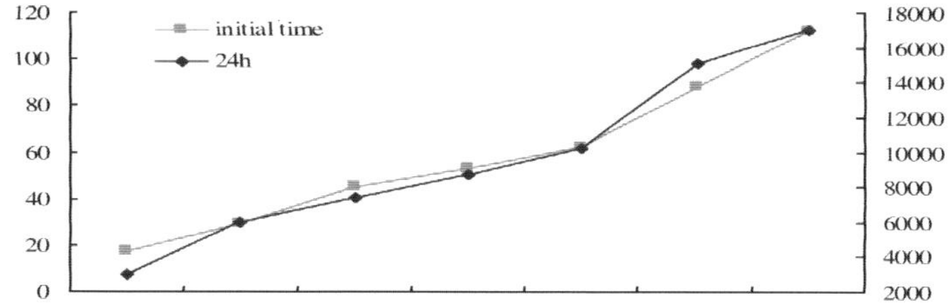

Figure 1. The relationship between concentration and viscosity of Cr^{3+} gel system.

3.1.2. Gel time of Cr^{3+} gel system.

Using Brookfield viscometer under the condition of 45℃, the viscosity of Cr^{3+} gel system which mass concentration is 1800mg/L and the concentration ratio of polymer to crosslinking agent is 10:1 at certain time interval is measured. The change of gel viscosity with time is observed, and the curve of Cr^{3+} gel stability is plotted in figure 2.

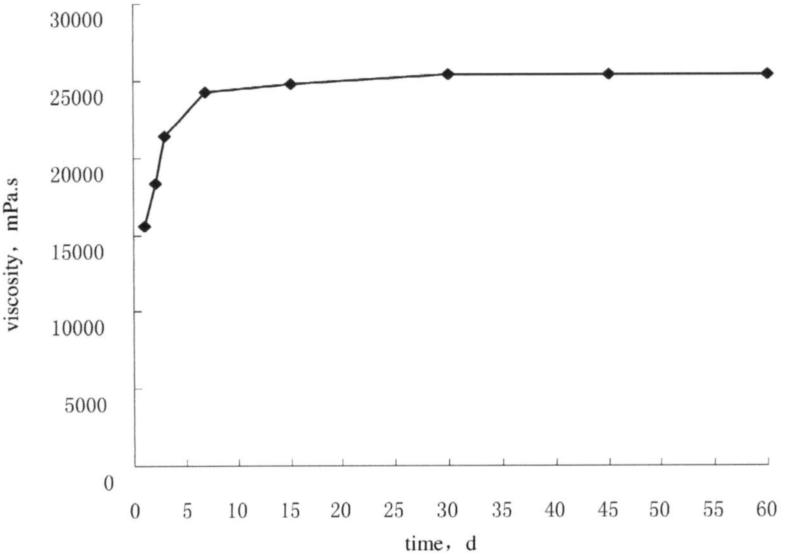

Figure 2. Curve of Cr^{3+} gel stability.

The gel viscosity increases rapidly at first. There're gels after preparing 11h, which the gel viscosity has risen to 2639mPa·s, and then the gel viscosity has risen to more than 15000mPa·s after 24h. It has risen to 25000mPa·s after 7d, and then remain stable at more than 20000mPa·s, which shows that gel performance is very good under the condition of 45℃.

3.1.3. Concentration ratio of polymer to crosslinking agent.

The influence of the concentration ratio to the viscosity relates to the gel time. The polymer solution whose mass concentration is 1800 mg/L is sheared into a dilute solution that viscosity is about 20mPa·s. The gel solutions are prepared according to the concentration ratio is 10:1, 20:1, 30:1, 40:1, 50:1, 60:1 respectively. Then their viscosities are measured when the time is 0h, 6h, 12h, 24h and 48h. The relationship between the concentration ratio and the viscosity can be known after measuring experiment, and it is shown in table 1.

Table 1. Viscosity (mPa·s) data tables of Cr^{3+} gel system of different concentration ratios.

concentration ratios	time, h				
	0	6	12	24	48
10:1	18	163.94	2963.28	7248.45	7222.39
20:1	17	141.95	2409.49	6318.65	6639.01
30:1	19.5	119.43	335.93	4479.04	4569.23
40:1	21.5	101.95	209.46	2819.32	3196.23
50:1	21.5	96.97	194.46	1986.23	2569.38
55:1	20.5	55.99	83.98	1789.45	2389.36

60:1	19.5	39.99	48.99	124.97	1985.89

The regularity can be seen that the greater the concentration ratio of polymer to crosslinking agent, the longer the gel time, and the weaker the gel strength. When the concentration ratio is relatively small, such as 10:1, the gel time is shorter and the gel viscosity has been more than 2000 mPa.s after 10h. If the gel system which concentration ratio is equal to 10:1 is used to displace oil, it may lead to gel in the near wellbore zone and even block, and the injection pressure may rise sharply, which is bad for deep profile control and flooding. Therefore, choosing the appropriate concentration ratio of polymer to crosslinking agent can generate gel in the depth of high permeability layers to be effective for the deep profile control and flooding. So that the subsequent injection of binary complex system could sweep more residual oil and further enhance oil recovery.

3.2. Compatibility of gel with surfactant

3.2.1. Compatibility of Cr^{3+} gel system with surfactant.

The first solution formula: the concentration of polymer (molecular weight is 2.5×10^7) is 1800mg/L, the concentration of Cr^{3+} crosslinking agent is 180mg/L.

The second solution formula: the concentration of polymer(molecular weight is 2.5×10^7) is 1800mg/L, the concentration of Cr^{3+} crosslinking agent is 180mg/L, the concentration of heavy alkylbenzene sulfonate is 0.3 percent.

The third solution formula: the concentration of polymer(molecular weight is 2.5×10^7) is 1800mg/L, the concentration of Cr^{3+} crosslinking agent is 180mg/L, the concentration of RMA-1 type surfactant is 0.3 percent.

According to the above three formulas, solutions were prepared, then were sealed and put in a thermostat set at 45°C. Changes of viscosity within a certain time can be observed, and correlation curve can be made a graph (Figure 3).

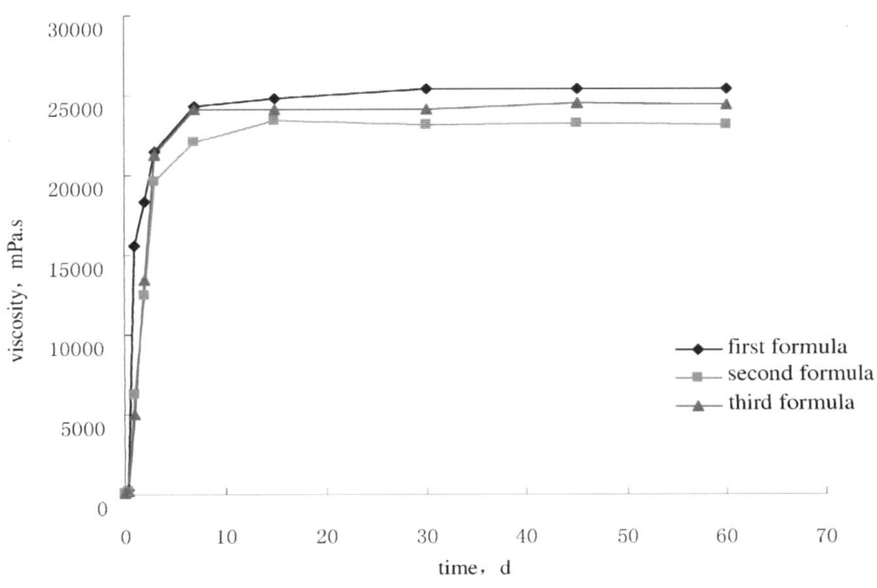

Figure 3. Correlation curve of viscosity with three compounded solutions of Cr^{3+} gel system.

By the three curves of figure 3, the growth trend of viscosity can be seen that increase rapidly at first and then remain stable at more than 20000mPa·s, which shows that the gel stability is very good.

By contrast with three solutions of viscosity, the compatibility of Cr^{3+} gel system with RMA-1 type surfactant is better than with heavy alkylbenzene sulfonate. Different forms of three formulas before and after the gelling process were shown in figure 4.

Figure 4. (a) Before Figure 4. (b) After

Figure 4. Before and after the gelling process about three formulas.

3.2.2. Compatibility of composite ion gel system with surfactant.

The first solution formula: the concentration of polymer (molecular weight is 2.5×10^7) is 1800mg/L, the concentration of Cr^{3+} crosslinking agent is 180mg/L.

The second solution formula: the concentration of polymer(molecular weight is 1.9×10^7) is 3500mg/L, the concentration of organic resin crosslinking agent A is 200mg/L, the concentration of organic resin crosslinking agent B is 1300mg/L, the concentration of heavy alkylbenzene sulfonate is 0.3 percent.

The third solution formula: the concentration of polymer(molecular weight is 1.9×10^7) is 3500mg/L, the concentration of organic resin crosslinking agent A is 200mg/L, the concentration of organic resin crosslinking agent B is 1300mg/L, the concentration of RMA-1 type surfactant is 0.3 percent.

According to the above three formulas, solutions were prepared, then were sealed and put in a thermostat set at 45℃. Changes of viscosity within a certain time can be observed, and correlation curve can be made a graph (Figure 5).

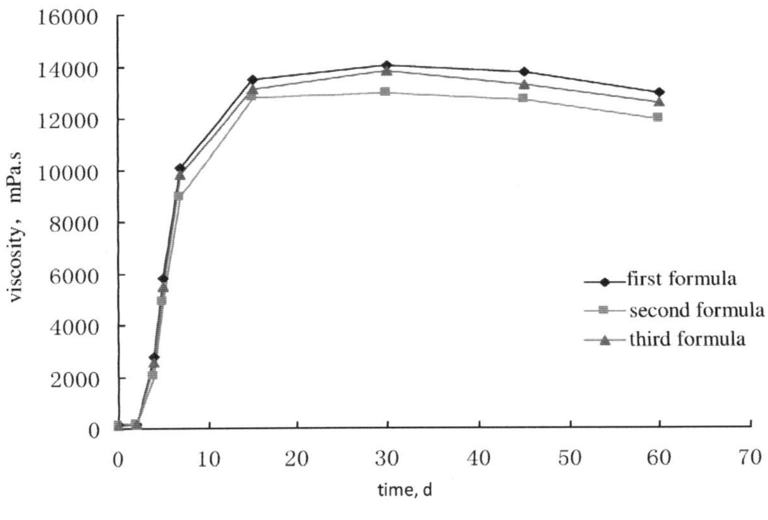

Figure 5. Correlation curve of viscosity with three compounded solutions of composite ion gel system

The curves shown in figure 5 indicates that there're not gels until three days later and their viscosity increase to more than 2000mPa·s. By contrast with three solutions of viscosity, the compatibility of composite ion gel system with RMA-1 type surfactant is better than with heavy alkylbenzene sulfonate. But from the viscosity variation tendency of composite ion gel it will be gel breaking. Compared with Cr^{3+} gel system, its stability is worse and its gel viscosity is less. It shows that gel performance of Cr^{3+} gel system is better than composite ion gel system.

3.2.3. The interface compatibility between gel system and surfactant.
The interfacial tension was measured by using interfacial tensiometer for two kinds of surfactants solution and the combination solutions of surfactants solution and gel system. The measuring results are shown in table 2.

Table 2. The measuring results of surfactant's interfacial tension

Sample	Interfacial tension mN/m
RMA-1 surfactant	1.43×10^{-3}
heavy alkylbenzene sulfonate	1.39×10^{-1}
Cr^{3+} gel + heavy alkylbenzene sulfonate	1.58
Cr^{3+} gel + RMA-1 surfactant	6.91×10^{-3}
composite ion gel + heavy alkylbenzene sulfonate	1.76
composite ion gel + RMA-1 surfactant	7.52×10^{-3}

The interfacial tension of RMA-1 surfactant has reached 10^{-3}mN/m, but the interfacial tension of alkybenzene sulfonate surfactants is only 10^{-1}mN/m. The heavy alkylbenzene sulfonate solution is added into prepared solution of Cr^{3+} gel system, and its interfacial tension is measured equal to 1.58mN/m, but the interfacial tension of solution that added RMA-1 surfactant has reached 6.91×10^{-3}mN/m. This shows that compared with heavy alkylbenzene sulfonate, RMA-1 surfactant has better interface compatibility with Cr^{3+} gel system, which can form ultra-low interfacial tension.

Similarly, heavy alkylbenzene sulfonate and RMA-1 surfactant solution is respectively added into prepared solution of composite ion gel system. The interfacial tension sequence is the same as that of Cr^{3+} gel system. It also shows that gel system has better compatibility with the RMA-1 type surfactant than heavy alkylbenzene sulfonate.

3.2.4. Compatibility of gel system and surfactant after gelling.
The equal amount solution of RMA-1 surfactant or heavy alkylbenzene sulfonate surfactants is respectively added into the gelled solutions, and then the changes of interface between gel and surfactant solution are observed regularly whether gel breaking or not. The observed result indicate that there is no obvious change on the interface between gel and surfactant, and no turbid matter and surfactant have better compatibility in static situation.

The author and others use the slug of the gel system (molecular weight of polymer is 2.5×10^{7}, mass concentration is 1800mg/L, the concentration ratio of polymer to crosslinking agent is 40:1) and polymer/surfactant compound system (molecular weight of polymer is 2.5×10^{7}, RMA-1 surfactant) to displace residual oil after polymer flooding in artificial large flat-panel model, and achieve success that enhanced oil recovery has been reached more than 10 percent after polymer flooding[13]. It can be a technical reserve after polymer flooding which can further enhance oil recovery.

4. Conclusions

- The influence factors of Cr^{3+} gel system viscosity are analyzed by experiment. The result shows that the viscosity of gel system increases with the increase of mass concentration, increases firstly and then remain stable with time, and the greater the concentration ratio of polymer to crosslinking agent, the longer the gel time, and the weaker the gel strength.

- Based on experimental analysis, the compatibility of gel with two types of surfactant compared with composite ion gel system has been studied. The experimental result shows that the compatibility of gel with RMA-1 type surfactant is very well. It can produce an ultra low interfacial tension to oil so that enhanced oil recovery has been reached more than 10 percent by using the gel system to displace residual oil after polymer flooding in artificial large flat-panel model.

Acknowledgments
This research was financially supported by the National Natural Science Foundation of China (No.51274070).

References
[1] Xia H F, Jiang Y and Wang G 2010 Study on improving residual oil recovery of polymer/surfactant system after polymer flooding *J. Journal of Xi'an Shiyou University (Natural Science)* **25** 45-9
[2] Zhao F L, Wang Y F and Dai C L 2006 Study on technology of improving oil recovery after polymer flooding *J. Journal of China University of Petroleum* **30** 86-8
[3] Wang D M, Lu H and Nu J G 1998 Application results and understanding of several problems of industral scale polymer flooding in Daqing oil field
[4] Yang Puhua 1999 Study on present situation of improving recovery and its development direction in the near future *J. Petroleum Geology and Recovery Efficiency* **6** 4-5
[5] Liu Y J and Wang X Z 1999 Technology of profile control and water plugging *M. Petroleum Industry Press* 127
[6] Liu X G 1998 Technical proceedings of water plugging used in oilfield *M. Petroleum Industry Press* 94
[7] Zhang J H, Liu M J and Zhang G 2008 Study of the rheological property of low temperature gelatinous plugging agent solution *J. Journal of Daqing Petroleum Institute* **32** 60-2
[8] Zheng B, Hou J R and Zhang M 2016 Research on oil displacement mechanism of crosslinked polymer using CT technique *J. Xinjiang Petroleum Geology* **1** 97-101
[9] Niu L W, Lu X G and Wang X Y 2014 Differences in molecular configuration and seepage properties among polymer, active polymer and Cr^{3+} polymer gel *J. Journal of China University of Petroleum* **6** 186-91
[10] Zhang J H, Dong X and Ye Y Z 2010 Effect of the gel and surfactant alternating injection after polymer flooding *J. Daqing Petroleum Institute* **34** 85-8
[11] Vargas-Vasquez S M, Romero-Zerón L B and Macmillan Bryce 2007 1H NMR characterization of HPAM/Cr(III) acetate Polymer Gel Components *International Journal of Polymer Analysis and Characterization* **12** 115-29
[12] G.X. Shen, J.H. Lee and K.S.Lee 2013 Performance of silicate gel and polymer/Cr(III) gel treatments in high temperature reservoirs *Applied Mechanics and Materials* **316-317** 782-5
[13] Wang Zhongguo, Zhang Jihong and Zhang Zhiming 2012 Oil Displacement by Alternately Injecting Slug of Gel and Polymer/Surfactant Compound System after Polymer Flooding *J. Northeast Petroleum University* **36** 49-53

IOP Conf. Series: Earth and Environmental Science **40** (2016) 012020 doi:10.1088/1755-1315/40/1/012020

The performance evaluation of nano-micron microsphere for profile control and displacement agent in low permeability reservoir

J H Zhang[1], Y N Wang [1, 3], X Y Wang [2], X J Zhuo[1], X Q Zhang[1] and C Liu [1]

[1]Key Laboratory for Enhancing Oil/Gas Recovery of Ministry of Education in Northeast Petroleum University, Daqing, Heilongjiang, 163318,China
[2]Oil Recovery Plant No.1, Daqing Oilfield Corp. Ltd., Daqing, Heilongjiang, 163318, China

E-mail: xiaoying_2323@163.com

Abstract. At present the waterflood efficiency is lower in most low permeable heterogeneous reservoirs, and the effect is poor using common polymer as profile control and displacement agent. As a new agent for profile control and displacement, the flow behavior of the nano-micron microsphere is evaluated in laboratory in this paper. The experimental result shows that it has well injectivity and flowability when flowing through the low permeability core. Because the nano-micron microsphere has a behavior of gradual expansion, it can be able to decrease permeability of porous medium very well. Especially in the process of subsequent water injection, the injection pressure increases firstly and then appears a small fluctuation declining, which shows that nano-micron microsphere solution has better antiscour performance and ability of gradual controlling and sealing.

1. Introduction

During waterflood process of low-permeability oilfield, the factors including oil-water viscosity difference, reservoir plane and vertical heterogeneity and others result in water breakthrough along high permeable layers, so that low permeable layers can not or poorly be developed, and the waterflood efficiency is seriously affected [1-4]. Therefore, the technology of the deep displacement used in low permeability oilfield is put forward [5-6] in order to weaken the heterogeneity of reservoir, enlarge the swept volume and improve waterflood effect in low permeability oilfield at high water-cut late stage [7-8].

However, the effect of the profile control and displacement agent that used commonly in the low-permeability oilfield is poor [9]. For example, the crosslinking controllability of movable weak gel as profile control and flooding is poor, and the cost is higher. LPS flooding control system is easy to be influenced by wastewater quality. Swellable particulate gel as profile control and flooding agent is so large that there exists contradiction between the injection depth and sealing strength, resulting in losing efficacy quickly in low permeable layers. Inorganic sealing agent is easy to precipitate, and can not enter into the indepth layer to plug [10-13]. In view of these problems, a new type of profile control and flooding agent with nano-micron microsphere is applied to achieve better effects [14-18].

[3] Address for correspondence: Y N Wang, Key Laboratory for Enhancing Oil/Gas Recovery of Ministry of Education in Northeast Petroleum University, Daqing, Heilongjiang, 163318, China. E-mail: xiaoying_2323@163.com

In this paper, the flowability and antiscour performance in low permeability core have been studied in the laboratory by the measurement of resistance factor and residual resistance factor of nano-micron microsphere solution and subsequent injection pressure in this flowing test. This research result can provide a reference for the application of this new profile control and flooding agent with nano-micron microsphere in low-permeability oilfield.

2. Experimental

2.1. Experimental conditions

The effective permeability of the rectangular homogeneous artificial core used in the experiment is $20\times10^{-3}\mu m^2$, $50\times10^{-3}\mu m^2$, $80\times10^{-3}\mu m^2$, $100\times10^{-3}\mu m^2$ and $200\times10^{-3}\mu m^2$ respectively. The concentration of nano-micron microsphere solution is 2000mg/L, 3000mg/L and 4500mg/L respectively, and viscosity is 0.91mPa·s, 0.92mPa·s and 0.96mPa·s respectively. The water used in saturating core is synthetic simulated formation brine that degree of mineralization is 8230.2mg/L, and the water used in core displacement test is advanced treated sewage that degree of mineralization is 6495.15mg/L. The experimental temperature is at 47.5°C (reservoir temperature).Experiment equipment includes thermostat, constant pressure and speed pump, high-pressure vessel, manual measuring pump, produced liquid meterrun, vacuum pump, pressure gauge, etc.

2.2. Experimental procedure

①Put rectangular cores with different permeability in core holder, add 5 MPa ring pressure, vacuumize cores and continue for 6h, saturate synthetic brine, and then measure porosity; ②Put rectangular cores saturated synthetic brine in an thermostat of 47.5°C for more than 12h, and then measure its permeability to water, the measurements are shown in Table 1; ③Inject the advanced treated sewage at the rate of 0.3ml/min before the pressure difference has been stable, then record the pressure difference (ΔPwi) at the two side of the core; ④ Respectively inject the different concentrations(2000mg/L、3000mg/L and 4500mg/L) of nano-micron microsphere solution at the rate of 0.3ml/min(calculated according to the frontal velocity of Daqing oilfield) before the pressure difference has been stable, then record the pressure difference (ΔPm) at the two side of the core; ⑤ Repeat the third step, and then record the pressure difference (ΔPwa) at the two side of the core; ⑥ Continue to inject the advanced treated sewage, observe the pressure changes constantly, and then record the stable pressure difference.

Table 1. Basic parameters and experimental scheme of rectangular homogeneous core.

Core number	length (cm)	area (cm^2)	porosity (%)	permeability ($10^{-3}\mu m^2$)	Experimental scheme (Concentration of injected nano-micron microsphere solution)
JZA-1	30.26	20.39	19.18	20.13	
JZA-2	30.39	20.13	19.48	48.53	
JZA-3	30.15	20.18	19.25	78.14	2000mg/L
JZA-4	30.30	20.28	17.93	100.32	
JZA-5	30.26	20.13	17.46	201.25	
JZB-1	30.25	20.14	17.76	21.24	
JZB-1	30.25	20.26	26.38	49.26	
JZB-1	30.28	20.17	26.42	83.42	3000mg/L
JZB-1	30.20	20.16	26.15	105.38	
JZB-1	30.22	20.10	27.42	210.2	
JZC-1	30.22	20.26	27.34	19.8	
JZC-1	30.18	20.16	27.35	48.94	
JZC-1	30.24	20.19	20.12	85.52	4000mg/L
JZC-1	30.26	20.24	20.34	102.64	
JZC-1	30.18	20.20	20.48	201.52	

3. Results and Discussion

According to the experiment results, a relationship curve is established between resistance factor and permeability (Figure 1.), and a relationship curve is also established between residual resistance factor and permeability (Figure 2.).

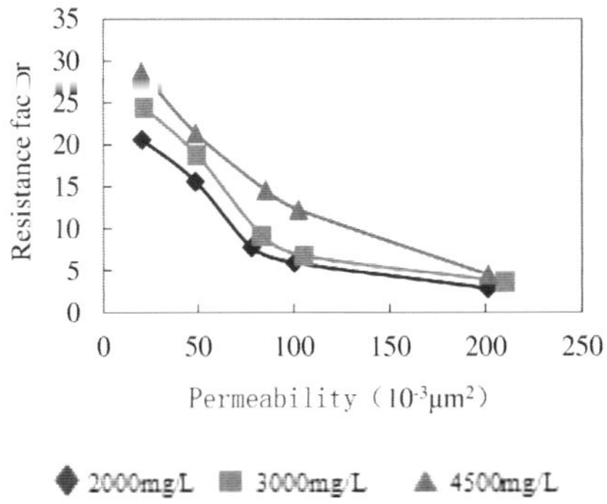

Figure 1. Curves of resistance factor with permeability.

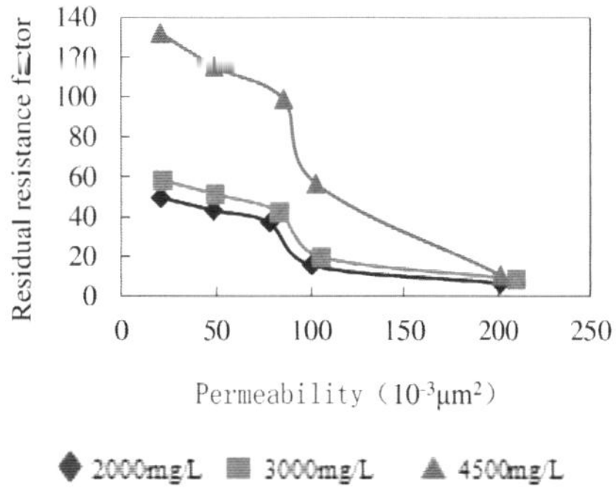

Figure 2. Curves of Residual resistance factor with permeability.

There are three curves of resistance factor with permeability in Figure 1, which depict three concentrations of nano-micron microsphere solution (2000mg/L, 3000mg/L, 4500mg/L), respectively. According to the three curves of Figure 1, the regularity can be seen that the higher the concentration of nano-micron microsphere solution the greater the resistance factor when the permeability is kept constant. This shows that the injectivity of high concentration of nano-micron microsphere solution is worse than that of low concentration in low permeability core. When low concentration of nano-micron microsphere solution (2000mg/L) is respectively injected into core samples with permeability from low to high, the resistance factor varies from 20 to 3. This shows that it has good injectivity and flowability not only in medium to high permeability core but also in low permeability core ($20\times10^{-3}\mu m^2$).

There are three curves of residual resistance factor with permeability in Figure 2, which depict three concentrations of nano-micron microsphere solution (2000mg/L, 3000mg/L, 4500mg/L), respectively. According to the three curves of Figure 2, the regularity can be seen that residual resistance factor decreases with the increase of the permeability. This shows that the sealing effect in low permeability cores is better than medium to high permeability cores. In low permeability cores, the residual resistance factor of high concentration of nano-micron microsphere solution (4500mg/L) is much greater than that of low concentration(2000mg/L or 3000mg/L), it is because that the nano-micron microsphere has gradual expansibility in solution, which can be seen in Figure 3. In the injection process, single-particle is continuous swelling and there is continuous mutual adsorption and accumulation among many particles, which is called hydration expansion of nano-micron microsphere. The higher the concentration is, the more possibility of sealing caused by more aggregate and adsorbate there is, so it can decrease the permeability of porous medium.

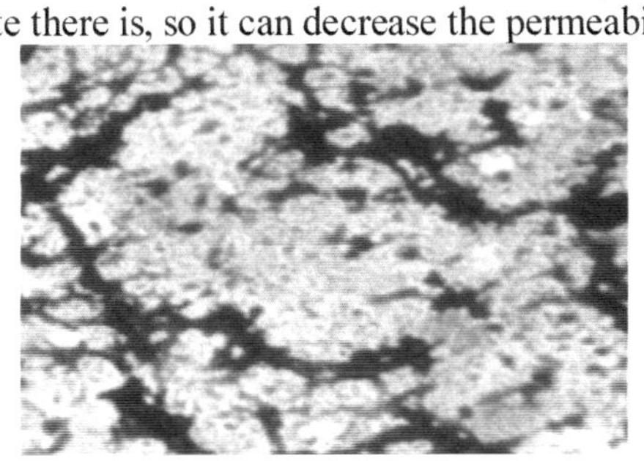

(a)1d

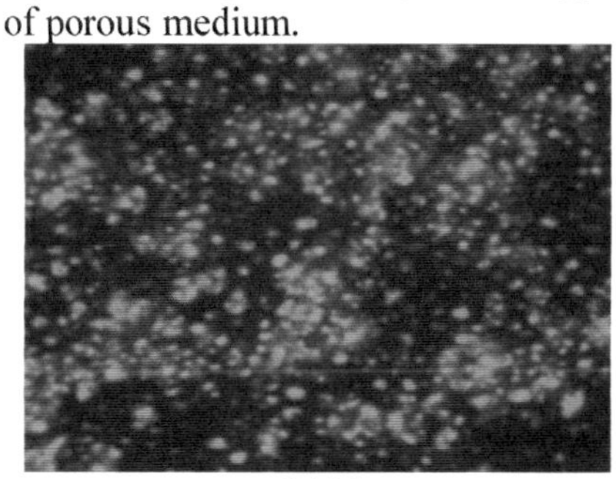

(b)5d

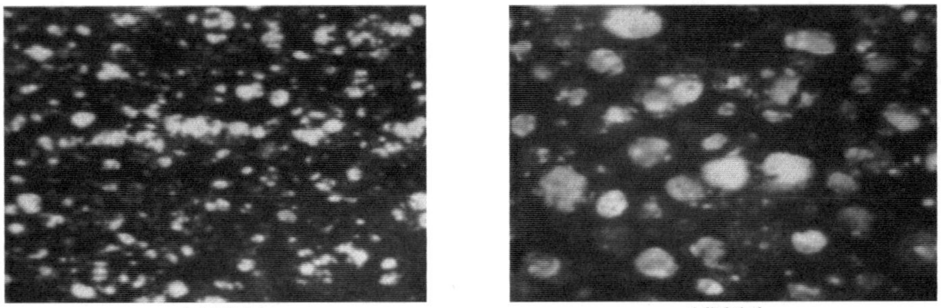

(c)10d (d)20d

Figure 3. Expansion variation of Nano-micron microsphere solution (2000mg/L) within 20 days under microscope.

After measuring residual resistance factor, water injection is continued to enter into the core and that pressure differences vary with injection pore volume multiple is measured, so as to evaluate the antiscour performance of nano-micron microsphere. With a low-permeability core called JZA-1 as example, the change of pressure with injection volume can be seen in Figure 4. Because of continuous hydration expansion in the injection process, part of porous channel is plugged by aggregate and adsorbate, which results in gradual rising of pressure. In the subsequent injection process, there are so many particles of nano-micron microsphere adsorbing and accumulating in the pore throat that the pressure increases further. After the pressure reaches a certain value, the nano-micron microspheres can pass the pore throat because of the compressive deformation. When it move to the deeper under the pushing of subsequent water, the pressure decreases. There are so many pore throats in the process of the following migration that the phenomena including the above process can appear continuously. The pressure increases firstly and then decreases in the process of experiment that subsequent water injection volume is 4.5PV. The experimental results showed that the nano-micron microsphere solution has better antiscour performance as well as the ability of gradual controlling and sealing in the low permeability layer.

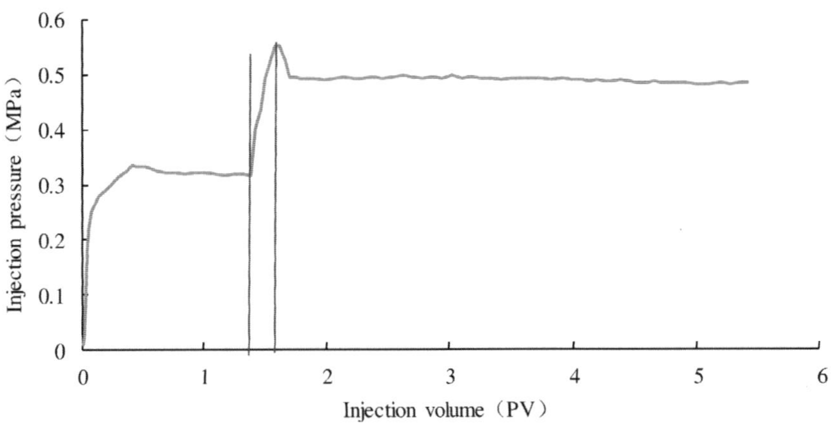

Figure 4. Curve of pressure with injection volume of Nano-micron microsphere (2000mg/L) through JZA-1 core.

4. Conclusions

- The resistance factor of nano-micron microsphere solution is low (20 or so) in low permeability cores, which shows that it has better injectivity and flowability in low permeability cores.
- Because of continuous hydration expansion in the injection process, part of porous channel is plugged by aggregate and adsorbate, which results in gradual rising of the pressure and residual

resistance factor becoming bigger. This shows that nano-micron microsphere solution has better ability to decrease permeability of porous medium.

- After injecting nano-micron microsphere solution in low permeability core ($20\times10^{-3}\mu m^2$), the pressure increases firstly and then appears a small fluctuation declining in the subsequent water injection. This shows that the nano-micron microsphere solution has better antiscour performance as well as the ability of gradual controlling and sealing in the low permeable medium.

Acknowledgments
This research was financially supported by the National Natural Science Foundation of China (No.51274070).

References

[1] Yao C J, Lei G L and Gao X M 2014 Rheological properties of pore-scale elastic microspheres profile controland flooding system *J. Petroleum Geology and Recovery Efficiency* **21** 55-8
[2] Sun H Q, Wang T and Xiao J H 2006 Novel technique of in-depth profile control step by step by polymer microsphers *J. Petroleum Geology and Recovery Efficiency* **13** 77-9
[3] Yin Y L and Zhang G C 2003 A summary of chemical oil well water sealing and profile control agent *J. Petroleum Geology and Recovery Efficiency* **10** 64-6
[4] Zhou H and Tang C Y 2006 Applying the dynamic two-step method to forcast remaining oil distribution of lower series xiaermen oilfield *J. China University of Geosciences* **17** 65-70
[5] Lei G L and Zheng J P 2007 Composing of pore-scale polymer microsphere and its application in improving oil recovery by profile control *J. China University of Petroleum* **31** 87-8
[6] Hirasaki G J, Miller C A and Puerto M 2011 Recent advances in sur-factant EOR *A. Oil and Gas Technology Expo & Conference* **16** 889-907
[7] Xiong C M and Tang X F 2007 Technologies of water shut-off and profile control: An overview *J. Petroleum exploration and development* **34** 83-8
[8] Gu X K, Du F Y and Wang X Q 2009 Current situation and development of chemical deep oil displacement regulation technology *J. Petrochemical Inoustry Application* **28** 4
[9] Yao C J, Lei G L and Gao X M 2014 Rheological properties of pore-scale elastic microspheres profile control and flooding system *J. Petroleum Geology and Recovery Efficiency* **21** 55-8
[10] Zhang X H 2011 Influence of resistance coefficient on polymer flooding effect *J. Special Oil and Gas Reserviors* **18** 82
[11] Eoff L, Dalrymple D and Everett D 2006 Globle field results of a polymeric gel system in conformance application *Society of Petroleum Engineers* **2** 1393- 8
[12] Shun H Q, Wang T and Xiao J H 2006 Novel technique of in-depth profile control step by step by polymer microspheres *J. Petroleum Geology And Recovery Efficiency* **13** 77- 9
[13] Bai B J, Tang X F and Liu G H 2000 Weak gel RNJ- 1 as flooding fluid for indepth permeability]control in low tempera-ture reservoirs *J. Oilfield Chemistry* **17** 343-5
[14] Li M Y, Dong Z X and Wu Z L 2001 Classfication of partially hydrolyzed polyacrylamide aluminum citrate systems *J. Oilfield Chemistry* **18** 347- 51
[15] Gao T Y, Dai B and Dai J Y 1995 Syntheses and appilcations of monodisperse polymer microsphere with large size *J. Acta Polymerica Sinica* **23** 174- 80
[16] Wang D M, Cheng J C and Xia H F 2001 Viscous-elastic fluids can mobilize oil remaining after waterflood by force parallel to the oil water interface *Spe Asia Pacific Improved Oil Recovery*

Rules of pulverized coal output under different components of coal petrography and different coal structure in Hancheng Block, China

S G Liu, K Tu[1], Z G Peng, Y Shao, Y Y Liu and Y Fu

School of Mechanics & Civil Engineering, China University of Mining and Technology, Beijing 100083, P.R. China

E-mail: tukun2014@163.com

Abstract. In order to study the output mechanism and influencing factors of pulverized coal under different components of coal petrography and different coal structures during the process of drainage, the physical simulation experiments were conducted under the state of single-phase water flow displacement. The results of this experiment for different coal petrography show the weight of pulverized coal output is normally 11# coal > 5# coal > 3# coal with different displacement velocities, and the increasing ratio of pulverized coal output is 5# coal > 11# coal with the different confining stress in the constant displacement velocity. For different coal structures the pulverized coal output weight of fragmented coal is much larger than the primary structure of coal. The particle size distribution curve shows 3#, 5# and 11# primary structure of coal have a double-peak, and the grain size of primary pulverized coal is relatively small and the secondary pulverized coal is relatively large. However, the grain size distribution of fragmented coal is a double-peak distribution, and the distribution scope is relatively concentrated and the average grain size is small. Therefore, the characteristics of pulverized coal were found to be related to its coal different coal petrography components and coal structure.

1. Introduction

China has abundant coalbed methane resources [4]. The exploration and exploitation of coalbed methane in China have made great achievement in recent years; meanwhile coalbed methane commercial development also has gained success and several breakthroughs during past 20 years [2, 17]. The problem of pulverized coal output is a common phenomenon in

---

[1] Address for correspondence: K Tu, School of Mechanics & Civil Engineering, China University of Mining and Technology, Beijing 100083, P.R. China. E-mail: tukun2014@163.com.

China during the process of coalbed methane drainage due to the complex tectonic-thermal evolution of coal basins and the particularity of coal reservoir [6, 8].

The elastic modulus of Coal petrography is little and the tensile strength is weak, and it is also fragile. Compared to the common sandstone coal is more prone to damage and produce pulverized coal under the same conditions, which can cause the blocking of coal petrography fracture and decreasing of the coal petrography permeability [5, 12, 16]. The large amount of pulverized coal output which aggregate in the bottoms of CBM wells and drainage system can lead to the pumps sticking and becoming buried. The pumps would then need to be checked frequently which would have a negative impact on the coalbed methane yields [9-10, 15]. For all of the above reasons, researching the characteristics and origin of the pulverized coal output is the point to keep the CBM wells which have high and stable coalbed methane yields. However, the most challenge is the pulverized coal in the seepage channel that is invisible. It is hard to identify the source of pulverized coal. Various factors can affect the output of pulverized coal during the drainage in the CBM wells. The properties of coal petrography are the basis of the pulverized coal output and the engineering disturbed can induce pulverized coal output. The geological factors include the components and relative content of the coal petrography, coal seam structure, and deformation and metamorphism and other factors. Mechanical damage of coal reservoir in the process of drilling and fracturing, abrasion damage caused by the gas-liquid-solid flowing during the drainage, and complex changes in the coal reservoir stress state are classified as engineering disturbed.

Increasing attention has been focused on the harm which is caused by output of the pulverized coal. Previous research has paid much attention to the control measures, the damage to the reservoir and management in the process of drainage [3, 6-7, 9, 12-14]. However, there are few papers on the rules of pulverized coal output for different components of coal petrography and different coal structure. Some scholars have discussed the mechanism and characteristics of the pulverized coal output [1, 2, 9].

2. Experimental method

2.1. Experimental principle

The coal reservoir fracture diversion system in this experiment is the main part. In this study the author used the quartz sand to support the existing cracks and simulated the process of CBM well drainage by single-phase water flow displacement, and simulated the change of the strata stress through changing confining pressure in different conditions. In order to identify the law of pulverized coal output, the dynamic output rules of pulverized coal was tried to be concluded by means of changing the size of displacement velocity of single-phase water flow and confining pressure between coal briquettes.

2.2 Sample preparation

The experiment was performed using the HXDL-II type acid fracture flow diversion meter. Sample preparation include coal briquette, proppant pack and displacement fluid.

2.2.1. Preparation of the coal briquette.

Coal samples were collected from 3#, 5#, and 11# coal seams in the Xiangshan coal mine, Hancheng area. The coal samples were crushed using the crushing machine and selected by

100 mesh standard sieve. The pulverized coal which is less than 100 mesh standard sieve was made of coal briquette. Using electronic scaled 35g pulverized coal which was used to make the coal briquette under the confining stress of 20MPa. The shape of the coal briquette by acid etching fracture diverting instrument diversion chamber specification shall prevail. It covers an area of 64.5 cm^2, straight line length is 139.7 mm, and the diameters of the semicircle on both ends are 38.1 mm.

2.2.2. The production of proppant pack

The proppant was the quartz sand with the standard of 20~40 mesh which were tiled in the middle of two pieces of coal briquette to simulate the coal reservoir with the pressed proppant.

2.2.3. Configuration of the displacement fluid

The formation water in the coal reservoir was simulated using the displacement fluid with salinity of 8% standard saline. The formula (mass ratio) of NaCl： $CaCl_2$： $MgCl_2{\cdot}H_2O$=7: 0.6: 0.4.

2.3 Experimental approach

The experiment was designed two groups of variables: the velocity of displacement fluid and confining stress. The method was to keep one variable stable and change another variable to study the rule of the pulverized coal output. The experimental device process and conditions are shown in Figure 1 and table 1, respectively. The experiment used the coal briquette with the same specification, and each displacement fluid flow velocity and confining stress condition lasted 2 hours in the experiment.

Table 1. The experimental method of different components of coal petrography

Coal samples	Experimental conditions	
3# coal	Confining stress for 5MPa, Flow velocity for 5ml/min	Flow velocity for 10ml/min, Confining stress for 3MPa
	Confining stress for 5MPa, Flow velocity for 15ml/min	Flow velocity for 10ml/min, Confining stress for 5MPa
	Confining stress for 5MPa, Flow velocity for 25ml/min	Flow velocity for 10ml/min, Confining stress for 7MPa
5# coal	Confining stress for 5MPa, Flow velocity for 5ml/min	Flow velocity for 10ml/min, Confining stress for 3MPa
	Confining stress for 5MPa, Flow velocity for 15ml/min	Flow velocity for 10ml/min, Confining stress for 5MPa
	Confining stress for 5MPa, Flow velocity for 25ml/min	Flow velocity for 10ml/min, Confining stress for 7MPa
11# coal	Confining stress for 5MPa, Flow velocity for 5ml/min	Flow velocity for 10ml/min, Confining stress for 3MPa
	Confining stress for 5MPa, Flow velocity for 15ml/min	Flow velocity for 10ml/min, Confining stress for 5MPa
	Confining stress for 5MPa, Flow velocity for 25ml/min	Flow velocity for 10ml/min, Confining stress for 7MPa

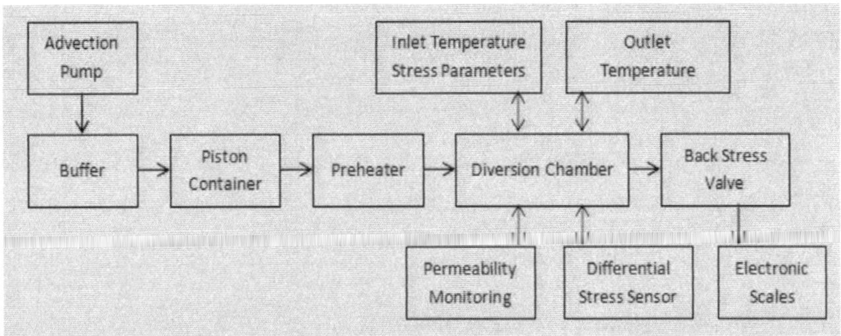

Figure 1. The physical simulation experimental device process of single-phase water of flow displacement

3. The effects of different coal petrography on the pulverized coal output

In this experiment for coal samples of 3#, 5#, and 11#, we kept the confining stress constant which was 5MPa and changed the displacement fluid flow velocity from 5ml/min to 15ml/min, 25ml/min gradually.

Table 2. The analysis of coal samples

Coal petrography samples	Microscope composition/%			Relative contents of vitrinite and inertinite/%		Moisture/%	Volatile/%	Ash/%
	Vitrinite	Inertinite	Clay minerals	Vitrinite	Inertinite			
3# coal	83.69	10.09	7.86	89.24	10.76	1.15	12.25	8.92
5# coal	75.38	13.16	11.45	85.14	14.86	1.21	13.12	11.31
11# coal	69.07	13.62	16.54	83.53	6.47	0.80	13.36	12.76

The analysis of pulverized coal output

We used the electronic scale to weigh the pulverized coal output which was collected in the experiment. The data on pulverized coal output are shown in Figure 2. We can see from the Figure 2, the weight of pulverized coal output in the experiment was 11# coal > 5# coal >3# coal with the increasing of the displacement fluid flow velocity.

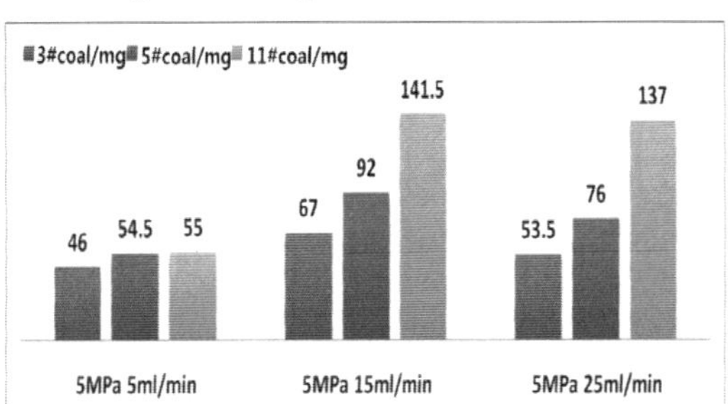

Figure 2. The pulverized coal output of different coal petrography under constant confining stress and variable displacement fluid velocity

This was due to the content of clay mineral was 11# coal > 5# coal > 3# coal (as shown in table 2) and the clay mineral more easily shed from coal skeleton particles under the shearing stress of the high-speed displacement fluid, which break into particles moved out with the displacement fluid.

When we kept the displacement fluid velocity constant and changed the confining stress (as shown in table 2), The rate of increase on pulverized coal output was 5# coal > 11# coal with the rising of the confining stress. The results are shown in Figure 3.

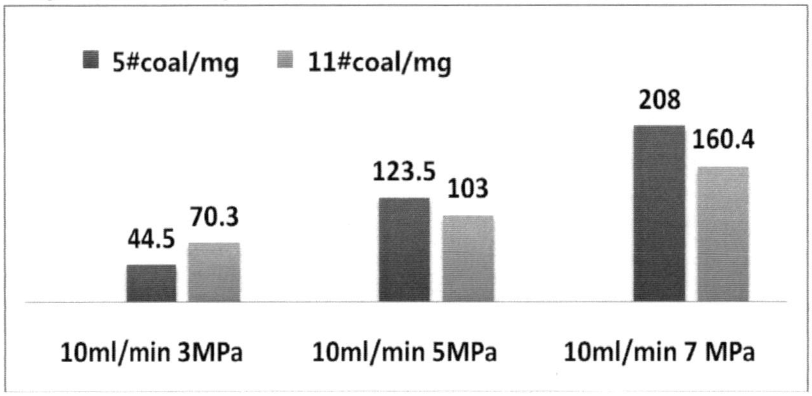

Figure 3. The pulverized coal output of different coal petrography under constant displacement fluid velocity and variable confining stress

The reason is that the content of vitrinite of 5# coal is more than 11# coal (as shown in table 1), microhardness of vitrinite is less than inetrinite and mineral, and microscopic brittleness is more than inetrinite and mineral. The vitrinite is more easily to break into coal particles.

4. The effect of different coal structure on pulverized coal output

Table 3. The experimental scheme of different coal structure

Coal samples	Coal structure	Experimental conditions	
3# coal	Primary structure of coal	Constant confining stress and variable displacement fluid flow velocity	Confining stress for 5MPa, flow velocity for 5ml/min
	Fragmented coal		Confining stress for 5MPa, flow velocity for 15ml/min
5# coal	Primary structure of coal		Confining stress for 5MPa, flow velocity for 25ml/min
	Fragmented coal	Constant displacement fluid flow velocity and variable confining stress	Flow velocity for 10ml/min, confining stress for 3MPa
11# coal	Primary structure of coal		Flow velocity for 5ml/min, confining stress for 5MPa
	Fragmented coal		Flow velocity for 5ml/min, confining stress for 7MPa

All equipment and coal samples and displacement fluid and other conditions used in this experiment were the same to the experiment which studied the effect of different coal petrography on pulverized coal output. Using displacement fluid and confining stress simulated the intensity of water production and the coal reservoir stress to study the effect of primary structure of coal and the fragmented coal on pulverized coal output. The experimental scheme was shown in table 3.

4.1. Analysis of pulverized coal output

As shown in table 4, the output of pulverized coal of the fragmented coal was much larger than the primary structure of coal under the same experimental conditions.

Table 4. The weight of pulverized coal output for different coal structure

Coal samples conditions / Experimental conditions	5MPa-5ml/min /mg	5MPa-15ml/min /mg	5MPa-25ml/min /mg	10ml/min-3MPa /mg	10ml/min-5MPa /mg	10ml/min-7MPa /mg	Percentage of pulverized coal output /%
3# primary structure of coal	46	67	53.5	80.5	66.5	103	0.06-0.14
3# fragmented coal	142.2	755	1227	1500	740	700	0.19-2.0
5# primary structure of coal	54.5	92	72	44.5	123.5	208	0.06-0.28
5# fragmented coal	460	567	930	720	390	410	0.52-1.24
11# primary structure of coal	55	141.5	137	70.3	103	160.4	0.07-0.21
11# fragmented coal	150	500	550	510	260	350	0.2-0.73

We can see in table 4 that the output of pulverized coal of 3#, 5#, 11# primary structure of coal was 46~208mg, making up 0.06~0.28% of the total weight of coal samples before conducting experiment. Compared to the fragmented coal, the output of pulverized coal was 142.2~1500mg and accounted for 0.19~2.0%, respectively. Due to the effect of tectonic stress, the coal structure of the fragmented coal was more broken and coal developed many tiny cracks within coal matrix. As a result of the fluid erosion and confining stress failure, the fragmented coal is more susceptible to form fine coal particles under the same experimental conditions causing the output increasing of pulverized output immediately.

4.2 Granularity characteristics of pulverized coal output

4.2.1. Research method

The size of the pulverized coal particles is an important factor to determine the stuck pumps. By studying the characteristics of different particles size of pulverized coal output, we can recognize and resolve the problem of stuck pumps. We used the China University of Mining and Technology (Beijing) State Key Laboratory of Coal Resources and Safe Mining's XY-P polarizing microscope and the CBM national Engineering Center' Mastersizer 2000 to analyze the sizes of coal particles.

4.2.2. Analysis of pulverized coal output of primary structure of coal

In this experiment the grain size distribution curves are shown in Figure 4. The pulverized coal output particle size distribution curve of 3#, 5#, 11# primary structure of coal showed a double-peak, and the grain size of primary pulverized coal was relatively small and the grain size of secondary pulverized coal was relatively large except for Figure 4(c). The grain size distribution had a wide scope with range from 0.02~2000μm.

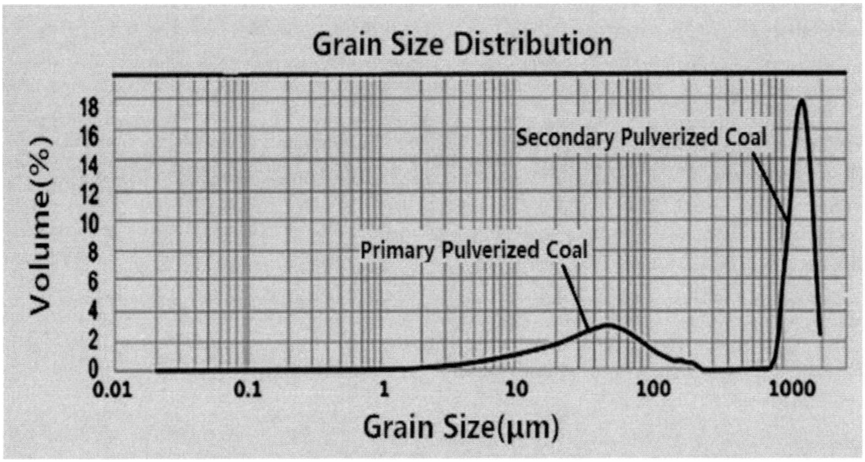

Figure 4. (a) 3# primary structure of coal

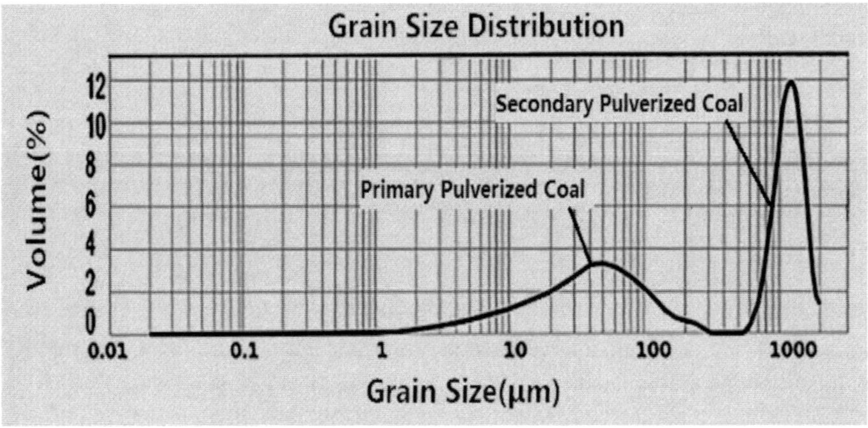

Figure 4. (b) 5# primary structure of coal

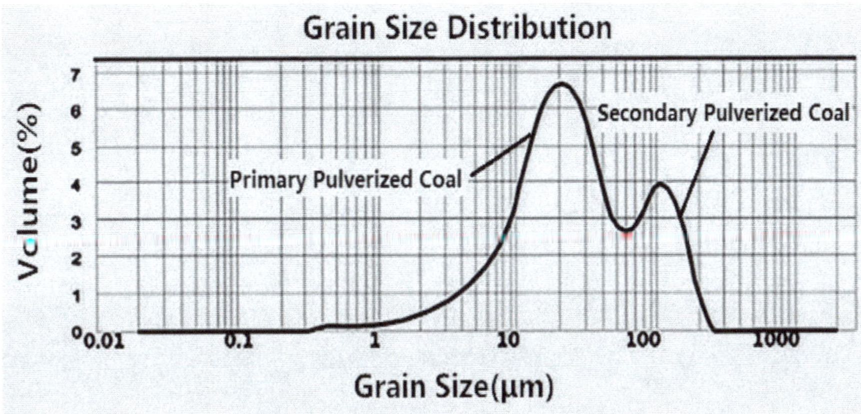

Figure 4. (c) 11# primary structure of coal

Figure 4. The grain size distribution of different primary structure of coal

The range of the grain size was 0.02~2000μm in Figure 4. We can see the grain size distribution of 3# primary structure of coal in Figure 4. (a), which showed a double-peak distribution curve. The primary pulverized coal appeared from 20μm to 90μm, and the second pulverized coal was from 900μm to 1600μm. In Figure 4. (b), the 5# primary structure of coal for the grain size distribution had a double-peak distribution curve. The size of the primary pulverized coal ranged between 20μm and 100μm, and the second pulverized coal was from 700μm to1700 μm. As shown in Figure 4. (c), the grain size distribution of 11# primary structure of coal presented a double-peak distribution. The primary pulverized coal was between 8 μm and 40 μm, and the second pulverized coal varied from 50μm to 180μm. Due to the effect of tectonic stress, the coal structure of the fragmented coal was more broken and coal developed many tiny cracks within coal matrix. As a result of the fluid erosion and confining stress failure, the fragmented coal is more susceptible to form fine coal particles under the same experimental conditions causing the pulverized coal output to increase immediately.

4.2.3. Analysis of pulverized coal output of fragmented coal

As shown in Figure 5, the grain size distribution of fragmented coal was a unimodal distribution, and the distribution scope was relatively concentrated and the average grain size was small. The particle size of fragmented coal ranged from 0.02~2000 μm in Figure 5.

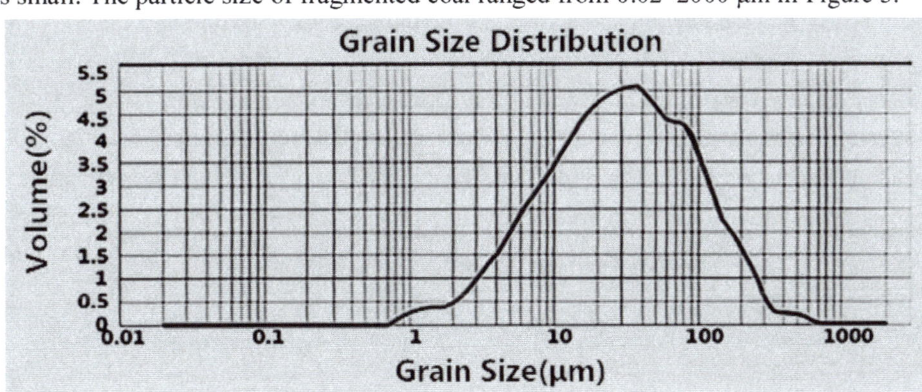

Figure 5. (a) 3# fragmented coal

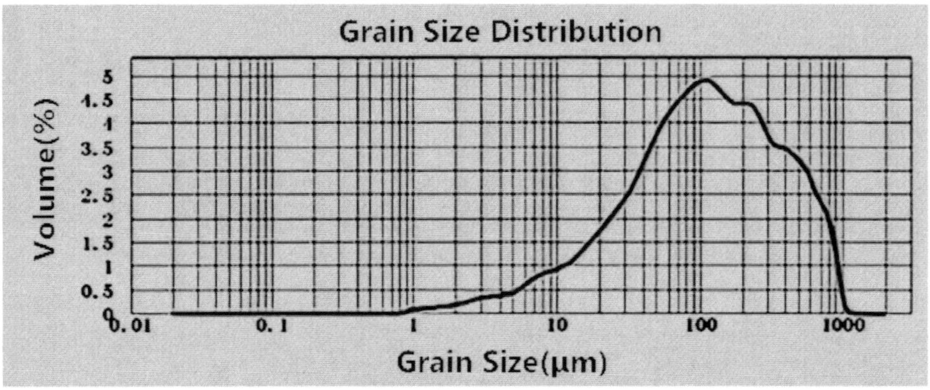

Figure 5. (b) 5# fragmented coal

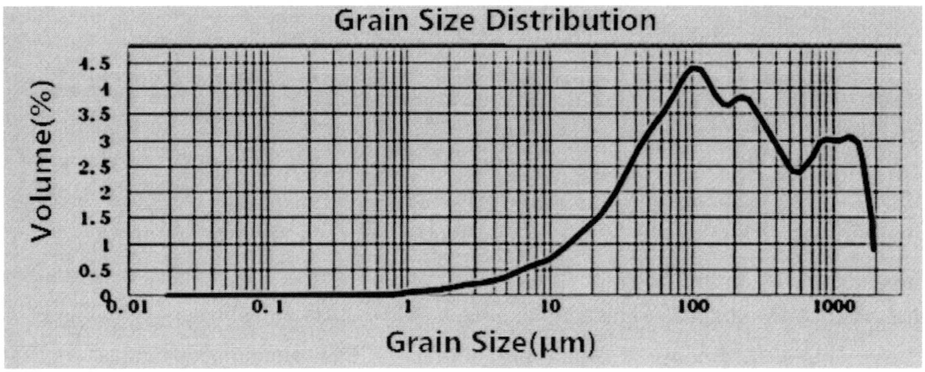

Figure 5. (c) 11# fragmented coal

Figure 5. The grain size distribution of different fragmented coal

The 3# fragmented coal showed a pulverized coal particle size distribution curve with a single peak. The grain size ranged from 0.02~2000 μm, and the main peak appeared between 10 μm and 100 μm. As shown in Figure 5, the pulverized coal particle size distribution curve of 5# fragmented coal was a single peak withe the range from 0.02~2000 μm and the main peak was between 50μm and 300 μm. The pulverized coal grain size distribution curve of 11# fragmented coal was shown in Figure (c). The particle size ranged from 0.02~2000μm and the main peak appeared between 60μm and 1300μm. The reason for the single peak was that the original fine pulverized coal accounted for large proportion of pulverized coal output of fragmented coal and the size range of pulverized coal output was concentrated.

4.3 Morphological characteristics of pulverized coal output in the experiment

The morphological characteristics of the pulverized coal were found to be related to the source of different coal body structure. Examinations of the morphological characteristics of pulverized coal output may lay a foundation for the development of the control measures. Special analysis was performed at the China University of Mining and Technology (Oxford) IE 250 S-3400 SEM.

Figure 6. (a) 3# primary structure of coal Figure 6. (b) 3# fragmented coal

Figure 6. Morphological characteristics of pulverized coal output under the condition of 10ml/min, 3MPa

As shown in Figure 6. (a) and Figure 7. (a), The microscopic morphology for the pulverized coal output of primary structure of coal was angular and roundness commonly sorted edge angle, and the shape characteristic was predominantly cylindrical shape and block shape, and the surface characteristics of pulverized coal was smooth.

Figure 7. (a) 5# primary structure of coal Figure 7. (b) 5# fragmented coal

Figure 7. Morphological characteristics of pulverized coal output under the condition of 15ml/min, 5MPa

We can see in Figure 6. (b) and Figure 7. (b) that the microscopic morphology for the pulverized coal output of fragmented coal was roundness, which was predominantly sub-angular shape and sub-round shape, and round shape. The surface characteristic of pulverized coal of fragmented coal was relatively rough and the shape characteristic sorted mainly granular.

5. **Conclusion**

The weight of pulverized coal output is related with displacement fluid velocity and confining stress. When the confining stress keeps stable, the weight of pulverized coal output increase with rising of the displacement fluid velocity. This shows that the weight of pulverized coal output is positively correlated with the displacement fluid velocity of single-phase water flow. When the velocity of displacement fluid keeps constant and changes the confining stress, the weight of pulverized decrease relatively. This shows that the weight of pulverized coal output is negatively correlated with the confining stress.

Coal body structure was found to be one of primary factors of pulverized coal output. The clay mineral more easily shed from coal skeleton particles under the shearing stress of the high-speed displacement fluid, which caused to break into particles with the displacement fluid moving out. Due to the effect of tectonic stress, the coal structure of the fragmented coal was more broken and coal developed many tiny cracks within coal matrix. The morphology of pulverized coal output particles was found to be related to the nature of the coal, coal body structure and the late mechanical damage.

References

[1] Bai J M, Chen H, Zu S Q, Cui J B and Wang W Y 2010 Coalbed Methane Exploration and Development of Theory and Technology, 2010 Coalbed Methane Symposium, ed Sun F.J and Feng S.L (Petroleum Industry Press, Beijing) 425-431 (in Chinese)

[2] Cao D Y, Yuan Y, Wei Y C, Liu S G, Li X M and Wang Q W 2012 A comprehensive classification study of coal fines genetic mechanism and origin site *Coal Geology of China* **24** 10-12

[3] Chen Z H, Wang Y B and Sun P 2009 Destructive influences and effectively treatments of pulverized coal to high rank coalbed methane production *J. China Coal Society* **34** 229-32

[4] Cui C C 2008 The promising prospect for developing and utilizing coal-bed gas *Natural Gas Chemical Industry* **3** 71-2

[5] Huang W A, Qiu Z S, Yang L, Wang Y Q, Ma Y L and Wang Z M 2013 Instability mechanism of sidewall and anti-sloughing drilling fluid technique for coalbed methane well drilling *Coal Geology & Exploration* **41** 37-41

[6] Li X Z, Long S X, Wu Q 2010 Coalbed Methane Exploration and Development of Theory and Technology, 2010 Coalbed Methane Symposium, ed Sun F.J and Feng S.L (Petroleum Industry Press, Beijing) 405-10 (in Chinese)

[7] Li Y M, Wang L L, Liu G W, Yan L and Zou Z G 2010 Study on coal reservoir damage mechanism in dewatering and extraction process of CBM wells *China Coalbed Methane* **7** 39-43 (in Chinese)

[8] Liu S G, He X H and Li H F 2011 Production mechanism and control measures of coal powder in coalbed methane horizontal well *J. Liaoning Technical University (Natural Science)* **30** 508-12 (in Chinese)

[9] Liu S G, Hu A M, Song B and Li H F 2012 Coal powder concentration warning and control measure during CBM well drainage *J. China Coal Society* **37** 86-90 in Chinese)

[10] Wang Q W, Cao D and Wang T 2013 Relationship between the stress-concentrated zone caused by cavity-completed wells and the coal powder output *Coal Geology & Exploration* **41** 35-37 (in Chinese)

[11] Wang H X and Lan W J 2012 Discussion on formation mechanism of coal powder in coalbed methane well *China Coal* **2** 95-7 (in Chinese)

[12] Yan H W 2010 *Study on condition of discharging coal dust from annulus between rod and tube in coalbed methane Wells* (Qingdao: China University of Petroleum (East China)) 75 (in Chinese)

[13] Yang Y, Guo C H and Luo T T 2011 The adsorption properties of viscoelastic surfactant

on coal powder *J. China Coal Society* **36** 110-113 (in Chinese)

[14] Zhang G S, Tian W T, Tao S, Bai J M, Shi H N and Cheng H 2011 Experimental research of coal grain migration rules of coalbed methane *J. Oil and Gas Technology* **33** 105-08 (in Chinese)

[15] Zhao J and Zhang S A 2013 Production dynamics of CBM bottom hole pressure *Coal Geology & Exploration* **41** 21-8 (in Chinese)

[16] Zhao J F, Wang S W, Qin Y, Zhao W X, Chen W W, Yang J and Li R 2013 Characteristics and origin of coal powder in coalbed methane well *Natural Gas Geoscience* **24** 1316-20 (in Chinese)

[17] Zhou F D, Yao G Q, Tang Z H and Oyinkepreye D O 2011 Influence and sensitivity study of matrix shrinkage and swelling on enhanced coalbed methane production and CO2 sequestration with mixed gas injection *Energy Exploration and Exploitation* **29** 759-76

A novel fault diagnosis method of PV based-on power loss and I-V characteristics

Y H Chen[2], R Liang[1,2,3], Y Tian[2] and F Wang[2]

[1]IOT Perception Mine Research Center, China University Of Mining And Technology, Xuzhou, China

[2]School of Information and Electrical Engineering, China University of Mining and Technology, Xuzhou, China

E-mail: liangrui@cumt.edu.cn

Abstract. The power loss and the changes of internal I-V output characteristics of photovoltaic (PV) module in the typical fault condition were analyzed. We proposed an on-line real time fault diagnosis method for PV module, which takes into account the power loss and the internal I-V characteristics. Taking into account the changes of temperature and irradiation, the running status of the PV module were simulated in real time. Firstly, by comparing the simulated power with the measured power, it could determine whether the abnormal power loss has occurred. Then based on the change of output voltage, it could decide if short-circuit fault has occurred and estimate the number of short circuited cells roughly. Further, the value of fill factor (FF) can be utilized to determine whether aging fault has occurred and to acquire the remaining service life of the module. The results of simulation and experiment show that this method can effectively detect the partial shadow short-circuit fault and aging fault. It proves the feasibility and accuracy of the fault diagnosis method.

1. Introduction

With the growing of global economy, primary energy resources such as coal, oil and natural gas have become increasingly scarce. Solar power is promoting and applying rapidly with the advantages of clean, pollution-free, unlimited and sustainability [1]. Generally, the working condition of PV power plant is poor, resulting in the continuous occurrence of different kinds of fault. Therefore, the monitoring of the operating states and fault diagnosis of the PV systems become more and more

[3] Address for correspondence: R Liang, IOT Perception Mine Research Center, China University Of Mining And Technology, Xuzhou, China. School of Information and Electrical Engineering, China University of Mining and Technology, Xuzhou, China. E-mail: liangrui@cumt.edu.cn.

important. For the moment, domestic and overseas scholars mostly were committed to the photovoltaic systems for large-scale power generation which is equipped with centralized inverter or series inverter. However, for small or medium power photovoltaic power generation system connected by micro inverter, there is no effective online fault detection system. In the small power photovoltaic system, each PV module is respectively connected with a micro inverter. When a module in the system cannot work normally, only this one will be affected, however, the other modules will work normally which makes system more efficient. Detecting and replacing module accurately in the running process of PV system can further improve the efficiency of the system.

Currently, there are several common fault diagnoses based on infrared picture analysis, based on electrical measurement, based on intelligent detection and based on monitoring system，respectively. Some authors took advantage of the thermic effect of photovoltaic cells to detect PV module by means of infrared technique [2-4]. By analyzing the characteristics of infrared images, the status of the PV module can be obtained. However the infrared images are easy to be affected by environment. A direct earth capacitance detection method was proposed to determine whether there is an open circuit fault and navigate to the cut point by measuring the earth capacitance of PV modules [5]. But this method can only detect the circuit fault. TDR was proposed to judge fault type and fault location by analyzing the signal reflected from the fault point [6]. A method for multi-sensor was proposed to judge fault type and fault location through analysis of voltage and current collected by the sensors [7]. By analyzing the power loss which is collected from the DC-side, it can realize fault diagnosis of PV module [8].An online fault diagnosis strategy was proposed by Wang and Li [9], who use BP neural network to detect the fault accurately. This method has abilities of self-study and adaptive, associative memory and nonlinear pattern recognition. And it is suitable for multi fault and complex pattern recognition. At present, the application of intelligent detection method in fault diagnosis of PV system is not much. Bruno Andò [10] presented an intelligent monitoring system with the aim of detecting the causes of efficiency losses, which equipped with voltage, current, irradiance, temperature, and inertial sensors. Because of the high cost of the monitoring system, method based on the monitoring system isn't suitable for small PV systems.

We propose a novel on-line real time fault diagnosis method for PV module, which takes into account the power loss and the internal I-V characteristics. Firstly, by comparing the difference between the simulated power and measured power, it could determine whether the abnormal power loss has occurred. Then identify the source of power loss based on the changes of radiation of the measuring points. What following is judging if short-circuit fault has occurred and estimating the number of short circuited cells roughly by the change of output voltage. If the aging fault occurs, analyze the fill factor and judge the degree of aging fault considering the temperature and solar light. If serious aging fault has occurred, replace the new photovoltaic module.

2. Analysis of common faults

Module failures happen easily with different kinds of faults because of the badly equipping environment of PV system. The fault criterion is determined by analyzing the output properties of different kinds of faults in detail and comparing with the output properties of normal module.

2.1 Fault type

PV module usually consists of multiple solar cells, as shown in figure 1. When one or more cells of PV module were shaded by surrounding buildings or trees, partial shadow will be formed. As shown in figure 1 (a). In theory each cell in PV module should be in parallel with a bypass diode. The main role of the bypass diode is to provide an additional current path so that the output power of PV module will not be affected when the cells are under the shadow or there is dust on the surface. But in practice, because of heating and the restriction of economy, usually a string which consists of 12 cells will be equipped with an anti-parallel diode. When the cells are in the shadow of occlusion, they show resistance properties. The diode will turn on and insulate the cell which fails to run from the PV module under forward bias voltage. Therefore partial shadow of PV module will cause a great loss of power. Faulty soldering and over welding in the process of production may cause the cells be short-circuited or disconnected, as shown in figure 1(b), figure 1(c).

The average lifespan for a PV module is 20 to 30 years. However, in practical application, because of the manufacturing processes and installation environment, PV module will appear abnormal aging fault after five to six years of operation, as shown in figure 1(d).

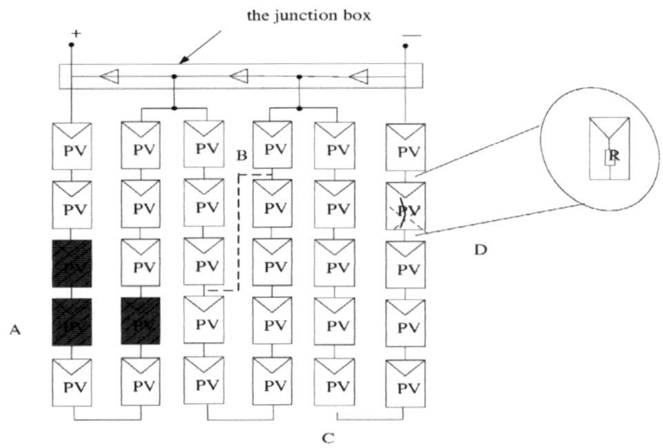

Figure1. Common faults of PV module.

2.2 Effects of faults on the power loss and the I-V characteristics of PV module

For the purpose of studying the power loss of PV module with different shadow rate, an experiment was carrying out on the JHX250P60 PV module. The change of the power loss of module with different shadow rate is shown in figure 2 and table 1.

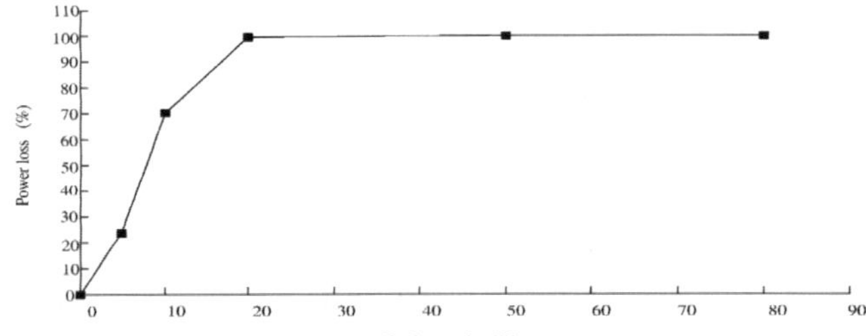

Figure 2. Power loss of module with different shadow rate.

Table 1. Power loss of module with different shadow rate.

shadow ratio /%	Uoc/V	Isc/A	Pm/W	Power loss rate/%
0	37.64	8.412	243.7948	0
5	37.05	6.391	186.4207	23.533
10	36.90	2.443	71.6658	70.604
20	33.25	0.031	0.4577	99.812
50	27.58	0.004	0.0130	99.998
80	14.46	0.001	0.0021	99.999

As is shown in figure 2 and table 1, the power loss shows a nonlinear increase with the increase of partial shadow rate. And the power loss rate of PV module is far larger than the partial shadow rate. When the shadow rate is up to 5%, the power loss rate exceed 20%; When the shadow rate is up to 10%, the power loss rate exceed 70%; When the shadow rate is up to 20%, the power loss rate exceed 99%, namely the output power approaches zero. The output power of PV module is affected by partial shadow seriously. When the shaded area is too large to meet the requirements of micro inverter, PV module will not continue to output power.

In order to study the influence of different types of faults on the output characteristics, we simulate and analyze in the case of faultless, partial shadow, short circuit and aging fault. The I-V characteristics of different faults are shown in figure 3.

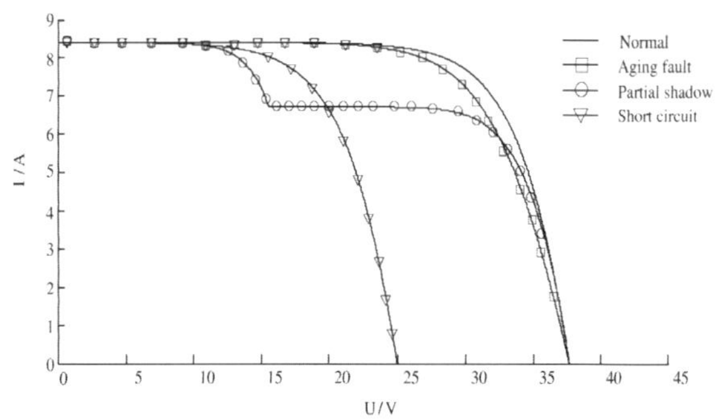

Figure 3. U-I curves under different faults.

PV module which equipped with three bypass diodes consists of 36 cells in simulation. As is shown in figure 3, the I-V curve shows "multiple knees phenomenon" because of partial shadow. However the short-circuit current and the open-circuit voltage of module will not change. Short circuit fault causes the reduction of the open circuit voltage and the maximum power point voltage. The shape of the curve is as basically the same as the I-V curve in the absence of fault. In the case of aging fault, the maximum power point voltage (U_m) and the maximum power point current (I_m) reduce accordingly, but the open circuit voltage and short circuit current remain basically unchanged.

As is shown in figure 4, simulate different degree of aging by changing the internal series resistance of PV module. When the series resistance increases, namely the degree of aging increase, the I-V curve shape of PV module remained unchanged, but shows a tendency of contraction.

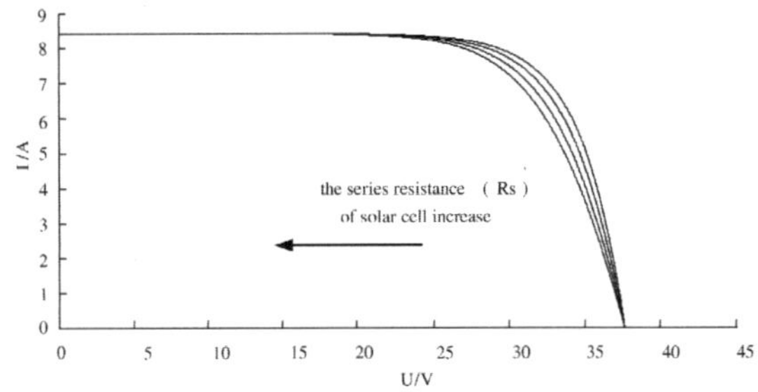

Figure 4. U-I curves under aging fault.

3. Fault diagnosis strategy

By analyzing the output power of faulty module, we learn that the abnormal power loss is likely to come from module failures. It is difficult to identify the fault simply by analyzing the power loss. Especially the changes of irradiation and temperature can also cause a high power loss which leads to misjudgment of the diagnosis system. We can identify some types of faults according to the internal characteristics changes of different types of faults. But it is difficult to identify the other faults because the changes of their internal characteristics are not so obvious. Therefore taking into integrative consideration the output power loss and the internal characteristics is helpful to improve the accuracy of fault diagnosis.

3.1 Diagnosis strategy of partial shadow

The micro inverter with a function of power recording plays an important role in timely detecting abnormal power loss and avoiding long-term work in fault state for PV module. An effective operation of a PV module is difficult to avoid power loss which includes the power loss caused by sun light incidence angle, the power loss caused by the temperature, the power loss cause by the inaccuracy of maximum power point tracking and the micro inverter switching loss etc. A brief description of the main factors in the conventional power loss of PV module is shown in figure 6. In the first part of power loss, the impacts of the incident angle, the reflecting of PV module and the dust make the irradiation of PV module different from the irradiation of simulation input, which causes the difference between the simulated power and measured power. Considering the general frequency of the cleanup, adopt 8% as the power loss rate. In the second part of power loss, the output voltage and the output current change with the change of the temperature which leads the reduction of output power. Therefore, the power loss caused by temperature is an important factor which should be considered. Generally, the average power loss rate of temperature is 3%. In the third part of power loss, the maximum power point tracking is accurate by default in the simulation model, namely the output power is the maximum power of the module. In practice, the most ideal tracking precision can reach 99.5% that result in some difference in the maximum power point power. In addition, there is a problem of the conversion efficiency of micro inverter. At present, the conversion efficiency of high quality micro inverse can reach 97.5% so that the power loss rate of the inverter efficiency is about 3%.

In order to determine whether the power loss is in the normal range, the boundary of daily power loss should be established firstly. As is shown in figure 7, the curve A is the actual output power curve of the experimental module without fault and the data were recorded during a sunny day in the summer of

2015. The curve B is the theoretical output power curve obtained from the simulation module which take the measured irradiation and measured temperature as its input variables. By comparing the measured power value of each measurement point with the theoretical value of the simulation, we can know that the normal power loss rate of the PV module without fault is generally between 13%-15%. If the power loss rate is more than 15%, it can be defined as abnormal power loss. Abnormal power loss may result from short circuit fault, aging fault and other faults, but also may come from partial shadow or the sudden decline of the irradiation caused by the cloud cover.

Therefore, it is also needed to determine whether the irradiation of the measuring point has been decreased abnormally in order to distinguish different faults. In the case of fine weather, the irradiations of two adjacent measuring moments are similar. So the irradiations of the first few measuring points can be used to get the irradiation of the next measuring point. $S_n = min\{S_i, i = 1,2,3\}$, where S_n is the constructed irradiation of the current measuring point, S_i are the irradiations of the first few measuring points. And S'_n is the measured irradiation of the current measuring point. If $S_n - S_0 \le S'_n \le S_n + S_0$ (S_0 is the margin for the variation), it can be judged as the normal irradiation change. If $S'_n < S_n - S_0$ is tenable within consecutive 30 minutes, it can be judged that partial shadow has occurred.

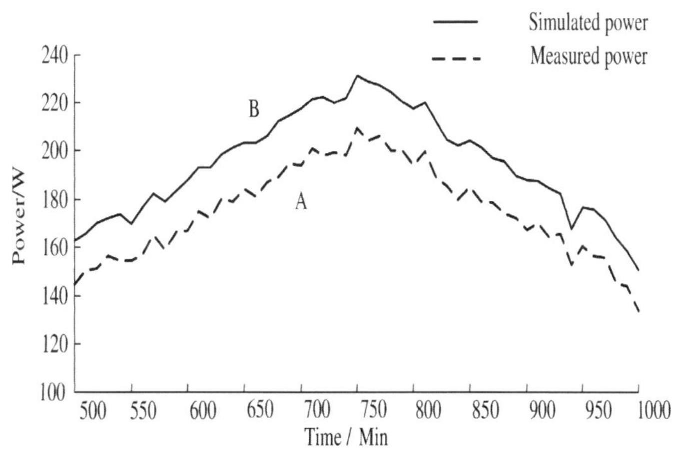

Figure 5. Comparison of the simulated power and the measured power.

3.2 Diagnosis strategy of short circuit fault

The output voltage of PV module will reduce when the circuit fault occurs. So it can decide if the short circuit fault has occurred by comparing the simulated output voltage with the measured output voltage. A variable α is introduced as shown in equation (1).

$$\alpha = \frac{V'_{oc} - V_{oc}}{u_{oc}} \tag{1}$$

Where V'_{oc} and V_{oc} are, respectively, the simulated open circuit voltage and the measured open circuit voltage, u_{oc} is the average open circuit voltage of the cells in PV module. If $\alpha > 1$, it can be determined that the short circuit fault has occurred. The number of the cells which are short circuited is determined by equation (2).

$$n_d = \left| \frac{V'oc - Voc}{u_{oc}} \right|_{\text{rounding}} \quad (2)$$

3.3 Diagnosis strategy of aging fault

In the classification of fault types, it is mentioned that the aging fault will result in the increase of series resistance. Direct measurement for internal series resistance of PV module is difficult and the accuracy is difficult to guarantee, so the fill factor which is easier to access to is introduced. The fill factor (FF) is a measure of the quality of solar cells (series resistance and shunt resistance). The fill factor (FF) is defined as the maximum output power divided by output power of ideal target ($I_{sc} \times V_{oc}$) as shown in equation (3).

$$FF = \frac{I_m V_m}{I_{sc} V_{oc}} \quad (3)$$

It is clear that FF should be as close as possible to 1 (i.e.100%). But the characteristic of the exponential function p-n junction prevents it from reaching 1. The value of FF is affected by the series resistance. We can get the curve of the filling factor (FF) with the change of the series resistance from the simulation. The fill factor decreases with the increase of the series resistance non-linearly as shown in figure 7. In this paper, the value of the fill factor is used to judge the occurrence of aging fault and to determine the degree of aging.

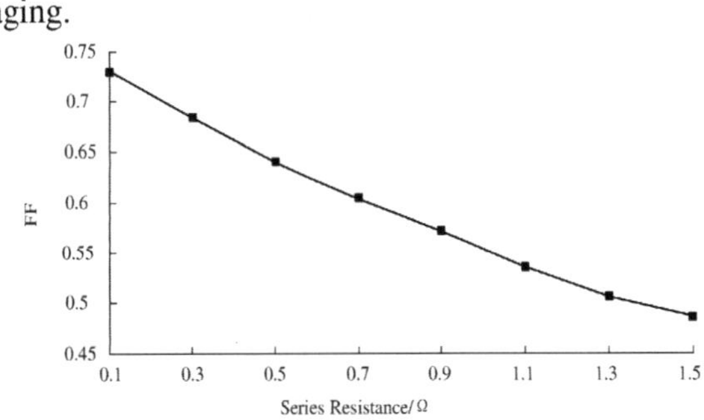

Figure 6. Curves of the fill factor with the change of resistance.

In order to calculate the FF of normal module, the value of series resistance of the module without aging fault need to be estimated under standard test conditions.

$$I = I_{ph} - I_0 \{ \exp[\frac{q(V + IR_s)}{AKT}] - 1 \} - \frac{V + IR_s}{R_{sh}} \quad (4)$$

The output characteristic of PV cells is shown in equation (4). Where I_{ph} is the photocurrent of the cell, I_0 is the reverse saturation current of the cell, V is the output voltage of the cell, q is the charge constant, which general is $1.6 \times 10^{-19} C$, A is the diode factor, K is the Boltzmann's constant, which is $1.38 \times 10^{-23} J/K$, T is absolute temperature in K, R_s and R_{sh} are, respectively, the series resistance and the parallel resistance of PV cells. Generally the parallel resistance (R_{sh}) of PV cells is very large which makes the value of $\frac{V+IR_s}{R_{sh}}$ far less than the photocurrent. So the value of $\frac{V+IR_s}{R_{sh}}$ can be neglected. Equation (4) can be simplified as equation (5).

$$I = I_{ph} - I_0\{\exp[\frac{q(V+IR_s)}{AKT}]-1\} \quad (5)$$

The "-1" can be neglected in equation (5) because the value of $\exp[\frac{q(V+IR_s)}{AKT}]$ is far large than 1. The equation (6) can be obtained by setting conditions of $I_{ph}\approx I_{sc}$ and $\lambda = q / AkT$.

$$I = I_{sc}[1-(\frac{I_0}{I_{sc}})\exp\lambda(V+IR_s)] \quad (6)$$

When the PV cells is open circuit (I=0), we can get that V=Voc. The equation (7) can be obtained by putting V= V_{oc} into the equation (6).

$$\lambda = \frac{1}{V_{OC}}\ln(\frac{I_{SC}}{I_0}) \quad (7)$$

The equation (8) can be obtained by putting the equation (7) into the equation (6).

$$I = I_{sc}\{1-\frac{I_0}{I_{sc}}\exp[\frac{1}{V_{oc}}\ln(\frac{I_{sc}}{I_0})(V+IR_s)]\} \quad (8)$$

We can obtain the conditions that I=Im and V=Vm at the maximum power point. The equation (9) can be obtained by putting the above conditions into the equation (8) and simplifying it.

$$R_s = [\frac{V_{oc}}{\ln(\frac{I_{sc}}{I_0})} g \ln(\frac{I_{sc}-I_m}{I_0}) - V_m] / I_m \quad (9)$$

Because I_0 is related to the nature of the solar cells' materials, it reflects the maximum composite ability of photon-generated which comes from the solar cells. It is generally considered that it is approximately a constant and not affected by the light intensity. Under standard testing conditions, $I_0 / I_{sc} \approx 10^{-10} \sim 10^{-8}$. In order to simplify the analysis, usually $I_0 / I_{sc} \approx 10^{-9}$ is adopted. As is shown in equation (9), based on the four technical parameters (Isc, Voc, Im, Vm) of PV module under the corresponding standard test condition which are provided by the manufacturer, it is easy to calculate the value of series resistance under the corresponding standard test condition. By putting the known data (V_{oc} =37.64V, I_{sc} =8.41A, V_m =30.87V, I_m =8.1A) of PV module into the equation (9), we can obtain that Rs= 0.0957Ω under standard test conditions. So taking R_s =0.1Ω as the value of the series resistance of PV module without aging fault under standard test condition. Under standard test conditions, the FF of module without fault is generally between 0.7 and 0.75. Therefore, it can be judged that when the fill factor FF is less than 0.7, aging fault has occurred. When the value of fill factor (FF) is below 0.6, it can determine that serious aging fault has occurred. In addition, the fill factor (FF) of PV module is affected by the irradiation and temperature. The value of the fill factor changes with irradiation, as shown in figure 7.

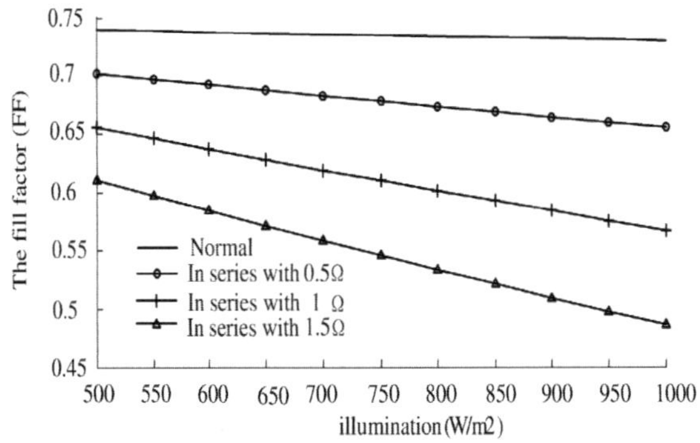

Figure 7. Values of the fill factor under different irradiation.

The fill factor is also related to the temperature of the solar cell. And the curve of current-voltage relationships will soften up because the leakage current of p-n junction increase with the temperature. And it makes FF generally decreases with the increase of temperature.

Calculating the fill factor (FF) by considering the factors of the irradiation and temperature. When the value of FF is less than 0.7, it can be judged that the module is in a failed state of aging. According to the fill factor FF at this time, the degree of aging fault can be obtained. Further, the remaining useful life can be calculated based on the using time of the module, so that the failure module can be replaced in time. Assuming that the module has been used for N years, the fill factor calculated at this time is FF. When the value of FF is between 0.6 and 0.7, the annual deterioration rate (x) of PV module can be calculated by equation (10).

$$x = \frac{0.7 - FF}{(0.7 - 0.6) \times N} \times 100\% \tag{10}$$

The remaining useful life of PV module can be calculated by equation (11).

$$M = \frac{1}{x} - N \tag{11}$$

3.4 Steps of fault diagnosis

The on-line fault diagnosis proposed in this paper can be roughly divided into the following steps: (1) Comparing the simulated output power with the measured output power, if the output power loss rate exceeds 15%, it can be judged that an abnormal power loss has occurred; (2) Taking into account the irradiation of several measurement points where the abnormal power loss occurred, if the irradiation drops suddenly, it can be judged that the power loss comes from the shadow, elsewise it comes from the faults; (3) Using the open circuit voltage of the module to determine whether the short circuit fault has occurred and roughly estimate the number of cells which are short-circuited; (4) Calculating the fill factor (FF) based on the condition of irradiation and temperature. It can determine whether the aging fault has occurred and obtain the degree of the aging fault based on the value of FF. The basic process of fault diagnosis is shown in figure 8.

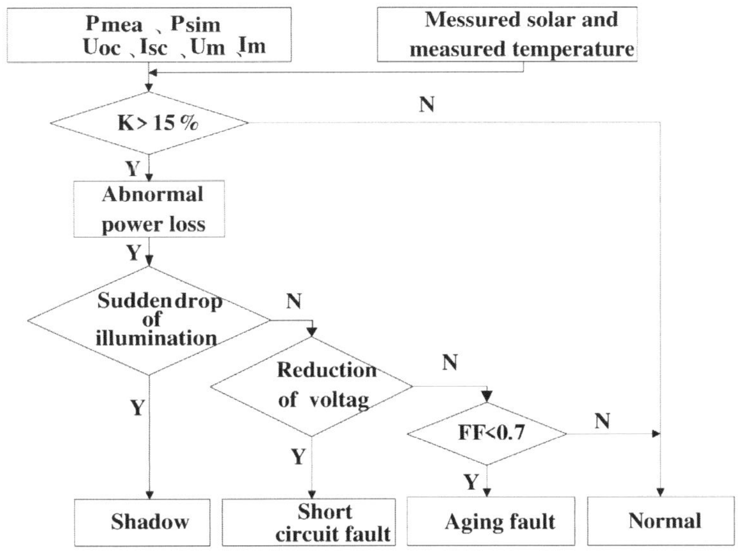

Figure 8. Flow chart of faults diagnosis.

When the short circuit fault happens, calculate the number of cells which are short-circuited based on the open circuit voltage. When the state of the module is judged to be between normal and severe aging, the remaining useful life of the module can be calculated roughly based on the annual aging rate.

4. Simulation and experimental validation

In order to verify the validity and accuracy of the fault diagnosis method proposed in this paper, the simulation and experiment are carried out. The parameters of simulation model come from the PV module which models for JHX250P60. Parameters under standard test conditions are shown in table 2.

Table 2. Parameters of experimental module.

Parameters	values
Rated power (Wp)	250W
Open circuit voltage （Voc）	37.64V
Short-circuit current (Isc)	8.41A
Voltage of Maximum power point (Vm)	30.87V
Current of maximum power point (Im)	8.10A

The model of micro inverter used in the experiment is ANI-250, which has the functions of DC-DC and the maximum power point tracking (MPPT). The micro inverter can record the maximum output power of the module at different times and sample the working voltage and the working current of the module at the same time. The open circuit voltage, the short circuit current, the voltage of maximum power point and the current of maximum power point are scanned and sampled every 10 minutes. The micro inverter has a built-in communications module which uploads the data to the server terminal in real time.

In order to verify the fault diagnosis method, the experiment should be carried out under the conditions of normal working, different shadow ratio, different degree of short circuit fault and different degree of aging fault. Considering the effects of irradiation and temperature on the fault diagnosis results, the 7 different time-points of the day were randomly selected. The module works properly at point A. Three cells of PV module are covered by the shadow at point B. At point C, twelve cells of PV module are covered by the shadow. At point D, 6 cells of the module are short-circuited. At point E, 9 cells of the module are short-circuited. At point F, the module is in series with a resistance of 0.5 ohm. At point G, the module is in series with a resistance of 1.1 ohm.

Table 3. Simulated power and measured power at different time-points.

Measurement points	Simulated power	Measured power	Power loss rate(k)	Sudden decline of irradiation
A	122.75	105.79	13.8%	None
B	140.90	78.55	44.3%	Yes
C	173.66	1.18	99.9%	Yes
D	223.72	175.84	21.4%	None
E	197.26	130.78	33.7%	None
F	171.01	143.31	16.2%	None
G	155.36	119.32	23.2%	None

By comparing the simulated power with the measured power of each measurement point, it can be obtained that abnormal power loss occurred at all time-points except point A. Combined with the situation of sudden decline of irradiation, it can be concluded that partial shadow occurred at point B and point C. And it can be initially judged as short circuit or aging fault occurred at point D, E, F and G.

Further, the characteristic parameters of PV module at point D, E, F and G are shown in table 4 and table 5.

Table 4. Test data of time-point D and E.

Measurement points	V_{oc}	V'_{oc}	α
D	34.78	31.63	6.3
E	33.66	28.74	9.84

Table 5. Test data of time-point F and G.

Measurement points	Isc	Voc	Pm	FF
F	5.87	37.14	143.31	0.66
G	5.46	36.85	119.32	0.59

By comparing the open circuit voltage of the module, it can be judged that the short circuit fault has occurred at point D and point E. And the open circuit voltage is used to estimate the number of short circuited cells roughly. Diagnostic results are shown in table 6. It can be judged that aging fault has occurred at measurement point F, G by calculating the value of FF. The remaining useful life of PV module can be roughly calculated based on the years that the module had been used for and the value of FF. The PV modules used in the experiment have been used for 3 years, namely N=3. Diagnostic results are shown in table 7.

Table 6. Diagnosis of time-point D and E.

Measurement points	Fault type	The number of short circuit cells
D	Short circuit	6
E	Short circuit	9

Table 7. Diagnosis of time-point F and G.

Measurement points	Fault type	Remaining useful life
F	Little aging	4.5
G	Serious aging	/

The experimental results show that the proposed method is effective and accurate. By using this method, we can make full use of real-time capability of the method based on power comparison and the accuracy of the diagnosis based on characteristic parameters. Experimental results show that the accuracy of this method is more than 90%. For erroneous judgments the possible reasons are as follows:

The measured data is limited by the precision of the sensor and the acquisition device, which leads to the errors between the input data of simulation model and the experimental data;

The maximum power point tracking of the PV module used in the experiment cannot track the maximum power point accurately, which leads to the error in the calculation of FF;

A variety of faults could occur at the same time during the working of PV module. The method proposed in this paper only uses a single criterion to diagnose a class of faults, which may result in a false diagnosis.

Compared with existing methods in literature, the method proposed in this paper has made the following improvements. First, compared with infrared picture analysis, this diagnosis result is not affect by the surrounding environment. Second, this method can detect three different types of faults. Finally, this method is more accurate and can reach more than 90%.

5. Conclusions

A real-time on-line fault diagnosis method based on power loss and the I-V output characteristics has been presented. We drew the following conclusion:

- Based on the comparison of the simulated power with the measured power of PV module, the power loss rate (K) can be obtained. By analyzing the value of K, it can be known whether abnormal power loss has occurred.
- If K is greater than 15% and the irradiation of the module is normal, it can determine whether short circuit has occurred by comparing the measured open circuit voltage and the simulated open circuit voltage.
- When a module has short circuit fault, the number of short circuited cells can be calculated on the basis of the open circuit voltage.
- The value of FF will decrease when aging fault occurs. If K is less than 0.7, it can be judged that the aging fault has occurred.
- When a module is short circuit, the degree of aging can be obtained by calculating the value of FF. Then the remaining useful life of PV module can be roughly estimated based on the years the module had been used for.

References

[1] Milligan M, Frew B, Schuerger B K M, Clark K, Lew D, Denholm P, Zavadil B, O'Malley M and Tsuchida B 2015 Alternatives no more: wind and solar power are mainstays of a clean, reliable, affordable grid *IEEE Power and Energy Magazine* **13** 78-87

[2] Wang P Z and Zheng S C 2010 Fault analysis of photovoltaic array based on infrared image *Acta Energiae Solaris Sinca* **3** 197-202

[3] Tsanakasa J A, Chrysostomoub D, Botsarisa P N and Gasteratosb A 2013 Fault diagnosis of photovoltaic modules through image processing and canny edge detection on field thermographic measurements *International Journal of Sustainable Energy* 1-22

[4] Tsai D M, Wu S C and Chiu W Y 2013 Defect detection in solar modules using ICA basis images *IEEE Transactions on Industrial Informatics* **9** 122-31

[5] Takumi T, Junji Y and Masayoshi I 2008 Disconnection detection using earth capacitance measurement in photovoltaic module string *Progress in Photovoltaics* **16** 669-77

[6] Takumi T, Junji Y and Masayoshi I. 2008 Fault detection by signal response in PV module strings *Conference Record of the IEEE Photovoltaic Specialists Conference* 1-5

[7] Hu Y H, Chen H and Xu R D 2011 Photovoltaic array fault diagnosis strategy based on optimal sensor placement *Proceedings of the CSEE* **31** 19-30

[8] Chouder A and Silvestre S 2010 Automatic supervision and fault detection of PV systems based on power losses analysis *Energy Conversion and Management* **51** 1929-37

[9] Wang Y Z, Li Z H and Wu C H Zhou D Q and Fu L 2013 A survey of online fault diagnosis for PV module based on BP neural network *Power System Technology* **37** 2094-100

[10] Bruno A, Salvatore B, Antonio P, Giuseppe M T and Cristina V 2015 Sentinella: Smart monitoring of photovoltaic systems at panel level *IEEE Transactions on Instrumentation and Measurement* **64** 2188-99

[11] Lan Q L and Wu Y C 2008 Reviews on the intelligent diagnostic system solar photovoltaic power generation *Journal of Wuhan University of Technology* **21** 23-6

[12] Chao K H and Wang M H 2007 Modeling and fault diagnosis of a photovoltaic system *Electric Power Systems Research* **78** 97-105

[13] Mao M Q, Yu S J and Su J H 2005 Versatile matlab simulation model for photovoltaic array with MPPT function *Journal of System Simulation* **17** 1248-51

[14] Takashima T, Yamaguchi J J and Otani K J 2009 Experimental studies of fault location in PV module strings *Solar Energy Materials and Solar Cells* **93** 1079-82

[15] Zhai Z T and Cheng X F 2008 Analytic solutions of solar cell model parameters *Acta Energiae Solaris Sinica* **30** 1078-82

[16] Ducange P, Fazzolari M and Lazzerini B 2011 An intelligent system for detecting faults in photovoltaic fields *ISDA Cordoba Argentina* 1341-6

[17] Zhai Z T, Cheng X F and Ding J L 2007 Determination of serial solar panels current equation based on maximum power *Proceedings of the CSEE* **27** 87-90

[18] Deceglie M G, Silverman T J, Marion B and Kurtz S R 2015 Real-time series resistance monitoring in PV systems without the need for I–V curves *IEEE Journal of Photovoltaics* **5** 1706-9

[19] Zhao Y, De Palma J F, Mosesian J, Lyons R and Lehman B 2012 Line–line fault analysis and protection challenges in solar photovoltaic arrays *IEEE Transactions on Industrial Electronics* **60** 3784-95

[20] Ropp M E and Gonzalez S 2009 Development of a MATLAB/simulink model of a single-phase grid-connected photovoltaic system *IEEE Transactions on Energy Conversion* **24** 195-202

[21] Zhu Y Q and Wang W S 2011 Fault diagnosis method and simulation analysis for photovoltaic array *ICECE Yichang China* 1569-73

[22] Villalva M G, Gazoli J R and Filho E R 2009 Comprehensive approach to modeling and simulation of photovoltaic arrays *IEEE Transactions on Power Electronics* **24** 1198-208

[23] Villa L F L, Picault D, Raison B, Bacha S and Labonne A 2012 Maximizing the power output of partially shaded photovoltaic plants through optimization of the interconnections among its modules *IEEE Journal of Photovoltaics* **2** 154-63

[24] Benghanem M and Maafi A 1998 Data acquisition system for photovoltaic systems performance monitoring *IEEE Transactions on Instrumentation and Measurement* **47** 30-3

[25] Hu Y H, Zhang J F, Cao W P, Wu J D, Tian G Y, Finney S J and Kirtley J L 2015 Online two-section PV array fault diagnosis with optimized voltage sensor locations *IEEE Transactions on Industrial Electronics* **62** 7237-46

Optimization design and compare of different solar-ground source heat pump system of office building in cold regions

H Xie, S He[1] and Y Fu

School of Mechanical Engineering, University of Science and Technology Beijing, Beijing, China

E-mail: hsdemi@163.com

Abstract. This paper presents two different operation modes of Solar-Ground Source Heat Pump System (SGSHP(S)). With the simulation tool TRNSYS, two different SGSHP system models were built to taking simulation. After making analysis and compare of different simulation results, series operation mode was believed to be better than parallel in the target building.

1. Introduction

Since the industry revolution, the shortage of energy and the deterioration of environment have been the two serious problems in this era. If we can take full use of earth or solar energy could reduce the consumption of primary energy source greatly in building. The solar-ground source heat pump system (SGSHP(S)) with ground heat exchanger being as heat storage can effectively overcome the declining efficiency and stability of system which caused by intermittent and unstable solar energy. The solar collector in system can apply solar energy to retrieve the soil temperature decrease caused by longtime operation of ground heat exchanger. So SGSHP system may be a great potential system contributing to our destination of sustainable development.

Research of solar-ground source heat pump system started from 1950s by Penrod[1][2]. After the energy crisis in 1977, some countries like American put a lot of research fund into research of ground source heat pump and solar-ground source heat pump system [3-5]. During 1979-1981, solar-ground source heat pump system has been simulated by many researchers (Bose J E, Klein, Parker, etc) [7-9]. In 1996, M Inalli applied Matlab to take calculations of soil temperature and researched the relationships between soil temperature and factors such as solar collector area [12]. In 2003, Andrew D.Chiasson and Cenk Yavuzturk used TRNSYS to take simulations of (SGSHP(S)) for the first time. They took 20 years as the operating period to research the law of soil temperature variation, system cost, etc. in their simulation [14]. In 2006, Trillat tried to establish a solar-ground source heat pump system and the result of it seems good with COP being higher than 3.75 [17]. In 2009, Kjellsson established different operation modes of (SGSHP(S)) to research their performance based on real projects [20, 23]. In 2013, FarzinM.Rad proved that (SGSHP(S)) is suitable for cold region, but there

[1] Address for correspondence: S He, School of Mechanical Engineering, University of Science and Technology Beijing, Beijing, China. E-mail: hsdemi@163.com.

exists a problem of imbalance soil temperature which should be solved by applying different operation modes [28].

SGSHP system has many kinds of operation mode, different modes lead to different COP and performance. This paper compared two kinds of main operation mode, in series and in parallel, to pick up the suitable mode for small office building in Beijing area.

(1) In series: heat-exchange fluid takes circular flow through the solar collector, ground heat exchanger, and finally the heat pump.

(2) In parallel: heat-exchange fluid flow through the solar collector and ground heat exchanger simultaneously, then converge into the heat pump.

Taking a small office building as target to research the SGSHP(S) optimization under different coupling-mode conditions based on the building type, features of loads, local climate, and solar/geothermal energy resources, this paper uses TRNSYS to simulate the operation of SGSHP(S) in two operation modes in consideration of the coupling relationship between system design and operation mode. Based on the simulation results, get the final conclusions.

2. Preliminary design of SGSHP(S)

2.1 Object building and load calculation

2.1.1. Introduction of object building and region

The object building researched in this paper is located in north third ring road of Beijing which belongs to cold region. Dry and cold in winter, rainy and hot in summer, along with short transition season, are the characters of this region. The object building is a new small office building with construction area of 4509 $m^2$and heating area of 3800m^2. The main building has four layers on the ground with height of 17.3m. The podium building has two layers on the ground with height of 8.05m. In main building, there are hall, conference room, control room and office, etc. on the ground floor; conference room and office, etc. on the first floor; office, etc. on the second floor; meeting room and office, etc. on the third floor. In podium building, the ground floor serves for refrigeration machine room and the first floor serves for multi-function room.

2.1.2. Load calculation

Energy simulation software DeST-C is applied to get the full-year heating load of this building as 238834.88kWh and cooling load as 155118.74 kWh. Because the type and characters of office building, it is required to keep 5℃ indoor in winter night but no need of air conditioning in summer night. As the result of this operation mode, the annual accumulative cooling load takes only 65% of heating load. Based on the calculation results of load, we take the following indicators to make preliminary design of air conditioning system in this building: maximum heating load as 250kW and maximum cooling load as 210kW.

2.2 Equipment selection and TRNSYS model building

TRNSYS simulation software is a dynamic simulation program based on modules. A complete system is composed of many small modules with different functions. Simulation of the system can be performed by invoking modules required and inputting the required parameters [21, 24, 26].

We choose the Flat-plate solar collector (Type 1 in TRNSYS) with liquid propellant working medium to collect solar energy which exchange heat indirectly (According to *Technical code for solar heating system GB 50495-2009*) [11]. It will be put on the roof of main building and podium building, facing south with fixed angle of 50°. We choose heat storage water tank (Type 4) as the seasonal heat storage facility.

Fan coil plus fresh air system is used as the end form of system combined with consideration of the building type [16, 25].

Comprehensively consider about the climate condition, geological condition and useable area of this project, we select the ground heat exchanger with single U-shaped vertical buried tube (Type 557) instead of horizontal buried tube to take the heat exchange work of ground in order to get a better performance on heat-exchange. Because the vertical buried tube can get more stable heat-exchange progress away from the influence of environment. After taking an estimated calculation and empirical estimation, we decided that tubes to be buried 100 meters deep with 5 meters far from each other. And external diameter of the buried tubes is 32mm.

According to the calculation results of load, we choose water -water single-stage pumps (Type927) to carry the loads [16, 25]. There are two kinds of control signal of this module which controls heating and cooling. The heating control is prior to the cooling control. TRNSYS will active the subroutine of heating and power consuming automatically when the heat pump starts to work.

To avoid frequent start-stop problems of water pump and heat pump, this project use the differential controller (Type2) to control temperature. To take control of time, we choose the time step function module (Type14). And we also take the Manifold/ Siamese module (Type11) to help finish our work.

3. Establishment of series and parallel modes of SGSHP(S)

There are many kinds of operating modes of Solar—Ground Source heat pump system. From the point of system connection, series and parallel connection types are mainly be used. In series connection mode, heat-exchange fluid takes circular flow through the solar collector, ground heat exchanger, and finally the heat pump. In parallel connection mode, heat-exchange fluid flow through the solar collector and ground heat exchanger simultaneously, then converge into the heat pump.

3.1 Series operation mode

In series operation mode: heat-exchange fluid takes circular flow through the solar collector, ground heat exchanger, and finally the heat pump (Shown as Figure 1). Under series operation mode, solar energy can be stored into ground heat accumulator to keep balance in absorbing and releasing of soil heat when the solar energy being sufficient.

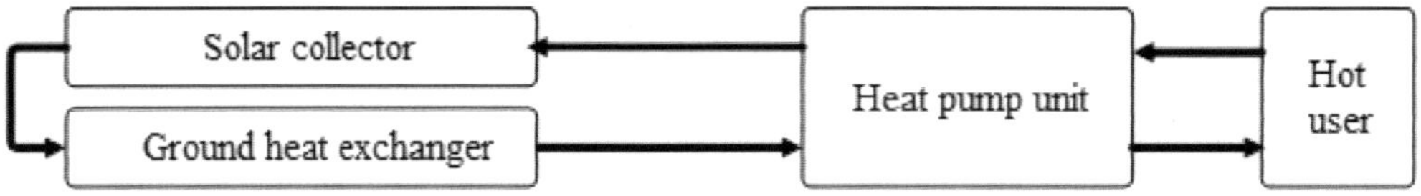

Figure 1 Function diagram of series mode

3.1.1. Operating conditions in different seasons

(1) In winter

Heat-exchange fluid flow through the solar collector first, then ground heat exchanger, and finally provides heat for heat pump as heat source. (In the right parts of the following Figure 2, takes the red lines marked No.1 in winter instead of the blue lines marked No.2.)

(2) In summer

Solar collector only absorbs heat in summer so it is always being closed during summer. The ground heat exchanger plays the role of the only cold source for heat pump. The left side parts of TANK (Founded in the following Figure 2) will be closed in summer.(In the right parts of the following Figure 2, takes the red lines marked No.1 in summer instead of the blue lines marked No.2.)

(3) In transition seasons

Heat load in transition seasons is small so it can be removed off only by the output water of ground heat exchanger without opening heat pump. The left side parts of TANK (Founded in the following Figure 2) will be closed in transition seasons. (In the right parts of the following Figure 2, takes the blue lines marked No.2 in transition seasons instead of the red line marked No.1.)

3.1.2. Main module connection of series operation mode

The main modules in TRNSYS as well as their connection form in series operation mode are shown in Figure 2. The main used modules are as below: meteorological parameter module(Type15); simple calculation module of soil temperature(Type77); flat-plate solar collector module(Type1); single-stage water pump module (Type3); convection heat exchanger module (Type5); heat storage water tank module (Type4); ground heat exchanger of single U-shaped buried tube module (Type557); Manifold/Siamese module (Type11); water-water source heat pump unit module (Type927); calculation module; date file reader (Type9).From the illustration diagram of series mode constructed in TRNSYS, there are five cycles in the system operating process yearly: first cycle of solar collection, second cycle of solar collection, buried tubes-heat pump cycle, heat pump-user cycle, buried tubes-user cycle.

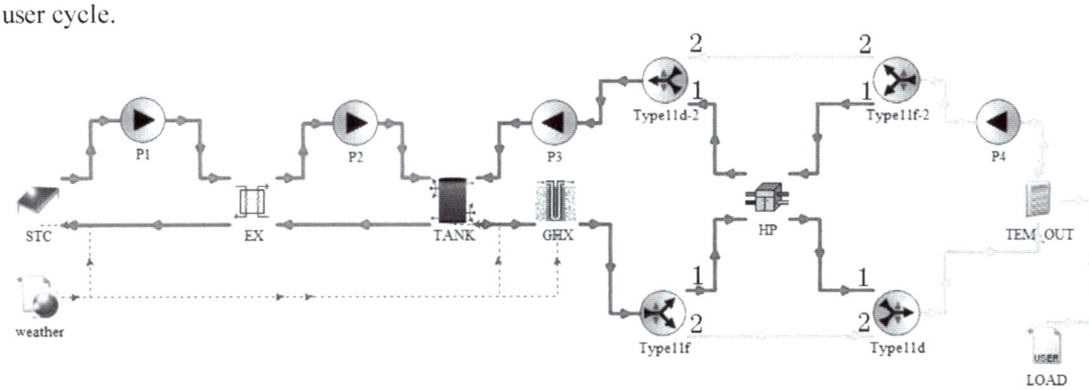

Notes: Solid lines in figure represent fluid cycle; Dotted lines in figure represent the input of external weather conditions.

Figure 2 Illustration diagram of series mode constructed in TRNSYS

3.2 Parallel operation mode

In parallel operation mode: heat-exchange fluid flow through the solar collector and ground heat exchanger simultaneously, then mixes in Siamese and flow into the heat pump finally (Shown as Figure 3). In the daytime with sufficient solar energy, we can increase the flow of heat-exchange fluid in solar collector if the water temperature from the output of solar collector is high. It can reduce the heat load carried by ground heat system which does benefits for the soil temperature maintenance. Conversely, in the operating time without sun, we can close the solar collector system to improve the operating efficiency of the whole system.

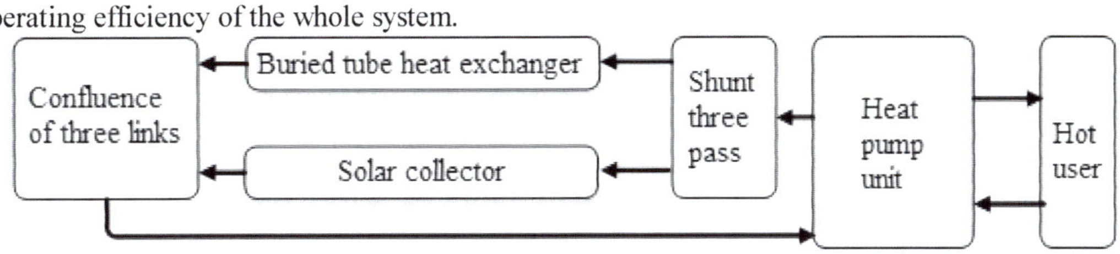

Figure 3 Function diagram of parallel mode

3.2.1. Operating conditions in different seasons

(1) In winter

When the temperature of output water of solar collector higher than that of ground heat exchanger, heat-exchange fluid flow through the solar collector and ground heat exchanger simultaneously, and finally provides heat for heat pump as heat source (In the right parts of the following Figure 4, takes the red lines marked No.1 in winter instead of the blue lines marked No.2.). When the temperature of output water of solar collector lower than that of ground heat exchanger, solar collector will be shut down, heat-exchange fluid flow through the ground heat exchanger only, and finally provides heat for heat pump as heat source (In the right parts of the following Figure 4, takes the red lines marked No.1 in winter instead of the blue lines marked No.2. In the left parts of the following Figure 3.4, the purple red lines marked No.2 will be closed in this case).

(2) In summer

Parallel operation mode in summer is as same as the series mode (refer to *3.1.1*). Solar collector will be shut down. All the left side parts represented by purple red lines (Founded in the following Figure 4) will be closed in summer. (In the right parts of the following Figure 4, takes the red lines marked No.1 in summer instead of the blue lines marked No.2.)

(3) In transition seasons

Parallel operation mode in transition seasons is as same as the series mode (refer to *3.1.1*). All the left side parts represented by purple red lines (Founded in the following Figure 4) will be closed in transition seasons. (In the right parts of the following Figure 4, takes the blue lines marked No.2 in transition seasons instead of the red lines marked No.1.)

3.2.2. The parallel operation mode main module connection

The main modules in TRNSYS as well as their connection form in parallel operation mode are shown in Figure 4. The main modules and five operating process cycles in parallel mode are the same as those in series mode.

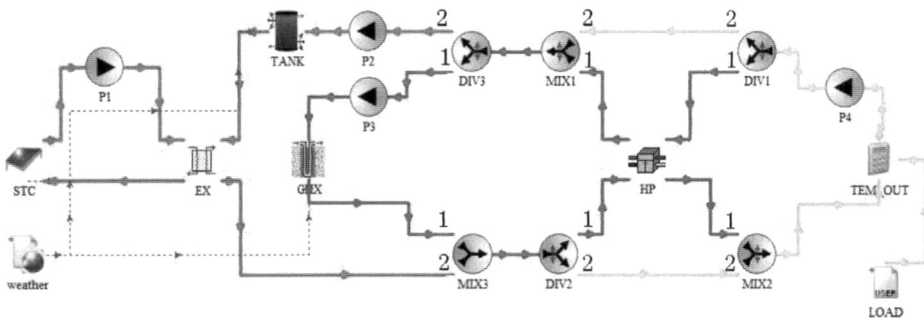

Notes: Solid lines in figure represent fluid cycle; Dotted lines in figure represent the input of external weather conditions.

Figure 4 Illustration diagram of parallel mode constructed in TRNSYS

4. Indicators and variables

In order to get the optimal operation mode of SGSHP(S), this section takes simulations of two different operation modes in series and in parallel connection separately based on the fore part. In each mode, we will evaluate the system performance from two indicators (accumulated system operation COP and temperature of soil) by changing the area of the solar thermal collector and the number of vertical ground heat exchanger tubes. Besides of these two indicators, we also take an economic indicator called LCC (Life Cycle Cost) into consideration. Finally, we will conclude the optimal operation mode by comparing the results.

4.1 Accumulated system operation COP

COP (Coefficient of Performance) refers to the ratio of operational heat (cold) load and operational input power. Accumulated system operation COP in this paper refers to the ratio of accumulated operational heat (cold) load and operational input power from the system beginning to the time point of making calculation.

$$Accumulated\ system\ operation\ COP = \frac{Operational\ Heat(Cold)\ Load(accumulated\ from beginning\ to\ end)}{Operational\ Input\ Power(accumulated\ from beginning\ to\ end)}$$

4.2 Average temperature of the ground heat accumulator

Ground heat accumulator refers to the cylindrical soil at the place of vertical ground heat exchanger including the tubes. Temperature can be considered as a uniformly distribution inside the ground heat accumulator but being as a radial-gradient distribution outside due to the effect of ground heat exchange. Computational method of the ground heat accumulator's volume shows as below:

Volume of ground heat accumulator=π×number of tubes×depth of tubes×(0.525×distance between tubes)^2 (1)

In the equation:

Depth of tubes——100m;

Distance between tubes——5m;

The average temperature of the ground heat accumulator derives from the output 3 of module Type557 in TRNSYS.

4.3 Life cycle cost LCC

LCC means the whole cost during the whole life of an equipment or a system. Initial investment (IC), operation cost (OC) and disposal cost (DC) are all included. The mathematical model of LCC in this system would be expressed as below:

$$LCC = IC + \sum_{k=1}^{n} OC(1+i)^{-k} + DC(1+i)^{-n} \quad (2)$$

In the equation:

IC——Initial investment (equipment fee, installation fee, bore-hole fee)

OC——Operation cost (maintenance and operating fee)

DC——Disposal cost(ignored in this case because of only 20-year operating simulation)

k——Time of equipment operating (year)

n——Economic life period of system (year)

i——Discount rate (estimated as 8%)

Maintenance fee (MF) here is estimated as 10% of depreciation expense (DE)

$$MF = 10\% \times DE \quad (3)$$

$$DE = IC \times ADR \quad (4)$$

$$ADR = [(1 - ENRV)/DL] \times 100\% \quad (5)$$

In the equations:

ADR——Annual depreciation rate

ENRV——Estimated net residual value (estimated as 4%)

DL——Depreciable life (20 years in this case)

Economic indicator LCC is a subsidiary indicator in this paper to support the analysis, so we do not expand introduction of detailed specific calculation values here.

4.4 Variables settings

Optimal design of the SGSHP(S) combined with the operation mode analysis is the final purpose of this research. Based on the overseas study work, area of the solar thermal collector and length of vertical ground heat exchanger as well as their coupled relationship exert the most influence on the performance of system. Now being convenient for calculation, we take the area of solar thermal collector and the number of vertical ground heat exchanger tubes as variables to make optimal system design. The number of vertical ground heat exchanger tubes ranges from 40 to 80 at step of one tube. The area of solar thermal collector changes to make sure that solar fraction ranges from 0% to 50% at step of ten percents.

Table 1 Solar collector system variables settings

Solar fraction (%)	0	10	20	30	40	50
Area of solar thermal collector (m^2)	0	163	325	490	650	815
Volume of water tank (m^3)	0	8	16	24	33	41
Area of plate heat exchanger (m^2)	0	1.80	3.59	5.39	7.18	8.98
Flow of cycle pump (m^3/h)	0	1.63	3.30	4.90	6.50	8.15

5. Results and discussion

In this section, we will take twenty-year simulations of SGSHP(S) in series and parallel connection modes separately by TRNSYS. Based on the results of accumulated system operation COP and temperature change of soil (two kinds of results: after 20-year operating period and during 20-year operating period), we make the analysis and discussion.

5.1 Analysis of series operation mode

In this section, we will make simulations to figure out the accumulated system operation COP of the SGSHP(S) and soil temperature after operating system in series mode for 20 years, results from TRNSYS shown as below.

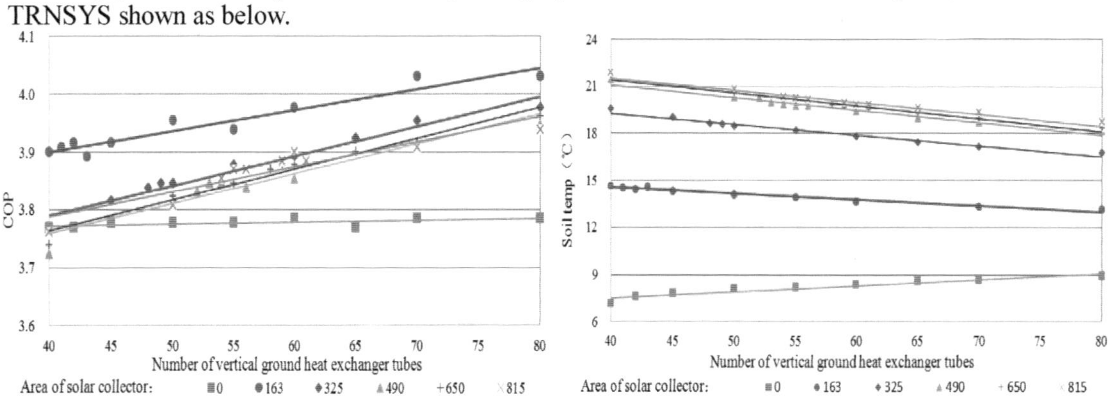

Figure 5. (a)Accumulated system operation COP Figure 5. (b)Soil temperature

Figure 5 Accumulated system operation COP and soil temperature after 20-year operating in series mode

From the Figure 5.(a) we can see that functioned solar collector in system can improve the COP up to 7% under the chosen condition of solar collector area and ground heat exchanger tubes, but increase the area of solar thermal collector larger than 163m^2 does not bring improvement of the accumulated system operation COP in series operation mode efficiently for a longtime. So, 163m^2 would be the most preferable area of solar collector to get the peak accumulated COP. To make analysis of it, if the area of solar collector being larger than 163m^2, the temperature of heat-exchange water must be higher than that of 163m^2 which may influence the soil temperature more heavily. And the increase of soil temperature with increasing area of solar collector has been proved in Figure 5. (b). If the soil temperature being higher than that one at area of solar collector being 163m^2, the effect on cooling process may be stronger than that on heating process throughout the whole year which will result in the decrease of system COP. Another conclusion from Figure 5. (a), accumulated system operation COP observed to rise in a long operating time with increasing number of ground heat exchanger tubes.

The average soil temperature of Beijing is 13.1℃. We can see that the soil temperature increases with the tubes number increasing without solar collector. It can also be considered as that the soil temperature decreases not so much compared to the average soil temperature of Beijing with the increasing tubes number. To make analysis of it, we should note that the heating load is far more than cooling load which has been calculated before. So the soil temperature will decrease in a long time without solar energy supply due to the difference between the releasing heat in summer and absorbing heat in winter. With the increasing number of ground heat exchanger tubes, the tubes become more and more crowded, so the heat cannot be released easily which will result in the increase of soil temperature (Same reason for Figure 9. (b)). And from the Figure 5. (b) we can conclude that increasing the area of solar collector will improve the soil temperature. But it will not cause an obviously ascending of the average temperature of ground heat accumulator in a longtime when the solar collector being larger than 490m^2. It may be due to the limitation of the heat conduction ability of soil.

Another conclusion from Figure 5. (b), amplitude of temperature variation reduces gradually with increasing number of ground heat exchanger tubes because that more tubes lead to larger volume of ground heat accumulator. And this conclusion may also be the reason for the rising of accumulated system operation COP in long operating time which mentioned before.

Taking economic factors into consideration, the LCC of system after 20-year operating in series is shown as below. From the figure, we can see that the LCC of system without solar collector is higher than that with solar collector area in 163m^2.The key reason of it is OC. OC of system without solar collector is higher due to the annually decrease of soil temperature which result from the lack of solar heat complement. By observing the variation of LCC in different solar collector areas, 60 seem to be a key number which should be noted to select the optimal number of vertical ground heat exchanger tubes.

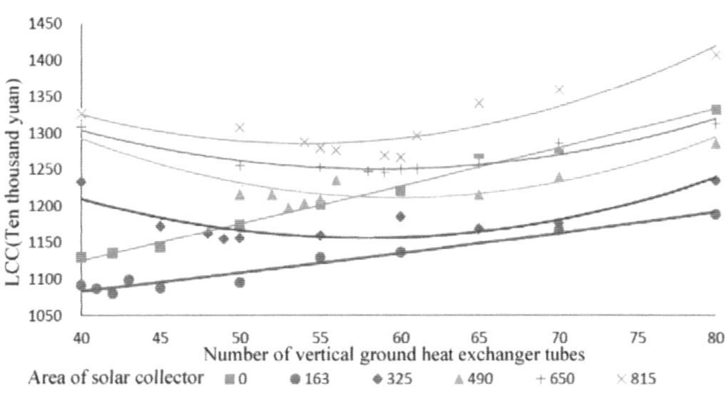

Figure 6 LCC of system after 20-year operating in series

On the basis of above analysis, area of solar thermal collector ranges from 0 to 325m^{2}with the number of vertical ground heat exchanger tubes ranges from 40 to 60 may be considered as suitable choice for the system design. We will following choose 9 groups of variable settings which vary in different solar collector areas in 0m^2, 163m^2, 325m^2 along with different numbers of vertical ground heat exchanger tubes of 40, 50, 60 to take simulations and observe the results of accumulated system operation COP as well as change of soil temperature in 20-year period.

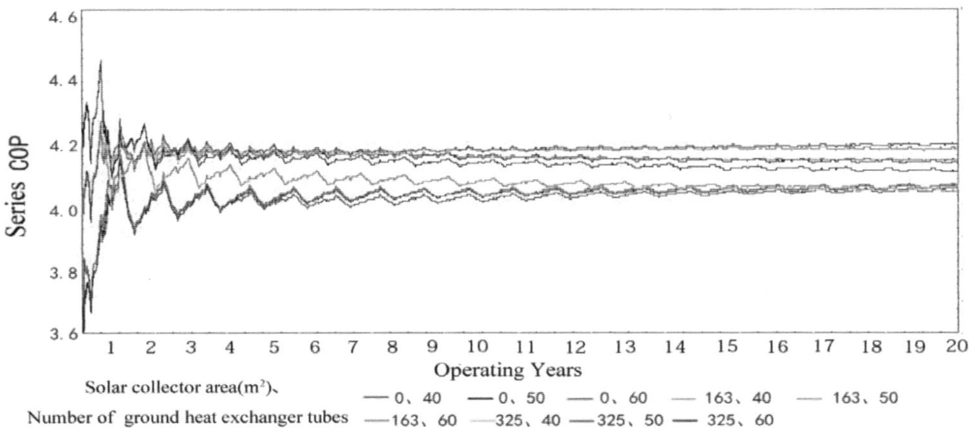

Figure 7 Fluctuation of accumulated system operation COP in series during 20-year operating period

In Figure 7, the accumulated system operation COP seems to change periodically by year which being lower in summer and winter. And in transition seasons, it will be higher due to the direct using of cycle water in ground heat exchanger tubes without operating the heat pump.

Operating in series connection mode, the accumulated system operation COP observed to be less fluctuant during the operating period and more stable after 20 years while the solar collector area being 163m^2. In the same time, we can see that the accumulated system operation COP being increased with increasing number of vertical ground heat exchanger tubes.

Notably the solar collector area being 325m^2 with the number of vertical ground heat exchanger tubes being40, accumulated system operation COP decreases sharply. It is probably resulted from significantly increase of soil temperature due to the excess solar heat input. If the solar heat input so much, the temperature of heat-exchange fluid throughout the heat exchanger tubes will be lifted up and this may result in the increase of soil temperature. This presage will be validated in the following part of the analysis about soil temperature variation. If the soil temperature increases, the useable temperature difference for heat pump system will be reduced in summer, it will decrease the COP of system in summer. And due to the releasing heat in summer is more than absorbing heat in winter based on the analysis before, COP of system throughout the whole year must be decreased for the increasing soil temperature.

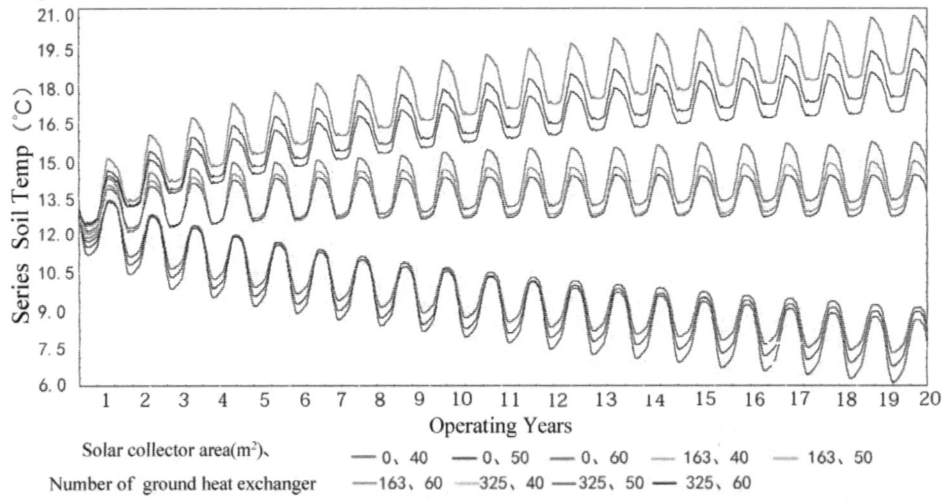

Figure 8 Variation of soil temperature in series during 20-year operating period

In Figure 8, the soil temperature seems to change periodically by year and being less fluctuant during the operating period while the solar collector area being 163m^2. If we do not operate solar system, soil temperature will decrease each year. If we choose the solar collector area as 325m^2, soil temperature will increases year by year owing to the overheat from sun. Especially with the number of ground heat exchanger tubes as 40, soil temperature increases sharply which can prove the presage of accumulated COP mentioned above.

We can also conclude from Figure 5.4 that soil temperature tends to be more stable when the number of ground heat exchanger tubes increase. It would also be a result of growing volume of ground heat accumulator.

5.2 Analysis of parallel operation mode

In this section, we will make simulations to figure out the accumulated system operation COP of the SGSHP(S) and soil temperature after operating system in parallel mode for 20 years, shown as below.

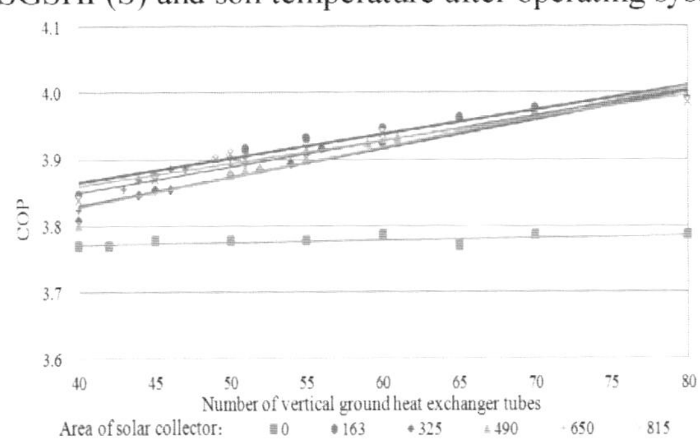

Figure 9. (a)Accumulated system operation COP

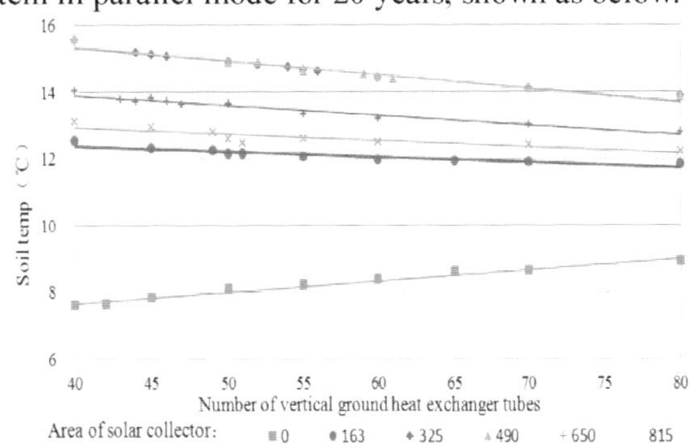

Figure 9. (b)Soil temperature

Figure 9 Accumulated system operation COP and soil temperature after 20-year operating in parallel mode

From the Figure 9. (a) we can see that functioned solar collector in system can improve the COP up to 5.8% under the chosen condition of solar collector area and ground heat exchanger tubes, but increasing the area of solar thermal collector does not result in improvement of the accumulated system operation COP in parallel operation model obviously for a longtime. But 163m^2 still be the most preferable area of solar collector to get the peak accumulated COP which can be observed.

Same with conclusion from series, accumulated system operation COP observed to raise in a long operating time with increasing number of ground heat exchanger tubes. And the values of accumulated system operation COP in different solar collector areas tend to focus on 4.02 finally.

The average soil temperature of Beijing is 13.1℃. From the Figure 9. (b), we get different conclusions from series mode. The soil temperature increases sharply with the solar collector area change from 0 to 325m^2, but increases no more than the peak amplitude (solar collector area in 325m^2) with the solar collector area increases from 490m^2 to 815m^2. The amplitude of calescence becomes smaller with the increasing area of solar collector. To make analysis of it, the reason should be the flow difference from series mode. Due to the increasing area of solar collector, the heat-exchange flow in solar collector will increase with the heat-exchange flow in ground heat exchanger decrease as a result. From the area 0 to 325m^2, the effect of rising temperature of heat-exchange fluid in ground heat exchanger tubes on soil temperature increasing is stronger than that effect of heat-exchange fluid decreasing in ground heat exchanger tubes. So the amplitude of calescence increases with the increasing area of solar collector. From the area 325m^2 to 490m^2, the effects on soil temperature variation from these two reasons may be offset with the increasing area of solar collector. Larger than 490m^2, the effect of heat-exchange fluid decreasing in ground heat exchanger tubes on soil temperature variation may be stronger, although the temperature of heat-exchange fluid is being rising all the time. So the tendency of soil temperature variation becomes weak with the area of solar collector increasing larger than 490m^2.

Same with conclusion from series, amplitude of temperature variation reduces gradually with increasing number of ground heat exchanger tubes because that more tubes lead to larger volume of ground heat accumulator. And this conclusion may also be the reason for the rising of accumulated system operation COP in long operating time which mentioned before.

Taking economic factors into consideration, the LCC of system after 20-year operating in parallel is shown as below. From the figure, we can see the same phenomenon that the LCC of system without solar collector is higher than that with solar collector area in 163m^2 which due to the same reason as analyzed before in serious. By observing the variation of LCC in different solar collector areas, there seems no key number which should be noted to select the optimal number of vertical ground heat exchanger tubes that differs from serious mode. But we can see the LCC of system seems to increase more quickly after the number 55 of ground heat exchanger tubes. Considering of being matched with serious mode, we also choose the number 60 to be the maximum tubes number to make the following simulations.

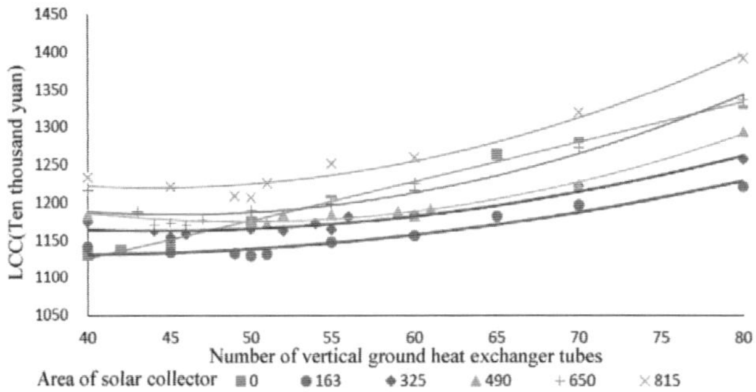

Figure 10 LCC of system after 20-year operating in parallel

On the basis of above analysis, area of solar thermal collector ranges from 0 to 325m^2 with the number of vertical ground heat exchanger tubes ranges from 40 to 60 may be considered as suitable choice for the system design. We will following choose 9 groups of variable settings which vary in different solar collector areas in 0m^2, 163m^2, 325m^2 along with different numbers of vertical ground heat exchangers tubes of 40, 50, 60 to take simulations and observe the results of accumulated system operation COP as well as change of temperature of soil in 20-year period.

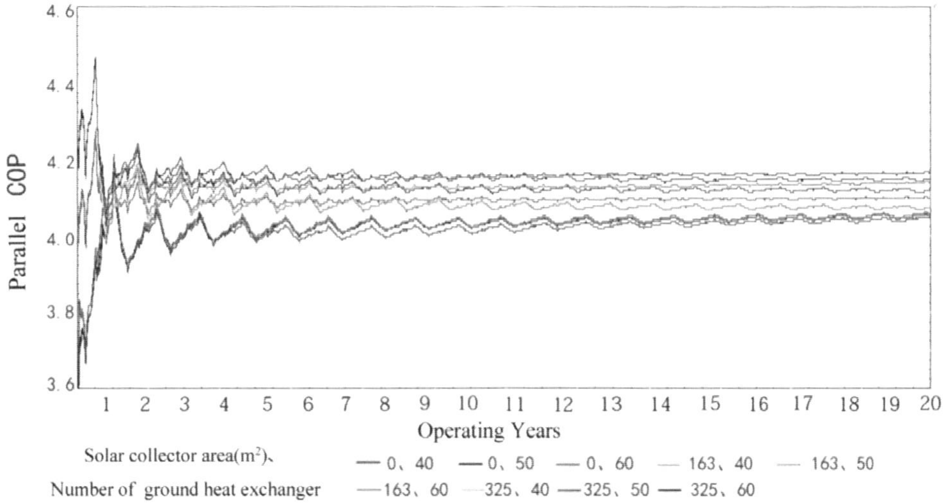

Figure 11 Fluctuation of accumulated system operation COP in parallel during 20-year operating period

In Figure 11, the accumulated system operation COP seems to change periodically by year which being lower in summer and winter. And in transition seasons, it will be higher due to the direct using of cycle water in ground heat exchanger tubes without operating the heat pump.

Operating in parallel connection mode, the accumulated system operation COP observed to be less fluctuant during the operating period and more stable after 20 years while the solar collector area being 163m^2. In the same time, we can see that the accumulated system operation COP increased with increasing number of vertical ground heat exchanger tubes.

Notably the same phenomenon as series connection mode, accumulated system operation COP decreases sharply under the condition of solar collector area being 325m^2 with the number of vertical ground heat exchanger tubes being 40. The reason of it is the same as series mode and can be validated in the following part of the analysis about soil temperature variation.

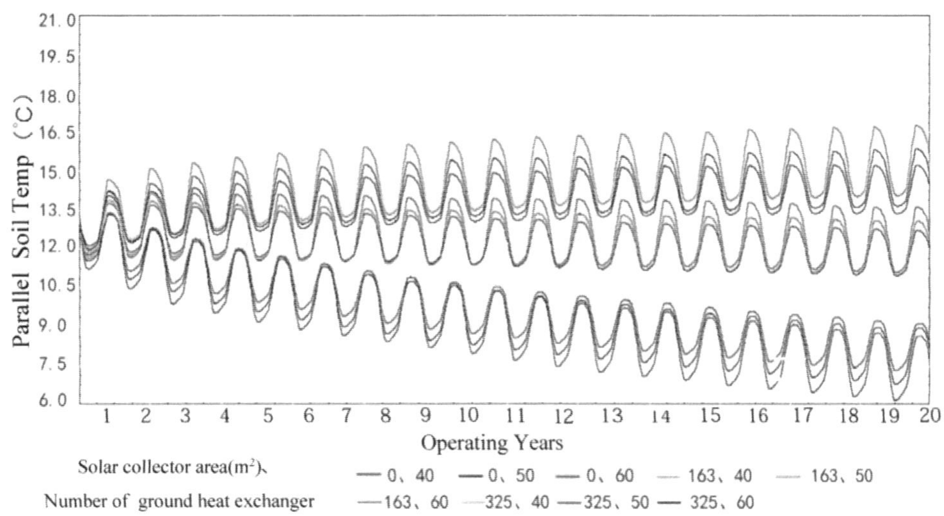

Figure 12 Variation of soil temperature in parallel during 20-year operating period

In Figure 12, the soil temperature seems to change periodically by year and being less fluctuant during the operating period while the solar collector area being 163m^2. Its variation pattern is the same as that in series. But different from series mode, the soil temperature tends to be more stable than that in series because that there is no direct heat exchange between solar collector system and ground heat exchanger system.

5.3 Discussion and compare of series and parallel operation modes.

Based on an overall analysis and conclusion above, 163m^2 should be the most preferable solar collector area in Beijing for both of series and parallel mode from all aspects of COP, soil temperature and LCC of system. Peak COP of system, lowest increase of soil temperature, most stable soil temperature variation and smallest cost of system, all of these will be realized in the choice of 163m^2 compared to the other area choices.

And the best number of vertical ground heat exchanger tubes in series would be 42 while 50 in parallel. It can be find the proof from the economic analysis. In the analysis of COP and soil temperature, system perform better with the increasing number of the ground heat exchanger tubes, so we get two optimal choices of 42 in serious while 50 in parallel from the aspect of economy.

So we choose these two competitive modes in optimal area of solar collector and number of ground heat exchanger tubes to take simulations and figure out the optimal mode finally. The simulation results are as below:

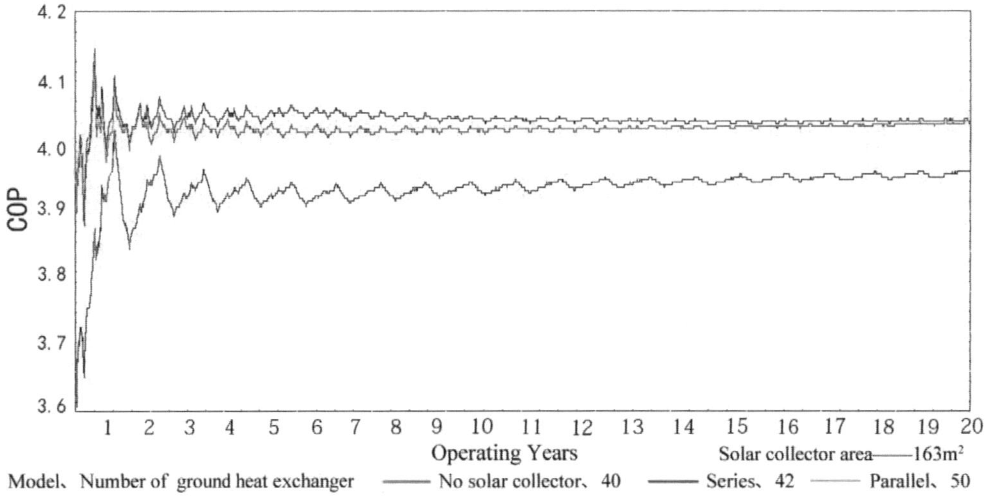

Figure 13 Compare of accumulated system operation COP in series and parallel during 20-year operating period

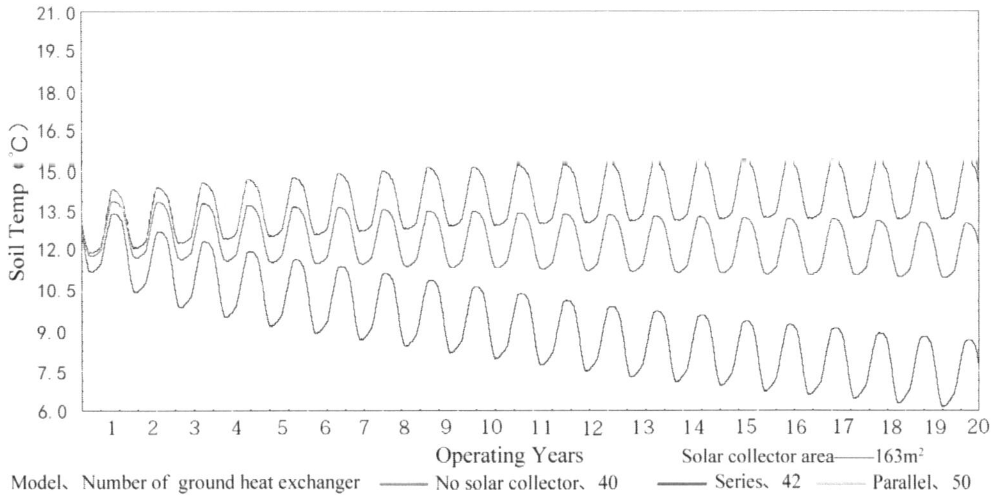

Figure 14 Compare of soil temperature in series and parallel during 20-year operating period

From the Figure 13, we can see the accumulated system operation COP in series is almost 5% higher than that in parallel in the first ten years. But in the later ten years, two of them come closer to each other.

From the Figure 14, we can see that the soil temperature in both of these two modes could make thermal equilibrium with environment. In series mode, soil temperature increases about 1℃ than the average soil temperature in Beijing. In parallel mode, soil temperature decreases about 1℃ than the average soil temperature.

Taking all of these conclusions into consideration, series mode performs better than parallel mode that would be the suitable mode of SGSHP(S) for the type of small office building in cold region which this paper researched in.

6. Conclusions

Two modes of SGSHP system, in series and parallel connection, were modeled and simulated using TRNSYS software in this paper. We took the area of solar collector and number of ground heat exchanger tubes as the variables to make 20-year simulations to get the following conclusions.

- Heat pump system with only heat (cold) source based geothermal energy is not suitable for the office building in Beijing area. It is necessary to take some actions to ensure the recovery of ground heat resource, or the soil temperature would decrease after system taking a longtime operation.
- In series connection mode, accumulated system operation COP increases first and then decreases while soil temperature increases all the time with the increasing area of solar collector. And both of accumulated system operation COP and soil temperature vary linearly with the increasing number of ground heat exchanger tubes.
- In parallel connection mode, there is no apparent change in accumulated system operation COP when the solar collector area increases with the soil temperature increasing first and then decreases. And both of accumulated system operation COP and soil temperature increase linearly with the increasing number of ground heat exchanger tubes.
- For all of the SGSHP system in office building in Beijing area, solar collector system should be subsidiary system in the whole system with appropriate solar fraction on 10%.

- Taking all of the indicators into consideration, the series mode performs better than parallel mode with the solar collector area in 163m^2 (10%solar fraction) and 42 ground heat exchanger tubes under the condition of office building in Beijing.

Acknowledgment
The work presented here is funded by School of Mechanical Engineering, University of Science and Technology Beijing.

References

[1] Penrod E B and Prasanna K V 1962 Design of a flat-Plate collector for a solar-earth heatpump *Solar Energy* **6** 9-22
[2] Penrod E B and Prasanna K V 1969 A procedure for designing solar-earth heat pumps *Heating, Piping & Air Conditioning* **41** 97-100
[3] Metz P D 1978 The potential for ground coupled storage with series solar assisted heat pump system *Proc. of the 1978 Annual Meeting. America Section of the International Solar Energy Society, Inc.* 331-4
[4] Metz P D 1978 Development of avalidated model of ground-coupling *Proc. the 1978 annual Section of the International Solar Society, Inc.* 344-8
[5] Metz P D 1982 The use of ground couple tanks in solar assisted heat pump system transaction of ASME *J. Solar Energy Engineering* **104** 366-72
[6] Andrews J W, Kush E A and Metz P D 1978 *Solar-assisted heat pump system for cost-effective space heating and cooling* (New York: Brookhaven NationalLaboratory) 138
[7] Parker J D and Frierson B 1981 *Performance monitoring of ground-coupled solar-assisted heat pump systems* (Stillwater: Oklahoma State University) 37
[8] Bose J E, Ledbetter C W and Partin J R 1979 Construction and evaluation of a heal pump system using earth coil *Proc. of the 3rd Annual Heat Pump Technology Conference* (Stillwater: OklahomaState University)
[9] Bose J E, Ledbetter C W and Partin J R 1980 Earth coupled and solar assisted heat pump system *Proc. of the 5th Annual Heat Pump Technology Conference* (Stillwater: Oklahoma StateUniversity)
[10] Lund P D and Ostman M B 1985 A numerical model for seasonal storage of solar heat in the ground by vertical pipes *Solar Energy* **34** 351-66
[11] Ministry of Construction of the People's Republc of China 1993 *Thermal design code for civil building* (Beijing: China Planning Press)
[12] Inalli M, Ünsal M and Tanyildizi V 1997 A computational model of a domestic solar heating system with underground spherical thermal storage *Energy* **22** 1163-72
[13] Yumrutas R and Ünsal M 2000 Analysis of solar aided heat pump systems with seasonal thermal energy storage in surface tanks *Energy* **25** 1231–43
[14] Chiasson A D and Yavuzturk C 2003 Assessment of the viability of hybrid geothermal heat pump systems with solar thermal collector *ASHRAE Transactions* **109** 487-500
[15] Qu Y X, Fang Z H, Zhang L H and Li A G 2003 The economic analysis of ground-coupled heat pump by using the solar energy as an assisted source *Renewable Energy* **01** 8-10
[16] Ministry of Construction of the People's Republc of China 2005 *Design standard for energy efficiency of public buildings* (Beijing: China Architecture&Building Press)
[17] Trillat-Berdal V, Souyri B and Fraisse G 2006 Experimental study of aground-coupled heat pump combined with thermal solar collectors *Energy and Buildings* **38** 1477-84
[18] Wu X H 2008 *Research and simulation analysis of combing solar collector and ground source heat pump heating system* (China: Jilin University)
[19] Rad F M, Fung A S and Leong W H 2009 Combined solar thermal and ground source heat pump system *Eleventh International IBPSAC onference* 2297-395
[20] Kjellsson E 2009 *Solar collectors combined with ground-sourceheat pumps in dwellings*

(Sweden: University of Lund)

[21] Guan Q L 2009 *Study on solar-ground coupled heat pump with seasonal storage by TRNSYS* (China: Tianjin University)

[22] Ministry of Construction of the People's Republc of China 2009 *Technical code for solar heating system* (Beijing: China Architecture&Building Press)

[23] Kjellsson E, Hellström G and Perers B 2010 Optimization of systems with the combinationof ground-source heat pump and solar collectors in dwellings *Energy* **35** 2667-73

[24] Qi H B 2012 Design analysis of a ground-coupled heat exchanger system based on trnsys simulation *Power Generation&Air condition* **02** 57-60

[25] China Academy of Building Research 2012 *Design code for heating ventilation and air conditioning of civil buildings* (Beijing: China Architecture&Building Press)

[26] Yin L Y 2013 *Simulation study on TRNSYS based solar-earth source heat pump system* (China: Taiyuan University of Technology)

[27] Garber D, Choudhary R and Leong K S 2013 Risk based lifetime costs assessment of a ground source heat pump (GSHP) systemdesign: Methodology and case study *Building and Environment* **60** 66-80

[28] Rad F M, Fung A S and Leong W H 2013 Feasibility of combined solar thermal and ground source heat pump systemsin cold climate, Canada *Energy and Buildings* **61** 224-32

The method of determining the insulation parameters in three-phase electrical networks with isolated neutral with voltages up to and above 1000 V

B B Utegulov[1]

Energy Department, S. Seifullin Kazakh Agro-Technical University, 62 Pobeda avenue, Astana, Republic of Kazakhstan

E-mail: bolatu@mail.ru

Abstract. In work the new method of definition of isolation's parameters in a three-phase symmetric electrical network with isolated neutral which is based on measurement of modules sizes of a linear voltage, voltages of phases C and A concerning to ground, after connection of active additional conductivity between phase A of electrical network and ground is shown and the analysis of a blunder of the developed method is made. The analysis of a blunder has shown, that the developed method provides satisfactory accuracy at definition of isolation's parameters, and also simplicity and the safety of operations in working electro installations of up to above 1000 V voltage.

1. Introduction

The implementation of modern technological processes in quarries, which are directly related to the growth of unit capacity of stripping and mining machines, are making increasingly stringent requirements to ensure the safety and security of the operation of electrical system [1, 2].

However, the increase in the length of electrical networks, which feed stripping and mining machines, increases the likelihood of single-phase ground fault, which, as a rule, are the main cause of the interruption of power supply. Relay protection and automation actions allow maintaining the continuity of power supply only if it is possible to control periodic insulation parameters mains phases with respect to ground.

Among the range of issues related to ensuring security of supply of electric power electrified mining equipment and the safety of its operation there is a development of methodology for determining the insulation parameters, which occupies a special place, as the results of the use of the method used are derived the main provisions of organizational and technical measures that promote a culture of service of the internal power supply of mining enterprises. The importance of developing the method to determine the insulation value is also determined by the fact that it can be used in other industries where there is three-phase electrical networks with isolated neutral with voltages up to and above 1000 V.

For experimental studies of the state of insulation of three-phase electrical networks with isolated neutral with voltages up to and above 1000, was proposed a number of methods [1, 2, 4 ÷ 5] with taking into account the specific characteristics of inherent in the internal power supply of open cast

[1] Address for correspondence: Energy Department, S. Seifullin Kazakh Agro-Technical University, 62 Pobeda avenue, Astana, Republic of Kazakhstan. Email: bolatu@mail.ru.

mining. For the insulation parameters measuring methods a number of requirements are presented, namely:

- Measurements should be carried out without interruption in the supply of electricity to consumers.
- The process of measurement should not cause damage to the insulation of electrical networks and electrical accidents.
- The measurements must be carried out using a small amount of electrical equipment and appliances.
- Execution of determining the insulation parameters should be safe both for researchers and for personnel servicing electrical systems.
- Measurements of baseline values should be sufficiently accurate and if possible have a short duration of works on measurement.
- Accuracy of the method must not exceed 10%.

Based on the analysis of existing methods [1, 2], considering the above requirements for experimental research of insulation parameters of three-phase electrical networks with isolated neutral was obtained the conclusion that the methods developed earlier is not fully meet the essential requirements. Therefore, at present in mines previously proposed methods were not used as a primary means of prevention, ensuring of uninterrupted power supply and operational safety of electrical installations.

In this regard, there are problems of further improvement of means of controlling insulation parameters of electrical networks in conjunction with the implementation of preventive measures and periodic measurements in different operating conditions. Method for determining the phase insulation parameters to earth of electrical networks must not affect the operation of the electricity system, and the calculation of insulation parameters must contain a minimum of computation.

In practice of operation of electrical networks with isolated neutral with voltage up to 1000 V and above it is necessary to know the value of the insulation parameters by which are developed the organizational and technical measures to ensure the safety and security of electricity supply of mining enterprises.

2. Method for determining the insulation parameters in an electrical network with insulated neutral

One of the factors of electric shock is the weakening of insulation condition of a three-phase electrical network with insulated neutral voltages up to and above 1000 V. In order to ensure the increase of efficiency of power supply system is necessary to develop a method of determining the parameters of isolation under operating voltage. Under the effectiveness we accept the ensuring of growth of electrical safety and reliability in the operation of electrical installations with voltage up to and above 1000 V. The known [1] method of determining the parameters of isolation, "Ammeter-voltmeter" is a classical method, as it provides a satisfactory accuracy of the unknown quantities, but it does not ensure work safety in electrical installations production works and reduces the reliability of power supply of industrial machinery and equipment. Reduction of electrical installations work reliability and level of electrical safety in the operation of three-phase power networks up to and above 1000 determined that by using the method "Ammeter-voltmeter" it is necessary to make the metal circuit of a mains phase to earth and measuring the total current single-phase fault ground. Since during a metal closure of any phase to earth phase voltage of the two other phases of the mains with respect to ground reaches linear values and can thus lead to a short circuit in a multi-phase mains operated, which determines the reliability of power decrease production machinery. A reduction in electrical safety determined by that in the metal closure of any phase of electrical network and ground, contact voltage and step voltage will have the maximum value, and thereby provides maximum increase the probability electric shock to persons.

The presented in work [3] method of determining the insulation parameters in three-phase electrical network with insulated neutral voltages above 1000 V, based on the measurement values of the modules of the line voltage, zero-sequence voltage and phase voltage with respect to ground when connected known active extra conduction between electrical network of the measured phase and ground, has a significant error. A significant error determined by that in determining the insulation parameters using the value of zero sequence voltage module, and thus it is necessary to use a voltage transformer windings, allowing to allocate the residual voltage

On the basis of the foregoing methods for determining the insulation parameters in three-phase mains with insulated neutral voltages up to and above 1000 V, which provides a satisfactory accuracy of the unknown quantities by eliminating the measurement of the modulus of the residual voltage, the operational safety of electrical installations and the reliability of the electricity system, in connection excluding the measurements of the total current of the module for single-phase earth fault between a mains phase with respect to ground.

A method for determining the insulation parameters in three-phase balanced networks with voltage up to and above 1000 V, based on the measurement values of the modules of the line voltage, the phase voltages A and C relative to the ground after connecting additional active conductivity between the phase A and the mains ground was developed.

As a result of the measurement values of the modules of the line voltage and phase voltage C and A with respect to the ground, taking into account the magnitude of the additional active conductivity by mathematical formulas are defined:

– the total conductance of network insulation

$$y = \frac{1.73 U_1 U_A}{U_C^2 - U_A^2} g_o, \tag{1}$$

– the active conductance of network insulation

$$g = \left(\frac{3U_l^2 (U_1^2 - 3U_A^2)}{(U_C^2 - U_A^2)^2} - 1 \right) 0.5 g_o, \tag{2}$$

– capacitive conductance of network insulation

$$b = \sqrt{y^2 - g^2}, \tag{3}$$

where U_l – line voltage; U_A – A phase voltage with respect to the ground; U_C – C phase voltage with respect to the ground; g_o – additional active conductance.

The method developed in the implementation does not require the creation of a special measuring device, since the measuring devices, i.e. voltmeters, available in the service manual. The PE-200 resistance is used as an active additional conductivity with R = 1000 ohms, where by means of parallel and serial connection provides the required power dissipation. To switch the active standby is used more conductivity cell load switch.

The developed method provides satisfactory accuracy, simple and safe in its implementation in the three-phase electrical networks with isolated neutral voltages up to and above 1000 V.

3. Analysis of error of method determining the insulation parameters in an electrical network with isolated neutral

The obtained mathematical dependences for determining the total and active conductances of electrical network insulation provide easy and safe work of electrical installations with voltage up to and above 1000 V.

Perform error analysis of the developed method of determining the insulation parameters in symmetrical three-phase electrical networks with isolated neutral which is based on measurement of unit line voltage, phase voltage C and A relative to the earth, after the active connection of additional conduction between phase And the electric network and earth.

To improve the efficiency of the developed method for determining the parameters of isolation in a symmetrical three-phase network with isolated neutral, based on error analysis for each specific network is selected additional active conductivity, in order to ensure satisfactory accuracy of required quantities.

Random relative error in determining the total conductivity of insulation and its components in three-phase balanced networks with voltage up to and beyond 1000, based on the measurement values of the modules of the line voltage, phase voltage C and A with respect to ground, after connecting the active additional conduction between the phase and the electric network and earth, determined according to equation (1), (2) and (3).

Random relative error in determining the total conductance of mains phase insulation relative to the ground is determined from the formula (1):

$$y = \frac{1.73 U_l U_A}{U_C^2 - U_A^2} g_o,$$

where U_l, U_A, U_C, g_o – values that define the total conductance of network insulation and obtained by direct measurement. The relative mean square error in determining the total conductance of mains phase insulation relative to the ground is determined from the expression [3]:

$$\Delta y = \frac{1}{y}\left[\left(\frac{\partial y}{\partial U_A}\Delta U_A\right)^2 + \left(\frac{\partial y}{\partial U_C}\Delta U_C\right)^2 + \left(\frac{\partial y}{\partial U_l}\Delta U_l\right)^2 + \left(\frac{\partial y}{\partial g_o}\Delta g_o\right)^2\right]^{1/2}, \quad (4)$$

where $\frac{\partial y}{\partial U_A}$, $\frac{\partial y}{\partial U_C}$, $\frac{\partial y}{\partial U_l}$, $\frac{\partial y}{\partial g_o}$ – partial derivatives $y = f(U_l, U_A, U_C, g_o)$.

Here ΔU_l, ΔU_A, ΔU_C, Δg_o – absolute errors of direct measurement values U_l, U_A, U_C, g_o which are defined by the following expressions:

$$\begin{aligned} \Delta U_l &= U_l \cdot \Delta U_{l*}; \\ \Delta U_C &= U_C \cdot \Delta U_{C*}; \\ \Delta U_A &= U_A \cdot \Delta U_{A*}; \\ \Delta g_o &= g_o \cdot \Delta g_{o*}. \end{aligned} \quad (5)$$

To determine the errors of measuring devices accept that $\Delta U_{l*} = \Delta U_{A*} = \Delta U_{C*} = \Delta U_*$, where: ΔU_* - the relative error of voltage measurement circuits; $\Delta g_{o*} = \Delta R_*$ - the relative error of the measuring instrument, which measures the resistance which is connected between the phase A electrical and ground. Determine the partial derivative functions $y = f(U_1, U_A, U_C, g_o)$ by the variables U_1, U_A, U_C, g_o:

$$\begin{aligned} \frac{\partial y}{\partial U_1} &= \frac{1.73 U_A}{U_C^2 - U_A^2} g_o; \\ \frac{\partial y}{\partial U_A} &= \frac{1.73 U_1 (U_C^2 + U_A^2)}{(U_C^2 - U_A^2)^2} g_o; \\ \frac{\partial y}{\partial U_C} &= -\frac{3.46 U_1 U_A U_C}{(U_C^2 - U_A^2)^2} g_o; \\ \frac{\partial y}{\partial g_o} &= \frac{1.73 U_1 U_A}{U_C^2 - U_A^2}. \end{aligned} \quad (6)$$

Solving the equation (4), substituting the values of the partial derivatives of equation (6) and private values of absolute errors (5), at the same time, assuming that $\Delta U_* = \Delta R_* = \Delta$, we obtain:

$$\varepsilon_y = \frac{\Delta y}{\Delta} = \frac{1.73 U_l U_A g_o}{U_C^2 - U_A^2}\left(2 + \frac{4U_C^4 + (U_C^2 + U_A^2)^2}{(U_C^2 - U_A^2)^2}\right)^{1/2}. \quad (7)$$

The obtained equation (7) is divided into the equation (1):

$$\varepsilon_y = \frac{\Delta y}{\Delta} = \left(2 + \frac{4U_C^4 + (U_C^2 + U_A^2)^2}{(U_C^2 - U_A^2)^2}\right)^{1/2}. \quad (8)$$

The obtained equation (8) we express in relative units, and after the conversion obtain:

$$\varepsilon_y = \frac{\Delta y}{\Delta} = \left(2 + \frac{4 + (1 + U_*^2)^2}{(1 - U_*^2)^2}\right)^{1/2}, \quad (9)$$

where $U_* = \dfrac{U_A}{U_C}$.

Random error in determining the active conductance of mains phase insulation relative to the ground is determined from the formula (2):

$$g = \left(\frac{3U_l^2 (U_1^2 - 3U_A^2)}{(U_C^2 - U_A^2)^2} - 1\right) 0.5 g_o,$$

where U_1, U_A, U_C, g_o – values that define the active conductance of network isolation and obtained by direct measurement.

Relative mean square error of the method when determining the active conductivity of phase insulation of electrical network relative to the ground is determined from the expression:

$$\Delta g = \frac{1}{g}\left[\left(\frac{\partial g}{\partial U_A}\Delta U_A\right)^2 + \left(\frac{\partial g}{\partial U_C}\Delta U_C\right)^2 + \left(\frac{\partial g}{\partial U_1}\Delta U_l\right)^2 + \left(\frac{\partial g}{\partial g_o}\Delta g_o\right)^2\right]^{1/2}, \quad (10)$$

where $\dfrac{\partial g}{\partial U_A}$, $\dfrac{\partial g}{\partial U_C}$, $\dfrac{\partial g}{\partial U_1}$, $\dfrac{\partial g}{\partial g_o}$ – partial derivatives, $g = f\ (U_1, U_A, U_C, g_o)$.

Here ΔU_1, ΔU_A, ΔU_C, Δg_o – absolute errors of direct measurement values U_1, U_A, U_C, g_o which are defined by the following expressions:

$$\begin{aligned} \Delta U_1 &= U_1 \cdot \Delta U_{1*}; \\ \Delta U_C &= U_C \cdot \Delta U_{C*}; \\ \Delta U_A &= U_A \cdot \Delta U_{A*}; \\ \Delta g_o &= g_o \cdot \Delta g_{o*}. \end{aligned} \quad (11)$$

To determine the accuracy of measuring devices accept that $\Delta U_{1*} = \Delta U_{A*} = \Delta U_{C*} = \Delta U_*$, where: ΔU_* - the relative error of voltage measurement circuits; $\Delta g_{o*} = \Delta R_*$ - the relative error of a measuring instrument that measures resistance which is connected between the phase A electrical and the ground.

Determine the partial derivatives $g = f(U_1, U_A, U_C, g_o)$ by the variables U_1, U_A, U_C, g_o:

$$\frac{\partial g}{\partial U_l}=\frac{3U_l(2U_l^2-3U_A^2)}{2(U_C^2-U_A^2)^2}g_o;$$
$$\frac{\partial g}{\partial U_A}=-\frac{3U_l^2U_A(3U_C^2+3U_A^2-2U_l^2)}{(U_C^2-U_A^2)^3}g_o;$$
$$\frac{\partial g}{\partial U_C}=-\frac{6U_l^2U_C(U_l^2-3U_A^2)}{(U_C^2-U_A^2)^3}g_o; \qquad (12)$$
$$\frac{\partial g}{\partial g_o}=\frac{3U_l^2(U_l^2-3U_A^2)}{2(U_C^2-U_A^2)}-0.5.$$

Solve equation (10), substituting the values of the partial derivatives of equation (12) and the values of the partial absolute errors (11), at the same time, assuming that $\Delta U_*=\Delta R_*=\Delta$, we obtain:

$$\frac{\Delta g}{\Delta}=\frac{3g_o}{(U_C^2-U_A^2)^3}\begin{pmatrix}(U_C^2-U_A^2)^2[2U_l^4(U_l^2-3U_A^2)^2-(U_C^2-U_A^2)^4]+\\+U_l^4\{U_A^4[3(U_C^2-U_A^2)-2U_l^2]^2+U_C^4(U_l^2-3U_A^2)^2\}\end{pmatrix}^{1/2}. \qquad (13)$$

Obtained equation (13) divide by the equation (2):

$$\varepsilon_g=\frac{\Delta g}{\Delta}=\begin{pmatrix}\dfrac{2U_l^4(U_l^2-3U_A^2)^2-(U_C^2-U_A^2)^4}{[3U_l^2(U_l^2-3U_A^2)-(U_C^2-U_A^2)^2]^2}+\\+\dfrac{U_l^4\{U_A^4[3(U_C^2-U_A^2)-2U_l^2]^2+U_C^4(U_l^2-3U_A^2)^2\}}{(U_C^2-U_A^2)^2[3U_l^2(U_l^2-3U_A^2)-(U_C^2-U_A^2)^2]^2}\end{pmatrix}^{1/2}. \qquad (14)$$

The resulting equation (14) the value of the line voltage is expressed in terms of the phase voltages in accordance with the fact that $U_l=\sqrt{3}U_\phi$:

$$\varepsilon_g=\frac{\Delta g}{\Delta}=3\begin{pmatrix}\dfrac{18U_{ph}^4(U_{ph}^2-U_A^2)^2-(U_C^2-U_A^2)^4}{[27U_{ph}^2(U_{ph}^2-U_A^2)-(U_C^2-U_A^2)^2]^2}+\\+\dfrac{3U_{ph}^4U_A^4(U_C^2-U_A^2-2U_{ph}^2)^2+U_C^4(U_{ph}^2-U_A^2)^2}{(U_C^2-U_A^2)^2[27U_{ph}^2(U_{ph}^2-U_A^2)-(U_C^2-U_A^2)^2]^2}\end{pmatrix}^{1/2}. \qquad (15)$$

Simplifying the formula (15) we obtain the equation (16):

$$\varepsilon_g=\frac{3}{27U_{ph}^2(U_{ph}^2-U_A^2)-(U_C^2-U_A^2)^2}\begin{pmatrix}18U_{ph}^4(U_{ph}^2-U_A^2)^2-(U_C^2-U_A^2)^4+\\+\dfrac{3U_{ph}^4U_A^4(U_C^2-U_A^2-2U_{ph}^2)^2}{(U_C^2-U_A^2)^2}+\\+\dfrac{U_C^4(U_{ph}^2-U_A^2)^2}{(U_C^2-U_A^2)^2}\end{pmatrix}^{1/2}. \qquad (16)$$

Obtained equation (16) express in relative units, and after the conversion obtain:

$$\varepsilon_g = \frac{\Delta g}{\Delta} = \frac{3}{27(1-U_{A*}^2)-(U_{C*}^2-U_{A*}^2)^2}\left(\begin{array}{l} 18(1-U_{A*}^2)^2-(U_{C*}^2-U_{A*}^2)^4+ \\ +\dfrac{3U_{A*}^4(U_{C*}^2-U_{A*}^2-2)^2}{(U_{C*}^2-U_{A*}^2)^2}+ \\ +\dfrac{U_{C*}^4(1-U_{A*}^2)^2}{(U_C^2-U_A^2)^2} \end{array}\right)^{1/2}, \quad (17)$$

where $U_{A*}=\dfrac{U_A}{U_{ph}}$, $U_{C*}=\dfrac{U_C}{U_{ph}}$.

Relative mean square error method for determining the conductivity of the capacitive isolation mains phases relative to the ground is determined by the expression (3):

$$\Delta b = \frac{1}{b}\left[\left(\frac{\partial b}{\partial y}\Delta y\right)^2+\left(\frac{\partial b}{\partial g}\Delta g\right)^2\right]^{1/2}, \quad (18)$$

or

$$\varepsilon_b = \frac{\Delta b}{\Delta} = \frac{\left[(1-\tan^2\delta)^2\left(\dfrac{\Delta y}{\Delta}\right)^2+\left(\dfrac{\Delta g}{\Delta}\right)^2\right]^{1/2}}{\tan^2\delta}. \quad (19)$$

Solving the equation (19) substituting the values of mathematical descriptions of the relative rms dependences of total (8) and active (16) conductivities of electrical installations phase insulation relative to the ground phase we get the following equation:

$$\varepsilon_b = \frac{\Delta b}{\Delta} = \frac{\left(\begin{array}{l} (1-\tan^2\delta)^2\left[2+\dfrac{4U_C^4+(U_C^2+U_A^2)^2}{(U_C^2-U_A^2)^2}\right]+ \\ +\dfrac{9}{[27U_{ph}^2(U_{ph}^2-U_A^2)-(U_C^2-U_A^2)^2]^2}\times \\ \times\left[\begin{array}{l} 18U_{ph}^2(U_{ph}^2-U_A^2)^2-(U_C^2-U_A^2)^4+ \\ +\dfrac{3U_{ph}^4U_A^4(U_C^2-U_A^2-2U_{ph}^2)^2+U_C^4(U_{ph}^2-U_A^2)^2}{(U_C^2-U_A^2)^2} \end{array}\right] \end{array}\right)^{1/2}}{\tan^2\delta}. \quad (20)$$

Obtained equation (21) express in relative units, and after the conversion obtain:

$$\varepsilon_b = \frac{\Delta b}{\Delta} = \frac{\left(\begin{array}{l} (1-\tan^2\delta)^2 \left[2 + \dfrac{4U_{C*}^4 + (U_{C*}^2 + U_{A*}^2)^2}{(U_{C*}^2 - U_{A*}^2)^2} \right] + \\ + \dfrac{9}{[27(1-U_{A*}^2) - (U_{C*}^2 - U_{A*}^2)^2]^2} \times \\ \times \left[\begin{array}{l} 18(1-U_{A*}^2)^2 - (U_{C*}^2 - U_{A*}^2)^4 + \\ + \dfrac{3U_{A*}^4 (U_{C*}^2 - U_{A*}^2 - 2)^2 + U_{C*}^4 (1-U_{A*}^2)^2}{(U_{C*}^2 - U_{A*}^2)^2} \end{array} \right] \end{array} \right)^{1/2}}{\tan^2\delta} . \qquad (21)$$

Based on the results of random relative mean square errors in determining the active, capacitive and total conductivities of mains phase isolation relative to the ground, build the dependence:

$$\varepsilon_y = \frac{\Delta y_*}{\Delta} = f(U_*);$$

$$\varepsilon_g = \frac{\Delta g_*}{\Delta} = f(U_{A*}, U_{C*});$$

$$\varepsilon_b = \frac{\Delta b_*}{\Delta} = f(U_{A*},\ U_{C*},\ \tan\delta),$$

shown in figure 1, figure 2, figure 3. Mathematical dependence of the relative mean square errors of the total – ε_y, active – ε_g and capacitive – ε_b conductivities of phase insulation of electrical network with insulated neutral on graphic illustrations (figure 1, figure 2, figure 3) characterize the change in error depending on the amount of additional active conduction g_o, which is inserted between the A-phase of electrical network and earth.

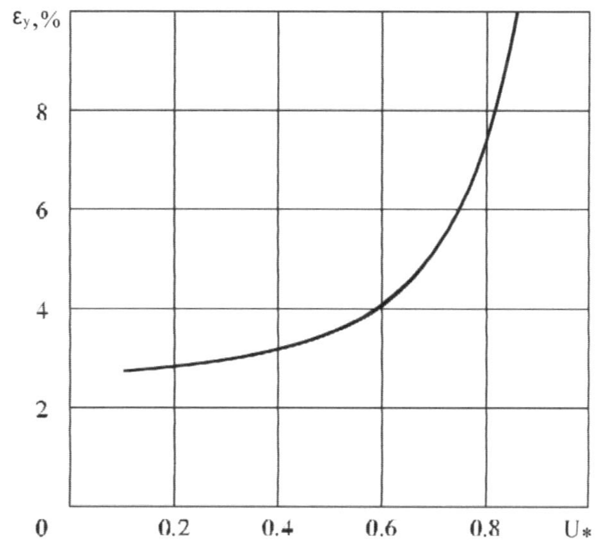

Figure 1. Analysis of the error in determining the total conductance of the network insulation.

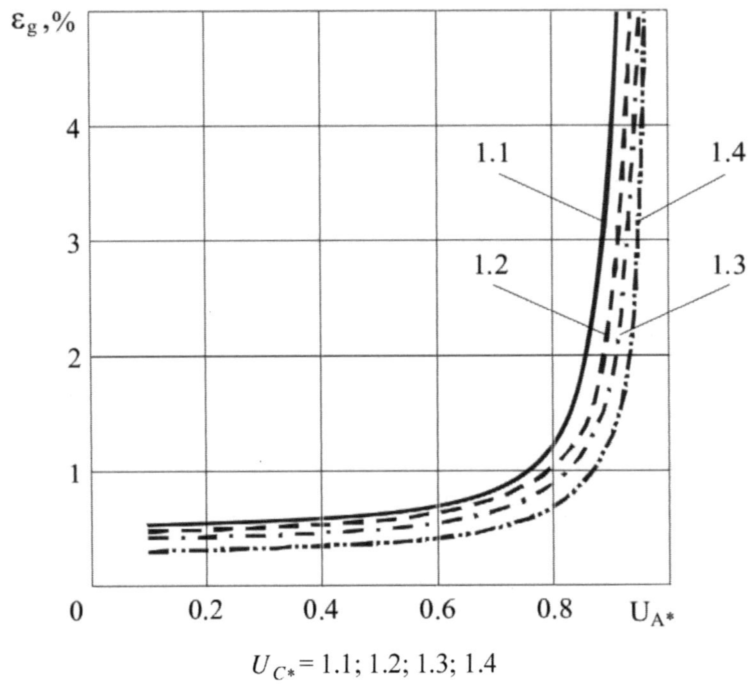

Figure 2. Analysis of the error in determining the active conductance of the network insulation.

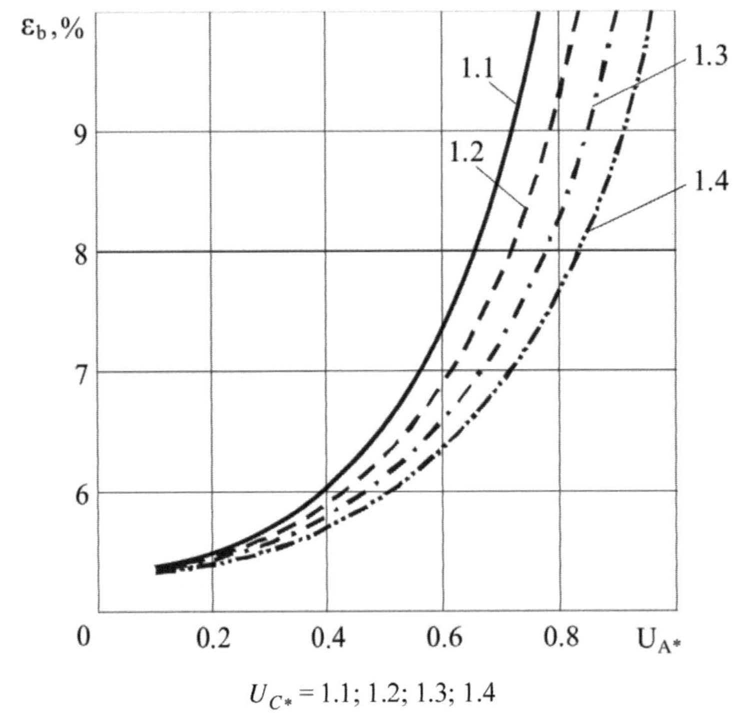

Figure 3. Analysis of the error in determining the capacitive conductance of the network insulation when $\tan\delta = 1.0$.

In determining the parameters of isolation in a symmetrical three-phase electrical network with isolated neutral on the basis of the method of analysis of error for each specific network select additional active conduction, so as to ensure satisfactory accuracy required.

In determining the total conductance of mains phases isolation relative to the ground is chosen such additional active conductivity, the values were within $U_* = 0.2 \div 0.8$, at the same time as shown in figure 1, the error does not exceed 5% when using measuring devices with accuracy class 1.0, and 2.5% when using measuring devices with accuracy class 0.5.

In determining the value of the active conductance in the three-phase electrical network with insulated neutral voltage up to 1000 V and above, select this additional g_o, so that $U_{A*} = 0.2 \div 0.8$, when $U_{C*} = 1.1 \div 1.6$, , then on the basis of graphic illustrations of figure 2 error does not exceed 3.5% when using measuring devices with accuracy class 1.0.

In determining the capacitive conductance mains phase isolation relative to the ground selection of additional active conductance g_o based on a graphic illustrations of figure 3 so that $U_{A*} = 0.2 \div 0.8$, when $U_{C*} = 1.1 \div 1.6$, when $\tan\delta = 1.0$, to provide error to 4 % when using measuring devices with accuracy class 1.0.

It should be noted that when using measuring instruments with an accuracy class of 0.5, errors of ε_y – total, ε_g – active, ε_b – capacitive admittances of isolation is reduced by half, to provide more reliable data when determining the insulation parameters developed method.

The developed method provides satisfactory accuracy when determining the parameters of isolation, as well as the ease and safety of production work in existing electrical installations voltages up to and above 1000 V.

4. Conclusion

In work was obtained the following results:

- A method for determining the parameters in three-phase networks with isolated neutral voltage up to 1000 V and above, is to measure the modulus of the line voltage, phase voltage with respect to ground and A, after you connect an additional active conductivity between the A-phase mains and earth was developed.
- Method for determining the parameters of isolation in three-phase electrical network with isolated neutral, through error analysis for each particular network is selected additional active conductance, so as to ensure satisfactory accuracy required when determining the:

The total conductance of mains phase insulation relative to the ground is chosen such additional active conductance, the values were within = 0.2 ÷ 0.8, with the error does not exceed 5% when using measuring devices with accuracy class 1.0, and 2.5% using measuring devices with accuracy class 0.5;

Active conductance in three-phase electrical network with isolated neutral voltages up to and above 1000 V select such active additional conductance g_o, so that $U_{A*} = 0.2 \div 0.8$, when $U_{C*} = 0.2 \div 0.8$, then the error does not exceed 3.5 % when using the measuring devices with accuracy class 1.0;

Capacitive conductance of electrical network phase insulation relative to the ground select such additional active conductance g_o, so that $U_* = 0.2 \div 0.8$, with the change $\tan\delta = 0.6 \div 1.6$, then the error does not exceed 5 % when using the measuring devices with accuracy class 1.0, and 2.5 % when using the measuring devices with accuracy class 0.5.

- The developed methods provide satisfactory accuracy simplicity, and security in its implementation in the three-phase electrical networks with isolated neutral voltages up to and above 1000 V.

References

[1] Gladilin L V, Shchutskiy V I, Batsezhev Y G and Chebotayev N I 1977 *Electrical Safety in the Mining Industry* (Moscow: Nedra) 327
[2] Utegulov B B and Zhumadirova A 2005 *Microprocessor Means for Monitoring of Insulation*

Condition and Protection from a Single-phase Earth Fault in Electrical Networks of 6-10 kV ed Uteguloy B B (Pavlodar: S.Toraighyrov Pavlodar State University) 151

[3] Gladilin L V and Utegulov B B 1980 Accuracy analysis of method for determining insulation parameters in a three-phase electrical networks with isolated neutral by voltage above 1000 V *Proceedings of the Higher Educational Institutions. Mining Journal* 94–97

[4] Utegulov B, Utegulov A, Begentaev M, Begentaev B, Uakhitova A, Zhumazhanov S and Zhakipov N 2011 Method for determining parameters of isolation network voltage up to 1000 V in mining enterprises *Source of the Document Proc. of the IASTED Int. Conf. on Power and Energy Systems and Applications (PESA)* 50–53

[5] Utegulov B, Utegulov A, Begentaev M, Begentaev B, Uakhitova A, Zhakipov N and Sadvakasov T 2011 Method for determining the insulation in asymmetric networks with voltage up to 1000 V in mining enterprises *Source of the Document Proc. of the IASTED Int. Conf. on Power and Energy Systems and Applications (PESA)* 54–57

Improved control strategy for PI-R current of DFIG considering voltage and current harmonics compensation

S Y Song[1], Q H Liu, Y N Zhao and S Y Liu

North China Electric Power University, State Key Laboratory of Alternate Electrical Power System with Renewable Energy Sources, Beijing 102206, China

E-mail: shiyusong209@126.com

Abstract. With the rapid development of wind power generation, the related research of wind power control and integration issues has attracted much attention, and the focus of the research are shifting away from the ideal power grid environment to the actual power grid environment. As the main stream wind turbine generator, a doubly-fed induction generator (DFIG) is connected to the power grid directly by its stator, so it is particularly sensitive to the power grid. This paper studies the improvement of DFIG control technology in the power grid harmonic environment. Based on the DFIG dynamic model considering the power grid harmonic environment, this paper introduces the shortcomings of the common control strategy of DFIG, and puts forward the enhanced method. The decoupling control of the system is realized by compensating the coupling between the rotor harmonic voltage and harmonic current, improving the control performance. In addition, the simulation experiments on PSCAD/EMTDC are carried out to verify the correctness and effectiveness of the improved scheme.

1. Introduction

In recent years, the research on DFIG high-performance control is gradually increased under the condition of unbalanced grid voltage according to papers [1-6].However, the research on the high-performance control strategy of DFIG wind turbine is relatively few in the case of harmonic. The research of paper [7] shows that, harmonic voltage will create the ripple of electromagnetic torque of the motor, increase the copper loss and iron loss, lead to stator and rotor current distortion and so on. What's worse, electric generator unit may split from the grid. At present, there have been some papers involved in the operation and control methods of DFIG wind power system in harmonic environment at home and abroad. For instance, the stator current harmonic compensation method based on the multiple synchronous rotating coordinate systems from papers [8-10], the multi-objective control method added harmonic calculation instruction according to papers [11-13], control method of harmonic current based on multiple rotating coordinate systems from papers [14-16], and selected harmonic elimination method based on resonant controller according to papers [17-21].In conclusion, the above methods can be divided into two categories. One can be named as the control strategy of the PI current controller based on multiple rotating *dq* coordinate systems, which means that PI controller

[1] Address for correspondence: S Y Song, North China Electric Power University, State Key Laboratory of Alternate Electrical Power System with Renewable Energy Sources, Beijing 102206, China. E-mail: shiyusong209@126.com.

is used to control the rotor current after selecting the control target in $(dq)^{+}$, $(dq)^{5-}$, $(dq)^{7+}$ reference frames. The other is the control strategy of the PI-R current controller based on *dq* coordinate system, which means using PI-R controller to control the rotor current in unified synchronous speed rotating *dq* coordinate system. But most of the methods mentioned above ignore the cross coupling characteristics between rotor voltage and current. Some of them only consider the coupling between the fundamental voltage and current without considering the coupling between the harmonic voltage and current. If there is no effective compensation for the coupling in the design of *d*, *q* axis rotor current controller, it will affect the speed and accuracy of rotor current control, thereby significantly reducing the dynamic characteristics of the control strategy.

Based on the above reasons, this paper takes the rotor voltage and current coupling characteristics into consideration in order to improve the existing control strategies. The harmonic coupling relationship between voltage and current are concluded after introducing the mathematical model of DFIG in harmonic environment. By increasing the rotor harmonic voltage and current coupling compensation, this paper proposes an improved control strategy of PI-R current controller based on *dq* coordinate system. A simulation is carried out in PSCAD/EMTDC to prove the effectiveness of the control strategy.

2. DFIG dynamic modelling in harmonic environment

The mathematical model of DFIG in vector form of synchronous speed rotating *dq* coordinate system is as follows

(1)Voltage equations

$$\begin{cases} \boldsymbol{u}_{\mathrm{dq1}} = R_1\boldsymbol{i}_{\mathrm{dq1}} + \dfrac{d\boldsymbol{\psi}_{\mathrm{dq1}}}{dt} + j\omega_1\boldsymbol{\psi}_{\mathrm{dq1}} \\ \boldsymbol{u}_{\mathrm{dq2}} = R_2\boldsymbol{i}_{\mathrm{dq2}} + \dfrac{d\boldsymbol{\psi}_{\mathrm{dq2}}}{dt} + j\omega_{\mathrm{s}}\boldsymbol{\psi}_{\mathrm{dq2}} \end{cases} \tag{1}$$

(2)Flux linkage equations

$$\begin{cases} \boldsymbol{\psi}_{\mathrm{dq1}} = L_1\boldsymbol{i}_{\mathrm{dq1}} + L_{\mathrm{m}}\boldsymbol{i}_{\mathrm{dq2}} \\ \boldsymbol{\psi}_{\mathrm{dq2}} = L_{\mathrm{m}}\boldsymbol{i}_{\mathrm{dq1}} + L_2\boldsymbol{i}_{\mathrm{dq2}} \end{cases} \tag{2}$$

Formula (2) can be expressed as follows

$$\begin{cases} \boldsymbol{i}_{\mathrm{dq1}} = \dfrac{\boldsymbol{\psi}_{\mathrm{dq1}} - L_{\mathrm{m}}\boldsymbol{i}_{\mathrm{dq2}}}{L_1} \\ \boldsymbol{\psi}_{\mathrm{dq2}} = \dfrac{L_{\mathrm{m}}\boldsymbol{\psi}_{\mathrm{dq1}}}{L_1} + \sigma L_2\boldsymbol{i}_{\mathrm{dq2}} \end{cases} \tag{3}$$

Where $\sigma = 1 - L_m^2/(L_1L_2)$ is the leakage factor.

Substituting (3) into (1) yields the relationship between rotor voltage and rotor current as follows

$$\boldsymbol{u}_{\mathrm{dq2}} = R_2\boldsymbol{i}_{\mathrm{dq2}} + \sigma L_2\frac{d\boldsymbol{i}_{\mathrm{dq2}}}{dt} + j\omega_{\mathrm{s}}\sigma L_2\boldsymbol{i}_{\mathrm{dq2}} + \frac{L_{\mathrm{m}}}{L_1}\frac{d\boldsymbol{\psi}_{\mathrm{dq1}}}{dt} + j\omega_{\mathrm{s}}\frac{L_{\mathrm{m}}}{L_1}\boldsymbol{\psi}_{\mathrm{dq1}} \tag{4}$$

In the consideration of the three-phase power grid voltage distortion of harmonics at the frequencies of $-5\omega_1$ and $7\omega_1$, DFIG modelling will become more complicated. In the stationary $\alpha\beta$ coordinate system, the space vector of DFIG (such as voltage, current, flux etc.) can be generally expressed as (5) according to the paper [22]

$$\begin{aligned} \boldsymbol{F}_{\alpha\beta}(t) &= \boldsymbol{F}_{\alpha\beta}^{(1)}(t) + \boldsymbol{F}_{\alpha\beta}^{(5)}(t) + \boldsymbol{F}_{\alpha\beta}^{(7)}(t) \\ &= \left|\boldsymbol{F}_{\alpha\beta}^{(1)}(t)\right|\boldsymbol{e}^{j(\omega_1 t+\phi^{(1)})} + \left|\boldsymbol{F}_{\alpha\beta}^{(5)}(t)\right|\boldsymbol{e}^{-j(5\omega_1 t+\phi^{(5)})} + \left|\boldsymbol{F}_{\alpha\beta}^{(7)}(t)\right|\boldsymbol{e}^{j(7\omega_1 t+\phi^{(7)})} \end{aligned} \tag{5}$$

where superscript (1),(5),(7) respectively on behalf of fundamental positive-sequence, 5th harmonic, 7th harmonic. $\left|\boldsymbol{F}_{\alpha\beta}^{(1)}(t)\right|$, $\left|\boldsymbol{F}_{\alpha\beta}^{(5)}(t)\right|$, $\left|\boldsymbol{F}_{\alpha\beta}^{(7)}(t)\right|$ means the amplitude of 1st,5th,7th harmonic vector. $\varphi^{(1)}$, $\varphi^{(5)}$,

$\varphi^{(7)}$ are the respective initial phase shift for the fundamental positive-sequence and harmonic components of $-5\omega_1$ and $7\omega_1$ according to the paper [23].

In addition to stationary $\alpha\beta$ coordinates, there are forward synchronous speed $(dq)^+$ coordinate system, 5 times the reversal synchronous speed $(dq)^{5-}$ coordinate system and 7 times the synchronous speed $(dq)^{7+}$ coordinate system. The vector transformation between the above coordinate systems and the stationary $\alpha\beta$ coordinates is as follows

$$\begin{cases} \boldsymbol{F}_{\mathrm{dq}} = \boldsymbol{F}_{\alpha\beta}\boldsymbol{e}^{-j\omega_1 t} \\ \boldsymbol{F}_{\text{-5dq}} = \boldsymbol{F}_{\alpha\beta}\boldsymbol{e}^{j5\omega_1 t} \\ \boldsymbol{F}_{\text{+7dq}} = \boldsymbol{F}_{\alpha\beta}\boldsymbol{e}^{-j7\omega_1 t} \end{cases} \tag{6}$$

Substituting (6) into (5) yields

$$\begin{cases} \boldsymbol{F}_{\mathrm{dq}}^{(1)} = \left|\boldsymbol{F}_{\alpha\beta}^{(1)}(t)\right|\boldsymbol{e}^{j\phi^{(1)}} \\ \boldsymbol{F}_{\text{-5dq}}^{(5)} = \left|\boldsymbol{F}_{\alpha\beta}^{(5)}(t)\right|\boldsymbol{e}^{-j\phi^{(5)}} \\ \boldsymbol{F}_{\text{+7dq}}^{(7)} = \left|\boldsymbol{F}_{\alpha\beta}^{(7)}(t)\right|\boldsymbol{e}^{j\phi^{(7)}} \end{cases} \tag{7}$$

Different harmonic vectors will be transformed into straight flow after transformation to the corresponding rotational coordinate system so that the model can be simplified to a certain extent.

2.1. Voltage and flux linkage equations of DFIG considering harmonic

When grid voltage contains 5th and 7th harmonic components, it can be expressed in the form of formula (3) as follows

$$\boldsymbol{u}_{\alpha\beta 1} = U_{\mathrm{sm}}^{(1)}\boldsymbol{e}^{j\left(\omega_1 t+\phi_0^{(1)}\right)} + U_{\mathrm{sm}}^{(5)}\boldsymbol{e}^{-j\left(5\omega_1 t+\phi_0^{(5)}\right)} + U_{\mathrm{sm}}^{(7)}\boldsymbol{e}^{j\left(7\omega_1 t+\phi_0^{(7)}\right)} \tag{8}$$

where $U_{\mathrm{sm}}^{(1)}, U_{\mathrm{sm}}^{(5)}, U_{\mathrm{sm}}^{(7)}$ are the amplitude of 1st ,5th ,7th harmonic voltage. $\phi_0^{(1)}$, $\phi_0^{(5)}$, $\phi_0^{(7)}$ means 1st ,5th ,7th harmonic vector angle.

Grid voltage transforms from stationary $\alpha\beta$ coordinate system to synchronous speed rotating dq coordinate system can be expressed as follows

$$\boldsymbol{u}_{\mathrm{dq}1} = \boldsymbol{u}_{\alpha\beta 1}\boldsymbol{e}^{-j\omega_1 t} = \boldsymbol{u}_{\mathrm{dq}1}^{(1)} + \boldsymbol{u}_{\text{-5dq1}}^{(5)}\boldsymbol{e}^{-j6\omega_1 t} + \boldsymbol{u}_{\text{+7dq1}}^{(7)}\boldsymbol{e}^{j6\omega_1 t} \tag{9}$$

Formula (9) shows that the fundamental component of the grid voltage is presented as a direct flow and 5th, 7th harmonic component is presented respectively as $+6\omega_1$, $-6\omega_1$ rotary alternating quantity in the dq coordinate system.

Similarly, expressions of stator current and flux in the dq coordinate system are available. Substituting them into (1) and (2), extracting common factor $\boldsymbol{e}^{-j6\omega_1 t}$ and $\boldsymbol{e}^{j6\omega_1 t}$, stator and rotor voltage equations from paper [13] are as follows

$$\begin{cases} \boldsymbol{u}_{\mathrm{dq}1}^{(1)} = R_1\boldsymbol{i}_{\mathrm{dq}1}^{(1)} + p\boldsymbol{\psi}_{\mathrm{dq}1}^{(1)} + j\omega_1\boldsymbol{\psi}_{\mathrm{dq}1}^{(1)} \\ \boldsymbol{u}_{\mathrm{dq}2}^{(1)} = R_2\boldsymbol{i}_{\mathrm{dq}2}^{(1)} + p\boldsymbol{\psi}_{\mathrm{dq}2}^{(1)} + j\omega_s\boldsymbol{\psi}_{\mathrm{dq}2}^{(1)} \\ \boldsymbol{u}_{\text{-5dq1}}^{(5)} = R_1\boldsymbol{i}_{\text{-5dq1}}^{(5)} + p\boldsymbol{\psi}_{\text{-5dq1}}^{(5)} - j5\omega_1\boldsymbol{\psi}_{\text{-5dq1}}^{(5)} \\ \boldsymbol{u}_{\text{-5dq2}}^{(5)} = R_2\boldsymbol{i}_{\text{-5dq2}}^{(5)} + p\boldsymbol{\psi}_{\text{-5dq2}}^{(5)} - j\omega_{5s}\boldsymbol{\psi}_{\text{-5dq2}}^{(5)} \\ \boldsymbol{u}_{\text{+7dq1}}^{(7)} = R_1\boldsymbol{i}_{\text{+7dq1}}^{(7)} + p\boldsymbol{\psi}_{\text{+7dq1}}^{(7)} + j7\omega_1\boldsymbol{\psi}_{\text{+7dq1}}^{(7)} \\ \boldsymbol{u}_{\text{+7dq2}}^{(7)} = R_2\boldsymbol{i}_{\text{+7dq2}}^{(7)} + p\boldsymbol{\psi}_{\text{+7dq2}}^{(7)} + j\omega_{7s}\boldsymbol{\psi}_{\text{+7dq2}}^{(7)} \end{cases} \tag{10}$$

where $\omega_s = \omega_1 - \omega_r$, $\omega_{5s} = 5\omega_1 + \omega_r$ and $\omega_{7s} = 7\omega_1 - \omega_r$.

The stator and rotor flux linkage equations are as follows

$$\begin{cases} \boldsymbol{\psi}_{\text{dq1}}^{(1)} = L_1 \boldsymbol{i}_{\text{dq1}}^{(1)} + L_{\text{m}} \boldsymbol{i}_{\text{dq2}}^{(1)} \\ \boldsymbol{\psi}_{\text{dq2}}^{(1)} = L_2 \boldsymbol{i}_{\text{dq2}}^{(1)} + L_{\text{m}} \boldsymbol{i}_{\text{dq1}}^{(1)} \\ \boldsymbol{\psi}_{\text{-5dq1}}^{(5)} = L_1 \boldsymbol{i}_{\text{-5dq1}}^{(5)} + L_{\text{m}} \boldsymbol{i}_{\text{-5dq2}}^{(5)} \\ \boldsymbol{\psi}_{\text{-5dq2}}^{(5)} = L_2 \boldsymbol{i}_{\text{-5dq2}}^{(5)} + L_{\text{m}} \boldsymbol{i}_{\text{-5dq1}}^{(5)} \\ \boldsymbol{\psi}_{\text{+7dq1}}^{(7)} = L_1 \boldsymbol{i}_{\text{+7dq1}}^{(7)} + L_{\text{m}} \boldsymbol{i}_{\text{+7dq2}}^{(7)} \\ \boldsymbol{\psi}_{\text{+7dq2}}^{(7)} = L_2 \boldsymbol{i}_{\text{+7dq2}}^{(7)} + L_{\text{m}} \boldsymbol{i}_{\text{+7dq1}}^{(7)} \end{cases} \tag{11}$$

2.2. *Power and electromagnetic torque equations of DFIG considering harmonic*

The equations of active power P_1 and reactive power Q_1 output from stator side are as follows

$$\begin{cases} P_1 = -\frac{3}{2}\text{Re}\left(\boldsymbol{u}_{dq1}\hat{\boldsymbol{i}}_{dq1}\right) \\ Q_1 = -\frac{3}{2}\text{Im}\left(\boldsymbol{u}_{dq1}\hat{\boldsymbol{i}}_{dq1}\right) \end{cases} \tag{12}$$

Where $\hat{\boldsymbol{i}}_{dq1}$ is the conjugate complex of $\boldsymbol{i}_{dq1}$.

Substituting (9), (10) into (1) and ignoring stator resistance will result the formulas of stator voltage and current considering harmonic. According to them, (12) could be rewritten as follows according to the paper [24]

$$\begin{cases} P_1 = P_{1,\text{dc}} + P_{1,\sin 6}\sin 6\omega_1 t + P_{1,\cos 6}\cos 6\omega_1 t + P_{1,\sin 12}\sin 12\omega_1 t + P_{1,\cos 12}\cos 12\omega_1 t \\ Q_1 = Q_{1,\text{dc}} + Q_{1,\sin 6}\sin 6\omega_1 t + Q_{1,\cos 6}\cos 6\omega_1 t + Q_{1,\sin 12}\sin 12\omega_1 t + Q_{1,\cos 12}\cos 12\omega_1 t \end{cases} \tag{13}$$

where $P_{1,dc}$, $Q_{1,dc}$ are the direct current component of stator active and reactive power. $P_{1,\sin 6}$, $P_{1,\cos 6}$, $Q_{1,\sin 6}$, $Q_{1,\cos 6}$ are the peak value of 6th harmonic components (sine and cosine) of stator active and reactive power. $P_{1,\sin 12}$, $P_{1,\cos 12}$, $Q_{1,\sin 12}$, $Q_{1,\cos 12}$ are the peak value of 12th harmonic components (sine and cosine) of stator active and reactive power. The above expressions refer to the paper [25].

The electromagnetic power of DFIG is

$$P_{\text{e}} = -\frac{3}{2}\text{Re}\left(j\omega_1\boldsymbol{\psi}_{\text{dq1}}\hat{\boldsymbol{i}}_{\text{dq1}} + j\omega_{\text{s}}\boldsymbol{\psi}_{\text{dq2}}\hat{\boldsymbol{i}}_{\text{dq2}}\right) \tag{14}$$

Simplify the electromagnetic power into function only related to stator flux and rotor current

$$P_{\text{e}} = -\frac{3\omega_{\text{r}}L_{\text{m}}}{2L_1}\text{Im}\left(\psi_{\text{dq1}}\hat{i}_{\text{dq2}}\right) = P_{\text{e,dc}} + P_{\text{e,sin6}}\sin 6\omega_1 t + P_{\text{e,cos6}}\cos 6\omega_1 t + P_{\text{e,sin12}}\sin 12\omega_1 t + P_{\text{e,cos12}}\cos 12\omega_1 t \tag{15}$$

where $P_{e,dc}$ is the direct current component of electromagnetic power. $P_{e,\sin 6}$, $P_{e,\cos 6}$ are the peak value of the 6th harmonic component of electromagnetic power (sine and cosine). $P_{e,\sin 12}$, $P_{e,\cos 12}$ are the peak value of the 12th harmonic component of electromagnetic power (sine and cosine). The above expressions also refer to the paper [20].

The relationship between electromagnetic torque T_e and electromagnetic power P_e is

$$T_{\text{e}} = p_{\text{n}}\frac{P_{\text{e}}}{\omega_{\text{r}}} \tag{16}$$

where means p_n pole pairs of motor.

The derivation and analysis above shows when the power grid contains 5th and 7th voltage harmonic components, not only the stator current will appear 5th and 7th harmonic but stator active power, reactive power and electromagnetic torque will also contain 6th and 12th harmonic components which is adverse to the operation of DFIG.

3. Improved control strategy of DFIG in power grid harmonic environment

DFIG is a complex mechanical and electrical system with a large number of cross coupling and the coupling between rotor voltage and current will weaken the performance of the control system. If the coupling between the rotor voltage and current is effectively compensated, the control performance could be significantly improved.

Although control strategy of PI-R current controller based on *dq* coordinate system needs resonant controller, it can greatly reducing the number of PI controllers. In addition to this, the control strategy doesn't need to extract the fundamental and harmonic components of the rotor current in rotor current control loop. It will reduce the control delay due to harmonic extraction and improve the dynamic performance of control.

3.1. Improved control strategy of PI-R current controller based on dq coordinate system

In order to clearly show the coupling relationship of DFIG, equation (4) could be rewritten as follows

$$\begin{cases} u_{d2}^{(1)}+u_{-5d2}^{(5)}e^{-j6\omega_1 t}+u_{+7d2}^{(7)}e^{j6\omega_1 t}=\left(R_2 i_{d2}^{(1)}+\sigma L_2\dfrac{di_{d2}^{(1)}}{dt}+\omega_s\sigma L_2 i_{q2}^{(1)}+\dfrac{L_m}{L_1}\dfrac{d\psi_{d1}^{(1)}}{dt}+\omega_s\dfrac{L_m}{L_1}\psi_{q1}^{(1)}\right) \\ +\left[R_2 i_{-5d2}^{(5)}+\sigma L_2\dfrac{di_{-5d2}^{(5)}}{dt}-(5\omega_1+\omega_r)\sigma L_2 i_{-5q2}^{(5)}+\dfrac{L_m}{L_1}\dfrac{d\psi_{-5d1}^{(5)}}{dt}-(5\omega_1+\omega_r)\dfrac{L_m}{L_1}\psi_{-5q1}^{(5)}\right]e^{-j6\omega_1 t} \\ +\left[R_2 i_{+7d2}^{(7)}+\sigma L_2\dfrac{di_{+7d2}^{(7)}}{dt}+(7\omega_1-\omega_r)\sigma L_2 i_{+7q2}^{(7)}+\dfrac{L_m}{L_1}\dfrac{d\psi_{+7d1}^{(7)}}{dt}+(7\omega_1-\omega_r)\dfrac{L_m}{L_1}\psi_{+7q1}^{(7)}\right]e^{j6\omega_1 t}=u_{d2}'+\nabla u_{d2} \\ u_{q2}^{(1)}+u_{-5q2}^{(5)}e^{-j6\omega_1 t}+u_{+7q2}^{(7)}e^{j6\omega_1 t}=\left(R_2 i_{q2}^{(1)}+\sigma L_2\dfrac{di_{q2}^{(1)}}{dt}-\omega_s\sigma L_2 i_{d2}^{(1)}+\dfrac{L_m}{L_1}\dfrac{d\psi_{q1}^{(1)}}{dt}-\omega_s\dfrac{L_m}{L_1}\psi_{d1}^{(1)}\right) \\ +\left[R_2 i_{-5q2}^{(5)}+\sigma L_2\dfrac{di_{-5q2}^{(5)}}{dt}+(5\omega_1+\omega_r)\sigma L_2 i_{-5d2}^{(5)}+\dfrac{L_m}{L_1}\dfrac{d\psi_{-5q1}^{(5)}}{dt}+(5\omega_1+\omega_r)\dfrac{L_m}{L_1}\psi_{-5d1}^{(5)}\right]e^{-j6\omega_1 t} \\ +\left[R_2 i_{+7q2}^{(7)}+\sigma L_2\dfrac{di_{+7q2}^{(7)}}{dt}-(7\omega_1-\omega_r)\sigma L_2 i_{+7d2}^{(7)}+\dfrac{L_m}{L_1}\dfrac{d\psi_{+7q1}^{(7)}}{dt}-(7\omega_1-\omega_r)\dfrac{L_m}{L_1}\psi_{+7d1}^{(7)}\right]e^{j6\omega_1 t}=u_{q2}'+\nabla u_{q2} \end{cases} \tag{17}$$

where $u_{d2}'=G_{PIR}(s)(i_{d2}^*-i_{d2}),\ u_{q2}'=G_{PIR}(s)(i_{q2}^*-i_{q2})$ are rotor voltage decoupling term in *dq* coordinate system and can be output through regulation of PI-R controller by adjusting the difference between the rotor current command value and the actual value. $G_{PIR}(s)$ is the transfer function of PI-R controller. ∇u_{d2}, ∇u_{q2} are rotor voltage compensation term. Considering the steady state operation, the amplitude of fundamental and harmonic wave of stator flux is constant (assuming the differential term is zero). According to (17), full compensation term of rotor voltage could be rewritten as follows (including the fundamental and harmonic compensation)

$$\begin{cases} \nabla u_{d2}=\omega_s\sigma L_2 i_{q2}^{(1)}-\sigma L_2\left(\omega_{5s}i_{-5q2}^{(5)}e^{-j6\omega_1 t}-\omega_{7s}i_{+7q2}^{(7)}e^{j6\omega_1 t}\right)+\dfrac{L_m}{L_1}\left(\omega_s\psi_{q1}^{(1)}-\omega_{5s}\psi_{-5q1}^{(5)}e^{-j6\omega_1 t}+\omega_{7s}\psi_{+7q1}^{(7)}e^{j6\omega_1 t}\right) \\ \nabla u_{q2}=-\omega_s\sigma L_2 i_{d2}^{(1)}+\sigma L_2\left(\omega_{5s}i_{-5d2}^{(5)}e^{-j6\omega_1 t}-\omega_{7s}i_{+7d2}^{(7)}e^{j6\omega_1 t}\right)-\dfrac{L_m}{L_1}\left(\omega_s\psi_{d1}^{(1)}-\omega_{5s}\psi_{-5d1}^{(5)}e^{-j6\omega_1 t}+\omega_{7s}\psi_{+7d1}^{(7)}e^{j6\omega_1 t}\right) \end{cases} \tag{18}$$

The improved control strategy of the PI-R current controller based on the *dq* coordinate system considering the rotor voltage and current coupling is shown in Figure 1.

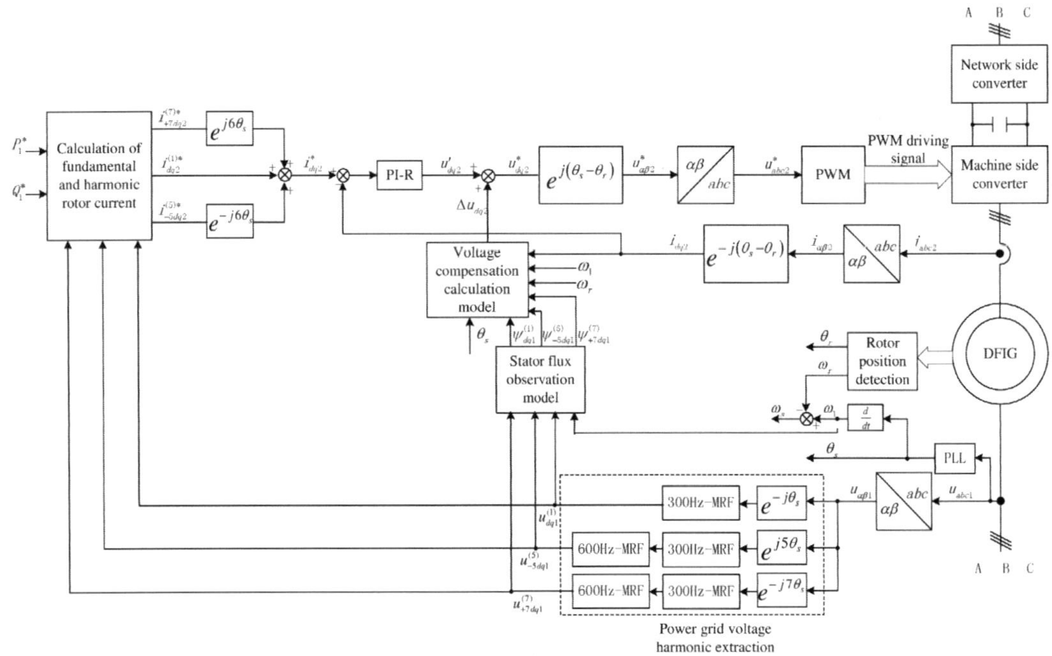

Figure 1. Improved control strategy of PI-R current controller based on *dq* coordinate system

The fundamental order of rotor current output by instruction calculation module in $(dq)^{+}$ reference frames, 5th harmonic order in $(dq)^{5-}$ reference frames and 7th harmonic order in $(dq)^{7+}$ reference frames are transformed into $(dq)^{+}$ reference frames and combined into total current command value. The detected rotor current can be directly transformed into *dq* coordinate system as feedback value no need the separation between the fundamental and harmonic. The pre compensation rotor voltage decoupling term is obtained by adjusting the PI-R current controller. Adding decoupling term to rotor voltage coupling compensation which is output by voltage compensation calculation model results the directive value of rotor voltage in *dq* coordinate system. PWM pulse signal of the turn-on and turn-off of the IGBT device of the generator side converter can be got through the PWM modulation link in order to realize power and harmonic control under the specified control target of DFIG.

The stator flux is needed when calculate the compensation voltage. Therefore, the stator flux linkage observation model is added in the control strategy, as shown in figure 2.

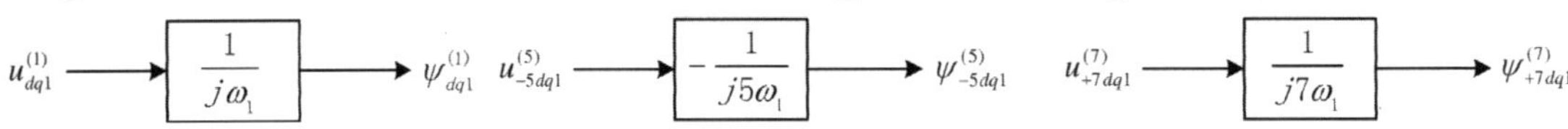

Figure 2. Stator flux observation model

Calculation model of compensation voltage based on the dq coordinate system can be obtained by formula (19), as shown in figure 3.

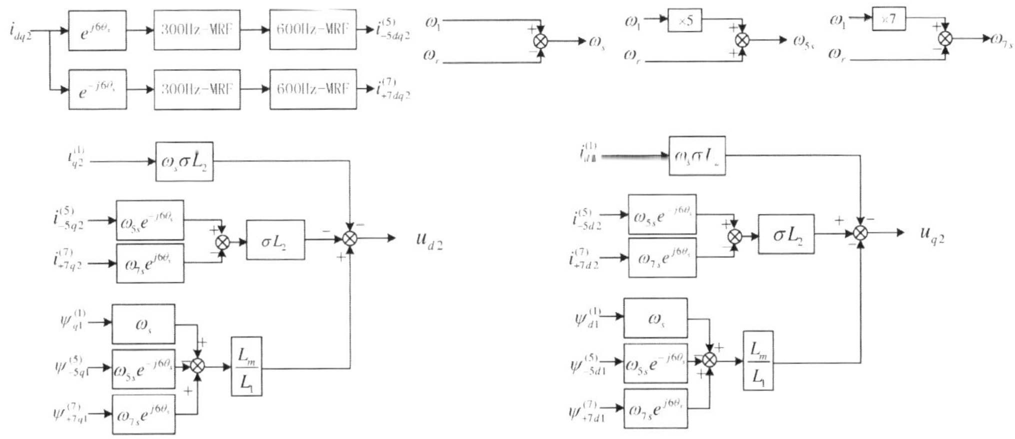

Figure 3. Compensation voltage calculation model。

3.2. *Design method of PI-R current controller.*

Based on (1), excluding the stator resistance results

$$\psi_{\rm dq1} = \frac{u_{\rm dq1}}{\frac{d}{dt} + j\omega_1} \tag{19}$$

Using Laplace transform to (3), (4) and (19) will obtain the transfer function of the DFIG control model as follows

$$\begin{cases} G_1(s) = 1/(s + j\omega_1) \\ G_2(s) = (s + j\omega_{\rm s})L_{\rm m}/L_{\rm s} \\ G_{\rm p}(s) = 1/(\sigma L_2 s + R_2) \end{cases} \tag{20}$$

The control structure diagram of DFIG control system under *dq* coordinate system can be obtained according to the mathematical model and control strategy above, as shown in figure 4.

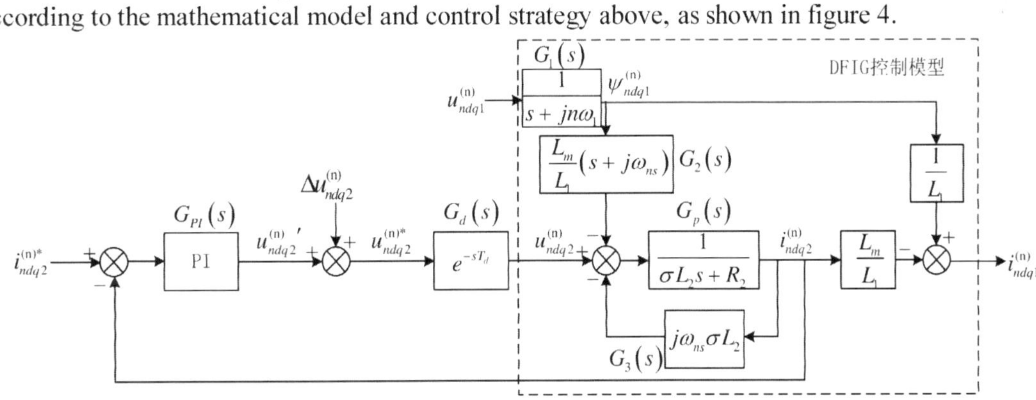

Figure 4. Control structure diagram of DFIG under dq coordinate system

The dotted line in the figure represents the DFIG model. $G_{\rm d}(s)$ represents the transfer function of the converter action delay according to the paper [26]. Generally, the time delay $T_{\rm d}$ is equal to the sampling period and equal to 1/2 of the switching period. From the converter parameters in Table 1, the sampling period is 250 μs.

The transfer function of the PI-R controller is

$$G_{\mathrm{PIR}}(s) = K_{\mathrm{P}} + \frac{K_{\mathrm{I}}}{s} + \frac{2K_{\mathrm{R}}\omega_{\mathrm{c}}s}{s^2 + 2\omega_{\mathrm{c}}s + \omega_0^2} \tag{21}$$

where K_{P} is proportional gain. K_{I} is integral gain. K_{R} is the transmission gain of resonant controller at the resonant frequency ω_0. ω_{c} is the open loop cutoff frequency of resonant controller when $K_{\mathrm{R}}=1$.

Excluding the compensation items and coupling terms, the structure of the current loop is shown in figure 5.

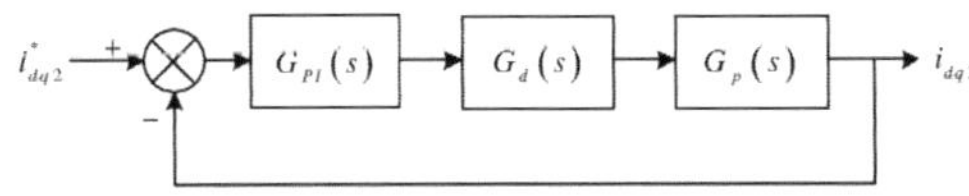

Figure 5. Current loop structure diagram of DFIG control system

ω_{cr} means the open loop crossing frequency. Making $s=j\omega_{\mathrm{cr}}$, rotor side time constant $\tau=(\sigma L_2)/R_2$, can get the frequency characteristic expression of rotor current open loop transfer function at the crossover frequency ω_{cr}.

$$G_{i2o}(j\omega_{\mathrm{cr}}) = \frac{K_{\mathrm{I}} - \dfrac{\left[K_{\mathrm{I}} + 2\omega_{\mathrm{c}}(K_{\mathrm{P}} + K_{\mathrm{R}})\right]\omega_{\mathrm{cr}}^2}{\omega_0^2} + j\left(K_{\mathrm{P}}\omega_{\mathrm{cr}} - K_{\mathrm{P}}\dfrac{\omega_{\mathrm{cr}}^3}{\omega_0^2} + K_{\mathrm{I}}\dfrac{2\omega_{\mathrm{c}}\omega_{\mathrm{cr}}}{\omega_0^2}\right)}{jR_2\omega_{\mathrm{cr}}(j\omega_{\mathrm{cr}}\tau + 1)\left[1 - \left(\dfrac{\omega_{\mathrm{cr}}}{\omega_0}\right)^2\right]} e^{-j\omega_{\mathrm{cr}}T_{\mathrm{d}}} \tag{22}$$

Frequently, the crossing frequency ω_{cr} is slightly larger than the resonant frequency ω_0 of the PI-R controller, and it can be considered that the two are approximately equal according to the paper [27]. Therefore, the real part of the upper numerators is $K_{\mathrm{I}}-K_{\mathrm{I}}\omega_{\mathrm{cr}}^2/\omega_0^2\approx0$.Because of $2\omega_{\mathrm{c}}\ll\omega_0$, the imaginary part of the upper numerators is $K_{\mathrm{I}}(2\omega_{\mathrm{c}}\omega_{\mathrm{cr}})/\omega_0^2\approx0$. The time constant of the rotor τ can be calculated by the DFIG simulation experiment parameter in Table 1 which is 11 more than 1. So the denominator $(j\omega_{\mathrm{cr}}\tau+1)$can be simplified to $j\omega_{\mathrm{cr}}\tau$. In summary, the frequency characteristic expression of rotor current open-loop transfer function at the crossing frequency ω_{cr} can be simplified to

$$G_{i2o}(j\omega_{\mathrm{cr}}) \approx \frac{2\omega_{\mathrm{c}}(K_{\mathrm{P}} + K_{\mathrm{R}})\left(\dfrac{\omega_{\mathrm{cr}}}{\omega_0}\right)^2 + jK_{\mathrm{P}}\omega_{\mathrm{cr}}\left[\left(\dfrac{\omega_{\mathrm{cr}}}{\omega_0}\right)^2 - 1\right]}{\omega_{\mathrm{cr}}^2\sigma L_2\left[1 - \left(\dfrac{\omega_{\mathrm{cr}}}{\omega_0}\right)^2\right]} e^{-j\omega_{\mathrm{cr}}T_{\mathrm{d}}} \tag{23}$$

The phase margin ϕ expression of rotor current open-loop transfer function at the crossing frequency ω_{cr} is

$$\phi = \pi + \angle G_{i2o}(j\omega_{\mathrm{cr}}) \approx \arctan\left\{\frac{K_{\mathrm{P}}\omega_{\mathrm{cr}}}{2\omega_{\mathrm{c}}(K_{\mathrm{P}} + K_{\mathrm{R}})}\left[1 - \left(\frac{\omega_0}{\omega_{\mathrm{cr}}}\right)^2\right]\right\} - \omega_{\mathrm{cr}}T_{\mathrm{d}} \tag{24}$$

The open loop transfer function gain at the open loop crossing frequency should be 1, that is

$$\left|G_{i2o}(j\omega_{\mathrm{cr}})\right| \approx \left|\frac{2\omega_{\mathrm{c}}(K_{\mathrm{P}} + K_{\mathrm{R}})\left(\dfrac{\omega_{\mathrm{cr}}}{\omega_0}\right)^2 + jK_{\mathrm{P}}\omega_{\mathrm{cr}}\left[\left(\dfrac{\omega_{\mathrm{cr}}}{\omega_0}\right)^2 - 1\right]}{\omega_{\mathrm{cr}}^2\sigma L_2\left[1 - \left(\dfrac{\omega_{\mathrm{cr}}}{\omega_0}\right)^2\right]} e^{-j\omega_{\mathrm{cr}}T_{\mathrm{d}}}\right| = \left|\frac{2\omega_{\mathrm{c}}(K_{\mathrm{P}} + K_{\mathrm{R}})\left(\dfrac{\omega_{\mathrm{cr}}}{\omega_0}\right)^2 + jK_{\mathrm{P}}\omega_{\mathrm{cr}}\left[\left(\dfrac{\omega_{\mathrm{cr}}}{\omega_0}\right)^2 - 1\right]}{\omega_{\mathrm{cr}}^2\sigma L_2\left[1 - \left(\dfrac{\omega_{\mathrm{cr}}}{\omega_0}\right)^2\right]}\right| = 1 \tag{25}$$

and
$$\omega_c\left(K_P+K_R\right)\approx 0.5\omega_{cr}\left[1-\left(\frac{\omega_0}{\omega_{cr}}\right)^2\right]\sqrt{\left(\omega_{cr}\sigma L_2\right)^2-K_P^2} \tag{26}$$

Combine (24) and (26) results
$$K_P\approx\frac{\sigma L_2\omega_{cr}}{\sqrt{1+\dfrac{1}{\tan^2\left(\phi+\omega_{cr}T_d\right)}}} \tag{27}$$

Based on the above derivation, the design steps of PI-R current controller in *dq* coordinates are as follows.

1) Select the appropriate phase margin and open-loop crossover frequency. Phase margin ϕ is usually recommended between 30 degrees to 60 degrees according to the paper [28]. The recommended open-loop crossover frequency ω_{cr} is between 1/10 and 1/5 of the switching frequency.

2) Put predetermined phase margin ϕ and open-loop crossover frequency ω_{cr} into (27) so that the parameter K_P of PI-R controller can be solved.

3) Put K_P,ϕ, ω_{cr} into (24) and the value of $(K_P+K_R)\omega_c$ will be obtained.

4) According to the accuracy of the digital processor, select a smaller value of ω_c. The usual recommended PI-R controller cutoff frequency is between 2rad/s and 10rad/s in the paper [27].

5) The value of K_R can be calculated from the known value of K_P, ω_c and $(K_P+K_R)\omega_c$.

6) In PI-R controller, the parameters of the PI controller mainly control the fundamental component of *dq* coordinate system through no steady-state error control. The impact of parameters of R controller is very small at the fundamental frequency. So the value of K_I in PI-R controller can be approximated the same as it in PI controller.

7) Check whether the solution of the PI-R parameters meet the control system's steady-state and dynamic performance requirements. If not, re-adjust the value of phase margin, open loop cross frequency and cut off frequency. Repeat the previous steps until the system's steady-state and dynamic performance is optimal.

According to the above steps, phase margin is selected as 30 degree in this paper. Open loop crossing frequency ω_{cr} is selected as 1/6 switching frequency which is 333×2πrad/s. Cutoff frequency is selected as 5rad/s. So the parameters of the PI-R controller are K_P=18、K_I=405、K_R=942.

4. Simulation

The model of DFIG wind power generation system is built in PSCAD simulation platform, and the system parameters are shown in Table 1 in Appendices according to the paper [29].

In order to verify whether the compensation term optimizes the generator control performance considering DFIG rotor voltage and current coupling, different control strategies are used for the same model, such as the traditional control strategy, control strategy of PI-R current controller without compensation and improved control strategy of PI-R current controller with compensation. Through simulation and comparison, not only can verify the superiority of the control performance of PI-R current controller under the *dq* coordinate which is better than the traditional control strategy in harmonic environment, but also can verify the superiority of the improved compensation control strategy of PI-R current controller under the *dq* coordinate which is better than that without compensation.

In the simulation, maintain a constant wind speed 6m/s. The total length of the simulation is 20s. DFIG runs in the state of sub synchronous. The rotor speed is about 0.8p.u. The stator reactive power instruction is 0Var. The grid voltage distortion includes 4% of the 5th negative sequence harmonic components and 4% of the 7th positive sequence harmonic components. The harmonic voltage distortion rate of THD is about 5.66%. The parameters of the PI-R current controller is calculated according to the design value in 2.2.The simulation control target is to balance the rotor current. The simulation waveform and its spectrogram under different control strategies is shown in Figure 6 and Figure 7. In Figure 6, each simulation waveform from left to right are respectively on behalf of the results of using the traditional PI control strategy, no compensation PI-R control strategy and PI-R control strategy with compensation. In Figure 7, each spectrum analysis diagram from top to bottom are also respectively on behalf of the results of using the traditional PI control strategy, no compensation PI-R control strategy and PI-R control strategy with compensation.

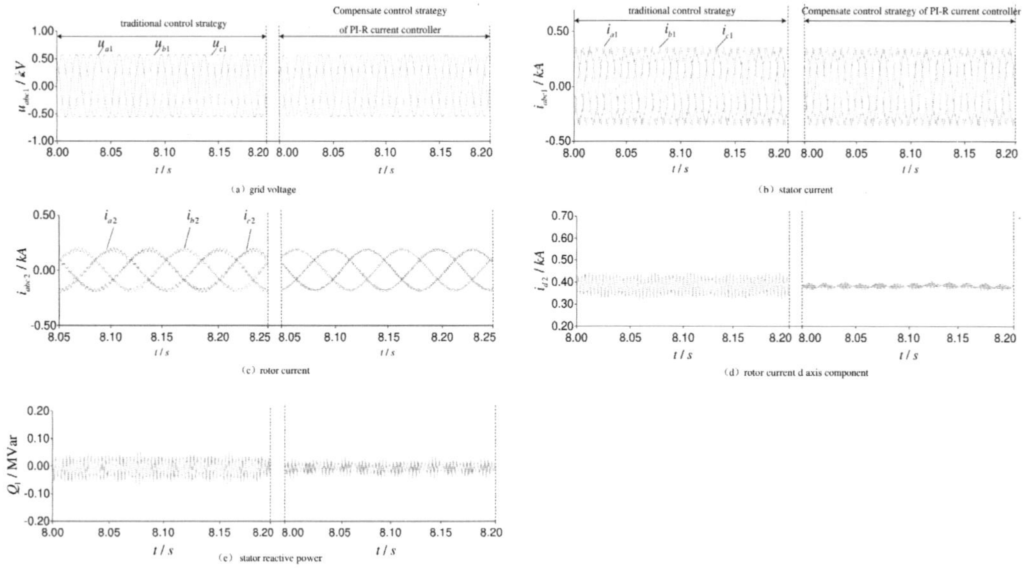

Figure 6. Waveform of different control strategies for balancing the rotor current as the control target

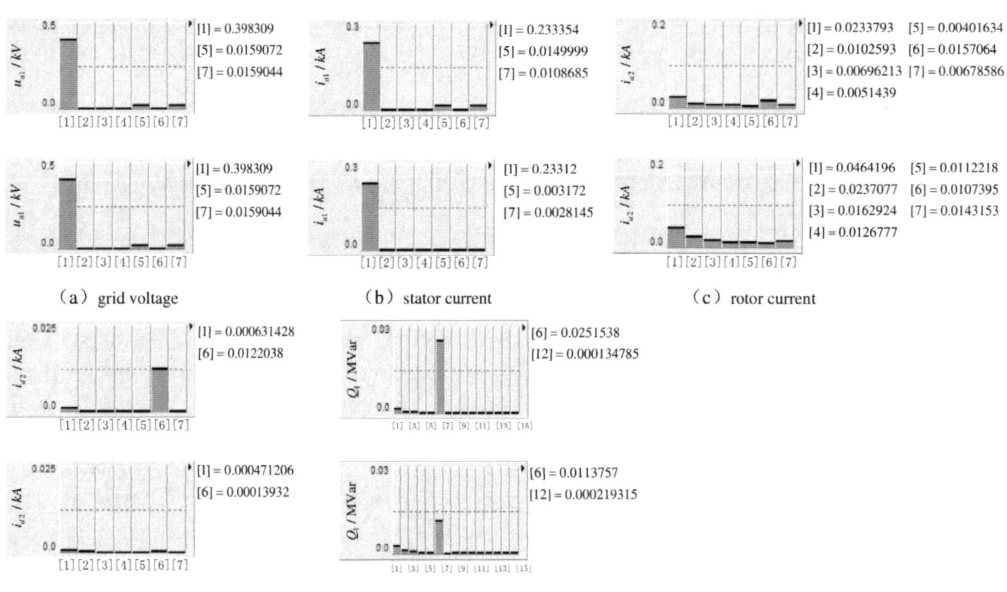

Figure 7. Spectrum analysis chart of different control strategies for balancing the rotor current as the control target

The simulation results of different control strategies can be seen qualitatively according to the waveform. Spectrum analysis chart can be used to quantify the control effect of different control strategies. The numerical value of the spectrum analysis chart is the effective value of the harmonic.

Choosing to balance the rotor current as the control target, the following conclusions can be drawn from figure 6 and Figure 7. (The defined current harmonic suppression effect of a control strategy here means the ratio of the square root of the d, q axis harmonic current in a certain control strategy to the ratio in the traditional control strategy.)

(1) After adding compensation, the control strategy of PI-R current controller based on *dq* coordinates system can reduce the content of the 5th and 7th harmonic currents in stator current better than the traditional control strategy, making the stator current waveform more tend to be good sine waveform. The harmonic suppression effect of the stator current with compensation is 22.9% and the effect is better after adding compensation.

(2) After adding compensation, the control strategy of PI-R current controller based on *dq* coordinates system can reduce the content of the 6th harmonic currents in rotor current better than the traditional control strategy, making the rotor current waveform more tend to be good sine waveform. The harmonic suppression effect of the rotor current with compensation is 1.5%. Obviously, the effect is better after adding compensation. Choosing to balance the rotor current as the control target, the harmonic suppression effect of the rotor current is more significant than that of the stator current.

(3) After adding compensation, the control strategy of PI-R current controller based on *dq* coordinates system can reduce the content of the 6th and 12th harmonic currents in stator reactive power better than the traditional control strategy, making the stator reactive power waveform more and more flat. The harmonic suppression effect of the rotor current with compensation is 45.2% and the effect is better after adding compensation.

5. Conclusion

In this paper, the dynamic modeling of DFIG is carried out under the harmonic environment of power grid. Based on that, the coupling relationship between rotor voltage and current is presented, including the fundamental wave and harmonic wave. The improved control strategy of PI-R current controller based on *dq* coordinate system is put forward through the full compensation of coupling. In addition, the DFIG control model is established and the method and process of designing the PI-R controller parameters according to the phase margin of the open loop transfer function and the crossing frequency are presented. Theoretical analysis and simulation results show that in the harmonic environment of power grid, the improved control strategy is more effective than the traditional and the unimproved control strategy to suppress the influence of the grid harmonics considering the coupling between DFIG rotor voltage and current. Choosing to balance the rotor current as the control target, the harmonic component of rotor current is effectively controlled and the harmonic component of stator current, stator active and reactive power is restrained to a certain extent along with the suppression of rotor current harmonics at the same time.

Acknowledgments

I wish to thank my supervisor Professor Liu whose continual guidance and support during my research process. Thanks for financial support from project support by State Grid Corporation of China (SGSDDK00KJJS1500155).

A. 1

NOMENCLATURE	
$\boldsymbol{u}_1,\boldsymbol{u}_2$	Stator and rotor voltage vectors
$\boldsymbol{i}_1,\boldsymbol{i}_2$	Stator and rotor current vectors
ψ_1,ψ_2	Stator and rotor flux linkage vectors

ω_1, ω_r, ω_s	Stator, rotor, and slip angular frequencies
P_1, Q_2	Stator output active and reactive powers
L_1, L_2	Stator and rotor self-inductances in *dq* coordinates
L_m	Mutual inductance
Subscripts	
α, β	Stationary α- and β-axis
d, *q*	Synchronous *d*- and *q*-axis
1, *2*	Stator and rotor
+, 5−, 7+	Fundamental, fifth-order and seventh-order components
dq、-5*dq*、+7*dq*	Vector corresponding coordinate system
Superscripts	
+, 5−, 7+	Positive $(dq)^{+}$, $(dq)^{5-}$, $(dq)^{7+}$ reference frames

Table A.1 Simulation parameters of DFIG

Wind turbine parameters	Rated power: 1.5MW	Blade radius: 35m	Air density: 1.225kg/m^3
	Rated wind speed: 12m/s	Cut in wind speed: 3m/s	Cut off wind speed: 25m/s
DFIG parameters	Rated power: 1.5MVA	Rotor stator ratio: 2.5	Pole-pairs:2
	Inertia constant: 1.14s	Stator resistance: $R_1 = 0.00698\Omega$	
	Rotor resistance: $R_2 = 0.008252\Omega$	Stator leakage reactance: $X_{1\sigma} = 0.05618\Omega$	
	Rotor leakage reactance: $X_{2\sigma} = 0.036818\Omega$		
	Mutual inductance between stator and rotor: $X_m = 1.485432\Omega$		
Converter parameters	DC Bus Voltage: 1100V	Switching frequency: 2kHz	Grid side wire inductance: 2mH
	Grid side wire resistance: 0.0003Ω	DC storage capacity: 8640μF	

References

[1] Wang Y and Xu L 2010 Coordinated control of DFIG and FSIG-Based wind farms under unbalanced grid conditions *Power Systems, IEEE Power & Energy Society* 367-77

[2] Yao J, Li H, Chen Z, Xia X, Chen X, Li Q and Liao Y 2012 Enhanced control of a DFIG-Based wind-power generation system with se-ries grid-side converter under unbalanced grid voltage conditions *IEEE Power Electronics Society* 3167-81

[3] Kearney J, Conlon M F and Coyle E 2009 The Integrated control of the rotor side and grid side converters in a dfig to reduce both pow-er and torque pulsations during network voltage unbalance condi-tions *Universities Power Engineering Conference (UPEC) Proc. the 44th International* 1-5

[4] Oshaba A S and Ali E S 2014 Bacteria foraging: a new technique for speed control of dc series motor supplied by photovoltaic system *WSEAS Transactions on Power Systems* **9** 185-195

[5] Oshaba A S, Ali E S and Abd-Elazim S M 2015 MPPT control design of pv system supplied srm using bat search algorithm *Sustainable Energy, Grids and Networks (SEGAN Elsevier)*

2C 51-60
[6] Oshaba A S, Ali E S and Abd-Elazim S M 2015 ACO Based Speed Control of SRM Fed by Photovoltaic System *Int. J. of Electrical Power and Energy Systems (IJEPES Elsevier)* **67** 529-36
[7] Emanuel A E 1991 Estimating the effects of harmonic voltage fluc-tuations on the temperature rise of squirrel-cage motors *IEEE Trans of Energy Conversions* **6** 161-8
[8] Ramos C J, Martins A P and Carvalho A S 2007 Rotor current controller with voltage harmonics compensation for a dfig operating under unbalanced and distorted stator voltage:industrial electronics society *IECON 2007. 33rd Annual Conference of the IEEE* 5-8
[9] Ali E S 2015 Speed control of induction motor supplied by wind turbine via imperialist competitive algorithm *Energy (Elsevier)* **89** 593-600
[10] Xu J, Chen W J, Xu D H and Chen M 2011 Stator harmonic current suppression of DFIG in wind power system under low order harmonic voltage of power grid *Automation of Electric Power Systems* **08** 87-92
[11] Yu Q, Heng N, Hu J Band Li J W 2010 Improved control of the grid-connected converter under the harmonically distorted grid voltage conditions *2010 International Conference on Electrical Machines and Systems (ICEMS)*
[12] Hu J B, Heng N, Hailiang Xand He Y K 2011 Dynamic modeling and improved control of dfig under distorted grid voltage conditions *IEEE Transantions on Energy Conversion* **26** 163-75
[13] Hu J B, He Y K, Guo X M and Nian H 2007 The unbalanced voltage of DFIG in wind power system modeling and control *Automation of Electric Power Systems* **14** 47-56
[14] Blaabjerg F, Teodorescu R, Liserre Mand Timbus A V Overview of control and grid synchronization distributed power generation system *Industrial Electronics , IEEE Transantions* **53** 1398-409
[15] Zhang S Q, Dai K, Xie B, Yu M and Kang Y 2010 Synchronous rotating coordinate system specified harmonic current control *Proc. the CSEE* **03** 55-62
[16] Yao J, Xia X F, Chen X Y and Liao Y 2012 Study on harmonic current suppression of full power converter for wind power grid connected power converter *Proc. of the CSEE* **16** 17-25
[17] Xu H L 2014 *Research on resonant control technology of double fed wind power converter* (Hangzhou: Zhejiang University)
[18] Liserre M, Teodorescu R and Blaabjerg F 2006 Multiple harmonics control for three-phase grid converter system with the use of PI-RES current con-troller on a rotating frame *IEEE Transantions on Power Electronics* **21** 836-41
[19] Trinh Q N and Lee H H 2011 Improvement of current performance foe grid connected converter under distorted grid condition *IET Conference on Renewable Power Generation (RPG 2011)* 233
[20] He Y K and Xu H L 2014 Adaptive problems of the power system of DFIG and its resonant control solution *Proc. the CSEE* **29** 5188-203
[21] Oshaba A S and Ali E S 2013 Speed control of induction motor fed from wind turbine via particle swarm optimization based pi controller *Research Journal of Applied Sciences, Engineering and Technology (Maxwell Science Publication)* **5** 4594-606
[22] Hu J B and He Y K 2011 Under the condition of harmonic in the power grid of DFIG wind generator modeling and control *Automation of Electric Power Systems* **11** 20-6
[23] Lingling F, Yuvarajan S and Kavasseri R 2010 Harmonic analysis of a DFIG for a wind energy conversion system *IEEE Transantions on Energy Conversion* **25** 181-90
[24] Nian H and Song Y P 2013 Based on vector proportional integral current regulator of DFIG operation control technology with harmonic in power grid *Proc. the CSEE* **06** 101-11
[25] Xu H L, Hu J B, Nian H and He Y K 2012 Dynamic model-ing and improved control of DFIG under unbalanced and distorted grid voltage conditions," in Industrial Electronics (ISIE) *2012 IEEE International Symposium* 1579-84
[26] Hongrae K, Degner M W, Guerrero J M, Fernando B and Robert D L 2010 Discrete-Time

IOP Conf. Series: Earth and Environmental Science **40** (2016) 012025 doi:10.1088/1755-1315/40/1/012025

current regualtor design for ac machine drive IEEE Transactions on Industry Applications **46** 1425-35
[27] Liu C J 2012 *The resonant control of DFIG in the grid environment* (Hanzhou: Zhejiang University)
[28] Miret J, Castilla M, Matas J, Guerrero J M and Vasquez J C 2009 Selective harmonic-compensation control for single-phase active power filter with high harmonic rejection *IEEE Transactions on Industry Electronics* **56** 3117-27
[29] Qi H R 2015 *Research on control technology of doubly fed wind power generator in power system harmonic environment* (Beijing: North China Electric Power University)

Review of optimization techniques of polygeneration systems for building applications

Rong A Y[1,3], Su Y[2] and Lahdelma R[1]

[1]Department of Mechanical Engineering, Aalto University, 00076, Aalto, Finland

[2]Department of Electromechanical Engineering, Macau University, Macau, China

Email: aiying.rong@aalto.fi

Abstract. Polygeneration means simultaneous production of two or more energy products in a single integrated process. Polygeneration is an energy-efficient technology and plays an important role in transition into future low-carbon energy systems. It can find wide applications in utilities, different types of industrial sectors and building sectors. This paper mainly focus on polygeneration applications in building sectors. The scales of polygeneration systems in building sectors range from the micro-level for a single home building to the large-level for residential districts. Also the development of polygeneration microgrid is related to building applications. The paper aims at giving a comprehensive review for optimization techniques for designing, synthesizing and operating different types of polygeneration systems for building applications.

1. Introduction

According to IEA [1], buildings are the largest energy consuming sector in the world, accounting for 32% of final energy use and equally important sources of CO_2 emissions. Polygeneration means simultaneous production of two or more energy products in a single integrated process. The typical energy products include electric power, heat and cooling. A larger potential of energy and emission savings have been recognized for polygeneration in building applications [2]. One viable way of exploiting the potential is through system optimization.

However, there are several challenges for polygeneration optimization for building applications. First, similar to applications in other areas such as utilities and industrial sectors, interdependence between different energy products makes system operations more complicated and production of different energy products need to be coordinated in an optimal manner [3]. Second, the scales of the systems range from micro-level to large-level [4]. There will be a significant difference between optimizing micro-scale systems and large-scale systems. Building integration of micro-generation systems is challenging because the loads are characterized by high diversity, stochastic nature and small quantity [5]. Third, driven by the development of sustainable buildings [1], utilization of renewable energy sources (RES) such as solar energy, wind power and biomass will be increased. Consequently, the system configuration varies significantly [6]. Finally, the concept of Smart Grid for power-only systems [7] has been extended to polygeneration systems, called Smart polygeneration microgrid (SPM) [8]. SPM can be applied in buildings. The techniques for optimizing polygeneration

[3] Address for correspondence: Rong A Y, Department of Mechanical Engineering, Aalto University, 00076, Aalto, Finland. E-mail: aiying.rong@aalto.fi.

microgrids (PM) or SPM should be different from those for optimizing traditional polygeneration systems.

In literature, there are various surveys for polygeneration systems from different perspectives [3], [4], [9-11]. There are also surveys that address specifically the development of micro-polygeneration technologies and systems [12-14]. In this paper, we attempt to give a review for optimization techniques for polygeneration systems for building applications, focusing on micro-, small- scale as well as PM. The typical applications of medium- and large-scale systems are district heating and cooling (DHC) systems. Reference [15] reviewed the optimization techniques for such systems. Reference [16] reviews the design and optimization of microgrids in residential application from the viewpoints of distributed energy systems including both power-only and polygeneration systems.

The paper is organized as follows. Section 2 describes the system configuration of polygeneration systems for building applications. Section 3 describes PM for building applications. Section 4 describes the development of optimization techniques for polygeneration systems and classifies optimization techniques for joint design, synthesis and operation of both traditional polygeneration system and PM.

2. System configuration

Figure 1 shows a schematic diagram for a typical polygeneration system that generates electricity, cooling and heat for buildings. A typical polygeneration system consists of a power generation unit (PGU) or a prime mover, heat recovery system (HRS) and thermal activated technology (TAT) such as absorb chiller, adsorb chiller and desiccant dehumidifier and auxiliary devices (not shown in Figure 1) that increase the flexibility of the system such as energy storage, electric chiller, fuel driven chiller and heat pump.

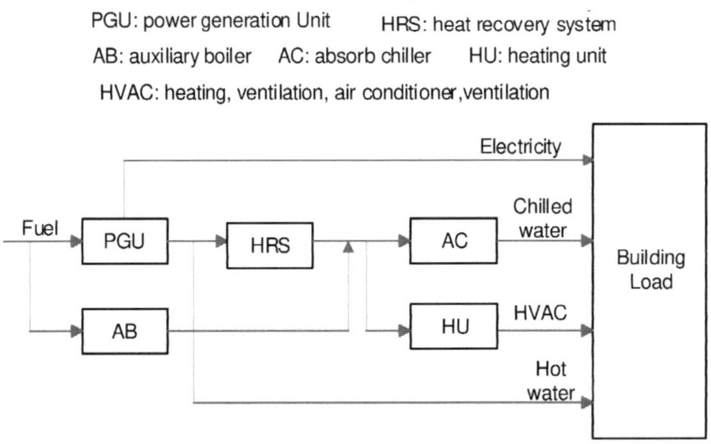

Figure 1. A schematic diagram of polygeneration system for building applications

The high energy efficiency of a polygeneration system comes from utilizing otherwise waste energy in the process. From Figure.1, the waste heat in the process was recovered twice. First, the exhaust heat from PGU (power generation unit) can be directly used for heating water in the pipe and providing hot water for buildings. Second, the waste heat can be recovered by HRS and used to drive TAT to achieve further use of waste heat. The auxiliary boiler (AB) is used to provide additional heat for heating and cooling purpose if the excess heat in the system is not sufficient. The selection of the heating unit is associated with the design of the HVAC (heating, ventilation, air conditioner) system. It can be seen that TAT technologies play an important role in utilizing the low grade heat in the process and thus improve the energy efficiency of the system.

The design of a polygeneration system needs considering the types of technologies and capacity of devices in the system. A polygeneration system can be classified according to (a) prime mover, (b) TAT technology, (c) auxiliary device, (d) primary energy sources (solar, fossil fuel, biomass) and (e) energy products (cooling, heat, power, hydrogen, and syngas) in the system [6]. In terms of scales of the system [4], the polygeneration system can be classified according to the capacity of electricity generation: micro-scale (under 20kW), small-scale (20kW-1MW), medium-scale (1MW-10MW) and large-scale (above 10MW). This classification is based on the type of prime movers. Fuel cell (FC), Stirling engine (SE) and micro gas turbine (MT) are suitable for micro-scale and small-scale systems. Traditionally, steam turbine (ST) and gas (combustion) turbine (GT) are usually for medium- and large-scale systems. However, they can also be applied for small-scale systems. Reciprocating internal combustion (IC) engine are applicable from micro-scale to medium- scale systems. Among all prime movers, fuel cell is an environment friendly prime mover where the chemical energy from the fuel is converted into electricity and water is the by-product. For remaining fuel based prime movers, CO_2 is the byproduct if fossil fuel is used. All of these fuel based types can be RES based, e.g., powered by solar energy or directly fueled by biomass or through biomass gasification [17]. RES can also be used directly to generate heat such as solar heat collector or power such as wind turbine and solar photovoltaics (PV).

The system configuration can be determined in the design stage by selecting proper types of prime movers and other technologies as well as the capacity of prime movers and other technologies based on the scale of the system. Once the system configuration is determined, operational strategies have impact on the efficiency of the system. In addition, load profile also affect the efficient operation of the system. To increase operational flexibilities and achieve the balance between supply and demand for different energy products, one option is to install different types of storages as a buffer for energy products. For traditional fuel based polygeneration systems, usually thermal energy storage (TES) are used [18].

With integration of RES such as wind and solar power into the energy systems, in additional to thermal storage, different types of electric storages (ES) such as battery and plug-in electrical vehicle (PEV) charging stations can also be considered [19]. The other option is to introduce electric heat pumps into the system to meet heat and cooling demands [20]. The efficiency of heat pump is higher than electric heater or chiller (one unit electricity can produce 3 units of thermal energy). An electric heat pump can work in both heating and cooling mode and thus can balance the load through years (summer: cooling mode, winter: heating mode). Electric heat pump in conjunction with thermal energy storage can increase additional flexibility to utilize RES and achieve power grid balance by avoiding starting up thermal demand-driven polygeneration plants when there is large RES based power production [21-23].

3. Polygeneration microgrid

Microgrids are small distributed energy systems integrating different distributed generation technologies such as reciprocating IC engine, micro-turbine, gas turbine, small-scale steam turbine, and fuel cell, photovoltaics (PV) and wind turbine, which can utilize local energy sources such as natural gas, biomass, solar and wind, to produce power, heat and cooling for satisfying local energy demands. When polygeneration technology is involved, the microgrid is called polygeneration microgrid (PM) as shown in Figure 2. This configuration is similar to [24].

PM can operate either in a stand-alone mode or a grid-connected (utility grid in Figure 2) mode. The polygeneration system in Fig 1 can be treated as a module connected to the microgrid in Figure 2. It means that higher efficiency of polygeneration can be carried to microgrids [25] because the overall electricity consumption of the absorb chiller is much lower than the compression (electric) chiller in the conventional system [25]. Thermal energy and electric power are distributed separately via district heating and cooling (DHC) network and local power network respectively and production of thermal energy and electric power need to be coordinated. This means that the structure of PM is the same as the microgrid for the power-only system in terms of power delivery. PM integrates DHC and power distribution network and supplies both electric power and thermal energy to customers by a mix of generator systems.

With integration of intermittent RES such as wind power and solar energy into microgrid, it is natural to extend the concept of Smart Grids for power delivery to PM for power and thermal energy delivery. On one hand, the higher efficiency of polygeneration can address sustainability issue more reasonably by considering demands of both electricity and thermal energy simultaneously [3]. On the other hand, polygeneration systems can provide additional flexibility to enhance the effectiveness and efficiency for power delivery and power utilization by exploiting the interdependence between different energy products in the polygeneration system [22], [23], as well as by utilizing auxiliary devices such as electric boiler and heat pump [21], [26]. This extension is called Smart PM (SPM) [8]. Infrastructure of PM and SPM may be the same. However, SPM is equipped with more advanced control and communication devices to facilitate the coordination of production for different energy products. More specifically, in SPM, the potential of thermal energy needs to be explored in depth. "Smartness" can be interpreted from different perspectives. For example, automatic adjustment, advanced control, accurate estimation, proactive coordination and integrated design.

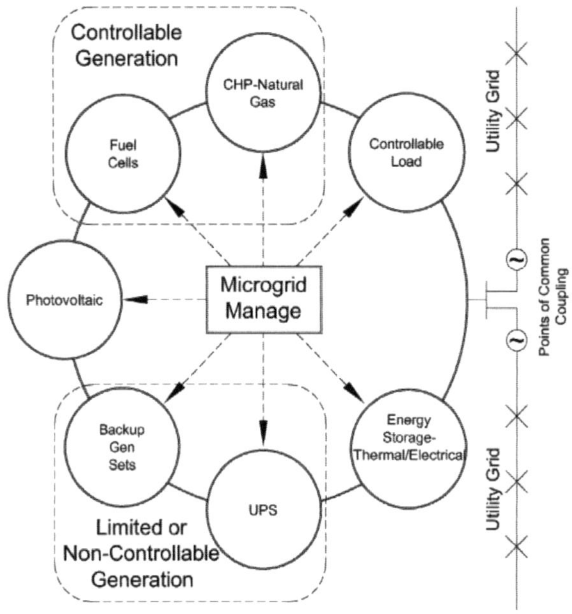

Figure 2. (Smart) Polygeneration microgrids

4. System optimization

A polygeneration system is more complicated than a power-only system regardless of the scale of the system due to interdependence between different energy products in the polygeneration plant and production scheduling need to be done in coordination through different production, storage and auxiliary facilities in the system [3]. On the top of this, there are different technical, legal [27] such as the capacity, efficiency and self-consumption percentage of polygeneration installations and social considerations that need to be respected in practice. Moreover, to promote polygeneration technology for building applications, the eligibility criteria for introducing technology needs to be modified or carefully designed [28] because it is not easy for a micro- and small-scale polygeneration plant to operate as efficiently as a large-scale plant. Furthermore, the system operates in a dynamic environment full of uncertainty. For example, the power price or the fuel price as well as load profile for different energy products varies from hour to hour in each day. Finally, the output of some intermittent RES such as wind and solar energy is hard to control or predict due to the stochasticity caused by wind speed or solar irradiance.

4.1 Methodology development

Cost and energy efficiency is the driver behind the development of micro- and small-scale polygeneration system. The system can utilize local energy resources to produce different energy products to meet the need of local customers. It can alleviate the dependency on imported energy resources to improve energy security. It can also reduce the energy loss due to the short distance between generation and the end user. Furthermore, it is suitable for serving the remote community which has no connection with the utility grid.

As compared with large- scale thermal systems where units for power, heat and cooling generation are subject to minimum up and down time constraints [29-32], micro- and small-scale systems can respond quickly and flexibly to load changes and market changes because generating units can be started up and shut down easily and quickly. Usually it is not necessary to consider such constraints [33]. However, it depends on the choice of the period length. If the period length is chosen to be 15 minutes, it is also necessary to consider minimum up and down time constraints [34]. Minimum down time (MDT) constraints indicate the necessary maintenance time after the unit is shut down while minimum up time (MUT) constraints indicate the need to minimize thermal stresses in the equipment which could otherwise arise.

However, such systems are prohibitively expensive [35] for building applications mainly due to the load profiles for buildings [36]. The cost and energy savings of the system can be achieved when thermal loads are sufficiently high and fully utilized [18]. The operating hours in the buildings are not long enough to recover the initial investment quickly. For example, thermal loads for residential buildings happen in the evening through early morning while the thermal loads for office buildings happen in the day time at office hours. In addition, the efficiency of the micro- and small-scale system may not competitive enough to gain market share as compared with domestic and commercial boilers in certain situations. Therefore, it is important to investigate different system configurations and operating strategies to make the system feasible, attractive and profitable. Different approaches such as simulation, thermodynamic analysis and optimization can be applied in designing, synthesizing and operating polygeneration systems. In practice, lots of effort is devoted to evaluating the performance of the system based on field tests, simulation and thermodynamic analysis due to complexity of the system. Reference [11] summarized development related to field test, simulation and thermodynamic analysis. It also touched the issues of optimization. The literature related to all these developments continues to grow. In the subsequent sections, we mainly focus on optimization approaches.

4.3 Overview of modeling and optimization based solution approaches

Optimization is one of the typical ways to deal with complicated systems. Optimization is a way to achieve the best possible performance criteria under certain conditions. Usually different types of mathematical models need to be constructed first to facilitate developing efficient algorithms or utilizing various standard optimization algorithms. The operating characteristics of a component can be either convex [37], [38] or non-convex [39]-[41]. Here the component refers to a device in the system. Convex characteristic means that if a component can operate on two points in a region, then it can also operate on the line segment connecting these two points. The convex characteristic represents a category of the "easy" optimization problem that is not difficult to find the optimal solution because the local optima is the global optima.

In terms of modelling techniques, reference [3] summarized three approaches. The first one is to apply general mathematical programming approach, including multi-objective optimization approach [42], [43] or thermodynamic modelling approach [44]. Thermodynamic models are used to describe or predict chemical and physical equilibria, which usually take nonlinear forms. The second one is to explicitly describe interaction between different components and the relationship between different energy products using input-output matrix [45-47]. The last one is to reformulate a general mathematical programming model of the component into the model with a special structure [48] according to extremal formulation [49]. Based on this modelling framework, power-only, heat-only and cooling-only components can be treated as a special case of polygeneration component model. Using this as basis, different types of efficient algorithms for single objective [50] and for multi-objective optimization [51], [52] can be developed for different system configurations.

Unit commitment (UC) is an important operational optimization problem in energy and power systems. UC determines the most efficient combinations of generation units to satisfy the demands for different energy products. UC problem can be modelled on the top of individual component (unit) model by introducing binary variables indicating whether the plant is ON or OFF. This will result in a mixed integer linear [33], [34] or non-linear [53] model. The operational flexibility of the system can be achieved by coordinating with heat pump, energy storage with the schedules of UC to increase the share of RES while considering other uncertainties such as price and demand [26]. In terms of dealing with uncertainty, fuzzy programming [54], [55], Monte Carlo simulation [56], [57], robust optimization [58], stochastic programming [43], sensitivity analysis [27], Markov method [59], [60] and forecast [61] have been applied.

4.4 Optimization techniques for micro and small-scale polygeneration systems

Reference [62] provided a paradigm for design, synthesis and operation of the energy system based on optimization approaches. The first step is to determine the system superstructure. The superstructure for a polygeneration system (as shown in Figure 1.) can be interpreted as the layout of the components in the system as well as the technology choices and capacity for the corresponding component. The second step is to build an optimization model that combine techno-economic evaluation with life cycle analysis while considering uncertainty. This is equivalent to building a mathematical model that optimizes some performance criteria while respecting technical constraints and accommodating uncertainties over a planning horizon. Then life cycle assessment is conducted based on the model. The third step is to apply or develop efficient computational algorithms for solving the resulting mathematical programming model.

[nc]**Table 1.** Optimization technique for single objective design and operation problems

Prime mover	Auxiliary device	Modelling	Solution	Reference
[c]GT,[d]IC,[e]MT,	[l]TES	[m]MILP	Robust	[58]
[c]GT	[i]EC	[m]MILP	[z]Standard	[74]
[d]IC	[l]TES	[m]MILP	[z]Standard	[75]
[c]GT, [d]IC		[m]MILP	[z]Standard	[76]
[a]FC, [c]GT, [d]IC,[e]MT	[k]HP	[n]MINLP	Bi-level	[77]
[a]FC,[d]IC,[e]MT,[g]SE	[l]TES	[m]MILP	[z]Standard	[79]
[b]GE		[m]MILP	[z]Standard +[u]LCA	[80]
[c]GT	[i]EC	[m]MILP	[z]Standard	[82]
[c]GT		[m]MILP	[z]Standard	[83]
[c]GT	[l]TES	[m]MILP	[z]Standard	[27]
[c]IC	[i]EC	[m]MILP	[z]Standard	[56]
[a]FC,[c]IC,[g]SE	[l]TES	[m]MILP	[z]Standard	[35]
[a]FC		Thermodynamic	Parametric study	[86]
[a]FC	[k]HP, [l]TES	[m]MILP	[z]Standard	[87]
[a]FC,[c]GT,[d]IC,[e]MT,[g]SE	[i]EC, [l]TES	[n]MINLP	Sequential	[88]
[a]FC,[c]GT,[d]IC,[e]MT,[g]SE	[i]EC, [l]TES	[n]MINLP	[z]Standard	[89]
[b]GE	[k]HP, [l]TES	[m]MILP	[z]Standard	[90]
[b]GE	[l]TES	[n]MINLP	[y]SA	[93]
[b]GE	[i]EC	[n]MINLP	[s]GA	[95],[96]
[c]GT	[i]EC, [l]TES	[n]MINLP	[x]PSO	[97]
[b]GE	[k]HP	[n]MINLP	[v]MPGA	[98]
[b]GE	[i]EC, [l]TES	[m]MILP	Graphic method	[100]
[a]FC		[m]MILP/[n]MINLP	[t]HO+[y]SA	[102]
[b]GE, [c]GT		[n]MINLP	Convexfication	[104]
[c]GT		Thermodynamic	[s]GA	[105]
[a]FC, [b]GE,[g]SE,[f]PV		[m]MILP	[z]Standard	**[106]**
[a]FC, [b]GE,[g]SE,[f]PV	[l]TES	[m]MILP	[z]Standard	**[107]**
[d]IC,[e]MT,[f]PV	[l]TES	[n]MINLP	[t]HO	**[109]**
[a]FC,[f]PV	[j]ES	[m]MILP	[t]PSO+[o]FCM	**[110]**

[nc] Tables 1 and 2 share the same notes. The notes appear at the end of Table 2.

Table 2. Optimization techniques for multi-objective design and operation problems

Prime mover	Auxiliary device	Modelling	Solution	Reference
[a]FC		Thermodynamic	[q]EA	[73]
[e]MT	[i]EC	Thermodynamic	[s]GA	[101]
[b]GE	[k]HP,[l]TES	[n]MINLP	[s]GA	[99]
[c]GT	[k]HP,[l]TES	[n]MINLP	[s]GA	[85]
[c]GT,[h]ST	[i]EC,[l]TES	[n]MINLP	[s]GA	[103]
[b]GE	[i]EC	[n]MINLP	[u]LCA	[94]
[a]FC, [d]IC,[e]MT,[h]ST	[l]TES	[n]MINLP	[p]Epsilon-c+forecast	[83]
[c]GT, [d]IC	[i]EC,[k]HP,[l]TES	[m]MILP	Robust	[91]
[c]GT,[d]IC	[k]HP	[m]MILP	Heuristic	[92]
[d]IC	[i]EC	[m]MILP	[o]BB+fuzzy	[54]
[d]IC	[i]EC	[m]MILP	[p]Epsilon-c+fuzzy	[55]
[b]GE		[m]MILP	[p]Epsilon-c+[u]LCA	[81]
[b]GE, [c]GT		[m]MILP	[p]Epsilon-c	[78]
[d]IC, [e]MT, [f]PV	[i]EC, [j]ES, [l]TES,	[m]MILP	[w]NSGA-II	[**108**]

[a] Fuel Cell.
[b] Gas Engine.
[c] Gas Turbine.
[d] reciprocating Internal Combustion engine.
[e] Micro gas Turbine.
[f] solar Photo Voltaics panel
[g] Stirling Engine.
[h] Steam Turbine.
[i] Electric Chiller.
[j] Electric Storage.
[k] Heat Pump.
[l] Thermal Energy Storage.
[m] Mixed Integer Linear Programming.
[n] Mixed Integer Non Linear Programming.
[o] Branch and Bound.
[p] Epsilon-constraint.
[q] Evolutionary Algorithm.
[r] Fuzzy Cognitive Map.
[s] Genetic Algorithm.
[t] Hierarchical Optimization.
[u] Life Cycle Analysis.
[v] Multiple Population Genetic Algorithm.
[w] Non-dominated Sorting Genetic Algorithm-II.
[x] Particle Swarm Optimization
[y] Sensitivity Analysis.
[z] Standard solver, e.g. CPLEX, LINGO, Excel solver

The overall problem can be formulated as a mixed integer linear or non-linear programming (MILP or MINLP) model. Reference [63] decomposed the problem at the three levels. Each level is associated with a subproblem, called synthesis, design and operation subproblems, respectively. The synthesis subproblem aims at identifying the components to be installed or to be excluded in the optimal lay-out. The problem can be formulated by an integer programming (IP) model by associating each component with a binary (0-1) variable indicating whether the component to be excluded or to be included in the optimal lay-out. The design subproblem aims at determining the optimal size/capacity of each component. Operation subproblem aims at optimization of the operating condition for each component based on the hourly load profile. For any energy system, these three subproblems should be solved jointly to identify the system configuration (in terms of the layout of the component and the size of the component) and the corresponding operational strategy to achieve the best results where the components can be operated in a convenient manner [64].

In literature, some research addresses optimization of joint design (synthesis) and operation problem while other research focuses on operation optimization. Operation optimization can be handled separately when the system configuration is determined. At the same time, operation optimization also has effect on sizing of the component. For the traditional system, operation optimization is related to determining operational strategy. Two commonly used operational strategies are FEL (following electric load) and FTL (following thermal load). The primary concern of economic dispatch (ED) in FEL is to satisfy electric demand while the primary concern of ED in FTL is to meet thermal demand. However, both of them suffers from energy waste in certain conditions [4]. Then some advanced operational strategies need to be developed based on the system configuration and the objective of the system. References [4] and [11] summarized the development of different operational strategies.

For the PM, operation optimization is associated with determining control strategy. The most commonly used control strategy is model predictive control (MPC) strategy [65]. MPC uses a dynamic model to estimate future outputs over a prediction horizon based on preceding and future actuator commands. It attempts to compute the best sequence of future control commands to minimize some performance criteria, subject to a set of physical or policy constraints. It uses the principle of closed loop control theory to penalize the error between output and its reference value and the deviation of mean values from their nominal values as well as supress rapid changes in input. The MPC is one of effective strategies to address the concerns of grid stability and load balance caused by the penetration of intermittent RES [66]. There are also other control strategies that use different predictors such as fuzzy logic based [67] and grey prediction based [68]. Real-time optimization and online control are major research themes related to operation optimization of the PM. The research follows this direction has started [69-72] and will continue in the future.

Here we mainly focus on joint design (synthesis) and operation problems. For the traditional polygeneration system, the system configuration more or less follows the structure of Fig 1, the major components include prime movers, HRS and TAT, i.e., prime movers associated with PGU and certain types of TAT are mandatory components. Auxiliary devices such as electric chiller, heat pump and thermal energy storage are optional. As we mentioned in Section 3, the traditional system is treated as a module in the PM. The PM should have more options for generators and storage facilities besides those mentioned in the traditional system because it needs to accommodate the penetration of intermittent RES. The additional components include PV panels, wind engine, electric storage such as battery and PEV. Therefore, we do not distinguish between PM and traditional system explicitly. The subtle difference can be reflected in the choices of prime movers and auxiliary devices in Table 1 and 2. Table 1 and 2 give single and multi-objective optimization techniques for design and operation problems respectively. In the reference column of two tables, the reference number appears bold if the problem is related to PM. The typical economic objective function is total cost which consists of investment cost and operational cost over a year. The investment cost includes the cost for purchasing equipment. The operational cost includes fuel cost and energy cost if there is a power exchange with main grid over the planning horizon. The additional objective may be related to CO_2 emissions and environmental impacts. In addition, the system energy efficiency and system reliability can also be used as an objective function. Multi-objective optimization approaches can address multiple conflicting objectives simultaneously and provide more realistic performance evaluation of the system.

5. Conclusion

Polygeneration is a promising technology to address efficient use of primary energy sources and CO_2 emissions for providing increasing energy demand requirements stimulated by economic development and population increase [111] due to its higher efficiency. Building sector accounts for one third of global energy consumption. The energy consumption can be reduced by increasing the efficiency of energy supply and by implementing efficient demand side management (DSM) [112]. DSM include two major activities, called demand response (load-shifting) and energy efficiency and conservation.

SPM plays a greater role in making DSM cost effective and convenient by providing two-way communication capabilities between customers and energy providers and real time information.

The sustainable development of the energy system drives the increasing use of local energy sources and local technologies to satisfy the energy demand. In this way, the environmental impact of energy production and use can be decreased. This leads to the paradigm shift for the energy system from the centralized production to distributed generation [113]. This is the reason why both micro- and small-scale polygeneration systems and RES based generation prosper. Main challenges for optimal design, synthesis and operation of polygeneration systems come from the requirements to increase the flexibility of integrating intermittent RES by selecting suitable system configurations, operating strategies and control strategies. It is expected that the development of SPM can enhance such capabilities effectively.

Acknowledgments
The research was funded by the STEEM (Sustainable Transitions of European Energy Markets) project (project No. 91002171) from Aalto University, Finland. Yan Su's work was supported by Solar Energy Lab of University of Macau with Project No. MYRG2014-00014-FST and FDCT/060/2014/A2.

References

[1] IEA (International Energy Agency) 2013 *Transition to sustainable buildings by 2050.*
[2] IEA (International Energy Agency) *Cogeneration and district energy –sustainable technologies for today and tomorrow* Report. http://www.iea.org/publications/freepublications/publication/name, 805, en. html
[3] Rong A and Lahdelma R 2016 Role of polygeneration in sustainable energy system development—challenges and opportunities from optimization viewpoints *Renew Sustain Energy Rev.* **53** 363-72.
[4] Liu M, Shi Y and Fang F 2014 Combined cooling, heating and power systems: A survey *Renew. Sustain. Energy Rev.* **35** 1-22.
[5] IEA EBC (Energy in Buildings and Communities Programme) 2014 *Integration of microgeneration and related technologies in building.*
[6] Calise F 2011 Design of a hybrid polygeneration system with solar collectors and a solid oxide fuel cell: dynamic simulation and economic assessment *Int. J of Hydrogen Energy* **36** 6128-50.
[7] Tuballa M L and Abundo M L 2016 A review of the development of Smart Grid technologies *Renew. Sustain. Energy Rev.* **59** 710-25.
[8] Bracco S, Delfino F, Pampararo F, Robba M and Rossi M 2013 The University of Genoa smart polygeneration microgrid test-bed facility: The overall system, the technologies and the research challenges *Renew. Sustain. Energy Rev.* **18** 442-59.
[9] Chicco G and Mancarella P 2009 Distributed multi-generation: a comprehensive review *Renew. Sustain. Energy Rev.* **13** 535-51.
[10] Murugan S and Horak B 2016 Tri and polygeneration systems—A review *Renew. Sustain. Energy Rev.* **60** 1032-51.
[11] Cho H, Smith AD and Mago P 2014 Combined cooling, heating and power: A review of performance improvement and optimization *Appl. Energy* **136** 168-85.
[12] Dong L, Liu H and Riffat S 2009 Development of small-scale and micro-scale biomass-fuelled CHP systems – A literature review *Appl. Therm. Eng.* **29** 2119–26.
[13] Maghanki M M, Ghobadian B, Najafi G and Galogah RJ 2013 Micro combined heat and power (MCHP) technologies and applications *Renew. Sustain. Energy Rev.* **28** 510-24
[14] Sonar D, Soni S L and Sharma D 2014 Micro-trigeneration for energy sustainability: Technologies, tools and trends *Appl. Therm. Eng.* **71** 790-6.
[15] Ortiga J, Bruno J C, Coronas A and Grossman I E 2007 Review of optimization models for the design of polygeneration systems in district heating and cooling networks *Comput Aided Chem Eng* **24** 1121-26.

[16] Liu P, Georgiadis M C and Pistikopoulosc E N 2013 An energy systems engineering approach for the design and operation of microgrids in residential applications *Chem Eng Res Des* **91** 2054-69.
[17] Raj N T, Iniyan S and Goic R 2011 A review of renewable energy based cogeneration technologies *Renew. Sustain. Energy Rev.* **15** 3640-80.
[18] Rong A, Lahdelma R and Luh P B 2008 Lagragian relaxation based algorithm for trigeneration planning with storages *Eur J. Oper Res.* **188** 240-57.
[19] Bracco S, Delfino F, Pampararo F, Robba M and Rossi M 2015 A dynamic optimization-based architecture for polygeneration microgrids with tri-generation, renewables, storage systems and electrical vehicles *Energy Convers Manag.* **96** 511-20.
[20] Danny Harvey L D 2006 *A Handbook on Low-energy Buildings and District-energy Systems: Fundamentals, Techniques and Examples* (UK: James and James).
[21] Meibom P, Kiviluoma J, Barth R, Barth H, Weber C and Larsen H V 2007 Value of electric boiler and heat pump for wind power integration *Wind Energy* **10** 321-37.
[22] Rong A and Lahdelma R 2016 An efficient model and algorithm for transmission-constrained multi-site combined heat and power system *Eur J. Oper Res* (minor revision)
[23] Rong A, Luh P B and Lahdelma R 2016 Dynamic programming based algorithm for the transmission-constrained multi-site combined heat and power system *ICPT 2016: 18th Int. Conf. Power Transmission* (accepted).
[24] Sellwood V 2015 Micorgrid and role of storage
[25] Bracco S and Delfino F 2015 The role of high efficiency trigeneration plants within sustainable smart microgrids: performance analysis and experimental tests *2015 AEIT Int. Annu.Conf. 1-6*
[26] Blarke M B 2012 Towards an intermittency-friendly energy system: Comparing electric boiler and heat bump in distributed cogeneration *Appl. Energy* **91** 348-65.
[27] Lozano M A, Ramos J C and Serra L M 2010 Cost optimization of the design of CHCP (combined heat, cooling and power) systems under legal constraints *Energy* **35** 794-805.
[28] Piacentino A, Barbaro C and Cardona F 2014 Promotion of polygeneration for buildings applications through sector- and user-oriented "high efficiency CHP" eligibility criteria *Appl. Therm. Eng.* **71** 882-94.
[29] Rong A, Hakonen H and Lahdelma R 2008 A variant of the dynamic programming algorithm for unit commitment of combined heat and power systems *Eur. J. Oper. Res* **190** 741-55.
[30] Rong A, Hakonen H and Lahdelma R 2009 A dynamic regrouping based sequential dynamic programming algorithm for unit commitment of combined heat and power systems *Energy Convers Manag.* **50** 1108-15.
[31] Rong A, Lahdelma R and Grunow M 2009 An improved unit decommitment algorithms for combined heat and power system *Eur. J. Oper. Res* **195** 552-62
[32] Bragin M A, Luh P B, Yan J H and Stern G A 2016 An efficient approach for solving mixed-integer programming problems under the monotonic condition *J Contr. Deci.* **3** 44-67.
[33] Collazos A, Marechal F and Gahler C 2009 Predictive optimal management method for the control of polygeneration systems *Comput. Chem. Eng.* **33** 1584-92.
[34] Bosman M G C, Bakker V, Molderink A, Hurink J L and Smit G J M 2012 Planning the production of a fleet of domestic combined heat and power generators *Eur. J. Oper. Res* **216** 140–51.
[35] Merkel E, McKenna R and Fichtner W 2015 Optimisation of the capacity and the dispatch of decentralized micro-CHP systems: A case study for the UK *Appl. Energy* **140** 120-34.
[36] Entchev E, Yang L, Ghorab M and Lee E J 2014 Performance analysis of a hybrid renewable microgeneration system in load sharing applications *Appl Therm Eng.* **71** 697-704.
[37] Lahdelma R and Hakonen H 2003 An efficient linear programming algorithm for combined heat and power production *Eur. J. Oper. Res.* **148** 141-51.
[38] Rong A and Lahdelma R 2005 An efficient linear programming model and optimization algorithm for trigeneration *Appl. Energy* **82** 40-63.
[39] Makkonen S and Lahdelma R 2006 Non-convex power plant modeling in energy optimisation *Eur. J. Oper. Res.* **171** 1113-26.

[40] Rong A and Lahdelma R 2007 An efficient envelope-based Branch and Bound algorithm for nonconvex combined heat and power production planning. *Eur. J. Oper. Res.* **183** 412-31.
[41] Liu P, Pistikopoulos EN and Li Z 2009 Global multi-objective optimization of a nonconvex MINLP problem and its application on polygeneration energy systems design *CMU-IBM Cyber-Infrastructure for MINLP.* http://www.minlp.org/library/index.php (accessed Dec 2009)
[42] Liu P 2009 *Modeling and optimization of polygeneration energy systems* (London, United Kingdom: Department of Chemical Engineering, Imperial College).
[43] Chen Y 2013 *Optimal design and operation of energy polygeneration systems* (United States of America: Department of Chemical Engineering, Massachusetts Institute of Technology).
[44] Yu Z, Han J and Cao X 2011 Investigation on performance of an integrated solid oxide fuel cell and absorb chiller tri-generation system *Int. J of Hydrogen Energy* **36** 12561-73.
[45] Chicco G and Mancarella P 2009 Matrix modelling of small-scale trigeneration systems and application to operational optimization. *Energy* **34** 261–73.
[46] Geidl M and Andersson G 2007 Optimal power flow of multiple energy carriers *IEEE Trans Power Syst* **22** 145-55.
[47] Liu M, Shi Y and Fang F 2013 Optimal power flow and PGU capacity of CCHP systems using a matrix modeling approach *Appl. Energy* **102** 794–802.
[48] Lahdelma R and Rong A 2005 Efficient re-fromulation of linear cogeneration planning model ed. M.H. Hamza *Proc. of the 24th IASTED Int. Conf. Modeling, Identifi. Contr* 1-6.
[49] Dantzig G 1963 *Linear programming and extensions* (NJ, USA : Princeton University Press)
[50] Rong A 2006 *Cogeneration planning under the deregulated power market and emissions trading scheme* (Finland: Department of Information Technology, University of Turku).
[51] Rong A, Figueira J R and Lahdelma R 2014 An efficient algorithm for bi-objective combined heat and power production planning under the emission trading scheme *Energy Convers Manag* **88** 525-34.
[52] Rong A, Figueira J R and Lahdelma R 2015 A two phase approach for the bi-objective non-convex combined heat and power production planning problem *Eur. J. Oper. Res* **245** 296-308.
[53] Hawkes A D, Brett D J L and Brandon N P 2009 Fuel cell micro-CHP techno-economics: Part 1 – model concept and formulation *Int. J. of Hydrogen Energy* **34** 9545-57.
[54] Mavrotas G, Demertzis H, Meintani A and Diakoulaki D 2003 Energy planning in buildings under uncertainty in fuel costs: The case of a hotel unit in Greece *Energy Convers Manag.* **44** 1303-21.
[55] Mavrotas G, Diakoulaki D, Florios K and Georgiou P 2008 A mathematical programming framework for energy planning in services' sector buildings under uncertainty in load demand: The case of a hospital in Athens *Energy Policy* **36** 2415-29.
[56] Mavrotas G, Florios K and Vlachou D 2010 Energy planning of a hospital using Mathematical Programming and Monte Carlo simulation for dealing with uncertainty in the economic parameters *Energy Convers Manag.* **51** 772-31.
[57] Li C, Shi Y, Liu S, Zheng Z, Liu Y 2010 Uncertain programming of building cooling heating and power (BCHP) system based on Monte-Carlo method *Energ Buildings* **42** 1369–75.
[58] Akbari K, Nasiri M M, Jolai F and Ghaderi SF 2014 Optimal investment and unit sizing of distributed energy systems under uncertainty: A robust optimization approach *Energ Buildings* **85** 275-86.
[59] Aval S M M, Ahadi A and Hayati H 2015 Adequacy assessment of Power systems incorporating building cooling, heating and power plants *Energ Buildings* **105** 236-46.
[60] Luh P B, Yu Y, Zhang B, Litvinov E, Zheng T, Zhao F, Zhao J and Wang C 2014 Grid integration of intermittent wind generation: A Markovian approach *IEEE T Smart Grid* **5** 732-41
[61] Fuentes-Corte L F, Santibanez-Aguilar J E and Ponce-Ortega J M 2016 Optimal design of residential cogeneration systems under uncertainty *Comput Chem Eng* **88** 86-102.
[62] Gong J and You F 2015 Sustainable design and synthesis of energy systems *Curr Opin Chem Eng* **10** 77-86

[63] Piacentino A, Barbaro C, Cardona F, Gallea R, and Cardona E 2013 A comprehensive tool for efficient design and operation of polygeneration-based energy microgrids serving a cluster of buildings. Part I: Description of the method *Appl Energy* **111** 1204–21.
[64] Piacentino A and Cardona F 2007 On thermoeconomics of energy systems at variable load conditions: integrated optimization of plant design and operation *Energy Convers Manag* **48** 2341–55.
[65] Banta L, Rossi I, Traverso A and Traverso AN 2014 Advanced control of a real smart polygeneration microgrid *Energ Proc* **61** 274-7.
[66] Rossi I, Banta L, Cuneo A, Luigi Ferrari ML, Traverso AN and Traverso 2016 A Real-time management solutions for a smart polygeneration microgrid *Energ Convers Manag* 112 11–20
[67] [Kyriakarakos G, Dounis A I, Arvanitis K G and Papadakis G 2012 A fuzzy logic energy management system for polygeneration microgrids *Renew Energ* **41** 315-27.
[68] Kyriakarakos G, Dounis A I, Arvanitis K G and Papadakis G 2013 Intelligent demand side energy management system for autonomous polygeneration microgrids *Appl Energ* **103** 39-51.
[69] Matics J and Krost G 2008 Micro combined heat and power home supply: Prospective and adaptive management achieved by computational intelligence techniques *Appl Therm Eng* **28** 2055-6.
[70] Basu A K, Bhattacharya A, Chowdhury S, and Chowdhury S P 2012 Planned scheduling for economic power sharing in a CHP-Based Micro-Grid *IEEE Trans on Power Syst* **27** 30-8.
[71] Gu W, Wu Z , Bo R, Liu W , Zhou G, Chen W and Wu Z 2014 Modeling, planning and optimal energy management of combined cooling, heating and power microgrid: A review *Int J Elec Power* **54** 26-37.
[72] Zhang Y, Zhang T, Wang R, Liu Y and Guo B 2015 Optimal operation of a smart residential microgrid based on model predictive control by considering uncertainties and storage impacts *Solar Energy* **122** 1052-65.
[73] Ahmadi P, Dincer I and Rosen MA 2013 Performance assessment and optimization of a novel integrated multigeneration system for residential buildings *Energ Buildings* **67** 568-78.
[74] Ameri M and Besharati Z 2016 Optimal design and operation of district heating and cooling networks with CCHP systems in a residential complex *Energ Buildings* **110** 135-48.
[75] Anatone M and Panone V 2015 A model for the optimal management of a CCHP plant *Energy Proc.* **81** 399-411
[76] Arcuri P, Florio G and Fragiacomo P 2007 A mixed integer programming model for optimal design of trigeneration in a hospital complex *Energy* **32** 1430-47.
[77] Arcuri P, Beraldi P, Florio G and Fragiacomo P 2015 Optimal design of a small size trigeneration plant in civil users: A MINLP (Mixed Integer Non Linear Programming Model) *Energy* **80** 628-41.
[78] Bracco S, Gabriele Dentici G and Siri S 2013 Economic and environmental optimization model for the design and the operation of a combined heat and power distributed generation system in an urban area *Energy* **55** 2014-24.
[79] Brandoni C and Renzi M 2015 Optimal sizing of hybrid solar micro-CHP systems for the household sector *Appl Therm Eng* **75** 896-907.
[80] Carvalho M, Serra L M and Lozano M A 2011 Optimal synthesis of trigeneration systems subject to environmental constraints *Energy* **36** 3779-90.
[81] Carvalho M, Lozano M A and Serra L M 2012 Multicriteria synthesis of trigeneration systems considering economic and environmental aspect *Appl Energy* **91** 245-54.
[82] Romero A, Carvalho M and Millar D L 2014 Application of polygeneration optimization techniques for a hospital in North Ontario *Trans Canad Soci Mech Eng* **38** 45-62.
[83] Fuentes-Corte L F, Santibanez-Aguilar J E and Ponce-Ortega J M 2016 Optimal design of residential cogeneration systems under uncertainty *Comput Chem Eng* **88** 86-102.
[84] Lozano M A, Ramos J C, Carvalho M and Serra L M 2009 Structure optimization of energy supply systems in tertiary sector buildings *Energy Buildings* **41** 1063-75
[85] Lythcke-Jørgensen C, Adriano Viana Ensinas AV, Münster M and Haglind F 2016 A methodology for designing flexible multi-generation systems *Energy*

http://dx.doi.org/doi: 10.1016/j.energy.2016.01.084
[86] Najafi B, Mamaghani A H, Baricci A, Rinaldi F and Casalegno A 2015 Mathematical modelling and parametric study on a 30 kWel high temperature PEM fuel cell based residential micro cogeneration plant *Int. J. Hydrogen Energy* **40** 1569-83.
[87] Ren H and Gao W 2010 A MILP model for integrated plan and evaluation of distributed energy systems *Appl Energy* **87** 1001–14.
[88] Rubio-Maya C, Uche J and Martínez A 2011 Sequential optimization of a polygeneration plant *Energy Convers Manag* **52** 2861–69.
[89] Rubio-Maya C, Uche-Marcuello J, Martínez-Gracia A and Bayod-Rújula A A 2011 Design optimization of a polygeneration plant fueled by natural gas and renewable energy sources *Appl Energy* **88** 449–57.
[90] Wu Q , Ren H, Gao W and Ren J 2016 Modeling and optimization of distributed energy supply network with power and hot water interchanges *Appl Therm Eng* **94** 635-43.
[91] Piacentino A and Cardona F 2008 EABOT – Energetic analysis as a basis for robust optimization of trigeneration systems by linear programming *Energy Convers Manag* **49** 3006–16.
[92] Piacentino A and Cardona F 2008 An original multi-objective criterion for the design of small-scale polygeneration systems based on realistic operating conditions *Appl Therm Eng* **28** 2391–404.
[93] Ren H, Gao W and Ruan Y 2008 Optimal sizing for residential CHP system *Appl Therm Eng* **28** 514-23.
[94] Jing Y, Bai H and Wang J 2012 Multi-objective optimization design and operation strategy analysis of BCHP system based on life cycle assessment *Energy* **37** 405-16.
[95] Wang J, Jing Y and Zhang C 2010 Optimization of capacity and operation for CCHP system by genetic algorithm *Appl Energ* **87** 1325–35.
[96] Wang J, Zhai Z, Jing Y and Zhang C 2010 Optimization design of BCHP system to maximize to save energy and reduce environmental impact *Energy* **35** 3388-98.
[97] Wang J, Zhai Z, Jing Y and Zhang C 2010 Particle swarm optimization for redundant building cooling heating and power system *Appl Energ* **87** 3668–79.
[98] Zeng R, Li H, Jiang R, Liu L, and Zhang G 2015 A novel method based on multi-population genetic algorithm for CCHP–GSHP coupling system optimization *Energy Convers Manag* **105** 1138–48.
[99] Zeng R, Li H, Jiang R, Liu L, and Zhang G 2016 A novel multi-objective optimization method for CCHP–GSHP coupling systems *Energ Buildings* **112** 149–58.
[100] Qin C, Tang J and Zhang Y 2015 An efficient algorithm for CCHP system sizing and an operational optimization model based on LP *J Nat Gas Sci Eng* **25** 189-96.
[101] Abdollahi G and Sayyaadi H 2013 Application of the multi-objective optimization and risk analysis for the sizing of a residential small-scale CCHP system *Energ Buildings* **60** 330–44.
[102] Gamou S, Yokoyama R and Ito K 2002 Optimal unit sizing of cogeneration systems in consideration of uncertain energy demands as continuous random variables *Energ Convers Manag* **43** 1349–61.
[103] Kavvadias KC and Maroulis ZB 2010 Multi-objective optimization of a trigeneration plant *Energ Policy* **38** 945–54.
[104] Savola T and Fogelholm CJ 2007 MINLP optimisation model for increased power production in small-scale CHP plants *Appl Therm Eng* **27** 89–99.
[105] Ghaebi H, Saidi MH and Ahmadi P 2012 Exergoeconomic optimization of a trigeneration system for heating, cooling and power production purpose based on TRR method and using evolutionary algorithm *Appl Therm Eng* **36** 113-25.
[106] Mehleri E D, Sarimveis H, Markatos N C and Papageorgiou L G 2012 A mathematical programming approach for optimal design of distributed energy systems at the neighbourhood level *Energy* **44** 96-104.
[107] Mehleri E D, Sarimveis H, Markatos N C and Papageorgiou L G 2013 Optimal design and operation of distributed energy systems: Application to Greek residential sector *Energy* **51** 331-42.

[108] Guo L, Liu W, Cai J, Hong B and Wang C 2013 A two-stage optimal planning and design method for combined cooling, heat and power microgrid system *Energ Convers Manag* **74** 433–45.

[109] Rivarolo M, Cuneo A, Traverso A and Massardo A F 2016 Design optimization of smart poly-generation energy districts through a model based approach *Appl Therm Eng* **99** 291-301.

[110] Karavas C P, Kyriakarakos G, Arvanitis K G and Papadakis G 2015 A multi-agent decentralized energy management system based on distributed intelligence for the design and control of autonomous polygeneration microgrids *Energ Convers Manag* **103** 166–79.

[111] Serra L M, Lozano M A, Ramos J, Ensinas A V and Nebra S A 2009 Polygeneration and efficient use of natural resources *Energy* **34** 575–86

[112] Marechal F, Favrat D and Jochem E 2005 Energy in the perspective of the sustainable development: The 2000 W society challenge *Resour Conserv Recy* **44** 245-62.

[113] Manfren M, Caputo P and Costa G 2011 Paradigm shift in urban energy systems through distributed generation: Methods and models *Appl Energ* **88** 1032-48.

Experimental investigation of static ice refrigeration air conditioning system driven by distributed photovoltaic energy system

Y F Xu[1, 2], M Li[14], X Luo[1], Y F Wang[1], Q F Yu[1], R H E Hassanien[3]
[1]Solar Energy Research Institute, Yunnan Normal University, Yunnan, China
[2]Zhejiang Solar Energy Product Quality Inspection Center, Zhejiang, China
[3]Agricultural Engineering Department, Faculty of Agriculture, Cairo University, Cairo, Egypt

E-mail: lmllldy@126.com

Abstract. The static ice refrigeration air conditioning system (SIRACS) driven by distributed photovoltaic energy system (DPES) was proposed and the test experiment have been investigated in this paper. Results revealed that system energy utilization efficiency is low because energy losses were high in ice making process of ice slide maker. So the immersed evaporator and co-integrated exchanger were suggested in system structure optimization analysis and the system COP was improved nearly 40%. At the same time, we have researched that ice thickness and ice super-cooled temperature changed along with time and the relationship between system COP and ice thickness was obtained.

1. Introduction

With the dramatic climate changes, the cooling demand has been increased and led to a rapid growth of energy consumption, which causes traditional fossil fuel energy shortage and great damage to climate and environment with the emissions of CO_2 and harmful particles by extensive use of traditional fossil energy. Furthermore, using the electric air conditioning increases the electrical pressure in the peak time and increases the tense situation between power supply and demands. Therefore, refrigeration driven by solar energy becomes one of the promising approaches to reduce or partially replace the conventional refrigeration systems. There are two main working modes of solar

[4] Address for correspondence: Solar Energy Research Institute, Yunnan Normal University, Yunnan, China.
Email: lmllldy@126.com

refrigeration namely solar thermal refrigeration and solar photovoltaic refrigeration [1, 2].

Solar photovoltaic (PV) refrigeration has more advantages on refrigerating effect, stable operation and energy utilization rate compared to solar thermal refrigeration. Aktacir [3] designed a multifunctional PV refrigerator and found that when indoor and outdoor average temperatures were 26.3 ℃ and 24.9 ℃, the minimum temperature of the refrigerator reached -10.6 ℃, but the system COP should be improved. To improve system efficiency, Bilgili [4] studied the performance of PV refrigerator. It was reported that the coefficient of the performance variation of the cooling system decreases with the decreasing of evaporating temperature. So a reasonable cooling temperature can manually defined is beneficial to the refrigeration efficiency. Ekren *et al.* [5] studied the photovoltaic DC refrigerator system and found that PV module conversion efficiency has a greater impact on the system exergy efficiency, so the increasing photo-electric conversion efficiency of PV module day by day is good for PV refrigeration. And the researched results also showed that the energy matching and coupling between photovoltaic arrays and refrigeration systems is a key point in PV refrigeration research. So in order to improve the energy matching and coupling characteristics, Kaplanis & Papanastasiou [6] improved the performance of a traditional refrigerator which driven by PV. The energy conversion, management and operation performance for PV refrigerator system powered on three conditions such as, photovoltaic components, battery and outage showed that the system COP gradually decreases from morning till night and the energy coupling between PV modules and refrigerator was very good when DC variable frequency compressor was adopted. And Mba *et al.* [7] used MATLAB software to simulate PV refrigeration system operating process and analysis system operating characteristics in different conditions. The energy conversion, management and operation performance for PV refrigerator system powered on three conditions such as, photovoltaic components, battery and outage showed that the system COP gradually decreases from morning till night [8]. Axaopoulos & Theodoridis [9] designed a PV Ice-maker without battery and studied its performance when the compressor operating efficiency was 9.2%, they found that this prototype have a good ice-making capability and reliable operation as well as a great improvement in the startup characteristics of the compressors, which remain working even during days with low solar irradiation of 150W/m^2.

Recently, the PV refrigeration system has been improved and developed in the aspects of product structure, operating efficiency and refrigeration performance. According to comprehensive analysis, PV refrigeration system research is currently mainly concentrated on ice generator driven by PV. Batteries are essential component to store energy and to solve the intermittent of solar energy in PV refrigeration system. Thus, static ice refrigeration conditioning system (SIRACS) driven by distributed photovoltaic energy system (DPES) was established based on our previous research results [10]. At first, experimental work has been done to analyze the energy coupling and transferring characteristics in light-electricity-cold conversion process. And then, the system structure optimization analysis was also investigated and optimized another SIRACS was put forward and experimental research has been done. The higher energy utilization efficiency and better financial rewards were achieved.

2. Materials and methods

2.1 Configuration of SIRACS driven by DPES

Static ice refrigeration conditioning system (SIRACS) driven by distributed photovoltaic energy

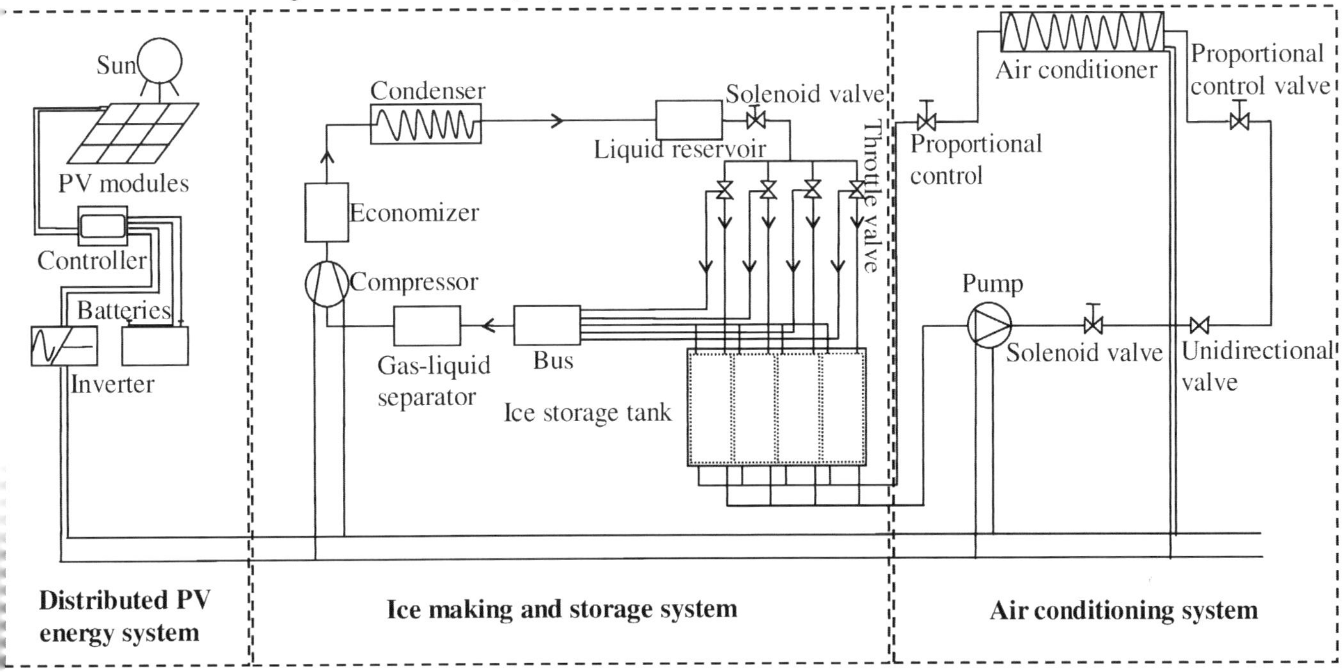

system (DPES) is mainly configured by DPES, ice generator, static ice storage system and air conditioning system. The system structure and work diagram are shown in Figure 1.

Figure 1. Structure and Work diagram of SIRACS driven by DPES

In daytime, DPES receives solar energy and turns it into direct-current (DC) electric power which can be converted to alternating current (AC) electric power by inverter to drive AC compressor, water pump, ethylene glycol pump and fan coil. To maintain the stability of electric energy supply, batteries are adopted and connected with controller to maintain the energy conversion and supply in the best way. Ice generator and storage system is made up with AC compressor, condenser, expansion valve, disc evaporator and static ice storage tank. Circulating water can be frozen in the disc evaporator and then the ice fallen into static ice storage tank. Thereby, the ice generator works as vapor compression refrigeration. In AC compressor, refrigerant R134a is compressed into high temperature and high pressure gas, filtered in gas-liquid separator and to release heat in condenser. Refrigerant is condensed to mild temperature and high pressure gas. When the gas inflows into throttle valve, it can be throttled to low temperature and low pressure liquid then feeds into plat evaporator. And then, refrigerant flows into the other gas-liquid separator to be sucked into the compressor. Thereby, the refrigeration cycle will be completed. Air conditioning system is mainly made up of coiled pipe cold exchanger fixed in static ice storage tank, Ethylene glycol pump, solenoid valve, proportional control valve and fan coil. Ethylene glycol is an adopted cold exchanging medium. According to the working principle diagram, a 0.2 kW SIRACS driven by DPES was established as shown in figure 2 and the main component parameters are shown in Table 1.

Table 1. Main component parameters of SIRACS driven by DPES

Component	Model	Parameters

DPES	PV module	JN-245	P_m: 245 W, V_m: 34.5 V, I_m: 7.10 A, V_{OC}: 43.5 V, I_{SC}: 8.18 A
	Controller	PL60	12-48 V 60 A charge,30 A load
	Inverter	solar 48 V	P: 3 kW, DC input voltage: 48 V, Output voltage: 220 V
	Batteries	SP12-65	Battery capacity: 12 V65 Ah, four batteries in series
SIRACS	Refrigerant	R134a	Boiling point: -26.1 ℃, Critical temperature: 101.1 ℃.
	Ice generator	IM50	Ice production: 2.08 kg/h, P: 380 W.
	Static ice storage tank	/	Capacity: $0.008m^3$
	Cold exchanging medium	Ethylene glycol	Melting point: -12.6 ℃, Viscosity: 25.66 mPa·s.
	Pump	RS15-6	Power: 46-93 W, Life: 6 m, Maximum flow rate: 3.4 m^3/h.
	Fan coil	/	Speed: 2800 r/min, number of fins: 95, Size: 23 cm*8 cm*20 cm, Coil numbers: 26, Coil inner diameter: 6 mm.

Figure 2. Pictures of 0.2 kW SIRACS driven by DPES

DPES were made up of two 245 W polycrystalline silicon PV in series. Four batteries in series were used to store electric power and the controller and inverter were adopted in DPES. SIRACS were made up with ice generator, static ice storage tank and fan coil.

Refrigerant (R134a) temperatures and pressure were measured by T-type thermocouples and pressure transducers, respectively. Voltages and currents of PV modules were measured by a digital multi-meter. The wind speed was measured by the wind speed transducer. Solar irradiation was measured by Pyranometer. Compressor input power was measured by a wattmeter. Electromagnetic flow meter was used to measure refrigerant flow and Ethylene glycol flow. The parameters of all instruments were shown in Table 2.

Table 2. Measuring instrument parameters

Instrument	Parameters and accuracy
Digital multi-meter	FLUKE F-179, DC voltage measurement accuracy: ±0.9%+2, DC current value measurement accuracy: ±1%+3.
Thermocouples	Type: T, range: -200 to 350 ℃, accuracy: ±0.4% of full scale
Wind speed transducer	EC-9S, range: 0-70 m/s, accuracy: ±0.3 m/s, resolution ratio: 0.1 m/s
Electromagnetic flow meter	SIEMENS FUP 1010, range: -12 m/s~12 m/s, accuracy: <0.3 m/s (1 ft/s), ±0.5% ~ ±2.0%,
Electronic balance	AHW-3, range: 0-3 kg, accuracy: ±0.05 g
Pressure transducer	Range: 0-0.6 MPa low side and 0-1.8 MPa high side; accuracy: ±0.25% of full scale at 25 ℃
Pyranometer	Kipp & zonen CMP6, Range: 0-2000 W/m^2, accuracy: ±5%
Wattmeter	DELIXI DDS607, Type: 220 V 20 A 50 Hz 3200 imp/kW, accuracy: 0.01 kW·h
Data acquisition	Agilent34972A data acquisition unit

2.2 Theoretical models of SIRACS driven by DPES

Solar energy absorbed by PV modules can be estimated as:

$$Q_{PV} = \alpha \tau G S_P \tag{1}$$

α is solar cells absorption coefficient and τ is PV module cover glass transmittance. G is the solar irradiance, W·m^{-2}. S_P is the total areas of PV modules, m^2.

The photo-electric transformation instantaneous efficiency of PV modules is given by:

$$\eta_{PV} = \frac{I_P \cdot V_P}{Q_{PV}} \tag{2}$$

I_P and V_P are respectively the PV modules output current and voltage values can be tested with Fluke digital multi-meter, A and V.

Batteries were used in order to provide the stability for system operation. The transient energy balance equation of batteries is expressed as:

$$Q_b = \pi\left(I_P \cdot V_P\right) - I_b \cdot V_b \tag{3}$$

Q_b is batteries storage electricity per time, W. I_b and V_b are respectively the batteries output current and voltage values, A and V. π is defined as switching constant in this equation. When DPES is connected with load, π equates 1 and π equates 0 when load is disconnected.

In ice making process, the energy balance equation is given by:

$$W_p = \dot{m}_r C_r (T_{con,in} - T_{con,out}) + \dot{m}_i h + Q_{loss} \tag{4}$$

W_p is ice generator compressor operation power, W. $\dot{m}_r$ is refrigerant mass flow tested by electromagnetic flow meter, kg·s^{-1}. C_r is average specific heat capacity of refrigerant flowing in the condenser, $J{\cdot}kg^{-1}{\cdot}K^{-1}$. $T_{con,in}$ is refrigerant temperature at condenser inlet, K. $T_{con,out}$ is refrigerant temperature at condenser outlet, K. $\dot{m}_i$ is the ice production per time, $kg{\cdot}s^{-1}$. h is ice latent heat, $J{\cdot}kg^{-1}$. Q_{loss} is energy loss per time in static ice making process, W.

And the instantaneous COP of static ice generator is shown as:

$$COP = \frac{\dot{m}_i h}{W_p} \tag{5}$$

Ice produced by static ice generator are stored in the tank when the fan and ethylene glycol pump are turn on, the cold stored by ice can be exchanged and transferred to cooling indoor air. The instantaneous efficiency of static ice storage air conditioner can be expressed by:

$$\eta_a = \frac{C_a \dot{m}_a \Delta T_a}{\dot{m}_r h} \tag{6}$$

C_a is specific heat capacity of air indoor, $kJ{\cdot}kg^{-1}{\cdot}K^{-1}$. $\dot{m}_a$ is air mass flow produced by fan, $kg{\cdot}s^{-1}$. ΔT_a is the temperature change of air through fan coil, K.

And system instantaneous energy efficiency is written:

$$\eta = \eta_{PV} \cdot COP \cdot \eta_a \tag{7}$$

3. Results and discussion

0.2kW SIRACS driven by DPES was tested on October 22 Kunming. Firstly, global solar radiation and ambient temperature were tested, as shown in Figure 3.

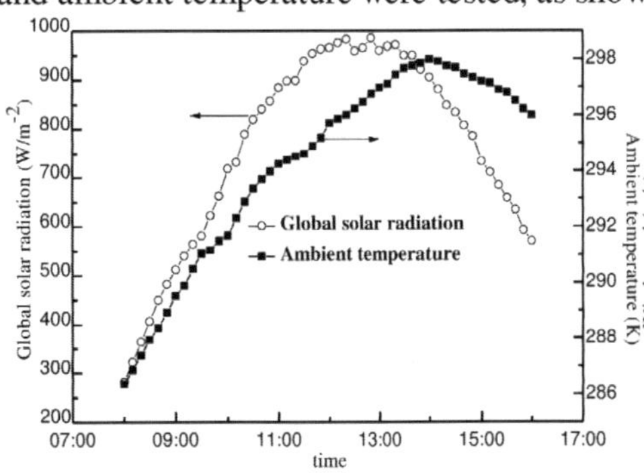

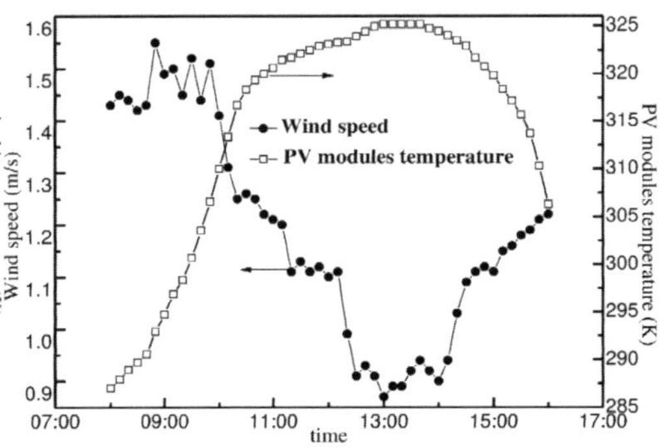

Figure 3. Variations of global solar radiation and ambient temperature on 22th NOV, 2015 in Kunming

Figure 4. Variations of wind speed and PV modules temperature on 22th NOV, 2015 in Kunming

PV modules temperature is a decisive component of output performance, which can be affected by wind speed changing at any time. Thus, wind speed and PV modules temperature can only be obtained by experimental measurement, as shown in Fig. 4. When SIRACS was driven by DPES, the output current and voltage of PV modules are tested and the change curves are shown in Fig. 5. Batteries were adopted to ensure power output stability and store excess power generated by PV modules. And so the output current and voltage of batteries changed with that of PV modules and the output characteristics of batteries were given by Figure 6.

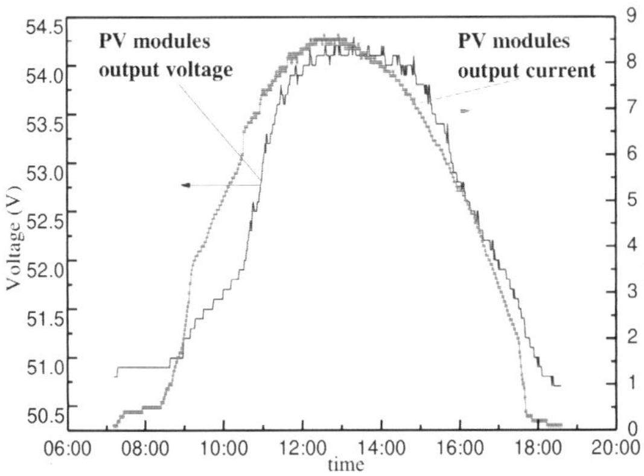

Figure 5. Voltage and current variations of PV modules

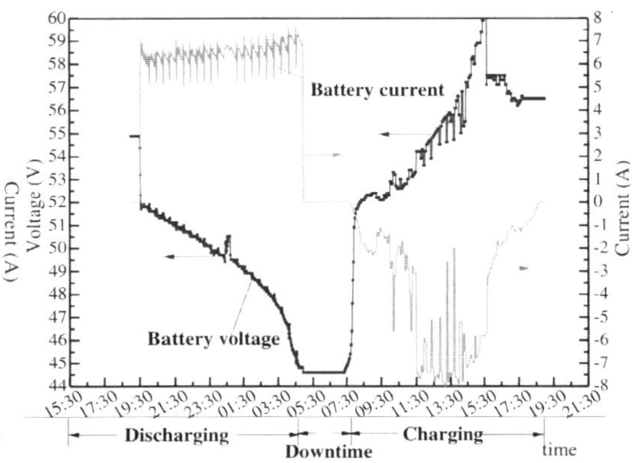

Figure 6. Voltage and current variations of batteries

It was observed that in figure 5 the maximum current value was 8.6 A at 12:35 and output voltage changes were 50.8 V-54.2 V and the maximum voltage appears at 12:36. Research results for batteries charging and batteries discharging are shown in figure 6. In batteries discharging, statics ice generator can be driven for 10 hours by batteries only which were in full power state and disconnected with PV modules until the batteries electricity decreased and cannot drive the statics ice maker. The voltage declined from 54.88 V to 44.6 V step by step and output current maintain about 7 A with a small fluctuations between 5.4 A and 7.4 A. In charging process for batteries, batteries disconnect with ice generator and connected with PV modules. Batteries current changed with solar irradiance and voltage increased gradually until batteries were in floating state.

The static ice generator operation cycle time was 10 min and the ice production was about 0.35 kg every cycle. Ice making efficiency was 2.08 kg/h. The temperatures of ice generator with every component in thermodynamic cycle have experimental tested, as shown in figure 7.

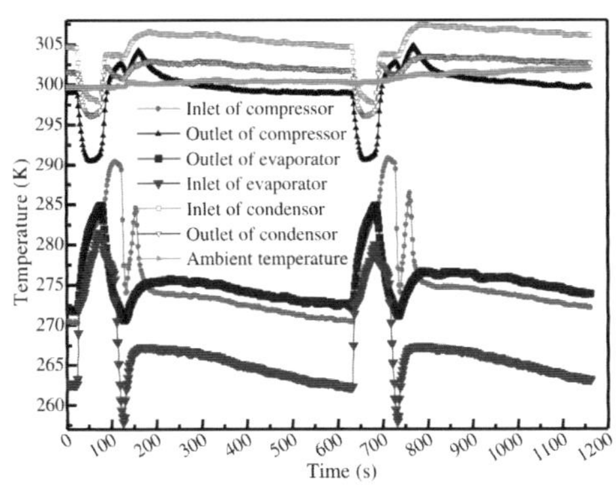

Figure 7. Temperature variations of ice generator components in thermodynamic cycle process

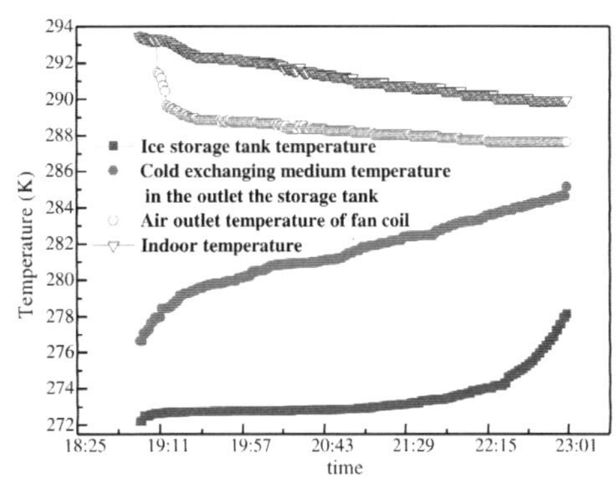

Figure 8. Temperatures change of exchanging cold and supplying cold process

According to the above experimental results and calculation formulas, the maximum value of PV modules photoelectric conversion efficiency can be calculated. At the same time, batteries storing or releasing electricity, the minimum energy loss and COP maximum value of static ice generator and the maximum energy utilization efficiency of air conditioner made up with fan coil also were calculated and shown in Table 3.

Table 3. Experimental and computational results of SIRAS driven by DPES

PV modules				Batteries		Static ice maker				Air conditioner	
α	0.92	S_p	3.25	I_b	-7.1~ 7.3	W_p	380	$T_{con,out}$	296.0~ 303.6	C_a	1
τ	0.90	I_p	0.1~ 8.6	V_b	44.6~ 56.5	$\dot{m}_r$	0.0127	$\dot{m}_i$	2.08	$\dot{m}_a$	0.12
G	282-985	V_p	50.8~ 54.2			$T_{con,in}$	297.7~ 307.5	h	335	ΔT_a	0~1.5
						C_r	650				
$\eta_{PV,max}$		17.58%		Q_b	-380~ 401	$Q_{loss,min}$	128.66	COP_{max}	0.51	$\eta_{a,max}$	93.00%
η_{max}								8.34%			

It was observed that in Table 3 the maximum value of system energy utilization efficiency was only 8.34%. The low photoelectric conversion efficiency of PV modules is one of the reasons leading the low system efficiency and static ice generator with a 0.51COP low refrigeration coefficient caused lost of energy loss. Because in ice making process, ice pasted well together with five solid walls of grid plate evaporator and cannot be separated off by ice gravity. Thus, compressor must be shut down and a special solenoid valve as shown in Figure 13 must be opened which used for adjust refrigerant flow and introduced the high temperature and high pressure refrigerant steam expelled from compressor into evaporator. And then, the evaporator plays the role of condenser to release heat to ice through solid walls and ice interfaces begin melting. Ice was divorced from evaporator and fall into tank. Meanwhile, evaporator inlet temperature and outlet temperature increased sharply until the solenoid valve was closed and compressor was turned on and a new ice making process begins. The refrigerant flowed out evaporator into throttle valve and become low temperature and low pressure liquid. And then, the refrigerant flow into condenser to absorb heat from outside and condenser inlet and outlet temperature decline sharply when low temperature refrigerant flows into it. Condenser outlet temperature was higher than condenser inlet because refrigerant absorb heat form outside through condenser, as shown in Figure 12. Refrigerant next flow into compressor and compressor inlet temperature increase a lot compared to ice making process, which can be changed with inlet and outlet temperature variation of evaporator and condenser. Compressor stops running in ice melting abscission process and the refrigerant gas cannot be compressed in compressor. Therefore, the compressor outlet temperature drops sharply when the solenoid valve opens and it returns smoothly when ice making process opened. Ice melting abscission process was extremely unfavorable to SIRAS for a number of reasons as the following:

- Ice generator operation period extended for 200s; furthermore, the ice melting abscission time was one third time of an ice making cycle.

- Service life of compressor will be shortened by frequent start and stop compressor, which has a great impact on ice making process and was extremely unfavorable for energy supply process of DPES. Electricity consumption of solenoid valve was 3.0×10^{-4} kW h in ice melting abscission process and the total electricity consumption was 0.024 kW h from 7:30 to 19:00.

Therefore, in order to improve the performance of SIRAS driven by DPES, some optimization and improvement measurements were proposed as follows:

- Photoelectric conversion efficiency of PV modules must be improved. But there are some difficult to greatly improve the efficiency limited the current technology level and material processing technology. It is very expensive to optimize the efficiency.
- In another way, the COP of static ice generator can be improved easily through structural improvement and performance optimization. Firstly, Evaporator immersion static refrigeration mode was adopted to replace ice harvester refrigeration mode. The optimized coil evaporator was immersed into water to absorb heat and produce ice and then all the energy could be utilized, as shown in figure 9.

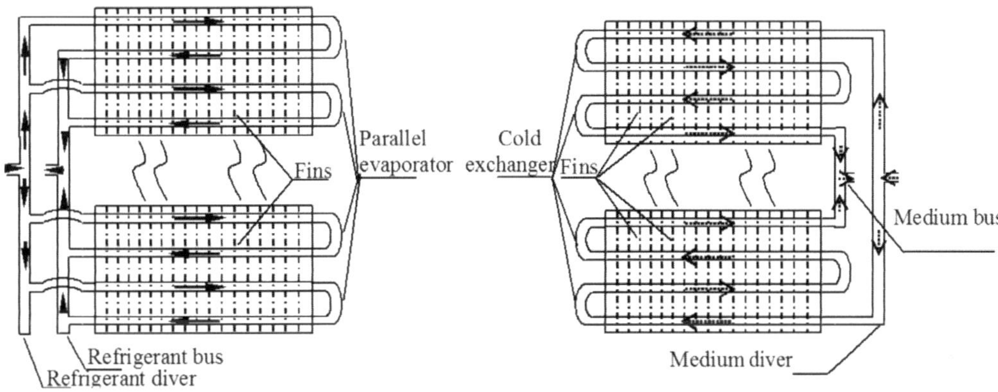

Figure 9. Profile of evaporators and cold exchanger immersed in the static ice storage tank

And then, The coiled pipe cold exchanger was co-integrated with coil evaporator. In refrigeration process, coiled pipe cold exchanger has the priority to get cold transferred from the coil evaporator next to the cold exchanger to supply cold for user. Surplus cold is used to make ice to store cold. Consequently, SIRACS not only has the out-of-the-box function of ordinary air conditioning, but also effectively improve the appearing phenomenon of over cooling and remedy the disadvantage of cold supply after ice making process in traditional submerged ice making system. The top view of co-integration evaporators and cold exchanger immersed in the static ice storage tank is shown in figure 10.

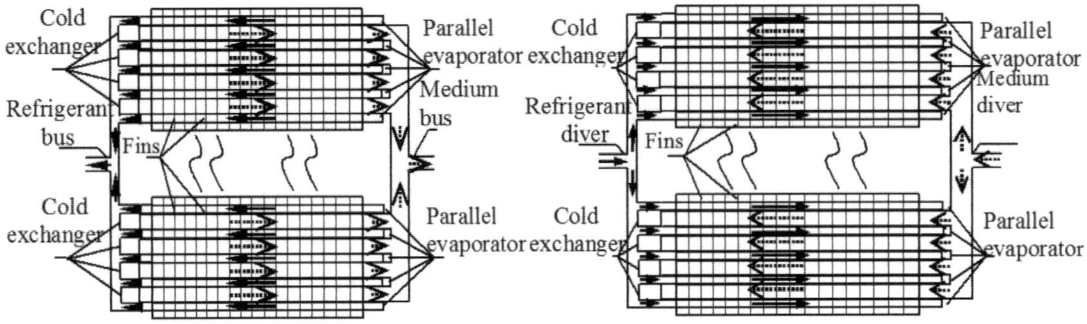

Figure 10. Top view of co-integration evaporators and cold exchanger immersed in the static ice storage tank.

Another evaporator immersion static ice generator was constructed according to the optimization design, shown as figure 11.

Figure 11. Photo of evaporator immersion static ice maker

In figure 11, the evaporator was co-integrated with cold exchanger with fins, which were immersed in the water stored with tank. The evaporator was connected with a 735 W refrigerator. Experimental test work has been done and the results are shown in Table 4.

Table 4. Experimental and computational results of evaporator immersion static ice maker

Compressor power /W	Runtime /s	The weight of water /kg	Initial water temperature /K	Final water temperature /K	The quantity of ice /kg	COP
735	7200	31	290.15	275.15	5.4	0.71

Experimental results showed that the COP of evaporator immersion static ice generator improved nearly 40% than that of ice slide static ice maker. But it is very obviously observed in experiment that the ice thickness increasing rate gradually decreased. For low heat conductivity, when ice thickness increased the cold in evaporator could not be delivered to the ice surface in time. So the cold could not be fully absorbed by water, which will cause that too much liquid refrigerant flowed into the gas-liquid separator and gas refrigerant was not enough sucked by compressor leading to compressor shutdown for too low air pressure. So the experimental research on ice thickness and ice super-cooled temperature must be carried out for system performance optimization and ensuring compressor operating safely and stably. The results were shown in figure 12. At the same time, the relationship between system COP constantly changing and ice thickness was studied with experiment also and the result was shown in figure 13.

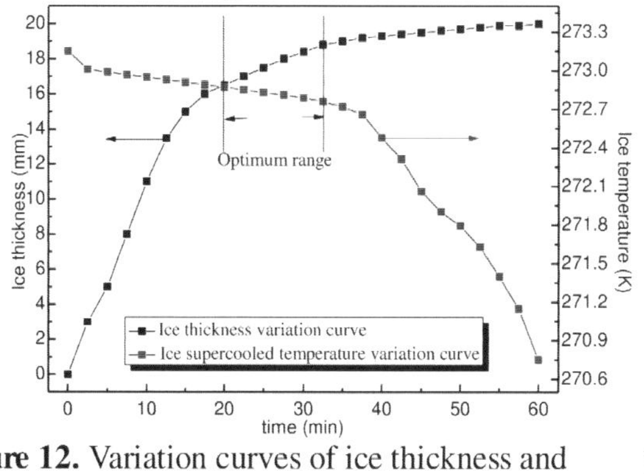

Figure 12. Variation curves of ice thickness and subcooled temperature

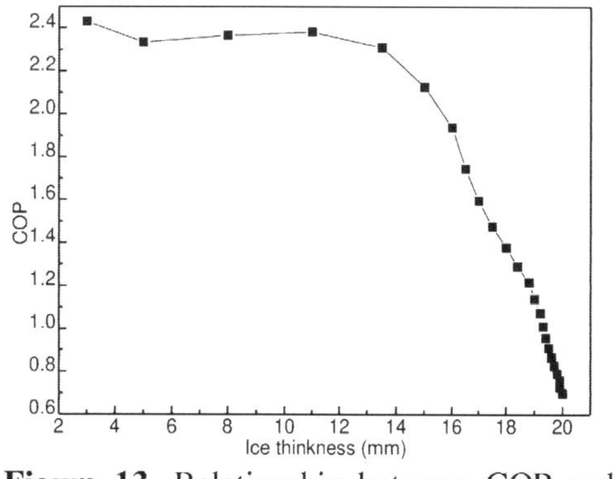

Figure 13. Relationship between COP and ice thickness

In figure 12, it was observed that the ice thickness on the evaporator gradually increased along with increasing time but the growth rate gradually reduced. At the same the ice subcooled temperature gradually declined and the decline rate was gradually increased. So to ensure the optimum coupling between ice thickness and ice temperature, the best working time range was given, which was 20min~32.5min when ice making process was begun. Both of ice thickness and ice temperature were in best value. Among them, the best ice thickness range was 16.5mm~18.8mm and the best ice super-cooled temperature was 272.76K~272.87K. From Figure 13, the system COP decreased along with ice thickness increasing and the COP decreased rapidly when ice thickness was more than 14mm. the COP range is 1.75~1.21 corresponding with the best ice thickness given by Fig. 12. All of the research Results can provide reference and technical support for immersion static ice generator working efficiently and the application and popularization of SIRAC driven by DPES.

4. Conclusion

In this paper, a 0.2kW SIRACS driven by DPES was constructed. The system characteristic parameters were tested and the energy conversion efficiency of each component was calculated according to the experimental data. And the optimal design scheme was proposed based on the analysis of each component energy loss. So another evaporator immersion static ice generator test platform was built after optimization and relevant experimental research work has been carried out. All of the research results are shown as:

- The maximum value of system energy utilization efficiency was only 8.34%. The main reasons is the COP of ice generator was only 0.51. Because the ice slide work model of ice generator include ice making process and ice melting abscission process, which not only leads to much of energy has been lost but also the ice generator operation period extended for 200s and ice melting abscission time was one third time of an ice making cycle.

- Optimal design scheme about ice making model was proposed. Evaporator immersion static refrigeration mode was adopted to replace ice harvester refrigeration mode and the coil cold exchanger was co-integrated with coil evaporator.
- Experimental test work on evaporator immersion static ice generator has been carried out. The COP of evaporator immersion static ice generator was 0.71 improved nearly 40% than that of ice slide static ice generator. The best ice thickness range was 16.5mm~18.8mm with the system COP changed among 1.75~1.21. The best ice super-cooled temperature was 272.87K~272.76K.

5. Acknowledgement

The authors gratefully acknowledge the financial support provided by National International Scientific and Technological Cooperation Program (2011DFA62380). The authors are also grateful to Renewable Energy Research and Innovation Development Center in Southwest China (05300205020516009), Research and Innovation Team of Renewable Energy in Yunnan Province and Yunnan Provincial Renewable Energy Engineer Key Laboratory and Yunnan Provincial Department of Education Science Research Fund Project (2015J035).

References

[1] Allouhi A, Kousksou T, Jamil A, Bruel P, Mourad Y and Zeraouli Y 2015 Solar driven cooling systems: An updated review *Renewable Sustainable Energy Reviews.* **44** 159-81

[2] Ghafoor A and Munir A 2015 Worldwide overview of solar thermal cooling technologies *Renewable Sustainable Energy Reviews.* **43** 763-74

[3] Aktacir M A 2011 Experimental study of a mufti-purpose PV-refrigerator system *Int Journal of Physical Sciences.* **6** 746-57

[4] Bilgili M 2011 Hourly simulation and performance of solar electric-vapor compression refrigeration system *Solar Energy.* **85** 2720-31

[5] Ekren O, Yilanci A, Cetin E and Ozturk H K 2011 Experimental performance evaluation of a PV-Powered refrigeration system *Electronics &Electrical Engineering.* **114** 7-10

[6] Kaplanis S and Papanastasiou N 2006 The study and performance of a modified conventional refrigerator to serve as a PV powered one *Renewable Energy.* **31** 771-80

[7] Mba E F, Chukwuneke J L and Achebe C H 2012 Modeling and simulation of a photo-voltaic Powered vapor compression refrigeration system *Journal of Information Engineering & Applications.* **2** 1-15

[8] Modi A, Chaudhuri A, Vijay B and Mathur J 2009 Performance analysis of a solar photovoltaic operated domestic refrigerator *Applied Energy.* **86** 2583-91

[9] Axaopoulos P J and Theodoridis M P 2009 Design and experimental performance of a PV Ice-maker without battery *Solar Energy.* **83** 1360-9

[10] Li Y, Zhou S H, Lin W D, Ji X, Luo X and Li M 2015 Performance analysis of photovoltaic refrigerator system *Acta Energiae Solaris Sinica.* **36** 422-9

IOP Conf. Series: Earth and Environmental Science **40** (2016) 012028 doi:10.1088/1755-1315/40/1/012028

A new cascaded control strategy for paralleled line-interactive UPS with LCL filter

X Y Zhang[1], X H Zhang[1, 3], L Li[1], F Luo[2] and Y S Zhang[2]

[1]School of Automation, Guangdong Polytechnic Normal University, Guangzhou, 510665, China

[2]Foshan Baykee New Energy Technology Co., Ltd. Foshan 528000, China

E-mail: 2740624927@qq.com

Abstract. Traditional uninterrupted power supply (UPS) is difficult to meet the output voltage quality and grid-side power quality requirements at the same time, and usually has some disadvantage, such as multi-stage conversion, complex structure, or harmonic current pollution to the utility grid and so on. A three-phase three-level paralleled line-interactive UPS with LCL filter is presented in this paper. It can achieve the output voltage quality and grid-side power quality control simultaneously with only single-conversion power stage, but the multi-objective control strategy design is difficult. Based on the detailed analysis of the circuit structure and operation mechanism, a new cascaded control strategy for the power, voltage, and current is proposed. An outer current control loop based on the resonant control theory is designed to ensure the grid-side power quality. An inner voltage control loop based on the capacitance voltage and capacitance current feedback is designed to ensure the output voltage quality and avoid the resonance peak of the LCL filter. Improved repetitive controller is added to reduce the distortion of the output voltage. The setting of the controller parameters is detailed discussed. A 100kVA UPS prototype is built and experiments under the unbalanced resistive load and nonlinear load are carried out. Theoretical analysis and experimental results show the effectiveness of the control strategy. The paralleled line-interactive UPS can not only remain constant three-phase balanced output voltage, but also has the comprehensive power quality management functions with three-phase balanced grid active power input, low THD of output voltage and grid current, and reactive power compensation. The UPS is a green friendly load to the utility.

1. Introduction

The International Electrical Commission defines three standardized UPS systems, i.e., on-line, off-line, and line-interactive type [1]. The Off-line UPS is also called passive stand-by UPS, and the load is supplied by the utility grid when the utility grid is normal, and transferred to the UPS when the utility grid fails. The output voltage accuracy of the UPS is poor. The on-line UPS is usually double conversion UPS with the rectifier and the inverter. The rectifier and the inverter both have to deal with the rated load power, which leads to the low efficiency. The rectifier generally adopts the thyristor phase-controlled method, which leads to the distorted current drained from the utility grid and deteriorates the power quality at the point of common coupling. So Power quality conditioner such as active power filter is needed [2-4].

[3] Address for correspondence: X H Zhang, School of Automation, Guangdong Polytechnic Normal University, Guangzhou, 510665, China. E-mail: 2740624927@qq.com.

In the 1990's, the series-parallel compensated line-interactive UPS was introduced, which is also often called as delta-conversion UPS. The UPS has two inverters connected to the common battery. One is the main inverter, which is fully rated at 100% of the load power, and the other is the delta inverter, which is typically rated at 20% of the load power. The delta-conversion UPS has the breakthrough progress in the structure and technical performances, which can realize an effective power factor correction, load harmonic current suppression and output voltage regulation. However, due to the need of two sets of inverters, as well as a bulky transformer, etc., the delta-conversion UPS still has some drawbacks, such as complex system structure, extra size and cost and so on [5-8].

In order to overcome these drawbacks, a paralleled line-interactive UPS with LCL filter was proposed [9]. The UPS is connected in shunt between the utility grid and the load. In normal operation, the utility grid supplies the active power of the load, and the inverter supplies only a small amount of reactive power to regulate the output voltage. The UPS with only one-stage conversion can achieve the output voltage quality and grid-side power quality control simultaneously, but the multi-objective control strategy design is difficult. The active power and power factor are usually regulated by the phase and the amplitude of the output voltage individually in the literatures available [10-11]. The inherent resonance characteristics of LCL filter are not considered, and the grid current is not directly controlled on the assumption that the grid current is sinusoidal if the grid voltage and the output voltage are sinusoidal. Actually the distortion of the grid voltage and output voltage often leads to the distortion of the grid current.

With the development of solar power and other renewable energy technologies, LCL filter in the grid-connected inverter application had been comprehensively studied [12]. LCL filter can significantly suppress the switching frequency harmonics, but it also increases the order of the system. In some frequency point it will show zero impedance to cause the system oscillation. In order to suppress resonance, the passive damping and active damping methods are introduced. Capacitance current active damping method is widely used, and the capacitance current and grid current dual loop control strategy is better to solve the grid current control problem. Work has been mostly concentrated in the current control of the single-phase or three-phase three-wire system [13-14]. With the increase of renewable energy penetration, grid inverter is required to have the inertia characteristics as the traditional synchronous generators to participate in the regulation of the grid, which requires the inverter to run in the voltage-source mode. The voltage mode control strategy, especially in the case of the three-phase four-wire system and unbalanced load, has been studied rarely.

The UPS with LCL filter is a three-phase four-wire system, and requires the constant three-phase balanced output voltage control. When the load is unbalanced and the grid active power is balanced, each phase inverter may work in the special bi-directional energy flow state. Due to the potential good characteristics of the UPS with LCL filter and the research results of the LCL filter, the study on the multi-objective optimization control strategy has significant meaning according to the modern control theory and digital control technology. This paper presents a paralleled three-phase three-level line-interactive UPS with LCL filter. Base on the circuit topology and operation mechanism analysis, a cascaded control strategy for the power, voltage, and current is designed. The setting of the controller parameters is also discussed. A 100kVA UPS prototype is built and experiments verification is carried out.

2. Description of the UPS topology

Figure 1 shows the circuit diagram of the paralleled three-phase three-level line-interactive UPS. It mainly consists of the battery sets and the three-phase inverter with LCL filter. Two battery sets Ubat1 and Ubat2 are connected in series to provide the DC bus. The middle point O of the DC bus is connected to the neutral points of the utility grid and the load together, which forms the three-phase four-wire power supply system to meet the power demand of the balanced or unbalanced load.

The inverter consists of three T-type half bridges, and the LCL filter with inductors L1, L2, and capacitor C. The UPS has three operation modes, namely bypass mode, standalone mode and norm mode. The load is powered by the utility grid with static switch K2 on，K1 and K3 off in the bypass mode. In the standalone mode, the load is powered by the battery sets with switch K3 on, K1 and K2 off. In the normal mode, the load is powered by the UPS and the utility grid together with the switch

K1 and K3 on, K2 off. The link inductor L2 not only plays a filtering function, but also endures the voltage difference between the utility grid and the output. The choice of L2 depends on the allowable voltage fluctuation.

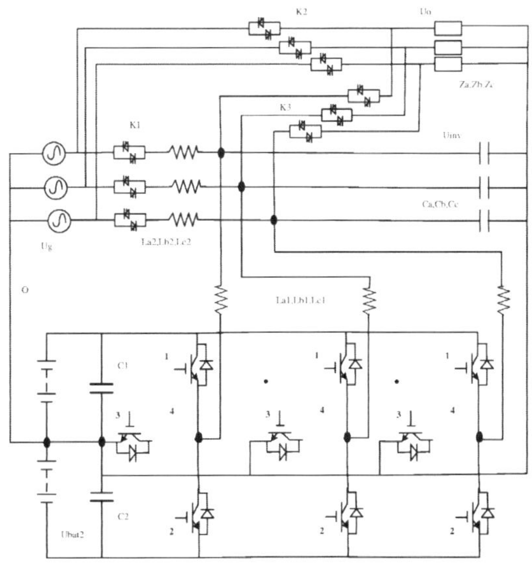

Figure 1. Circuit diagram of the paralleled three-phase three-level line-interactive UPS

3. Normal operation mode analysis

In the standalone mode, the UPS works as the conventional inverter, the operation mechanism is very clear and not explained here. The operation mechanism analysis is detailed only for the normal mode in the paper.

Due to the three-phase four-wire system, each phase can be controlled with the same control structure independently. Here only phase a is illustrated as an example. Figure 2 shows the voltage vector diagram of the normal operation. The case that the grid voltage is lower than the output voltage is shown in figure 2. (a). The case that the grid voltage is higher than the output voltage is shown in figure 2. (b).

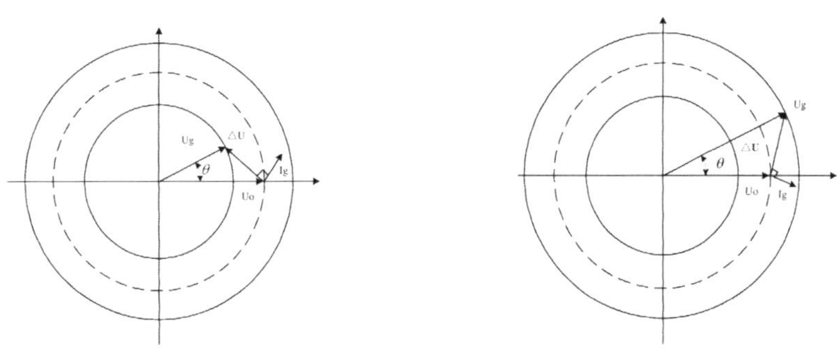

Figure 2. (a) Figure 2. (b)

(a) Utility grid voltage is lower than the output voltage.

(b) Utility grid voltage is higher than the output voltage.

Figure 2. Voltage vector diagram of normal operation.

The output voltage is desired to be constant, so the output voltage is chosen as the reference $\overrightarrow{U_o} = U_o\angle 0^o$. The grid voltage is expressed as $\overrightarrow{U_g} = Ug\angle\beta^o$ and β is the phase angle difference. The active and reactive power P_g and Q_g from the utility grid can be calculated as

$$P_g = \frac{U_o U_g \sin\beta}{\omega L_2} \quad (1)$$

$$Q_g = \frac{U_O(U_g COS\beta - U_o)}{\omega L_2} \quad (2)$$

Where ω is the measured grid frequency, and I_g is the grid current.

The active power of the load should be three-phase balanced supplied by the utility grid, so the active power reference P_g^* of each phase should be

$$P_g^* = \frac{P_{loada} + P_{loadb} + P_{loadc}}{3} \quad (3)$$

Where P_{loada}, P_{loadb}, and P_{loadc} are the actual period-averaged load active power of phase a, b, and c. Because Uo remains 220V, and Ug is uncontrollable, Equations (1) and (3) show that β is determined by the total load active power and $\beta \geqslant 0$. Equation (2) shows that the capacitive reactive power is outputted to compensate for the decreased grid voltage in figure 2(a) and the inductive reactive power is outputted to compensate for the increased grid voltage in figure 2(b). Therefore, the UPS has certain reactive power compensation capacity when the grid voltage fluctuates.

4. Cascaded control strategy design

The control objective of the UPS is to maintain balanced three-phase output voltage with small THD and balanced three-phase active power input from the grid with small-distorted grid current. Thus, the UPS is a "green" load to the utility grid and not a polluter. The UPS only has single-stage conversion, so the multi-objective optimization control of voltage, current and power is difficult. The a-phase cascaded control block diagram of the UPS is shown in figure 3. The entire control system mainly consists of the effective value control loops and the instantaneous value control loops. The effective value control loops include active power and output voltage control. The instantaneous value control loops include capacitance voltage and capacitance current control. When the grid voltage is unbalanced, the digital phase-locked control is used to extract the positive sequence component [15]. Then signals are expressed in the dq rotating coordinate frame. A closed-loop control system which regulates u_{gq} to zero is capable of getting the actual positive sequence angle θ.

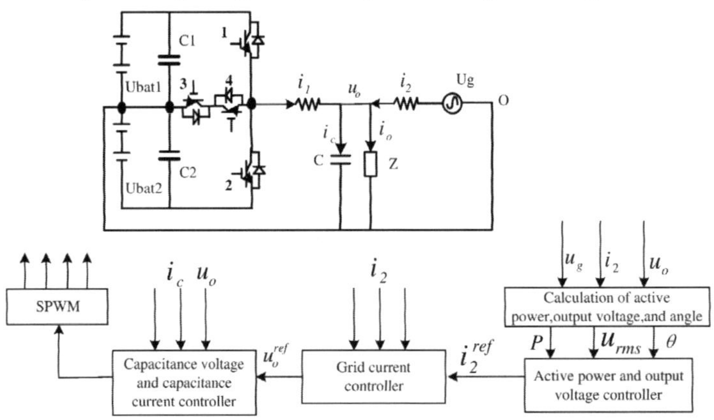

Figure 3. Control block diagram of the UPS

4.1. Effective value controls

The effective value control loop is shown in figure 4. The active power of the grid is controlled by regulating the grid active current component. The amplitude of the output voltage is controlled by regulating the grid reactive current component. $cos\theta$ and $sin\theta$ present the angle information of grid active current and reactive current individually.

The active power of the utility grid is regulated by a proportional integral controller PI1. The active power reference P_g^* is obtained from (3). The actual active power of the grid is calculated using the periodic mean method as follows.

$$P_{ga} = \frac{\int_0^T u_{ga} i_{ga} dt}{T} \tag{4}$$

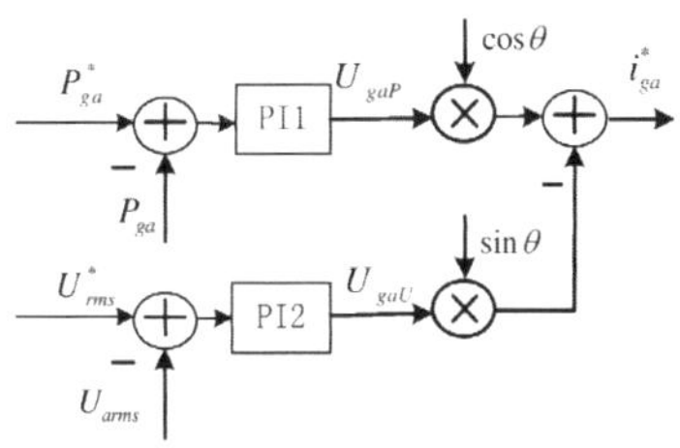

Figure 4. Effective value control block diagram

Where u_{ga} and i_{ga} are the a-phase instantaneous voltage and current of the grid. T is the system period time.

The effective value reference of the output voltage U_{rms}^* is 220V. The effective value of the output voltage is regulated by a proportional and integral controller PI2. The output U_{gaP} of the active power controller multiplied by $cos\theta$ equals to the grid active current reference. The output U_{gaU} of the output voltage controller multiplied by $sin\theta$ equals to the grid reactive current reference. The sum of the active and reactive current references is the instantaneous gird current reference.

4.2. Instantaneous voltage and current control

Instantaneous voltage and current control block diagram is shown in figure 5. The controller is composed of three parts. ① is the grid current control loop, and② is the output voltage control loop, and③ is the output voltage waveform quality control loop.

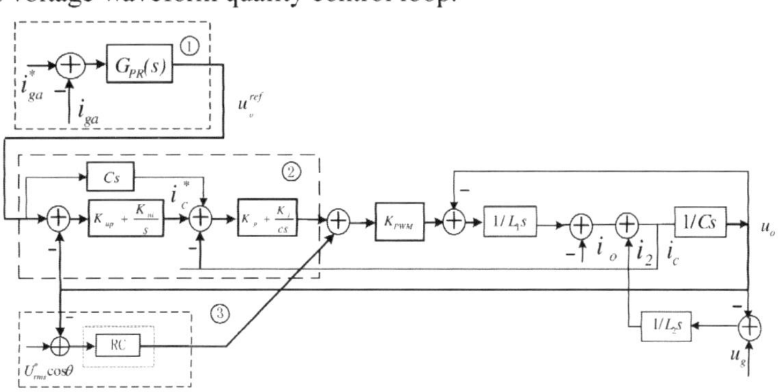

Figure 5. Control block diagram of instantaneous voltage and current

4.2.1. Grid current control

From the above effective value control loops, we can get the grid current reference.

$$i_{ga}^{*} = U_{gaP}\cos\theta + U_{gaU}\sin\theta \tag{5}$$

Equation (5) shows that the current reference is sinusoidal and the current control loop should be fast. The proportional-resonant controller is an appropriate choice [16-17]. The resonant controller is essentially an AC PI regulator with the working frequency. The gain at working frequency is infinite, while in the non-fundamental frequency the gain is very small, so in theory the controller can achieve zero static error tracking of the sinusoidal signal with working frequency. When the digital method is adopted, because of that the digital system accuracy and digital operation truncation error, the working frequency is easily shifted away from the specified system frequency. Therefore, the actual used controller is the quasi-resonant regulator. So the proportional-resonant controller (PR) shown in figure 3 is described as

$$G_{PR}(s) = K_i + \frac{2K_r\omega_c s}{s^2 + 2\omega_c s + \omega_o^2} \tag{6}$$

Where Ki is the proportional coefficient, ω_o is the grid frequency, Kr is the gain coefficient, ω_c is the cut-off frequency, taking ω_c=8 in this case. In this paper, the bilinear transformation is used to the discretization of equation (6).

4.2.2 Instantaneous output voltage control

From figure 3, we can get the state equation as following.

$$\begin{aligned} L_1 di_1/dt &= u_{inv} - u_o \\ L_2 di_2/dt &= u_g - u_o \\ Cdu_o/dt + i_o &= i_1 + i_2 \end{aligned} \tag{8}$$

The output voltage transfer function can be obtained.

$$[L_1L_2Cs^2 + (L_1 + L_2)]u_o = L_1u_g + L_2u_{inv} - L_1L_2si_o \tag{9}$$

The characteristic root of the system $s=\pm j\sqrt{\frac{L_1+L_2}{L_1L_2C}}$, which is critical stable. The damping of the system can be injected by using the capacitance current feedback, so the capacitance current PI controller is the inner loop. The capacitance current transfer function is

$$i_c = \frac{L_2(K_p cs+k_i)k_{PWM}\, i_c^{*} - L_1L_2cs^2 i_o + L_1csu_g}{(L_1L_2cs^2 + k_pk_{PWM}\, L_2cs + L_2 + L_1 + k_ik_{PWM}\, L_2)} \tag{10}$$

Where Kp is the proportional coefficient and Ki is the integral coefficient. The damping coefficient of the characteristic equation from (10) $\xi = \frac{k_pk_{PWM}\, L_2C}{2}\sqrt{\frac{L_1L_2C}{L_1+L_2+k_ik_{PWM}\, L_2}}$. It shows that the larger the Kp, the greater the system damping, and the better the resonance suppression effect. However, if the damping coefficient is too large, the dynamic characteristic becomes worse and the regulating time becomes longer. In order to take into account the damping effects and dynamic performance, usually ξ is greater than 0.6 and less than 0.8. The parameters of L1, L2, and C usually exist a certain degree of changes in the actual operation, so ξ = 0.707 is a preferred value [18]. The cut-off frequency of the current loop is generally selected within the one-tenth of the switching frequency. The switching frequency is 9.6kHz, so the cut-off frequency fc is 900Hz in this paper. From the above comprehensive consideration, the controller parameters Kp = 2.3, Ki = 0.95.

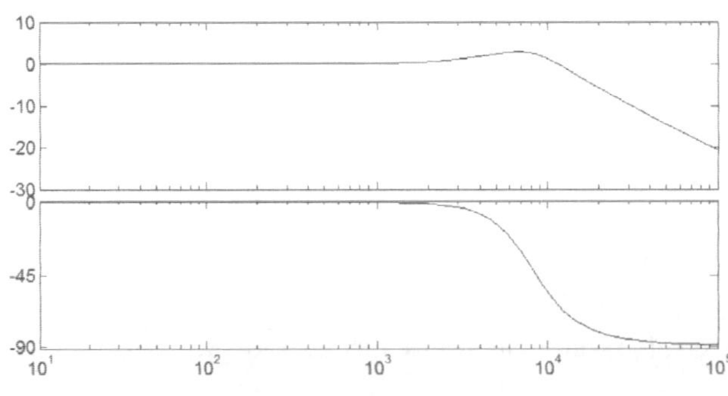

Figure 6. Bode diagram of capacitance current

Figure 6 is the bode diagram of capacitance current. When the outer voltage control loop is designed, the inner current control loop can be treated as a proportional coefficient. The outer voltage controller is a PI controller, and the output voltage transfer function can be obtained as follows.

$$u_o = \frac{Ck_w s^2 + k_{up} k_w s + k_{ui} k_w}{Cs^2 + k_{up} k_w s + k_{ui} k_w} u_o^{ref} \quad (11)$$

The damping coefficient $\xi = \frac{k_{up} k_w}{2\omega_n}$, and the natural frequency $\omega_n = \sqrt{\frac{k_{ui} k_w}{C}}$. When the damping coefficient is given, the greater the natural frequency, the worse the dynamic responses is. In this paper,$\xi = 0.707$ and the natural frequency is 10Hz. So the control parameters Kui=0.317,Kup=46.7. The parameters of the UPS is shown in Table 1.

Table 1. Parameters of the UPS

Vdc1/V	Vdc2/V	fs/kHz	Kpwm	L1/mH	L2/mH	C/uf
400	400	9.6	0.8	0.2	1.6	75

4.2.3. Output voltage waveform control

To ensure the sinusoidal output voltage with small THD, repetitive control (RC) is introduced as shown in figure 3. Repetitive control is a control strategy based on the internal model principle. The internal model principle means that if the signal generator for the reference signal is contained in the stable closed loop system, the controlled output can track the reference instruction without the steady error. If the system requires zero steady-state error for the sinusoidal reference signal, the model of the sine signal should be included in the stable closed-loop transfer function. Figure 7 shows the block diagram of the repetitive controller [18-19].

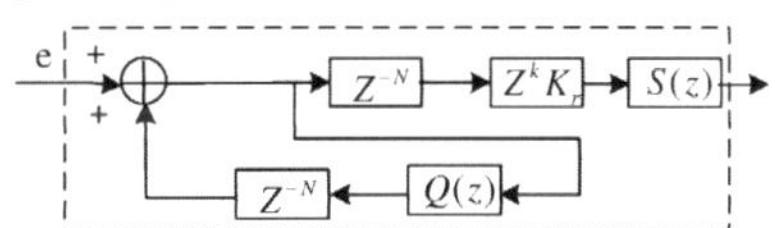

Figure 7. Block diagram of the repetitive controller

The transfer function is given as

$$H(z) = \frac{K_r S(z) Z^{k-N}}{1 - Q(z) Z^{-N}} \quad (11)$$

Where $e = U_{rms}^* \cos\theta - u_{inva}$, z^{-N} with N=192 is the periodical delay. Kr = 0.2 is the proportional factor, ultimately determining the amplitude of the output. Q(z) = 0.95 is designed to enhance the robustness of the system. z^{-K} with K = 5 is the lead loop, which make the controller generate correction signal ahead of K step in the next period according to the error signal of the past period. S(z) is the compensator to shape characteristics of the controlled objective to ensure the system stability.

$$S(z) = \frac{0.1813z + 0.1551}{z\text{^}2 - 1.294z + 0.63} \quad (12)$$

The output of the repetitive controller is directly fed to the SPWM modulator.

5. Experimental results

In order to verify the effectiveness of the cascaded control strategy of the UPS shown in figure 1, the 100kVA three-phase three-level paralleled line-interactive UPS prototype was built, shown in figure 8. The controller was implemented using the Texas Instrument float-point DSP TMS320F28335 and the three-level power module type F3L300R12PT4. A large number of experiments are carried out on the prototype.

Figure 8. Experimental prototype of the three-phase three-level paralleled line-interactive UPS

5.1 Resistive load test

Figure 9 shows the experimental waveforms under the single-phase unbalanced resistive load. Figure 9. (a) shows the grid voltage is 230V. Figure 9. (b) shows the output voltage remains three-phase balanced 220V. Figure 9. (c) shows the load current waveform. The a-phase load is 10kW, and b-phase and c-phase loads are zero. Figure 9. (d) shows that the grid current is three-phase balanced. Although the load is unbalanced, but the grid active power is three-phase balanced with each phase 4.2kW shown in figure 9. (e). Figure 9. (f) shows the THD of the output voltage is 1.5. The load experiments show that under the unbalanced single-phase load, the UPS achieves three-phase balanced grid active power input and three-phase balanced output voltage with small THD. The UPS has good performance under resistive loads.

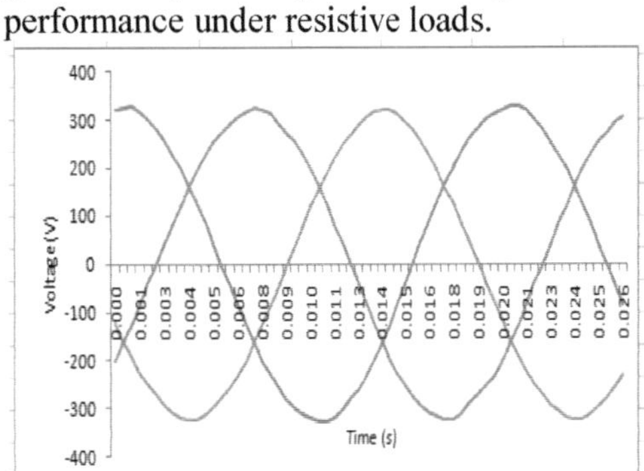

Figure 9. (a) Grid voltage.

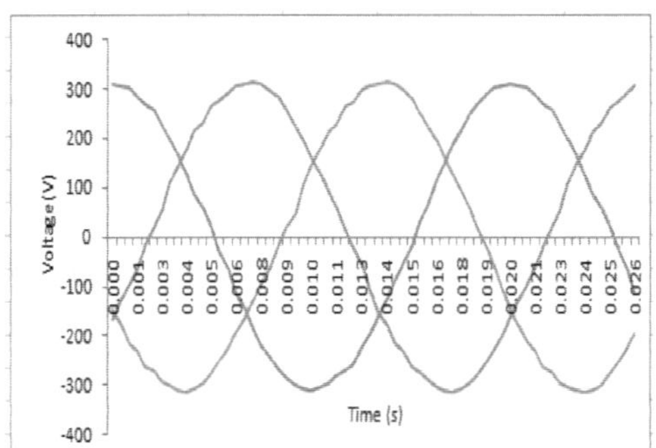

Figure 9. (b) Output voltage.

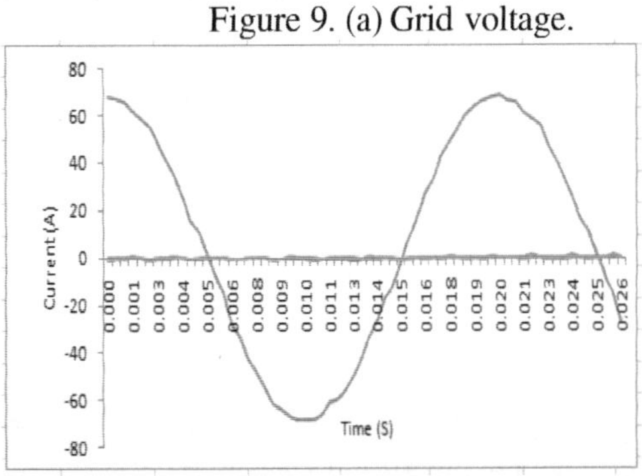

Figure 9. (c) Load current.

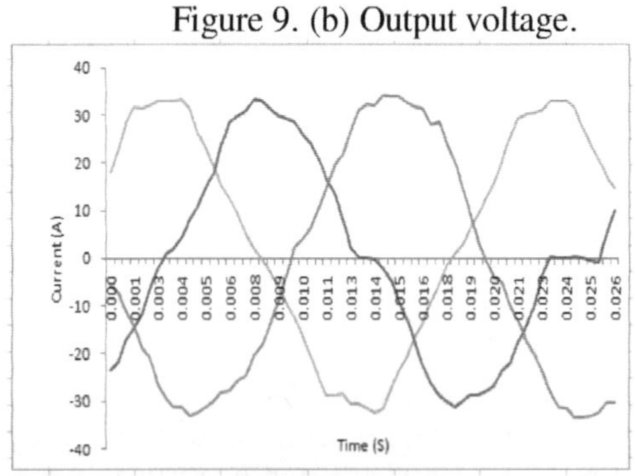

Figure 9. (d) Grid current.

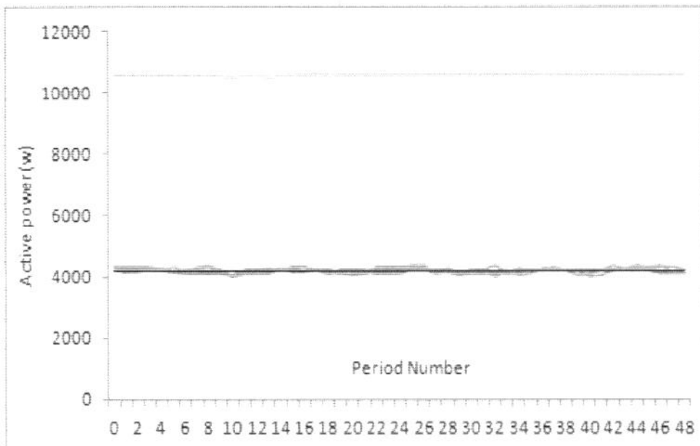

Figure 9. (e) Grid active Power.

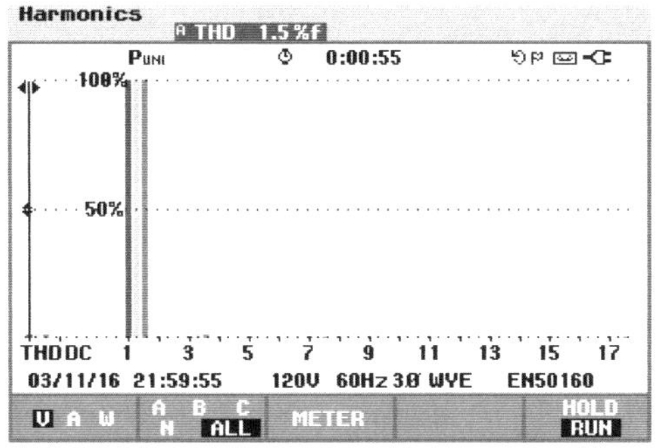

Figure 9. (f) THD of output voltage.

Figure 9. Experimental waveforms under the unbalanced single-phase resistive load.

5.1. Nonlinear load test

Figure 10 shows the experimental waveforms under the nonlinear load. The actual nonlinear load is a frequency converter. Figure 10. (a) shows the load current. The load is serious nonlinear, and the THD is 95%. Figure 10. (b) shows the grid current. The grid current is three-phase balanced and has the THD of 4.1%. The output voltage in figure 10. (c) is three-phase balanced with THD of 3.5%. Figure 10. (d) shows that the grid active power is three-phase balanced with each phase 11.6kW. The RMS of the output voltage remains 220V in figure 10. (e). Figure 10. (f) shows the output voltage when the grid fails. Because the UPS is voltage-mode control, the transition process is very smooth and the output voltage exist no sudden change. Figure 10 shows that the UPS realizes three-phase balanced grid active power input and three-phase balanced output voltage with small THD under nonlinear load. The UPS has good steady and dynamic performance under nonlinear loads.

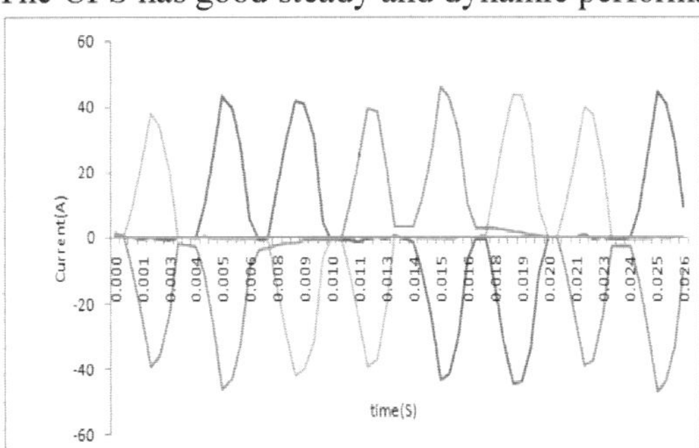

Figure 10. (a) Load current.

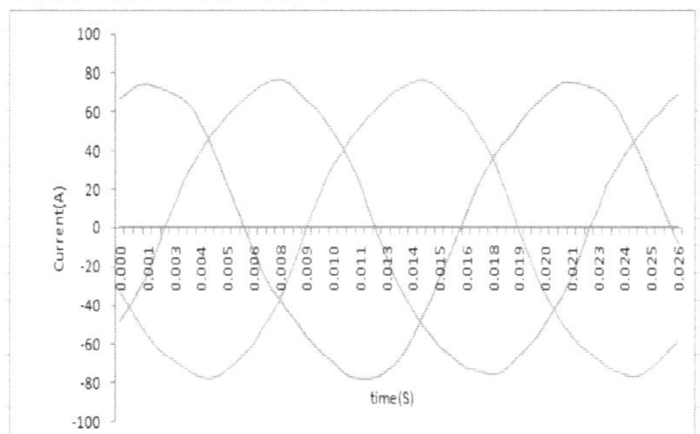

Figure 10. (b) Grid Current.

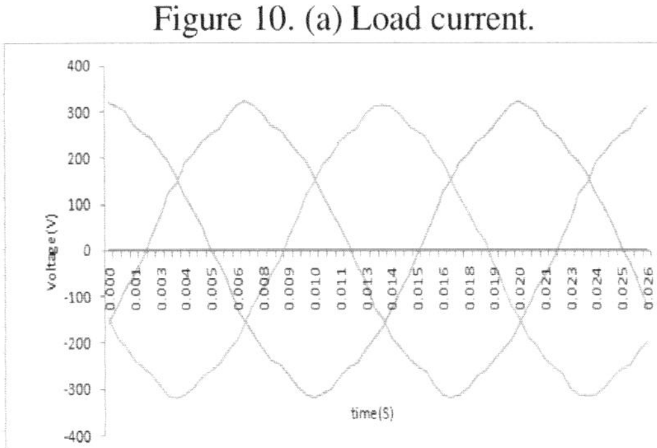

Figure 10. (c) Output voltage

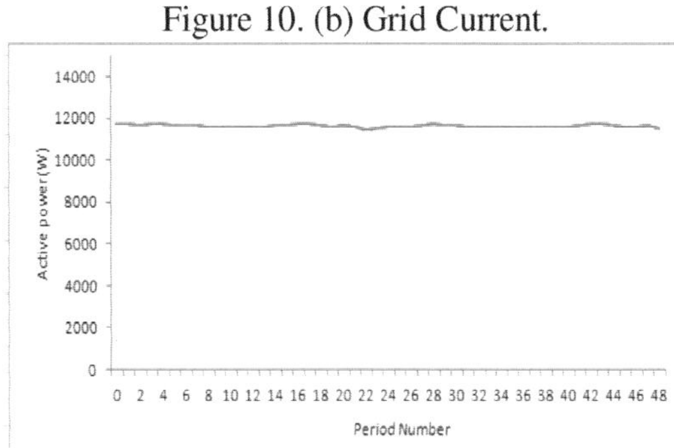

Figure 10. (d) Grid active power

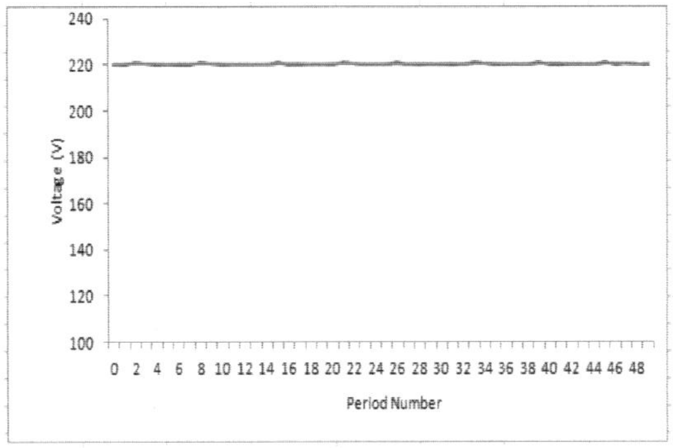

Figure 10. (e) RMS of Output voltage

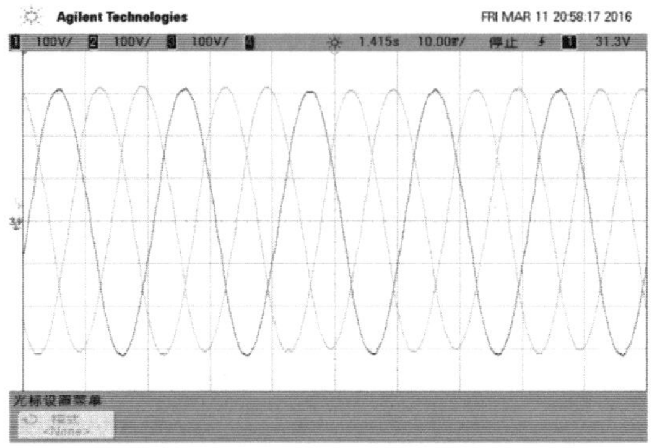

Figure 10. (f) Output voltage when grid fails.

Figure 10. Experimental waveforms under the nonlinear load.

6. Conclusions

A three-phase three-level paralleled line-interactive UPS with LCL filter is presented in this paper. Based on the detailed analysis of the circuit structure and operation mechanism, a new cascaded control strategy for the multi-objective control of power, voltage, and current is discussed. The inner control loop is the output voltage control, and the outer control loop is the grid current control. The grid active current is regulated to meet the active power demand of the load and the grid reactive current is used to regulate the amplitude of the output voltage. The controller parameters design is also detailed discussed.

Experiments results under the unbalanced resistive load and the nonlinear load show the effectiveness of the cascaded control strategy. The capacitance voltage and capacitance current feedback control achieves the system stable and the fast dynamic response. The proportional-resonant controller realizes small THD of the grid current. The repetitive controller realizes small THD of the output voltage. The UPS can not only realize three-phase balanced output voltage with small THD, but also realize three-phase balanced grid active power input. The UPS also has certain reactive power compensation ability when gird voltage varies. The UPS becomes a green load to the grid. Although the UPS has only single-conversion power stage, but it has good comprehensive performances. If the battery is connected to the renewable sources such as PV, it can realize the flexible power output to the grid. It deserves further discussions.

Acknowledgments

This work was supported by the Natural Science Foundation of Guangdong Province under Grant S2013010015007, and the Guangdong Province Science and Technology Plan under Grant 2015A030401099.

References

[1] Solter, W 2002 A new international UPS classification by IEC 62040-3 *Proc. IEEE Int. Conf. Telecommunications Energy Conference* 541-5
[2] Kamran F and Habetler T G 1998 A novel on-line UPS with universal filtering capabilities *IEEE Trans. Power Electronics* **13** 410-8
[3] Chand S and Chawla K 2002 EMC evaluation and analysis of UPS *Proc. Int. Conf. Electromagnetic Interference and Compatibility* 37-42.
[4] Z Giovanni, C Emiliano C. F Andrea, R Carlo 2009 New control techniques for UPS dynamic efficiency optimisation *Int. Conf. Telecommunications Energy* 1-5
[5] Jeon S J and Cho G H 1997 A series-parallel compensated uninterruptible power supply with sinusoidal input current and sinusoidal output voltage *Proc IEEE Power Electrnics Specialists Conf.* 297-303
[6] Nasiri A, Amac A E and Emadi A 2004 Series-parallel active filter/uninterruptible power supply system *Electric Power Components and Systems Journal* **32** 1151-63

[7] RA Modesto, R Barriviera, dSSA Oliveira, AA Oliveira 2013 A simplified strategy used to control the output voltage and the input current of a single-phase line-interactive UPS system *Power Electronics Conference Brazilian* 420-6
[8] Kwon B H, Jin-Ha Choi, and Tae-Won Kim. 2001 Improved Single-phase Line-interactive UPS *IEEE. Trans. Power Electron* **48** 804-11
[9] Wu J C and Jou H L 1995 A new UPS scheme provides harmonic suppression and input power factor correction *IEEE Trans. Ind. Electron.* **42** 629–35
[10] Abusara M A, Guerrero J M and Sharkh S M 2014 Line-interactive UPS for microgrids *IEEE Trans. Ind. Electronics* 61 1292-300
[11] Vasquez J C, Guerrero J M, Savaghebi J M and Eloy-Garcia R 2013 Modeling, analysis, and design of stationary-reference-frame droop-controlled parallel three-phase voltage source inverters *IEEE Trans. Ind. Electron.* **60** 1271–80
[12] J Dannehl, FW Fuchs, S Hansen, PB Thogersen 2010 Investigation of active damping approaches for PI-based current control of grid-connected pulse with modulation converters with LCL filter *IEEE Trans. Industry Applications* **46** 1509-17
[13] J Zhao, Y Zhang, K Qu, F Li, F Kurokawa 2013 Overview of control strategy for single-phase grid-connected inverter with LCL filter *Transaction of china electrotechnical society* **28** 134-42
[14] D Pan, R Xinbo, X Wang, C Bao, LI Weiwei 2013 A capacitor-current real-time feedbackactive damping method for improving robustness of the LCL-type Grid-connected inverter *Proc. the Chinese Society for Electrical Engine* **33** 1-10
[15] Masoud K G and Iravani M R 2004 A method for synchronization of Power electronic converters in Polluted and Variable-frequency Environments *IEEE Trans. Power Systems* **19** 1263-70
[16] Jang J, Joung M, Choi B, Hong S and Lee S 2012 Dynamic analysis and control design of optocouplerisolated LLC series resonant converters with wide input and load variations *IET Power Electronics* 5 755-64
[17] Freijedo F D, Yepes A G, Malvar J 2011 Frequency tracking of digital resonant filters for control of power converters connected to public distribution systems *IET Power electronics* **4** 454-62
[18] Ramos G A and Costa C 2012 Power factor correction and harmonic compensation using second-order odd-harmonic repetitive control *IET Control Theory & Applications* **6** 1633-44
[19] Loh P C, Tang Y and Blaabjerg F 2011 Mixed-frame and stationary-frame repetitive control schemes for compensating typical load and grid harmonics *IET Power Electronics* **4** 218-26

IOP Conf. Series: Earth and Environmental Science **40** (2016) 012029 doi:10.1088/1755-1315/40/1/012029

A new metal detection method based on balanced coil for mobile phone wireless charging system

B Zhou[1], Z Z Liu, H X Chen, H Zeng, T Hei

School of Electrical Engineering, Shandong University, Jinan, 250061, China

E-mail: zhoubo_0524@163.com

Abstract. The wireless charging time of mobile phone will increase greatly if the metal objects mix in the magnetic field coupling area. In addition, the fire may be caused as for the high temperature of metal objects. The paper proposed an improved detecting method based on balance coil for mobile phone wireless charging system according to comparing the advantages and disadvantages of traditional metal detection methods. The circuit model was established, and hardware and software were optimized. At last, experimental results verified the theoretical analysis.

1. The introduction

In recent years, with the popularity of smart phones, our mobile phones possess more and more functions and bigger screen, which leads to larger power consumption and charging frequency. Messy wires, frequent plug, along with interface damage, disgust us deeply for the charging process [1]. We need a more convenient and reliable charging way. As a new charging way, wireless charging gets the favor of most people. Through the space magnetic field coupling, electricity was transmitted to batteries. It enables us to make full use of fragments of time to charge phones, which greatly improve the customer experience.

However, with the development of wireless electricity transmission technology, new problem has emerged. When metal objects enter the power transmission area, part of the energy is consumed on the metal because of the eddy current effect. This may lead to the increase of charging time, and more seriously, the sharply rising temperature of metal may cause a fire. Above all, the metal detection problem needs to be solved urgently.

In this paper, a new detection method based on balance coil was proposed, which can be used in mobile phone wireless charging system to detect metal foreign body.

2. The influence of different metal object in mobile phone wireless charging system

2.1. The influence of metal foreign bodies on mobile phone wireless charging time

As shown in Figure 1, this experiment adopts the mobile phone wireless charging device. It includes transmitting part and receiving part. There are three kinds of metal mediums: ¥1 coins (steel material plated with nickel), ¥0.1coins (aluminum-magnesium alloy), and a piece of iron.

[1] Address for correspondence: B Zhou, School of Electrical Engineering, Shandong University, Jinan, 250061, China. E-mail: zhoubo_0524@163.com.

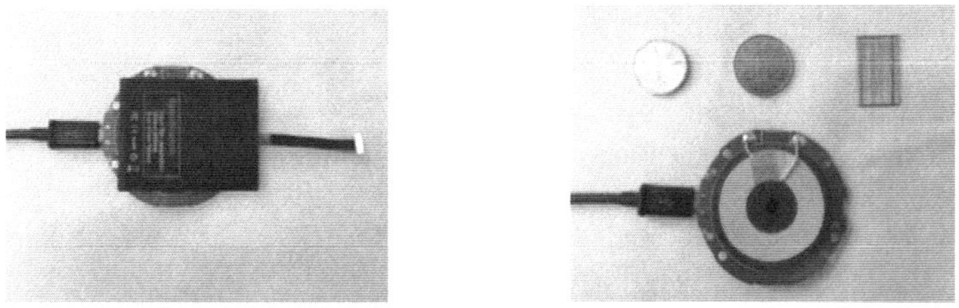

Figure 1. Transmitting and receiving part of the charging device

In order to ensure that the distance of receiving coil and transmitting coil is consistent for each test, we put a cardboard between the receiving coil and the transmitting coil, which is one centimeter thick. We set the initial charging capacity as 63%, and the end of charge capacity as 73%. Then the charging times of the air medium, mixed with ¥ 1 coin, mixed with ¥0.1coin and mixed with iron medium are recorded relatively. The physical map of testing system mixed with metal foreign body is shown in the Figure 2.

Figure 2. The physical map of metal foreign body in charging system

Then, Figure3 shows the the device of wireless charging system.

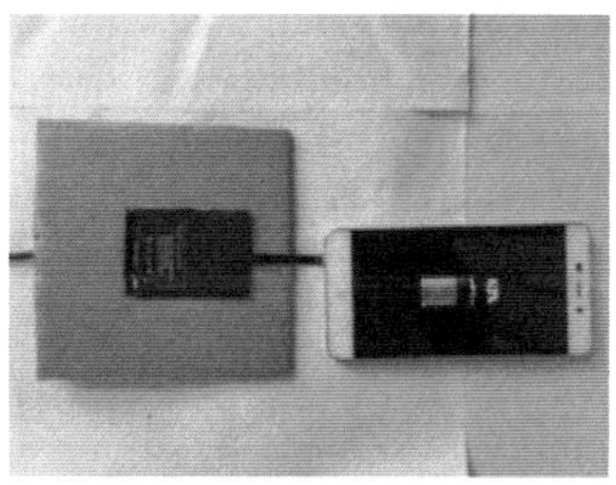

Figure 3. The device of charging system

Finally, the results of experiment are shown in table 1.

Table1. The statistics of charging time

Coupling area media	Charging time(min)
Air medium	38
Mixed with ¥1 coins	66
Mixed with ¥0.1 coins	46
Mixed with iron	54

According to the experimental data, the charging time will increase greatly, and the charging efficiency will decrease obviously when there is metal foreign body in the wireless charging system.

2.2. The introduce of metal heating

Metal objects can not only lead to the increase of charging time, also aggravate its heating situation. Because of the eddy current effect, the temperature of the metal rises sharply, which is easy to cause a fire. The ¥1 coin was placed in the energy exchange area. When the charging time reaches five minutes, the temperature of the coin is shown in Figure 4, which is caught by the infrared temperature measuring instrument.

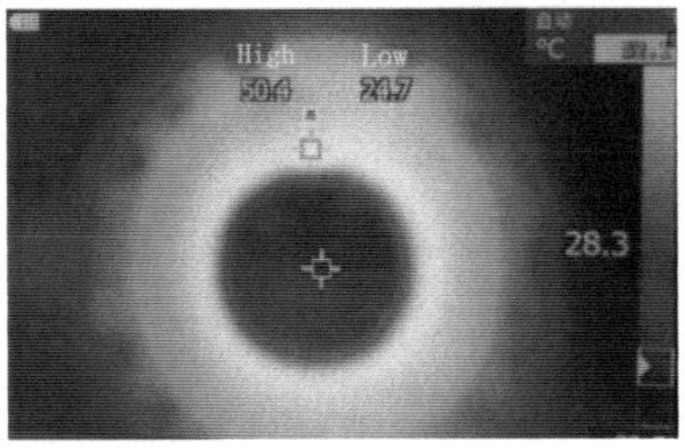

Figure 4. The temperature of the coin caught by the infrared temperature measuring instrument

The temperature is raised to 50 degrees after five minutes, so it is indeed a risk for charging system. Above all, it is necessary to carry out the metal detection for the mobile phone wireless charging system.

3. The comparison of traditional metal detection methods

At present, there are three kinds of metal detection methods in practice, which include the LC resonance detection technology, the differential frequency detection technology and the balanced coil detection technology.

The core of the LC resonance detection technology is the LC oscillation circuit. The operating frequency will deviate from the resonance point when metal objects mix in the system. By detecting the output voltage, we can judge the existence of objects. However, the LC oscillation circuit can be greatly influenced by environmental changes, which is easy to cause the circuit out of order [2].

The differential frequency metal detector is composed of two groups of oscillators with the same frequency. One is the reference oscillator, and the other is the detecting oscillator. When there is metal object, the frequency of detection oscillator may change, therefore, the value of frequency may be different from the reference detector. According to the difference, we can detect the presence of metal objects. The obvious advantage is the simplification of its circuit. However, as its unsteadiness of overall performance, the method has eliminated since the 1970's.

The traditional balanced coil metal detection technology contains three circular coils, which possess equal diameter. And also, their central axes keep in a straight line, as shown in Figure5. With the transmitting coil is in middle, two receiving coils are symmetrically placed on both sides. When the transmitter is coupled with sinusoidal signal, the two receiving coils induce the same electromotive force. However, when there are metal objects, voltages of the two receiving coils are different[3]. By comparing the difference of these two voltages, we can detect the presence of undesirable metal objects. In view of the excellent anti-interference and high sensitivity, the balance coil detection technology has widely used in precise metal detection at home and abroad.

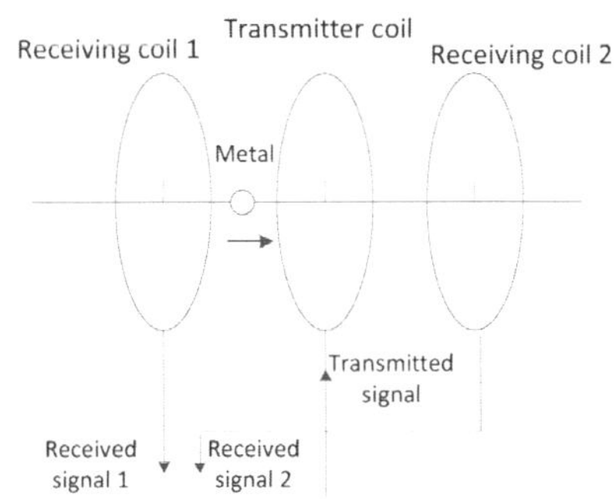

Figure 5. The structure of traditional balanced coil metal detection method

4. A new balance coil metal detection method used for mobile phone wireless charging system

4.1. The introduction of the improved balanced coil

Taking into account the magnetic field of mobile phone charging system, this paper proposes an improved balance coil structure. As is shown in Figure 6, the receiving coil is composed of two parallel symmetrical half circle coils.

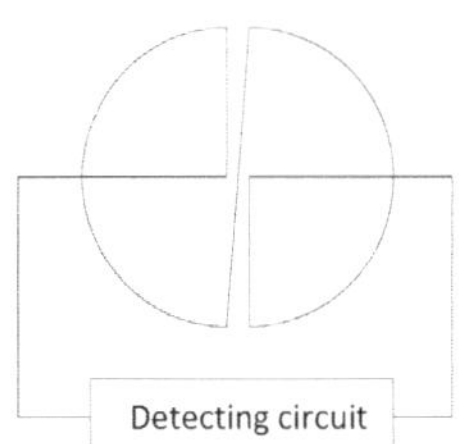

Figure 6. An improved balanced coil structure

Different from the traditional structure, the detection coil is placed close to the launch coil. In order to increase the signal variation caused by metal object, we increase the turn number of the coil. The position of the detecting coil in the charging system is shown in Figure 7.

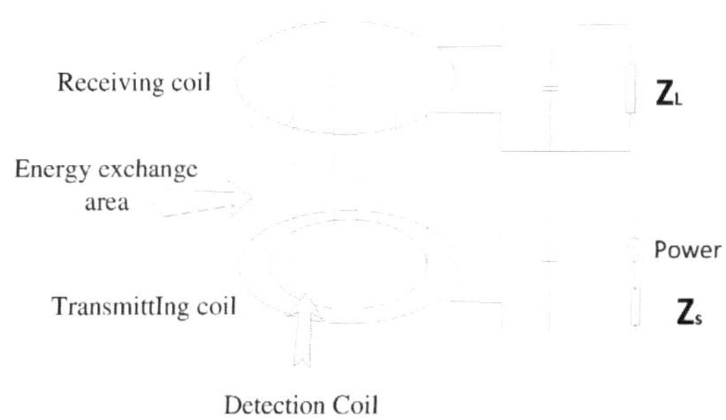

Figure 7. The diagram of detection system

Ideally, the electromagnetic field of the left and the right semicircle are the same when there is no metal. According to the principle of electromagnetic induction, the electromotive force induced in the left and right coils are equal, but the direction is opposite [4]. So the voltage of the balanced coil is almost zero after offsetting. However, when the charging system is mixed with metal object, the original magnetic field can be changed, which leads to the changes of coil terminal voltage. Then we can determine the existence of metal object in the magnetic field coupling region.

4.2 The equivalent circuit model of the system

The eddy current effect of the metal in magnetic field can be equivalent to a circuit that consists of an inductance and a resistance [5]. The equivalent circuit model of wireless charging system after mixing with metal foreign body is shown in Figure 8.

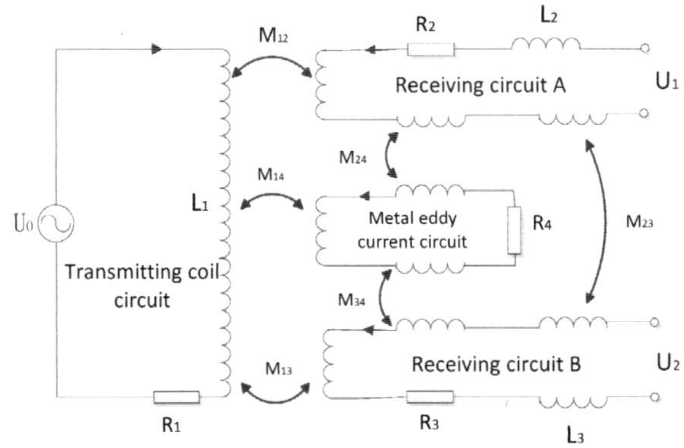

Figure 8. The equivalent circuit model of wireless charging system

The system's equivalent circuit is composed of four circuits, including the launching loop, the receiving loop A, the receiving loop B, and the eddy current loop. R_1, R_2, R_3, R_4 are the equivalent resistance; L_1, L_2, L_3, L_4 are the coefficient of self-inductance; $M_{12}, M_{13}, M_{14}, M_{23}, M_{24}, M_{34}$ are the coefficient of mutual induction; I_1 , I_2, I_3, I_4 are the current of the four loops; U_0 is the excitation voltage of the transmitting coil; U_A , U_B are the terminal voltage of the receiving coil A and B. According to the Kirchhoff's law, the voltage equations of circuits are established as equation (1).

$$\begin{aligned}
&(R_1 + j\omega L_1)\, I_1 - j\omega M_{12} I_2 - j\omega M_{13} I_3 - j\omega M_{14} I_4 = U_0 \\
&-j\omega M_{12} I_1 + (R_2 + j\omega L_2)\, I_2 - j\omega M_{23} I_3 - j\omega M_{24} I_4 = U_A \\
&-j\omega M_{13} I_1 - j\omega L_2 I_2 + (R_3 + j\omega L_3)\, I_3 - j\omega M_{34} I_4 = U_B \\
&-j\omega M_{14} I_1 - j\omega M_{24} I_2 - j\omega M_{34} I_3 + (R_4 + j\omega L_4)\, I_4 = 0
\end{aligned} \tag{1}$$

According to the special structure of the balance coil in this paper, it has the following equation (2).

$$I_1 = -I_2 \tag{2}$$

The output voltage of the balanced coil is:

$$U = U_A - U_B \tag{3}$$

4.3 Hardware components of the detection circuit

Usually, the voltage signals obtained from the balanced coil are relatively weak, and are always mixed with harmonic component. Firstly, we use the amplifier to amplify the signal, using low-pass filter to eliminate the higher harmonics. The frequency of detection signal is so high that the MCU can't deal with. By using peak detection circuit, the peak value of AC signal is obtained [6]. At last, the A/D collecting system sends the signal to the microcontroller. Figure 9 is the structure of the hardware circuit.

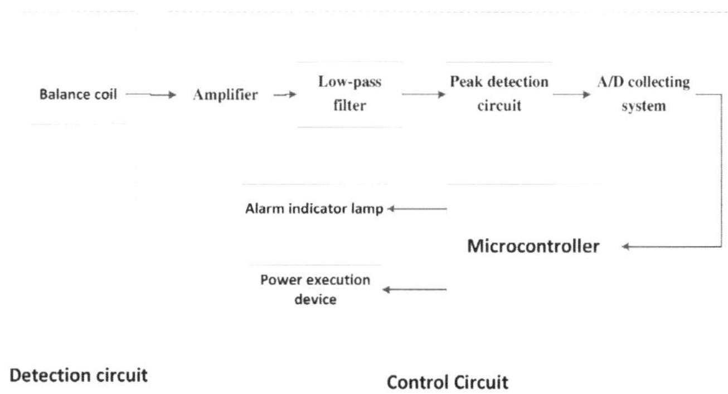

Figure 9. The structure of hardware circuit

4.4 Software Process

The system software is responsible for the setting of discharge time of the peak detection circuit, the sampling frequency of the A/D conversion and the warming voltage.

Actually, the balance coil can't be completely symmetrical, and there are some factors such as temperature drift and external vibration [7]. All above may lead to a fluctuation in the terminal voltage of the balance coil, which has great influence on the fine metal detection. In order to eliminate the influence, a self-balancing algorithm is adopted in the program. When the power is on, the microcomputer records the initial voltage of the balance coil. Then, the subsequent detection voltage value is modified with the initial voltage value, which can effectively reduce the influence of external factors, and improve the accuracy of detection.

In addition, the charging process of mobile phone may emerge power fluctuations, which easily lead to false operation of the detection mechanism. In order to eliminate the influence of power fluctuation, we continuously collect five voltage signals in the experiment and obtain the average value. Comparing the mean value with the threshold voltage, we can judge whether there is a metal foreign body. This method can avoid the error caused by accidental power fluctuations, and increase the reliability of the detection system. The flow chart of software system is shown in Figure 10.

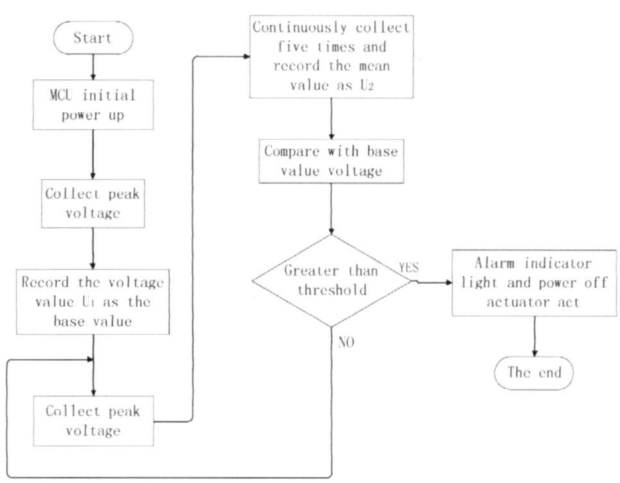

Figure 10. The flow chart of software system

5. Experiment

In order to verify that the proposed balance coil can be used to detect metal, we use a wireless charging device with a coil. The radius of transmitting coil is 25mm as well as the number of turns is 5. The physical diagram of the balance coil is shown in Figure 11.

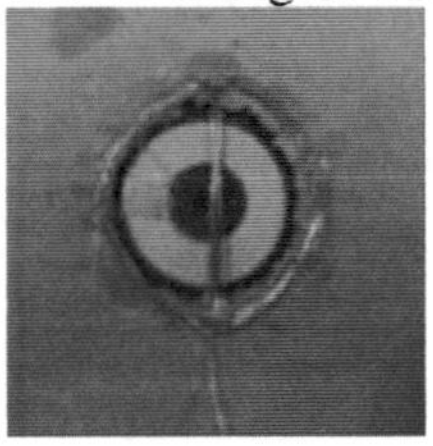

Figure 11. The physical diagram of the balance coil

The experimental device comprises a transmitting coil, a receiving coil and a balanced detecting coil. The transmitting coil is a litz wire with the radius of 25mm, the wire diameter of 0.5mm and the turn number of 10. The receiving coil is a copper enameled wire with the radius of 20mm, the wire diameter of 0.1mm, the turn number of 20. The height of the energy exchange area is 5mm. The input signals of transmission coil in the frequency of 155 kHz, and the peak value of voltage is 10.4V. Figure 12 shows the waveform of terminal voltage.

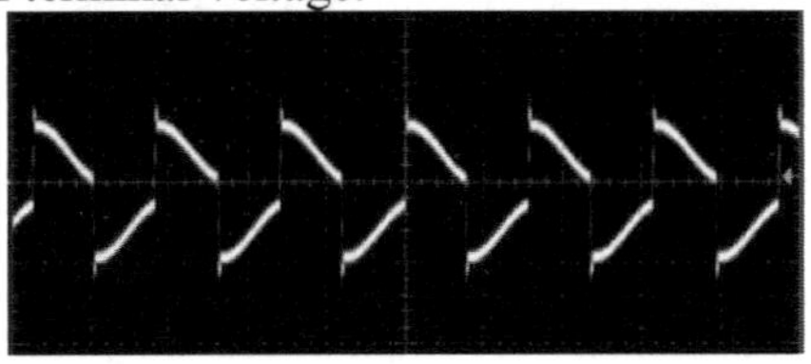

Figure 12. The terminal voltage waveform of the transmitting coil

5.1. The measured voltage waveform of each link in the experiment

In this experiment, we use ¥1 coin as a sample of metal foreign body. As shown in Figure 13, the paper measures the voltage waveform of balanced coil when there is a metal object or not.

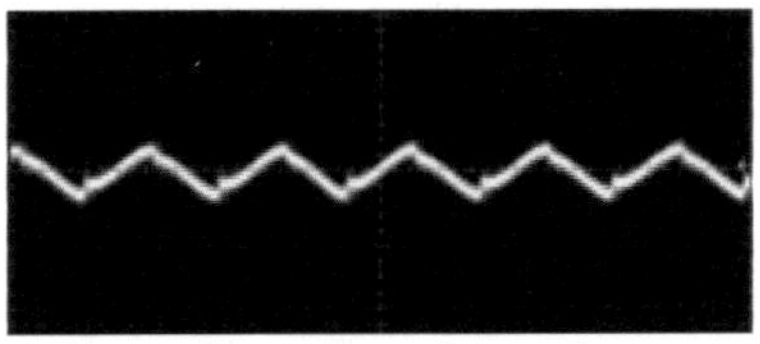

Figure 13. (a) The voltage waveform of balance coil

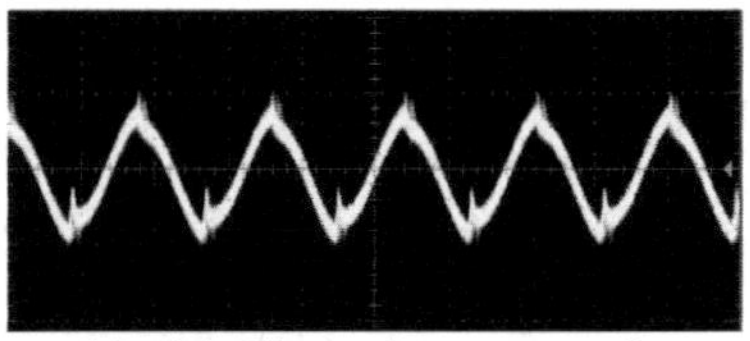

Figure 13. (b) The voltage waveform of balance coil

before the metal is placed after the metal is placed

Figure 13. The voltage waveform of balance coil

Although the signal amplitude of output voltage increased significantly after the metal was placed in energy exchanging area, it's still only 400 mV. In order to detect the changes of signal, we add the amplifying and filtering element. The waveform is shown in Figure 14.

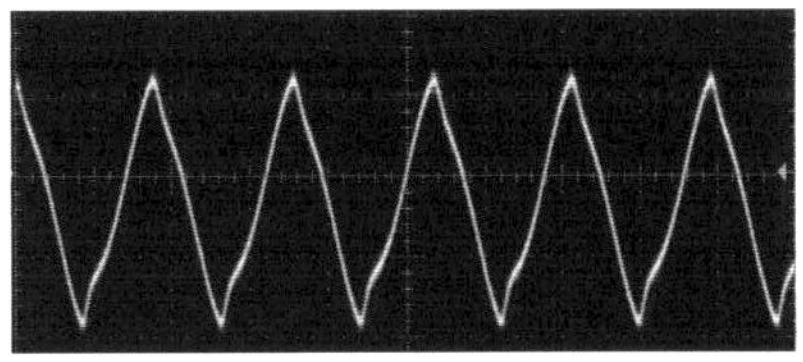

Figure 14. The voltage waveform obtained after the amplifier and filter circuit

The AC signal frequency is so high that the microcontroller can't handle with. So, we add the peak detection circuit. The voltage waveform obtained is shown in Figure 15. The magnitude of DC voltage is 3.2V.

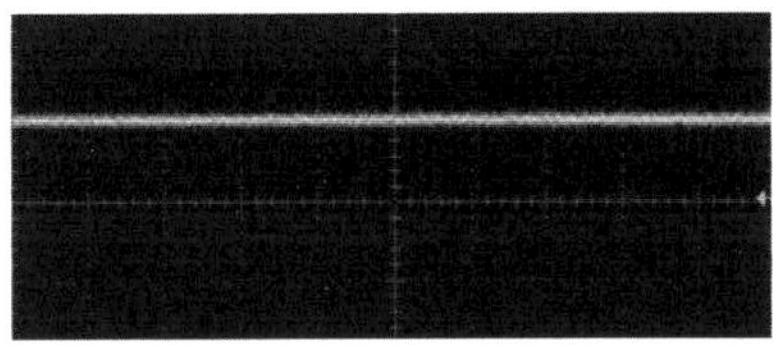

Figure 15. The voltage waveform obtained after the peak detection circuit

5.2. *The experimental results*

To test the performance of the system, we use the metal foreign bodies of different materials, including ¥ 1 coin, ¥ 0.1 coin and iron medium. The experimental data are shown in table 2.

Table 2. The data of experiment

Metal material	Balance coil voltage /mV	Peak detection voltage /V	Testing number	Alarming number
¥ 1 coin	400	3.2	10	10
¥ 0.1 coin	310	2.3	10	9
Iron medium	360	2.6	10	10

From table 2, we can see that the detection system successfully detects metal object of different media and raises the alarm. It is indicated that the balance coil detection system is safe and reliable.

6. Conclusion

The improved metal detection technology introduced in this paper, excellently accomplishes the metal detection task. The experiment proves that this scheme is feasible, and possesses high application values. The metal detection device based on balance coil is high precision and high sensitivity. And, the software algorithm adopted in the experiment, perfectly solves the detection problem caused by the unsymmetrical coil and the power fluctuation of wireless charging.

In addition, there are some problems remained to be solved. As we all know, the main obstacle hindering the development of mobile phone wireless charging is the economy. So we should try to reduce the cost of the raw materials of the metal detection system. Then, there is still room for further optimization of the detecting method, including the way to insure the stability of hardware and the sensitivity of software. At last, the security will be more and more important along with the marketable launch of the wireless charging system. So the metal detection method introduced in this paper will get a wide range of applications in future.

References

[1] Zhou L Y 2014 *Design of the Wireless Charging System for Mobile Phones* (Chongqing: Chongqing University)
[2] Yamazaki S, Nakane H and Tanaka A 2002 Basic analysis of a metal detector *IEEE*

Transactions on Instrumentation and Measurement **51** 810-4

[3] Wang Q L 2010 *Design of metal detector based on balance coil technique* (Jinan: Shandong University)

[4] Qu X D,Yang Y, Liu Z Z and Hou Y J 2014 Metal Detection Technology Based on Balance Coil in Application of Wireless Power Transmission *Power Electronics* 10 61-3

[5] Lan T Y 2009 *Design of metal detection system based on DSP* (Chengdu: Southwest Jiaotong University)

[6] Fan L Z and Li S H 2006 Design of an intelligent metal detector based on at89s52 *Acta Scienti arum Naturalium Universitatis NeiMongol* **02** 185-9

[7] Zhang Z X 2009 *The Research of Metal Detector Based on ATmega8515 Single-chip* (Jinan: Shandong University)

A fault diagnosis system for PV power station based on global partitioned gradually approximation method

S Wang, X N Zhang, D D Gao[1], H X Liu, J Ye and L R Li

School of Mechanical Engineering, Qinghai University, Xining, Qinghai Province, 810016, China

E-mail: gaodd@qhu.edu.cn, 1185725268@qq.com

Abstract. As the solar photovoltaic (PV) power is applied extensively, more attentions are paid to the maintenance and fault diagnosis of PV power plants. Based on analysis of the structure of PV power station, the global partitioned gradually approximation method is proposed as a fault diagnosis algorithm to determine and locate the fault of PV panels. The PV array is divided into 16×16 blocks and numbered. On the basis of modularly processing of the PV array, the current values of each block are analyzed. The mean current value of each block is used for calculating the fault weigh factor. The fault threshold is defined to determine the fault, and the shade is considered to reduce the probability of misjudgments. A fault diagnosis system is designed and implemented with LabVIEW. And it has some functions including the data real-time display, online check, statistics, real-time prediction and fault diagnosis. Through the data from PV plants, the algorithm is verified. The results show that the fault diagnosis results are accurate, and the system works well. The validity and the possibility of the system are verified by the results as well. The developed system will be benefit for the maintenance and management of large scale PV array.

1. Introduction

In the modern society of energy shortage and environmental pollution, the solar energy, an ideal type of new energy, is widely developed and utilized [1]. With the strong development of PV industry, the number of PV power generation system is also increasing. Over the past ten years, the PV industry has been developed rapidly in China. The PV installed capacity get to 43GW until 2015, which take up about 25% of the world. However, China is considered as the mostly world's largest PV manufacturer with unlocked its PV market potential. The country is planning to meet a goal of 20 GW of solar power installed by 2020, 5 GW of this by 2015 which is of course negligible considering its huge potential. The PV power plant's life enters a recession, more and more problem has cropped up. There are two main aspects need attention. Firstly, it is required to improve the efficiency of the system and reduce costs. Secondly, the failure of PV panels limits the normal run of the power station. The fault can be attributed to the electrical disturbances. So it is necessary to design a fault diagnosis method for PV array for improving the power generating efficiency.

At present, the fault detection method of the PV array is mainly divided into traditional diagnostic methods, intelligent algorithm and model algorithm [2-4]. The traditional diagnostics include online

[1]Address for correspondence: D D Gao, School of Mechanical Engineering, Qinghai University, Xining, Qinghai Province, 810016, China. E-mail: gaodd@qhu.edu.cn, 1185725268@qq.com

diagnostics and offline diagnostics [5]. Multi-sensor method is the most widely used for online diagnosis [6-7]. Offline diagnostics require downtime checking. The real-time online fault detecting and location can't be implemented, and diagnostic accuracy is limited by equipments [8]. Circuit model [9] and power model [10] are used to judge the fault, but the model algorithm is fail in fault location. BP neural network [11], data fusion theory [12-13], expert system, et al. are utilized in fault diagnosis to detect and locate the fault. However, the fault detection accuracy depends on the parameter settings and samples, which leads to low determinacy of the intelligent algorithm.

There are many methods about the design of the upper computer operating interface for the monitoring system, such as VB, VC++, web browsing, MCGS configuration software or LabVIEW development environment [14-19]. LabVIEW is widely used in industrial measurement and control platform due to the characteristic of powerful and simple programming.

In this paper, a new fault diagnosis algorithm, Global partitioned gradually approximation (GPGA) method, is proposed to realize fault detection and location for PV array. And a fault diagnosis monitoring simulation system for 20MW large-scale PV power station combined with the algorithm is implemented based on LabVIEW. The simulation system has the following functions: 1) data real-time display, 2) online check, 3) data statistic, 4) data save and print, 5) real-time power prediction, 6) fault diagnosis.

This paper is organized as follows: section 2 presents the algorithm for PV array. At first, the PV power station structure is analyzed. Then the fault diagnosis algorithm is designed according to the structure. In section 3, the simulation system is implemented based on LabVIEW, and the experiments are designed to verify the global partitioned gradually approximation method. In section 4, the simulation results and discussion are provided to illustrate the feasibility and validity of the fault diagnosis system. Finally, conclusions and future work are given in Section 5.

2. Algorithm of fault diagnosis for PV array

2.1. Photovoltaic power station structure analysis

In order to study the composition of PV power station system, a number of PV power stations are investigated in Golmud City, Qinghai Province. Figure 1 diagrams the structure of a 20MW PV power station. 20 single PV panels with the same specification (240W) in series form a branch. 16 branches go into a combiner box. 8 combiner boxes are connected to a single inverter, 2 inverters with the same specification (500kW) form an inverter square. There are 20 inverter squares in total. 4 inverter squares in parallel go into a box type transformer. 5 box type transformers go into 10kV bus to boost, and the main transformer feeds into the grid with 35kV high voltage output.

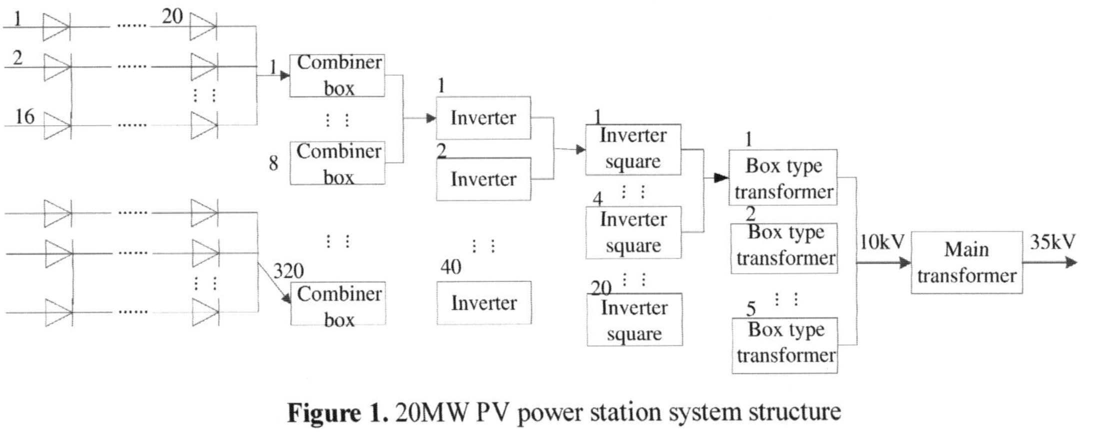

Figure 1. 20MW PV power station system structure

IOP Conf. Series: Earth and Environmental Science **40** (2016) 012030 doi:10.1088/1755-1315/40/1/012030

In the PV power station, the electric signal (including voltage and current) of combiner boxes and inverters can be collected and recorded. Through the analysis and processing of the history data, it's helpful for us to determine whether there is a fault in the PV plants.

2.2. *Global partitioned gradually approximation method*

When a fault occurs on PV panels, the collected voltage and current will become smaller than that of normal operation. Through the analysis of data, the disturbance of this electrical signal can provide some important information about the fault and its location [20]. And the fault diagnosis algorithm is implemented for large scale PV array according to the the disturbance.

By analyzing the basic structure of PV power station, the total number of PV modules in a 20MW PV power plant is over 10 million. When the data acquisition is completed and sent to the computer to determine the failure, all the data is processed at the same time, which will slow down the speed of the system and make the fault localization more difficult. Therefore, the matrix block processing method is employed to implement the modular processing of data. The data is divided into 20 modules with an inverter square as the unit. Each unit contains all the current data of the 16 combiner boxes. Figure 2 shows the collecting data information of a block region.

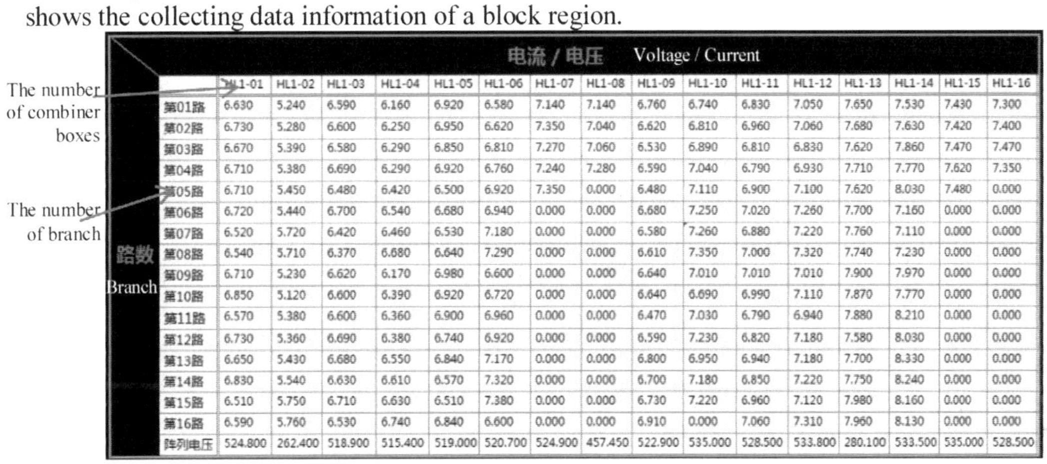

	HL1-01	HL1-02	HL1-03	HL1-04	HL1-05	HL1-06	HL1-07	HL1-08	HL1-09	HL1-10	HL1-11	HL1-12	HL1-13	HL1-14	HL1-15	HL1-16
第01路	6.630	5.240	6.590	6.160	6.920	6.580	7.140	7.140	6.760	6.740	6.830	7.050	7.650	7.530	7.430	7.300
第02路	6.730	5.280	6.600	6.250	6.950	6.620	7.350	7.040	6.620	6.810	6.960	7.060	7.680	7.630	7.420	7.400
第03路	6.670	5.390	6.580	6.290	6.850	6.810	7.270	7.060	6.530	6.890	6.810	6.830	7.620	7.860	7.470	7.470
第04路	6.710	5.380	6.690	6.290	6.920	6.760	7.240	7.280	6.590	7.040	6.790	6.930	7.710	7.770	7.620	7.350
第05路	6.710	5.450	6.480	6.420	6.500	6.920	7.350	0.000	6.480	7.110	6.900	7.100	7.620	8.030	7.480	0.000
第06路	6.720	5.440	6.700	6.540	6.680	6.940	0.000	0.000	6.680	7.250	7.020	7.260	7.700	7.160	0.000	0.000
第07路	6.520	5.720	6.420	6.460	6.530	7.180	0.000	0.000	6.580	7.260	6.880	7.220	7.760	7.110	0.000	0.000
第08路	6.540	5.710	6.370	6.680	6.640	7.290	0.000	0.000	6.610	7.350	7.000	7.320	7.740	7.230	0.000	0.000
第09路	6.710	5.230	6.620	6.170	6.980	6.600	0.000	0.000	6.640	7.010	7.010	7.010	7.900	7.970	0.000	0.000
第10路	6.850	5.120	6.600	6.390	6.920	6.720	0.000	0.000	6.640	6.690	6.990	7.110	7.870	7.770	0.000	0.000
第11路	6.570	5.380	6.600	6.360	6.900	6.960	0.000	0.000	6.470	7.030	6.790	6.940	7.880	8.210	0.000	0.000
第12路	6.730	5.360	6.690	6.380	6.740	6.920	0.000	0.000	6.590	7.230	6.820	7.180	7.580	8.030	0.000	0.000
第13路	6.650	5.430	6.680	6.550	6.840	7.170	0.000	0.000	6.800	6.950	6.940	7.180	7.700	8.330	0.000	0.000
第14路	6.830	5.540	6.630	6.610	6.570	7.320	0.000	0.000	6.700	7.180	6.850	7.220	7.750	8.240	0.000	0.000
第15路	6.510	5.750	6.710	6.630	6.510	7.380	0.000	0.000	6.730	7.220	6.960	7.120	7.980	8.160	0.000	0.000
第16路	6.590	5.760	6.530	6.740	6.840	6.600	0.000	0.000	6.910	0.000	7.060	7.310	7.960	8.130	0.000	0.000
阵列电压	524.800	262.400	518.900	515.400	519.000	520.700	524.900	457.450	522.900	535.000	528.500	533.800	280.100	533.500	535.000	528.500

Figure 2. Data information of a block region

According to the above analysis, all the collected data is treated as a M × N matrix A, and the data of each inverter is treated as 16×16 sub-matrix A_i (i=1, 2, 3,...... , 20). A and A_i are written as Equation (1) and (2).

$$A=\begin{bmatrix} [A_1]_{16\times16} & \cdots & [A_5]_{16\times16} \\ \cdots & \cdots & \cdots \\ [A_{11}]_{16\times16} & \cdots & [A_{20}]_{16\times16} \end{bmatrix}_{m\times n} \tag{1}$$

$$A_i=\begin{bmatrix} A_{pv11} & A_{pv12} & \cdots & A_{pv1c} \\ A_{pv21} & A_{pv12} & \cdots & A_{pv2c} \\ \cdots & \cdots & A_{pvrl} & \cdots \\ A_{pvb1} & A_{pvb2} & \cdots & A_{pvbc} \end{bmatrix}_{b\times c} \tag{2}$$

In Equation (2), b=c=16, r=1~b, l=1~c.

In this method, the relative error between the average value of the A_i and all data A_{pvrl} ($r=1,2,\ldots,b$, $l=1,2,\ldots,c$) in A_i is calculated in Equation (3). The relative error is the important parameter for judging whether there is a fault, which is called Fault weight factor δ.

$$\delta = \left| \frac{A_{\arg} - A_{pvrl}}{A_{\arg}} \right| \tag{3}$$

In Equation (3),

$A_{\arg} = A_{bc} / (b \times c)$;

$A_{bc} = \sum A_{pv11} + A_{pv12} + ... + A_{pv1c} + ... + A_{pvb1} + A_{pvb2} + ... + A_{pvbc}$.

The data A_{pvrl} has different properties for the different states. If there is no fault in the array, the branch current data A_{pvrl} is allowed to fluctuate within a small range. The small range is artificially set value, which is called fault threshold δ_0. If in a failure state, the deviation between the data A_{pvrl} in the fault branch and the average value A_{arg} is greater than fault threshold δ_0.

A fault function is f. $\delta \geq \delta_0$, which means the system has a fault, f=1. $\delta < \delta_0$, which means the system is normal, f=0. The conditions for determining the fault is written as Equation (4).

$$f = \begin{cases} 1 & \delta \geq \delta_0 \\ 0 & \delta < \delta_0 \end{cases} \tag{4}$$

The PV plant is divided into finite units, and the fault of each unit can be determined by Equation (4). The fault positions are located by searching all the collected data from the PV plant. The global partitioned gradually approximation algorithm is implemented with Matlab software, and its flow chart is shown in Figure 3.

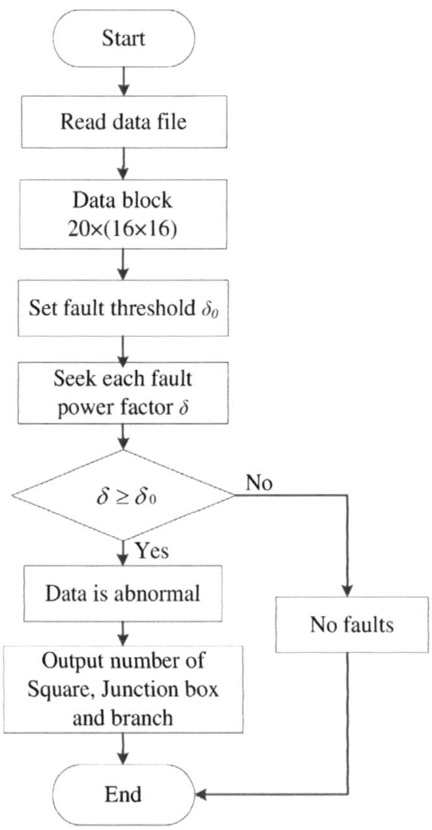

Figure 3. Flow chart of programmed algorithm

2.3. Shade determination

For large PV power plants, the shade is a very common phenomenon. The shade makes data measured of the shaded panels smaller than the normal state, which is similar to fault panels. But it is not a fault type [21]. Thus this phenomenon possibly leads to the mistake of fault detection with the proposed algorithm. So the shade situation should be considered for improving the validity of the GPGA algorithm.

As shown in Figure 4, four types of shade phenomenon are diagramed to illustrate the shade arrangement. Through the shade arrangement, the shade phenomenon has the following characteristics: (1) there are a larger number of shaded PV modules within the region; (2) most of the obscured PV modules are adjacent within the region.

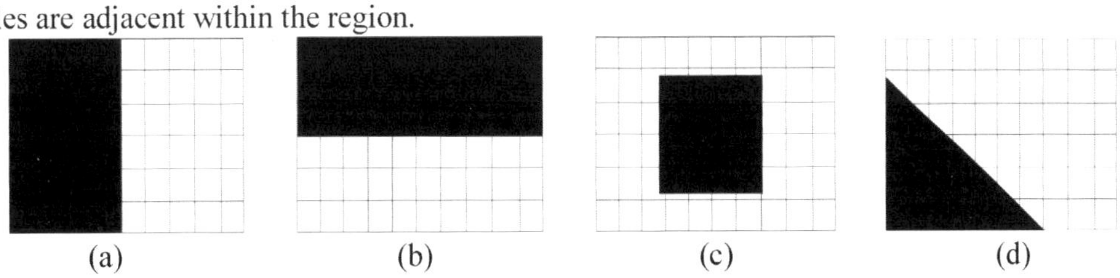

Figure 4. Shade situation analysis

The total number of PV units in an area is u, the number of shaded units is v, the number of adjacent shaded modules is w. The algorithm for determining the shaded panels is established as shown in Figure 5.

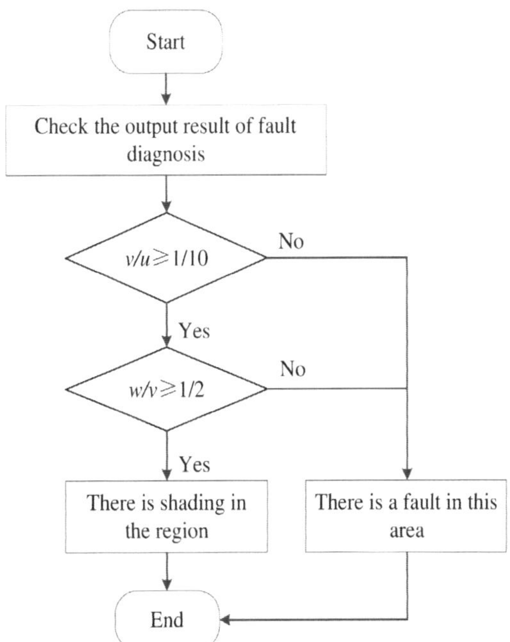

Figure 5. Flow chart of shading decision algorithm

The GPGA method combined with the shading decision algorithm is used to implement fault detecting and locating when the PV panels are shaded by the clouds. The shade-determining conditions are summarized as follows:

(1) $v/u \geq 1/10$ in the shading decision diagram.

(2) $w/v \geq 1/2$ in the shading decision diagram.

A shade function is y, when $v/u \geq 1/10 \cap w/v \geq 1/2$, the system has a shade, y=1. In other conditions, the system has no a fault, y=0. The shade-determining conditions are written as

$$y=\begin{cases}1 & v/u\geq 1/10 \text{ I } w/v\geq 1/2\\ 0 & other \quad conditions\end{cases} \quad . \tag{5}$$

2.4. *Determination algorithm of fault diagnosis*

The GPGA method is used to judge whether there is a fault, and the shading decision algorithm is employed to judge whether there is a shade. When f=1∩y=1, there is a shade in the region. When f=1∩y=0, there is a fault in the region.The flow chart of fault diagnosis algorithm is shown in Figure 6.

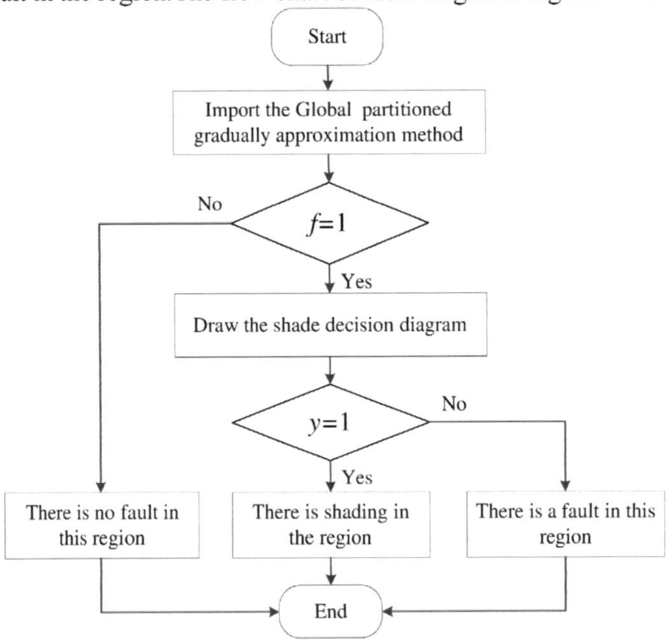

Figure 6. Flow chart of fault diagnosis algorithm

3. Simulation system design

3.1. *System function analysis for PV power station*

This simulation system is designed combing with the actual demand for PV power station. It has the following major functions.

1) Overall structure of PV power plant. The simulation system shows the overall structure block diagram of 20MW PV power station. The parameters, including the voltage, current, active power and the real-time power, are displayed with simulation system. The smallest unit in the structure block diagram is inverter square.

2) Real-time displaying for link data. The displayed data include voltage and current of each branch of the combiner box, basic data amount of the inverter power generation (the phase current, the phase voltage, total power, active power, reactive power, power factor, accumulated generated energy), the daily generated energy, temperature, solar irradiation, wind speed and so on.

3) History data viewing online. The daily power, daily accumulated energy, the daily inverter power, the environmental parameters are recorded and viewed in the simulation system.

4) Data statistics. This function is to analyze the data collected from the sensors, and plot the curves for the inverter and combiner box under different time periods.

5) Report generation. This function is to get the history data information. Through the choice of date, the data information of the inverter can be saved. Data font, font size, format, save type, save path and other basic settings also can be set in the function.

6) Power prediction. Based on the analysis of the history big data, daily power electricity is predicted. In order to view clearly, the system has the function of zoom in and out about power prediction curve, as well as the upper limit curve and lower limit curve.

7) Fault diagnosis. The global partitioned gradually approximation method is embedded into the simulation system, with which the fault detection and location of large-scale PV array is implemented.

The overall structure of the monitoring simulation system is shown in Figure 7.

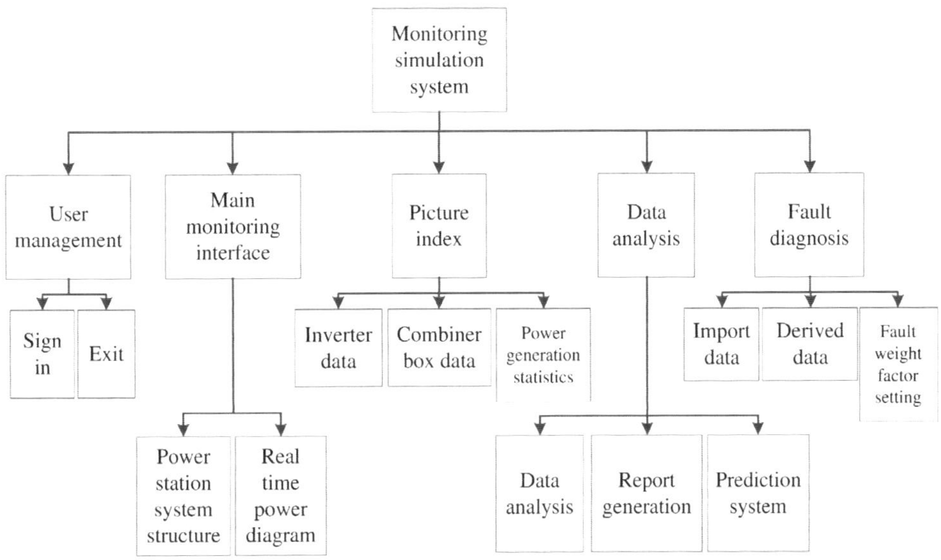

Figure 7. The overall structure of the monitoring simulation system

3.2. Simulation system design

The Qinghai-Tibet Plateau is one of the most abundant areas in the world. Because of the high altitude, the sunshine is sufficient. The amount of solar radiation in Qaidam Basin is more than 6800MJ/m^2, in which the desertification land used for PV power station gets to100 thousand square kilometers. It is the best area of solar photovoltaic power generation in China, because its average annual sunshine gets to 3000 hours. The Golmud City is located in the hinterland of Qinghai-Tibet Plateau, and the city is located in Qaidam Basin at an altitude of 2780 meters. The actual data of a Golmud 20MW PV power station is used as the data source of the simulation system.

According to the requirement of the user and the function of the PV power station system, the development and design of the fault diagnosis simulation system based on GPGA method is carried out on LabVIEW. The simulation system mainly includes user management, data real-time display, storage management, data statistics, data prediction and fault diagnosis.

The following figures are a part of the application interface of fault diagnosis simulation system. Figure 8 shows the user login screen. The correct user name and password is necessary to enter the system main interface for system security.

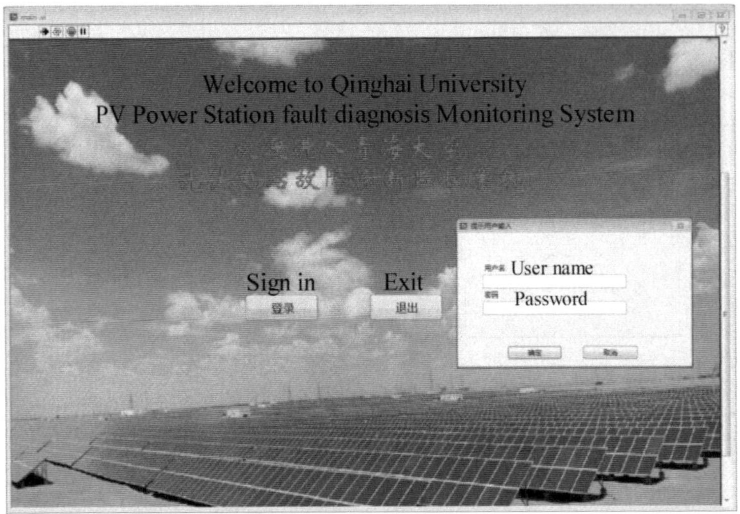

Figure 8. The user login interface

The system main interface is shown in Figure 9. The main interface is composed of three parts including the menu switch control, the tree structure of PV power station and the real-time power display window. The system menu includes the main interface display, the interface index, data analysis and fault diagnosis. The data could be acquired in real-time, and the data includes the current, voltage and active power. The frequency of the acquitted data is $1.67\times10^{-3}s^{-1}$. The real-time data power curve is displayed in the display window, and the history data information can be observed at the same time. The data information in display window includes power, solar irradiation, ambient temperature and wind speed.

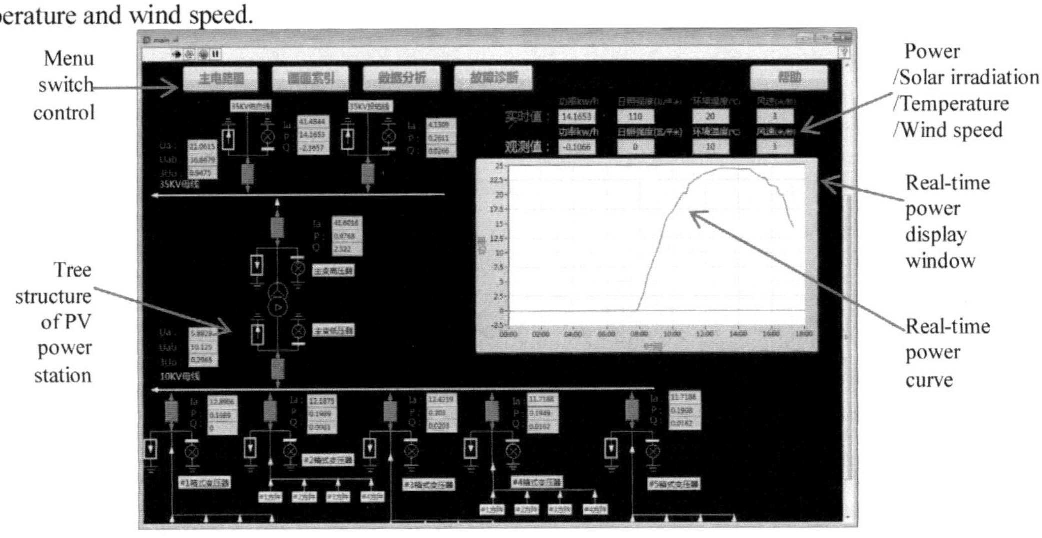

Figure 9. System main interface

Figure 10 gives the specific fault diagnosis interface. The interface has the following functions: 1) setting the fault threshold, 2) importing and exporting the data, 3) fault display. The data is calculated by LabVIEW program which is based on Global partitioned gradually approximation method. Its results show as "There is a fault in inverter No. ***, combiner box No. ***, branch No. ***." .

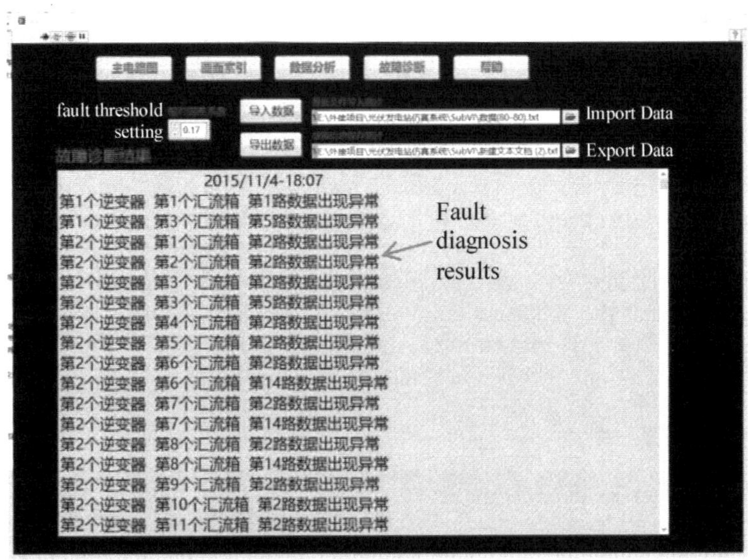

Figure 10. Fault diagnosis interface

4. Result and discussion

4.1. Simulation result of fault diagnosis algorithm

Some collected data is employed to validate the proposed fault diagnosis algorithm. The data is acquired from the PV power station in Glomud City. The data includes voltage and current of all the combiner boxes in July 11 to 13. And some data is modified as the faults to test and verify the proposed fault diagnosis algorithm. In this experiment, the fault threshold is set as 6 types, 5%, 7%, 10%, 12%, 15%, and 20%. Through calculating, the result of fault diagnosis is seen at the interface. Table 1 shows the accuracy of fault detection under different fault threshold.

Table 1. Fault detection accuracy rate under different fault threshold (δ_0)

	The original data fault points/a	Detect fault points/a	accuracy rate /%
δ_0=5%	100	210	---
δ_0=7%	100	203	---
δ_0=10%	100	100	100
δ_0=12%	100	100	100
δ_0=15%	100	94	94.0
δ_0=20%	100	88	88.0

The simulation results show that the number of abnormal data points is much larger than the actual fault points when the fault threshold is less than 10%. The normal data can be regarded as the fault easily. When the fault threshold is in 10% - 12%, the fault detection accuracy rate is 100%. When the fault threshold continues to increase, the accuracy rate of fault detection will decrease.

Finally the results of fault diagnosis are used for judging whether there is a fault occurred in some regions.

(1)If the abnormal data does exist in a region, f=1, and its number is more than 1/10 of the total number, shading decision algorithm is used for determining whether there is a shade or a fault. If more than 1/2 of the abnormal data are adjacent, y=1, it can be determined that there is shade in the region, not a fault.

(2) If the abnormal data does exist in a region, f=1, and its number is less than 1/10 of the total number, and most of the abnormal data are nonadjacent, y=0, we can determine that there is fault in the region.

Taking the 2nd square inverter as an example, the shade decision algorithm is used for detecting the fault and the results are showed in Figure 10. The Figure 11 displays the shading decision algorithm (In Figure 11, A denotes the number of combiner box, and B denotes the number of breach). The branch number of an inverter is 256, and the abnormal data branch number is 52 (v/u=1/5). Then we found that more than 1/2 of the abnormal data points are adjacent, so we can determine the shade in the 2nd inverter square. But the 5th branch of the 3rd combiner box, the 6th branch of the 14th combiner box and 4th branch of the 7th combiner box are in fault.

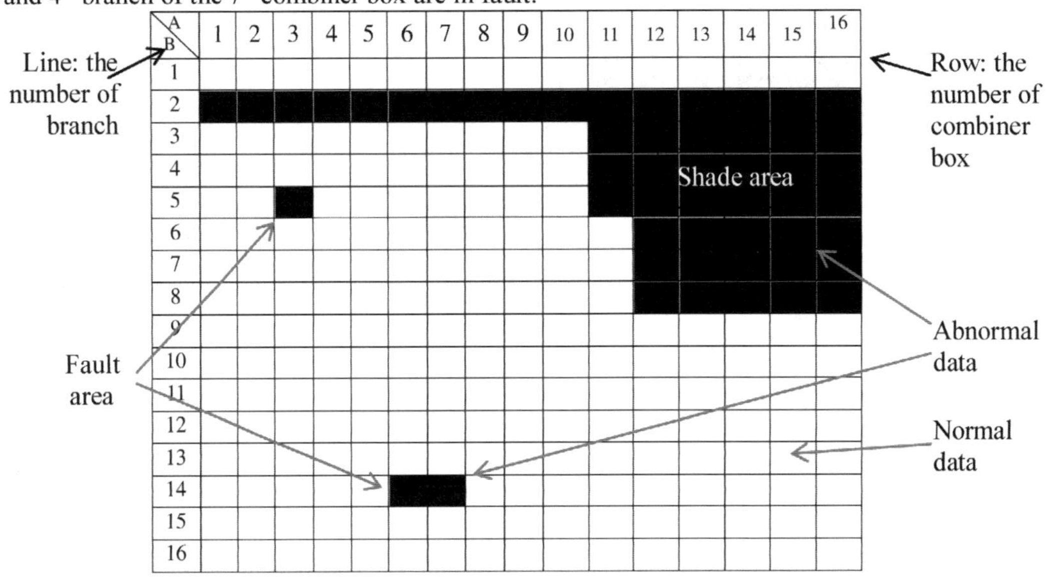

Figure 11. Shade decision diagram of the 2nd square inverter

4.2. Losses measure of PV power station

The power losses occur on the PV power station due to the low fault diagnosis accuracy rate and slower troubleshooting. If the fault is found and maintained in time, the power losses mainly depend on the generated energy of fault branches which are not detected.

20 same (240W) PV panels in series form a branch, and the power of one branch is measured about 4KW (it less than the rated power in normal operation). The total power of this PV power station is 20MW. The different power losses are calculated at different fault detection accuracy rate. The calculated results are shown in Table 2.

Table 2. Power losses under different fault detection accuracy rate

accuracy rate /%	The number of un-detected fault branch	Power losses/KW	Percentage/%
100	0	0	0
94	6	24	0.12
88	12	48	0.24
82	18	72	0.36
76	24	96	0.48

Through the calculating and contrast, the result shows that power losses increases with the loss fault detection accuracy rate. So, fault diagnosis algorithm of high accuracy is helpful to reduce power losses.

5. Conclusions

A new fault diagnosis algorithm, global partitioned gradually approximation method, is proposed for the large-scale PV arrays in high-altitude desert area. The proposed fault diagnosis algorithm takes account for the shade affects, which reduces the misjudgment probability of the algorithm. A system for PV power station monitoring and fault detection is designed and implemented with LabVIEW software based on the new fault diagnosis algorithm. The function of the system includes real time data display, online-browse, data statistics, and fault diagnosis. The proposed fault diagnosis algorithm can predict and locate the fault arrays through the experiments. The developed system can be applied in the large-scale PV plants without changing any original hardware and infrastructure.

Acknowledgments

This research work is sponsored in part by the National Natural Science Foundation of China (No. 51566015), and the Projects from Qinghai Science &Technology Department (No. 2012-G-Y25A-4 and 2014-GX-212).

References

[1] Kelly N A and Gibson T L 2009 Improved PV energy output for cloudy conditions with a solar tracking system *Solar Energy* **83** 2092-2102
[2] Wang Y, Li Z and Wu C 2013 Fault diagnosis technologies for photovoltaic system *Chinese Journal of Power Sources* **37** 1700-05
[3] Cheng Z, Li B F, Liu L and Liu Y L 2010 A fault detection method for new PV module strufure *Journal of Electronic Measurement and Instrument* **24** 131-6
[4] Liu Y and Li B 2010 Research on domestic PV module structure based on fault detection *Intelligent Control and Automation* **20** 171-5
[5] Wang Y Z, Wu C H, Zhou D Q, Fu L, Li Z H. 2013 A survey of fault diagnosis for PV array based on BP neural network *Power System Protection and Control* 41 108-14
[6] Hu Y H, Chen H, Xu R D, Li R. 2011 Photovoltaic(PV) array fault diagnosis strategy based on optimal sensor placement *Proceedings of the Chinese Society of Electrical Engineering* **31** 19-30
[7] Hu Y, Chen H and Xu R 2010 A type of PV array fault diagnosis strategy based on voltage scan *Proceedings of the Chinese Society of Electrical Engineering* **30** 185-91
[8] Hu Y, Deng Y and He X 2013 A summary on PV array fault diagnosis method *Power Electronics* **47** 21-3
[9] Takashima T, Yamaguchi J, Otani K, Oozeki T, Kato K, Ishida M. 2009 Experimental studies of fault location in PV module strings *Solar Energy Materials and Solar Cells* **93** 1079-82
[10] Patel H and Agarwal V 2008 MATLAB-based modeling to study the effects of partial shading on PV array characteristics *IEEE Transactions on Energy Conversion* **23** 302-10
[11] Lan Q and Zhang Y 2011 Application of BP neural network in fault diagnosis of solar photovoltaic system *Telecom Power Technology* **28** 38-40
[12] Zhong D 2011 *Research on fault detection of PV array based on composite data fusion* (Tianjin: Tian Jin University)
[13] Tang J 2012 *The fault detection and simulation research of solar array* (Beijing: Noah China Electric Power University)
[14] Zhang H 2006 *The design and research of distant monitoring system for the PV plant* (Qinhuangdao: Yan Shan University)
[15] Wang K 2013 *Design and implementation of embedded monitoring system for PV power plant* (Bejing: Beijing Jiaotong University)

[16] Zhang X and Zheng J 2010 Design of monitoring system for photovoltaic power station *Electrotechnics Electric* **9** 12-6

[17] Miao J 2012 *The research and implementation of monitoring system for PV power plant* (Beijing: Beijing Jiaotong University)

[18] Zhao Y, Meng C, Chen Y, Sun Y. 2012 Photovoltaic Power Station Monitoring System Based on LabVIEW *Instrument Technique and Sensor* 50-3

[19] Zhang W 2014 *The research and implementation of PV's remote monitoring system* (Yangzhou: Yang Zhou Univesity)

[20] Zhang X, Gao D, Liu H, Ye J, Wang S. 2016 A noval method for multi-sensor fault detection and positioning of PV array *Renewable Energy Resources* **34** 166-72

[21] Chowdhury S R and Saha H 2010 Maximum power point tracking of partially shaded solar photovoltaic arrays *Solar Energy Materials & Solar Cells* **94** 1441-7

Feasibility of drying system using waste heat as the heating source

M N Xie[1], Y L Shi and L X Chen

Quanzhou Institute of Equipment Manufacturing, Fujian Institute of Research on the Structure of Matter Chinese Academy of Sciences

Email: xmn@fjirsm.ac.cn

Abstract. In this study, a wastewater heat pump system was proposed and its thermal performance was analyzed. The proposed system includes two evaporators: an air-source evaporator and a water-source evaporator. The air-source evaporator absorbs heat from the moist hot air which exhaust from the drying oven. The water-source evaporator absorbs heat from the waste water, while the waste water recovers heat from the mechanical energy, which was produced by cutting and polishing in stone production. The thermodynamic model was developed to evaluate the performance of the proposed system. The energetic analysis was carried out to investigate the influences of the temperature of fresh air. The results show significantly higher energy efficiency, compact-sized and energy-saving compared with the system which uses air as the heat source. Among the seven of alternative refrigerants (R152a, R123, R1234yf, R1234ze, R600a, R22 and R600) investigated, R123 was suggested to be used in this heat pump for its high heating efficiency, inflammable, very low ODP(Ozone Depletion Potential) and GWP(Global warming potential).

1. Introduction

The industrial drying energy consumption is responsible for 14-35% of the china industrial energy use. Large amounts of low-grade heat were wasted, and cannot be used reasonably. By analysing the existing heat pump drying device, the results show that the problems drying device mainly existed are as follows. Firstly, the mainly heating modes of industrial dryers are electric heating and steam heating. The energy consumption of these drying modes is relatively large, which would cause negative effect on environment. And the drying effect is not as ideal as expected. Secondly, heat pump drying technology used in the domestic and heat pump dryers used in industry are still in the stage of laboratory research [1-4].

The industrial energy is lost as waste heat in the form of hot exhaust gases, cooling water, and the heat lost from hot equipment surfaces. There are two main works in the process of stone production: cutting and polishing. In order to improve the service life of the cutter, the production efficiency and product quality, cooling cutting tools and removing abrasive dust by water during processing are need. Since the stone processing consumes large amounts of water, lots of electric energy was consumed during cutting and polishing, much of the energy was absorbed by wastewater and finally dissipated in the environment. At present, the wastewater treatment equipment of the enterprise is simplified. A

[1] Address for correspondence: M N Xie, Quanzhou Institute of Equipment Manufacturing, Fujian Institute of Research on the Structure of Matter Chinese Academy of Sciences. Email: xmn@fjirsm.ac.cn.

common way to deal with wastewater flowing from the plant is to pass through several simple sedimentation tanks, or directly discharge into the environment without any treatment. This is not only a waste of valuable water resources, but also causing great damage to the environment and the waste heat energy in the water absorbed from the process of cutting and grinding. The traditional stone drying mainly takes the form of electric heating drying, where all heat is from electricity consumption, and the hot air is discharged into the atmosphere directly after drying. This approach is not only a waste of energy, but also causes serious environmental pollution.

Typically, the electricity power consumed by cutting and grinding, is much higher than the thermal energy required for drying. Analysis and demonstration of the industrial waste heat recovery system, recycling the heat from wastewater and exhaust presented in this article, has higher economy, energy saving and better effect on environment. At the same time, to encourage enterprises recycling wastewater, supporting the construction of settling ponds, reservoirs and circular pool, minimize the consumption of non-renewable energy, and to save operating cost. The main contents are as follows. A wastewater heat pump system was proposed and introduced. Its performance was theoretical analyzed and its feasibility was proved by comparing with conventional air cooling systems. The superior thermal performance with highest COP is found by comparing different substances.

2. The composition and working principles of the system

2.1. The fundamental composition of the system

The system is consisted of three subsystems, namely a wastewater recycling system, a refrigerant cycle system and a hot air heating system, as shown in figure 1.

Refrigerant cycle system evaporator contains two parts: air-source evaporator and water-source evaporator. Air-source evaporator heat transfer process is mainly carried out by the moist hot air at the end of the drying oven, and water-source evaporator is heated by recovering heat from waste water. Low-grade waste heat is converted into high-grate thermal energy by the utilization of wastewater from recycling system, and the energy of the wastewater is recovered by absorbing the mechanical energy from the cutting and grinding. Hot air heating system is an open-loop system, with the fresh air

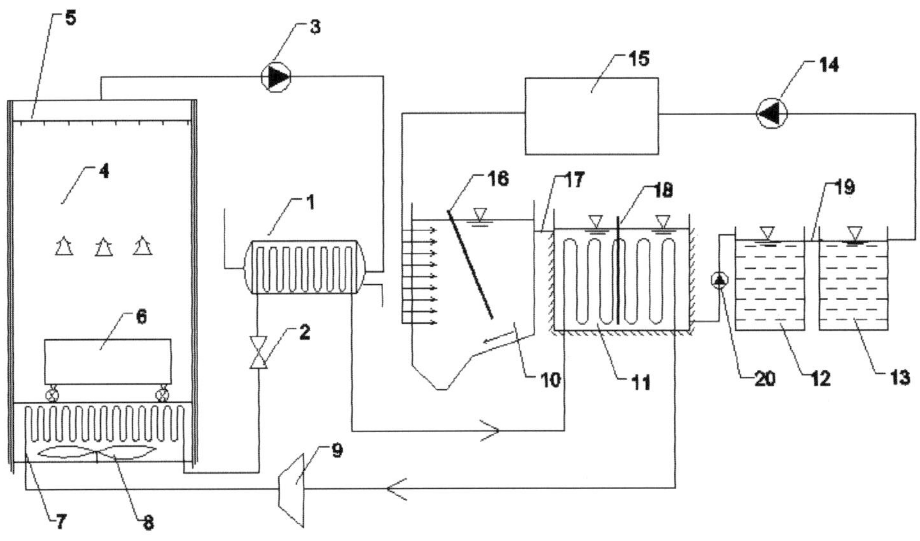

and exhaust recycling unit.

Figure 1. Schematic illustration of the principal design of the heat recovery system.

Table 1. Illustration of the serial number in figure 1.

Number	Component	Number	Component
1	Air-Source Evaporator	11	Immersed Water-Source Evaporator
2	Expansion Valve	12	Second Settlement Tank
3	Air Pump	13	Cleanwater Storage Tank
4	Drying Oven	14	Water Pump
5	Sieve Plate	15	Cutters And Sanders
6	Vehicle For Material Transportation	16	First Plate
7	Condenser	17	First Overflow Pipe
8	Fan	18	Second Plate
9	Compressor	19	Second Overflow Pipe
10	First Settlement Tank	20	Wastewater Pump

2.2. The working principles of the system

Water temperature rise of stone production is generally 5-8℃ with tremendous heat. The wastewater was collected through the irrigation plant in the region, and then pulled into the first settlement tank, so most of the suspended solids sink to the bottom and the supernatant directly flows into immersed water-source evaporator as heat source. The wastewater of lower temperature was pulled into the next settler under the traction of water pump, and its supernatant flowed into the Clearwater storage tank. Eventually, clean water entered the production plant for recycling by water pump.

In the system, the heat pump cycle is consists of a compressor, a condenser, an expansion valve, two evaporators and auxiliary system equipment. The working fluid is refrigerant. The refrigerant enters the condenser as a superheated vapor after compressed, and gives heat to the inlet fresh air which providing the main energy source for the oven dried until it becomes saturated liquid and leaves the condenser. The refrigerant is then throttled to the evaporator pressure while its temperature drops below the wastewater temperature. After it absorbs heat in the evaporator, it leaves as saturated vapor.

The fresh air enters the drying oven after absorbed heat in the condenser, and then heating stones and taking away the water in the materials, so the air temperature reduce and the humidity increase at the same time. Then, the air is discharged into the atmosphere after recovery heat from the oven heat and humid air in the evaporator. Recovering the latent heat in the exhaust need the gas temperature cooled below the dew-point temperature. The air pass through the evaporator for heat recovery to provide the necessary heat for the refrigerant cycle, realization of heat recycling, and make full use of the limited heat.

2.3. The calculation formula

2.3.1. The energy for the drying of stone material.

$$Q = Q_1 + Q_2 + Q_3 \tag{1}$$

$$Q_2 = Cm\Delta T \tag{2}$$

$$Q_3 = mI \tag{3}$$

Here, *I, C, Q*and *m* represent the latent heat of vaporization, the specific heat of the water, the heat transfer rate and the mass flow rate of the water, respectively. The subscripts 1, 2 and 3 stand for the heat absorbed by dry stone matter, water heated to the desired temperature and water evaporation heat, respectively. Assume that $Q_1 = (Q_2 + Q_3) \times 0.5$.

2.3.2. The calculation formula of humid air.

Humid air can be regarded as the ideal gas mixture when it is at atmospheric pressure [5].

$$d = 0.622 \times \frac{\varphi \times P_s}{P - \varphi \times P_s}$$

(4)

$$h = 1.005t + d \times (2501 + 1.86t) \quad (5)$$

where *d* is the moisture content, φ is the relative humidity, P_s is the saturated vapor pressure, *h* is the enthalpy and *t* is the temperature.

2.3.3. The heat pump system.

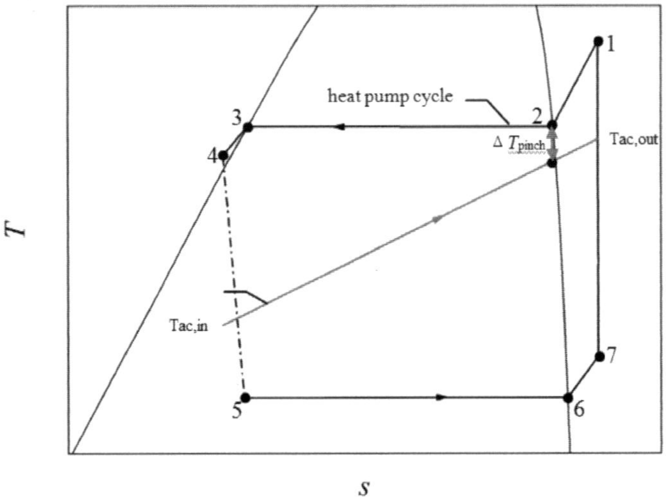

Figure 2. T-S diagram of heat pump cycle[6].

T-s diagram of heat pump cycle is shown in figure 2. The low-pressure sub-cooled steam point 7 is compressed to point 1, which is an adiabatic compression. It becomes saturated liquid point 3 after release heat at the same pressure, and sub-cooled to point 4. The refrigerant absorbs heat to point 7 after it is expanded from point 4 to point 5. The theoretical cycle shall be deemed reversible process except the constant-enthalpy process; it is possible to calculate by the equilibrium state parameters of the refrigerant. The heat balance equation of fresh air and refrigerant in condenser is as in equation (6).

$$Q_c = m_f(h_1 - h_4) = m_a(h_{ac,in} - h_{ac,out}) \quad (6)$$

Where $h_{ac,in}$ and $h_{ac,out}$ are the enthalpy of the fresh air at the inlet and outlet of the condenser, m_f is the mass flow rate of the refrigerant of heat pump, m_a is the mass flow rate of the fresh air, *h* is the enthalpy of refrigerant.

The heat balance equations in evaporators are as follows.

$$Q_e = m_f(h_7 - h_5) = m_f(h_7 - h_4) = Q_{ww} + Q_a \quad (7)$$

$$Q_{ww} = m_{ww} C_{ww} \Delta T = m_{ww} C_{ww} (T_{ww,in} - T_{ww,out}) \quad (8)$$

$$Q_a = m_a(h_{ac,in} - h_{ac,out}) \quad (9)$$

Where the subscripts "ww", "a", "ww,in", "ww,out", "ae,in" and "ae,out" stand for the wastewater, air, the inlet side of the wastewater, the outlet side of the wastewater, the inlet side of the air in the evaporator and the outlet side of the air in the evaporator, respectively.

The pinch point temperature difference (PPTD) of condenser represents the narrowest point between the fresh air temperature curve and refrigerant temperature curve as shown in Figure 2. It generally occurs at the beginning of the condensation process and can be expressed as follows:

$$\Delta T_{pinc\,h} = T_2 - T_P \quad (10)$$

Where T_2 is the condensing temperature; $T_P$represents the temperature of fresh air at the outlet of the condensing section in condenser.

The PPTD$\Delta T_{pinc\,h}$ is 15℃, For condensing temperature T_2, it is given an assumed value first, then all the state points of the cycle can be simulated, after that T_P of the fresh air is determined according the follows:

$$T_P = T_{ac,out} - \frac{h_1 - h_2}{h_1 - h_4}\left(T_{ac,out} - T_{ac,in}\right) \tag{11}$$

If the difference of T_P and T_2 is not equal to $\Delta T_{pinc\,h}$, the value of T_2 is adjusted and iterated to meet the PPTD $\Delta T_{pinc\,h}$.

Assuming that part of the refrigerant m_{f1} in the air-source evaporator is heated to the saturated steam, and the rest of the refrigerant m_{f2} is overheated to superheated steam in the water-source evaporator.

$$m_f = m_{f1} + m_{f2} \tag{12}$$

$$m_{f1}(h_6 - h_4) = Q_a \tag{13}$$

The theoretical compression work shows as:

$$W = m_f(h_1 - h_7) \tag{14}$$

The heat of the fresh air get can be calculated as following.

$$Q = Q_e + W = m_a\left(h_{ac,in} - h_{ac,out}\right) \tag{15}$$

2.4. The performance evaluation indices

Heat pump dryers have a lot of performance indicators. Commonly used to evaluate the performance of the heat pump drying device are the specific power consumption and the thermodynamic coefficient of the heat pump.

The COP of the overall system can be defined as in equation (16).

$$COP = \frac{h_7 - h_4}{h_1 - h_7} \tag{16}$$

Dry evaporate 1kg of water consumed energy in the process of known as the unit energy consumption of SPC [7]. The energy dissipations of the 1kg of water evaporation in the drying process are called the unit energy consumption of SPC, which reflects the relationship between the dehumidification and energy consumption. This SPC will be defined in equation (17).

$$SPC = \frac{Q}{COP} \tag{17}$$

2.5. The Feasibility Evidences

Under the assumptions of the operating system, the basic parameters as follows:

1. The pressure losses of refrigerant in condenser and evaporator are ignored.
2. For the heat pump cycle, R123 is employed as the working fluid. The evaporator outlet superheat degree is 5℃ to ensure the operation security of the compressor. The condenser outlet subcooling degree is 5℃. The PPTD $\Delta T_{pinc\,h}$ is 15℃. It is assumed that h_4 equals h_5.
3. Setup parameter：The inlet temperature of new air is 25℃，the relative humidity of fresh air is 0.5，the aimed temperature of drying air is $T_{ar,in} = 70$℃，the inlet temperature of evaporator is $T_{ae,in} = 70$℃。
4. The circulating water temperature is $T_{ww,in} = 25$℃ after stone polished and cut, with an outlet temperature $T_{ww,out} = 20$℃ away from water-source evaporator.

Theoretical cycle analysis is based on the thermodynamic properties to evaluate the cycle performance of working fluid, the results list in table 2.

Table 2. The R123 parameter values of heat pump drying system at various points.

Each state point	Temperature T (°C)	Pressure P (MPa)	Enthalpy h (kJ/kg)	Entropy s (kJ/kg*K)
1	-	0.4139	424.66	1.67
2	73.5	0.4139	-	-
3	73.5	0.4139	-	-
4	68.5	0.4139	270.81	-
5	-	-	270.81	-
6	20.0	0.0756	393.49	-
7	25.0	0.0756	396.93	1.67

As the moisture content of exhaust is higher, affecting the drying effect, the best option is discharge into the atmosphere directly. Hot air exchanges the heat with refrigerant in the evaporator after living the drying oven and then discharge into the atmosphere directly. The results of each state moist air list in the table 3.

Table 3. The parameter values of humid air in the heat pump drying system.

Each state point	Condenser		Drying Oven		Air-Source Evaporator	
	$T_{ac,in}$	$T_{ac,out}$	$T_{ar,in}$	$T_{ar,out}$	$T_{ae,in}$	$T_{ae,out}$
Temperature (°C)	25	70	70	40	40	30
Moisture content (kg)	0.0098	0.0098	0.0098	0.0181	0.0181	0.0181
Enthalpy (kJ/kg)	50.31	96.36	96.36	86.71	86.71	76.32

Because of the open type heat pump drying is influenced by the environment, the waste water recycling system as the auxiliary heating source. The reason is that Water has a higher specific heat and better heat transfer performance, temperature and flow stability. Temperature of wastewater is constant all over the year and its temperature is almost 20 higher than the ambient air temperature and the amount of it is significantly high and its flow rate is almost constant all over the year. For the sake of reaching the same drying efficiency, the calculation results of this system are as follows, it can be seen from the result that the value are in a reasonable range.

Table 4. The parameter values of the heat pump drying system.

Parameter	COP	Q (kJ/kgmin)	m_{ww} (kg/min)	SPC (kJ/kg)
Value	5.55	3756.69	159.99	676.88

3. Results and discussion

Since the huge of waste water quantity and heat excess supply of water source, the parameters passes through the refrigerant cycle system set to a fixed value.

3.1. Contrast with single air source heat pump

To highlight the advantages of the heat pump system that take stone processing plant's waste water as heat source, compared with traditional air source heat pump. Replacing the existing wastewater recycling system with air heat exchanger, the required quantity of heat m_{f2} is provided by air. The parameters of the air inlet temperature of the evaporator is $T_{ae2,in} = 20°C$, relative humidity 0.5 and outlet temperature of the evaporator is $T_{ae2,out} = 10°C$. The comparison of both results shows in the table 5.

Table 5. The results contrast of the wasterwater source heat pump and the air source heat pump.

	Parameter Value	m_{f2} for air source	m_{f2} for water source
Inlet	Temperature Of Evaporator (°C)	20.00	20.00
	Relative Humidity	0.50	-
	Enthalpy (kJ/kg)	38.54	-
Outlet	Temperature Of Evaporator (°C)	10.00	15.00
	Relative Humidity	1.00	-
	Enthalpy (kJ/kg)	28.35	-
	m_{a2} (kg)	311.96	151.81
Volume	m_{a2} (m^3/min)	241.83	0.10

Based on the data in this table, a pure air source heat pump source volume is 2418 times as much as the source volume of the wastewater source heat pump system for making the same quantity of heat. Since the density of air is far less than the water, the volume of air heat exchanger is much larger than the water-cooled heat exchanger's, equipment cost and covering area increase. Instead, the heat tubes of waste water sources is embedded in the heat recovery pool, which needs small covering area and the waste heat recovery will be conducive to wastewater treatment. Reducing the temperature of the wastewater accelerate the velocity of sludge sedimentation to a certain extent. Silt suction machine are adopted regularly to remove the sludge at the bottom of the first settlement tank and second settlement tank. The fresh water supply is needed in order to maintain the water level.

3.2. Influence of different factors

Since the wide range of temperature in southern summer and winter especially in Fujian. Summer temperatures up to 30 ℃, and the winter temperature is not below 9 ℃. The influences of air temperature on the unit operation were considered. To illustrate the feasibility of the system under different air parameters, the temperature change range is set from 5 to 30℃.

Seven of alternative refrigerants (R152a, R123, R1234yf, R1234ze, R600a, R22 and R600) were invested in these case, with the fresh air temperature is set within [5-30℃], and the changes of the COP as shown at figure 3. When the refrigerant is R123, m_a changed as shown in figure 4. According to figure 4, the higher temperature of the fresh air, the moisture absorption of the air is stronger at the same relative humidity. In the case of the same amount of drying stone, summer need less air than winter and the higher temperature air has a higher COP value.

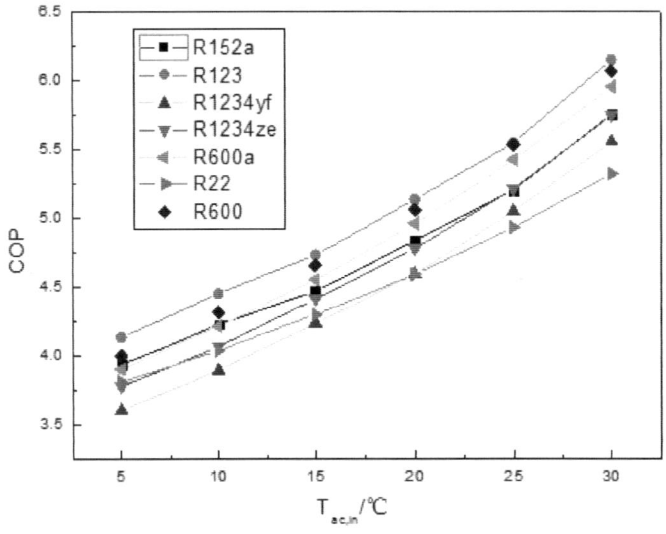

Figure 3. COP of different refrigerants.

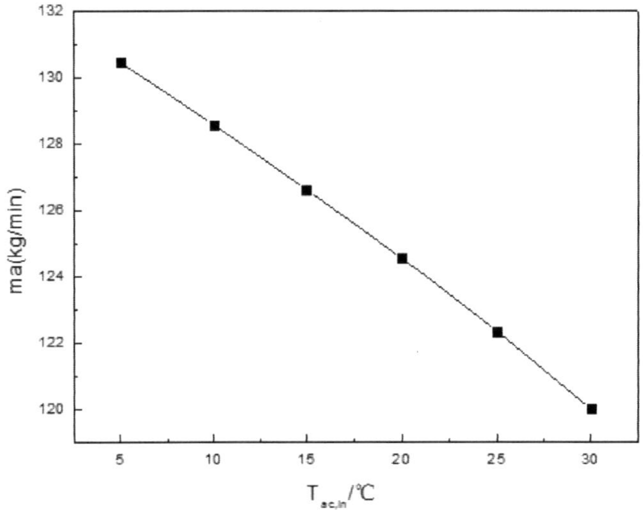

Figure 4. m_a of different refrigerants.

Table 6. All the properties of the refrigerants[8].

Refrigerator	Safety Level	Boiling Point (°C)	P_{cond} (MPa)	T_{cond} (°C)	ODP	GWP	COP
R152a	A2	-25.0	2.0599	74.0	0	0.02	5.75
R123	B1	27.7	0.4193	74.0	0.012	120.00	6.15
R1234yf	A2	-29.4	2.0753	70.7	0	4.00	5.55
R1234ze	A1	-30.0	1.6852	72.0	0	6.00	5.75
R600a	A3	-11.8	1.1429	72.3	0	0.10	5.95
R22	A1	-40.8	3.2911	74.6	0.034	1700.00	5.32
R600	A3	-0.5	0.8624	72.8	0	20.00	6.06

All the "Properties" of the refrigerants, including: Safety level, boiling point, T_{cond}, P_{cond}, ODP and COP, were included in the table.6. Combining with the above three figures, it can be seen from that the change of environmental temperature has great influence on operation of device. The temperature difference is decrease with the increase of environmental temperature, so the value of COP increased. R123 has the highest efficiency with the maximum value of COP at the different working conditions. R22 is normally used in the heat pump as a work fluid. While both R123 (ODP=0.012) and R22 (ODP=0.034) have a value of ODP, which will be limit usage by Montreal Protocol [9]. Although, R152a, R600a and R600 have ODP=0, and with lower GWP [10,11], they are highly flammable. R152a must be equipped with the independent secondary loop device making an additional cost. R1234yf and R1234ze was proposed as environmentally friendly refrigerants in recent years [12, 13], for their zero ODP and lower GWP and mild flammable, while have the lowest value of COP in this work. Comprehensive environmental benefits and the value of COP, the system select the refrigerant R123, which has high heating efficiency and environment friendly under a negative pressure working condition.

4. Conclusions

The heat pump system was studied in this paper, which provides necessary energy for the stone drying by recovery heat from the circulating water working in the processes of cutting and polishing. The energy of the circulating water is supplemented by the thermal energy produced during cutting and polishing. The wastewater is discharged from the stone processing work as stable heat source; the recovery technology for exhausting steam realize reliable operation; with obvious energy-saving effect, a small part of the power consumption can complete drying process; the proposed system is energy-saving and eco-friendly with reducing primary energy consumption and improving energy efficiency.

Compared with conventional air cooling systems, the required heat exchanger has many advantages such as: smaller size, more compact structure, lower investment cost, larger temperature range and so on. The R123 shows the superior thermal performance with the highest COP in the selected seven promising refrigerants. Therefore, this paper recommends R123 as the working refrigerant. The circulating water system is particularly suitable for the areas where water resources are scarce and the annual average temperature are over 10℃, which provides a reference for practical engineering application.

References

[1] Guo J 2013 *Research on optimization and operation strategy of air source heat pump water heater* (Shanghai: Shanghai Jiao Tong University)
[2] Hepbasli A, Biyik E and Ekren O 2014 A key review of wastewater source heat pump (WWSHP) systems *Energy Conversion and Management* **88** 700-22
[3] Shen C, Jiang Y and Yao Y 2012 A field study of a wastewater source heat pump for domestic hot water heating *J Build Services Eng Res Technol* **34** 433-48
[4] Hu P and Zhu W B 2015 Performance analysis of Heat Pump Water Heating System with Cascade Utilization of Waste Heat from Wastewater *Proc. the 24th iir international congress of refrigeration* ed Chen Z S (Japan: Youkohoma)
[5] Guo X 2012 *Design and research on an air low temperature drying cycle system* (Nanjing: university of Aeronautics and astronautics)
[6] Lemmon E W, Huber M L and Mclinden M O 2010 *Reference Fluid Thermodynamic and Transport Properties-REFPROP. 9.0.* (Gaithersburg: NIST Standard Reference Database 23)
[7] Pan Y K 1998 Modern drying technology (Beijing: chemical Industry Press) p 12
[8] Calm J M and Hourahan G C 2011 *Physical, safety and environmental data for current and alternative refrigerants* (Prague: Czech Republic) P 915
[9] Li M 2014 *Comparative study on heat pump technology applied in waste heat recovery* (Tianjin: Tianjin University)

[10] Garimella S 2003 Innovations in energy efficient and environmentally friendly space-conditioning systems *Energy* **28** 1593-614
[11] Mcculloch A 1999 CFC and Halon replacements in the environment *J. Fluorine Chemistry* **100** 163-73
[12] Akasaka R, Tanaka K and Higashi Y 2010 Thermodynamic property modeling for 2,3,3,3-tetrafluoropropene (HFO-1234yf) *International Journal of Refrigeration* **33** 52-60
[13] Brown J S, Zilio C and Cavallini A 2009 The fluorinated olefin R-1234ze(Z) as a high-temperature heat pumping refrigerant *International Journal of Refrigeration* **32** 1412-22

Establishment of key grid-connected performance index system for integrated PV-ES system

Q Li[1 3], X D Yuan[1], Q Qi[2] and H M Liu[2]

[1]Jiangsu Electric Power Research Institute, Nanjing 211103, China
[2]College of Energy and Electrical Engineering, Hohai University, Nanjing 211100, China

E-mail: 35830342@qq.com

Abstract. In order to further promote integrated optimization operation of distributed new energy/ energy storage/ active load, this paper studies the integrated photovoltaic-energy storage (PV-ES) system which is connected with the distribution network, and analyzes typical structure and configuration selection for integrated PV-ES generation system. By combining practical grid-connected characteristics requirements and technology standard specification of photovoltaic generation system, this paper takes full account of energy storage system, and then proposes several new grid-connected performance indexes such as paralleled current sharing characteristic, parallel response consistency, adjusting characteristic, virtual moment of inertia characteristic, on-grid/off-grid switch characteristic, and so on. A comprehensive and feasible grid-connected performance index system is then established to support grid-connected performance testing on integrated PV-ES system.

1. Introduction

In recent years, solar energy and wind energy as the representative for new energy have been increasingly developed and applied all over the world. Because grid-connected power generation system of new energy with intermittent and random has great impact on stable operation of power grid, new energy power generation system with energy storage devices has attracted more and more attention, whose development and scale of production have been gradually expanded. While the corresponding testing method and testing standard for their grid-connected performance are relatively lagging, which has become a barrier to new energy and energy storage integrated power generation system for industrial production and scale application.

The existing technologies on inverter unit for grid-connected new energy are relative mature, such as maximum power point tracking (MPPT) efficiency, power quality, low voltage ride through and so on. In Southern California of United States, Edison power company proposed improvement suggestions about performance of photovoltaic (PV) inverter from the point of view of high penetration adapting to PV power generation system through analysis on the test data of three-phase 480 V AC commercial photovoltaic inverters [1]. NTUA laboratory gave performance test methods of PV inverters for current

[3] Address for correspondence: Jiangsu Electric Power Research Institute, Nanjing 211103, China. Email: 35830342@qq.com.

test standard of PV inverters by testing their characteristics under different operating conditions in Greece and Cyprus [2]. Dutch Energy Research Association gave test results and overall evaluation of six commercial photovoltaic inverters whose capacity was up to 5 kW in terms of energy conversion efficiency, power point tracking efficiency, power quality and electromagnetic compatibility and so on [3]. In order to meet the needs of the development of new energy and power system, IEC 61727-2004 [4], IEEE 1373-1999 [5] and IEEE 1547-2003 [6] were published one after another. These technical standards are proposed mainly from the system level based on the new energy power generation system, while the test standards and evaluation system based on both the new energy power generation and energy storage system (ESS) have not been completed.

To promote coordination development between solar power and power grid, the State Grid Corporation of China had unveiled the "State Grid Corporation Technical Regulations of PV Power Plants Connection to the Grid (Trial)" in July 2009, and had formally released Q-GDW 617-2011 [7] and Q-GDW 618-2011 [8] successively in 2011, which made clear requirements of technical indexes and test regulations about response characteristic, anti-islanding protection characteristic, power output characteristic, power quality indexes as well as other general performance of middle and small scale grid-connected PV power stations, and made detailed technical regulations of all aspects about grid-connected characteristic of ESS in distribution network, whose voltage is 35 kV and below, in 2010 [9]. After that, Electric Power Research Institute of China (CEPRI) had developed a test platform of grid-connected photovoltaic inverters [10], and analyzed power quality of photovoltaic power generation system in terms of test conditions, test methods and test results [11]. Meanwhile, CEPRI went on studying and simulating by means of two test methods on the PV inverter low voltage ride through [12], and analyzed the basic structure and operation mode of PV inverter [13]. Tianjin Electric Power Research Institute of China inspected performance of grid-connected inverter equipments through the construction of test platform of grid-connected inverters and improved related test standards [14]. State Grid Electric Power Research Institute of China (SGEPRI) had developed mobile test platform of small PV power plant, which can test grid-connected performance and power station's general performance of grid-connected PV power station, whose rated capacity does not exceed 200 kW and grid-connected voltage level is 380 V [15].

Different from wind power, photovoltaic and other new energy, the integrated distributed photovoltaic and energy storage (PV-ES) system is weakly connected with the distribution network, and its grid-connected performance is relevant lower. On the basis of grid-connected characteristic of current photovoltaic generation system, ESS and active load is bound to affect the dynamic grid-connected characteristic of the whole system. The following issue is how to detect grid-connected performance of distributed PV-ES.

The existing grid-connected performance indexes on distributed generation, PV station and energy storage system cover a lot of aspects, such as in-grid principle, electric power quality, power control, voltage regulation, frequency characteristics, safety and reliability, relay protection, communication, electricity measurement and grid-connected detection, etc. Several novel indexes suitable for distributed integrated PV-ES system are proposed in this paper, such as "paralleled current sharing characteristic", "parallel response consistency", "adjusting characteristic", "virtual moment of inertia characteristic", "on-grid/off-grid switch characteristic", and so on. These novel indexes will effectively develop the grid-connected performance indexes of the PV-ES system, and have important significance for the scientific evaluation of the grid-connected characteristics of the integrated PV- ES system.

2. Configuration determination of integrated PV-ES system

In order to improve conversion efficiency of PV-ES and voltage stability of the DC buses, and optimize charging and discharging control of energy storage, four typical topological structures of integrated distributed PV-ES system, shown in Figure 1, are comparatively analyzed, and their main advantages and disadvantages are assembled in Table 1.

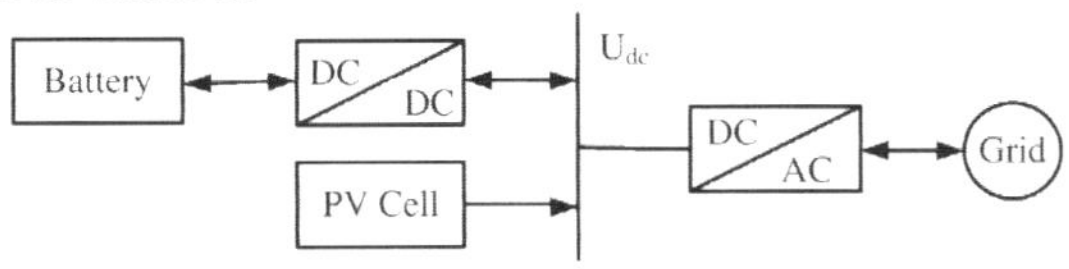

Figure 1. (a)

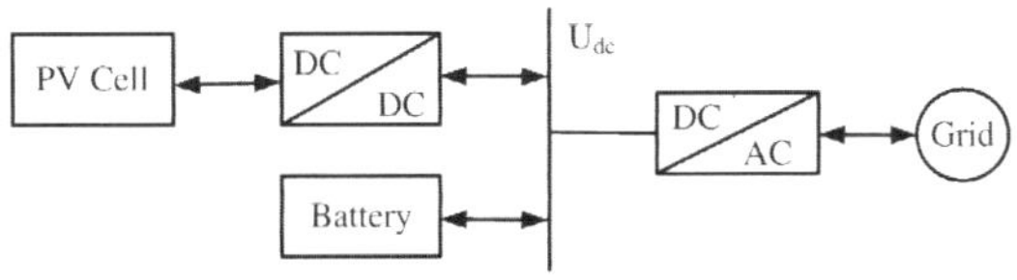

Figure 1. (b)

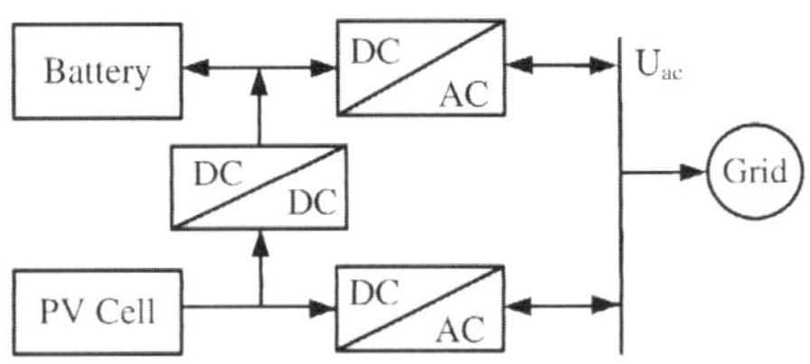

Figure 1. (c)

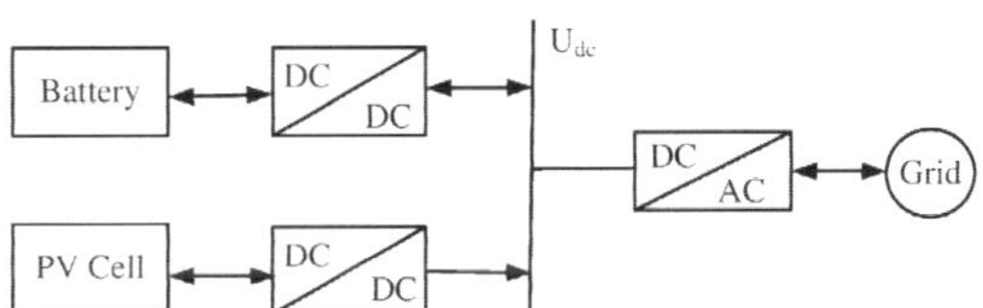

Figure 1. (d)

Figure 1. Four typical configurations of PV-ES

Table 1. Comparatively analysis of four typical configurations of PV-ES

Configuration Types	(a)	(b)	(c)	(d)
PV Utilization Ratio	high	lower	high	high
PV Grid-Connected Efficiency	high	low	high	lower
ES Charging Efficiency	high	high	high	low
ES Discharging Efficiency	low	high	high	lower
Complexity	low	low	high	low
Controllability	high	low	high	high
Whether to Recommend	yes	no	no	yes

A typical demonstration pilot in China selected type (b), it contains 250 kVA voltage-source inverters, 150 kVA inverter with dual-mode and 150 kW charging controller. This configuration has the following advantages:

- The maximum charging efficiency of PV-ES is up to 98.5%;
- The maximum discharging efficiency of energy storage is up to 95.7%;

- It has the technologies of centralized parallel operation control, mode smooth switching and protection.

But this topology has the following problems:

- In order to control voltage of battery, charging control of battery and MPPT control of PV will not be taken into account under some certain conditions, which results in a waste of photovoltaic energy;
- Energy management systems (EMS) cannot dispatch and control directly the energy storage devices, which increases the coupling degree of system and causes the system to more difficultly realize optimization on efficiency of PV and economic operation of energy storage.

It can be seen in table 1 that four topologies have their own advantages and disadvantages in terms of the utilization ratio of PV, charging/discharging efficiency of PV-ES, system complexity and controllability. There is a big disadvantage on efficiency of PV in type (b), and EMS cannot dispatch and control directly the energy storage devices, which decreases the coupling degree of system, it is difficult to realize the aim about economical operation control of energy storage; type (c) has the highest utilization ratio of PV and charging/discharging efficiency of PV-ES, but it is difficult to form standardized products and the operation condition is complex. By contrast, there are only some disadvantages in type (a) and (d) in terms of discharging efficiency of energy storage, but utilization ratio of PV and controllability of system is higher. Therefore, the comprehensive performances of type (a) and (d) are better.

3. The key performance indexes on the integrated PV-ES system

Grid-connected performance and output power quality of PV generation system are the key factors for scale application. Grid-connected performance indexes on the integrated PV-ES system remain the same with the basic requirements of the traditional PV generation system after it accesses ESS. Meanwhile, new problems about grid-connected characteristic are also proposed.

3.1. Grid-connected performance indexes of traditional PV system

Existing grid-connected performance evaluation indexes of PV system mainly contain principles of accessing system, power quality, power control, voltage regulation, related characteristic of voltage, current and frequency, security, relay protection and security automatic equipment, communication and information, energy metering and grid-connected detection and so on. Among them,

- <u>power quality indexes</u>: voltage deviation, frequency deviation, voltage fluctuation and flicker, unbalance of three-phase voltage, harmonic, DC component, and electromagnetic compatibility, and so on;
- <u>power control and voltage regulation indexes</u>: active power control, voltage/reactive power regulation, and start-stop, and so on;
- <u>characteristic of voltage, current and frequency indexes</u>: voltage response, frequency response, over-current response, and maximum permissible short-circuit current, and so on;
- <u>relay protection and security automatic equipment indexes</u>: element protection, system protection, anti-island protection, fault information, recovery grid-connected protection, and so on.

The indexes of power quality are selected as example in this paper.

(1) Voltage Deviation

According to the GB/T 12325-2008 "Power Quality Power Supply Voltage Deviation" [16], below 20 kV three-phase voltage deviation is ±7% of nominal voltage, and 35 kV and above voltage deviation does not exceed ±10% of nominal voltage.

(2) Frequency Deviation

According to the GB/T 15945-2008 "Power Quality Power System Frequency Deviation" [17], normal frequency deviation is ±0.2 Hz, and especially, frequency deviation is ±0.5 Hz when the system capacity is small.

(3) Voltage Fluctuation and Flicker

According to the GB/T 12326-2008 "Power Quality Voltage Fluctuation and Flicker" [18], voltage fluctuation, denoted by d, is the difference between the adjacent two extreme voltages in the curve $V_u(t)$, the square root value of voltage, which is expressed as a percentage of system voltage in table 2.

Table 2. Voltage fluctuation limit

r (times/ h)	$d\%$	
	LV[a] or MV[b]	HV[c]
$r \leq 1$	4	3
$1 \leq r \leq 10$	3	2.5
$10 < r \leq 100$	2	1.5
$100 < r \leq 1000$	1.25	1

According to the GB/T 156-2007, grades of nominal voltage U_N are divided as follows:

[a] Low Voltage (LV): $U_N \leq 1$ kV

[b] Medium Voltage (MV): 1 kV $< U_N \leq 35$ kV

[c] High Voltage (HV): 35 kV $< U_N \leq 220$ kV

(4) Unbalance of Three-Phase Voltage

According to the GB/T 15543-2008 "Power Quality Three-Phase Voltage Unbalance" [19], at the common connection point between PV-ES and power grid, unbalance of negative sequence voltage does not exceed 2% and especially, it does not exceed 4% if time is short.

(5) Harmonic Current

According to the GB/T 14549-1993 "Power Quality Public Power Grid Harmonic" [20], the limit of the Total Harmonic Distortion (THD) of current is THD ≤ 5% and the limits of odd harmonic and even harmonic current ratio are shown in table 3.

Table 3. Limits of harmonic current ratio

	Harmonic Number	Limits of Ratio (%)
Odd Harmonic	3th- 9th	4.0
	11th- 15th	2.0
	17th- 21th	1.5
	23th- 33th	0.6
	above 35th	0.3
Even Harmonic	2th- 10th	1.0
	12th- 16th	0.5
	18th- 22th	0.375
	24th- 34th	0.15
	above 36th	0.075

(6) DC Current Component

According to Q/GDW 617-2011 [7], Q/GDW 564-2011 [9] and Q/GDW 480-2010 [21], for converter type distributed power resources such as PV and ESS, the DC current feeding into the grid should not be more than 0.5% of its AC current value.

3.2. The novel grid-connected performance indexes

Addition of ESS has greatly improved the power quality issues, such as harmonic, voltage deviation, voltage fluctuations and flicker and so on. Several new indexes are proposed for grid-connected performance of the integrated PV-ES system.

3.2.1. Paralleled current sharing characteristic

Paralleled current sharing characteristic refers to automatic balance and sharing of currents withstood by all the distributed power branches. In order to improve reliability of the system, do not add any additional control of external current sharing as far as possible. When the input of voltage or load current changes, voltage should be kept stable, transient response of current sharing is good. Numerically, paralleled current sharing *PCS* is as reciprocal of unbalance *S*.

$$PCS = S^{-1} = \left| \frac{I_k}{I_{k\max}} - \frac{I_{PCC}}{\sum_k I_{k\max}} \right|^{-1} \times 100\% \tag{1}$$

where, k=1,...,K, k and K are the number of the distributed power supply branch and the total number of its branch respectively, I_{PCC} is current of public connection point, I_k and $I_{k\max}$ are the current and its maximum value of the kth distributed power supply branch respectively.

3.2.2. Parallel response consistency

Parallel response consistency *PRC* of distributed PV-ES integration device refers to when other technical parameters are same, response deviation of each parallel branch should be kept within a certain small range under the same excitation. Its value can be expressed as the reciprocal of response deviation δ of each branch.

$$PRC = \delta^{-1} = \left(\sum_{k=1}^{K} \left| R_k - \frac{\sum_{k=1}^{K} R_k}{N} \right| \right)^{-1} \tag{2}$$

where, R_k is response of the kth distributed power supply branch when excitation of public connection point is given.

3.2.3. Adjusting characteristic

When active or reactive power of PV-ES integration device is above 20% total rated output, the device should be able to achieve continuous smooth regulation of active or reactive power, and participate in controlling active or reactive power of the system. The differential coefficient of active power k_P and the differential coefficient of reactive power k_Q are as follows.

$$k_P = \frac{\Delta f}{\Delta P} \times 100\% \tag{3}$$

$$k_Q = \frac{U_0 - U_R}{U_N} \times 100\% \tag{4}$$

where, Δf is the variation of frequency of power grid, ΔP is the variation of active power of distributed power supply, U_0 is the terminal voltage of the distributed power supply without any load, U_R is the terminal voltage when reactive current was rated, U_N is rated terminal voltage of the distributed power supply.

3.2.4 Virtual Moment of Inertia Characteristic

In order to improve the stability of total power grid, the control strategy of energy storage converter is designed to simulate the droop characteristic of the synchronous generator and the moment of inertia of the generator rotor based on virtual synchronous machine technology, which produces the virtual inertia to participate in grid frequency regulation, so as to realize the steady support of the grid voltage and frequency.

Once the system frequency deviated from the rated value due to the system power supply or load demand changes, the integrated PV-ES system changes the power output following the frequency modulation auxiliary power value. The output variation of PV and ESS will result in the frequency regulation ability of the integrated PV- ES system, that's virtual inertia response capability. The virtual inertia index is noted as $H_{PV\text{-}ES}$, which is kept as a constant in converter control system, and it can be described as

$$H_{PV\text{-}ES} = \frac{\Delta P_{PV} + \Delta P_{ES}}{\Delta f} \tag{5}$$

where, Δf is the frequency deviation, ΔP_{PV} and ΔP_{ES} are the corresponding output change of PV and ESS respectively. If the MPPT strategy is adopted in PV's converter, ΔP_{PV} was kept zero, the total virtual moment of inertia is only decided by the response performance of ESS's converter.

3.2.5 On-Grid/off-Grid Switch Characteristic

In normal in-grid operating mode, the distributed integrated PV-ES system connected into the grid in parallel. When some fault occurs in power grid or power quality is terrible, the distributed integrated PV-ES system is switched into isolating operating mode. After the fault restoration, the integrated PV-ES system would re-connect into the grid. In normal in-grid operating mode, the P/Q control strategy is selected to inject active power to grid. While in isolating operating mode, the U/f control strategy comes into operation to maintain an acceptable level of voltage and frequency for the distributed integrated PV-ES system. Some key parameters of the converter can be used to picture the On-Grid/off-Grid Switch Characteristic, such as rated input voltage of DC bus, rated input voltage of AC side, rated grid-connected frequency, efficiency of the whole machine, overload ratio, and so on.

3.3. Grid-connected performance index system

According to the above analysis, this paper puts forward two layers index system on grid-connected performance evaluation of the integrated PV-ES system. Among which, the upper layer contains power characteristic S1, power quality S2, response characteristic S3, security and protection S4 as well as system characteristic S5, the comprehensive index system is shown in table 4.

Table 4. Grid-connected performance evaluation index system of the integrated PV-ES system

First Grade Index	Second Grade Index
S1: Power Characteristic	S11: Active Power Output S12: Power Factor S13: Active Power Control S14: Reactive Power Control
S2: Power Quality	S21: Harmonics in Public Supply Networks S22: Admissible Deviation of Supply Voltage S23: Three Phase Voltage Unbalance Factor S24: Voltage Fluctuation and Flicker S25: Permissible Deviation of Frequency S26: Temporary and Transient Overvoltage
S3: Response Characteristic	S31: Voltage Response S32: Frequency Response S33: Over-current Response S34: Maximum Permissible Short Circuit Current

	S35: Sensitivity of Voltage Adjustment S36: Paralleled Current Sharing Characteristic S37: Parallel Response Consistency S38: Adjusting Characteristic
S4: Security and Protection	S41: General Requirements S42: Component Protection S43: System Protection S44: Anti- islanding Protection S45: Fault Information S46: Grid-connected Recovery
S5: System Characteristic	S51: Grid Connected/ Off Switch Characteristic S52: Virtual Moment of Inertia Characteristic

The above index system is more comprehensive. Based on the index system, some methods can be used to go on stratified evaluation, such as Analytic Hierarchy Process (AHP) and fuzzy evaluation and so on. AHP is a decision-making methods combined with qualitative and quantitative analysis, and it takes elements decomposed into goals, guidelines, programs and other levels. The fuzzy comprehensive evaluation uses the fuzzy mathematics method based on certain criteria, gives the possibility of each item which is usually affected by multiple factors. In this paper, the possibility refers to the pros and cons of one index or overall performance of the distributed integrated PV- ES system. Therefore, real grid-connected performance of the integrated PV-ES power generation system can be evaluated comprehensively.

4. Conclusion

At present, the integrated PV-ES power generation system is very popular all over the world, but it still lacks grid-connected performance testing and evaluation criteria. By comparing and analyzing the existing schema of PV-ES system, this paper further carries on improving utilization ratio of PV, and charging and discharging efficiency of PV-ES and system configuration with reducing complexity of system. In the basis of existing grid-connected performance testing index of PV, several novel performance indexes are proposed to evaluate the integrated PV-ES power generation system, which effectively develops existing evaluation indexes, and finally a more comprehensive grid-connected performance index system on the integrated PV-ES power generation system has formed.

References

[1] Brave R J, Yinger R, Robles S and Tamae W 2011 Solar PV Inverter Testing for Model Validation *IEEE Power and Energy Society General Meeting*

[2] Romanosl P, Hatziargyriou N, V La Rocca, G Tsolakos, K Pantazis, N Daniil and S Bonias 2010 PV inverters supporting the grid first experiences testing and qualification according to the grid interconnection guideline in Greece and Cyprus *National Technical University of Athens* **186** 7 - 10

[3] Wildenbeest E J, De-Haan S W H and Van Der Borg N C 1994 Recent test results of 6 commercial inverters for grid connected PV system *IEEE Netherlands Energy Research Foundation ECN Renewable Energy* **12** 5 - 9

[4] IEC 61727- 2004 *PV System-Grid Interface Characteristic Standard* 2004

[5] IEEE 1373-1999 *Grid-Connected PV System Test Methods and Procedure Recommendation Standard* 1999

[6] IEEE 1547-2003 *Grid-Connected Interface Standard of Distributed Generation* 2003
[7] Q-GDW 617-2011 *Technical Regulations of PV Power Station Connection to the Grid* 2011
[8] Q-GDW 618-2011 *Test Procedures of PV Power Station Connection to the Grid* 2011
[9] Q-GDW 564-2011 *Technical regulations of Storage Energy System Connection to Distribution Network* 2011
[10] Yan H G, Zhang X, Yang X J, Wang H, Fan Y, Jiang L M and Li T Y 2011 Development of Detection Platform for Grid-Connected Photovoltaic Inverters *Power System Technology* **35** 139-43
[11] Dong Y H, Liu M Y and Huang J S 2013 *Comparison and Analysis of Power Quality Test Method in PV Grid System* (Beijing: 2nd IET Renewable Power Generation Conference China electric power research institute)
[12] Qu L N, T Shi, Sun M M and Zhu L Z 2012 *Study on low voltage ride through test methods for PV inverter* (Shanghai: China International Conference on Electricity Distribution)
[13] Jiang N and Xu H H 2010 *Detection platform design of solar energy grid-connected photovoltaic system* (Hangzhou: International Conference on Power System Technology)
[14] Cui R J, Li S W, Li X H and Wang Z 2013 Preliminary Study on Testing Technology of Photovoltaic Grid-Connected Inverter *Proc. the CSU-EPSA* **25** 103-7
[15] Li H T, Zhang J J, Bao S J and Lin X J 2011 Design and Development of a Mobile Detection Platform for Grid Connected of Small Photovoltaic Power Station *Automation of Electric Power Systems* **19** 39- 42
[16] GB/T 12325-2008 *Power Quality-Deviation of Supply Voltage* 2008
[17] GB/T 15945-2008 *Power Quality-Frequency Deviation for Power System* 2008
[18] GB/T 12326-2008 *Power Quality-Voltage Fluctuation and Flicker* 2008
[19] GB/T 15543-2008 *Power Quality-Unbalance of Three-Phase Voltage* 2008
[20] GB/T 14549-1993 *Power Quality-Harmonic of Public Grid* 1993
[21] Q-GDW 480-2010 *Technical Regulations of Distributed Power Connection to the Grid* 2010

IOP Conf. Series: Earth and Environmental Science **40** (2016) 012033 doi:10.1088/1755-1315/40/1/012033

Fuzzy comprehensive evaluation for grid-connected performance of integrated distributed PV-ES systems

Z H Lv[1,3], Q Li[1], R W Huang[2], H M Liu[2] and D Liu[1]

[1]Jiangsu Electric Power Research Institute, Nanjing 211103, China
[2]College of Energy and Electrical Engineering, Hohai University, Nanjing 211100, China

E-mail: 517480859@qq.com

Abstract. Based on the discussion about topology structure of integrated distributed photovoltaic (PV) power generation system and energy storage (ES) in single or mixed type, this paper focuses on analyzing grid-connected performance of integrated distributed photovoltaic and energy storage (PV-ES) systems, and proposes a comprehensive evaluation index system. Then a multi-level fuzzy comprehensive evaluation method based on grey correlation degree is proposed, and the calculations for weight matrix and fuzzy matrix are presented step by step. Finally, a distributed integrated PV-ES power generation system connected to a 380 V low voltage distribution network is taken as the example, and some suggestions are made based on the evaluation results.

1. Introduction

Photovoltaic (PV) power generation, one of the widespread new energy generations, offers great convenience to remote areas or rural areas, where extension of power grid is infeasible or uneconomic. At the same time, with natural unpredictability and randomness, PV vastly accesses to the grid brings greater challenges for safe, reliable and efficient operation of the power grid. Energy storage (ES) technologies can improve instability of PV generation output, whose principle is to restrain fluctuation of voltage and frequency in grid-connected operation mode through rapid active and reactive control, so that the output power of new generations remains stable within a certain range of time [1][2]. Energy storage technologies can be divided into mechanical energy storage, electromagnetic energy storage and chemical energy storage [3][4]. At present, there are more and more extensive and in-depth researches on ES technologies all over the world.

The random fluctuation of distributed PV generation output distributes on wide-area time scale from millisecond to hour, it is hard to meet the demand of multiple time scale power restrain of distributed energy for single energy storage technology [5]. The battery energy storage is used to restrain fluctuation of PV in [6], which results in too large charging current of battery, polarization and abscission of active material in the electrode. If combination of energy storage technologies about super capacitor and battery with strong complementary in performance, that is to take advantage of large rated power of super

[3] Address for correspondence: Jiangsu Electric Power Research Institute, Nanjing 211103, China. Email: 517480859@qq.com

capacitor and large rated capacity of battery ES, and optimize charging and discharging strategy of them to extends their life, then the total life-cycle costs would be decreased obviously.

Study on evaluation for grid-connected performance of PV is rather mature. Many aspects of the impacts on load characteristics, power grid scheduling, voltage and reactive power, those come from the large-scale PV power accessed to the grid, was analyzed in [7]. Voltage offset and power flow change with different capacity PV generation connected to the grid were studied in [8][9]. The impacts of grid-connected PV systems in large-scale and high-voltage on security and stability of power grid was analyzed in [10].

Although these well-done studies have certain guiding significance for the evaluation of the grid-connected performance of integrated PV-ES systems, it does not consider all the impacts of PV-ES systems, distributed power and new energy generation systems, and the testing standard and evaluation system of PV with ES have not been established.

This paper gives two typical topologies of distributed integrated PV-ES systems, and proposes a more comprehensive performance index evaluation system combining with current ES technologies. Grey correlation is introduced to form a multi-level fuzzy comprehensive evaluation model together with the ambiguity function theory [11], which combines both advantages of grey correlation and fuzzy comprehensive evaluation.

2. Anslysis on typical topologies of distributed integrated PV-ES systems

The integrated PV-ES system is usually composed of the PV array, ES device, DC/DC converter and loads. ES is mainly divided into four categories: electrochemical energy, physical energy, electromagnetic energy and phase-change energy. The typical technical parameters of them are listed in Table 1. According to the table, battery is not suitable for high power and frequent charging/discharging due to its big energy density, small power density, short cycle life, low charge/discharge efficiency and sensitive to the charge/discharge process. Super capacitor is not suitable for large-capacity situation due to its lower energy density, which is 30% of the battery, but it can be applied to frequent charge and discharge cycle due to their high power density, long cycle life and high charge/discharge efficiency. The two ES technologies have strong complementarities. Lots of literatures state that the hybrid energy storage system can make use of the complementary characteristics of battery and super capacitor, so as to improve power output capability, reduce losses, increase charge and discharge times, and extend the life of the battery.

Table 1. Typical technical parameters of different energy storage technologies

ES Types	Energy Density /kW·h·m^{-3}	Power Density /kW·m^{-3}	Conversion Efficiency %	Smallest Unit Capacity /kW·h	Cycle Life Year /a
Superconducting Magnetic ES	178	1000.0	96	500	30
Super Capacitor	53	176700	94	1.0	30
Flywheel ES	424	1766.8	90	4.0	30
Lead-acid Battery	7.07	106	92	0.5	8
Lithium Battery	212	212	88	5.0	7
Acid Battery	243.7	530.0	88	5.0	7

Two kinds of typical topologies of integrated PV-ES system including single ES and hybrid ES technologies are analyzed here, and the topologies are show in Figure 1.

(1) Single Integrated PV-ES System

Typical integrated PV-ES system is composed of PV array, battery, bidirectional DC-DC converters, DC-AC converter, photovoltaic power control system and ES control system and so on. The DC-DC converter followed by PV array is a boost converter and its main function is to take Maximum Power Point Tracking (MPPT) control of PV output power and keep voltage of DC bus stable.

(2) Hybrid Integrated PV-ES System

Compared with single integrated PV-ES system, hybrid integrated PV-ES system has other type ES units apart from battery. Among which the more representative one is composed of battery and super capacitor (SC). PV power control system mainly controls output voltage to maximize output power. Hybrid energy storage control system regulates active power output of PV power station, and regulates two bi-directional DC-DC transform devices for distribution of active power between battery and SC, then achieves the optimal solution of charging and discharging process.

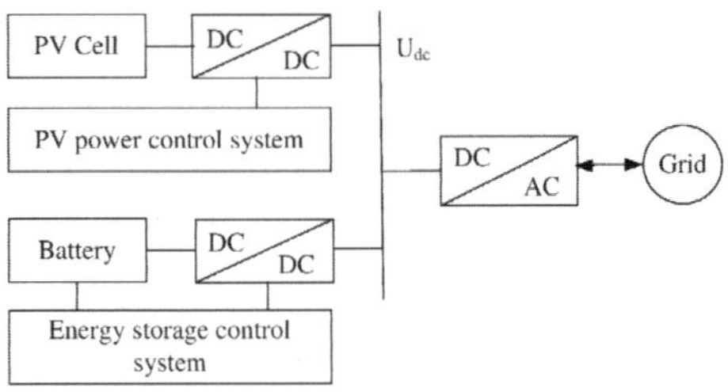

Figure 1. (a) single ES system

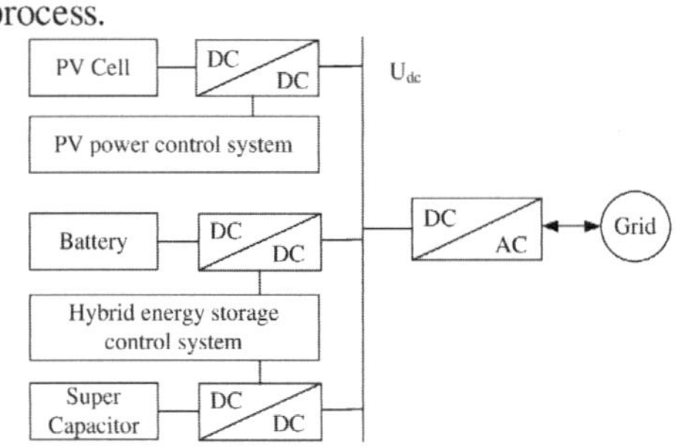

Figure 1. (b) hybrid ES system

Figure 1. Topology of integrated PV-ES system

3. Grid-connected performance index system of distributed integrated PV-ES systems

Grid-connected performance, especially output power quality, is the key to scale application of solar PV generation system. Existing grid-connected performance evaluation indexes of PV power system mainly include: principles accessing the system, power quality, power control, voltage regulation, characteristics of voltage, current and frequency, security, relaying protection and safety automatic equipments, communication and information, energy metering and grid-connected detection and so on. Among them,

- Indexes of power quality mainly includes: harmonics, voltage deviation, voltage fluctuation and flicker, unbalanced three-phase voltage, electromagnetic compatibility, etc.
- Indexes of power control and voltage regulation mainly includes: active power control, voltage/reactive power regulation, starting and stopping, etc.
- Indexes of characteristics of voltage, current and frequency mainly includes: voltage response, frequency response, overflow response, maximum permissible short-circuit current, etc.
- Indexes of relaying protection and safety automatic equipment mainly includes: component protection, system protection, anti-islanding protection, fault information, grid-connected recovery, etc.

Addition of ESS has greatly improved power quality, such as harmonic, voltage deviation, voltage fluctuation and flicker and so on. Based on the above grid-connected performance indexes, a novel grid-connected performance indexes of integrated PV-ES system is proposed, as shown in table 2.

Table 2. Grid-connected performance index system of the integrated PV-ES system

First Grade Index	Second Grade Index
S1: Power Characteristic	S11: Active Power Output
	S12: Power Factor
	S13: Active Power Control
	S14: Reactive Power Control

S2: Power Quality	S21: Harmonics in Public Supply Networks S22: Admissible Deviation of Supply Voltage S23: Three Phase Voltage Unbalance Factor S24: Voltage Fluctuation and Flicker S25: Permissible Deviation of Frequency S26: Temporary and Transient Overvoltage
S3: Response Characteristic	S31: Voltage Response Characteristic S32: Frequency Response Characteristic S33: Over-current Response Characteristic S34: Maximum Permissible Short Circuit Current S35: Sensitivity of Voltage Adjustment S36: Paralleled Current Sharing Characteristic S37: Parallel Response Consistency S38: Adjusting Characteristic
S4: Security and Protection	S41: General Requirements S42: Component Protection S43: System Protection S44: Anti- islanding Protection S45: Fault Information S46: Grid-connected Recovery
S5: System Characteristic	S51: Grid Connected/ Off Switch Characteristic S52: Virtual Moment of Inertia Characteristic

4. Grid-connected performance comprehensive evaluation

Comprehensive evaluation is a multi-disciplinary technology. The performance factors of distributed PV-ES system are more and miscellaneous, and it is hard to determine existing interaction between different factors and the index values quantitatively. This paper proposes a comprehensive evaluation method based on grey correlation degree and fuzzy evaluation theory. The grey correlation degree is applied in fuzzy comprehensive evaluation to achieve weights of indexes about distributed grid-connected PV-ES, which effectively overcomes the respective defects, and enjoys the advantages of both at the same time.

4.1. Method and process of comprehensive evaluation

Fuzzy evaluation is a method that gives the possibility of a certain specific comment according to some special criteria. Considering factors about grid-connected performance of distributed PV-ES system, this paper quantifies the original qualitative evaluation for some important factors based on their importance degrees.

The improved grey correlation method is introduced here to calculate the weights of indexes. The weight experience judgment values from experts is considered as the original data to carry on pure further numerical calculation, and it makes full use of sub-index information of experts experience judgment, which causes the weight coefficients impartially. The whole evaluation process is divided into two steps, the first step is to obtain the fuzzy matrix T, and then to obtain the weight matrix A, finally the fuzzy comprehensive evaluation result S is obtained by fuzzy matrix T multiplying weight matrix A. The comprehensive evaluation process is shown in Figure 2.

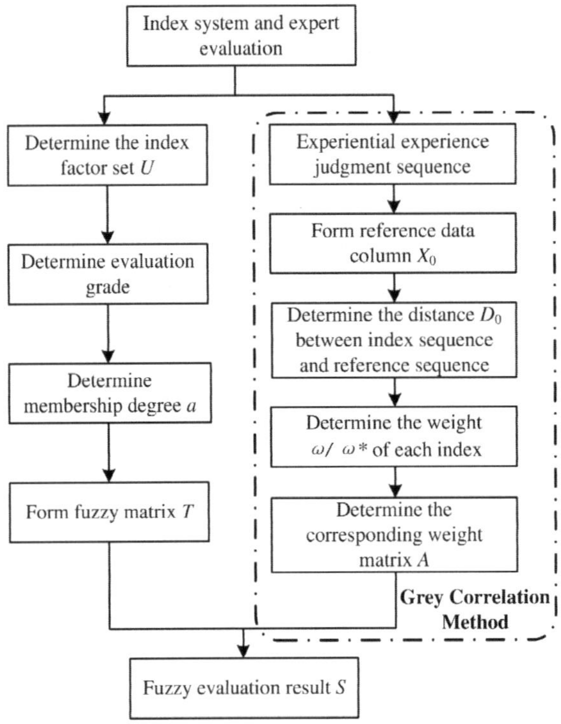

Figure 2. The comprehensive evaluation process

4.2. Calculation of weight matrix A

(1) Suppose that m experts make experience judgment for n evaluation index weights, and to form data columns as follows:

$$X=[X_1,X_2,\ldots,X_n]^T=\begin{bmatrix} x_1(1) & x_1(2) & \cdots & x_1(m) \\ x_2(1) & x_2(2) & \cdots & x_2(m) \\ \vdots & \ddots & \cdots & \vdots \\ x_n(1) & x_n(2) & \cdots & x_n(m) \end{bmatrix} \tag{1}$$

(2) Select the biggest weight from X as the reference one X_0.

$$X_0=[x_0(1),x_0(2),\ldots,x_0(m)] \tag{2}$$

(3) Calculate the distance between each row weight index X_i, i=1,…,n, and the reference one X_0.

$$D_{0i}=\sum_{k=1}^{m}[x_0(k)-x_i(k)]^2 \tag{3}$$

(4) Calculate the weight of each index ω_i.

$$\omega_i=\frac{1}{1+D_{0i}} \tag{4}$$

(5) Calculate the normalization weight of each index ω_i^*, and $A=[\omega_1^*,\omega_2^*,\cdots,\omega_n^*]^T$.

$$\omega_i^*=\frac{\omega_i}{\sum_{i=1}^{n}\omega_i} \tag{5}$$

4.3. Calculation of fuzzy matrix T and comprehensive evaluation result S

(1) Determine each index factor set *U*;

(2) Determine the evaluation grade *P*, *P*={Excellent, Good, Average, Barely Passed, Unqualified} = {E, G, A, P, U};

(3) Determine the membership degree of every grade evaluation index quantitatively, and to form fuzzy matrix $T_{5\times n}$;

(4) Calculate the evaluation value *S* as follows,

$$S = T \cdot A = \begin{bmatrix} s_1 & s_2 & s_3 & s_4 & s_5 \end{bmatrix}^T \tag{6}$$

The value *S* is the indexes in single layer. If there is more than one level in the index system, it should be calculated from the base level to the top level one by one, and the evaluation results of all grade indexes can be obtained.

5. Analysis of example

An integrated PV-ES system accessed to 380 V low voltage distribution network is selected as the simulation case to analyze grid-connected performance.

5.1. Calculation of weight matrix A

According to the statistics with experience judgment given by 23 experts, the weight of each index by improved grey correlation degree is achieved level by level according to equation (1) to (5), the indexes in first level include: power characteristic, power quality, response characteristic, security and protection, and system characteristic, shown in table 3.

Table 3. The indexes weights results

First level index	Weights	Second level index	Weights
S1	0.275	S11	0.348
		S12	0.278
		S13	0.237
		S14	0.137
S2	0.247	S21	0.18
		S22	0.18
		S23	0.171
		S24	0.157
		S25	0.15
		S26	0.162
S3	0.194	S31	0.132
		S32	0.125
		S33	0.128
		S34	0.109
		S35	0.108
		S36	0.109
		S37	0.105
		S38	0.185
S4	0.143	S41	0.18
		S42	0.18
		S43	0.171
		S44	0.159
		S45	0.149
		S46	0.161

S5	0.141	S51	0.473
		S52	0.527

5.2. Calculation of Fuzzy Evaluation

According to industry standards, measuring calculation data and experts evaluation statistics, the assessment grades {E, G, A, P, U} can be determined. Here a single example, the membership statistic results of power quality are shown in table 4.

Table 4. Membership statistic result of power quality

Second Level Indexes	Measuring/Calculation Data	Industry Standards	Grade membership				
			E	G	A	P	U
S21	1.8	[0, 4.0][a]	0.5	0.3	0.2	0	0
S22	5%	[-7%, +7%]	0.2	0.5	0.3	0	0
S23	0.8%	[0, 1.3%]	0.3	0.4	0.2	0.1	0
S24	2.5	[0, 4][b]	0.2	0.2	0.3	0.3	0
S25	0.8	[0, 1]	0.1	0.2	0.3	0.4	0
S26	0.3%	[0, 0.5%]	0.2	0.3	0.3	0.2	0

[a] the allowed distortion rate of harmonic current when the number of odd harmonic ≤ 11;
[b] the limit of voltage fluctuation when the voltage variable frequency ≤ 1.

Then the fuzzy matrix of power quality is

$$T_2 = \begin{bmatrix} 0.5 & 0.2 & 0.3 & 0.2 & 0.1 & 0.2 \\ 0.3 & 0.5 & 0.4 & 0.2 & 0.2 & 0.3 \\ 0.2 & 0.3 & 0.2 & 0.3 & 0.3 & 0.3 \\ 0 & 0 & 0.1 & 0.3 & 0.4 & 0.2 \\ 0 & 0 & 0 & 0 & 0 & 0 \end{bmatrix}$$

Combined with improved grey correlation analysis presented above, we can calculate the corresponding index weight and fuzzy evaluation:

$$S_2 = T_2 \cdot A_2 = \left[0.2561\ 0.3224\ 0.2649\ 0.1566\ 0\right]^T$$

According to the principle of the maximum membership, evaluation result of power quality characteristic about grid-connected performance of distributed integrated PV-ES generation system is good.

Similarly, fuzzy evaluation results of power characteristic, response characteristic, security and protection, system characteristic on distributed integrated PV-ES generation system are obtained as follows,

$$S_1 = \left[0.3745\ 0.4374\ 0.1644\ 0.0237\ 0\right]^T$$

$$S_3 = \left[0.2237\ 0.3089\ 0.3227\ 0.1457\ 0\right]^T$$

$$S_4 = \left[0.6931\ 0.2749\ 0.0320\ 0\ 0\right]^T$$

$$S_5 = \left[0.3473\ 0.3946\ 0.2054\ 0.0527\ 0\right]^T$$

Evaluation results of the second grade index are {G, G, A, E, G}. Similarly, the first level indexes are carried out as follows,

$$S = T \cdot A = \left[0.3577\ 0.3548\ 0.2068\ 0.0809\ 0\right]^T$$

According to the principle of the maximum membership, fuzzy comprehensive evaluation result about grid-connected characteristic of the distributed integrated PV-ES generation system is excellent, the comprehensive performance is nice and it is recommended.

In order to observe the effect of ES bringing to the grid-connected performance of integrated PV-ES systems, the capacity of ES is designed as half. The specific measurement data and membership degrees are as shown in table 5.

Table 5. Membership statistic result of power quality in low ES configuration

Second Level Indexes	Measuring/Calculation Data	Grade membership				
		E	G	A	P	U
S21	3.2	0	0.1	0.5	0.4	0
S22	6%	0	0.1	0.4	0.5	0
S23	1.1%	0	0	0.5	0.5	0
S24	3.2	0	0.1	0.3	0.6	0
S25	0.85	0	0.2	0.5	0.3	0
S26	0.4%	0.1	0.1	0.4	0.4	0

The calculation process is similar to the previous. Due to the limitation of length, no more tautology will be here. The final evaluation result S' is as follows,

$$S' = T' \cdot A' = [0.0162 \quad 0.0979 \quad 0.4344 \quad 0.4515 \quad 0]^T$$

And the comprehensive evaluation result is Barely Passed, it shows that the configuration of ES is not optimal, and it should not be recommended.

6. Conclusion

As for the distributed integrated PV-ES generation system, the grid-connected performance evaluation index system was established, and the comprehensive assessment method based on fuzzy evaluation theory and grey correlation degree was proposed. The simulation results show that ES could greatly improve the grid-connected performances of distributed PV power generation system. How to determine the optimal configuration of hybrid ES and realize the control strategy of PV-ES system need to be further studied.

References

[1] Mutale J, Strbac G, Curcic S and Jenkins N 2000 Allocation of losses in distribution systems with embedded generation *IEE Proceedings-Generation, Transmission and Distribution* **147** 7-14
[2] Brown R E and Freeman L A A 2001 *Analyzing the reliability impact of distributed generation* Power Engineering Society Summer Meeting
[3] Blasko V and Kaura V 1997 A new mathematical model and control of a three-phase ac- dc voltage source converter *IEEE Transactions on Power Electronics* **12** 1231-8
[4] Thompson G L, Wall D L 1981 A branch and bound model for choosing optimal substation locations *IEEE Transaction on Power Apparatus and Systems* **100** 2683-8
[5] Wang C S 2006 *Modeling and control of hybrid wind/photovoltaic/ fuel cell distributed generation system* (Bozeman: Montana State University)
[6] Dougal R A, Liu S and White R E 2002 Power and life extension of battery- ultra capacitor hybrids *IEEE Transactions on Components and Packaging Technologies* **25** 120-31
[7] Zhao P and Yan Y T 2009 Research on Effect of Grid-connected photovoltaic System on Power Grid *Electrical Engineering*

[8] Hu B, Nonaka Y and Yokoyama R 2012 Influence of Large-scale Grid-connected Photovoltaic System on Distribution Networks *Automation of Electric Power Systems* **36** 34-8

[9] Liu y, Bebic J and Ren W 2008 *Distribution system voltage performance analysis for high-penetration PV* Proceedings of the IEEE Energy 2030 Conference

[10] Yang W D, Zhou X and Xue F 2010 *Impacts of large scale and high voltage level photovoltaic penetration on the security and stability of power system* Proceedings of the Power and Energy Engineering Conference (APPEEC)

[11] Choobineh F and Li H 1993 An index for ordering fuzzy numbers *Fuzzy Set and Systems* **54** 287-94

A monitoring system based on electric vehicle three-stage wireless charging

T Hei[1, 3], Z Z Liu[1], Y Yang[2], CHEN Hongxing [1], B Zhou[1] and H Zeng[1]

[1]School of Electrical Engineering, Shandong University, Jinan, 250061, China
[2]Hefei Power Supply Corporation, Hefei 230022, China

E-mail: heitong93@163.com

Abstract. An monitoring system for three-stage wireless charging was designed. The vehicle terminal contained the core board which was used for battery information collection and charging control and the power measurement and charging control core board was provided at the transmitting terminal which communicated with receiver by Bluetooth. A touch-screen display unit was designed based on MCGS (Monitor and Control Generated System) to simulate charging behavior and to debug the system conveniently. The practical application shown that the system could be stable and reliable, and had a favorable application foreground.

1. Introduction

Considering the increasingly prominent fuel supply contradiction and environmental pollution, energy-saving and emission-reduction have become a trend all the world. In recent years, under the guidance of national policy and the efforts of all parties, new energy EV(electric vehicle) industry has developed vigorously in our country. EV is playing a more and more significant role in development of national economy industry, but charging diversification and convenience are still bottlenecks, which restrict its development. To solve this problem, thus, wireless charging technology emerges. As WCT(Wireless Charging Transfer) doesn't contain exposed ports, doesn't need manual operation, doesn't occupy a large of space, relative to wired charging, it provides a more comfortable and safe travelling environment for drivers[1-3].

At present, most of the researches have been devoted to the realization and optimization of wireless energy transfer system. When it comes to monitoring and management system, reference [4] proposed an online monitoring system for EV charging stations, optimizing the operation management of charging station. While reference [5] designed a wireless charging system for small power electronic devices, based on the theory of electromagnetic propagation and magnetic induction, but it didn't discuss on the large power equipment.

Currently there have been some EV surveillance applications. Most of them are based on BMS (Battery Management System)to collect the information of battery, and the communication between EV and charging control apparatus is based on the CAN(Controller Area Network) bus. However there are some defects in domestic battery management system, such as limited industrialization, unstandardized systems, verified standards and different reliability, which have restricted the large-

[3]Address for correspondence: T Hei, School of Electrical Engineering, Shandong University, Jinan, 250061, China. E-mail: heitong93@163.com

scale applications of it. And additionally, CAN Bus communication doesn't apply to non-contacting wireless charging system.

Considering the above problems, this paper proposed the monitor system of EV wireless charging, including independently designed acquisition circuit of voltage and current signals as well as a software development platform based on ARM. And the system was independent of SOC(State of Charging) and battery information collected by EV BMS system. Taking charging and discharging characteristics of Li-Ion into consideration, the system adopted the three-stage charging method to achieve safe and efficient charging management. And a touch-screen display unit was designed based on MCGS to debug the system conveniently.

2. The three-stage charging method and control algorithms

In order to balance the safety and rapidness of charging process, and considering the efficient use of the battery, wireless charging system adopted three-stage charging method. The method divided the process into three parts based on fitting the characteristic curve of the battery charge and discharge, containing current limiting, balanced charging and float charging.

Take the rated voltage 4.1V battery as an example. Firstly, in the early stage of charging, if the battery voltage is less than 2.9V, then charging circuit enter the current limiting stage, using a constant rate of 0.1C to charge, with 1C represents the battery is full of 1 hours in the ideal state. The current limiting stage is mainly to avoid the impact of current when the battery voltage is too low. And then the charging mode is converted to balanced charging, adopting constant current to speed up the charging, when the monomer battery voltage rises to saturation voltage(4.1V/4.2V), the capacity of the charge is close to 40% -70% of total capacity at this point. Next, adopting the constant voltage to charge with charge current gradually decreasing. When the current down to 0.1C, it is therefore rational to consider that the battery is full. After that, charging process will enter the floating stage, charging with 0.01C until the end of the charge. Figure 1 shows the three-stage curve diagram of the battery.

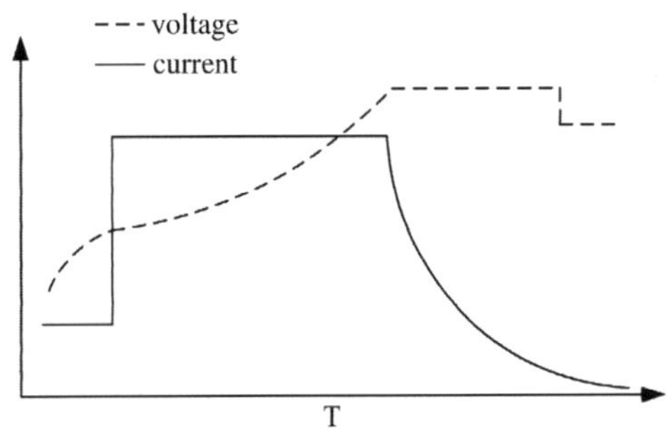

Figure 1. The three-stage curve diagram of the battery diagram

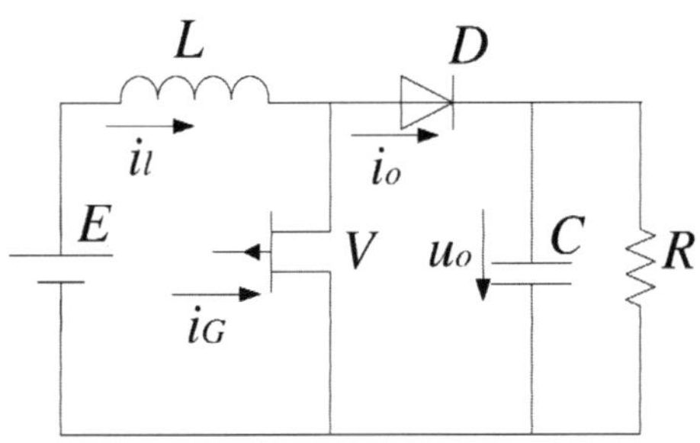

Figure 2. Chopping circuit principle diagram

The system was designed mainly based on the boost chopper circuit in wireless charging to achieve charging current and voltage control. As shown in Figure. 2 is the boost chopper circuit.

When the switch V is on the state, power-E charge to inductive-L, at this time current and output voltage is constant, while when the V-Switch is off, power-E and nductive-L charge to capacitance-C, providing electric energy to load at the same time. The relationship between conduction duty cycle and voltage as well as the average value of the output current are shown in (1)and (2).

$$U_o = \frac{1}{\beta} E = \frac{1}{1-\alpha} E \tag{1}$$

$$I_o = \frac{U_o}{R} = \frac{1}{\beta}\frac{E}{R} \tag{2}$$

In the formula (1) and (2), α represents the PWM duty cycle, while the β represents reciprocal of boost ratio. It is obvious that the output current and voltage value can be regulated through controlling the PWM duty cycle. So we could use the relationship between the duty cycle and the output voltage and current to achieve the three stage charging. By examinating and judging the parameters such as battery voltage and charge current to realize charging mode switching: when the charging voltage and current is higher than the set value, then increasing the duty factor, and vice-versa.

In order to achieve stable and rapid charging, this paper applied double closed loop control system. First of all, substracting the reference voltage value and the measured value can obtain the deviation, next the current reference value can be got after PI adjusting and limiting operation. The duty ratio of the chopper circuit can be real time regulated from the deviation.of current operated by PI and limiting. Figure.3 shows the algorithm block diagram.

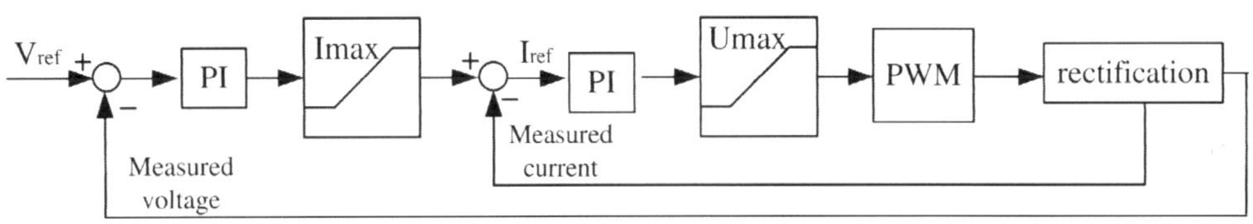

Figure 3. The algorithm block diagram

3. The composition of the monitoring system

Considering that currently we can't collect a large amount of user data to establish a database, thus, this paper proposed a charge monitoring system based on the thought of modularization, which is aimed at a single EV. It specifically included transmitting control module, receiving module, measuring module, display module and so on.

Transmitting control and receiving module adopt STM32F103 as core board, using Bluetooth hands-free communications to realize real-time connection as well as ensure the reliability of the three stage charging. The receiving module includes a current voltage measurement circuit, and the data is transmitted to the transmitting end core board via the receiving side, to achieve real-time control of the charging and discharging. In addition, as the battery has the working temperature limit, the temperature detection also has great significance during charge-discharge process. Based on this, the receiver sets the temperature sampling sensor, communicating with the core board through the serial port.

It's necessary to collect the real-time voltage and current during charging process, sending the collected data to the receiving core board through ADC conversion. The signal is transmitted to the control core board via Bluetooth to implement feedback control. By comparing the real time voltage current with the standard voltage and current, the PWM waveform is continuously adjusted.

As shown in Figure 4, the R1 is a standard resistor, and the charge current is transmitted back to the control panel through a differential amplifier circuit. Differential amplifier circuit has a strong ability of common mode rejection and small output drift voltage, which can accurately measure the magnitude of the R1 voltage, thus the size of the charge current is obtained. What was more, there's a voltage acquisition, as shown in Figure 5. This paper adopted the method of precision resistance voltage divider to regulate the voltage signal. The collected battery terminal voltage was divided into the resistance voltage through the resistance, using a filter capacitor and a voltage stabilizing diode as the subsequent circuit to meet the requirements of ADC input voltage.

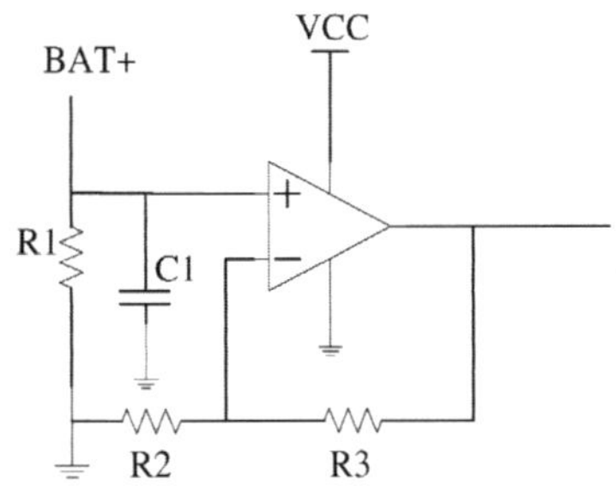

Figure 4. The current acquisition circuit

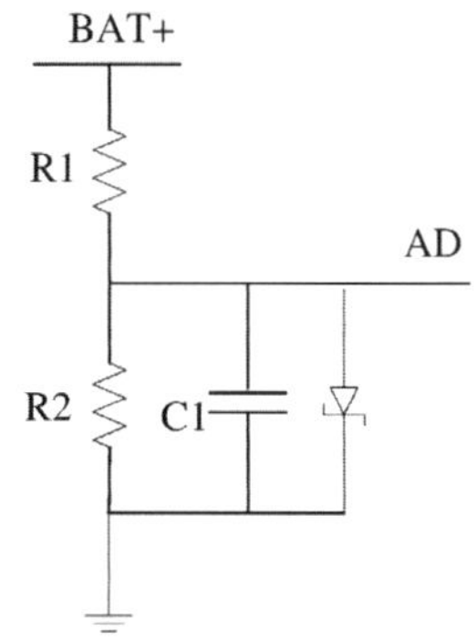

Figure 5. The voltage acquisition circuit

In the system, we adopted LT-111 as the core of metering module, which could accurately measure the electrical parameters of the line, to provide voltage and current, power, energy and power factor, communicating with the transmitting terminal core board through the RS232. The display module chosen 7 inch MSCG display, to simulate the user side charging control and display parameter in system debugging, which could communicate with the core board though RS485.

As shown in Figure 6 is the communication process.

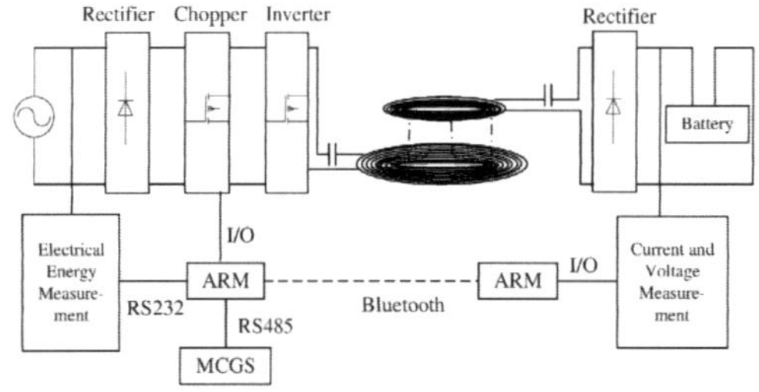

Figure 6. The communication process diagram

3. Monitoring system implementation and function display

After the system design was completed, choosing "the Taiun 100" EV as a reference to test. Its battery rated voltage is 74V, rated current is 20A.

As the internal battery pack of the EV is composed of series connected cells, based on the laboratory environment, we adopt a series of lithium battery with rated voltage of 12V to simulate the voltage of the electric vehicle. The experimental apparatus is shown in the following Figure 7.

Figure 7. The experimental apparatus

In the three-stage charging process, the trickle threshold was set at 54V, constant current setting value was set at 1A and the constant pressure value is set at 74V to make the charging and discharging completely. When the start key of the operations panel MCGS was pressed, the charging command can be sent to the control board via RS232/485 networks. After that transmitter and receiver Bluetooth chip perform handshake, the current and voltage information collected by the receiver is sent to the transmitting end and the display screen at the same time. The three stage charging process controlled by ARM is displayed on the screen. The designed interface includes voltage and current power display in real time, as shown in Figure 8.

Drawing from measured data, the charging voltage curves were obtained, as shown in Figure 9. The experimental data shown that the wireless charging monitoring system could complete information acquisition and real-time control in the process of charging. Besides the system was reliable on security, expansibility and universality, and could be used in different models.

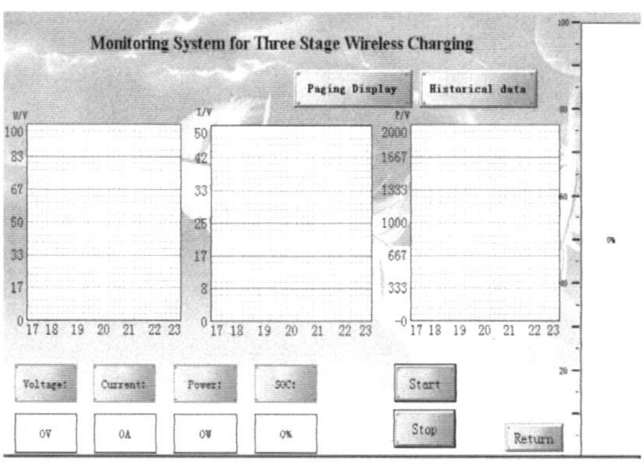

Figure 8. The display interface of MCGS

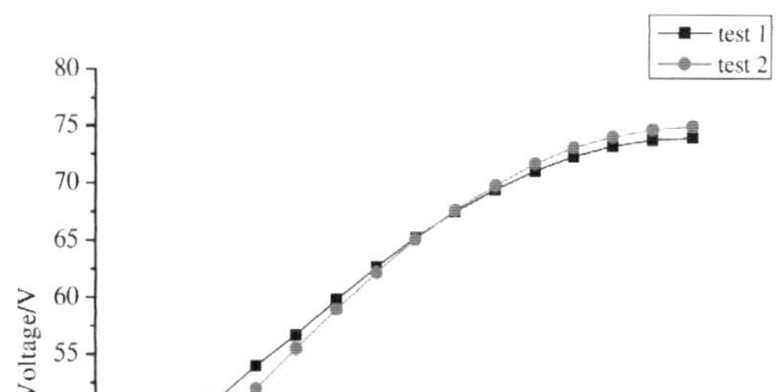

Figure 9. The charging voltage curves

4. Conclusion

Based on the introduction of the three section type EV battery charge and discharge characteristics, a wireless charging monitoring system was designed, which achieved an organic unity of rapidity and efficiency. Besides, a current & voltage monitoring circuit is specially designed to make the system more universal and a tangibly screen is adopted to facilitate debugging and control, which makes the system more intelligent.

Experiments showed that the system can complete the real-time charging monitoring reliably and its charging process meets the battery characteristics. That is to say, it prolongs battery life and has a high cost performance ratio. In addition, the system is not for a given vehicle segment. It avoids drastic alteration to the car and it applies to a variety of application scenarios. Therefore, it has good market prospects. But it must be noted that this system is designed for the changing process of a single EV, without considering the scheduling problem of large scale EVs charging. The direction of research is to optimise the system to achieve coordinate charging, and to make the charging more efficient

References

[1] Wang Z Y, Wang X M, Zhang B and Qiu D Y 2014 Advances of wireless charging technology in electric vehicle *J. Power Supply* **3** 27-32
[2] Ni G Q and Wang L N 2013 Design scheme of wireless charging system *Electronic Engineer* **39** 51-4
[3] Ai M H and Qi L H Electric vehicle charging stations' on-line monitoring system based on ordered charging *Shanxi Electric Power* **42** 6-11
[4] Du C Q and Du G 2015 Remote monitoring technology research and platform development for electric vehicle *Automotive Engineering* **37** 1071-6
[5] Yang Y, Liu Z Z, Qu D M and Wang J 2015 *Research on coordinated charging for electric vehicles* International Conference on Electrical Machines and Systems (ICEMS) 525-30

The design of a vehicle-mounted test system for the thermal performance of solar collector

S R Wen[1], X H Wu, L Zhou, W Zheng, L Liu and J C Yan

National Solar Water Heater Supervision and Inspection Center of Product Quality (Wuhan), No.6 Gong Ping Road, Wuchang District , Wuhan 430061, China

Email: 24148424@qq.com

Abstract. To increase the test efficiency of thermal performance of solar collector, a vehicle-mounted test system with high automation, simple operation, good mobility and stability is proposed in this paper. By refitting a medium bus, design of mechanical system and test loop, and using PC control technology, we implemented the vehicle-mounted system and realized effective integration between vehicle and test equipment. A number of tests have been done, and the results show that the vehicle-mounted test system has good parameters and performance and can be widely used to provide door-to-door testing services in the field of solar thermal application.

1. Introduction

Solar collector is the key component of domestic and industrial thermal solar systems, which absorbs solar radiation and converts it into heat and be widely used around the world. Whether the quality of solar collector is good or not, is directly relating to the quality of thermal solar systems. Thermal performance test is most important to evaluate the quality of solar collectors, but the test method is very complex because of the following factors [1]:

- According to GB/T 4271-2007 and ISO 9806: 2013, performance tests (steady-state efficiency, angle modifier, time constant) required for the collector be mounted outdoors where the hemispherical solar irradiance shall be greater than $700 w/m^2$ [2].
- Temperature and flow of the heat transfer fluid need to be controlled precisely and kept stable at set value.
- The feature of solar collector is big in size, heavy in weight.
- The solar collector belongs to the glass products, sample transport costs are relatively high.

With the haze weather increasing in recent years, effective test days decrease significantly, the existing fixed test equipments are unable to meet the growing test requirements [3]. The vehicle-mounted test system designed in this paper can realize mobile tests in different places; the laboratory can provide a door-to-door service, which is convenient to the clients by improving the test efficiency.

2. System design

[1] Address for correspondence: S R Wen, National Solar Water Heater Supervision and Inspection Center of Product Quality (Wuhan), E-mail: 24148424@qq.com.

2.1. General design

The general design from initial design to final design has experienced several stages, such as data collection, spot investigation, vehicle selection, scheme discussion, final scheme demonstration and determination.

After taking into account all the requirements such as system functionality, vehicle safety, technical parameters, technology advancement, integral aesthetics and anti-leakage, the final design scheme was determined. The core of its content includes:

- To choose medium bus as the basis of vehicle system and refit to meet the need of design.
- When the vehicle-mounted test system is moving or at non-working state, all devices are fixed inside the vehicle, as show in figure 1.
- When working, a few devices, mainly including the collector tracking frame, automatically be moved out of Vehicle, and then the collector sample will be installed and prepared for testing, as show in figure 2. After the test, the tracking frame will be taken back into Vehicle.

Above all, this design scheme has the following advantages:

- During the test tracking frame is supported on the ground, installing or removing the test sample is more convenient, and that can ensure its high load-bearing capacity and good stability.
- Medium bus has properly payload, moreover, its mobility and flexibility are better than heavy vehicles.

The vehicle-mounted test system consists mainly of five parts: modified vehicle, mobile platform, tracking frameclosed test loop, measurement and control system, etc.

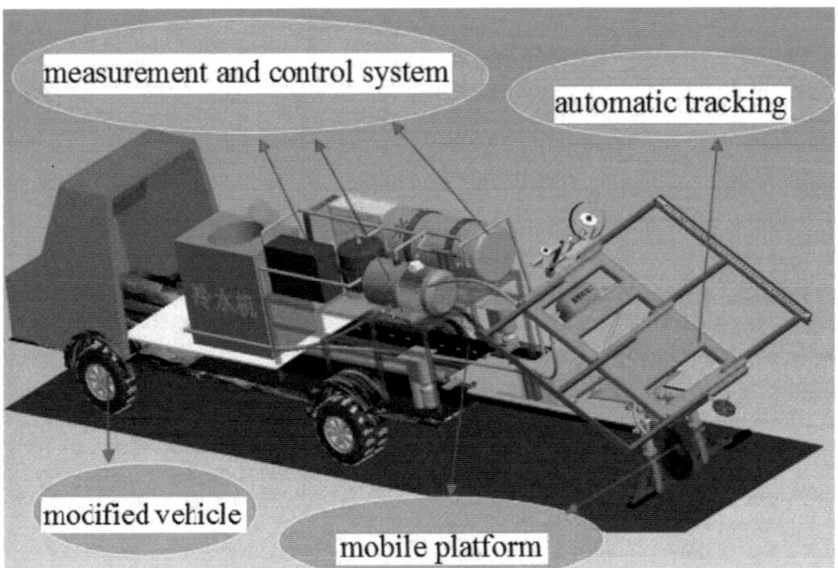

Figure 1. All devices in the vehicle.

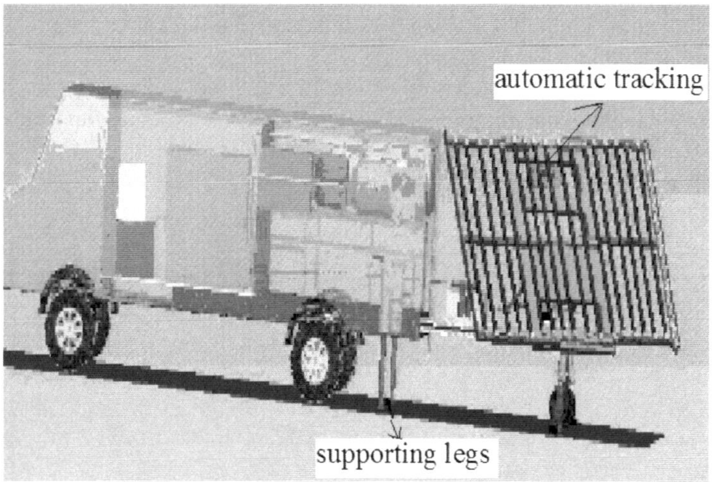

Figure 2. Tracking frame out of the vehicle.

2.2. Vehicle refit

Vehicle is the carrier of the test system. A commercial vehicle was chosen, its parameters are shown in the table 1 below.

Table 1. Parameters of vehicle.

Manufacture	JMC transit
Wagon / Van	Transit Wagon
Engine	Diesel inline, 2.8T
Overall Dimensions (length * Width * Height)	6503*2000*2615 (mm)
Roof (Low, Medium, High)	High
Rear wheels	dual
Wheelbase	Long Extended, 3750mm
Maximum Payload	1400 kg
Gross Vehicle Weight	3500kg
Maximum speed	120 km/h

According to system layout and the load requirement, the corresponding refit for the vehicle has been made:

- Interior layout adjustment: remain the driver seat and copilot seat, remove other seats; divide interior space into seat area, operating area and equipment area by detachable clapboards.

- Top refit: The air-conditioning has been installed on the top of the vehicle, whose outlet is in the operating area.
- Bottom refit: in order to support and balance the whole vehicle, at the bottom of the vehicle, the crossbeam has been strengthened and another four electric supporting legs have been added.
- The whole floors of the vehicle have been strengthened.

2.3. Mechanical system design

The mechanical system mainly includes the mobile platform and the collector mounting frame. As the key of the effective integration of the test system and the vehicle, the mobile platform system is responsible for removing and taking back the mounting frame by chain transmission mode. Positioning and fixing have been designed for the mobile platform to ensure the system stability. The collector mounting frame has been designed as a two-axis tracking system[4], can move automatically to follow the sun in azimuth and altitude, the support surface on which can stretch out and draw back automatically for the different sizes of solar collectors, as show in figure 3.

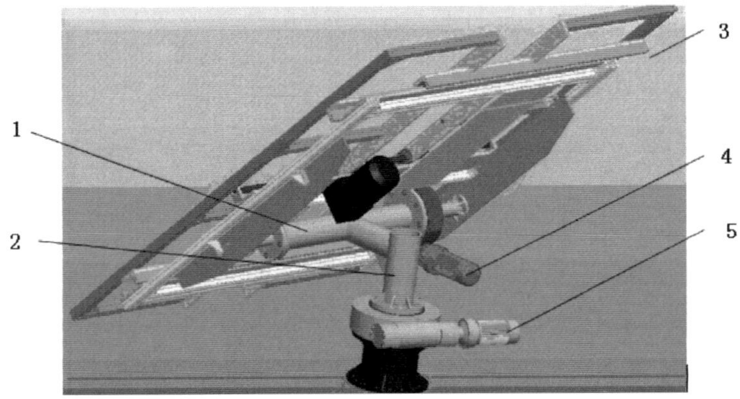

Figure 3. Collector frame using automatic tracking.

1. Altitude angle tracking axis; 2.Azimuth angle tracking axis; 3.Collector supporting surface; 4. Altitude angle motor; 5.Azimuth angle motor

2.4. Design of test loop

The closed test loop [5], based on the closed test loop diagram in ISO 9806-2013 [1]and GB/T 6424-2007, designed for the self-control and self-regulation of the temperature and flow of working fluids, is mainly composed of cooler, the heat insulating water tank, heater, the mixing pump, flow meter, temperature sensors, the pipe, etc.

2.5. Design of measurement and control subsystem

2.5.1. Hardware design

The measurement and control system, based on PC for the data acquisition and electrical control as shown in figure 4, is mainly composed of different kinds of sensors, signal conditioning circuit board, 16 channel 16 bit A/D conversion board, industrial PC, 8 channel 16 bits D/A conversion board, isolated I/O module and the actuators, etc. Data and status of various sensors can be collected in real-time by motion control cards and data acquisition cards, and then used to control the operation of system after program processing and output.

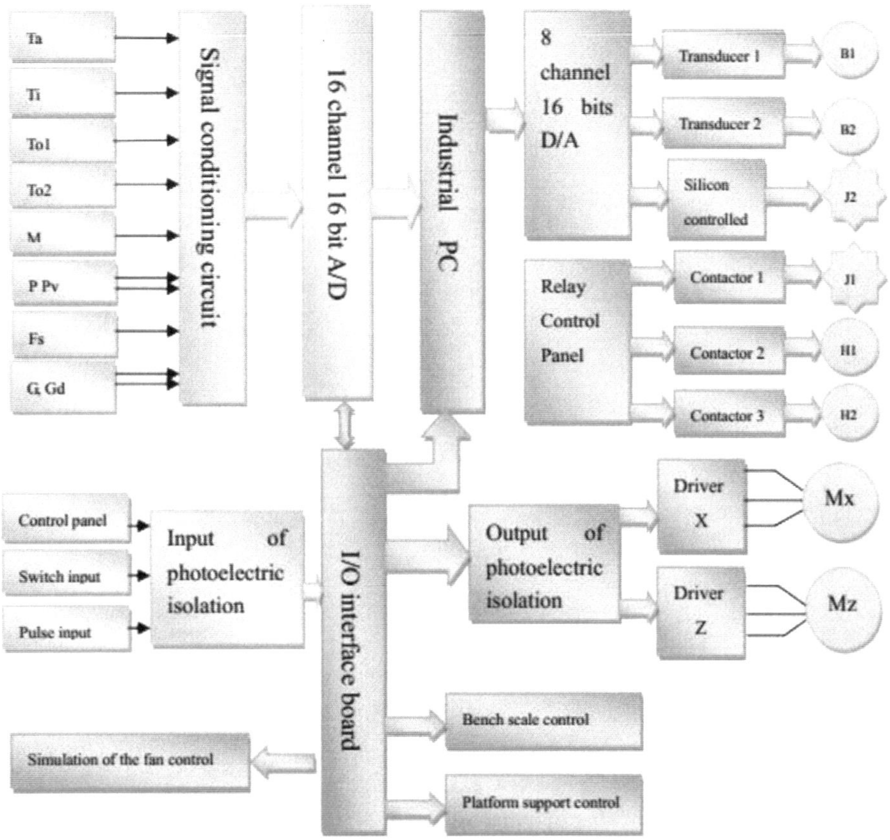

Figure 4. Hardware structure of measurement and control system

Ta: environment temperature; Ti: Inflow temperature; M: Quantity of flow; To1: Effluent temperature 1; To2: Effluent temperature 2; P: Pressure; Pv: Differential pressure; Fs: Air speed; G, Gd: Solar irradiation;

2.5.2. Software design

The integrated software uses Windows as the operating system, the friendly man-machine way for interaction, the advanced graphical programming language LabVIEW for programming [6], and the multi-user, multi-threaded technology as well as distributed architecture for the task processing. The system software includes modules such as the state machine with LabVIEW as its main body structure, the signal acquisition module of limit signals and servo motor encoder, the movement module of horizontal mobile platform and two-axis solar tracking frame, the real-time tracking test module and data processing module, etc. Figure 5 shows the software structure based on the standard state machine.

During the signal acquisition, all signals emitting from various sensors and servo motor encoders can be mainly acquainted by motion control cards and data acquisition cards. In the motion control module, servo motor movement parameters are set up, such as servo motor pulse mode, encoder feedback, speed direction setting, sport mode settings, and so on. In the data analyzing and processing module, it mainly processes the data analysis, curve fitting, visual display and reports printing. According to user's selection and setting, the integrated software system can automatically control the two-axis tracking frame, test procedure, temperature and flow of the heat transfer fluid. It also can do the self-check and calibration to ensure reliability of the system.

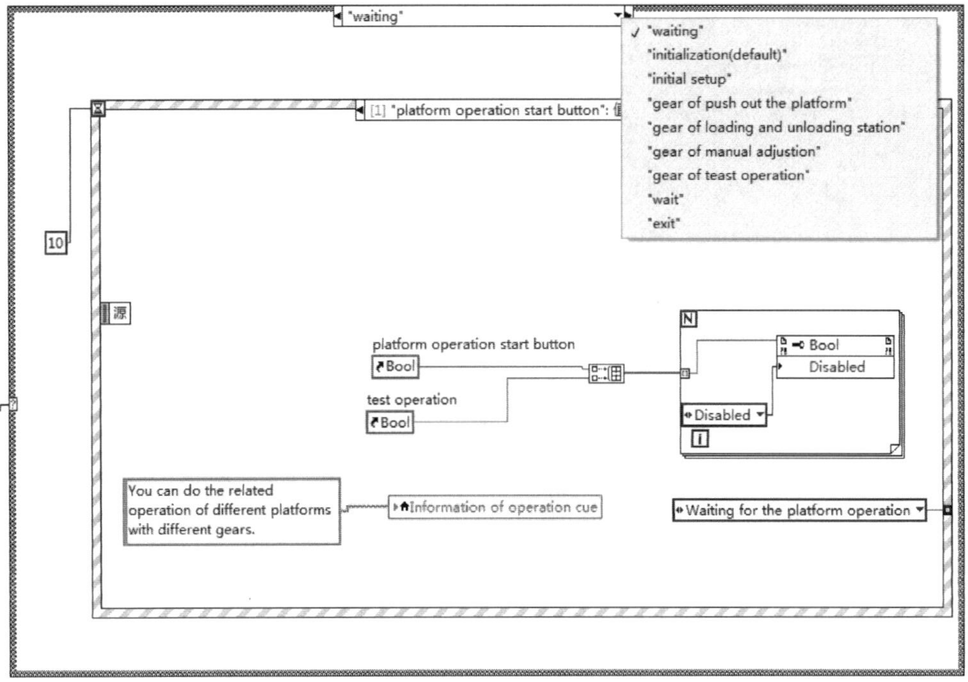

Figure 5. Software structure based on the standard state machine

3. Implementation and testing

According to the design scheme of the previous chapters, a Vehicle-mounted test system was produced successfully, as shown in Figure 6-9. Unlike fixed test system, this vehicle-mounted test system has the following operation principles:

- Non-working status: The back door of the vehicle is closed. Not only the electrical cabinets, water tank, table, computer and printer, but also the automatic tracking frame, the whole test system are placed inside the vehicle.
- Working status: After the test Vehicle arriving at destination, the back door shall be opened first, then the automatic tracking frame automatically be moved out of Vehicle and supported well. Next, the solar collector may be mounted on the tracking frame and be prepared for tests. When the test ends, the tracking frame will be taken back into Vehicle.

Figure 6. Non-working status

Figure 7. Working status

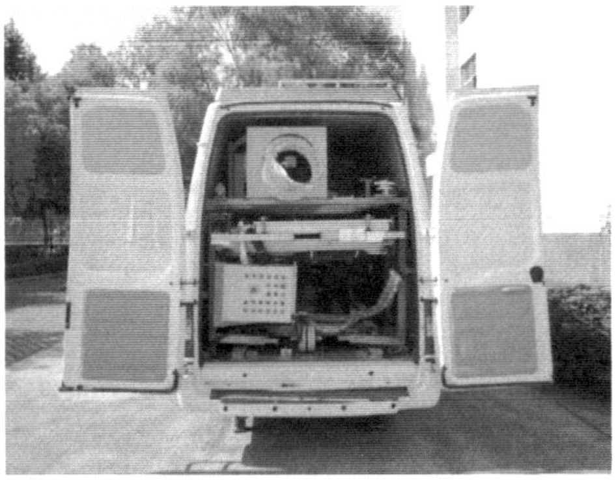

Figure 8. Non-working status

Figure 9. Working status

In order to prove the feasibility and advancement of the vehicle-mounted test system, a number of tests have been done, such as repetition test, calibration, contrastive test, electro-magnetic compatibility test, safety test for vehicle operating on road and so on. These tests have shown that the design scheme is feasible and the vehicle-mounted test system has the following properties:

- It can be applied on thermal performance testing of solar collectors according to the standards of ISO 9806:2013 and GB/T 6424-2007, including outdoor steady-state efficiency test, time constant, incident angle modifier and pressure drop measurement.
- It has high automation: the mounting frame can automatically move out form the vehicle or take back, and track the sun; it can automatically control test procedure and conditions; it can automatically record and analyze data, then print the test report.
- Parameters and performance of the vehicle-mounted test system has many advantages in many aspects compared with Standard requirements, as shown in table 2.

Table 2. Overall comparisons

Parameter/ characteristic		deviation /mode	
		Permitted deviation (ISO 9806:2013)	The Vehicle-mounted Test System
Test solar irradiance		± 50 W/m^2	± 50 W/m^2
Fluid temperature Measurement		±0.1K(inlet), ±0.5K(outlet)	±0.1K(inlet), ±0.1K(outlet)
Fluid inlet temperature control	control accuracy	stable within ±1K	stable within ±0.2K
	Control mode	Two stages	Two stages
Fluid flow rate	Measurement	±1%	±1%
	control accuracy	stable within ±2% of the set value	stable within ±1% of the set value
	response time	/	about 1min
	Stable time	/	maintain more than 15 min
	Control mode	Using a simple bypass loop and manually controlled needle valve	automatic control by Frequency Converter

Data logger	a 12-bit digital system	a 16-bit digital system
Track the sun	manually or Single-dimension tracking	two-dimensional automatic tracking control accuracy :0.2 mm

4. Conclusions

As a movable device, the vehicle-mounted test system can provide door-to-door testing services, which can increase efficiency and shorten test cycle, also has other advantages such as high automation, friendly man-machine interface, simple operation, good mobility and stability. It can be widely used in testing laboratories and manufacturers for thermal performance test of solar collectors to meet the rising demand, and that will obtain great economic and social benefits.

The design of effective integration between vehicle and test device, especially part of devices can be removed out from vehicle and taken back safety and reliably, provides a good reference for the mobile test equipment.

Future research work should be oriented toward increasing the range of temperature control to satisfy the demands of medium-temperature collector test.

References

[1] 2013 Solar energy –solar thermal collectors-Test methods ISO 9806:2013 117
[2] 2007 Test methods for the thermal performance of solar collectors GB/T 4271-2007
[3] RudićVranić M 2011 Development of a Solar Collector for Thermal Applications. *Exploitaion of Renewable Energy Sources(EXPRES)*
[4] Zhong Y N, Zeng L X, Ding S T, Zhang X L 2011 Optimized design of sun position detection sensor *Applied Mechanics and Materials* **120** 499-503
[5] Cao H F, Yang L G, Cheng J B and Cheng X 2013 A semi-instantaneous heat exchanger for mobile solar collectors test system *Telkomnika* **11** 4069-76
[6] Johnson G and Jennings R 2006 *LabVIEW graphical programming* (NEW York: McGraw-Hill Education)

Analysis on using biomass lean syngas in micro gas turbines

C Mărculescu[1], V E Cenuşă and F N Alexe

Power Engineering Faculty, University Politehnica of Bucharest, Romania 313, Splaiul Independentei, Bucharest 060042, Romania

E-mail: cosminmarcul@yahoo.uk.com

Abstract. The paper presents an analysis on small systems for converting biomass/wastes into power using Micro Gas Turbines (MGT) fed with gaseous bio-fuels produced by air-gasification. The MGT is designed for burning various fossil liquid and gas fuels, having catalogue data related to natural gas use. Fuel switch changes their performances. The present work is focused on adapting the MGT for burning alternative low quality gas fuel produced by biomass air gasification. The heating values of these gas fuels are 3 to 5 times lower than the methane ones, leading to different air demand for the stoichiometric burning. Validated numerical computation procedures were used to model the MGT thermodynamic process. Our purpose was to analyze the influence of fuel change on thermodynamic cycle performances.

1. Introduction

Biomass and organic residues can be used as renewable energy source for the production of gaseous derived fuels. The gasification process and technologies involve a decrease in process energy efficiency compared to combustion because of the endothermic characteristic that limits this efficiency to about 0.6 – 0.7 (cold gas efficiency) significantly below the 0.85 – 0.98 specific to combustion processes [1]. Biomass source size and time availability limit the power generation to small scale applications. Moreover, the thermodynamic cycles with good energy conversion efficiency in the low power range (up to 15-20 MWel) are Brayton and Otto-Diesel [2, 3, 4, 5]. The lower energy efficiency of the gasifying process is balanced by the superior energy efficiency of the gas turbines and reciprocating spark ignition engines in the scale range of the applications. The society evolution involves the increase of food industry as well as of its residues to be disposed. Nevertheless the waste streams have flows significant lower than fossil fuels ones. Consequently, waste to energy applications demand small power units in the range of 40 kW - 500 kW [1]. In this range the steam turbines efficiency is usually limited to 0.12 – 0.18 (with power output starting from 150–200 kW) while the gas turbines and engines generally maintain their conversion efficiency of about 0.3 – 0.4 (with power output starting from 20–200 kW) [3, 6, 7, 8].

The gas turbines directory performances are reached by using standard gas fuel – the methane. The shift from natural gas to lean gases, that are usually produced through gasification processes, conduct to significant changes in turbines behavior that are subject to current investigations. This research area is developing and the results provide information both to equipment manufacturers and system engineers for a better assessment of biomass to energy conversion solutions using gasification process.

[1] Address for correspondence: C Mărculescu , Power Engineering Faculty, University Politehnica of Bucharest, Romania 313, Splaiul Independentei, Bucharest 060042, Romania. Email: cosminmarcul@yahoo.uk.com.

A MGT unit, using natural gas, up to 500kW$_{el}$ that can be used also in cogeneration was analyzed for the case of burning alternative lean syngas from air gasification of organic waste. The syngas composition (specific energy content) was varied according to different feedstock (wood biomass and organic residues from food industry), and the thermo-chemical process (atmospheric pressure air gasification – the most common type) [9, 10, 11]. Consequently a range of syngas type was used in our computation from low to higher quality gas fuel using as reference the methane.

The paper presents the effect of syngas physical-chemical properties on Micro GT performances.

2. Materials and methods.

2.1. Gas fuel properties

The syngas fuel used in our research comes from the air gasification of wood biomass and organic residues from meat processing industry. Two different feedstock were considered as variation range for the produced syngas quality. The syngas produced by wood gasification has a higher specific energy content related to feedstock properties (high combustible fraction, average humidity and low inert content). The food industry waste consists in meat and bones residues (low specific energy content due to water content –45%, and inert fraction – 30%) [9]. Consequently, the syngas energy properties will be inferior to the first one. Feedstock and syngas detailed properties are presented in previous works of the authors [9]. The common bibliographic sources report the syngas composition on dry basis that is a theoretical situation only. The syngas humidity used in the analysis was chosen based on condensing the H_2O vapors in surface heat exchangers cooled with river water at average temperature in temperate climate. By cooling the syngas nearly to 31.4°C, the water vapors molar participation in the syngas, with a relative humidity near to 100%, will be about 4.6%.

The range of the syngas components are: CO (17.9÷33.5%); H_2 (11.1÷20.8%), N_2 (33.3÷57.9%), CO_2 (4.4÷7%), C_nH_m (0.5÷1.%). Syngas molar masses (22÷26 kg/kmol) are higher compared to methane. The volumetric Higher Heating Value of these gases (HHV varies between 4.3 MJ/Nm^3 symbol D and 8 MJ/Nm^3 symbol A) are about 5 to 15 times lower compared to CH_4 [9].

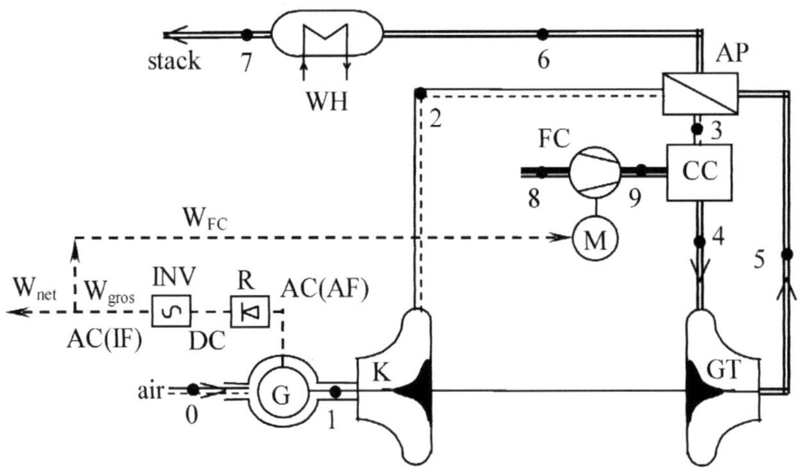

Figure 1. Typical micro GT design in single shaft arrangement.

Note: G-generator; K-air compressor; GT-gas turbine; M-engine; FC-fuel compressor; CC-combustion chamber; AP-air pre-heater; WH-water heater; AC-alternative current; R-rectifier; DC-direct current; INV-inverter; W-power.

Almost all MGT employs an internal recovery Air Preheater (AP). This consists into a

IOP Conf. Series: Earth and Environmental Science **40** (2016) 012036 doi:10.1088/1755-1315/40/1/012036

regenerative gas-air surface heat exchanger. The compressed air is pre-heated by recycling the thermal potential of turbines exhaust gases.

The syngas obtained by air gasification contents a reduced fraction of combustible gases. Consequently the process performances will be affected if this new gas fuel will be used and some components of the MGT will need redesigning [12].

2.2. Problem formulation

The two non-dimensional factors used to analyze the MGT thermodynamic process are: **1)** the compression ratio (Π=p_{max}/p_{min}), and **2)** the ratio of extreme absolutes temperatures (θ=T_{max}/T_{min}). When Π is fixed and θ increases, the electrical efficiency (η_{el}) and the specific work relative to flue gas mass flow rate (W_{sp} - kJ/kg), are increasing too. Consequently becomes interesting to raise T_{max}, with minimum intervention on the MGT components.

It is generally known [13] that η_{el} and W_{sp} can be maximized at a specific value of Π. For classic Brayton cycle, without AP, at a given θ value:

$$\Pi\ (W_{sp\ max}) < \Pi\ (\eta_{el\ max}) \quad (1)$$

The influence of the AP on MGT process leads to [14, 15]:

$$\Pi\ (\eta_{el\ max}) < \Pi\ (W_{sp\ max}) \quad (2)$$

These characteristics ensure good electrical efficiency for low Π. The MGT are equipped with compressor for the fuel. Therefore the unit power and its efficiency given in directories consider the compressor energy consumption and the values listed are net.

The engineering computations necessary to analyze the MGT cycle use lots of variables and nonlinear equations. The analytical solving becomes unusable. Therefore, validated procedures [14,15] were used in the numerical computation customized to cycle.

The main parameters (temperatures of compressed air and exhaust gases, excess air at Combustion Chamber, α_{CC}) and indicators were computed: AP $_{efficiencies}$ (ε), η_{el}, W_{sp}, depending on: **1) Π** ($\Pi\in$[2 to 5], in geometric progression with $10^{1/50}$ ratio); **2) θ** ($\theta\in$[3.9 to 4.9], corresponding to $T_{max}\in$[1095 to 1441] K, or $t_{max}\in$[822 to 1168]°C); **3)** different fuels compositions. As reference fuel we used methane "**M**", while the other 4 fuel gas properties (**A**, **B**, **C**, and **D**) are presented in previous work of the authors [9], and the range of syngas components and HHV is presented above.

The following assumptions were used:

- ISO parameters were used for the compressor air inlet (T_{atm}=T_{ref}=288.15 K, relative humidity φ=60%, and p_{atm}=101.324 kPa).
- Fuel gas pressures at FC inlet are considered equals to the atmospheric one, while the temperatures at FC inlet are different: $T_{8\ methane}$=283.15 K, and $T_{8\ BS}$=304.55 K. $T_{8\ BS}$ is the temperature of the syngas after cooling and water vapors removal (the condensing temperature of the water vapors at 4.6 kPa).
- The AF, CC, AP, and WH pressure losses are between 1.5 to 3 % from the specific inlet pressures.
- The energy losses in CC, AP, and WH were computed based on η_{CC}=η_{AP}=η_{WH}=0.99.
- The heat losses in the compressors and turbines were neglected. The isentropic efficiencies of this equipment are: 0.81 for air compressor, 0.77 for fuel gas compressors, respectively 0.82 for turbine.
- The pressure ratios of fuel gas compressors are superior to the air compressor ratios, according to equipment manufacturer specifications.
- Electric gross efficiency, $\eta_{el\ g}$, was calculated based on the AC power at inverter terminals. It includes the losses in rotating machines, Rectifier, and Inverter. We considered that $\eta_{mec\ K\&T}\ \eta_{gen}\ \eta_{rectifier}\ \eta_{inverter}$ = 0.915.
- The electricity net efficiency is calculated taking into consideration the fuel gas compressor power consumption that is 1.12 times higher than the real work transmitted to the fuel gas into compression process.

- The mass flow rates were computed based on combustion and energy balance equations of CC.
- The compositions of fuel, air, and flue gases are mixtures of real gases (CO, H_2, CO_2, O_2, N_2, and others) considering their quota and thermodynamic properties.

Additionally we introduced a condition for AP, by fixing its exergetic efficiency:

$$\eta_{ex\ AP}=\{m_{flue\ gas}\ [h_5-h_6-T_{ref}\ (s_5-s_6)]\}/\{m_{air}\ [h_3-h_2-T_{ref}\ (s_3-s_2)]\}=0.9 \quad (3)$$

m – mass flow rate [kg/s]; h – specific enthalpy [kJ/kg]; s – specific entropy [kJ/kg/K] in the points related to the schema.

The exergy degradation into AP, given mainly by differences of temperature has a limiting effect on the Number of Transfer Units (NTU). The ratio is chosen assuming that AP is designed according to the turbo-machineries, limiting the AP size and cost.

3. Results and their analyses

By using the computations results we built diagrams of parameter's variations and completed analyses on their evolutions.

Compressed fuel temperatures, t_9 and the compressed air ones, t_2, depends mainly on Π, increasing with this ones. Assuming that compressors have the same inlet's temperatures ($t_1=t_8$), pressures ($p_1=p_8$), compression ratios ($p_2=p_9$), and isentropic efficiencies, the outlet K and FC temperatures should be different due to the properties of gases ($t_2 \geq t_{9\ BS} > t_{9\ Methane}$).

For the same Π and θ, the flue gas temperatures at the gas turbine output (t_5) depend only on the molar participation of the components in the flue gases. Our results show that there are no significant differences between the studied cases. The variations of this parameter with Π and θ are: t_5 decreases with Π increase, at constant θ, and diminish with θ, at constant Π.

As previous mentioned, the peculiarity of Micro GT design is given by the AP. It recover a quota of heat from flue gas to the air, increasing the air temperature at CC inlet (reducing the fuel consumption) and decreasing the heat losses associated to flue gas flow. Consequently, all other parameters in the schema depend on AP performances.

The AP energy balance equation is:

$$\eta_{en\ AP} = [m_{fg}\ (h_5-h_6)]/[m_{air}\ (h_3-h_2)] = [m_{fg}\ c_{p\ av\ fg}\ (t_5-t_6)]/[m_{air}\ c_{p\ av\ air}\ (t_3-t_2)] \quad (4.1)$$

c_p – specific heat of working gases at constant pressure kJ/kg/K.

It results the following equation between the temperature differences:

$$(t_3-t_2) = (t_5-t_6)\ \eta_{en\ AP}\ (m_{fg}/m_{air})(c_{p\ av\ fg}/c_{p\ av\ air}) \quad (4.2)$$

$\eta_{en\ AP}$ – energy efficiency of the air preheater.

Knowing t_2 and t_5, the temperatures t_3 and t_6 could be theoretically computed from a system with two equations the imposed exergetic efficiency (3), and the AP energy balance equation (4.2). In reality there is a strong dependence between these parameters, processes evolutions and substance properties. Consequently all the parameters must be established by iterative computation. The iterations stop when the imposed exergetic efficiency of the AP is achieved. Thus we also computed the ratio $m_{fuel}/m_{flue\ gas}$ depending on Π, θ, and fuel type. Note that the lean syngas fuel mass flow rate is considerably bigger than the methane one. When the HHV decreases, the fuel mass flow rate increases. Therefore, per 1 kg air, the flue gas flow rate increase substantially from methane to syngas A, and moderate from syngas A (high quality) to D (lean). The wide range of the results conducted to the use of an average bio-syngas for representing in Figure 2 the variation of the ratio $m_{fuel}/m_{flue\ gas}$ for methane and the average bio-syngas (BS).

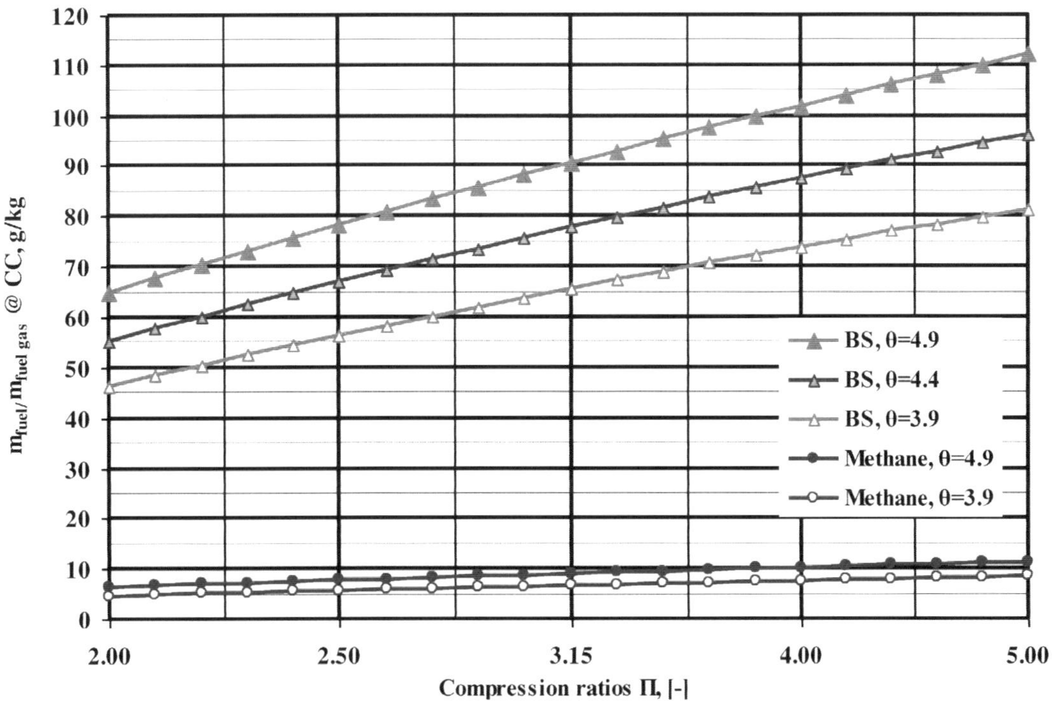

Figure 2. $m_{fuel}/m_{flue\ gas}$·for methane, and the average for all BS $m_{fuel}/m_{flue\ gas}$ values, in g/kg, vs. Π, for θ_{ratios}=3.9 to 4.9

Rising Π, at constant θ, increase t_2, which pushup t_6 (t_6>t_2) (Figure 3). For given Π and θ, rising flue gas mass flow rate, decrease the temperature difference on the flue gas flow part and t_6 increase because t_5 has insignificant variation with the fuel type.

As mentioned before, rising Π, at constant θ, decrease t_5. It pushes down t_3 (t_5>t_3). As a result the temperature differences on both AP flows, (t_3-t_2) and (t_5-t_6), diminish. Consequently t_3 decrease with Π and increase with θ. Because the flue gas mass flow rate per 1 kg air is higher for syngas than for methane, the value of t_3 is also higher for syngas preheating compared to methane (Figure 4).

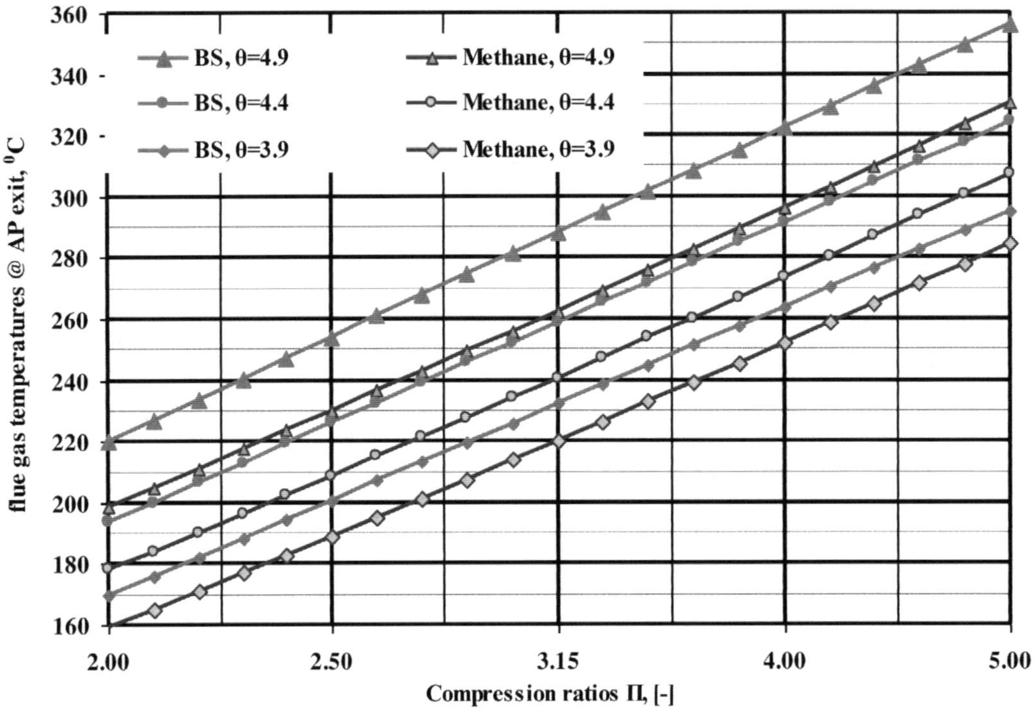

Figure 3. Variation of t_6 for CH_4 and the average t_6 values for all BS, vs. Π, for $\theta \in 3.9 \div 4.9$

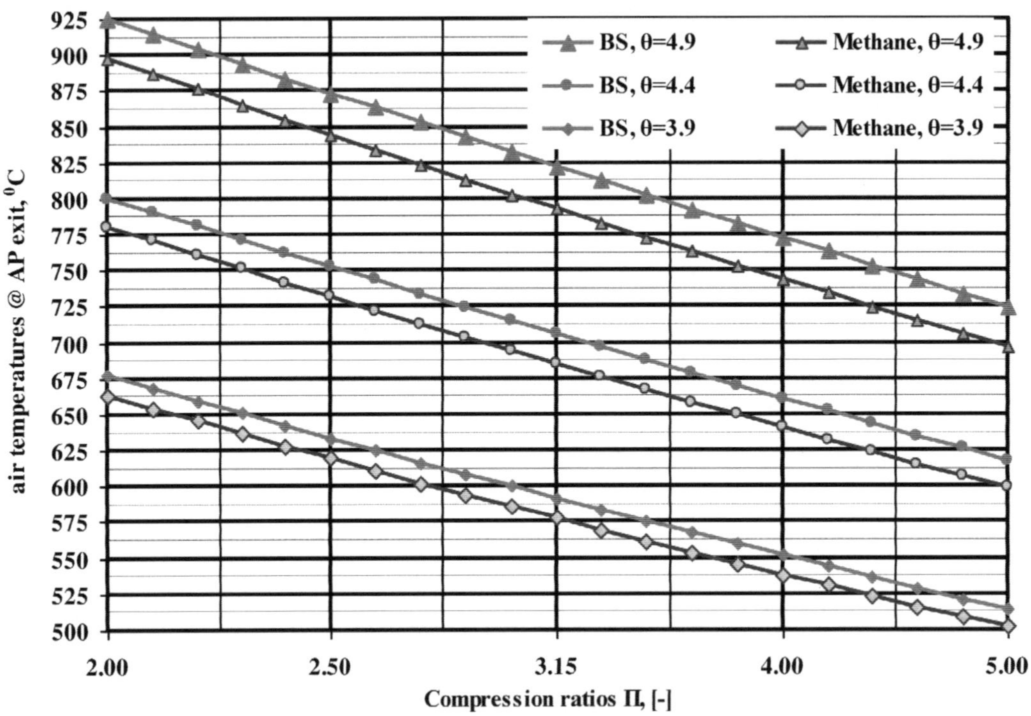

Figure 4. Variation of t_3 for CH_4 and the average t_3 values for all BS, vs. Π, for $\theta \in 3.9 \div 4.9$

The average AP heating effectiveness $\varepsilon_{AP}=(t_3-t_2)/(t_5-t_2)$ diminishes with Π, being higher for lean syngas and minimal for CH_4 (Figure 5). For methane the values are in range with available data [14].

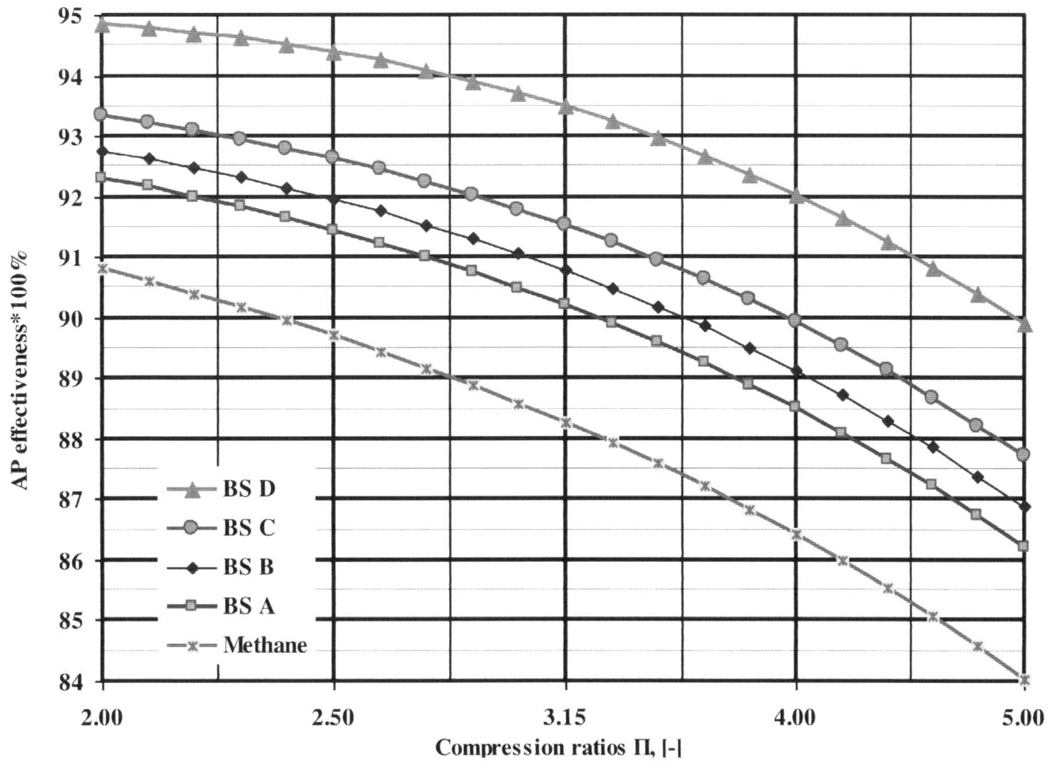

Figure 5. Average AP effectiveness, ε_{AP}, at θ_{ratios}=3.9 to 4.9, for different fuels, vs. Π

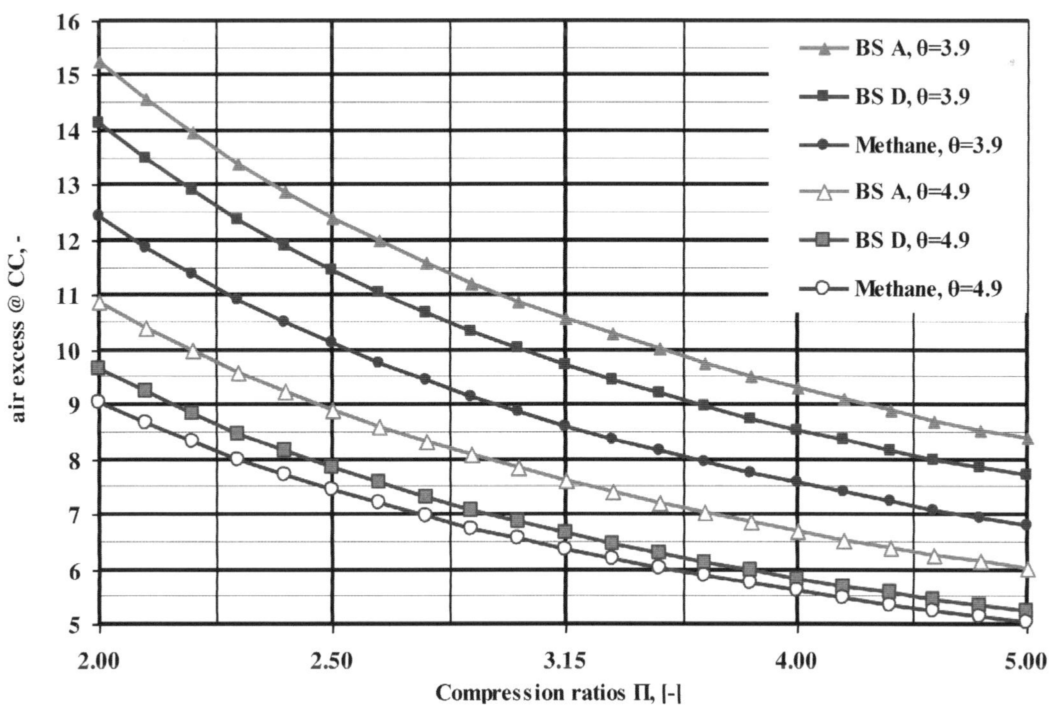

Figure 6. Variation of α_{CC} for CH_4 and different BS vs. Π, for θ =3.9 to 4.9

Figure 6 shows the air excess coefficient in combustion chamber (α_{CC}) variation depending on Π, θ, and fuel type. It reveals the differences between syngases (BS D and BS A) and methane. For the same Π and θ values α increase when the fuel gas heat values are higher. For all analyzed fuels the evolution of air excess is similar to the conventional GT: for a given θ it diminishes with Π growth, while for a given Π it diminishes with θ growth. The differences are given by the numerical values of α_{CC}, higher than for classical GT without AP, because at micro GT the air temperature at CC input is higher.

Figure 7 shows the variation of ratio between $P_{th\ AP}$ (the thermal flow rate recovered by AP) and $P_{fuel@CC}$ (the heat input by burning fuel into CC) for an average syngas and methane. For all θ values these ratios decrease when Π increase. The values obtained when burning methane are higher compared to syngas use. If compression ratio increases, the difference between the temperatures at turbine and compressor exhaust decreases and the possibilities for internal heat recovery decrease.

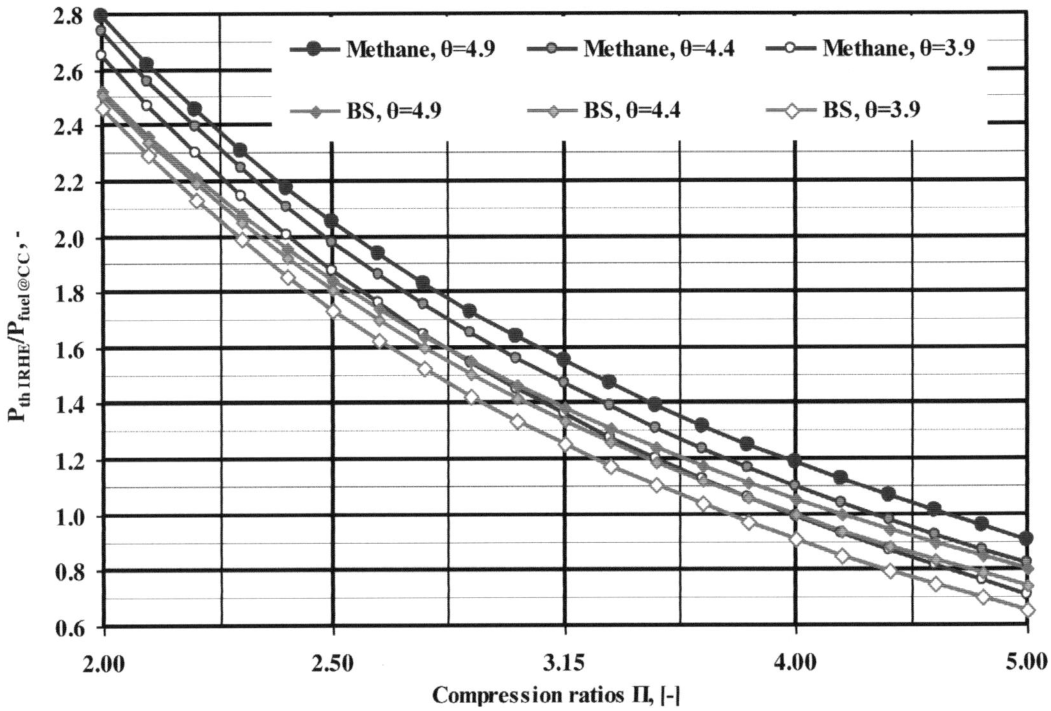

Figure 7. Variation of $P_{th\ AP}/P_{fuel@CC}$ for methane and the average values of $P_{AP}/P_{fuel@CC}$ for BS, vs. Π, at θ=3.9 to 4.9

The generated power per mass gas flow, P_g/M_{fg}, increases with θ. Nevertheless this parameter does not reach a maximum value for Π range. P_g/M_{fg} for syngas is higher than for methane, but the fuel switch has a low effect.

For the end user it is important the effect of fuel change on the net power flow, respectively the net electrical efficiency. To establish the equipment performances were computed and represented in following diagrams the evolutions of: gross electric efficiency $\eta_{el\ gross}$ (Figure 8); fuel compressor relative power consumption $P_{el\ fuel\ compressor}/P_{el\ gross}$ (Figure 9); net electrical energy per flue gas unit $W_{el\ net}/m_{flue\ gas}$ (Figure 10) and net electric efficiency $\eta_{el\ net}$ (Figure 11).

The electric gross efficiency, $\eta_{el\ gross}$, has its maximum value in the Π range, for the used θ ratios (Figure 8). We noticed that fuel calorific value has a low effect on this parameter. When θ varies in the considered range the electric gross efficiency increases with about 5 percentage points.

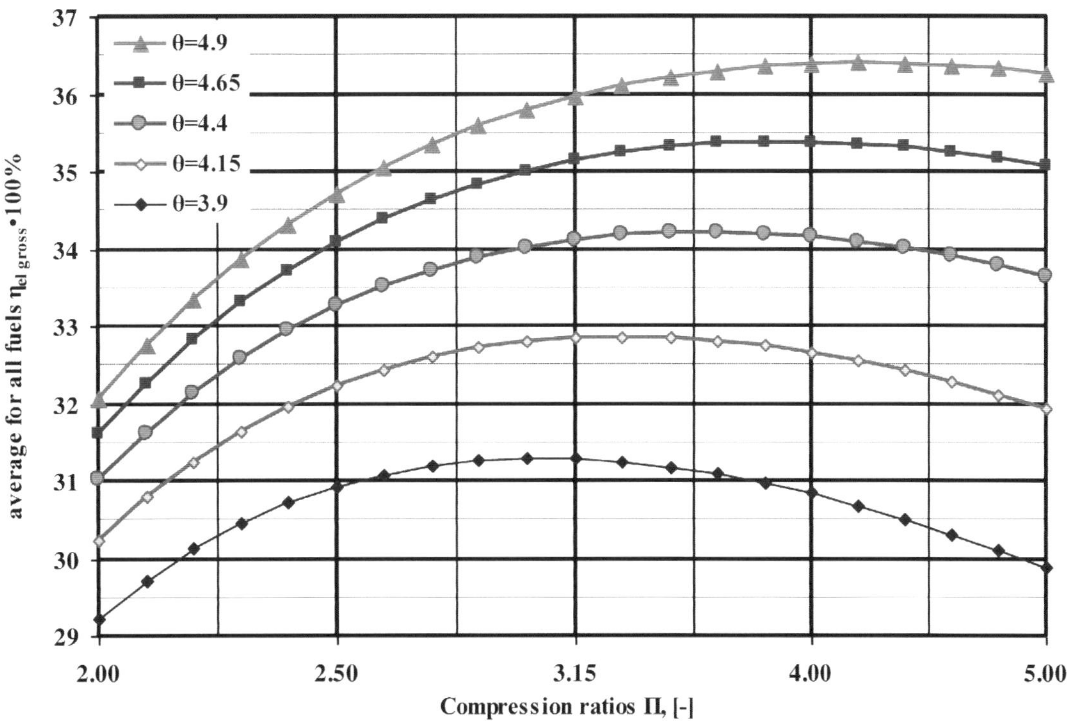

Figure 8. Average for all fuels $\eta_{el\ gross}$ vs. Π, for θ_{ratios}=3.9 to 4.9

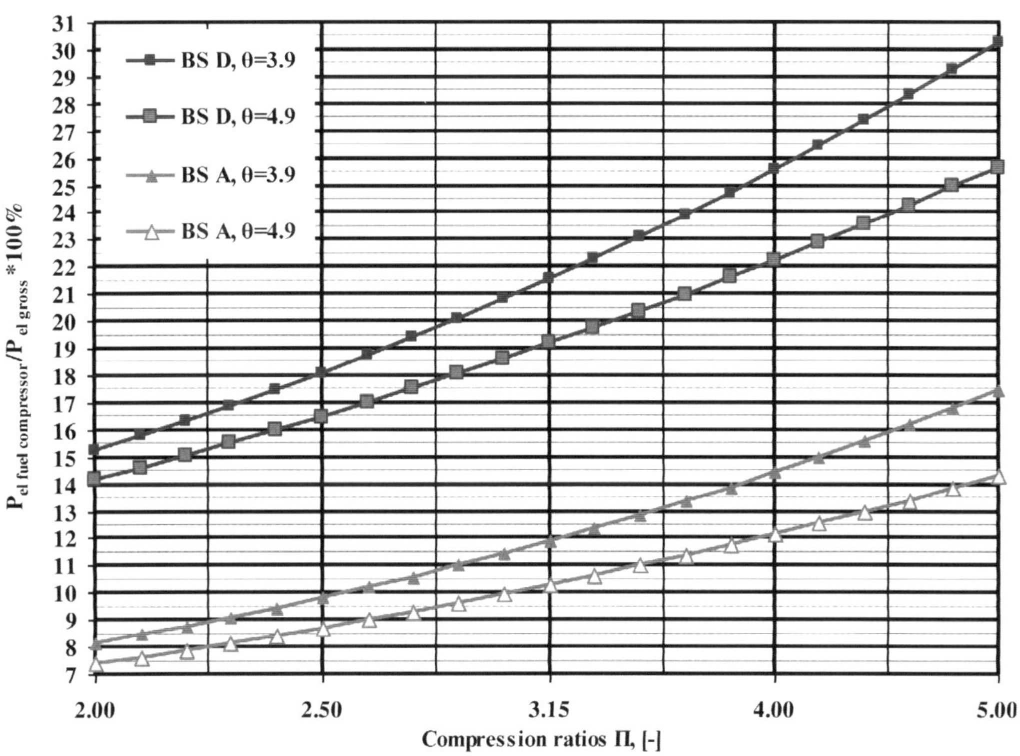

Figure 9. Variation of $P_{el\ fuel\ compressor}/P_{el\ gross}$, vs. Π, for BS "A" and "D", at θ =3.9 to 4.9

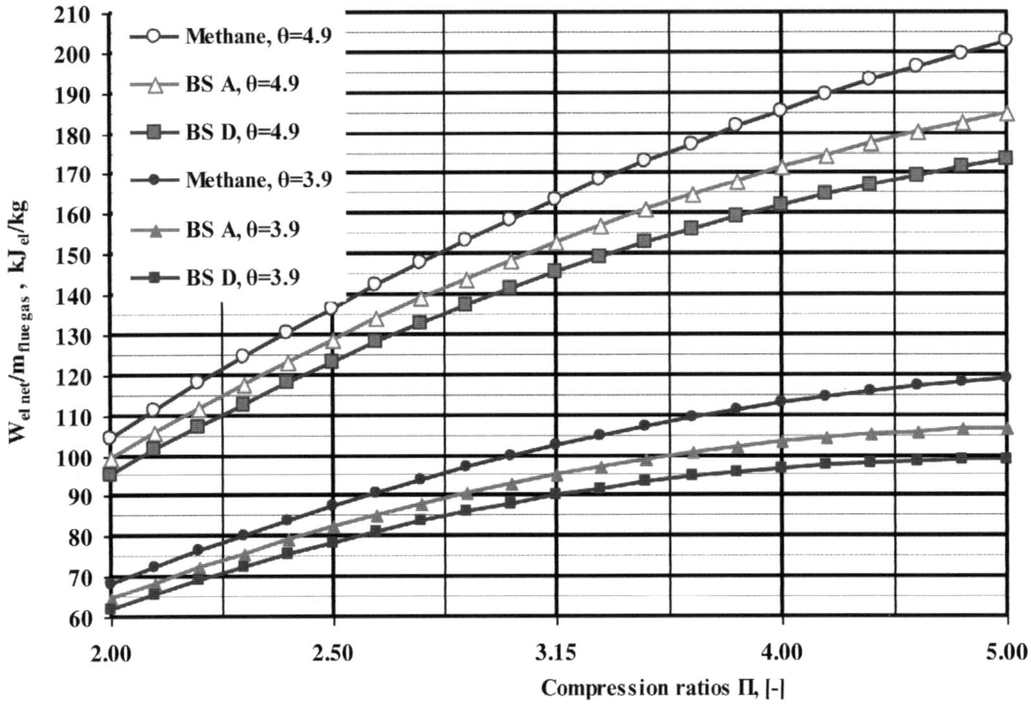

Figure 10. Variation of $W_{el\ net}/m_{flue\ gas}$ ratio, vs. Π, for methane, BS "A", and BS "D", at θ=3.9 to 4.9

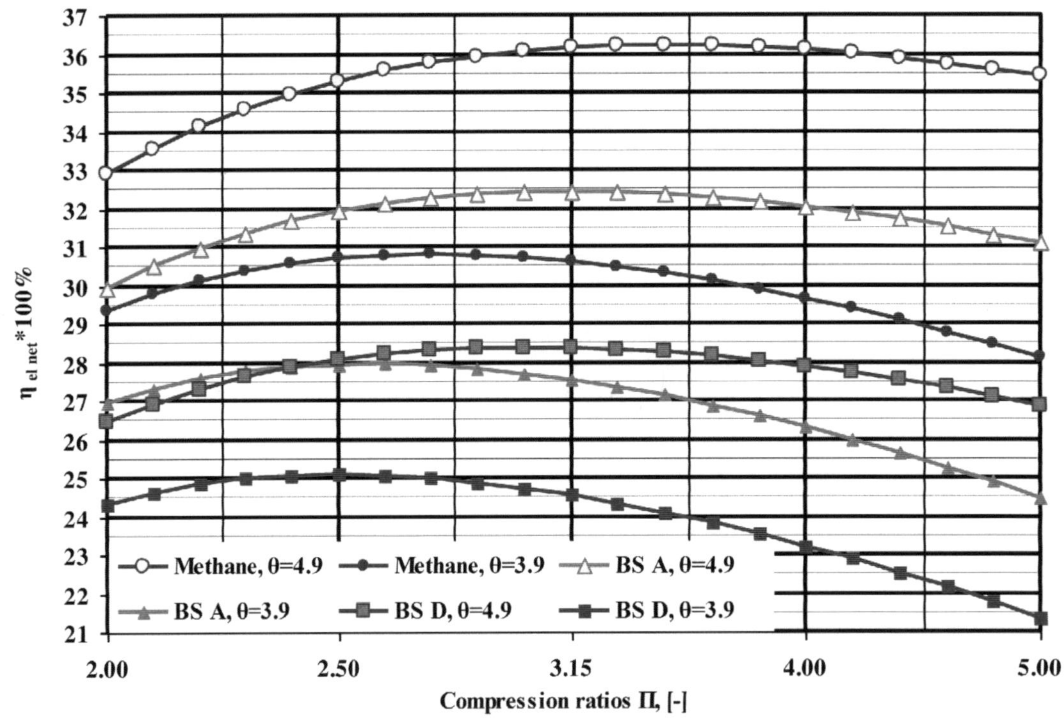

Figure 11. $\eta_{el\ net}$ vs. Π, for methane, BS "A", and BS "D" at different θ =3.9 to 4.9

When using standard gas fuel (methane) the values of fuel compressor power consumption relative to gross electric power ($P_{el\ fuel\ compressor}/P_{el\ gross}$) are in the range of 1.4 (for θ=4.9; Π=2) to 3.45 (for θ=3.9; Π=5). The power consumption varies directly with Π, and increases when fuel quality decreases. Figure 9 presents this parameter variation for syngas use highlighting the major influence of syngas quality on fuel compressor power consumption. For the BS A the gas compressor consumes starting from 7.4% to 8% (for Π=2) up to 14.2% to 17.4% (for Π=5) from the generated power. For the BS D these quota increase at 14% to 15.2% (for Π=2) up to 25.7% to 30.3% (for Π=5). That's why changing the fuel leads to a contrary effect on $W_{el\ net}/m_{flue\ gas}$ compared to $W_{el\ gross}/m_{flue\ gas}$. The $W_{el\ net}/m_{flue\ gas}$ is bigger when using CH_4 than syngas (Figure 10).

The higher flue gas compressor consumption when using syngas than for methane use has two effects (see Figure 11): 1) for the same heat flow rate input the net power and consequently the electrical net efficiencies decreases; 2) for the same θ, the values of Π that maximize $\eta_{el\ net}$ are smaller than the values of Π that maximize $\eta_{el\ gross}$.

4. Conclusions

The micro gas turbines can be modified to burn syngas obtained from air gasification of biomass and waste keeping the main equipment (gas turbine and air compressor). The paper focused on the thermodynamic aspects of the process. There are, of course, necessaries some changes for adapting the Micro GT to syngas use, to the burner of the combustion chamber and the fuel compressor. The results of the study revealed an important influence of fuels composition on micro GT performances. At lower HHV of gas fuel and constant heat flow rate in the combustion chamber, the gas flow rate increases to keep the temperature of flue gases constant. For the analyzed case (Π =4 and θ=4.5) the thermal power from fuel combustion increases in the range of 1.1% - 3.4 %. Even if the gross electric power increases with 0.9 to 2.9 %, the net electric output will decrease with about 0.15% - 0.35 % compared to methane use. The net energy efficiency will decrease with 0.9 to 3.75 % compared to the reference value (directory specifications of the unit).

The paper results can be used in design engineering for waste to derived fuels conversion chains using gasification conversion and power generation using micro gas turbines.

Acknowledgements
This work was supported by a grant of the Romanian National Authority for Scientific Research, CNDI – UEFISCDI, project number PN-II-PT-PCCA-2011-3.2- 1687 (62/2012).

References

[1] Mărculescu C 2012 Comparative analysis on waste to energy conversion chains using thermal-chemical processes *Energy Procedia* **18** 604–11
[2] Gupta K K, Rehman A and Sarviya R M 2010 Bio-fuels for the gas turbine: A review *Renew. Sustain. Energy Rev.* **14** 2946-55
[3] Cameretti M C and Tuccillo R 2015 Combustion features of a bio-fuelled micro-gas turbine *Appl. Therm. Eng.* **89** 280-90
[4] Arroyo J, Moreno F, Munoz M and Monne C 2013 Efficiency and emissions of a spark ignition engine fuelled with synthetic gases obtained from catalytic decomposition of biogas *Int. J. Hydrogen Energy* **38** 3784–92
[5] Arroyo J, Moreno F, Munoz M, Monne C, *et al.* 2014 Combustion behavior of a spark ignition engine fueled with synthetic gases derived from biogas *Fuel* **117** 50-58
[6] Delattin F, Lorenzo G D, Rizzo S, Bram S and Ruyck De J 2010 Combustion of syngas in a pressurized microturbine-like combustor: Experimental results *Appl. Eng.* **87** 1441-52
[7] Barsali S, Marco De A, Giglioli R, Ludovici G and Possenti A 2015 Dynamic modelling of biomass power plant using micro gas turbine *Renew. Energy* **80** 806-18
[8] Cavarzere A, Morini M, Pinelli M, Spina P R, Vaccari A and Venturini M 2014 Experimental Analysis of a Micro Gas Turbine Fuelled with Vegetable Oils from Energy Crops *Energy Procedia* **45** 91-100
[9] Mărculescu C, Cenuș ă V and Alexe F 2016 Analysis of biomass and waste gasification lean

syngases combustion for power generation using spark ignition engines *Waste Manag.* **47** 133-40

[10] Cascarosa E and Gasco L 2012 Meat and bone meal and coal co-gasification: Environmental advantages *Resources, Conservation and Recycling* **59** 32-37

[11] Loha C, Chatterjee P K and Chattopadhyay H 2011 Performance of fluidized bed steam gasification of biomass – Modeling and experiment *Energy Convers. Manage.* **52** 1583–88

[12] Neilson C 1998 LM 2500 gas turbine modifications for biomass fuel operation *Biomass and Bioenergy* **15** 269-73

[13] Romier A 2004 Small gas turbine technology *Appl. Therm. Eng.* **24** 1709-23

[14] Cenuşă V and Alexe F 2007 Comparative Analysis on Performances of Micro Gas Turbines Burning Biogas vs. Natural Gas *WSEAS Trans. Environ. Dev.* **3** 72-80

[15] Cenuşă V, Benelmir R and Merabtine A 2012 Reference Calculations for Gas Turbine Installations *International Journal of Energy, Environment and Economics* **20** 247-65

IOP Conf. Series: Earth and Environmental Science **40** (2016) 012037 doi:10.1088/1755-1315/40/1/012037

Wind farm micro-siting optimization using novel cell membrane approach

W Huang[1], W X Che[1,3], R S Tan[2], M X Li[1], Q Liu[1]

[1]State Key Laboratory of Power Systems on New Energy, North China Electric Power University, Beijing 102206, China
[2]China Energy Engineering Group Guangdong Electric Power Design Institute Co., Ltd., Guangzhou 510663, China

E-mail: Che0wenxue@163.com

Abstract. Micro-siting aims to determine every wind turbine's position to reduce velocity deficits caused by the wake effect. The Novel CMO (cell membrane optimization) approach is proposed to overcome this weakness. It plays a vital role to utilize more wind resources while the type of wind turbine and the area to build a wind farm have been determined. The work is based on the Jensen wake model, and the hypothetical situations are the same as those used by the former researchers. There are three wind cases: constant speed with one direction, constant speed with variable directions and variable speeds with variable directions. The area of wind farm is assumed to be a plane 2km×2km square. The numbers of the wind turbines is 26, 19 and 15 in three cases respectively. Compared with Gene Algorithm introduced by G. Mosetti, CMO's results are acceptable and the velocity deficit is smaller, which results from that CMO's variables is continuous and can make the most of the area the wind turbines can be placed. Moreover, it performs well to avoid the local optimal solutions by dividing the searching particles into different types which move according to different rules.

1. Introduction

The use of renewable energy has been paid attention to due to the energy crisis and environmental pollution in recent years. Power generation with wind energy, solar energy and biomass has been developing rapidly, of which the wind power technology is relatively mature[1]. Because wind resources of middle and western China are plentiful, but far away from load centers which located in the eastern, wind farms in China are often large-scale and the wind power is transported by long-distance transmission network. The layout optimization of wind turbines plays a significant role in reducing the wake deficit in large-scale wind farms. The wind turbines are arranged in accordance with the prevailing wind direction, and are kept a distance from each other to avoid wake deficit nowadays [2]. The wake effect means that the turbine absorbs the wind energy from the wind flows, and converted the energy to mechanical energy, the speed behind the wind turbine decreases, which would obviously affect the downstream wind turbine. But with the distance along the wind direction increasing, the wake effect dies away [3]. Therefore, micro-siting issue, that is to arrange each wind turbine reasonably in the wind farm, is especially important in large-scale wind farm because of the

[3]Address for correspondence: W X Che, State Key Laboratory of Power Systems on New Energy, North China Electric Power University, Beijing 102206, China. E-mail: Che0wenxue@163.com.

large number of wind turbines. Although reasonable arrangement cannot improve the power generation efficiency of every wind turbine, the wake effect of the whole wind farm is reduced by making wind turbine capture more wind energy relatively.

N. O. Jensen proposed a simple wake model which could be used to optimize the wind turbine layout in 1983[4], furthermore the model was developed by Katic in the year of 1987 [5], and the model was accepted by the researchers and even the commercial software. Before G. Mosetti [6] brought the intelligence method (Gene Algorithm, GA) in solving the micro-siting problem, researchers used classic method to solve this complex problem, which would take much time. Many intelligent approaches are introduced to solve the problem after GA, including developed GA [7], particle filtering approach [8], cyber swarm algorithm [9] and so on, and most of them used discrete variables to represent wind turbines' locations. In parallel, intelligent approaches are easy to trap in local optima. This paper uses a novel approach which uses continuous variables and has less chance to get the local optima, which is compared with other algorithm and the results are acceptable.

2. Micro-siting models

The micro-siting problem is to arrange the layout of the wind turbines to minimize the wake effect of the whole wind farm, and is based on the following assumptions:

- The rated capacity is known because it is determined when the project gets the government approval, and so is the wind turbine type. In this way, the number and the power curve of the wind turbines are determined.
- The wind farm's scale and the area which the wind turbine can be arranged is known.
- The condition of wind resources is known, including direction, speed and their frequency.

This paper uses the same model as the one in the paper which G. Mosetti wrote. The model is shown as follows, which is divided into three parts, the wake model, objective function and the cases of wind. G. Mosetti divided the wind farm into small squares which wind turbines can be situated, and used Genetic Algorithm to solve the problem. The squares which wind turbines located were represented by '1' and the others were '0'. The original numbers were chosen at random, and then new numbers generated by the old ones with crossovers and mutations, the best solution could be found after iterations.

2.1. Jensen wake model

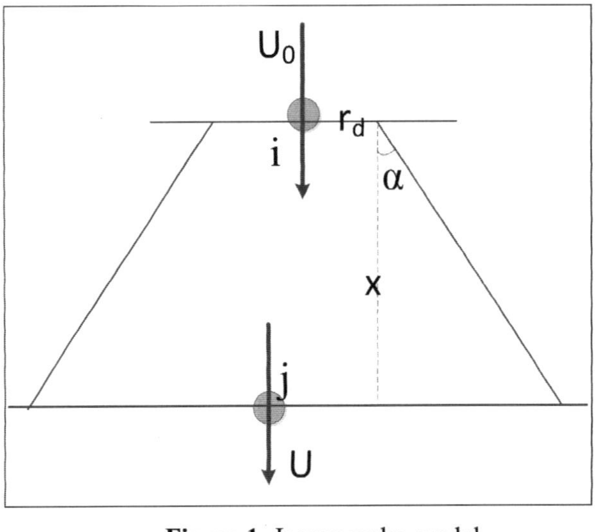

Figure 1. Jensen wake model.

As shown in Figure 1, $\mathbf{u_0}$ represents the free wind speed without wake effect, $\mathbf{u}$ represents the wind speed affected by wind turbine **i** at the distance of **x** along the wind direction, r_d is the rotor radius

$$u = u_0 \left[1 - 2a \times \left(1 + \alpha \frac{x}{r_d} \right)^{-2} \right] \tag{1}$$

where α represents entrainment constant, a is axial induction factor.

Entrainment constant α can also be expressed as Equation (2):

$$a = 0.5 \times (1 - (1 - c_t)^{1/2}) \tag{2}$$

where c_t is rotor thrust coefficient.

Axial induction factor a can be expressed as Equation (3):

$$\alpha = 0.5 \times (\ln \frac{z}{z_0})^{-1} \tag{3}$$

where z represents the hub height of the wind turbine and z_0 represents the surface roughness, in this paper it is 0.3m.

2.2. Objective function

The parameters of the wind turbine are shown by Table 1 below.

Table 1. Parameters of the wind turbine.

parameter	value
Hub height(m)	60
Roter diameter(m)	40
c_t	0.88

The power curve can be expressed as Equation (4):

$$P = 0.5 \times C_p \times \rho \times \pi \times R^2 \times u^3 \tag{4}$$

Assuming the value of ρ is 1.2kg/m^3, the power coefficient C_p is 40%, and R is 20m,and the unit of P is changed from watt to kilowatt. Equation (4) can be simplified as follows:

$$P = 0.001 \times 0.5 \times 0.4 \times 1.2 \times 3.14 \times 20^2 \times u^3 \approx 0.3 \times u^3 \tag{5}$$

And the power the whole wind farm generates is

$$P_{tot} = \sum_{i=1}^{N} 0.3 \times u_i^3 \tag{6}$$

where **N** is the number of the wind turbines in the wind farm.

G. Mosetti supposed that if the number of the wind turbine is large enough the cost of the wind farm can be reduced by one third.

$$Cost = N \times (\frac{2}{3} + \frac{1}{3} e^{-0.00174 N^2}) \tag{7}$$

where **N** is the number of the wind turbines in the wind farm.

And the objective function is:

$$Obj = \omega_1 \times P_{tot}^{-1} + \omega_2 \times Cost \times P_{tot}^{-1} \tag{8}$$

where ω_1 and ω_2 are weights and ω_1 is kept small relatively, the ω_2/ω_1 value is 9 in this paper .

2.3. Cases of wind

- Case A: The wind speed is 12m/s and the direction is constant.
- Case B: The wind speed is 12m/s, the direction is divided into 36 parts and every direction is of equal frequency.
- Case C: There are three speeds, which are 8m/s, 12m/s and 17m/s, and still 36 directions. But their frequency is different, which can be seen from Figure 2.

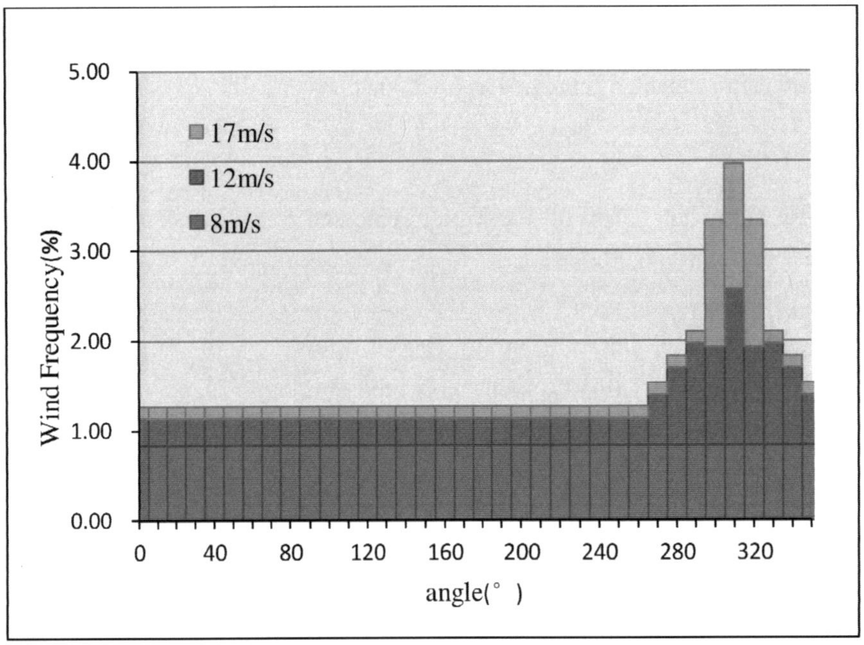

Figure 2. Frequency distribution of wind angles and speeds.

3. CMO algorithm based micro-siting

Micro-siting is a multi-dimensional nonlinear optimization problem and is difficult to solve with traditional optimization methods. Since genetic algorithm was used to solve this problem by in 1993, heuristic algorithms have been paid more attention to. But heuristic algorithm is easy to fall into local optima, so this paper introduces a new optimization algorithm - membrane optimization algorithm to avoid this problem. Based on the researches of DNA molecule computing, Cell Membrane Computing was proposed Gheorghe Paun in the year of 1998 [10]. Then a number of scholars established different membrane computing models according to different rules. In the year of 2007, membrane computing and optimization methods are combined for the first time by Huang Liang, and was called Cell Membrane Optimization [11]. And then Tanshi Heng proposed a simple CMO according to the material transport inside and outside the cell membrane [12], which is used in this paper.

3.1. Basic steps of the CMO algorithm

3.1.1. Step 1: Substance initialization

Randomly generate m n-dimensional materials in the solution space, with each material randomly distributed in the space. Calculate their function values, and place the substance whose function value is the smallest in X^{best}.

3.1.2. Step 2: Classify the substances

Calculate the concentration of each position where the substance stays based on Equation (9). Sort the substances based on the order of their concentration, from small to large, and sort the top 50% of the substances into the low-density material (LS). Consider the others of higher concentration as the high-density material (HS). Divide HS into FS and NS. The substance in HS whose number is odd is classified as fat-soluble, and the even number is classified as non-fat-soluble.

$$con = \frac{s}{m} \tag{9}$$

where s represents the number of the substances whose distance from the set substance is smaller than a set length; m is the number of all the substances.

3.1.3. Step 3: Free diffusion of fat-soluble HS

Each fat-soluble HS diffuses toward the direction of the LS group lcon times. Choose the best location as the final result of diffusion. The diffusion method is as follows: choose any substance from the LS group to determine the direction vector of the movement, and unitize the vector.

$$\vec{F} = (LS^{randi} - FS^{i})\times(\left\|LS^{randi} - FS^{i}\right\|)^{-1} = (LS^{randi} - FS^{i})\times(\sum_{k=1}^{n}(LS_k^{randi} - FS_k^{i})^2)^{-\frac{1}{2}} \tag{10}$$

The fat-soluble HS then moves to the LS group based on this vector to obtain a new material.

$$newFS_k^i = \begin{cases} FS_k^i + rand()*F_k*(u_k - LS_k^{randi}) & F_k \geq 0 \\ FS_k^i + rand()*F_k*(LS_k^{randi} - l_k) & F_k < 0 \end{cases}$$
$$k = 1,L\ ,n \tag{11}$$

3.1.4. Step 4: Movement of non-fat-soluble HS

Non-fat-soluble HS cannot freely diffuse the same as the fat-soluble FS because it requires a carrier when it moves to the LS group. This process is known as facilitated diffusion. Non-fat-soluble HS moves toward the global optimum material when no carrier is required.

The movement of non-fat-soluble HS can be described into three steps:

- Distinguish whether a carrier exists.
- The non-fat-soluble FS that has a carrier moves in support of the carrier, just as free diffusion.

$$\vec{F} = (LS^{randi} - NS^{i})\times(\left\|LS^{randi} - NS^{i}\right\|)^{-1} = (LS^{randi} - NS^{i})\times(\sum_{k=1}^{n}(LS_k^{randi} - NS_k^{i})^2)^{-\frac{1}{2}} \tag{12}$$

$$newNS_k^i = \begin{cases} NS_k^i + rand()*F_k*(u_k - NS_k^{randi}) & F_k \geq 0 \\ NS_k^i + rand()*F_k*(NS_k^{randi} - l_k) & F_k < 0 \end{cases}$$
$$k = 1,L\ ,n \tag{13}$$

- The non-fat-soluble FS without a carrier moves toward the global optimum material.

$$newNS_k^i = NS_k^i + rand()*(X_k^{best} - NS_k^i) \quad k = 1,L\ ,n \tag{14}$$

3.1.5. Step 5: Movement of LS

The movement of LS requires a carrier and energy so that the two judgments should be conducted.

The LS that cannot meet the conditions of energy moves in the searching area at random lcon times, and the best position is chosen as the final result.

$$newLS_k^i = l_k + rand()*(u_k - l_k) \quad k = 1,L\ ,n \tag{15}$$

The LS that meets the energy condition but not the carrier condition moves toward the current optimum material.

$$newLS_k^i = LS_k^i + rand()*(X_k^{best} - LS_k^i)$$
$$k = 1,L\ ,n \tag{16}$$

The LS that has both energy and carrier can move toward the HS group.

$$\overset{V}{F} = (HS^{randi} - LS^{i}) \times (\left\| HS^{randi} - LS^{i} \right\|)^{-1} = (HS^{randi} - LS^{i}) \times (\sum_{k=1}^{n} (HS_{k}^{randi} - LS_{k}^{i})^{2})^{-\frac{1}{2}} \quad (17)$$

$$newLS_{k}^{i} = \begin{cases} LS_{k}^{i} + rand()*F_{k}*(u_{k} - LS_{k}^{randi}) & F_{k} \geq 0 \\ LS_{k}^{i} + rand()*F_{k}*(LS_{k}^{randi} - l_{k}) & F_{k} < 0 \end{cases} \quad k = 1, \ldots, n \quad (18)$$

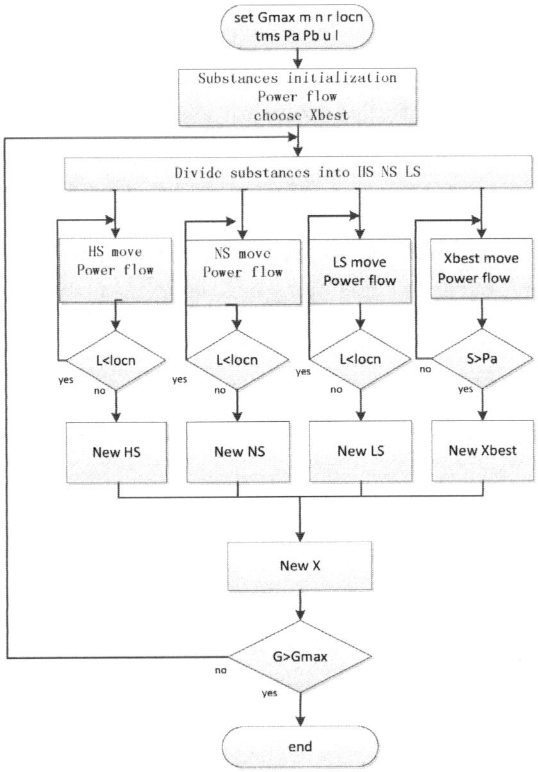

Figure 3. Flow diagram of CMO algorithm.

3.1.6. Step 6: Movements of the current optimal material

A better-quality material always exists in the vicinity of a good-quality material. Thus, searching around the current optimum substance to improve the accuracy of the material is necessary. This study sets the current optimum material as the centre and in a radius that gradually shrinks to search for a better substance and then uses it to replace the current best material.

3.1.7. Step 7: Update substance

Use new X^{best}, new fat-soluble FS, new non-fat-soluble HS, and new LS to replace the old ones relatively after all the movements, and retain the old X^{best} to replace any one of the new substances.

3.2. Optimization steps

Assuming that the land is 2km×2km, Figures. 4-10 describe the 2km×2km square land, the little squares and circles in the land are wind turbines respectively. And the target is to optimize the layout of the wind farm in the three cases which was mentioned previously.

- Setting the values of X-axis to be the first half of the materials, and the other are the value of Y-axis. For example, the wind turbines' number is 30, the dimension n should be 60, and 1~30 are X-axis values of the 30 wind turbines and 31~60 are Y-axis values.
- Finding the wind turbines whose wake affect its speed for each wind turbine, and calculate the actual speed.
- Calculating the power generated by each wind turbine and whole wind farm.

- Calculating the objective function value. Move the materials towards the best one. For the one which cannot meet the distance need, set its function value to be a large enough number to delete it.

4. Results and Discussion

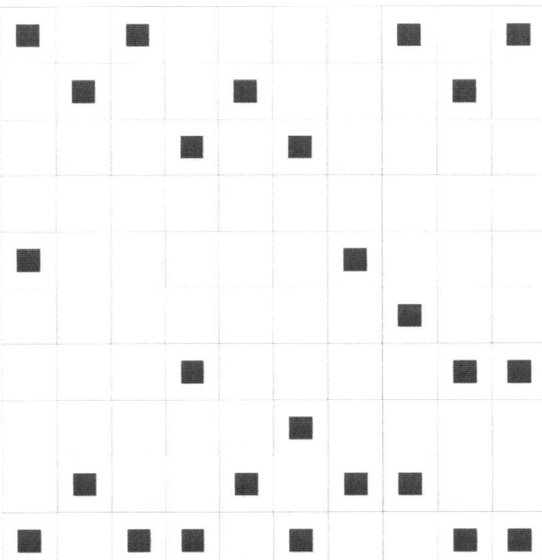

Figure 4. Layout optimized by GA in case A.

Figure 5. Layout optimized by CMO in case A.

In case A, the wind blows from one direction, which is from the upper part to the lower half of figure 3and figure 4. In this case, the wind turbines' number is set to be 26, so the dimension of the variable is 52, and the optimal layout obtained by CMO is shown in figure 4 and the one obtained by GA method in G. Mosetti's paper is shown in figure 3. The efficiency of the CMO is 92.4%, which increases in this paper compared with 91.645 % in G. Mosetti's paper and 92.015% [6] in Grady's paper [7]. The reason why the CMO gets better results may be that the wind turbines cannot be placed arbitrarily but can only at the centre of the divided squares with GA method. The wind turbines can be placed closer to the boundary and the land can be made better use of in this way.

As the wind turbine's wake will affect the downstream wind turbine, the distance along the wind direction is longer relatively, that is the vertical distance is longer than the horizontal distance of the wind turbines in the picture. And the two pictures are similar in this way, and this is easy to understand because the only direction of the wind is blowing top to bottom in the picture.

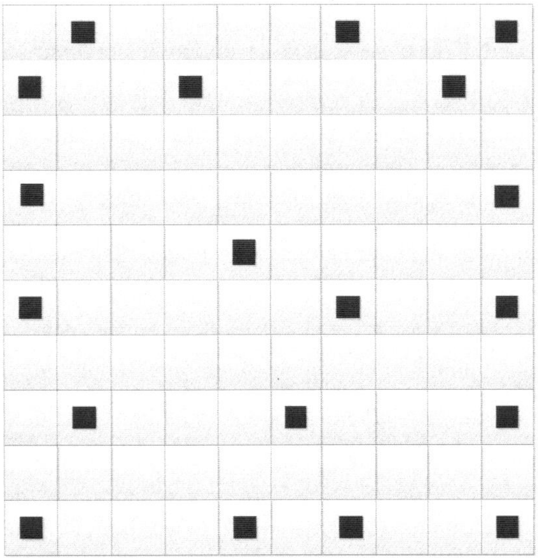

Figure 6. Layout optimized by GA in case B.

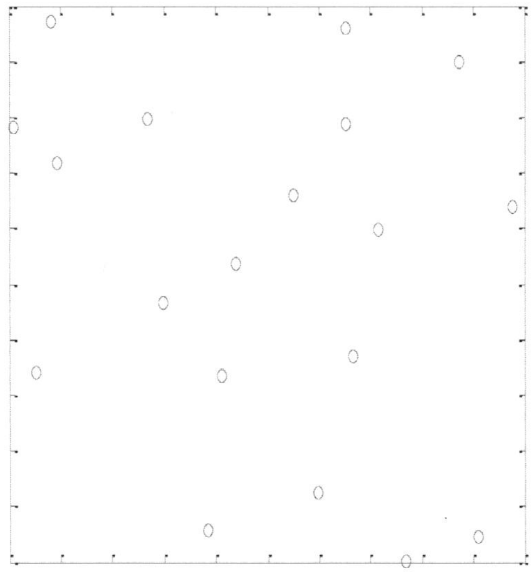

Figure 7. Layout optimized by CMO in case B.

In case B, the number of wind turbines is 19, the wind of 12m/s blows from 36 directions, therefore, the wind turbines distribute along the edge of the land as closer as possible to make the distance longer from each other, which can be seen from figure 5 and figure 6. Furthermore, the frequency of every direction is equal so the wind turbines distribute around evenly. And the efficiency of the layout shown in figure 5 and figure 6 is 93.859% [6] and 95.15%, which is improved obviously on the condition that the numbers of wind turbines are equal in both of the two layouts.

As this paper tries to justify the effectiveness of the CMO method, so the number of the wind turbines isn't set to be a variable to be optimized. The objective function values are different when the weights ω_1 and ω_2 changes, so this paper doesn't compare the objective function values in the paper. The generations of the CMO is different as we can see in the explanation of it, in every generation each kind of materials moves many times, so the generations may less than ten times, but the number of the iterations may be a hundred or even thousand. That's because the materials are divided into different kinds and moves according to different rules, which is the reason why the CMO has little chance to trap in the local optima.

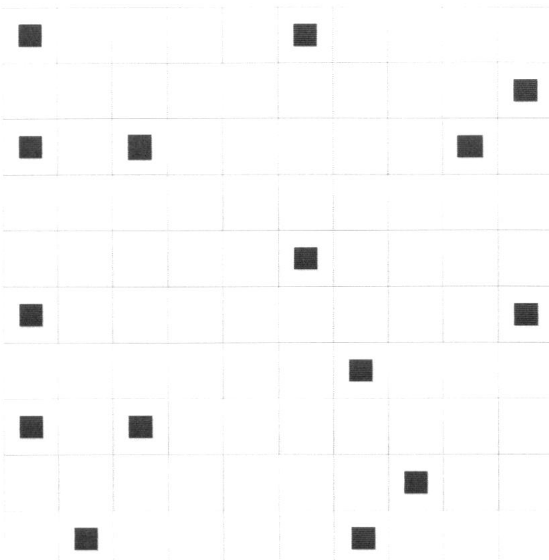

Figure 8. Layout optimized by GA in case C.

Figure 9. Layout optimized by CMO in case C.

In case C, the wind condition is more complicated and the number of wind turbines is only 15. There are 36 directions as in case B, the wind speeds are 8m/s, 12m/s and 17m/s and the frequency of different directions and speed may vary in a large range. As in case B, the wind turbines scatter in the area. But as the frequency of the wind around 300° is larger than the others, so the distance along this direction is longer. And the efficiency of GA conducted by G. Mosetti is 94.62% [6] and CMO's is 97.12%. But the improvement of the efficiency may be owe to the number's decrease of the wind turbines compared to case B. Set the number to be 20 and the optimal layout is shown in Figure 9 and the efficiency is reduced to 94.48%. So the number of the wind farm is still needed to be optimized in the future, the wake effect increases as the number increases, but the rated power of the wind farm also increases.

Figure 10. Layout of 20 wind turbines optimized by CMO in case C.

5. Conclusions

- The novel optimization method, which is called CMO method, can be used to solve the micro-siting problem. This paper carries out the progress to optimize a wind farm of 2km×2km in three cases, and compare the results with other methods. The reason why the efficiency of the wind farm is improved can be that the discrete values of the positions are converted into continuous ones, which makes better use of the land.
- The CMO method has little chance to get the local minima by dividing the searching area into two parts and the material can move from one area to another.
- It turns out that this method can be applied into micro-siting problem, but it needs to be applied in more complex situations in which the number of wind turbine is larger and the model may be more complicated.

Acknowledgments
This work was funded by China Energy Engineering Group Guangdong Electric Power Design Institute Co., Ltd. with project number EV02861W.

References

[1] Wang Z, Jiang C, Q Ai amd Wang C 2009 The key technology of offshore wind farm and its new development in China *Renewable & Sustainable Energy Reviews* **13** 216-22
[2] Gao X, Yang H and Lu L 2014 Investigation into the optimal wind turbine layout patterns for a Hong Kong offshore wind farm *Energy* **73** 430-42
[3] Wan C, Wang J, Yang G, Gu H and Zhang X 2012 Wind farm micro-siting by Gaussian particle swarm optimization with local search strategy *Renewable Energy* **48** 276-86
[4] Jensen N O 1983 *A Note on Wind Generator Interaction* (Roskilde, Denmark: Risø National Laboratory for Sustainable Energy, Technical University of Denmark)
[5] Katic I, Jensen N O 1987 *A simple model for cluster efficiency* (Rome: EWEC'86. Proceedings) 407-10
[6] Mosetti G, Poloni C and Diviacco B 1994 Optimization of wind turbine positioning in large windfarms by means of a genetic algorithm *J. Wind Engineering and Industrial Aerodynamics* **51** 105-16
[7] Grady S A, Hussaini M Y and Abdullah M M 2005 Placement of wind turbines using genetic algorithms *Renewable Energy* **30** 259-70
[8] Eroglu Y and Seckiner S U 2013 Wind farm layout optimization using particle filtering approach *Renewable Energy* **58** 95-107
[9] Yin P Y, Wu T H and Hsu P Y 2016 A power-deficiency and risk-management model for wind farm micro-siting using cyber swarm algorithm *Applied Mathematical Modelling* **40** 2177-89
[10] Paun G 1998 Turku center for computer science-tucs report
[11] Liang H 2007 Research on Membrane Computing Optimization Methods (Hangzhou: Zhejiang University)
[12] Shiheng T 2011 A New Swarm Intellingent Optimization Algortithm: Cell Membrance Optimization and Its Applications (Guangzhou: South China University of Technology)

IOP Conf. Series: Earth and Environmental Science **40** (2016) 012038 doi:10.1088/1755-1315/40/1/012038

Transmission line icing prediction based on DWT feature extraction

T N Ma, D X Niu and Y L Huang[1]

Research Institute of Technology Economics Forecasting and Assessment, North China Electric Power University, Beijing 102206, China

E-mail: huangyali5210@163.com

Abstract. Transmission line icing prediction is the premise of ensuring the safe operation of the network as well as the very important basis for the prevention of freezing disasters. In order to improve the prediction accuracy of icing, a transmission line icing prediction model based on discrete wavelet transform (DWT) feature extraction was built. In this method, a group of high and low frequency signals were obtained by DWT decomposition, and were fitted and predicted by using partial least squares regression model (PLS) and wavelet least square support vector model (w-LSSVM). Finally, the final result of the icing prediction was obtained by adding the predicted values of the high and low frequency signals. The results showed that the method is effective and feasible in the prediction of transmission line icing.

1. Introduction

In recent years, frequent occurrences of all kinds of extreme weather caused by global climate abnormal variation have resulted in many transmission line icing disasters, which have caused power outages, leading to the destruction of a large number of fixed assets and huge economic losses [1-4]. Especially the freezing disaster happened in 2008 in southern China, caused great damage to the power system. It led directly to the collapse of 8709 base of 110 kV and above lines, the break of more than 2.07 million lines, the outage of 1497 substations, resulting in direct property loss of the State Grid Corporation of 10.45 billion and reconstruction funds of 390 million [5]. Through the lessons learned from the previous power grid icing disasters and the current grid development situation and global cli-mate changes, the power grid in the future will still withstand the tests of freezing disaster. So we must discover the disaster in advance and prevent it as soon as possible. Therefore, the research on trans-mission line icing prediction has important practical application value.

At present, the research on transmission line icing prediction mainly according to the movement laws of fluids and heat transfer mechanism of transmission icing, focuses on establishing all kinds of icing prediction models combined with meteorological factors, environmental factors and line parameters. In general, the models can be divided into three categories: mathematical-physical models, statistical models and intelligent prediction models. Mathematical-physical models are built on the icing growth process by observing the physical process of ice formation to make prediction, including the Goodwin model [6] and the Makkonen model [7]. These models are put forward on the basis of experiments. However, their prediction accuracy is not ideal. Statistical models are based on the

[1] Address for correspondence: Research Institute of Technology Economics Forecasting and Assessment, North China Electric Power University, Beijing 102206, China. Email: huangyali5210@163.com.

statistical theory and traditional statistical methods to process the historical data, not considering the physical process of the icing formation, such as the icing growth rate - temperature model, the icing growth rate - wind speed model, the icing growth rate - wind direction model, the extreme value model and the multiple linear regression model [8]. Statistical models have simple calculation and easy realization, but can't fit well on the nonlinear characteristics of icing. When the samples are complex, the prediction precision of statistical models will be seriously affected. Intelligent prediction models are established combined with modern computer technology and Mathematical Sciences. This kind of models can deal with high-dimensional nonlinear problems well and have high accuracy of prediction due to their powerful studying and generalizing ability. In the research of icing prediction, the most common intelligent models include support vector machine (SVM) model [9], BP neural network (BPNN) model [10], fuzzy theoretical model [11] and so on. The intelligent prediction models have achieved good effect in icing forecasting, but always have the phenomenon of over-fitting, slow convergence speed and the local optimum because of the defects of the model itself. Besides, they can't explain the nonlinear features of icing effectively, thus their prediction accuracy is affected.

For the sake of those problems, a method for icing prediction based on the discrete wavelet transform (DWT) feature extraction was proposed in this paper. DWT model can make infinite de-composition of the input signal to get smooth low-frequency signals and large wave of high frequency signals respectively. According to different frequency signals, the partial least squares regression model (PLS) and wavelet least squares support vector machine (w-LSSVM) were adopted to learn and fit them, so as to make accurate prediction. The method has the advantages of both the PLS model and the w-LSSVM model, which improves the ac-curacy of icing prediction. The simulation results show that the method proposed in this paper not only has higher prediction accuracy, but also has better prediction performance than other single algorithms and is more suitable for icing prediction in transmission line.

2. Feature extraction and prediction method

2.1. Feature extraction based on DWT

Discrete wavelet transform (DWT) is always used in computer image processing because it can decompose the image signals in different scale according to the different goals. The wavelet function usually can be expressed as follow [12].

$$\Psi_{a,t}(t) = a^{-\frac{1}{2}}\psi\left(\frac{t-\tau}{a}\right) a > 0, \tau \in R \tag{1}$$

Where a is the wavelet scale stretch factor, τ is the wavelet shift factor. Generally, wavelet discretization aims at the parameters a, τ . Let $a = 2^j$, $\tau = 2^j \cdot T_s \cdot n$, the wavelet base function is here.

$$\Psi_{j,n}(t) = 2^{-\frac{j}{2}}\psi\left(\frac{t}{2^j} - n \cdot T_s\right) \quad j, n \in N \tag{2}$$

For any function $f(t)$, the discrete wavelet coefficient is as follow.

$$WT_f(j,n) = \int_R f(t)\overline{\Psi_{j,n}(t)}dt \tag{3}$$

Where, $\overline{\Psi_{j,n}(t)}$ is the dual function of $\Psi_{j,n}(t)$. The discrete wavelet coefficient WT_f can be used to build the function $f(t)$.

$$f(t) = \frac{1}{A}\sum WT_f(j,n) \cdot \psi_{j,n}(t) \tag{4}$$

Where, A is constant. The discrete wavelet transformation is shown in figure 1.

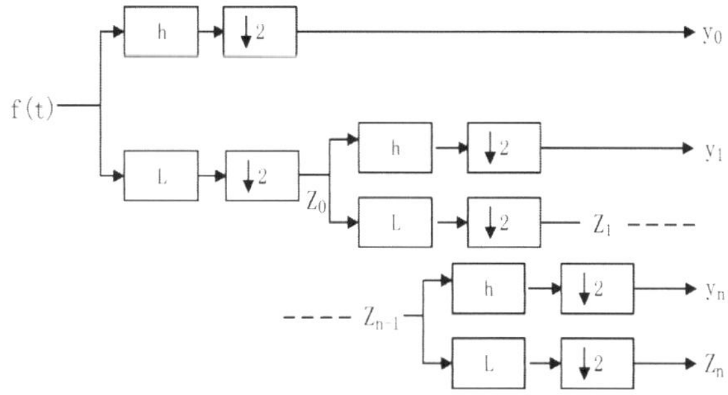

Figure 1. Discrete wavelet transformation.

The process of discrete wavelet transform can be represented as a decomposition tree composed of low frequency signals and high frequency signals. And the process of signal decomposition is iteration process. The signals will be decomposed into multi-level wavelet to obtain the lower frequency signals and higher frequency signals. In the practical application, the decomposition frequency is determined by the coefficient WT_f.

In icing forecasting, a group of high frequency signals ($y_0 \rightarrow y_n$) and a group of low frequency signals (Z_n) will be obtained during the discrete wavelet transform on the icing historical data. The low frequency signals are steady and smooth relatively, which are good for making linear regression according to the regression model. On the contrary, the high frequency signals generally show high nonlinear characteristics, which can be for learning and fitting according to the intelligent prediction model.

2.2. *Icing prediction based on PLS*

The low frequency signal curve obtained by DWT is relatively smooth and stable, so can be fitted by adopting the partial least squares linear regression model (PLS) [14] to make accurate prediction.

Assuming the meteorological factors, x_j, $j = 1,2,\cdots,m$, are independent variables and the icing thickness, y, is the dependent variable. And suppose that there are n samples.

The first step is to standardize the independent variables by using the following formulas:

$$\begin{cases} \boldsymbol{Y} = (\boldsymbol{Y_0})_n \\ \boldsymbol{Y_0} = [\boldsymbol{y} - \boldsymbol{E}(\boldsymbol{y})]/\boldsymbol{S}_y \end{cases} \tag{5}$$

$$\begin{cases} \boldsymbol{X} = (\boldsymbol{X_0})_{n\times m} \\ (\boldsymbol{X_0})_j = [\boldsymbol{x}_j - \boldsymbol{E}(\boldsymbol{x}_j)]/\boldsymbol{S}_{x_j} \end{cases} \tag{6}$$

Where, $\boldsymbol{X}$ is the standard matrix of the meteorological factors while $\boldsymbol{Y}$ is the standard matrix of the icing thickness. $E(y)$ is the mean value of y while $E(x_j)$ is the mean value of x. S_y and S_x are both mean square errors.

The second step is to extract the first principal component t_1 from the matrix X : $t_1 = XW_1$, $\|W_1\| = 1$. And extract the first principal component u_1 from the matrix Y : $u_1 = YC_1$, $\|C_1\| = 1$.

The regression equations are showed as follow.

$$X = t_1 P_1^T + X_1 \tag{7}$$

$$Y = t_1 r_1 + Y_1 \tag{8}$$

Where, $r(x_i, y)$ is the correlation coefficient of x_i and y. $P_1 = X^T t_1 / \|t_1\|^2$, $r_1 = Y^T t_1 / \|t_1\|^2$. X_1 and Y_1 are residual matrixes of the regression equation.

The third step is to get the second principal component variable t_2 using the same method as above and replacing X and Y with X_1 and Y_1. Then the following equations are got.

$$X_1 = t_2 P_2^T + X_2 \tag{9}$$

$$Y_1 = t_2 r_2 + Y_2 \tag{10}$$

Cycle the process above until A principal components, $t_1, t_2, \cdots, t_A$, are extracted. Then the order is determined by the cross validation test.

$$Q_A^2 = 1 - \frac{PRESS_A}{SS_{A-1}} \tag{11}$$

$$PRESS_A = \sum_{i=1}^{n} (y_i - \hat{y}_{A(-i)})^2 \tag{12}$$

$$SS_A = \sum_{i=1}^{n} (y_i - \hat{y}_{Ai})^2 \tag{13}$$

Generally, when $Q_A^2 < 0.0975$, the prediction ability of the model has no obvious improvement, so the algorithm stops and the order of the composite variables is A.

The fourth step is the multiple linear regression for Y to $(t_1, t_2, \cdots, t_A)$ and the following formula is got.

$$\hat{Y} = r_1 t_1 + r_2 t_2 + \cdots + r_A t_A \tag{14}$$

Because $t_1, t_2, \cdots, t_A$ are the linear combinations of X, the following formulas are obtained.

$$t_i = XW_i^* \tag{15}$$

$$W_i^* = \prod_{k=1}^{i-1} (I - W_k P_k^T) W_i \tag{16}$$

Then from the formula above, the formula got is here.

$$\hat{Y} = r_1 X W_1^* + r_2 X_2^* + \cdots + r_A X_A^* \tag{17}$$

Suppose that,

$$y^* = Y \tag{18}$$

$$x_i^* = X_i \tag{19}$$

$$\alpha_i = \sum_{k=1}^{A} r_k W_{ki}^* (i = 1, 2, \cdots, m) \tag{20}$$

The formula above is changed.

$$\hat{y}^* = \alpha_1 x_1^* + \alpha_2 x_2^* + \cdots + \alpha_m x_m^* \tag{21}$$

The formula also can be written as follow further.

$$\hat{y} = \left[E(y) - \sum_{i=1}^{m} \alpha_i \frac{S_y}{S_{x_i}} E(x_i) \right] + \alpha_1 \frac{S_y}{S_{x_1}} x_1 + \cdots + a_m \frac{S_y}{S_{x_m}} x_m \tag{22}$$

In summary, the prediction results $\hat{y}$ can be calculated by the least square regression equation according to the meteorological data and the raw icing thickness data.

2.3. *Icing prediction based on w-LSSVM*

The high frequency signals got by DWT have strong nonlinear characteristics, which makes it hard to fit the data accurately using the common regression methods. Therefore, the least squares wavelet support vector machine (w-LSSVM) regression model was used to train and learn to make an accurate prediction for the icing thickness on the basis of establishing the relationship between meteorological factors and icing thickness in this paper.

LSSVM is a extension of SVM [13]. It cast the input vectors into the high dimension space by nonlinear mapping and constructed the optimal decision surface. Then the inequality operation of SVM was transformed to the operation of solving equations by applying the risk minimization principle, which reduced the computational complexity and accelerated the operation speed.

Set a given sample set $T = \{(x_i, y_i)\}_{i=1}^{N}$, the number of samples N. The regression model of the sample is here.

$$y(x) = w^T \bullet \varphi(x) + b \tag{23}$$

Where, $\varphi(*)$ is a high dimensional space that the training samples are projected onto. w is the weight vector, b is the biased.

For LSSVM, the optimization problem was showed as follow.

$$min \frac{1}{2} w^T w + \frac{1}{2} \gamma \sum_{i=1}^{N} \xi_i^2 \tag{24}$$

$$s \cdot t \quad y_i = w^T \varphi(x_i) + b + \xi_i, i = 1,2,3,\cdots N; \tag{25}$$

To solve the problem above, the Lagrangian function was built.

$$L(w,b,\xi_i,\alpha_i) = \frac{1}{2} w^T w + \frac{1}{2} \gamma \sum_{i=1}^{N} \xi_i^2 - \sum_{i=1}^{N} \alpha_i [w^T \varphi(x_i) + b + \xi_i - y_i] \tag{26}$$

Where, α_i is the Lagrange multiplier. The following can be got by performing derivatives on each variable.

$$\begin{cases} \dfrac{\partial L}{\partial w} = 0 \rightarrow w = \sum\limits_{i=1}^{N} \alpha_i \varphi(x_i) \\ \dfrac{\partial L}{\partial b} = 0 \rightarrow \sum\limits_{i=1}^{N} \alpha_i = 0 \\ \dfrac{\partial L}{\partial \xi} = 0 \rightarrow \alpha_i = \gamma \xi_i \\ \dfrac{\partial L}{\partial \alpha} = 0 \rightarrow w^T + b + \xi_i - y_i = 0 \end{cases} \tag{27}$$

The following problem can be got after eliminating w and ξ_i.

$$\begin{bmatrix} 0 & e_n^T \\ e_n & \Omega + \gamma^{-1} \cdot I \end{bmatrix} \cdot \begin{bmatrix} b \\ a \end{bmatrix} = \begin{bmatrix} 0 \\ y \end{bmatrix} \tag{28}$$

Where, $\Omega = \varphi^T(x_i)\varphi(x_i)$, $e_n = [1,1,\ldots,1]^T$, $\alpha = [\alpha_1, \alpha_2, \ldots, \alpha_n]$, $y = [y_1, y_2, \ldots, y_n]^T$.

The following could be obtained when the linear equations above were solved.

$$y(x) = \sum_{i=1}^{N} \alpha_i K(x_i, x) + b \tag{29}$$

Where, $K(x_i, x)$ is the kernel function.

$$K(x_i, x) = \prod_{i=1}^{N} \psi(\frac{x_i - x_i'}{\sigma_i}) \tag{30}$$

Where, $\psi(x) = \cos(1.75x) \cdot \exp(\frac{-x^2}{2})$. And the wavelet least squares support vector machine regression prediction model is followed [15].

$$y(x) = \sum_{i=1}^{N} a_i \prod_{i=1}^{N} \{\cos[\frac{1.75(x_i - x_i')}{\sigma_i}] \cdot \exp[\frac{-(x_i - x_i')^2}{2}]\} + b \tag{31}$$

3. Experimental setup and analysis

The icing thickness data and micro meteorological data of one southern route of January 10, 2008 to March 10, 2008 were selected to carry on the simulation computation in the paper. The meteorological factors mainly include temperature, humidity, wind speed, light intensity and wind direction, as shown in figure 2.

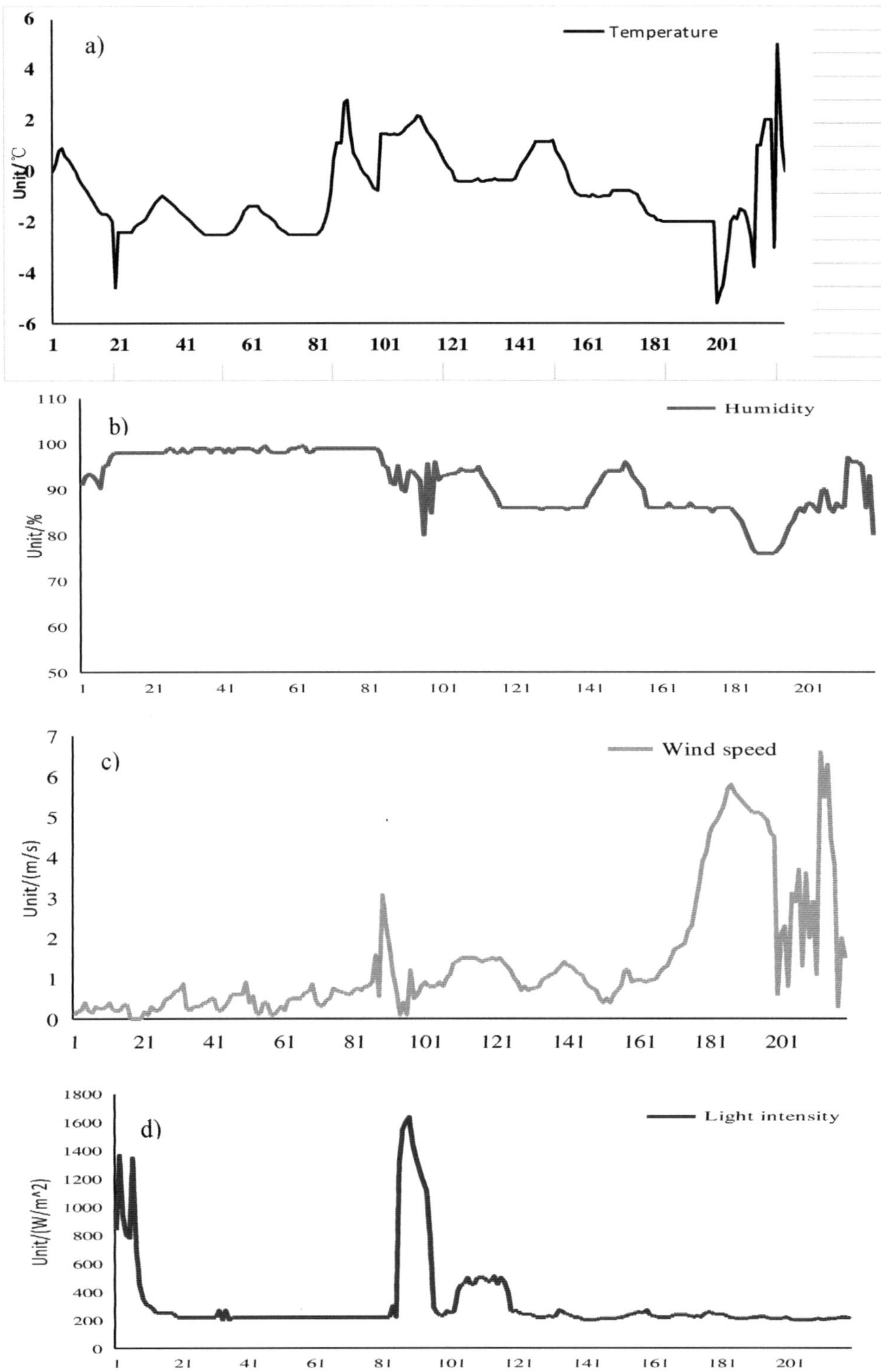
a)
Temperature
Unit/℃
b)
Humidity
Unit/%
c)
Wind speed
Unit/(m/s)
d)
Light intensity
Unit/(W/m^2)

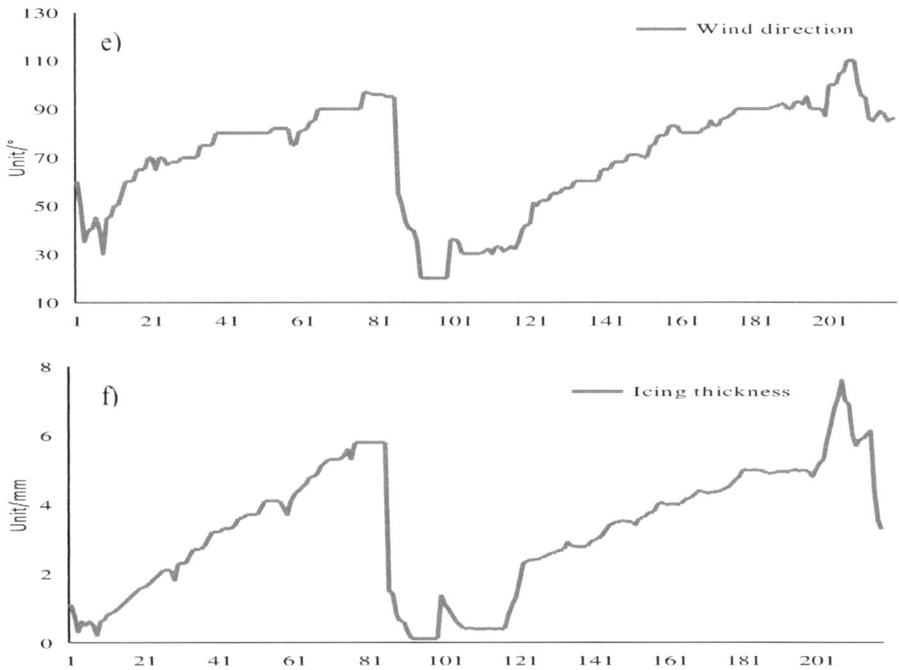

Figure 2. Meteorological data and ice thickness data graph.

The results were shown in figure 3 by using DWT decomposition and reconstruction on the icing thickness data above. The low frequency results reconstructed shown as curve D1 were smooth, the tendency of which could roughly reflect the original icing thickness data. For the high-frequency D1-D3 curves, they obviously showed stronger volatility. And the volatility gradually decreased from D1 to D3. Besides, the volatilities of the high frequency curves, D4 and D5, were not particularly different compared with other phases, indicating that five layers decomposition tree structure could meet the need of icing prediction well. Figure 4 included the original signal, reconstructed signal and the error between the two. Figure 4 showed that the error between the reconstructed signal and the original signal was very small after DWT processing, which fully met the expected accuracy.

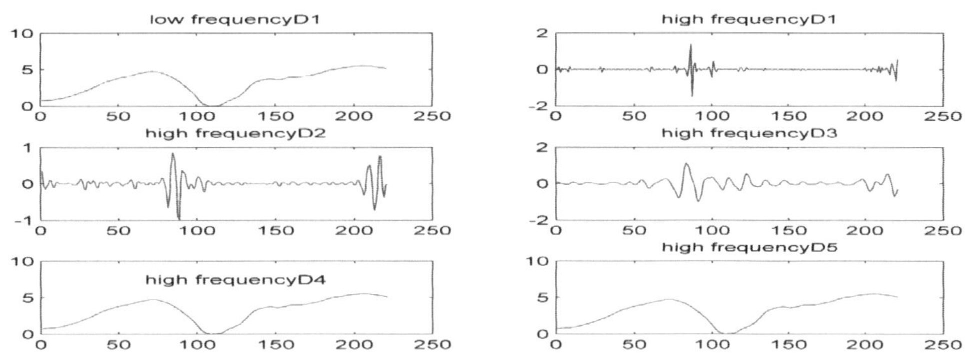

Figure 3. Decomposed frequency signal figures.

Furthermore, the first 200 sets of data were selected as the training sample data and the remaining 21 sets of data were as testing sample data. The meteorological factors were taken as the input while the ice thickness data as output. 21 groups of icing thickness prediction value $Z_0(i)$ were obtained by adopting PLS model to fit the low-frequency curve D1. And 21 groups of icing thickness prediction value $Y_j(i)$ were obtained by adopting w-LSSVM model to fit the high-frequency curves D1-D5. Then the final icing prediction result was got by adding all prediction results.

$$f(i)=\sum_{i=1}^{21}\sum_{j=1}^{5}\left(Z_0(i)+Y_j(i)\right) \tag{32}$$

In addition, in order to compare with other models, the PLS model, SVM model and BP neural network model were also used to predict the icing thickness of the 21 sets of data and the results were shown in table 1 and figure 5.

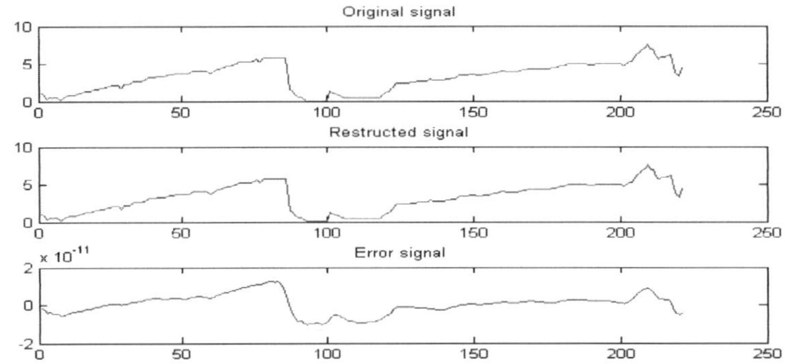

Figure 4. The original signal、reconstructed signal and error figures.

Table 1. Forecast results and error values of each model.

	Actual Value(cm)	The proposed method	Error (%)	PLS	Error (%)	SVM	Error (%)	BP	Error (%)
1	4.8	4.71	-1.88	4.57	-4.79	4.7	-2.08	4.65	-3.12
2	5	5.1	2.00	4.67	-6.60	4.82	-3.60	4.77	-4.60
3	5.2	5.32	2.31	4.93	-5.19	4.95	-4.81	5.01	-3.65
4	5.3	5.21	-1.70	5.14	-3.02	5.19	-2.08	5.42	2.26
5	5.9	5.8	-1.69	5.76	-2.37	5.77	-2.20	5.69	-3.56
6	6.3	6.17	-2.06	6.03	-4.29	6.12	-2.86	6.09	-3.33
7	6.8	6.7	-1.47	6.62	-2.65	6.67	-1.91	6.92	1.76
8	7.1	7.25	2.11	6.88	-3.10	6.92	-2.54	7.26	2.25
9	7.6	7.48	-1.58	7.26	-4.47	7.4	-2.63	7.37	-3.03
10	7	7.13	1.86	6.76	-3.43	7.29	4.14	6.82	-2.57
11	6.9	6.85	-0.72	7.21	4.49	6.78	-1.74	6.68	-3.19
12	6	6.14	2.33	6.34	5.67	5.92	-1.33	6.25	4.17
13	5.7	5.76	1.05	5.61	-1.58	5.81	1.93	5.69	-0.18
14	5.9	5.82	-1.36	5.68	-3.73	5.8	-1.69	5.75	-2.54
15	5.9	5.75	-2.54	5.56	-5.76	5.73	-2.88	5.66	-4.07
16	6.04	6.17	2.15	5.83	-3.48	5.91	-2.15	5.9	-2.32
17	6.14	6.04	-1.63	5.88	-4.23	5.94	-3.26	5.85	-4.72
18	4.42	4.33	-2.04	4.57	3.39	4.31	-2.49	4.25	-3.85
19	3.53	3.41	-3.40	3.29	-6.80	3.42	-3.12	3.65	3.40
20	3.27	3.36	2.75	3.1	-5.20	3.37	3.06	3.43	4.89
21	4.48	4.52	0.89	4.66	4.02	4.57	2.01	4.25	-5.13

In order to evaluate the predictions of the models above accurately, the maximum absolute percentage error (MaxAPE), the mean absolute percentage error (MAPE), the median absolute percentage error (MdAPE) and relative error (RE) were adopted in the paper [11].

$$MaxAPE = \max\left(\left|\frac{f(i)-\hat{f}(i)}{f(i)}\right|\right)\times 100\% \tag{33}$$

$$MAPE = \frac{1}{N}\sum_{i=1}^{N}\left|\frac{f(i)-\hat{f}(i)}{f(i)}\right|\times 100\% \tag{34}$$

$$MdAPE = \underset{i}{Med}\left(\left|\frac{f(i)-\hat{f}(i)}{f(i)}\right|\right)\times 100\% \tag{35}$$

$$RE = \frac{f(i)-\hat{f}(i)}{f(i)}\times 100\% \tag{36}$$

Where, $f(i)$ is the original value, $\hat{f}(i)$ is the predicted value, $i = 1,2,\cdots,N$.

The calculation results of relative error (RE) were shown in table 1 and figure 6. And the calculation results of other indicators were shown in table 2.

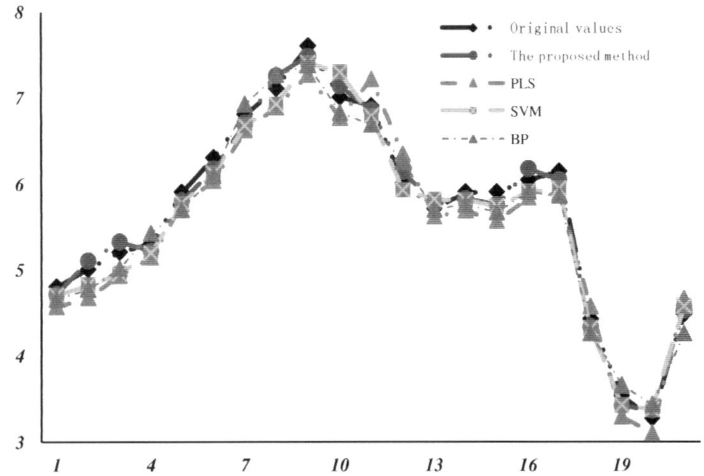

Figure 5. The prediction results line chart of each model.

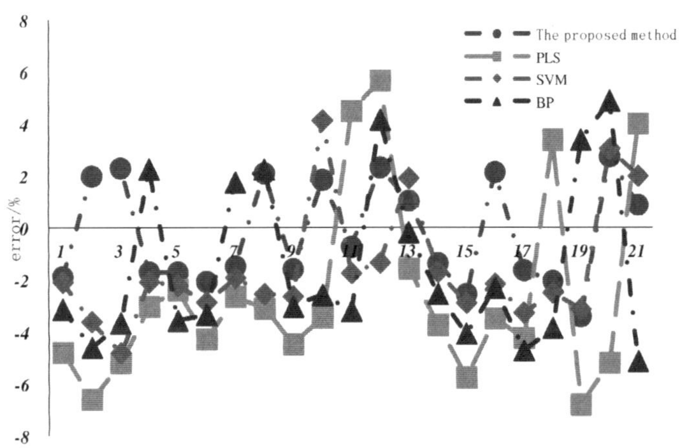

Figure 6. The relative error line chart of each model.

From table 1 and figure 6, the relative error line of the method put forward in the paper is much closer to the X axis than that of the other three methods, indicating that it is better to fit the icing nonlinear characteristics after DWT decomposition and improve the accuracy of the prediction of icing thickness. In addition, the relative error line of SVM is closer to the X axis than that of BP and PLS, meaning that the overall error of SVM is less than that of BP and PLS, so the nonlinear fitting ability of SVM is stronger than that of BP and PLS. However, the relative error line of PLS has the biggest volatility and is the farthest away from the X axis, indicating that it is not suitable to learn and fit the high-dimensional, nonlinear characteristics of icing data using PLS model only, which is more suitable for the application of icing prediction for low-frequency signal after DWT decomposition.

As shown in table 2, the MaxAPE value of the method in the paper was 3.4%, the MAPE value was 1.88% and the MdAPE value was 1.875%. They were obviously smaller than the evaluation index values of other methods in the table. It showed that it can significantly improve the prediction accuracy by using different methods to predict the icing thickness for different frequencies signals and add the predicted results after DWT decomposition. The prediction of high frequency signal could reflect the robustness and learning performance of w-LSSVM model while the prediction of low frequency signal could avoid the problems occurred when using PLS model only and fit the trend of curve better.

Table 2. Evaluation index values of each model.

	The proposed method	PLS	SVM	BP
MaxAPE/%	3.40	6.8	4.81	5.13
MAPE/%	1.88	4.2	2.6	3.27
MdAPE/%	1.875	4.235	2.489	3.333

4. Conclusion

In this paper, the icing prediction model based on DWT feature extraction was proposed aiming at the nonlinear characteristics of icing thickness of transmission lines. First of all, the historical icing data was decomposed by using DWT and a set of high frequency signals ($y_0 \rightarrow y_n$) and a set of low frequency signals (Z_n) were obtained. Secondly, PLS model and w-LSSVM model proposed in the paper were used to fit and forecast for the low frequency signal that was smooth and the high frequency signal that has strong volatility respectively. Then the final predicted result of icing thickness was got by adding the predicted value of PLS and the predicted value of w-LSSVM. Finally, the algorithm was verified by the example of one south line and the results showed that the icing prediction accuracy of the method proposed in this paper was much higher than that of PLS, SVM and BP. It indicated that it could make full use of PLS model and w-LSSVM model respectively after DWT decomposition, which could improve the robustness of the algorithm so as to improve the accuracy of the icing prediction.

References

[1] Hu Y 2008 Analysis and Countermeasures of large area ice disaster in power network *High Voltage Technology* **34** 215-19

[2] Huang X B, Ou Y, Li S, Wang Y N, *et al.* 2011 Analysis of key influence factors on transmission line icing *High Voltage Technology* **37** 1677-82

[3] Yang L, Hao Y P, Li L C, *et al.* 2011 Fuzzy expert system for the assessment of icing state of overhead transmission line *High Voltage Technology* **37** 3028-35

[4] Yao C G, Zhang L, Li C X, *et al.* 2013 Measurement method of wire icing thickness based on mechanical analysis and sag measurement *High Voltage Technology* **39**

[5] Lu J Z, Peng J W, Zhang H X, Li B and Fang Z 2009 Icing meteorological genetic analysis of Hunan Power grid in 2008 *Electric Power Construction* **30** 29-32

[6] Goodwin E J I, Mozer J D, Digioia A M J, *et al.* 1983 Predicting ice and snow loads for transmission line design *Predicting Ice & Snow Loads for Transmission Line Design*

[7] Makkonen L 1998 Modeling power line icing in freezing precipitation *Atmospheric Research* **46** 131-42

[8] Liao Y F and Duan L J 2010 Study on estimation model of wire icing thickness in Hunan Province *Transactions of Atmospheric Sciences* **33** 395-400

[9] Dai D, Huang X T, Dai Z, *et al.* 2013 Transmission line icing regression model based on support vector machine *High Voltage Technology* **39**

[10] Sheng C, Dong D, Xiao T H, *et al.* 2012 Short-term Prediction for Transmission Lines Icing Based on BP Neural Network *Power and Energy Engineering Conference (APPEEC)* Asia-Pacific IEEE 1-5

[11] Liu J, Li A J and Zhao L P 2012 Ice thickness prediction model based on T-S fuzzy neural network *Hunan electric power* **32** 1-4

[12] Yu X B and Dong T 2002 Fast decomposition and reconstruction algorithm for discrete wavelet transform *Journal of Southeast University: Natural Science Edition* **32** 564-68

[13] Mao L F, Jiang Y C, Long R H, *et al.* 2008 Mid long term electric load forecasting based on partial least squares regression analysis *Grid technology* **32** 71-7

[14] Huang X N, Xu J H, Yang C S, *et al.* 2014 Transmission line icing forecasting based on data driven algorithm and LS-SVM *Automation of Electric Power Systems*

[15] Liao G C 2014 Hybrid Improved Differential Evolution and Wavelet Neural Network with load forecasting problem of air conditioning *International Journal of Electrical Power & Energy Systems* **61** 673-82

IOP Conf. Series: Earth and Environmental Science **40** (2016) 012039 doi:10.1088/1755-1315/40/1/012039

Design and development of a solar powered mobile laboratory

L Jiao[1], A Simon, H Barrera, V Acharya and W Repke

School of Engineering, Grand Valley State University, Grand Rapids, Mi, 49504 USA

E-mail: jiaoh@gvsu.edu

Abstract. This paper describes the design and development of a solar powered mobile laboratory (SPML) system. The SPML provides a mobile platform that schools, universities, and communities can use to give students and staff access to laboratory environments where dedicated laboratories are not available. The lab includes equipment like 3D printers, computers, and soldering stations. The primary power source of the system is solar PV which allows the laboratory to be operated in places where the grid power is not readily available or not sufficient to power all the equipment. The main system components include PV panels, junction box, battery, charge controller, and inverter. Not only is it used to teach students and staff how to use the lab equipment, but it is also a great tool to educate the public about solar PV technologies.

1. Introduction

Our climate is changing. According to the data collected at NASA's Goddard Institute for Space Studies (GISS), the average global temperature has increased by 0.8°C since 1880, two thirds of which has occurred since 1975 at a rate of 0.15°C-0.2°C per decade [1]. This temperature increase is mainly caused by the emission of greenhouse gasses, 25% of which are accounted for by energy production [2]. Despite the fact that increasing awareness of climate change has led to initiatives from governments, businesses, and the society to tackle this problem, greenhouse gas emissions have not been reduced to safe levels. The global temperature is predicted to continue to increase between 3.7°C and 4.8°C by 100 [3], which would lead to serious consequences for food security, fresh water availability, and the frequency and intensity of storms [4]. To reach the goal of keeping global warming below 2°C, greenhouse gas emissions need to be reduced by 40%-70% by 2050 compared to the levels in 2010 [5]. This is an ambitious goal. To make this happen, seeking alternative renewable energy sources and further increasing the awareness of the importance of adapting to renewable energy is necessary in preventing climate change.

Solar energy is the most clean, abundant, and renewable alternative energy resource available. Solar photovoltaic (PV) is viewed as a quick, scalable, and sustainable solution to mitigate climate change. In many countries, the initiative of solar installation is one of the most prevalent issues that their governments are reaching out to solve. According to Green Tech Media research, global solar PV installations have grown 34% in 2015 compared to that in 2014 [6]. The US installed 7.26 GW of PV in 2015, 16% above that in 2014[7]. In Germany, PV-generated energy amounted to 34.9 TWh in 2014 and covered approximately 5.7% of Germany's net electricity consumption8. Studies have also

[1] Address for correspondence: L Jiao, School of Engineering, Grand Valley State University, Grand Rapids, Mi, 49504 USA. E-mail: jiaoh@gvsu.edu

been carried out on the stand-alone PV systems. Benghanem et al. reported a methodology for setting up several PV water pumping systems for irrigation purposes in Algeria [9]. Song *et al.* designed stand-alone PV systems to power aerators for natural purification of acid mine drainage [10]. Sakiliba et al. assessed the stand-alone residential solar PV applications in sub-saharan Africa [11]. Most of these studies focused on the applications of the solar PV not on educating the public about PV technologies. In addition, most installations occurring in remote regions hinder the public knowledge of the renewables used. We have developed a solar powered mobile laboratory (SPML) that can be used to showcase PV technologies and help educate both the public and students on the importance of solar PV.

The SPML is a mobile platform that provides an alternative and flexible method for schools, universities, and communities to give students and staff access to laboratory environments. The primary power source of the SPML is solar PV which allows the laboratory to be operated in places where the grid power is not readily available or is not sufficient to power all the equipment. In the SPML, the grid electricity is used to supplement the solar power.

In this paper, detailed design and development of the solar powered mobile laboratory will be presented.

2. System Requirements and Functional Specifications

After studying a list of typical laboratory equipment, the system requirements and functional specifications were determined and listed in Table 1.

Table 1. The SPML system requirements and functional specifications

Alternative Power System requirements	
001	The primary power of the SPML system shall be supplied by the solar PV.
002	The SPML system shall be capable of receiving power from grid through two NEMA 5-15 receptacles when: the solar power system is not capable of supplying power AND two NEMA 5-15 receptacles are available on the site where the system is deployed.
Circuit requirements	
003	Two power receptacles shall be provided for charging two laptop computers.
004	Two power receptacles shall be provided for charging two cellular telephones.
005	Two power receptacle shall be provided for two soldering station.
006	One power receptacle shall be provided for one laser engraver.
007	One power receptacle shall be provided for one ventilation system dedicated to the laser engraver.
008	One power receptacle shall be provided for one projector.
009	Two power receptacles shall be provided for two 3D Printers.
010	One power receptacle shall be provided for charging one tablet computer.
011	Two power receptacles shall be provided for two desktop computers.
Monitoring and Control requirements	
012	The SPML system shall be capable of operating when exposed to environmental conditions typical to Pennsylvania for the months of February through October.
Solar Power and Storage System requirements	
013	The SPML system shall have an array of three to five solar panels capable of charging a battery bank.
014	The solar panel array shall be mounted to the roof of a vehicle and fold flat for transport.
015	The SPML system shall have a bank of batteries capable of powering the LED lighting, two Cube 3 3D printers, one iMac desktop PC, and two soldering stations for three hours

IOP Conf. Series: Earth and Environmental Science **40** (2016) 012039 doi:10.1088/1755-1315/40/1/012039

	per day.
016	The SPML system shall have an inverter capable of converting the DC power of the solar panels to AC power.

3. System Architecture

The functional block diagram of the SPML is shown in Figure 1. Solar PV modules are used to generate direct electric current (DC) upon exposure to solar radiation. The outputs of the PV modules are connected through the junction box. The battery is used to store and regulate energy generated by the PV modules. The charge controller provides the maximum power point tracking (MPPT) function and controls the charging and discharging of the battery. To be compatible with the load power requirements, the inverter converts the DC power to AC power. The switch is used to automatically switch the grid connection to the inverter if sufficient solar energy is available.

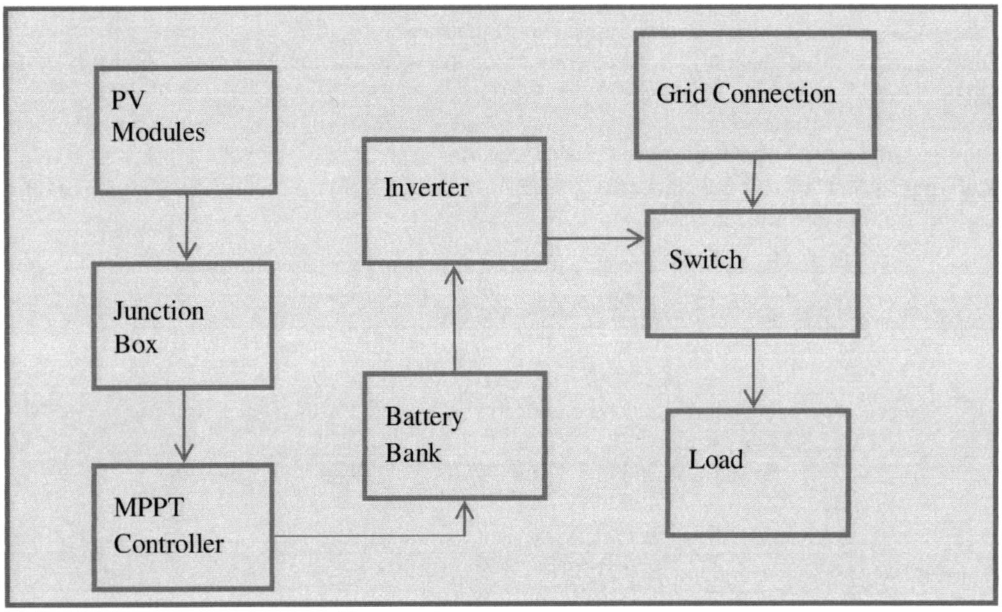

Figure 1. Functional block diagram of the SPML

4. System Design

This section provides the detailed design of each component of the solar powered mobile laboratory system. The design is based on the IEEE standards for sizing PV components in stand-alone PV systems [12].

4.1 Determination of Load Energy Requirements

The devices that were selected to be powered by the solar power are the LED lighting, two 3D printers, an iMac computer that controls the 3D printers, and two 40W soldering stations. These were chosen because they were identified as important, high-use loads and the power consumption is low enough to ensure proper operation for three hours with a small number of PV modules. The average power consumption is 402W and the average energy consumption for a three-hour operation is 1206Wh.

4.2 Sizing the Inverter

With an average power consumption of 402W, an inverter of 2000W with an efficiency of 94% was selected. It is sufficient to cover any starting power surge and leave room for future expansion.

4.3 Sizing the Battery

Since solar energy inherently varies with time, most of the stand-alone PV systems incorporate batteries to store the generated energy and to provide sufficient energy for the required load. The daily energy needed by the load of the battery consists of the load energy consumption and the energy losses of the inverter and wiring. Considering a typical 2% loss in the wires, the daily load on the battery is calculated to be

$1206W / 94\% / (1 - 2\%) = 1309Wh / day$.

The required capacity of the battery can be computed from Equation 1 where factors of days of autonomy, battery charge/discharge losses, depth of discharge, and losses due to the lower operation temperature are taken into consideration[13].

$$Capacity(Ah) = \frac{Wh / day}{V_b} \frac{DOA}{D_T D_{ch} DOD} \qquad \text{---------------------} \quad (1)$$

Where V_b is the battery voltage, D_T is the temperature derating factor, D_{ch} is the charge/discharge derating factor, DOA is the number of days of autonomy, and DOD is the depth of discharge. Assuming a combined 12% loss from low temperature operation and battery charge/discharge, one day of autonomy, and 70% of depth of discharge, the capacity required for a 12V battery is 177Ah. A 12V 200Ah deep cycle AGM battery was selected for this system.

4.4 Sizing the PV Panel

Proper sizing of the PV panel ensures that the system is recharged and able to deliver the required energy in the specified days of recovery. The required panel power is determined by the total daily energy need, solar irradiation, solar panel losses, and losses in wiring and the charge controller. Since solar irradiation changes with time, monthly average daily irradiation is often used. The SPML system will be deployed in the Pittsburgh region; hence, the solar irradiation in Pittsburgh is referenced and can be found in Table 2, where the monthly average daily irradiation and the yearly average daily irradiation are shown for the panel tilt angles of 0°, latitude – 15°, latitude, latitude + 15°, and 90°, respectively[14].

Table 2. Solar Radiation for Flat-Plate Collectors Facing South at a Fixed Tilt (KWh/m^2/day)

Tilt (°)	**Jan**	**Feb**	**Mar**	**Apr**	**May**	**Jun**	**Jul**	**Aug**	**Sep**	**Oct**	**Nov**	**Dec**	**Year**
0	1.7	2.5	3.5	4.6	5.5	6.1	5.9	5.2	4.2	3.0	1.8	1.4	3.8
Lat. -15	2.4	3.2	4.1	4.9	5.5	5.9	5.9	5.5	4.8	3.8	2.4	1.9	4.2
Lat.	2.6	3.4	4.2	4.8	5.2	5.4	5.5	5.3	4.8	4.1	2.6	2.1	4.2
Lat. +15	2.7	3.5	4.1	4.4	4.6	4.7	4.8	4.8	4.6	4.1	2.7	2.2	3.9
90	2.5	3.0	3.1	3.1	2.9	2.5	2.6	2.9	3.2	3.3	2.3	2.0	2.7

It can be seen from Table 2 that in the time frame of February to October, the lowest average daily irradiation is 3.5 PSH when the panel is tilted at an angle equal to (latitude + 15°= 55°). The decision to go with the fixed tilt angle instead of one- or two- axis tracking is based on the available space for the tracking mount and cost. When deployed in a site, the system will be parked to ensure that the panels are facing south and tilted at an angle of 55° from the horizontal.

Assuming a 12% loss in wiring and the charge controller, the required panel power is determined to be $\frac{1309Wh/day}{88\% \times 88\% \times 3.5PSH} = 483W/day$. Three 160W PV panels would meet the daily energy needs of the required loads.

4.5 Sizing the Charge Controller

The main requirements of the charge controller are: (1) it can charge a 12V battery bank; (2) it can handle the combined current output of the 3 solar panels in parallel; (3) it has the ability to disconnect the battery bank from connected loads if the batteries drop below a certain voltage. The total current from the three parallel-connected panels is 40A. The Conext MPPT 60-150 charge controller, which has a rated input current of 60A, was selected for the SPML system. This MPPT controller has separate wire terminals for the array, battery bank, and a low power auxiliary output which enables the controller to disconnect the load when the batteries discharge to a programmable voltage. This auxiliary output will be used to switch a relay that disconnects the inverter from the battery bank when the batteries reach a programmed low voltage limit. As the current output of the auxiliary output is limited, a small pre-stage relay is required to drive the coil of a larger relay, which will be the one to disconnect the inverter from the batteries.

4.6 Balance of System (BOS) Design

The required current capacity, minimum required wire gauge, and the number of conductors for the solar powered mobile laboratory are summarized in Table 3. The DC wiring and AC wiring are sized to handle 156% and 125% of the peak current, respectively, per the National Electrical Code (NEC), except where an equipment manufacture specifies a current limit. All ground wires are 6AWG bare stranded wire.

Table 3. Wire Current Capacity and Size for the SPML

Cable Position	Max Current (A)	NEC (A)	Min Wire Size (AWG)	Conductors
1. Solar Modules to Combiner	9.60	14.98	14	10
2. Combiner to Charge Controller	48.00	74.88	6	2
3. Charge Controller to Battery Breaker	48.00	74.88	6	2
4. Battery Breaker to Battery Bank	52.83	82.41	4	2
5. Battery bank to Inverter	52.83	82.41	4	2
6. Inverter to Transfer Switch	16.67	20.84	10	2
7. Transfer Switch to Load Center	16.67	20.84	10	2
All Ground Wires			6	1

4.7 Junction Box, Disconnect, and Load Center Enclosure designs

A Junction box was selected to combine the power from the three solar panels. The short circuit current of each module is about 9.6A. Each string is connected to a 15A DC breaker with 14AWG conductors in the junction box. An enclosure was designed to host the various DC circuit breakers and bus bars that are used to protect and disconnect the equipment in the system. One such breaker is a 60A circuit breaker that is used to disconnect the solar array. In order to handle the 48A current from the solar array, the size of the conductors between this breaker and the charge controller is 6AWG. The load disconnect relays are mounted in a small enclosure that can be mounted close to the inverter. The system also has a circuit load center for the connection of AC load. Since the peak continuous AC current that the inverter can supply is 16.67A, all AC circuits downstream of the inverter were sized with 10AWG conductors. All specified loads will be located on the same circuit and will be protected by a 15A breaker. The cabling from the AC load center panel to the specific devices can be standard 14AWG indoor wiring.

5. System Schematic

The detailed circuit schematic of the SPML is illustrated in Figure 2. Five 160W solar panels were incorporated in the schematic to provide room for future expansion.

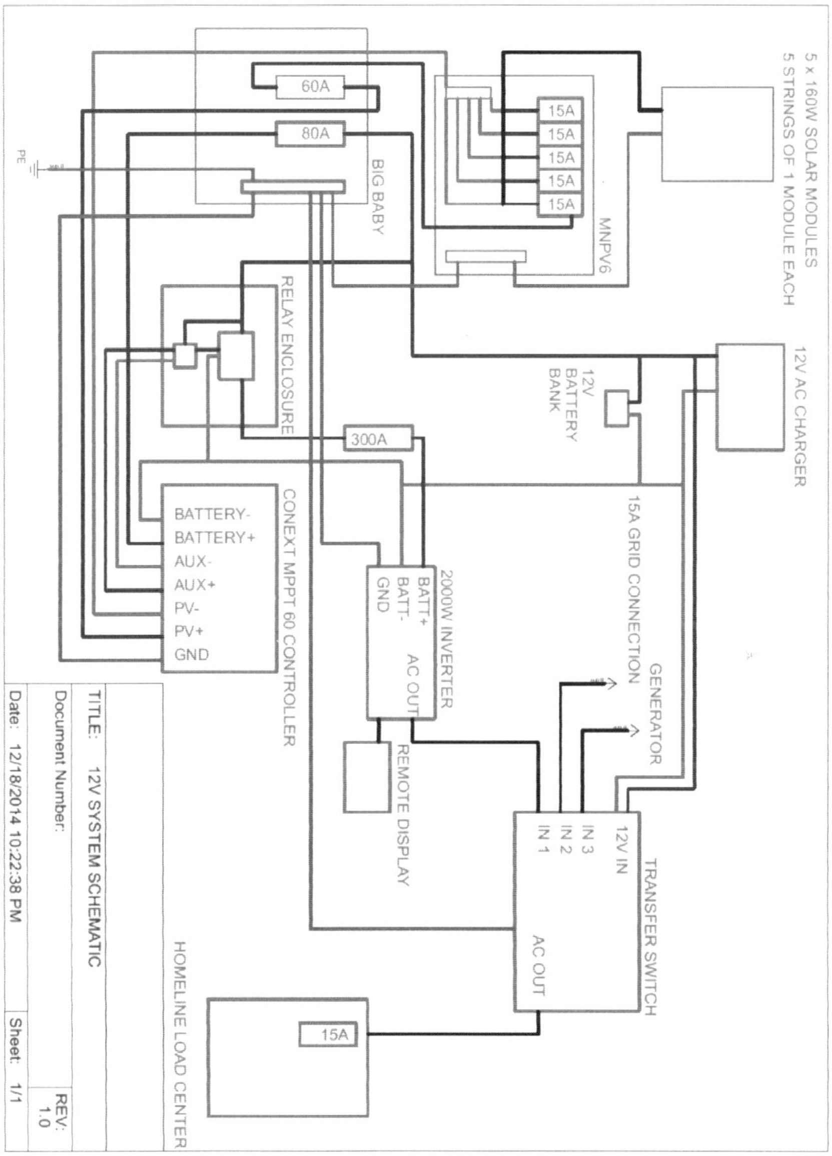

Figure 2. Circuit schematic of the SPML

6. System Testing

The system was built and tested according to the IEEE Standards[15]. The tested parameters were load capacities under different load conditions, inverter efficiency, battery charging currents, and maximum power under different light conditions. The test results warranted that the system met all the design requirements.

7. Conclusions

A 480W solar powered mobile laboratory was successfully designed and built. The SPML is a mobile platform that provides an alternative and flexible method for schools, universities, and communities to grant students and staff access to laboratory environments. The lab provides equipment like 3D printers, computers, and soldering stations. The primary power of the system was supplied by three 160W solar PV modules connected in parallel. The grid electricity is used to supplement the solar power. To provide consistent power to the required loads, a 200Ah battery was included. To properly charge/discharge the battery, a 60A MPPT charge controller was selected. A 2000W inverter was sized to convert the DC power from the battery to the AC power required by the lab equipment. A relay is designed to disconnect the inverter when the battery voltage reaches its low voltage limit. In addition to teaching students and staff how to use the lab equipment, the SPML is a great tool to educate the public about solar PV technologies.

References

[1] Hansen J, Ruedy R, Sato M, and Lo K 2010 Global surface temperature change *Reviews of Geophysics,* 48 (RG4004)

[2] Edenhofer O *et al* 2014 *Climate change 2014 mitigation of climate change* Cambridge University Press, New York

[3] Elzen M *et al* 2010 *The emissions gap report: are the copenhagen accord pledges sufficient to limit global warming to 2 ℃ or 1.5 ℃?* The United Nations Enviornmental Programme (UNEP)

[4] Field C B *et al* 2014 *Climate change 2014: impacts, adaptation and vulnerability* Cambridge University Press, New York

[5] European Commission 2015 *The Paris protocal – a blueprint for tackling global climate change beyond 2020* Brussels COM (2015)81

[6] Munsell M 2016 *GTM research: global solar PV installations grew 34% in 2015* Greentech Media

[7] Kann S *et al* 2015 *U.S. solar market insight* SEIA/GTM Research

[8] Burger B 2015 *Electricity production from solar and wind in Germany in 2014* Fraunhofer ISE

[9] Benghanem M and Arab A H 2007 Photovoltaic water pumping systems for Algeria *Desalination* **209** 50–7

[10] Song J Y and Choi Y 2015 Design of photovoltaic systems to power aerators for natural purification of acid mine drainage *Renewable Energy* **83** 759-66

[11] Sakiliba S K, Hassan A S, Wu J Z, Sanneh E S and Ademi S 2015 Assessment of Stand-Alone Residential Solar Photovoltaic Application in Sub-Saharan Africa: A Case Study of Gambia *J. Renewable Energy*

[12] IEEE Standards Coordinating Committee 21 2008 IEEE guide for array and battery sizing in stand-alone photovoltaic (PV) systems *IEEE Std 1562-2007*

[13] Messenger R and Ventre J 2010 Photovoltaic Systems (FL, USA: Engineering CRC Press) 294

[14] Marion W and Wilcox S 1994 *Solar Radiation Data Manual for Flat-Plate and Concentrating Collectors* National Reviewable Energy Laboratory

[15] IEEE Standards Coordinating Committee 21 2004 1526-2003-IEEE recommended practice for testing the performance of stand-alone photovoltaic systems *IEEE Std 1526-2003*

IOP Conf. Series: Earth and Environmental Science **40** (2016) 012040 doi:10.1088/1755-1315/40/1/012040

Difference between the maximum empirical and field measured peak Watt values of thermal power system under highly sufficient solar conditions

O A Elhussain[1] and T I M Abdel-Magid[2]

[1]Faculty of Engineering Sciences, Omdurman Islamic University, 382, Khartoum, Sudan
[2]Faculty of Engineering, International University of Africa, Khartoum, Sudan

Email: elhafiz999@yahoo.com

Abstract. Mono-Crystalline solar cell module is experimentally conducted in Khartoum, Sudan to study the difference between maximum empirical value of peak Watt and maximum value of thermal power produced in field under highly sufficient solar conditions. Field measurements are recorded for incident solar radiation, produced voltage, current and temperature at several time intervals during sun shine period. The thermal power system has been calculated using fundamental principles of heat transfer. The study shows that solar power for considered module could not attain the empirical peak power irrespective to maximum value of direct incident solar radiation and maximum temperature gained. A loss of about 6% of power can be considered as the difference between field measurements and the manufacturer's indicated empirical value. Solar cell exhibits 94% efficiency in comparison with manufacturer's provided data, and is 31% more efficient in thermal energy production than in electrical power extraction for hot-dry climate conditions.

1. Introduction

The climatic conditions in Sudan are favorable for utilization of solar energy, as reviewed in figure 1. Sudan is exposed to solar incident of more than world's average radiation. Such potential is highly favorable in both thermal and photovoltaic energy extraction. This leads to receive many more hours of bright sunshine during the course of the year within the continent of the Earth. However, this will draw the attention to explain how the photovoltaic principle works to transform sunlight directly into electricity, what sort of hardware does the job, how the source is hitched to the load, and how to go through designing an appropriate system to specific needs. But before it has been immersed in the details of fascinating the power technology, it needs to be paused and consider how truly revolutionary begin to use photovoltaic devices to grow accustomed to their presence in our lives. This attitude could be happened in the respect of concept that should reflect on how marvelous it to be able to get electricity directly from the sun. It is worth mentioning that, Sudan as geographically located in the Sahara desert, with less cloud cover and a better solar angle, one could ideally obtain closer to 8.3kWh/m^2/day provided the nearly ever present wind would not blow sand onto the units. The area of

[1] Address for correspondence: Faculty of Engineering Sciences, Omdurman Islamic University, 382, Khartoum, Sudan. Email: elhafiz999@yahoo.com.

the Sahara desert is over 9 million km². 90,600 km², or about 1%, could generate as much electricity as all of the world's power plants combined.

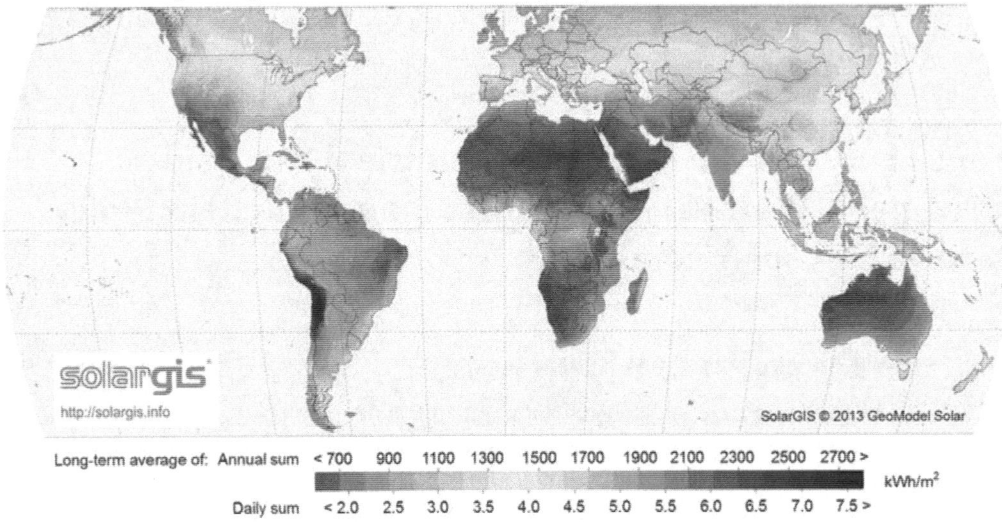

Figure 1. World Solar Map

2. Theory

Principally, the output of a photovoltaic module or array is directly proportional to the amount of sunlight it receives, in terms of both intensity and duration. Seasonal and daily variations in available sunlight are a function of the earth's rotation and its angle of tilt toward the sun. In simplest terms, this explains why sunlight becomes more abundant towards noon and tapers off in the afternoon and why it falls upon a given area with less intensity during hourly time of the day. Thus, for solar radiation to be most direct there are optimum tilt angles at which collecting devices (either photovoltaic or thermal) ought to be placed.

Due to the incident solar radiation on module surface area, an amount of heat transfer exists between exposed surface area and photo voltage cells. The rate of steady state heat transfer in a radial direction through a homogeneous material plane wall of Mono-crystalline cell is defined by the following equation of heat transfer fundamental:

$$Q = (Tsr - Ti) \sum A/R \quad (1)$$

Where:
Q = Amount of heat transfer, W
T_i = Temperature following the cell uniform thickness wall, ^0C
A = Radiated cell surface area, m^2
R = Thermal resistance of cell material, $m^2.^0C/W$
T_{sr} = Incident solar air temperature, ^0C
Value of T_{sr} during maximum value of radiation is calculated by:

$$T_{sr} = T_{amb} + (\alpha\, I / h_o) \quad (2)$$

Where:
T_{amb}= Ambient air temperature, ^0C
α = Solar absorptivity of outside film surface, dimensionless value. (For Mono-crystalline silicon, α= 0.5)
I = Incident solar radiation, W/m^2

IOP Conf. Series: Earth and Environmental Science **40** (2016) 012040 doi:10.1088/1755-1315/40/1/012040

h_o = Convective film coefficient of outside surface, W/ $m^2.^0C$. (For direct Incident solar radiation this value depends on wind velocity, and is equal to 22.7W/$m^2.^0C$ at wind velocity of 3.4 m/s)
Thermal resistance of cell material can be found through:

$$R = (\Delta x / k) \tag{3}$$

Where:
Δx = Cell Thickness, m
k = Thermal conductivity, W/ $m.^0C$
Electrical power produced is measured via:

$$P = (I \times V) \tag{4}$$

Where:
I = Current, A
V = Voltage, V

3. Results and analysis

This study was conducted in Khartoum, Sudan (Latitude: 15° 35' 53" North) for a photovoltaic module directed to incident solar radiation during the month of June, which is considered as the mostly hottest month during the year as averaged for 40 years based on the Metrological Climatic Data of the country.

A crystalline solar cell produced by RAGGIE Manufacturer (Model: RGM-130W ISO9001 CE) having the following technical specifications: maximum Peak Power (Pmax=130W), Voltage (Vmp=17.5 V), Current (Imp=7.42 A), surface area (1.495m x 0.670m), cell thickness (2.5cm), and thermal conductivity (149 W/$m.^0C$) was used to conduct the experimental data. The cell is directed towards the sun and exposed to the North direction by an angle of 15^0. Data is recorded at different time intervals during sun shine, and panel was considered as both solar cell and solar collector. Measurements of average interval values of both solar cell and collector are obtained in table 1, which indicates that for 11 hours of sun shine exposure total radiation received was measured to be 7,003Whr/m^2 and the average voltage, current and surface temperature were 12.05V, 1.89A and 44.45^{0}C respectively.

Table 1. Experimental data.

Time Interval, O'clock	Radiation (Whr/m^2)	Voltage (V)	Current (A)	Temperature (^{0}C)
07	146.75	11.45	0.42	25.65
08	381.50	11.66	1.17	33.33
09	603.50	11.85	1.94	42.23
10	775.25	11.99	2.53	47.90
11	883.00	12.10	2.96	52.90
12	931.00	12.17	3.13	53.98
01	903.50	12.23	3.06	54.03
02	817.50	12.26	2.78	53.00
03	692.25	12.27	2.33	50.23
04	493.50	12.24	1.50	45.70
05	285.00	12.19	0.74	40.05
06	090.25	12.13	0.18	34.45

Figure 1 shows the relationships of voltage and current in accordance to the hourly change of incident solar radiation. It clearly shows that for the herein considered module voltage increases as the incident solar radiation increases until it reaches its maximum value and dramatically falls with the decreased amount of solar radiation. While current exhibits a relatively similar behavior as it falls smoothly with the decrease of incident solar radiation.

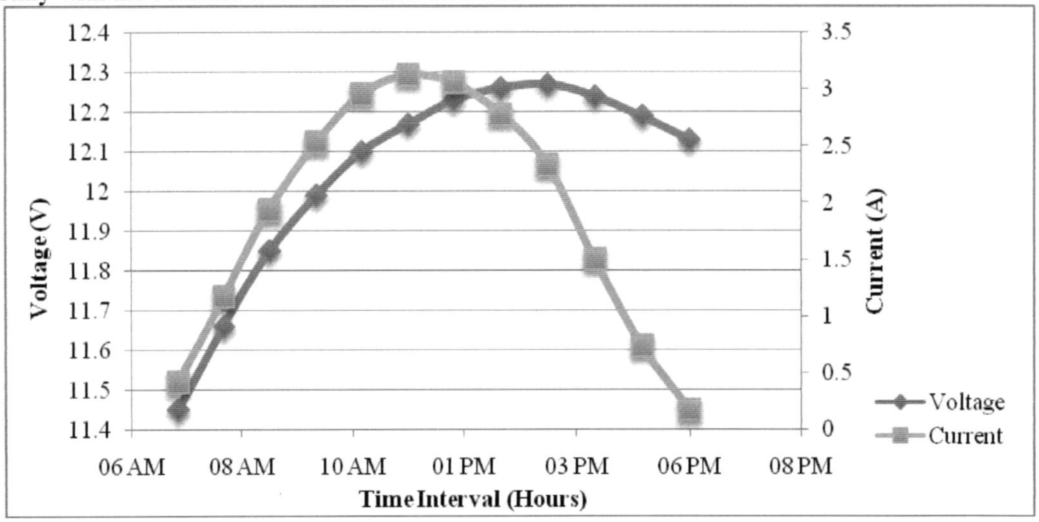

Figure 2. Voltage and current through time intervals of sun shine.

Figure 2 represents the relationships of direct incident solar radiation and temperature versus different time intervals. The figure indicates that, as the temperature increases along sun shine period the direct incident solar radiation proportionally increases and vice versa. However, temperature and radiation reach their maximum values at mid day hours, mainly at 12 and 1 O'clock PM.

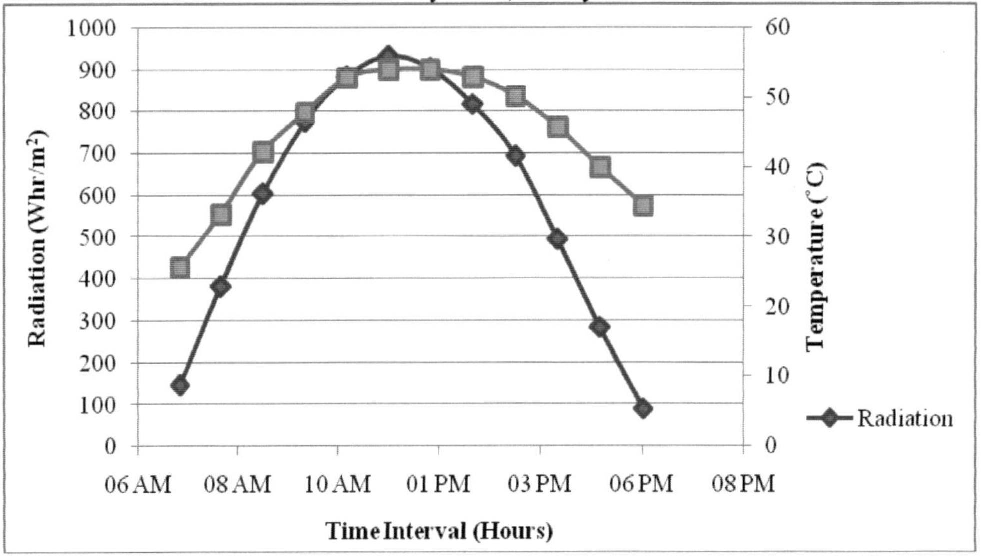

Figure 3. Incident solar radiation and temperature through sun shine time intervals.

Observation of incident radiation, voltage and current behaviors are illustrated in figure 3, as it clearly shows the direct relation between solar cell outputs and the amount of collected solar radiation.

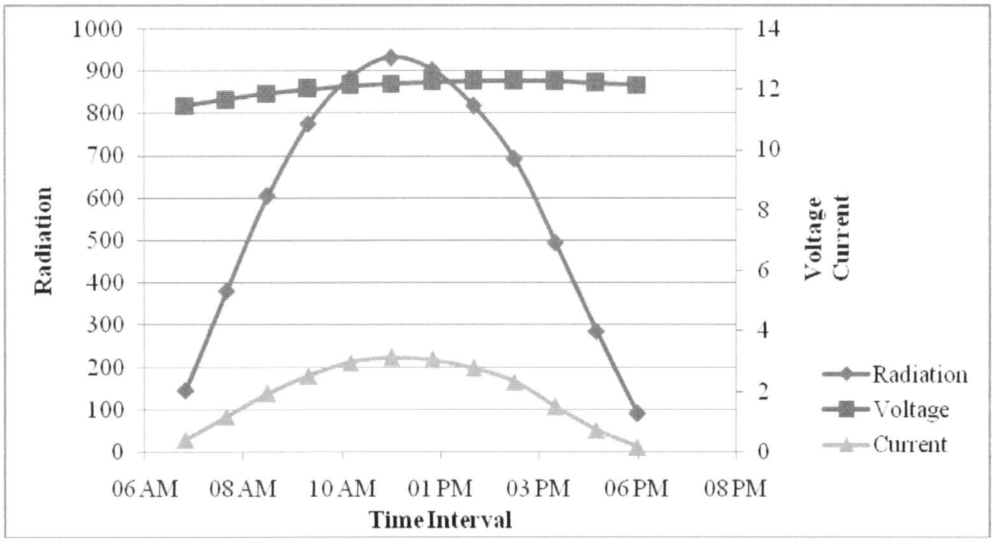

Figure 4. Voltage and current change with incident solar radiation.

As targeting for maximum thermal energy to be produced by the solar cell, maximum values of data mentioned in table 1 are substituted into equations (2) and (3). Solar air temperature and thermal resistance of cell material were found to be respectively equal to 74.5^{0}C and 0.1678 m2.^{0}C/ W. When substituting these values into equation (1) it gives an amount of heat transfer equal to 122.3 W. As for electricity production, using maximum values induced into equation (4), maximum electrical power produced is 38.4 W, which is found to be 31% of thermal potential extracted power from the solar cell.

The difference in Watts between the thermal power produced by solar cell and peak power of the same solar module measured empirically by the manufacturer was found to be equal to 7.6 W, which indicates solar cell efficiency of 94.1%.

Figure 5 compares between thermal and electrical power outputs as functions of both time intervals and solar radiation. Solar radiation is demonstrated in a curve as a function of time, while power values are represented as a variation of bubble size. The figure indicates that potential thermal power delivered is greater than extracted electrical power at all values of incident solar radiation.

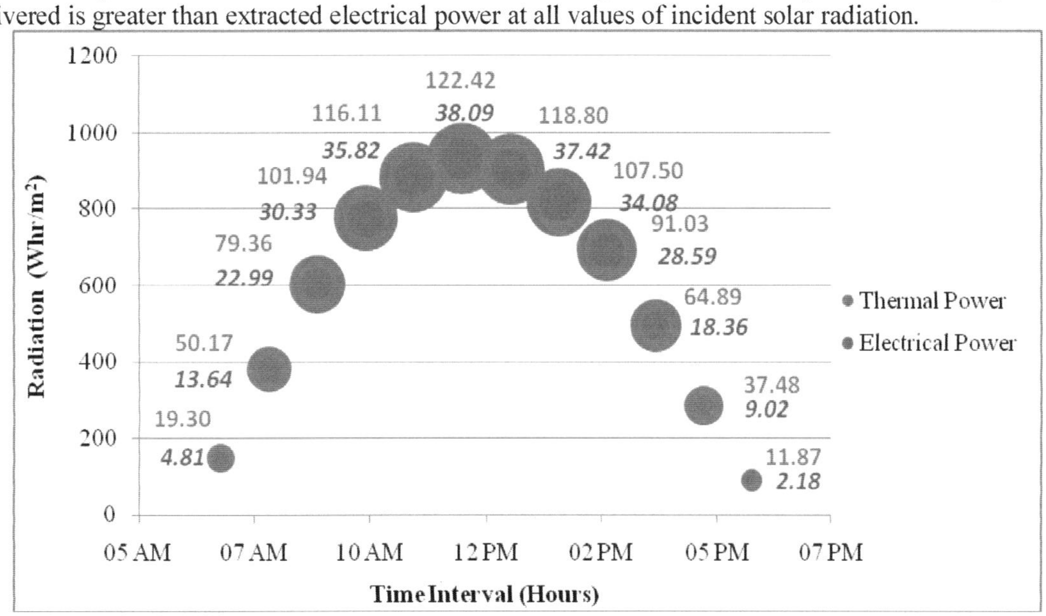

Figure 5. Thermal and electrical power produced with incident solar radiation.

4. Conclusion

Solar power for such module could not attain peak value of its power irrespective to the maximum value of direct incident solar radiation and maximum temperature attained. A loss of about 6% can be considered as the difference between the field measurements to be obtained at maximum value and that indicated by the manufacturer for empirical maximum value. These losses encountered on the system may be caused due to many factors which related to in accuracy of the measurements, instruments and devices error, in appropriate inclination of module and other circumstances. Whereas over 68% of incident solar energy is dissipated into losses for electricity extraction when compared to thermal energy gained.

Acknowledgments

The authors would like to thank Dr. Eng. Osman Merqhani for his support, and for providing necessary information and measuring the field data hence giving the principals of scientific experience in this area which positively enhance this paper to be produced in this form.

References

[1] Green M A 2003 Crystalline and thin-film silicon solar cells: state of the art and future potential *Solar Energy* **74** 181–92

[2] Nieuwenhout F *et al.* 2001 Experience with solar home systems in developing countries: a review *Progress in Photovoltaic's: Research and Applications* **9** 455–74

[3] Schroeder D P, Aksamija Z, Rath A, Voyles P M, Lagally M G and Eriksson M A Thermal Resistance of Transferred-Silicon-Nanomembrane Interfaces *Physical Review Letters* **115**

[4] Green M A 2008 Self-consistent optical parameters of intrinsic silicon at 300K including temperature coefficients *Solar Energy Materials and Solar Cells* 1305–10

[5] Santbergen R and Van Zolingen R J C 2007 The absorption factor of crystalline silicon PV cells: A numerical and experimental study (The Netherlands: Department of Mechanical Engineering, Eindhoven University of Technology)

[6] Strong S J and Scheller W G 1993 *The solar electric house: energy for the environmentally-responsive, energy-independent home*: (Still River: Sustainability Press)

[7] A.M. Abdalla 1997 *Environmental Control of Agricultural Structure under Climatic Conditions of Desert* (Khartoum: Khartoum University Press) supply filing No.408 (in Arabic).

The basic approach to the performance test of the ALIP for GENIV sodium fast reactor

H R Kim[1, 3] and Y B Lee[2]

[1] Ulsan National Institute of Science and Technology, 50 UNIST-gil, Uljoo-gun, Ulsan, Republic of Korea
[2] Korea Atomic Energy Research Institute, 150 Dugjin-dong, Yuseong-gu, Daejeon, Republic of Korea

E-mail: kimhr@unist.ac.kr

Abstract. The annular linear induction electromagnetic pump (ALIP) with the flowrate of 900 L/min and the developed pressure of 4 bar has been designed by electric equivalent circuit analysis. It was fabricated by the consideration of materials compatible to the sodium environment of high temperature. Basic characteristic test of the ALIP was carried out in advance for its installation to the integral effect TEst Loop for safety simuLation and Assessment (STELLA) loop to confirm the sodium-thermo-hydraulic components. The test showed that the magnetic field had been linearly increased when the input current was increased, where input current and voltage had represented linear relation each other. The generated electromagnetic force was proportionate to the square of the applied current. The velocity of the aluminium pipe was proportionally increased when the input current was increased. It was verified that the basic characteristic of the ALIP showed a good accordance with the theoretical calculation.

1. Introduction

An annular linear induction electromagnetic pump (ALIP) has been used for transportation of electrically conducting liquid metal [1, 2]. Advantages of no rotating parts which diminish the corrosion due to contact with liquid metal and of no maintenance for a long time, have made the EM pump possible to be alternative for the circulation of sodium coolant in the prototype GENIV sodium fast reactor (PGSFR). The concept for the design of linear induction motors is used for designation of the EM pump. An electrical equivalent circuit with primary and secondary resistances and reactances given by pump geometrical and electromagnetic variables is used for the design analysis of the EM pump. The EM pump with the flowrate of 900 L/min and developed pressure of 4 bar is designed and fabricated considering material for the environment of high temperature sodium [3]. In the present study, the basic operation tests for the pump are carried out on the change of input current before installing it in the sodium loop for verifying the elements of the sodium-thermo-hydraulic experimental system.

2. Design of ALIP

2.1. Basic Concept

[3] Address for correspondence: H R Kim, Ulsan National Institute of Science and Technology, 50 UNIST-gil, Uljoo-gun, Ulsan, Republic of Korea. E-mail: kimhr@unist.ac.kr.

Figure 1 shows the cross sectional view of the annular linear induction EM pump. It has two parts of primary and secondary ones. The electromagnet composed of the stacked silicon-iron cores and driving copper coils forms primary part. The ducts of a narrow annular gap for the sodium flow makes secondary part. First of all, the pump has geometrical variables including inner core diameter, inter-core gap and the outer core length and electromagnetic ones of frequency, input voltage and current. The optimized design variables are drawn by analysis on the developed pressure and the efficiency using the equivalent circuit.

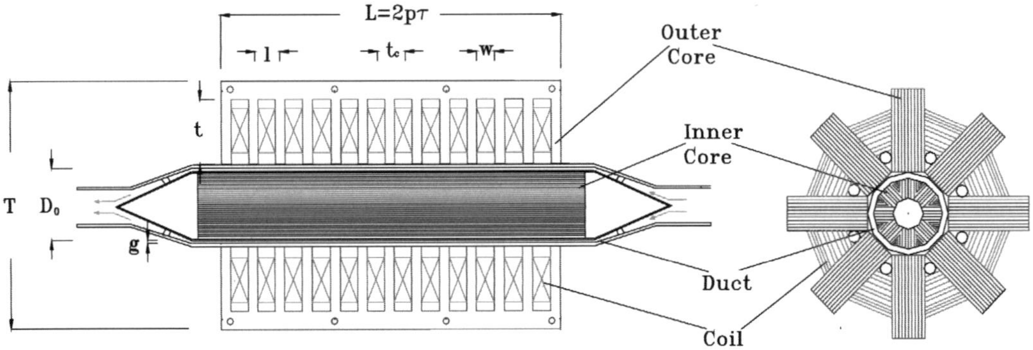

Figure 1. Cross-sectional view of the EM pump.

2.2. Circuit Analysis

The EM pump connected by 3-phase AC power has the resistance by copper winding and liquid sodium, magnetization and leakage reactance. The pump is analyzed on one phase because it has the balanced three-phase inputs. In Figure 2 is represented the electric equivalent circuit for one phase with primary part composed of the coils and cores, and the secondary one of the sodium flowing along the channel of the pump [3].

From balance between input and output power using the electric equivalent circuit variables from the pump geometrical and electromagnetic ones in figure 2, the developed pressure is derived as a function of flowrate and other variables as seen in Equation (1), where slip, s is defined as the ratio of difference between synchronized flowrate and fluid flowrate to the synchronized flowrate [3], where the resistances and reactances of the circuit are represented in Equation (2) ~ (5) given by Laithwaite's standard design formula [3].

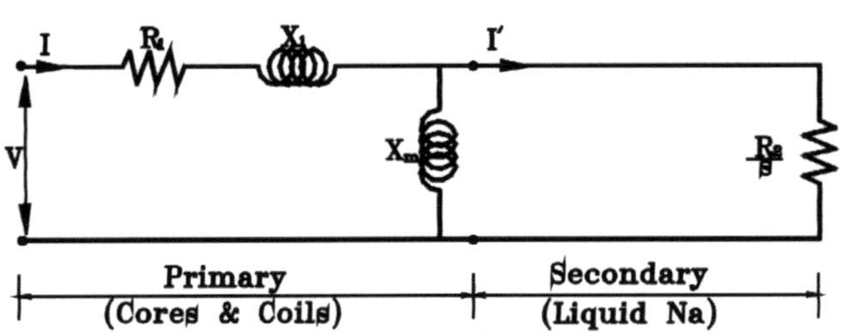

Figure 2. Electric equivalent circuit of the EM pump

$$\Delta P = \frac{3I^2}{Q}\frac{R_2(1-s)}{s(R_2^{\,2}/X_m^{\,2}s^2+1)} \tag{1}$$

$$R_1 = \frac{\pi\rho_c q k_p^{\,2} m^2 D_0 N^2}{k_f k_d p\tau^2} \tag{2}$$

$$X_1 \cong \frac{2\pi\mu_0\omega D_0\lambda_c N^2}{pq} \tag{3}$$

$$X_m = \frac{6\mu_0 \omega \tau \pi D_0 (k_w N)^2}{\pi^2 p g_e} \tag{4}$$

$$R_2 = \frac{6\pi D \rho_r '(k_w N)^2}{\tau p} \tag{5}$$

Substituting Equation (2) ~(5) into Equation (1), the developed pressure of the pump and its efficiency are derived as functions of electromagnetic and geometric variables in Equation (6) and (7).

$$\Delta P = \frac{36\sigma s^2 f \tau^2 (\mu_0 k_w N I)^2}{p g_e^2 \{\pi^2 + (2\mu_0 \sigma s f \tau^2)^2\}} \tag{6}$$

$$\varepsilon = \frac{\Delta P \bullet Q}{\sqrt{3} V I \cos\varphi} = \frac{6 k_w^2 (1-s)}{\left\{ \frac{\rho_c q k_p^2 m^2 \sigma g_e}{k_f k_d \tau} \{1 + (\frac{\pi}{2\mu_0 f s \sigma \tau^2})^2\} + \frac{6 k_w^2}{s} \right\} \cos\varphi} \tag{7}$$

The developed pressure and efficiency are calculated for the different geometrical and electromagnetic variables to be optimized.

3. Fabrication and test of the EM pump

3.1. Fabrication of ALIP

The designed ALIP was fabricated for Sodium integral effect TEst Loop for safety simuLation and Assessment (STELLA-1) for testing components such as sodium-air heat exchanger (AHX) and sodium to sodium decay heat exchanger (DHX) in the STELLA-1 loop. In Figure 3, the completed 3-phase ALIP with a flowrate of 900 L/min and a developing pressure of 4 bar was shown [3], where the ALIP would be installed in a vertical direction to meet full filling of liquid sodium in the pump as seen in figure 3.

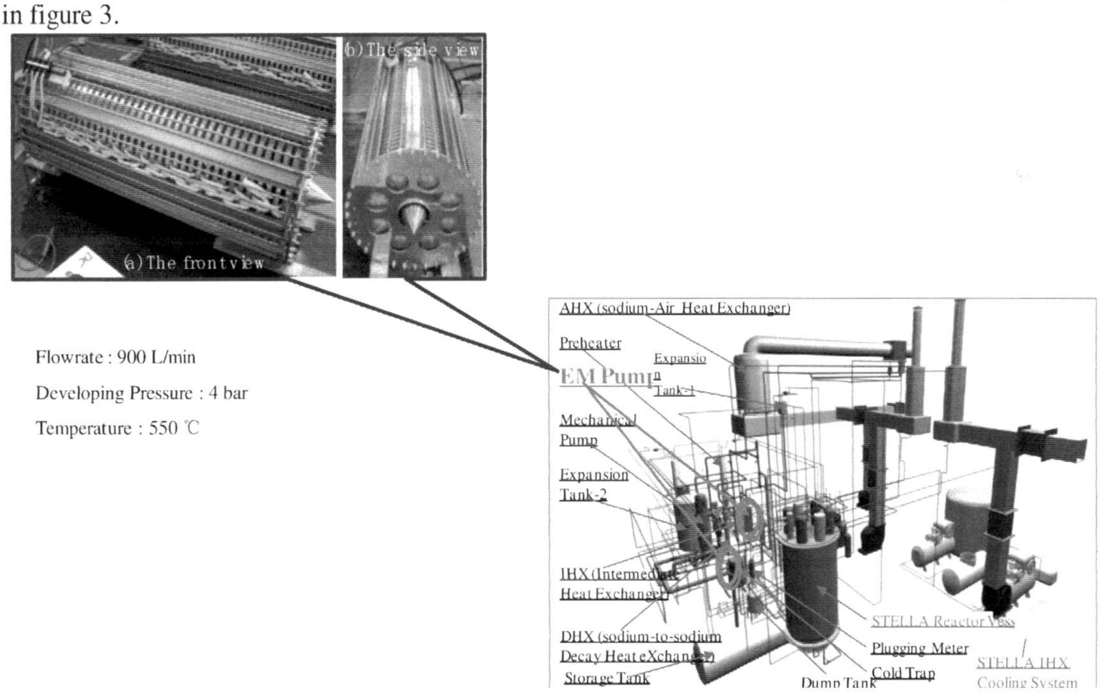

Figure 3. The completed ALIP for sodium circulation in STELLA loop

Figure 4 represented the system for the basic operation test of the ALIP where input voltage was controlled by using 3-phase AC slidacs with range of 0~380 V. Solid aluminum pipe with outer and inner diameter of 130 mm and 122 mm each was inserted into the narrow annular gap of the pump.

The force generated by pump and velocity of the aluminum pipe, which were equivalent to the developed pressure and flow velocity of sodium flow, respectively, were measured when the voltage and current were changed in the room temperature. The electrical conductivity of the aluminum was order of 10^7 /(ohm m), similar to that of sodium while the density and viscosity are different each other, and the phases of the aluminum and liquid sodium are solid and liquid. However, the operation of the pump can be verified due to the same principle of electromagnetic force generation on the aluminum and sodium with high electrical conductivity.

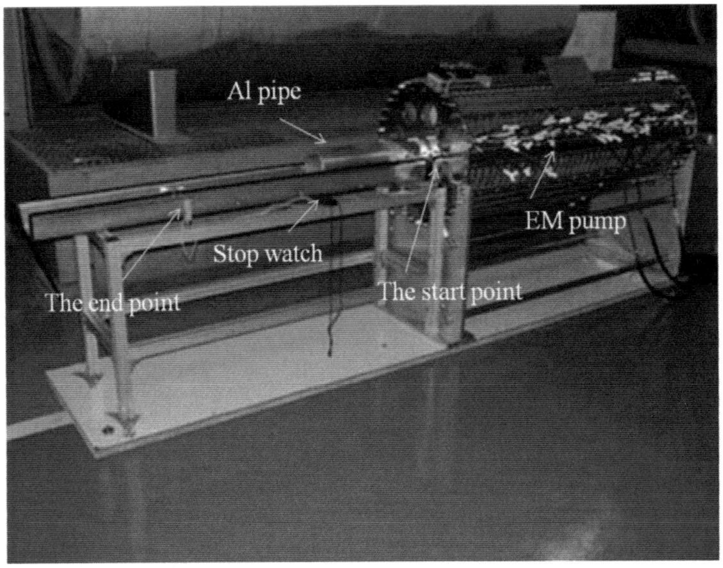

Figure 4. The basic operation test device of the EM pump

3.2. The basic function test of the EM pump

The input current of the pump was linearly increased as input voltage of each phase (R, S and T) of three phase power is increased as represented in figure 5.

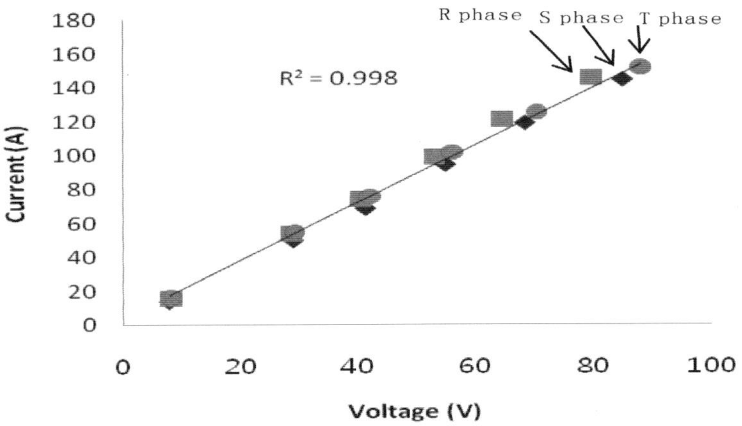

Figure 5. The input current and voltage applied to the EM pump

The magnetic flux density in the gap of the pump, where the liquid sodium flows, was linearly increased when the input current was increased as understood by Ampere's law [4] in figure 6 (a). It was twice higher in case of 2–parallel connection than in case of 6-parallel connection where the pump had six pole pairs. Figure 6 (b) represented that the stand-still force was increased in proportion to the square of input current as it was increased. It was thought because the developed pressure corresponding to the electromagnetic force generated in the pump was given by the function of square of the current in Eqn. (8), where the ratio of secondary resistance to magnetizing reactance multiplied by slip, $R_2/(X_m s)$, of the pump was negligible.

$$\Delta P = \frac{3I^2}{Q}\frac{R_2(1-s)}{s(R_2^2/X_m^2 s^2+1)} = \frac{36\sigma s^2 f\tau^2(\mu_0 k_w NI)^2}{pg_e^2\{\pi^2+(2\mu_0\sigma sf\tau^2)^2\}} \quad (8)$$

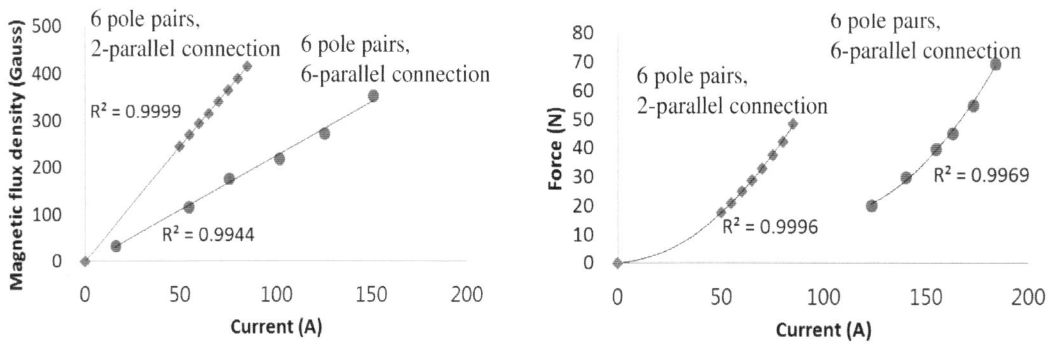

Figure 6. (a) The magnetic induction Figure 6. (b) Standstill force

Figure 6. (a) The magnetic induction and (b) standstill force of the ALIP

The magnitude of the generated force was much larger when the pump was connected in the 2–parallel method than in the 6-parallel method. As a result, the pump was understood to generate larger pumping power when it was connected 2–parallelly.

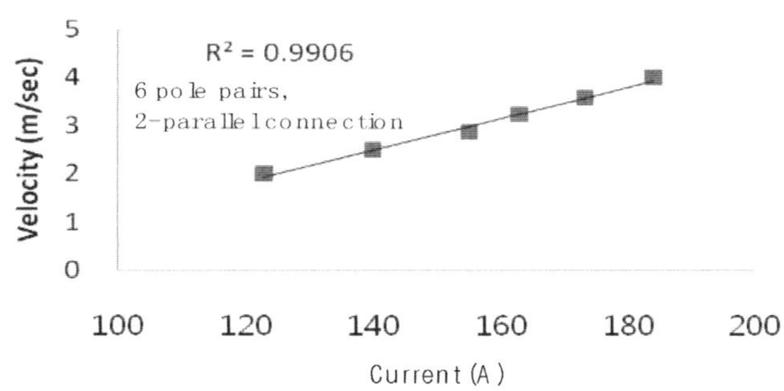

Figure 7. The measurement of velocity of the aluminium pipe in the annular channel of the EM pump

In figure 7, the higher the current was, the larger the velocity of the aluminium pipe was. The measurement showed that its velocity and current were linear. Actually, the factor, $(2\mu_0\sigma sf\tau^2)^2$, in Equation (8) is numerically much small by calculation for the designed ALIP and slip is proportionate to velocity. Therefore, the relation between the velocity and current is found to be linear when the developed pressure is fixed in Equation (8).

4. Conclusion

The ALIP for sodium circulation in the STELLA loop was designed and completed using equivalent circuit analysis for the requirement of the flowrate of 900 L/min and developed pressure of 4 bar. It was understood the basic characteristic experiment including force and velocity showed a good accordance with the theoretical prediction.

Nomenclature

$\cos\varphi$	power factor
$\boldsymbol{D}$	outer duct diameter [m]
$\boldsymbol{D_0}$	inner core diameter [m]
f	input frequency [Hz]
g	gap between outer and inner cores [m]
g_e	effective gap [m]
I	applied current [A]
K	hydraulic pressure loss coefficient
k_d	slot depth (t) / slot width (w)
k_f	slot filling factor (0.5-0.6)
k_p	slot pitch (t_c) / slot width (w)
k_w	winding factor
l	tooth pitch [m]
L	pump core length
m	number of phase
N	coil turns
p	number of pole pairs
q	number of slot per phase per pole pair
R_1	primary equivalent resistance [ ]
R_2	Secondary equivalent resistance [ ]
Q	flowrate of the sodium [m^3/s]
Q_s	synchronized flowrate [m^3/s]
s	slip (=1-Q/Q_s)
t	slot depth [m]
T	pump outer core diameter [m]
t_c	slot pitch [m]
w	slot width [m]
X_m	magnetizing reactance [ ]
ΔP	developed pressure of the pump [N/m^2]
ΔP_L	head loss in the loop [N/m^2]
ε	pump efficiency
μ_0	magnetic permeability of vacuum [H/m]
φ	phase angle between voltage and current
τ	pole pitch [m]
ρ	sodium density [kg/m^3]
σ	electrical conductivity of the fluid [1/]
ω	angular frequency (=$2\pi f$)

References

[1] Blake L R 1959 Electro-magnetic pumps for liquid metals *Journal of Nuclear Energy Part B Reactor Technology* **1** (2) 65-76
[2] Carroll D G and Boardman C E 2002 The super-PRISM reactor system *Nuclear Engineer* **43** (6) 165–168
[3] Kim H R 2010 The design of the annular linear induction em pump with a sodium flowrate of 35 kg/sec *Korean Nuclear Society Spring Meeting* Pyeonhchang Korea
[4] John D K 1983 *Electromagnetics* (New York : McGraw-Hill Book Company)
[5] Merle C P and David C W 1991 *Mechanics of Fluids* (Tokyo: Prentice-Hall International Inc.)

Co-hydrogasification of lignocellulosic biomass and swelling coal

J Zhang, N Zheng and J Wang[1]

Department of Chemical Engineering for Energy, East China University of Science and Technology, 130# Meilong Road, Shanghai 200237, PR China

E-mail: jwang2006@ecust.edu.cn

Abstract. The hydrogasification of pine wood (PW) and rice husk (RH) was carried out in a two-stage fixed-bed reactor to investigate the effects of hydrogen pressure and hydrocracking temperature on the yields of gas and tar compositions. The elevation in hydrogen pressure promoted the conversion of two biomasses, leading to the improvement in gaseous hydrocarbons but resulted in a decrease in the yield of BTX (benzene, toluene and xylene). The increased severity of hydrocracking boosted the yield of methane, ethane and BTX mainly at the expense of heavy compounds in tar for PW under 1 MPa. The co-hydrogasification of biomass and DWG swelling coal chiefly showed a synergistic effect on the yields of BTX and PCX (phenol, cresol and xylenol) at 500 °C hydrocracking temperature under 5 MPa.

1. Introduction

Lignocellulosic biomass, as a relatively abundant and renewable carbon source alternative to the traditional fossil resources, has recently received great interest to produce bio-fuels and bio-chemicals. The hydrogasification of biomass may be an effective way to produce energy-dense oil enriched with light aromatics as well as methane rich gases [1]. Pindoria *et al.* [2] used sugarcane bagasse for pyrolysis under different atmosphere of hydrogen and helium, and found that the existence of hydrogen inhibited the repolymerization of tar. Pütün *et al.* [3] conducted pyrolysis and hydropyrolysis of sunflower bagasse at different temperatures, heating rates and pressures, and observed that pressurized hydrogen (> 50 bar) increased the oil yields by up to about 10 wt.%.

The availability of biomass largely relies on seasons and geographical environment. During the last few years, the co-pyrolysis of coal and biomass has become a technique that can manage seasonal fluctuations in biomass availability and increase energy density of the feedstock [4]. Collot *et al.* [5] used coal and biomass samples for co-pyrolysis and co-gasification with the final temperature of 850 °C or 1000 °C and pressure up to 25 bar, but the gas and liquid compositions were not determined. However, so far, there are seldom any literature reports referring to the effect of hydrogen atmosphere on product distribution and bio-oil composition, nor the co-hydrogasification of biomass and coal.

The aim of this work is to investigate the formation of gaseous and oil products from a pinewood (PW) and a rick husk (RH) under pressurized hydrogen atmosphere. The co-hydrogasification of RH

[1]Address for correspondence: J Wang, Department of Chemical Engineering for Energy, East China University of Science and Technology, 130# Meilong Road, Shanghai 200237, PR China. E-mail: jwang2006@ecust.edu.cn.

and DWG subbituminous swelling coal was also checked to provide insights into the mutual effects between biomass and coal.

2. Experimental

The proximate and ultimate analyses of two biomass samples (pine wood and rice husk) and a subbituminous coal featured in strong swelling are shown in Table 1. The process flow diagram of the pressurized reactor had been shown elsewhere [6]. In this case, the vertical tubular reactor was heated by the electric furnace by two separate temperature controllers. The upper zone of reactor was preheated at a heating rate of 20 °C min^{-1} and held at 500 °C or 700 °C, and the sample (2.5 or 4 g) or the mixture of biomass and coal (mass ratio=1:1) at the bottom of reactor was then heated from ambient temperature to 700 °C at a heating rate of 15 °C min^{-1}. The temperature in upper zone is called the hydrocracking temperature, and the temperature in lower zone is referred to the devolatilization temperature. The hydrogen gas flowed from the inlet at the bottom of the reactor to the outlet at the top. The gas flow rate was controlled at 100 ml min^{-1} under 1.0 MPa and 500 ml min^{-1} under 5.0 MPa to ensure the same residence time. To prevent the unexpected condensation of volatile matter and preheat the hydrogen flow, the heating tape twined around the tubular lines which were connected to the inlet and outlet of reactor and was kept at the temperature of 260 °C. A series of stainless steel traps were dipped into the cooled salt water (−12 to 6 °C) to collect the liquid product and expect that the last section was cooled by liquid nitrogen (~−196 °C) to condensate liquid products as much as possible. The yield of liquid product was determined by analytical balance. Char product was collected after the experiment. The incondensable gases were gathered in the gas bags for compositional analysis. The analytical method of these products can be seen elsewhere [6].

Table1. Proximate and ultimate analyses of the biomass and coal samples.

	Proximate analysis (wt.%, ad basis)				Ultimate analysis (wt.%, daf basis)				
	Moisture	Ash	Volatile	Fixed carbon	C	H	N	S	O[a]
Pine wood	14.14	0.52	72.32	13.02	49.96	5.78	0.10	0.00	44.16
Rice husk	1.55	9.86	72.58	16.01	51.20	6.17	0.36	0.10	42.17
DWG	0.43	15.97	28.88	54.72	85.29	5.19	1.50	2.76	5.26

[a] by difference.

3. Results and discussions

3.1. Overall distribution of products

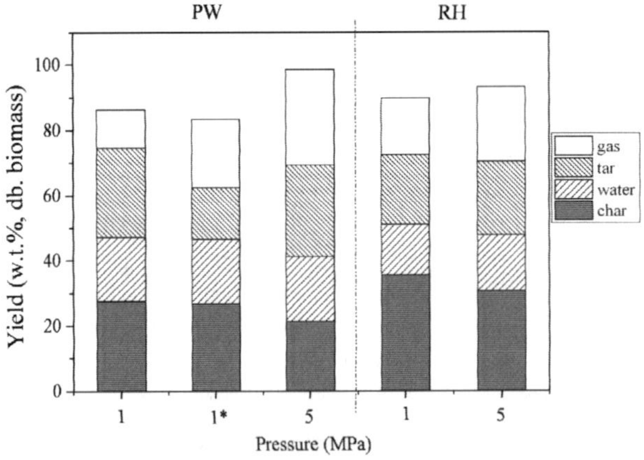

Figure 1. Yields of gas, tar, water and char products from hydrogasification at the different hydrocracking temperature of 500 °C and 700 °C under 1 MPa and 5 MPa for PW and RH biomass. * cracking temperature: 700 °C and in all other cases, 500 °C.

Figure 1 shows the gross distributions of gaseous, tar, water and char products from two biomass samples at 1 MPa and 5 MPa where all yields are counted in mass percent on a dry biomass basis. The yield of gas contains the yields of seven gases (CH_4, CO_2, CO, C_2H_6, C_2H_4, C_3H_8 and C_3H_6) regardless of other gaseous products, and the yield of water deducted the amount of moisture used in the air-dried sample.

With the increasing hydrogen pressure from 1 MPa to 5 MPa at the same hydrocracking temperature of 500 °C, the yield of gaseous products derived from PW increased at the expense of the char products. The vaporization of volatile matter may be retarded by high hydrogen pressure, and thus the secondary reaction of volatile matter was strengthened to produce more gases. The water yield remained almost constant at a level of 19.5%. The tar yield increased slightly from 27.4% to 28.12%. The hydrogasification behavior of RH was similar to that of PW.

With the increasing hydrocracking temperature from 500 °C to 700 °C under 1 MPa hydrogen pressure, the char yield and water yield obtained from PW almost remained unchanged. However, the tar yield decreased from 27.4% to 15.8% and this corresponded to the increase of gaseous products. It is because the secondary reaction of volatile matter was promoted by the increasing hydrocracking temperature to release more gases.

3.2. Yields of main liquid products

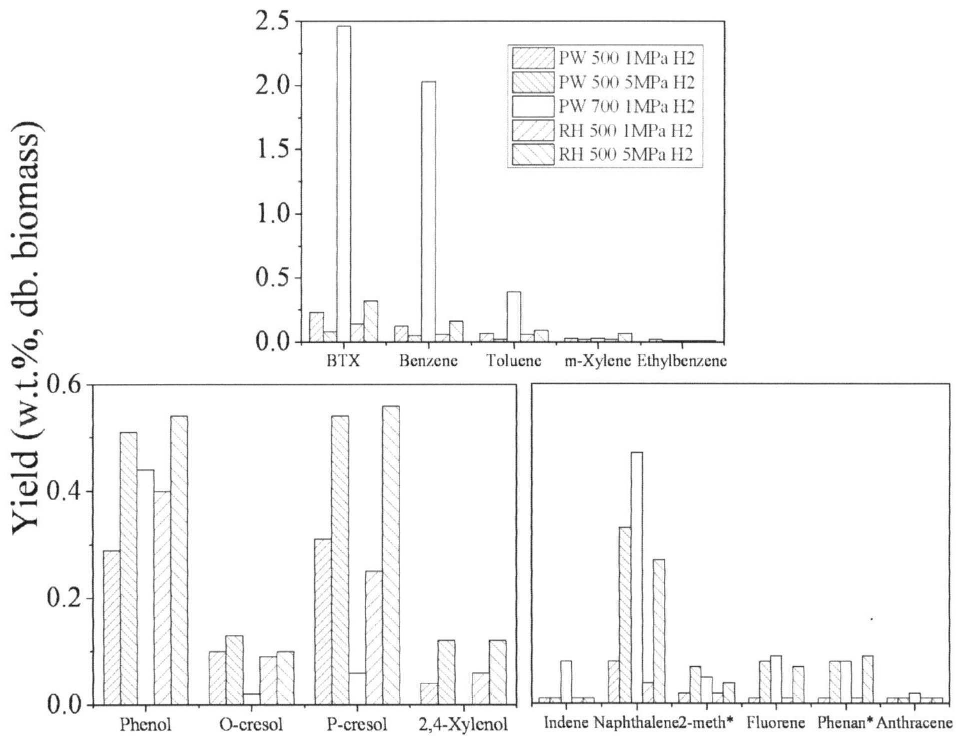

Figure 2. Quantitative analysis of 14 compounds in tar obtained from hydrogasification of two biomasses at two-stage reactor under 1 MPa and 5 MPa. 2-meth*, 2-methyl-naphthalene. Phenan*, Phenanthrene.

Figure 2 shows the yields of bio-oil products from two biomass samples at different hydrocracking temperatures under different pressures. With the hydrogen pressure increasing from 1 MPa to 5 MPa at the same hydrocracking temperature of 500 °C, the yield of BTX (benzene, toluene, m-xylene) derived from PW decreased slightly, but the yield of phenolic and polycyclic aromatics increased, of which phenol, p-cresol, 2, 4-xylenol, naphthalene, 2-methyl-naphthalene, fluorene and phenanthrene increased obviously and o-cresol, indene and anthracene showed a minor increase. When comparing the yields of compounds obtained from PW by increasing hydrocracking temperature from 500 °C to 700 °C under 1 MPa, it was observed that the yield of BTX was soundly enhanced, especially for benzene and toluene. The yield of phenol and polycyclic aromatics also showed pronounced increase, while the o-cresol, p-cresol and 2,4-xylenol presented a decline trend, indicating that the high hydrocracking temperature facilitated the secondary reactions of volatile matter to produce more light aromatics such as BTX and phenol. The elevation off hydrogen pressure from 1 MPa to 5 MPa showed a similar impact on changes of main compounds yields from RH bio-oil in disregard of the decline in 2, 4-xylenol yield.

3.3. Yields of individual gas products

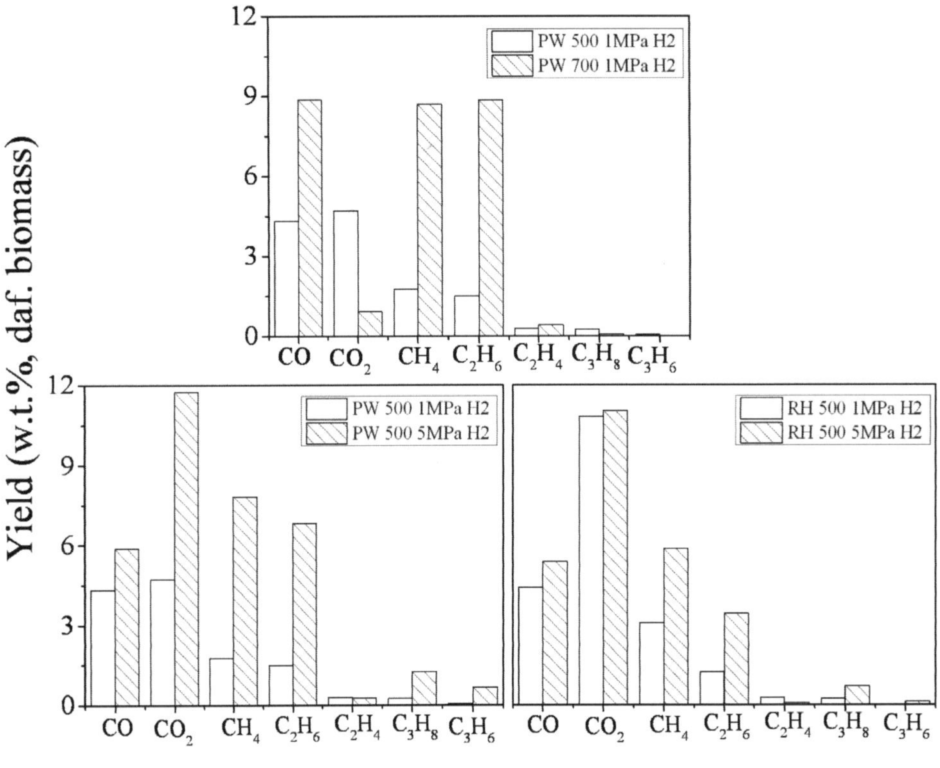

Figure 3. Quantitative analysis of main gases derived from hydrogasification of two biomass at the hydrocracking temperatures of 500 °C and 700 °C under 1 MPa and 5 MPa.

Figure 3 shows the yield of gaseous products of PW and RH under hydrogen pressures of 1 MPa and 5 MPa at the hydrocracking temperatures of 500 °C and 700 °C. Firstly, the high hydrogen pressure led to the increase in yields of CO, CO_2, CH_4, C_2H_6, C_3H_8 and C_3H_6. Meanwhile, the yield of C_2H_4 remained almost constant. The yield of CO_2 decreased while the yield of CO increased with the increasing hydrocracking temperature. This was reasonable because the reversal water-gas shift reaction was promoted by the high hydrocracking temperature thermodynamically:

$$CO_2 + H_2 \rightleftarrows CO + H_2O \quad \Delta H = +41\ KJ/mol \tag{1}$$

The yield of CH_4 and C_2H_6 increased substantially while the yield of C_2H_4 merely changed at 700 °C hydrocracking temperature. However, C_3H_8 disappeared due to its decomposition at high hydrocracking temperature. The pressurized hydrogen facilitated the yields of CO, CH_4 and C2~C3 hydrocarbons for RH, but the change of CO_2 was unremarkable.

3.4. Co-hydrogasification of RH and DWG

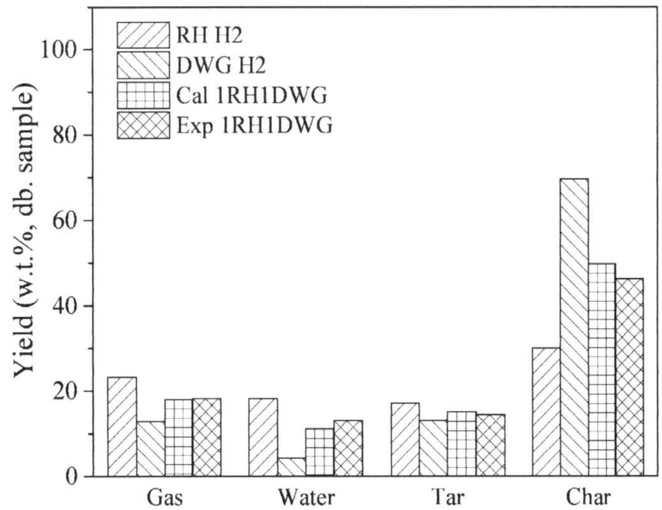

Figure 4. Yields of gas, water, tar and char products produced by hydrogasification at 700 °C for individual sample or mixture of RH and DWG. Hydrocracking temperature, 500 °C; heating rate, 15 °C min^{-1}. Holding time, 0h.

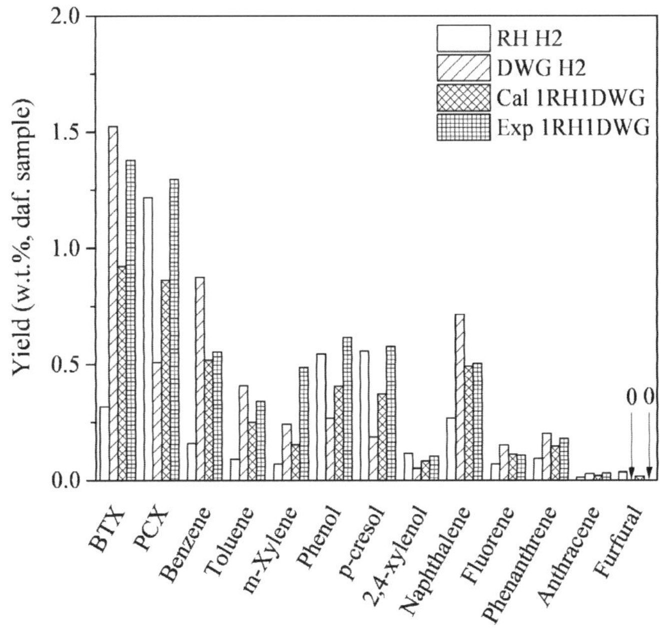

Figure 5. Yields of 11 light aromatics obtained under 5 MPa without holding time at the final temperature of 700 °C for the hydrogasification of individual sample or mixture of RH and DWG.

The co-hydrogasification of RH and swelling DWG coal was investigated to check whether the mutual effects existed by adding fossil fuel into renewable sources under H_2 atmosphere. Figure 4 shows the distributions of gas, water, tar and char products obtained by the hydrogasification of RH/DWG or co-hydrogasification of RH and DWG mixture. Compared with the distribution of RH products, the hydrogasification of DWG produced exhibited much higher char yield as well as lower water and gas yield. This is reasonable because RH contained higher organic oxygen content (Table 1) to produce water and carbon oxides. The co-hydrogasification of RH and DWG mixture showed lower char yield than calculated char yield assuming no mutual effect. It indicated that the co-hydrogasification of RH and DWG can promote the total conversion of mixture. The gas and tar yield almost stayed constant, despite a slight increase in water yield.

Table 2 shows the yields of CO, CH_4, CO_2 and C2~C3 hydrocarbons obtained by hydrogasification or co-hydrogasification process. It can be seen that methane mainly derived from DWG coal and carbon oxides chiefly produced by RH during hydrogasification. But those two sample showed similar patterns in gas yields of C2~C3. The co-hydrogasification of RH and DWG merely showed mutual effect on alterations of final gas compositions. It demonstrated that no synthetic effect existed in gas phase.

Figure 5 presents the yields of 11 main light aromatics quantitatively determined under 5 MPa at the final temperature of 700 °C for hydrogasification or co-hydrogasification process. DWG produced higher yield of BTX while RH harvested more PCX. It is interesting to found that the co-hydrogasification significantly enhanced the yield of both BTX and PCX, but showed minor influences on other aromatics. The strengthened yields of BTX and PCX could be probably due to the addition of biomass into coal. It destroyed the swelling behaviour of DWG during heating-up period, which benefit the escaping of volatile matter from coal particles. On the other hand, the biomass contained appreciable amount of potassium (0.42wt.%, dry sample basis) which has been widely regarded as a catalyst during thermal conversion of coal. It is possible that this kind of potassium would act as active sites which promote the hydrocracking of primary tar compositions into mono-cyclic enriched aromatics during co-hydrogasification.

Table 2. Yields of CO, CH_4, CO_2 and C2~C3 hydrocarbons (wt. %- daf.) at 700 °C hydrogasification temperature for individual sample or mixture of RH and DWG. Hydrocracking temperature, 500 °C.

Sample	CO	CH_4	CO_2	C_2H_6	C_2H_4	C_3H_8	C_3H_6
RH	4.3	5.9	12.5	2.4	0.1	0.5	0.1
DWG	1.1	10.8	0.7	2.3	0.0	0.4	0.0
1RH1DWG[a]	2.7	8.3	6.6	2.4	0.1	0.5	0.1
1RH1DWG[b]	3.0	8.4	6.8	2.3	0.1	0.5	0.0

[a] calculated value.
[b] experimental value.

4. Conclusions

At a lower cracking temperature of 500 °C, the production of phenolic and polycyclic aromatics was enhanced with increasing hydrogen pressure, and BTX decreased slightly for hydrogasification of lignocellulosic biomass. Increasing the cracking temperature up to 700 °C significantly promoted the formation of BTX. Co-hydrogasification significantly enhanced the yield of both BTX and PCX. More studies are needed to investigate the co-effect of temperature and hydrogen pressure.

A. 1 Nomenclature

Symbol	Definition
PW	Pine wood
RH	Rice husk
BTX	Benzene, toluene and xylene
PCX	Phenol, cresol and xylenol
DGW	The subbituminous coal from the place of origin named DWG
FID	Flame ionization detector
TCD	Thermal conduct detector
H_2	Hydrogen gas
CH_4	Methane
CO	Carbon monoxide
CO_2	Carbon dioxide
C_2H_6	Ethane
C_2H_4	Ethylene
C_3H_8	Propane
C_3H_6	Propylene
C2	Ethane and Ethylene
C3	Propylene and Propane

Acknowledgements
This work is funded by the National "863" Scientific Research Program (2011AA05A201) and the National Natural Science Foundation of China (Grand No. 21376080).

References
[1] Balagurumurthy B and Bhaskar T 2013 Hydropyrolysis of lignocellulosic biomass: state of the art review *Biomass Conversion and Biorefinery* **4** 67-75
[2] Pindoria R V, Lim J Y, Hawkes J E, Lazaro M J, Herod A A and Kandiyoti R 1997 Structural characterization of biomass pyrolysis tars/oils from eucalyptus wood waste: effect of H2 pressure and sample configuration *Fuel* **76** 1013-23
[3] Pütün A E, Koçkar Ö M, Yorgun S, Gerçel H F, Andresen J, Snape C E and Pütün E 1996 Fixed-bed pyrolysis and hydropyrolysis of sunflower bagasse: Product yields and compositions *Fuel Processing Technology* **46** 49-62
[4] Ellis N, Masnadi M S, Roberts D G, Kochanek M A and Ilyushechkin A Y 2015 Mineral matter interactions during co-pyrolysis of coal and biomass and their impact on intrinsic char co-gasification reactivity *Chemical Engineering Journal* **279** 402-8
[5] Collot A G, Zhuo Y, Dugwell D R and Kandiyoti R 1999 Co-pyrolysis and co-gasification of coal and biomass in bench-scale fixed-bed and fluidised bed reactors *Fuel* 78 667-79
[6] Zhang J, Wang X, Wang F and Wang J 2014 Investigation of hydrogasification of low-rank coals to produce methane and light aromatics in a fixed-bed reactor *Fuel Processing Technology* **127** 124-32

IOP Conf. Series: Earth and Environmental Science **40** (2016) 012043 doi:10.1088/1755-1315/40/1/012043

Design of multi-energy fields coupling testing system of vertical axis wind power system

Q Chen, Z X Yang[1], G S Li, L Song and C Ma

Institute of Systems Science and Engineering, Henan Engineering Laboratory of Wind Power Systems, Henan University of Science and Technology, No.48 Xiyuan Road, Luoyang 471003, China

E-mail: zxyang@haust.edu.cn

Abstract. The conversion efficiency of wind energy is the focus of researches and concerns as one of the renewable energy. The present methods of enhancing the conversion efficiency are mostly improving the wind rotor structure, optimizing the generator parameters and energy storage controller and so on. Because the conversion process involves in energy conversion of multi-energy fields such as wind energy, mechanical energy and electrical energy, the coupling effect between them will influence the overall conversion efficiency. In this paper, using system integration analysis technology, a testing system based on multi-energy field coupling (MEFC) of vertical axis wind power system is proposed. When the maximum efficiency of wind rotor is satisfied, it can match to the generator function parameters according to the output performance of wind rotor. The voltage controller can transform the unstable electric power to the battery on the basis of optimizing the parameters such as charging times, charging voltage. Through the communication connection and regulation of the upper computer system (UCS), it can make the coupling parameters configure to an optimal state, and it improves the overall conversion efficiency. This method can test the whole wind turbine (WT) performance systematically and evaluate the design parameters effectively. It not only provides a testing method for system structure design and parameter optimization of wind rotor, generator and voltage controller, but also provides a new testing method for the whole performance optimization of vertical axis wind energy conversion system (WECS).

1. Introduction

The clean and renewable energy sources such as solar and wind energy are gaining importance due to their less influences to environmental. The rapidly increasing demand of electrical energy, the issues associated with limited reserves and the rising cost of fossil fuels such as oil, coal, and natural gas are also responsible for the growth and rise of renewable energy application [1]. Although the capacities of WT units have increased from a few tens of kW power capacity to multi-MW level over the past few decades, the conversion efficiency of WECS is also the focus of researches and concerns [2].

The WECS is mainly composed of wind rotor, generator, storage battery and voltage controller. Now more advanced generators, improved wind rotor, power converter systems and optimal control

[1] Address for correspondence: Z X Yang, Institute of Systems Science and Engineering, Henan Engineering Laboratory of Wind Power Systems, Henan University of Science and Technology, No.48 Xiyuan Road, Luoyang 471003, China. E –mail: zxyang@haust.edu.cn.

solutions have developed to enhance the conversion efficiency of wind energy, such as Daroczy and so on used two-dimensional CFD simulation and parameter optimization for H style darrieus rotor turbulence aerodynamic modeling [3]. McTavish and so on proposed a new type of composite structure of vertical shaft wind turbines, and conducted two-dimensional steady and three-dimensional dynamic CFD simulation analysis [4]. Considering two different vertical shaft WT, Li yan team attempted to explore the torque, power characteristics and the influence of real parameters through performance simulation and wind tunnel experiment [5,6]. Aubree proposed a sensorless maximum power point tracking method for small wind power permanent magnet synchronous generator with dc load [7]. Qi ZY proposed an integrated power control method by involving the maximum power point tracking, load power tracking control, overspeed protection control and battery management optimization [8]. The method above are mostly by improving the wind rotor structure, optimizing the generator parameter and energy storage controller to enhance the conversion efficiency separately. Though the conversion efficiency has improved in varying degree after using different methods, the effect is not particularly evident. While when different energy conversion occurs, such as wind rotors convert air kinetic energy into mechanical energy, generators convert mechanical energy into electrical energy, batteries convert electrical energy into chemical energy, the coupling effect between them will occur and influence the overall conversion efficiency. At present, the research of wind power MEFC technology is still in blank, therefore this research has a certain practical significance.

The purpose of this paper is to provide a MEFC testing method for the vertical axis WECS, the aim is to solve the existing technology question that can only test one part of the system separately. This method can test the whole WECS performance systematically, when the coupling parameters are allocated to an optimal state, it can improve the overall conversion efficiency of WECS.

2. Composition of MEFC system

The main WECS equipment designed in this paper includes the alternating current (AC) frequency converter, the open jet return flow wind tunnel, the vertical axis wind rotor, the permanent magnet generator, the storage battery and the load box. To keep it simple in following, these equipment are simplified as converter, wind tunnel, wind rotor, generator, battery and load respectively.

The testing and auxiliary equipment of MEFC system include the anemometer, disc coupling sensor, UCS, three-phase electric parameter tester, voltage controller, inverter and power management device. The composition block diagram of this system is shown in figure1.

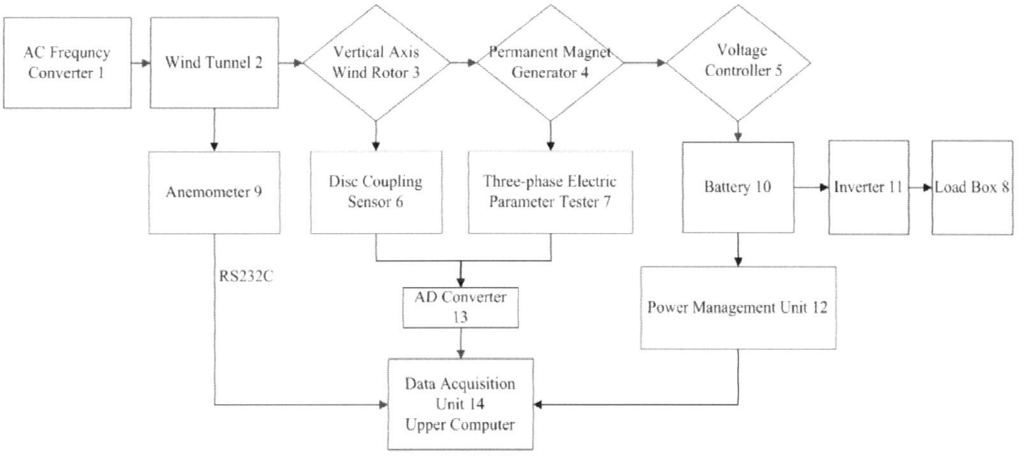

Figure 1. The block diagram of MEFC testing system.

In the wind tunnel there is a ventilator which speed can be controlled by the converter. There are three test-beds in the wind field of the wind tunnel outlet. The first test-bed is used to install the vertical axis wind rotor specimen, it equips an anemometer which is used to measure the wind speed.

The second test-bed is used to install the generator specimen which connects to the vertical axis wind rotor specimen, it equips an three-phase electric parameters monitoring device which is used to measure the output power of the generator specimen. Between the first test-bed and the second test-bed there is a torque sensor which is used to measure the torque of wind rotor specimen, here the torque sensor is a disc coupling sensor. The third test-bed is used to install the voltage controller specimen which can convert the three-phase alternating current from the generator specimen into the direct current, and it also equips an power management device which is used to measure the output power of voltage controller specimen.

The UCS is used to collect and analyze the data from measuring instruments, the data acquisition unit of the UCS is used to keep communication connection to the anemometer, torque sensor, three-phase electric parameters monitor device and power management device. The output of torque sensor and three-phase electric parameters monitoring device all connect to the data acquisition unit through A/D conversion module. The output of voltage controller specimen connects to the battery, and the battery connects to load through the inverter.

The working process of this system is as follows. The size of the flow velocity of wind tunnel is controlled by the converter, when change the power frequency, the flow velocity is changed accordingly. The rotor specimen converts wind energy into mechanical energy. The torque sensor detects the performance parameters of wind rotor specimen, such as output rotational speed, torque, power and so on, and it converts those analog signals into digital signal through A/D module. Generator converts the mechanical energy into electrical energy, three-phased electric parameter tester detects the performance parameters of generator, such as the output current, voltage, power and so on, and converts those analog signals into digital signals through A/D module, then we can obtain the conversion efficiency characteristics of generator through the values detected by the torque sensor and three-phased electric parameter tester. The voltage controller specimen converts the unstable output from generator into the charging voltage and current of the battery. Through the analysis and calculation of the electricity parameters detected by power management device and the test values from three-phase electric parameter tester, we can obtain the efficiency characteristics of voltage controller specimens. The stored energy of battery is released through the load box. The UCS conducts the data collection for MEFC analysis.

3. The basic mode of MEFC

The performance parameters of MEFC testing system designed in this paper include the wind energy E_W, the output power E_M of wind rotor, the output power E_G of generator and the electricity power parameter E_C of voltage controller.

As shown in equation (1), the power of wind energy can be calculated according to the air density ρ, the swept area A of the wind rotor and the wind speed V.

$$E_W = \frac{1}{2} \times \rho \times A \times V^3 \tag{1}$$

The output power of wind rotor E_M can be calculated according to the rotor torque T and rotor speed n which are detected by the disc coupling sensor, as shown in equation (2).

$$E_M = \frac{2\pi}{60} \times T \times n \tag{2}$$

The output power of the generator E_G can be calculated according to the single output voltage U and current I which are detected by the disc coupling sensor. When the load is the resistive, the computation formula of the generator output power E_G is equation (3).

$$E_G = 3 \times U \times I \tag{3}$$

The conversion energy E_C of the voltage controller can be calculated by the output voltage U and current I which are detected by the power management system.

$$E_C = U \times I \tag{4}$$

In the WECS, The rotor transformed the wind energy into mechanical energy, the utilization of wind energy is η_M. The generator transformed the mechanical energy into the electric power, the utilization of generator is η_G. The working efficiency of the voltage controller is η_C. Those corresponding calculating formulas are as follows.

$$\eta_M = \frac{E_M}{E_W} \times 100\% \tag{5}$$

$$\eta_G = \frac{E_G}{E_M} \times 100\% \tag{6}$$

$$\eta_C = \frac{E_C}{E_G} \times 100\% \tag{7}$$

Here, the output parameters (rotor torque, rotational speed and power) of wind rotor are the coupling design parameters which are determined by the magnetic circuit calculation and structure of generator, while the wind rotor output parameters correspond to the maximum wind energy utilization rate which is captured by the wind energy utilization curve (n-η_M characteristic curve of wind speed as shown in figure 2(a) [9]. The generator output parameters (voltage, current and power) are the input design parameters of voltage controller [10], while The generator output parameters correspond to the maximum efficiency which is obtained from the generator characteristic curve (n-η_G characteristic curve of rotational speed as shown in figure 2(b). The MEFC testing method proposed in this paper can measure the comprehensive performance of WECS. The calculation formula of the overall efficiency of this system is equation (8).

$$\eta = \eta_M \times \eta_G \times \eta_C \tag{8}$$

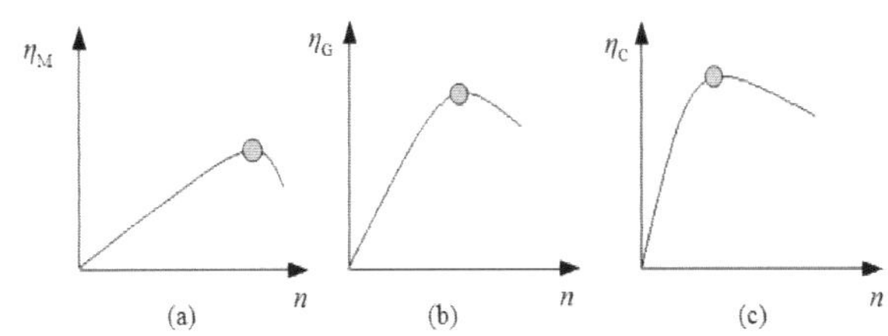

Figure 2. The characteristic curves-(a) is the wind rotor efficiency curve, (b) is the generator efficiency curve, (c) is the battery controller efficiency curve.

Through the characteristics of these parameters, we can get the coupling relationship between the properties which are used to coupling performance test.

When design the whole coupling controller, we can use the following state equation to describe it.

$$\begin{cases} \dot{x} = A(t)x + B(t)u \\ y = C(t)x + D(t)e \end{cases} \tag{9}$$

Here, A is the state matrix which is determined by the structure parameters of WECS. B is the input matrix which is determined by the controller parameters of energy transformation and storage system. C is the output matrix which determined the output information y. D is the direct transmission matrix which represents the error of the testing process.

$x = (x_1, x_2, x_3, x_4)^T$ is the state vector, x_1 and x_2 are the speed and output torque of wind wheel, x_3 and x_4 are the charging current and SOC of battery.

$u = (u_1, u_2, u_3, u_4)^T$ is the controlled variable, u_1 is the wind speed, u_2 is the charging voltage, u_3 and u_4 are the internal resistance and temperature of battery.

According the control equation and the corresponding coupling theory, we can design the controller which we needs.

4. Component description and fundamental principle of testing system

As shown in figure 1, the converter 1 provides the variable AC frequency power source for ventilator of wind tunnel 2, the wind tunnel 2 provides wind field simulation parameters (uniform laminar flow velocity *V*) for MEFC testing. The wind rotor testing specimen 3 converts the wind energy into mechanical energy under the action of the wind speed *V*, the generator specimen 4 converts the mechanical energy into electrical energy, the voltage controller specimen 5 makes the unstable electric energy into stable power source for battery 10. The electric energy stored in the battery 10 is absorbed by the load 8 after the inverter 11. The power management device refers to the battery management system.

The output characteristic of wind tunnel 2 is detected by the anemometer 9, the testing parameters (wind speed *V*) are transmitted to the data acquisition unit 14 through RS232C communication mode. The output characteristic parameters obtained from the disc coupling sensor 6 of the wind rotor specimen 3, such as the rotational speed *n*, torque *T* and mechanical power E_M and so on, are transmitted into the data acquisition unit 14 through AD converter 13. The output characteristics of the generator specimen 4 are detected by three-phase electric parameter tester 7, its characteristic parameters including the voltage *U*, current *I*, electric power E_G and so on are transmitted into the data acquisition unit 14 through AD converter 13. Through the power management unit 12, the output characteristics of the voltage controller specimen 5 including the DC voltage and current parameters and so on are transmitted to the data acquisition unit 14 through RS232C communication mode. Data acquisition unit 14 completes the data analysis of MEFS performance parameters such as wind energy, mechanical energy, electric energy and so on, and it creates the coupling test report.

Figure 3 is the input and output characteristics calculation model of wind rotor specimen 3. The wind rotor swept area *A* can be ascertained according to the size of the diameter and height of the testing wind rotor. The input wind power E_W of the wind rotor test specimen 3 can be calculated by equation (1), while the output mechanical energy E_M can be calculated by the equation (2) with the parameters of the rotor torque *T* and rotor speed *n* which obtained from the disc coupling sensor 6. The wind energy utilization η_M which corresponds to the system input parameters under the condition of different frequencies and different wind speed can be derived from the equation (5), thus we can draw out the wind energy utilization curve (V-η_M characteristic curve figure 2(a) of the vertical axis wind turbines specimen 3. The extreme value at the inflection point of the V-η_M characteristic curve is the maximum wind energy utilization rate value, its corresponding parameters including the rotor torque *T*, rotational speed n and the power E_M are the coupling design parameters which are determined from the magnetic circuit calculation and structure of generator 4.

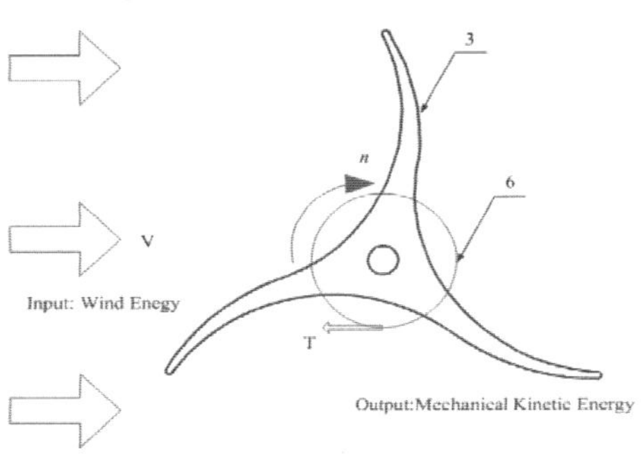

Figure 3. The energy transmission model of vertical axis wind rotor.

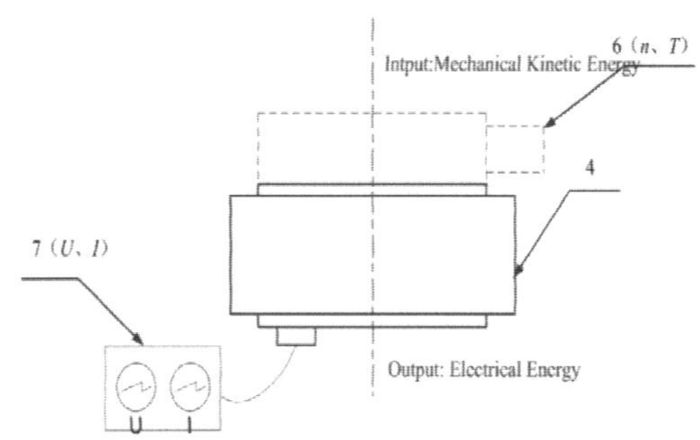

Figure 4. The energy transmission model of generator.

Figure 4 is the input and output characteristics calculation model of the generator 4, the input energy is the output mechanical energy E_M of the wind rotor 3, while the output electrical energy E_G can be derived from the equation (3) with the parameters of single-phase output voltage U and current I which detected from the three-phase intelligent electric parameters monitor 11. When the generator efficiency η_G is calculated by equation (6), we can draw out the n-η_G characteristic curve (Figure 2(b)) of generator 4. The extreme value at inflection point of the n-η_G characteristic curve is the maximum power output value of the generator, its corresponding parameters including the voltage U, the current I and electric power E_G are the circuit design and performance simulation coupling parameters of voltage controller 5.

Figure 5 is the input and output characteristic calculation model of voltage controller 5, its input is the three phase unstable power source of generator 4, its output is the charging voltage and charging current of battery. Its input electric power is the output energy E_G of voltage controller 5, while its output energy E_C can be obtained from the equation (4) according to the output voltage U and current I which are detected by the power management system. Then through the equation (7) we can calculate the work efficiency η_C of voltage controller 5.

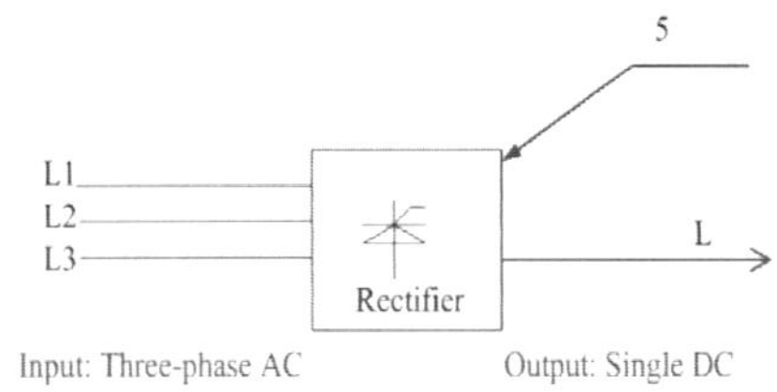

Figure 5. The voltage controller model.

The comprehensive performance of the vertical axis WECS can be obtained from the overall efficiency of the testing system, while the overall efficiency can be calculated from equation (8) according to the results of MEFC test.

5. The implementation process of testing system

When the design of each part component completes, we can enter into the following testing steps.

Step 1. Wind rotor test. It tests the wind energy conversion efficiency of wind rotor by coupling sensor. The input detecting parameter is wind speed, through the wind speed it can calculate the wind energy. The output detecting parameter is torque and rotational speed which can calculate the mechanical kinetic energy.

This test can evaluate the wind rotor efficiency curve, at the same time it can make the start torque, rotating speed (corresponding to the wind speed at 10 m/s of the commonly used small-sized rotor) and the power corresponding to the rotating speed as the original design parameters of the generator. If there exists a larger difference between the best efficiency point of the wind rotor efficiency curve and the rotating speed, it needs to adjust the wind rotor geometry parameters to the best efficiency point. The equipments required here are the wind tunnel and rotor testbed, the test instruments needed are anemometer and coupling sensor. The coupling sensors is connected to the RS232C interface of the UCS, then the value of testing rotating speed, torque and power parameters can be transmitted into the UCS directly [11].

Step 2. Generator test. It tests the input parameters (torque, speed and power) of generator through coupling sensor and the output parameters (current, voltage and electric power) through three-phase integrated electric parameter tester. This test can evaluate the efficiency curve of the generator. The equipment required here are the converter, driving motor, generator specimen and resistive load. The test instruments needed are the coupling sensor and three-phase integrated electric parameter tester. This test can share the whole equipment with wind rotor test, also can build it respectively.

Step 3. Voltage controller test. It tests the input parameters (current, voltage) of the controller through three-phase integrated electric parameter tester and the output parameters (current, voltage and electric power) from the battery tester. This test can evaluate the efficiency curve of the voltage controller. Note that it requires to do this test cooperated with the generator test.

Step 4. Parameter analysis. This step needs to analysis the component performance of the wind rotor, generator and controller synthetically from a whole perspective of the system, and to evaluate the overall performance of the generator according to their respective efficiency curve, then improve the design parameters, and provide technical preparation for the whole system coupling test at last.

Step 5. Overall test. Assemble the wind rotor, generator, controller and battery to be a WT prototype, do the simulation experiment of this prototype in the wind tunnel testing site. The input detecting parameter is the wind speed, through wind speed we can work out the wind energy, the output detecting parameter is the electric power. This test can evaluate the efficiency curve of WECS, from the characteristic curve we can determine the best performance of the overall coupling system.

Note that only when the component performance analysis from step 1 to step 4 achieve to the optimal state, we can enter into system integration stage of step 5 effectively.

The design solution of the best efficiency point mentioned above is that through the analysis of the aerodynamic of wind rotor structure and the experimental test we can get the efficiency curve of wind rotor, according to the best efficiency point we can estimate the starting torque, rated speed and power of generator then we can design it, and make the best efficiency point of generator consistent with the wind rotor as far as possible. Because the influence factors are much more in practice, it is difficult to realize their optimal synchronously, at this time we needs the coupling test to verification it, to change the design parameters of the generator according to the test results and then to improve the structure and performance. As for the voltage controller, its low cost makes the adjustment easy relatively [12]. At last the final performance of WECS will needs the coupling test equipment proposed in this paper to detect.

It needs to firstly point out that the wind speed is a random variable, the wind tunnel is just a simulation test equipment only. The wind speed, as the input variable for theoretical calculation, its error can reach to 50%. Therefore, the commonly used method is CFD analysis combined with the experimental at present, the performance of the wind rotor depends on its geometry structure and shape, the big difference will exist in the output characteristics [13]. The test results of the single wind rotor provide the coupling design parameters for generator, its result still needs the overall coupling test to reach the overall optimal performance. Secondly, the wind field site selection and evaluation are the first link of wind power generation, its purpose is mainly to investigate the stability and sustainability of wind energy. For the WT structure and wind field, there is only a little relationship between them, for the outdoor WT, even though its rigidity and strength are satisfied (the limited wind speed required in Japan is 60 m/s, the British is 100 m/s, China has not clear regulation about this at present) [14], the safety is the first factor also need to consider. When the specific meteorological data is known, the range of the wind speed will be not restricted to the wind field. As for the rare and powerful storm, it must to take the stall control measures for the WT. At present, the main consideration is to choose the suitable size of the WT for the different wind field.

Acknowledgments
We most gratefully acknowledge the financial support of the National Natural Science Foundation of China (NSFC) under grant No.71071078 & No.70671035, and the Doctoral Scientific Research Foundation of Henan University of Science and Technology.

References

[1] Dipesh K, Kalyan C 2016 A review of conventional and advanced MPPT algorithms for wind

energy systems *Renewable and Sustainable Energy Reviews* **55** 957-70

[2] Tripathi S M, Tiwari A N and Deependra S 2015 Grid-integrated permanent magnet synchronous generator based windenergy conversion systems: A technology review *Renewable and Sustainable Energy Reviews* **51** 1288-305

[3] Daroczy L, Janiga G, Petrasch K, Webner M and Thevenin D 2015 Comparative analysis of turbulence models for the aerodynamic simulation of H-Darrieus rotors *Energy* **90** 680-90

[4] McTavish S, Feszty D and Sankar T 2012 Steady and rotating computational fluid dynamics simulations of a novel vertical axis wind turbine for small-scale power generation *Renewable Energy* **41** 171-9

[5] Feng F, Li Y, Chen L X, Tian W Q and Zhang Y H 2014 The simulation and experimental research of combined vertical axis wind turbine aerodynamic characteristics *Acta Energiae Solaris Sinica* **35** 855-60

[6] Ding G Q, Li Y, Wang S L, Xie W and Tian W Q 2013 The solidity impaction to the vertical axis wind turbine performance of wind tunnel test *Renewable Energy Sources* **31** 58-62

[7] Aubree R, Auger F, Mace M and Loron L 2016 Design of an efficient small wind-energy conversion system with an adaptive sensorless MPPT strategy *Renewable Energy* **86** 280-91

[8] Qi Z Y and Lin E 2012 Integrated power control for small wind power system *J. Power Sources* **217** 322-8

[9] Aubree R, Auger F, Mace M and Loron L 2016 Design of an efficient small wind-energy conversion system with an adaptive sensorless MPPT strategy *Renewable Energy* **86** 280-91

[10] Yang B, Jiang L, Wang L, Yao W and Wu Q H 2015 Nonlinear maximum power point tracking control and modal analysis of DFIG based wind turbine *Int. Journal of Electrical Power & Energy Systems* **74** 429-36

[11] Qi Z Y and Lin E 2012 Integrated power control for small wind power system *Journal of Power Sources* **217** 322-8

[12] Saeidi D, Sedaghat A, Alamdari P and Alemrajabi A A 2013 Aerodynamic design and economical evaluation of site specific small vertical axis wind turbi *Applied Energy* **101** 765-75

[13] Almohammadi K M, Ingham D B, Ma L and Pourkashan M 2013 Computational fluid dynamics (CFD) mesh independency techniques for a straight blade vertical axis wind turbine *Energy* **58** 483-93

[14] Beiter P 2015 2014 *Renewable Energy Data Book* (Washington: U. S. Department of Energy) 52-61

Design of wide-area time-delay supplementary controller for interconnected Network based on Hamilton function method

G Hailati[1] and Z H Hu

Shanghai Polytechnic University Rm513, #16, Lane 2360, Jinhai Road, Pudong, Shanghai (201209), China

E-mail: glzt128@qq.com

Abstract. The transient stability of interconnected network with supplementary time-delay controller for generator excitations and static var compensator (SVC) has been investigated in this paper. Firstly, a delay-dependent stability criterion based on Hamilton function method is derived, and the criterion is in term of matrix inequalities. Secondly, a nonlinear time-delay Hamilton function model of interconnected network with SVCs is constructed. Thirdly, the wide-area time-delay supplementary controller (WATSC) for the interconnected network is designed and converted into the form of Hamiltonian system. The delay-dependent stability of the closed-loop power system is analysed. The gains of the WATSC are determined by using the theoretical analysis results. It is effective for the designed WATSC installed in the 16-machine, 68-bus power system for damping the inter-area modes. Then simulation results show that the method of the controller is effective.

1. Introduction

Collection and transmission of terminal data, and the remote control technology is widely used in intelligent interconnected grid [1-2]. Damping of inter-area oscillation can be controlled not only rely on local feedback signal generator. Therefore, the main function of the controller is designed to obtain the remote data signal through the device, in order to coordinate the control interval ringing problem [3].

As we all know, the transmission delay of dynamic data is the actual problems existing in the wide-area control [4]. And it will make worse the effect of the controller, and may lead to destabilization of the whole system. Thus, the design of the controller should consider the delays of the transmission signal [5]. Such a method can effectively improve the effect of the wide area controller. Calculate the maximum delay time process is a necessary step of the processing of the controller design. In [6] and [7], proposed a damping controller attached to the SVC device, and it used the data of steady-state voltage of the grid, current and power. In [8], designed damping controller attached to the STATCOM equipment, the use of the data signal from the entire system. A similar method is used in [9], and damping controller attached to the TCSC device is designed. Robust control is a popular method in the nonlinear control system.

Such as a multi-agent H_∞ controller [10] and a mixed H_2 / H_∞ controller [11] are used in wide-area damping controller design. Generator excitation control, while additional control device SVC method

[1] Address for correspondence: G Hailati, Shanghai Polytechnic University Rm513, #16, Lane 2360, Jinhai Road, Pudong, Shanghai (201209), China. E-mail: glzt128@qq.com.

can effectively control the range of problems caused by the ringing signal propagation delay caused by [12]. Wide-area delay controller is predicted based on the prediction method in [13]. Power vibration controller is proposed based on adaptive approach in [14]. Linear controller is designed used by LMI equations in [15]. Practical experience used in the damping controller for HVDC system in [16]. Wide area damping controller is presented based on Lyapunov theory and technical design model simplification, and taking into account the time delay of two different forms: the constant and time-varying in [17]. Factors wide area power controller is based on the prediction method highlight, while using variable delay in [18]. The controller design using varying norm - bounded uncertainty method in [19].

Most wide-area damping controller method is basically a simplified linear model of the power system. However, according to the characteristics of the power system, which are nonlinear and dynamic characteristics. Controller design by using nonlinear dynamics method is proposed in [20-21]. Hamilton function method is an important method for analyzing nonlinear systems [22-23]. This method first used in ordinary nonlinear system, then extended to the study of nonlinear time-delay system. Hamilton is an energy-based approach, to analyze the energy distribution problems and the stability of nonlinear systems. It is significance for design wide-area delay damping controller by using energy stability analysis methods.

A wide area damping controller WATSC is designed in this paper Consider using Hamilton function method and the time delay. Firstly, nonlinear systems Hamilton stable condition is derived. Secondly, Hamilton model is established of the power system, including generators, AVR (auto voltage regulator), PSS (power system stabilizer) and SVC and other major electrical equipment. Finally, calculate the relationship between the delay time and the controller parameters according to the stability condition, and to determine the optimal parameters of the controller.

The remainder of the paper is organized as follows: nonlinear systems Hamilton stable condition is derived in Sec. II. Model of interconnected network based on Hamilton function is established in Sec. III. The designed methodology of WATSC is described in Sec. IV. The simulation results of the test system and the effectiveness of the controller with and without time-delay are presented in Sec.V. Finally, the conclusions are drawn in Sec. VI.

2. Stable condition of Hamilton system

Consider the following Hamilton systems:

$$\dot{x}_1(t) = (J_{1t} - R_{1t})\nabla_{x_1} H_t + g_1 u_1(t)\,; \tag{1a}$$

$$\dot{x}_2(t) = (J_{2t} - R_{2t})\nabla_{x_2} H_t + g_2 u_2(t)\,, \tag{1b}$$

$$\dot{x}_c(t) = (J_q - R_q)\nabla_{x_c} H_t + g_c u_c(t)\,, \tag{1c}$$

let $x := \left(x_1^T \;\; x_2^T \;\; x_c^T\right)^T$, and $\nabla_{x_i} H_t = \dfrac{\partial H(x(t))}{\partial x_i}$ for $i = 1, 2$, where $x_i(t) \in R^n$ is the state vector of the system, $x_c(t) \in R^n$ is the state vector of SVC installed in the system, $\tilde{x}_i(t) = x_i(t-\tau)$, for $i = 1,2$, τ is the signal transmission delay, $u_i(t) \in R^m$ is the control input of the system, $u_c(t) \in R^m$ is the control input of the SVC, H is a Hamilton function, $g_i \in R^{n \times m}$ and $g_c \in R^{n \times m}$ is the port matrix, the anti-symmetric matrix J_{it} and J_q are the interconnection matrix, the symmetric positive semi-definite matrix R_{it} and R_q is the damping matrix.

Consider the problem of dynamic variable delay, state feedback input is expression as:

$$u_1(t) = -K_{11}\nabla_{x_1} H_t - K_{12}\nabla_{x_2} H_{t-\tau}\,, \tag{2a}$$

$$u_2(t) = -K_{21}\nabla_{x_1} H_{t-\tau} - K_{22}\nabla_{x_2} H_t\,, \tag{2b}$$

$$u_c(t) = -K_{c1}\nabla_{x_1} H_{t-\tau} - K_{c2}\nabla_{x_2} H_{t-\tau}\,, \tag{2c}$$

where τ is the time-delay from generator dynamically variable feedback to the controller. Suppose there is a limit of time delay, so that they meet the following criteria:

$$0 \le \tau \le h\,. \tag{3}$$

Form Equation (1)-(3) we can get the following closed loop system:

$$\begin{pmatrix} \dot{x}_1(t) \\ \dot{x}_2(t) \\ \dot{x}_c(t) \end{pmatrix} = \begin{pmatrix} J_{1t}-R_{1t} & 0 & 0 \\ 0 & J_{2t}-R_{2t} & 0 \\ 0 & 0 & J_q-R_q \end{pmatrix} \begin{pmatrix} \nabla_{x_1} H_t \\ \nabla_{x_2} H_t \\ \nabla_{x_c} H_t \end{pmatrix} + \begin{pmatrix} -g_1K_{11} & 0 & 0 \\ 0 & -g_2K_{22} & 0 \\ 0 & 0 & -g_3K_c \end{pmatrix} \begin{pmatrix} \nabla_{x_1} H_{x_1(t),x_{20},x_{c0}} \\ \nabla_{x_2} H_{x_{10},x_2(t),x_{c0}} \\ \nabla_{x_c} H_{x_{10},x_{20},x_c(t)} \end{pmatrix}$$

$$+\begin{pmatrix} 0 & -g_1K_{12} & 0 \\ -g_2K_{21} & 0 & 0 \\ -g_3K_{c1} & -g_3K_{c2} & 0 \end{pmatrix} \begin{pmatrix} \nabla_{x_1} H_{x_{1t}(t-\tau),x_{20},x_{c0}} \\ \nabla_{x_2} H_{x_{10},x_2(t-\tau),x_{c0}} \\ \nabla_{x_c} H_{x_{10},x_{20},x_c(t-\tau)} \end{pmatrix}, \tag{4}$$

where $\nabla_{x_i} H_{t-\tau} = \dfrac{\partial H(x(t-\tau))}{\partial x_i}$ for $i=1,2$.

Then Equation (4) can be expressed in the following form:

$$\dot{x}(t) = (J-R)_t \nabla_x H_t + T_1 \nabla_x H_t + T_2 \nabla_{x_\tau} H_{t-\tau}. \tag{5}$$

We can write (5) in a transformed form as

$$\dot{x}(t) = (\bar{J}-\bar{R})_t \nabla_x H_t + T_2 \nabla_{x_\tau} H_{t-\tau}. \tag{6}$$

where $J-R+T_1 = \bar{J}-\bar{R}$. $\bar{J}$ and $\bar{R}$ are symmetric matrix which satisfies $\bar{R} \ge 0$.

Proposition 1: Denote $A := \nabla_{xx} H(x) = \dfrac{\partial \nabla_x H(x)}{\partial x}$. For all x in an open neighbourhood of the origin, and the system described by Equ. (6) is globally asymptotically stable, and matrices exist as $0 \le P^T = P \in R^{n\times m}$, $0 \le Q^T = Q \in R^{n\times n}$, $0 \le S^T = S \in R^{n\times n}$ and for any $0 \le \tau \le h$, such that

$$\Xi = \begin{pmatrix} M_{11}+Q-S & M_{12} \\ M_{12}^T & -Q-S+h^2T_2^TA^TSAT_2 \end{pmatrix} < 0 \tag{7}$$

holds for all $x \in R^n$, where

$$M_{11} = -\bar{R} + (\bar{J}-\bar{R})^T A^T P + PA(\bar{J}-\bar{R}) + h^2 (\bar{J}-\bar{R})^T A^T SA(\bar{J}-\bar{R});$$

$$M_{12} = T_2 + PAT_2 + S + h^2 (\bar{J}-\bar{R})^T A^T SAT_2,$$

then the stability of system (6) is local.

Proof: Let

$$V_1(x(t)) = H_t + \nabla_x H_t^T P \nabla_x H_t;$$

$$V_2(x(t)) = \int_{t-\tau}^{t} \nabla_x H_s^T(x(s)) Q \nabla_x H_s(x(s)) ds;$$

$$V_3(x(t)) = h\int_{t-\tau}^{t}\int_{v}^{t} \left(\frac{d}{ds}\{\nabla_x H_s^T(x(s))\}\right) S \left(\frac{d}{ds}\{\nabla_x H_s(x(s))\}\right) ds dv,$$

then we have

$$\frac{d}{dt}V_1 = \nabla_x H_t^T (\bar{J}-\bar{R}) \nabla_x H_t + \nabla_x H_t^T T_2 \nabla_{x_\tau} H_{t-\tau} + \nabla_x H_t^T \left\{PA(\bar{J}-\bar{R}) + (\bar{J}-\bar{R})^T A^T P\right\} \nabla_x H_t$$
$$+ \nabla_x H_t^T PAT_2 \nabla_{x_\tau} H_{t-\tau} + \nabla_{x_\tau} H_{t-\tau}^T T_2^T A^T P \nabla_x H_t, \tag{8}$$

$$\frac{d}{dt}V_2 = \nabla_x H_t^T Q \nabla_x H_t - \nabla_{x_\tau} H_{t-\tau}^T Q \nabla_{x_\tau} H_{t-\tau}, \tag{9}$$

$$\frac{d}{dt}V_3 = h^2 \left\{\nabla_x H_t^T (\bar{J}-\bar{R})^T A^T SA(\bar{J}-\bar{R}) \nabla_x H_t + \nabla_x H_t^T (\bar{J}-\bar{R})^T A^T SAT_2 \nabla_{x_\tau} H_{t-\tau}\right.$$
$$\left. + \nabla_{x_\tau} H_{t-\tau}^T T_2^T A^T SA(\bar{J}-\bar{R}) \nabla_x H_t + \nabla_{x_\tau} H_{t-\tau}^T T_2^T A^T SAT_2 \nabla_{x_\tau} H_{t-\tau}\right\} - \nabla_x H_t^T S \nabla_x H_t + \nabla_x H_t^T S \nabla_{x_\tau} H_{t-\tau}$$
$$+ \nabla_{x_\tau} H_{t-\tau}^T S \nabla_x H_t - \nabla_{x_\tau} H_{t-\tau}^T S \nabla_{x_\tau} H_{t-\tau}. \tag{10}$$

Now, choose the following Lyapunov-Krasovskii functional

$$V(x(t)) = V_1(x(t)) + V_2(x(t)) + V_3(x(t)),$$

and compute the time derivative along the orbit of system (6), by using the relations (8), (9) and (10), we get

$$\frac{d}{dt}V = \begin{pmatrix} \nabla_x H_t \\ \nabla_{x_\tau} H_{t-\tau} \end{pmatrix}^T \Xi \begin{pmatrix} \nabla_x H_t \\ \nabla_{x_\tau} H_{t-\tau} \end{pmatrix}. \tag{11}$$

From the inequality (7), we can find that $\dot{V}(x(t)) < 0$ for all nonzero $x(t)$ which proves the system (6) is global asymptotic stable.

3. Model of interconnected network

The interconnected network with n machines, AVR, PSS and SVC equipment can be described as follows:

$$\dot{\delta}_i = \omega_i - \omega_0 ; \tag{12a}$$

$$\dot{\omega}_i = \frac{\omega_0}{M_i} P_{mi} - \frac{D_i}{M_i}(\omega_i - \omega_0) - \frac{\omega_0}{M_i} \sum_{j=1}^{n} E'_{qi} E'_{qj} B_{ij} S_{\delta_{ij}} ; \tag{12b}$$

$$\dot{E}'_{qi} = -\frac{1}{T'_{d0i}} \left(E'_{qi} + \left(x_{di} - x'_{di} \right) \sum_{j=1}^{n} E'_{qj} (G_{ij} S_{\delta_{ij}} - B_{ij} C_{\delta_{ij}}) \right) + \frac{E_{fi} + u^c_{fi}}{T'_{d0i}} ; \tag{12c}$$

$$\dot{v}_{1i} = -\frac{1}{T_{Ri}}(v_{1i} - v_{1i0}) + \frac{1}{T_{Ri}}(u_{ti} - u_{ti0}) ; \tag{12d}$$

$$\dot{v}_{2i} = -\frac{1}{T_{\omega i}} v_{2i} + K_{si}(\omega_i - \omega_0) ; \tag{12e}$$

$$\dot{v}_{si} = -\frac{1}{T_{2i}} v_{si} + \left(\frac{1}{T_{2i}} - \frac{T_{1i}}{T_{2i}} \frac{1}{T_\omega}\right) v_{si} + \frac{T_{1i}}{T_{2i}} K_{si}(\omega_i - \omega_0) \tag{12f}$$

$$\dot{B}_{svck} = \frac{1}{T_{svck}} \left[K_{svck}(V_{ref\_svck} - V_{svck} - u^c_{svck}) - B_{svck} \right], \tag{12g}$$

respectively, n and m is the number of generators and SVC, $i = 1, \cdots, n$ $k = 1, 2, \cdots, m$, δ_i means the rotor angle, ω_i means the angular velocity, ω_0 means synchronous speed; x_{di} means d-axis synchronous reactance, x'_{di} means the d-axis transient reactance ($x'_{di} < x_{di}$), T'_{d0i} means d-axis transient open-circuit time constant, E'_{qi} means q-axis transient potential, M_i and D_i means inertia time constant and damping, P_{mi} means mechanical input power, supposed as constant; u^c_{fi} means excitation input, u_{ti} means generator terminal voltage. Notice $S_{\delta_{ij}} = \sin(\delta_i - \delta_j)$, $C_{\delta_{ij}} = \cos(\delta_i - \delta_j)$.

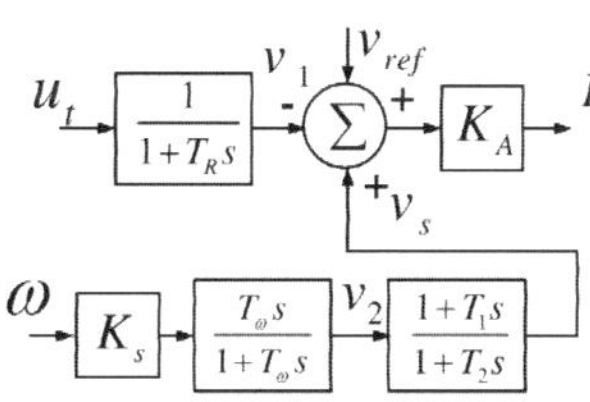

Figure1. Structure of excitation system with AVR and PSS

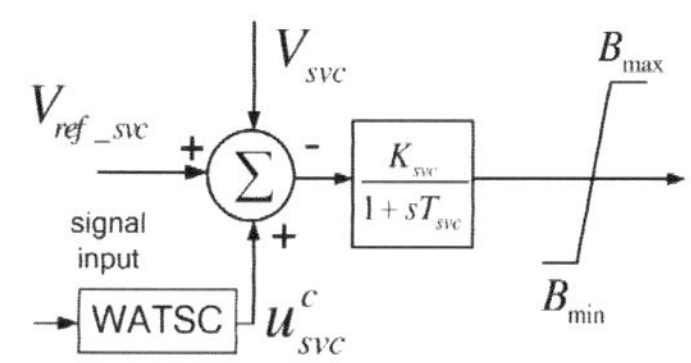

Figure2. Structure of SVC control block diagram with WATSC

The Structure of the system with AVR and PSS is shown in Figure 1 see [28]. u_{ti} means terminal voltage of generator, $v_{refi} = const.$, field voltage $E_{fi} = K_{Ai}(v_{refi} + v_{si} - v_{1i})$, variable v_{1i} means per unit value, v_{1i0} and u_{ti0} means corresponding initial values. K_{ai} means the gain coefficient, the AVR means a simple PI regulator, the PSS includes gain link, K_{si} and K_{svck} means the gain coefficients, T_{svck}, $T_{\omega i}$, T_{Ri}, T_{1i} and T_{2i} means the time constants in seconds, B_{svck} represents the equivalent susceptance of the SVC, V_{svck} means the voltage magnitude of the SVC bus, V_{ref_svck} means the SVC reference voltage and u^c_{svck} means the input of auxiliary control WATSC. The Figure 2 is the SVC control block diagram with WATSC, where $T_1 = R_s C_s = (R_{s1} + R_{s2}) C_s$, $T_2 = R_{s2} C_s$, $T_\omega = R_\omega C_\omega$, $T_{svc} = R_{svc} C_{svc}$.

The system with Hamilton function (12) is constructed as follows:

$$H(x)=\frac{1}{2}\sum_{i=1}^{n}\frac{M_i}{\omega_0}(\omega_i-\omega_0)^2-\sum_{i=1}^{n}P_{mi}\delta_i+\sum_{i=1}^{n}G_{ii}\delta_i E'^2_{qi}+\sum_{1\le i<j\le n}E'_{qi}E'_{qj}(G_{ij}S_{\delta_{ij}}-B_{ij}C_{\delta_{ij}})-\frac{1}{2}\sum_{i=1}^{n}B_{ii}E'^2_{qi}$$
$$+\sum_{i=1}^{n}\frac{E'^2_{qi}}{2(x_{di}-x'_{di})}+\frac{1}{2}\sum_{i=1}^{n}(C_{1i}v_{1i}^2+C_{\omega i}v_{2i}^2+C_{si}v_{si}^2)+\frac{1}{2}\sum_{k=1}^{m}C_{svck}B_{svck}^2, \tag{13}$$

where $H(x)$ means total storage energy of system. The (13) includes the kinetic energy and the other term define the summation of the potential energy.

By calculation, the partial derivatives of Hamilton function are:

$$\partial_{\delta_i}H=-P_{mi}+G_{ii}E'^2_{qi}+\sum_{j\ne i}^{n}E'_{qi}(G_{ij}C_{\delta_{ij}}+B_{ij}S_{\delta_{ij}})E'_{qj}; \tag{14a}$$

$$\partial_{\omega_i}H=\frac{M_i}{\omega_0}(\omega_i-\omega_0); \tag{14b}$$

$$\partial_{E'_{qi}}H=2G_{ii}\delta_i E'_{qi}-\sum_{j\ne i}^{n}E'_{qj}(G_{ij}S_{\delta_{ij}}-B_{ij}C_{\delta_{ij}})-B_{ii}E'_{qi}+\frac{E'_{qi}}{x_{di}-x'_{di}}; \tag{14c}$$

$$\partial_{v_{1i}}H=C_{1i}v_{1i}; \tag{14d}$$

$$\partial_{v_{2i}}H=C_{\omega i}v_{2i}; \tag{14e}$$

$$\partial_{v_{si}}H=C_{si}v_{si}; \tag{14f}$$

$$\partial_{B_{svck}}H=C_{svck}B_{svck}. \tag{14g}$$

Substitute the equations above into Equ. (12), and then the equations above can be simplified as follows:

$$\dot{\delta}_i=\frac{\omega_0}{M_i}\partial_{\omega_i}H; \tag{15a}$$

$$\dot{\omega}_i=-\frac{\omega_0}{M_i}\partial_{\delta_i}H-\frac{D_i\omega_0}{M_i^2}\partial_{\omega_i}H; \tag{15b}$$

$$\dot{E}'_{qi}=-\frac{x_{di}-x'_{di}}{T'_{d0i}}\partial_{E'_{qi}}H-\frac{K_{Ai}}{T'_{d0i}C_{1i}}\partial_{v_{1i}}H+\frac{K_{Ai}}{T'_{d0i}C_{si}}\partial_{v_{si}}H+\frac{u^c_{fi}}{T'_{d0i}}+\frac{K_{Ai}}{T'_{d0i}}v_{refi}; \tag{15c}$$

$$\dot{v}_{1i}=-\frac{1}{T_{Ri}C_{1i}^2}\partial_{v_{1i}}H-C_{1i}u_{ti}; \qquad \text{b} \tag{15d}$$

$$\dot{v}_{2i}=-\frac{K_{si}\omega_0}{M_i}\partial_{\delta_i}H-\frac{K_{si}D_i\omega_0}{M_i^2}\partial_{\omega_i}H-\frac{1}{T_{\omega i}C_{\omega i}^2}\partial_{v_{2i}}H; \tag{15e}$$

$$\dot{v}_{si}=-\frac{T_{1i}K_{si}\omega_0}{T_{2i}M_i}\partial_{\delta_i}H-\frac{T_{1i}K_{si}D_i\omega_0}{T_{2i}M_i^2}\partial_{\omega_i}H-\frac{1}{T_{2i}C_{si}}\partial_{v_{si}}H+\left(\frac{1}{T_{2i}C_{\omega i}}-\frac{T_{1i}}{T_{2i}T_{\omega i}C_{\omega i}}\right)\partial_{v_{2i}}H. \tag{15f}$$

$$\dot{B}_{svck}=-\frac{1}{T_{svck}C_{svck}}\partial_{B_{svck}}H+\frac{K_{svck}}{T_{svck}}(V_{ref\_svck}-V_{svck}-u^c_{svck}). \tag{15g}$$

Defining $x_i=(\delta_i,\omega_i,E'_{qi},v_{1i},v_{2i},v_{si},B_{svck})^T$, the system (15) rewritten as the following form:

$$\dot{x}_i=(J_i-R_i)\nabla_{x_i}H_t+g_i(x)u_i. \tag{16}$$

where

$$J_i(x)=\begin{pmatrix}0&-F_1&0&0&-F_3&-F_5&0\\F1&0&0&0&-F_4&-F_6&0\\0&0&0&-F_2&0&-F_7&0\\0&0&F_2&0&0&0&0\\F_3&F_4&0&0&0&-F_8&0\\F_5&F_6&F_7&0&F_8&0&0\\0&0&0&0&0&0&0\end{pmatrix},\quad R_i(x)=\begin{pmatrix}0&0&0&0&N_2&N_4&0\\0&\frac{D_i\omega_0}{M_i^2}&0&0&N_3&N_5&0\\0&0&\frac{x_{di}-x'_{di}}{T'_{d0i}}&N_1&0&N_6&0\\0&0&N_1&\frac{1}{T_{Ri}C_{1i}^2}&0&0&0\\N_2&N_3&0&0&\frac{1}{T_{\omega i}C_{\omega}^2}&N_7&0\\N_4&N_5&N_6&0&N_7&\frac{1}{T_{2i}C_{si}}&0\\0&0&0&0&0&0&\frac{1}{T_{svck}C_{svck}}\end{pmatrix},$$

$$g_i = \begin{pmatrix} 0 & 0 & K_{Ai}/T'_{d0i} & 0 & 0 & 0 & 0 \\ 0 & 0 & 0 & -C_{1i} & 0 & 0 & 0 \\ 0 & 0 & T'^{-1}_{d0i} & 0 & 0 & 0 & 0 \\ 0 & 0 & 0 & 0 & 0 & 0 & K_{svck}/T_{svck} \\ 0 & 0 & 0 & 0 & 0 & 0 & -K_{svck}/T_{svck} \\ 0 & 0 & 0 & 0 & 0 & 0 & -K_{svck}/T_{svck} \end{pmatrix}^T, \quad u_i = (v_{refi} \quad u_{ti} \quad u^c_{fi} \quad v_{ref\_svck} \quad v_{svck} \quad u^c_{svck})^T,$$

and

$$F_1 = -\frac{\omega_0}{M_i},\ F_2 = \frac{K_{Ai}}{2T'_{d0i}C_{1i}},\ F_3 = -\frac{K_{si}\omega_0}{2M_i},\ F_4 = -\frac{K_{si}D_i\omega_0}{2M_i^2},\ F_5 = -\frac{T_{1i}K_{si}\omega_0}{2T_{2i}M_i},\ F_6 = -\frac{T_{1i}K_{si}D_i\omega_0}{2T_{2i}M_i^2},\ F_7 = -\frac{K_{Ai}}{2T'_{d0i}C_{si}},$$

$$F_8 = -\left(\frac{1}{2T_{2i}C_{\omega i}} - \frac{T_{1i}}{2T_{2i}T_{\omega i}C_{\omega i}}\right),\ N_1 = \frac{K_{Ai}}{2T'_{d0i}C_{1i}},\ N_2 = \frac{K_{si}\omega_0}{2M_i},\ N_3 = \frac{K_{si}D_i\omega_0}{2M_i^2},\ N_4 = \frac{T_{1i}K_{si}\omega_0}{2T_{2i}M_i},\ N_5 = \frac{T_{1i}K_{si}D_i\omega_0}{2T_{2i}M_i^2},$$

$$N_6 = -\frac{K_{Ai}}{2T'_{d0i}C_{si}},\ N_7 = -\left(\frac{1}{2T_{2i}C_{\omega i}} - \frac{T_{1i}}{2T_{2i}T_{\omega i}C_{\omega i}}\right).$$

where $g_i \in R^{n \times m}$ means the port matrix, $u_i(t) \in R^m$ means the control input $x_i(t) \in R^n$ means the state vector of the system, the anti-symmetric matrix $J_i(x)$ means the interconnection matrix, $J_i(x) = -J_i^T(x)$, the symmetric positive semi-definite matrix $R_i(x)$ means the damping matrix, $R_i(x) = R_i^T(x) \geq 0$. $\boldsymbol{x} := \left[\boldsymbol{x}_1^T \ \boldsymbol{x}_2^T \cdots \boldsymbol{x}_n^T\right]^T$, and $\nabla_{x_i} \boldsymbol{H}_i = \{\partial \boldsymbol{H}(\boldsymbol{x}(t))\}/\partial \boldsymbol{x}_i$, $i = 1, \cdots, n$.

Let x_0 be the equilibrium point of the system (16), and it is satisfied $(J_i - R_i)\nabla_{x_i} H_i\big|_{x=x_0} = 0$. It is obvious that $H(x_0) = 0$ and $\nabla_{x_i} H_i\big|_{x=x_0} = 0$. If the equilibrium point x_0 is an extremum of $H(x)$, then the equally sufficient condition is

$$Hess(H(x_0)) = \left.\frac{\partial \nabla_x H(x)}{\partial x}\right|_{x=x_0} > 0.$$

4. Wide-Area damping coordination controller design

4.1. Coordination Control of Excitation and SVC with delay

To illustrate the idea of coordination controller, the block diagram of coordinated controller WATSC shown in Figure3. This additional controller includes two outputs: first, an additional input to generator excitation system is u^c_{fi}, and the other is u^c_{svck} which an additional input to SVC. The inputs signals of WATSC are dynamic remote measurement information from the other regions or areas. The three parts of WATSC input including: local signal, wide-area signal and the time delay τ of wide-area signal.

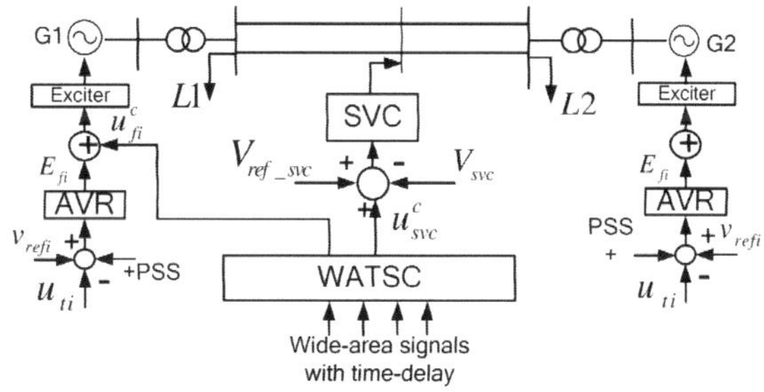

Figure 3. Structure of wide-area damping control scheme with time-delay

There are two control outputs of WATSC, u_{fi}^c and u_{svck}^c, that is to say there are two parts of WATSC, and each part includes a washout links, a lead-lag links and a gain links (K^c). The part of structure of wide-area damping coordination controller in this study is illustrated at Figure 4.

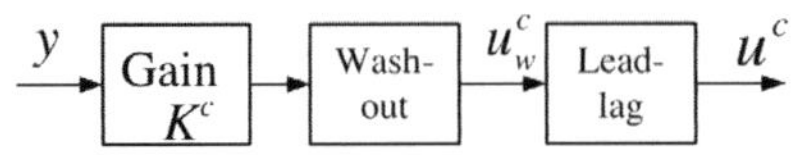

Figure 4. The part of structure of wide-area damping coordination controller

Selection of the input signal y usually is the speed difference signal or tie-line power signal, and the signal contains time delays. When $i = j$, take $\tau_{ii} = 0$. And when $i \neq j$, it is a wide-area signal , take $\tau_{ij} > 0$. The output signals of the WATSC are used as an additional inputs, which u_f^c to the excitation system block and u_{svc}^c to the SVC device. The main factor affecting the controller effect, here is the relationship between time delay and the parameters of controller. So, it is necessary to determine the relationship between delay margin and the parameters of controller.

Selection of the installation position and the feedback signal are main problems for the design of the WATSC. The geometric controllability/observability analysis method [29] has been used to select the suitable installation position and the feedback signals.

The washout link is $T_w s/(1+T_w s)$, T_w is the washout constant and chosen as 10s in this paper. The lead-lag link and the gain of the controller is designed by using *sisotool* function in MATLAB.

4.2. *Convert the controller into the form of Hamiltonian system*

The transfer function of the WATSC is easily obtained by the method mentioned in the previous section. As we all known, the transfer function model is equivalent to the circuit model. Thus, the energy functions of WATSC are obtained by using the energy of the equivalent circuit [30].

Assuming that the transfer function of the lead-lag link of WATSC is described as

$$\frac{u^c(s)}{u_w^c(s)} = \frac{b_0 s^2 + b_1 s + b_2}{a_0 s^2 + a_1 s + a_2} = d + \frac{\beta_1 s + \beta_1}{a_0 s^2 + a_1 s + a_2}, \ (d = b_0). \tag{17}$$

The differential equations of each link of WATSC branch can be written as follows:

$$\dot{u}_w^c = -\frac{1}{T_w} u_w^c + K^c y ; \tag{18a}$$

$$\dot{u}_1 = u_2 ; \tag{18b}$$

$$\dot{u}_2 = -a_2 x_1 - a_1 x_2 + u_w^c ; \tag{18c}$$

$$\dot{u}^c = -\beta_1 a_2 x_1 + (\beta_2 - \beta_1 a_1) x_2 + (\beta_1 - d / T_w) u_w^c + d K^c y ; \tag{18d}$$

The circuit energy of controller can be expressed as

$$H^c = \frac{1}{2} C_{u1}^c u_1^2 + \frac{1}{2} C_{u2}^c u_{2i}^2 + \frac{1}{2} C_w^c (u_w^c)^2 + \frac{1}{2} C_u^c (u^c)^2 ; \tag{19}$$

The energy storage components of each branch of controller are capacitors C_{u1}^c, C_{u2}^c, C_w^c, C_u^c.

Then the differential equations of each link of WATSC can be converted into the form of Hamiltonian system as

$$\dot{u}_w^c = -\frac{1}{T_w C_w^c} \frac{\partial H^c}{\partial u_w^c} + K^c y ; \tag{20a}$$

$$\dot{u}_1 = \frac{1}{C_{u2}^c} \frac{\partial H^c}{\partial u_2} ; \tag{20b}$$

$$\dot{u}_2 = -a_2 \frac{1}{C_{u1}^c} \frac{\partial H^c}{\partial u_1} - a_1 \frac{1}{C_{u2}^c} \frac{\partial H^c}{\partial u_2} + \frac{1}{C_w^c} \frac{\partial H^c}{\partial u_w^c} ; \tag{20c}$$

$$\dot{u}^c = -\beta_1 a_2 \frac{1}{C_{u1}^c} \frac{\partial H^c}{\partial u_1} + (\beta_2 - \beta_1 a_1) \frac{1}{C_{u2}^c} \frac{\partial H^c}{\partial u_2} + \frac{(\beta_1 - d / T_w)}{C_w^c} \frac{\partial H^c}{\partial u_w^c} + d K^c y . \tag{20d}$$

Let $x_t^c = \left(u_w^c \quad u_1 \quad u_2 \quad u^c\right)^T$, thus the system (20) can be rewritten as the following form:

$$\dot{x}^c = \left(J^c - R^c\right)\nabla_{x^c} H_t^c + g^c u , \tag{21}$$

where

$$J^c = \begin{pmatrix} 0 & 0 & \frac{-1}{2C_w^c} & \frac{d - T_w\beta_1}{2C_w^c T_w} \\ 0 & 0 & \frac{C_{u1}^c + a_2 C_{u2}^c}{2C_{u1}^c C_{u2}^c} & \frac{a_2\beta_1}{2C_{u1}^c} \\ \frac{1}{2C_w^c} & \frac{C_{u1}^c - a_2 C_{u2}^c}{2C_{u1}^c C_{u2}^c} & 0 & \frac{\beta_1 a_1 - \beta_2}{2C_{u2}^c} \\ \frac{T_w\beta_1 - d}{2C_w^c T_w} & -\frac{a_2\beta_1}{2C_{u1}^c} & \frac{\beta_2 - \beta_1 a_1}{2C_{u2}^c} & 0 \end{pmatrix}, \quad R^c = \begin{pmatrix} \frac{1}{T_w C_w^c} & 0 & \frac{-1}{2C_w^c} & \frac{d - T_w\beta_1}{2C_w^c T_w} \\ 0 & 0 & -\frac{C_{u1}^c + a_2 C_{u2}^c}{2C_{u1}^c C_{u2}^c} & \frac{a_2\beta_1}{2C_{u1}^c} \\ \frac{-1}{2C_w^c} & -\frac{C_{u1}^c + a_2 C_{u2}^c}{2C_{u1}^c C_{u2}^c} & \frac{a_1}{C_{u2}^c} & \frac{\beta_1 a_1 - \beta_2}{2C_{u2}^c} \\ \frac{d - T_w\beta_1}{2C_w^c T_w} & \frac{a_2\beta_1}{2C_{u1}^c} & \frac{\beta_1 a_1 - \beta_2}{2C_{u2}^c} & 0 \end{pmatrix},$$

$g_i = \left(K^c \quad 0 \quad 0 \quad dK^c\right)^T$, $u = y$.

As mentioned above ,the WATSC has two control outputs, u_{fi}^c and u_{svck}^c, so the control gain is K_f and K_s, respectively, they are the main adjustable parameters to achieve the effect of suppressing the interval damping. The control gain of WATSC is described as $K_f = k_{fii} = -k_{fij}$, $K_s = k_{sii} = -k_{sij}$, i, $j = 1, \cdots, n$.

Then, the Hamiltonian system of controller (21) is added to the original Hamiltonian system (16) of interconnected network to obtain complicate time-delay Hamiltonian system. The delay margin is found for the tolerated capability of power system by using the criterion Equ. (7). The function *feasp* and *gevp* in MATLAB [31] is used to calculate the feasibility problems of matrix inequality. Finally, the relationship between delay margin and the parameters of controller will be found.

4.3. *Summary of steps of the controller design*

To design the controller proposed in this paper has the six steps as following.

Step 1) Model the system in the toolbox of MATLAB/Sim Power system. To linearize the nonlinear model by using the Linear Analysis Toolbox of MATLAB. And then get the state space model.

Step 2) Calculate the eigenvalues, damping ratios and frequencies of state matrix of the system to determine the inter-area low frequency oscillation mode. Select the installation sites and the feedback input signals of the WATSC by calculating the participation factor of each mode and analysing the controllability and observability [32] of linear model of the system.

Step 3) According to the characteristics of the dominant eigenvalues to design the lead-lag link and determine the gain range of the controller by using the interactive MATLAB GUI called *sisotool* function.

Step 4) Convert the differential equations of the complicate power system into the form of Hamiltonian system. Obtain the relationships between the delay margins and the gains of WATSC based on the method described in Proposition 1.

Step 5) Choose the gains of WATSC by a trade-off between damping ratios of the inter-area modes and delay margins.

Step 6) Model the WATSC by using MATLAB/Sim Power system and add it on the original nonlinear details models of the studied power system to verify the effectiveness of the designed wide-area controller.

5. Simulation Studies

5.1. *System description and WATSC design*

The 16-machine, 68-bus power system consists of five areas linked together, shown in Figure 5, and the all data illustrated in [33]. Each generator is equipped with excitation system. This is a prototype power system using a SVC. The Area#2 is required to import 889MW from Area#5. To facilitate this large amount of power transfer, a 200 Mvar SVC is installed in the middle of the tie-line connecting bues #51 and #52, and the parameters of SVC as $K_{svc}=20$ and $T_{svc}=0.05$. Under nominal operating condition, the tie-line power flow from Area#1 to Area#2 is 701.3MW. And the eigenvalues of the state matrix as show in Table 1.

It is obvious that Mode 1 (vibration frequency is 0.6743Hz) shows the swing of the opposite phase between the G1 and the G13. The largest swing is released on the G1 (participation factor is 0.2). Mode 2 (vibration frequency is 0.8216Hz) is a inter-area mode, and the swing of the opposite phase formed between the G10, G14 and the G16. The largest swing of mode 2 is released on the G16 (participation factor is 0.3). Mode 1 and Mode 2 have the smallest damping ratio. Thus, the intention to design WATSC provides additional damping to Mode 1 and Mode 2 in order to achieve coordinated control.

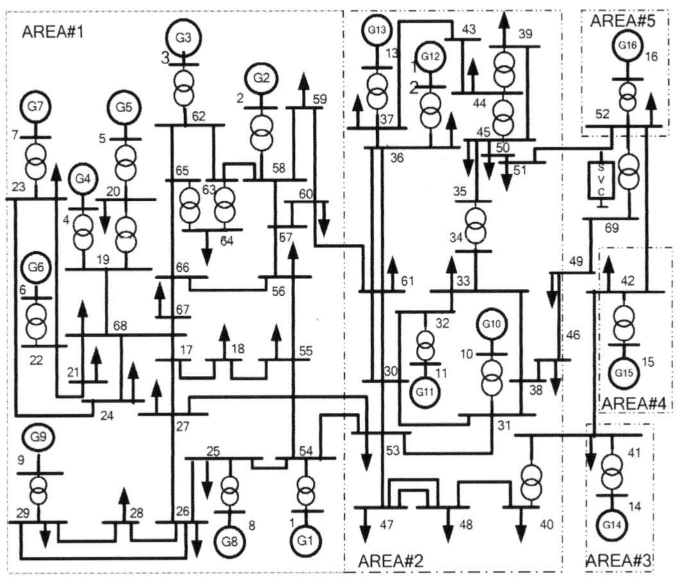

Figure 5. The16-machine, 68-bus power system with SVC

Table 1. Inter-area modes of the 16-machine 68-bus system

Mode No.	Eigenvalue	Damping Ratio ς	Frequency f (Hz)
1	-0.0408+j0.6731	**0.0604**	**0.6743**
2	-0.0539+j0.8198	**0.0657**	**0.8216**
3	-0.5239+j0.5812	0.6696	0.7825
4	-0.5778+j0.5703	0.7117	0.8118
5	-0.6168+j0.6304	0.6994	0.8820

Table 2. Analysis result of maximal geometric controllability/ observability of system

No.	u_{fi}^{c}	Geometric Controllab -ility measure	Wide -area Input signal	Geometric Observab -ility measure	Wide-area Input signal	Geometric Observab -ility measure

		Mode 1		Mode 1		Mode 2
1	G1	**0.1083**	$\Delta\omega_{1-10}$	0.0022	$\Delta\omega_{10-13}$	0.0020
2	G2	0.0349	$\Delta\omega_{1-11}$	0.0029	$\Delta\omega_{10-14}$	0.0020
3	G3	0.0938	$\Delta\omega_{1-12}$	0.0023	$\Delta\omega_{10-15}$	0.0025
4	G4	0.0254	$\Delta\omega_{1-13}$	**0.0038**	$\Delta\omega_{1-16}$	0.0010
5	G5	0.0301	$\Delta\omega_{1-14}$	0.0010	$\Delta\omega_{16-10}$	**0.0033**
6	G6	0.0692	$\Delta\omega_{1-15}$	0.0026	$\Delta\omega_{16-11}$	0.0022
7	G7	0.0586	$\Delta\omega_{10-12}$	0.0023	$\Delta\omega_{16-12}$	0.0023
8	G8	0.0736	ΔI_{27-53}	0.0028	$\Delta\omega_{16-13}$	0.0014
9	G9	0.0368	ΔI_{53-54}	0.0015	$\Delta\omega_{16-14}$	0.0030
10	G10	0.0189	ΔI_{60-61}	0.0016	$\Delta\omega_{16-15}$	0.0028
11	G11	0.0274	ΔP_{27-53}	0.0030	$\Delta\omega_{13-15}$	0.0022
12	G12	0.0499	ΔP_{60-61}	0.0036	$\Delta\omega_{14-15}$	0.0015
13	G13	0.0163	ΔP_{53-54}	0.0024	ΔP_{42-52}	0.0029
14	G14	0.0209	ΔP_{49-52}	0.0024	ΔP_{41-42}	0.0025
15	G15	0.0523	ΔP_{50-52}	0.0019	ΔP_{40-41}	0.0012

The control scheme proposed in this paper is illustrated at Figure 3. According to the analysis of eigenvalues and eigenvectors of the test system, we can determine the candidate input signal and the role of place of WATSC by the participation factor of each mode. The installation sites and the feedback input signals of the WATSC are selected by calculating the participation factor of each mode and analysing the controllability and observability of linear model of system. The geometric controllability/observability measures associated with Mode1 and mode 2 are calculated and then used to choose the feedback signal of the WATSC, as shown in Table 2. As the G1 has the largest geometric joint controllability index, it is chosen as a suitable location to install wide-area feedback control signal u_f^c . As the rotor angle difference between G1 and G13 has the largest geometric observability index, it is chosen as one of the remote feedback signal of the WATSC. The speed difference signal $\Delta\omega_{16-10}$ has the largest geometric observability index of mode 2, as shown in Table 2, it is chosen as the other one of the remote feedback signal of the WATSC.

The WATSC has two output signals, one is u_{svc}^c (acting on SVC), and the other is u_f^c (acting on the generator excitation of G1). The WATSC uses remote signals with time-delay to supplement excitation for generator and SVC if we assuming that control time-delay is negligible.

According to the characteristics of the dominant eigenvalues to design the lead-lag link and determine the range of gain of the controller without considering time-delay by using the interactive MATLAB GUI called *sisotool* function. The control parameters of WATSC with different feedback signals (speed difference signal $\Delta\omega_{16-10}$, $\Delta\omega_{16-14}$ and tie line active power ΔP_{42-52} have the larger geometric observability index, as shown in Table 2) are computed and given in Table 3 , where K^c means the gain range of WATSC which changes from 0 to K^c .

Table 3. The control parameters of WATSC

Output Of WATSC	Gain name	Feedback signal	Parameters of WATSC				
			K^c	a0	a1	b0	b1
u_f^c	K_f	$\Delta\omega_{1-13}$	86	0.86	1.1	0.36	0.47
u_{svc}^c	K_{sI}	$\Delta\omega_{16-10}$	8469	1.96	1.4	0.81	0.49
	K_{sII}	$\Delta\omega_{16-14}$	8520	0.58	1.5	0.66	0.38
	K_{sIII}	ΔP_{42-52}	0.0134	4.41	3.8	0.88	0.02

Adding WATSC to the selected machine G1 and SVC as shown in the Figure 3, and we can obtain the gains K_f and K_s changed from 0 to K^c of WATSC in Table. 3, where K_s represents K_{sI}, K_{sII} and K_{sIII}. It is clear that the damping ratio is significantly improved as shown in Table 4. And the gain K_{sI} has better damping ratio and the largest frequency than the gains K_{sII} and K_{sIII} for the Mode 2. Thus, the speed difference signals ω_{1-13} and speed difference signals ω_{16-10} are selected as input signals of WATSC.

Convert the WATSC into the form of Hamiltonian system by using the method described in section 4.2. And Hamiltonian system is establishing of the system, then calculate the delay margin h by using the stability criterion. Calculating the matrix inequalities (6) described in the previous section, and using MATLAB Control Toolbox, the results are shown in Table 3. It can be found that h decreases with the increase of the gains K_f and K_{sI}.

Table 4. Inter-area modes and delay margin with different gain of WASC (f /Hz)

No.	h /ms	K_f	Mode 1 ς	Mode 1 f	K_s		Mode 2 ς	Mode 2 f
1	—	0	0.060	0.674		0	0.0657	0.8216
2	473.12	10	0.0938	0.682	K_{sI}	1000	0.117	0.849
3	—	—	—	—	K_{sII}	1000	0.112	0.826
4	—	—	—	—	K_{sIII}	0.001	0.0759	0.823
5	383.32	20	0.128	0.69	K_{sI}	2000	0.14	0.887
6	—	—	—	—	K_{sII}	2000	0.156	0.832
7	—	—	—	—	K_{sIII}	0.002	0.0878	0.825
8	290.18	30	0.165	0.698	K_{sI}	3000	0.149	0.917
9	—	—	—	—	K_{sII}	3000	0.197	0.841
10	—	—	—	—	K_{sIII}	0.003	0.101	0.828
11	176.94	40	0.203	0.706	K_{sI}	4000	0.154	0.94
12	—	—	—	—	K_{sII}	4000	0.235	0.853
13	—	—	—	—	K_{sIII}	0.004	0.111	0.832
14	99.36	50	0.242	0.714	K_{sI}	5000	0.157	0.959
15	—	—	—	—	K_{sII}	5000	0.269	0.868
16	—	—	—	—	K_{sIII}	0.005	0.123	0.838
17	85.74	60	0.285	0.725	K_{sI}	6000	0.16	0.975
18	—	—	—	—	K_{sII}	6000	0.298	0.887
19	—	—	—	—	K_{sIII}	0.006	0.134	0.844
20	77.61	70	0.33	0.737	K_{sI}	7000	0.163	0.988
21	—	—	—	—	K_{sII}	7000	0.323	0.907
22	—	—	—	—	K_{sIII}	0.007	0.145	0.853
23	54.26	80	0.384	0.756	K_{sI}	8000	0.166	1.0
24	—	—	—	—	K_{sII}	8000	0.342	0.927

IOP Conf. Series: Earth and Environmental Science **40** (2016) 012044 doi:10.1088/1755-1315/40/1/012044

25	—	—	—	—	K_{sm}	0.009	0.16	0.874
26	36.98	86	0.43	0.773	K_{sI}	8469	0.167	1.01
27	—	—	—	—	K_{sII}	8520	0.35	0.938
28	—	—	—	—	K_{sIII}	0.013	0.175	0.92

Generally, the time delay of wide-area signals between tens to hundreds of milliseconds, therefore, the time delay is afforded by the wide-area controller should be as large as possible. Thus, the picking of the gains of WATSC should be satisfied the time margin and damping ratio as large as possible (15). Form the Table 4, the number 8 with gains K_f =30 and K_{sI} =3000 should be selected to be as the gains of WATSC.

5.2. *Simulation results*

The fault which is a 3-phase line to ground occurs at bus 61 followed by tripping one of the lines connecting buses 60 and 61 at $t = 1$ s, and the fault is cleared at $t = 1.1$ s. A time-domain simulation was carried out based on detailed nonlinear model to check the effectiveness of WATSC. And the output of the u_f^c is limited by ± 0.05 p.u. and the output of the u_{svc}^c is limited by ± 0.1 p.u.

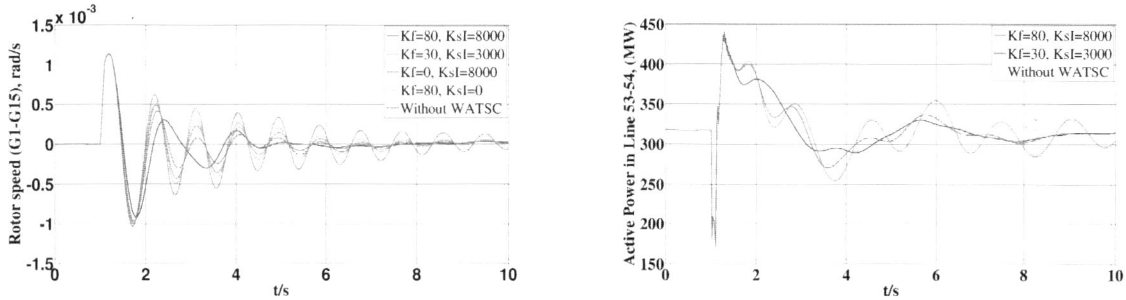

Figure 6. (a) Figure 6. (b)

Figure 6. Dynamic response of the system with different gains of WATSC, without considering time-delay

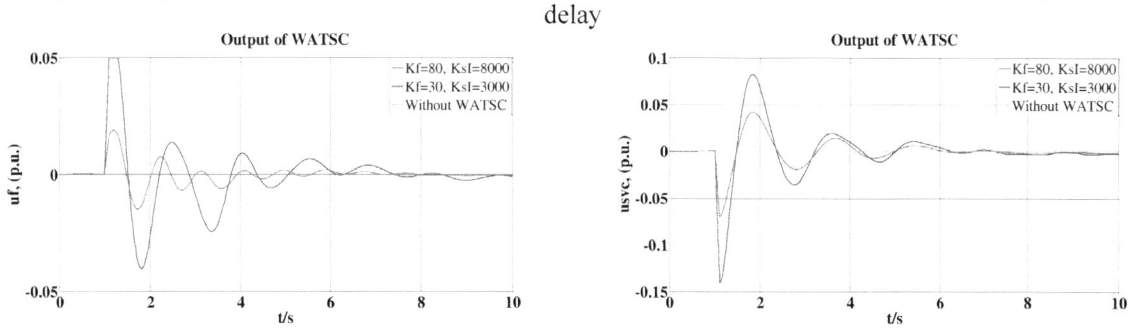

Figure 7. (a) Figure 7. (b)

Figure 7. Output of the WATSC without time-delay
(uf = u_f^c, usvc = u_{svc}^c)

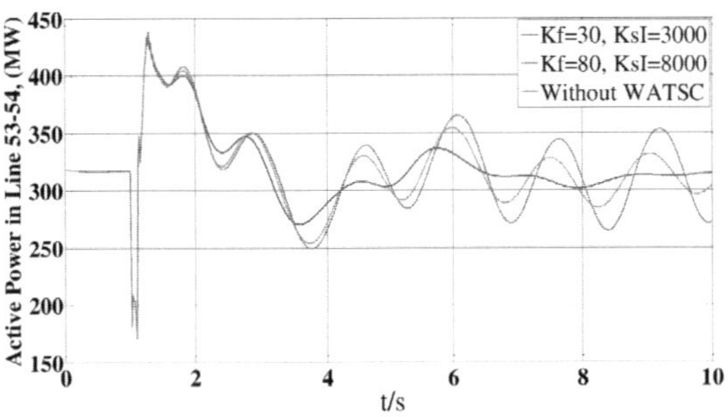

Figure 8. Dynamic response of the system with different gains of WATSC (τ =100ms).

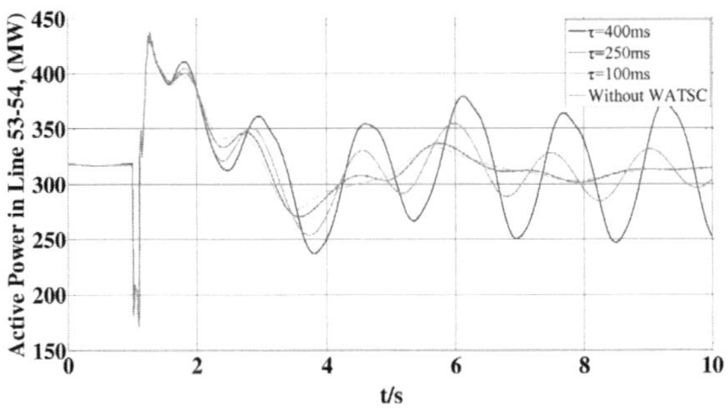

Figure 9. Dynamic response of the system with different time-delay (K_f =30 and K_{sI}=3000)

As shown in Figure 6 and Figure 7, the response of the system is compared with and without the WATSC. The Figure 6 (a) shows the rotor speed difference of G1-G15 that the WATSC with different gains are applied to the case. It is means that the inter-area oscillation is caused by the proposed controller. And the damping performance better with the increase of gains of WATSC. It is confirmed that the result shown in Table 3.

The responses of the system with different time delay by adding WATSC are shown in Figure 8 and Figure 9. It is clear that the system performance deteriorates considerably with increasing delays. It is mean that the WATSC (K_f =30 and K_{sI}=3000) is effectively when the time-delay as large as 100ms. Results show that the WATSC (K_f =30 and K_{sI}=3000) has a big delay margin of 290.18ms, and it is better than the WATSC (K_f =80 and K_{sI}=8000) has the delay margin 54.26ms. Thus, the gains K_f =30 and K_{sI}=3000 of the WATSC should be selected to provide damping for the Mode 1 and Mode2 of the test system.

From the simulation results above, the delay margin can provide an effective guide for the design of supplementary controller, and the gain of supplementary controller can be chosen to achieve trade-off between the delay margin and damping performance. It is mean that the wide-area controller mentioned in this paper is effective.

6. Simulation Studies

Wide-area damping controller is designed by using the theory of Hamilton and the affection of time delay from remote signals. The optimal controller parameters are selected of the power system. Coordinated control between the generator excitation and SVC device has certain rationality. Simulation results show that the ringing interval has been effectively suppressed. And the relationship of the delay margin and the gains of WATSC are analyzed. The delay margin can be used as design specifications to guide the selection of different gain of the wide-area damping controller.

Acknowledgments

This work is supported by the Shanghai Young Teachers' Training financed projects (No.ZZZZEGD15001), School Foundation of Shanghai Polytechnic University (No.EGD15XQD02).

References

[1] Hauer A J F, Trudnowski D J and DeSteese J G 2007 A perspective on wams analysis tools for tracking of oscillatory dynamics *IEEE/Power Engineering Society General Meeting(PES)* **1-10** 4324-33

[2] Far H G, Banakar H, Li P, Luo C L and Ooi B T 2009 Damping inter-area oscillations by multiple modal selectivity method *IEEE Trans.On Power Systems* **24** 766-75

[3] Aboul-Ela M E, Sallam A A, McCalley J D and Fouad A A 1996 Damping controller design for

power system oscillations using global signals *IEEE Trans. On Power Systems* **11** 767-73

[4] Wu H X, Ni H and Heydt G T 2002 The impact of time delay on robust control design in power systems *Proc. IEEE PES Winter Meeting* **2** 27-31

[5] Pagola F L, Perez-arriaga I J and Verghese G C 1989 On sensitivities, residues and participation application to oscillation stability analysis and control *IEEE Trans. on Power Systems* **4** 278-85

[6] Chang Y, Xu Z, Cheng G and Xie J 2007 A novel SVC supplementary controller based on wide area signals *Electric Power Systems Research* **77** 1569-74

[7] Ray S and Venayagamoorthy G K 2008 Wide-area signal-based optimal neurocontroller for a UPFC *IEEE Transactions on Power Delivery* **23** 1597-605

[8] Xie X R, Li J, Xiao J Y and Han Y D 2004 Inter-area damping control of STATCOM using wide-area measurements *2004 IEEE Int. Conf. on Electric Utility Deregulation, Restructuring and Power Technologies* 222-7

[9] Chaudhuri B and Pal B C 2004 Robust damping of multiple swing modes employing global stabilizing signals with a TCSC *IEEE Transactions on Power System* **19** 499-506

[10] Ni H, Heydt G T and Mili L 2002 Power system stability agents using robust wide area control *IEEE Trans Power Systems* **17** 1123-31

[11] Zhang Y and Bose A 2008 Design of wide-area damping controllers for interarea oscillations *IEEE Trans Power Systems* **23** 1136-43

[12] Zabaiou T, Dessaint L A, Okou F A and Grondin R 2008 Wide-area measurements based coordination of SVCs and synchronous generators *Power and Energy Society General Meeting Conversion and Delivery of Electrical Energy in the 21st Century, 2008 IEEE* 1-8

[13] Chaudhuri B, Majumder R and Pal B C 2004 Wide-area measurement based stabilizing control of power system considering signal transmission delay *IEEE Trans Power Systems* **19** 1971-79

[14] Chaudhuri N R, Ray S, Majumder R and Chaudhuri B 2010 A new approach to continuous latency compensation with adaptive phasor power oscillation damping controller (POD) *IEEE Trans. Power Syst.* **25** 939-46

[15] Yao W, Jiang L, Wen J Y and Cheng S J 2014 Wide-area damping controller of facts devices for inter-area oscillations considering communication time delays *IEEE Trans. Power Syst.* **29** 318-28

[16] He J, Lu C, Wu X, Li P and Wu J 2009 Design and experiment of wide area HVDC supplementary damping controller considering time delay in China southern power grid *IET Gen. Transm. Distrib.* **3** 17-25

[17] Dotta D, Silva A S and Decker I C 2009 Wide-area measurement -based two-level control design considering signals transmission delay *IEEE Trans. Power Syst.* **24** 1 208-16

[18] Wang S, Meng X and Chen T 2012 Wide-area control of power systems through delayed network communication *IEEE Trans Control Syst. Technol.* **20** 495-503

[19] Xu S and Lam J 2005 Improved delay-dependent stability criteria for time-delay systems *IEEE Trans. Autom. Control* **50** 384-7

[20] Zarghami M, Crow M L and Sarangapani J 2010 Nonlinear control of FACTS controllers for damping inter-area oscillations in power systems *IEEE Trans.on Power Del.* **25** 3113-21

[21] M. Mokhtari, F. Aminifar,et al. 2013 Wide-Area power oscillation damping with a fuzzy controller compensating the continuous communication delays *IEEE Trans. Power Syst.* **28** 1997-2005

[22] Cheng D Z, Xi Z R, Hong Y G and Qin H S 2000 Energy-based stabilization of forced Hamiltonian systems and its application to power systems *Control Theory & Applications*, **17** 798-802

[23] Shi F and Wang J 2012 Stabilizing control of multi-machine power systems with transmission losses based on pseudo-generalized Hamiltonian theory *Control Theory & Applications*, IET, **6** 173-81

[24] Yang R and Wang Y 2012 Stability analysis for a class of nonlinear time-delay systems via Hamiltonian functional method *Science China: Information Sciences* **55** 1218-28

[25] Xi Z R, Cheng D Z, Lu Q and Mei S W 2002 Nonlinear decentralized controller design for multi-machine power systems using Hamiltonian function method *Automatica* **38** 527-34

[26] Pasumarthy R and Kao C Y 2009 On stability of time-delay Hamiltonian systems". Proceedings of the 2009 American Control Conference 4909-14

[27] Sun W and Wang Y 2007 Stability analysis for some class of time-delay nonlinear Hamiltonian systems *J. Shandong University (Natural Science)* **42** 1-9

[28] Jovcic D and Pillai G N 2005 Analytical Modelling of TCSC Dynamics *IEEE Transactions on Power Delivery* **20** 1097-104

[29] Zhang Y and Bose A 2008 Design of wide-area damping controllers for inter-area oscillations *IEEE Trans. Power Syst.* **23** 1136-43

[30] Jing Q, Zeng Y and Zhang L 2012 High order generator hamiltonian model included AVR and PSS energy function *Power and Energy Engineering Conference (APPEEC) 2012 Asia-Pacific* 1-4

[31] Gahine P, Nemirovski A, Laub A J and Chilali M 2000 *LMI Control Toolbox User's Guide* Natick, MA: MathWorks

[32] Heniche A and Kamwa 2008 Assessment of two method to select wide-area signals for power systems damping control *IEEE Trans. power Syst.* **23** 572-81

[33] Chaudhuri B and Pal B 2005 *Robust control in power systems* (New York: Springer-Verlag New York Inc.)170-8

IOP Conf. Series: Earth and Environmental Science **40** (2016) 012045 doi:10.1088/1755-1315/40/1/012045

Experimental investigation of thermal characteristics of lithium ion battery using phase change materials combined with metallic foams and fins

Y C Deng, H Y Zhang[1], X Xia

College of Automotive Engineering, Shanghai University of Engineering Science 333 Longteng Road, Songjiang, Shanghai, China

E-mail: zhanghengyun@sues.edu.cn

Abstract. Phase change materials are of great interest in energy storage and energy management applications due to their high latent heat and excellent cycling stability. In this paper, the thermal characteristics of phase change materials (PCM) for thermal management of cylindrical 18650 lithium-ion battery (LIB) were experimentally investigated. A commercial paraffin wax with a melting temperaturerange between 47 – 53.8°C was used in this study. A metal cylinder with a heater was used to emulate the heat generation from a battery, which was surrounded with the paraffin PCM and containted in a metal housing. The experiment was conducted in an environmental test chamber with controlled ambient temperatures and power inputs. Both the battery temperature and the housing wall temperature were measured during steady-state heating and cyclic heating conditions. Since PCM has low thermal conductivity, thermal enhancement techniques were investigated by adding metal foams (MFs) or combining metallic foam and fins into the PCM to enhance the thermal conductivity. The battery temperatures were measured for all the cases and the results were analyzed and discussed.

1. Introduction

Lithium-ion batteries generate a lot of heat during discharge, which can raise their temperature and affect their operational performance [1]. High temperatures could cause thermal runaway, which is a catastrophic self-accelerated degradation of the Li-ion battery as described in [2], leading to spontaneous combustion or even explosion. Due to the high latent heat and chemical stability, paraffin wax based phase change material (PCM) can be used to maintain the battery at the appropriate temperature level[3-5]. When the heating temperature reaches the melting point of the PCM during the discharge stage of high C-rates (the discharge current to the nominal current ratio [6]), PCM can absorb the heat and maintain temperature of battery at a constant level, reducing the temperature rise of the battery itself as well as temperature nonuniformity among the cells. When the battery is at idle state or operating at low work load, the absorbed heat by the PCM can be released during the solidification process, which helps to maintain the battery temperature at a smaller range of temperature swing. Among the available PCMs, paraffin has been found with the advantages of high

[1] Address for correspondence: H Y Zhang, College of Automotive Engineering, Shanghai University of Engineering Science 333 Longteng Road, Songjiang, Shanghai, China. E-mail: zhanghengyun@sues.edu.cn.

latent heat, small volume change during melting, good chemical stability, low cost and least subcooling phenomenon, which is the most suitable PCM candidate for the thermal management of lithium ion batteries.

In the past, the phase change material used in electric vehicles (EV) battery thermal management has been studied [7]. However, the shortcoming of low thermal conductivity inherent in paraffin affects its heat transfer performance. The improvement of the thermal conductivity for paraffin can be categorized into different types such as adding fins [8-9], graphite[10-11], metallic foams[12] and thermally conductive fillers[13], etc. Metallic foam with a large number of randomly distributed but similar-sized pores form a high thermal conductivity matrix and conduct the heat to the PCM filled in the matrix effectively. In addition, the porosity of open-cell foam metal reach more than 95%, which does not reduce the PCM amount and thus the heat storage capacity significantly [14]. Zhou et al.[15]experimentally investigated the heat transfer performance of paraffin wax with open-cell metallic foams and compared with the case of embedded expanded graphite. The results indicated that both foam materials improved the heat transfer performance of PCM, and metallic foams can provide better heat transfer performance than expanded graphite due to the continuous inter-connected structures. Duan et al. [16] implemented PCM surrounding the cylindrical heater which was used to simulate the heat source of power battery at low heating power. The experimental results showed that the addition of phase change material, in comparison with natural convection cooling, can significantly reduce the temperature of the heat source. Khateeb et. al[17]applied paraffin wax with embedded aluminum foam to the battery pack in a scooter and measured the temperature rise of battery under low discharge C-rates (1/5-1C). In comparison with the case of PCM only, the addition of aluminum foam further reduced the battery temperature rise. However, the aluminum foam in [17] had a low porosity (porosity80%) which may affect the heat storage capacity of the PCM. As will be shown in this paper, the highly porous copper foam (97.5%) will be used in combination with fin structure, which would be appropriate for high discharge rate applications.

In this paper, the thermal characteristics of the battery cell with PCM and thermal enhancements were examined experimentally at a high power dissipation level. The mock-up battery was made of aluminium cylinder (18mm in diameter, 65mm in length) with inserted heater to simulate the heat generation of 18650-size lithium-ion battery. The cases with pure PCM, PCM combined with metallic foam, and PCM combined with fins and metallic foams were designed and tested. The heating power was set to 6.6W, 8.8W, 13.2W, respectively, corresponding to high discharge rates 8-11C [18], which is to represent the high power dissipation from power batteries during conditions such as uphill and acceleration. The cycling operation conditions were also tested to examine the responses of the battery. Finally, the heat transfer characteristic of the battery by combining the metallic foams and fins in the PCM was also investigated.

2. Experimental setup

Three PCM configurations were designed to surround the battery in the metal housing to examine the thermal characteristics of the battery. The schematic diagram for the three configurations is given in Figure 1. The housing was made of aluminium with the inside diameter of 31mm and outside diameter of 41mm.In the first configuration, pure paraffin wax approximately 25 grams was put into the space between the battery and the housing, as shown in Figure 1.(a), Figure 1. (b) shows the second configuration, the copper foams with pore density of 20 PPI and porosity of 97.5% was first put into the housing contained with the battery, and then the paraffin of the same mass of 25 grams was filled into the foam. In the third design, four aluminium fins with the size of $65\times6.8\times1mm^3$ were first mounted into the pre-fabricated grooves on the surface of the mock-up battery, and then filled with the melted PCM and the metallic foam as shown in Figure 1.(c). Care must be taken to minimize voids in filling the PCM into the metallic foam.

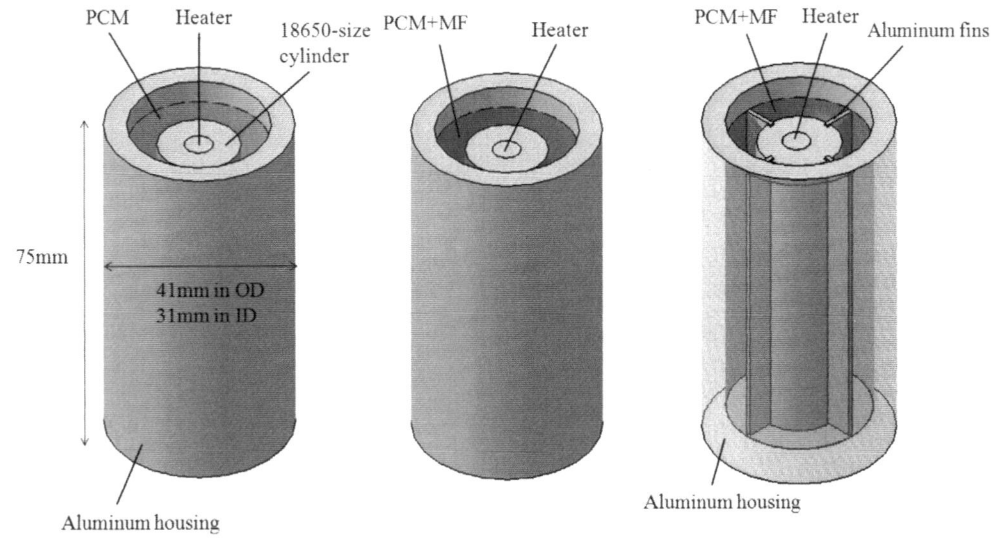

Figure 1. (a) Pure PCM Figure 1. (b) PCM-MF Figure 1. (c) PCM-MF-Fins

Figure 1. Three PCM configurations for battery thermal management in the present experiment

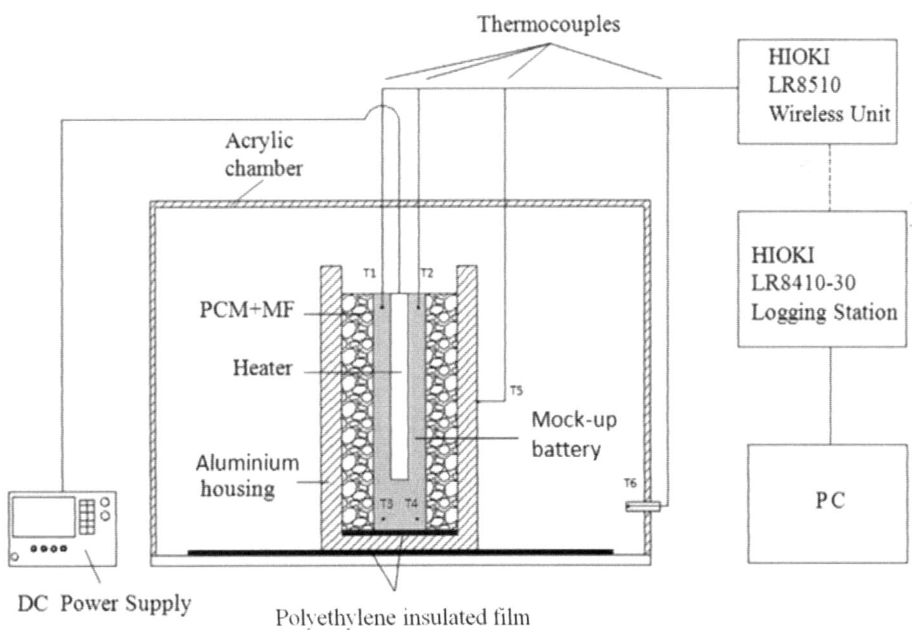

Figure 2. Schematic diagram of the experimental system

All the experiments were tested in an acrylic chamber with the outer dimension of 410mm×410mm×410mm and wall thickness of 5mm to avoid the adverse effects of ambient temperature fluctuations. The whole testing system was placed in the environmental test chamber and the chamber temperature was set to 25°C. The experimental setup is schematically showed in Fig.2.In this experiment, the phase change material was supplied as paraffin sections from Shanghai Yi Yang

Instrument Co.Ltd. Differential Scanning Calorimeter (DSC) was used to obtain the melting and solidification temperatures and heat release of PCM during phase transition. The onset melting temperature was taken as the intersection of the heat flow line w/o phase change and the tangent line drawn at the turning point near the peak temperature. According to the above method, the pure paraffin wax started melting at the intersecting temperature of 47°C, and completed at the peak temperature 53.8°C. The calculated latent heat during melting is 204.5 J/g based on the DSC test. The porosity of copper foam can be calculated as follows [14].

$$\varepsilon = \left(1 - \frac{M}{\rho_1 V}\right) \times 100 \quad (1)$$

Where, M is the mass of the composite sample; ρ_1 is the density of copper ρ_1=8.92g/cm^3, V is the volume of the composite sample.

The porosity was experimentally found to be 97.5%, and the pore density is 20 PPI (pore per inch) which was given by the supplier. The conductivity of paraffin wax with embedded copper foams was measured using LFA457MicroFlash system, which has the measurement range 0.1-2000W/mK, accuracy ±5%, repeatability ±3%. Due to the limitation of instrument sample holder, composite material sample was prepared in the size of 6.1 × 14.07 × 13.9mm^3.The specific preparation procedure is described as follows. The copper foam was cut into specified size, and then put into a polyethylene (PE) bag. The melted paraffin was poured directly into copper foam in PE bag and compressed to the desired shape. The whole sample was cut into specific size after it was solidified, and the surface was polished with fine sandpaper to improve the accuracy in the measurement of the thermal diffusion. Thus, the thermal diffusion rate was measured by the flash method. The value should be transferred into thermal conductivity by the following formula [19]:

$$k = \alpha \rho C_p \quad (2)$$

Where, αis the thermal diffusivity which was measured by flash method$\alpha = 1.2\ \mathrm{mm}^2/\mathrm{s}$, ρ is the density of sample, Cp is the specific heat of composite sample which can be calculated as

$$C_p = \frac{\varepsilon V C_{p2} \rho_2 + (1-\varepsilon) V C_{p1} \rho_1}{M} \quad (3)$$

where, ε is the porosity of copper foam ε=97.5%, V is the volume of composite sample,$C_{p2}$is the specific heat of paraffin wax C_{p2}=2.1J/gK, ρ_2 is the density of paraffin ρ_2=0.85g/cm^3, $C_{p1}$is the specific heat of copper, M is the mass of composite sample.

Based on above equations, the conductivity of the composite PCM can be obtained as k=2.2W/mK. The relative deviation was found to be -2.3% compared with the two-dimensional model of the periodic hexagonal structure of metallic foam matrix given in Calmidi's work [20]. The aluminium cylinder with the diameter of 18mm, and the length of 65mm inserted with heater inserted was fabricated to simulate the lithium-ion battery heating for heat transfer study. A Watlow catridge 1/4'' heater was used to connect to a DC power supply. The aluminium foil was wrapped tightly to the heater to strengthen the heat transfer between heater and aluminium cylinder. HIOKI LR8410-30 thermometer was used to measure the temperature. Four thermocouples (T1, T2, T3, T4) were inserted into the corresponding holes on the top and bottom surface of the battery. In order to improve the temperature measuring accuracy, aluminium foils were used to fix the four K type thermocouples, securing the thermocouples in good contact on the bottom of the holes. Thermocouple T5 was fixed by Kapton tape on the surface of aluminium housing in the middle height. The ambient temperature inside the acrylic chamber was measured by thermocouple T6 which was held with a sleeve tube connected to the measurement unit and filled with an insulating material to prevent from outside influences. The relative error of temperature measurement is estimated to be 2%. The error of the heating amount can be calculated based on error propagation formula ΔP/P=ΔI/I+ΔU/U, which is within 2%. Overall the error is estimated around 4%. After completion of experimental apparatus connection, the temperature inside the test chamber in Fig.2 was set to 25°C till steady-state was achieved. In the experiments, the heating power was set to 6.6W, 8.8W, 13.2W, and the heating time for each corresponding powers was controlled so that the housing temperature reached the peak temperature of 53.8°C. The temperature evolution for the mock-up battery was obtained by averaging four thermocouples, whereas the housing temperature was recorded at each heating power, and the temperature profiles were plotted and analyzed.

3. Results and discussion

3.1. Temperature reduction of mock-up battery using metallic foam-PCM composite materials

In this paper, heating powers of 6.6W, 8.8W, 13.2Wwere implemented to examine the limit state of the battery cooling capacity for the three PCM cases illustrated in Figure 1. Figure 3 shows the comparison of temperatures for the battery and housing for the case of pure paraffin and the composite materials with paraffin embedded copper foam at the heating power of 6.6W. The temperature rise of the mock-up battery can be divided into three stages. The first stage corresponds to the sensible heating stage 0 ~ 380s, and the battery temperature (Tb) reached 47°C at 380s. At this stage, the paraffin remained unmelted and the PCM heat transfer was through conduction only, thus the profile of temperature rise was steep. The second stage was the melting process which started at 380s and ended at 2700s. At this stage, the battery surface temperature went up beyond 47°C which corresponded to the onset of melting as obtained by the DSC test. Paraffin near to the battery surface started to melt, absorbed the heat and slowed down the temperature rise of battery. When the temperature of housing (Tw) reached 53.8°C at around 2700s, the melting was considered completed. The temperature plateau was observed obviously and the battery temperature kept almost unchanged. This could be explained by the effect of natural convection induced inside the melting PCM, which maximized the heat transfer rate from battery to the PCM melting interface. The third stage started at 2700s, the temperature of the aluminium housing reached 53.8°C, the PCM was considered fully melted, and the battery temperature started to rise rapidly again after the completion of PCM melting.

In the next experiment, metallic foam was added to increase the effective thermal conductivity of paraffin. Heat generated from battery could be conducted to the paraffin by copper foam and battery temperature and internal temperature gradient within paraffin were reduced. But on the other hand, higher flow resistance posed by metallic foams suppresses natural convection of PCM [21].Thus, there is no clear-cut temperature plateau as shown in Fig. 3. The temperature rise of battery gently increased during the melting process. However, the improvement of heat conduction overwhelmed the adverse effect on natural convection in paraffin. The distinction of each stage was not as obvious as the case of pure paraffin, the temperature difference between battery and the housing was narrowed, and the battery temperature was lower than the case of paraffin without copper foam. When the temperature of housing reached 53.8°C, the battery temperature was measured 65.9°C which dropped by 3.4°C compared to pure paraffin as shown in Fig.3 at 6.6W heat input. The similar battery temperature decreased with the addition of copper foam can be found in Fig. 4, Fig. 5 at increased heating power of 8.8W, 13.2W, respectively. On one hand, the battery temperature and the temperature difference between battery and housing further increased with the increasing heating power. On the other hand, the composite of paraffin combined with copper foam showed better performance than the pure PCM case. The larger the heat input was, the larger the battery temperature drop was between the composite PCM and pure PCM cases. The battery temperatures upon the completion of the melting are listed in Table 1 for the different heating powers.

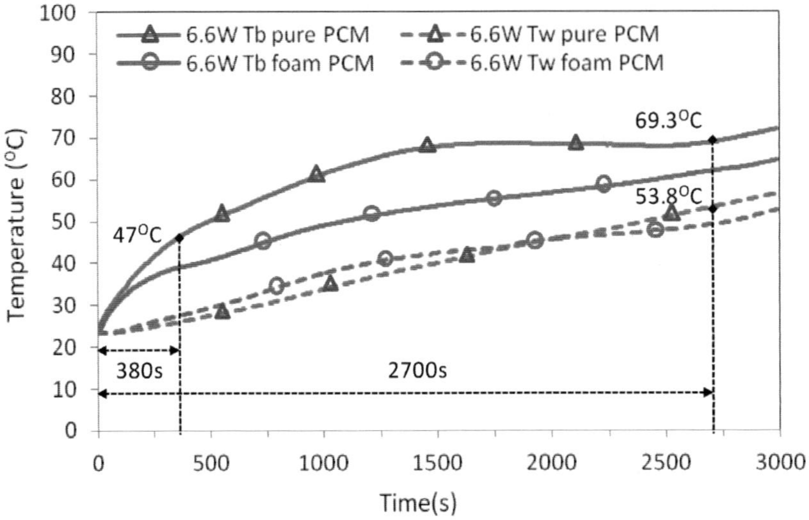

Figure 3. Temperature profiles for battery and housing wall for pure PCM and PCM-MF cases at heating power of 6.6W

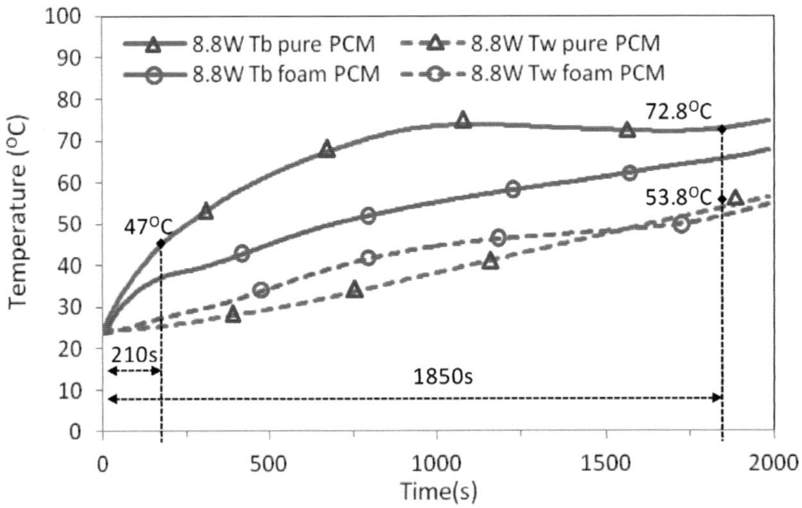

Figure 4. Temperature profiles for battery and housing wall for the pure PCM and PCM-MF cases at heating power of 8.8W

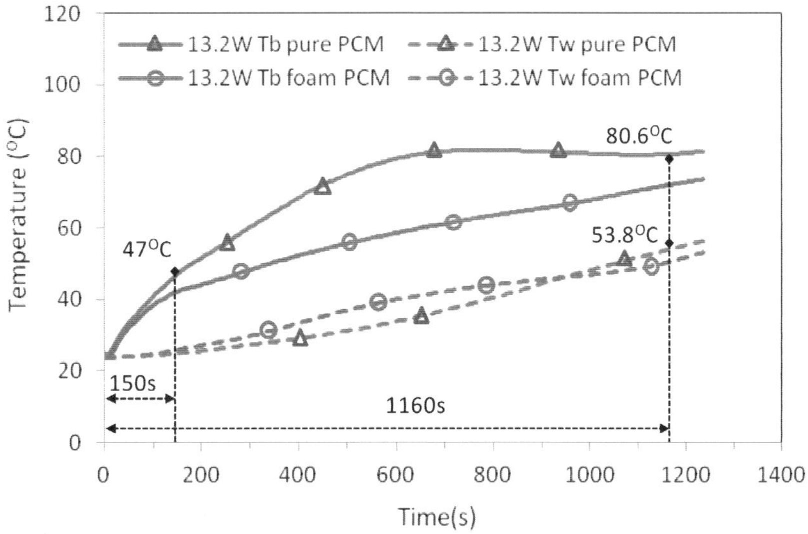

Figure 5. Temperature profiles for battery and housing wall for pure PCM and PCM-MF cases at heating power of 13.2W.

Table 1. Comparison of battery temperature between the three investigated cases at completion of melting at the three investigated heating powers

	Pure PCM		PCM - MF		PCM-MF-Fins	
P (W)	Tb (°C)	Tw (°C)	Tb (°C)	Tw (°C)	Tb (°C)	Tw (°C)
6.6	69.3	53.8	65.9	53.8	60.8	53.8
8.8	72.8	53.8	68.7	53.8	62.2	53.8
13.2	80.6	53.8	74	53.8	65.1	53.8

3.2. The temperature evolution for battery with foam-paraffin at heating and cooling cycles

The above study compared the experiment case of pure paraffin and paraffin combined with copper foam under steady-state heating conditions .To further examine the effectiveness of the configuration of paraffin with copper foam, the cyclic heating test was conducted by switching on and off the power supply to simulate the dynamic electric vehicle operation. . The ambient temperature was set to 25°C, and the heating power was set to 4.4W. Both cases were heated in 3000s and cooled in 3000s under natural convection, and then the same cycles were repeated. The temperature curves of battery for cycles were plotted in Fig.6. The battery temperature with pure paraffin reached 60°C at the end of second heating, beyond the allowable working temperature for lithium ion battery which was within the range of 0-55°C. In contrast, the battery temperature was maintained below 53°C for the case of paraffin combined with copper foam during heating period. The temperature rise was obviously lower than the case of pure paraffin, indicated that battery surrounded with the copper foam-paraffin composite materials was more effectively maintained within the proper operating temperature at the cyclic heating condition.

IOP Conf. Series: Earth and Environmental Science **40** (2016) 012045 doi:10.1088/1755-1315/40/1/012045

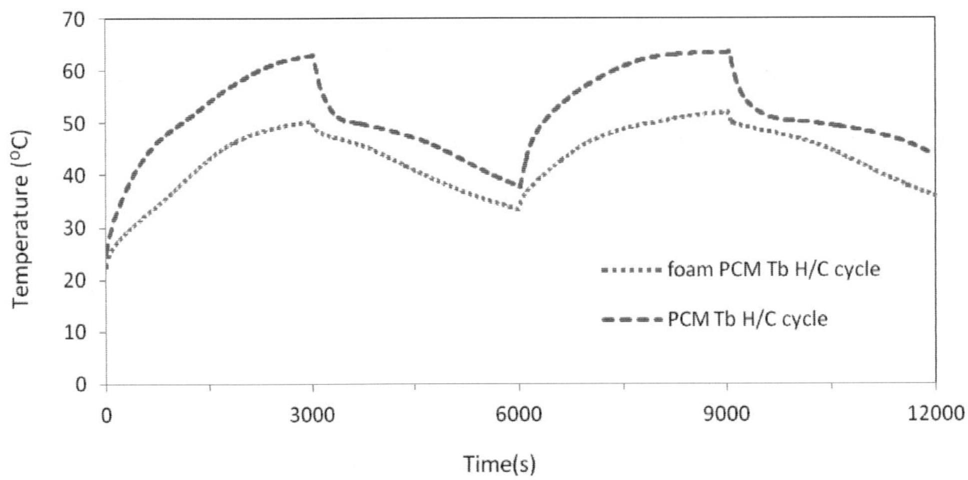

Figure 6. Comparison of effect on battery temperature for pure paraffin and foam-paraffin in heating and cooling cycles

3.3. Effect of fin inserts into foam-paraffin composite materials

In this section, the combination of fins and copper foam in the PCM was examined as the thermal enhancement technique. Four aluminium fins of 1mm thick were symmetrically inserted into foam-paraffin composite materials as indicated in Figure. 1(c). The experimental condition was controlled in the same way as mentioned above. The temperature rise of battery and housing are shown in Figures.7-9 for the different heating power. The shape of curves of battery and housing temperature evolution for the foam-paraffin-fins was similar to the case in foam-paraffin, but both the absolute temperature and the battery to housing temperature difference in the former case were lower than the latter case. Adding fins, on the one hand, improved the effective conductivity in paraffin and, on the other hand, reduced the thermal contact resistance by extending the contact area between battery and the copper foam, and enhanced the heat transfer performance of metallic foam [22]. At the heating power of 6.6W, battery temperature reached 60.8°C when the housing temperature reached 53.8°C, dropped by 5.1°C compared to the case of foam-paraffin, 8.5°C compared to the pure paraffin. The temperature drop was more significant for the higher heating power. The battery temperature of three experimental cases at the fully melted time at different heating power of 6.6W, 8.8W, 13.2W were listed in Table 1.

IOP Conf. Series: Earth and Environmental Science **40** (2016) 012045 doi:10.1088/1755-1315/40/1/012045

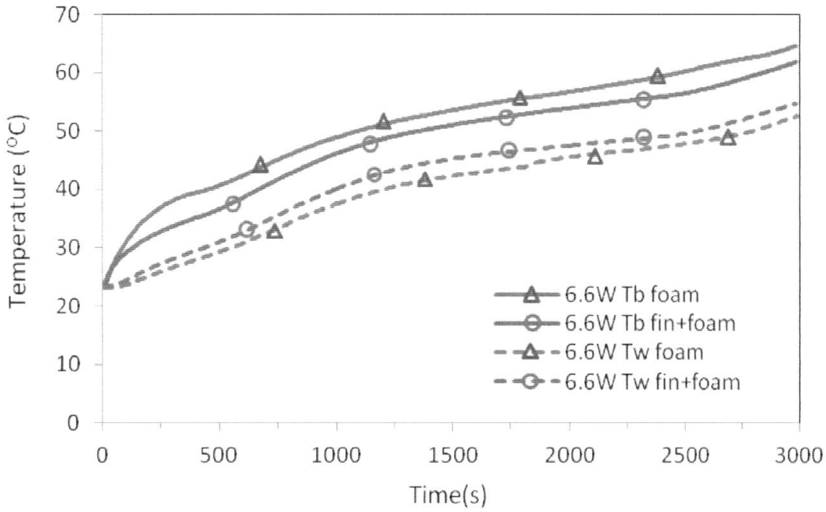

Figure 7. Temperature profiles on battery and housing for the cases of PCM-MF and PCM-MF-fins at the heating power of 6.6W

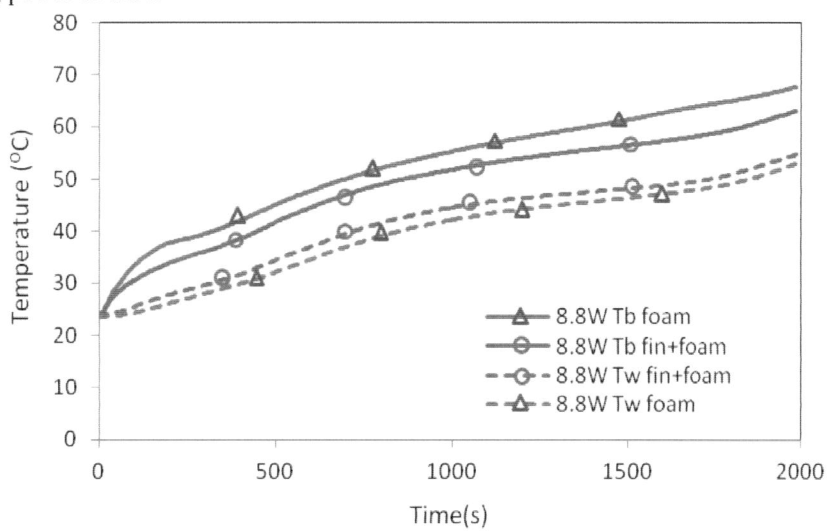

Figure 8. Temperature profiles on battery and housing for the cases of PCM-MF and PCM-MF-fins at the heating power of 8.8W

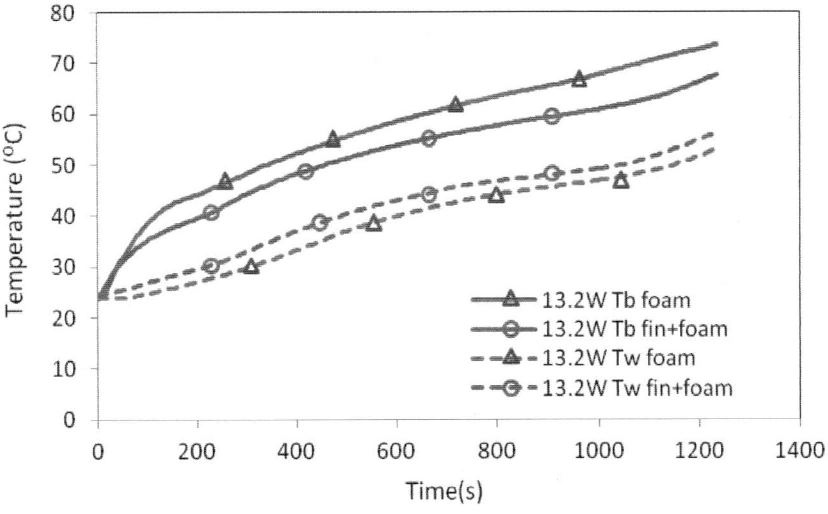

Figure 9. Temperature profiles on battery and housing for the cases of PCM-MF and PCM-MF-fins at the heating power of 13.2W

The temperature difference between battery and the aluminium housing at different time steps for different heating powers were listed in Table 2 to compare the thermal performance of the three PCM cases. Better thermal performances were achieved in both the foam-paraffin and foam-paraffin-fin cases than that the pure paraffin case. Among them, the case of foam-paraffin-fin exhibited the lowest battery to housing temperature difference with time evolution. Similar trend was found for the cases of heating power 8.8W and 13.2W. Nonetheless, adding fins can slightly affect the capacity of heat storage of the composite materials since the volume of four fins in the present study is more significant than the metallic foam, which is the unfavourable factor. The increase in volume by adding four fins of 1mm is $1.56cm^3$, whereas the actual volume of the copper foam is $0.735\ cm^3$. In contrast, adding fins could reduce volume fraction of PCM by 5.3%, which may adversely affect the capacity of the heat storage.

Table 2. Comparison of the battery to housing temperature differences for the three cases, PCM, MF-PCM and MF-PCM-fin at the different heating power.

P(W)	Time(s)	Tb – Tw [°C]		
		PCM	MF-PCM	MF-PCM-fins
6.6	600	23.8	11.5	5.8
	1100	28.0	11.2	6
	1600	27.3	11.5	5.9
	2500	17	12.3	6.9
8.8	600	34	13.1	7.5
	1100	33.9	13.2	7.3
	1600	23.2	15.5	8.3
	1900	18.6	15.4	8.3
13.2	600	45.5	18.6	10.9
	900	37.3	19.8	11.4
	1100	29	21.7	11.7

4. Conclusions

In this paper, an experimental study of the thermal characteristics of mock-up LIB was conducted with PCM and the thermal enhancement techniques. The conclusions can be drawn as follows.

- In the PCM case, the battery temperature evolution can be divided into three phases depending on the melting condition of the pure PCM: temperature rise with unmelted PCM, temperature plateau with partially melted PCM, and temperature rise with fully melted PCM.
- The addition of metallic foam can enhance the heat conduction and reduce the battery temperature obviously. The battery temperature at the time PCM fully melted is 3.4-6.6°C lower than that with pure PCM for heating power 6.6W-13.2W.
- Under cycling condition, the temperature swing with metallic foam and PCM can be reduced by ~ 10°C in comparison with the case with the PCM only.
- The addition of four fins can further reduce the Tb by 4.1-8.9°C compared with the case with metallic foam and PCM for heating power 6.6W-13.2W.
- The addition of fins can further reduce the battery temperature, which also reduce the PCM volume and may have an adverse effect on capacity of heat storage.

Acknowledgements
We acknowledge the financial supports from the Eastern Scholar program by Shanghai Higher Education Institution and the project 14520501100 granted by Shanghai Scientific and Technological Commission.

A. 1

Nomenclature

C_p specific heat capacity of sample (J/gK)
C_{P1} specific heat capacity of copper (J/gK)
C_{P2} specific heat capacity of paraffin (J/gK)
I current (A)
k thermal conductivity (W/mK)
M mass of sample (g)
P heating power (W)
Tb battery temperature (°C)
Tw housing wall temperature (°C)
U voltage (V)
V volume of sample (cm^3)
α thermal diffusivity (cm^2/s)
ε porosity
ρ_1 density of copper (g/cm^3)
ρ_2 density of paraffin (g/cm^3)

References

[1] Rao Z H and Wang S F 2011 A review of power battery thermal energy management *Renewable and Sustainable Energy Reviews* 9 4554-71
[2] W. Golubkov A, Fuchs D, Wagner J, Wiltsche H, Stangl C, Fauler G, Voitic G, Thaler A and Hacker V 2014 Thermal-runaway experiments on consumer Li-ion batteries with metal-oxide and olivin-type cathodes *Royal Society of Chemistry Advances* **4** 3633-42
[3] Sharma A, Tyagi V V, Chen C R and Buddhi D 2009 Review on thermal energy storage with phase change materials and applications *Renewable and Sustainable Energy Reviews* 13 318-45
[4] Ling Z Y *et al.* 2014 Review on thermal management systems using phase change materials for electronic components, Li-ion batteries and photovoltaic modules *Renewable and Sustainable Energy Reviews* **31** 427–38

[5] Mills A and Al-Hallaj S 2005 Simulation of passive thermal management system for lithium-ion battery packs *Journal of Power Sources* 141 307-15
[6] US Department Of Energy, 1996, USABC Electric Vehicle Battery Test Procedures ManualRevision 2 (Washington DC: Office of Scientific & Technical Information Technical Reports)
[7] Rao Z H, Huo Y T, Liu X J and Zhang G Q 2014 Experimental investigation of battery thermal management system for electric vehicle based on paraffin/copper foam *Journal of the Energy Institute* 1-6
[8] Mahmoud S, Tang A, Toh C, Al-Dadah R and Soo S L 2013 Experimental investigation of inserts configurations and PCM type on the thermal performance of PCM based heat sinks *Applied Energy* **112** 1349-56
[9] Fok S C, Shen W and Tan F L 2010 Cooling of portable hand-held electronic devices using phase change materials in finned heat sinks *Int. Journal of Thermal Sciences* **149** 109-17
[10] Zhang Z, Zhang N, Peng J, Fang X, Gao X and Fang Y 2012 Preparation and thermal energy storage properties of paraffin/expanded graphite composite phase change material **91** 426-31
[11] Wu W X, Zhang G Q, Ke X F,Yang X Q, Wang Z Y and Liu C Z 2015 Preparation and thermal conductivity enhancement of composite phase change materials for electronicthermal management *Energy Conversion and Management* **101** 278–84
[12] Zhao C Y, Lu W and Tian Y 2010 Heat transfer enhancement for thermal energy storage using metal foams embedded within phase change materials(PCMs) *Solar Energy* **84** 1402–12
[13] Zhang Y L, Rao Z H, Wang S F, Zhang Z and Li X P 2012 Experimental evaluation on natural convection heat transfer of microencapsulated phase change materials slurry in a rectangular heat storage tank *Energy Conversion and Management* **59** 33-9
[14] Liu P S and Chen X 2012 *Metal Foams 2012* (Changsha: Central South University Press)
[15] Zhou D and Zhao C Y 2011 Experimental investigationson heat transfer in phase change materials(PCMs) embedded in porous materials *Applied Thermal Engineering.* **31** 970-7
[16] Duan X and Naterer G F 2010 Heat transfer in phase change materials for thermal management of electric vehicle battery modules *Int. J. Heat Mass Transfer* **53** 5176-82
[17] Khateeb S A, Amiruddin S, Farid M, Selman J R and Al-Hallaj S 2005 Thermal management of Li-ion battery with phase change material for electric scooters: experimental validation *J. Power Sources* **142** 345-53
[18] Drake S J and Wetz D A 2014 Measurement of anisotropic properties of cylindrical Li-ion cells *J. Power Sources* **252** 298-304
[19] Incropera F P, DeWitt D P, Bergman T L and Lavine A S 2007 *Fundamentals of Heat and Mass Transfer, Sixth Edition* John Wiley & Sons, Inc.
[20] Calmidi V V and Mahajan R L 1999 The effective thermal conductivity of high porosity fibrous metal foams *Journal of Heat Transfer* **121** 166-71
[21] Tian Y and Zhao C Y 2011 A numerical investigation of heat transfer in phase change materials (PCMs) embedded in porous metals Energy **36** 5539-46
[22] Sadeghi E, Hsieh S and Bahrami M 2011 Thermal conductivity and contact resistance of metal foams *Journal of Physics D Applied Physics* 44 355-65

IOP Conf. Series: Earth and Environmental Science **40** (2016) 012046 doi:10.1088/1755-1315/40/1/012046

Numerical analysis of phase change materials for thermal control of power battery of high power dissipations

X Xia, H Y Zhang[1] and Y C Deng

College of Automotive Engineering, Shanghai University of Engineering Science 333 Longteng Road, Songjiang, Shanghai, China

Email: zhanghengyun@sues.edu.cn

Abstract. Solid-fluid phase change materials have been of increasing interest in various applications due to their high latent heat with minimum volume change. In this work, numerical analysis of phase change materials is carried out for the purpose of thermal control of the cylindrical power battery cells for applications in electric vehicles. Uniform heat density is applied at the battery cell, which is surrounded by phase change material (PCM) of paraffin wax type and contained in a metal housing. A two-dimensional geometry model is considered due to the model symmetry. The effects of power densities, heat transfer coefficients and onset melting temperatures are examined for the battery temperature evolution. Temperature plateaus can be observed from the present numerical analysis for the pure PCM cases, with the temperature level depending on the power densities, heat transfer coefficients, and melting temperatures. In addition, the copper foam of high thermal conductivity is inserted into the copper foam to enhance the heat transfer. In the modeling, the local thermal non-equilibrium between the metal foam and the PCM is taken into account and the temperatures for the metal foam and PCM are obtained respectively.

1. Introduction

Since the phase change materials (PCMs) can store large amounts of heat with chemical stability and many other advantages [1-3], they have has been widely used in various fields such as electronics cooling and battery thermal management. Duan & *Natere* [4] experimentally studied two types of PCM for thermal management of battery cell for application in electric vehicles. The first type of PCM had a melting temperature 18°C, and the second type had a melting temperature of 50°C. The results showed that the temperature of heater can be better maintained by PCM in comparison with natural convection. Kandasamy *et al.* [5] investigated the application of PCM package for thermal management of electronic components experimentally. The effects of power input, orientation of package, and various melting/freezing times under cyclic conditions were investigated. The results showed increased power input would accelerate melting rate, but the thermal properties of the package has minimal impact at different orientation to the gravity. Wu *et al.* [6] prepared a type of phase change material board (PCMB) with thermal storage capacity of 141.74 J/g and thermal conductivity of 7.654 W/ (m K). PCMB displayed much better cooling effect in comparison to natural air cooling in their investigation. Compared with natural air cooling and liquid cooling, PCM could keep

[1] Address for correspondence: H Y Zhang, College of Automotive Engineering, Shanghai University of Engineering Science 333 Longteng Road, Songjiang, Shanghai, China. E-mail: zhanghengyun@sues.edu.cn.

temperature of electronics at a smaller range of temperature swing and minimize the temperature difference among the cells.

The above studies shows that phase change materials have improved performance in heat storage and temperature control at low heating rate. However, PCM would not perform well for applications with high heating rate and narrow temperature range such as power potteries in electric vehicles, due to the disadvantage of low thermal conductivity of the phase change material, which may not dissipate the heat in a timely manner. To solve this problem, the researchers have proposed several methods. [7-11]. Mills *et al.* [10] combined expanded graphite with paraffin wax, and the results showed the thermal conductivity of this new composite material was two orders of magnitude larger than that of pure paraffin. Fleming *et al.* [11] fabricated a shell and tube latent heat storage unit using aluminium foam for heat transfer enhancement. The experimental measurements showed that during melting and solidification process of phase change material, the metal foam significantly improved the heat transfer performance. Literature survey showed that the phase change material has excellent performance in thermal management for battery cells, but most of work is focused on experimental studies. Though some numerical study on the thermal enhancement has been conducted for PCM inserted with metal foam, the corresponding numerical studies are still limited for cylindrical batteries, and the local thermal equilibrium is often used for the composite PCM with foam structure.

In this paper, numerical simulation method is applied to study battery thermal management system with phase change material either with or without metal foam enhancement, which is targeted to the application in electric vehicles. The battery was located in the center of a concentric PCM container, dissipating heat to the ambient through the PCM layer and the convection at the outer wall of the container. High power densities are applied of 399223W/m^3, 532297W/m^3, and 798446W/m^3, respectively. Both the heat conduction and natural convection of PCM will be numerically simulated and analyzed based on the battery temperature evolution. Thermal profiling of battery is also studied with the addition of copper metal foam in the phase change material.

2. Numerical model

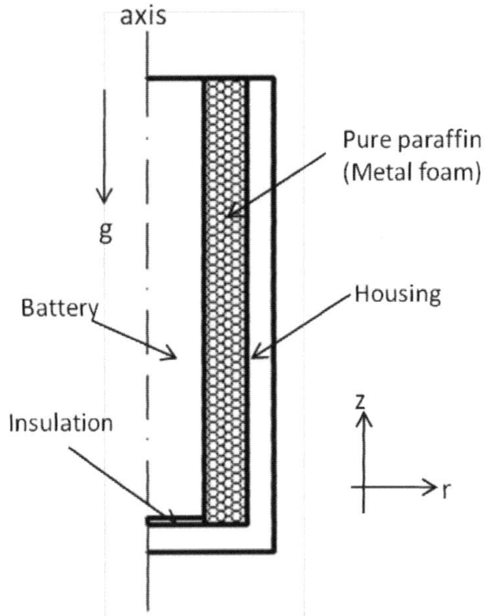

Figure 1. The two-dimensional geometrical model in the present study

The physical problem was depicted in Figure 1, which consisted of the battery, the pure paraffin PCM or PCM composite made of paraffin and copper foam (25PPI, 97%porosity), insulation and cylindrical enclosure. The diameter and height of 18650 lithium ion battery were 18mm and 65mm, respectively. The thickness of insulating spacers, made of polycarbonate, was 1mm, the housing had an inner diameter of 31mm and depth of 66mm, with a wall thickness of 4mm. Due to symmetry, the two-dimensional computational model was developed. The battery generated a constant heat source q. The boundary at the bottom surface of the housing was adiabatic. Outer wall surface of cylindrical housing was set to the boundary of constant heat transfer coefficient h. The initial temperature of computational domain was 25°C. Table1 showed thermo-physical properties of PCM, whereas Table2 showed thermo-physical properties of the other materials in the model. Surface area weighted average temperatures of the battery outer wall and the housing outer wall and liquid fraction of PCM were monitored during simulation.

Table 1. Thermo-physical properties of paraffin PCM

Parameters	Value
Density of paraffin (kg/m^3)	920
Specific heat capacity(J/kgK)	2400
Latent heat of paraffin(J/kg)	195000
Melting temperature of paraffin (℃)	47-53
Thermal conductivity of paraffin (W/mK)	0.2
Thermal expansion coefficient (K^{-1})	0.00085

Table2. Thermo-physical properties of aluminum, copper foam and polycarbonate

	Density (kg/m^3)	Thermal conductivity(W/mK)	specific heat (J/kgK)
Polycarbonate	1200	0.19	1260
Copper foam	8978	387.6	381
Aluminum	2719	202.4	871

The following assumptions were made in the present numerical simulation:

- The liquid paraffin was considered as incompressible Newtonian fluid and the buoyancy was subjected to the Boussinesq approximation.
- The local thermal non-equilibrium was assumed between the copper foam and the PCM.
- The effect of radiation heat transfer was ignored.
- The specific heat and thermal conductivity of battery and PCM were constant.
- The copper foam was assumed homogeneous and isotropic.

Based on the above assumptions, the Darcy-Brinkman-Forcheimer Momentum equations were used in numerical calculations. Two energy equations model was applied to describe the local thermal non-equilibrium between paraffin and copper foam [12].

Continuity equations,

$$\frac{\partial \rho_f}{\partial \tau}+\frac{\partial}{\partial z}(\rho_f \mu)+\frac{1}{r}\frac{\partial}{\partial r}\left(r\rho_f \upsilon\right)=0 \tag{1}$$

Momentum equations,

$$\frac{\rho_f}{\varepsilon}\frac{\partial u}{\partial \tau}+\frac{\partial}{\partial z}\left(\frac{\rho_f}{\varepsilon}uu\right)+\frac{1}{r}\frac{\partial}{\partial r}\left(r\frac{\rho_f}{\varepsilon}\upsilon u\right)=-\frac{\partial p}{\partial r}+\frac{\partial}{\partial z}\left(\frac{\mu_f}{\varepsilon}\frac{\partial u}{\partial z}\right)+\frac{1}{r}\frac{\partial}{\partial r}\left(r\frac{\mu_f}{\varepsilon}\frac{\partial u}{\partial r}\right)$$
$$-\frac{\mu_f u}{K}-\frac{\rho C\varepsilon}{\sqrt{K}}|u|u-\frac{(1-\beta)^2}{\beta^3+\delta}A_m u+\rho g\alpha(T_f-T_{ref}) \tag{2}$$

$$\frac{\rho_f}{\varepsilon}\frac{\partial v}{\partial \tau}+\frac{\partial}{\partial z}\left(\frac{\rho_f}{\varepsilon}uv\right)+\frac{1}{r}\frac{\partial}{\partial r}\left(r\frac{\rho_f}{\varepsilon}\upsilon\upsilon\right)$$
$$=-\frac{\partial p}{\partial r}+\frac{\partial}{\partial z}\left(\frac{\mu_f}{\varepsilon}\frac{\partial v}{\partial z}\right)+\frac{1}{r}\frac{\partial}{\partial r}\left(r\frac{\mu_f}{\varepsilon}\frac{\partial v}{\partial r}\right)-\frac{\mu_f v}{K}-\frac{\rho_f C\varepsilon}{\sqrt{K}}|v|v-\frac{\mu_f v}{\varepsilon r^2}-\frac{(1-\beta)^2}{\beta^3+\delta}A_m v \quad (3)$$

The energy equations are developed with the two-temperature energy model.
For paraffin,

$$\varepsilon\rho_f\left(c_f+H\frac{d\beta}{dT_f}\right)\frac{\partial T_f}{\partial \tau}+\rho_f c_f\left(u\frac{\partial T_f}{\partial z}+\frac{1}{r}\frac{\partial}{\partial r}(vT_f)\right)=\left(k_{td}+k_{fe}\right)\nabla^2 T_f+h_{sf}a_{sf}\left(T_s-T_f\right) \quad (4)$$

For copper foam,

$$\left(1-\varepsilon\right)\rho_s c_s\frac{\partial T_s}{\partial \tau}=k_{se}\nabla^2 T_s-h_{sf}a_{sf}\left(T_s-T_f\right) \quad (5)$$

$$\beta=\begin{cases}0, T_f<T_{m1}\\(T_f-T_{m1})/(T_{m2}-T_{m1}), T_{m1}\le T_f\le T_{m2}\\1,\ T_f>T_{m2}\end{cases} \quad (6)$$

In the numerical investigation, the liquid fraction β for the PCM was denoted in the followings. $\beta=0$ represents the solid phase, $\beta=1$ represents the liquid phase and $\beta=0\sim1$ represents that the paraffin was in the mushy region. T_{m1} and T_{m2} were the onset melting temperature (MT) and the peak temperature of the PCM according to the DSC test curve.

Permeability K and inertial coefficient C were determined by the following equations,

$$K=\frac{\varepsilon^2 d_k^{\,2}}{36\chi(\chi-1)} \quad (7)$$

$$C=\frac{0.00212\left(1-\varepsilon\right)^{-0.132}\left(d_f/d_p\right)^{-1.63}}{\sqrt{K}} \quad (8)$$

Where χ was the tortuosity coefficient of the copper foam relating to the porosity,

$$\chi=2+2\cos\left(\frac{4\pi}{3}+\frac{1}{3}\cos^{-1}(2\varepsilon-1)\right) \quad (9)$$

The interfacial surface area, a_{sf}, can be determined by,

$$a_{sf}=\frac{3\pi d_f}{d_p^{\,2}} \quad (10)$$

The interfacial heat transfer coefficient, h_{sf}, was formulated as,

$$h_{sf}=\begin{cases}0.76\,\mathrm{Re}^{0.4}\,\mathrm{Pr}^{0.37}\,k_f/d_f, 0<\mathrm{Re}\le 40\\0.52\,\mathrm{Re}^{0.5}\,\mathrm{Pr}^{0.37}\,k_f/d_f, 40<\mathrm{Re}\le 1000\\0.26\,\mathrm{Re}^{0.6}\,\mathrm{Pr}^{0.37}\,k_f/d_f, 1000<\mathrm{Re}\le 20000\end{cases} \quad (11)$$

3. Numerical method

The commercial computational fluid dynamics software Ansys Fluent was used as the solver in the present study. Due to symmetry, the two-dimensional geometry was considered, which could reduce the number of grids and decrease computing cost. Quadrilateral surface mesh was utilized. To better capture internal characteristics of paraffin during melting, refined grids were applied at the PCM zone. In total 17260 grids was applied, and the time step of 0.1s were used to balance the accuracy of results and the computational time. The second-order-upwind scheme was used to compute convection term and pressure-implicit with splitting of operation (PISO) algorithm for pressure-velocity coupling, and PRESTO for pressure correction. Since the dynamic viscosity of liquid paraffin was sensitive to temperature [13], it was determined by the following equation,

$$\mu = 0.11994 - 6\times10^{-4}T + 9\times10^{-7}T^2 \tag{12}$$

The other thermo-physical parameters of paraffin wax have been shown in Table 1.

4. Results and discussion

4.1. *Effects of battery power density*

Figure 2 showed the numerical results of the temperature distributions of pure paraffin at different time steps τ=500s, 1000s, 1500s and 2000s, respectively. At τ=500s, temperature distribution decreased from the battery on the left to the right almost uniformly in paraffin zone. At this time, heat dissipation of battery was mainly through heat conduction, because paraffin just began to melt as showed in Fig. 3. Temperature distribution of paraffin was no longer uniform as the heating time evolved. Temperature was higher at the upper left corner and lower at the lower right corner since the melted paraffin, heated by the battery, moved to the upper left region under the effect of buoyancy. Paraffin remains the solid phase or mushy phase at the lower right corner and the heat transfer thereof was mainly by heat conduction.

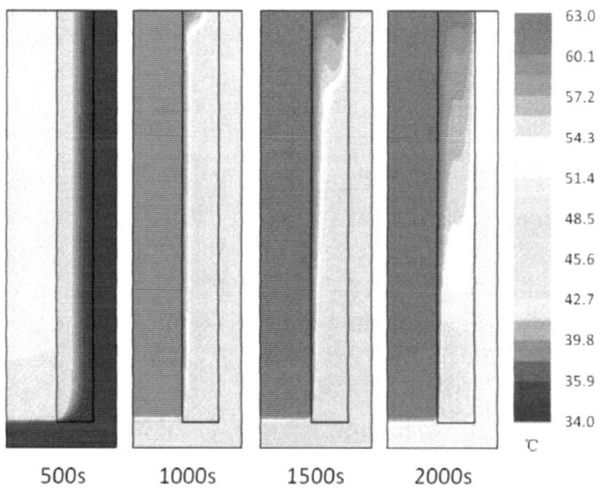

Figure 2. Temperature distributions for pure PCM at different time steps

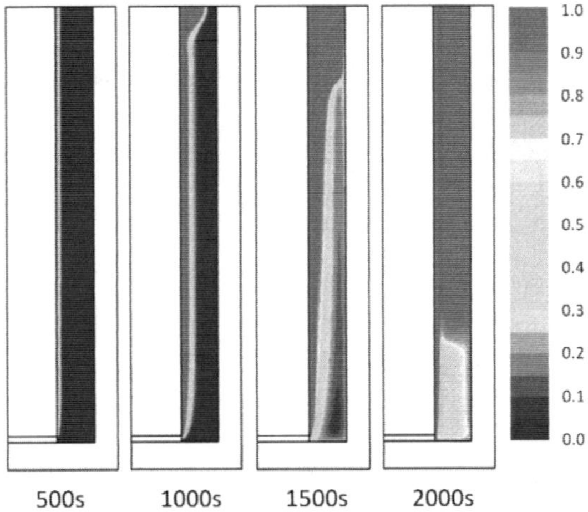

Figure 3. Liquid fraction distributions for pure PCM at different time steps

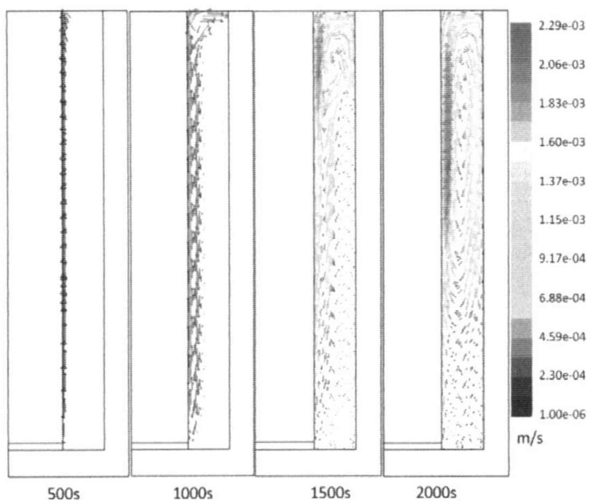

Figure 4. Velocity fields for pure PCM at different time steps

Figure 3 showed the numerical results of evolution of the solid liquid interface of pure paraffin at the time steps of 500s, 1000s, 1500s and 2000s, respectively, at the power density of 399223W/m^3, the heat transfer coefficient of the housing outer wall 4W/m^{2}K, and the onset melting temperature of paraffin wax was 47°C. The red-colour area denoted the liquid phase, blue-colour area was the solid phase and the mushy region had the colour between red and blue. Solid liquid interface of pure paraffin become inclined, as indicated in liquid fraction contour from 500s to 1000s since the density of liquid paraffin decreased with the increasing temperature, which generated buoyancy force and thereby resulted in natural convection as showed in Fig.4. At first, liquid paraffin near the high temperature battery wall flowed upward to the top of the paraffin zone due to the buoyancy effect and then moved downward along the solid liquid interface as a result of heat absorbing of PCM at the melting front. The induced natural convection of the liquid paraffin accelerated the heat transfer, which would maintain the battery temperature near a constant level of temperature plateau. As the time evolves, the natural convection was intensified as indicated in the velocity fields in Fig. 4. Thus the heat transfer was also intensified on the upper part of the paraffin zone, further melting the PCM.

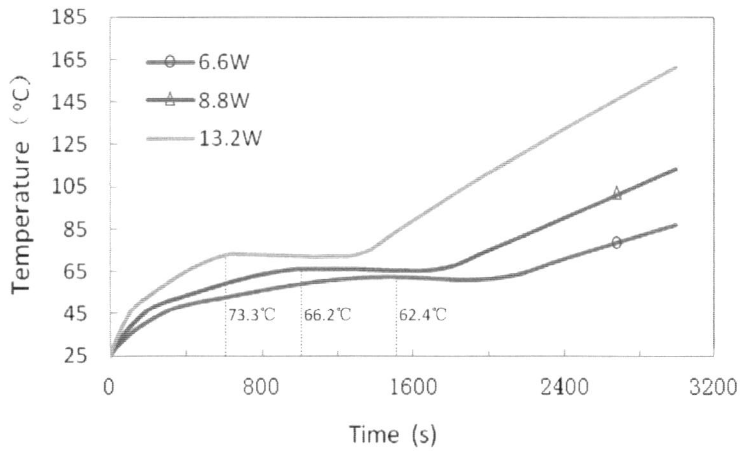

Figure 5. Temperature profile at simulate battery wall

Figure 5. showed the numerical results of the battery wall temperature variation at the power density of 399223W/m^3, 532297W/m^3, and 798446W/m^3, respectively. Here the heat transfer coefficient of housing outer wall was 4W/m^{2}K, and the onset melting temperature of paraffin wax was 47°C. It can be seen that paraffin entered stage of phase change when the battery temperature reached melting point, the battery temperature curves began to level off, forming temperature plateaus. With the maximum temperatures of 62.4^{o}C, 66.2^{o}C, and 73.3^{o}C, respectively for the three different power densities 399223W/m^3, 532297W/m^3, and 798446W/m^3.It was also noted that the battery temperature reduced by about 1.2^{o}C, 0.8^{o}C and 1.4^{o}C at the end of the plateau. This was mainly because the liquid paraffin began to contact the top inner wall of the metallic housing during the melting process, so heat transfer between paraffin and housing was through solid phase heat conduction as well as the convection heat transfer.

4.2. Effects of the convective heat transfer coefficient of housing wall

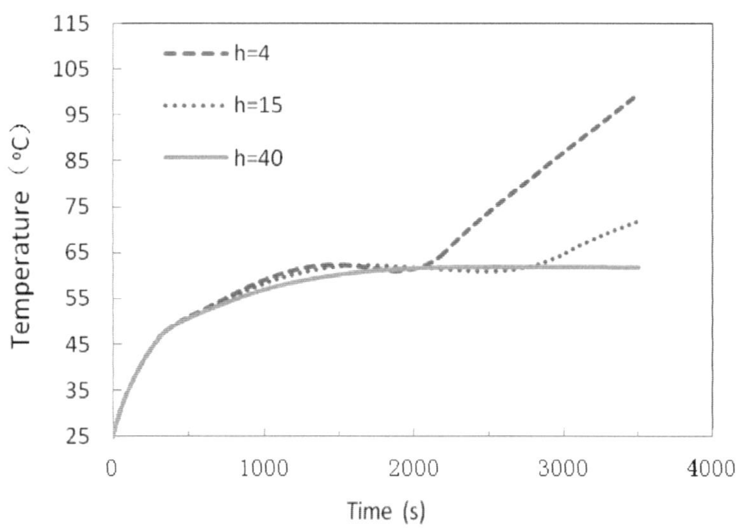

Figure 6. Battery temperature profile at different h in W/m^{2}K

The convective heat transfer coefficient had an impact on the battery temperature in battery thermal management. Here the convective heat transfer coefficient of the housing outer wall of 4W/m^{2}K, 15W/m^{2}K, 40W/m^{2}K were considered, the onset melting temperature of paraffin wax was 47°C, and power density was 399223W/m^3. Fig.6 shows the numerical results of simulate battery wall temperature variation. The temperature evolutions of battery outer wall before 2000s were basically the same with the improvement of convective heat transfer coefficient. The plateau temperatures formed owing to natural convection in PCM at around 63°C for different heat transfer coefficients. However, there was a great difference in the time points required for paraffin fully melted, which were 2244s and 3064s, when convective heat transfer coefficient of the housing outer wall was 4W/m^{2}K and 15W/m^{2}K. At the highest heat transfer coefficient h=40W/m^{2}K, the melting of paraffin wax was not completed even after 6000s elapsed. It demonstrated that an enhancement of the convective heat transfer coefficient would maintain the battery around plateau temperature for a longer time.

4.3. Effects of onset melting temperatures of paraffin

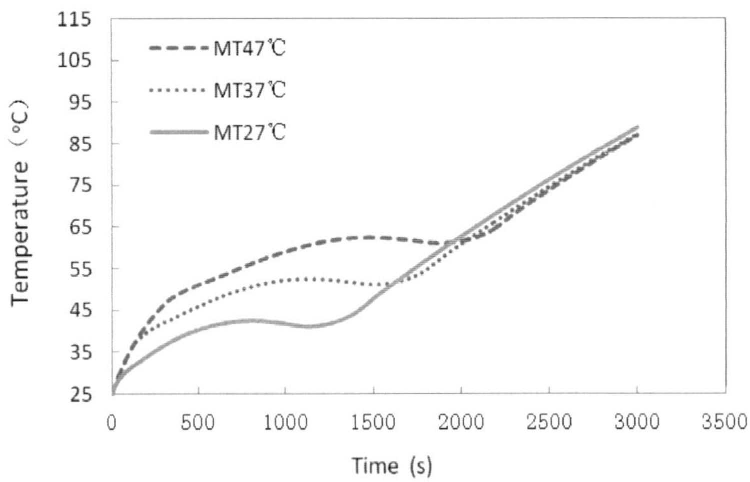

Figure 7. The evolution of the battery temperature with different onset melting temperatures (MT).

In this section, the effect of the onset melting temperatures of the PCM was examined. Three onset melting temperature (MT) 27°C, 37°C and 47°C, were considered with the power density of 399223W/m^3. The heat transfer coefficient was fixed to be 4W/m^{2}K for the housing outer wall. The numerical results of the battery wall temperature were illustrated in Fig.7. The time required for onset melting of paraffin wax become shorter with the decrease of the onset melting point of paraffin wax. The paraffin with the lowest melting point of 27°C first began to melt and then entered the stage of temperature plateau, at which the temperature of battery wall was kept around 42°C. The paraffin fully melted at τ=1468s. The paraffin of higher melting points exhibited similar trends, though the plateau temperatures were higher, around 52°C for the onset melting temperature of 37°C and 62°C for the onset temperature of 47°C. It is seen that the temperature difference of battery and the onset melting temperature remains almost constant in spite of different onset melting temperature.

4.4. Heat transfer characteristics of pure paraffin and paraffin/metal foam composite

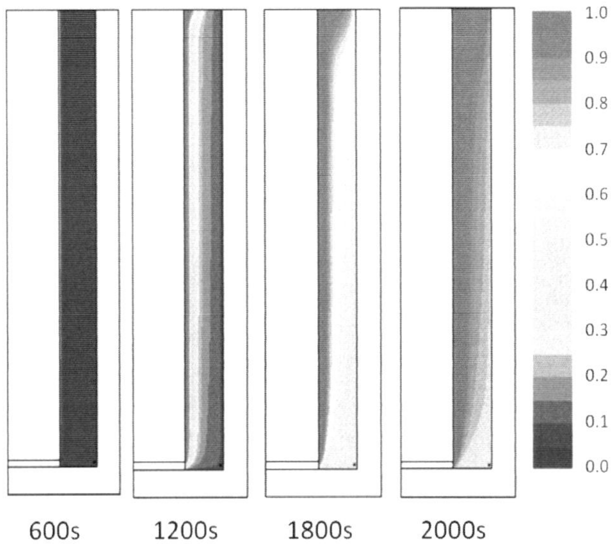

Figure 8. Liquid fraction distribution for paraffin/copper foam composite

Figure 8 showed the liquid fraction distribution for paraffin/copper foam composite. Comparing the liquid fraction of pure paraffin and paraffin/copper foam composite shown in Fig. 3 and Fig. 8, it can be found that the slope of solid liquid interface of PCM composite was much smaller than that in pure paraffin. This was due to the existence of metal foam, so that the liquid paraffin flow was hampered by the ligaments of copper foam. But at τ=1200s, the majority of the solid phase paraffin had changed into the mushy region, which was quite different from pure paraffin. This was due to the fact that the metal foam composite had a much higher high thermal conductivity and can spread the heat more rapidly within the paraffin. Therefore most of paraffin became mushy before it was fully melted into the liquid. The generation of liquid paraffin in the left PCM area was due to the fact that paraffin in this area was heated by both battery wall and metal foam simultaneously.

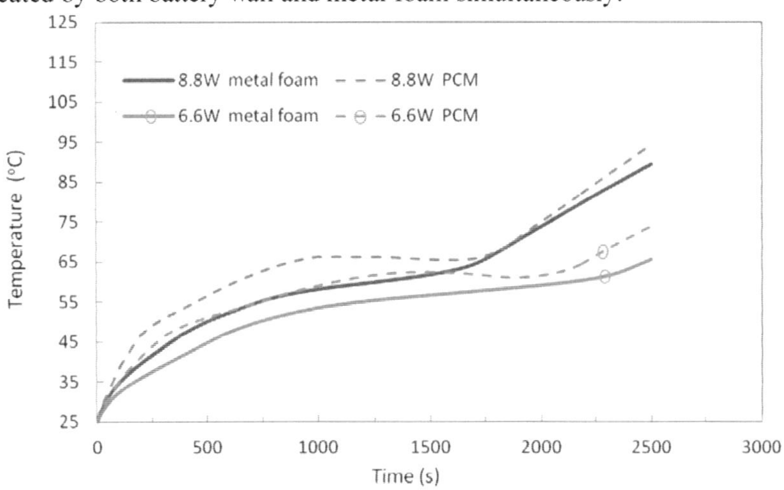

Figure 9. Temperature profile at simulate battery wall for pure PCM and paraffin/copper foam composite

The battery temperature evolutions were exhibited in Figure 9 for both the pure PCM and PCM-copper foam composite materials. It was seen that the overall temperature levels of the PCM composites were much lower than the pure PCM case. At the onset of the melting, the temperature difference between the pure PCM and PCM composite reached the highest temperature, which was up to 4.8°C,7.4°C for the heat density 399223W/m^3,532297W/m^3. The temperature difference between the two heat densities reduced to the minimum when the PCM was fully melted. In comparison, the temperature plateau was more obviously seen in the pure PCM case which helped maintain the battery temperature at a constant level. For the PCM composite case, the addition of copper foam suppressed the natural convection but enhanced the heat conduction greatly, which exhibited a slight slope in the battery temperature evolution.

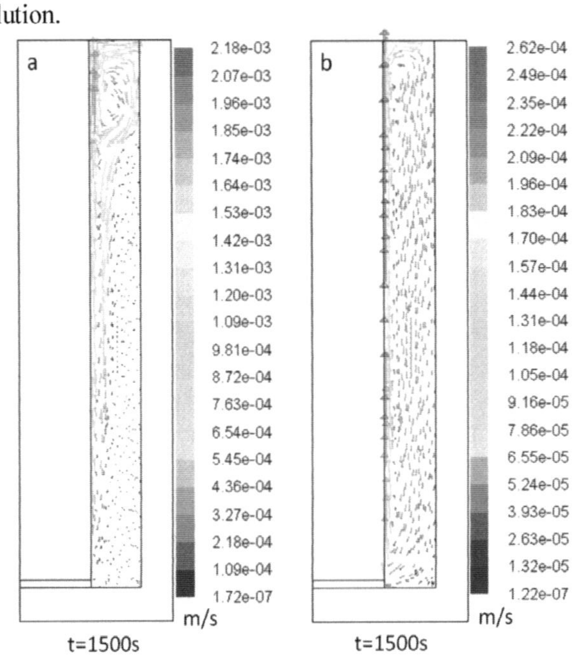

Figure 10. Velocity field of the numerical simulation for (a) pure PCM; (b) paraffin/copper foam composite

Figure 10 compared the velocity field of the numerical simulation for pure PCM(left) and paraffin/copper foam composite(right) recorded at the heat density, 399223W/m^3, and heating time of 1500s. In the pure paraffin case, stronger velocity magnitude was found at the upper left part of the liquid paraffin region. This was due to the high temperature wall surface of the battery continued to heat up the paraffin in its vicinities. In comparison, the velocity magnitude in PCM composite was two orders of magnitude lower. This mainly resulted from the effects of viscous and inertial resistance due to the existence of the ligament of copper foam, which suppressed the natural convection in the PCM composite. Another difference was the mushy regions. As shown in the case of pure paraffin, there were still large portions of solid phase (Figure 3). However, in the case of paraffin/copper foam, due to the high thermal conductivity and volumetric heating effects of the copper foam, all the regions had the velocity magnitude of similar levels (Figure 8). Through the present numerical analysis, the fluid flow and heat transfer for the pure PCM and PCM composite can be well understood.

5. Conclusions

A numerical study had been carried out in using PCM for thermal control of the Lithium ion battery power battery cell. The conclusions can be drawn as follows.

- The temperature plateau phenomena were observed for all the heating power densities. Nonetheless, the plateau temperatures were higher than the maximum melting temperature of

PCM by 9°C, 12°C, 20°C at the power densities of 399223W/m^3, 532297W/m^3, and 798446W/m^3, respectively.

- The effect of the heat transfer coefficient from the housing outer wall was examined. It was shown that the battery temperature increases at the same rate before melting. Nonetheless, longer melting time was observed with higher heat transfer coefficient. If the heat transfer coefficient was larger enough, 40W/m^{2}K in this case, PCM would not be fully melted even after 6000s.
- The effect of PCM onset melting temperature was examined. The lower the onset melting temperature was, the lower the plateau temperature was, whereas the difference of the plateau temperature and the onset melting temperature kept constant. On the other hand, it took longer time for the PCM to melt fully for higher onset melting temperature due to larger sensible heat before melting.
- The copper foam was inserted into the PCM to overcome the shortcoming of the low thermal conductivity. Numerical results showed that there was no obvious temperature plateau for the PCM composite with the copper foam case. In comparison with the pure PCM case, the battery temperature with PCM in copper foam was lower. The natural convection of the melted PCM in copper foam was also weaker than the pure PCM case, leading to a less inclined melting front.

Acknowledgments
Financial supports from the Eastern Scholar program by Shanghai Higher Education Institution and the project 14520501100 granted by Shanghai Scientific and Technological Commission are acknowledged.

A. 1

Nomenclature

- a_{sf} interfacial surface area (m^{-1})
- C inertial coefficient(m^{-1})
- C_P specific heat capacity(J/kg K)
- d_f fiber diameter(m)
- d_p pore size (m)
- d_k characteristic length (m)
- g gravitational acceleration(m/s^2)
- H latent heat of paraffin (J/kg)
- h heat transfer coefficient (W/m^{2}K)
- h_{sf} Interfacial heat transfer coefficient (W/m^{2}K)
- K permeability (m^{-2})
- k thermal conductivity(W/m K)
- k_{fe} effective thermal conductivity of paraffin (W/m K)
- k_{se} effective thermal conductivity of copper foam(W/m K)
- P_r prandtl Number
- p pressure(pa)
- q heat density(W/m^3)
- R_e Reynolds number
- T temperature (°C)

Greek symbols

- α thermal expansion coefficient(K^{-1})
- β liquid fraction
- ω pore density(PPI)
- ε porosity
- μ dynamic viscosity (kg/ms)
- ρ density(kg/m^3)
- τ time(s)
- χ tortuosity coefficient

subscripts

- f paraffin
- ref reference
- s copper foam
- td thermal dispersion

References

[1] Sharma A, Tyagi V V, Chen C R and Buddhi D 2009 Review on thermal energy storage with phase change materials and applications *Renewable and Sustainable Energy Reviews* **13** 318-45

[2] Ling Z Y *et al.* 2014 Review on thermal management systems using phase change materials for electronic components, Li-ion batteries and photovoltaic modules *Renewable and Sustainable Energy Reviews* 31 427–38

[3] Mills A and Al-Hallaj S 2005 Simulation of passive thermal management system forlithium-ion battery packs *J. Power Sources* **141** 307-15

[4] Duan X and Naterer G F 2010 Heat transfer in phase change materials for thermal management of electric vehicle battery modules *Int. J. Heat Mass Transfer* **53** 5176-82

[5] Kandasamy R, Wang X Q, Mujumdar A S 2007 Application of phase change materials in thermal management of electronics *Applied Thermal Engineering* **27** 2822-32

[6] Wu W X, Zhang G Q, Ke X F, Yang X Q, Wang Z Y and Liu C Z 2015 Preparation and thermal conductivity enhancement of composite phase change materials for electronic thermal management *Energy Conversion and Management* **101** 278–84

[7] Zhang Z, Zhang N, Peng J, Fang X, Gao X and Fang Y 2012 Preparation and thermal energy storage properties of paraffin/expanded graphite composite phase change material *Applied Energy* **91** 426-31

[8] Zhao C Y, Lu W and Tian Y 2010 Heat transfer enhancement for thermal energy storage using metal foams embedded within phase change materials *Solar Energy* **84** 1402–12

[9] Zhang Y L, Rao Z H, Wang S F， Zhang Z, Li X P 2012 Experimental evaluation on natural convection heat transfer of microencapsulated phase change materials slurry in a rectangular heat storage tank *Energy Conversion and Management* **59** 33-9

[10] Mills A, Farid M, Selman J R and Al-Hallaj S 2006 Thermal conductivity enhancement of phase change materials using a graphite matrix *Applied Thermal Engineering* **26** 1652-61

[11] Fleming E, Wen S, Shi L and Silva A K D 2015 Experimental and theoretical analysis of an aluminum foam enhanced phase change thermal storage unit *Int. J. Heat Mass Transfer* **82** 273-81

[12] Liu Z, Yao Y and Wu H 2013 Numerical modeling for solid–liquid phase change phenomena in porous media: Shell-and-tube type latent heat thermal energy storage *Applied Energy* **112** 1222-32

[13] Muhammad M D and Yeung O B H Validation of a CFD Melting and Solidification Model for Phase Change in Vertical Cylinders *Numerical Heat Transfer Part A: Applications* **68** 501-11

IOP Conf. Series: Earth and Environmental Science **40** (2016) 012047 doi:10.1088/1755-1315/40/1/012047

Regulation for Optimal Liquid Products during Biomass Pyrolysis: A Review

F Wang[1,2], L J Hu[3], Y W Zheng[1,2,4], Y B Huang[1,2,4], X Q Yang[1,2], C Liu[1,2], J Kang[1,2] and Z F Zheng[1,2,5]

[1]College of Materials Engineering, Southwest Forestry University, Kunming, 650224, China
[2]University Key Laboratory of Biomass Chemical Refinery & Synthesis, Yunnan Province, Kunming, 650224, China
[3]College of Forestry, Nanjing Forestry University, Nanjing, 210037, China
[4]College of Materials Science and Engineering, Northeast Forestry University, Harbin, 150040, China

E-mail: 18208809424@163.com

Abstract. The liquid product obtained from biomass pyrolysis is very valuable that it could be used for extraction of chemicals as well as for liquid fuel. The desire goal is to obtain the most bio-oil with desired higher heating value (HHV), high physicochemical stability. The yields and chemical composition of products from biomass pyrolysis are closely related to the feedstock, pyrolysis parameters and catalysts. Current researches mainly concentrated on the co-pyrolysis of different biomass and introduce of novel catalysts as well as the combined effect of catalysts and pyrolysis parameters. This review starts with the chemical composition of biomass and the fundamental parameters and focuses on the influence of catalysts on bio-oil. What is more, the pyrolysis facilities at commercial scales were also involved. The classic researches and the current literature about the yield and composition of products (mainly liquid products) are summarized.

1. Introduction

At present, the fossil fuels including coal, petroleum and natural gas, dominate the energy which we are depending on. However, the fossil fuels are estimated that they might be used up until 2050[1]. Confronting the increasing energy crisis and the severe environmental pollution resulted from abuse of fossil fuels, we have to explore the renewable substitutes for present fossil fuels. As the only one renewable organic resource, bio-energy has attracted more and more attention because of its abundance and non-pollution to environment [2][3][4].

Biomass pyrolysis is the most attractive one among thermochemical process[5]. In the absence of air or oxygen, at the temperature range of 350°C-700°C biomass can be converted into three kinds of substances: bio-oil, char and the non-condensate gas. Especially, the bio-oil is a kind of valuable oxygenic liquid that it consists of anhydrosugars, C2-C6 light oxygenates along with phenolics and

[4] Address for correspondence: Z F Zheng, College of Materials Engineering, Southwest Forestry University, Kunming, 650224, China; University Key Laboratory of Biomass Chemical Refinery & Synthesis, Yunnan Province, Kunming, 650224, China. E-mail: 18208809424@163.com.

aromatics hydrocarbons, so that it can be used for combustion as well as for extracting valuable chemicals[6]. The phenols and sugars in bio-oil could be extracted from bio-oil for chemicals. Hydrocarbons as the best substitution for fossil fuel can be obtained from biomass catalytic pyrolysis. Biomass pyrolysis deeply depends on the parameters, by which the yield and composition of products can be controlled. Catalysts used in biomass pyrolysis could increase the content of hydrocarbons and increase the stability of bio-oil. Javaid reviewed operating parameters for optimum liquid oil yield during biomass pyrolysis[7]. Furthermore, in this review the control of feed, pyrolysis parameters as well as catalysts for optimal bio-oil is overviewed.

2. Properties of Bio-oil

Physical and chemical properties of bio-oil are necessary to determine the potential of bio-oil used for liquid fuel and extraction of chemistry. Gas chromatography and mass spectrometry detector (GC-MS), gas chromatogram(GC) and high performance liquid chromatography (HPLC) are always used for analysis of chemical components in bio-oil[8][9][10]. FTIR and NMR are also applied for the analysis of characterization of bio-oils[9]. Among physical properties of bio-oil, including higher heating value (HHV), lower heating value (LHV), water content, pH value, viscosity and stability, HHV and LHV are considered the most important factors. The HHV means the total energy released from fuel when it burns in oxygen, while the LHV presents the energy which doesn't include the vaporization energy contained in the water vapour[11]. The HHV of bio-oil is always measured with oxygen bomb calorimeter[12][13]. However, the HHV and LHV of bio-oil could also be estimated empirically from the ultimate and proximate analysis by the following formulas[14][15]:

$$HHV = 0.3491 \times C + 1.1783 \times H + 0.1005 \times S - 0.1034 \times O - 0.0151 \times N - 0.0211\ A$$

In the formula the unit of HHV is MJ/kg. C, H, S, O, N mean the mass percent of carbon, hydrogen, sulphur and nitrogen respectivly and A is the weight percent of ash.

$$LHV = HHV - 218.3 \times H$$

In this formula the unit of LHV and HHV is KJ/kg. H means the mass percent of hydrogen.

In the Table 1, the HHV of bio-oil from various biomass pyrolysis is listed according to the previous researches.

Table 1. HHV of bio-oil under different conditions.

Feedstock	Condition	HHV(MJ/kg)	Analysis method of HHV	Ref
Bio-oil from pine sawdust	Feeding rate of 1.47 $kg{\cdot}h^{-1}$; Pyrolysis temperature: 700°C; Solid residence time: 6 min	18.76	Oxygen bomb calorimeter	[12]
Bio-oil of pine sawdust	Feeding rate of 1.47 $kg{\cdot}h^{-1}$; Pyrolysis temperature: 800°C; Solid residence time: 6 min	17.13	Oxygen bomb calorimeter	[12]
Bio-oil of pine sawdust	Feeding rate of 1.47 $kg{\cdot}h^{-1}$; Pyrolysis temperature: 700°C; Solid residence time: 3 min	13.76	Oxygen bomb calorimeter	[12]
Bio-oil of coconut core	Pyrolysis temperature: 550°C; Grain size: 1.18-1.8 mm; Heating rate: 60 $°C{\cdot}min^{-1}$	14.75-10.88	Calorimeter pump	[16]
Bio-oil of hazelnut cupulae	Pyrolysis temperature: 600°C; Sweep gas atmosphere;	35.70	Calculation	[11]

Bio-oil of hazelnut cupulae	Fixed-bed reactor Pyrolysis temperature: 600°C; Sweep gas atmosphere; Fixed-bed reactor; ZnO catalyst	40.03	Calculation	[11]
Bio-oil of willow	Pyrolysis temperature: 450°C; Nitrogen gas;	16.10	Calorimeter	[17]
Bio-oil of 1:1 of willow/PHB	Pyrolysis temperature: 400°C; Nitrogen gas;	20.22	Calorimeter	[17]
Linseed residue	Pyrolysis temperature: 450°C; Particle size: 1.75-2.36 mm	33.35	Calorimete	[18]
Nutshel	Pyrolysis temperature: 450°C; Particle size: 300-600 μm	19.34	Calorimete	[18]
Switchgrass with moisture contents of 5%	Pyrolysis temperature: 500°C; Fluidized-bed pyrolysis reactor	17.4	Calorimeter	[19]
Switchgrass with moisture contents of 15%	Pyrolysis temperature: 500°C; Fluidized-bed pyrolysis reactor	18.0	Calorimeter	[19]

3. The Effect of Feedstock

3.1. Composition of biomass

Lignocellulosic biomass is mainly composed of cellulose, hemicelluloses and lignin. In addition, a little of mineral is also present in biomass. Every type of biomass has its specific chemical composition depending on proportions of cellulose, hemicellulose, lignin as well as minerals.

Cellulose, hemicellulose, lignin perform in different ways during pyrolysis process due to their entirely different chemical compositions and structures. Every component of biomass including cellulose, hemicellulose, lignin has its optimal thermal cracking temperature range. Hemicellulose decomposes easily at about 220°C-320°C because of amorphous structure with many branched units so that it has low activation energy. Cellulose cracks in higher temperature range about 300°C-400°C while lignin in broad temperature range about 150°C-600°C[20]. What is more, the yield and compositions of the products from each component are distinctly different. Cellulose yields more bio-oil, hemicellulose prefers vast gaseous product while lignin products the most char. Stefanidis *et al* studied the effect of cellulose, hemicellulose and lignin on bio-oil and found that through pyrolysis, cellulose mainly yielded dehydrated sugar and phenols, while hemicellulose produced ketones and phenolic compounds, and simultaneously the main products from lignin were exclusively complex phenols which were different from those derived from cellulose and hemicelluloses[21].

The products from various biomass are different due to difference of contents of cellulose, hemicellulose, lignin and mineral. Herbaceous biomass is mainly composed of cellulose and hemicellulose, while woody biomass consists of lignin. Moreover, woody biomass has low ash content compared with herbaceous biomass[22]. Burhenne *et al* compared the difference between woody biomass and herbaceous biomass during pyrolysis process and the result indicated that herbaceous biomass produced a larger fraction of gaseous products, while the woody biomass led to higher yield of char[23]. In the study of Părpăriț ă *et al*, the more amounts of carboxylic acids, ketones and furans and a lower phenols were obtained from forestry biomass compared with agricultural biomass[24]. Considering various composition of biomass, co-pyroysis were conducted. Zhang *et al* conducted the catalytic pyrolysis using the mixture of corn stalk and food waste and the result indicated that apparent synergistic effect between corn stalk and food waste promoted the production of aromatics dramatically[25]. In another research, sewage sludge and rice straw were used for co-pyrolysis, the result showed calorific value of the pyrolyzed biomass increased with the addition of rice straw[26].

The minerals in the biomass mainly contain K, Na, Mg, Ca, Si (Table 2). Minerals act as catalysts for biomass decomposition and charing reactions.

Table 2. Composition of switchgrass ash.[27]

Compound	SiO_2	Al_2O_3	Fe_2O_3	SO_3	CaO	MgO	Na_2O	K_2O	P_2O_5	TiO_2	SrO	BaO
Content (wt%)	72.2	3.74	1.28	0.39	6.19	2.33	0.77	6.17	2.63	0.18	0.02	0.01

Pretreatment of biomass could really eliminate the minerals from biomass. Compared with raw biomass, the biomass after pretreatment leads to higher liquid product, less char and gaseous product[28]. The properties of bio-oil obtained from pretreated biomass are showed in Table 3. Furthermore, Richards & Zheng[29] studied the influence of metal ions and salts on pyrolysis of wood and found that Fe or Cu which were incorporated into the biomass resulted in the high content of levoglucosan and levoglucosenone.

Table 3. The properties of bio-oil obtained from pretreated biomass.

T/°C	Feedstock	Pretreatment	Ash content/%	Organic yield /wt%	Char yield /wt%	Gas yield/ wt%	Water yield /wt%	Composition of Bio-oil	Ref
530	Pine wood	90°C, 120min, Synthetic condenser	0.04	58.0	8.6	24.9	8.6	Levoglucosan concentration increased to 35.9 wt%	[29]
530	Pine wood	90°C, 120min, Real condenser 2 liquid	0.02	56.2	8.7	29.8	6.5	Levoglucosan concentration increased to 37.0wt%	[29]
500	Northern red oak	0.40 wt%, sulphuric acid		decrease	increase	decrease		Sugar yield increased by 8.1wt%; Light oxyge-nates yield decreased.	[30]
500	Switchgrs	2.0 wt%, sulphuric acid		slightly increase	increase	decrease		Sugar yield increased by 11.7wt%; Light oxyge-nates yield decreased.	[30]
500	Hardwood trembling aspen	Hot Water Treatment, 90°C						Anhydrosugars, Fatty Acids increased;	[31]

500	Softwood white spruce	Hot Water Treatment, 90°C	Phenols decreased Phenols, Anhydrosugars, Furans increased	[31]

3.2. Particle size of feedstock

The finer particle has a lager surface areas compared with the larger so that it has a high heating rate, less mass and heat limitation and cracks at lower temperature. The apparent activation energy values reduced with the particle size decreased so that the small sample cracked easier[32]. Moreover, Bridgeman *et al* analyzed two size fractions of switch grass and reed canary grass, the result indicated that smaller particles have a higher concentration of inorganic matter and moisture content as well as higher cellulose concentration than larger particles[33].

In general, the decrease of size of feedstock resulted in higher liquid product, less char and gaseous product. However, when the particle is very small, the biomass particle would splice each other and lead to inadequate cracking of biomass, so that the content of char increases with the decrease of bio-oil and gaseous products. Therefore, the particle size of biomass has an optimal value for the content of products. From the research of Pütün *et al*[34], in the pyrolysis of soybean-cake with particle size range of 0.425 mm to D_p (>1.800mm), 0.850 mm< D_p <1.250 mm particle size led to the highest yield of bio-oil. The size of particle also makes a difference in the compositions of bio-oil. Zhou *et al* studied the effects of the particle size on bio-oil during fast pyrolysis of beech wood and found that when the wood particle size increased from 0.3 mm to 3 mm, the yields of pyrolytic lignin decreased drastically[35]. However, when the particle size was more than 3 mm, the yield of pyrolytic lignin was constant. Shen *et al* proved that with the increase of particle size, the light bio-oil fraction increased and heavy bio-oil fraction decreased. Meanwhile, the content of water improved and the viscosity of bio-oil reduced[36]. However, in the research of Septien *et al*, during fast pyrolysis of woody biomass at high temperature (1000-1400°C), the size of particle had less effect on the final pyrolysis products due to the thorough decomposition at high temperature[37]. It can be deduced that the influence of size of particle on compositions of bio-oil is closely related to the decomposition degree of biomass.

4. Pyrolysis Parameters

4.1. The final pyrolysis temperature

Among all parameters, the final pyrolysis temperature and catalysts are the most effective factors affecting the products during pyrolysis of biomass[38]. The degree of heating influences the decomposition of biomass directly. When the final pyrolysis temperature is below 250°C, the biomass has no obvious change almost in either structure or composition. However, when the temperature exceeds 250°C, the chemical structure of biomass changes drastically. In general, the pyrolysis of biomass starts at 350°C. Under high temperature, the abundant energy from outside ruptures the active chemical bonds in biomass, such as O-H, C-H, olefinic C=C, C-O and C=O.

Plentiful researches have studied the role of the final pyrolysis temperature on the yields of products[39][40][41]. With the increase of temperature, the yield of char decreases while the yield of gas increases markedly. In the temperature range of 500°C-600°C, the most liquid oil is obtained. Depending on difference of the others parameters, the temperatures which result in the highest yield of liquid are various. As the final temperature increases, the cracking degree of the feedstock improves so that leads to the increasing amount of bio-oil and gas. At high pyrolysis temperature, the second cracking reaction takes place in the char and bio-oil to generate more gaseous product. This process leads to further decrease of char and slight degree of bio-oil, conversely, the increase of gas.

The high final temperature ruptures structure of biomass thoroughly, thus obtains the products which are different from that at low temperature. The liquid product from pyrolysis of biomass is a very complex mixture of oxygenated compounds with water content about 20 wt%-30 wt%. The water in bio-oil which is harmful to the property of bio-oil in aspects of lower calorific value and physical stability, derives from not only the initial water in biomass but also dehydration reaction. Amutio *et al* operated flash pyrolysis of Pinewood sawdust with conical spouted bed reactor and found that with improvement of the final temperature, the water content increased slightly[42]. This change was related to the intense dehydration reaction in high temperature. What is more, the amount of phenols was next to water and it peaked at 500°C. However, when the temperature exceeded 600°C, the yield of phenols decreased notably. What is more, the yields of ketones and saccharide decreased with the increase of temperature because of their instability in high temperature. In the research of Horne & Williams, the relation between temperature and concentrations of the polycyclic aromatic hydrocarbons (PAH) as well as high-value oxygenated aromatic compounds in bio-oil were studied[39].

4.2. Heating rate
The heating rate also performs a critical role on the yield and composition of products during biomass pyrolysis. High heating rate promotes vast outflow of primary volatile from solid feed, thus leads to high yield of liquid and gaseous production leaving low char amount. The results from study of Sensöz & Angin indicated that only the high heating rate above 50°C/min could make a great difference in yield of bio-oil. As the heating rate changed from 10 to 50°C/min, the temperature for maximum yield of bio-oil had no change[43]. Nevertheless, from result of the research of Demirbas, when the heating rate varied from 50 to 250°C/min, the temperature for maximum yield of bio-oil changed significantly[44]. The heating rate could also upgrade the bio-oil by change its composition. Debdoubi *et al* studied the effect of heating rate on oil products from esparto pyrolysis and found that with the increase of heating rate, H/C ratio increase and O/C decrease that contributed to improved energy content of pyrolysis oil[45]. In another study, when the heating rate increased, the improved higher heating value of bio-oil was detected[46].

4.3. Sweep gases
4.3.1. The types of sweep gases.
During biomass pyrolysis, the sweep gases are necessary to remove the vapors instantly and restrain the reactions of feedstock with oxidizing medium as well as provide hydrogen. The inert gases including nitrogen, helium, argon gas and active gases containing steam, methanol can be used as the sweep gases. In general, the use of nitrogen gas and steam are common. The inert gases used in biomass pyrolysis not only restrict the reactions between the biomass and oxidizing medium, but also remove the vapors out reactors timely. However, the steam not only as vehicle sweeps the volatile, but also reacts with the products by providing hydrogen. At the same time, the water vapor permeates through solid and results in desorption, distillation and removal of volatile. The steam as sweep gas drastically increased the yield of liquid and decreased the yield of gaseous and solid products[46]. The steam also influences the composition of bio-oil significantly. Pütün *et al* compared the effects of steam with inert gases and static atmosphere on the composition of bio-oil, and found that polar fractions dominated during the steam pyrolysis while inert gases and static atmospheres favored aliphatic and aromatic compounds[34]. In the research of Minkova *et al*, the solid from steam pyrolysis had a higher absorption compared with inert gases and static atmospheres[47].

4.3.2. The rate of sweep gases.
The rate of sweep gas influences the biomass pyrolysis because it determines the residence time of vapor. The vapors from biomass in pyrolysis would experience the secondary reactions, for instance, decomposition and repolymerization if they are not instantly removed out with long residence time. Through the decomposition reaction, preliminary vapors are converted into permanent gases, while the repolymerization leads to high yield of char. The short residence time yields more liquid product, but the quality of bio-oil decreases simultaneously because of its heavy molecular weight[48]. Rapid

sweep gas sweeps the volatile instantly so that avoids the secondary reactions. If the volatiles are quenched timely, the high amount of liquid products is obtained with the expense of char and gaseous products. The researches in regard to this respect were conducted[46][49]. Jae *et al* studied the influence of velocity of sweep gas on catalytic fast pyrolysis of wood using ZSM-5 catalyst and found that increase of the velocity led to decreased aromatic and increased olefin due to the reduced vapor residence time[50]. What is more, with the increase of velocity of sweep gas, the selectivity of benzene, toluene and C4 olefins was improved.

5. Catalysts

The composition of rude bio-oil derived from biomass pyrolysis is complex highly. There are vast oxygenated group in bio-oil, such as carboxyl, ketone group, methoxyl groups, hydroxyl so that the rude bio-oil has some negative factors (difficult ignition, corrosiveness, coking) to use directly[51]. The catalysts applied in biomass pyrolysis can efficiently upgrade the products. Abundant catalysts with high catalytic performance of cracking, aromatization, deoxygenation as well as esterification have been tested. Depending on the various reaction conditions, the specific catalysts could be chosen.

5.1. The catalysts for hydrocarbons

The aromatics from biomass pyrolysis have high motor octane number so that they can be used directly as fuels and a high octane fuel additive. To yield more aromatic hydrocarbon as well as less oxygenated compounds is beneficial to improve the bio-oil quality by the catalysts which possess excellent catalytic activity. The aromatics obtained by catalysts in two routes: either from depolymerization and dehydration of phenols or condensation-dehydrocyclization-aromatization reaction of light olefins on ZSM-5 with strong acid sites[52]. A mass of catalysts have been tested for production of aromatics in biomass pyrolysis, the related research results are summarized in Table 4.

Table 4. The influence of catalysts on bio-oil.

Feed	Catalysts	Temp /°C	Process types	Bio-oil yield /wt%	Hydrocarbon yield /wt% of biomass	Change of bio-oil composition	Ref
White pine	$Na_{0.2}H_{0.8}Y$	500	ex-situ	50	9.5	Hydrocarbon, phenols increased drastically; Sugar and furans decreased	[51]
White pine	HY	500	ex-situ	45	3.4	Hydrocarbon, phenols increased drastically; Sugar and furans decreased	[51]
Beech wood	Al-MCM-41	500	in-situ	45.34	1.53	Phenols decreased;	[52]
Beech wood	H-ZSM-5	500	in-situ	43.95	7.49	Mainly hydrocarbon; Phenols decreased;	[52]
Beech wood	Silicalite	500	in-situ	47.58	7.47	Ether increased;	[52]
Food waste	BEA zeolite	500	in-situ		43(of bio-oil)	Aromatics including polyaromatic hydrocarbons increases	[53]
Food waste	Al-SBA-15	500	in-situ		30(of bio-oil)	Reduction of polymeric fatty acid species; Increase of hydrocarbons	[53]
Rice	ZSM-5	550	in-situ		12.8		[54]

talk							
Rice stalk	LOSA-1	550	in-situ		10.0		[54]
Rice stalk	spent FCC catalysts	550	in-situ		6		[54]
Rice stalk	γ-Al_2O_3	550	in-situ		5	Mainly unidentified compounds, lower molecule-weight oxygenating compounds	[54]
Lignin	Beta	650	in-situ	2.0×10^9 (peak area)	90 (of bio-oil)	Oxygenates decreased considerably; Aromatics increased; substantially3-ring and 4-ring polyaromatics increased	[55]
Lignin	Mordenite,	650	in-situ	1.0×10^9 (peak area)	30 (of bio-oil)	Oxygenates decreased considerably, aromatics increased	[55]
Lignin	Y zeolites	650	in-situ	0.5×10^9 (peak area)	90 (of bio-oil)	Oxygenates decreased considerably; Aromatics increased	[55]
Lignin	ZSM-5	500	in-situ	4×10^9 (peak area)	69 (of bio-oil)	Mianly aromatics; Oxygenates decreased considerably, aromatics increased;	[55]
Radiata pine	Meso-MFI	500	ex-situ	42.9	4.1	Oxygenates reduced significantly; Aromatic hydrocarbons increased	[56]
Radiata pine	1wt%Ga/ Meso-MFI	500	ex-situ	45.9	6.2	Oxygenates reduced significantly; Aromatic hydrocarbons increased, mainly mono aromatic.	[56]
Radiata pine	HZSM-5	500	ex-situ	46.6	3.7	Oxygenates reduced significantly; Aromatic hydrocarbons increased	[56]
Radiata pine	Mesoporous ZSM-5	500	ex-situ	50.6	0.3	Oxygenates reduced significantly; Mainly phenolics;	[56]
Beech wood	ZSM-5	650	in-situ		23.7	Aromatic hydrocarbons increased; Oxygenates decreased significantly	[61]
Beech wood	Mesoporous ZSM-5	650	in-situ		30.1	Aromatic hydrocarbons increased; Oxygenates decreased significantly	[61]

Various catalysts were tested in biomass pyrolysis for upgrading bio-oil. In the research of Nguyen & Williams, H-faujasite (Na-FAU, $Na_{0.2}H_{0.8}$-FAU and H-FAU) were carried out in pyrolysis of pinewood and $Na_{0.2}H_{0.8}$-FAU led to the high-quality bio-oil by decreasing the acid content, aldehydes and ketones as well as increasing content of the phenols and hydrocarbon (mainly aromatics)[39]. Park *et al* studied the effect of β zeolite and Al-SBA-15 catalyst on the composition of bio-oil and the result indicated that β zeolite converted polymeric acid compounds into aromatics and

PAHs due to its strong acidic sites, while Al-SBA-15 tended to hydrocarbons and oxygenates[52][53]. By now ZSM-5 is considered as the optimal catalysts for optimizing aromatic hydrocarbons because of its remarkable effect of shape-selectivity. Zhang *et al* compared the ZSM-5 with various catalysts for catalytic pyrolysis of biomass and ZSM-5 resulted in maximum yields of aromatics (12.8%) and olefins (10.5%)[54]. ZSM-5 and LOSA-1 exhibited higher deoxygenation and cyclization than γ-Al_2O_3 and FCC catalysts. In another researches, ZSM-5 showed similar property[55].

In catalysts, Brønsted acid sites are critical to convert oxygenates to aromatic hydrocarbon, while the pore structure possess effect of shape-selective[56]. Carlson *et al* tested the relationship between catalytic parameters and catalytic activity using ZSM-5, silicalite and SiO_2-Al_2O_3. Silicalite which has the same pore structure with ZSM-5 produced primarily coke, whereas, SiO_2-Al_2O_3 containing Brønsted acid sites also yielded mainly coke[57]. The high selectivity for aromatic of ZSM-5 attributes to its proper density of Brønsted acid sites and unique pore structure.

ZSM-5 catalysts can be optimized by changing its acid sites, pore structure as well as other characters. The density of Brønsted acid sites determines catalytic activity of ZSM-5 for the deoxygenation and aromatic formation. Changing the acid sites on ZSM-5 catalysts through adjusting SiO_2/Al_2O_3 can apparently increase the content of aromatics at the cost of oxygenates[55][58]. However, if the numbers of acid sites further increase, two acid sites are close exceedingly so that the repolymerization dominates to lead to more char[59]. From the previous researches, the lignin-derived oxygenates in the CFP (catalytic fast pyrolysis) of biomass can not enter into the micropores of conventional ZSM-5 because the size of reactants is larger than ZSM-5[47]. However, enhancing the external areas of ZSM-5 and introducing the mesopores into the bulk can improve evidently the conversion of heavy lignin-derived oxygenates. Zheng *et al* studied the effect of crystal size (2 μm, 200 nm and 50 nm) of ZSM-5 on the aromatic yield and selectivity and the result indicated that ZSM-5 with crystal size 200 nm produced the maximum aromatic yield, minimum BTX selectivity and char yield. This attributed to the highest micropore surface area, amount of weak acid sites and maximum B/L ratio[60]. Creating mesopores permits the larger oxygenated molecules into the ZSM-5 and converts them to aromatics afterwards through reaction of cracking, deoxygenation and aromatization. What is more, the intraparticle mesopores in the bulk of ZSM-5 shorten micropore diffusion path length inhibiting secondary reaction of monoaromatics for char. Compared with regulating the particle size of catalysts introducing mesopore in ZSM-5 is more useful to convert the large volatile intermediate into light products. Li *et al* carried out pyrolysis of biomass with mesoporous ZSM-5 zeolites prepared by desilication with NaOH and found that mesoporous ZSM-5 has less influence on the products distribution in the process of cellulose pyrolysis, but increases the yield of aromatic yield and decreases the coke yield pronouncedly in pyrolysis of lignin[61]. The similar result also came about in another research[59]. What is more, in the research of Rownaghi *et al*, the mesoporous ZSM-5 obviously reduced the coke formation and enhanced production of alkyl aromatics from methanol[62]. In addition, the effect of other important parameters such as pyrolysis temperature, ratio of catalyst/lignin as well as biomass weight hourly space velocity on products from catalytic pyrolysis of biomass was also studied[58][63].

5.2. The catalysts for ester

The raw bio-oil from biomass during pyrolysis is acid and corrosive because of presence of a mass of organic acids such as formic, acetic and propionic acids. It is necessary to upgrade the bio-oil to convert organic acids and aldehydes to stable and valuable esters. With the help of catalysts, organic acids are firstly converted into aldehydes and water through deoxygenation and then the aldehydes are transformed into alcohols which can esterify with the residual acids. Xu *et al* studied the effect of Co which merged with Ru/γ-Al_2O_3 on the property of bio-oil and the result exhibited that Co-Ru/γ-Al_2O_3 increased esters in upgrading bio-oil by 2-fold than that in raw bio-oil. In the catalytic process, Co-Ru/γ-Al_2O_3 slightly improved the property of bio-oil by hydrotreating as well as esterifying reactions[64]. In another research, Mo-Ni/γ-Al_2O_3 catalysts were tested[65]. However, considering the catalysts with limitation of hydrotreating reaction, to add ethanol into bio-oil can pronouncedly increases the degree of esterification. In the research of Song *et al*, after adding ethanol/acetic acid into the bio-oil, SO_4^{2-}/SiO_2-TiO_2 catalyst increased drastically catalytic performance by increasing the yield of esters and alkane compounds, as well as decreasing organic acids, phenolic, aldehyde compounds. Interaction between SiO_2 and Ti promoted the formation of firm bond of Ti-O-Si so that improved the activity of catalysts[66]. What is more, from another study, when ethanol was added, the solid acid catalysts Amberlyst-70 not only transformed the organic acids into stable esters, but also converted sugar oligomers into stabilized methyl-α-D-glucopyranoside[67].

6. State-of-art of Biomass Pyrolysis

For the sake of biomass pyrolysis products which could be used for fuels and chemicals, biomass pyrolysis has been conducted at commercial scales. By now the reactor of biomass pyrolysis mainly includes fixed bed, fluidized bed, rotating cone, scew feeder and vacuum pyrolyser. Kan *et al* reviewed the characteristics and current status of pyrolysis reactors and bubbling, fluidized bed, rotating cone have been applied in market commercially[15]. Table 5 lists pyrolysis facilities at commercial scales.

Table 5. Pyrolysis facilities at commercial scales.

Company	Location	Pyrolysis reactor	Feed	Capacity (t/day)	Primary product	Ref
Dynamotive	West Lorne, Ontario, Canada	Bubbling fluidized bed	Waste sawdust and woodchips	100-130 (dry)	Bio-oil	[68]
Mitsubishi Heavy Industries/ Mie Chuo Kaihatsu	Mie Prefecture, Japan	Indirect heating rotary kiln	Woodchips	100	Syngas	[68]
Ensyn	Rhinelander, Wisconsin	Circulating fluidised bed	Hardwood wastes	40(dry)	Bio-oil	[68]
BTG	Malaysia	Rotating cone	Palm oil waste	50(dry)	Bio-oil	[68]
Choren	Germany	Slow pyrolysis and gasification	Woody biomass and agricultural residue	180(dry)	Syngas	[68]
Power China Hubei Electric Engineering Corporation	Hubei province, China		Cotton stalk, wheat straw, rape stalk	132	Syngas	[69]
Anyang biomass carbonization Co	Henan province, China		Straw	49	Syngas; Bio-char	[70]
Xinjiang development of agriculture and forestry renewable resources company	Xinjiang province, China		Cotton stalk, branch	41	Bio-char	[71]

7. Conclusion and Prospect

The whole world consumes vast energy and chemicals mainly derived from fossil fuel. Biomass pyrolysis is a promising way to yield bio-oil which contains phenols, sugars, furans and acids. The phenols, sugars and furans are valuable chemicals that could be extracted from bio-oil. The yield and composition of bio-oil both are strongly depended on the feed, pyrolysis parameters including pyrolysis temperature, heating rate and sweep gases. By controlling the various parameters the desire bio-oil is obtained.

The raw bio-oil is incompetent to used directly as liquid fuels because of its high oxygen content and high acidity. Catalysts could change the composition of bio-oil highly. Metals supported on the Al_2O_3 increase the content of esters, improving the stability of bio-oil. Hydrocarbons could replace a large percentage of fossil fuels in a variety of applications[72]. So far ZSM-5 is considered as the most efficient catalyst to convert oxygenated chemicals in bio-oil to hydrocarbons. By loading metal on the ZSM-5 or enlarging the pore size of ZSM-5 could improve the catalytic performance.

However, by now the mechanism of pyrolysis and catalytic pyrolysis is not clear because of the complicate composition of biomass. For the better using the bio-oil for chemicals and fuels, further researches should focus on the following aspects:

- For the regulation for optimal liquid products during biomass pyrolysis: pretreatment of feedstock containing more cellulose and hemicelluloses, pyrolysis temperature within 450-600°C, high heating rate, high-speed sweeping gas decreasing residence time of vapor in reactor, steam as sweeping gas, medium particle size of feedstock and rapid quenching of the vapor could increase the content of liquid, while the catalysts, especially ZSM-5, upgrade the quality of bio-oil. Therefore, both aspects should be considered simultaneously during biomass pyrolysis.
- The single component and model compound should be used for the studies of mechanism of pyrolysis and catalytic pyrolysis;
- The others biomass or chemicals could be introduced for biomass pyrolysis and the co-pyrolysis is recommended;
- The novel catalysts should be explored, for instance, metal-supported ZSM-5 and enlarging pore size ZSM-5.

Acknowledgments
Authors thank the 948 project (No.2013-4-08), State Forestry Administration of the People's Republic of China and the Major Project of Scientific Research Foundation (No.ZD2014012), the Department of Education, Yunnan Province, China for the financial support. Thanks also to professor Zoran S. Petrovic and Chan Mo for polishing this article.

References

[1] Gozal D and Saxena R C 2008 Bio-fuels from thermochemical conversion of renewable resources A review 2008 *Renew. Sust. Energ. Rev* **12** 504-17
[2] Klass D L 2006 *Biomass for Renewable Energy, Fuels and Chemicals* (San Diego: Academic Press) pp 29-32
[3] Huber G W, Iborra S and Corma A 2001 Synthesis of transportation fuels from biomass chemistry, catalysts and engineering *Chem. Rev* **106** 4044-98
[4] Demirba A 2001 Biomass resource facilities and biomass conversion processing for fuels and chemicals *Energ. Convers. Manage* **4** 1357-78
[5] Mohan D, Pittman C U and Steele P H 2006 Pyrolysis of wood/biomass for bio-oil: a critical review *Energ. Fuel* **20** 848-89
[6] Park K H, Park H J, Jeongnam K, Ryoo R, Jeon J K, Park J and Park Y K 2010 Application of Hierarchical MFI Zeolite for the Catalytic Pyrolysis of Japanese Larch *J. Nanosci. Nanotechno* **10** 355-9
[7] Akhtar J and Amin N S 2012 A review on operating parameters for optimum liquid oil yield in biomass pyrolysis *Renew. Sust. Energ. Rev* 2012 **16** 5101-09
[8] Aysu T 2015 Catalytic pyrolysis of Eremurus spectabilis for bio-oil production in a fixed-bed reactor Effects of pyrolysis parameters on product yields and character *Fuel. Process.*

Technol **129** 24-38
[9] Özbay N, Esin A V, Başak B U and Eren P A 2008 Characterization of bio-oil obtained from fruit pulp pyrolysis *Energy* **33** 1233-40
[10] Choia Y S, Johnston P A, Brown R S, Shanksa B H and Leec K H 2014 Detailed characterization of red oak-derived pyrolysis oil Integrated use of GC, HPLC, IC, GPC and Karl-Fischer *J. Anal. Appl. Pyrol* **110** 147-54
[11] Bilgen S, Keles S, Kaygusuz K 2012 Calculation of higher and lower heating values and chemical exergy values of liquid products obtained from pyrolysis of hazelnut cupulae *Energy* **41** 380-85
[12] Gao N B, Liu B L, Li A M and Li J J 2015 Continuous pyrolysis of pine sawdust at different pyrolysis temperatures and solid residence times *J. Anal, Appl. Pyrol* **114** 155-62
[13] Yakubmohammed I, Kabirkazi F, Abdallaabakr Y, Yusuf S and Razzaque A 2015 Novel method for the determination of water content and higher heating value of pyrolysis oil *Bioresources* **10** 2681-90
[14] Channiwala S A and Parikh P P 2002 A unified correlation for estimating HHV of solid, liquid and gaseous fuels *Fuel* **81** 1051-63
[15] Kan T, Strezov V and Jevans T 2016 Lignocellulosic biomass pyrolysis A review of product properties and effects of pyrolysis parameters *Renew. Sust. Energ. Rev* **57** 1126-40
[16] Shirley J D, Jorge L, Dario A and Rolón J C 2016 Effect of temperature and particle size on the yield of bio-oil, produced from conventional coconut core pyrolysis *Int. J. Chem. Eng. Appl* 7 102-8
[17] Cornelissen T, Jans M, Yperman J, Reggers G, Schreurs S and Carleer R 2008 Flash co-pyrolysis of biomass with polyhydroxybutyrate Part 1 Influence on bio-oil yield, water content, heating value and the production of chemicals *Fuel* **87** 2523-32
[18] Mohammad N I, Mustaifzu R and Sheikh Mohamma N H 2013 A comparative study on the fixed bed pyrolysis of different biomass solid wastes *Proc. The 4th BSME-ASME International Conference on Thermal Engineering* (Dhaka) 796-801
[19] He R, Philipye X, Cenglish B and Satrio J A Influence of pyrolysis condition on switchgrass bio-oil yield and physicochemical properties *Bioresource. Technol* 2009 **100** 5305-11
[20] Yang H P, Yan R, Chen H P, Ho L D and Zheng C G 2007 Characteristics of hemicellulose, cellulose and lignin pyrolysis *Fuel* **86** 1781-88
[21] Stefanidis S D, Kalogiannis K G, Iliopoulou E F, Michailof C M, Pilavachi P A and Lappas A A 2014 A study of lignocellulosic biomass pyrolysis via the pyrolysis of cellulose, hemicellulose and lignin *J. Anal, Appl. Pyrol* **105** 143-50
[22] Carpenter D, Westover T L, Whitney C S and Jablonski W 2014 Biomass feedstocks for renewable fuel production a review of the impacts of feedstock and pretreatment on the yield and product distribution of fast pyrolysis bio-oils and vapors *Green Chem* **16** 384-406
[23] Burhenne L, Messmer J, Aicher T and Laborie M P 2013 The effect of the biomass components lignin, cellulose and hemicellulose on TGA and fixed bed pyrolysis *J. Anal, Appl. Pyrol* **101** 177-84
[24] Părpăriț ă E, Brebu M, Uddin M A, Yanik J and Vasile C 2014 Pyrolysis behaviors of various biomasses *Polym. Degrad. Stabil* **100** 1-9
[25] Zhang B, Zhong Z P, Min M, Ding K, Xie Q L and Ruan R 2015 Catalytic fast co-pyrolysis of biomass and food waste to produce aromatics Analytical Py-GC/MS study *Bioresour. Technol* **189** 30-5
[26] Huang Y F, Shi C H, Chi P T and Lo S L 2015 Microwave co-pyrolysis of sewage sludge and rice straw *Energy* **87** 638-44
[27] Patwardhan P R, Satrio J A, Brown R C and Shanks B H 2010 Influence of inorganic salts on the primary pyrolysis products of cellulose *Bioresource Technol* **101** 4646-55
[28] Lou R, Wu S B, Lv G J and Zhang A L 2013 Factors related to minerals and ingredients influencing the distribution of pyrolysates derived from herbaceous biomass *BioResources* **8**

1345-60
[29] Richards G N and Zheng G C 1991 Influence of metal ions and of salts on products from pyrolysis of wood applications to thermochemical processing of newsprint and biomass *J. Anal, Appl. Pyrol* **21** 133-46
[30] Dalluge D L, Daugaard T, Johnston P, Kuzhiyil N, Wright M M and Brown R C 2014 Continuous production of sugars from pyrolysis of acid-infused lignocellulosic biomass *Green Chem* **16** 4144-55
[31] Roux E L, Diouf P N and Stevanovic T 2015 Analytical pyrolysis of hot water pretreated forest biomass *J. Anal, Appl. Pyrol* **111** 121-31
[32] Haykiri A H 2006 The role of particle size in the non-isothermal pyrolysis of hazelnut shell *J. Anal, Appl. Pyrol* **75** 211-6
[33] Bridgeman T G, Darvell L I and Jone J M 2007 Influence of particle size on the analytical and chemical properties of two energy crops *Fuel* **86** 60-72
[34] Pütün A E, Apaydin E and Pütün E 2002 Bio-oil production from pyrolysis and steam pyrolysis of soybean-cake product yields and composition *Energy* **27** 703-13
[35] Zhou S, Manuel G P, Pecha B, McDonald A G and Westerhof Roel J M 2014 Effect of particle size on the composition of lignin derived oligomers obtained by fast pyrolysis of beech wood *Fuel* **125** 15-9
[36] Shen J, Wang X S, Manuel G P, Daniel M, Rhodes M J and Li C Z 2009 Effects of particle size on the fast pyrolysis of oil mallee woody biomass *Fuel* **88** 1810-17
[37] Septien S, Valin S, Dupont C, Marine P and Sylvain S 2012 Effect of particle size and temperature on woody biomass fast pyrolysis at high temperature (1000-1400°C) *Fuel* **97** 202-10
[38] Aysu T and Küç ük M M 2014 Biomass pyrolysis in a fixed-bed reactor Effects of pyrolysis parameters on product yields and characterization of products *Energy* **64** 1002-05
[39] Horne P A and Williams P T 1996 Influence of temperature on the products from the flash pyrolysis of biomass *Fuel* **75** 1051-59
[40] Onay O 2007 Influence of pyrolysis temperature and heating rate on the production of bio-oil and char from safflower seed by pyrolysis, using a well-swept fixed-bed reactor *Fuel. Process. Technol* **88** 523-31
[41] Xiao R R and Wang W 2013 Influence of temperature on organic structure of biomass pyrolysis products *Renew. Energ* **50** 136-41
[42] Amutio M, Lopez G, Artetxe M, Elordi G, Olazar M and Bilbao J 2012 Influence of temperature on biomass pyrolysis in a conical spouted bed reactor *Resour. Conserv. Recy* **59** 23-31
[43] Sensöz S and Angin D 2008 Pyrolysis of safflower (Charthamus tinctorius L) seed press cake part 1 The effects of pyrolysis parameters on the product yields *Bioresource. Technol* **99** 5492-97
[44] Demirbas A 2004 Determination of calorific values of bio-chars and pyro-oils from pyrolysis of beech trunkbarks *J. Anal. Appl. Pyrol* **72** 215-9
[45] Debdoubi A, Elamarti A, Colacio E, Blesa M J and Hajjaj L H 2006 The effect of heating rate on yields and compositions of oil products from esparto pyrolysis *Int. J. Energ. Res* **30** 1243-50
[46] Pütün A E, Apaydin E and Pütün E 2004 Rice straw as a bio-oil source via pyrolysis and steam pyrolysis *Energy* **29** 2171-80
[47] Minkova V, Razvigorova M, Bjornbom E, Zanzi R, Budinova T and Petrova N 2001 Effect of water vapor and biomass nature on the yield and quality of the pyrolysis products from biomass *Fuel. Process. Technol* **70** 53-61
[48] Shen L and Zhang D K 2005 Low-temperature pyrolysis of sewage sludge and putrescible garbage for fuel oil production *Fuel* **84** 809-15
[49] Özbay N, Apaydin V E, Uzunb B B and Ayşe E P 2008 Characterization of bio-oil obtained

from fruit pulp pyrolysis *Energy* **33** 1233-40

[50] Jae J, Coolman R, Mountziaris T J and Huber G W 2014 Catalytic fast pyrolysis of lignocellulosic biomass in a process development unit with continual catalyst addition and removal *Chem. Eng. Sci.* **108** 33-46

[51] Nguyen T S, Zabeti M, Lefferts L and Seshan G B 2013 Catalytic upgrading of biomass pyrolysis vapours using faujasite zeolite catalysts *Biomass. Bioenerg* **48** 100-10

[52] Stephanidis S, Nitsos C, Kalogiannis K, Iliopoulou E F, Lappas A A and Triantafyllidis K S 2011 Catalytic upgrading of lignocellulosic biomass pyrolysis vapours *Catal. Today* **167** 37-45

[53] Park Y K, Lee H W, Jeon J K, Kim S S, Ryu C, Kim J M, Chae H J and Jeong K E 2011 The application of Py-GC/MS for the catalytic upgrading of oil separated from summer food waste leachate *Res. Chem. Intermediat* **37** 1283-91

[54] Zhang H Y, Xiao R, Jin B S, Shena D, Chena R and Xiao G M 2013 Catalytic fast pyrolysis of straw biomass in an internally interconnected fluidized bed to produce aromatics and olefins *Bioresour. Technol* **137** 82-7

[55] Yu Y Q, Li X Y, Su L, Zhang Y, WangY J and Zhang H Z 2012 The role of shape selectivity in catalytic fast pyrolysis of lignin with zeolite catalysts *Appl. Catal. A-Gen* **447-8** 115-23

[56] Park H J, Heo H S, Jeon J K, Kim J, Ryoo R, Jeong K E and Park Y K 2010 Highly valuable chemicals production from catalytic upgrading of radiata pine sawdust-derived pyrolytic vapors over mesoporous MFI zeolites *Appl. Catal. B-Environ* **95** 365-73

[57] Carlson T R, Vispute T P and Huber G W 2008 Green Gasoline by Catalytic Fast Pyrolysis of Solid Biomass Derived Compounds *ChemSusChem* **1** 397-400

[58] Li X Y, Su L, Wang Y J, Yu Y Q, Wang C W, Li X L and Wang Z H 2012 Catalytic fast pyrolysis of Kraft lignin with HZSM-5 zeolite for producing aromatic hydrocarbons *Front. Env. Sci. Eng* **6** 295-303

[59] Foster A J, Jae J, Cheng Y T, Huber G W and Lobo R F 2012 Optimizing the aromatic yield and distribution from catalytic fast pyrolysis of biomass over ZSM-5 *Appl. Catal. A-Gen* **423-424** 153-61

[60] Zheng A Q, Zhao Z L, Chang S, Huang Z, Wu H X, Wang X B, He F and Li H B 2014 Effect of crystal size of ZSM-5 on the aromatic yield and selectivity from catalytic fast pyrolysis of biomass *J. Mol. Catal. A-Chem* **383-4** 23-30

[61] Li J, Li X Y, Zhou G Q, Wang W, Wang C W, Komarneni S and Wang Y J 2014 Catalytic fast pyrolysis of biomass with mesoporous ZSM-5 zeolites prepared by desilication with NaOH solutions *Appl. Catal. A-Gen* **470** 115-22

[62] Rownaghi A A, Rezaei F and Hedlund J 2012 Uniform mesoporous ZSM-5 single crystals catalyst with high resistance to coke formation for methanol deoxygenation *Micropor. Mesopor. Mat* **151** 26-33

[63] Jae J, Coolman R, Mountziaris T J and Huberb G W 2014 Catalytic fast pyrolysis of lignocellulosic biomass in a process development unit with continual catalyst addition and removal *Chem. Eng. Sci* **108** 33-46

[64] Xu Y, Wang T J, Ma L L and Chen G Y 2012 Upgrading of fast pyrolysis liquid fuel from biomass over Ru/γ-Al_2O_3 catalyst *Energ. Convers. Manage* **55** 172-7

[65] Xu Y, Wang T J, Ma L L, Zhang Q and Liang W 2010 Upgrading of the liquid fuel from fast pyrolysis of biomass over MoNi/γ-Al_2O_3 catalysts *Appl. Energ* **87** 2886-91

[66] Song M, Zhong Z and Pand Dai J J 2010 Different solid acid catalysts influence on properties and chemical composition change of upgrading bio-oil *J. Anal. Appl. Pyrol* **89** 166-70

[67] Gunawan R, Li X, Larcher A, Hu X, Daniel M, Weerawut C, Wu H W and Li C Z 2012 Hydrolysis and glycosidation of sugars during the esterification of fast pyrolysis bio-oil *Fuel* **95** 146-51

[68] Mulligan C J, Strezov L, and Strezov V 2014 *Biomass Processing Technologies* (Boca Raton, FL CRC Press) pp 123-53

[69] Report of biomass polygeneration project Power China Hubei Electric Engineering Corporation 2013 [In Chinese]
[70] Report of biomass gasification and carbonization project Anyang biomass carbonization Co LTD 2010 [In Chinese]
[71] Report of comprehensive utilization of biochar and soil improvement projects Xinjiang development of agriculture and forestry renewable resources Co LTD 2011 [In Chinese]
[72] Dickerson T and Soria J 2013 Catalytic Fast Pyrolysis: A Review *Energies* **6** 514-38

Relation between quality and production cost for pure biodiesel bases on the mixes of raw materials

C G Tsanaktsidis[1,6], K G Spinthiropoulos[2], Fariz Guliyev[5], D Dimitriou[3], K Euthaltsidou[4] and G T Tzilantonis[1]

[1]Department of Environmental Engineering and Pollution Control and Technologies Laboratory of Qualitative Fuel Control, Technological Education Institute of Western Macedonia, 501 00, Kila, Kozani, Greece
[2]Department of European Studies, University of Macedonia, Egnatias Street, Thessoloniki, Greece
[3]ELIN OIL A.C. Company, Industrial Area, Volos, Greece
[4]Department of Accounting and Finance, Technological Education Institute of Western Macedonia, 501 00, Kila, Kozani, Greece
[5] Department of Finance, Azerbaijan State University of Economics-Economics, Baku, Azerbaijan

E-mail: tsanaktsidis@teiwm.gr, k.tsanaktsidis@gmail.com

Abstract. Nowadays biodiesel has become more attractive because it is made from renewable resources. The main ingredients of industrial biodiesel are rap oil, sun oil, fat acid, olive oil cooked. In this study we verify that, the proportion of these components sets the qualitative composition and energy efficiency of the final product. Essential we link the raw materials (rap oil, sun oil, fat acid, olive oil cooked) used in the manufacture of industrial biodiesel the proportion of mixes, with the variation of physicochemical properties of biodiesel produced. According to the quantitative analysis we notice that the physiochemical properties which alter the value for example humidity, acidity, while a large number of physicochemical properties do not change their value depending on the ratio of raw materials in each mixture. The analysis of these changes seems that the presence of fat acids is negative for the quality of the mixture. From the analysis of the cost of the final mixtures that lower cost is achieved in the mixture was 10 and the highest cost was in the mixture 3. Based on a study of the cost of the mixtures can determine a basic relation between the quality and the cost of the final product.

1. Introduction

It is generally accepted that biodiesel as a renewable energy product is clean, non-toxic and biodegradable, contains no aromatic compounds and emissions of the pollutants sulfur oxides, carbon monoxide, unburned hydrocarbons and soot from the burning of the diesel engines are very low. The presence of sulfur in fuels is responsible oxides of sulfur in the exhaust gas which are one of the main pollutants of diesel. In biodiesel the sulfur density is very small, almost zero.

[6] Address for correspondence: C G Tsanaktsidis, Department of Environmental Engineering and Pollution Control and Technologies Laboratory of Qualitative Fuel Control, Technological Education Institute of Western Macedonia, 501 00, Kila, Kozani, Greece. E-mail: tsanaktsidis@teiwm.gr, k.tsanaktsidis@gmail.com.

The biodiesel is commonly produced by the transesterification of the vegetable oil or fat acid feedstock. The material used for the production of fatty acid methelester is oil. This oil comes from various kinds of oils such us olive oil cooked and fat acids, vegetable oils which content sunflower, cottonseed, rapeseed, soybean and vegetable oil who comes from energy crops and is already prepared when they arrive at the plant as a raw material.

Nevertheless the problems which have to be solved, in relation to the production route, are many but the most important are the cost and the quality of the final products. Biodiesel is categorized as one of the liquid biofuels, an alternative diesel fuel, derived from renewable biological sources such as virgin oil feedstock, waste vegetable oil, algae, oil from halophytes and fat acids [1–3]. It is generally known that this approach comes from the period when Rudolf Diesel tested vegetable oil as fuel for his engine, more than one century ago[1].

During the decades of 1930s and 1940s vegetable oils were used as diesel fuels but usually only in emergency situations. Because of the continuous growing demand for energy, the scientific community, the oil companies as wells as the governments worldwide, are looking for new energy sources including the biofuels. In recent years up to 16 % of global final energy consumption comes from renewable sources [4] and this figure is continuously increasing.

Because of the way of the production of biodiesel inevitably the final mixture contain free fatty acids, phospholipids, sterols, water, odorants and other impurities. Even refined oils and fats contain small amounts of free fatty acids and water [5]. The diesel fuels, both conventional diesel fuels as well as biodiesel/diesel blends, need to satisfy a wide range of engine types. Alongside it must be on top for different operating conditions and duty cycles, as well as in variations of fuel system technology, at the engine temperatures and fuel system pressures. It must also be suitable for a variety of climates. The properties of each grade of diesel fuel must be balanced to provide satisfactory performance over an extremely wide range of circumstances. In some respects, the prevailing quality standards represent certain compromises so that all the performance requirements may be satisfied. By controlling specifications and properties, it is possible to satisfy the requirements of millions of compression ignition engines with a single grade of diesel fuel.

The study of physicochemical properties of biodiesel has become much written lately [6-13], but there are not references to the effect of mixing the raw materials on properties of biodiesel produced as well as the correct mixture that gives the balance between quality and production cost of the final product.

In this study it is easily understood that the values of the properties of pure biodiesel are dependent on the mix of raw materials. Specifically the quality of the final biodiesel is directly affected by the analogy of raw material such us fat acid and olive oil cooked. When the values are increased the final product is not good from qualitative point of view.

With the methodology developed in this work is given the possibility to determine the blending limits of raw materials (rap oil, sun oil, fat acid, olive oil cooked) for the production of industrial biodiesel before the production process, identified and costs in relation to raw materials, the produced biodiesel.

The production of the mixtures made • by•mixing the raw materials which came from the energy plant oils, oils from fat acids and oils cooked. The finished biodiesel depends on the value of the acidity and the water present in the raw materials used for mixing.

2. Materials and methods

2.1. Materials

All the experiments were performed employing fuel samples from commercial sources. Especially, the vegetable and fat oil based biodiesel was procured from ELIN OIL A.C.company.

Oils from energy crops for the production of mixtures used, rap, sun oil, and oils from fat acid, fat acid, and olive oil cooked. Especially vegetable blends are derived from rapeseed and soybean oil. In particular were generated 10 mixtures of raw materials (rap oil, sun oil, fat acid, olive oil cooked) in various proportions as shown in table 1.

Table 1. Mixing of raw materials (rap oil, sun oil, fat acid, olive oil cooked) to create mixtures produced pure biodiesel

Mix	1	2	3	4	5	6	7	8	9	10
Olive oil cooked	50%-	40%-	37%	50%	30%	50%	40%	30%	60%	40%
Rap oil	20%	_	42%	–	20%	25%	35%	–	–	5%
Sun oil	–	25%	16%	10%	–	25%	–	40%	5%	50%
Fat acid	30%	35%	5%	40%	50%	–	25%	30%	35%	5%

Then, after appropriate treatment of the mixtures were determined the values • of the physicochemical properties of biodiesel produced by the method EN ISO [14-42]. These values • are also the properties of the produced industrial pure biodiesel.

3. Results and discussion

After the creation of the mixtures were determined values • of physicochemical properties and forming properties of the produced pure biodiesel industry, as shown in table 2 and table 3.

3.1. Physicochemical properties of mixtures.

In table 2: Prices physicochemical properties of raw material mixes (rap oil, sun oil, fat acid, olive oil cooked) in various ratios to prepare pure industrial biodiesel.

Table 2.A. Values physicochemical properties of raw material mixes (rap oil, sun oil, fat acid, olive oil cooked) in various ratios to prepare pure industrial biodiesel (remains stable during the experiment).

					Mix 1	Mix 2	Mix 3	Mix 4	Mix 5
Properties	**Unit**	**MIN**	**MAX**	**Method**	**50% Olive oil cooked 20% Rape 30% Fat acid**	**40% Olive oil cooked 25% Sun Oil 35% Fat acid**	**37% Olive oil cooked 16% Sun Oil 5% Fat acid 42% Rape**	**50% Olive oil cooked 10% Sun Oil 40% Fat acid**	**30% Olive oil cooked 20% Rape 50% Fat acid**
Density	Kg/m3	860	900	ENISO 12185	881,5	881,5	881,5	881,5	881,5
Viscosity	mm2/s	3.5	5	ENISO 3104	4,5	4,5	4,5	4,5	4,5
Total sulfur	mg/kg	-	10	ENISO 20846	4,8	4,8	4,8	4,8	4,8
Carbon residue	%m/m	-	0,3	ENISO 10370	-	-	-	-	-
Sulphated ash	%m/m	-	0,02	ENISO 3987	<0,01	<0,01	<0,01	<0,01	<0,01
Solid	mg/kg	-	24	ENISO	10	10	10	10	10

				12662					
Phosphorus	mg/kg	-	4	EN14107	1,2	1,2	1,2	1,2	1,2
Na	mg/kg	-	5	EN14108	0,1	0,1	0,1	0,1	0,1
K	mg/kg	-	5	EN14108	0,3	0,3	0,3	0,3	0,3
Ca	mg/kg	-	5	EN14538	0,2	0,2	0,2	0,2	0,2
Mg	mg/kg	-	5	EN141538	0,2	0,2	0,2	0,2	0,2
Methanol	%min	-	0,2	EN14110	0,03	0,03	0,03	0,03	0,03

Table 2.B. Values physicochemical properties of raw material mixes (rap oil, sun oil, fat acid, olive oil cooked) in various ratios to prepare pure industrial biodiesel (remains stable during the experiment).

					Mix 6	Mix 7	Mix 8	Mix 9	Mix 10
Properties	**Unit**	**MIN**	**MAX**	**Method**	**50% Olive oil cooked 25% Rape 25% Sun Oil**	**40% Olive oil cooked 35% Rape 25% Fat acid**	**30% Olive oil cooked 40% Sun Oil 30% Fat acid**	**60% Olive oil cooked 5% Sun Oil 35% Fat acid**	**40% Olive oil cooked 50% Sun Oil 5% Fat acid 5%Rape**
Density	Kg/m3	860	900	ENISO 12185	881,2	881,5	881,5	881,1	881,1
Viscosity	mm2/s	3.5	5	ENISO 3104	4,5	4,5	4,5	4,5	4,5
Total sulfur	mg/kg	-	10	ENISO 20846	4,8	4,8	4,8	4,8	4,8
Carbon residue	%m/m	-	0,3	ENISO 10370	-	-	-	-	-
Sulphated ash	%m/m	-	0,02	ENISO 3987	<0,01	<0,01	<0,01	<0,01	<0,01
Solid	mg/kg	-	24	ENISO 12662	10	10	10	10	10
Phosphorus	mg/kg	-	4	EN14107	1,2	1,2	1,2	1,2	1,2
Na	mg/kg	-	5	EN14108	0,1	0,1	0,1	0,1	0,1
K	mg/kg	-	5	EN14108	0,3	0,3	0,3	0,3	0,3
Ca	mg/kg	-	5	EN14538	0,2	0,2	0,2	0,2	0,2
Mg	mg/kg	-	5	EN141538	0,2	0,2	0,2	0,2	0,2
Methanol	%min	-	0,2	EN14110	0,03	0,03	0,03	0,03	0,03

Table 3.A. Values physicochemical properties of raw material mixes (rap oil, sun oil, fat acid, olive oil cooked) in various ratios to prepare pure industrial biodiesel (change during the experiment).

					Mix 1	Mix 2	Mix 3	Mix 4	Mix 5
Properties	**Unit**	**MIN**	**MAX**	**Method**	**50% olive oil cooked 20% Rape 30% Fat acid**	**40% olive oil cooked 25% Sun Oil 35% Fat acid**	**37% olive oil cooked 16% Sun Oil 5% Fat acid 42% Rape**	**50% olive oil cooked 10% Sun Oil 40% Fat acid**	**30% olive oil cooked 20% Rape 50% Fat acid**
Ignition point	°C	101	-	ENISO 3679	180	164	172	176	178
Water	mg/kg	-	0,02	ENISO 12937	289,7	269	230	300,1	250,1
CFPP	°C	-	=-5_/_5	EN116	0	-1	0	0	0
Methylester	%m/m	96,5	-	EN1403	97,08	96,98	97,05	96,03	97
Acid methyl. His linelonikou acid	%m/m	-	12	EN14103	2,61	3,18	3,08	3,08	3,01
Acidity	mgKOH/g	-	0,5	EN14104	0,12	0,13	0,08	0,13	0,12
Iodine Value	g iodine / 100g	-	120	EN14111	99,5	99,7	99,7	99,4	99,6
Monoglycerides	%m/m	-	0,8	EN14105	0,285	0,329	0,305	0,3	0,299
Diglycerides	%m/m	-	0,2	EN14105	0,079	0,079	0,079	0,077	0,076
Triglycerides	%m/m	-	0,2	EN14105	0,087	0,087	0,088	0,088	0,087
Free glycerol	%m/m	-	0,02	EN14106	0,008	0,004	0,001	0,001	0,001
Total glycerol	%m/m	-	0,25	EN14105	0,101	0,108	0,1	0,1	0,101

Table 3.B. Values physicochemical properties of raw material mixes (rap oil, sun oil, fat acid, olive oil cooked) in various ratios to prepare pure industrial biodiesel (change during the experiment).

					Mix 6	Mix 7	Mix 8	Mix 9	Mix 10
Properties	**Unit**	**MIN**	**MAX**	**Method**	**50% olive oil cooked 25% Rape 25% Sun Oil**	**40% olive oil cooked 35% Rape 25% Fat acid**	**30% olive oil cooked 40% Sun Oil 30% Fat acid**	**60% olive oil cooked 5% Sun Oil 35% Fat acid**	**40% olive oil cooked 50% Sun Oil 5% Fat acid 5%Rape**
Ignition point	°C	101	-	ENISO 3679	181	180	180	175	180
Water	mg/kg	-	0,02	ENISO 12937	257,9	279,9	249,9	305,9	232,9
CFPP	°C	-	=-5_/5	EN116	0	0	0	0	0
Methylester	%m/m	96,5	-	EN1403	96,09	97,01	98,01	96	98
Acid methyl. His linelonikou acid	%m/m	-	12	EN14103	3,03	3,01	3,01	3	3,01
Acidity	mgKOH/g	-	0,5	EN14104	0,14	0,14	0,13	0,14	0,07
Iodine Value	g iodine / 100g	-	120	EN14111	98,6	98,2	98,7	97,7	98,8
Monoglycerides	%m/m	-	0,8	EN14105	0,382	0,389	0,229	0,399	0,231
Diglycerides	%m/m	-	0,2	EN14105	0,091	0,094	0,094	0,294	0,097
Triglycerides	%m/m	-	0,2	EN14105	0,087	0,087	0,087	0,087	0,087
Free glycerol	%m/m	-	0,02	EN14106	0,001	0,001	0,001	0,003	0,001
Total glycerol	%m/m	-	0,25	EN14105	0,101	0,101	0,101	0,102	0,101

From analysis of the above tables it is apparent that the physicochemical properties of table 3 are those which are altered by the creation of mixtures of blending raw materials. From these properties we observe that the greatest change occurs, humidity and acidity. These properties are expected to be altered because it is directly associated with the percentage of animal fatty acids that contained into fat acid (fat assets). The result of those that mentioned above is that with the increasing of the percentage of animal fats in the mixture we have a comparable change in the cost of the final product. This is a first conclusion from the analysis of the tables on the qualitative analysis as far as the final product.

4. Presentation of cost mixtures

Taking as a base the table 2A, table 2B and table 3A and table 3B we can observe the properties of the final product the values that remain constant or vary depending on the mix.

It will be very useful to investigate the effects on the total cost of the final product when specific properties change their values because we believe that this variation, based on the cost of raw materials, will define the price of the final product.

Alongside the purpose of this study is to ascertain what mixing ratio of raw materials will produce a final product of biodiesel based on the features of maximum quality and minimum cost. To do that we take the Greek Market in order to investigate the reaction of the minimum and maximum values in raw materials but it is obvious that this approach can be adapted to any country.

An analysis of table 2 and table 3 shows that mixing of raw materials for the creation of mixtures affect the physicochemical properties of ignition point, water, methylester and acidity which constitute essential characteristics of the suitability of the produced net biodiesel.

According with our methodology there is a relationship among growth limits of animal oils in order not to adversely affect the properties of mixtures and final cost. We know that animal oils are cheaper for biodiesel production but at the same time reduce the quality of biodiesel produced.

For the vegetable blends the production cost is a priori very high. The problem occurs in animal mixes, as we already mentioned above, where the product is cheaper but low quality product produced is observed. Depending on the prices of raw materials and the area where it operates the biodiesel industry may be, making use of the conclusions of this work and of the methodology that proposed, to carve the biodiesel production policy accordingly. We can observe from the Scatter plot in figure 1 the relation of between the mixtures and the production cost. At the same time by the diagrammatic illustration it is obvious that the increasing of % of fat acid gives cheaper final product but short on quality. We should mention that the use of vegetable oil meaning sun oil gives quality in product, but make it more expensive for the producer and consequently for the final consumer. In conclusion it is easily understood that with this resolution by changing the % of oils in the mixtures we can receive each time a different product from qualitatively point of view. Alongside we could know the cost of the final product by choosing the quality of our blends.

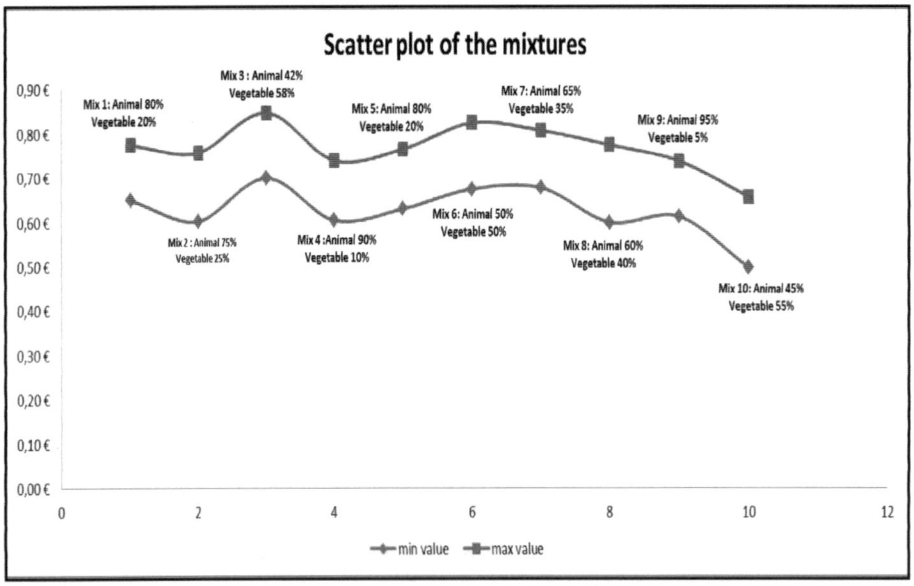

Figure 1. Scatter Plot. Representation of the Mixtures mixtures costs in Euro/liters

Based on the above values investigation table 4 shows, with regard to the final produced biodiesel, the relation between quality and cost.

Table 4. Relationship between quality and cost based on the mixing ratio of raw materials (in liters)

	Mix 1		Mix 2		Mix 3		Mix 4		Mix 5	
Composition measured on liters	50% Olive oil cooked 20% Rape 30% Fat acid		40% Olive oil cooked 25% Sun Oil 35% Fat acid		37% Olive oil cooked 16% Sun Oil 5% Fat acid 42% Rape		50% Olive oil cooked 10% Sun Oil 40% Fat acid		30% Olive oil cooked 20% Rape 50% Fat acid	
					Cost (€) / lit					
	Min	Max	Min	Max	Min	Max	Min	Max	Min	Max
Olive oil cooked	0,325 €	0,375 €	0,260 €	0,195 €	0,241 €	0,278 €	0.330 0 €	0,380 €	0,195 €	0,225 €
Rape	0,160 €	0,190 €	0,000 €	0,000 €	0,336 €	0,399 €	0,000 €	0,000 €	0,160 €	0,190 €
Fat acid	0,165 €	0,210 €	0,193 €	0,135 €	0,028 €	0,035 €	0,220 €	0,280 €	0,275 €	0,350 €
Sun Oil	0,000 €	0,000 €	0,150 €	0,128 €	0,096 €	0,136 €	0,060 €	0,085 €	0,000 €	0,000 €
Total Cost	0,650 €	0,775 €	0,603 €	0,457 €	0,700 €	0,848 €	0,610 €	0,740 €	0,630 €	0,765 €
Composition measured on liters	Mix 6		Mix 7		Mix 8		Mix 9		Mix 10	
	50% Olive oil cooked 25% Rape 25% Sun Oil		40% Olive oil cooked 35% Rape 25% Fat acid		30% Olive oil cooked 40% Sun Oil 30% Fat acid		60% Olive oil cooked 5% Sun Oil 35% Fat acid		40% Olive oil cooked 50% Sun Oil 5% Fat acid 5%Rape	
					Cost (€) / lit					
	Min	Max	Min	Max	Min	Max	Min	Max	Min	Max
Olive oil cooked	0,325 €	0,375 €	0,260 €	0,300 €	0,195 €	0,225 €	0,390 €	0,450 €	0,130 €	0,150 €
Rape	0,200 €	0,238 €	0,280 €	0,333 €	0,000 €	0,000 €	0,000 €	0,000 €	0,040 €	0,048 €
Fat acid	0,000 €	0,000 €	0,138 €	0,175 €	0,165 €	0,210 €	0,193 €	0,245 €	0,028 €	0,035 €
Sun Oil	0,150 €	0,213 €	0,000 €	0,000 €	0,240 €	0,340 €	0,030 €	0,043 €	0,300 €	0,425 €
Total Cost	0,675 €	0,825 €	0,678 €	0,808 €	0,600 €	0,775 €	0,613 €	0,738 €	0,498 €	0,658 €

In the basis of the methodology was developed it becomes evident-clear that the presence of raw materials, fat acid, olive oil cooked, determines the change of physicochemical properties of mixtures and quality in the produced pure biodiesel.

While we are using larger precedence of raw materials the properties that determine the quality of the finished biodiesel are low acidity values for example water but we have large production cost as we can see on the mixes 1, 3 and 6 from the table 4. In contrast when we are using higher precedence of fat acids then we lower cost and at the same time low quality biodiesel produced because the values of the examined properties are undesirable high. This can be found on the mixes 2,4,8 and 10 from the table 4.This encourages, in general, finding the best mixing ratio of raw materials to ensure a balance between cost and quality of raw materials. An explanation it can be given an explanation in the result of low quality in the final product of biodiesel when we are using higher precedence of animal raw materials because of the composition of raw materials, fat acid, olive oil cooked, since they contain a large proportion of unsaturated fatty organic acids with negatively affect to the quality of the fuel.

Table 5. Chemical composition of biodiesel

Caprylic	C8/0	$CH_3\ (CH_2)_6COOH$
Capric	C10/0	$CH_3(CH_2)_8COOH$
Lauric	C12/0	$CH_3(CH_2)_{10}COOH$
Myristic	C14/0	$CH_3(CH_2)_{12}COOH$
Palmitic	C16/0	$CH_3(CH_2)_{14}COOH$
Palmitoleic	C16/2	$CH_3(CH_2)_5CH{=}CH(CH_2)_7COOH$
Stearic	C18/1	$CH_3(CH_2)_{16}COOH$
Oleic	C18/1	$CH_3(CH_2)_7CH{=}CH(CH_2)_7COOH$
Linoleic	C18/2	$CH_3(CH_2)_4CH{=}CHCH_2CH{=}CH(CH_2)_7COOH$
Linolenic	C18/3	$CH_3(CH_2)CH{=}CHCH_2CH{=}CHCH_2CH{=}CH(CH_2)_7COOH$
Arachidic	C20/0	$CH_3(CH_2)_{18}COOH$
Eicosenoic	C20/1	$CH_3(CH_2)_7CH{=}CH(CH_2)_9COOH$
Behenic	C22/0	$CH_3(CH_2)_{20}COOH$
Eurcic	C22/1	$CH_3(CH_2)_7CH{=}CH(CH_2)_{11}COOH$

In general we know that the presence of unsaturated carbon chains imparts non-stability in the fuels used. It is also known that conventional fuels such as diesel, there are unsaturated hydrocarbons in the final product.

Their presence is detected only on biodiesel and thus the possible mixtures diesel-biodiesel. Thus, the higher rate will have to biodiesel, the higher rate will have the final mix and so/therefore the low stability of fuel should I use with corresponding negative effects on the environment.

Since the unsaturated hydrocarbons are concentrated in the fat acid and cooking oil, we understand that as they are used as starting material, the above problems will be presented.

The main criterion for the quality of biodiesel is the agreement regarding the corresponding specifications (standards). Generally the quality of the fuel can be influenced by several factors including the quality of the raw materials, the content of fatty acids, oils or fats, the production process, other materials used in the process and the post-production parameters. Specifically these parameters are the physicochemical properties that certify the suitability of the product and directly related to the type of raw materials, for example, animal or vegetable biodiesel. Furthermore with this study we know which of the properties altered and therefore affects the production and the cost of biodiesel production.

Concluding our analysis we know that animal oils are cheaper for biodiesel production but at the same time reduce the quality of biodiesel produced. With the equations that we mention in this study we can know the percentage of animal biodiesel that can be used in order to reduce the production cost and alongside to have a suitable product produced.

5. Conclusions

The analysis in these tables is proved, that the results of each mixture biodiesel regarding the properties affected by the quality of raw materials purchased by the factory, as in laboratory biodiesel. Soybean oil, rapeseed oil and sunflower oil is of very good quality oil and therefore reason the biodiesel containing a large percentage of them have excellent results compared to those containing cooked oil and fat acid, which is lower class raw materials. In particular as far as the quality of the pure biodiesel the presence of raw materials derived from fats acids affect the quality because increase properties such as acidity, triglycerides, methylester and monoglyceride. Similarly, the cost analysis have showed that the raw materials derived from fats acids are cheaper but do not contribute to the quality of the finished product. Contrasting the use of vegetable raw materials enhances the quality but charged to the final cost. So on the

IOP Conf. Series: Earth and Environmental Science **40** (2016) 012048 doi:10.1088/1755-1315/40/1/012048

basis of cost analysis table we must find the mixture to be equilibrium between quality and cost to a finished biodiesel industry. Generally, the above methodology can be a manual production process in this direction.

Alongside biodiesel could be used with all the other forms of energy meaning supplementary energy for electricity generation, household applications and transportation modes in accordance with the European Directive 2020. According with this Directive by 2020 the use of biodiesel must get to 20%.Currently on the market the Industries are using 8% in order to produce biodiesel . Consequently the proposed application can be used as tool in order to cover the remaining 12% according with the European Directive.

References

[1] McKeon Thomas, Hayes Douglas, Hilderbrad David and Weselake Randall 2008 *Industrial Oil Crops* AOCS Press Published by Elsevier Inc , Director Janet Brown

[2] Gerpen J V 2005 Biodiesel processing and production *Fuel Processing Technology.* **86** 1097–107

[3] Balat M and Balat H 2010 Progress in biodiesel processing Applied Energy. **87** 1815–35

[4] REN21 2011 *Renewables : Global Status Report* 17–8

[5] Ma F and Hanna M A 1999 Biodiesel production: a review *Bioresource Technology* **70** 1–15

[6] Naik S N, Goud V V, Rout P K and Dalai A K 2010 Production of first and second generation biofuels: A comprehensive review *Renewable and Sustainable Energy Reviews* **14** 78–97

[7] Gomez L D, Steele-King C G, McQueen-Mason S J 2008 Sustainable liquid biofuels from biomass: the writing's on the walls *New Phytologist* **178** 473–85

[8] Kamm B and Kamm M 2004 Principles of biorefineries *Appllied Microbiology & Biotechnology* **64** 37–45

[9] Tyson K S, Bozell J, Wallace R, Petersen E and Moens L 2004 Biomass oil analysis: research needs and recommendations *NREL\Technical Report*

[10] Tsanaktsidis C G, Spinthiropoulos K G, Christidis S G, Basileiadis V M and Garefalakis A E 2012 A mathematic equation using statistical data for the determination of kinematic viscosity in blends of diesel fuel with biodiesel *Computer Technology and Application* **3** 393-399

[11] Tsanaktsidis, C. G, Spinthiropoulos, K. G, Christidis, S. G, Sariannidis, N 2013 Mathematical Models for Calculating the Density of Petroleum Diesel Fuel/Biodiesel Blends *Chemistry and Technology of Fuels and Oils* **49** 399-405.

[12] Tsanaktsidis CG, Scaltsoyiannes AV, Katsidi EX, Christidis SG, Tzilantonis GT 2014 Using resin, a natural product, in order to reduce humidity in diesel fuel, *Chemistry and Technology of Fuels and Oils* **49** 497-501

[13] Tsanaktsidis CG, Vasiliadis V, Itziou A, Petrakis LA, Moisiadis SA 2013 Application of factor analysis for the study of physicochemical properties in different blends of diesel fuel with biodiesel *International Journal of Soft Computing and Engineering* **3** 42-46.

[14] ENISO 12185:1996 Crude petroleum and petroleum products -- Determination of density -- Oscillating U-tube method.

[15] ENISO 3104:1994 Petroleum products -- Transparent and opaque liquids -Determination of kinematic viscosity and calculation of dynamic viscosity.

[16] ENISO 3679:2004 Determination of flash point -- Rapid equilibrium closed cup method.

[17] ENISO 20846:2004 Petroleum products -- Determination of sulfur content of automotive fuels -- Ultraviolet fluorescence method.

[18] ENISO 10370:1993 Petroleum products -- Determination of carbon residue -- Micro method.

[19] ENISO 3987:1994 Petroleum products -- Lubricating oils and additives -- Determination of sulfated ash.

[20] ENISO 12937:2000 Petroleum products -- Determination of water -- Coulometric Karl Fischer titration method.

[21] ENISO/TR 12662:1997 Certification scheme for prestressing steels.
[22] BS EN 116:1998, BS 2000-309:1998 Diesel and domestic heating fuels. Determination of cold filter plugging point.
[23] EN-14103 Determination of Total FAME and Linolenic Acid Methyl Esters in biodiesel.
[24] EN-14103 Determination of Total FAME and Linolenic Acid Methyl Esters in biodiesel.
[25] BS EN 14112:2003 Fat and oil derivatives. Fatty acid methyl esters (FAME). Determination of oxidation stability (accelerated oxidation test).
[26] BS EN 14104:2003 Fat and oil derivatives. Fatty acid methyl esters (FAME). Determination of acid value .
[27] BS EN 14111:2003 Fat and oil derivatives. Fatty acid methyl esters (FAME). Determination of iodine value.
[28] EN 14105 Determination of Free and Total Glycerin in Pure biodiesel (B100) by GC in Compliance.
[29] EN 14105 Determination of Free and Total Glycerin in Pure biodiesel (B100) by GC in Compliance.
[30] EN 14105 Determination of Free and Total Glycerin in Pure biodiesel (B100) by GC in Compliance.
[31] BS EN 14106:2003 Fat and oil derivatives. Fatty acid methyl esters (FAME). Determination of free glycerol content.
[32] EN 14105 Determination of Free and Total Glycerin in Pure biodiesel (B100) by GC in Compliance.
[33] BS EN 14107:2003Fat and oil derivatives. Fatty acid methyl esters (FAME). Determination of phosphorous content by inductively coupled plasma (ICP) emission spectrometry.
[34] BS EN 14108:2003 Fat and oil derivatives. Fatty acid methyl esters (FAME). Determination of sodium content by atomic absorption spectrometry.
[35] BS EN 14108:2003 Fat and oil derivatives. Fatty acid methyl esters (FAME). Determination of sodium content by atomic absorption spectrometry.
[36] BS EN 14538:2006, BS 2000-547:2006 Fat and oil derivatives. Fatty acid methyl ester (FAME). Determination of Ca, K, Mg and Na content by optical emission spectral analysis with inductively coupled plasma (ICP OES).
[37] BS EN 14538:2006, BS 2000-547:2006 Fat and oil derivatives. Fatty acid methyl ester (FAME). Determination of Ca, K, Mg and Na content by optical emission spectral analysis with inductively coupled plasma (ICP OES).
[38] Duvekot C and Technologies A Determination of methanol content in biodiesel using agilent select biodiesel for methanol with headspace sampling to EN-14110 *Agilent Technologies*
[39] Burton R 2008 An overview of ASTM D6751: biodiesel standards and testing methods, alternative fuels consortium.
[40] Lee S.Y, Hubbe M.A and Saka S 2006 Prospects for biodiesel as byproduct of wood pulping – a review. *BioResource*s **1** 150–171.
[41] Gerpen J V, Shanks B, Pruszko, Clements D and Knothe G 2004 Biodiesel production technology NREL/SR-510-36244 *National Renewable Energy Laboratory*, Colorado, USA.
[42] Stanfel C 2009 Fuel filtration: Protecting the diesel engine *Filtration & Separation* **46** 22–5

A new anti-slough drilling fluid study and application

X Y Zhao[1, 2], Y F Meng[1, 3], X T Hou[2], C Xiao[2], G Li[1] and C C Niu[2]

[1]State Key Laboratory of Oil and Gas Reservoir Geology and Exploitation, Southwest Petroleum University, Chengdu 610500, China
[2]Research Institute of Petroleum Engineering, SINOPEC Beijing 100101, China

E-mail: swpuzxy@163.com

Abstract. Currently the BIOCAT drilling fluid system adopted in oil field results in wellbore collapsing frequently, borehole enlargement rate over 100%. In order to settle these problems, the Impermeable anti-collapsing inhibitive Polymer Drilling Fluids System is developed by optimizing the inhibitors with high anti-collapsing inhibitive instead of BIOCAT inhibitor, adding impermeable agent SMT3, sealing materials such as graphite, calcium carbonate, etc. Laboratory experiments results show that the developed drilling fluid system has stronger inhibitive and anti-collapse capacity. The swelling increment decreases by over 10% and recovery rate increases by 60%-70%, also, it can efficiently prolong the borehole wall stable period by over 100%. 24 wells applications with the Impermeable anti-collapsing inhibitive Polymer Drilling Fluids System achieved good results, comparing to adjacent wells with BIOCAT drilling fluid system, wellbore collapsing decreasing significantly and borehole diameter enlargement of unstable hole section in wells decreased from 100% to 10%, well drilling time decreased by 15%, drilling cost decreased by 10%. It shows that the developed Impermeable anti-collapsing inhibitive Polymer Drilling Fluids System KPAM-NH4PAN can provide strong and long-term inhibition, close to oil-based drilling fluid. so it is suggested that it be used in subsequent drilling operations to enhance the wellbore stability in an oilfield.

1. Introduction

SJ Basin was influenced by compression and nappe (fault) cutting during the tectonic evolution. The formation has large dip, with fragmented strata, micro-fractures developed, and average width of core fractures is 4.3 micron. The upper formation is mainly mudstone and tuff, and the lower mainly tuff and mudstone. Formation has high content of montmorillonite and quartz, strongly water-sensitive. The well drilling process is prone to collapse, lost circulation and formation damage.

1.1. Borehole collapse

Serious borehole collapse occurred during drilling Well PCC3027 in PCC Block, which led to sticking of logging tool twice, sticking of casing, when density of drilling fluid was increased from 1.2 g/cm^3 to 1.26g/cm^3, more serious borehole sloughing occurred. For other wells in PCC Block, borehole collapse was even more serious at the interval of 300-800m, with relatively big hole diameter.

[3] Address for correspondence: Y F Meng, State Key Laboratory of Oil and Gas Reservoir Geology and Exploitation, Southwest Petroleum University, Chengdu 610500, China. Email: swpuzxy@163.com.

Figure 1. Well PCC3027 Collapsing and Breaking

The hole diameter for second spud-in of Well EHH4103 in EHH Block was irregular, with the average hole diameter enlargement rate of almost 30%. Baker Hughes analyzed and believed that the action of drilling fluids to clay resulted in hole instability, so this problem should be solved by using a proper drilling fluid system that could reduce the reaction between mud and clay formation.

1.2. Soft and thick mud cake

The well log data indicated that the thickness of mud cake at the reservoir interval was up to 25.4mm, and tight hole resulted in serious sticking of logging tool. Besides, the cake thickness also indicated serious water loss of drilling fluid under the downhole environment and deep formation damage might be resulted from filtration into the formation.

1.3. Poor inhibition performance of drilling fluids

The bentonite content of drilling fluids for the stage of second spuding was 25kg/m^3, however, it became over 85kg/m^3 after finishing drilling, with bit balling happened during drilling, which indicated that drilling fluid could not inhibit mud making.

2. Borehole instability mechanisms study

2.1. Core Analysis

The purpose of the experiment was to analyze the mineral composition of cores, the clay mineral content in the cavy interval, so as to provide basis for optimization of new drilling fluids and some inhibited treating agents.

Table 1. Whole rock analysis of samples taken from well EHH3111

Core No.	Well No.	Depth (m)	Quartz	Potassium Feldspar	Albite	Calcite	Barite	Total Content of Clay Minerals (%)
13	EHH3111	2212	38.7	/	26.1	3.0	/	32.2
25	EHH3111	1733	64.8	/	12.0	0.1	/	23.1
28	EHH3111	1391	25.5	1.2	19.7	6.8	7.9	38.9

Table 2. Whole rock analysis of samples taken from well EHH3111

Core No.	Well No.	Depth (m)	Relative Content of Clay Minerals (%)					Mixed-layer Ratio（%S）
			S	I/S	I	K	C	I/S

13	EHH3111	2212	/	58	2	4	36	30
25	EHH3111	1733	58	/	1	23	18	
28	EHH3111	1391	51	/	/	34	15	

From the experiment, the content of clay minerals in cavey interval is up to 30%, dominated by water-sensitive montmorillonite and illite-montmorillonite layers, formation strength reduced by the hydration of clay minerals, leading to borehole collapse. Besides, the samples have high contents of feldspar and quartz with strong hardness and brittleness, and the inhomogeneous stress distribution resulted from expansion of clay minerals, water absorption may also lead to borehole collapse.

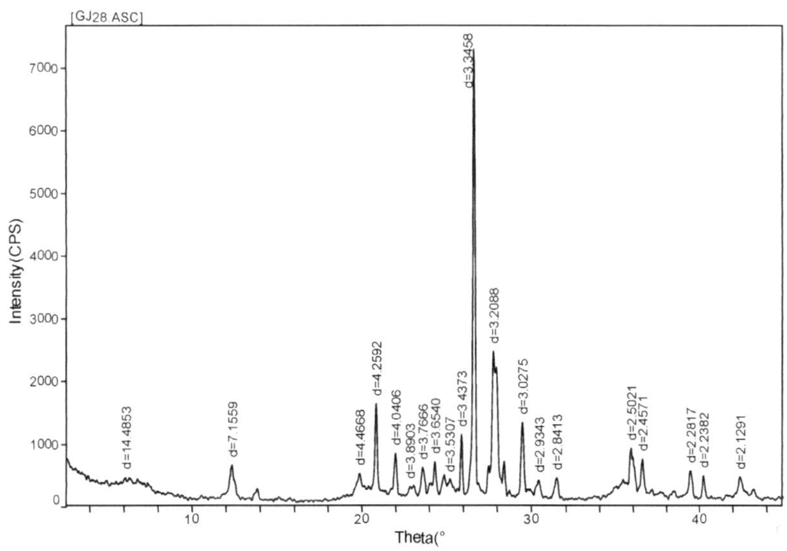

Figure 2. X-ray diffraction diagrams for rock mineral analysis with sample 28

2.2. Electron microscope picture of the core taken from sloughing formation

Microscope photos show that the sample is unconsolidated with developed fracture. During drilling process, drilling fluids and filtrate could invade the formation through these fractures, which could cause clay hydration swelling and lead to borehole collapse.

Figure 3. Microscope photo 1209-0725

Figure 4. Microscope photo 1209-0726

2.3 Water-Sensitive experiment with cuttings from sloughing formation

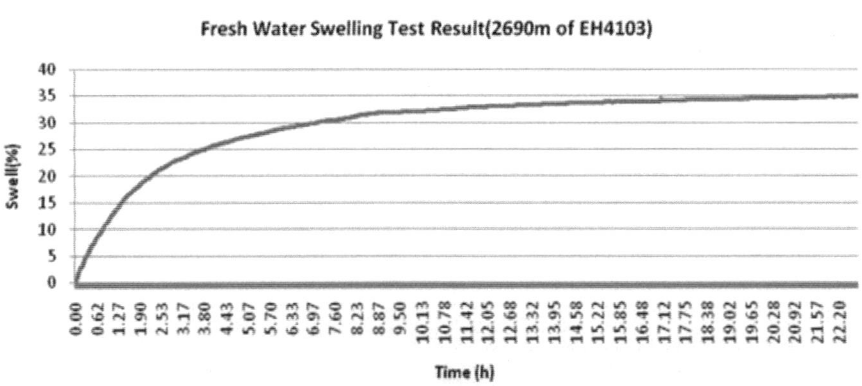

Figure 5. Fresh Water Swelling (2690m)

It can be seen from the picture that the formation water sensitivity of the EHH block in an Oilfield is strong, with swell capacity varying in different depths.it is the strong water absorbent or swelling formation at 1690m, with fresh water swelling amount up to 70%; while at 2670 and 2690m, it is the easy-swelling formation, with the fresh water swell capacity of 35%. During drilling, the cuttings are easy to disperse in the drilling fluid, and cause increasing of bentonite content; consequently, the formation tends to be hydrated, which causes formation strength reduction and makes formation lose stability.

1.4.2.4 Recovery rate experiment

Table 3. Roller recovery rate with fresh water

Depth(m)	Cuttings quality before rolling (g)	Cuttings quality after rolling (g)	Recovery rate (%)
1532	19.9	2.04	10.3
1735	20.1	1.98	9.9
1895	19.8	2.1	10.6
2070	20.2	2.37	11.7
2410	20.2	1.94	9.6
2630	20.2	3.57	17.7

Seen from the result of roller recovery rate experiment, the recovery rate of Well EHH4103 is relatively low, indicating that the cuttings are strong in dispersibility, easy to cause cutting solids contamination. The borehole wall tends to lose stability when soaked in fresh water or drilling fluids for a long time.

3. New anti-slough drilling fluid system study

The effects to drilling fluid rheology caused by polyamine are large; drilling fluid inhibitive capability can't fully meet the needs of anti-sloughing, the HTHP filtrate loss of drilling fluid is larger at lower section in EHH block. The initial idea is to change polyamine into KCl as clay anti-swelling agent, in higher temperature EHH block, use SPNH to improve high temperature inhibitive capability of the drilling fluids**[1-5]**.The drilling fluid formula and test results for PCC and EHH blocks(well depth 1800 m) are as follows:

Table 4. The drilling fluid prescription in 300m-1800m well section of PCC and EHH blocks

Treating agents	Concentration (kg/m^3)	Treating Chemical	Concentration (kg/m^3)
Water	855.54	BARITINA (3,8 Pe)	100.00
BENTONITA	25.00	SOLTEX	18.00
NaOH	3.00	MARPAC GOLD LV / HV	6.00
MARPOL 507	2.00	KCl	50
MAR VIS XCD	1.50		

Table 4. The drilling fluid formula at well section 1800m-2700m in EHH block

Treating Chemical	Concentration (kg/m^3)	Treating Chemical	Concentration (kg/m^3)
Water	855.54	BARITINA (3,8 Pe)	100.00
BENTONITA	25.00	SOLTEX	18.00
NaOH	3.00	MARPAC GOLD LV / HV	6.00
MARPOL 507	2.00	KCl	50
MAR VIS XCD	1.50	SPNH	20

3.1 The drilling fluid inhibitive capability experiments and research

From the immersing experiments in optimized drilling fluid, we can find that the drilling cuttings remain hard after 24 hours immersion, which illustrates the drilling fluid, has a strong inhibitive capability, and is able to prevent bit balling and borehole instability**[6-10].**

3.2 Drilling cuttings swelling experiments

The experiment procedures can be seen in "water swelling experiment", the optimized drilling fluid is employed and drilling cuttings are taken from well EHH4103. The experiment results are as follows:

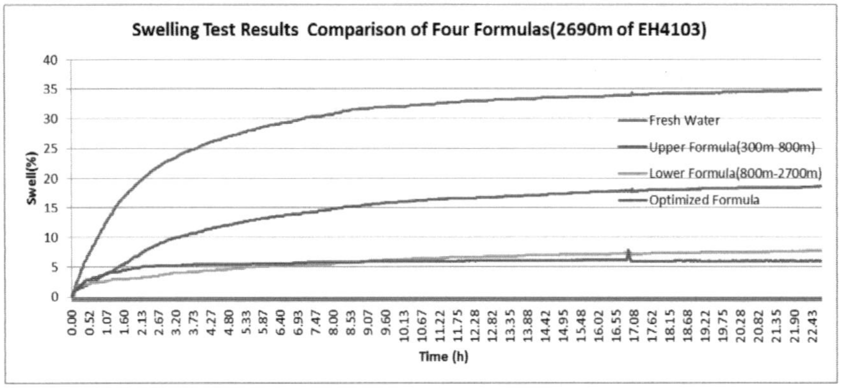

Figure 6. Comparison of swelling test results with four drilling fluids (cuttings taken from the depth of 2690m in EHH4103)

The above picture show that the cuttings swelling amount drops from 35% in fresh water to 6% in optimized drilling fluids, which has stronger anti-swelling capability than on-site drilling fluids.

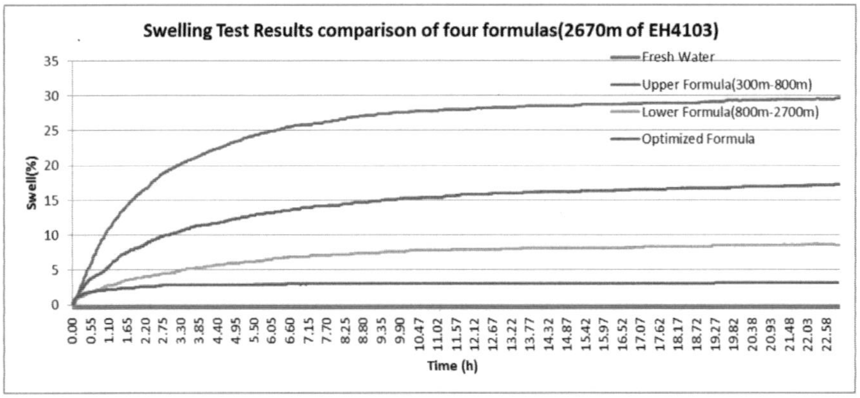

Figure 7. Comparison of swelling test results with four drilling fluids (cuttings taken from the depth of 2670m in EHH4103)

The above picture show that the cuttings swelling amount drops from 30% in fresh water to 3% in optimized drilling fluids, and the anti-swelling capability enhances by 60% compared to on-site drilling fluids.

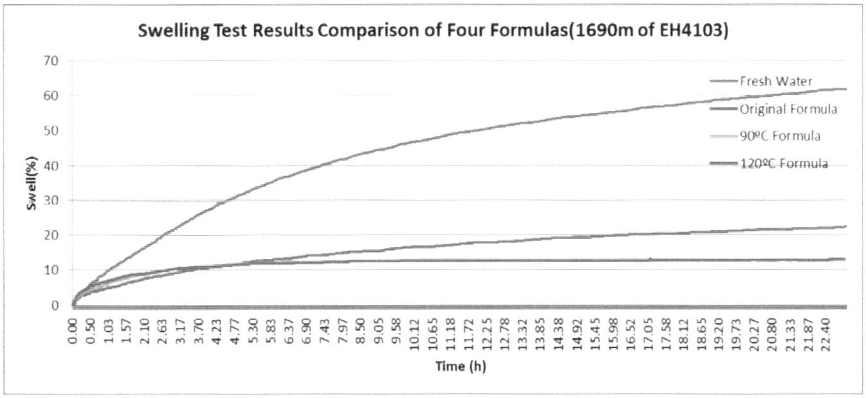

Figure 8. Swelling test results comparison of four formulas(1690m of EHH4103)

The picture show swelling test results respectively in fresh water, on-site drilling fluids (800-2700m), optimized drilling fluids (special for PCC block, resistance temperature of 90 ℃), and in optimized drilling fluids (special for EHH block, resistance temperature of 120℃).

The test results show that swelling amount is 61% in fresh water, it is 23% in on-site drilling fluids, but the swelling amount is only 13% in optimized drilling fluid, which means that the optimized drilling fluid has stronger anti-swelling capability.

3.3 Drilling cuttings rolling recovery test

The test procedure of drilling cuttings rolling recovery test in optimized drilling fluid is the same with that in fresh water. The drilling cuttings of well EHH4103 rolling recovery test results using optimized drilling fluid are as follows:

Table 4. Drilling cuttings rolling recovery test result of optimized drilling fluid

Well Depth (m)	Drilling cuttings quality before rolling (g)	Drilling cuttings quality after rolling (g)	Drilling cuttings recovery rate (%)
1532	20.1	7.32	36.4
1735	20	6.89	34.5
1895	20	7.98	39.9
2070	20.7	8.14	39.3
2410	19.9	5.9	29.6
2630	20	5.37	26.9

The experimental data show that drilling cuttings recovery rate in optimized drilling fluid is 26%-40%, enhancing by 5%-26% comparing with on-site drilling fluid, which illustrates the optimized drilling fluid can prevent drilling cuttings dispersion **[11-14]**.

4. Field Application

The optimized drilling fluid system was used in two wells with good effect; the wellbore enlargement in sloughing intervals in wells PCC3033 and LH3018 have reduced from 100% to 10%, and the average wellbore enlargement has reduced from 30% to 5%. In Well PCC3033, two layer blowing during testing, and the production met the anticipated requirements, so fracturing job was eliminated, and the target on wellbore stability and reservoir protection were obtained.

For the caliper of the whole well, the maximum borehole diameter enlargement rate of test well Lh-3018 is less than 10% and that of the contract wells LHH2077 and LHH2078 is greater than 100%.

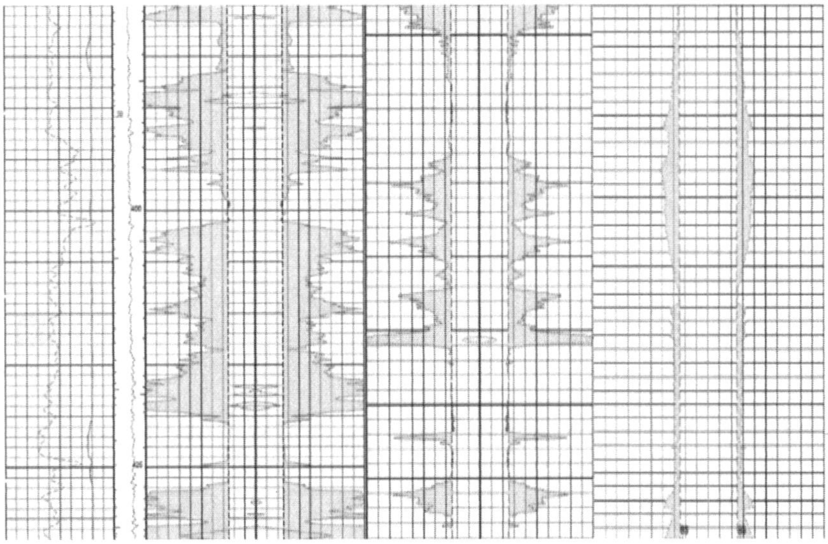

Figure 9. Caliper comparisons of collapsing section of 375-425m among Wells LH2078 and LH2077 and test wells

5. Study Conclusions

- The main reasons that cause wellbore unstable include micro-fractures, high clay mineral content and strong water sensitivity as well as crushing formation. It is suggested that enough drilling fluids and gel liquids should be stored at well site, so as to supplement drilling fluid or maintain its property
- The optimized drilling fluid system KPAM-NH4PAN can provide strong and long-term inhibition, close to oil-ased drilling fluid. This fluid system has good rheology and filtration performance, so it is suggested that it can be used in other oilfield with similar formation to enhance the wellbore stability.
- Test results show that KCl polymer drilling fluid has stronger inhibitive and anti-collapse capability than drilling fluid used before, and has good rheology as well as less filtration. It is suggested to choose KCl polymer drilling fluid for drilling deep formation. In order to better use the system and have good performance in oilfield, drilling fluid properties maintenance measures need to be developed scientifically and implemented strictly.

References

[1] Chesser B G 1987 Design considerations for an inhibitive, stable water-based mud system *Spe Driling Engineering* **2**

[2] Dearing H L and Ali S A 1996 Drill-In fluid selection crucial to well productivity *Petroleum Engineer International*

[3] Durand C, Lecourtier J, Rosenberg and Loeber L 1996 Relationship between composition, structure, and permeability of drilling filter cakes *Oil & Gas Science & Technology* **51** 777-8

[4] Fjær E, Holt R M, Nes O M and Sønstebø E F 2002 *Mud chemistry effects on time-delayed borehole stability problems in shales* SPE 78163

[5] Li Y, Rosenberg E, Argillier J F, Durrieu J and Montes J 1995 *correlation between filter cake structure and filtration properties of model drilling fluids* (Richarson, TX: Society of Petroleum Engineers, Inc.)

[6] Zhang J L 2013 Anti-sloughing drilling fluid technology of deep formation in offshore of shengli oilfield *Drilling Fluid & Completion Fluid* **30** 40-2

[7] Xu J F, Qiu Z S, Wang W G, Li S J, Huang C J and Qiao G W 2006 Study and application of a strong inhibitive drilling fluid KCC-TH *Petroleum Drilling Techniques* **34** 36-8
[8] Li F, Yu Y L and Pu X L 2013 Research on novel anti-caving drilling fluid system *Drilling Fluid & Completion Fluid* **30** 1-3
[9] Peden J M, Arthur K G and Avalos M 2013 *The analysis of filtration under dynamic and static conditions* (Bakersfield, California: 1984 SPE Formation Damage Control Symposium) SPE 12503
[10] Fordham E J and Ladva H K J 1989 and 1992 Cross flow filtration of bentonite suspension I and II *J. Colloid Interface Science* **148** 29-34
[11] Altun G, Osgouei A E and Ozyurtkan M H 2014 An alternate mud proposal to minimise borehole instability (Kuala Lumpur, Malaysia: International Petroleum Technology Conference)
[12] Aadnoy B S 1987 A complete elastic model for fluid-induced and in-situ generated stresses with the presence of a borehole *Energy Sources* **9** 239–59
[13] Aadnoy B S and Chenevert M E 1987 Stability of highly inclined boreholes *Spe Drilling Engineering* **2** 364–74
[14] Fraser L J, Reid P I, Williamson D and Enriquez J 1995 Mechanistic investigation of the formation damaging characteristics of mixed metal hydroxide drill-in fluids and comparison with polymer base fluids *Spe Driling & Completion* **14**

Research of Hybrid Three-phase equilibrium Technology

K Xu[1,3], **Z Z Liu**[1], **G Z Qi**[1] and **Y J Hou**[2]

[1] School of Electrical Engineering, Shandong University, Jinan 250061, China
[2] Shandong Academy of Sciences, LiXia District, Jinan 250061, China

Email: xuzuokai@foxmail.com

Abstract. This paper puts forward a kind of managerial method based on the combination of PPF (passive power filter) and APF (active power filter) for the problem of three-phase current balance in three-phase four-wire system. This method uses two special reactors to filter zero-sequence current and uses APF to filter negative-sequence fundamental current, positive-sequence and negative-sequence harmonic current. It is more effective, reliable and economic. This paper proves feasibility of the method by the simulation results.

1. Introduction

Now, electricity load is bigger and bigger. Proportions of nonlinear load like LED lights, computers are increasing in building low voltage power distribution system. Load distribution between the three phases changes over time, so three-phase unbalance degree is variable. Even if three-phase load balancing, on account of nonlinear load, neutral current is often too large, even more than the phase current. Therefore, building low voltage power distribution systems often have serious problems of three-phase current unbalance [1].

The problems of three-phase current unbalance seriously affect the power quality and safety [2], [3]. Governance methods of three-phase current unbalance mainly include: 1) Load management of power station area like [4],[5], it can make three-phase unbalance degree to reduce, but it doesn't take nonlinear load into consider; 2) reactive compensation between the three phases like [6], to a certain extent, it can adjust the asymmetric three-phase load, but its dynamic performance is poor.
APF is power electronic device to dynamically suppress harmonic and compensate reactive power. Compared with PPF, it has a better response speed and compensation characteristics. But its capacity of compensation of zero-sequence current is smaller, about 25% of the rated capacity. When APF compensates zero-sequence current, the dc side capacitor voltage fluctuation is increscent, which cause degradation of APF performance. So this paper uses two special reactors to filter zero-sequence current. Compared to the traditional LC filter, it can filter out all the number of zero-sequence current and its dependence on the parameters of the system is tiny. APF only is used to filter negative-sequence fundamental current, positive-sequence and negative-sequence harmonic current. Therefore, this method is more effective, reliable and economic.

2. The basic principle and system structure

[3] Address for correspondence: K Xu, School of Electrical Engineering, Shandong University, Jinan, 250061, China. E-mail: xuzuokai@foxmail.com.

System's basic structure is shown in figure 1. ① is PWM voltage source converter. It can generate offset current according negative-sequence fundamental current, positive-sequence and negative-sequence harmonic current, equivalents to controlled current source. ② is ZSBT (Zero-Sequence Blocking Transformer) [7]-[9]. It is not impedance for positive-sequence, negative-sequence current and is impedance for zero-sequence current. ③ is Zig-Zag TR (Zig-Zag transformer). It allows only zero-sequence current to pass and is about 0.005Ωon the impedance of the zero-sequence current.

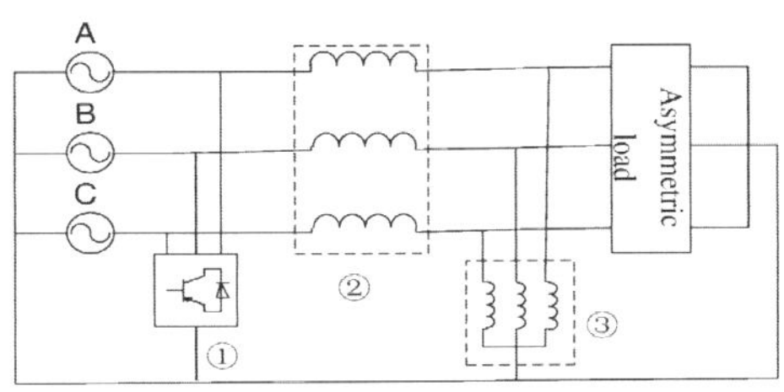

Figure 1. System's basic structure

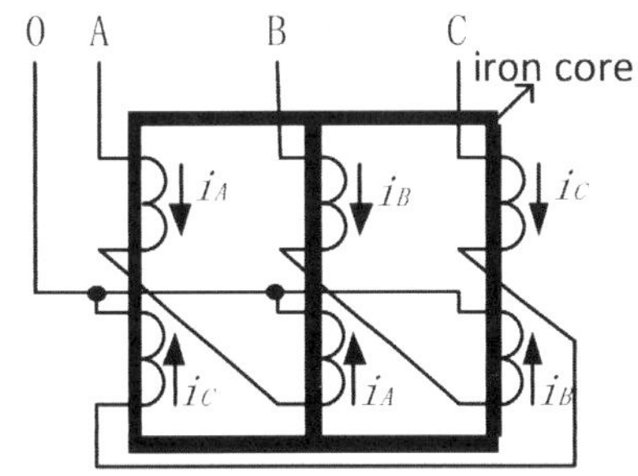

Figure 4. Structure of Zig-Zag TR

Three phase unbalance current can be divided into positive-sequence fundamental current component, negative-sequence fundamental current component, zero-sequence fundamental current component, positive-sequence secondary harmonic current component, negative-sequence secondary harmonic current component, zero-sequence secondary harmonic current component, and so on [10]. With the exception of positive-sequence fundamental current component, others need to filter out. Equivalent circuit of system is shown in figure 2 for zero-sequence current. X_0 is short circuit impedance of system. As a general rule, X_0 is far less than X_1 and can be ignored. X_1 is zero-sequence impedance of ZSBT. X_2 is short circuit impedance of Zig-Zag TR. The figure 2 can be seen that system filter efficiency is . Choose impedance of ZSBT to meet the requirements of zero-sequence filter. Equivalent circuit of system is shown in figure 3 for positive-sequence and negative-sequence current. i_h is the current to be compensated. i_f is positive-sequence fundamental current. The figure 3 can be seen that only i_f is retained.

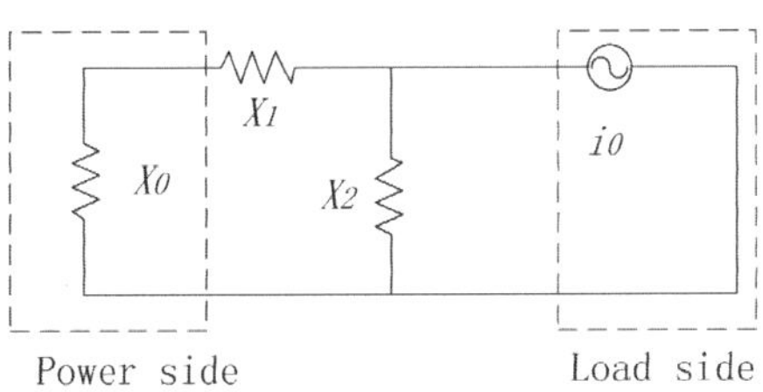

Figure 2. Equivalent circuit for zero-sequence current

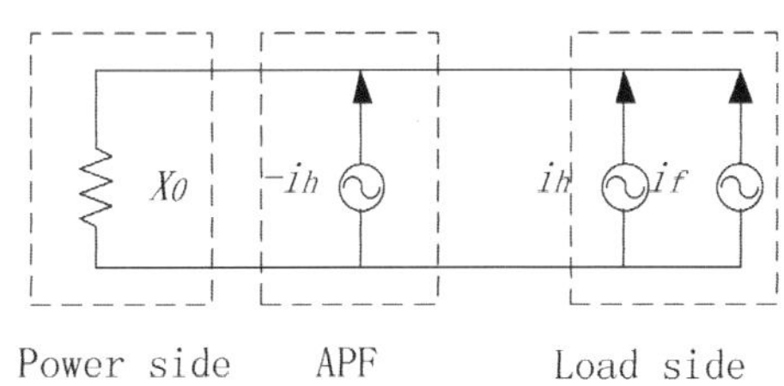

Figure 3. Equivalent circuit for positive - sequence current

3. The principle and the design of reactor

3.1 The principle and the design of Zig-Zag TR

Zig-Zag TR only has one side of the winding. It's a special reactor. In each phase, there are two of the same winding, which roll on iron core column, as shown in figure 4.

Flux linkage of iron core column can reflect the reactance according to definition of inductance [11]. Assuming that the flux linkage Ψ_A, Ψ_B, Ψ_C respectively is generated by the phase current i_A, i_B, i_C. Ψ_{Ae}, Ψ_{Be}, Ψ_{Ce} respectively is total flux linkage in iron core column A,B,C. Total flux linkage of iron core column that is generated by positive-sequence current is three-phase symmetrical, as shown in figure 5. The sizes of Ψ_A, Ψ_B and Ψ_C are equal. Zig-Zag TR has big impedance to positive-sequence current. Leakage current is about 0.01A and can be ignored. Total flux linkage of iron core column that is generated by negative-sequence current is similar to that of positive-sequence current. So Zig-Zag TR is equivalent to open circuit for positive-sequence and negative-sequence current. Total flux linkage of iron core column that is generated by zero-sequence current is 0 Wb because the sizes of Ψ_A, Ψ_B, Ψ_C are equal, as shown in figure 6. Ignore leakage flux, Zig-Zag TR is equivalent to short circuit for zero-sequence current. Design of Zig-Zag TR can use the traditional design method of transformer [12].

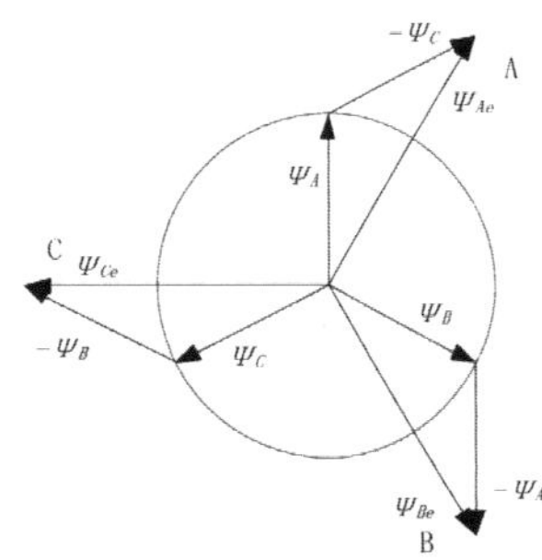

Figure 5. Flux linkage of iron core column for positive-sequence current

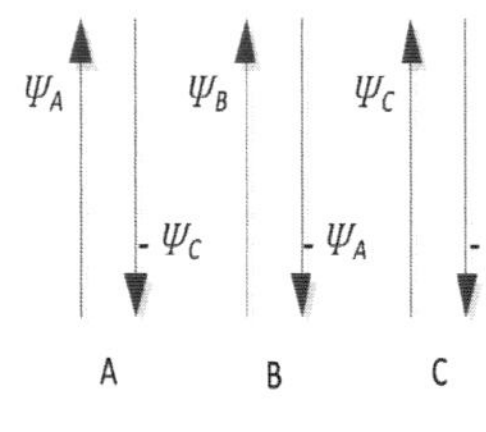

Figure 6. Flux linkage of iron core column for zero-sequence current

3.2 The principle and the design of ZSBT

ZSBT is a special reactor. Three identical windings of ZSBT roll on the same iron core column [8], as shown in figure 7.

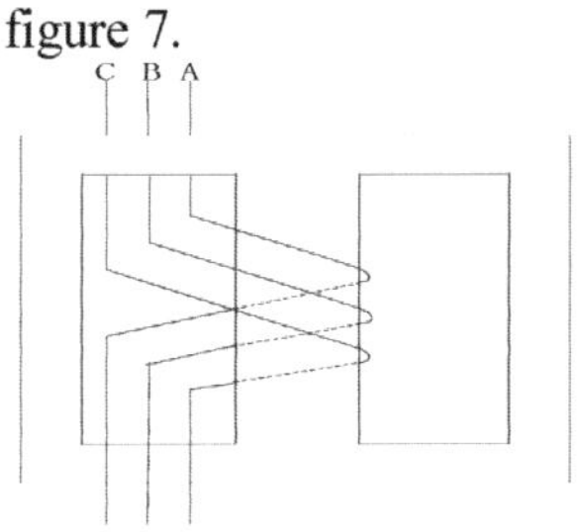

Figure 7. Structure of ZSBT

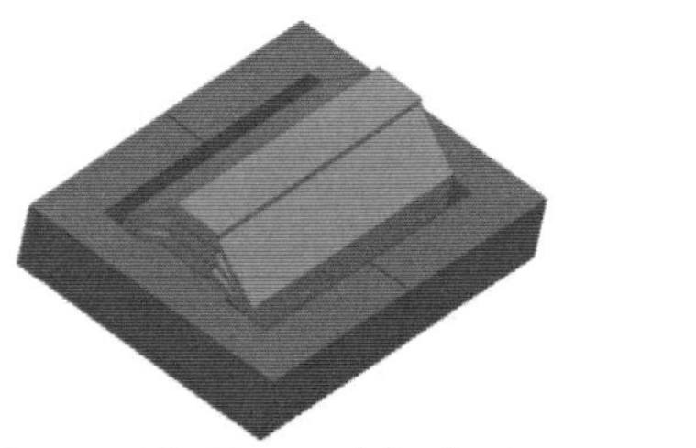

Figure 10. 3D model of ZSBT

Total flux linkage of iron core column that is generated by positive-sequence current is 0 Wb, as shown in figure 8. Ignore leakage flux, ZSBT is equivalent to short circuit for positive-sequence current. Total flux linkage of iron core column that is generated by negative-sequence current is similar to that of positive-sequence current. So ZSBT is equivalent to short circuit for positive-sequence and negative-sequence current. Total flux linkage of iron core column that is generated by zero-sequence current is a sum of Ψ_A, Ψ_B and Ψ_C, as shown in figure 8. Because the sizes of Ψ_A, Ψ_B, Ψ_C are equal, it is 3 times as single phase flux linkage. So impedance of ZSBT is 3 times as single phase impedance.

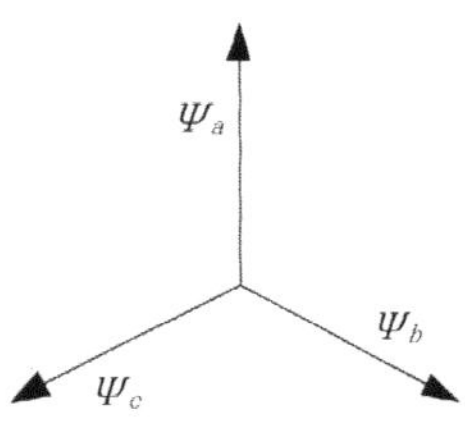

Figure 8. Flux linkage of ZSBT for positive-sequence current

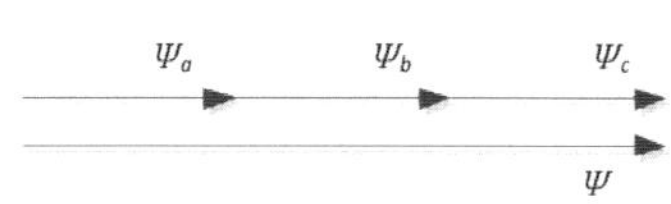

Figure 9. Flux linkage of ZSBT for zero-sequence current

IOP Conf. Series: Earth and Environmental Science **40** (2016) 012050 doi:10.1088/1755-1315/40/1/012050

There is a big error by traditional design method for special structure of ZSBT. The paper design ZSBT by the magnetic coupling method. Core cross-sectional area, turns of winding, Wire diameter and estimate of air gap are defined by traditional design method. 3D model of ZSBT is set up in Ansoft, as shown in figure 10. Air gap is defined by simulation with different air gap length. electromagnetic properties of ZSBT can be analyze in Ansoft.

4. Design of APF

In three-phase four-wire system, a sum of positive-sequence and negative-sequence phase current is 0 A; a sum of zero-sequence phase current is 3 times of zero-sequence phase current. So output current of DC side is 0 A when APF compensates positive-sequence and negative-sequence phase current; it's 3 times of zero-sequence phase current when APF compensates zero-sequence current. Therefore, fluctuation of DC side voltage is increased when APF compensates zero-sequence current, which astricts the ability of APF compensation.

In this paper, main circuit of APF uses diode-clamped three-level inverter [13] that is the most widely used at present; harmonic detection uses instantaneous reactive power theory [14],[15]; current regulator uses VR (vector resonant regulator) [16]-[18]. Control chart is shown in figure 11. Directive current i_C* is composed of current of voltage regulator i_u and harmonic current *ih*. Output current of APF i_C follows i_C*.

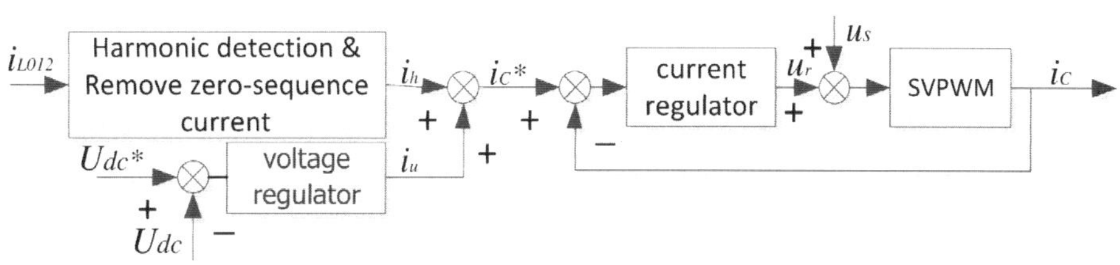

Figure 11. Control chart of APF

5. Simulation experiment

In Simulink, the model was set up to prove the feasibility of this method, as shown in figure 12. The load is shown in table 1. System currents are shown in figure 13 when the load access to the system. It shows the phase current is not balanced and neutral current is mainly fundamental current. A little harmonic current of neutral line is generated by Single-phase uncontrollable rectifier of A-phase. Parameters of mixed mode are shown in table 2.

Table 1. Compositions of the system load

the load in parallel	Phase of load	Structure of load	R(Ω)/L(mH)
1	A&B&C	three-phase uncontrolled rectifier	5.0/2.0
2	A	Single-phase uncontrollable rectifier	1.5/1.0
	B	resistance-inductance load	5.0/1.0
	C	resistance-inductance load	10.0/1.0

Table 2. Parameters of mixed mode

Parameters	value	Parameters	value
zero-sequence impedance of ZSBT	0.03140Ω	Voltage of DC side	800V
short-circuit impedance of zag-zig TR	0.00628Ω	system voltage	220V
inlet wire inductance of APF	1.0mH	Capacitor of DC side	3300uF

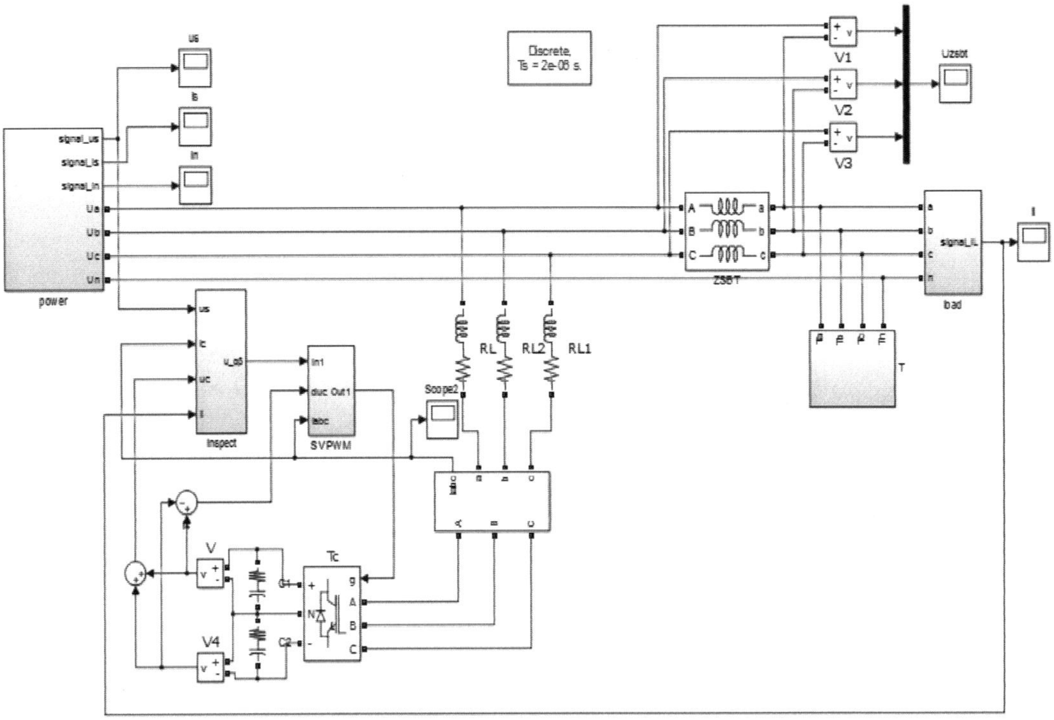

Figure 12. Mode of three-phase hybrid power balancing technique

After using methods of this paper, THD (distortion rate) of phase current respectively drop to 2.3%, 2.3%, 2.3%, 2.3% from 14.3%, 15.6%, 17.8%; neutral current drop to 2.26A from 115.26A; three-phase unbalance degree drop to 1.4% from 51.28%; system currents are shown in figure 14. Output current of APF and current that is filtered by Zag-Zig TR are shown in figure 15. A sum of APF output phase current is 0 A and Zag-Zig TR can filter out neutral current of each time. The contrast of capacitor voltage fluctuations that the system only uses APF to governance unbalanced three-phase current and that it uses mixed mode is shown in figure 16. Capacitor voltage is more volatile with APF. Voltages of two capacitors aren't equal, which has big effects on performance of APF.

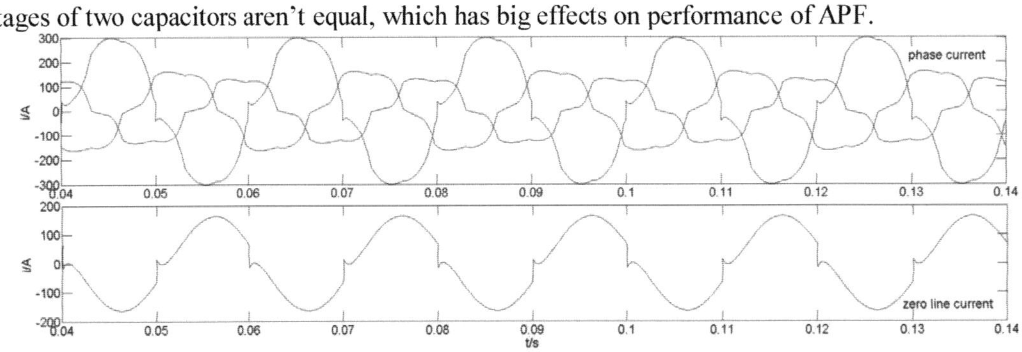

Figure 13. System currents without three-phase balancer

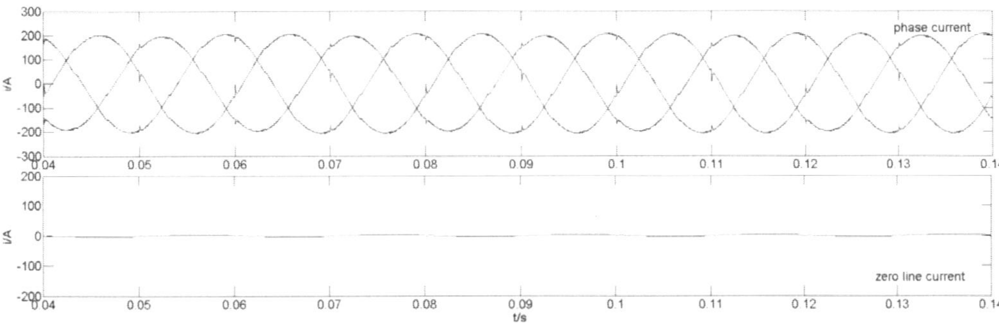

Figure 14. System currents with three-phase balancer

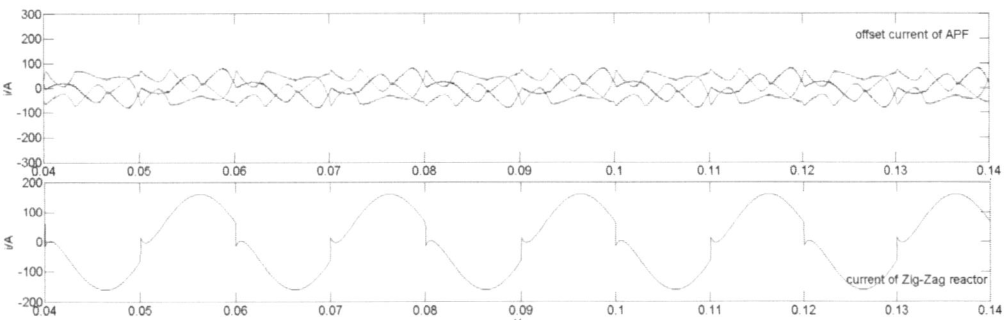

Figure 15. Output current of APF and current that is filtered by Zag-Zig TR

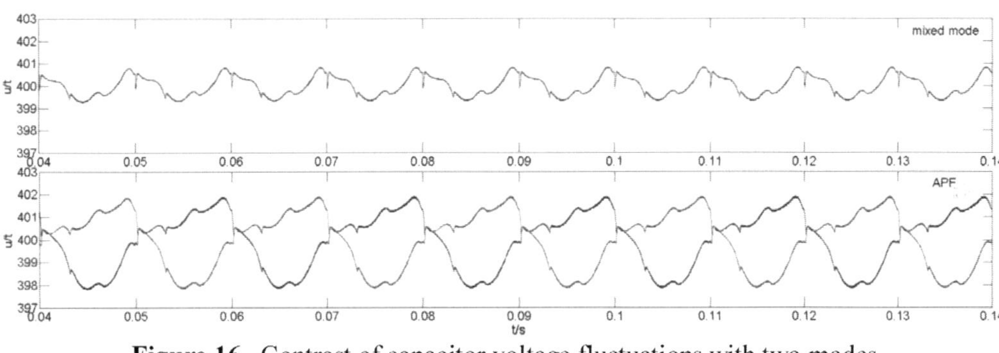

Figure 16. Contrast of capacitor voltage fluctuations with two modes

6. Conclusion

The technology of hybrid three-phase current balance that is researched by this paper is suitable for three-phase four-wire system. Compared with traditional technology of three-phase current balance, it can better play the advantage of PPF and APF. With the deepening of research, it will have a public concern in the building voltage power distribution system.

References

[1] Ping Y, Chen A, Du C Sand Zhang C H 2011 Review of shunt active power filters based on trilevel inverters in three-phase four-wired system *Chinese electrotechnical society* 483-9

[2] Lin Z X, Chen Y, Cai J D and Li T Y 2009 Impact and countermeasure of three-phase unbalanced operration for low voltage distribution systems *J. Electric Power Science And Technology* **24** 63-6

[3] Pan B R, Yu S S and Chen S K 2012 Review of analysis and research about distribution

transformer three-phase imbalance *Power System Technology* **10** 63-5

[4] Fang H F and Sheng W X 2015 Research on the method for real-time online control of three-phase unbalanced load in distribution area *Proc. the CSEE* **35** 2185-93

[5] Zhang Z K, Zhang Y D and Shi L 2016 Based on the load transfer between the distribution area of three-phase unbalanced load management technology *Rural Electrification* **11** 29-30

[6] Yang Y L and Wang F Q 2004 Additional loss and voltage deviation caused by unbalanced operation of disribution transformer and countermeasures *Power System Technology* 28 73-6

[7] Syafrudin M, Hadzer C M and Jusmin S 2002 Zero-sequence harmonics current minimization using zero-blocking. Transformer and Shunt LC Passive Filters *IEEE Trans on Industry Applications* **2** 116-20

[8] Laka A, Barrena J A, Chivite-Zabalza J, Vidal M A R and Calvo G 2014 IP-ZSBT magnetic configuration for parallelization–serialization of three-phase high power converters *IEEE Transactions on Power Electronics* **29** 366-74

[9] Laka A, Barrena J A, Chivite-Zabalza J, Vidal M A R and Calvo G 2013 Novel zero-sequence blocking transformer (zsbt)using three single-phase transformers *IEEE Transactions on Power Electronics* **28** 234-42

[10] Yang J, Wang Z A and Qiu G Y 1996 A study on real-time detecting method for harmonic and fundamental negative sequence currents in unbalanced three-phase circuits *J. Xi An Jiao Tong University* **30** 94-9

[11] Cui L J 1995 *Special transformer theory and design* (Beijing: Scientific and Technical Documents Publishing House) 917-22

[12] Wang X H 2009 *Electrical Machinery* (Beijing: China Machine Press) 2-15 63-4

[13] Tan G J, Zhang J L and Han Y F 2009 Research of three-level shunt active power filter based on SVPWM *Industry and Mine Automation* **26** 27-30

[14] Zhang S Q, Dai K and Xie B 2010 Selective harmonic current control based on multiple synchronous rotating coordinates *Proc. the CSEE* **30** 55-61

[15] Liang Y Y, Liu J Z and Xu X T 2015 Feedforward-feedback control strategy based on detection of power current and load current for three-phase four-wire APF *Electric Power Automation Equipment* **35** 94-100

[16] Yi H, Zhuo F and Yao Y 2015 Research on source current detection type apf control scheme based on vector resonant regulator *Transactions of china electrotechnical society* **30** 72-9

[17] Yi H, Zhuo F and Yan W D 2014 Comparison analysis on selectivity of resonant controllers for parallel active power filters *Proc. the CSEE* **34** 5320-8

[18] Lascu C and Asiminoaei L 2007 High performance current controller for selective harmonic compensation in active power filters *IEEE Transactions On Power Electronics* **22** 1826-35

Study on impact of energy price comparison on energy saving benefits of heat pump in North China

X Dong[1,4], Q Tian[1], Y G Zhang[2] and H F Bai[3]

[1]School of Environmental Science and Engineering, Taiyuan University of Technology, Jinzhong 030600, China
[2]School of Civil Engineering and Mechanics, Yanshan University, Qinhuangdao 066004, China
[3]Shanxi Zhonglv Environmental Protection Group Co., Ltd., Taiyuan 030024, China

E-mail: 784786056@qq.com

Abstract. As to the heat pump technology applying in the HVAC engineering, the relationship between energy saving rate (ESR) and electricity cost saving rate (ECSR) of heat pump should be a positive correlation in theory. But in the actual energy price system, due to the fluctuating energy price comparison, the relationship between them is of less coordination. Moreover, despite the high ESR, the economic benefit of ECSR is lost. In this paper, via the case analysis under the condition of average technical and economic parameters in North China, the critical point rate of economic benefit of ECSR in energy price comparison among prices of residential electricity, steam-coal, and residential natural gas is found, which is about 2:3:8. Also, a viewpoint as well as method is suggested to promote the wide usage of heat pump, balance energy supply structure, save energy consumption, and reduce emissions by optimizing the energy price comparison, which is feasible and desirable to raise the price comparison between residential electricity and natural gas, and reduce the price comparison between residential electricity and steam-coal in a certain extent.

1. Introduction

Energy is the important material basis and driving force of the survival then development of human society. So far, the energy structure has not been essentially changed, which is still primarily based on fossil fuels, including the auxiliary application of clean and renewable energy in many fields. Threatening the sustainable energy development, energy structure problems mainly have two aspects: One is the serious harm to environment such as acid rain and greenhouse effect caused by the utilization of fossil fuels; the other is the contradiction between the growing energy demand in human society and the increasing difficulty of exploiting finite fossil fuels in terms of technology and policy.

In order to solve the above troubles, many studies on the techno-economic performance of heat pump have been conducted in recent years. Sahraie *et al.* [1] have simulated and optimized the thermo-economic and thermodynamic properties of a two-stage irreversible heat pump from the perspective of employing Fuzzy and LINMAP and TOPSIS decision-making approaches to obtain the Pareto frontier. Lazzarin and Noro [2,3] implemented an economic analysis of a gas engine heat pump

[4] Address for correspondence: X Dong, School of Environmental Science and Engineering, Taiyuan University of Technology, Jinzhong 030600, China. E-mail: 784786056@qq.com

coupled to two condensing boilers, which was connected to the district heating grid in Vicenza, Italy. Lambert and Beyene [4,5] made a thermo-economic comparative analysis between a solar thermal cooling system and a solar electric cooling system for 24 cities across US Sunbelt (south of 37°N). Thygesen and Karlsson [6] have simulated and analyzed three solar-assisted ground source heat pump systems with regards to economics and energy. Blum *et al.* [7] made the techno-economic and spatial analysis of vertical ground source heat pump systems in the state of Baden-Württemberg, South Germany. Yazdi *et al.* [8] have researched the exergy, economic and environmental performances of a heat pump to cool the input air from a gas turbine compressor both in Yazd whose climate is arid and Tehran whose climate is temperate, Iran. Li H and Yang HX [9,10] conducted experiments then energy and exergy analysis on a multi-functional solar-assisted heat pump system proposed to supply domestic hot water and space cooling in Hong Kong. Wu *et al.* [11] have carried on a year-round simulation and techno-economic analysis of heating system with air source absorption heat pump under weather conditions of four cities in China. Wu *et al.* [12] have also studied annual economic performance of heating system with compression-assisted air source absorption heat pump.

The application of heat pump technology can reduce the energy consumption during cooling and heating, relieve the tense energy situation, and decrease environmental pollution, which has significant social and economic benefits. Even so, there are still possibilities existing in the discordant and fluctuating energy price system to cause the loss of economic benefits of heat pump, from the artificial other than technical factors perspective. It means that even though the application of heat pump is energy-saving, this kind of behavior may not be money-saving. The purpose of this paper is to find reasons and explore effective measures to eliminate the adverse impact of energy price comparison and promote the usage of heat pump widely as well as quickly in North China.

2. Primary energy rate of heat pump

Heat pump is a kind of high-efficiency and energy-saving equipment used to improve heat grade for heating. Heat pump uses the low-grade heat from nature or manufacturing process emissions as a low-temperature heat source, through a small amount of high-grade power consumption of compressor or other forms, thus improves the heat grade of low-temperature heat to apply to heating users of high-temperature heat source. Taking electric vapor compression heat pump most widely used as an example, its energy balance model is shown in Figure 1.

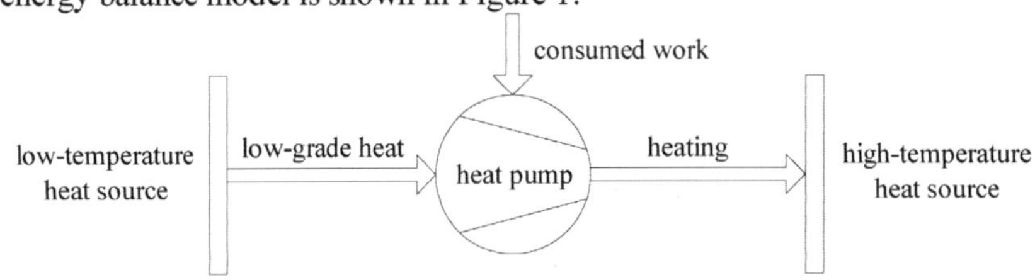

Figure 1. Heat pump in the energy balance model

Electric vapor compression heat pump is an equipment transforming electricity into heat. The relative scale between benefits and costs of heat pump is evaluated by COP in the thermodynamics as follows:

$$COP = \frac{Q_H}{W} \tag{1}$$

where Q_H is the heating capacity and W the quantity of consumed work.

The expression of COP is the comparison rate between thermal energy and mechanical work, so the economic feasibility of transforming mechanical work into thermal energy must be considered in certain conditions of energy price comparison and device availability. From the thermal power point of view, the COP expression cannot explain the rationality of heat energy utilization in theory. To this issue, the introduction of primary energy rate (PER) of heat pump is essential. The meaning of PER is that the quantity of consumed work W in the COP expression can be converted into equivalent primary energy consumption of thermal power generation as follows:

$$PER = \frac{Q_H}{W}\eta_p \tag{2}$$

where η_p is the efficiency of thermal power generation.

Analyzing the Formula (2), it is feasible to compare PER with the thermal efficiency of boiler. Both the techno-economic analysis of heat pump and boiler can be based on the conversion and utilization efficiency within transforming the chemical energy of fossil fuel combustion into heat energy. Their specific energy utilization processes are shown in Figure 2.

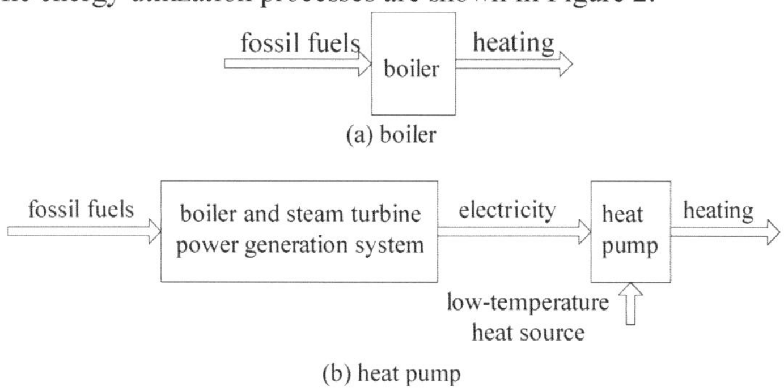

Figure 2. Energy utilization process comparison between (a) boiler and (b) heat pump

Illustrated in Figure 2, the chemical energy of fossil fuel combustion is directly used for heating in the energy utilization process of boiler, thus the thermal efficiency must be less than one. Concerning the energy utilization process of heat pump, the chemical energy of fossil fuel combustion is W/η_p, which is used for thermal power generation through boiler and steam turbine power generation system. Electricity generated from thermal power generation system drives heat pump to consume the work W and improve the low-grade heat. The heat pump process ends up with the heat Q_H for heating. Because the energy utilization process of heat pump is by way of electricity transmission and driving the compressor to use the low-grade heat from low-temperature heat source, W/η_p is just an idealized expression of primary energy consumption quantity of heat pump to facilitate comparative calculation, which has ignored the influence of efficiency of electricity transmission and utilization. There is no unified evaluation method whether the usage of low-temperature heat source is calculated on the energy consumption of heat pump or not, and how to estimate the low-temperature waste heat discharged from steam turbine. But crucially in terms of the conversion process from chemical energy to heat energy, the PER of heat pump and thermal efficiency of boiler are comparable [13].

3. Energy saving rate of heat pump

Energy saving benefits of heat pump are generally evaluated by two indicators which are energy saving rate (ESR) and electricity cost saving rate (ECSR). The calculation of ESR refers to the comparison of primary energy consumption between different equipments. Meanwhile, the count of ECSR is the comparison of electricity economic efficiency of end users. The ESR of electric vapor compression heat pump shows the comparison of primary energy consumption between heat pump and traditional coal-fired boiler, as well as between heat pump and gas-wall boiler, under the same thermal load condition. Under the condition of the same heating quantity Q_H of heat pump as boiler, the ESR of heat pump is calculated as:

$$\lambda_h = \frac{E_b - E_h}{E_b} \tag{3}$$

where λ_h is the ESR of heat pump; E_h and E_b are the quantity of primary energy consumption of heat pump and boiler respectively.

The thermal efficiency of boiler η_b is estimated as:

$$\eta_b = \frac{Q_H}{E_b} \tag{4}$$

Using the PER of heat pump and the thermal efficiency of boiler, the ESR of heat pump is shown as follows:

$$\lambda_h = 1 - \frac{\eta_b}{PER} \tag{5}$$

The average thermal power efficiency is 39% in 2015, China [14]; the thermal efficiency of coal-fired boiler is about 60% [15] and for gas-wall boiler is 80-96% [16]; the average COP can reach more than three for air source heat pump coupled with low-temperature hot water floor-heating system in the whole heating season [17]. What have been mentioned above are the average technical parameters in North China. For certain and calculation, the thermal power efficiency is 39%, the thermal efficiency of coal-fired and gas-wall boiler is respectively 60% and 88% average, and the COP of heat pump is three. Results calculated in Formula (5) and gained by these parameters are that, the ESR of heat pump comparing with primary energy consumption of coal-fired boiler is about 48.7%, and the ESR comparing with gas-wall boiler is about 24.8%. Results also show that under the same thermal load condition, the application of heat pump is effectively able to save the corresponding rate of non-renewable energy consumption, and reduce the corresponding rate of carbon emissions.

4. Electricity cost saving rate of heat pump

As one kind of heating equipment energy-saving and emissions-reducing, heat pump will replace boiler for heating in some conditions. Its economic benefit is affected by many factors, such as the change of investment and operating cost, the change of life cycle and depreciation of equipment. Only from the aspect of electricity cost saving rate (ECSR) of heat pump comparing with the heating operating cost of coal-fired and gas-wall boiler, this article discusses the economic feasibility of heat pump as a substitute for energy-saving and environment-protecting technology.

When the heating load is same, the economic feasibility of heat pump can be computed according to Formula (6):

$$\eta_h = \frac{m_b - m_h}{m_b} \tag{6}$$

where η_h is the ECSR of heat pump; m_h is the electricity cost of heat pump and m_b the fuels cost of boiler.

Combining with the thermal efficiency η_b of boiler, the ECSR η_{hc} of heat pump comparing with fuels cost of coal-fired boiler and the ECSR η_{hg} comparing with gas-wall boiler can be got respectively as follows:

$$\eta_{hc} = 1 - \frac{Q_c \eta_b}{3600COP} \cdot \frac{c_p}{c_c} \tag{7}$$

$$\eta_{hg} = 1 - \frac{Q_g \eta_b}{3600COP} \cdot \frac{c_p}{c_g} \tag{8}$$

where Q_c is the calorific value of steam-coal, kJ/kg, and Q_g for residential natural gas, kJ/Nm3; c_p is the price of residential electricity, CNY/kW·h, and c_c for steam-coal, CNY/kg, and c_g for residential natural gas, CNY/Nm^3.

The average technical parameters in North China above used for calculation are also given here. Namely, the thermal efficiency of coal-fired and gas-wall boiler is respectively 60% and 88% average, and the COP of heat pump is three. Here more adds the calorific value Q_c of steam-coal Q5500 commonly used in thermal power plant, whose national general standard is 5,500 kcal/kg or 23,022 kJ/kg [18], and the calorific value Q_g of residential natural gas whose national general standard is 3.8×10^4 kJ/Nm^3 [19]. Moreover, the average economic parameters in North China include the average Free-On-Board price c_c of steam-coal Q5500 in the Qinhuangdao Port, which is 0.415 CNY/kg [20], the national average price c_g of residential natural gas 2.48 CNY/Nm^3 [21], and the average price c_p of residential electricity in North China 0.485 CNY/kW·h [22].

According to Formula (7) and (8), results show that comparing with gas-wall boiler, the ESR of heat pump is 24.8% and the ECSR is 39.5%, while comparing with coal-fired boiler, the ESR is 48.7% and the ECSR is -49.5%. The results reveal that the operation of heat pump is more energy-saving and money-saving than utilization of gas-wall boiler. Even though the operation of heat pump is more energy-saving than that of coal-fired boiler, the former is much more expensive than the latter. Namely, heat pump has to spend 49.5% more money on heating operating cost to save 48.7% energy consumption. Energy saving loses electricity cost saving for heat pump in this case which severely lowers the energy-saving benefits and market value of heat pump.

In Formula (7) and (8), the thermal efficiency of boiler, COP of heat pump, and calorific value of fuels are all relatively stable technical parameters in a certain region and during a certain period, which are not primary causes directly affecting ECSR of heat pump. Average prices of residential electricity, steam-coal, and residential natural gas are crucial economic parameters within spontaneous economy market regulation and political active intervention both at home and abroad. Their energy price comparison c_p/c_b between residential electricity price relative to heat pump and fuels price relative to boiler is decisive for ECSR, including c_p/c_g between the price of residential electricity and natural gas, and c_p/c_c between the price of residential electricity and steam-coal. What has been shown in Figure 3 is the impact of energy price comparison c_p/c_b on ECSR η_h of heat pump in North China. Under the aforementioned computational condition of average technical and economic parameters in North China, the critical point rate of economic benefit of ECSR in energy price comparison among prices of residential electricity, steam-coal, and residential natural gas is found, whose $c_p:c_c:c_g$ is about 2:3:8. On the basis of this critical point rate, the lower the rate is, the higher ECSR is, and η_h is in the profit interval, which is favorable for popularizing heat pump; while the higher the rate is, the lower ECSR is, and η_h is in the deficit interval, which is unfavorable for heat pump. The reason why ECSR comparing with coal-fired boiler is negative is because c_p/c_c is 1.17 more than 0.67, whose η_{hc} is within the deficit interval, so heat pump loses economic benefit. The c_p/c_c should be controlled beneath 0.67 in order to popularize heat pump heating technology of energy-saving benefits in the central heating areas using coal-fired boiler.

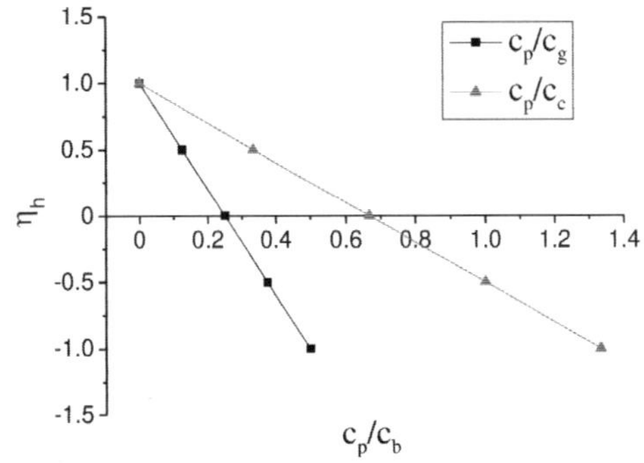

Figure 3. Impact of energy price comparison c_p/c_b on ECSR η_h of heat pump in North China

Although the energy price comparison c_p/c_b between residential electricity price relative to heat pump and fuels price relative to boiler is lower, the ECSR η_h is higher. The low c_p/c_b within the profit interval must be controlled within a certain range to ensure a safe and orderly economic market. Because when c_p/c_b is low to a certain extent, economic market would spontaneously transform heating behavior from fossil fuel combustion into electrical heating entirely. It would lead to unbalancing the structure of energy supply as well as energy import and export, and threatening energy economy security of China. It would also threaten other social stability factors such as stable average employment rate in related fields. Nowadays, the overall situation of electricity supply in China is still relatively tense. Through the reasonable behavior of market economy and government regulation, it is feasible and desirable to raise the price comparison c_p/c_g between residential electricity and natural gas, and reduce the price comparison c_p/c_c between residential electricity and steam-coal in a certain extent, in order to balance energy supply structure and promote the sustainable utilization of heat pump which is energy-saving and environmentally friendly. However, meeting the demands of residents, enterprises, government and other various interests farthest, the best energy price comparison solution remains to be determined via the hearing finally.

5. Conclusions

The application of heat pump can obtain certain ESR, reduce emissions and protect environment. It is quite significant for solving energy and environment problems in China.

Under the condition of average technical and economic parameters in North China, the example shows the electricity cost comparison among the air source heat pump coupled with low-temperature hot water floor-heating system, traditional coal-fired boiler and gas-wall boiler. The results explain that the critical point rate of economic benefit of ECSR in energy price comparison among prices of residential electricity, steam-coal, and residential natural gas is about 2:3:8. Based on it, the lower the rate is, the higher ECSR is, and ECSR is in the profit interval, vice versa. Even if heat pump has energy-saving advantages, the corresponding economic benefit may not come along with the energy price system which is not harmonious.

Through the reasonable behavior of market economy and government regulation, it is feasible and desirable to raise the price comparison c_p/c_g between residential electricity and natural gas, and reduce the price comparison c_p/c_c between residential electricity and steam-coal in a certain extent, in order to balance energy supply structure and promote the sustainable utilization of heat pump which is energy-saving and environmentally friendly. However, meeting the demands of residents, enterprises, government and other various interests farthest, the best energy price comparison solution remains to be determined via the hearing finally.

Acknowledgements
The work described in this paper was supported by the International Cooperation Project (No. 2013DFA61580) from the Ministry of Science and Technology of China (MOST).

References

[1] Sahraie H, Mirani M R, Ahmadi M H and Ashouri M 2015 Thermo-economic and thermodynamic analysis and optimization of a two-stage irreversible heat pump *Energy Conversion and Management* **99** 81-91
[2] Lazzarin R and Noro M 2006 District heating and gas engine heat pump: Economic analysis based on a case study *J. Applied Thermal Engineering* **26** 193-9
[3] Lazzarin R and Noro M 2006 Local or district heating by natural gas: Which is better from energetic, environmental and economic point of views? *J. Applied Thermal Engineering* **26** 244-50
[4] Lambert M A 2007 Design of solar powered adsorption heat pump with ice storage *J. Applied Thermal Engineering* **27** 1612-28
[5] Lambert M A and Beyene A 2007 Thermo-economic analysis of solar powered adsorption heat pump *J. Applied Thermal Engineering* **27** 1593-611
[6] Thygesen R and Karlsson B 2013 Economic and energy analysis of three solar assisted heat pump systems in near zero energy buildings *Energy and Buildings* **66** 77-87
[7] Blum P, Campillo G and Kölbel T 2011 Techno-economic and spatial analysis of vertical ground source heat pump systems in Germany *Energy* **36** 3002-11
[8] Reza M, Yazdi M, Aliehyaei M and Rosen M A 2015 Exergy, economic and environmental analyses of gas turbine inlet air cooling with a heat pump using a novel system configuration *Sustainability* **7** 14259-86
[9] Li H and Yang H X 2010 Study on performance of solar assisted air source heat pump systems for hot water production in Hong Kong *Applied Energy* **87** 2818-25
[10] Li H and Yang H X 2010 Energy and exergy analysis of multi-functional solar-assisted heat pump system *Int. J. Low-Carbon Technologies* **5** 130-6
[11] Wu W, Wang B L, Shi W X and Li X T 2014 Techno-economic analysis of air source absorption heat pump: Improving economy from a design perspective *Energy and Buildings* **81** 200-10
[12] Wu W, Shi W X, Wang B L and Li X T 2015 Annual performance investigation and economic analysis of heating systems with a compression-assisted air source absorption heat pump *Energy Conversion and Management* **98** 290-302
[13] Zhang Y G 2003 Technical and economic evaluation on heat pump heat-supply system *Gas & Heat* **23** 364-7
[14] NEA of China 2015 Technical supervision report of electric power construction engineering quality. Available online: http://zfxxgk.nea.gov.cn/auto93/201504/t20150420_1905.htm?keywords= 施 工 单 位 (accessed on 10/04/2015)
[15] Branch Station of Bureau of Economic Operation Regulation of NDRC of China 2015 State Grid promotes energy instead of electricity over 50 billion kilowatt-hours in 2014. Available online: http://ceppc.chinapower.com.cn/newsarticle/1228/new1228815.asp (accessed on 28/01/2015)
[16] Guangzhou Diessen Household Boiler Manufacturing Co.,Ltd. and National Gas Appliances Quality Supervision and Inspection Center of China 2010 Gas-fired heating and hot water combi-boilers: GB 25034-2010 (*Beijing: China Standards Press* **10**)
[17] Li Z, Tian Q and Guo W Q 2014 Research of ventilating performance in North China of air

source heat pump direct radiant floor heating system *Science Technology and Engineering* **14** 228-32

[18] Beijing Coal-chemical Industry Research Branch of Coal Science Research Institute, Shanxi Guoyang New Energy Co.,Ltd. 2006 Coal Product Varieties and Hierarchies: GB/T 17608-2006 China Standards Press **2**

[19] Natural Gas Technology Research Institute of Southwest Oil and Gas Field Company of PetroChina, Southwest Branch of Engineering Design Co.,Ltd. of PetroChina. 2012 Natural gas: GB 17820-2012 China Standards Press **1**

[20] Qinhuangdao Seaborne Coal Market. BSPI – FOB price of Qinhuangdao Port. Available online: http://www.cctcw.cn/news/2014-09-18/new_20302_s.html (accessed on 27/05/2015)

[21] Price Department of NDRC of China. Notification about adjusting natural gas price (NDRC price [2013] No.1246).
Available online: http://money.163.com/13/0628/18/92FOA28D00253B0H.html (accessed on 28/06/2013)

[22] Price Department of NDRC of China. Notification about adjusting electricity price of power grid in North China (NDRC price [2011] No.2619).
Available online: http://www.sdpc.gov.cn/zcfb/zcfbtz/201112/t20111201_448582.html (accessed on 01/12/2011)

Development of economic and environmental metrics for forest-based biomass harvesting

F L Zhang, J J Wang[1], S H Liu[1] and S M Zhang

Department of Mechanical and Transportation Engineering, China University of Petroleum, Beijing, 102249, China

E-mail: jwang@cup.edu.cn, liu_shu_hai@163.com

Abstract. An assessment of the economic, energy consumption, and greenhouse gas (GHG) emission dimensions of forest-based biomass harvest stage in the state of Michigan, U.S. through gathering data from literature, database, and other relevant sources, was performed. The assessment differentiates harvesting systems (cut-to-length harvesting, whole tree harvesting, and motor-manual harvesting), harvest types (30%, 70%, and 100% cut) and forest types (hardwoods, softwoods, mixed hardwood/softwood, and softwood plantations) that characterize Michigan's logging industry. Machine rate methods were employed to determine unit harvesting cost. A life cycle inventory was applied to calculating energy demand and GHG emissions of different harvesting scenarios, considering energy and material inputs (diesel, machinery, etc.) and outputs (emissions) for each process (cutting, forwarding/skidding, etc.). A sensitivity analysis was performed for selected input variables for the harvesting operation in order to explore their relative importance. The results indicated that productivity had the largest impact on harvesting cost followed by machinery purchase price, yearly scheduled hours, and expected utilization. Productivity and fuel use, as well as fuel factors, are the most influential environmental impacts of harvesting operations.

1. Introduction

Increasing concerns associated with global warming have caused regulatory agencies and nongovernmental organizations (NGOs) to require a sharp reduction in greenhouse gas (GHG) emissions from the transportation sector, U.S. [1]. Depletion of fossil fuels supports the need for alternative renewable energy resources for transportation [1]. Ethanol produced from renewable feedstocks (e.g. woody biomass, agricultural residues, energy crops) could serve as a substitute for gasoline and therefore help reduce GHG emissions and contribute to achieving environmental goals.

To ensure economically feasible and environmentally beneficial biofuel production, decreasing cost, energy consumption and GHG emissions from cradle to grave life cycle of the biofuel is necessary [2]. The harvesting operations of biomass represent the largest component of cost associated with biomass energy production and generation [2]. The economic and environmental assessment of the supply chain focused on forest-based biomass in Michigan. Forest-based biomass is a major feedstock source of woody biomass from forestlands in the U.S. [2]. The statewide timberland acreage data from U.S. Forest Service indicates that it is the seventh largest in the U.S. [2]. And the current

[1] Address for correspondence: J J Wang, S H Liu, Department of Mechanical and Transportation Engineering, China University of Petroleum, Beijing, 102249, China. E-mail: jwang@cup.edu.cn, liu_shu_hai@163.com.

annual growth exceeds annual mortality and removals in most types of forest [2-3]. In addition, decline in other traditional roundwood industries (pulp and lumber) in recent years have opened up new prospects for the sustainable use of forest resources [4].

The goal of this study is to develop economic and environmental impact performance metrics for forest-based biomass harvesting cost, energy demand and GHG emissions within the state of Michigan. The three factors are considered as measures of economic growth, energy security, and environmental protection, respectively [1]. To this end, the following were conducted: (1) a harvesting cost analysis using the machine rate costing method, and (2) a life-cycle assessment (LCA) procedure for energy demand and GHG emissions estimation. Both methods were used to evaluate a series of forest biomass harvesting scenarios differentiating harvesting systems, harvest types, and forest types that characterize Michigan's logging industry.

A forest harvesting system is defined as the method of organizing forestry equipment and duties among loggers to harvest the particular forest in the most effective way possible [5]. Three main harvesting systems of forest in the Michigan's logging industry are [6-7]: (1) cut-to-length (CTL) harvesting with harvester/forwarder; (2) whole tree harvesting (WTH) with feller-buncher/grapple skidder/slasher; and (3) motor-manual harvesting (MMH) with chainsaws/cable skidder, as illustrated in Figure 1. After being felled with a feller in the cut-to-length harvesting solution, timber is delimbed and bucked into, for example, pulpwood logs for biofuels, at the stump [8]. Logs are then transported via a forwarder to the roadside [9]. With the WTH method, trees are felled then transported to the roadside with branches and the top intact. The whole trees are processed at the roadside into pulpwood logs [9]. After trees are felled and bucked with a chainsaw in a MMH system, they are skidded to roadside by a ground-based cable skidder [5].

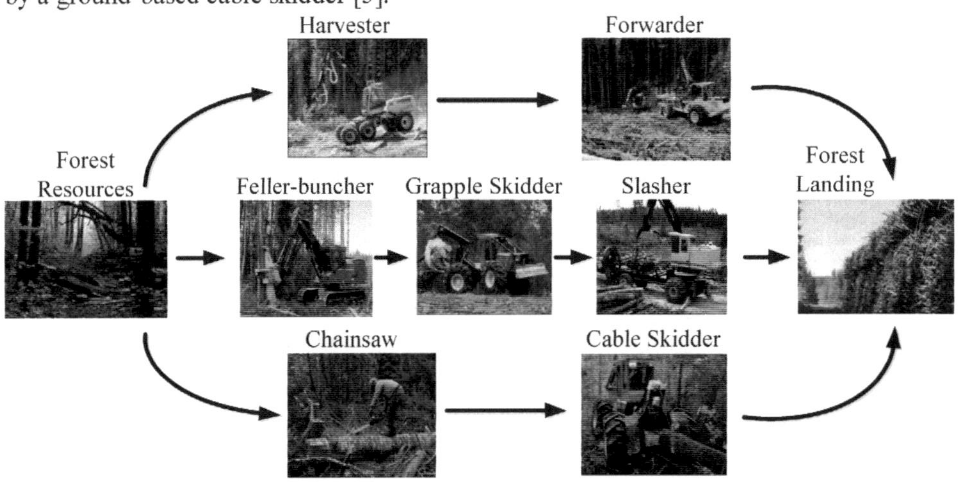

Figure 1. Three forest harvesting systems (Image courtesy of Google)

The three dominant harvest types evaluated were (1) removal of all merchantable trees, (2) a 70% removal of all merchantable trees, and (3) a 30% removal of all merchantable trees [6-7]. Forest resources are differentiated by natural hardwoods, natural softwoods, mixed hard and softwood, and softwood plantations [6-7].

This study is the first to run an explicit assessment of the three indicators (cost, energy and emissions) for Michigan's harvesting activities differentiating harvesting systems, harvest types, and forest types. Previous LCAs of biofuel produced from various biomass feedstocks focused on environmental performance of producing and supplying feedstocks [7, 10-14], without paying much attention on economic viability. Moreover, the evaluation of harvesting/forwarding activities was mostly based on the assumption of employing cut-to-length harvester/forwarder only [12] or whole tree harvesting system only [13]. Other studies did not differentiate harvest type (assuming 100%

clear-cutting) [12] or forest type for environmental impact assessment [14]. This research expands on the scope of prior work and adds to the existing body of knowledge in this research area.

The rest of this paper is structured as follows. Section 2 describes the research methods along with the goals and the scope definition, functional unit, and life cycle input data. The results are presented and discussed in Section 3. Sensitive analysis on key input variables is performed and presented in Section 4. Section 5 concludes the discussions to include research findings and major contribution.

2. Material and methods

2.1. LCA method

2.1.1. Goal and scope.

LCA is an effective tool to identify environmental implications but does not account for the economic implications [1]. The primary goal of the LCA is to evaluate energy consumption and GHG emissions from forest harvesting operations in Michigan, U.S. System boundary defines the scope for LCA and is illustrated in Figure 2. Harvesting activities include cutting standing trees from the stump, processing them into typical log length (e.g., 2.54 m), and then transporting the processed logs to forest landings [14] (Figure 2). Activities from "upstream" inputs of the biomass feedstock production are excluded from this study. These "upstream" activities may include reforestation and landscape carbon stock changes due to direct and indirect land use change [14]. We also excluded inputs for "downstream" activities of producing, distributing and end use of biofuels.

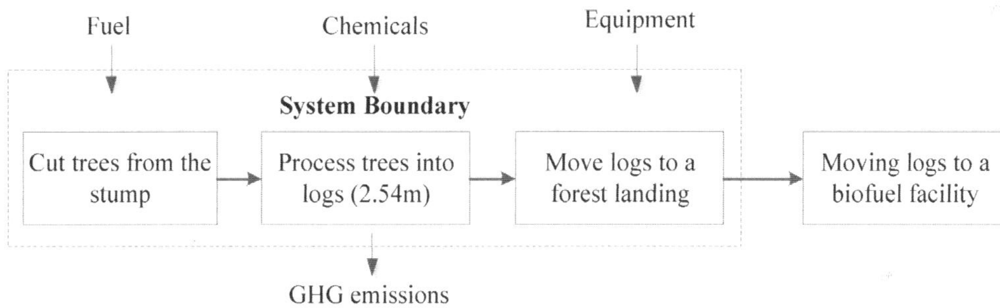

Figure 2. Diagram of the system boundary

2.1.2. Functional unit.

One green tonne of pulpwood is the functional unit which includes the harvesting and collection using forestry machinery commonly found in Michigan. For the purpose of this study, a green tonne is at 50% moisture content wet basis [14]. Since no byproducts were considered, the environmental impacts were assigned to bioethanol, the primary product.

2.1.3. Life cycle input data.

Environmental impacts of producing energy and material inputs as well as the direct utilization of them in the forest harvesting activities [15] were included as part of the LCA using the GREET (Greenhouse gases, Regulated Emissions, and Energy use in Transportation) model [16], peer-reviewed scientific articles, reports, or other sources. These material and energy inputs included fossil fuel, major machinery used to harvest wood (e.g. harvesters, forwarders), and lubricants and other inputs during machine fabrication, maintenance and large process equipment replacing [14]. The global warming potential (GWP) was established to measure a particular chemical's propensity to cause global climate change represented by carbon dioxide equivalents (CO_{2eq}) over different time horizons [17]. For this application, emissions of greenhouse gases, such as CO_2, N_2O, CH_4, etc, were normalized using the IPCC GWP 100-year average and then aggregated for the overall impact estimation of a process or product.

2.2. Harvesting cost analysis method

Two key methods to establish harvesting costs are the machine rate and cash flow methods [18-19]. The cash flow method applies past data of loggers to develop a harvesting cost ($/tonne), by adding business related costs and for the tonnes of timber delivered for the last year. The actual average harvesting cost for the prior year is calculated by dividing the harvesting cost ($/tonne) by the tonnes of timber delivered [19]. It should be noted that this method performs poorly on distributing larger costs associate with, e.g., equipment purchases and capital equipment maintenance except that all equipment is leased [19]. The machine rate method combines hourly system costs and productivity to predict a harvest cost [19], which is useful when predicting the approximate hourly costs of a machine and for cost comparison between two machines [18]. It has been identified as the most common approach of estimating the cost of individual machines and forest harvesting operations [20-21], and thus was employed for this study.

The machine rate method divides hourly system cost by estimated productivity to derive harvesting cost [18]. Productivities were collected for the three harvest types in Michigan as stated above. The hourly system cost was calculated by adding up machine cost per hour for each piece of equipment in a harvesting system [19]. Hourly machine cost was calculated using the machine rate model of Machine Costing Spreadsheet from Virginia Polytechnic Institute [19]. The spreadsheet allows users to enter basic machine costing values to calculate fixed, operating, and labor costs for the machine. Basic machine costing values were collected from different sources like manufactures, peer-reviewed research papers, reports, and Bureau of Labor Statistics [22]. In particular, purchase price was collected by contacting forestry equipment manufacturers and local dealers, such as Forestry Equipment Sales [23], and Machinery Trader [24]. Different equipment models from brands like Caterpillar, John Deere, and Timberjack were investigated. The machine purchase prices presented in Table 1 are averaged values without indicating specific brands or models. As expected, the purchase price of the CTL forest machines is higher than that of WTH machines. This may be because CTL machinery is technically more advanced. On the other hand, fuel consumption is lower with CTL harvesting machines than with WTH machines, which may be a result of the CTL machines lighter weight. These conclusions are consistent with those drawn by Ponsse specialises in harvesting solutions for CTL logging [25]. Fuel consumption with chainsaws system is significantly lower compared to the more mechanized systems.

Table 1. Collected machine information.

Harvesting System	Equipment	Purchase Price ($)	Fuel use[a](l/hr)
CTL harvesting system	CTL harvester	400,000	18.55
	Forwarder	350,000	12.11
WTH system	Feller-buncher	200,000	23.85
	Grapple skidder	260,000	19.31
	Slasher	150,000	14.76
MMH system	Chainsaw	1,500	4.16
	Cable skidder	250,000	9.08

Source: [a] Collected from Abbas *et al.* [6]

The remaining basic machine costing values were assumed as follows:

- Depreciation period was assumed to be 20 years [26-27] for equipment and 5 years for chainsaws. The straight line method was employed for calculating depreciation.
- Salvage value was assumed as 10% of the initial investment [26].
- The full time utilization rate of equipment was assumed to be 2,000 hours per year. The calculation was based on 8 hours/day for a 5 day per week. For 52 weeks/year and a two week breakup for vacation, it would be 8 hours x 5 days x 50 weeks = 2,000 hours/year.
- An interest rate of 8% and an insurance/taxes rate of 7% were applied [28].

- Maintenance and repair costs were assumed as 50% of depreciation [29].
- Machine utilization rate was assumed as 65% [29], and
- A labor rate of $16.17 per hour (2012 Median pay, including fringe benefits) [22] was assumed for logging workers.

3. Results and discussion

3.1. Harvesting costs

The resultant cost estimate (Table 2.) is per Scheduled Machine Hour (SMH), that is, for every working hour of the machine is scheduled plus downtime [19]. Based on the machine rates in Table 2, the system cost for the three harvesting systems of CTL, WTH, and MMH are $114.33 SMH, $138.01 SMH, and $63.07 SMH, respectively.

Table 2. Calculated machine cost (all cost are expressed in $/SMH).

Equipment	Fixed operating cost	Running cost	Labor cost	Hourly machine cost
CTL harvester	26.18	18.85	16.17	61.20
Forwarder	22.90	14.06	16.17	53.13
Feller-buncher	13.09	20.96	16.17	50.22
Grapple skidder	17.01	18.33	16.17	51.51
Slasher	9.82	10.29	16.17	36.28
Chainsaw	0.21	2.88	16.17	19.26
Cable skidder	16.36	11.28	16.17	43.81

Average harvesting cost ($/tonne) was calculated for each harvesting system (CTL, WTH, or MMH) in each harvest type. The results are illustrated in Figure 3. As expected, average harvesting cost for the MMH system is higher than the more mechanized systems (CTL and WTH) across most forest and harvest types. For both fully-mechanized systems of CTL and WTH, average harvesting cost decreased as removal intensity increased from 30% to 100%, with WTH slightly higher than CTL in most cases except for shelterwood cutting on natural hardwoods and clearcutting operations on natural hardwoods and mixed hardwood/softwood. This may be because more machines are required for the WTH system, and higher fuel consumption with the WTH system machines than with the CTL harvesting system machines (Table 1). In addition, labor costs are higher for WTH because more operators are needed.

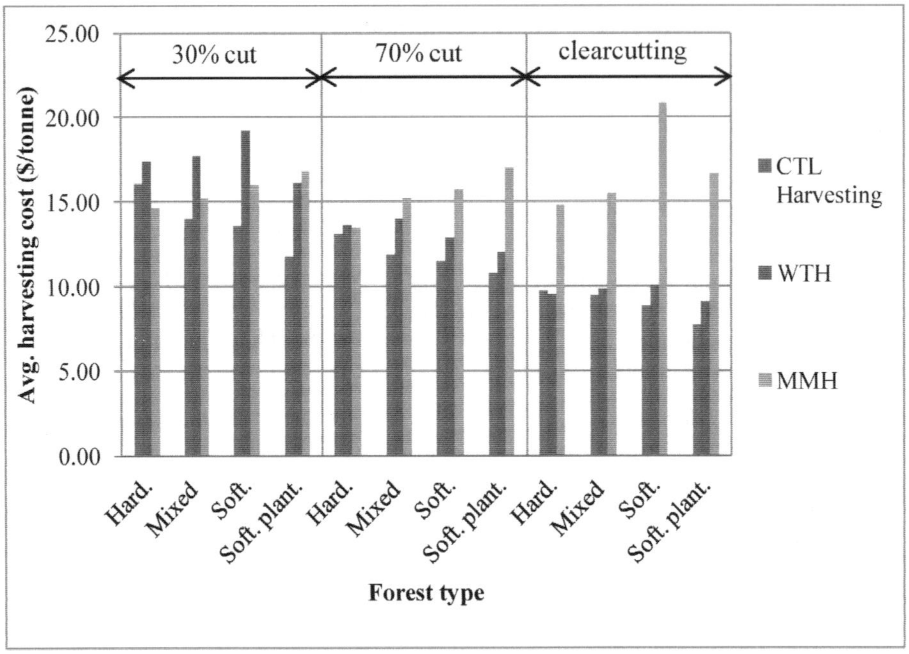

Figure 3. Average harvesting cost for all natural stands for each harvest system in each harvest type

3.2. Environmental impact assessment

Based on the life cycle inventory data, calculated energy consumption and GHG emissions are presented in Tables 3-5. Compared with more mechanized systems (CTL and WTH), chainsaws/cable skidder system leads to higher energy consumption and GHG emissions because of significantly lower productivities. The feller-buncher system consumes more fossil fuels, resulting in high levels of environmental impacts.

Table 3. Environmental impacts of the CTL harvesting system in Michigan.

Treatment	Forest type	Emissions (kg CO_{2eq}/tonne)	Energy (MJ/tonne)
30% cut (selective)	Natural hardwoods	14.58	211.32
	Mixed hardwood/softwood	12.76	188.82
	Natural softwoods	12.41	184.47
	Softwood plantations	10.78	164.23
70% cut (shelterwood)	Natural hardwoods	11.98	179.20
	Mixed hardwood/softwood	10.90	165.82
	Natural softwoods	10.58	161.78
	Softwood plantations	9.95	154.07
Clearcutting	Natural hardwoods	9.01	142.34
	Mixed hardwood/softwood	8.77	139.35
	Natural softwoods	8.21	132.51
	Softwood plantations	7.21	120.14

Table 4. Environmental impacts of the WTH system in Michigan.

Treatment	Forest type	Emissions (kg CO_{2eq}/tonne)	Energy (MJ/tonne)
30% cut (selective)	Natural hardwoods	23.47	313.44
	Mixed hardwood/softwood	23.88	318.54

	Natural softwoods	25.83	342.63
	Softwood plantations	21.81	292.87
70% cut (shelterwood)	Natural hardwoods	18.53	252.28
	Mixed hardwood/softwood	18.98	257.92
	Natural softwoods	17.52	239.82
	Softwood plantations	16.37	225.56
Clearcutting	Natural hardwoods	13.06	184.60
	Mixed hardwood/softwood	13.50	190.00
	Natural softwoods	13.83	194.10
	Softwood plantations	12.56	178.46

Table 5. Environmental impacts of the MMH system in Michigan.

Treatment	Forest type	Emissions (kg CO_{2eq}/tonne)	Energy (MJ/tonne)
30% cut (selective)	Natural hardwoods	20.88	264.14
	Mixed hardwood/softwood	21.68	274.02
	Natural softwoods	22.88	288.91
	Softwood plantations	23.97	302.40
70% cut (shelterwood)	Natural hardwoods	19.24	243.88
	Mixed hardwood/softwood	21.68	274.02
	Natural softwoods	22.43	283.31
	Softwood plantations	24.29	306.38
Clearcutting	Natural hardwoods	21.12	267.15
	Mixed hardwood/softwood	22.10	279.25
	Natural softwoods	29.62	372.38
	Softwood plantations	23.72	299.29

4. Sensitivity analysis and discussion

A sensitivity analysis was conducted to explore the importance of input variables to harvesting cost determination, energy demand and GHG emissions estimation Productivity estimate for different harvesting equipment configurations is a key input for all three indicators. Additional key inputs to harvesting cost include machine purchase price, salvage value, economic life in years, scheduled hours per year, expected utilization rate, repair and maintenance cost, interest insurance and taxes, labor rate and fringe, fuel and lubricants, etc. Other key inputs to environmental impact include a series of intensity factors, such as GHG intensity of fossil fuel consumption, lubricants, machine construction, repair and maintenance. The input variables are increased or decreased by 10%. Harvesting cost, energy demand and GHG emissions were recalculated and compared to the base case.

4.1. Sensitivity analysis of harvesting cost

The sensitivity analysis was performed for the CTL harvesting system. The key inputs to harvesting cost were classified into three categories: system data (productivity and labor rate), data for CTL harvester, and data for forwarder. The analysis of results is illustrated in Figure 4. Note that harvest and forest types were not differentiated because the percent changes in harvesting cost are not affected by harvest types as well as forest types. Figure 4 shows that productivity had the largest impact on harvesting cost, while large machinery repair and maintenance had the smallest, as would be expected. Besides machinery purchase price, yearly scheduled hours and expected utilization (%) also had large impacts on the results.

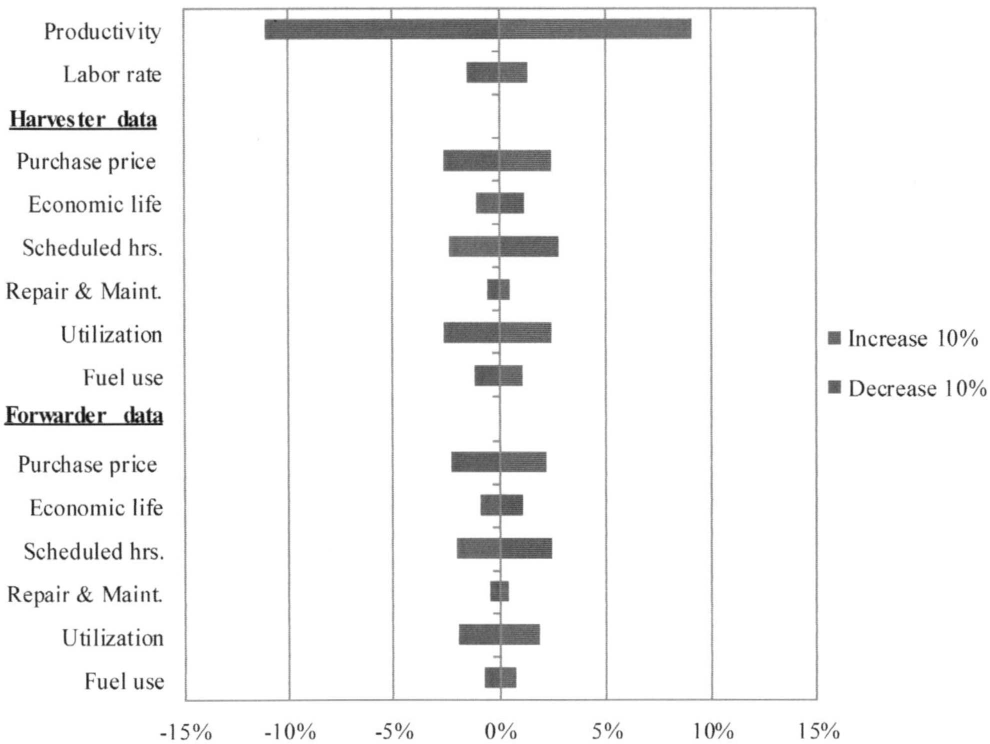

Figure 4. Sensitivity analysis for significant inputs to harvesting cost

4.2. Sensitivity Analysis of Environmental Impacts

The sensitivity analysis of environmental impacts was conducted for the CTL harvesting system. The input variables considered were productivity, fuel and chemical inputs, and the corresponding GHG emissions or energy demand intensity factors. The analysis results are illustrated in Figure 5. Note that harvest and forest types were not differentiated because the percent changes in energy demand or GHG emissions was not affected by these dimensions. Figure 5 highlights how productivity, fuel use as well as fuel factors had greater impacts on energy consumption and GHG emissions. Additionally, productivity had a positive impact while fuel use and its intensity factor had negative impacts. All other input variables (e.g. oil/lubes, grease, and large machinery repair and maintenance) had smaller impacts, as would be expected. Note that in Figure 5, fuels or chemicals use were placed in the same group as their corresponding intensity factors (e.g., fuel use and its emission factor, oil/lubes and its energy demand factor). This is because the two items are multiplied in the model for calculating energy demand or GHG emissions, and thus they have the same degree of impacts on the results.

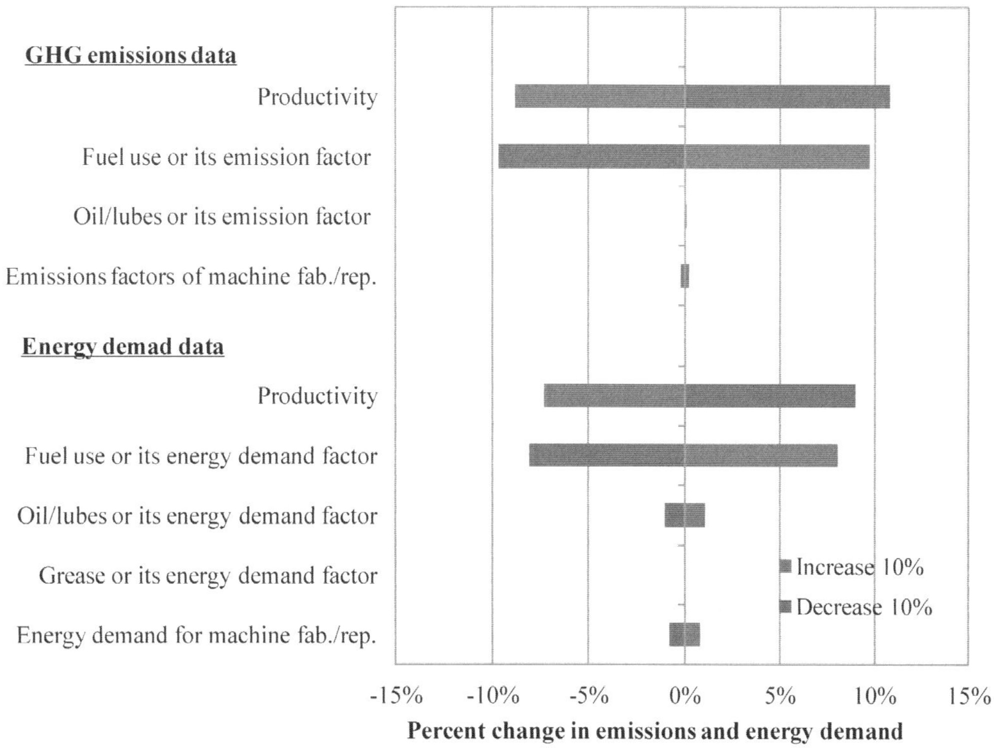

Figure 5. Sensitivity analysis for significant inputs to emissions and energy demand

5. Summary and conclusions

By collecting data from literature and database sources, an assessment of economic, energy demand and GHG emissions was performed for forest-based biomass harvesting in Michigan. The assessment differentiates harvesting systems (CTL, WTH, or MMH), harvest types (30%, 70%, or 100% cut), and forest types (hardwoods, softwoods, mixed or softwood plantations) that characterized Michigan's logging industry. It is believed that this is the first time explicit analysis of these three indicators for Michigan's harvesting activities has been conducted. The machine rate method was employed to determine harvesting cost. A life cycle inventory was applied to estimate energy demand and GHG emissions for different scenarios, considering energy and material inputs (diesel, machinery, etc.) and outputs (emissions) in each process (e.g., cutting, skidding/forwarding, slashing, etc.). Sensitivity analysis was performed for significant input variables to the harvesting operation to explore their importance. The input variables were increased or decreased by 10%. Harvesting cost, energy demand and GHG emissions and compared to the baseline. Several conclusions were drawn as follows:

- Harvesting cost for MMH system is higher than the more mechanized systems (CTL and WTH) across most forest and harvest types.
- For CTL and WTH mechanized systems, average harvesting cost decreased as removal intensity increased from 30% to 100%, with WTH slightly higher than CTL in most cases except for shelterwood cutting on natural hardwoods and clearcutting operations on natural hardwoods and mixed hardwood/softwood.
- Compared with more mechanized systems of CTL and WTH, the MMH system leads to higher GHG emissions and energy demand due to significantly lower productivities.
- The WTH system consumes more fossil fuels, resulting in high levels of environmental impacts.
- Productivity had the largest impact on harvesting cost, while large machinery repair and maintenance had the smallest. Besides, machinery purchase price, yearly scheduled hours and expected utilization (%) also had large impact on the results.

- Productivity, fuel use as well as fuel factors had huge impacts on energy consumption and GHG emissions, except that productivity had a positive impact while fuel use and its intensity factor had negative impacts. All other input variables (e.g. oil/lubes, grease, and large machinery repair and maintenance) had small impact.
- Percent changes in harvesting cost, energy demand and GHG emissions are regardless of harvest types as well as forest types.
- Fuels or chemicals use and their corresponding intensity factors (e.g., fuel use and its emission factor, oil/lubes and its energy demand factor) have the same degree of impacts on energy demand and GHG emissions.

This study is believed to be the first to conduct an explicit assessment of the three indicators (cost, energy and emissions) for Michigan's harvesting activities differentiating harvesting systems, harvest types, and forest types. Most previous LCAs of biofuels produced from various biomass feedstocks focused on environmental performance of producing and supplying feedstocks [7,10-14], without paying much attention on economic efficiency. Moreover, the estimation of harvesting/forwarding activities were based on the assumption of using cut-to-length harvester/forwarder only [12], or whole tree harvesting system only [13]. Other studies did not differentiate harvest types (assuming 100% clear-cutting) [12] or forest types for environmental impact assessment [14]. This research increases the scope of prior work and expands the body of knowledge in this area.

For future work, a life cycle cost, energy efficiency and GHG emissions would be performed for other stages of forest-based biomass supply chain, including the upstream of biomass transport and storage and the downstream of biofuel production and distribution. An integrated analysis of the entire supply chain may be also conducted.

Acknowledgments
This research was supported by National Natural Science Foundation of China (No. 71502173), the Program for New Century Excellent Talents in University (No. NCET-13-1028), and Science Foundation of China University of Petroleum, Beijing (No. 2462014YJRC040 and 2462014YJRC039). The Project-sponsored by SRF for ROCS, SEM.

References

[1] Kannan R, Leong K C, Osman R and Ho H K 2007 Life cycle energy, emissions and cost inventory of power generation technologies in Singapore *Renewable and Sustainable Energy Reviews* **11** 702-15

[2] Department of Energy. U.S. 2011 *Billion-Ton update: biomass supply for a bioenergy and bioproducts industry* (TN, U.S.: Oak Ridge National Laboratory)

[3] Van Deusen P C and Roesch F A 2008 Alternative definitions of growth and removals and implications for forest sustainability *Forestry* **81** 176-82

[4] Leefers L A and Vasievich J M 2010 *Timber Resources and Factors Affecting Timber Availability and Sustainability for Kinross, Michigan.* (East Lansing, MI: Michigan State University, Department of Forestry)

[5] Fennell K 2013 *Economic and energy efficiency assessment of biomass harvesting at a northern off-grid community: a case study of Barren Lands First Nation at Brochet, Manitoba, Canada* Natural Resources Institute Clayton H. Riddell Faculty of Environment, Earth, and Resources, University of Manitoba

[6] Abbas D, Handler R, Lautala P, Hartsough B, Dykstra D and Hembroff L 2014 A Survey analysis of forest harvesting and transportation operations in michigan *Croatian Journal of Forest Engineering* **35** 179-92

[7] Handler R M, Shonnard D R, Lautala P, Abbas D and Srivastava A 2014 Environmental impacts of roundwood supply chain options in Michigan: life-cycle assessment of harvest and transport stages *Journal of Cleaner Production* **76** 64-73

[8] Proto A R and Zimbalatti G 2016 Firewood cable extraction in the southern Mediterranean area of Italy *Forest Science and Technology* **12** 16-23
[9] Manavakun N 2014 *Harvesting operations in eucalyptus plantations in Thailand* (Helsinki: University of Helsinki, Faculty of Agriculture and Forestry, Department of Forest Sciences)
[10] Fan J, Kalnes T N, Alward M, Klinger J, Sadehvandi A and Shonnard D R 2011 Life cycle assessment of electricity generation using fast pyrolysis bio-oil *Renewable Energy* **36** 632-41
[11] Shonnard D R, Williams L and Kalnes T N 2010 Camelina-derived jet fuel and diesel: sustainable advanced biofuels *Environmental Progress & Sustainable Energy* **29** 382-92
[12] Zhang F, Handler R, Johnson D M, Shonnard D R 2011 *Comparative Analysis of Life Cycle Greenhouse Gas Emissions of Supply Chains for Biofuel and Fossil Fuel Production* (Reno, Nevada, USA: Proc. of the Production and Operations Management Society (POMS) 22nd Annual Conference)
[13] Prada M, Martínez-Alonso C, Sánchez-García S and Canga E 2015 Analysis of three forest chippers: productivity, costs and GHG emissions in Northern Spain *Journal of Cleaner Production* **101** 238-44
[14] Zhang F, Johnson D M and Wang J 2015 Life-Cycle energy and ghg emissions of forest biomass harvest and transport for biofuel production in Michigan *Energies* **8** 3258-71
[15] Michigan Forest Biofuels Research
Available from: http://michiganforestbiofuels.org/ (visited Feb. 20, 2015).
[16] Transportation Technology R&D Center. Argonne National Laboratory, GREET.net 2014. Available from: https://greet.es.anl.gov/
[17] Houghton J T, Ding Y, Griggs D J, Noguer M, van der Linden P J., Dai X., Maskell K., Johnson C.A. Climate Change 2001: The Scientific Basis. IPCC Third Assessment Report. Geneva, Switzerland: Intergovernmental Panel on Climate Change. 2001.
Available from: http://www.grida.no/climate/ipcc_tar/wg1/pdf/wg1_tar-front.pdf
[18] Tenyah T M 2009 *Environmental impact of different logging methods in the bankhead national forest, alabama: a comparative analysis* (Alabama : Alabama A&M University)
[19] Technology and Development at the USDA Forest Service.
Available from: http://www.fs.fed.us/t-d/programs/forest_mgmt/projects/textbook/cost/ (visited April 13, 2015).
[20] Brinker R W, Kinard J, Rummer R and Lanford B 2002 Machine rates for selected forest harvesting machines (Auburn Alabama: Alabama Experiment Station. Auburn University) 32
[21] Bilek E M (Ted) 2009 Machine cost analysis using the traditional machine-rate method and chargeout! (Lake Tahoe: Council on Forest Engineering (COFE) Conference Proceedings: Environmentally Sound Forest Operations)
[22] Bureau of Labor Statistics, U.S. Department of Labor, Occupational Outlook Handbook, 2014-15 Edition, Logging Workers,
on the Internet at http://www.bls.gov/ooh/farming-fishing-and-forestry/logging-workers.htm (visited May 02, 2015).
[23] Forestry Equipment Sales. Available from: https://forestryequipmentsales.com/
[24] Machinery Trader. Available from:
http://www.machinerytrader.com/list/list.aspx?Manu=CATERPILLAR&Mdltxt=501+HD
[25] Ponsse specialises in harvesting solutions for cut-to-length logging (CTL). 2013 Environmentally friendly and efficient cut-to-length logging *Wood Southern Africa & Timber Times* **38** 10-11.
[26] Parsakhoo A, Hosseini S A, Lotfalian M and Jalilvand H 2009 Efficiency and cost analysis of forestry machinery usage in Hyrcanian Forests of Iran *World Applied Sciences Journal* **6** 227-33
[27] LaHood R and Jackson L O 2017 later model year light-duty vehicle greenhouse gas emissions and corporate average fuel economy standards *Federal Register*
[28] Cavalli R, Zimbalatti G, Grigolato S and Proto A R 2008 *Forest wood chips supply chain in Southern Italy* Formec, Schmallenberg, Germany

[29] Akay A E 1998 Estimating machine rates and production for selected forest harvesting machines operating in the western united states and determining the most economical machine combinations under representative conditions in turkey (Oregon State : Oregon State University)

Study on load forecasting to data centers of high power density based on power usage effectiveness

C C Zhou[1], F Zhang[2], Z Yuan[1], L M Zhou[1], F M Wang[2], W Li[2] and J H Yang[1,3]

[1]College of Information and Electrical Engineering, China Agricultural University, Beijing 100083, China
[2]Beijing Electric Power Economic Research Institute Beijing 100055, China

Email: yang.haag@163.com

Abstract. There is usually considerable energy consumption in data centers. Load forecasting to data centers is in favor of formulating regional load density indexes and of great benefit to getting regional spatial load forecasting more accurately. The building structure and the other influential factors, i.e. equipment, geographic and climatic conditions, are considered for the data centers, and a method to forecast the load of the data centers based on power usage effectiveness is proposed. The cooling capacity of a data center and the index of the power usage effectiveness are used to forecast the power load of the data center in the method. The cooling capacity is obtained by calculating the heat load of the data center. The index is estimated using the group decision-making method of mixed language information. An example is given to prove the applicability and accuracy of this method.

1. Introduction

Currently, data centers are becoming more and more important and their load density is much more centralized than ever with the rapid innovation of Information Technology (IT) and the development trend of energy Internet [1]. The accuracy of load forecasting to the data centers has a significant impact on Spatial Load Forecasting (SLF). There are some methods of SLF [2-8], which can be divided into 4 groups in accordance with the principle of prediction [9]: local simulation method, load density index method, the method of multivariable variants, and trend method. The methods mostly focus on a wide range of power supply area and need a large amount of historical data or sample data. However, there are little methods to forecast the load of the data center for the power density being very concentrated.

A load forecasting method of the data center of high power density on the basis of Power Usage Effectiveness (PUE) is proposed using a combination of energy consumption indexes and group decision-making of mixed-language information for SLF in the paper. The load density index of the data center can be established according to the method with data center architectures, equipment condition, geographic and climatic factors.

2. PUE Index

[3] Address for correspondence: College of Information and Electrical Engineering, China Agricultural University, Beijing 100083, China. Email: yang.haag@163.com.

The concept of PUE can be indicated in Equation (1) [10], the lower a data center's PUE is, the better green environment performance it is.

$$PUE=P_{TF}/P_{IT} \tag{1}$$

where P_{TF} refers to Total Facility Power (TFP) and P_{IT} is IT Equipment Power.

TFP consists of IT Equipment Power and non-IT equipment power (P_{NIT}), and non-IT equipment is usually made up of infrastructures[11]. Therefore, Equation(1) can be changed into Equation (2).

$$PUE=1+P_{NIT}/P_{IT} \tag{2}$$

3. Thermal load calculation of data centers

The computer room at a data center often includes more than one thermal source, which are as follows[12]: 1) Thermal conduction of building envelopes; 2) Solar heat; 3) Thermal dissipation of IT and other equipment; 4) Heat and moisture dissipation of human bodies; 5) Thermal dissipation of illumination device; 6) Ventilation loads.

The requirements for air conditioning's capacity can be estimated after the above thermal loads are given. Aside from eliminating the heat in a data center, the air conditioning also controls humidity. The rise of humidity adds an extra heat load to the air conditioning, and then the refrigerating capacity need to be expanded. The expansion range of the refrigeration system is typically not more than 30%, 30% mainly refers to the systems containing a huge number of mixed gases in a data center, while the smaller range is mainly for the small systems to return gas through the pipeline.

3.1. Thermal conduction of building envelopes

There are 3 ways of thermal transfer: conduction, convection and radiation. Heat that goes into a data center by its building envelops such as roofs, walls and partitions has a lot to do with climatic factors like the season, time, locations and irradiation angle of the sun.

Heat conducted to the computer room by planar walls Q_C (kcal/h) is calculated as in Equation (3) when air temperatures inside and outside the door are stable.

$$Q_C = K_C F_C(t_{zp} - t_n) \tag{3}$$

where K_C is thermal conduction factor, kcal/(m^2•h•℃); F_C is the building envelopes area, m^2; t_{zp} is the temperature indoor, ℃; t_n is the temperature outdoor, ℃.

3.2. Solar heat going through windows

Some of the heat, Q_s (kcal/h), absorbed penetrated into from window glass becomes the thermal loads of the computer room because of convection, which can be calculated as in Equation (4).

$$Q_s = K_s F_s q_s \tag{4}$$

where K_s is the penetration factor of solar heat that depends on the types of the glass, and it goes between 0.36~0.4; F_s is the area of the glass, m^2; q_s is the level of solar heat, kcal/(m^2•h).

3.3. Heat of servers

Servers are the major equipment in a data center, and their power consumption can be equal to the maximum power consumption provided in the server brochures.

3.4. Heat and moisture dissipation of human bodies

Human bodies release heat, which contains moisture. And corresponding thermal loads include real and potential heat. When the indoor temperature is 24℃, the real thermal load is 70W, and the potential thermal load is 112W; or if the indoor temperature is 21℃, the real thermal load is 87W, and the potential thermal load is 94W. In both occasions, the overall thermal loads are near the same. Thermal loads from human bodies can't be overlooked if there are a lot of persons in the room.

3.5. Thermal dissipation of illumination device

Electric power consumed by the illumination device in the computer room transfers into light as well as heat. Light will eventually turn into heat after it is absorbed by the building and the equipment.

Consequently, light needs to be taken into account as a thermal load, too. The heat generated by the illumination device is generally calculated according to the floor area of the room.

3.6. Thermal loads brought in by outdoor air

The requirements of new air for a computer room in a data center are 40m^3/h for per person and air volumes required to maintain the positive pressure of the room. The thermal load of ventilation, Q_v (kcal/h), can be calculated as in Equation(5).

$$Q_v = G_v(h_{nw} - h_{in}) \tag{5}$$

where G_v refers to fresh air, kg/h, while per cubic meter of air is regarded 1.2 kg; h_{nw} and h_{in} are respectively outdoor and indoor air enthalpy.

4. Load Forecasting Method Based on PUE

4.1. Group decision-making of mixed language information

After the investigations to several data centers of a city in northern China, it is found that four main kinds of factors influence PUE of a data center such as geographical, structural, equipment and management factors. Every type of factors involves in several indexes, as shown in Figure 1.

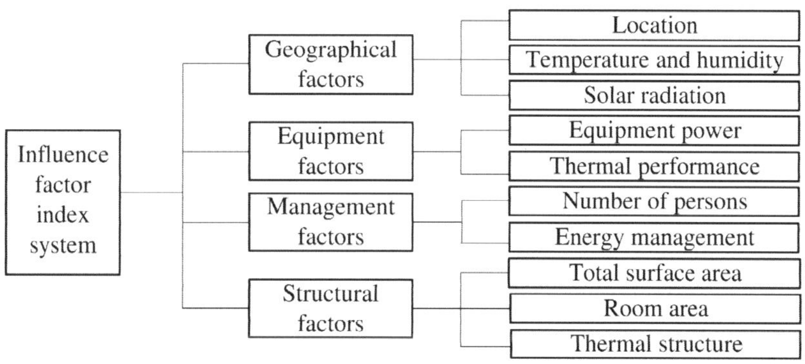

Figure 1. Influence factor index system of PUE

In Figure 1, some indexes are numerical such as solar radiation, equipment power, number of persons and so on, which are convenient for calculation. However, some indexes are qualitative such as location, energy management, thermal structure and so on. These indexes need to be quantized.

A zero-symmetrical assessment scale is used as shown in Equation (6) [13].

$$S_1 = \left\{ S_\alpha \middle| \alpha = -(\tau - 1), -\frac{2}{3}(\tau - 2), \cdots, \frac{2}{3}(\tau - 2), (\tau - 1) \right\} \tag{6}$$

where S_α stands for language term, and when its subscript is above 0, it can be calculated as in Equation (7).When subscript comes down 0, S_α can be calculated as in Equation (8).

$$S_1^+ = \left\{ S_\alpha \middle| \alpha = \frac{2(i-1)}{\tau + 2 - i}, \mathrm{i} = 2, \mathrm{L}, \tau - 1, \tau \right\} \tag{7}$$

$$S_1^- = \left\{ S_\alpha \middle| \alpha = -\frac{2(i-1)}{\tau + 2 - i}, \mathrm{i} = \tau, \tau - 1, \cdots, 2 \right\} \tag{8}$$

Besides, there are some calculation rules: 1) if $\alpha>\beta$, then $S_\alpha>S_\beta$. 2) $S_\alpha \oplus S_\beta = S_{\alpha+\beta}$. 3) $\lambda S_\alpha = S_{\lambda\alpha}$.

A specific situation in Figure 2 shows when τ stands for 4 where S_{-3} stands for "extremely bad", $S_{-4/3}$ stands for "very bad", $S_{-1/2}$ stands for "bad", S_0 stands for "not bad", $S_{1/2}$ stands for "good", $S_{4/3}$ stands for "very good", and S_3 stands for "extremely good".

It can be seen in Figure 2 that this language assessment scale is not balanced, and the closer to S_0, the more intensive scales.

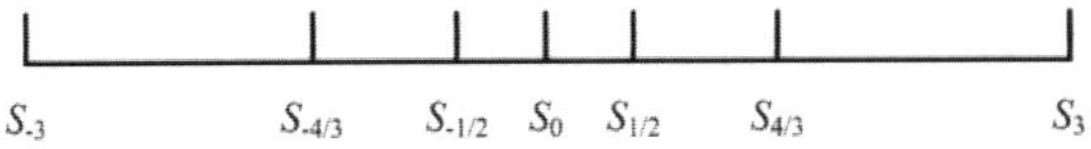

Figure 2. Language assessment scale of τ=4

There are two integrated operators, OWA operator and LHA operator. OWA operator is defined as in Equation (9).

$$\text{OWA}(S_{\alpha 1}, S_{\alpha 2}, \cdots, S_{\alpha n}) = v_1 S_{\beta 1} \oplus v_2 S_{\beta 2} \oplus \cdots \oplus v_n S_{\beta n} \tag{9}$$

where $\boldsymbol{v}=(v_1,v_2,\cdots,v_n)$ stands for weighted vector, and it satisfies that $v_i \geq 0$ (i=1, 2, $\cdots$, n), $\sum_{i=1}^{n} v_i = 1$; $S_{\beta j}$ (j=1,2,$\cdots$,n) stands for the jth number in language information; n stands for the number of attributes.

The essence of OWA operator is to arrange data from large to small, and the weight v_i reflects the importance of data position. The effects of subjective evaluations can be eliminated using OWA operator to assess language information, but the importance of data themselves may be ignored at the same time. Then LHA operator is defined as in Eq.(10) for this reason.

$$\text{LHA}(S_{\alpha 1}, S_{\alpha 2}, \cdots, S_{\alpha n}) = v_1 S_{\chi 1} \oplus v_2 S_{\chi 2} \oplus \cdots \oplus v_n S_{\chi n} \tag{10}$$

where $S_{\chi j}$ (j=1, 2, $\cdots$, n) stands for the jth number among ($nw_1 S_{\alpha 1}$, $nw_2 S_{\alpha 2}$, $\cdots$, $nw_n S_{\alpha n}$); $\boldsymbol{w}=(w_1,w_2,\cdots,w_n)$ stands for another weighted vector, which reflects the importance of data themselves and is independent of their location, for which it is called as the data weight.

4.2. Load forecasting of data centers

The process of load forecasting to a data center is as follows: 1) Data collecting; 2) To build up factor index system; 3) To calculate the two kinds of weight, $\boldsymbol{v}$ and $\boldsymbol{w}$, and then group decision-making of mixed language information is applied to PUE estimation [5]; 4) To calculate heat load of the data center; 5) To forecast the total power load of the data center according to PUE.

5. Case study

5.1. Case 1

A data center named A in northern China for example is taken to prove the applicability and accuracy of this method. Data center A mainly serves for science research and education, and the construction area is 17000 m^2, which includes 8500 m^2 of the server room. There are three experts who take part in the evaluation. The equipment factors are taken as an example to explain the detail process. The experts take the τ=4 language assessment scale. The results of the evaluation are shown in Table 1.

Table 1. Equipments factors type indexes evaluation of 3 experts

Expert	Equipment power capacity	Thermal performance	Function
1	S_3	S_0	S_3
2	S_3	$S_{1/2}$	S_3
3	S_3	$S_{4/3}$	S_3

Table 2. Final evaluation on the targets of equipment factors by 3 experts

Expert	Equipment factors
1	$S_{1.50}$
2	$S_{2.13}$
3	$S_{2.20}$

Using Equation (10) and Equation (11), the two kinds of weight can be calculated: $\boldsymbol{v}_1=\boldsymbol{v}_2=\boldsymbol{v}_3$= (0.25, 0.5, 0.25), $\boldsymbol{w}_1$= (0.3333, 0, 0.3333), $\boldsymbol{w}_2$= (0.4615, 0.0769, 0.4616), $\boldsymbol{w}_3$= (0.4091, 0.1818, 0.4091). Then LHA operator is used, the two weights and evaluation result in Table 2, the final evaluation of equipment factors comes out as shown in Table 2. As the process above, other evaluations can be assessed in this way. After getting all evaluations of each kind of factor, the PUE can be computed using OWA operator. PUE of data center A is 2.06.

The total heat load of data center A is 4111kW. When 30% margin is considered, the cooling capacity of data center A is 5344.3kW. According to PUE, the total power load of data center A is 10386.1kW. The actual maximum load of this data center is 10513.5kW on June 25th, 2015, which means the error is 1.21% and it can be accepted.

5.2. *Case 2*

Another data center named B is also in northern China, but it is an industrial data center whose scale is smaller than data center A. The construction area of data center B is 2500 m^2, which includes 1300 m^2 of the server room. Calculating process is just like that in subsection 5.1, and as a result, PUE of data center B is 2.58.

The total heat load of data center B is 847 kW. When 30% margin is considered, the cooling capacity of data center B is 1101 kW. According to PUE, the total power load of data center A is 1797kW. The actual maximum load of this data center is 1823.1 kW on July 13th, 2015, which means the error is 1.45% and it is acceptable.

5.3. *Comparison and Discussion*

Comparing the two cases above, it can be seen that two data centers, A and B, have similar climatic conditions but different scales. For one thing, the equipment in data center A is newer than B. Therefore the radiating performance of the equipment in A is much better. Moreover, the building structure of A is better designed than B, and the larger scale allows more scientific arrangement for cooling system in A. These features lead to the result that PUE of A is better. It can be easily calculated that load density in data center A is smaller than B. Besides the different functions, the development of the two data centers also contributes to this result. Data center A is a new one, which means it can contain more severs in the future and the power load of A may increase rapidly. On the other hand, data center B has operated for 8 years, and its power load has already been saturated.

6. Conclusion

A load forecasting method of data centers based on PUE is proposed. The scientific basis for power network planning can be provided by quantifying qualitative language information with geographical factors, structural factors, equipment factors, and so on. And the applicability and accuracy of this method is proved according to the application and the comparison of two cases.

Acknowledgments

This work is supported by Science and Technology Project of SGCC.

References

[1] Tian S M, Luan W P, Zhang D X, Liang C H and Sun Y J 2015 Technical forms and key technologies on energy internet *Proc. the CSEE* **35** 3483-94

[2] Yi G L and Zhang Y 2004 Land usage-based spatial load forecasting in distribution system *Electric Power Automation Equipment* **24** 20-3

[3] Heydt G, Khotanzad A and Farahbakhshian N 1981 A method for the forecasting of the probability density function of power system loads *IEEE Transactions on Power Apparatus and Systems* **100** 5002-10

[4] Xiao B, Yang X Y and Mu G 2014 A load density index method based on historical data of cell load *Power System Technology* **38** 1014-9

[5] Zhou S Y, Zhou R J, Li H Y, Kang X W 2014 Power load density prediction method of using group decision-making of mixed language information *Power System Protection and Control* **42** 15-22

[6] Liu Z F, Pang C C, Wei J W and He J T 2012 Index Calculation of Load Density Based on IAHP and TOPSIS Methods *Automation of Electric Power System* **36** 56 -60
[7] Gu C J, Yang D Z, Walsh W M and Reindl T 2014 Spatial load forecasting with communication failure using time-forward kriging *IEEE Transactions on Power System* **29** 2875-82
[8] Wang T H, Wang P Y and Fan M T 1998 Spatial load forecasting, gis platform and layout framework for distribution planning system *Power System Technology* **22** 6-10
[9] Xiao B, Mu G, Yang Y H, Yan H, Yang C J and Chen H W 2010 R&D of spatial load forecasting platform based on GIS *Electric Power* **43** 53-7
[10] Barroso L Z and Hoelzle U 2009 The DataCenter as a compute *Morgan and Claypool Publishers* 70-5
[11] Ma Y S 2012 Interpretation of PUE of Green Data Center *Building Electricity* **8** 54-8
[12] Zhang G M, Chen B and Zhang Y H 2012 Datacenter Infrastructure *Publishing House of Electronics Industry* 260-73
[13] Dai Y Q, Xu Z S, Li Y and Da Q L 2008 New evaluation scale of linguistic information and its application *Chinese J. Management Science* **16** 145-8

Analysis on influencing factors of EV charging station planning based on AHP

F Yan and X F Ma[1]

School of electrical and electronic engineering, North China Electric Power University, Baoding, Hebei 071003, China

Email: maxiaof888@163.com

Abstract. As a new means of transport, electric vehicle (EV) is of great significance to alleviate the energy crisis. EV charging station planning has a far-reaching significance for the development of EV industry. This paper analyzes the impact factors of EV charging station planning, and then uses the analytic hierarchy process (AHP) to carry on the further analysis to the influencing factors, finally it gets the weight of each influence factor, and provides the basis for the evaluation scheme of the planning of charging stations for EV.

1. Introduction

With the rapid progress of battery technology, EVs get more and more attention in recent years, and EVs have many positive effects on solving the present crisis of environment pollution and resources shortage.

At present, the development of the EV charging station is still in the demonstration and operational stage, the EV technology is increasingly developing, and the EV battery technology has improved greatly, with its unique advantages of high efficiency of charging and discharging, large output power and long service life, lithium-ion batteries are the most widely used in the electric car battery system now. However, the common charging time of EV batteries is too long (5 to 8 hours), and the mileage is limited after each charging, which is always the crux of the EV industry development [1]. EVs are not only a kind of load, but so can serve as energy storage for power grid, except that they have strong randomness, which is associated with the moving characteristic as a means of transport.

EVs can not only bring low carbon social economic prosperity, but also face many challenges, such as the social acceptability of EV launch in fossil-fuels-oriented energy economies, the cost of EVs: sales, long-life, and its decommission specifications/features-in the supplementary choices of recycling and reuse, etc. For the protection of EV battery systems, and lengthen the service life of the battery as far as possible, EVs should generally adopt the practice of conventional slow charging, in order to ensure the user for a long time and emergency driving demand, quick charge will be as a public service facilities and play an important complementary role in the road[2].

EV charging station planning problem directly determines the driving experience of EV users, and that also affect people's purchase intention to EV in a large extent, thus it affects the development of the popularity of EV and EV industry. The construction of the planning index system of EV charging

[1] Address for correspondence: X F Ma, School of electrical and electronic engineering, North China Electric Power University, Baoding, Hebei 071003. Email: maxiaof888@163.com.

station is very urgent, and it deeply affects the evaluation of the planning scheme of EV charging station [3].

The area of EV charging station planning has become a research focus, the literature [4] analyzed several factors that affecting electric vehicle charging station planning, and put forward the principle of its layout proposal, but it is short for quantitative analysis of the influence factors. Literature [5-7] put forward the planning method of EV charging station, but they are lack of systematic research for the influence factors that affecting the EV charging station planning. In conclusion, the related literature for systemic study on the influencing factors of EV charging station planning is less, and the related literature that considering the weight of every factor is much scarcer. This paper made a systemic in-depth research for the influencing factors of EV charging station planning with the application of AHP method, and studied their weights and provided the theoretical basis of the planning for EV charging stations in future.

2. Analysis on Influencing factors of EV charging station planning

The factors that affect the layout planning of EV can be summarized as the power system, transportation system, economic cost, user convenience and other factors [8].

2.1. Power system factors

Grid capacity factor is the main cause of capacity limit of the distribution network, and the charging current of general EV fast charging stations charger is large (up to 150A – 400A), and a charging stations is often equipped with more than one charger, so it is a huge load for the original grid, the grid capacity is one of the important factors that affect the layout planning of EV charging station.

Grid planning alignment refers to the blending degree of EV charging station planning and the network layout, especially the short to medium term planning of grid, can largely affect the EV charging stations within a period of time of planning, it can largely effect the EV charging station planning within a period of time[9].

In addition, a large number of EV that connected to distribution network will affect the power quality, reduce power supply reliability and cause a series of problems, so the consideration of these factors in the process of charging station layout of EV is also unavoidable.

2.2. Traffic system factors

Road congestion is an important factor restricting the layout of the charging station, and excellent EV charging station planning program should be selected that traffic congestion situation is relatively light zone.

Road network density affects the convenience of charging to a large extent, but also it can expand the scope of services to a certain extent.

2.3. Economic cost factors

EV charging station is an infrastructure construction. There is a large economic investment from the planning and construction to operation and maintenance, which also directly affect the choice of charging station planning. In this paper, the main economic cost factors are land cost, power grid reconstruction cost, operation and maintenance cost, etc.

2.4. Users' convenience factors

EV charging station is a kind of public service facilities, the users' charging convenience is particularly important, and this article mainly focuses on the service coverage and traffic convenience.

2.5. Other factors

The market share of EV industry is still very limited at present, and the present layout of the EV in the future will certainly experience reconstruction problems over a period of time, taking this factor into consideration within planning period will save costs for subsequent modification.

Safety is very important to the long-term safe operation of the charging station[10], and it is an important factor to influence the selection of the planning scheme of the EV charging station.

EV technology maturity is also one of the key factors that largely affecting EV charging station planning, especially with the constant breakthrough of EV battery technology, the electric vehicle charging station layout should also adapt to the requirement of the EV battery.

3. Model construction

Analytic hierarchy process method (AHP) was proposed by the operational research experts T L Satty in the end of the 70s of the 20th century. As a kind of multi-objective decision method, AHP with a systematic and hierarchical features has been widely applied, especially some multi-objective, multi-criteria and lack of structural characteristics of relatively complicated problems [11].

3.1. Index system

First of all, the AHP logic level construction was made for the influence factors of the EV charging station planning, and the index system is built.

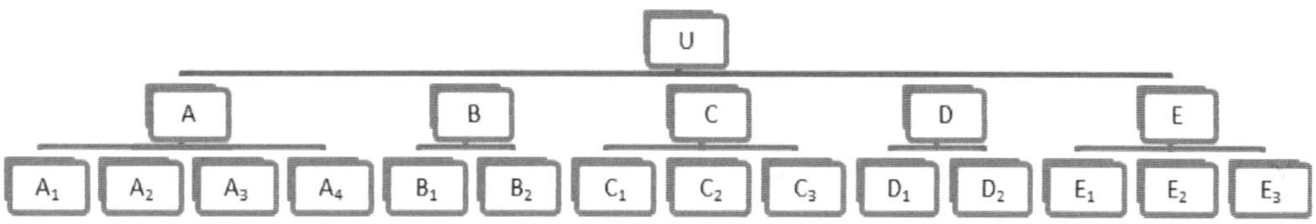

Figure 1. EV charging station planning impact factors analysis index system

Note: U-Influence factor analysis and others can be seen in table 4 and table 5.

3.2. Judgment matrix structure

After setting up the index system, a judgment matrix is constructed based on the index system, the values determined by contrast a hierarchy index important degree of influence on the upper indexes of every element in matrix[12], $b_{ij}=b_i/\ b_j$ said, the b_{ij} can use 0-9 said, see AHP relative importance of standard table.

Table 1. AHP relative importance of the standard

Scale	Meaning
1	b_i and b_j are equally important
3	b_i is slightly more important than b_j
5	b_i is more important than b_j significantly
7	b_i is more important than b_j strongly
9	b_i is extremely important than b_j
2、4、6、8	intermediate value of the judgment

3.3. Level sorting and consistency checking

After the establishment of the judgment matrix for each layer, the judgment matrix of each layer with the largest eigenvalue λmax and the corresponding eigenvector W is obtained, then the feature vector normalization, obtained in the same layer of each evaluation index compared with the rank weights of the upper one index importance. This process is single level sequencing. Random consistency index (CR) was used to measure the consistency of the decision maker's ideas.

$$CR = \frac{CI}{RI} \tag{1}$$

Where CR is the consistency ratio, CI is the consistency index of the judgment matrix, CI= (λmax-n) / (n-1), RI is the average immediately consistency index, and its value can be used to refer to the RI index table.

Table 2. Mean random consistency index

Order n	1	2	3	4	5	6	7
RI	0	0	0.52	0.89	1.12	1.26	1.36

For the matrix whose order number is 1 or 2, always has a complete consistency, therefore the RI value is 0, if matrix order number greater than 2, when the CR < 0.1 is considered the consistency of the judgment matrix meet the requirements, and it is in the reasonable scope, otherwise, it must be properly adjusted[13], and related formulas are as follows:

Each line element of the corresponding judgment matrix B is multiplied, and Mi is obtained:

$$M_i = \prod_{j=1}^{n} b_{ij} \qquad i = 1,2,3 \dots \dots n \tag{2}$$

Wi is Mi's n times square root obtained by calculation

$$w_i = \sqrt[n]{M_i} = \sqrt[n]{\prod_{j=1}^{n} b_{ij}} \tag{3}$$

Square root unit, we get the sort weight W_i:

$$W_i = \frac{w_i}{\sum_{i=1}^{n} w_i} \tag{4}$$

The largest eigenvalue of the B matrix is calculated:

$$\lambda_{max} = \frac{1}{n}\sum_{i=1}^{n} \frac{(BW)_i}{W_i} \tag{5}$$

In which, (BW)i means the first i element of vector BW.

4. A numerical example

The influence factors of EV charging station planning was analyzed by AHP, first of all, the judgment matrix for influence factors of sub reference layer is obtained. Hierarchical ordering and consistency checking were made according to the judgment matrix, finally we got the ranking of the reference layer and the corresponding weight of each layer. Results are shown in Table 3 shows.

Table 3. Sub reference layer factor weight table

Reference layer	Sub reference layer	Weight
A	A_1	0.439
	A_2	0.07
	A_3	0.23
	A_4	0.261
B	B_1	0.675
	B_2	0.325
C	C_1	0.594
	C_2	0.157
	C_3	0.249
D	D_1	0.683
	D_2	0.317
E	E_1	0.163
	E_2	0.54

E_3	0.297

In the same way, we calculated the influence factors and the weight of the reference layer, and the results are shown in the following table.

Table 4. Ranking factors of the reference layer

Target layer	Reference layer	Weight	Ranking	Factor name
	A	0.448	1	power system factors
	B	0.219	2	traffic system factors
U	C	0.109	3	economic cost factors
	D	0.083	4	user convenience factors
	E	0.141	5	Other factors

We got the overall order of the overall objectives for various factors combined with the above two parts, it can be seen that the grid capacity and road congestion are the most two important aspects. In addition in the selection of electric vehicle planning program, we should pay attention to grid capacity, road congestion, power supply reliability and power quality, etc. Rankings can be more intuitive to get from Figure 2.

Table 5. Weight table of influence factors

Index	Weight	Factor name
A_1	0.197	Grid capacity
A_2	0.031	Grid planning alignment
A_3	0.103	power quality
A_4	0.117	power supply reliability
B_1	0.148	Road congestion
B_2	0.071	Road network density
C_1	0.065	land cost
C_2	0.017	power grid reconstruction cost
C_3	0.027	operation and maintenance cost
D_1	0.057	service coverage
D_2	0.026	traffic convenience
E_1	0.023	Reconstruction and extension problems
E_2	0.076	Safety
E_3	0.042	EV technology maturity

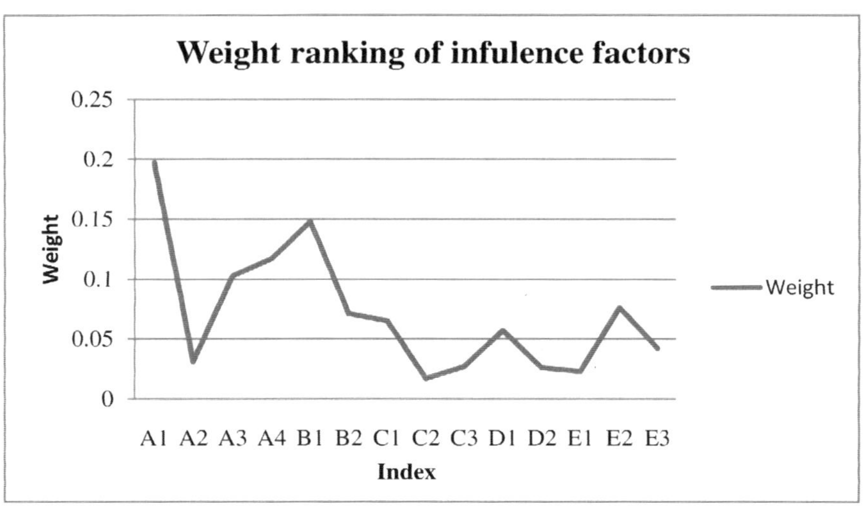

Figure 2. Weight ranking of influence factors

5. Conclusions

Reasonable EV charging station layout planning can provide electrical energy supplement better for users of EVs, and it is good for the safe operation of power grid, meanwhile it provides a strong guarantee for the smooth traffic network, it is of great significance for the popularization and promotion of EVs. There are many factors that affect the planning of EV charging station. In this paper, the main influencing factors are sorted by AHP, and the corresponding weights are obtained. It provides theoretical basis for the follow-up of the evaluation method of EV charging station.

References

[1] Gomez J C and Morcos M M 2003 Impact of EV battery chargers on the power quality of distribution systems *IEEE Trans. on Power Delivery* 18 975-81

[2] Mehdi E A, Kent C and Jason S 2010 Rapid charge electric vehicle stations *IEEE Trans. on Power Delivery* **25** 1883-7

[3] Wang J X, Cui Y L and Zhu Minghui 2014 Probabilistic harmonic calculation in distribution networks with electric vehicle charging stations *J. Applied Mathematics*

[4] Xu F, Yu G Q, Gu L F and Zhang H 2009 Tentative analysis of layout of electrical vehicle charging stations *East China Electric Power* **37** 1678-82

[5] Nakul S and Scott K 2013 A approach for the optimal planning of electric vehicle infrastructure for highway corridors *Transportation Research Part E* **59**15-33

[6] Ren Y L, Shi L F, Zhang Q, Han W J and Huang S J 2011 Optimal distribution and scale of charging stations for electric vehicles *Automation of Electric Power Systems* **35** 53-7

[7] Li R Q and Su H Y 2011 Optimal allocation of charging facilities for electric vehicles based on queuing theory *Automation of Electric Power Systems* **35** 58-61

[8] Ahn Y and Yeo H 2015 An analytical planning model to estimate the optimal density of charging stations for electric vehicles *Plos One* 1011

[9] Pan Z J and Zhang Y 2016 A novel centralized charging station planning strategy considering urban power network structure strength *Electric Power Systems Research*

[10] Li X P, Ma J Q, Cui J X, Ghiasi A and Zhou F 2016 Design framework of large-scale one-way electric vehicle sharing systems: A continuum approximation model *Transportation Research Part B*

[11] Karleuša B, Ožanić N and Deluka-Tibljaš A 2014 Improving decision making in defining priorities for implementation of irrigation plans using AHP methodology *Technical Gazette* 213

[12] Singh R P and Nachtnebel H P 2016 Analytical hierarchy process (AHP) application for reinforcement of hydropower strategy in Nepal *Renewable and Sustainable Energy Reviews* 55-8

[13] Riahi A and Moharrampour M 2016 Evaluation of strategic management in business with AHP case study: Pars house Appliance *Procedia Economics and Finance* 36-8

IOP Conf. Series: Earth and Environmental Science **40** (2016) 012055 doi:10.1088/1755-1315/40/1/012055

Objective criteria ranking framework for renewable energy policy decisions in Nigeria

Nwofor O K[1, 2, 4] and Dike V N[1, 2, 3]

[1]Atmospheric Physics Group, Department of Physics, Imo State University, PMB 2000, Owerri, Nigeria
[2]International Center for Climate and Environment Sciences, Institute of Atmospheric Physics, Chinese Academy of Science, Beijing 100029, China
3Centre for Renewable Energy Research and Development, Imo State Polytechnic Umuagwo, P.M.B 1472 Owerri, Nigeria

Email: okeynwofor@yahoo.com

Abstract. We present a framework that seeks to improve the objectivity of renewable energy policy decisions in Nigeria. It consists of expert ranking of resource abundance, resource efficiency and resource environmental comfort in the choice of renewable energy options for large scale power generation. The rankings are converted to a more objective function called Resource Appraisal Function (*RAF*) using dependence operators derived from logical relationships amongst the various criteria. The preferred option is that with the highest average *RAF* coupled with the least *RAF* variance. The method can be extended to more options, more criteria, and more opinions and can be adapted for similar decisions in education, environment and health sectors.

1. Introduction

Energy is very important in the economic development of Nigeria as well as in the country's diplomatic relations with other countries. Nigeria is currently witnessing a near energy crisis as deduced from current energy generation in the country. At present there are 11 power stations; 2 hydro-powers and 9 thermal, with total generating capacity of about 5,486 mW. There are also about 14 thermal stations under construction with projected capacity of about 4832mW [1]. This means that with a population of about 169,000,000 people, the energy *per capita* will only be $61x10^{-6}$mW when the new power projects are realized; which is abysmally low when compared to the present energy *per capita* in other large economies of Africa such as South Africa (energy *per capita* = $650x10^{-6}$mW) and Egypt (*per capita* = $262x10^{-6}$mW) [1].

The abysmally low public energy generation in Nigeria is the reason for the current over reliance of the populace on petroleum products largely for generator-based electricity production. This is coupled with the use of petroleum as the traditional energy resource for transportation. The domestic price of petroleum products has been increased recently by the government causing a lot of social tension in the country. The diminishing public and household access to energy from the non-sustainable

[4] Address for correspondence: Nwofor O K, Atmospheric Physics Group, Department of Physics, Imo State University, PMB 2000, Owerri, Nigeria. International Center for Climate and Environment Sciences, Institute of Atmospheric Physics, Chinese Academy of Science, Beijing 100029, China. Email: okeynwofor@yahoo.com

petroleum-based public energy supply system is affecting economic activities in the country drastically [2, 3, 4], and also contributes to dwindling revenue generation in the country as most people are not willing to pay for unavailable, inadequate or unstable power supply which is presently the case [5].

The precarious energy situation in Nigeria had precipitated several calls for serious diversification of the country's energy generation especially to the renewable options. Critical resource surveys show substantial renewable energy potentials across the country: A total of 278 unexplored sites for small hydropower plants that would each average between 2 mW and 10 mW when developed has been identified in the country; An average solar energy potential of about 5.5 $kWhm^{-2}day^{-1}$ is estimated for the country, while a total biomass abundance of over 56 tones with energy value put at about $7.1x10^5$ MJ is estimated [6]. Agbetuyi *et al.* [7], in a study of 22 stations across the entire country deduced an annual average wind energy of 5458 $kWhr^{-1}$ using the 10 meter blade rotor and 35,934 $kWhr^{-1}$ using the 25 meter blade rotor (also see Dike *et al.* [8] and Okeniyi *et al.*[9]).

In addition to the major renewable energy resources (i.e. hydro, solar, biomass and wind) which have received very elaborate evaluations in the country, some less fancied renewable energy sources such as geothermal and biogas from various sources, have equally been appraised for integration into the renewable energy mix. Amoo [10] for instance evaluated the energy potential of geothermal sites in Nigeria and concluded that although the geothermal fields were generally of low energy values; their cumulative energy value was potentially considerable with the right technology. In the case of biogas, an appraisal of land-fill gas (LFG) potentials of government-operated landfills in Lagos by Idehai and Akujieze [11], yielded a yearly energy resource of ~ 646,663.2 mWh, while Mohammed *et al.* [12], estimated in Lagos alone, a total renewable energy potential of 697.15TJ from crop residue, 455.80TJ from animal waste and 442mW from municipal waste. It has also been shown that the extensive aquatic macrophyte (*water hyacinth*) ; a plant that covers much of the water body in the Niger Delta area has huge renewable energy potential as a methane biogas resource [13]; the same applies to another plant known as *Nipa Palm* [14].

The reasonably large renewable energy resource potentials as considered above had served as basis for the articulation of a Renewable Energy Master Plan (REMP) by the Energy Commission of Nigeria (ECN) [15]. Actual development however is still within this policy stage due to lack of political will on the part of the government to do the needful [16]. Although several components which are expected to give boosts to policy implementation especially from the legal perspective have been suggested [17, 18], there are for sure, obvious environmental and economic challenges of executing the REMP in an extensive and all-inclusive form as can be deduced from the study by Okedu [19].

Government seeming inertia for large scale development of renewable energy in the country could then be persisting due partly to the obvious scarcity of technical resources to achieve elaborate diversification. This problem can largely be addressed by optimizing the logic of choice [20, 21], through the construction of objective frameworks that draw from both science and economics to enhance renewable energy development policy decisions.

From basic renewable energy economics [22], such an objective framework for renewable energy choices must be interfaced with the following major issues as previously noted [21], which in a more recent study, such as [23], are described in terms of ecological, economic and social sustainability indicators:

- Resource abundance or availability
- Financial and technical capacity
- Environmental concerns

2. The criteria framework

Following the above issues, the development of methodologies for energy decision framework by criteria ranking had been introduced [20, 21]. The present method is specific to renewable energy options and introduces more parameters geared towards raising the objectivity and reliability of the method.

The following underlying parameters and functions are required in the new framework:

2.1 Resource Appraisal Criteria (RAC)

The Resource Appraisal Criteria (*RAC*) are the key factors identified as the basis for making choices on the use of a renewable energy option for large scale energy power generation (or for small scale generation as the case may be) in a country. For clarity these are:

- Resource Abundance (represented by symbol; α) -referred to as the level of natural abundance of the renewable energy resource.
- Resource Economy (represented by symbol; β) -referred to as the cost efficiency of the resource energy conversion process i.e. energy output generated from the invested cost. This criterion is obviously determined by technological capacity.
- Resource Environmental Sustainability or Resource Environmental Comfort (represented by symbol; ϵ)-referred to as the aggregated environmental impacts of utilizing the resource for energy generation.

2.2 Resource Appraisal Parameter (RAP)

As previously proposed in Nwofor et al. [21], we define an objective parameter called Resource Appraisal Parameter (Ƞ), a function of the energy resource criteria (μ_i) i=1, 2, 3…*n criteria* as;

$$Ƞ = f(\mu_i) \quad (1)$$

For the three *RAC* considered in this work, we assign for Resource Abundance (μ_1 = α), for Resource Economy (μ_2= β) and for Resource Environmental Sustainability (Resource Comfort) (μ_1 = ϵ), Equation 1 can then be written as;

$$Ƞ = f(\alpha, \beta, \epsilon) \quad (2)$$

2.3 Criteria Weighting Parameters (CWP) and Aggregated Weighted Parameter (AWP)

RAP is essentially weighted in the various criteria with Criteria Weighting Parameters (*CWP*) of weights, $\omega_i, \ldots\ldots \omega_n$, The *CWP* is presented in the form of number ranking between 1- 5 as follows; Very Low ranking (VL) - 1, Moderately Low ranking (ML)-2, Average ranking (A) – 3, Moderately High ranking (MH) – 4, and Very High ranking (VH) – 5, specific to a given locality such as state, region or country reflecting expert perceptions, such that it's magnitude increases with expert opinion preferences as given from the weights assigned.

In addition we expect the appraisal parameter for an energy option such as solar, tide, biomass or nuclear to be additive for all criteria, hence by applying the weights, we have the Aggregated Weighted Parameter *(AWP)* for an option as;

$$Ƞi = \sum_i^n \omega_i \mu_i \quad (3)$$

Or in the present case as;

$$Ƞi = \omega_\alpha \alpha + \omega_\beta \beta + \omega_\epsilon \epsilon \quad (4)$$

2.4 Logical Dependence Operator (LDO)

The Logical Dependence Operator (*LDO*) denoted by Ω, is introduced to infuse objectivity to the *CWP*, by a simple process of "interdependence" mapping analogous to logical mapping, which in general terms, translates a temporal relation based on mutually causal dependencies [24]. The *LDO* allows for the synchronization of the expert-assigned weights to more global expectations thereby checking the inherent subjective nature of the preference ranking method.

One key outcome of the dependence mapping framework is the fact that although renewable energy resources are said to be environmental friendly, they are not exploited and utilized at zero impact to the environment. *LDO* is therefore derived by exploiting the logical interdependence of the different criteria (see figure 1).

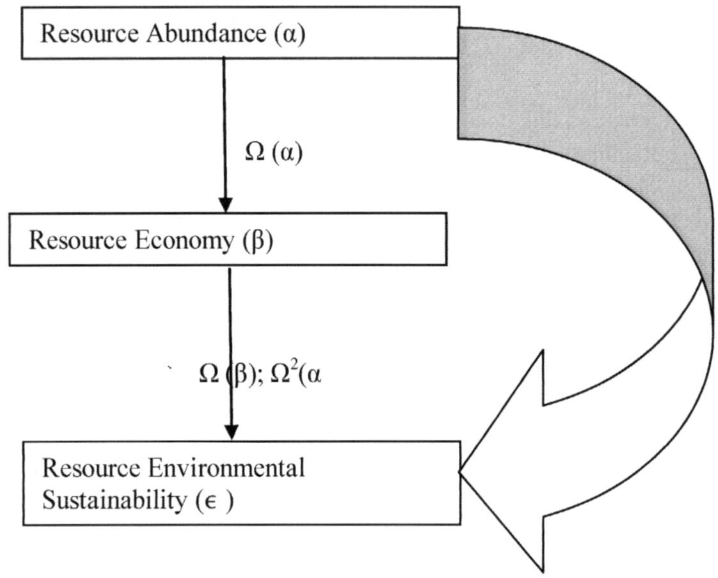

Figure 1. Schematic of the interdependence of the Resource Appraisal Criteria (*RAC*) and their operator mapping unto one another using Logical Dependence Operator *(LDO)*.

From figure 1, economy ~ efficiency (β) is directly derived from abundance (α), i.e.

$$\Omega\,(\alpha) \rightarrow \beta \qquad (5)$$

This is obvious from first principles since efficiency; $\beta = \frac{Energy\ Output}{Cost\ Input}$, and energy is a function of mass (abundance), β has a first order dependence on α; and resource abundance determines resource efficiency and not vice versa.

On the other hand, improved efficiency of energy utilization -from improved technology- reduces stress on the environment. Miniaturization for instance which these days comes with improved technology reduces the space used up for solar and wind installations. In the case of tidal power generation, improved technology may reduce coastal flooding, and would reduce geological failures in the case of geothermal reservoirs, ϵ therefore depends on β in the first order, i.e.;

$$\Omega\,(\beta) \rightarrow \epsilon \qquad (6)$$

And by using equation 5 in equation 6, ϵ depends on α in the second order, i.e.

$$\Omega^2(\alpha) \rightarrow \epsilon \qquad (7)$$

Although therefore we may see environmental friendliness as the basis for renewable energy development, abundance and economy of utilization (technical capacity) may be much more important criteria for accepting a given renewable energy resource.

2.5 Resource Appraisal Function (*RAF*)

Using ϵ as basis criteria, and assigning the highest power of the weighting parameter corresponding to the given criteria interdependence mapping, the transformation: *CWP*→ *LDO* will yield: $\omega_\alpha \rightarrow \Omega^2$, $\omega_\beta \rightarrow \Omega$, $\omega_\epsilon \rightarrow 1$

LDO converts the weighted Resource Appraisal Parameter (*RAP*) to a function, which we call the *Resource Appraisal Function* (*RAF*).

Hence equation 4 becomes:

$$\eta_f = \Omega^2 \alpha + \Omega \beta + \epsilon \tag{8}$$

3. Sample implementation of the framework

Figure 2 is a program implementation scheme of the proposed criteria ranking framework. It begins with identification of the resource appraisal criteria (*RAC*) for the problem. Then the initial ratings of experts are assimilated into a resource appraisal parameter (*RAP*). The operational flexibility of the method permits at these two stages the imputation of more options, ranking criteria and expert opinions. The criteria weighting parameters (*CWP*) i.e. weights are applied and the results summed up in the aggregated weighted parameters (*AWP*). Without any established logical interdependence of the criteria the *AWP* might be averaged at this stage for the number of opinions and used to inform decision; otherwise, the logical dependence operator (*LDO*) is derived and imputed to generate the resource appraisal functions (*RAF*) which are then used to arrive at the decisions.

Table 1 is a sample decision response table from three experts on each of the 4 renewable energy options being considered for Nigeria. The table shows the preferences in the opinions in the assigned weights (α, β and ϵ), the evaluated Resource Appraisal Parameter (η_i) the Average ($<\eta>$) and the Variance (VAR)$^{1/2}$) for the three opinions. The opinion with the highest series maximum for the *RAF* is the preferred option. The significance of the variance in the method is that it signifies the level of agreement in the opinions. Hence lower series variance accompanying large *RAF* is important.

In figure 2, the plots of the *RAF* derived using the operator function on *RAP* is shown for the same energy options. Fig 2 shows similar trends in the maximum value of the *RAP* and *RAF*. However, a slight modulation of the *RAP* is obvious in the plotted *RAF* reflecting a more universal logic in the appraisal. The variance can easily be evaluated from the plots.

4. Conclusion and outlook

The "Objective Criteria Ranking (*OCR*) Framework" has been introduced. The method aims at enhancing the objectivity of renewable energy decisions in Nigeria by ranking the important choice criteria for renewable energy options objectively. It basically employs a generating function called the *Resource Appraisal Function* with all the parameters and functions used for its determination fully explained and a sample graphics result using three opinions on four renewable energy options have been provided.

The framework introduced in this paper is inherently robust since it can be expanded for more energy options, more criteria, and more expert opinions and can also with suitable modifications be used for decision making in other sectors such as education, health and environment. This later possibility requires one to appreciate the tenets of the general decision problem, to which the presented methodology belongs.

A *Decision Problem* generally involves making of choices from a number of options or alternatives for the purpose of prompting action on a problem. It draws from the truism that whereas theoretically, the existence of a "best" decision might be argued, a "most objective" decision would not be in contention once the logic is established. One of the earliest formalizations of the Decision Problem is found in Participatory *Appraisal* which as a bottom-up framework [25, 26], utilizes community-based solutions to decision matters as had been used extensively in health as in [27], and in rural agricultural decision problems as in [28]. For decision problems in specialized areas such as education, health and the environment, expert opinion is however preferred and decisions can be reached by aggregating expert opinions ranking; analogous to having a team of doctors in an operating theatre as against just one [29, 30].

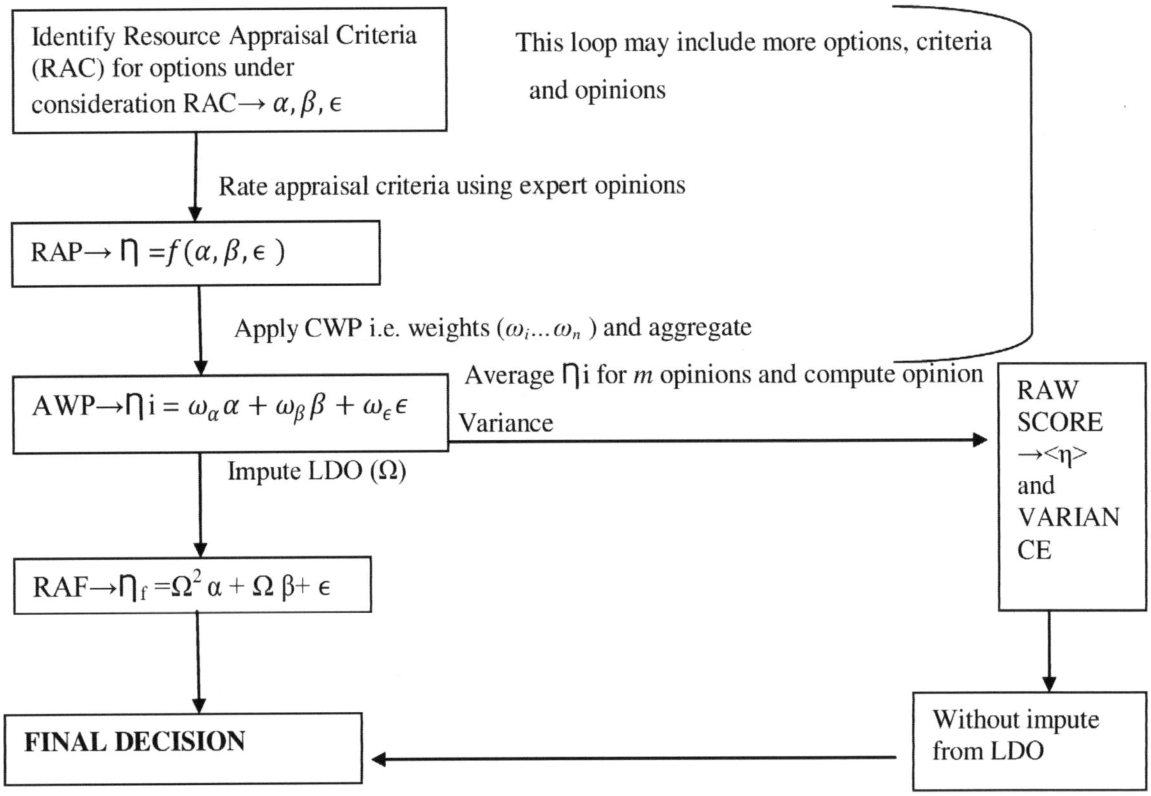

Figure 2. Program implementation scheme of the Objective Ranking Framework

Table 1. Decision Response Table from three experts for 4 renewable energy options in Nigeria (adapted from Nwofor *et al.* [21])

Resource Option	Expert Opinion	α	β	ϵ	η_i	$<\eta>$	$(VAR)^{1/2}$
Biomass	1	5	4	5	14		2.10
	2	4	1	3	8	10.3	1.32
	3	4	2	3	9		0.75
Hydro	1	3	4	4	11		0.23
	2	2	4	5	11	10.6	0.23
	3	3	3	4	10		0.34
Solar	1	5	3	5	13		0.80
	2	5	1	5	11	11.6	0.30
	3	4	2	5	11		0.30
Wind	1	2	1	3	6		0.80
	2	3	1	5	9	7.3	0.90
	3	3	2	2	7		0.20

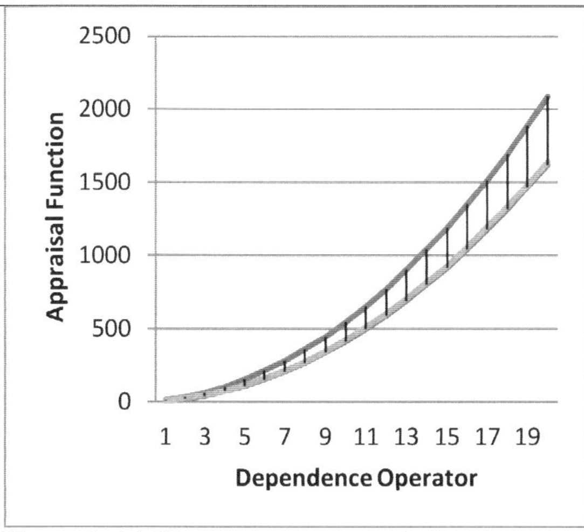

Figure 3. (a) Biomass resource option

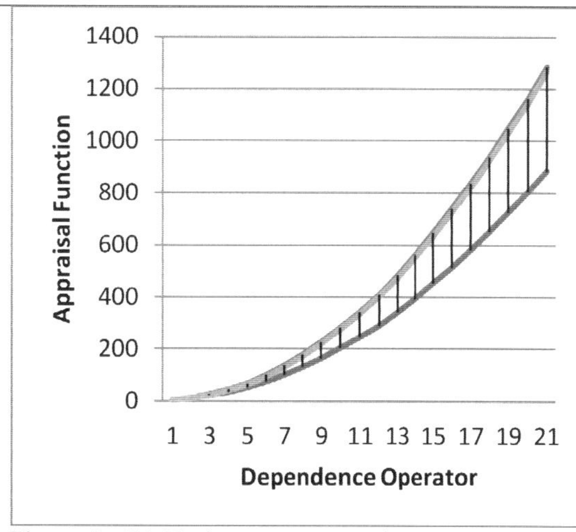

Figure 3. (b) Hydro resource option

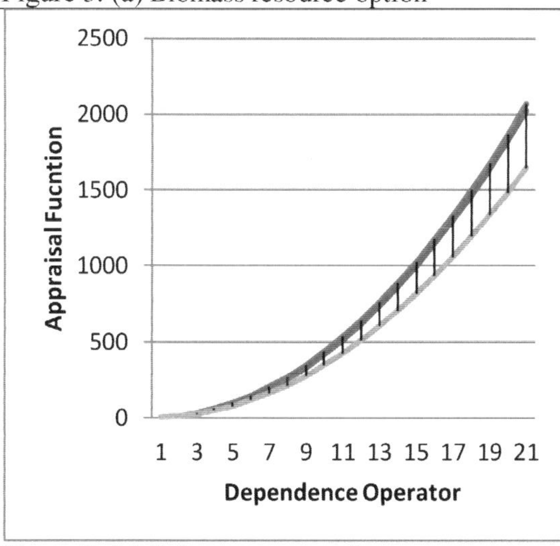

Figure 3. (c) Solar resource option

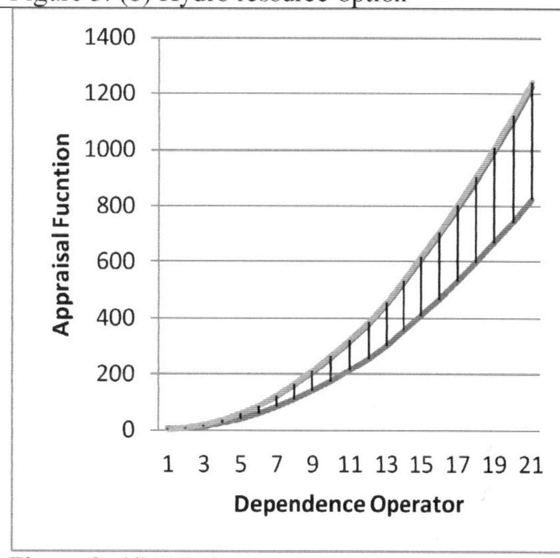

Figure 3. (d) Wind resource option

Figure 3. Graphics results of the implementation of the framework using the data of Table 1

Data Plots: blue-opinion series 1; red-opinion series 2; green- opinion series 3

The following possible applications of *OCR* in education, health and environment, may then suffice:

- Education: *OCR* can be applied in educational decision making in developing countries. In Nigeria, a good example exists in early childhood education in which Sooter [31], had identified access (availability of affordable education), content (curriculum provisions), and teacher competence as major issues. Given the scarce resources in the country, *OCR* can be usefully applied in appraising the three criteria for the purpose of prioritizing intervention in an objective manner.
- Health: The health sector like education is highly specialized and so many issues are decided on the basis of decision optimization. An example is found in Gnanlet and Gilland, [32], which discussed the optimal resource levels required to meet health care demands in a hospital

setting within sequential and simultaneous criteria paradigms- with very useful conclusions. In Nigeria where the issue of health financing has become highly vexed especially with regard to sourcing of required funds (see for instance, [33]), *OCR* will come handy in choosing whether to anchor funding on government, private enterprises or on external aid agencies. *OCR* might equally be adopted by the political class in choosing their target social group for health care insurance financing.

- Environment: With regard to environmental issues, *OCR* is easily applicable. One area of need in a country like Nigeria is waste management where the cost of effectively managing waste is far above the resources committed for the process as had been argued by Fadare and Afon [34]. There is a range of policy choices available to individuals, groups and governments with regard to waste management. These include disposal by incineration, landfill, or recycling. But the various waste disposal choices are commonly constrained by social conditions such as poverty. Hence the search for sustainability in the sector in the country should not just be based on environmental concerns alone but be linked with purchasing power as well [35, 36]. Decisions on waste disposal policy in various areas in the country can therefore be optimized using *OCR*.

Acknowledgement
The authors are grateful to the anonymous reviewer for very useful comments and for making available, recent related literatures that improved the work considerably.

References

[1] Akujor C E , Ogungbenro O A, Alozie G A and Nwofor O K 2008 A nuclear energy option for Nigeria *Advances in Science & Technology* **2** 71-7
[2] Ohunakin O S 2010 Energy utilization and renewable energy sources in Nigeria *Journal of Engineering and Applied Sciences* **5** 171-7
[3] Oseni M O 2012 Improving households' access to electricity and energy consumption pattern in Nigeria *Renewable and Sustainable Energy Reviews* **16** 3967-74
[4]. Shaaban M and Petinrin J O 2014 Renewable energy potentials in Nigeria: Meeting rural energy needs *Renewable and Sustainable Energy Reviews*, **29** 72-84
[5] Akinwale Y O , Ogundari I O , Ilevbare O E and Adepoju A O 2014 A descriptive analysis of public understanding and attitudes of renewable energy resources towards energy access and development in Nigeria *International Journal of Energy Economics and Policy* **4** 636-46
[6] Sambo A S 2009 Strategic developments in renewable energy in Nigeria *International Association for Energy Economics Third Quarter Publication* 15-9
[7] Agbetuyi A F , Akinbulire T O , Abdulkareem A and Awosope C O A 2012 *International Electrical Engineering Journal (IEEJ)* **3** 595-601
[8] Dike V N, Chineke T C, Nwofor O K and Okoro U K 2011 *Pacific Journal of Science and Technology* **12** 598 - 604
[9] Okeniyi J O, Ohunakin O S and Okeniyi E T 2015 *Scientific World Journal* Art. No. 581679
[10] Amoo O M A 2014 Thermodynamic based resource classification of renewable geothermal energy in Nigeria *Journal of Renewable and Sustainable Energy* **6** art. No. 033129
[11] Idehai I M and Akujieze C N 2015 Estimation of landfill gas and its renewable energy potential in Lagos, Nigeria *International Journal of Energy and Environmental Engineering* **6** 329-43
[12] Mohammed Y S, Mustafa M W, Bashir N and Mokhtar A S 2013 Renewable energy resources for distributed power generation in Nigeria: A review of the potential *Renewable and Sustainable Energy Reviews* **22** 257-68

[13] Adeleye B A , Adetunji A , and Bamgboye I 2013 Towards deriving renewable energy from aquatic macrophytes polluting water bodies in Niger delta region of Nigeria *Research Journal of Applied Sciences, Engineering and Technology* **5** 387-91
[14] Okugbo O T, Usunobun U, Adegbegi J A, and Okiemien CO 2012 A review of Nipa Palm as a renewable energy source in Nigeria *Research Journal of Applied Sciences, Engineering and Technology* **4** 2367-71
[15] Mas'ud A A , Vernyuy Wirba A , Muhammad-Sukki F, Mas'ud I A , Munir A B and Yunus N 2015 *Renewable and Sustainable Energy Reviews* **51** 775-84
[16] Chineke T C , Nwachukwu R , Nwofor O K , Ugboma E and Ndukwu O 2015 Much ado about little; renewable energy and policy *Materials, Methods & Technologies* **9** 123-31
[17] Oniemola P K 2015 *Journal of African Law* **27** (in press)
[18] Aliyu A S, Dada J O and Adam I K 2015 Current status and future prospects of renewable energy in Nigeria *Renewable and Sustainable Energy Reviews* **48** 336-46
[19] Okedu K E, Uhunmwangho R and Promise W 2015 Renewable energy in Nigeria: The challenges and opportunities in mountainous and riverine regions *International Journal of Renewable Energy Research* **5** 222-9
[20] Nwofor O K, Chineke T C and Akujor C E 2004 On the choice of energy options for Nigeria by preference ranking of selected criteria: Presentation at 27th annual conference of Nigerian Institute of Physics, Kaduna Polytechnic
[21] Nwofor O K , Chineke T C , Akujor C E and Okoro U O 2007 Energy diversification choices for Nigeria: Optimization by the criteria ranking method *Energy & Environment* **18** 577-89
[22] Timmons D , Harris J M and Roach B 2014 *The Economics of Renewable Energy, A GDAE Teaching Module on Social and Environmental Issues in Economics* (Medford, MA 02155: Global Development and Environment Institute, Tufts University)
[23] Abdullahi T 2015 Assessment of renewable energy sources for sustainable development in Nigeria *Jurnal Teknologi* **77** 9-16
[24] Pomares S E, Hernandez L A, Morales R, Jorge Ramirez J E and Gomez G R 2008 An intermedia synchronization model for multimedia distributed systems *Journal of Multimedia* **3** 33-41
[25] Chambers R 1992 Rural appraisal: rapid, relaxed and participatory *Discussion Paper* (East Sussex Brighton: Brighton Institute of Development Studies) **311**
[26] Chambers R 1994 The origins and practice of Participatory Rural Appraisal *World Development* **22** 953-69
[27] Alaszewski A and Crones M 1996 *Final report of a rapid appraisal study of primary health care* (Hull: The University of Hull, Institute of Health Studies)
[28] Cavestro L 2003 *Participatory Rural Appraisal: Concepts, Methodologies and Techniques*, Master in Cooperazione Allo Sviluppo Nelle Aree Rurali, Universita' Degli Studi di Padova Facolta' di Agraria, Dipartimento Territorio e Sistemi Agro-Forestali.
[29] Torres K , Sadovski A , Carl W and Steidley C W 2005 Program implementation of the rating methods of preference ranking *Revista de Matem´atica: Teor´ıa y Aplicaciones,* **12** 165–17
[30] Volkovs M N, Larochelle H and Zemel R S 2012 *Learning to rank by aggregating expert preferences CIKM'12* Maui, HI, USA
[31] Sooter T 2013 Early childhood education in Nigeria: issues and problems *Journal of Educational and Social Research* **3** 173-9
[32] Gnanlet A and Gilland W G 2009 Sequential and simultaneous decision making for optimizing health care resource flexibilities *Decision Sciences* **40** 295–326

[33] Uzochukwu B , Ughasoro M D, Etiaba E, Okwuosa C, Envuladu E and Onwujekwe O E 2015 Health care financing in Nigeria: implications for achieving universal health coverage *Niger J Clin Pract* **18** 437-44

[34] Fadare S O and Afon A O 2010 Waste Handling Practices at Abattoirs: Experience from Ile-Ife, Nigeria *The Lagos Journal of Environmental Studies* **7** 51-7

[35] Oyeniyi B A 2011 Waste management in contemporary Nigeria: the Abuja example *International Journal of Politics and Good Governance* **2** Quarter II

[36] Obidi A B and Adeoti A I 2015 Preference for improved solid waste management attributes among urban poor and non-poor households' in Delta-State, Nigeria *Asian Journal of Agricultural Extension, Economics & Sociology* **5** 46-56

Research on the competitiveness and development strategy of china's modern coal chemical industry

Q Wang[1, 3], Y J Han[2] and Z F Yu[2]

[1]China University of Mining & Technology, Beijing, China
[2]Shenhua Science and Technology Research Institute CO. LTD, Beijing, China

E-mail: wqchmc@sina.com

Abstract. China's modern coal chemical industry has grown into a certain scale after over a decade of development, and remarkable progress has been made in key technologies. But as oil price collapsed since 2015, the economic benefit of the industry also slumped, with loud controversies in China over the necessity of modern coal chemical industry. The research believes that the modern coal chemical industry plays a positive role in the clean and sustainable exploitation of coal in China. It makes profit when oil price is no lower than $60/bbl, and outperforms petrochemical in terms of cost effectiveness when the price is between $60/bbl and 80$/bbl. Given the low oil price and challenges posed by environmental protection and water restraints, we suggest that the state announce a guideline quickly, with adjusted tax policies and an encouragement to technological innovation, so that the modern coal chemical industry in China can grow sound and stable.

1. Introduction

A rich deposit of coal and a relatively small deposit of oil and gas are the characteristics of China's resource endowment. During the 11th Fifth-Year Plan and the 12th Fifth-Year Plan (2006-2015), as the global oil price remained high, the Chinese government started a series of modern coal chemical demonstration projects focusing on chemical products from energy commodities, with a view to ensure national energy security, accelerating the process of raw material diversification and fostering new growth areas. Currently a series of demonstration projects are already up and running, including coal to oil, coal to gas, coal to olefins and coal to ethylene glycol. Some of them are already in commercial operation, leading the global industry in terms of technological innovation and industrial production. The growth of the industry has become a major highlight of China's energy and chemical sector during the 12th Five-Year Plan (2010-2015).

Since 2015, global oil price collapsed while China's economic growth slowed down to a medium-to-high speed with less demand for energy and chemical products. This led to massive profit losses in China's modern coal chemical projects, as the heated industry cools down and pessimism rises up. Hence, it is crucial to understand the significance of the coal chemical industry, to realize its problem and analyze its economic benefits when determining the positioning and development path of the industry.

[3] Address for correspondence: Q Wang, China University of Mining & Technology, Beijing. E-mail: wqchmc@sina.com.

2. Progressing China's modern coal chemical industry

2.1. The Industry has formed scale

Modern coal chemical projects such as coal to oil, coal to olefins and coal to gas now have completed industrial procedure, with their scale continuing to grow and the industrial layout starting to form. The demonstration projects of coal to oil and coal to olefins are able to operate stably over a long term, manifesting superior technology and economic benefits to traditional coal chemical industries.

Table 1. Scale of China's modern coal chemical industry in 2015

Category	Coal Chemical Productivity	Coal Chemical Output	Petrochemical Output	Import	Apparent Consumption	Proportion of Coal Chemical Productivity
Refined Oil (1,000 t)	2630	1193	297141	4675	301816	0.54%
Natural Gas (10 million m3)	310	85	12077	5981	18059	1.72%
Olefins (1,000 t)	4680	1470	35790	27740	65000 equivalent	7.26%
Ethylene Glycol (1,000 t)	1500	580	3900	7870	12350	12.15%

(Investigated and calculated by Shenhua Science and Technology Research Institute)

2.2. Products not Easily Produced in the Petrochemical Industry are Obtained

Oil products of direct coal liquefaction has low condensation point, large specific gravity and high calorific value. It is suitable for military or aerospace use and is a special oil product with promising military prospects. In 2014, simulation testing and the first testing flight were held for coal-based jet fuel. In 2015, hot commissioning of missile diesel engine burning coal-based fuel was conducted, and diesel with a super-low condensation point was sent to cold areas like the South Pole and Mohe for trial use. Oil products of indirect coal liquefaction have a high cetane number (above 70) and could serve as a clean additive to oil. Once mixed with normal diesel, the additive could raise the grade of the oil (Guo V or Guo VI) and significantly cut pollutants in car exhausts, and it has already been used during the APEC summit in Beijing with good results. In addition, direct coal liquefaction and indirect coal liquefaction have a strong complementarity, forming a special category of coal-based oil products.

2.3. Constantly improving the level of clean and efficient utilization

Due to the characteristics of its raw materials, modern coal chemical projects consume relatively more resources and energy than the petrol approach. During the 12th Five-Year Plan", on the basis of complete procedures, the consumption of energy and water had been constantly reduced through technical and managerial improvement. (Figure 1, 2).

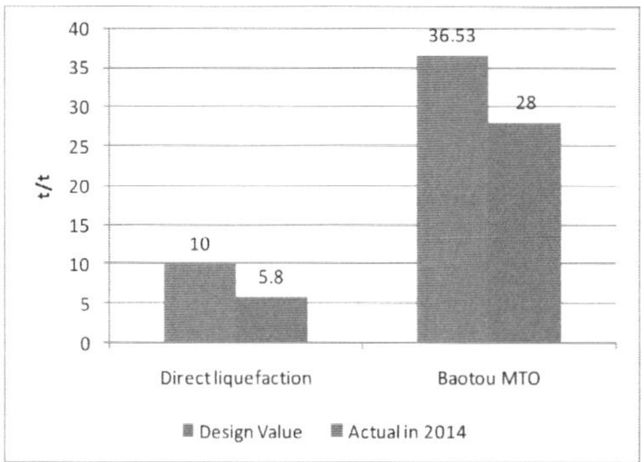

Figure 1. Consumption of water in coal chemical projects. (Provided by Shenhua Erdos and Shenhua Baotou CO. LTD.)

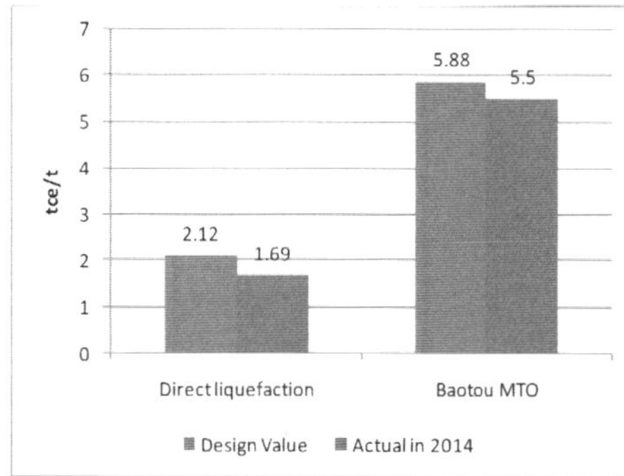

Figure 2. Consumption of energy in coal chemical projects. (Provided by Shenhua Erdos and Shenhua Baotou CO. LTD.)

3. Ten years of demonstration as a testimony to the necessity of developing modern coal chemical industry in china

3.1. Developing modern coal chemical industry to compensate for the lack of basic resources

China's rich-coal and poor-oil endowment results in a weak guarantee of oil and gas, unable to ensure sustainable supply, with obvious resource restraints (Table 2).

Table 2. Known reserves of major traditional primary energies (2014) [1]

Energy	Deposit	Global Share	Deposit-Production Ratio
Oil	18.5 billion bbl	1.1%	11.9
Gas	3.5 trillion m3	1.8%	25.7
Coal	114.5 billion t	12.8%	30

China's dependence on external oil and gas has increased over the years (Figure 3). 50% of imported oil is from the politically unstable Middle East, and 70% has to go through the Malacca Straights. Gas is principally imported from Asian countries with a high premium. Hence there is risk in the importation of oil and gas[2].

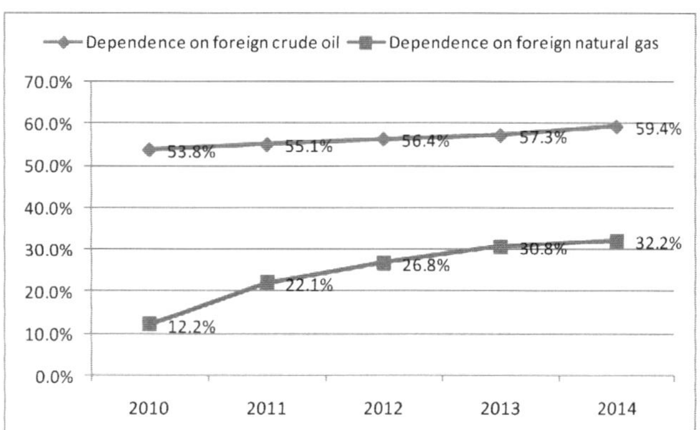

Figure 3. China's external dependence in oil and gas [2]

Modern Coal Chemical Industry led by chemical products from energy commodities aims at replacing oil and gas, directly or indirectly by coal, relatively rich in deposit in China. It is a practical and feasible energy development path that is in line with China's fundamentals.

3.2. Developing modern coal chemical industry for clean use of coal

Coal burning has a relatively low efficiency and creates high pollution. The real thermal efficiency of coal burning furnaces in China is 15% to 20% lower than that in advanced countries[3]. China's total emission of pollutants like SO_2 and NO_x, as well as CO_2 all rank the first in the world[4][5]. PM10 density is four to five times the WHO normal[6]. Among all the causes the burning of coal is certainly a principal one.

Modern Coal Chemistry turns complex and "dirty" raw material into high-quality and clean oil products and chemicals, and is an important approach to use coal both as energy and as resources. Based on the gasification of coal, modern coal chemical projects turn coal into gas by adding pure oxygen in a reducing environment, while the nitrogen largely becomes nitrogen gas and ammonia with virtually zero nitrogen oxides. Sulfur in coal is reduced to hydrogen sulfide, which then turns into sulfur or sulfuric acid through multiple-stage Claus reactions, recovering more than 99.9% of sulfur. Only some furnaces emit a small amount of sulfur dioxide, while the emission of air pollutants (SO_2, NO_x) in coal to oil plants is about one fifth to one tenth of that of traditional thermal plants. The process emits almost no dust or smoke with most of the dust captured in a liquid phase. During the process some carbon is fixed in the products while some turns into carbon dioxide. The waste air is dense with CO_2 (87%-99%) which is easy to capture and store. Most of the mercury in coal enters coal ash water, and the emission of mercury into the air is extremely low.

3.3. Developing modern coal chemical industry for the transformation of the coal industry

According to BP energy analysis, the share of China's coal consumption will decline from 68% in 2012 to 52% in 2035, with the actual consumption being largely equal to the present and coal remaining the dominant form of energy[7]. But due to the impact of shrinking downstream demand, China's coal industry has been declining for more than three years. By the end of 2015, coal production registered negative growth for the second year (Figure 4).

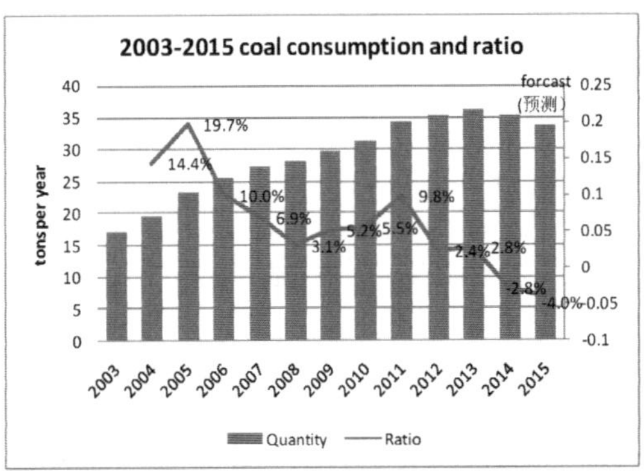

Figure 4. Dynamics of China's coal production (National Bureau of Statistics Department)

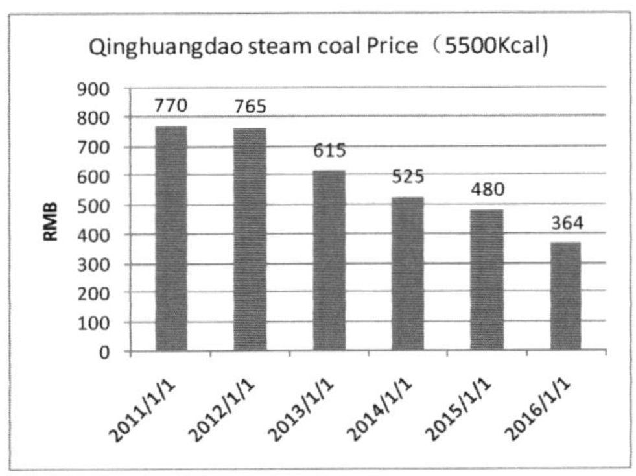

Figure 5. Dynamics of Qinghuangdao steam coal price (China Coal Resource Network)

Coal price has dropped significantly (Figure 5). 80% of the industry was in losses in the first ten months of 2015. The market condition has been so severe that it forces the industry to reform. Modern coal chemical projects are an effective way to extend the traditional industrial chain, with large investment, high value-added and high tax per ton of products. Developing modern coal chemical industry can effectively boost regional economic growth and help big producing provinces and coal companies improve and upgrade their structure. It has good economic benefits when oil price is moderate, and is a way of transformation and upgrading of coal companies.

3.4. Developing modern coal chemical to fill the structural shortage of the petrochemical industry

China has an oil refinery capacity of over 750 million tons with 200 million tons' glut, while ethylene, propylene, paraxylene and other chemical commodities are in short supply. In 2014, China relies on import for 51% of ethylene, 33% of propylene and 52% of paraxylene[8]. China became a net importer of naphtha[9] in 2009 with an increasingly sharp shortage, which amplified the lack of supply of commodities such as ethylene and propylene. By diversifying raw materials and approaches through modern coal-based chemical industry is an effective way of breaking the resources bottleneck and enhance the global competitiveness of primary products.

In addition, the improvement of oil quality also calls for diversified raw materials and approaches. With a higher content of alkenes and arenes, the current oil products in China do not compare well with Europe or the US. The upgrading of existing refineries is circumscribed by many factors[10][11][12], unable to fully realize oil upgrading to meet Euro VI standards.

Coal-based oil has a low content of sulfur, nitrogen, arenes and a low condensation point. It is a very clean fuel which can meaningfully reduce emissions of particle matters and promote high oil quality at a lower cost.

4. Economic competitiveness analysis of china's modern coal chemical industry

4.1. Breakeven analysis of modern coal chemical industry

The result of demonstration project operation suggests that when oil price is no lower than 60 dollars, modern coal chemical projects could generally stave off loses; when oil price is no less than 80 dollars, they could generally reach the benchmark earnings rate of the industry (11% pre-tax); if calculated using the lower limit for activating oil price adjustment set by the NDRC (40 dollars per barrel), the projects will suffer some losses. Below are the calculations.

Table 3. Economic competitiveness analysis of modern coal chemical projects

Item		Breakeven Point ($, Brent)	Benchmark Earnings Rate ($, Brent)	Losses Incurred at $40/bbl (Yuan)
Coal to Oil	DCL	55-60	75-80	500-700
	ICL	60-65	80-85	600-800
	Coal Tar Hydrogenation	60-65	75-80	600-800
Coal to Chemicals	Olefin	45-50	70-75	0-200
	Glycol	50-55	75-80	300-500
	Arenes	55-60	80-85	500-700

Note: On Jan 13, 2016, the NDRC of China finalized the bottom line of 40 dollars per barrel of oil, below which no further adjustment will be made to the price of oil products. (Calculated by Shenhua Science and Technology Research Institute CO. LTD.)

4.2. Competitiveness of modern coal chemical industry relative to petrochemical

In the area of 80 dollars per barrel, overall cost of coal oil would be lower than that of petroleum oil by 1300-1800 Yuan per tonne of products; profit margin would be higher by 1000-1300 Yuan per tonne under China's existing taxation system; and net profit would still be higher than that of petroleum oil by 400-700 Yuan per tonne. (Figure 6).

In the area of 60 dollars per barrel, overall cost of coal oil would be lower than that of petroleum oil by 200-600 Yuan per tonne of products; profit margin would be higher by 230-300 Yuan per tonne. Because coal oil pays more tax than petroleum oil, net profit of coal oil would be lower than that of petroleum oil by 130-310 Yuan per tonne. (Figure 7)

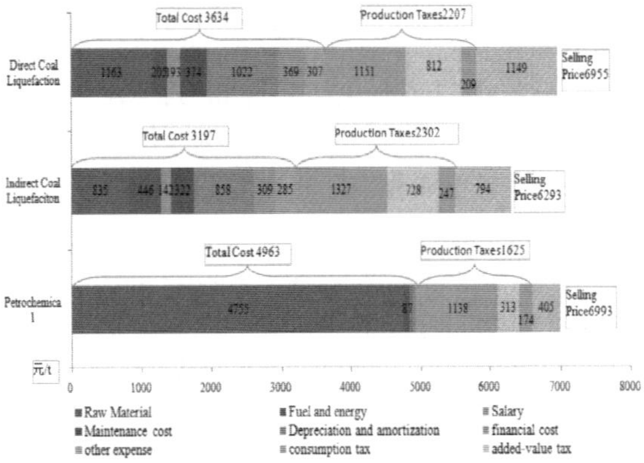

Figure 6. Coal to Oil vs. Petroleum Refinery in Economic Terms (at around \$80/bbl)

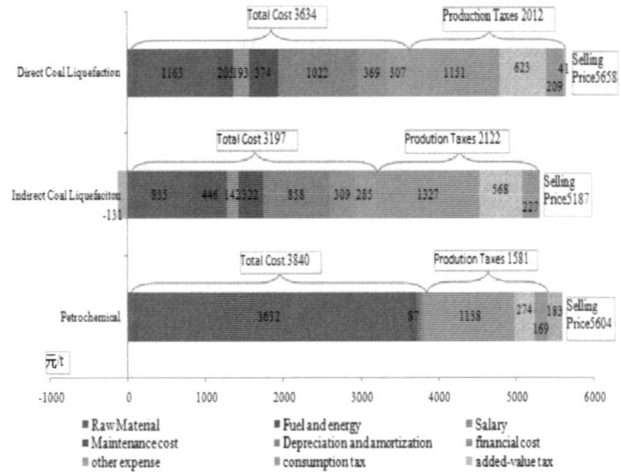

Figure 7. Coal to Oil vs. Petroleum Refinery in Economic Terms (at around \$60/bbl)

In the area of 80 dollars per barrel, overall cost of olefin production from coal would be lower than that from petroleum by 2,160 Yuan per tonne; profit margin higher by 2,000 Yuan per tonne; and net profit higher by 860 Yuan per tonne. (Figure 8)

In the area of 60 dollars per barrel, overall cost of olefin production from coal would be lower than that from petroleum by 260 Yuan per tonne. Because coal-derived olefins pay an extra 570 yuan per tonne in tax, net profit of olefin production form coal is lower than that from petroleum by 300 Yuan per tonne. (Figure 9)

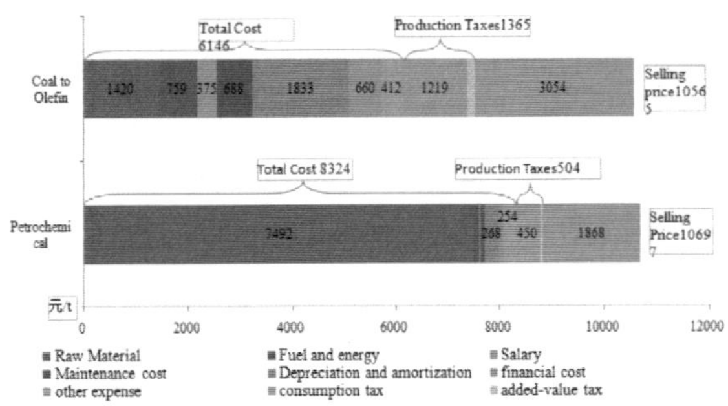

Figure 8. Coal vs. petroleum in production of olefins in economic terms(at around \$80/bbl)

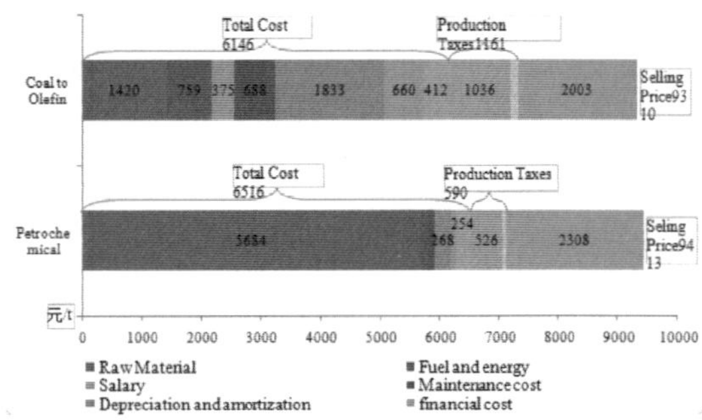

Figure 9. Coal vs. petroleum in production of olefins in economic terms(at around \$60/bbl)

4.3. Prediction of global oil price

Many factors influence global oil price, including global economic growth, balance of supply and demand of oil, oil reserves, exploitation of non-traditional oil and gas, development of new energy, balance of payments of oil producers and geopolitics. So far, many institutions have predicted the oil price during the 13th Five-Year Plan (2015-2020), as shown in Table 4.

Table 4. Estimates by different institutions of global oil price during the 13th five-year plan (\$/bbl)

Institution	Time of Announcement	2016-2017	2017-2020
IEA	December 2015	60-70	70-80
EIA	October 2015	53-62	80
BP	November 2015	<60	<60
Morgan Stanley	September 2015	45-65	75-85

In conclusion, oil price is likely to fluctuate at a low level during the 13th Five-Year Plan, which will probably move around 60 dollars per barrel in 2016-2017 and rise moderately to 70-80 dollars per barrel in 2017-2020.

5. China's modern coal chemical industry faces many challenges in development

5.1. Industrial competitiveness is affected by low oil prices and high taxes
The world economy is still in deep transition with deep uncertainty in growth. In 2015 Brent Oil fell to 57 dollars per barrel and below 40 dollars in the second half of the year. Coal to olefins fell below its breakeven point, and coal to oil and coal to ethylene glycol operate at losses. With basic energy price falling, the Middle-East and North American countries will move faster from oil and gas production to chemical export, adding to the risk of fluctuation in the energy and chemical markets and a fiercer competition. In addition, the overweight tax burden seriously impacts on production and operation of the enterprises (as shown in Table 5). At present, the consumption tax on diesel oil is up to ￥1411.20 / ton ($214.31), the consumption tax on naphtha is up to ￥2105.20 / ton ($319.70), the comprehensive tax burden is close to 50%.

Table 5. Effect of consumption tax on the economic benefit of coal to oil

Item	IRR（before tax）				
	40 $/bbl	50 $/bbl	60 $/bbl	70 $/bbl	80 $/bbl
Full amount	-4.0%	-0.5%	3.1%	9.5%	13.3%
Free of consumption tax	0.3%	4.2%	6.1%	12.4%	16.3%

Note: The benchmark IRR（before tax）of the project is calculated by 11%.

5.2. Controversies exist around environmental protection
China's Modern Coal Chemical Industry is largely located in the western regions, where ecology is fragile with prominent conflict between industrial development and environmental protection. Treatment of waste water is the key issue facing the industry. Many pilot projects are working hard on key technologies in waste water treatment and improving environmental facilities, but so far the treatment of waste water with a high concentration of salt still remains a problem, and the treatment cost of high-concentration organic waste is too high to make companies profitable.

5.3. Restraints of Water Resources Become Prominent
At present, China's modern coal chemical industry uses a huge amount of water, with 6 million tons of water use for million-tonne class oil projects, 20 million tons for 600 thousand-tonne olefin projects, and 10 million tons for 1.3 billion cubic-meter gas projects. The projects largely being located in Mid-West China where water resource is scarce and the ecology is weak further intensifies the restraints on the development of the industry.

6. Suggestions on the Development of China's Modern Coal Chemical Industry

6.1. Modern Coal Chemical Industry Urgently Needs a National Guideline
Over a decade, China's policy on modern chemical industry, while regulating the growth of the industry, had problems such as unclear planning, inconsistency, and multi-head administration. The underlying reason of the problems is that there has neither been a national guideline nor a strategic positioning for the industry.

I participated in the initial research of the *Plan on Deep Processing of Coal during the 13th Five Year for NEA* and I believe that the formulation and release of a plan on the development of modern coal chemical industry, which sheds light on the strategic positioning and development path for coal to oil, coal to gas and coal to chemical, will be conducive to gradually constructing a healthy, scientific and orderly environment in which the market takes the lead in allocating the resources, and will provide the necessary basis to competent authorities for drawing up specific industrial policy.

6.2. By exploring a reasonable tax system and reducing the cost to deal with low oil prices
First, The Coal-to-Oil Industry Needs an Appropriate Taxation Regime. Consumption tax levied with a reference to the petrochemical industry is not in accordance with the features of coal to oil, preventing the industry to grow in a sound and scientific way. The taxation structure at present aggravates the burden of companies and retards the growth and technological progress of the industry. Coal to oil industry should be allowed to enjoy differential low tax policies, namely reduced or exempted consumption tax for oil products and differential VAT and income tax support policies, so as to cushion the blow by oil price slump to the coal-to-oil industry and ensure sound growth. Subsequently, by optimizing the process design and utilities system, to further reduce the production cost. During the 13th Five Year period, the project breakeven point reduced 10 $/bbl compared to the current.

6.3. Industrial upgrading and demonstration are needed during the 13th five-year plan
The 13th Five-Year Plan (2016-2020) is a good window for solving the problems of competitiveness, environmental protection and water resource restraints in modern coal chemical industry through industrial upgrading and demonstration projects.

Processes need to be improved to integrate more technology. Ways to reduce cost and increase efficiency need to be explored. We should further lower project cost and increase competitiveness through systemic integration, process optimization and managerial improvement.

Water conservation should be strengthened in processes and systems. Water pinch technology and refined management of water systems are introduced to improve water balance in the whole system. We use coal mine drainage and recycled water first, separate different waste water in procedures, separate clean water from dirty water, and treat wastes separately and differently so as to maximize the use of water resources.

We should actively research and apply advanced processes and technology with a high energy transformation rate and low pollutant emission, ensuring that appropriate coal is used with specific technology and equipment, to improve response to irregular conditions and reduce pollutant emission at the source. Treatment of high-concentration organic waste water and recycling of different salts should be applied in demonstration projects to address the problems in waste water discharge.

Given the high concentration of carbon dioxide in gas emission from modern coal chemical industry, we should explore embracing IGFC advanced generation system, CCS and EOR, and engage in developing forest carbon sinks.

6.4 Appropriate demonstration projects can be started in eastern china
In China, energy chemical industry represented by petrochemicals is largely located in Eastern China (Figure 10), with a clear coastal layout[13].

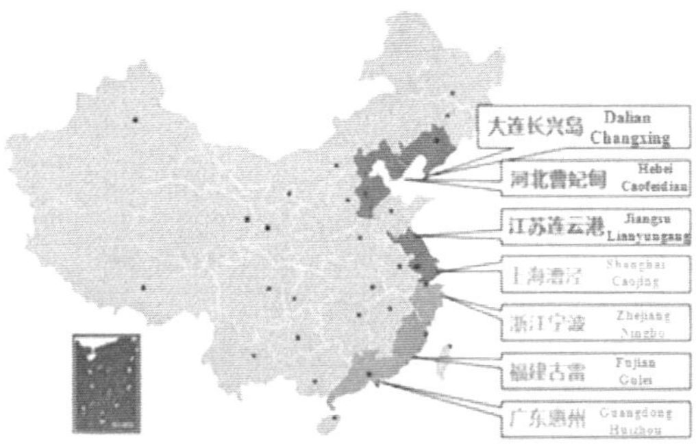

Figure 10. Seven petrochemical bases in China petro-chemical industry layout scheme

Figure 11. China's Modern Coal Chemical Bases *China Coal Chemical Industry 13th Five-Year Plan*

China's modern coal chemical industry is mainly concentrated in the mid-west (Figure 11), similar to the pattern of its coal resource distribution. During the 13th Five-Year Plan, we suggest that coal chemical demonstration projects could be tentatively started in eastern regions rich in oil and gas, water and chemical resources so as to circumvent the environmental and water restraints the industry facing in western regions.

Many problems facing the modern coal chemical industry are transitional, which could be gradually solved during the 13th Five-Year Plan through upgrading and demonstration.

References

[1] *Chinese Energy Statistical Yearbook* 2014 National Bureau of Statistics Department, China Statistics Press
[2] Tian Y J 2014 Opinions on the development of china's coal chemical industry—remarks drawn before the 13th five-year plan *Coal Chemical Industry* **6** 1-8
[3] Xu L L 2014 *Study on the international competitiveness of Chinese industrial boiler* (Kaifeng :Henan University) 1-2
[4] Gu J H 2012 *Analysis of the relationship between economic growth and sulfur dioxide emissions in China* (Changchun: Jilin University) 1-2
[5] Wu L J and Tian Q B 2016 Time trend and regional differences of carbon emission in China. *J. Finance and Economics for Shanxi University* **38** 25-35
[6] Chen J P and Li Z J 2013 The situation and problems of air pollution control in china and some policy suggestions *The Research on the Development* **10** 4-14
[7] BP *BP 2035 World Energy Outlook* 2014 (London: BP) 68-71
[8] China Petroleum and Chemical Industry Federation 2015 China Petroleum and Chemical Industry Economic Data Express Import and Export Volumes (Beijing: China Petroleum and Chemical Industry Federation)
[9] Luan H Y and Bai X S 2011 *Word Naphtha Market Analysis and Its Price Valuation*. Chemical Industry **29** 22-9
[10] Fang Z J, Ji C W and Wang S F 2008 Current situation and development trend of gasoline additives *Small Internal Combustion Engine and Motorcycle* **37** 91-4
[11] Ren Z M, Mu H Y and Li H 2015 The domestic oil upgrading technology comparison *Industrial Production* **7** 65-6
[12] Li Y and Li D F 2008 Technical progress of fcc light gasoline etherification process *Petrochemical Technology* **37** 528-33
[13] Zhang W Z, Dong K G, Tian S C 2009 Petrochemical industrial development and location changes in China *Geographical Research* **28** 1378-88

A maximum power point tracking algorithm for buoy-rope-drum wave energy converters

J Q Wang[1], X C Zhang[2], Y Zhou[1], Z C Cui[1] and L S Zhu[1, 3]

[1]Research Center of Mechanics and Mechatronic Equipment, Shandong University, Weihai 264209, China
[2]The center of Marine environment monitoring, Weihai 264209, China

E-mail: 18766317216@163.com

Abstract. The maximum power point tracking control is the key link to improve the energy conversion efficiency of wave energy converters (WEC). This paper presents a novel variable step size Perturb and Observe maximum power point tracking algorithm with a power classification standard for control of a buoy-rope-drum WEC. The algorithm and simulation model of the buoy-rope-drum WEC are presented in details, as well as simulation experiment results. The results show that the algorithm tracks the maximum power point of the WEC fast and accurately.

1. Introduction

Ocean contains abundant wave energy which has a promising prospect to exploit and utilize as a clean renewable energy [1-4]. In order to maximize the energy conversion efficiency of the buoy-rope-drum wave energy converters (WEC), using the maximum power point tracking (MPPT) control strategy is essential [5, 6].

There are many kinds of methods of MPPT control strategy including Perturb and Observe (P&O), Incremental Conductance, Fractional Open-Circuit Voltage, Fuzzy Logic Control, etc. Though the widely applied P&O method has the advantages of having relatively simple structure, less measured parameters and being easy to implement [7], it has the defect of being unable to satisfy the fast tracking speed of the MPPT and the high tracking precision at the same time. What is more, the wave energy is extremely unstable and the wave is nonlinear and stochastic, which makes it harder to track the maximum power point (MPP) of the WEC effectively. In order to overcome these disadvantages and maximize the energy conversion efficiency of the WEC, a novel variable step size P&O MPPT algorithm is presented. Compared with other P&O MPPT algorithms, the new one has a power classification standard and a variable step size standard. The results of simulation experiments show that the variable step size P&O MPPT algorithm tracks the MPP fast and accurately.

2. Buoy-rope-drum WEC simulation model overview

For the sake of testing the performance of the variable step size P&O MPPT algorithm presented in this paper, the buoy-rope-drum WEC simulation model using this algorithm is built in

[3] Address for correspondence: L S Zhu, Research Center of Mechanics and Mechatronic Equipment, Shandong University, Weihai 264209, China. E-mail: 18766317216@163.com.

MATLAB/Simulink which is shown in figure 1. This model mainly contains the following four modules: buoy hydrodynamic force response module, electric generator module, DC-DC boost converter module and MPPT algorithm module.

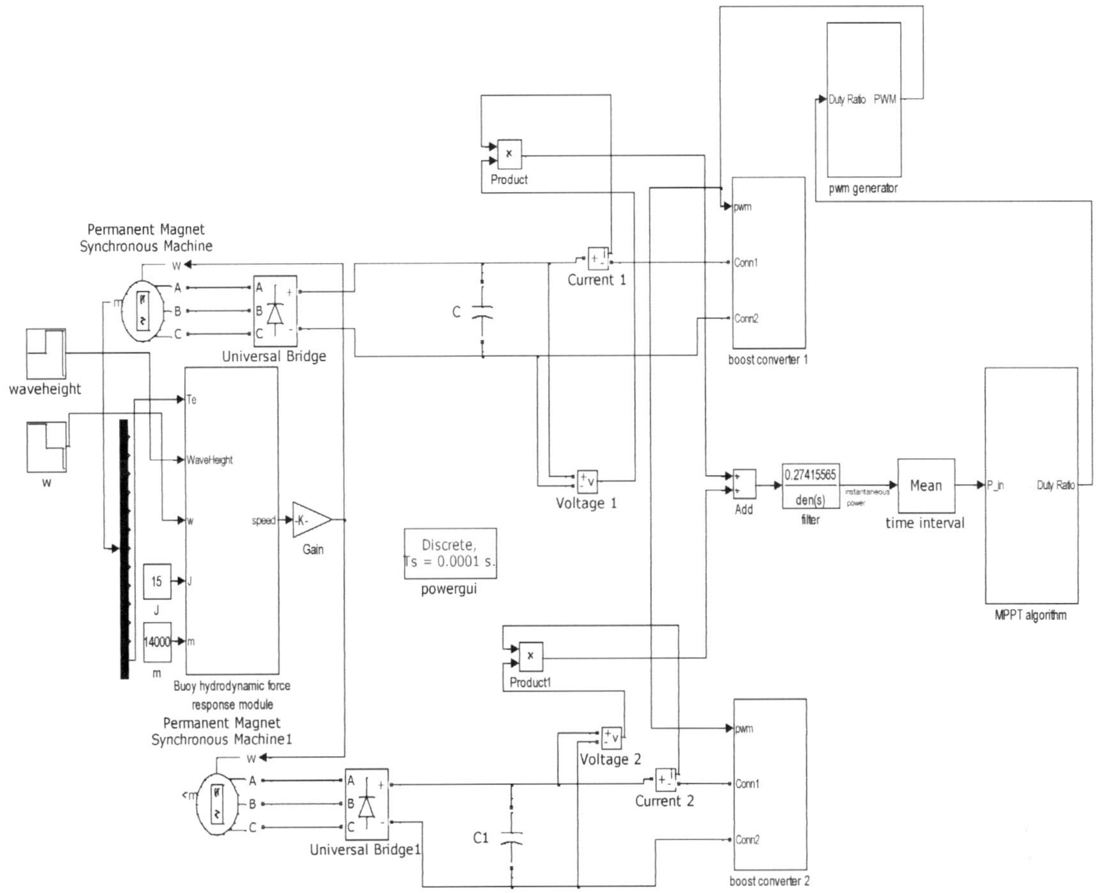

Figure 1. Simulink model of buoy-rope-drum wave energy converters

2.1. Buoy hydrodynamic force response module

In [8], authors derive the hydrodynamic force response equation of the WEC. In order to build the module conveniently and get the motion state of the buoy accurately, the equation is modified as

$$\ddot{x}+\frac{C_1}{m+\frac{J}{r^2}}\dot{x}+\frac{\gamma A}{m+\frac{J}{r^2}}x=\frac{\frac{H}{2}e^{-kD}\sqrt{(\gamma A)^2+(C_1\omega)^2}}{m+\frac{J}{r^2}}\cos(\omega t+\sigma)-\frac{T_e}{mr+\frac{J}{r}} \tag{1}$$

Where x is the displacement of the buoy, C_1 is the linear friction coefficient of sea water, m is the mass of the WEC, J is the rotational inertia of the drum, r is the radius of the drum, γ is the seawater specific gravity, A is the cross-sectional area of the buoy, H is the significant wave height, k is the wave number, D is the static draft depth of the floating body, ω is the wave radian frequency and T_e is the electromagnetic torque of the generator.

In real ocean environment, the buoy is affected by the loads of wave, wind and tide. Shandong University tested a buoy-rope-drum WEC in ocean whose buoy diameter is 4.5 m and height is 2 m. In [9], the loads of wave, wind and tide imposed on the WEC tested by Shandong University are analyzed and calculated. The results show that because the size of buoy is relatively small, the values of wind load and tidal load are small leading to little effects on the motion of the buoy. In order to simplify the mathematical model of the buoy motion, equation (1) ignores the effect of loads of wind and tide.

The buoy hydrodynamic force response module is built in MATLAB/Simulink according to the equation above. To accurately simulate the running state of the buoy-rope-drum WEC tested in the ocean before, the parameters of the buoy are set the same as the Shandong University's WEC.

2.2. Electric generator module

In this module, the permanent magnet synchronous generator (PMSG) block whose parameters are as same as the engineering prototype is selected. Because there are two generators in the WEC engineering prototype, two PMSG models are used in the module.

2.3. DC-DC boost converter module

Boost converters have the function of changing the impedance across the dc-link as given in

$$R_{eq} = (1 - D)^2 \cdot R \tag{2}$$

Where D is the boost converter duty ratio and R is the load resistance at the output [10]. In order to realize load matching when the WEC runs in the ocean, the variable step size P&O MPPT algorithm can be used to manipulate the boost converter duty ratio to maximize the output power. The DC-DC boost converter module used in this paper is shown in figure 2.

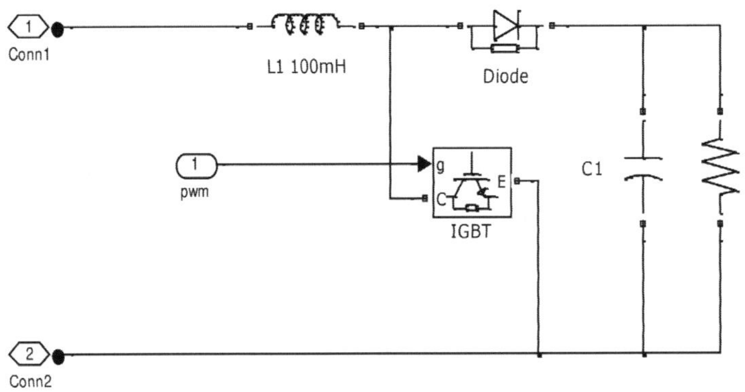

Figure 2. Simulink model of boost converter module

2.4. MPPT algorithm module

The variable step size P&O MPPT algorithm presented in this paper is written in the Embedded MATLAB Function block of Simulink.

3. Variable step size P&O MPPT algorithm

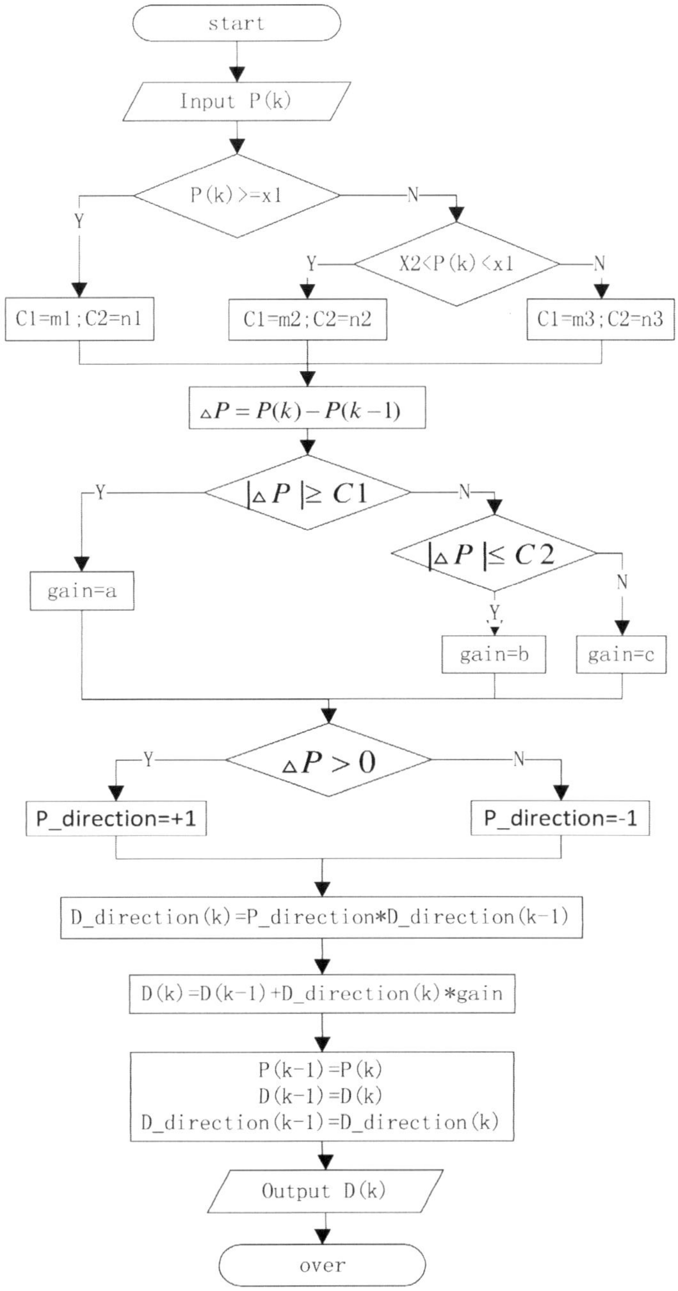

Figure 3. The flowchart of the variable step size P&O MPPT algorithm

Figure 3 shows the flow chart of the variable step size P&O MPPT algorithm. First, the algorithm divides the current average input power (i.e., output power of the WEC), P(k), into different grades which determine the corresponding variable step size criterion, C1, C2 (C1>C2). Second, the algorithm calculates the difference, ΔP, between P(k) and the average input power from the previous time interval, P(k-1). If |ΔP| is greater than or equal to C1, then the step size, *gain*, is set to a. If |ΔP| is less than or equal to C1, then the step size is set to b, otherwise *gain* is set to c, where $a > c > b$. If ΔP is positive, P_direction is set to +1. The assumption is that the direction of change in the boost converter duty ratio in last time interval, D_direction (k-1), has increased the power production from the WEC and therefore is left unchanged. On the contrary, if ΔP is negative, P_direction is set to -1, which means the direction of change in the boost converter duty ratio has decreased the power production and will change sign. The product of P_direction and D_direction (k-1) produces the direction of change in the boost converter duty ratio, D_direction (k). The output duty ratio, D (k), is then increased or decreased according to the product of $gain$ and D_direction (k). Executing the algorithm repeatedly will realize maximum power point tracking.

In the algorithm, the interval time, T, the variable step size criterion, C1, C2, and the step size, $gain$, are the key parameters which affect the performance of the algorithm significantly. Because the buoy-rope-drum WEC has the characteristics of having large inertia and only producing electric during the upward stroke of the buoy, the interval time, T, over which the input power is averaged and control updates are commanded need to be long enough to wait for the WEC getting into steady state. If T is too short, the algorithm will tune the boost converter duty ratio before the WEC is steady, which will cause misjudgment making it impossible to track the MPP. Wave energy is proportional to the product of the wave height squared and the wave period [11] and wave state is significantly different at different times, which makes the energy carried by different waves very different. If the input power is not classified before the step size is determined, the algorithm will only realize the MPPT in some wave conditions. In order to make further improvement on the performance of the algorithm used in the simulation experiments, P(k) is divided into six grades. The value of the step size affects the tracking speed and the steady state tracking precision of the algorithm. When the WEC works far away from the maximum power point, larger steps allow the algorithm to close more quickly on it. However, when the WEC works near the MPP, smaller steps would avoid excessive dithering and decrease energy loss. Determining the step size reasonably would improve the efficiency and the stability of the algorithm. In this paper, the parameters of the algorithm are determined after extensive test optimization using the buoy-rope-drum WEC simulation model presented above. Finally, in the algorithm, T=50s, a=0.05, b=0.01, c=0.03 and P(k) is divided in 6 grades where $P(k) \geq 7000w$, $5500w \leq P(k) < 7000w$, $4000w \leq P(k) < 5500w$, $2500w \leq P(k) < 4000w$, $1000w \leq P(k) < 2500w$, $P(k) < 1000w$.

4. Simulation experimental results and analysis

4.1. Simulation experimental results and analysis of the buoy-rope-drum WEC with resistive loads

The purpose of the simulation experiment is to confirm the maximum power point of the WEC at different wave conditions and provide the control group for subsequent experiments. First, the boost converter module in figure.1 is replaced by resistive loads, then the WEC operates under 12 wave conditions respectively where the resistive loads increase from 5Ω to 50Ω with 5Ω every increase. The experimental results are shown in figure.4. The wave heights of the first group's three wave conditions are all 0.5m and the wave periods are 5.5, 6, 6.5s respectively. The wave heights of the other three groups are 1, 1.5, 2m with periods as same as the group one. As shown in figure 4, the output power curves of the WEC are unimodal curves and the MPPs at different wave conditions are acquired.

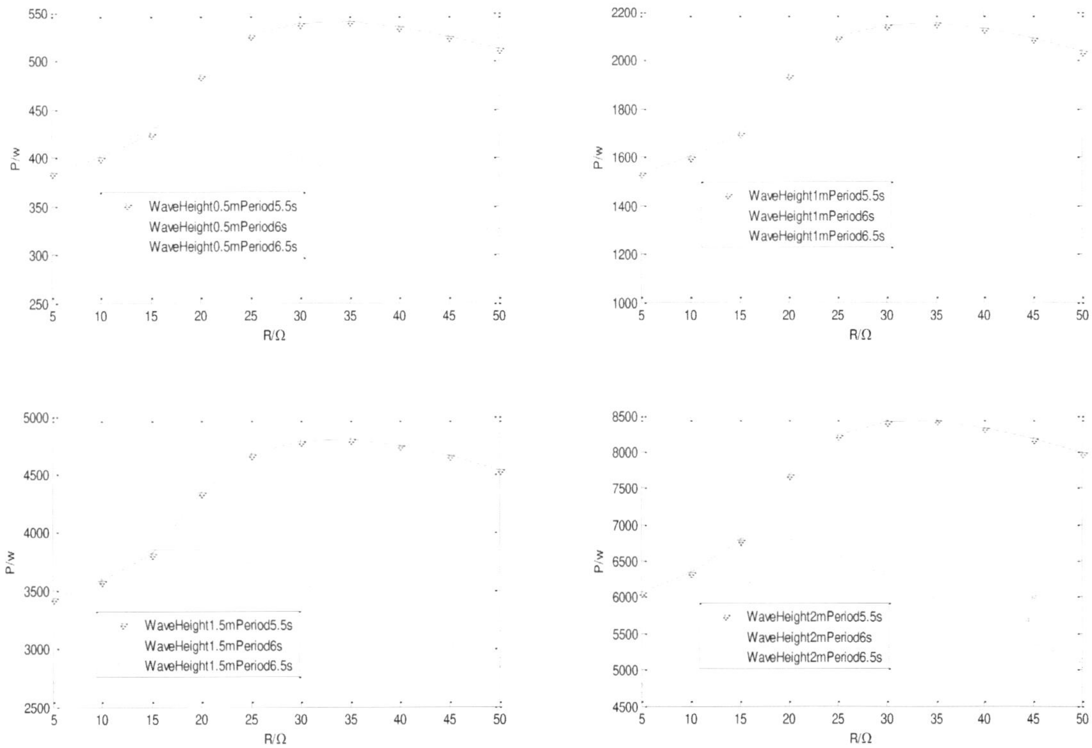

Figure 4. The output power curves of buoy-rope-drum wave energy converters in 12 different wave conditions

4.2. Simulation experimental results and analysis of the buoy-rope-drum WEC under MPPT algorithm control

The purpose of this simulation experiment is to testify that the variable step size P&O MPPT algorithm presented in this paper has good steady state tracking precision. The simulation model used in this experiment is shown in figure.1. The steady state output powers of the WEC and the power tracking error rates are obtained in the simulation experiment with the wave conditions as same as the experiment's operated in 4.1. Results are listed and compared with the results of the 4.1 experiment in table1. As shown in table1, the power tracking error rates are all less than 5%, which means the MPPT algorithm tracks the MPP nicely. There are two main reasons causing the existence of the power tracking error. First, the power electronic devices chosen for the simulation model like IGBTs, diodes are not ideal components so the forward voltage drop and the switching loss exist. Second, even if the novel variable step size P&O MPPT algorithm is used, the power oscillation near the MPP cannot be eliminated completely leading to the power waste. The power oscillation under steady state is the inherent defect of P&O methods which cannot be avoided.

Table 1. The output power and power tracking error rates of buoy-rope-drum wave energy converters in different wave conditions

Wave conditions（Wave height, H, Wave period, T,）	Maximum power point values/W	Steady state output power of WEC under MPPT control/W	power tracking error rates
H=2 m; T=5.5s	8430	8150	3.32%
H=2 m; T=6s	7780	7640	1.80%

H=2 m; T=6.5s	6810	6700	1.62%
H=1.5m;T=5.5s	4795	4620	3.65%
H=1.5m;T=6s	4415	4390	0.57%
H=1.5m;T=6.5s	3855	3800	1.43%
H=1 m; T=5.5s	2152	2055	4.51%
H=1 m; T=6s	1977	1930	2.38%
H=1 m; T=6.5s	1725	1700	1.45%
H=0.5m;T=5.5s	540.5	516	4.53%
H=0.5m; T=6s	496	490	1.21%
H=0.5m;T=6.5s	432	423	2.08%

4.3. Simulation experimental results and analysis of the buoy-rope-drum WEC under MPPT algorithm control with changing wave conditions

This experiment is to testify that the variable step size P&O MPPT algorithm can track the MPP fast and accurately under changing wave conditions. The simulation model used in this experiment is shown in figure 1. The experimental process is divided into three stages. First, the experiment is run with wave height set to 1.5m and wave period set to 5.5s. When simulation time is 300s, the wave height is adjusted to 2m and wave period to 6s. Keep this wave condition until simulation time is 600s then the wave height and wave period are adjusted to 1m and 6.5s respectively. The steady state output powers at different stages are listed in table 2 and the instantaneous output power curve is shown in figure 5. As observed, the algorithm tracks the maximum power points of different stages fast and accurately.

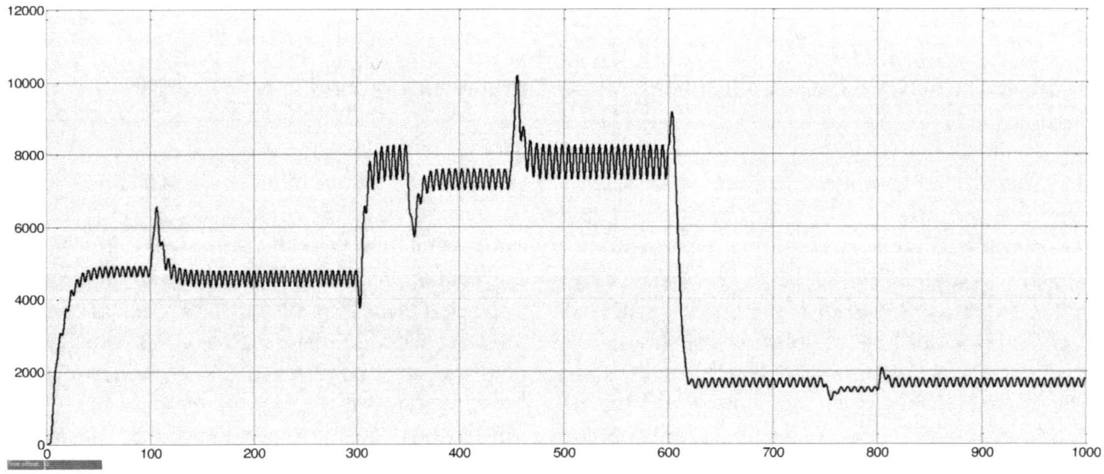

Figure 5. The instantaneous output power curve of the buoy-rope-drum wave energy converter

Table 2. The output power and power tracking error rates of buoy-rope-drum wave energy converters in changing wave conditions

Stages	Maximum power point values/W	Steady state output power of WEC under MPPT control/W	power tracking error rates
Stage 1: H=1.5m;T =5.5s	4795	4593	4.21%
Stage 2: H=2 m; T=6s	7780	7770	0.13%
Stage 3: H=1 m; T=6.5s	1725	1690	2.03%

5. Conclusions

This paper presents a novel variable step size P&O MPPT algorithm designed to maximize the energy conversion efficiency of the buoy-rope-drum wave energy converter. Simulation model of the WEC is built in MATLAB\Simulink environment and simulation experiments are done. The results show that the algorithm can track the maximum power point fast and accurately and maximize the output power during the whole operating process. The power classification standard and variable step size standard introduced in the algorithm enhance the robustness and adaptability of the algorithm significantly relieving the contradiction of the tracking speed and the tracking precision of the algorithm. This investigation also provides theoretical direction for the design of the DC-DC boost converters and the power control system of the buoy-rope-drum WEC.

References

[1] You Y G, Li W, Liu W M Li X Y and Wu F 2010 Development status and perspective of marine energy conversion system *Automation of Electric Power Systems* **34** 25-37
[2] Antonio F 2010 Wave energy utilization: A review of the technologies *Renewable and Sustainable Energy Reviews* **14** 899–918
[3] Mccormick M 1981 *Ocean Wave Energy Conversion* (NewYork: John & Sons Inc)
[4] Leijon M *et al.* 2006 An electrical approach to wave energy conversion *Renewable Energy* **31** 1309-19
[5] Amon E A, Schacher A A and Brekken T K A 2009 *A novel maximum power point tracking algorithm for ocean wave energy devices* (San Jose, CA, USA: IEEE Energy Conversion Congress and Exposition)
[6] Amon E A, Brekken T K A and Schacher A A 2012 Maximum power point tracking for ocean wave energy conversion *IEEE Transactions on Industry Applications* **48** 1079-86
[7] Zhou L, Wu J, Li Q H and Guo K 2008 Survey of maximum power point tracking techniques for photovoltaic array *High Voltage Engineering* **34** 1145-54
[8] Zhu L S, Qu Y M, Wang Y G and Yang Z 2014 Buoy-rope-drum Wave power efficiency *Acta Energiae Solaris Sinica* **35** 1381-6
[9] Liu B 2015 *Research on the Mooring System of Buoy-Rope-Drum Wave Power* (Jinan: Shandong University)
[10] Li W C 2011 *Research on maximum power capture technology for horizontal axis tidal current turbine* (Qingdao: Ocean University of China)
[11] Brekken T K A, Von Jouanne A and Hai Y H 2009 *Ocean wave energy overview and research at oregon state university* (Lincoln, NE: IEEE Power Electronics and Machines in Wind Applications)

Study on pyrolysis gas in thermal extraction of Bai Yinhua lignite with industrial washing oil

Y M Cui[1], X P Lian, F Y Zhao, Y Q Xu, Y Q Hu, Z K Yuan and X R Hao

Department of Chemical and Pharmaceutical Engineering, Hebei University of Science and Technology, Shijiazhuang 050018, China
E-mail: 1198635475@qq.com

Abstract. Industrial washing oil as solvent, pyrolysis gas produced from Bai Yinhua lignite during thermal extraction was studied by gas chromatography. The effects of temperature and solvent coal ration on coal pyrolysis gas were investigated. The results showed that: Pyrolysis gas produced mainly in CO, CO_2, O_2, H_2, CH_4, and so on, in which the total amount of oxygen containing compounds nearly 40%, significant effects of deoxidation was achieved. The increase of heat extraction temperature can significantly increase the degree of bond breaking and the gas formation rate, the gas yield and the rate of oxygen increase significantly, while the gas yield increases with the decrease of the solvent coal ration.

1. Introduction

China is rich in lignite resources, the reserves of which account for more than 13% of proven coal. Among them, the lignite reserves in the northeast of Inner Mongolia accounts for about 3/4. The southwest region is mainly dominated by Yunnan, which makes up about 1/5. Northeast, east and south-central region is about 5% [1-2]. Lignite is the lowest degree of metamorphism of coal, the degree of coalification between peat and bituminous coal. Lignite has the disadvantages of high inherent moisture content, high volatile, easy to self ignition and low heat [2-4]. Without any bond performance, most of the lignite is used for combustion to heat, so it cannot be directly used for coking. Super clean coal can be extracted from lignite by using solvent thermal extraction technology [5-6]. Because Super clean coal has strong bonding property, it can be used as a high-quality binder for coking coal blending which can achieve the purpose of expanding coking coal resources [7-8].

Coal-derived oils have similar structural properties with coal. Their thermo chemical properties are stable, and their sources are reliable, easy to recycle, and have good economy. Cui *et al* [6] reported that the extraction yield was 66.8% for Xiao-kang village 1/3 coking coal using thermal extraction technology with industrial washing oil at 380 °C. The recycling of industrial washing oil was studied. It shows that industry washing oil has the excellent extraction performance, and it is a preferred solvent for thermal extraction of coal.

In the process of thermal extraction, the organic matter with high activity in coal is extracted, ash is retained in the residue. During thermal extraction, the break of bonds in the coal molecular structure is accompanied with the pyrolysis of coal and gas was generation. Because of the presence of solvent, the gas produced in the pyrolysis process and the high temperature pyrolysis of the coal alone are different during thermal extraction. It reflects the change of coal molecular structure in a certain extent

[1]Address for correspondence: Y M Cui, Department of Chemical and Pharmaceutical Engineering, Hebei University of Science and Technology, Shijiazhuang 050018, China. E-mail: 1198635475@qq.com.

during thermal extraction. In addition, thermal extraction process is accompanied by reduction, condensation, hydrogenation reaction, which requires a qualitative and quantitative study on the pyrolysis gas [9].

With the popularization and development of gas chromatography technology, it possible to study the complex composition of gases produced in the process of thermal extraction. Either qualitative research or quantitative research will be better than the traditional methods. In this paper, pyrolysis gas produced from Bai Yinhua lignite during thermal extraction with wash oil as solvent will be studied by gas chromatography.

2. Experiment

2.1 Coal samples

In the experiments coal samples were used of the Inner Mongolia Bai Yinhua lignite. The sample preparation was according to the National Standard GB474 - 1996. They were crushed and divided less than 100 μm in 150min at 100 °C in drying and vacuum conditions. The coal samples were stored in a dryer, which were low temperature and avoiding light. Industrial analysis and elemental analysis data of coal was shown in table 1. The main functional groups of active components in Bai Yinhua lignite was shown in table 2.

Table 1. Proximate and ultimate analysis of samples.

	Proximate analysis / %[1]			Ultimate analysis / % [1], ad			
Mad	Aad	Vad	C	H	O[2]	N	S
16.29	14.61	36.42	52.98	4.06	41.44	0.99	0.53

[1]Percent of weight

[2]By defferent

Table 2. Main functional groups of active components in Bai Yinhua lignite.

Group		Structural formula
Core structure	Aliphatic hydrocarbon chain	—C—C— —C=C—
	Aromatic nucleu	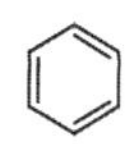
Oxygen group	Hydroxyl group	—OH
	Carboxyl group	—COOH
	Methoxyl group	—OC
	Carbonyl group	C=O
Alkyl side chain	Methyl side chain	—C
	Ethyl side chain	—C—C
	Propyl side chain	—C—C—C
Nitrogen groups	Pyridine	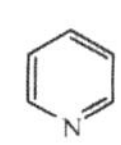

Pyrrole 

2.2 Experimental method

The experiments were carried out in a thermal extraction device which was designed by ourselves. A certain size of Baiyinhua lignite was added in an autoclave after full mixing with the industry washing oil. The coal was extracted at a certain temperature in N_2 atmosphere after staying for 10 min. After the extraction, when the reaction system was cooled to room temperature, The qualitative and quantitative analysis were carried out by gas chromatography.

2.3 Analysis method

Using GC-9790 gas chromatograph. The gas chromatograph was producted by Wenling Analytical Instrument Co. Ltd. Operating conditions:

- Carrier gas: Argon, 5A molecular sieve packed column;
- Carrier gas: Hydrogen, GDX-103 packed column.
- Column temperature: 50 °C;
- Gasification room temperature: 70℃;
- Detector temperature: 180 °C;
- Sample volume: 1 ml.

Standard gas was choosed by the known gas mixture componets which was producted by the Dalian Special Gas Co. Ltd, Qualitative analysis with retention time, quantified by an external standard method.

3. Results and discussions

3.1 Analysis of main gas composition and content

The gas content after extraction was shown in figure 1, which was at the condition of the solvent and the coal 10:1; the extraction temperature is 380 °C. It was shown that CO_2 and N_2 content were highest; CH_4, CO and C_2H_6 were low; H_2, O_2 and C_2H_4 were lowest. N_2 content is higher because the whole process of heat extraction was carried out in the N_2 atmosphere, so the reaction of the mixed gas, N_2 also occupies a considerable part of the content. The total amount of oxygen-containing compounds in the figure was nearly 40%, indicating that the oxygen removal effect of solvent extraction of Baiyinhua lignite was significant. In addition, the content of organic hydrocarbon gas is relatively large such as CH_4, this was because the thermal extraction temperature exceeds the thermal decomposition temperature of coal, which will lead to functional groups and carbon chain rupture. Figure 2 shows main ways to generate methane during Thermal Extraction.

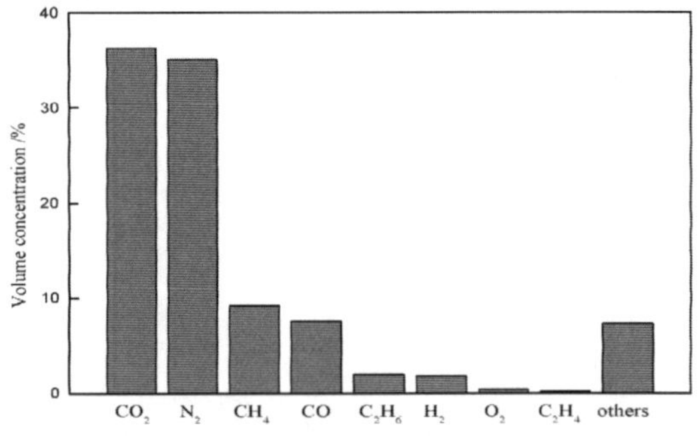

Figure 1. Distribution of gas content

IOP Conf. Series: Earth and Environmental Science **40** (2016) 012058 doi:10.1088/1755-1315/40/1/012058

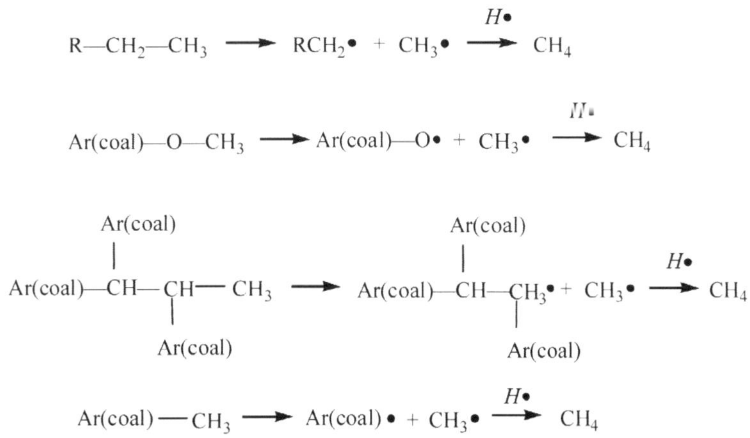

Figure 2. Main ways to generate methane during thermal extraction

3.2 Effect of temperature on pyrolysis gas

The effect of temperature on pyrolysis gas was investigated under the condition of the solvent and the coal 10:1. The change of gas yield with temperature was shown in figure 3. It can be seen that the gas yield increases with the increase of temperature. When the temperature was lower than 300 °C, the increase of gas yield was relatively slow, when the temperature reaches 300 ℃, the gas production rate increased rapidly, reaching 4.9% at 400 °C.

According to the theory of modern coal chemistry, coal was a complex macromolecular compound, it was composed of a three-dimensional cross-linked space network [10] These aromatic and hydrogenated aromatic structures were connected by methylene or longer aliphatic hydrocarbon chains. The periphery of the basic structure was connected with an alkyl side chain and a variety of functional groups. Alkyl side chain mainly has $-CH_2-$, $-CH_2-CH_2-$, the functional group mainly contains oxygen functional group, including phenolic hydroxyl group, carboxyl group, carbonyl group, carbonyl group and so on[11]. Figure 4 was a schematic diagram of two-phase model. The network structure of coal was demonstrated by the two-phase model.In the process of thermal extraction, with the increase of temperature, the side chains and functional groups in the coal molecules increased, the chain of the alkyl side chain would generate methane and other hydrocarbon gases. The fracture of the oxygen containing functional groups would generate CO, CO_2 and other gases. Small aromatic polycyclic aromatic molecules dissolved in the solvent. The bond breaking degree of coal molecules and the gas yield increased significantly when increasing temperature at the extraction temperature exceeding 300 °C [12].

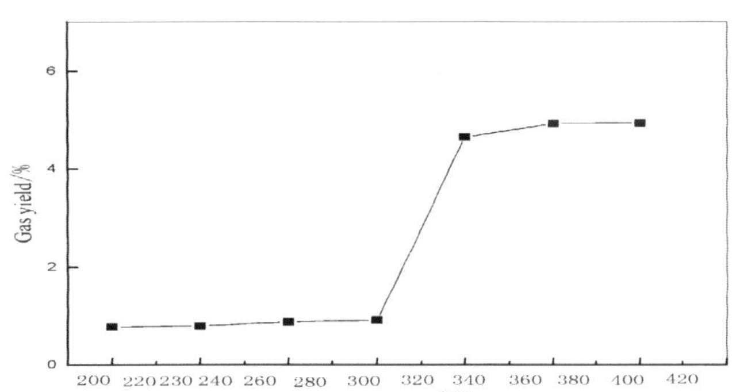

Figure 3. Change of gas yield with temperature

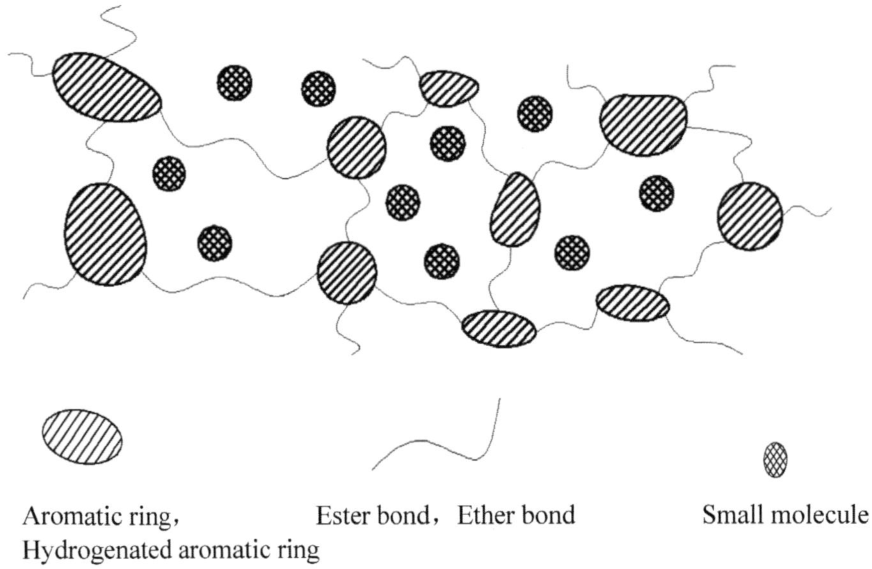

Figure 4. Schematic diagram of Two-phase model

The change of deoxidation rate with temperature was shown in figure 5. Before 300 °C the deoxidation rate increased unobviously with the increase of extraction temperature, After 300℃ the rapid increase to 340 °C deoxidation rate reached 7.1%, continue to heat up the deoxidation rate changed slowly. The removal of oxygen element in coal mainly due to the weak oxygen bond, which was easily broken by the thermal effect, and the CO and CO_2 gas could be formed quickly. The oxygen bond broken more because of temperature rising. When the temperature reached a certain degree, the deoxidation rate increased rapidly. At last the weak oxygen bond rupture was basically completed, the deoxidation rate did not increase.

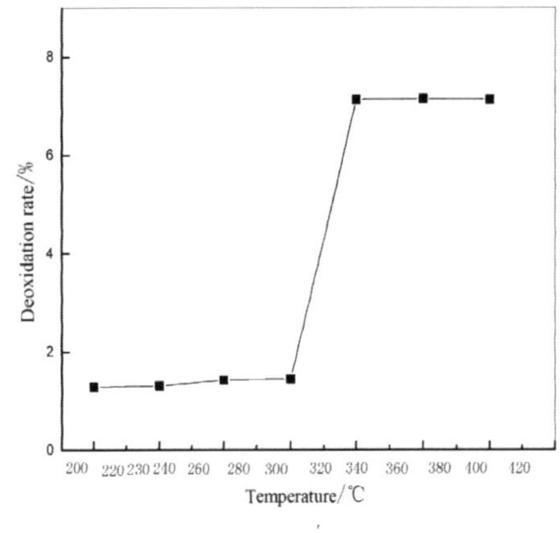

Figure 5. Change of deoxidation rate with temperature

3.3 Effect of the ratio of solvent to coal on pyrolysis gas

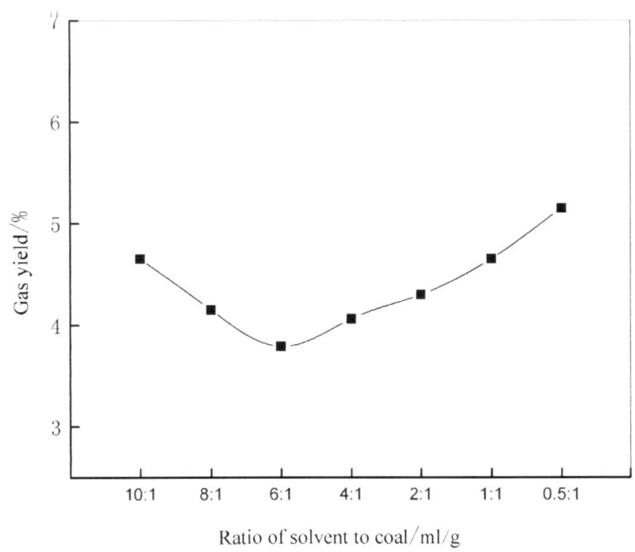

Figure 6. Effect of the ratio of solvent to coal on the pyrolysis gas

The effect of the ratio of solvent to coal on the pyrolysis gas was investigated at 380 °C. Figure 6 shown the change of gas yield with the ratio of solvent to coal. It was shown that the gas yield decreased firstly and then increased with the decrease of the ratio of coal to gas, which reached a turning point when the ratio of solvent to coal was 6:1. The results shown that the ratio of solvent to coal has a great influence on the formation of gas. The solvent can penetrate and swell the coal, which was helpful to the fracture of the bond in the coal, and the smaller molecular free radicals were generated, and the gas was produced by hydrogenation and condensation. Therefore, the larger soluble coal ratio was favorable for the formation of gas. But with the ratio of solvent to coal continued to decrease, the lack of adequate organic solvent components dissolved in the coal, the organic components of pyrolysis reaction under high temperature and gas generation. So with the further decrease of the ratio of coal to gas, the gas yield increased gradually.

4. Conclusion

- During the thermal extraction of Bai Yinhua lignite, the gas produced by pyrolysis mainly include CO, CO_2, O_2, H_2, CH_4, C_2H_4, C_2H_6, etc.. The total amount of oxygen-containing compound was nearly 40%, It indicated that the oxygen element removal of Baiyinhua lignite was significant by thermal extraction.
- The degree of bond breaking and the gas formation rate can significantly accelerated by increasing temperature. The effect of deoxidation rate tended to be gentle when the temperature touched 340 °C.
- The effect of the ratio of solvent to coal on the pyrolysis gas was significant, With the decrease of the ratio of solvent to coal, the number of small molecular organic residue increased, it was easier to form pyrolysis gases when no solvent protected at high temperature.

Acknowledgements

This work is supported by the natural science foundation of Hebei Province Item No. B2016208014 and Science and technology research project of Hebei Province Item NO. ZD2016088

References

[1] Li X H, Xue Y L, Feng J, Qun Y, Li W Y, Guo X F and Liu K 2015 Co-pyrolysis of lignite and shendong coal direct liquefaction residue *Fuel* **144** 345-8

[2] Zhao Y Y 2012 Study on the influence of High Moisture Lignite combustion on boiler *J. Northeast Dianli University* **32** 55-8

[3] Wang B 2011 Discussion on the present situation of lignite upgrading technology *Coal Quality Technology* **4** 9-13

[4] Ashida R, Morimoto M, Makino Y, Umemoto S, Nakagawa H, Miura K, Saito K and Kato K 2009 Fractional of brown coal by sequential high temperature solvent extraction *Fuel* **88** 003

[5] Cui Y M,Xu Y Q, Hu Y Q, Yuan Z K, Li L Y, Zhang Z X and Zhao F Y 2013 Preparation and characterization of super pure coal *Advanced Materials Research* **2657** 49-53

[6] Cui Y M, Zhao F Y, Zhang Z X, Duan X Y, Hu Y Q, Zhang X L, Huang S P and Li L Y 2014 Research progress on preparation of ultralow-ash coal bv solvent thermal extraction *Modern Chemical Industry* **04** 34-7

[7] Fu X H, Zhu S Q and Yang Q W 2006 Deep physical processing of coal and preparation of ultra clean coal *Coal Preparation Technology* **10** 46-57

[8] Luda R 2008 Low quality coal using high temperature solvent extraction of coal coking *Fuel & Chemical Processes* **39** 54-9

[9] Pan C X, Liu H L, Zhu W W, Li H P, Liu J R, Wei X Y, Shui H F and Wang Z C 2015 Characterization of the thermal dissolution products of a subbituminous coal at different temperatures *J. Fuel Chemistry and Technology* **43** 416-21

[10] Kawashima H, Koyano K and Takanohashi T 2013 Changes in nitrogen functionality due to solvent extraction of coal during HyperCoal production *Fuel Processing Technology***106** 275-80

[11] Murata S, Hosokawa M and Kidena K 2000 Analysis of oxygen-functional groups in bown coals *Fuel Processing Technology* **67** 213-43

[12] Han D H, Hu Y Q, Wang J S, Yao R and Shao J X 2011 Study and analysis of influence factors of coal pyrolysis *Coal Technology* **30** 164-5

Preparation and characterization of NiO photocathodes sensitized with PbS quantum dots

X W Li[1], R X Chen[1], Q Lu[1], K J Wei[1], Y P Du[1], X J Wang[1], T Hao[1], L Li[1, 3], X X Yuan[1] and M Zhang[2]

[1]Heibei Key Laboratory of Optic-electronic Information Materials, College of Physics Science and Technology, Heibei Unversity, Baoding 071002, China
[2]College of Chemistry and Environmental Science, Hebei University, Baoding 071002, China

E-mail: lilinghbu@163.com

Abstract. The nanosized tremella-like NiO was synthesized by a simple hydrothermal method at low temperature. A novelty modified $(CH_3)_4N)_2S/((CH_3)_4N)_2S_n$ electrolyte was introduced in solar cell successfully and NiS as the counter electrode. Moreover, PbS sensitized p-type NiO was synthesized by chemical bath deposition (CBD) in an precursor solution to ensure the nanosized tremella-like NiO films obtain a better uniformity and high penetration. At the same time, a series of comparative experiments were designed for studying the influence of mesoporous NiO film thickness on the photoelectron characteristic of the cells. The result indicated that when there were three layers of screen printing and the thickness of NiO film was approximately 3μm, the maximum power conversion efficiency of 0.87% was achieved, with 512 mV of V_{oc}, 0.33% of ff and 5.14 mA cm^{-2} of J_{sc}.

1. Introduction

Quantum dot sensitized solar cells (QDSSCs) exhibit many attractive features, including low production cost, high power conversion efficiency in theory and easy fabrication process, were considered as a good choice to be a substitute for dye-sensitized solar cells (DSSCs) [1, 2]. QDs are usually employed to construct n-type photoanodes where photocurrents owe to electrons injection into n-type semiconductor oxide [3, 4]. On the other hand, the possibility of using p-type photocathodes in solar cell was first reported in 1999 [5]. In addition, p-type photocathodes operate in an inverse mode, the holes are injected from a QDs into the valence band of the p-type semiconductor oxide when QDs are attached to hole acceptors [6, 7]. Nickel oxide (NiO) is attractive as electrode material in photovoltaic applications due to their good stability, a wide band gap and transparency [8, 9]. In the condition of solar cells, the major research of the p-type semiconductor materials that can be used as sensitized photocathodes is that they can be combined with n-type photoanodes in a tandem configuration. Such tandem solar cell, the maximum power conversion efficiency achievable increases from 30% for solar cell using a single junction to 42% when two semiconductors are employed [10]. Nevertheless, the photoelectron conversion efficiency of such tandem p-n solar cells have been

[3] Address for correspondence: L Li, Heibei Key Laboratory of Optic-electronic Information Materials, College of Physics Science and Technology, Heibei Unversity, Baoding 071002, China. E-mail: lilinghbu@163.com.

hindered by the poor performance of the p-type photocathodes to date, resulting from the fact that the lower photocurrent of n-type and p-type solar cell [11, 12]. As a result, the photoelectron performances of the solar cell with the p-type photocathodes are significantly enhanced as a promising strategy to improve the photo-electrochemical characteristic of tandem p-n type solar cells [13]. Safarialamuti *et al.* prepared CdS/CdSe deposition onto mesoporous NiO films by successive ionic layer adsorption and reaction method (SILAR). Finally, the photoelectron conversion efficiency of 0.02% was obtained [14]. Subsequently, the conversion efficiency for CdSe sensitized p-type NiO solar cells to a best report of 0.35% has been reached by Park *et al* [13]. In 2016, Raissi *et al.* [15] have reported the fabrication of NiO-based QDSSCs using PbS QDs with a cobalt electrolyte and Pt counter. Finally the best conversion efficiency of solar cell achieved 0.4%.

In this paper, we synthesised the nanosized tremella-like NiO by a simple hydrothermal method at low temperature, and a series of comparative experiments were designed for studying the influence of mesoporous NiO film thickness on the photoelectron characteristic of solar cells. We assembled A novelty modified $(CH_3)_4N)_2$ $S/((CH_3)_4N)_2S_n$ electrolyte was introduced in PbS sensitized p-type NiO solar cell successfully and NiS as counter electrode. The best efficiency of PbS sensitized NiO solar cells using polymer electrolyte and NiS counter electrode is 0.87% was obtained by the optimization of NiO film thickness. Moreover, a maximum IPCE value of solar cell was reached 32.7% at 340 nm.

2. Experimental Section

The tremella-like NiO were synthesized by simple hydrothermal method. In this process, 0.466 g of $Ni(NO_3)_2$ $6H_2O$ was dissolved in 50 mL N, N-dimethylform (DMF), under magnetic stirring at room temperature to obtain a transparent solution. Then the prepared solution was poured into a 200 ml Teflon-lined stainless autoclave, which was kept at 80°C for 20 h. A green-blue ptoduct was got, which was washed with deionized water and ethanol for several times, and dried in vacuum at 60°C for 5 h. Finally, the black NiO was obtained through annealing at 500°C for 3 h. Nanoporous NiO films were fabricated by screen printing an aqueous suspension of NiO nanoparticles on FTO conductive glass substrates and sintering at 500°C for 30 min. CBD is considered as assemble PbS QDs deposited to the NiO photocathode. NiO films were modified by soaking in 0.1 M thioglycolic acid (TGA) for 1 min. Subsequently, the TGA-modified NiO film was soaked into an ethanol solution including 0.5 M $Pb(NO_3)_2$ for 1 min before cleaned by ethanol, and dried. Then after soaked for 1 min into a 0.5 M NaS solution, the sample was cleaned again with methanol. NiS as working electrode. Cleaned FTO glass was soaked into 0.5 M Ni $(NO_3)_2$ ethanolic solution for 30 s, cleaned by ethanol and dried. After the sample was soaked into 0.5 M Na_2S methanolic solution for 30 s, cleaned by methanol and dried. This process was continued five times to gain the NiS electrode. The modified electrolyte be made up of 0.01 M $((CH_3)_4N)_2S$, 0.02 M 4-tert-butylpyridine (TBP), 0.02 M $LiClO_4$, 0.002 M S, 3-Methoxypro-pionitrile (MPN) was employed as a solvent for the polysulfide electrolyte. The $((CH_3)_4N)_2S$ was synthesised by heating $(NH_4)_2S$ and $((CH_3)_4N)OH$ at 100°C.

3. Results and Discussion

3.1 Analysis of SEM morphology

Figure 1 reveals the typical FESEM morphology of NiO nanoparticles, it is clear that the as-prepared NiO has a tremella-like architecture with a diameter ranging from 3 to 4μm. These tremella-like NiO are constructed by some thin nanosheets with smooth surface and about 8 nm in thickness, which are formed a multi layered structure.

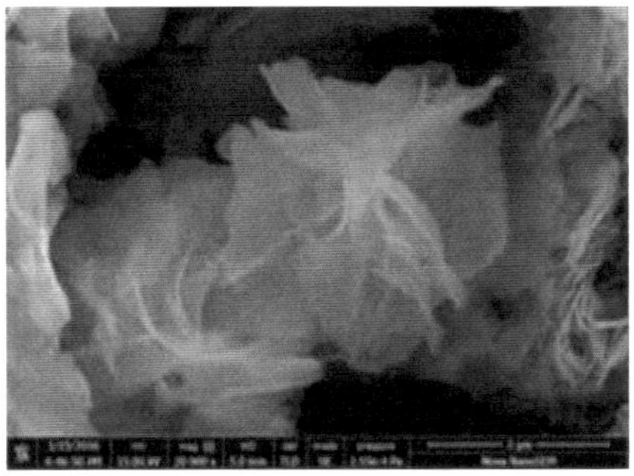

Figure 1. SEM images of tremella-like NiO

3.2 Analysis of XRD spectra

Figure 2 illustrates the typical XRD pattern of NiO/PbS composite and the pure NiO nanoparticles. As shown in Figure. 2a, the typical pattern peaks at 2θ values of NiO nanoparticles are approximately 37.24°, 43.15°, 62.58° and 75.26°, which correspond to (111), (200), (220) and (311) bunsenite planes of NiO, respectively(JCPDS card NO. 47-1049). In Figure. 2b, compared with the pattern of pure NiO nanoparticles, the diffraction peaks of the NiO/PbS are perfectly indexed as pure galena PbS phase (JCPDS card NO.05-0592). Furthermore, there are no clear peaks for impurity on the patterns, which indicates that the product is pure and other impurities were not generated during their processing.

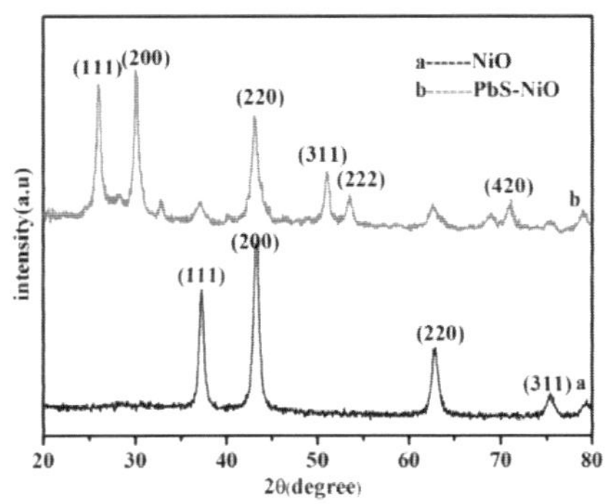

Figure 2. XRD spectra of NiO/PbS and pure NiO

3.3 J-V characteristics

The photocurrent-voltage (*J-V*) curves of the QDSSCs based on NiO with different thickness as shown in Figure 3. Key photoelectron characteristic parameters of solar cells are summarized in table 1. It can be obviously observed that with the increasing number of layers of screen printing, both J_{sc} and V_{oc} of the NiO/PbS solar cells slightly increase from 3.38 mA cm^{-2} and 191 mV for one layer to 5.14 mA cm^{-2} and 512 mV for three layers, respectively. A maximum efficiency of 0.87% are achieved for the solar cell with three layers of screen printing, and then decreases with a further increase in the number of layers. In present situation, the thickness of NiO film would increase after each layer. Such an increase layers of screen printing results in adsorbing more quantum dots, which is advantageous to increase the amount of collected holes and the efficiency of solar cell. However, the injection of the hole will be hampered with the sequentially increasing of film thickness, which is responsible for the reduce of J_{sc} and V_{oc} values.

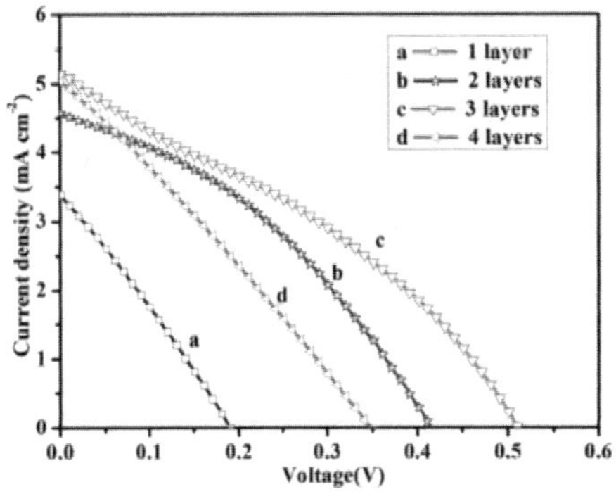

Figure 3. *J-V* spectrum of NiO with different layers

Table 1. Photoelectron characteristic of the QDSSCs measured under 1 sun illumination

Sample	V_{oc}(V)	J_{sc}(mA cm^{-2})	FF	Eff/%
1	0.191	3.38	0.27	0.18
2	0.416	4.57	0.37	0.70
3	0.512	5.14	0.33	0.87
4	0.349	5.04	0.27	0.48

3.4 IPCE spectrum

The incident photon to current efficiency (IPCE) spectra of a p-QDSSCs assembled from different thickness NiO thin film can be found in Figure 4. The IPCE spectra for three layers of screen printing displays much larger photo response at wavelengths below 420nm compared to one layer. The IPCE reached a maximum of 32.7% at 340 nm. The IPCE spectra was not enhanced with further increase in the NiO layer thickness. It can be reasonable as far as 30%-40% of the incident photons were absorbed by the NiO thin film over most of the visible wavelength range. Photocurrents were not observed due to photons were absorbed by the NiO thin film.

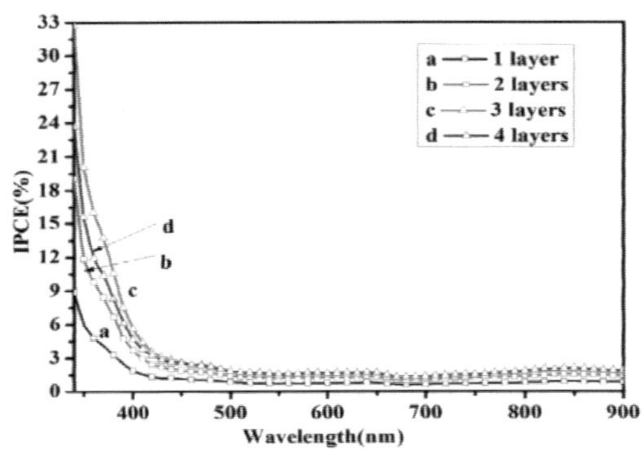

Figure 4. IPCE spectrum of NiO with different layers

4. Summary

In this paper, we have successfully synthesized tremella-like NiO by a simple hydrothermal method, and researched the influence of NiO thin film thickness on the photovoltaic performance of p-type QDSCs. We found that when the layers of screen printing were 3 and when the thickness of NiO thin film was about 3μm, the conversion efficiency of 0.87% was achieved, with 512 mV of V_{oc}, 0.33% of ff and 5.14 mA cm^{-2} of J_{sc}. At the same time, the IPCE reached a maximum of 32.7% at 340 nm. It was suggested that the p-type NiO/PbS structure could be successfully used to assemble p-type solar cell.

Acknowledgements
We gratefully acknowledge the financial support from the following sources: National Natural Science Foundation of China (NSFC) (Grants 21201053, 51077035), Specialized Research Fund for the Doctoral Program of Higher Education (20121301120005), Fund in Hebei Province Natural Science (F2014201078, A2015201050), Youth Fund in Hebei Province Department of Education China (QN2014057), Fund in Hebei Province Applied Basic Research Project (14964306 D) and Outstanding Youth Fund of Hebei University (2015JQ02). Hebei University College Students' Innovative Entrepreneurial Training Program (201410075014, 201410075082, CYE409, CYE410), Hebei University Innovation Entrepreneurship Training Program(2015066, 2015064, 026, 201510075047, 201610075066).

References

[1] Kamat P V 2008 *J. Phys. Chem. C* **112** 18737-53
[2] Wang H, Bai Y S, Zhang H, Zhang Z H, Li J H and Guo L J 2010 *J. Phys. Chem. C* **114** 16451-5
[3] Santra P K, Kamat P V 2012 *J. Am. Chem. Soc* **134** 2508-11.
[4] Vasko J, Victoria G P, Sixto G, Eneko A, Germán C, Ramon T Z, Iván M S, Juan B 2011 *J. Am. Chem. Soc* **133** 20156-9
[5] He J, Lindström H, Hagfeldt A, Lindquist S E 1999 *J. Phys. Chem. B* **103** 8940-3
[6] Barceló I, Guillén E, Lana-Villarreal T, Gómez R 2013 *J. Phys. Chem. C* **117** 22509-17
[7] Wang Z, Shakya A, Gu J, Lian S, Maldonado S 2013 *J. Am. Chem. Soc* **135** 9275-8
[8] Nattestad A, Mozer A J, Fischer M K R, Cheng Y B, Mishra A, B¨auerle P, Bach U 2010 *Nat. Mater* **9** 31-5
[9] Rhee J H, Lee Y H, Bera P, Seok S I 2009 *J. Chem. Phys. Lett* **477** 345-8
[10] Vos A D 1980 *J. Phys. D App. Phys* **13** 839-46
[11] Borgström M, Blart E, Boschloo G, Mukhtar E, Hagfeldt A, Hammarström L, Odobel F, 2006 *J. Phys. Chem. B* **109** 22928-34
[12] Morandeira A, Fortage J, Edvinsson T, Pleux L L, Blart E 2008 *J. Phys. Chem. C* **112** 1721-28
[13] Park M A, Lee S Y, Kim J H, Kang S H, Kim H 2014 *Mol. Cryst. Liq. Cryst* **598** 154-62
[14] Safarialamuti A, Jennings J R, Hossain M A, Yung L Y L, Wang Q 2013 *Phys. Chem. Chem. Phys* **15** 4767-74
[15] Raissi M, Pellegrin Y, Jobic S, Boujtita M, Odobel F 2016 *Sci. rep* **6**

Comparative study on the ventilation mode of the residential building in Beijing area in winter

G J Wang[1], Q P Li, Y Guo and Y H Hu

Beijing Institute of Residential Building Design & Research Co., Ltd, Beijing, China

E-mail: wgj86365413@126.com

Abstract. With the development of residential energy conservation technology, the air tightness requirement of the window is higher and higher. So in winter the cold penetration wind cannot satisfy the requirement of indoor personnel to fresh air. The common ventilation mode includes natural ventilation, natural inlet and mechanical exhaust, wall type ventilator with heat exchange, ventilation unit with heat exchange. Looking for energy saving, comfortable way of ventilation, the application effect of the way of ventilation is evaluated in air distribution and comfort performance and the initial investment by FLUENT software. The conclusion is that the mode of ventilation unit with heat exchange has higher superiority compared with the others.

1. Introduction

With the development of residential energy conservation technology, the air tightness requirement of the window is higher and higher. That is to say, the amount of air exchange between indoor and outdoor is greatly reduced, and the heat loss caused by cold air infiltration is also largely deceased. For the commonly used plastic steel window, when the test pressure is 10 Pa, its air infiltration is very little, less than $0.7 m^3/(m \cdot h)$, which is considered to be nearly enclosed [1]. The amount of air infiltration is largely less than the requirement of indoor personnel to fresh air, and indoor pollutants can't be timely ruled out. The indoor air quality problems are becoming more and more prominent. However, the basic requirement for healthy housing is give a good indoor air quality, and ventilation is considered as the most basic and effective way to reach this aim. Therefore, there is a conflict between the residential energy conservation technology and ventilation. In this case, how to operate the ventilation of the residential building has become a more prominent problem [2].

In this study, air distribution, energy consumption characteristics, and initial investment of different ventilation modes including natural ventilation, natural inlet and mechanical exhaust, wall type ventilator with heat exchange, ventilation unit with heat exchange are analyzed and evaluated to present their advantages and disadvantages. This study can guide to find a ventilation mode that can meet the thermal comfort, air quality and energy saving.

2. Evaluation method

2.1 Air distribution

The aim of ventilation is realized by the reasonable air distribution. A reasonable air distribution not only

[1] Address for correspondence: G J Wang, Beijing Institute of Residential Building Design & Research Co., Ltd, Beijing, China. E-mail: wgj86365413@126.com.

gives a healthy and comfortable environment, but also makes the initial investment and operating costs to be relatively low. Therefore, according to characteristics and requirements of the indoor environment, it is very important to adopt the most appropriate ventilation system and air distribution to achieve high quality and efficient operation. In this study, the indoor air velocity field is simulated by FLUENT software, then it is evaluated using the indicator, namely, airflow distribution uniformity.

2.2 Thermal comfort index

Thermal comfort index is considered to be a comprehensive response of human body to various factors of thermal environment, whose indicators includes effective temperature, thermal stress index, predicted mean vote (PMV) and so on. According to ISO7730 standard, the thermal environment can be described and evaluated using indicators, namely, PMV and PPD [3]. In this study, the indoor air temperature and velocity are simulated by FLUENT software, then PMV is calculated based on its definition, and it is used to evaluate thermal environment.

2.3 Initial investment

Initial investment is a barrier of technology promotion. Therefore, initial investment is an important indicator evaluating technical products.

3. Physical model and boundary conditions

In this study, the research object is a residential building with area of 120 m^2, which is shown in Fig.1. In order to make an effectively comparative study, air volume of different common ventilation mode including natural ventilation, natural inlet and mechanical exhaust, wall type ventilator with heat exchange, ventilation unit with heat exchange, are set to the same. The air velocity is calculated, according to their vent area. Their temperature boundary conditions also are set to be the same. Details of their boundary conditions as follows,

Figure 1. Plan of research object

Velocity boundary conditions

1) Air volume. According to indoor air quality standard [4], fresh air requirement should be 30m³/(h·p). There are 3.5 peoples in a room. So the total fresh air rate is 105m³/h. According to design standard for energy efficiency of residential buildings in severe cold and cold zones [5], air change rate should be $0.5h^{-1}$. So the

total fresh air rate is 168m³/h. According to design code for heating ventilation and air conditioning of civil buildings [6], air change rate should be $0.5h^{-1}$, when the room area is 20 m^2-50 m^2. So the total fresh air rate is 168m³/h. Therefore, the total fresh air rate is 168m³/h.

2) Vent area. The vent area of natural ventilation is equal to the area of open surface of window. The vent area of natural inlet and mechanical exhaust, wall type ventilator with heat exchange, ventilation unit with heat exchange is the same to its actual size.

3) Air velocity. Air velocity is calculated, according to their air volume and vent area. The direction of natural ventilation is the same as that of the leading wind in winter, according to design code for heating ventilation and air conditioning of civil buildings [6].

Temperature boundary conditions

1) External wall temperature. In winter, heat is transferred from indoor air to the inner surface of external wall by convection. Then, this heat is transferred from the inner surface to the outer surface by conduction. At last, this heat is transferred from the outer surface to outdoor air by convection [7]. According to the fact that the heat transferring from indoor air to outdoor air is equal to that transferring from indoor air to the inner surface of external wall, the temperature of inner surface of external wall can be calculated, which can be considered as the dirichlet boundary condition of external wall. The temperature of interior wall is set as same as the indoor design temperature.

2) External windows temperature. Similar to the external wall temperature, the temperature of the external windows is calculated. Of course, the heat transfer coefficient of the external window is different from that of the exterior wall.

3) Inlet air temperature. For these ventilation modes without heat recovery, the inlet air temperature is the same to the heating outdoor temperature. For these ventilation modes with heat recovery, the inlet air temperature is calculated, according to the fact that its heat recovery efficiency is 60%.

4. Comparative analysis of ventilation modes

4.1 Air velocity field

The air velocity field of different ventilation modes is simulated by FLUENT software, and they are shown in figure 2-5, which indicate the simulation results of room 1.5m.

From these figures, when the natural ventilation is operated, the air distribution of rooms except bathrooms and kitchens is relatively uniform. However, its comprehensive effect is not stable, due to the limitation of opening time and area of windows, the outdoor meteorological environment. When the natural inlet and mechanical exhaust is operated, the air distribution of rooms is relatively uniform, and the temperature distribution is also relatively reasonable. When the wall type ventilator with heat exchange is operated, the air distribution of rooms is not reasonable. The fresh air is concentrated in the area near the vents, and they have a little influence on the air inside the room. When the ventilation unit with heat exchange is operated, the air distribution of rooms is relatively reasonable.

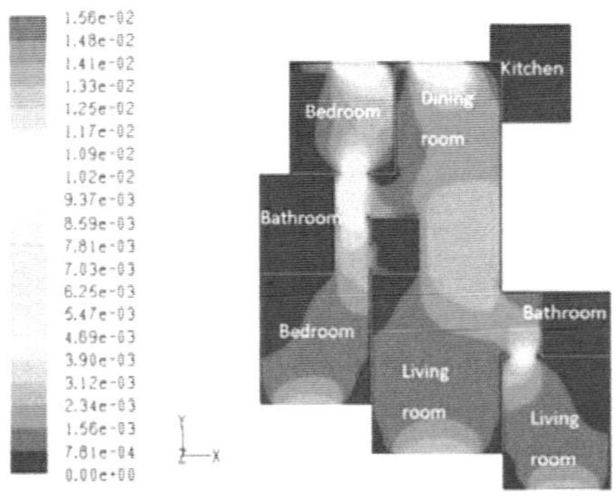

Figure 2. Air velocity field of natural ventilation

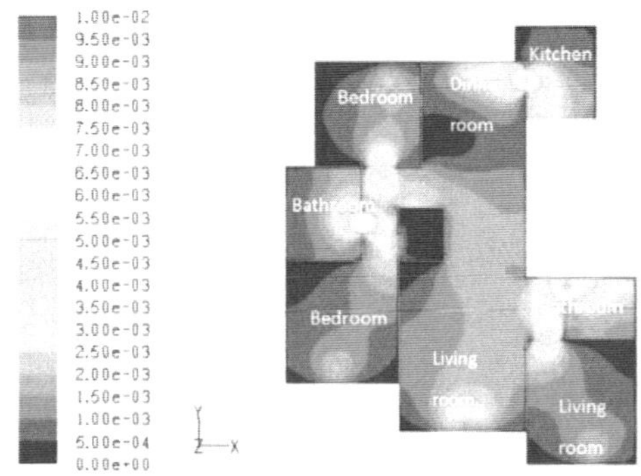

Figure 3. Air velocity field of natural inlet and mechanical exhaust

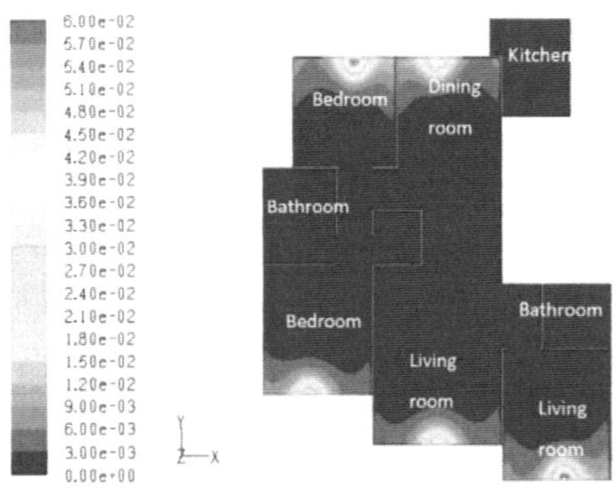

Figure 4. Air velocity field of wall type ventilator with heat exchange

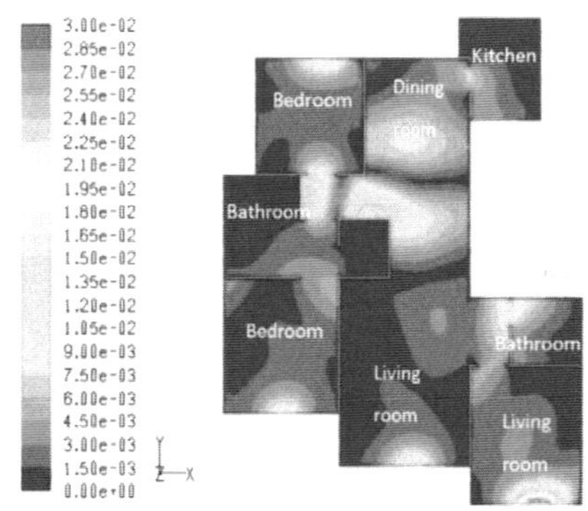

Figure 5. Air velocity field of ventilation unit with heat exchange

4.2 PMV

The air temperature and velocity of different ventilation modes are simulated by FLUENT software, then their PMV distributions are calculated and shown in Figure 6-9, whose height is 1.5m.

According to these figures, the feeling of the human body in bedroom and dining room with the natural ventilation is cold. Sometimes PMV is less than -1. The feeling in all rooms with the natural inlet and mechanical exhaust is comfortable, and PMV is between 0 and 0.5. PMV distribution of all rooms with the wall type ventilator with heat exchange is reasonable. But PMV in the area near the vents is low, less than -0.5. PMV of all rooms with the ventilation unit with heat exchange focuses on -0.5-0.5, where the feeling of the human body is comfortable.

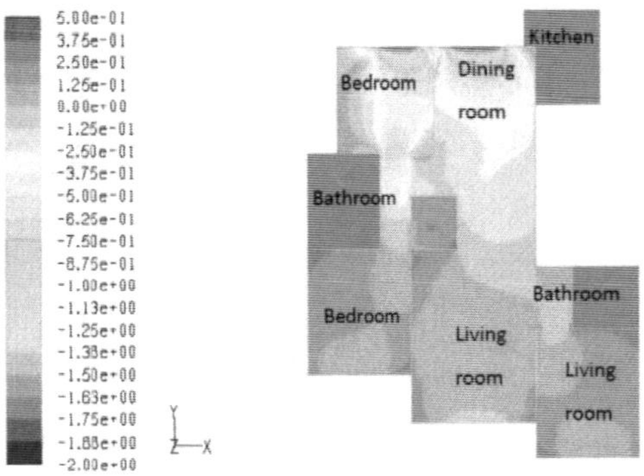

Figure 6. PMV distribution of natural ventilation

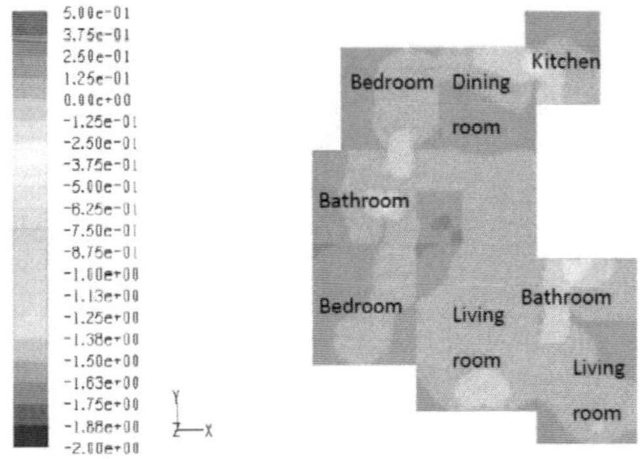

Figure 7. PMV distribution of natural inlet and mechanical exhaust

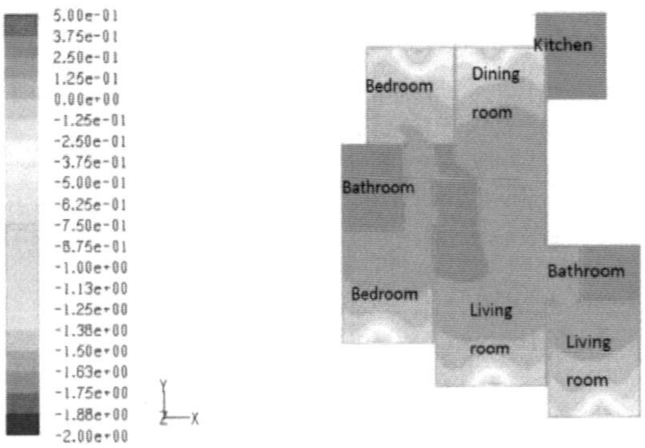

Figure 8. PMV distribution of wall type ventilator with heat exchange

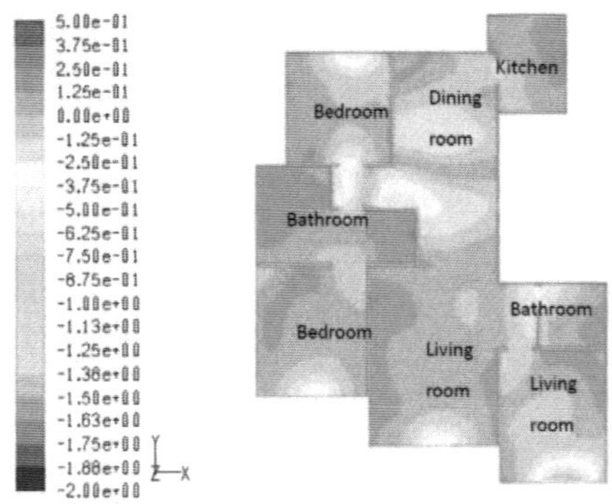

Figure 9. PMV distribution of ventilation unit with heat exchange

4.3 Initial investment

Initial investment of different ventilation modes are listed in Table 1.

Table 1. Initial investment of different ventilation modes

	Natural ventilation	Natural inlet and mechanical exhaust	Wall type ventilator with heat exchange	Ventilation unit with heat exchange
Required equipment	——	1 fan+6 inlets+4 outlets+ air duct	5 wall type ventilators	1 air exchange unit+5 inlets+4 outlets
Initial investment	——	3000-3600 RMB	20000-25000 RMB	5000~6000 RMB

5. Conclusions

In this study, air distribution, thermal comfort index, and initial investment of different ventilation modes are analyzed and compared. Therefore, their comprehensive evaluation index are listed in Table 2.

Table 2. Comprehensive evaluation of different ventilation modes

Ventilation modes	Air distribution uniformity	thermal comfort performance	Economic performance	Comprehensive performance
Natural ventilation	Common	Common	Best	Common

Natural inlet and mechanical exhaust	Better	Best	Best	Better
Wall type ventilator with heat exchange	Common	Better	Common	Better
Ventilation unit with heat exchange	Best	Best	Better	Best

References

[1] Qin O Y and Zhu Y X 2004 Study on winter ventilation scheme for residential buildings in northern China *China Housing Facilities* **6** 26-8
[2] Wang Z C 2011 *Residential ventilation design and evaluation* (Beijing: China Architecture& Building Press)
[3] Guan D H 1991 *Air conditioning* (Beijing: Tsinghua University Press)
[4] Indoor air quality standard GB/T 18883-2002
[5] Design standard for energy efficiency of residential buildings in severe cold and cold zones JGJ 26-2010
[6] Design code for heating ventilation and air conditioning of civil buildings GB 50736-2012
[7] Zhang X M, Ren Z P and Mei F H 2001 *Heat transfer* (Beijing: China Architecture& Building Press)

IOP Conf. Series: Earth and Environmental Science **40** (2016) 012061 doi:10.1088/1755-1315/40/1/012061

Research on feasibility of computational fluid dynamics (CFD) method for traffic signs board calculation

S Chao[1], C W Jiao and S Liu

Research Institute of Highway Ministry of Transport, 8 Xitucheng Rd. Haidian Dist., Beijing, China

E-mail: z.chao@rioh.cn

Abstract. At this stage of the development of China's highway, the quantity and size of traffic signs are growing with the guiding information increasing. In this paper, a calculation method is provided for special sign board with reducing wind load measures to save construction materials and cost. The empirical model widely used in China is introduced for normal sign structure design. After that, this paper shows a computational fluid dynamics method, which can calculate both normal and special sign structures. These two methods are compared and analyzed with examples to ensure the applicability and feasibility of CFD method.

1. Background

In recent years, with the continuous development of China's highway, the investment of traffic signs is huge to ensure the safety of the highway [1]. The study on low wind load traffic sign board can reduce the sign board support structure and materials, and greatly reduce the production cost of signs. However, the wind load calculation model for signs in China national standard can only be used for typical sign board, which cannot meet the calculation requirements for special sign boards, which take measures (including punching, slotting etc.) to reduce the wind load. In this paper, CFD simulation model is provided to calculate the wind load on sign board. It will be provided as a basis for traffic signs structure design to reduce construction cost. This method will have wide application prospect.

2. Calculation method of wind load on traffic signs

Wind load is the main external load of traffic sign structure. The components of the structure, e.g. the sign board, the post and the crossbeam exposure to the sign board, are all under the wind load. Because of its large area, the largest wind load is on the board of traffic sign [2][3]. This study mainly refers to the wind load on the traffic sign board.

At present, there are two methods for calculating the structure of traffic signs normally [4]. One is application of structural mechanics and mechanics of materials, simplifying the structure into a thin walled bar system, and using limit state design method to check the design. The other one is approximate calculation method of finite element. The first method is widely used in real design because of rigorous and simple. There two methods considering wind load are both based on the static for the method that the wind load are distributed on board of traffic sign [5].

---

[1] Address for correspondence: Research Institute of Highway Ministry of Transport, 8 Xitucheng Rd. Haidian Dist., Beijing, China. Email: z.chao@rioh.cn.

The calculation formula of the wind load on traffic signs based on design method based on limit state is provided in National Standards GB5768 [6]:

$$F_{wb} = \frac{1}{2}\gamma_0\gamma_Q\rho CV^2A$$

F_{wb} (unit N) is the wind load on the sign board. γ_0 is structure importance coefficient, equals 1.0, because traffic signs structure security level refers to level 2. γ_Q is variable load factor, normally equals 1.4. ρ is air density, commonly equals $1.2258N \cdot s^2 \cdot m^{-4}$. C is coefficient of wind load, equals 1.2 for traffic sign board. V (unit m/s) is wind speed. A is the area toward the wind.

In recent years, with the development of CFD and its software, the relevant research on the wind load on traffic signs has been analyzed using CFD. In 2008, a research on the method of determining the wind load on signs and traffic signal structures was carried out in the University of Iowa. In this research report, the normal calculation method for signs wind load in relevant standards is static design method considering only static wind load, is only for isolated rectangular board. But this method is not suitable for the adjacent and complex structure sign boards. The CFD method is applied in this research to assess the total wind load on sign boards and its actual distribution on, in front and behind the board. The research focus on the aspect ratio and spacing of sign board, the thickness of variable information sign, back-to-back signs, add-on exit sign, the wind tunnel effect in the formation of trucks underneath the signs potentially by using CFD. These research findings have affirmed the necessity of applying CFD to study on the wind load on traffic sign board [7].

3. Computational Fluid Dynamics (CFD) overview

CFD is the product of the combination of modern fluid mechanics, numerical mathematics and computer science [8]. It can apply various discrete mathematical methods by using computers to do Numerical experiments, simulation, and analysis on various problems in fluid mechanics [9][10].

Basic features of CFD are numerical simulation [11] and computer experiment. It has greatly replaced the expensive experimental equipment for fluid dynamics based on the basic physics theorem. Therefore, it has a tremendous impact on scientific research and engineering technology. At present, CFD is widely used in many engineering fields, e.g. space design, automobile design, biomedical industry, chemical industry, turbine design, etc.. CFD method has the advantages of low cost and can be simulated more complex or more ideal process, etc.. After a certain assessment, CFD software can broaden the scope of the experimental study, and reduce the cost of expensive experimental work. Using a computer to simulate phenomenon under given parameters is equivalent to a numerical experiment. CFD software is normally able to achieve a variety of optimization of the physical model, e.g. steady and unsteady flow, laminar flow, turbulent flow, incompressible and compressible flow, heat transfer, chemical reaction, etc.. CFD software can be used to find a suitable numerical solution according to the flow characteristics of each physical problem. Furthermore, it can achieve the best in terms of computational speed, stability and precision [12].

4. Problems and computing software selection

The fluid involved in the wind load on the sign board problem is the air that can be simplified as steady state, steady, compressible viscous fluid. This type of fluid can be solved using most of the main fluid mechanics software. But the wind load on the sign board problem has three following characteristics large calculation area, special shape of computing object and huge amount of computational grids [13][14].

For the first characteristic, in practical application traffic signs are always in open space where very open around and without structure higher than signs. For this kind of practical application simulation, the calculation area must be set large enough to avoid similar situation of pipe flow, and to avoid the influence of calculation regional boundary on the fluid motion around traffic signs. In addition, the flow field variation must be fully grasped around traffic signs.

For the second one, traffic signs have a specific shape, whose prominent feature is the characteristics of thin board. In other words, the sign board area is large, whose length and height can reach a few meters, but its thickness is quite thin. In China National Standard, the minimum thickness of the aluminum alloy board, thin steel board and synthetic resin board are 1.5mm, 1.0mm and 3.0mm. In practical use, the thickness of the sign board is generally no more than 5mm. Therefore, the area and the thickness of the signs are nearly 106 times.

For the last characteristic, the grids division of CFD simulation calculation for traffic signs can both reflect close to the sign board of the smaller size (mm level), but also can fill the whole simulation area (10m level). If all the grids are divided by the mm level, and the computational are is calculated according to the length of the 10m cube, the number of the mesh is 109. This order of magnitude grids number cannot be calculated by any existing commercial CFD software.

All above, CFD analysis of wind load on traffic sign board is faced with a huge amount of grids. There are two ways to solve this problem. One is to meet the computational area size, while to reduce the number of computing grids to the acceptable range. The other is to use large grids processing capacity of the CFD software. Accordingly, the CFX simulation software (from Ansys co.) was chosen in this study, and the icem-CFD in its software package was used as a grids partition tool.

Computational speed and stability are improved comparing the traditional method, because fully implicit multi grids coupled solution technique is used in CFX software. And because its convergence rate is fast, which is faster than other fluid software 1-2 magnitudes in the same conditions, it can calculate a million grids model.

Icem-CFD can generate a boundary layer mesh. According to the principle of fluid mechanics, dense mesh is generated in a certain region close to the simulation object to describe the details of the model and fluid. The gradually sparse mesh is generated in the calculation area far from the model. Thus, the number of the grids is reduced in the region.

5. Computational area and boundary conditions

According to the actual environmental characteristics of the sign setting, the computational area and boundary conditions of CFD simulation were set up. The size of the sign board is 6m length, 4m width and 0.003m thickness. For minimizing the number of grids, the fluid region is shown in Figure 1.

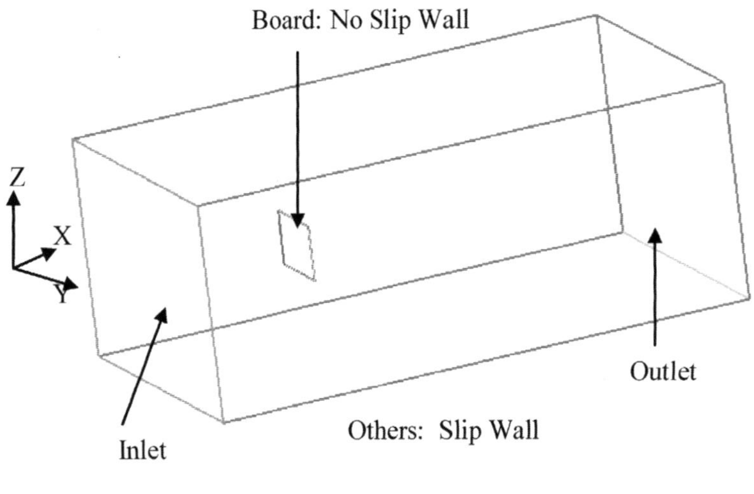

Figure 1. Computational area and boundary conditions from CFD software

In this condition, X=-12~30m, which is in front of and behind sign board 12m and 30m. Y= -12~12m, which is 12m on the left and right side of the median line of the sign board. Z=-8~8m, which is 5m and 7m from board lower and upper edge to bottom and top of computational region, and 4m high sign board. For fluid flow characteristics, considering the calculation of wind load on traffic signs the reynolds number is $Re = \rho VL/\mu$. In the formula, ρ=1.2258$N \cdot s^2 \cdot m^{-4}$stands for air density. V stands for the calculation wind speeds which are 20m/s, 25 m/s, 30 m/s, 35 m/s, 40 m/s. Feature length L is 6m the longest side of the sign board. μ=17.9×10-6 Pa·s is air dynamic viscosity. After that, Re is 9.6×106~1.92×107 in this condition. For the boundary conditions, the computational area import is speed imports, which are 20m/s, 25m/s, 30m/s, 35m/s, 40m/s. The export is free export, with 0Pa pressure. The four wall boundaries in the calculation area are all sliding wall boundaries. The effect of the air flow is neglected, because the calculation of CFD simulation is for cantilever or framed sign as an example whose height is 5m from board lower edge to bottom of computational region. In the test phase, non-slip and slip are compared for boundary conditions of fluid field bottom. The test results of wind load on sign board also show consistent in non-slip and slip conditions. Therefore, the bottom can be considered as a slip boundary.

6. Calculation for wind load on typical sign board

In the conditions above, CFD simulation for wind load on a typical sign board with 5 different wind speeds has been finished. The simulation results are consistent with the results of the empirical formula given by GB5768. Therefore, the method of CFD simulation calculation for wind load on typical sign board can be determined.

6.1 Computational model

6.1.1 Model set up

According to the above calculation region and boundary conditions, wind load on traffic sign board simulation model has been set up, and grids has been divided using icem-CFD software in this study.

6.1.2 Grid division

The essence of CFD is that the control equation is point or region dispersed in the required area to become defined algebraic equations at each grid point or in sub-region. Thus, the algebraic equations can be solved by using the method of linear algebra. The grid division is a geometric approximation of the physical component model. After grid division, the physical component model is divided into discrete aggregate of nodes and elements containing many simple geometric elements.

Grid division is one of the key steps in finite element calculation. When the object is very complex, the grid division technology often plays a vital role, which directly affects the accuracy of the subsequent numerical analysis results. The different grid divisions will cause the different solving method of numerical integration. Therefore, a reasonable unit must be used in this simulation solution.

The geometry of the traffic sign board is simple, but the thickness of the plate is only 3-5mm, which is far less than its height and width (generally more than 1m). Thus the grid division needs to reflect the details, but also to control the total number of grids. In addition, for the complex physical model structure (the different typical and special sign boards) the method of tetrahedral mesh and boundary layer was used in the computation phase to avoid the influence of hexahedral mesh block form the result.

Because of the large calculation area (length × width × height=42m×24m×16m) and the only 3mm thickness, the boundary layer mesh was set up to maximum reduce the number of grid. The grid density is increased in the area near the sign board, and the volume of a single grid is gradually increasing away from the signs to achieve the purpose of reducing the grids' number while meeting the computing requirement. The thickness of the first layer is 0.2mm, according to the 3mm thickness of the sign board.

Finally, the area was divided into 2,744,000 computational model grids, as shown in Figure 2.

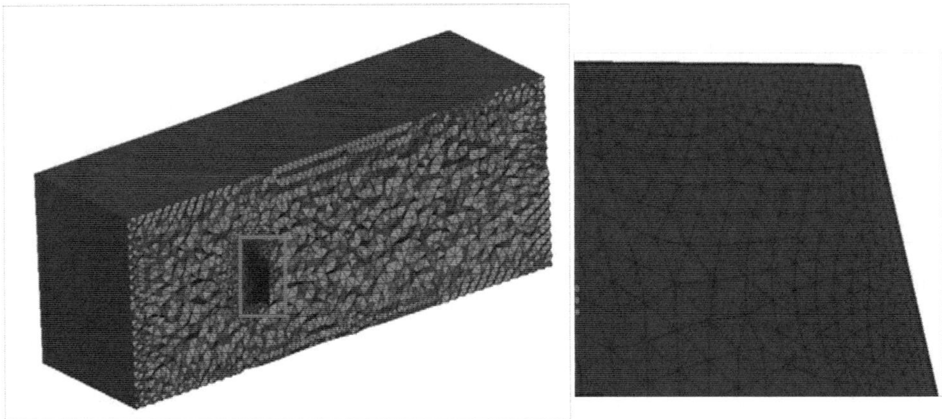

Figure 2. Computational area and boundary conditions for typical sign board from CFD software.

6.2 Calculation results and analysis

After the above settings, the 6m×4m sign board was simulated with the 5 different wind speeds, as the following results.

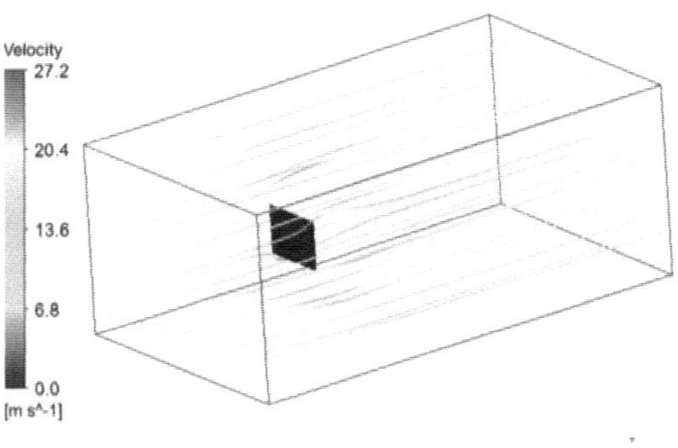

Figure 3. (a) v=20m/s

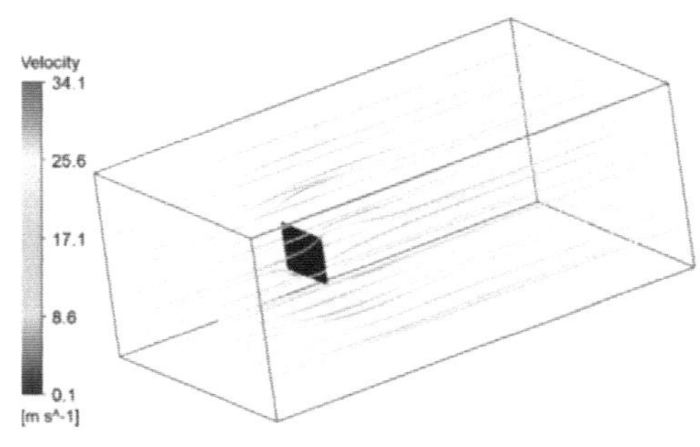

Figure 3 . (b) v=25m/s

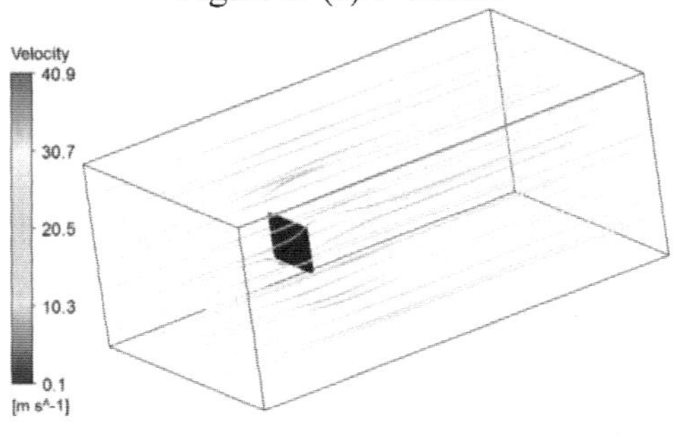

Figure 3. (c) v=30m/s

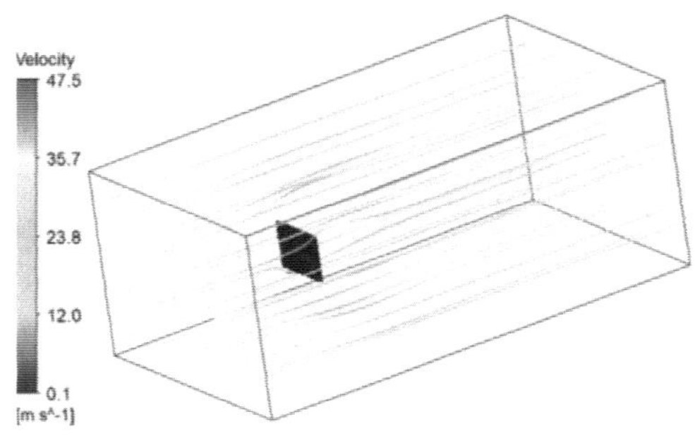

Figure 3. (d) v=35m/s

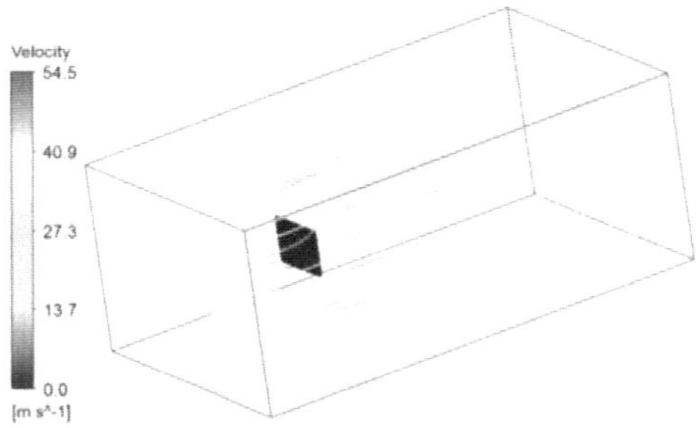

Figure 3. (e) v=40m/s

Figure 3. Calculation of watershed river flow diagram from CFD software

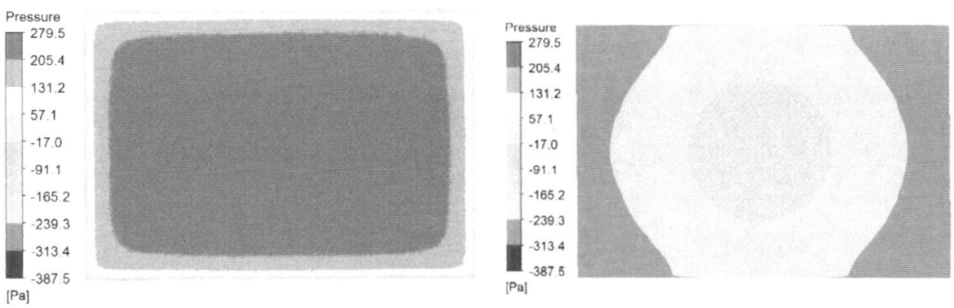

Figure 4. (a) v=20m/s

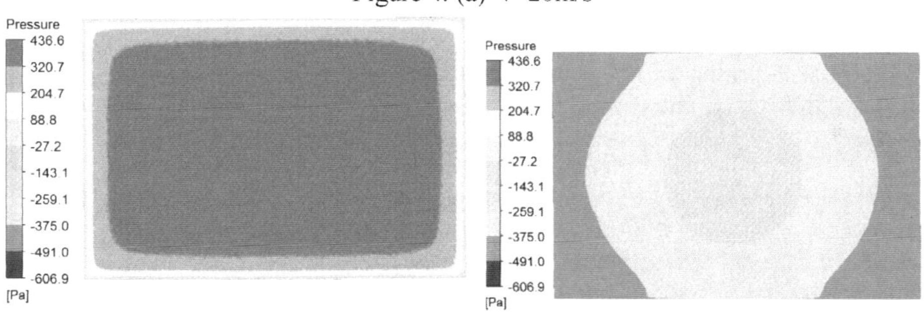

Figure 4. (b) v=25m/s

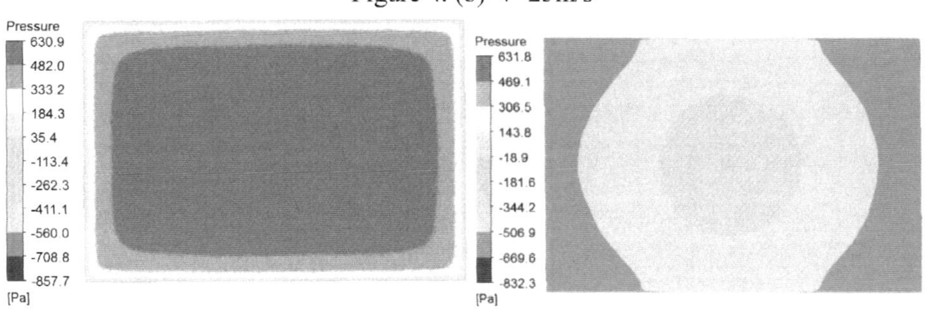

Figure 4. (c) v=30m/s

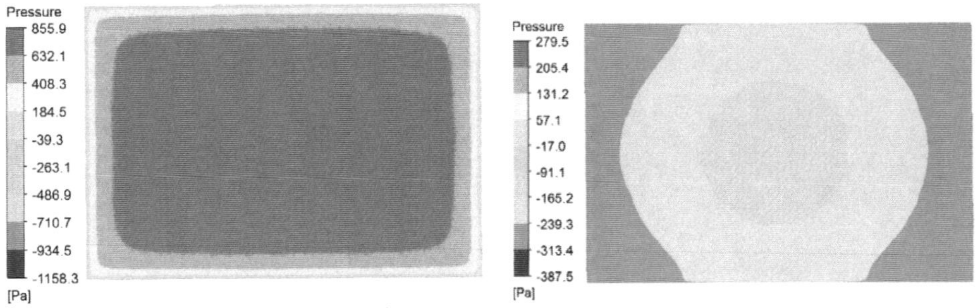

Figure 4. (d) v=35m/s

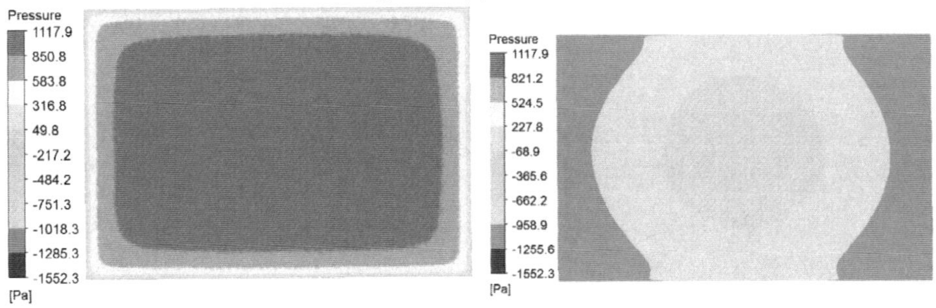

Figure 4. (e) v=40m/s

Figure 4. Calculate the watershed pressure distribution from CFD software.

In figure 4, the upwind and leeward sides pressure distribution of sign board decreases from inside to outside, but the decreasing trends are different. On upwind side, the pressure distribution is average with small gradient on central area. But the pressure gradient is large on the outer edge, where the speed increases significantly. On leeward side, pressure distribution is from the center to the outside similar to a ring.

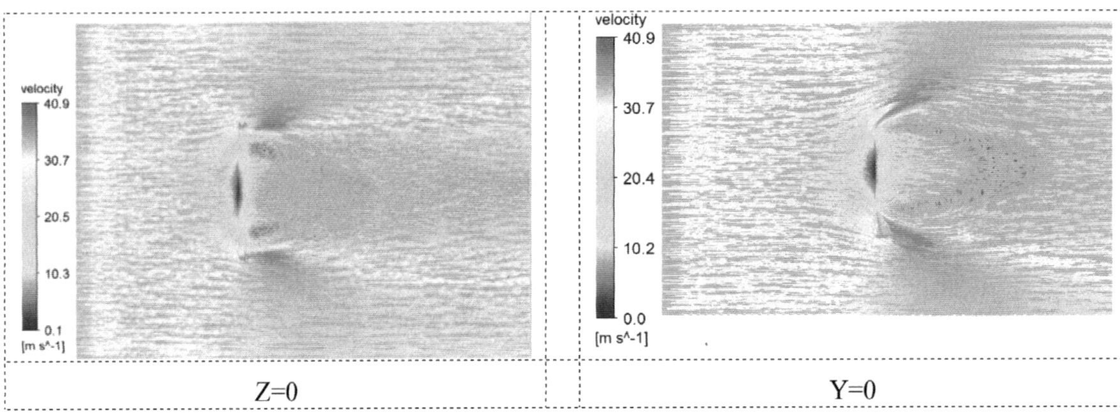

Figure 5. Calculate watershed velocity vector diagram from CFD software.

According to figure 5, the main reason of the pressure distribution on leeward side is the formation of the return zone to absorb fluid. At the same time, it pushes the fluid around the field to form a new distribution of velocity and pressure.

Figure 6 shows the velocity distribution on the central line. The recirculation zone can be seen within 7.2m behind the sign board, obviously.

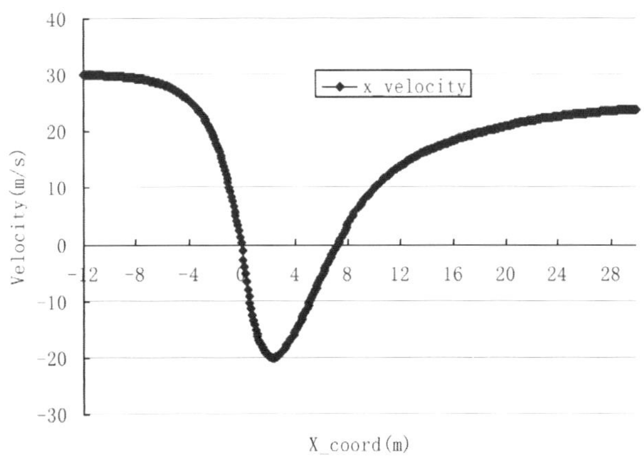

Figure 6. X direction speed on centre line: 30m/s from results.

Through integral on pressure on sign board surface, the pressure direction of wind direction was received. Then the wind load was compared with the empirical values. After comparison, the calculation results were very close, whose difference is about 1% shown in table 1. It can be considered that the grid is close to the different Re number. It shows that the simulation results are consistent with different wind speeds. Therefore, the computational model, grid division, computational area and boundary conditions are set reasonably.

Research Institute of Highway Ministry of Transport, 8 Xitucheng Rd. Haidian Dist., Beijing, China

Table 1. Wind load calculation accuracy analysis from results.

Wind speed(m/s)	Empirical formula results(N)	CFD calculation result(N)	Relative error
20	9885	9979	0.95%
25	15445	15594	0.96%
30	22241	22468	1.02%
35	30272	30600	1.08%
40	39539	39988	1.13%

7. Summary

In this study, two methods for calculating wind load on traffic sign board have been provided and compared. The empirical method, calculate the wind load based on the area of the sign board, cannot solve the wind load on sign board with complex structure. The other CFD method can calculate the finite element the air flow field of complex structure. And it can well describe and solve the problem of fluid mechanics, which cannot be solved by the method of arithmetic. Furthermore, a typical sign board has been calculated by using CFD simulation software with 5 different wind speeds. The results are no significant difference with the results by using the empirical model. Therefore, it can be proved that the method of CFD is feasible to calculate the wind load of the sign board. In addition, it can develop the wind load reduction measures and the study on traffic signs in the complex situation.

However, the method is too complex for common traffic engineers. And it takes huge time about 3 days at least (normally a week) to build model and run program on large computer in CFD software for every different signs. Therefore, most traffic engineers prefer to use traditional method in Chinese standard and specification, which will take only several minutes to calculate wind load [10][11][12].

But this simple method cannot make calculation for special sign board (e.g. punched or slotted). For widely using in real traffic situation, some selected typical sign boards (or most common sign board) can be simulated and operated wind tunnel test. The results for those boards will be used in other projects. Furthermore, the approach has been successfully practiced in a demonstration project to calculate reducing wind load measures (including punching, slotting etc.). And the total steel consumption, construction and maintenance costs were reduced.

Acknowledgments
This research is supported by "The National Key Technology R&D Program: Highway Safety Design Technology for Qinghai-Xizang Plateau Special Circumstances" (No.2014BAG05B02) and "special funding project of basic scientific research for central level public welfare research institute: Development and research of reducing wind load on traffic sign board".

References
[1] JTG B01-2014, Ministry of Transport of the People's Republic of China Technical standard of highway engineering. 2014
[2] British Columbia Ministry of Transport and Highways 2009 Manual of Standard Traffic Signs & Pavement Markings
[3] JTG/T D81-2006, Ministry of Transport of the People's Republic of China Specification for Design of Highway Safety Facilities. 2006
[4] Federal Highway Administration 2009 Manual on uniform traffic control devices 2009 edition
[5] Constantinescu G, Bhatti A and Tokyay T 2008 Improved method for determining wind load on highway sign and traffic-signal structures
[6] GB5768.2-2009 Ministry of Transport of the People's Republic of China 2009 *Road traffic signs and markings – Part 2: Road traffic signs.* GB5768.2-2009
[7] Simiu E and Seanlan R H 1995 Wind Effects on Structures - An Introduction to Wind Engineering.The3rd Edition. John Wiley & Sons, INC. **33-99** 135-88
[8] Roache P J 1980 Computational Fluid Dynamics Hermosa Publishers
[9] Krishna P 1995 Wind load on low rise buildings – a review *J. Wind Eng. hid. Aerodyn* **54/55** 383-96
[10] Hangan H, Roberts D and Xu Z 2004 Downburst simulations experimental and numerical challenges *Proc. the 2004 Structures Congress – Building on the Past: Securing the Future* 1657-64
[11] Murakami S and Mochida A 1987 Three dimensional numerical of airflow around a cubic mode by means of large eddy simulation *J. Wind Engr* **25** 291-305
[12] Tamura T, Kawai H, Kawamto S, Nozawa K, Sakamoto S, Ohkuma T, 1997 Numerical prediction of wind loadings and structures - Activities of AIJ Cooperative Project on CFD *J. Wind Eng. Ind. Aerodyn.* **67/68** 671-85
[13] BS EN 1991-1-4:2005 Eurocode 1: General Actions – Wind Acitons
[14] Kim J and Hangan H 2007 Numerical simulations of impinging jets with application to downbursts *J. Wind Engineering and Indus-trial Aerodynamics* **95** 279-98

Numerical simulation on a straight-bladed vertical axis wind turbine with auxiliary blade

Y Li[1 3], Y F Zheng[1], F Feng[2], Q B He[1] and N X Wang[1]

[1]Engineering College, Northeast Agricultural University, Harbin, China
[2]College of Science, Northeast Agricultural University, Harbin, China

Email: liyanneau@163.com

Abstract. To improve the starting performance of the straight-bladed vertical axis wind turbine (SB-VAWT) at low wind speed, and the output characteristics at high wind speed, a flexible, scalable auxiliary vane mechanism was designed and installed into the rotor of SB-VAWT in this study. This new vertical axis wind turbine is a kind of lift-to-drag combination wind turbine. The flexible blade expanded, and the driving force of the wind turbines comes mainly from drag at low rotational speed. On the other hand, the flexible blade is retracted at higher speed, and the driving force is primarily from a lift. To research the effects of the flexible, scalable auxiliary module on the performance of SB-VAWT and to find its best parameters, the computational fluid dynamics (CFD) numerical calculation was carried out. The calculation result shows that the flexible, scalable blades can automatic expand and retract with the rotational speed. The moment coefficient at low tip speed ratio increased substantially. Meanwhile, the moment coefficient has also been improved at high tip speed ratios in certain ranges.

1. Introduction

According to the relative position of the axis and the direction of the wind, wind turbines can be divided into horizontal axis wind turbine (HAWT) [1] and vertical axis wind turbines (VAWT).With the many attributes the straight-bladed vertical axis with the advancement of low vibration, low noise, high safety, and the simple structure [2-3] has become very popular. Meanwhile, the vertical axis wind turbines have an increasing share in the small and middle wind turbine market year by year, and it also gets more attention of experts and scholars [4-7].

Based on the working principle of the VAWT, it can be divided into the lift type and drag type. With them, the Darrieus wind turbine and the Savonius rotor were the typical representative of the VAWT by lift and drag type. Under low wind speeds, the performance does not start off well. However the blade does have high tip speed ratio, with high rotor power efficiency characterizes lift-type. Compared to a lift type wind turbine, drag type can start by itself quickly, but rotor power efficiency is low at high tip speed ratio. To solve the contradiction between the starting performance and high rotor power efficiency of the VAWT, a large number of researchers are studying it. For the Darrieus wind turbine, the installation of guide plates around the wind turbine to improve the rotor power efficiency and the starting performance was widely adopted by some scholars. For the condition

[3] Address for correspondence: Engineering College, Northeast Agricultural University, Harbin, China. Email: liyanneau@163.com.

of the VAWT with guide plate, CFD numerical calculation was carried on by Wang Xin [8-10], the result shows that guide plate can effectively improve the turbine inflow wind speed. Then the rotor power efficiency can be enhanced. B.D.A ltan [11] placed the curtain in front of the Savonius wind rotor, and W.T.C Hong [12] has designed a novel omni-direction-guide-vane (ODGV) around a vertical axis wind turbine, both the aims were to improve the inflow wind speed around the wind turbine. About the Savonius wind machine, the optimal design was taken with structure parameters by some scholars, including blade aspect ratio, blade overlap ratio, the number of stages, geometry of the blade, number of blades [13-15]. Also to improve the rotor power efficiency, front guide plate was proposed by Golechato, which could reduce the blade return resistance [16]. Mohamed [17] through numerical simulations discussed the impaction on performance from different guide plates.

To improve the condition of a drag-type wind turbine in high speed with low rotor power efficiency and lift-type wind turbine starting by itself, an innovative vertical axis wind turbine with a flexible, scalable auxiliary blade was put forward. As the low-speed wind turbine works, the flexible blade is open, creating a triangle edge, increasing the drag force. As the speed increases, due to the effect of centrifugal force, the flexible blade can be curled into by-blade; then the by-blade returns to the dynamic characteristics, becomes a kind lift blade completely: at this moment, the lift is the mainly driving force. By testing the torque output characteristics under different wind and rotating speeds, the two-dimensional wind turbines carried computational fluid dynamics (CFD), resulting in the effectiveness of the new type of wind turbine. In this study, the influence on the performance of this new wind turbine was discussed as a function of different tip speed ratios, which could provide the theoretical basis and guidance for practical applications.

2. Basic Theory

Figure 1 shows the three-dimensional structure of the new wind turbine. To express the work theory clearly, the figure of three-dimensional (Figure 1) and two-dimensional (Figure 2) depict two blades, while the model was with four blades in the calculation, thus consistent with the actual situation.

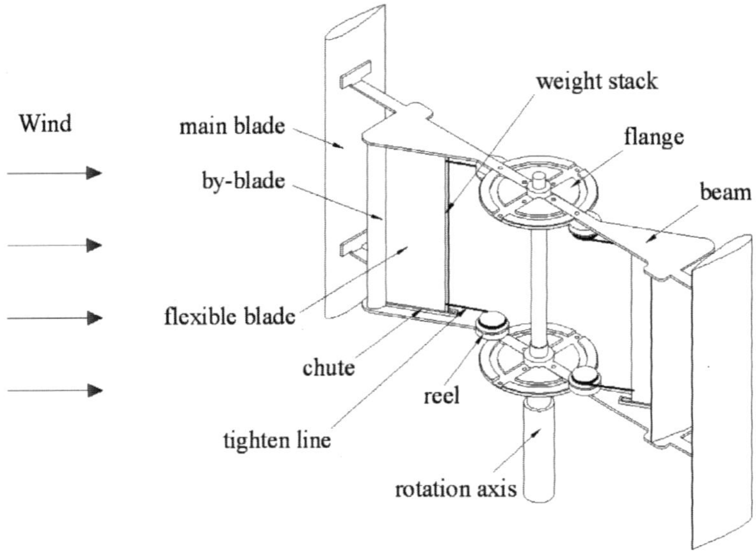

Figure 1. 3D view of the new type SB-VAWT

Main blades installed on the rotation axis by beams and flange; by-blades installed between the upper and lower beams which rotate around the rotation axis together with main blades. At the most thickness of by-blade, there is a round hole. The rotating drum was instrumented in round hole of the by-blade. One end of flexible blade connected to the drum through the bar hole on the by-blade, and the other end fixed weight stack. By attaching the lower and upper-end instrument to the chute, the weight stack could then be connected to the reel of the tightening line. The drum and the end of reels respectively, were equipped with torsion spring and spiral spring. In the initial state of wind turbines, the flexible blade was in a state of tension under the role of spring force (Figure 2).

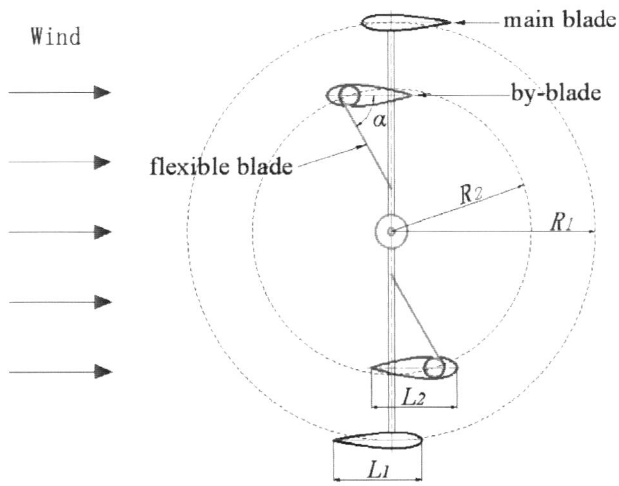

Figure2. 2D schematic of the new type SB-VAWT

At low speed of wind turbine, the strength of torsion spring and centrifugal force were smaller than the strength of spiral spring, so the flexible blade was open. Triangle drag blade could produce larger driving moment so that the new wind turbine can start by itself at low wind speed. With the rotational speed increase, the centrifugal force (provided by weight stack) increases gradually, at the same time, the flexible blade shrinks slowly under the action of centrifugal force. When the speed reached a certain value, the flexible blade curled into the by-blade completely. The by-blade recovered its original edge profile. At this state, the lift characteristics of the main blade and by-blade play gradually, rotating moment of wind turbine mainly produced by the main blade, and the by-blade also provided a certain amount of driving moment. At this time, the rotor power efficiency had been improved under the high rotational speed.

From the initial condition, obtaining the high rotor power coefficient and optimal aerodynamic performance of the blade, we needed to apply different pre-tightening forces on the torsion and spiral springs, choose different weight stacks, change the rate range from the beginning of the telescope; and control the curled by blade from the rotational speed within the weight stack.

3. Numerical simulations

To explore the working mechanism of the flexible blade deeply, and to research the characteristics of the rotor moment at different tip speed ratio, unsteady CFD calculation about this new wind turbine was carried out. Because the straight-bladed vertical axis wind turbine has the same cross section shape in the height direction, the influence of the length direction can be ignored, and the results of the two dimensional model can also have a high reliability according to the previous research results [18]. The unit length (1m) was selected for two-dimensional numerical simulation calculation. In the process of calculation, to reduce the amount of calculation, we removed the unnecessary structure characteristics like the beam, which does not much affect the calculation accuracy of numerical calculation; hence, we just retain the blade.

For comparing with the wind tunnel experimental results conveniently, blade number is 4, NACA0018 airfoil blade was selected by main blade, NACA0025 airfoil blade (drum are necessary added at the maximum thickness of the by-blade) was selected by by-blade. The main structural parameters of this numerical calculation were shown in Table 1. Compare the change of moment coefficient with the added auxiliary scalable blades, the operating conditions of the main blade, including the main blade and the auxiliary blade, and all the blades, calculated, and the moment coefficient curve of different combinations of blade obtained. Also, the wind speed in this calculation was 6 ms-^{1} and10 ms-^{1}.

Table 1. Main structural parameters of the wind turbine

Parameter	Value
Main blade radius, R_1	0.4m
By-blade radius, R_2	0.25m
Main blade chord, L_1	125mm
By-blade chord, L_2	120mm
Angles between by-blade and soft blade, α	60 degree

3.1 Computational Region and Mesh Division

To make the flow fully development, we avoided the calculation of the areas too narrow or too small, which may have an adverse impact on results, the calculation area was set to 10D×4D rectangular area. The calculation model put in the length direction of the symmetry axis and 3D distance to the inlet in calculation area, such as shown in Figure 3.

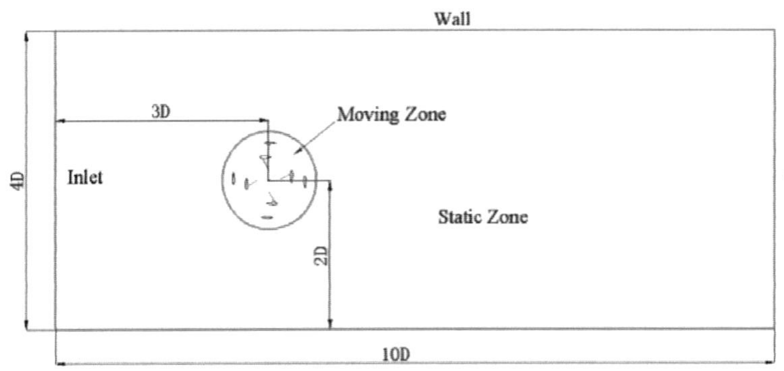

Figure 3. Calculation region and boundary conditions

Using sliding mesh technique, the computational area was divided into a static and a rotating areas, and the wind turbine was wrapped in the rotating area. The setting of the boundary in the computational area included: the boundary between static and rotational area set as slip boundary, outlet boundary set as pressure export, upper and lower boundary set as slip surface, blade surface set as no slip wall, inlet boundary and set as constant speed entrance.

The Gambit software generates the computation grid. The whole calculation area meshes. Around the blade, the grid is encrypted. The grid node density utilized in the rotor domain is higher than in the other domains. The growth rate of layer is 1.05, the maximum cell squish is 1.97407e-001 (0 is the best quality, 1 is the worst quality), so the calculation of boundary layer is very good. The number of elements is about two hundred to three hundred thousand cells, such as shown in Figure 4.

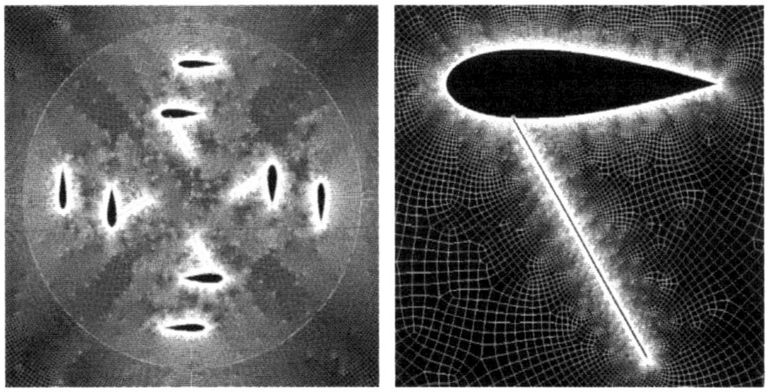

Figure 4. Moving zone mesh and local mesh

3.2 Calculation setting

The calculation based on the model of RNG k-ε used the standard wall function method to solve the problem of low Reynolds number. On the control equation of the discrete format, pressure interpolation adopted Standard method, pressure and velocity coupling used SIMPLE algorithm. To ensure the calculation precision, the momentum equations, turbulence kinetic energy k equation and dissipation rate ε equation all adopted the second order upwind format.

In the conventional VAWT CFD calculation, the state of blade was constant, but in this paper, the position of flexible blade was changing at different rotational speed, the model need to grid by time. For the purpose of simplified calculation and finding out the optimal tip speed ratio flexible blade completely into by-blade, fixed the position of the flexible blade, just calculate the moment coefficient at different tip speed ratio.

4. Results and discussion

The moment coefficient (C_m) and tip speed ratio (λ) defined as Equation (1) and (2). The two parameters are important parameters for measuring the performance of the wind turbine.

$$C_m = \frac{M}{\frac{1}{2}\rho A U^2 R} \tag{1}$$

$$\lambda = \frac{\omega \cdot R}{U} = \frac{\pi n R}{30 U} \tag{2}$$

Where n is the rotor speed rpm; R is the rotor radius, m; U is the flow speed, ms^{-1}; M is the wind turbine moment, N/m; ρ is the air density; A is the swept area of wind turbine, m^2.

Figure 5 shows the wind turbine moment curve between the moment coefficient and the tip speed ratio. Figure 5 (a) is for U = 6 ms^{-1} and Figure 5 (b) is for U =10 ms^{-1}. "All blade" includes flexible blade, by-blade and main blade in the figure.

By comparing the traditional SB-VAWT and adding the flexible blade, the low tip speed improved. However, the coefficient decreased. This result is in agreement with the Savonius rotor on the moment coefficient change trend. After completely curled into by-blade, the moment coefficient of the wind turbine which has double blades compared with the original almost has the same change trend, the biggest moment coefficient is almost the same, but the overall move towards low tip speed ratio. This provides favorable conditions for using the wind power at lower tip speed ratio.

Comparing the change of the moment coefficient curve under U = 6 ms^{-1} and U = 10 ms^{-1}, it can be seen that the change trend of the moment coefficient at different wind speeds is consistent, but with the increase of wind speed, the moment coefficient increases at the same tip speed ratio. The figures show that with the increase of wind speed, also wind energy efficiency undergoes a certain degree of increase.

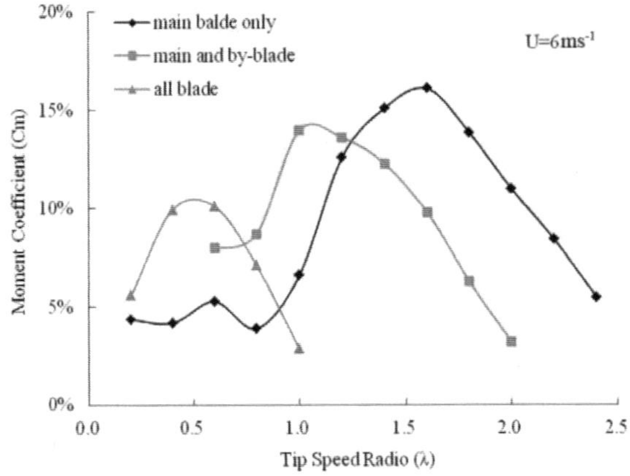

Figure 5. (a)

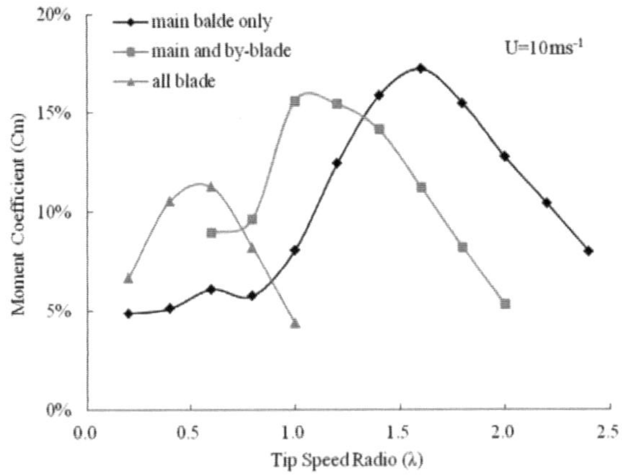

Figure 5. (b)

Figure 5. Moment coefficients of the rotor (a) U = 6 ms^{-1}; (b) U=10ms^{-1}

5. Conclusions

This paper puts forward a new type of straight-bladed vertical axis wind turbine with flexible, scalable blades in the rotor, and its aerodynamic characteristics were studied by means of numerical simulation. From the above results, we can conclude the following.

Flexible scalable blade plays an important role in the new type of vertical axis wind turbine. At low tip speed ratio, flexible auxiliary blade will open, with by-blade together assembling into the resistant wind wheel, to solve the straight-blade vertical axis wind turbine startup problems. At high tip speed ratio, flexible auxiliary blade retracted into by-blades, lift characteristics of a wind turbine is recovered then, which can improve the utilization of wind energy to some extent.

Acknowledgments

This study was sponsored by Scientific Research Fund of Heilongjiang Provincial Education Department (No.: 12541012). The authors would like to express thank to the support.

References

[1] Burton T, Sharpe D, Jenkins N, Ervin Bossanyi 2001 *Wind Energy Handbook*. West Sussex: John Wiley &Sons. http://dx.doi.org/10.1002/0470846062
[2] Islam M, Ting D S K and Fartaj A 2008 Aerodynamicmodels for Darrieus-type straight-

bladedvertical axis wind turbines *Renewable and SustainableEnergy Reviews* **12** 1087-109

[3] Dai G, Xu Z and Huangpu, K.L 2010 Recent Research Progress in the Vertical Axis Wind Turbine *Fluid Machinery* **38** 39-43

[4] Li Y and Tagawa K 2009 A visual experimentby smoke wire way on the static flow field of astraight-blade vertical axis wind turbine *Renewable Energy Resources* **27** 32-5

[5] Howell R, Qin N, Edwards J and Durrani N 2009 Wind tunnel and numerical study of asmall vertical axis wind turbine *Renewable Energy* **35** 412-22

[6] Rolland S, Newton W, Williams A J, Croft T N, Gethin D T and Cross M 2013 Simulations technique for the design of a vertical axis windturbine device with experimental validation *Applied Energy* **111** 1195-203

[7] Ohya Y, Shimatami T and Zhang X 2009 Distributed energy and clean energy technology (1) the research on the compact type gathering windtype wind turbine and large-scale. *shanghai electric power* **3** 210-2

[8] Wang X, Tong Z M and Wang Q K 2010 Numerical investigation on aerodynamic performanceof the vertical axis wind turbine with guiding impeller *J. University of Shanghai for Science and Technology* **32** 423-6

[9] Wang Q K 2011 Numerical investigation on aerodynamic performance for Guiding VAWT with Combined Blades *Chinese J. Mechanical Engineering* **47** 126-32

[10] Liu J, Chen K and Zhang X 2013 Numerical simulation of the wake flow of wind turbine witha flanged diffuser *J. Engineering Thermophysics* **34** 2140-3

[11] Altan B D and Atilgan M 2008 An experimentaland numerical study on the improvement of theperformance of savonius wind rotor *Energy Conversionand Management* **49** 3425-32

[12] Chong W T, Fazlizan A, Poh S C, Pan K C, Hew W P and Hsiao F B 2013 The design, simulationand testing of an urban vertical axiswind turbine with the omni-direction-guide-vane *Applied Energy* **112** 601-9

[13] Saha U, Thotla S and Maity D 2008 Optimum design configuration of savonius rotor throughwind tunnel experiments *J. Water Engineeringand Industrial Aerodynamics* **96** 1359-75

[14] Irabu K and Roy J 2011 Study of direct forcemeasurement and characteristics on blades of savoniusrotor at static state *Experimental Thermaland Fluid Science* **35** 653-9

[15] Akwa J, Junior G and Petry A 2012 Discussionon the verification of the overlap ratio influence onperformance coefficients of a savonius water rotorusing computational fluid dynamics *RenewableEnergy* **38** 141-9

[16] Golecha K, Eldho T and Prabhu S 2011 Influenceof the deflector plate on the performance ofmodified savonius water turbine *Applied Energy* **88** 3207-17

[17] Mohamed M H, Janiga G, Pap E and Thévenin D 2010 Optimization of savonius turbines using an obstacle shielding the returning blade *Renewable Energy* **35** 2618-26

[18] Louis Angelo Danao, Jonathan Edwards, Okeoghene Eboibi, Robert Howell 2014 A numerical investigation into the influence of unsteady windon the performance and aerodynamics of a vertical axis wind turbine *Applied Energy* **116** 111–24

IOP Conf. Series: Earth and Environmental Science **40** (2016) 012063 doi:10.1088/1755-1315/40/1/012063

Degradation analysis of photovoltaic modules based on operational data: effects of seasonal pattern and sensor drifting

X Y Li[1]

REC Solar PTE LTD, Singapore 637312

E-mail: ann.lixueyan@gmail.com

Abstract. Degradation analysis of photovoltaic (PV) modules based on real operational data is essential to the future development of the PV industry. Weather conditions and system drifting often lead to large uncontrollable fluctuations in operational data, which present great challenges for calculating degradation rates of PV modules. In this paper, we propose a new numerical two-step approach to overcome these difficulties. In particular, we will show that our method is able to eliminate effects of seasonal patterns and systematic sensor drifting in evaluating degradation rates of PV modules. The method is applied to the six-year operational data of a solar PV system installed at CA United States. We demonstrate that our approach can greatly improve the degradation calculations, compared with other widely used methods.

1. Introduction

The performance degradation of photovoltaic (PV) modules is one of the key factors that limit the future development of the PV industry. The degradation of modules is the main reason for the power output loss of PV systems over time. Reducing the module degradation rates is therefore technically essential for the purpose of improving the energy efficiency of solar PV systems. Because PV modules are usually guaranteed for more than 20 years by manufacturers, it is also important for financial reasons to track and control the long-term degradation of PV modules. It has been well known that the degradation of individual modules (in long or short term) can be attributed to the intrinsic properties change of the PV materials caused by either external potential --- potential induced degradation (PID) [1], or the light --- light induced degradation (LID) [2]. Since the module always operates in a system, some other mechanism external to the solar cells such as interconnect issues and solder bonds could also play a significant role in the performance degradation, which makes it necessary to determine the degradation rates under operation conditions instead of indoor testing of isolated modules. Consequently, a reliable numerical method for degradation analysis of real operational data is of great importance for not only the fundamental research but also the future development of the PV industry.

There are two major difficulties in evaluating degradation rates of PV modules from real operational data. One is the large fluctuations of the operational data due to uncontrollable external parameters such as weather conditions (raining, clouds movement, etc.) and unexpected changes of factors external to PV systems (unexpected shading, inverter problems, etc.). The other one is the

[1] Address for correspondence: X Y Li, REC Solar PTE LTD, Singapore 637312. E-mail: ann.lixueyan@gmail.com.

IOP Conf. Series: Earth and Environmental Science **40** (2016) 012063 doi:10.1088/1755-1315/40/1/012063

systematic 'degradation' in real operational data caused by the commonly existing seasonal patterns and/or the sensor drifting (for example, the drifting of irradiance sensors) instead of PV modules. The first problem has been intensively discussed. Many numerical filtering techniques that help people to effectively extract useful information from real operational data with large uncontrollable fluctuations have been proposed and widely used in degradation analysis of PV systems [3-10], a nice example of which can be found in Ref. 3. On the contrary, the second problem (caused by seasonal pattern and/or sensor drifting) has been rarely discussed in literature. It has been well known that energy output of PV systems sensitively depends on weather conditions [11] [37], but the seasonal patterns that often occurred in operational data have not been seriously considered in degradation analysis yet. Sensors are important parts of PV systems [12-19]. The sensor drifting is an inevitable issue in assessing the performance of systems. From a simple estimation, we can see how significant the sensor drifting could be for the degradation analysis of PV systems: The degradation rate of silicon based PV modules typically is around -0.7% per year [11], and the performance 'degradation' caused by sensor drifting could be easily more than -2% per year [20]. In this paper, we suggest a two-step approach (TSA) that takes into account effects of seasonal patterns and sensor drifting. From our calculations based on 6-year operational data of a PV system in California, we show that our TSA (together with the effective filtering techniques [3]) is able to significantly improve the quality of degradation analysis of PV systems, providing a reliable numerical tool for tracking module degradation for the PV industry.

2. Data collection and initial processing

The concrete roof top mounted system under study was installed at Chowchilla, CA, United States. Up to 1736 multi-crystalline solar PV modules were used with the peak power of 220W for each of them. After the commissioning of both the solar PV system and a weather station nearby (used for recording weather data), 6-year data (both solar power and weather data) is available from January 1st 2009 to December 31st 2014. In order to minimize the inverter effects, here we only present our analysis of DC data. Before the degradation analysis, the original data was filtered by irradiance limits, stability, and outlier filters to eliminate the uncertainty originated from uncontrollable fluctuations due to weather conditions, shading and system interrupts [3]. In figure 1, we show the filtered power and irradiance data. These data are used for the degradation analysis that will be discussed later. A periodic seasonal pattern can be clearly seen in both power and irradiance data. The linear fitting of the irradiance data suggests that the irradiance sensor has a drifting rate of -1.0% per year, which will cause a significant 'degradation' in calculating performance of modules.

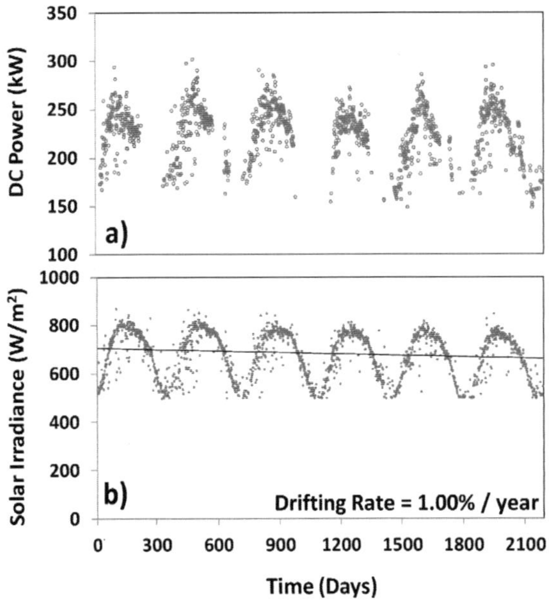

Figure 1. Filtered a) DC power of the solar system and b) irradiance data over 6 years.

The periodic seasonal pattern can be clearly seen in both figures. In b), the solid strait line is the linear fit of the irradiance data. The negative slope (-1.0% per year) of the line is caused by the drifting of the irradiance sensor.

3. Method and discussions

In conventional methods, the degradation rate of PV modules is obtained by linearly fitting either the performance ratio (PR) or the DC power of modules as a function of time. The linear fitting process eliminates the complicated periodic seasonal patterns in real operational data in a very simple manner. Also, the effects of the sensor drifting (Figure 1. b) are neglected, which may cause big problems as we shall see later. Currently, three different numerical methods of degradation analysis are widely used: 1) the PR-R_d method in which the degradation rate (Rd) is obtained by linearly fitting the PR as a function of time [7]. In this method, the PR is calculated by $PR = \frac{E_{output} \cdot P_0^{-1}}{H_I \cdot G_0^{-1}}$, where E_{output}, H_I are the measured energy output of the modules and solar irradiation, respectively. Two constants, P_0 and G_0, are the nametag rating of solar systems and the peak value of irradiance; 2) the so-called DC/ $G_{POA\text{-}corr}$ method [8] in which the degradation rate is calculated by linearly fitting the temperature corrected ratio of DC power over the irradiance as a function of time. The temperature corrected ratio ($DC/G_{POA\text{-}corr}$) is expressed as $DC/G_{POA-corr} = \frac{Power_{DC} \cdot [(T_{mod} - 25) * 0.43\% + 1] \cdot P_0^{-1}}{G_{POA} \cdot G_0^{-1}}$, where $Power_{DC}$ is the measured DC power of solar systems, G_{POA} is the plane-on-array irradiance, T_{mod} is the module temperature, and P_0, G_0 are two normalized constants as before; 3) the regression PVUSA method [9] in which the parameter R_d is computed by linearly fitting the DC power of the solar system as a function of time. In this method, the DC power is modeled as $P_{DC} = G_{POA} \cdot (A + B \cdot T_{amb} + C \cdot G_{POA} + D \cdot ws)$, where T_{amb} is the ambient temperature, ws is the wind speed, and A, B, C, D are four regression constants derived from the operational data.

We first apply the aforementioned three methods to the data shown in figure 1. The calculated PR or DC power and the linear fitting of them are shown in figure 2. The degradation rates obtained from these three calculations are +0.13%/year for the PR-R_d method, +0.84%/year for the $DC/G_{POA\text{-}corr}$, and -1.39%/year for the PVUSA. Obviously, the previous two methods (PR-R_d and $DC/G_{POA\text{-}corr}$) give the wrong sign of R_d, and the PVUSA greatly overestimate the degradation rate compared with the typical value of the degradation rate of silicon PV module (around -0.7% per year [11]). The problems of these three methods are caused by the neglect of significant irradiance sensor drifting effects (-1.0% per year for the system under study). For the PR-R_d and DC/ $G_{POA\text{-}corr}$ methods, the irradiance is in the denominator of the linearly fitted function, the neglect of the negative sensor drifting therefore leads to underestimated R_d, resulting in the wrong sign of R_d for the system under study. While, for the PVUSA method, the irradiance appears as a multiplier in the fitted function, causing an overestimated degradation rate.

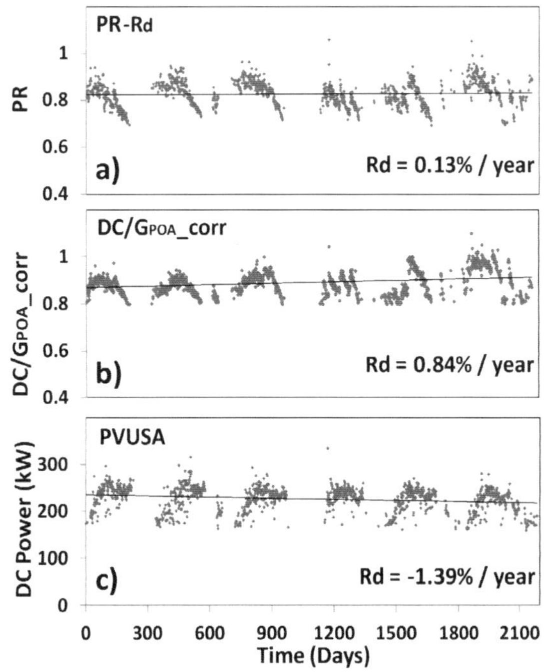

Figure 2. Degradation rates demined from linearly fitted functions with three different methods: a) PR-R_d, b) DC/$G_{POA\text{-}corr}$ and c) PVUSA. Blue dots are calculated functions from measured values, and solid lines are linear fitting.

In order to correctly take into account effects of the sensor drifting and also the seasonal pattern, we propose here a two-step approach for evaluating the degradation rate from real operational data. In the first step of the approach, the irradiance data is corrected by a linear time-varying factor so that the sensor-drifting effects can be eliminated. The corrected irradiance $\tilde{G}_{POA}$ is calculated by the following equation,

$$\tilde{G}_{POA} = G_{POA} \cdot (1 + \Delta \cdot Time) \quad (1)$$

where G_{POA} is the measured irradiance and Δ is the sensor drifting rate that can be determined by a linear fitting (figure 1b). For the system under study, Δ is -1.0% per year. The corrected irradiance is then used in the second step to determine the degradation rate. The simple linear fitting in conventional methods is a rough approximation, which may miss some important information of the fitted function that in general has a periodic seasonal pattern. In light of this, we use a new model as following to fit the DC power (not linear fitting), from which R_d can be determined,

$$\tilde{P}_{DC} = (R_d \cdot t + 1) \cdot \left(A \cdot T_{amb} + B \cdot \tilde{G}_{POA} + a_1 \cdot T_{amb}^2 + a_2 \cdot \tilde{G}_{POA}^2 + a_3 \cdot RH + a_4 \cdot ws + a_5\right) \quad (2)$$

where $\tilde{P}_{DC}$ is the fitted DC power, t is the time, RH is the relative humidity, A,B are initially set system parameters, and a_1 to a_5 are five constants that together with R_d can be determined by fitting $\tilde{P}_{DC}$ to measured DC power. For current study, the humidity and wind speed are not available, so a_3, a_4 are set to zero.

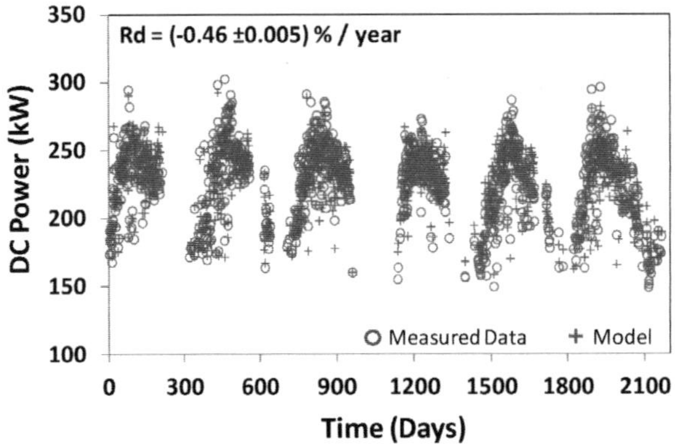

Figure 3. The degradation analysis from the two-step approach. Comparison of fitted $\tilde{P}_{DC}$ and the measured DC power. The periodic seasonal pattern is nicely captured.

We now use the two-step approach to evaluate the degradation rate of the system under study. In figure 3, we show the fitted $\tilde{P}_{DC}$ (by the standard least square method) together with the measured DC power. Clearly, the fitted data nicely captured the periodic seasonal pattern of the measured ones. The R_d is estimated to be -0.46% per year by this method, which agrees nicely with previous experimental values of similar Si modules [11]. To separate the effects of seasonal pattern and the sensor drifting, we also tried to use the equation (2) with the original irradiance data (not the corrected one in equation (1)) to fit the DC power. By doing this, we neglect the sensor drifting and obtain an underestimated R_d around 0.58% per year as expected, which is also significantly different from aforementioned linear-fitting methods.

4. Conclusion

Season patterns and sensor drifting are two common issues in the operation of solar systems [11-32,]. While, the effects of these two factors on degradation of PV systems are often neglected in conventional theoretical models [33-36].In this paper, we proposed a two-step approach to numerically evaluate the performance degradation rate of PV systems from real operational data, which takes into account effects of seasonal patterns and irradiance sensor drifting. The method is applied to the 6-year data of a PV system installed in Chowchilla, CA, United States. Results from our two-step model are compared with those based on three other widely used methods. We show that together with existing effective filtering techniques, the two-step model is able to yield reasonable estimations of degradation rates of PV modules under operating conditions, providing a reliable numerical tool for tracking and analyzing the performance of PV systems.

Acknowledgments
We acknowledge the support from Quality Assurance Department of REC Solar PTE LTD.

References

[1] Pingel S, Frank O, Winkler M, Daryan S, Geipel T, Hoehne H and Berghold J 2010 Potential induced degradation of solar cells and panels *Proc. 35th IEEE Photovolt. Spec. Conf.* 2817-22
[2] Sopori B, Basnyat P, Shet S, Mehta V, Binns J and Appel J 2012 Understanding light induced degradation of c-Si solar cells *Proc. 38th IEEE Photovolt. Spec. Conf.* 1115-20
[3] Jordan D C and Kurtz S R 2014 The dark horse of evaluating long-term field performance --- data filtering *IEEE J. Photovolt.* **4** 317-23
[4] Jordan D C, Smith R M, Osterwald C R, Gelak E and Kurtz S R 2010 Outdoor PV degradation comparison *Proc. 35th IEEE Photovolt. Spec. Conf.* 2694-97
[5] Jordan D C and Kurtz S R 2013 Photovoltaic degradation rates-an analytical review *Prog. Photovolt. Res. Appl.* **21** 12–29

[6] Osterwald C R, Adelstein J, del Cueto J A, Kroposki B, Trudell D and Moriarty T 2006 Comparison of degradation rates of individual modules held at maximum power *Proc. 4th IEEE Photovolt. Spec. Conf.* 2085-88
[7] Haeberlin H and Beutler Ch 1995 Normalized representation of energy and power for analysis of performance and online error detection in PV systems *Proc. 13th Eur. Photovolt. Sol. Energ. Conf.* 934–7
[8] Jordan D C and Kurtz S R 2012 PV degradation risk. *Technical Report in World Renewable Energy Forum*
[9] Jennings C 1988 PV module performance at PG&E *Proc. 20th Photovolt. Spec. Conf.* 1225-9
[10] [10] Ndiaye A, Charki A, Kobi A, Kebe C M F, Nadiaye P A and Sambou V 2013 Degradations of silicon photovoltaic modules: a literature review *Sol. Energ.* **96** 140-51
[11] [11] Osterwald C R, Anderberg A, Rummel S and Ottoson L 2002 Degradation analysis of weathered crystalline-silicon PV modules *Proc. 29th IEEE Photovolt. Spec. Conf.* 1392-5
[12] Guerriero P, Di Napoli F, Vallone G, Dalessandro V and Daliento S 2016 Monitoring and diagnostics of PV plants by a wireless self-powered sensor for individual panels *IEEE J. Photovolt.* **6** 286-94
[13] Driesse A and Zaaiman W 2015 Characterization of global irradiance sensors for use with PV Systems *Proc. 42nd IEEE Photovolt. Spec. Conf.* 1-5
[14] Campobello G, De Caro S, Giordano O, Russo A, Segreto A, Serrano S, Scimone T and Testa Efficiency assessment of MIC PV plants using Wireless Sensor Networks *Proc. 5th IEEE ICCEP* 793-9
[15] Tsuzuki S and Yamada Y 2015 Feasibility study of ubiquitous sensor networks by inductively coupled PLC over PV power systems *Proc. IEEE ISPLC 2015* 274-9
[16] Alfaris F, Alzahrani A and Kimball J W 2014 Stochastic model for PV sensor array data *Proc. IEEE ICRERA 2014* 798-803
[17] Guerriero P, Vallone G, Primato M, Di Napoli F, Di Nardo L, D'Alessandro V and Daliento S 2014 A wireless sensor network for the monitoring of large PV plants *Proc. IEEE SPEEDAM 2014* 960-5
[18] Pau M, Locci N and Muscas C 2014 A tool to define the position and the number of irradiance sensors in large PV plants *Proc. IEEE ENERGYCON 2014* 374-9
[19] Goma E, Orabi M, Hasaneen E-S and Qahouq J A 2014 Single-output-sensor on-chip integrated MPPT for PV solar system power management *Proc. IEEE APEC 2014* 655-61
[20] [20] Davis K O, Kurtz S R, Jordan D C, Wohlgemuth J H and Sorloaica-Hickman N 2013 Multi-pronged analysis of degradation rates of photovoltaic modules and arrays deployed in Florida *Prog. Photovolt: Res. Appl.* **21** 702-12
[21] Fu Z, Liu Z, Yuan Y, Zhao M and Li Q 2015 Construction of wireless sensor based monitoring network for distributed grid-connected PV system *Dianli Zidonghua Shebei/Electr. Pow. Automat. Equip.* **34** 36-41
[22] Karthikeyan R, Mahalakshmi P and Gowrishankar N 2014 Wireless sensor networks for efficient irrigation system using hybrid pv -wind system *Proc. IEEE ICAEE 2014* 1-5
[23] Dyreson A R, Morgan E R, Monger S H and Acker T L 2014 Modeling solar irradiance smoothing for large PV power plants using a 45-sensor network and the Wavelet Variability Model *Sol. Energ.* **110** 482-95
[24] Chen Y-T and Lai J-H 2014 A novel variable step-size MPPT method for PV system with single sensor *Proc. 9th IEEE ICIEA* 718-22
[25] Qahouq J A, Jiang Y and Huang W 2014 Load-voltage-based single-sensor MPPT controller for multi-channel PV systems *Proc. IEEE APEC 2014* 3451-4
[26] He Y T and Xiao L X 2013 A high efficiency solar PV micro-power supply integrated with wireless sensor nodes *Appl. Mech. Mater.* **401-403** 495-9
[27] Guerriero P, Vallone G, D'Alessandro V and Daliento S 2013 Innovative algorithm for true maximum detection based on individual PV panel sensor network *Proc. 4th IEEE ICCEP* 42-7

[28] Bosch J L and Kleissl J 2013 Deriving cloud motion vectors from a PV power plant on-site sensors *Proc. 42nd ASES Natl. Sol. Conf.* 567-71
[29] Dos Santos W M and Martins D C 2012 Digital MPPT technique for PV panels with a single voltage sensor *Proc. IEEE INTELEC 2012* 1-8
[30] Driesse A, Dirnberger D, Reise C and Reich N 2012 Spectrally selective sensors for PV system performance monitoring *Proc. 38th IEEE Photovolt. Spec. Conf.* 3294-9
[31] Barreiro C D, Bross A, Schmalzel J L and Jansson P M 2011 Sensor array for PV shading measurements *Proc.10th IEEE Sens.* 1889-92
[32] Kuszamaul S, Ellis A, Stein J and Johnson L 2010 Lanai High-Density Irradiance Sensor Network for characterizing SOLAR resource variability of MW-scale PV system *Proc. 35th IEEE Photovolt. Spec. Conf.* 283-88
[33] Makrides G, Zinsser B, Georghiou G E, Schubert M and Werner J H 2010 Degradation of different photovoltaic technologies under field conditions *Proc. 35th IEEE Photovolt. Spec. Conf.* 2332–7
[34] Dows R N 1995 PVUSA procurement, acceptance, and rating practices for photovoltaic power plants *PG&E Co. Report* # 95-30910000.1
[35] Rosenthal A L, Thomas M G and Durand S J 1993 A ten-year review of performance of photovoltaic systems *Proc. 23rd IEEE Photovolt. Spec. Conf.* 1289-91
[36] King D L, Quintana M A, Kratochvil J A, Ellibee D E and Hansen B R 2000 Photovoltaic module performance and durability following long-term field exposure *Prog. Photovolt: Res. Appl.* **8** 241–56
[37] Li X, Prawiradiraja TP, Battul D 2013 The role of humidity in energy output of solar panels in coastal regions *GSTF J. Eng. Tech. (JET)* **2** 72–6

IOP Conf. Series: Earth and Environmental Science **40** (2016) 012064 doi:10.1088/1755-1315/40/1/012064

An improved AVC strategy applied in distributed wind power system

Y N Zhao[1], Q H Liu, S Y Song and W Mao

North China Electric Power University, State Key Laboratory of Alternate Electrical Power System with Renewable Energy Sources, Beijing 102206, China

E-mail: zhaoyn10@163.com

Abstract. Traditional AVC strategy is mainly used in wind farm and only concerns about grid connection point, which is not suitable for distributed wind power system. Therefore, this paper comes up with an improved AVC strategy applied in distributed wind power system. The strategy takes all nodes of distribution network into consideration and chooses the node having the most serious voltage deviation as control point to calculate the reactive power reference. In addition, distribution principles can be divided into two conditions: when wind generators access to network on single node, the reactive power reference is distributed according to reactive power capacity; when wind generators access to network on multi-node, the reference is distributed according to sensitivity. Simulation results show the correctness and reliability of the strategy. Compared with traditional control strategy, the strategy described in this paper can make full use of generators reactive power output ability according to the distribution network voltage condition and improve the distribution network voltage level effectively.

1. Introduction

In the field of new energy power generation, wind power is one of the most competitive and efficient way of generating. At present, the main form developing and utilizing wind power is large-scale wind farm which is located at the end of grid and far away from load centre. Insufficient transmission capacity and limitations of the grid consumptive ability result in a wide range "wind abandoned" phenomenon, seriously reducing the economic benefits of wind power [1]. Compared with large-scale centralized power generation, small-scale distributed wind power generation can access to the distribution network directly and used locally, not requiring long-distance transmission. However, the access of distributed wind power generation will have impact on steady operation and power quality of distribution network with weak structure. Therefore, the study of distributed wind power key technology has a very important significance.

Current study about reactive power optimization control of distributed wind generator accessing to low voltage distribution network are mainly focused on three fields, which are reactive power optimization containing distributed generators, voltage control of reactive power compensation devices and wind farm hierarchical AVC (automatic voltage control) control system. Objective function and constraint conditions are needed in reactive power optimization [2-4]. However, this optimizing strategy can only be used in long time scale global optimization, not for short time rapid

[1] Address for correspondence: Y N Zhao, North China Electric Power University, State Key Laboratory of Alternate Electrical Power System with Renewable Energy Sources, Beijing 102206, China. E-mail: zhaoyn10@163.com.

voltage control. As for reactive power compensation device, most of papers pay attention to devices and ignore the reactive power supporting ability, which increase the investment and operation cost [5-8]. The AVC system of wind farm can be divided into reactive power setting layer and distributing layer [9-12]. Reference of reactive power is obtained in setting layer and is distributed into reactive power source in distributing layer. But most research literature about AVC system is for wind farm and only concern about grid connection point, which haven't take full account of features of distributed wind generator.

In order to solve above problems, this paper puts forward an improved AVC strategy applied in wind power system based on doubly-fed induction generator (DFIG). The strategy takes all nodes of distribution network into consideration and chooses the node having the most serious voltage deviation as control point to calculate the reactive power reference. In addition, distribution principles can be divided into two conditions: when wind generators access to network on single node, the reactive power reference is distributed according to reactive power capacity; when wind generators access to network on multi-node, the reference is distributed according to sensitivity. This strategy considers flexible features of distributed wind power generators, regulating reactive power control and strengthening the resist of distribution network to voltage fluctuation.

2. Characteristics of distribution network voltage considering the integration of wind power

Characteristics of distribution network voltage considering the integration of wind power. There are N nodes on the electricity line, and the load of each node is $P_n+jQ_n(n=1,2,\cdots,N)$. The voltage of initial node is U_0 and the voltage in note N is U_n. What's more, the impedance between two nodes is R_n+jX_n. Define the flow of active power and reactive power in the direction of load is positive, whereas is negative.

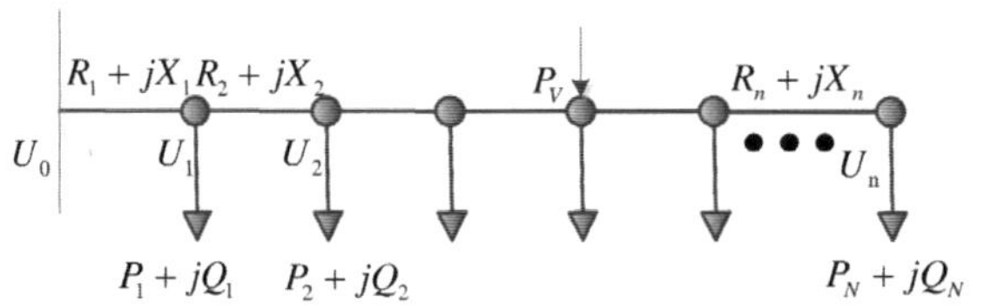

Figure 1. Typical radial distribution network.

Ignoring power loss, the loss of the voltage between nodes N and N-1 is [13]:

$$\Delta U_n = U_n - U_{n-1} = -\frac{\sum_{m=n}^{N} P_m R_n + \sum_{m=n}^{N} Q_m X_n}{U_{n-1}} \tag{1}$$

Due to the load of active power and reactive power are over 0, the loss of voltage between two nodes in equation (1) is negative constantly without considering reactive power compensation devices. According to equation (1), we can get the voltage on node N is:

$$U_n = U_0 + \sum_{k=1}^{n} \Delta U_k = U_0 - \sum_{k=1}^{n} \frac{\sum_{m=k}^{N} P_m R_k + \sum_{m=k}^{N} Q_m X_k}{U_{k-1}} \tag{2}$$

Considering that wind generators access to distribution network on multi-node, and the active power of generator on nodes M is P_{Vm}, voltage of node N is shown in equation.

$$U_n = \begin{cases} U_0 - \sum_{k=1}^{n} \dfrac{(\sum_{m=k}^{N} P_m - P_V) R_k + \sum_{m=k}^{N} Q_m X_k}{U_{k-1}} & (0 < n \le p) \\ U_p - \sum_{k=p+1}^{n} \dfrac{\sum_{m=k}^{N} P_m R_k + \sum_{m=k}^{N} Q_m X_k}{U_{k-1}} & (p < n \le N) \end{cases} \tag{3}$$

Considering that wind generators access to distribution network on multi-node, and the active power of generator on nodes M is P_{Vm}, voltage of node N is shown in equation.

$$U_n = U_0 - \sum_{k=1}^{n} \frac{\sum_{m=k}^{N}(P_m - P_{V_m})R_k + \sum_{m=k}^{N} Q_m X_k}{U_{k-1}} \tag{4}$$

If reactive power of generator Q_{Vm} is taken into consideration, the node voltage is:

$$U_n = U_0 - \sum_{k=1}^{n} \frac{\sum_{m=k}^{N}(P_m - P_{V_m})R_k + \sum_{m=k}^{N}(Q_m - Q_{V_m})X_k}{U_{k-1}} \tag{5}$$

As shown in equation (4), the change of generator active power will result in voltage fluctuation. The reactive power regulation of generator can strengthen the resist of distribution network to voltage fluctuation.

3. Reactive power limits of DFIG

The stator of DFIG is connected to the grid directly and the rotor is connected to the grid through rotor-side converter and grid-side converter which is showed in figure2. Where P_g and Q_g are active power and reactive power of stator, P_r and Q_r are active power and reactive power of rotor, P_c and Q_c are active power and reactive power of grid-side converter, P_g and Q_g are active power and reactive power of generator system.

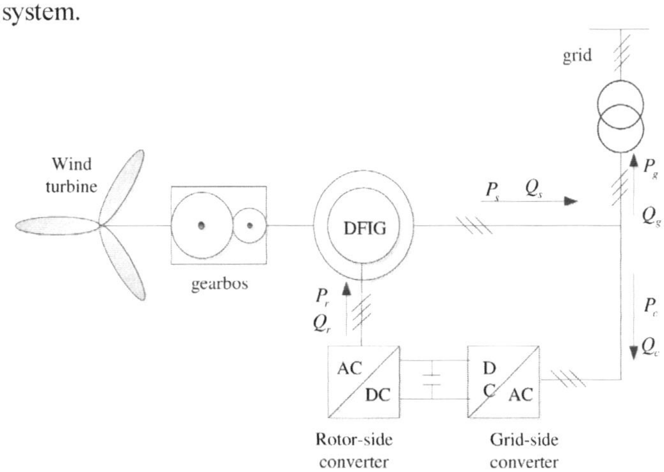

Figure 2. Power relationship of DFIG.

The stator of DFIG can send out active power and reactive power. The grid-side converter can operate at four quadrants. Thus, DFIG can not only be used as active power source, but also a reactive power source to regulate grid voltage.

3.1. *Reactive power limit of stator*

Stator power of DFIG is limited by stator winding, rotor winding and current of rotor-side converter, and the most important factor is the limit of rotor-side current from which the range of stator reactive power is obtained [14].

$$\begin{cases} Q_{s\max} = -\dfrac{U_1^2}{X_1} + \sqrt{\left(\dfrac{X_m U_1}{X_1} I_{r\max}\right)^2 - P_s^2} \\ Q_{s\min} = -\dfrac{U_1^2}{X_1} - \sqrt{\left(\dfrac{X_m U_1}{X_1} I_{r\max}\right)^2 - P_s^2} \end{cases} \tag{6}$$

Where Q_{smax} and Q_{smin} are reactive power maximum and minimum of stator, U_1 is vector amplitude of grid voltage, X_m is the magnetizing reactance, X_1 is stator reactance, I_{rmax} is the current maximum of grid-side converter.

3.2. Reactive power limit of grid-side converter

The grid-side converter can operate at four quadrants and its capacity generally depends on maximum slip power of DFIG. When the generator is operating at low wind speed, the grid-side converter does not fully play its power handling capability. Consequently, if the network has reactive power requirement, the grid-side converter can output reactive power.

Assuming the rated capacity of grid converter is S_c, reactive power limit is showing in equation (7) considering capacity limitation.

$$-(S_c^2 - P_c^2)^{\frac{1}{2}} \le Q_c \le (S_c^2 - P_c^2)^{\frac{1}{2}} \tag{7}$$

According to the power relationship between stator and rotor, equation (8) can be obtained.

$$\begin{cases} Q_{c\max} = \left[S_c^2 - (sP_s)^2 \right]^{\frac{1}{2}} \\ Q_{c\min} = -\left[S_c^2 - (sP_s)^2 \right]^{\frac{1}{2}} \end{cases} \tag{8}$$

Where s is the slip ration of generator, Q_{cmax} and Q_{cmin} are reactive power maximum and minimum of grid-side converter.

In conclusion, the reactive power limit of DFIG is obtained as follows:

$$\begin{cases} Q_{\max} = Q_{s\max} + Q_{c\max} \\ Q_{\min} = Q_{s\min} + Q_{c\min} \end{cases} \tag{9}$$

4. Improved AVC system

The AVC system can be divided into reactive power setting layer and distribution layer [9]. The setting layer detects real time data of DFIG and network to calculate reactive power reference. The distribution layer distributes the reference to each generator according to a certain principle. But traditional AVC system is used for large-scale wind farm and only concerns about grid connection point, which haven't take features of distributed wind generator into consideration. Therefore, this paper comes up with an improved AVC system, which calculates voltage of each node and choses the node with the most serious voltage deviation as control point. The scheme of improved AVC strategy is shown in figure 3.

Considering actual situation of distribution network, the voltage of all nodes cannot be obtained by measurement due to lack of measuring installation. In view of radiant network, node voltage can be calculated at the basis of measured data and grid data according to formula (5) when the request for control accuracy is not rigid.

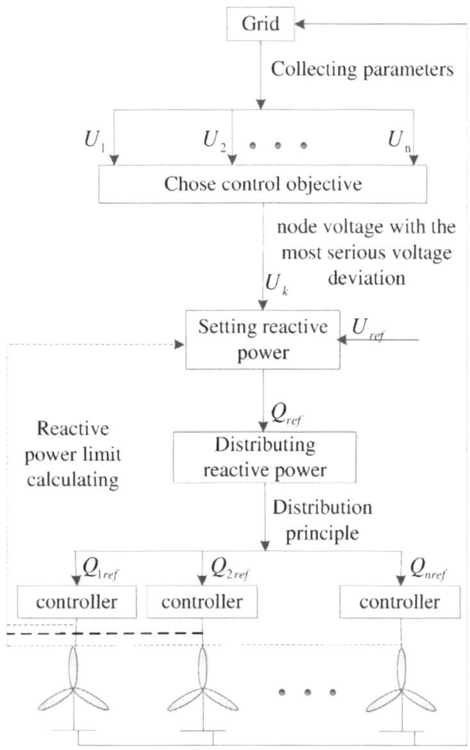

Figure 3. Scheme for AVC strategy.

4.1. Reactive power setting layer

Reactive power setting layer is to setting reactive reference. The control block diagram is shown in figure 4. The reference voltage and actual voltage of control point are compared to get the adjustment voltage. Then the adjustment voltage goes through dead zone and PI controller. Finally reactive power reference can be obtained while meeting reactive power limit. When calculating adjusting voltage, voltage standard of distribution network is taken into consideration that voltage fluctuation should not exceed the rated voltage to -7% ~ 7%.

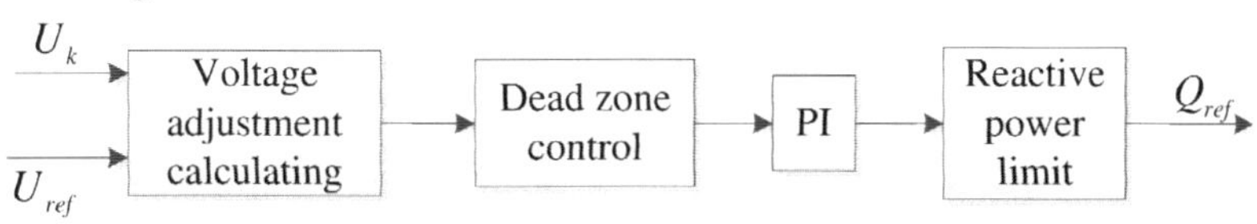

Figure 4. Control Block Diagram of Reactive power setting.

4.2. Reactive power distributing layer

4.2.1. Distributing principle between wind generators.

Reactive power reference is distributed to wind generators in distribution layer. Traditional distribution principles have equal power factor algorithm [11] and reactive power capacity proportion algorithm [12]. It is important to note that equal power factor algorithm may cause overload of wind generator when active power is very large. Therefore, reactive power capacity proportion algorithm is used in this paper when wind generators access to network on single node.

The reactive power reference is distributed according to reactive power capacity when wind generators access to network from on single node.

$$Q_{\text{iref}} = \left(Q_{\text{imax}} / Q_{\text{totalmax}}\right) Q_{\text{ref}} \tag{10}$$

Where Q_{iref} is the reactive power reference of each generator, Q_{imax} is the maximum of the reactive power that each generator can generate/absorb, $Q_{totalmax}$ is the maximum of the reactive power that all generators can generate/absorb, Q_{ref} is the reactive power reference obtained from setting layer.

When distributed wind generators access to network on multi-node, the scale and direction of power flow will change more frequently and complicatedly. Therefore, voltage regulation will have blindness and power loss of network will increase when reactive power capacity proportion algorithm is adopted. Paper [16] came up with a sensitivity algorithm to calculate reactive power reference of reactive power compensation equipment. This paper applies this method to distribute reactive power between wind generators accessing to network on different nodes.

Assuming that active power of generator is a constant, the relationship between control node voltage and reactive power of DFIG is shown in equation (11).

$$\sum_{i=1}^{N} \frac{\partial U_k}{\partial Q_{Vi}} \Delta Q_{Vi} = \Delta U_k \tag{11}$$

Where U_k is voltage of control point, Q_{Vi} is the reactive power of wind generator accessing to node i.

Define the sensitivity of node voltage U_j to wind generator reactive power output Q_{Vi} accessing to node i as K_{ij}.

$$K_{ij} = \frac{\partial U_j}{\partial Q_{Vi}} \tag{12}$$

The sensitivity is as follows from equation (5) and equation (11).

$$K_{ij} = \begin{cases} \sum_{m=1}^{i} \frac{X_m}{U_{m-1}} & (i > j) \\ \sum_{m=1}^{j} \frac{X_m}{U_{m-1}} & (i < j) \end{cases} \tag{13}$$

Thus, the reactive power reference of each wind generator accessing to node i is obtained.

$$Q_{iref} = \frac{k_i}{\sum_{i=1}^{N} k_i} Q_{ref} \tag{14}$$

4.2.2. Distributing principle between stator and grid-side converter.

The output of stator reactive power is realized by controlling the rotor side power, and the rotor side only needs to control slip power. Therefore, distributing reactive power should give priority to stator to reduce the power addressed by control system [15], which means when the reactive power reference is within the limit of stator, it will generate only by stator. Otherwise, grid-side converter will be used to generate reactive power.

5. Case study

In order to verify the validity of control strategy, radiant network with 7nodes has been established in PSCAD. The rated voltage of network is 10 kV, impedance between nodes is 1.12+j1.32Ω and load parameters of distribution network are in table1. There are wind generators in node3, node4 and node6. The number of generator is shown in figure5 and parameter is in table 2.

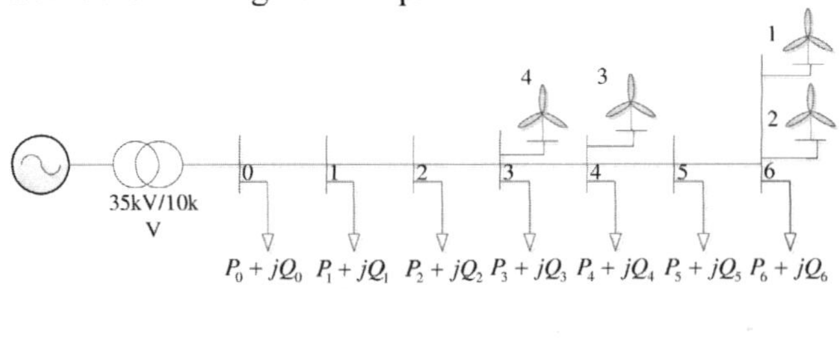

IOP Conf. Series: Earth and Environmental Science **40** (2016) 012064 doi:10.1088/1755-1315/40/1/012064

Figure 5. Distribution network topology.

Table 1. Load parameters of distribution network.

Node	Active load	Reactive load
node0	25MW	5Mvar
node1	0.45MW	0.09Mvar
node2	0.45MW	0.09Mvar
node3	0.45MW	0.09Mvar
node4	0.45MW	0.09Mvar
node5	0.45MW	0.09Mvar
node6	0.45MW	0.09Mvar

Table2. Parameters of DFIG.

Parameters	Value	Parameters	Value
Rated power/ MW	1.5	Rotor leakage inductance/H	0.0003982
Rated voltage/ V	690	Line resistance/Ω	0.0003
Mutual inductance/ H	0.00973	Line inductance/H	0.0008
Stator resistance/Ω	0.00579	Rotor radius/m	38.5
Stator leakage inductance/H	0.0002645	Gearbox ratio	104
Rotor resistance/Ω	0.00621	DC capacitor/uF	8640

Design two cases to analyze and verify the control strategy: power output change of wind generator and load change of network.

Case 1: The operation speed of wind generator3 and wind generator4 is 10m/s and the operation speed of wind generator1 and wind generator2 is by 9 m/s step to 11 m/s at 5s.

Active power of wind generators in node6 increases with increasing of wind speed, resulting voltage of node6 and node5 beyond the upper limit 1.07, just as figure (a). Figure (b) represents change of node voltage adopting traditional AVC strategy which only considering the voltage of grid connection point. After voltage regulation, voltage of nodes6 is 1.07pu and still greater than upper limit. This is due to that only generators accessing to node6 are involved in voltage control and largest reactive power absorption of generators is 0.38 MVar, when the wind speed is 11m/s shown in figure3, which limits the reactive power adjustment ability of wind generator. Figure (d) represents adopting control strategy described in this paper which develops reactive power adjustment ability of other generators to control the voltage within normal range. Figure (e) shows reactive power output of wind generators.

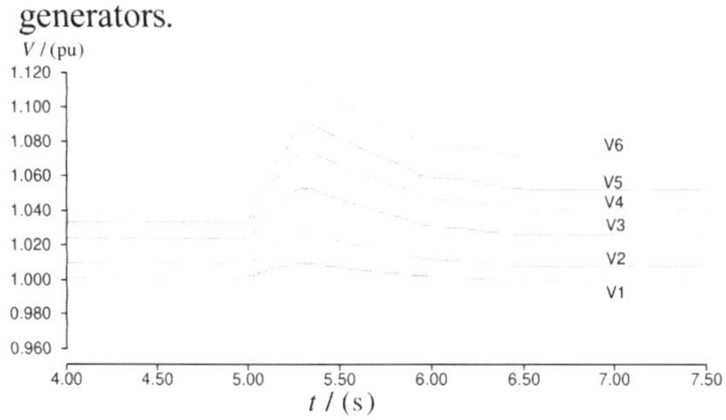

Figure 6. (a) Node voltage when the operation speed of generators is increasing.

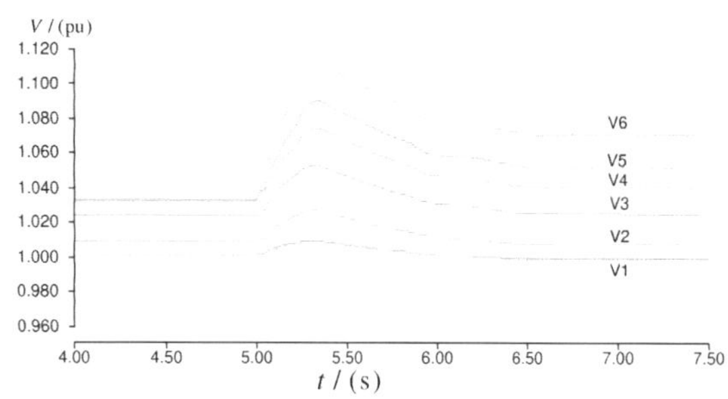

Figure 6. (b) Node voltage when adopting traditional AVC strategy.

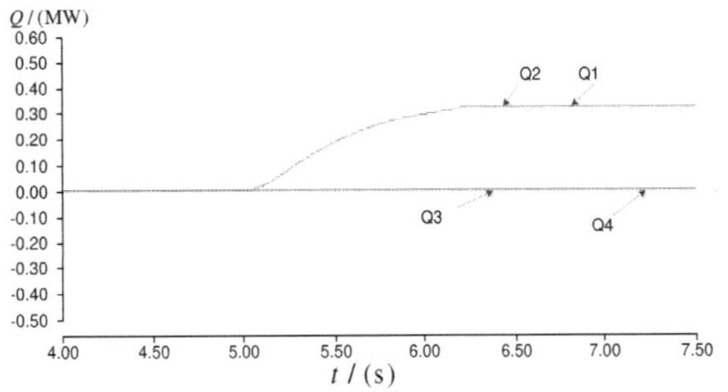

Figure 6. (c) Reactive power output of generators when adopting traditional AVC strategy.

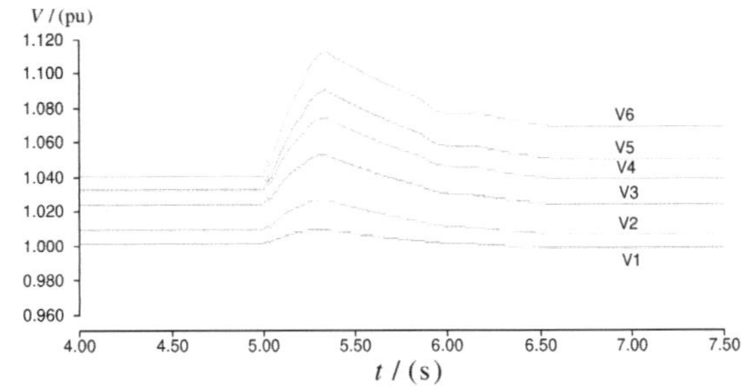

Figure 6. (d) Node voltage when adopting improved AVC strategy.

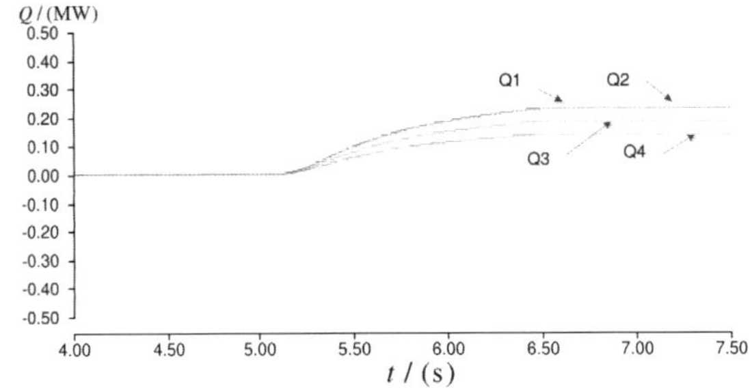

Figure 6. (e) Reactive power output of generators when adopting improved AVC strategy.

Figure 6. Simulation results for Case 1.

Case 2: The operation speed of 4 wind generators is 8m/s. The load in node5 is by 0.45MW+j0.09Mvar step to 1.65MW+j0.21Mvar at 5s.

Variation of load results voltage of node6 , node5 and node4 beyond the lower limit 0.93, just as figure (a). Figure (b) represents change of node voltage adopting traditional AVC strategy. At the moment, generator3 outputs reactive power to compensate the voltage of node4 and generator1 and generator2 outputs reactive power to compensate the voltage of node6. Figure (c) shows reactive power output of wind generators. At last, voltage of nodes5 is 0.9203pu and is still lower than lower limit 0.93. The control point will be node5 which has the most serious voltage exceeding when adopting control strategy described in this paper. At end, the node voltage is within the normal range as shown in figure (d). Figure (e) shows reactive power output of each wind generator.

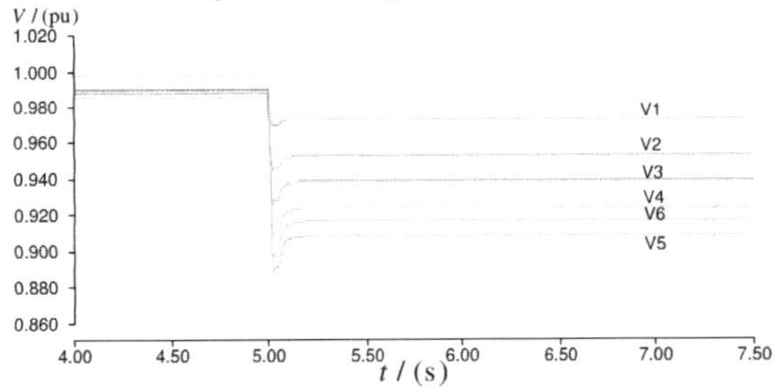

Figure 7. (a) Node voltage when the load increasing.

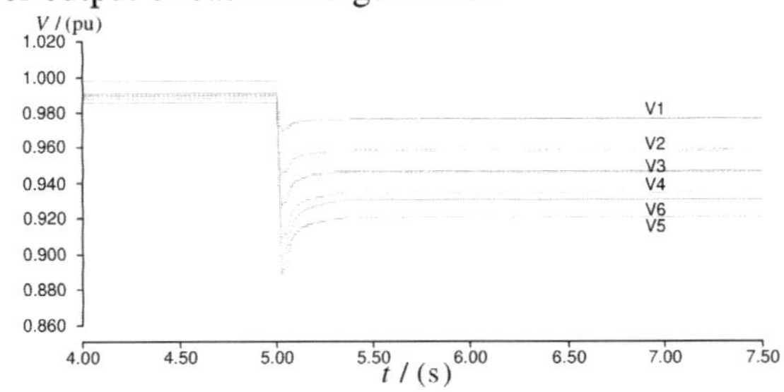

Figure 7. (b) Node voltage when adopting traditional AVC strategy.

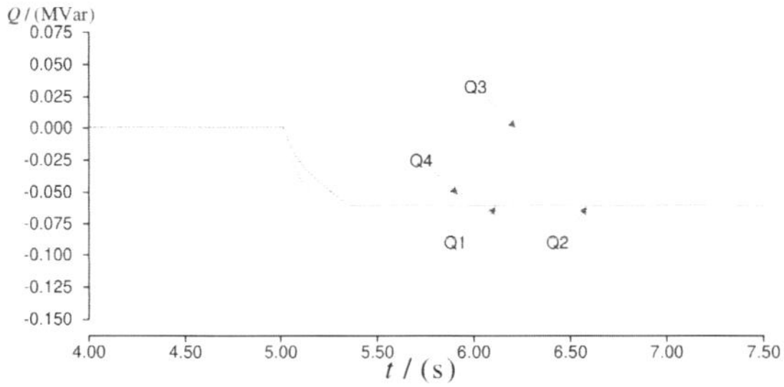

Figure 7. (c) Reactive power output of generators when adopting traditional AVC strategy.

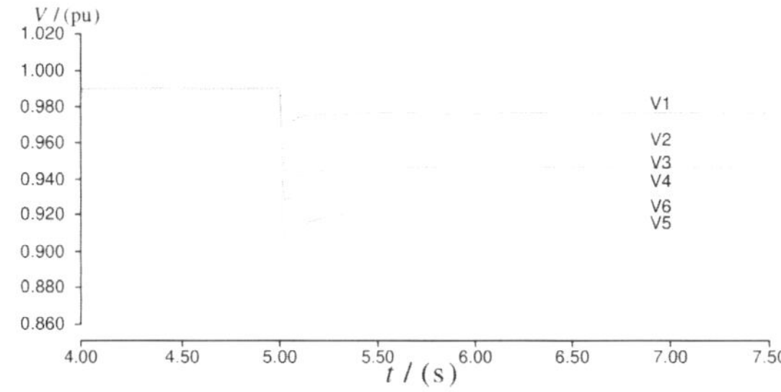

Figure 7. (d) Node voltage when adopting improved AVC strategy.

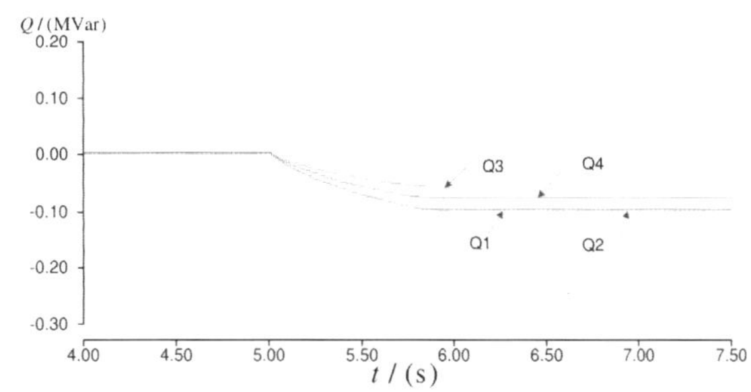

Figure 7. (e) Reactive power output of generators when adopting improved AVC strategy.

Figure 7. Simulation results for Case 2.

6. Conclusion

In this paper, the AVC strategy is improved and applied in distributed wind generator. First, the voltage variation mechanism is researched before and after the connection of wind power generator from the viewpoint of voltage drop. The influences of connecting single point and multi points are analyzed. Then the reactive power limit of DFIG is analyzed. At the basis of it, improved AVC strategy is used in reactive power control. The voltage variation of nodes in network is taken into consideration and he node having the most serious voltage deviation is chosen as the control point when setting the reactive power reference. After that, the reference is divided according to the reactive power capacity of generator system for single point and sensitivity for multi points. The improved strategy is suitable for distributed wind generator and gives full play to the reactive power ability of DFIG. In addition, network voltage is been improved effectively when adopting this strategy, also optimizing the power flow distribution and power quality. The research results of this paper look forward to having promoting significance on the rapid development of distributed wind power.

Acknowledgments

I wish to thank my supervisor Professor Liu whose continual guidance and support during my Research process. Thanks for financial support from project support by State Grid Corporation of China (SGSDDK00KJJS1500155).

References

[1] Yang J Y, Cui , Xing Z X, Wang H X and Wang G 2015 Reactive power control strategy for dispersed wind farm considering wind power forecasting *Automation of Electric Power Systems* **13** 8-15

[2] Senjyu T, Miyazato Y, Yona A, Urasaki N and Funabashi T 2008 Optimal distribution voltage control and coordination with distributed generation *IEEE Transactions on power delivery* **23** 1236-42
[3] Zhao J J, Fu Y and Li D D 2011 Reactive power optimization in distribution network considering reactive power regulation capability of dfig wind farm *Automation of Electric Power System* **11** 33-8
[4] Wei Q, Harley R G, Venayagamoorthy G K 2009 Coordinated reactive power control of a large wind farm and a statcom using heuristic dynamic programming *IEEE Transactions on Energy Conversion* **24** 493-503
[5] Wei C Z, Su J H, Wen A, Liu N and Huang H F 2015 Reactive voltage coordinated control for distribution network with DSTATCOMs and DGs *Automation of Electric Power Systems* **09** 132-7
[6] Tapia G, TapiaA and Ostolaza J X 2007 Proportional-integral regulator-based approach to wind farm reactivepower management for secondary voltage control *IEEE Transactions on Energy Conversion* **22** 488-98
[7] Zhang W Y, Li Y L, Sun G Y, Jin W and Li X Y 2015 Zonal-voltage control for active distribution network based on distributed staticsynchronous compensator P*roc. the CSEE* **35** 1644-54
[8] Jia J, Liu J and Zhang Y 2010 Novel reactive power optimization control strategy for doubly fed inductionwind power generation system *Proc. the CSEE* **30** 87-92
[9] Yang H, Liang H F and Li G Y 2011 A coordinated voltage control strategy for wind farm containing doubly fed induction generators *Power System Technology* **2** 121-6
[10] Chen N, Zhu L Z and Wang W 2009 Strategy for reactive power control of wind farm for improving voltage stability in wind power integrated region *Proc. the CSEE* **10** 102-8
[11] Wang S Y, Zhu L Z, Chen N, Yu J L 2009 A Reactive power control strategy for wind farm based on hierarchical layered principle *Automation of Electric Power Systems* **13** 83-8
[12] Liang Y Q, Zhang X G, Xu D G, Ma H F and Hadianmrei S R 2007 Reactive power analysis and control of doubly fed induction generator wind farm *Proc. the CSEE* **09** 77-82
[13] Xu X Y, Huang Y H, Liu C and Wang W S 2010 Influence of distributed photovoltaic generation on voltage in distribution network and solution of voltage beyond limits *Power System Technology* **10** 140-6
[14] Ammar M and Joos G 2013 Impact of distributed wind generators reactive power behavior on flicker severity IEEE Transactions on Energy Conversion **28** 425-33
[15] Shen H, Wang W S and Dai H Z 2003 Reactive power limit of variable-speed constant-frequency wind turbine *Power System Technology* **11** 60-3
[16] Qiao Y, Chen H F, Lu Z X, Xu F and Li J 2013 Design and application of automatic voltage control system in doubly-fed induction generator wind farms *Automation of Electric Power Systems* **05** 15-22

Study on application of capillary plane radiation air conditioning system based on the slope roof

Y G Li[1], T T Wang, X L Liu and X Z Dong

School of Thermal Energy Engineering, Shandong Jianzhu University, Jinan 250101, China

E mail: sdjglya@163.com

Abstract. In this paper, based on the principle of the capillary plane radiation air conditioning system, taking the slope roof as an example, the application of the capillary plane radiation air-conditioning system is studied and analysed. Then the numerical solution of differential equations is obtained by the technology of CFD. Finally, we analyze the distribution of indoor temperature of the slope roof and the predicted mean votes (PMV) using Airpak simulation software by establishing a physical model. The results show that the PMV of different sections ranges from 0 to 2.5, which meets the requirement of the comfort. These provide a theoretical basis for application and promotion of capillary plane in the slope roof.

1. Introduction

With the development of economy and pepple's living conditions, air conditions has been widely applied in our daily life. Capillary air-conditioning system, supplying cooling and heating mainly by radiation, which can control temperature and humidity individually[1]. The sensible heat load and the latent heat load of the air conditioning are borne by the capillary network and the independent fresh air system respectively. Thus, the temperature of chilled water is raised from 7 ℃ to about 18℃ in summer, so that the coefficient of function(COP) is improved by more than 40%, and energy-consumption of the air-conditioning system is reduced by more than 30% [2]. In winter, with "low temperature" (28-32 ℃) water heating,the effect of energy-saving is remarkable. In addition, because of ease silence and no feelings of air draft,it enjoys more obvious advantages among assorted systems[3]. So it is no wonder that it becomes more high-grade building designers' priority in the recent years.

Modern architects tend to introduce art and humanity into buildings, trying to highly reflect the local history background and cultural environment. Combining the large scale geometric figure with a large number of local materials makes the public feel friendly and encourages them to show their emotions and concerns. During the appreciating process, we realize the biomimetic and free concept from the exterior shape and transparent or translucent walls and roofs. They regard one construction as a creature, talking with her about the design thought. Seeing buildings in the light of their development, advocating the ideas of complexity and movement of the architectural space, and considering the requirement for multifunction architects manage to create brilliant architectures with simple geometric

[1] Address for correspondence: Y A Li, School of Thermal Energy Engineering, Shandong Jianzhu University, Jinan 250101, China. Email: sdjglya@163.com.

shapes. The city will be dressed in colorful styles of the buildings, meanwhile, it brings new topics for the design of capillary air-conditioning system. In the building of slope roof, sloving problems about layout of the capillary broadly, the effect of cooling in the summer and supplying heat in winter and the indoor comfort needs further research.

2. Establishment of model

2.1 System's schematic diagram

Capillary plane radiation air-conditioning system mainly consists of cooling-heating sources, heat exchangers, circulating pumps, temperature control system, desiccant system and capillary network [4]. These are shown in figure 1.

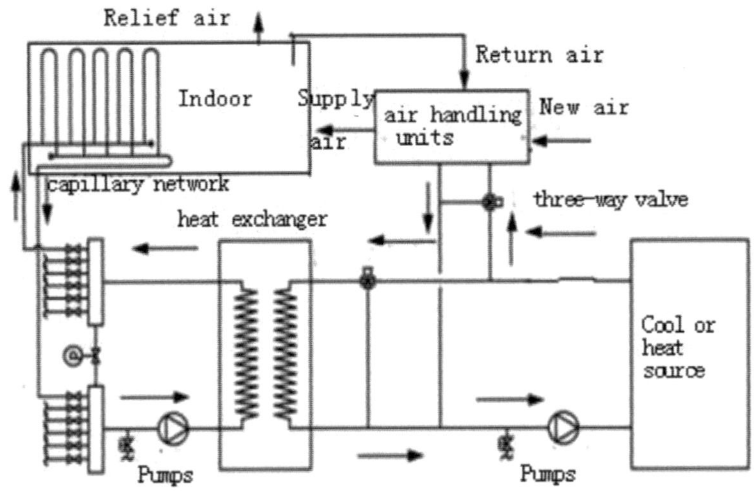

Figure 1. Schematic of capillary air-conditioning system

It is shown in the Figure 1 capillary plane radiation air conditioning system consists of capillary net terminal and fresh air system, which are two independent systems. Capillary network is mainly responsible for the indoor sensible heat load, and fresh air system is in charge of indoor humidity control.

2.2 Physical models

In this paper, taking a common room in Ji'nan as an example, a physical model is established to simulate the real condition. Room size (length * width * height) is 4.5 m× 3 m× 2.8 m and the angle of the slope roof is 30 ° with a building area of 13.50 m^2. The main equipments of the room include three persons, fluorescent lamp×6, computer×3, desk×3, south ex-windows (2.5 × 1.8) and a gate (2 × 1.1). The device parameters are listed in the table below.

Table 1. The main equipment parameters in the model room

Devices	Quantity/Size	Remakes
Human bodies	3	Quantity of body heat loss: 75W; Moisture scattering: 109g/h
Computers	3	100W
Desks	3	2×0.6m^2
fluorescent lamps	6	6×10W
external window	2.5×1.8 m^2	Thickness: 18mm
Gates	2×1.1 m^2	Thickness: 5cm

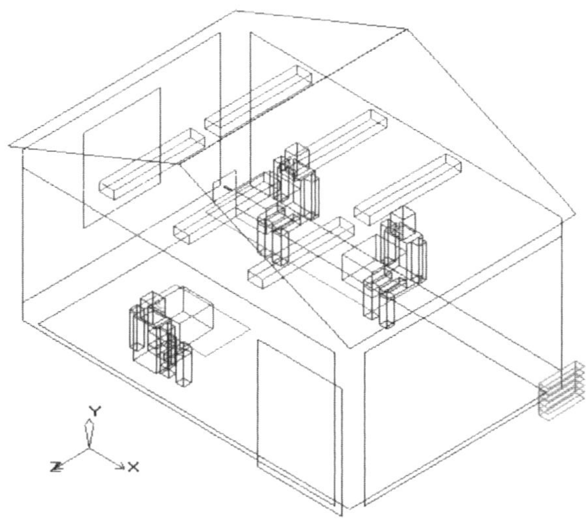

Figure 2. The model of slope roof

We conduct this experiment based on the following assumptions:
(1) In order to simplify the model, humans, computers and desks are set into cube block;
(2) The temperature of capillary network radiation surface is well-distributed. Except forcapillary network radiation surface, the external windows, exterior walls and doors the exterior protected construction are adiabatic;
(3) In the air outlet, the parameter of jet-flow is uniform, and the thermal parameters of indoor air are constant.

2.3 Mathematical models
The indoor air flow is relatively complex, unsteady and irregular vortex. Each physical parameter of fluid keeps randomly changeing along with the temporal and spatial conditions. It is a typical kind of turbulent flows, corresponding to turbulence models. The model of the 0 equation is used in this study. The model of the 0 equation does not require the solution of differential equation. However, a link between turbulent viscosity coefficient and average physical quantity is established by the algebraic relationship. The 0 equation do not need the solution of the K equation and the ε equation [5]. So this model is timesaving and the calculating speed is also much faster compared with one equation model and the standard model (K- ε model). Its formula is very simple, consisting of velocity V and the nearest distance (l) being away from the wall:

$$\mu_t = 0.03874\rho v l \tag{1}$$

According to the law of conservation of energy, the general form of governing equations is like Equation (2).

$$\frac{\partial(\rho\phi)}{\partial\tau} + div(\rho U\phi) = div(\Gamma_\phi grad\phi) + S_\phi \tag{2}$$

In Equation (2), $\frac{\partial(\rho\phi)}{\partial\tau}$ is non-steady state term; $div(\rho U\phi)$ is convection term; $div(\Gamma_\phi grad\phi)$ is diffusion term; Γ_ϕ is generalized diffusion coefficient; S_ϕ is generalized source term; ϕ is expressed general variable, which can be represented u, V, W, T waiting for the variable.

By reference to general form of the governing equation (2), the Reynolds Stress are assumed as a kind of algebraic function of average physical quantity according to the 0 equation model. The parameter Φ represents the velocity and the temperature of fluid flow. The corresponding air flow controlling equation is different due to representing different physical quantities, which are listed in table 2.

Table 2. Indoor air flow control equation of zero equation turbulent model

S_φ	φ	Γ_φ
1	0	0
u	μ_e	$-\frac{\partial p}{\partial x}+\frac{\partial}{\partial x}\left(\mu_e\frac{\partial u}{\partial x}\right)+\frac{\partial}{\partial y}\left(\mu_e\frac{\partial v}{\partial x}\right)+\frac{\partial}{\partial z}\left(\mu_e\frac{\partial w}{\partial x}\right)+g_x\left(\rho-\rho_{ref}\right)$
v	μ_e	$-\frac{\partial p}{\partial y}+\frac{\partial}{\partial x}\left(\mu_e\frac{\partial u}{\partial y}\right)+\frac{\partial}{\partial y}\left(\mu_e\frac{\partial v}{\partial y}\right)+\frac{\partial}{\partial z}\left(\mu_e\frac{\partial w}{\partial y}\right)+g_y\left(\rho-\rho_{ref}\right)$
w	μ_e	$-\frac{\partial p}{\partial z}+\frac{\partial}{\partial x}\left(\mu_e\frac{\partial u}{\partial z}\right)+\frac{\partial}{\partial y}\left(\mu_e\frac{\partial v}{\partial z}\right)+\frac{\partial}{\partial z}\left(\mu_e\frac{\partial w}{\partial z}\right)+g_z\left(\rho-\rho_{ref}\right)$
h	$\frac{\mu_e}{\sigma_h}$	S_h

In the formula $\mu_e=\mu_l+\mu_t$, μ_l is the dynamic viscosity of laminar air flow; μ_t is calculated by the Equation (1); u, v, w represent the velocity of x axis, y axis and z axis respectively; h and P represent the air enthalpy and pressure respectively; g_x, g_y, g_z refer to the acceleration of gravity in three dimensional directions; ρ_{ref} is the referenced air density; μ_e, μ_l, μ_t represent the effective viscosity coefficient, turbulent viscosity coefficient and laminar viscous coefficient, respectively; σ_h is equivalent Prandtl number; S_h is calorific power of unit volume [6].

3. The numerical solution of mathematical model

3.1 Set the boundary condition

It is crucial to set the boundary conditions in the numerical simulation. The boundary conditions are related to the proximity level of the simulation results and the actual situation. The temperature of water in capillary plane radiation system increased by 1℃ compared with traditional comfortable air conditioning [7]. In this paper, room temperature is 26℃and the relative humidity is 50% [8].

(1) Outdoor calculating parameters

The air temperature curve in Ji'nan in some years is shown in Figure 3. The design parameters of outdoor air conditioning in summer in Ji'nan is: the temperature of dry bulb is 34.4℃, wet bulb is 26.7℃ and the average wind speed is 2.8m/s.

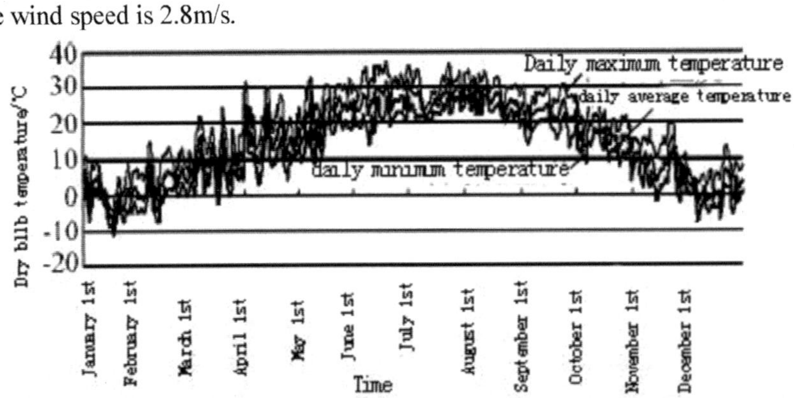

Figure 3. A curve of annual temperature change in Ji'nan

(2) Boundary conditions of retaining structure

Western and southern walls are all exterior wall in this model. Western Wall is $4.5\times2.8m^2$ and the south wall is $3\times2.8m^2$. There is a external window on the southern wall. In order to simplify the calculation, heat transfer of the exterior wall can be regarded as a boundary condition of constant heat flux with conduct heat density of $19.5W/m^2$ and heat transfer coefficient of $0.49W/(M^2 K)$. Southern exterior window is $2.5 \times 1.8m^2$ with thickness of 18mm, heat transfer coefficient of $3.38W/(m^2 K)$, solar radiation heat load of $59.05W/m^2$ and heat flux density of $81.23W/m^2$. Other retaining structure is adiabatic boundary without difference in temperature, and heat flow density is $0W/m^2$.

(3) Boundary conditions of capillary network radiation

Capillary net radiation air-conditioning system supplies cooling in summer. The temperature of the medium usually is 18℃ and the return water temperature is set up to 20℃ considering influence of difference in temperature. We assume that the temperature of radiation surface is uniform, and the radiation surface can be regarded as a steady-state boundary condition of constant temperature [9].

(4) Boundary conditions of air port

In this model, air diffuser is at the bottom. The guiding diffuser is simplified to rectangle diffuser, the temperature and velocity of the fresh air is 20 ℃, 0.5m/s, respectively [10]. Return air inlet is arranged on the top, and boundary conditions of air inlet are set for entrance velocity.

3.2 Comfort conditions and simulation results

Professor Fanger, who is a famous scholar in Denmark Industrial University has started to research on the comfort conditions of indoor thermal environment since 1960s.[11] He provided the PMV-PPD index system of thermal comfort to predict the PMV(predicted PMV mean vote) index according to the thermal comfort equation and ASHRAE 7-point scales (the PMV index of cold, cool, slightly cool, neutral, slightly warm and warm heat are -3, -2, -1, 0, +1, +2 and +3, respectively). This index has been widely used in Europe. The range of the PMV index is from -3 to +3 corresponding to the degree of human feeling from cold to hot. PMV index represents the feeling of most people in the same environment, so it can be used to predict the human's response to thermal environment. However, even if most people are satisfied with the thermal environment, there will be still some people are not comfortable due to physiological differences among humans. In order to illustrate this relationship, Fanger provided a standard of PPD (predicted percentage of dissatisfied), which measured the number of people being not satisfied with the thermal environment. In fact, when the value of PMV is 0, that is the feeling of human body to the environment is best, there are still 5% people being dissatisfied with the environment. Therefore, the PPD index affirms the most people's perception and retains the dissatisfaction of the few people with the environment. It is an index that most people can accept. The PMV-PPD index is based on thermal comfort equation. It is an environmental evaluation index, which is developed through the analysis of thermal sensation of 1396 American and Dane. It has been accepted by the world due to a big experimental data. In ISO Standard 7730 and ASHRAE Standard 55-2004, PMV and PPD index were used to evaluate the thermal environment respectively [12].

The PMV of indoor capillary network air-conditioning system of slope roof is simulated and analyzed The vertical section across the human and the computer is x=1.7m; the vertical section of the respiratory system corresponding to seating human body is y=1.2m and the vertical sections of human body at the position of z=0.2m, z=2.8m. Figure 4~9 describe the simulating results of capillary plane radiation air-conditioning system of slope roof.

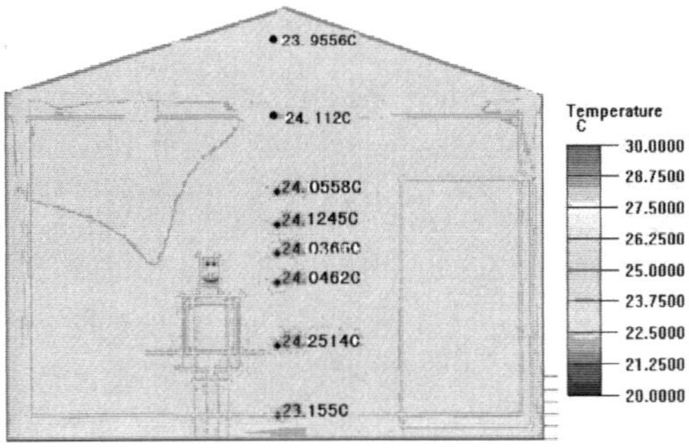

Figure 4. The vertical section of indoor temperature distribution (z=1.5m)

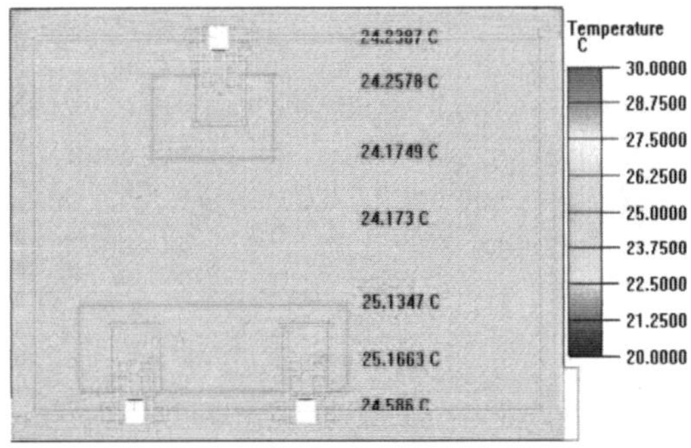

Figure 5. The plane section of indoor temperature distribution (y=1.25m)

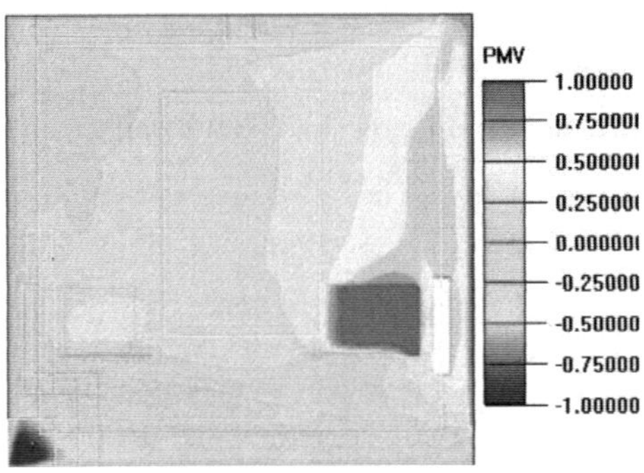

Figure 6. The section of indoor PMV distribution(x=1.7m)

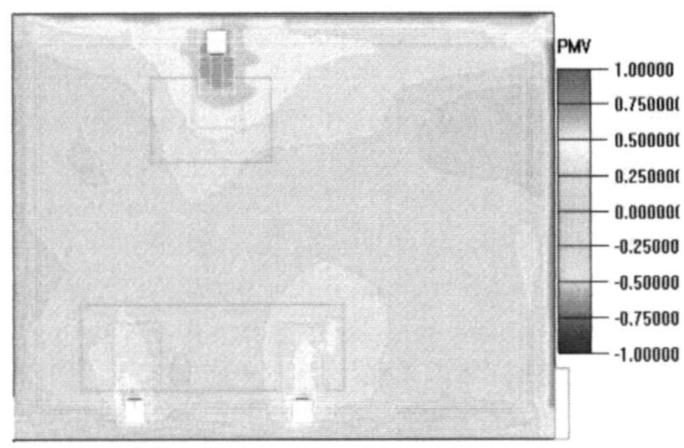

Figure 7. The section of indoor PMV distribution(y=1.2m)

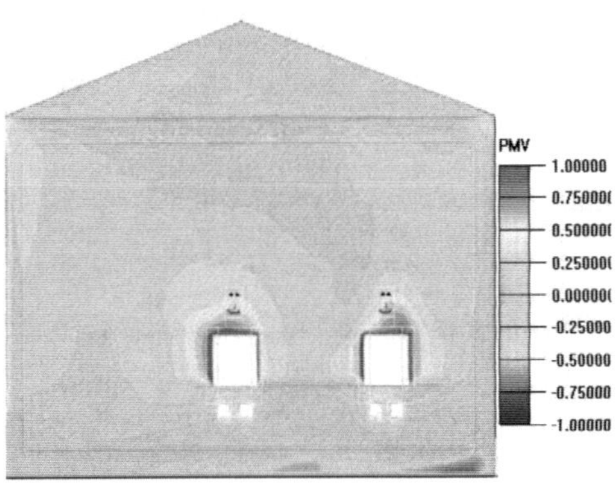

Figure 8. The section of indoor PMV distribution(z=0.2m)

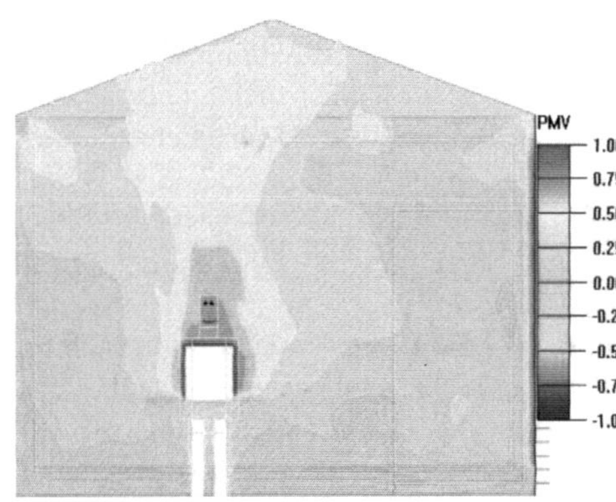

Figure 9. The section of indoor PMV distribution(z=2.8m)

From Figure 4 and Figure 5, it can be seen that the application of capillary radiation air-conditioning system in slope roof make the indoor temperature distribution uniform, which is maintained 24 ℃. This is because cooling load from heat and the radiation area of retaining structure is proportional to the temperature difference of supplying and returning water. The larger radiation area is conducive to supply cooling. Figure 6~9 show that the values of PMV are generally from 0 to 0.25, satisfying requirements of indoor comfort. Although the values of PMV surrounding human and computer are slightly higher, it does not affect the overall comfort.

4. Conclusions

In order to enrich the landscape of city and realize the goal of "better city, better life", new architectural shapes is emerging in an endless stream, which requires air-conditioning system keep pace with the times. The capillary plane radiant air conditioning system has outstanding performance in comfort, energy saving and so on. However, due to the various aspects of the comprehensive reasons, China's capillary plane radiation air-conditioning system has not been widely used.

In this paper, the application of the capillary plane radiation air conditioning system is studied based on the slope roof, so built a physical model by using a sloping roof construction in Ji'nan. Then by using Airpak software, our simulation experiment proves that it is feasible to apply capillary plane radiation air-conditioning system in the building of slope roof, for the value of PMV (predicted mean vote) of indoor in different sections is from 0 to 0.25, meeting the requirements of comfort. At the same time, the slope of the roof laying indoor temperature about 24 degrees, cooling capacity has been improved, as long as the layout of capillary plane radiation air-conditioning system is reasonable in the building of slope roof, the gradient of indoor temperature is not more than 1 ℃. It is conform to the requirement that the gradient of the temperature is not more than 3 ℃. so it is feasible to apply the capillary plane radiation air conditioning system in the building of slope roof. This provides a theoretical basis for the application and popularization of the capillary plane radiation air conditioning system in the slope roof construction. This is of great significance.

References:

[1] Jiang Y 2005 Temperature and Humidity Independent Control Air Conditioning System (Beijing: China Building Industry Press)
[2] Li Y G 2008 Huge Energy Saving Potential of New Air-conditioning Systems - capillary Plane Air-conditioning System *Heating and Cooling* **6** 24
[3] Mumma S A and Jeong J W 2005 Direct digital temperature, humidity, and condensate control for a dedicated outdoor air-ceiling radiant cooling panel system *ASHRAE Transactions* **111** 547-58
[4]Xue H X 2009 *Research on comfort and airflow in rooms with capillary air-conditioning systems* (Jinan: Shandong Jianzhu University)
[5] Tao W Q 1991 *Computational fluid and thermal dynamic* (Beijing: China Building Industry Press)
[6] Tao W Q 2003 *Numerical Heat Transfer* (Xi'an: Xi'an Jiaotong University Press)
[7] Jin F W, Yu M X and Jin G Y 2010 Experiments on Cooling Capillary Tube System HVAC **40** 102-6
[8] GB 50736,Design Code for Heating Ventilation and Air Conditioning of Civil Buildings[S]. 2012.
[9] Miriel J, Serres L and A. Trombe 2002 Radiant ceiling panel heating-cooling systems: experimental and silnulated study of the performances, thermal comfort and energy consumptions *Applied Thermal Engineering* **22** 1861-73
[10] Cui X Y, Li Y G and Yin G L 2009 Discussion of thermal parameter design of capillary air-conditioning systems *J. Shandong Jianzhu University* **24** 581-6
[11] Murakami S and Kato S 2001 Climate design based on cfd coupled simulation of convection, radiation and HVAC control for attaining a given PMV value *Building and Environment* **36** 701-9
[12] Wu S C 2013 A new specification learning experience *HVAC Expo.* **1** 19.

IOP Conf. Series: Earth and Environmental Science **40** (2016) 012066 doi:10.1088/1755-1315/40/1/012066

Study of a control strategy for grid side converter in doubly-fed wind power system

D J Zhu, Z L Tan[1], F Yuan, Q Y Wang and M Ding

School of Automation, China University of Geosciences, No. 388 Lumo Road, Wuhan, China

E-mail: tanzhili@cug.edu.cn

Abstract. The grid side converter is an important part of the excitation system of doubly-fed asynchronous generator used in wind power system. As a three-phase voltage source PWM converter, it can not only transfer slip power in the form of active power, but also adjust the reactive power of the grid. This paper proposed a control approach for improving its performance. In this control approach, the dc voltage is regulated by a sliding mode variable structure control scheme and current by a variable structure controller based on the input output linearization. The theoretical bases of the sliding mode variable structure control were introduced, and the stability proof was presented. Switching function of the system has been deduced, sliding mode voltage controller model has been established, and the output of the outer voltage loop is the instruction of the inner current loop. Affine nonlinear model of two input two output equations on d-q axis for current has been established its meeting conditions of exact linearization were proved. In order to improve the anti-jamming capability of the system, a variable structure control was added in the current controller, the control law was deduced. The dual-loop control with sliding mode control in outer voltage loop and linearization variable structure control in inner current loop was proposed. Simulation results demonstrate the effectiveness of the proposed control strategy even during the dc reference voltage and system load variation.

1. Introduction

Doubly-fed induction generator is one of the currently widely used in variable speed constant frequency (VSCF) wind power generation system. Its excitation system composed of the rotor side converter and grid side converter. When the wind speed changes caused by generator speed, by controlling the frequency of the rotor current, can make the output frequency constant on stator side. Due to the variable speed constant frequency control scheme are implemented by the rotor circuit, the slip power is only a small part of the generator rated power, so the capacity of converter for the excitation system is small with low voltage. In addition, the excitation control can adjust the amplitude of the excitation current, frequency, phase and phase sequence, so the doubly-fed induction generator with rotor excitation can control of active and reactive power flexibly, which means that it can compensate the reactive power for grid. As an important part of the excitation system of doubly-fed asynchronous generator, the grid side converter can not only transfer slip power in the form of active

[1] Address for correspondence: Z L Tan, School of Automation, China University of Geosciences, No. 388 Lumo Road, Wuhan, China. E-mail: tanzhili@cug.edu.cn.

power, but also adjust the reactive power of the grid, which means that the wind power system has reactive power adjustment ability. The grid side converter couples with the rotor side converter by the dc bus capacitance, the doubly-fed induction generator can be normal excited and good running characteristics, directly related to dynamic and static characteristics of it, so its performance has effect on the efficiency and stability of power system [1, 2].

The grid side converter is a three-phase voltage source PWM converter, the common control methods for which include: PID controller and its improvement method[3,4], Fast adaptive control [5], Model Predictive Current Control [6] ,Dead-Beat control [7,8], one-cycle control [9], intelligent control methods such as neural network [10], Hamiltonian control[11], etc. These control methods have their advantages respectively, but at the same time they have poor adaptability and is difficult to realize coordinated control. Sliding mode control (SMC) is a kind of effective nonlinear control method. It has strong robustness and implement simply, suitable for controlling nonlinear three-phase voltage source PWM converter. This paper aims at the control strategy for the grid side converter , a double-closed-loop control strategy is proposed, which includes outer voltage loop based on sliding mode variable structure control and inner current loop based on variable structure with input-output linearization control strategy .The simulation results prove its validity.

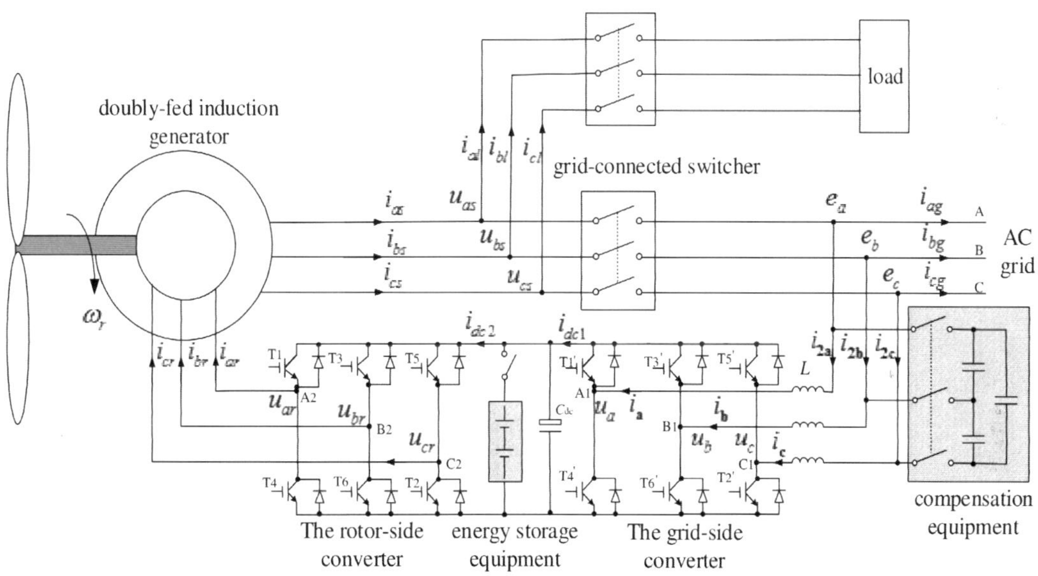

Figure 1. Doubly-fed induction generation in wind power system

2. The mathematical model of grid side converter

From the perspective of the grid side converter, the rotor side converter and dc capacitance can be considered as its loads. The grid side converter is a three-phase voltage source PWM rectifier [2]. Defined the switch function is

$$S_k = \begin{cases} 1 & \text{the upper bridge arm is on and the lower off} \\ 0 & \text{the upper bridge arm is off and the lower on} \end{cases} \quad (k = a,b,c) \tag{1}$$

According to the Kirchhoff's voltage law, the circuit equation for each phase of the three-phase PWM converter is

$$L\frac{di_k}{dt} + R_L i_k = e_k - u_k \quad (k = a,b,c) \tag{2}$$

In addition, using the Kirchhoff's current law on the positive node on the dc side capacitor, we get the Equation (3).

$$C\frac{dU_{dc}}{dt} = i_a S_a + i_b S_b + i_c S_c - i_{dc2} \tag{3}$$

Choose the d axis is consistent with the grid voltage vector $\boldsymbol{E}$, namely $e_d = E$, $e_q = 0$. The grid side converter mathematical model on d-q axis can be transformed from the three-phase static coordinate described as in Equation (2) and (3).

$$\begin{cases} L\frac{di_d}{dt} = -R_L i_d + \omega L i_q - u_d + e_d \\ L\frac{di_q}{dt} = -R_L i_q - \omega L i_d - u_q \end{cases} \tag{4}$$

$$C\frac{dU_{dc}}{dt} = S_d i_d + S_q i_q - i_{dc2} \tag{5}$$

The equivalent circuit of the grid side converter on d-q axis is shown in figure 2.

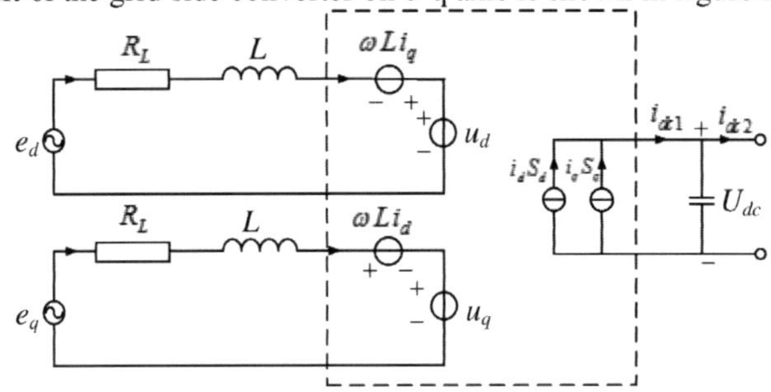

Figure 2. The equivalent circuit of the grid side converter on d-q axis

3. The voltage loop based on sliding mode variable structure control

It can be seen that the grid side converter is a mutual coupling nonlinear time-varying systems from the figure 2, so the conventional linear control strategy is difficult to achieve satisfactory control effect in the aspect of dynamic characteristics and robustness [5]. Therefore, to design a controller voltage with fast dynamic response and robustness for grid side converter, without relying on the switch devices and load parameters, the sliding mode variable structure control is definitely a good choice.

If the loss of inductance and resistance on ac side of the converter are ignored, equation (5) is multiplied by U_{dc} on both sides and then the output power on the dc side P can be obtained by:

$$P = u_d i_d + u_q i_q = (C\frac{dU_{dc}}{dt} + i_{dc2})U_{dc} = (S_d i_d + S_q i_q)U_{dc} \tag{6}$$

so

$$\begin{cases} S_d = u_d / U_{dc} \\ S_q = u_q / U_{dc} \end{cases} \tag{7}$$

Under the steady state conditions, ignoring the ac resistance R_L in equation (4), and considering that $di_d/dt = di_q/dt = 0$, u_d and u_q can be got from equation (4), as shown in equation (8)

$$\begin{cases} u_d = e_d + \omega L i_q \\ u_q = -\omega L i_d \end{cases} \tag{8}$$

Let $e_i = i_q^* - i_q$, $e_u = U_{dc}^* - U_{dc}$, define two sliding surfaces described by

$$\begin{cases} \sigma_i = c_1 e_i + \dot{e}_i \\ \sigma_u = c_2 e_u + \dot{e}_u \end{cases} \tag{9}$$

Where $\dot{e}_i$ and $\dot{e}_u$ are their derivative respectively, and c_1 and c_2 are positive constants.

To prove the stability of the proposed SM at the origin ($\sigma_i = 0$ and $\sigma_u = 0$), let the Lyapunov candidate be

$$\begin{cases} V(e_i) = \frac{1}{2} e_i^2 \\ V(e_u) = \frac{1}{2} e_u^2 \end{cases} \tag{10}$$

Therefore, its time derivative can be expressed as

$$\begin{cases} \dot{V}(e_i) = e_i \dot{e}_i = e_i(-c_1 e_i) = -c_1 e_i^2 < 0 \\ \dot{V}(e_u) = e_u \dot{e}_u = e_u(-c_2 e_u) = -c_2 e_u^2 < 0 \end{cases} \tag{11}$$

Since constant c_1 and c_2 are positive, the proposed control is asymptotically stable.

Ignore the resistance R_L, $e_i = i_q^* - i_q$ and $e_u = U_{dc}^* - U_{dc}$ are taken into (9), after simultaneous solution with equation (4) and(5),it can be got that

$$\begin{cases} \sigma_i = c_1(i_q^* - i_q) - \omega i_q + \frac{1}{L} u_d - \frac{1}{L} e_d \\ \sigma_u = c_2 e_u - \frac{S_d i_d + S_q i_q - i_{dc2}}{C} \end{cases} \tag{12}$$

Obviously, $dU_{dc}^* / dt = di_q^* / dt = 0$. The switching function of the proposed SM is as follows from equation (7), (8) and (12)

$$\begin{cases} \sigma_i = c_1(i_q^* - i_q) = 0 \\ \sigma_u = \frac{e_d}{CU_{dc}} \left(\frac{c_2 C U_{dc}}{e_d} e_u + \frac{U_{dc} i_{dc2}}{e_d} - i_d \right) = 0 \end{cases} \tag{13}$$

The two equations in (13) are divided by c_1 and e_d/CU_{dc} respectively, $c_m = c_2 C U_{dc} / e_d$ the equation can be simplified as

$$\begin{cases} \sigma_i = i_q^* - i_q = 0 \\ \sigma_u = i_d^* - i_d = 0 \end{cases} \tag{14}$$

in (14)

$$\begin{cases} i_q^* = 0 \\ i_d^* = c_m(U_{dc}^* - U_{dc}) + \frac{U_{dc} i_{dc2}}{e_d} \end{cases} \tag{15}$$

It can be seen that the model of voltage SMC controller contains only value of the output dc voltage and load current, dc side capacitor and the grid line voltage RMS e_d, so it is easy to realize digitally. The dc output of the voltage outer loop is the given current i_d^* for the inner current feedback loop.

4. Variable structure current controller based on the input output linearization

The response speed of current loop is more quickly than the voltage loop, so its current instruction i_d^* Variable structure current controller based on the input output linearization, namely the output of voltage outer loop can be considered as constant. Set $e_1 = i_d - i_d^*$, $e_2 = i_q - i_q^*$, equation (4) can be turned into

$$\frac{d}{dx}\begin{bmatrix} e_1 \\ e_2 \end{bmatrix} = \begin{bmatrix} -\frac{R_L}{L}e_1 - \frac{R_L}{L}i_d^* + \omega i_q + \frac{e_d}{L} - \frac{u_d}{L} \\ -\frac{R_L}{L}e_2 - \frac{R_L}{L}i_q^* - \omega i_d - \frac{u_q}{L} \end{bmatrix} \tag{16}$$

When not to consider system disturbance, the equation (16) can be expressed as the two input two output affine nonlinear models

$$\begin{cases} \dot{x} = f(x) + G(x)u \\ y_1 = h_1(x) \\ y_2 = h_2(x) \end{cases} \tag{17}$$

Where

$$f(x) = \begin{bmatrix} -\frac{R_L}{L}e_1 - \frac{R_L}{L}i_d^* + \omega i_q + \frac{e_d}{L} \\ -\frac{R_L}{L}e_2 - \frac{R_L}{L}i_q^* - \omega i_d \end{bmatrix}, \quad G(x) = \begin{bmatrix} -\frac{1}{L} & 0 \\ 0 & -\frac{1}{L} \end{bmatrix}$$

$x = [x_1, x_2] = [e_1, e_2] = [i_d - i_d^*, i_q - i_q^*], u = [u_d, u_q]$, $y = [y_1, y_2] = h(x) = [h_1(x), h_2(x)] = [e_1, e_2]$。

Affine nonlinear system shown in (17) meets the exact linearization condition. System vector relative degree $r = [r_1, r_2] = [1,1]$, Lie derivative of the function h to the function G is

$$L_g h(x) = \frac{\partial h(x)}{\partial x}G(x) = \begin{bmatrix} 1 & 0 \\ 0 & 1 \end{bmatrix}\begin{bmatrix} -\frac{1}{L} & 0 \\ 0 & -\frac{1}{L} \end{bmatrix} = \begin{bmatrix} -\frac{1}{L} & 0 \\ 0 & -\frac{1}{L} \end{bmatrix} \tag{18}$$

Defining a $m \times m$ matrix, $A(x)$

$$A(x) = \begin{bmatrix} L_{g_1}\left(L_f^{r_1-1}h_j\right) & \cdots & L_{g_m}\left(L_f^{r_1-1}h_j\right) \\ \vdots & \ddots & \vdots \\ L_{g_1}\left(L_f^{r_m-1}h_m\right) & \cdots & L_{g_m}\left(L_f^{r_m-1}h_m\right) \end{bmatrix} = \begin{bmatrix} -\frac{1}{L} & 0 \\ 0 & -\frac{1}{L} \end{bmatrix} \tag{19}$$

The Lie derivative of $h(x)$ to function f, $L_f h_1$ and $L_f h_2$ are shown in (20) and (21)

$$L_f h_1 = \frac{\partial h_1(x)}{\partial x}f(x) = [1 \quad 0]\begin{bmatrix} -\frac{R_L}{L}e_1 - \frac{R_L}{L}i_d^* + \omega i_q + \frac{e_d}{L} \\ -\frac{R_L}{L}e_2 - \frac{R_L}{L}i_q^* - \omega i_d \end{bmatrix} = -\frac{R_L}{L}e_1 - \frac{R_L}{L}i_d^* + \omega i_q + \frac{e_d}{L} \tag{20}$$

$$L_f h_2 = \frac{\partial h_2(x)}{\partial x}f(x) = [0 \quad 1]\begin{bmatrix} -\frac{R_L}{L}e_1 - \frac{R_L}{L}i_d^* + \omega i_q + \frac{e_d}{L} \\ -\frac{R_L}{L}e_2 - \frac{R_L}{L}i_q^* - \omega i_d \end{bmatrix} = -\frac{R_L}{L}e_2 - \frac{R_L}{L}i_q^* - \omega i_d \tag{21}$$

The system output derivatives of time can be expressed as:

$$\begin{bmatrix} \dot{y}_1 \\ \dot{y}_2 \end{bmatrix} = \begin{bmatrix} L_f h_1 \\ L_f h_2 \end{bmatrix} + A(x)\begin{bmatrix} u_1 \\ u_2 \end{bmatrix} \tag{22}$$

Obviously $A(x)$ is reversible, the feedback control law can be express as

$$u = -A^{-1}(x)\begin{bmatrix} L_f h_1 \\ L_f h_2 \end{bmatrix} + A^{-1}(x)v \tag{23}$$

The available input/output dynamic equations is

$$\begin{bmatrix} \dot{y}_1 \\ \dot{y}_2 \end{bmatrix} = v = \begin{bmatrix} v_1 \\ v_2 \end{bmatrix} \tag{24}$$

Comprehensive formula (4) (23) (40) are

$$\begin{cases} u_d = -R_L e_1 - R_L i_d^* + \omega L i_q + e_d - L v_1 \\ u_q = -R_L e_2 - R_L i_q^* - \omega L i_d - L v_2 \end{cases} \tag{25}$$

Take $v = [v_1, v_2] = [k_1 y_1, k_2 y_2] = [k_1 e_1, k_2 e_2]$, then the above equation can be translated into

$$\begin{cases} u_d = -(R_L + k_1 L) e_1 - R_L i_d^* + \omega L i_q + e_d \\ u_q = -(R_L + k_2 L) e_2 - R_L i_q^* - \omega L i_d \end{cases} \tag{26}$$

Select the appropriate feedback coefficient k_1,k_2 can obtain good dynamic and static performance. When the relative order of the uncertain disturbance system is lower than that of the system without perturbation, the zero dynamics of the system may become unstable due to disturbances. In order to reduce the possibility, when current loop controller is designed, variable structure control added into the input-output linearization above mentioned to enhance the anti-interference ability.

$$\dot{x} = f(x) + G(x)u + d(t,x) \tag{27}$$

where $d(t,x)$ is a uncertain perturbation.

Assuming uncertain disturbance $d(t,x)$ is within the scope of control, let $s = s(x)$ is a sliding surface, the following variable structure controller is designed

$$u = -(\frac{\partial s}{\partial x} G)^{-1} (\frac{\partial s}{\partial x} f + \lambda s + \mu \operatorname{sgn}(s)) \tag{28}$$

Where $\lambda \geq 0$, $\mu > 0$ are to be determined, and $\frac{\partial s}{\partial x} G$ is reversible. A corresponding nonlinear state transformation is as follows

$$z_i^j = T_i^j(x) = L_f^j h_i, i = 1, \cdots, m; j = 0, 1, \cdots, r_i - 1 \tag{29}$$

Namely

$$\begin{cases} z_1 = T_1(x) = L_f^0 h_1 = e_1 \\ z_2 = T_2(x) = L_f^0 h_2 = e_2 \end{cases} \tag{30}$$

then

$$z = y = [e_1, e_2]^T \tag{31}$$

Then the switching function of $s = s(x)$ can be expressed as

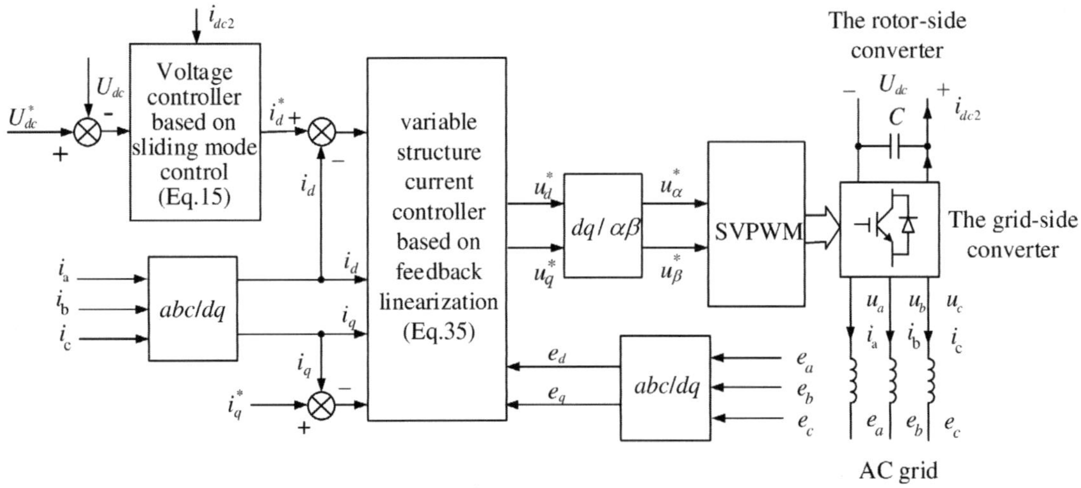

Figure 3. The dual-loop control diagram with sliding mode control in outer voltage loop and linearization variable structure control in inner current loop

$$s_i(z_i) = z_i \tag{32}$$

Namely
$$s(x) = z = [e_1, e_2]^T \tag{33}$$

Combination (28) with (33) can get control law

$$u = -(\frac{\partial s}{\partial x}G)^{-1}(\frac{\partial s}{\partial x}f + \lambda s + \mu \operatorname{sgn}(s))$$
$$= -\begin{bmatrix} -L & 0 \\ 0 & -L \end{bmatrix}\left(\begin{bmatrix} -\frac{R_L}{L}e_1 - \frac{R_L}{L}i_d^* + \omega i_q + \frac{e_d}{L} \\ -\frac{R_L}{L}e_2 - \frac{R_L}{L}i_q^* - \omega i_d \end{bmatrix} + \begin{bmatrix} \lambda_1 e_1 \\ \lambda_2 e_2 \end{bmatrix} + \begin{bmatrix} \mu_1 \operatorname{sgn}(e_1) \\ \mu_2 \operatorname{sgn}(e_2) \end{bmatrix}\right) \tag{34}$$

$$\begin{cases} u_d = (\lambda_1 L - R_L)e_1 + \mu_1 L\operatorname{sgn}(e_1) - R_L i_d^* + \omega L i_q + e_d \\ u_q = (\lambda_2 L - R_L)e_2 + \mu_1 L\operatorname{sgn}(e_1) - R_L i_q^* - \omega L i_d \end{cases} \tag{35}$$

Setting $[m_1, m_2] = [\lambda_1 L - R_L, \lambda_1 L - R_L]$ and then calculating the appropriate m_1, m_2, μ_1 and μ_2 assure of good static and dynamic performance. After u_d, u_q have gotten, transforming them into α-β axis, control single can be gotten by using SVPWM modulation .The overall control block diagram is shown in Figure 3.

5. Simulation

The proposed control system presented in figure 3 has been simulating evaluated by using a prototype shown in figure 1. The grid-side converter is connected to the AC grid by using input filter inductors L = 5.0mH with R_L=0.02Ω . The dc link is composed by capacitors of 2200 μF with a reference voltage of 300V. In the voltage sliding mode control controller module based on the formula (15), the coefficient c_m=1, as well as in the current controller according to formula (35), the coefficient $m_1 = m_2 = 50$, $\mu_1 = 50$, $\mu_2 = 25$.The rms value of AC side voltage is110V with frequency of 50Hz. The resistance R_L = 0.02, L = 5mH, switching frequency for power devices is 10 kHz, the rotor-side convert is instead by a resistance of 50Ω on DC side.

Figure 4 shows the simulation waveforms under different conditions. It can be seen that from figure 4 (a), the overshoot of dc voltage is 5%, with voltage peak of 315V. It reaches 300V in 0.03 seconds and then maintains stability. When the reference voltage suddenly increased from 300V to 400V at 0.1 seconds, as shown in figure 4 (c), the dc voltage dropped off about 15V at firstly. After 0.009 seconds, it reaches steady 400V without overshoot. If the dc load reduced to its half at 0.1 seconds, as shown in figure 4(e), a 1V voltage jump happened on dc side and back to the steady-state value of 300V after 0.02 seconds. If dc voltage of 400V generated by an inverter add to the dc side at 0.1 seconds, as shown in figure 4 (g), it can be seen that dc voltage return its steady state after 0.03 seconds. The simulation result approves that the control strategy this paper proposed has good dynamic characteristics and robustness.

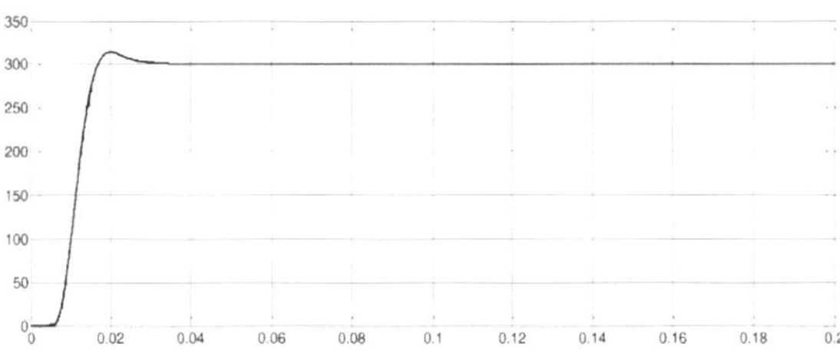

Figure 4. (a) dc voltage waveform

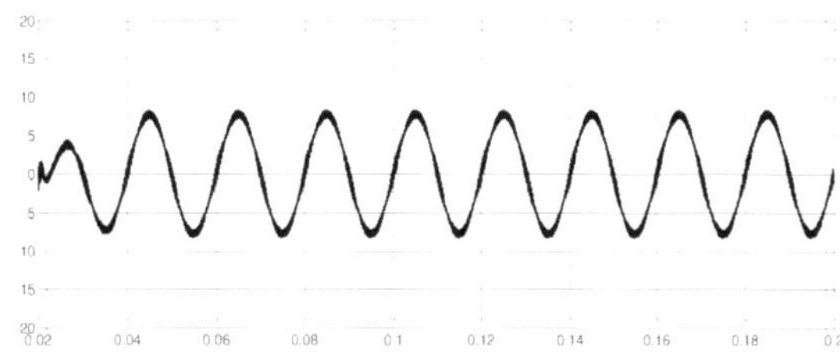

Figure 4. (b) a phase current waveform on ac side

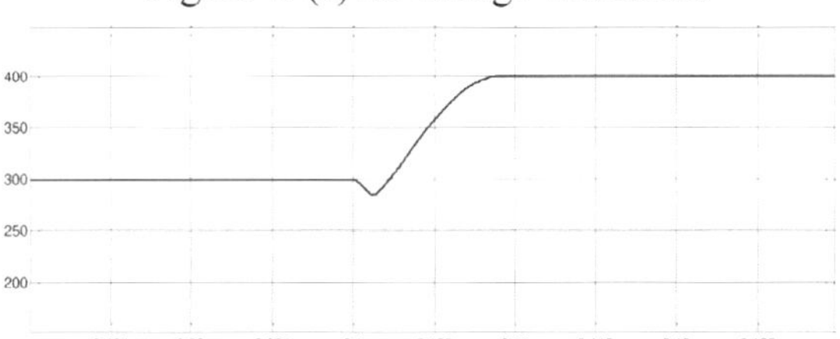

Figure 4. (c) DC voltage waveform when dc reference voltage changed suddenly

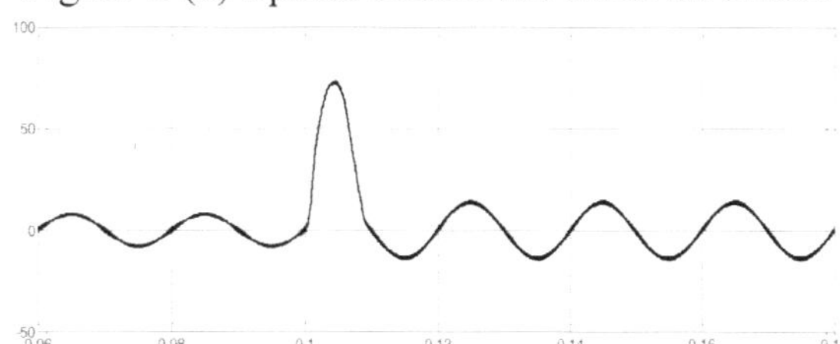

Figure 4. (d) A phase current waveform on ac side when dc reference voltage changed suddenly

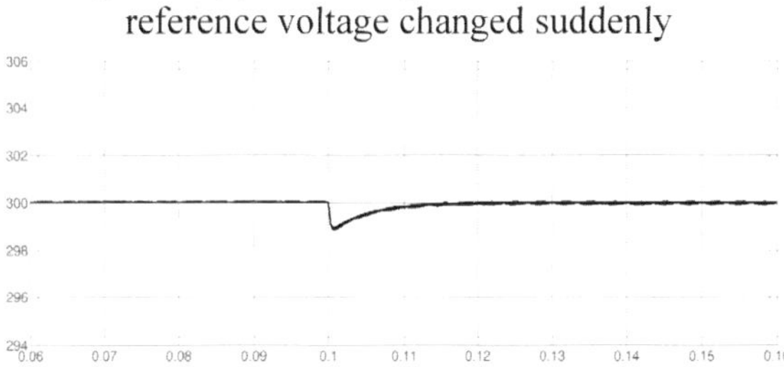

Figure 4. (e) DC voltage waveform when the load reduced to its half

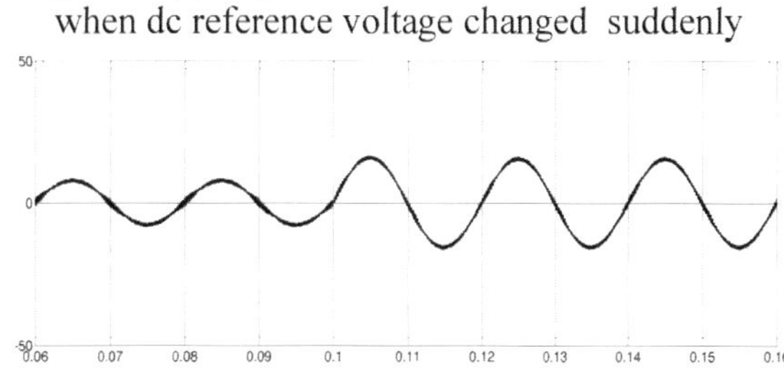

Figure 4. (f) A phase current waveform on ac side when the load reduced to its half

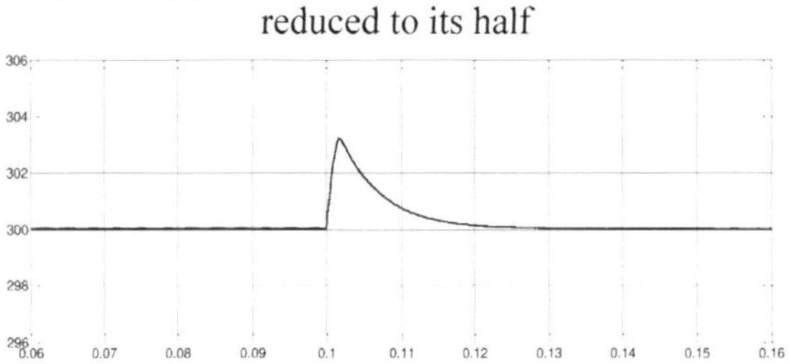

Figure 4. (g) dc voltage waveform when dc voltage source 400V added on dc side

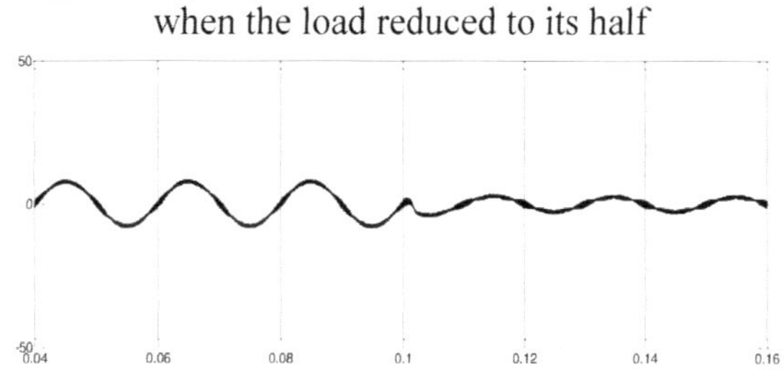

Figure 4. (h) A phase current waveform on ac side when dc voltage source of 400V added on dc side

Figure 4. Simulation waveforms under different conditions

6. Conclusion

This paper has proposed a control approach for improving the performance of the grid side converter in a doubly-fed induction generator used in the wind power system. In this control approach, the dcvoltage is regulated by a sliding mode variable structure control scheme and current by a variable structure controller based on the input output linearization. The theoretical bases of the sliding mode variable structure control was introduced, and the stability proof was presented. Switching function of the system has been deduced, sliding mode voltage controller model has been established, and the output of the outer voltage loop is the instruction of the inner current loop. Affine nonlinear model of two input two output equations on d-q axis for current has been established, its meeting conditions of exact linearization were proved. In order to improve the anti-jamming capability of the system, a variable structure control was added in the current controller, the control law was deduced. The dual-loop control diagram with sliding mode control in outer voltage loop and linearization variable structure control in inner current loop was proposed. Simulation results demonstrates the effectiveness of the proposed control strategy even during the dc reference voltage and system load variation.

Acknowledgments
The research work was supported by the Natural Science Foundation of Hubei Province, China, No. 2012FFB6413

References
[1] Ren J, Wang J , Hu Y and Ji Y 2014 *Electric Power Automation Equipment* **34** 108-13
[2] Song H and Qu Y 2012 *J. Control Theory and Applications* **10** 435-40
[3] Shao Z L, Shi M Q, Han Z F, Wang X D and Li Q M 2014 *Applied Decisions in Area of Mechanical Engineering and Industrial Manufacturing* **577** 321-4
[4] Tang H, Zhao R X, Tang S Q and Zeng Z 2012 *Ieee International Symposium on Industrial Electronics* 270-4
[5] Patnaik R K and Dash P K 2016 *IET Renewable Power Generation* **10** 598-610
[6] Shen Y, Zhang L, Chen H, Lu X F and Qian Z 2014 *26th Chinese Control and Decision Conference*(CCDC2014) 991-3
[7] Zhang Y C, Zhang Q, Li Z X and Zhang Y C 2013 *International Conference on Electrical Machines and Systems* 2207-12
[8] Chen W, Geng X J, Liu T and Xia C L 2014 *Electric Power Systems Research***108** 223-33
[9] Gao J L, Zheng T Q and Lin F 2011 *J. Power Electronics* **11** 64-73
[10] Setiawan I, Priyadi A, Miyauchi H and Purnomo M H 2015 *Ieej Transactions on Electrical and Electronic Engineering* **10** 674-82
[11] Wang C and Zhou J 2015 *Proc. Chinese Automation Congress*(CAC2015) 1252-7

IOP Conf. Series: Earth and Environmental Science **40** (2016) 012067 doi:10.1088/1755-1315/40/1/012067

Thermal performance and stress analyses of the cavity receiver tube in the parabolic trough solar collector

F Cao[1], Y Li, L Wang and T Y Zhu

College of Mechanical and Electrical Engineering, Hohai University, Changzhou, China

E-mail: fcao@hhu.edu.cn

Abstract. A light ray tracing model and a heat transfer model were built to analyse the heat flux distribution and heat transfer in a 1m cavity receiver tube with Parabolic Trough Collectors as the concentrator. The numerical methods were used to simulate the thermal stress and deformation of the receiver tube. The temperature fields of the receiver tube and the thermal stress distribution in the steel tube at the cross section and along the fluid flowing direction were presented. It is obtained from this study that non-uniform heat flux distribution is absorbed at the receiver tube outer surface due to the structure of the cavity receiver tube. Temperature fields in the steel receiver tube at the inlet and the outlet match well with the incident solar radiation. An eccentric circle temperature gradient is observed at cross section of the outlet fluid. The equivalent stress is a complex result of solar heating flux, energy transfer inside the PTC and the fluid and steel characteristics. Highest deformation is 3.1mm at 0.82m. On increasing the fluid mass flow rate, higher fluid mass flow rate results in higher equivalent stress along the absorber tube.

1. Introduction

Solar energy has been recognized as one of the most important energy sources at present and in the further energy consumption. Due to the discontinuous, low energy flux, periodicity and unsteady characteristics of solar energy, solar concentration is commonly utilized in solar engineering fields, among which the solar parabolic trough collectors (PTC) are the earliest and most widely accepted solar concentration style [1]. There have been many studies on the PTC, which can be generally classified as: the light ray tracing analysis under different structural [2-6], the collector tube thermal performance analysis [7-9], the thermal stress analysis [10-15] and the system performance evaluation [16-18].

Solar energy is reflected by the parabolic trough reflector, which generate nonlinear solar radiation distribution. There are also several types of solar receivers, such as the plat plate [2], the tube [3,14,15] and the cavity [19-21]. Solar radiation on the receiver is simulated through tracing each solar light [2-4] and the structural is also optimized according to the solar radiation distribution [5-6]. Serrano-Aguilera et al. proposed an Inverse Monte Carlo Ray Tracing method to calculate the numerical description of the PTC. Their method is expected to define continuous linear reflectors for flat plat receivers, where a quasi-planar concentrated flux distribution is required [2]. Cheng *et al.* made a three-dimensional

[1] Address for correspondence: F Cao, College of Mechanical and Electrical Engineering, Hohai University, Changzhou, China. E-mail: fcao@hhu.edu.cn.

numerical study of heat transfer characteristics in the receiver tube of PTC [3]. Based on the methodology, they then carried out a comparative analysis for PTC with a detailed Monte Carlo ray-tracing optical model [4]. They concluded that the ideal comprehensive characteristics and optical performance of the PTC systems were very different from some critical points determined by the divergence phenomenon of the non-parallel solar beam [4]. Liang *et al.* compared three optical models for the PTC, and optimized the geometric parameters according to their models [5]. Cheng *et al.* optimized the geometrical structure of the PTC based on the particle swarm optimization algorithm and the Monte Carlo ray-tracing method, which found a balance between the calculating speed and result accuracy [6]. As solar heating flux reaches the absorber, it will heat the fluid inside the absorber [7-9]. Kumar and Reddy carried out a 3-D numerical analysis of the porous disc line receiver for PTCs and investigated the influences of thermic fluid properties, receiver design and solar radiation concentration on overall heat collection [7]. Liang *et al.* summarized and compared the one-dimensional mathematical models under different assumptions and details for PTCs in the literature [8]. Huang et al. proposed a new analytical model for optical performance and a modified integration algorithm, with which the structural, operation and sun tracking parameters can be changed conveniently [9]. The solar heating flux and the temperature and pressure inside the absorbers can be utilized to analyze the thermal stress of the PTCs [10-15]. Giannuzzi *et al.* defined a guideline for steel structures design and assessment of components of PTC. The codes were developed for practical usage and were evaluated under some specific conditions [10]. Du et al. analyzed thermal stress and fatigue fracture of a single tube for the solar tower molten salt receiver and concluded that the distribution of stress was similar to that of inner/outer wall temperature difference [11]. Henshall *et al.* focused on the analysis of evacuated enclosures for flat-plate solar collectors and concluded that with a suitably low temperature sealing process vacuum the designed enclosure can successfully withstand imposed stresses [12]. Wu *et al.* derived the distribution of temperature field and stress field of a typical PTC and found that the maximum equivalent stress on the collector had close relationship with the convective heat transfer coefficient inside the collector [13]. Cui *et al.* studied thermal stress of DSG solar absorber tube in stratified two-phase regime and argued that the heat transfer difference of vapor and liquid phases are of high importance on analyzing the effective stress and deflection [14]. Wang *et al.* studied the effects of material selection on the thermal stresses of tube receiver and concluded that the stainless steel condition has the highest stress failure ratio and the copper condition has the lowest stress failure ratio [15]. As the thermal stress is analyzed and the strength of the receivers is satisfied, the PTC is coupled into the thermodynamic cycles to generate steam or power [16-18]. Zadeh *et al.* introduced the genetic algorithm and sequential quadratic programming into the thermal optimization of the PTC and analyzed the influences of the operation parameters like the nanoparticle concentration ratio and the operation temperature [16]. Kumaresan *et al.* reported an experimental study on performance of a PTC integrated with a storage unit and analyzed the parameters like the collector's useful heat gain and thermal efficiencies [17]. Reddy *et al.* carried out energetic and exergetic analyses for the components of the solar PTC power system and obtained the influences of the operation pressure and load conditions to the energetic and exergetic efficiencies [18].

Some special structural are also utilized for the PV or PV/T in PTC [3]. However, tube receivers are more commonly used in PTC circulation combined with Rankine Cycle [16-18]. In general, two kinds of tubes are utilized in the PTC, viz. the vacuum tube [2, 4-13] and the cavity tube [19-21]. Main advantages of the vacuum tube are the high thermal maintenances and low module cost, whereas the structural is frangible. Though a kind of metal inner and vacuum glass outer tube is proposed, the connection area of metal and glass is also frangible due to uninterrupted and periodical thermal stress from the inner and outer sides. The other kind of solar receiver is the cavity tube, whose schematic is shown in Figure 1. The outer cover of the receiver is elliptic, with an open inlet towards the parabolic trough reflector. The incoming solar lights are reflected by the parabolic trough reflector and entered into the absorber through the open inlet. Solar lights are then reflected for several times and finally reached the receiver tube at the ellipse focus. There are some studies on the thermal performance of the solar cavity receivers [19-21]. Very limited studies have been presented to analyze the thermal stress on the solar tube inside the ellipse cavity. Considering this, the thermal performance of the cavity receiver in the parabolic trough solar collectors is first analyzed. Then, the thermal stress of the receiver tube is presented through numerical simulation using STAR-CCM+ and Abaqus.

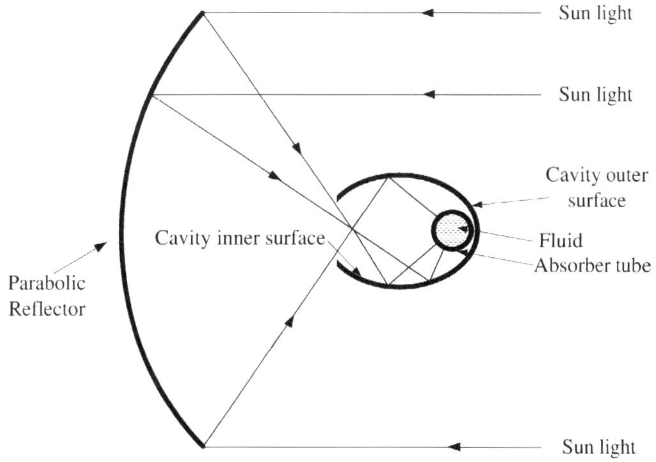

Figure 1. Schematic of the cavity receiver tube in the parabolic trough solar collecting system.

2. Mathematical model

2.1. Solar ray tracing

To track the solar ray transferring in the PTC, a coordinate system is built as shown in Figure 2. The major and minor axes and the focal distance of the elliptic are $2a$, $2b$ and $2c$ respectively. The focal distance and open length of the parabola are f and B respectively. According to the coordinate system, the elliptic cavity and the parabolic reflector can be expressed as:

$$\frac{(x-f-c)^2}{a^2}+\frac{y^2}{b^2}=1 \tag{1}$$

$$y^2=4fx \tag{2}$$

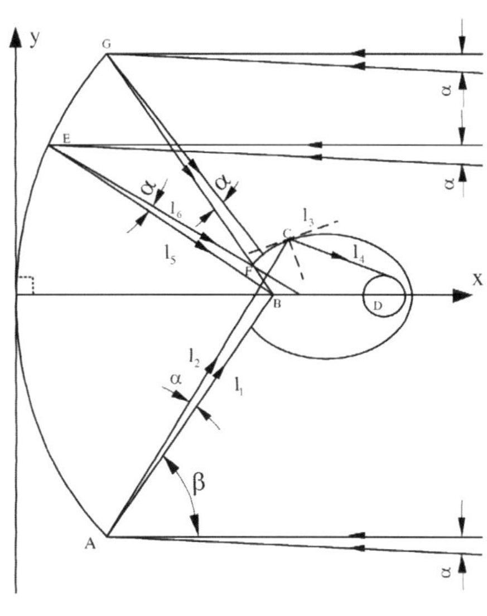

Figure 2. Sun light trance in the parabolic trough solar cavity collector

The marginal light BG can be expressed as:

$$y = -k_1(x - c) \tag{3}$$

where:

$$k_1 = \frac{-\frac{B}{2}}{\frac{B^2}{16} - f} \tag{4}$$

$$\beta = \arctan k_1 \tag{5}$$

The other marginal light is AC, which is reflected by the cavity inner surface and generates the tangent line of the absorber tube l_4. The slope for l_4 is

$$k_4 = \frac{k_2 - k_3 + k_3(1 + k_2 k_3)}{1 + k_2 k_3 - (k_2 - k_3)\ k_3} \tag{6}$$

where

$$k_3 = -\frac{b^2(m - f - c)}{a^2 n} \tag{7}$$

Correspondingly, in order to reach the absorber tube inside the cavity, the relationship between the absorber tube radius and the light slopes is:

$$\frac{|k_4(f + 2c) - k_4 m + n|}{\sqrt{k_4^2 + 1}} \le r_{tube} \tag{8}$$

2.2. Thermal balance in the cavity

The schematic of the thermal balance in the cavity is shown in Figure 3. The energy balance of the tube steel wall is:

$$Q_{3\mathrm{SolAbs}} = Q_{23\mathrm{cond}} + Q_{35\mathrm{rad}} + Q_{36\mathrm{rad}} + Q_{34\mathrm{conv}} \tag{9}$$

where $Q_{3SolAbs}$ is the solar radiation absorbed by the receiver tube, Q_{23cond} is the heat conduction through the steel wall, Q_{35rad} is the radiation exchange between the steel tube and the gall at the cavity inlet, Q_{36rad} is the radiation exchange between the cavity inner surface and the steel tube, Q_{34conv} is the heat convection between the steel tube and the fluid inside it.

IOP Conf. Series: Earth and Environmental Science **40** (2016) 012067 doi:10.1088/1755-1315/40/1/012067

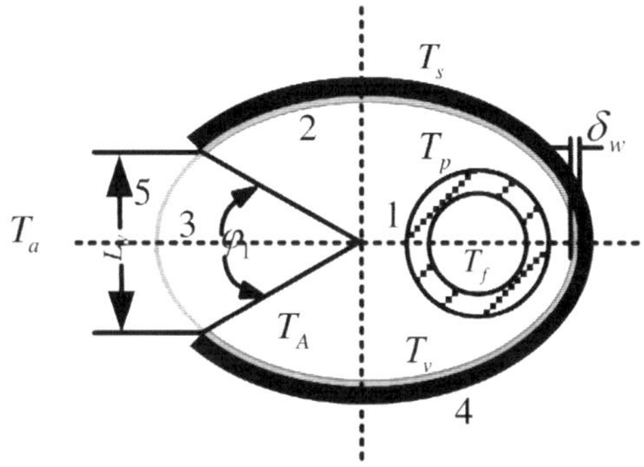

Figure 3. (a)

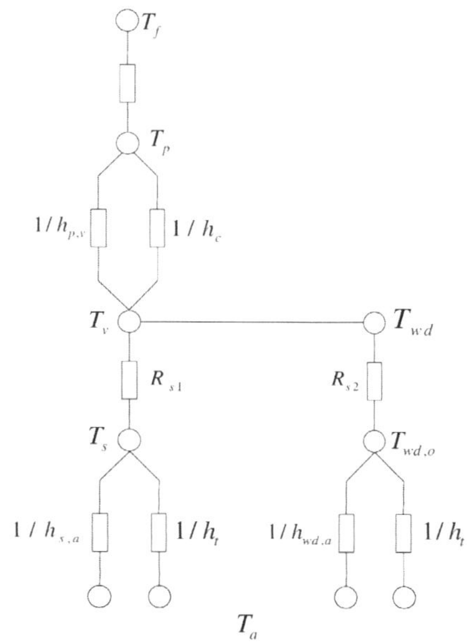

Figure 3. (b)

Figure 3. Thermal balance in the ellipse cavity.

The energy balance of the cavity glass is:

$$Q_{7\text{SolAbs}} + Q_{45\text{conv}} + Q_{56\text{cond}} + Q_{57\text{cond}} + Q_{35\text{rad}} = Q_{7,10\text{conv}} + Q_{7,10\text{rad}} \tag{10}$$

where $Q_{7SolAbs}$ is the heat absorbed by the cavity glass, Q_{56cond} is the heat conduction between the cavity inner surface and the glass, Q_{45conv} is the heat convection between the glass and the air inside the cavity, $Q_{7,10rad}$ is the radiation exchange between the glass and the atmosphere and Q_{57cond} is the heat conduction between the two surfaces of the glass.

The energy balance of the air inside the cavity is:

$$Q_{4\text{SolAbs}} + Q_{34\text{conv}} = Q_{45\text{conv}} \tag{11}$$

2.3. Solutions

Equation (1)-(8) are then converted into the codes to calculate the radiation distribution on the steel absorber tube. After obtaining the solar radiation absorbed by the receiver tube, $Q_{3SolAbs}$, the thermal balance inside the cavity is then calculated according to Equation (9)-(11). After that, the thermal stress of the receiver tube is presented using STAR-CCM+ and Abaqus.

3. Results and discussion

3.1. Thermal flux on the absorber tube

Dimensions of the cavity receiver tube are summarized in Table 1. Distributions of the light and heat flux around the receiver tube outer surface are shown in Figuation 4 (a) and (b) respectively. It is found from the figure that the sun light is sheltered by the cavity, which causes no light and heat flux from 0° to 10° at Figure 4. (a). The other light enters the cavity and reaches the absorber tube, leading to the first increase of the heat flux from 10° to 30°. A minor peak appears at 26° because of the circular receiving surface. After that, the direct light from the cavity open inlet and the reflected light from the elliptic inner surface are merged from 40° to 90°, generating a peak at 56°. The reflected sun light from the elliptic inner surface then enters the absorber surface from 90° to 125°. No light can reach the range between 125° and 180°.

Table 1. Parameters of the cavity and absorber tube

Parameter/Unit	Value
Concentration ratio /-	16.6
Major axis/mm	50
Minor axis/mm	30
PTC focal distance /mm	800
Elliptic focal distance /mm	40
PTC open width /mm	1000
Absorber tube radius /mm	18
Absorber tube length /m	1.0

3.2. Temperature distribution in the absorber tube

The temperature distributions in the absorber tube at the inlet and the outlet under fluid mass flow rate of 0.1 kg/s are shown in Figure 5. The temperature and pressure of the inlet fluid are 270 °C and 5MPa respectively. It is found from the figure that the highest temperature of the steel receiver tube increases from 284.86 °C at the inlet to 287.07 °C at the outlet. Temperature fields in the steel receiver tube at the inlet and the outlet are similar, which matches well with the incident solar radiation, following the principle that higher solar radiation generates higher steel receiver wall temperature. Kumar and Reddy also found the similar tendency of the radiation distribution and the wall temperature [7]. The fluid inside the receiver tube, however, differs a lot from the inlet to the outlet. The fluid inlet temperature is uniform at the inlet, which is settled by the boundary conditions. As the fluid is heated along the receiver tube, its temperature reaches 280 °C at the outlet. And its temperature distribution is not uniform, with higher fluid temperature at the steel wall with higher temperature. And an eccentric circle temperature gradient is observed in Figure 5 (b).

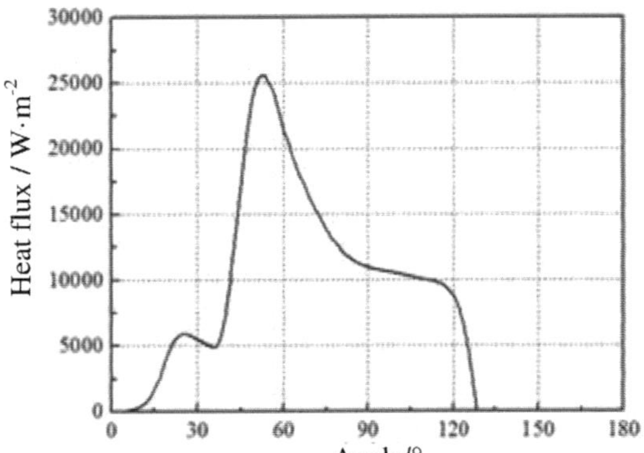

(a)

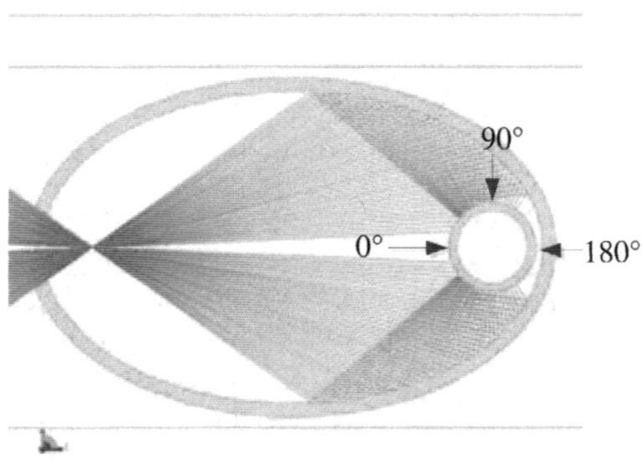

(b)

Figure 4. Distributions of the light and heat flux around the receiver tube outer surface.

IOP Conf. Series: Earth and Environmental Science **40** (2016) 012067 doi:10.1088/1755-1315/40/1/012067

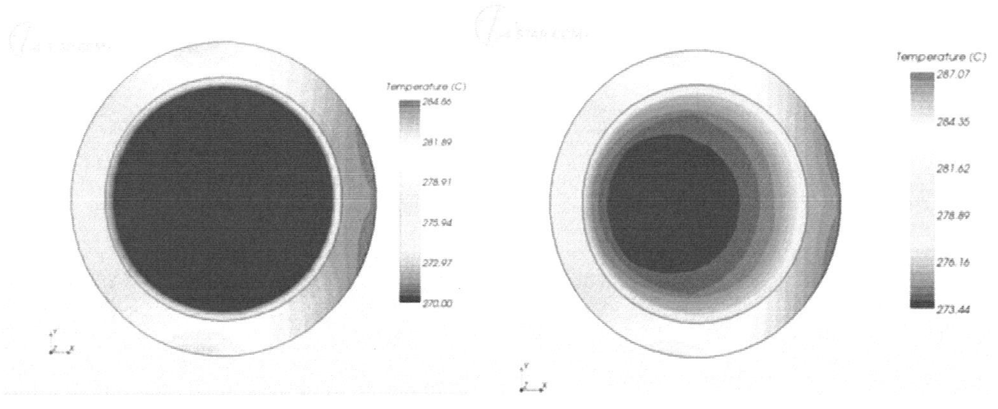

(a) (b)

Figure 5. Temperature distribution in the absorber tube
(a) temperature field at the inlet and (b) temperature field at the outlet.

3.3. Thermal stress in the absorber tube

After obtaining the temperature distributions at the steel tube receiver, the equivalent stresses and the deformation of the receiver tube can be simulated using STAR-CCM+ and Abaqus. The equivalent stress along the fluid direction, along the radial direction and at the tube outer surface and the deformation of the tube along the fluid direction are shown in Figure 6.

It is found from Figure 6 that the equivalent stress along the radial direction nearly remains at 16 MPa at the heating side and 7.1 MPa at the non-heating side. The equivalent stress at the cross section of the receiver tube decreases from 0° to 47°, then increases from 47° to 85°, then decreases till 142° and finally increases till 180°. The peak equivalent stress locates at 0°, with the value of 16.4 MPa. The equivalent stress in the wall of the cross section increases linearly. The deformation along the axis increases from the inlet, with the peak of 3.1mm at 0.82m. Then the defamation has a fluctuation, with a valley of 2.4mm at 0.89m.

The equivalent stress is a complex result of solar heating flux, energy transfer inside the PTC and the fluid and steel characteristics. First of all, the solar radiation heats the steel tube and is transferred into the tube through the steel wall, which generates the temperature difference between the inner and outer wall. Du *et al.* demonstrated the similar tendency between the thermal stress and the inner/outer wall temperature difference [11]. Then, the heating flux is transferred into the fluid inside the tube through heat convection, which generates the temperature difference between the inner wall and the fluid. Even though the fluid is in liquid or vapor phase, it still has effects on the steel walls. Wu *et al.* had concluded the close relationship between the thermal stress and the convective heat transfer coefficient inside the collector [13]. Finally, as fluid is continuously heated inside the tube, phase change will occur under suitable conditions, which leads to different thermal stresses onto the tube inner wall. Cui *et al.* argued that the heat transfer difference of vapor and liquid phases are of high importance on analyzing the effective stress and deflection [14]. In general, the thermal stress is generated by the temperature difference between the tube inner/outer wall and the fluid inside the tube, the fluid pressure and the fluid state.

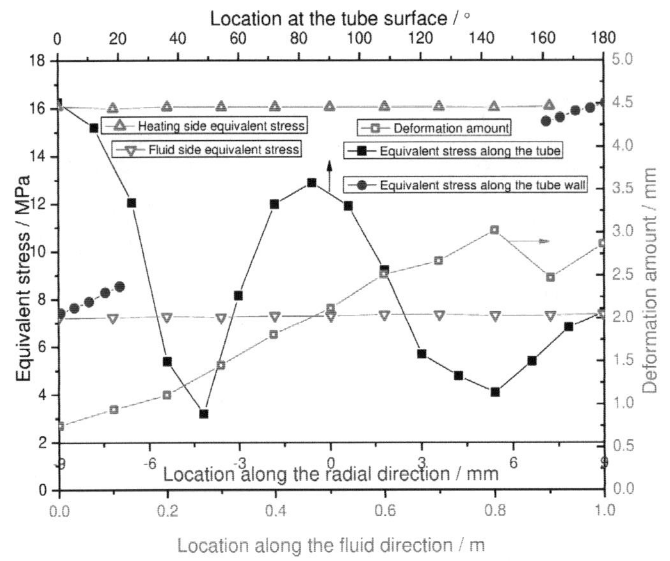

Figure 6. Equivalent stress along the fluid direction, along the radial direction and at the tube outer surface and the deformation of the tube along the fluid direction.

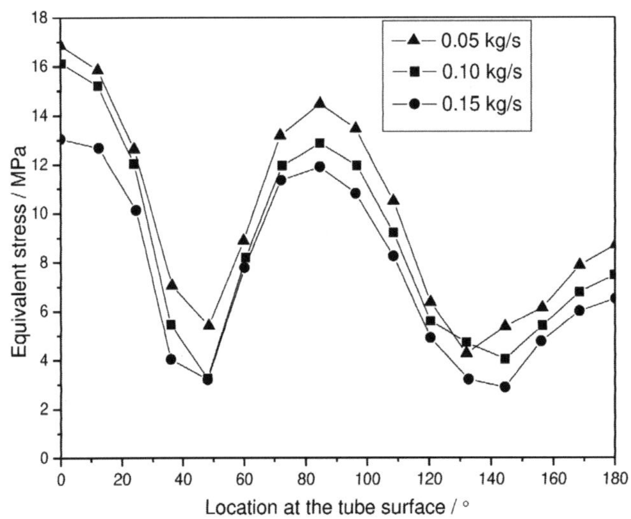

Figure 7. Equivalent stress at the outlet under different fluid mass flow rates.

3.4. Thermal stress under different water mass flow rates

As the inside fluid property affects the equivalent stress and deformation of the steel receiver tube, different fluid mass flow rates are considered in Figure 7. It is found that the equivalent stress distributions in the absorber tube along the fluid direction have the similar tendencies under different fluid mass flow rates, following the rule that higher fluid mass flow rate leading to higher equivalent stress.

4. Conclusions

Solar parabolic trough collectors (PTC) are the most widely accepted solar concentration style. The PTC with cavity receiver tuber has not been well discussed in the literature. In the present study, a ray tracing model and a heat transfer model were built to analyze the heat flux distribution and heat transfer in a 1m cavity receiver tube. The temperature fields of the receiver tube and the thermal stress distributions of the steel tube at the cross section and along the fluid flowing direction were obtained. The following conclusions are obtained through this study:

- A non-uniform heat flux distribution is absorbed at the receiver tube outer surface due to the structure of the cavity receiver tube.
- Temperature fields in the steel receiver tube at the inlet and the outlet match well with the incident solar heat flux. An eccentric circle temperature gradient is observed at cross section of the outlet fluid.
- The thermal stress is generated by the temperature difference between the tube inner/outer wall and the fluid inside the tube, the fluid pressure and the fluid phase, which are caused by solar heating flux, energy transfer inside the PTC and the fluid and steel characteristics. The highest deformation in this study is 3.1mm at 0.82m.
- On increasing the fluid mass flow rate, higher fluid mass flow rate results in higher equivalent stress along the absorber tube.

Acknowledgement
This research was funded by the National Natural Science Foundation of China (No.: 51506043) and the Fundamental Research Funds for the Central Universities (No.: 2014B19714).

References
[1] Jebasingh V K and Herbert G M J 2016 A review of solar parabolic trough collector *Renewable and Sustainalbe Energy Reviews* **54** 1085-91
[2] Serrano-Aguilera J J, Valenzuela L and Fernandez-Reche J 2016 Inverse monte carlo ray-tracing method (IMCRT) applied to line-focus reflectors *Solar Energy* **124** 184-97
[3] Cheng Z D, He Y L, Xiao J, Tao Y B and Xu R J 2010 Three-dimensional numerical study of heat transfer characteristics in the receiver tube of parabolic trough solar collector *International Communications in Heat and Mass Transfer* **37** 782-7
[4] Cheng Z D, He Y L, Cui F Q, Du B C, Zheng Z J and Xu Y 2014 Comparative and sensitive analysis for parabolic trough solar collectors with a detailed Monte Carlo ray-tracing optical model *Applied Energy* 559-72
[5] Liang H B, You S J and Zhang H 2016 Comparison of three optical models and analysis of geometric parameters for parabolic trough solar collectors *Energy* **96** 37-47
[6] Cheng Z D, He Y L and Du B C 2015 Geometric optimization on optical performance of parabolic trough solar collector systems using particle swarm optimization algorithm *Applied Energy* **148** 282-93
[7] Kumar K R and Reddy K S 2009 Thermal analysis of solar parabolic trough with porous disc receiver *Applied Energy* **86** 1804-12
[8] Liang H B, You S J and Zhang H 2015 Comparison of different heat transfer models for parabolic trough solar collectors *Applied Energy* **148** 105-14
[9] Huang W D, Hu P and Chen Z S 2012 Performance simulation of a parabolic trough solar collector *Solar Energy* **86** 746-55
[10] Giannuzzi G M, Majorana C E, Miliozzi A, Salomoni V A and Nicolini D 2007 Structural design criteria for steel components of parabolic-trough solar concentrators *J. Solar Energy Engineering* **129** 382-90
[11] Du B C, He Y L, Zheng Z J and Cheng Z D 2016 Analysis of thermal stress and fatigue fracture for the solar tower molten salt receiver *Applied Thermal Engineering* **99** 741-50
[12] Henshall P, Eames P, Arya F, Hyde T, Moss R and Shire S 2016 Constant temperature induced stresses in evacuated enclosures for high performance flat plate solar thermal collectors *Solar Energy* **127** 250-61
[13] Wu J J, Hou H J and Yang Y P 2015 Thermal stress analysis of parabolic trough solar receivers

Kung Cheng Je Wu Li Hsueh Pao/Journal of Engineering Thermophysics **36** 249-53

[14] Cui W, Li H, Li L and Liao Q 2013 Thermal stress analysis of DSG solar absorber tube in stratified two-phase regime *AIP Conference Proceedings* **1558** 2217-20

[15] Wang F, Shuai Y, Yuan Y and Liu B 2012 Effects of material selection on the thermal stresses of tube receiver under concentrated solar irradiation *Materials and Design* **33** 284-91

[16] Zadeh M, Sokhansefat T, Kasaeian A B, Kowsary F and Akbarzadeh A 2015 Hybrid optimization algorithm for thermal analysis in a solar parabolic trough collector based on nanofluid *Energy* **82** 857-64

[17] Kumaresan G, Sridhar R and Velraj R 2012 Performance studies of a solar parabolic trough collector with a thermal energy storage system *Energy* **47** 395-402

[18] Reddy V S, Kaushik S C and Tyagi S K 2012 Exergetic analysis and performance evaluation of parabolic trough concentrating solar thermal power plant (PTCSTPP) *Energy* **39** 258-73

[19] Chen F, Liu M, Zhang P and Luo X 2015 Thermal performance of a novel linear cavity absorber for parabolic trough solar concentrator *Energy Conversion and Management* **90** 292-9

[20] Chen F, Liu M, Hassanien R H E, Luo X, Hong Y R, Feng Z K, Ji M G and Zhang P 2015 Study on the optical properties of triangular cavity absorber for parabolic trough solar concentrator *Int. J. Photoenergy* 895946

[21] Gao W, Xu G Q, Li T T and Li H W 2015 Modeling and performance evaluation of parabolic trough solar cavity-type receivers *Int. J. Green Energy* **12** 1263-71

Identification to oil cracking in east of Sichuan Basin

Jian Ming Guo[1, 3] **and Yan Xia Li**[2]

[1]School of Petroleum Engineering, Xi'an Shiyou University, Shaanxi xi'an, 710065, China
[2]School of Earth Sciences & Engineering, Xi'an Shiyou University, Shaanxi xi'an 710065, China

E-mail: jmguo@xsyu.edu.cn

Abstract. In Sichuan basin the Lower Palaeozoic source rock has reached high to over-mature stage, but it is still considered that it has obvious genetic link with some gas reservoirs which were formed later. Therefore, the accumulation process between ancient high to over-mature source rock and the late formation of gas reservoirs has become one of the key factors to recognize the law of natural gas accumulation in Sichuan Basin. Based on some characteristics of Carboniferous gas reservoir in eastern basin, such as the gas compositions, stable carbon isotope, it is found that the gas reservoir is oil-cracking gas. The result reveals that the formation and evolution of present gas reservoir is that an ancient oil pool experienced the processes of forming, accumulating, destructing, and thermal maturating and cracking into gas reservoir.

1. Introduction

Similar to the process of kerogen that pyrolysed hydrocarbon, the thermally altered action (cracking) of crude oil in the oil reservoir is essentially a pyrolysis reaction at certain temperature, with the generation of gaseous hydrocarbon and residues (solid asphalt). Therefore, the course of reaction is consistent with the law of the chemical kinetics. It means that the cracking process of crude oil can be described by the equations of chemical kinetic. In recent years many scholars at home and abroad [1-11] have carried out the study of the reaction of pyrolysis crude oil using chemical kinetics and hermal history to quantitatively evaluate the degree of cracking of crude oil.

In the eastern Sichuan Basin the Lower Paleozoic Silurian source rock has reached high to over-mature stage. And vitrinite reflectance (Ro) reached more than 2% at present, with the distribution of 3.0 ~ 3.5% for majority of them, which illustrates that the hydrocarbon potential has been exhausted. However, a large number of gas reservoirs have been found in its overlying Carboniferous carbonate reservoir. Among the compositions in addition to the high concentration of methane (95.5 ~ 98%) in the gas reservoir, the surplus are in large quantities of solid carbon bitumen that remain in the carbonate reservoir [12].

The paper is based on the characteristics of chemical compositions and of stable carbon isotope of natural gas in Carboniferous gas pools in eastern Sichuan Basin, together with the study of the types of inclusions, homogenization temperature, composition, phase and solid asphalt, etc., to look for the evidence of thermal cracking of crude oil. At the same time, the chemical kinetics of cracking gas of

[3] Address for correspondence: Jian Ming Guo, School of Petroleum Engineering, Xi'an Shiyou University, Shaanxi xi'an, 710065, China. Email: jmguo@xsyu.edu.cn.

crude oil is used to calculate the efficiency of the pyrolysis of paleo-oil pools. All the above are necessary to reconstruct the process that the paleo-oil reservoir cracked into gas reservoir.

2. Geological background

East Sichuan area is a relatively tectonically active zone in stable Sichuan Basin Block. It is adjacent to Huaying Mountains - Shilong canyon from west, east to Qiyue Mountains, north from Daba Mountains, south to Nanchuan-Kailong line, with an area of about $5.5 \times 10^4 km^2$. And it is a parallel fold area, which is in control of the great fault, with the predominance of high anticline zone. There are 10 rows of high steep tectonic belts primarily extending toward the east north-east from west to east. The current structure is an ejective fold of north north-east and north-east; the anticline belts often have many high focuses which present slimly extending tens of hundreds of kilometers, many Triassic carbonate rocks exposed at the structure core, steep or vertical inverted formations are at the flank area. The Syncline is broad with red clastic Jurassic rocks; the physiognomy is gentle hilly area and the terrain does not fluctuate greatly. The width ratio of high steep anticline and syncline is 1: 3, which shows an association feature of ejective fold. And the main anticlines formed accompanied by the great faults (Figure 1).

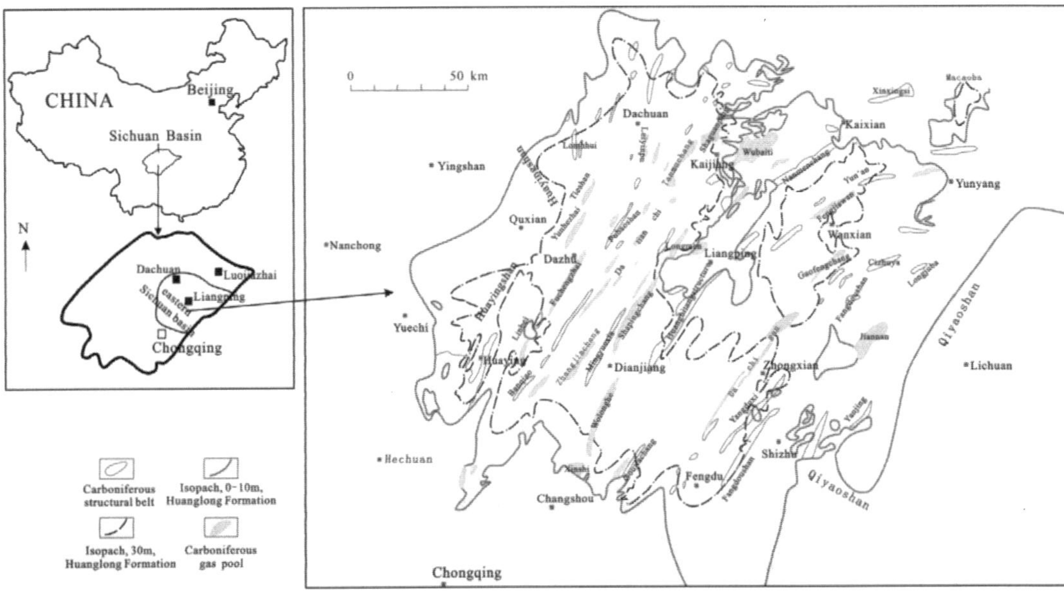

Figure 1. Distribution map of structure and large - medium size gas fields of east of Sichuan Basin.

In Sichuan Basin, as a major land-and-sea superimposed basin, there are multiple sets of source rock, multiple reservoir layers, changeable combination of reservoir and seal, multi-phase hydrocarbon generation, migration and accumulation process. In addition, Carboniferous gas reservoir is a relatively independent reservoir system which has its own source rock and reservoir layer. It is found that Chuandong Carboniferous gas reservoir sources is mainly from its underlying Silurian Longmaxi formation through comparison of gas source [13]; and the layer is the initial hydrocarbon source kitchen of Carboniferous reservoir.

Longmaxi formation is a black shale, dark gray mudstone and calcareous shale of deep water continental shelf facies, which is an anaerobic and reduction deposition. A few drillings reveal that the thickness of Longmaxi formation is about 300-672m in Chuandong area. The average thickness of graptolite shale facies of hydrocarbon source rock is about 400 m in the lower part, changing between 100-900m, and the thickness of high-quality hydrocarbon source rock of the black shale changes between 20-70m. More over, the content of organic carbon is high, with 0.5-7.5% in outcrop section. And the drillings reveal that the content of organic carbon is 0.07 ~ 2.77% (Well Wuke -1), which

IOP Conf. Series: Earth and Environmental Science **40** (2016) 012068 doi:10.1088/1755-1315/40/1/012068

indicates that it is a good set of source rocks that are the main gas source of the overlying Carboniferous gas.

3. Characteristics of gas composition

3.1. The chemical composition of gas

The hydrocarbon gas has an absolute advantage in gas components in Chuandong Carboniferous gas reservoir, with the content of the hydrocarbon gas 95.5 ~ 99.5%, mostly 97~98%. Methane is the main gas in total, even as high as 95.5 ~ 98% (figure 2a), which indicates that the gas is the typical dry gas, the drying coefficient 0.986 ~ 0.998. The content of heavy hydrocarbon gas is as low as 0.2-1.5%, generally including only ethane and propane, little butane. In addition, non-hydrocarbon gas accounts for only 1 ~ 4.5% (figure 2b).

3.2. Carbon isotope composition of gas

Methane carbon isotope values are between -37‰ ~ -29‰, at the top of -31‰ ~ -32‰ (figure 3), Ethane carbon isotope values are between -40 ~ -33%, Propane carbon isotope values between -39‰~ -32‰, $\delta^{13}C_1 > \delta^{13}C_2 < \delta^{13}C_3$; all of which suggest that carbon isotope value of ethane and methane reverses (Figure 4). Obviously, this shows that methane carbon isotope values increase with maturity. On the other hand, ethane carbon isotope values tend to increase along with methane carbon isotope values in Chuandong Carboniferous gas reservoir.

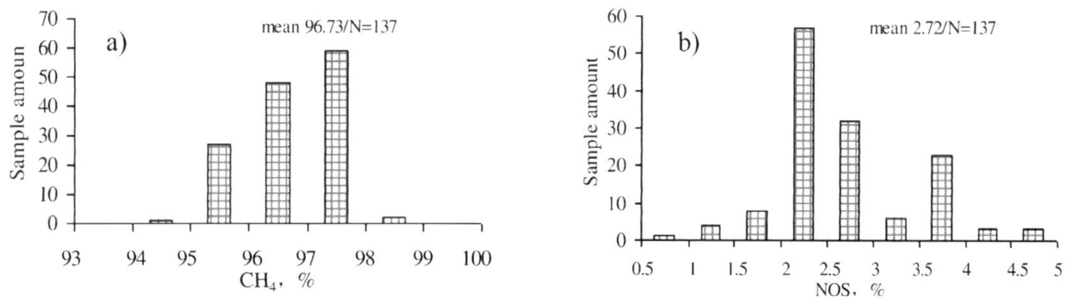

Figure 2. Frequency distribution of gas composition for the Carboniferous gas reservoir of east of Sichuan Basin.

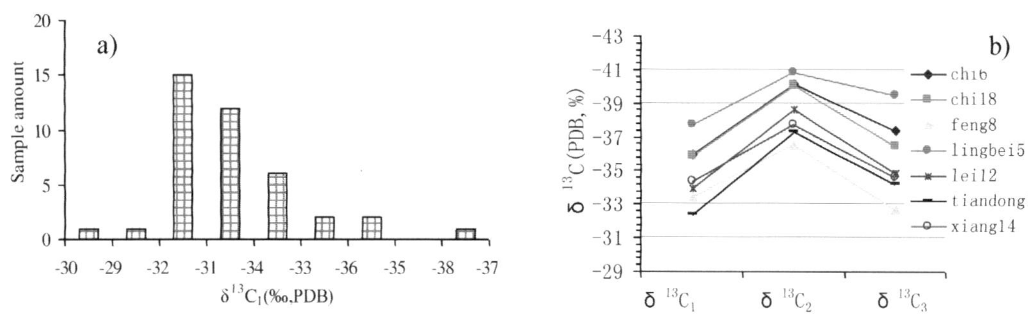

Figure 3. Distribution of Frequency and curve of carbon isotopes of Carboniferous natural gases.

4. According to gas composition to identify Crude oil cracking gas and kerogen pyrolysis gas

4.1. Identification mode of $ln(C_2/C_3) - ln(C_1/C_2)$

Prinzhofer & Huc [14]. suggested to use the relevant charts of $\ln(C_2/C_3)-\ln(C_1/C_2)$ and $\delta^{13}C_2-\delta^{13}C_3$ and $\ln(C_2/C_3)$ to distinguish the Primary Cracking gas of kerogen and the secondary cracking gas of crude oil. But the two relevant diagrams are based on two principal aspects. Firstly, Behar et al. carried out thermal simulation experiments of the different types of kerogen (type II and type III) in a closed system [15]. The conclusion is the ratio of C_1/C_2 and C_2/C_3 of the kerogen Cracking gas (Primary Cracking) and the oil cracking gas(secondary cracking) are completely different, and the ratio of C_1/C_2 increases with Primary Cracking of kerogen, but remains stable with secondary cracking. On the contrary, the ratio of C_2/C_3 is basically the same during the Primary Cracking of kerogen, but increases rapidly during the secondary cracking. Secondly, the difference of $\delta^{13}C_2-\delta^{13}C_3$ of gas gradually decreases with maturity, which tends to zero.

Prinzhofer & Huc analyzed gas resources in Kansas and thought that the gases were from crude oil craking gas (Figure 4a) [14]. The relevant mode of $\ln(C_1/C_2)$ and $\ln(C_2/C_3)$ of gases is applied in Chuandong Carboniferous gas reservoir group. The ratio of $\ln(C_2/C_3)$ changes obviously in single gas reservoir; with vertical distribution, it changes more greatly in the entire Eastern Sichuan region. In single gas reservoir the data points of $\ln(C_1/C_2)$ nearly coincide, with small changes laterally and small changes totally. The result suggests that the gases are characterized by oil cracking gas (figure 4b).

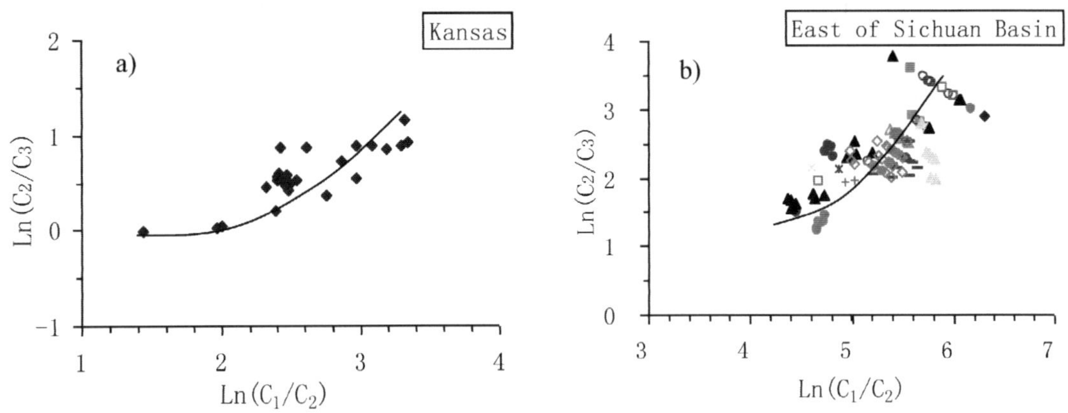

Figure 4. Correlation of C_1/C_2 and C_2/C_3 ratios for natural gases from from different area [16]

4.2. *Identification mode of $\delta^{13}C_2-\delta^{13}C_3$ and $\ln(C_2/C_3)$*

Prinzhofer & Huc established the diagram of $\delta^{13}C_2-\delta^{13}C_3$ vs. $\ln(C_2/C_3)$ in order to distinguish pyrolysis gases of kerogen and crude oil. The difference of$\delta^{13}C_2-\delta^{13}C_3$ of primary kerogen pyrolysis gas changes greatly totally, instead, $Ln(C_2/C_3)$ basically does not change. The difference of $\delta^{13}C_2-\delta^{13}C_3$ of crude oil-cracking gas tends to be zero, whereas the value of $Ln(C_2/C_3)$ increases. Prinzhofer & Huc applied the chart to conclude the gas source in Kansas was the crude oil-cracking gas (figure 5a). According to this chart, the gas source of Carboniferous gas reservoir is the cracking gas of crude oil in Eastern Sichuan region (figure 5b) with its $\delta^{13}C_2-\delta^{13}C_3$ tending to be zero.

5. Conclusion

In Lower Silurian Longmaxi Formation the high efficient hydrocarbon source rocks has come to high to over-mature (VRo as high as 2.5 ~ 3.5 %) in eastern Sichuan basin. This high efficient hydrocarbon source kitchen basically reached over the dead line of the oil generation between late Indosinian epoch and early Yanshannia epoch. The oil formed in Late Triassic - early Jurassic migrated upward and charged to the upper Middle Carboniferous carbonate reservoir to produce the Carboniferous ancient oil reservoir. Due to the increasing burial depth, the ancient oil reservoirs experienced crude oil cracking in the early Himalayan movement and turned into gas source kitchen. The gases generated from the cracking of crude oil were stored in the existing Carboniferous reservoirs to form large Carboniferous gas reservoirs.

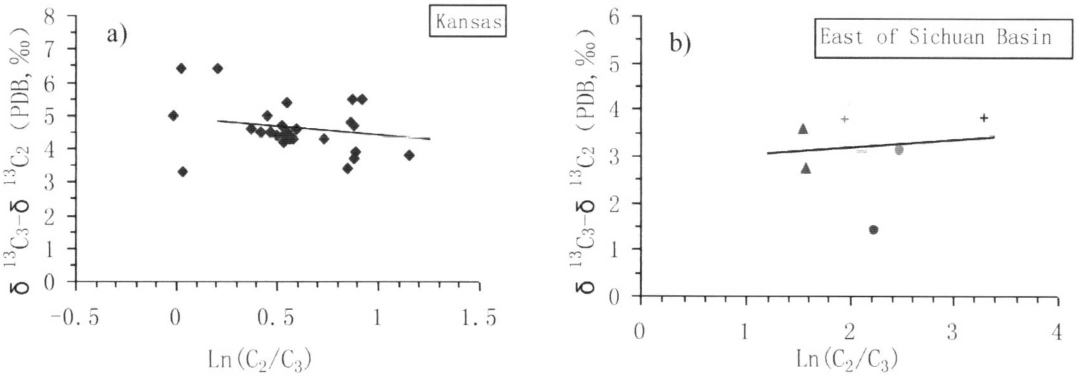

Figure 5. Plot of $\delta^{13}C_2-\delta^{13}C_3$ versus $\ln(C_2/C_3)$ for natural gases from different area.

Acknowledgments
In this paper, this research was sponsored by the Natural Science Foundation of Shaanxi Province (Project No. 2013JM5006).

References

[1] Braun R L and Burnham A K. 1987 A nalysis of chemical reaction kinetics using a distribution of activation energies and simp lermodels Energy and Fuels **1** 153
[2] Burnham A.K., Gregg, H.R, Ward, R L, et al. 1997 Decomposition kinetics and mechanism of n-hexadecane-1,2-$^{13}C_2$ and dodece-1-ene-1,2-$^{13}C_2$ doped in petroleum and n-hexadecane Geochimica et Cosmochimica Acta **61** 3725-37
[3] Cramer B, Kross B M and Littke R 1998 Modelling isotope fractionation during primary cracking of natural gas: A reation kinetic approach Chemical Geology **149** 235-50
[4] Cramer B, Faber E, Gerling P, et al. 2001 Reaction kinetics of stable carbon isotopes in natural gas-insights from dry, open system pyrolysis experiments Energy & Fuels 15 517-32.
[5] Domine F. 1989. Kinetics of hexane pyrolysis at very high pressures. 1. Experimental study Energy and Fuels **3** 89-96
[6] Domine F 1991 High-pressure pyrolysis of n-hexane, 2,4-dimethylpentane and 1-phenylbutaneIs pressure an important geochemical parameter Organic Geochemistry **17** 619-34
[7] Horsfield B, Schenk H J, Mills N, et al. 1992 An investigation of the in-reservoir conversion of oil to gas, compositional and kinetic findings from closed-system pro-grammed-temperature pyrolysis Organic Geochemistry **19** 191-204
[8] Lu S F, Fu X T, Wang Z P, et al. 1996 The thermal kinetic models of kerogen to oil and kerogen to gas as well as its calibration Scientia Geologica Sinica **31** 15-21
[9] Quigley T M, Mackenzie A S, et al. 1988 The temperatures of oil and gas formation in the subsurface Nature **333** 549-52
[10] Ungerer P, Behar F, Villalba M, et al. 1988 Kinetic modelling of oil cracking.Organic Geochemistry **13** 857-68
[11] Waples D W 2000 The kinetics of in-reservoir oil destruction and gas formation: constraints from experimental and empirical data, and from thermodynamics Organic Geochemistry **31** 553-75
[12] Li Y X and Zhong N N 2007 Approach To Solid Migrabitumen Mechanism Of Carboniferous Gas Reservoir In The East Of Sichuan Basin Petroleum Geology & Experiment **29** 402-05

[13] Wang L S, Chen S J, Yang J J, et al. 2002 Geochemical mode of gas reservoir formation in carboniferous in east Sichuan Natur.Gas Industry supplementaty issue **22** 102-06
[14] Prinzhofer A A and Huc A Y 1995 Genetic and post-genetic molecular and isotopic fractionations in natural gases Chemical Geology **126** 281-90
[15] Behar F., Ungerer P., Kressmann S., et al. 1991 Thermal evolution of crude oils in sedimentary basins, experimental simulation in a confined system and kinetic modeling Revue de l'Institut Francais du Petrole **46** 151-83
[16] Jenden P D, Newell K. D, Kaplan I R, et al. 1988 Composition and stable isotope geochemistry of natural gases from Kansas, Midcontinent, U. S. A. In, M. Schell. Origins of Methane in the Earth Chem. Geol. **71** 117-47 (special issue)

LPG gaseous phase electronic port injection on performance, emission and combustion characteristics of Lean Burn SI Engine

Pradeep Bhasker J and Porpatham E[1]

School of Mechanical Engineering, VIT University, Vellore, Tamil Nadu, INDIA

E-mail: porpatham.e@vit.ac.in

Abstract. Gaseous fuels have always been established as an assuring way to lessen emissions in Spark Ignition engines. In particular, LPG resolved to be an affirmative fuel for SI engines because of their efficient combustion properties, lower emissions and higher knock resistance. This paper investigates performance, emission and combustion characteristics of a microcontroller based electronic LPG gaseous phase port injection system. Experiments were carried out in a single cylinder diesel engine altered to behave as SI engine with LPG as fuel at a compression ratio of 10.5:1. The engine was regulated at 1500 rpm at a throttle position of 20% at diverse equivalence ratios. The test results were compared with that of the carburetion system. The results showed that there was an increase in brake power output and brake thermal efficiency with LPG gas phase injection. There was an appreciable extension in the lean limit of operation and maximum brake power output under lean conditions. LPG injection technique significantly reduces hydrocarbon and carbon monoxide emissions. Also, it extremely enhances the rate of combustion and helps in extending the lean limit of LPG. There was a minimal increase of NOx emissions over the lean operating range due to higher temperature. On the whole it is concluded that port injection of LPG is best suitable in terms of performance and emission for LPG fuelled lean burn SI engine.

1. Introduction

The increasing energy demand globally compels the researchers to go for more and more alternative cleaner energy source. It also emphasizes the need to study the technological alternatives and the emerging situations to provide a fresh look towards energy supply and environmental protection. Gaseous fuels have proved to be a promising way to reduce emissions in spark ignition engines. In particular, Liquefied Petroleum Gas (LPG) established to be a recommendatory fuel for Spark Ignition (SI) engines because of their higher octane rating, efficient combustion characteristics and lower emissions [1]. Table 1 shows the properties of LPG and it is clear that the properties of LPG like hydrogen to carbon ratio and wide ignition limit makes them favourable for lean burn operation [2,3]. Nitrogen oxides (NO_x) emission are less in LPG compared to gasoline counterparts [4,5]. LPG could be considered as a powerful alternative and sustainable energy source, but the influence of the fuel remains oppressed by means of conventional conversion applications from gasoline. Dedicated fuel systems should be manufactured for LPG fuel systems which can obtain better combustion, better

[1] Address for correspondence: Porpatham E, School of Mechanical Engineering, VIT University, Vellore, Tamil Nadu, INDIA. E-mail: porpatham.e@vit.ac.in.

thermal efficiency at higher compression ratio. LPG in carburetted engines is undoubtedly unresolved and in such cases the control of fuel flow is not possible. This greatly affects the efficacy of carburetted engines where the emissions increase due to improper combustion of the mixture. One of the major problem faced by gaseous fuels is power loss acted upon by its volumetric efficiency [6]. Also Hydrocarbon (HC) emissions with LPG in a premixed carburetion SI engine employing gas carburetors have been significant. The suggested solution to overcome these problems is the precise electronically controlled LPG fuel injection method.

Table 1. Properties of LPG.

Property	LPG
Composition (%vol)	Propane (30%), Butane (70%)
Lower heating value (kJ/kg)	46000
Flame speed (cm/s)	38.25
Stoichiometric AFR	15.5
Flammability limits (%vol in air)	2.15
Octane number (Research)	103-105
Auto ignition temperature (° C)	405-450

Several researchers have reviewed the applicability of injection system in LPG SI engines but only a few discusses about standalone systems. On the whole LPG fuel was injected in the engine manifold either in liquid or gaseous phase [3,4,7–10]. Because of the density play, LPG liquid phase injection would eliminate the problem of volumetric efficiency as mentioned earlier but requires complex hardware. Hence, numerous tryouts have been carried out to foster arrangement for electronic control of LPG injection for higher power output and thermal efficiency [11–13]. Though liquid phase injection would eliminate the problem of power loss, it requires a very complex setup which has attracted many researches on computerized control of liquid phase injection to improve efficiency and power. Pradeep and Ramesh [14,15] worked on an innovative twin injector system for a two stroke SI engine using cylinder barrel injection technique. The novel direct gaseous injection technique contributed to the reduction of hydrocarbon emissions around 80% by reducing short-circuiting losses as compared to that of manifold injection. The results show that the system helps in improving thermal efficiency and reducing HC emissions. Enhanced combustion rates were also observed when related to customary port injection technique. Owing to charge stratification and rich in-cylinder charge, lower peak NO emissions and higher Carbon monoxide (CO) emissions were observed. Ceviz *et al.* [16] experimented with different temperature of intake LPG fuel. Based on the test results, the temperature of the intake LPG fuel before manifold injection had a positive impact on the performance and NO emission properties. Baris *et al.* [17] observed experiments on LPG SI engine by varying the excess air coefficient (1.0 and 1.3) and ignition timing. Tests were carried out to analyze the performance of lean burn at a speed of 4300 rpm and at wide open throttle condition. It may be noted that as the air excess coefficient was higher than 0.8, advanced spark timing showed positive influence on performance of the engine. At an excess air coefficient of 1.3, the emissions reported lower. On the whole, HC and NOx emissions increased due to spark advance whereas the CO emissions were inconsequential due to the spark timing. Despite the fact that there are copious information about the study of LPG on engine performance and in reduction of emissions, there are very few analogous works on the consequences of carburetion and electronically controlled fuel injection control methods of LPG run SI engine and its operation under lean conditions.

Li *et al.* [18] established an electronic LPG gaseous phase injection technique for a 4-stroke water cooled SI engine. The study was carried out under injection pressure condition of 0.1 MPa and an orifice of 1mm. The emissions obtained from the injection technique were compared with the mechanical mixer system with LPG. Using electronic LPG injection, better emission control was reported for HC, CO and NOx. Smith *et al.* [19] observed experiments on a dual-fuel, four cylinder SI engine provided with a 3-way catalytic converter. Dual fuel involves single point LPG gas injection and multipoint gasoline injection. Over the studied speed range, the maximum Brake Mean Effective Pressure (BMEP) decreased with the use of LPG. On average the thermal efficiency under full load condition improved by 8%. The prime intention of the current research is to analyze the aftermath of injection and carburetion techniques on SI engine performance, emission and combustion characteristics. Moreover in the present study, the tests were executed using a dedicated LPG gaseous injection controller unit developed by the authors.

2. Experimental Setup

A four stroke, air cooled, single cylinder diesel engine with a displacement volume of 661 cc, with a compression ratio 17.5:1 and power of 4.4 kW at 1500 rpm was altered to run as SI engine. A diesel engine was altered because it can combat knock and higher compression ratio. The engine was operated in SI mode with gaseous fuels by replacing the fuel injection pump and the injector with carburettor system, a spark plug and an ignition system. The overall system was made favourable for LPG operation in SI mode.

Table 2. Engine specifications.

Type	Kirloskar TAF1, air cooled, single cylinder CI engine
Displacement	661 cc
Stroke	110 mm
Bore	87.5 mm
Connecting Rod	231 mm
Compression ratio	17.5:1 (CI version), 10.5:1 (SI version)
Rated Power	4.4 kW @ 1500 rpm
Inlet Valve Open	4.5° bTDC
Inlet Valve Close	35.5° aBDC
Exhaust Valve Open	35.5° bBDC
Exhaust Valve Close	4.5° aTDC

The spark timing could be varied precisely by 1° crank angle from 280° to 360° (0° to 80° CA before TDC) through a microcontroller based circuit with input from a proximity sensor. The compression ratio was modified by altering the clearance volume of a bowl-in type piston and was fixed at 10.5:1. Shim plates of standard thickness were placed between crank case and cylinder block to compensate the required clearance. The piston squish area was maintained at 30% of the total land area maintaining the a squish velocity of 4 m/s [20].

The engine used for experimentation was Kirloskar TAF1 engine as mentioned in Table 2.The intake manifold was especially designed to conduct experiments on carburetion and port-injection of gaseous phase LPG. The manifold was provided with throttle valve and a gas mixer unit for carburetion, while a 3-gas injector type rail was mounted for injection. The angle of injection was kept at 45° along the fuel line path. A dedicated LPG injection circuit using microcontroller ATMEGA 8A was developed and connected to the injectors. The pulse width and the delay of signals can be controlled precisely by 1ms. The injector rail has provision for mounting 4 injectors on it but only 2 were used according to the flow requirement. The injector works on a pressure range of 0-3 bar and voltage of 10-14 V.

Figure 1. Pictorial view of experimental setup.

Figure 1 shows the pictorial view of the experimental setup developed for the research work. The engine was connected to an eddy current dynamometer which was set to constant speed mode to carry out the experiments at 1500 rpm. The cylinder head surface was made to mount a KISTLER make piezoelectric pressure transducer in order to obtain the in-cylinder pressure. An AVL optical crank angle encoder was fixed to attain the corresponding pressure values. 100 cycle pressure data was recorded for each operating condition. The pressure values are corrected based on the intake manifold pressure obtained from the KISTLER make piezoresistive pressure sensor flush mounted in the engine intake manifold. The throttle position sensor was connected to a digital display in order to vary the exact throttle opening positions accordingly. A data acquisition system connected to a computer was used to acquire online in-cylinder pressure details from the flush mounted piezoelectric sensor and the intake manifold pressure from a piezoresistive sensor for every 1 ° resolution. The CO and CO_2 emissions were measured using NDIR analyzer and HC emissions were measured using FID analyzer and NO and NOx were measured using CLD analyzer.

2.1. Experimental Conditions

Experiments were conducted at constant throttle opening condition of 20% based on the ECU controller module connected to a throttle position sensor fixed in the intake manifold. The LPG gas cylinder was immersed in a hot water tank which maintained a temperature of 50^0C. A fuel surge tank was also made which was wrapped with tape heaters to ensure gaseous flow near the intake manifold as well. The fuel surge tank maintains a temperature of 50^0C and also helps in maintaining constant supply and reduces pressure fluctuations during high load conditions. The flow of LPG gas was measured using thermal mass flow meter. Airflow rate was obtained from a positive-displacement type flow meter (Make: DRESSER) connected with a surge tank on the suction side.

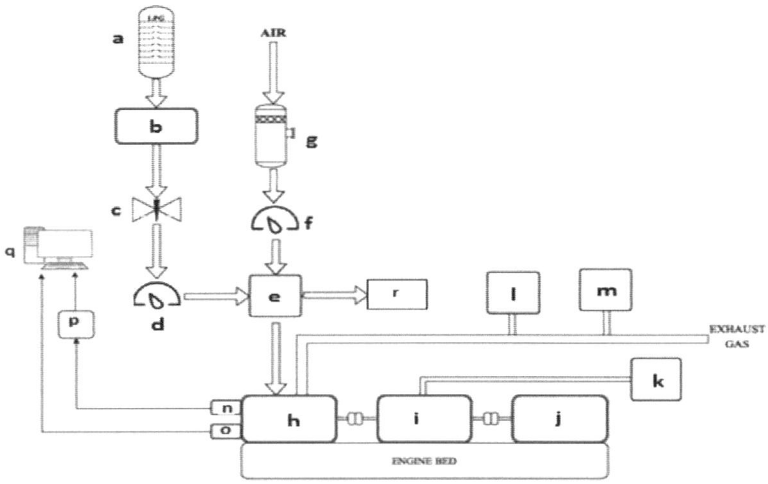

[a]LPG cylinder in hot water bath (50^0C)
[b]Surge Tank with electric heater
[c]Control Valve
[d]Thermal mass Flowmeter
[e]Injector unit
[f]Air flowmeter [g] Air surge tank
[h]Engine
[i]Eddy current dynamometer
[j]Motoring unit
[k]Dynamometer controller
[l]FID and CLD analyzer [m]NDIR analyzer
[n]Pressure pickup
[o]Crank angle encoder
[p]Data Acquisition system
[q]Computer
[r]Electronic injection circuit

Figure 2. Schematic diagram of experimental setup

The mixture strength was varied from richest to the leanest possible by controlling the amount of fuel using a fine tune adjuster. Extreme care was taken in maintaining the gaseous phase addition of LPG fuel to the primary flow line without causing hindrance in the desired flow rate. The leanest mixture i.e. lowest equivalence ratio limit was limited by unstable operation of the engine. At each of the equivalence ratio, MBT (Minimum spark advance for best torque) was set. However, under knocking conditions the spark timing was adjusted for knock free operation rather than MBT adjustment. Such an adjustment was required only during rich operating conditions which are of no significance in this research work. Readings were always taken on after the engine attains stability of operation. Torque, MBT timing, intake pressure, cylinder pressure, air flow rate, LPG fuel flow rate, injection pulse width, injection delay, inlet and exhaust temperatures, exhaust emission levels of HC,CO,NO and NOx were recorded for different equivalence ratios. The data acquisition system recorded the values of pressures at each crank angle degrees at regular time intervals for computing cycle pressure data. The pressure signals from the piezoelectric transducer were referenced based on the suction stroke cylinder pressure at BDC, which was equalled to the average manifold pressure obtained from the intake pressure piezoresistive transducer. Dedicated software was developed using LABVIEW to calculate the pressure average against crank angle for the obtained number of cycles.

Knowledge of the pressure variation in the combustion chamber will enable calculation of energy changes taking place in an engine by thermodynamic analysis. Also important aspects like peak pressure, indicated mean effective pressure, ignition delay, combustion duration and cyclic variations in pressure can be studied with the aid of cylinder pressure crank angle data. To obtain an in-depth view of combustion process, it is mandatory to correlate the above mentioned factors with in-cylinder pressure. This can be termed as Heat Release Analysis. This data provides a clear picture of what happens inside the cylinder during combustion process. First law of thermodynamics is used to interpret the pressure data to calculate the mass fraction of the fuel burnt and the rate of heat release [21, 22]. The analysis is done considering a closed cycle where the intake and exhaust valves are closed. Hence, a simple approach of single zone model is considered where the properties and the thermodynamic state are considered uniform throughout the cylinder and represented by the average values. The model is considered to be zero dimensional since no spatial variations are taken into account. For a closed cycle period, the first law can be written with suitable assumptions as follows,

$$\frac{dQ_{hr}}{d\theta} = \frac{\gamma}{\gamma-1}P\frac{dV}{d\theta} + \frac{1}{\gamma-1}V\frac{dP}{d\theta} + hA_s(T-T_w)\frac{dt}{d\theta}$$

where Q_{hr} is the rate of heat release and θ is crank angle in degrees, γ is the ratio of specific heats of the air, fuel and residual gas mixtures. The total mass of the mixture per cycle is computed as the addition of air mass, residual gas mass present in the clearance volume and the mass of fuel. Ideal gas equation is used to calculate the mass of residual gas trapped where the temperature and pressure are assumed equal to the exhaust manifold properties respectively [23]. The residual gas constant was assumed as that of the air. Based on the above information and assumptions, the molecular weight of the charge, mole fraction of fuel, residual gas and air and the charge gas constant are found. Using the expression of polynomials, C_P value of each of the LPG gas constituent was determined knowing the composition of the gas. Cylinder gas temperature T was obtained using ideal gas equation at every crank angle degree. Based on the engine speed, t is the time period and Tw being the temperature of combustion chamber wall. The coefficient of heat transfer was obtained using the Hohenberg's correlation as shown below [24]

$$h = C_1 V^{-0.06} P^{0.8} T^{-0.4} (\overline{Vp} + C_2)^{0.8}$$

where h is the coefficient of heat transfer in W/m^2 K, C_1 and C_2 the tuning constants (130 and 1.4), V volume of cylinder in m^3, P the in-cylinder pressure in bar, T the cylinder gas temperature in Kelvin

and $\overline{Vp}$ the piston mean speed in m/s. A_S is the area in m^2 where the heat transfer between the gas and combustion chamber wall occurs. This comprises top squish area, piston bowl surface area, cylinder wall and cylinder head. All the areas are considered constant excluding the cylinder wall area for a desired form. Along with the heat release, cumulative heat release and mass fraction burned at every crank angle degree was calculated.

3. Results and Discussions

All test results were obtained at a constant speed of 1500 rpm and throttle opening position of 20%. The load was varied based on the fuel inlet from rich to leanest possible limit. The performance, emissions and combustion characteristics were compared between LPG gas phase port injection and LPG carburetion system.

3.1. Performance parameters

Figure 3 shows the variation of brake power with equivalence ratio for carburetion and injection mode. The maximum brake power for carburetion was observed to be 2.5 kW at an equivalence ratio of 0.8 while for injection it was 3.1 kW. The values higher than equivalence ratio of 1 are not of interest in this research study hence maximum power output was considered for comparison at an equivalence ratio of 0.8. It may be noted that an increase of 24% higher power output is observed by injection technique. This clearly states that electronic control of fuel injection precisely meters the quantity of fuel and provides an enhanced breathing capacity when compared to that of carburetion which improves the brake power output and also brake thermal efficiency as shown in figure 4. The peak brake thermal efficiency for injection was found to be 34.1% at an equivalence ratio of 0.88 and was about 27% respectively for LPG carburetion. This shows an improvement of 22% higher brake thermal efficiency using LPG electronic injection technique as compared to that of carburetion. It can also be observed that the LPG electronic injection method helps in extending the lean limit from an equivalence ratio of 0.69 by carburetion to 0.63 by injection method. An increase of 8.6% of lean limit certainly reduces the fuel consumption of LPG using injection technique. On the whole, the maximum efficiency was observed at a much leaner mixture for injection on comparison and hence it can be stated that gas phase manifold electronic injection is much suitable for leaner operations for improving higher thermal efficiencies and brake power output.

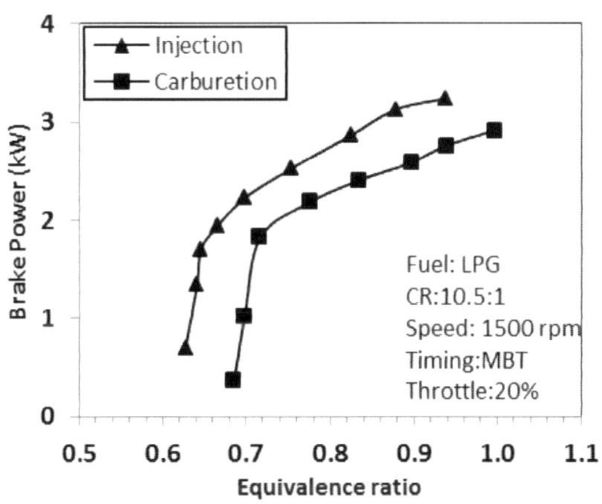

Figure 3. Variation of brake power.

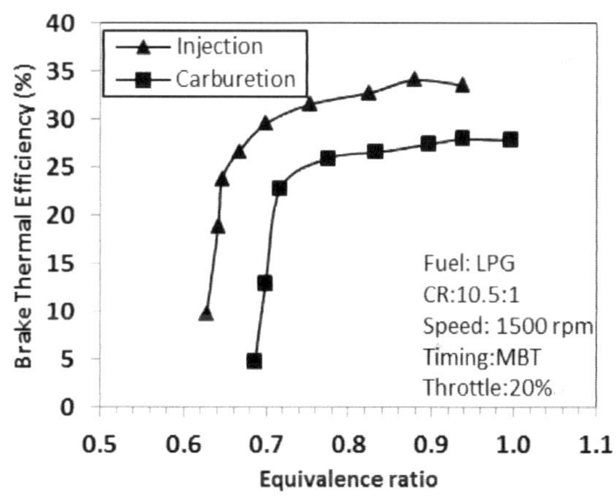

Figure 4. Variation of brake thermal efficiency.

3.2. Combustion Parameters

Figure 5 shows the rate of in-cylinder pressure rise, maximum cylinder pressure and occurrence of maximum pressure rise which was around 29.5 bar at 374 deg.CA at an equivalence ratio of 0.88 was favourable for gaseous injection method than carburetion which was around 24 bar at 376 deg.CA. This may also be noted in the rate of heat release curves as shown in figure 6. The HRR values are highest with about 42.3 J/deg.CA at 364 deg.CA for injection and were also better by 16.5% than carburetion which was around 36.3 J/deg.CA. The cumulative heat release which gives the total energy from combustion was around 850 Joules for carburetion which was also lower when compared to 960 Joules observed from electronic injection technique. This indicates that injection paves way for enhanced combustion. Since the charge is injected at a precise timing, the fuel is possibly rich at the vicinity of the spark plug which is comparably better than the homogenous charge inducted during carburetion. The charge stratification by injection and the turbulence created by the bowl type piston helps in flame propagation which results in enhanced combustion of LPG. LPG injection has the least ignition delay period i.e. 5% mass fraction burnt (5%MFB) of 52 deg.CA when compared to that of carburetion which can be clearly seen from figure 6. Figure 7 shows the variation of ignition delay and it can be seen that spark timing has been advanced earlier in LPG gaseous phase injection over carburetion from figure 8. The total burn duration as well as the segment wise combustion duration for the LPG port injection technique are less with minimum spark timing and combustion duration. This indicates enhanced combustion rate involving flame travel and turbulent kinetic energy and minimum spark advance is possible in lean burn condition by gaseous phase LPG injection method.

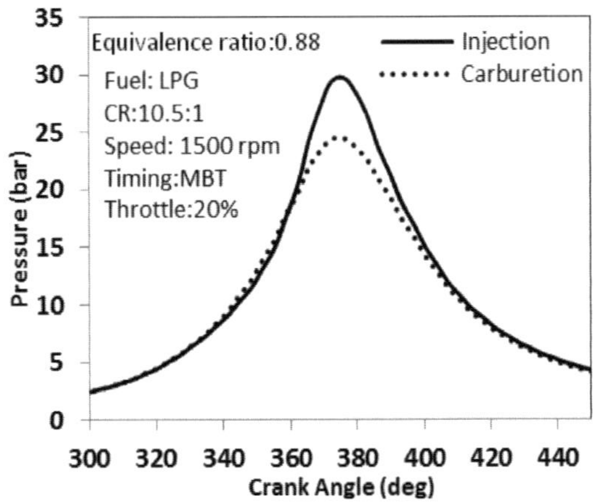

Figure 5. Variation of In-cylinder pressure.

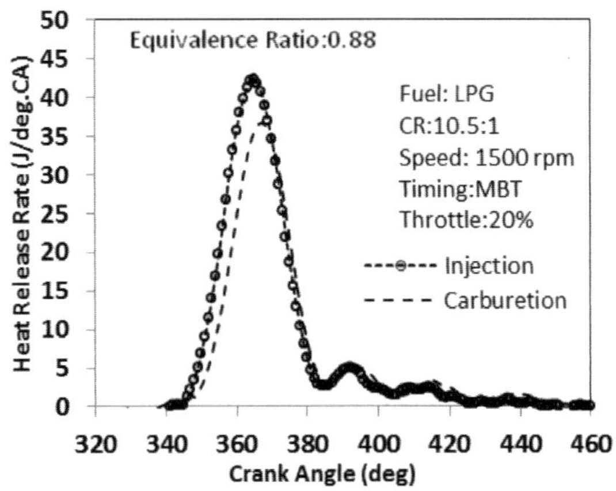

Figure 6. Variation of Heat Release Rate (HRR).

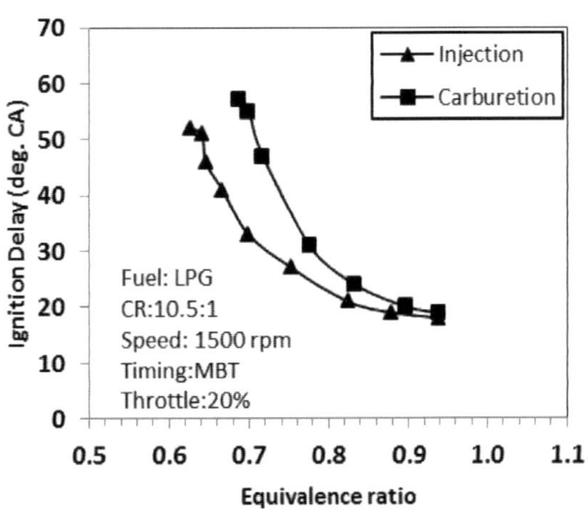

Figure 7. Variation of Ignition delay.

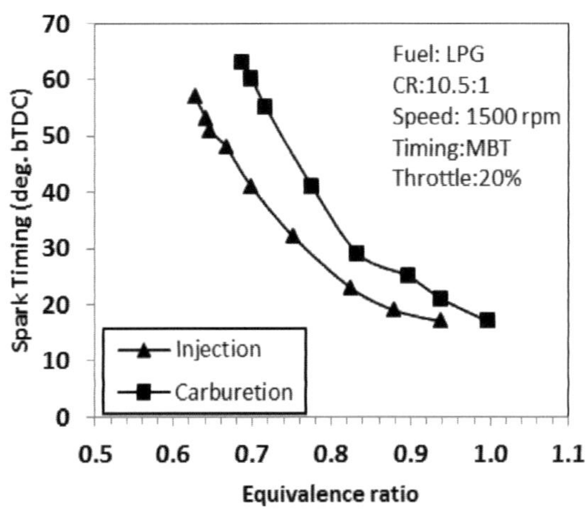

Figure 8. Variation of Spark timing.

3.3. Emissions

The variations of HC and CO emissions could be clearly seen from the lean equivalence ratios between 0.6 and 0.8 for carburetion and injection. Figure 9 shows that within the stable limit of combustion on the leaner side of 0.70 equivalence ratio HC emissions of carburetion was around 3720 ppm whereas for injection it reported as low as 300 ppm respectively. As the equivalence ratio reaches a particular lower limit there is a sudden rise in HC level due to misfiring. The decreasing trend of HC towards rich limit is primarily due to the influence of increase in the in-cylinder temperature. This can be attributed towards the enhanced combustion by injection and the combined effect of turbulence and laminar flame speed of the bowl-in-piston type. Figure 10 shows that at an equivalence ratio of 0.75 the CO emissions were 0.08% and 0.05% respectively for carburetion and injection. This can also be observed from Fig. 11 which shows the trend of CO_2 emissions of LPG injection which was higher because of complete combustion. The NO and NOx emission comparison from figure 12 and 13 clearly indicate that there was an increase of nearly 58% at an equivalence ratio of 0.82. The NOx emission were certainly higher for LPG injection than LPG carburetion because of higher combustion temperature and heat release rate, this can also be inferred from higher exhaust gas temperature obtained from injection technique compared with carburetion as shown in Figure 14. Hence, it can be stated that LPG electronic control manifold injection provides a better suitable way to reduce HC and CO emissions.

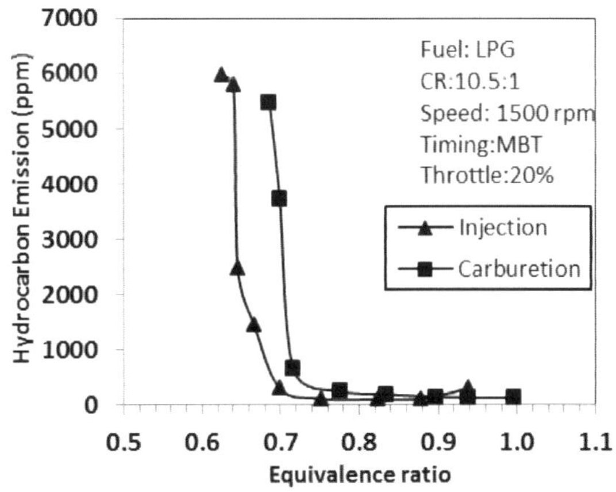

Figure 9. Variation of HC.

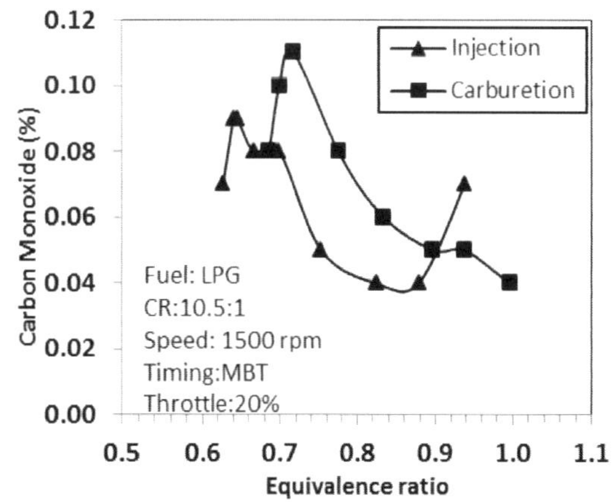

Figure 10. Variation of CO.

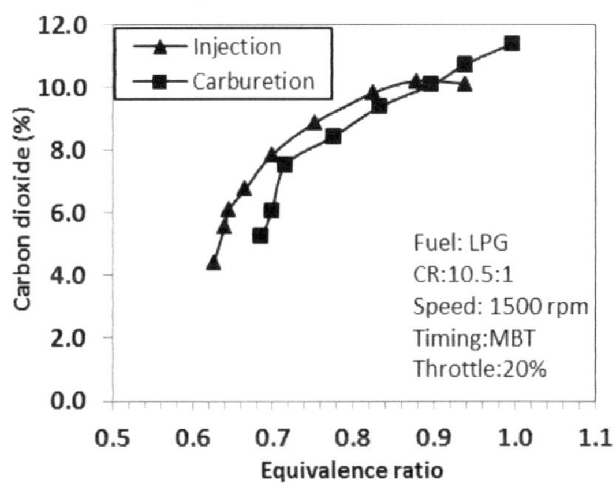

Figure 11. Variation of CO_2.

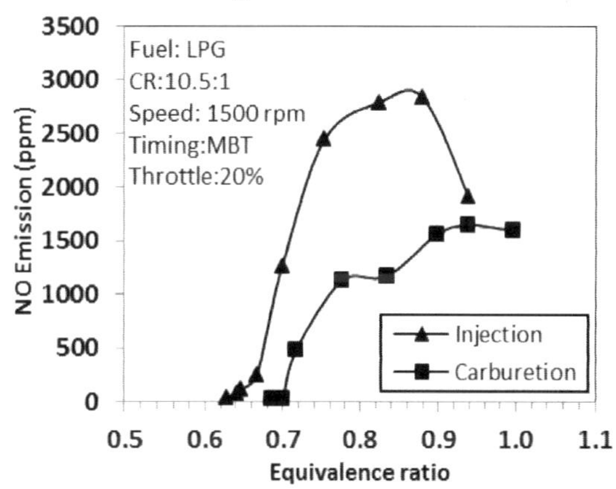

Figure 12. Variation of NO.

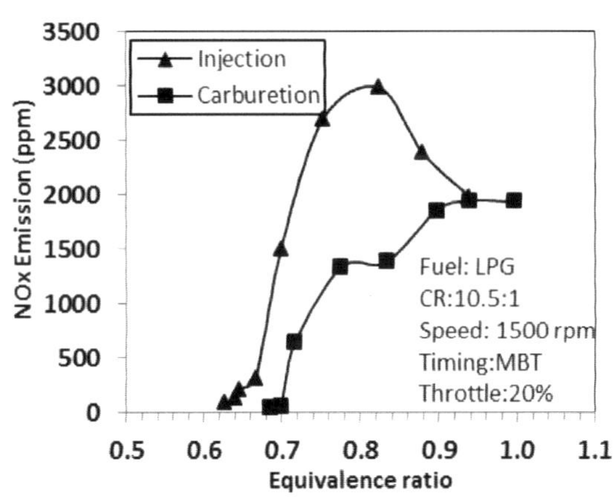

Figure 13. Variation of NOx.

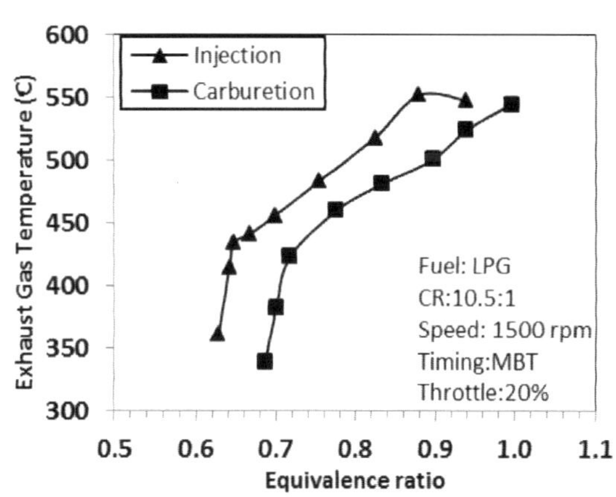

Figure 14. Variation of EGT.

3.4 Aspects of LPG electronic injection

From the above research, it is clear that LPG gaseous phase electronic injection technique paves way for improvements in terms of brake power, brake thermal efficiency, lean limit and also helps in enhanced combustion of LPG fuelled lean burn SI engine. Even on the emissions front, notable reductions in HC and CO were observed. Although, there have been a rise in NO and NOx emissions, this could well be reduced by further implications like Exhaust gas recirculation, hydrogen supplementation or by other simple techniques. If the heat release rate was analysed in the real condition the results would have been more realistic compared to the assumption of pressure and temperature being uniform. On the whole, LPG in SI engine can be established as an alternate fuel for ecological reasons and can be extended over agricultural sector for irrigation purposes as well as to other remote places in India. This definitely will promote a newer way of utilizing alternate energy source for sustainable growth and development.

4. Conclusion

An electronically controlled port-injection system was successfully developed and tested; comparison of its performance, combustion and emission characteristics with carburetion was also done. Based on the experimentation the following outcomes can be arrived.

- There was a significant improvement in port injection technique as compared to the carburetion in both power output and brake thermal efficiency. There was an improvement of 22% increase in brake thermal efficiency and 24% higher brake power output for injection system. The maximum brake thermal efficiency under lean operating range reduces the fuel consumption as well.
- Lean limit extension was possible from an equivalence ratio of 0.69 using carburetion to an equivalence ratio of 0.63, about 8% increase using gaseous port injection technique.
- In general, HC and CO emissions were less for port-injection as compared to carburetion but witnessed an increase of NO and NOx emission of about 58% due to higher in-cylinder temperature and enhanced combustion.
- The LPG electronic control port-injection has clear advantage over carburetion in terms of performance, combustion and emission characteristics of LPG lean burn SI engine.

The electronic LPG gas injection system in SI engine will promote a new way to develop cleaner sustainable power engines with reduced emission levels. The developed system provides a fresh look on lean burn combustion of LPG and helps in reduced fuel consumption.

Acknowledgement

The authors wish to thank The Department of Science and Technology (DST), Government of India and VIT University for the financial support extended for doing this work.

References

[1] Morganti K J, Foong T M, Brear M J, Da Silva G, Yang Y and Dryer F L 2013 The research and motor octane numbers of Liquefied Petroleum Gas (LPG) **Fuel 108** 797–811

[2] Li L, Wang Z and Wang H 2014 A study of lpg lean burn for a small SI engine *SAE Technical Paper* 2002-01-2844; 2002

[3] Bayraktar H and Durgun O 2005 Investigating the effects of LPG on spark ignition engine combustion and performance *Energy Conversion and Management* **46** 2317–33

[4] Ceviz M A and Yüksel F 2006 Cyclic variations on LPG and gasoline-fuelled lean burn SI engine *Renewable Energy* **31** 1950–60

[5] Akif Ceviz M, Sen A K, Küleri A K and Volkan Öner İ 2012 Engine performance, exhaust emissions, and cyclic variations in a lean-burn SI engine fueled by gasoline–hydrogen blends *Appl. Therm. Eng.* **36** 314–24

[6] Sunwoo M, Sim H and Lee K 1999 Design and development of an ECU and its air-fuel ratio control scheme for an LPG engine with a Bypass injector, Vehicle Electronics Conference,

(IVEC '99) Proceedings of the IEEE International page. 508 - 513 Suppl. IVEC.1999.830740; 1999

[7] Pourkhesalian A M, Shamekhi A H and Salimi F 2010 Alternative fuel and gasoline in an SI engine: A comparative study of performance and emissions characteristics *Fuel* **89** 1056–63

[8] Li L, Wang Z, Deng B, Han Y and Wang H 2002 Combustion and emissions characteristics of a small spark-ignited lpg engine *SAE Technical* **01** 1738

[9] Mistry, C., Comparative assessment on performance of multi cylinder engine using CNG , LPG and petrol as a fuel, *SAE Technical Paper* 2005-01-105; 2005

[10] Campbell M, Wyszy Ł P and Stone R 2004 Combustion of LPG in a spark-ignition engine *SAE Technical Paper* 2004-01-0974; 2004

[11] Watson H C and Phuong P X 2007 Why liquid phase LPG port injection has superior power and efficiency to gas phase port injection *SAE Technical Paper* **22** 48-52

[12] Wendeker M, Jakli P, Czarnigowski J, Boulet P and Breaban F 2007 Operational parameters of LPG Fueled SI Engine – Comparison of Simultaneous and Sequential Port Injection *SAE Technical Paper* 2007-01-2051;2007

[13] Sierens R 1992 *An* Experimental and theoretical study of liquid LPG injection *SAE Tech. SAE International* 922363; 1992

[14] Pradeep V and Ramesh A 2015 A novel twin injector strategy for direct cylinder barrel injection of LPG in a 2S-SI engine *Fuel* **162** 128–37

[15] Pradeep V, Bakshi S and Ramesh A 2015 Direct injection of gaseous LPG in a two-stroke SI engine for improved performance *Appl. Therm. Eng.* **89** 738–47

[16] Ceviz M A, Kaleli A and Güner E 2015 Controlling LPG temperature for SI engine applications *Appl. Therm. Eng.* **82** 298–305

[17] Baris Erkus, Ihsan M and Sürmen A 2015 Enhancing the heavy load performance of a gasoline engine converted for LPG use by modifying the ignition timings *Appl. Therm. Eng.* **85** 188-94

[18] Li L G, Liu Z M, Wang H P, Deng B Q, Wang Z S, Xiao Z C, Su Y and Jiang B 2003 Development of a gas-phase LPG injection system for a small SI engine *SAE Technical Paper* 2003-01-3260; 2003

[19] Smith W J, Timoney D J, Lynch D P 1997 Emissions and efficiency comparison of gasoline and LPG fuels in a 1.4 litre passenger car engine *SAE Technical, SAE International* 972970; 1997

[20] Krishnaiah R, Ekambaram P and Jayapaul P B Investigations on the effect of piston squish area on performance and emission characteristics of LPG fuelled lean burn SI engine *SAE Technical Paper* 2016-28-0123; 2016

[21] Gatowski J A, Balles E N, Chun K M, Nelson F E, Ekchian J A and Heywood J B 1984 Heat release analysis of engine pressure data *SAE Technical Paper* 841359; 1984

[22] Hayes T K, Savage L D and Sorenson S C 1986 Cylinder pressure data acquisition and heat release analysis on a personal computer *SAE Technical Paper* 860029; 1986

[23] Pedersen P S and Qvale B 1974 A Model for the physical part of the ignition delay in a diesel engine *SAE Technical Paper* 740716; 1974

[24] Hohenberg G F 1979 Advanced approaches for heat transfer calculations *SAE Technical Paper*790825 ;1979

Chlorine behavior during co-hydrothermal treatment of high alkali coal and PVC

N Huang[1], P T Zhao[1, 2, 3], Z Z Li[1] and T Li[1]

[1]School of Electric Power Engineering, China University of Mining and Technology, Xuzhou 221116, PR China
[2]National Engineering Research Center for Coal Preparation and Purification, China University of Mining and Technology, Xuzhou 221116, PR China

E-mail: pt.zhaoseu@gmail.com cerg.cumt@gmail.com

Abstract. The hydrothermal treatment (HT) is an effective method to remove chlorine from chlorinated wastes under mild conditions. However, the alkali was required to improve the dechlorination efficiency. Meanwhile, the alkalis contents removed was necessary to realize the clean and highly efficient utilization of coal. This work was trying to investigate the feasibility of simultaneously removal alkalis and chlorine by co-hydrothermal treatment of PVC and high-alkali coal. The effect of operating conditions including the HT temperature, the holding time and particle sizes of coal on the dechlorination efficiency (DE) of PVC during the HT process was experimentally studied in this work. The results show that the DE increased with the rise of: 1) particle sizes (0.054~0.22mm), 2) holding time (30~90 min) and 3) temperature (240~300 °C). In detail, under 300 °C and 60min of holding time, the 85.18% of DE with first PS was lower than the 93.93% of DE with second PS and 100% of DE with third PS. The organic chlorine mainly transferred into chloridion in aqueous solution in HT process. All the results indicate that it is a prospective way to simultaneously removal alkalis and chlorine by co-hydrothermal treatment of chlorinated wastes and high-alkali coal.

1. Introduction

PVC [poly (vinyl chloride)] has been widely used in many fields including agriculture, households, automobiles, packing, toys, construction and electronic applications. A report from Ceresana (2014) stated that the global demand for PVC is supposed to increase by around 3.2% per year until 2021 [1]. The Asia-Pacific region holds the largest sales market for PVC with a market share of about 56%.

PVC is a good material in many aspects because of its low cost and high performance. However, the wide application of PVC also caused environmental pollution, called "white pollution". With the development of the economy and society, the chlorine content of solid waste wastes keeps increasing and the decomposition of PVC has received more and more concern. As for the management of waste plastics, three types of methods including landfill, incineration and recycling are usually applied [2, 3]. Nevertheless, the landfilling of plastic wastes containing PVC has become an unsustainable option

[3] Address for correspondence: P T Zhao, School of Electric Power Engineering, China University of Mining and Technology, Xuzhou 221116, PR China; National Engineering Research Center for Coal Preparation and Purification, China University of Mining and Technology, Xuzhou 221116, PR China. E-mail: pt.zhaoseu@gmail.com; cerg.cumt@gmail.com.

which poses a considerable problem to future generations because of the difficulty to decomposition, the destruction of soil structure, the adverse effect on vegetation and the exceeded occupation of lands. The incineration also faces the emission of toxic products (TCDD, tetrachlorodibenzo-p-dioxin), resulting in detrimental effect on human's health. Consequently, it is of great importance to develop an innovative method to remove the chlorine in PVC to avoid the emission of toxic products, high temperature corrosion, the pollution of soil and water etc. [4]. Many authors have studied the dechlorination of PVC (virgin and waste) [5-9] and stated that the hydrothermal treatment (HT) is an effective way to remove chlorine under mild conditions.

Hydrothermal dechlorination of PVC refers to decompose PVC in aqueous solutions with high temperature and pressure. Akimoto *et al.* (2004) found that 98% of chlorine could be eluted when municipal waste plastics were pretreated with 0.2 M NaOH [10]. Lu *et al.* (2002) suggested alkali is an effective additives on the dechlorination of PVC during HT process [11]. Inoue *et al.* (2005) have investigated the dechlorination performance of KOH and NaOH during the co-grinding with PVC and found that grinding of the mixture lead to the size reduction of PVC, with causing its mechanochemical reaction with the additive to form partially dechlorinated PVC and chloride (KCl or NaCl)[12]. Yoshioka *et al.* (2000) reported the major products were oxalic acid, a mixture of benzenecarboxylic acids, and CO_2 when oxidized PVC in NaOH solutions at elevated temperatures [13]. Takeshita *et al.*, (2004) stated the chlorine in PVC dissolved in water as hydrochloric acid. No harmful chlorinated organic compounds were observed in the liquid and gas fractions after treatment at 300 °C [14]. Between 250 and 350 °C, polyene low-molecular weight aromatic and aliphatic compounds were observed in the solid, liquid and gas fractions respectively. According to the previous introduction, the HT is effective to remove chlorine from virgin/waste PVC, whereas the alkalis are required to improve the dechlorination efficiency.

The utilization of alkalis in HT dechlorination of PVC would also cause some environmental problem, e.g., the treatment of the wastewater. The coals reserved in Zhundong, Xinjiang, China are featured with the very high quality and very low levels of impurities and could meet the coal consumption of China for the next 100 years according to the current coal consumption[15]. Whereas, the high alkalis content of Zhundong coal would erode the boiler tubes and boiler furnace when burning. Content of Na and Ca are 0.2~1 wt% and 0.6~2.5 wt%, respectively. Yang *et al.* (2016) found that Na is mainly water soluble (Na_w/Na_{total}=50~85%) and Ca is mainly acid soluble (Ca_{ac}/Ca_{total}=60~90%) [16]. Li *et al.* (2015) found that the release rates of Na and K were higher than those of Ca and Mg in all types of gas environments [17]. Naruse *et al.* (2005) found that high alkali metal compounds existed in the fine particulates with size of less than 1 μm and these sticky particles become easy to adhere on heat exchanger tube surfaces [18]. Alkali metal compounds could cause the slagging, fouling and erosion/corrosion. Therefore, it is crucial to remove the alkalis (main Na) before using Zhundong coal. Accordingly, co-hydrothermal treatment means the processing of mixture (PVC and high-alkali coal) in water slurries at elevated temperature and pressure to facilitate the chemical (chlorine, alkalis) conversion of the organic structures in PVC into chlorine and eliminate the alkalis from Zhundong coal simultaneously. The aim of this work was to investigate the chlorine behavior during the co-hydrothermal treatment of PVC and Zhundong coal to check the feasibility of this technology.

2. Materials and methods

2.1. Materials

The polyvinyl chloride powder purchased from Shanghai Yangli Electromechanical Technology Co., Ltd with an average molecular weight ~90 000 Da and a chlorine content of 61.5% was utilized as raw material. The high-alkali coal was collected from Zhundong, China. The ultimate analysis of the materials was listed in Table 1. Before utilization, the coal was first dried in an electronic oven at a temperature of 65 °C and ground into three kinds of particle sizes, 0.054~0.075mm, 0075~0.1mm and

0.1~0.22mm. All the chemicals were analytical reagent and purchased from Sinopharm Chemical Reagent Co., Ltd (Shanghai, China).

2.2. Co-hydrothermal treatment of PVC and coal

The HT dechlorination experiments were performed in an autoclave. The experimental procedures and the schematic diagram of the autoclave are illustrated in Figure 1. Before the HT process, 2 g of PVC and 18 g of coal were well mixed and immerged into 150 ml deionized water in a 500 ml quartz tube. The glass tube was then placed in the reactor which was subsequently sealed by 8 bolts. After that, the argon with a purity of 99.999% supplied from an argon cylinder was transported into the reactor to replace the air in the reactor. Then the autoclave was heated to the pre-determined temperature (240, 260 and 300 °C) by an electric heater, preserving for pre-determined time (30, 60 and 90 min). After that, the reactor was removed to atmosphere environment to cool down. When the pressure was reduced to atmosphere and the temperature was below 100 °C, the gaseous product was collected by a gas-sampling bag. And the reactor was then opened and the slurry samples were taken out from the glass tube. The slurry was divided into two parts as separated liquid and solid products by filtration. During the whole HT process, the stirrer was kept stirring at a rotary speed of 100 rpm to ensure uniformity of temperature and mass transfer within the samples.

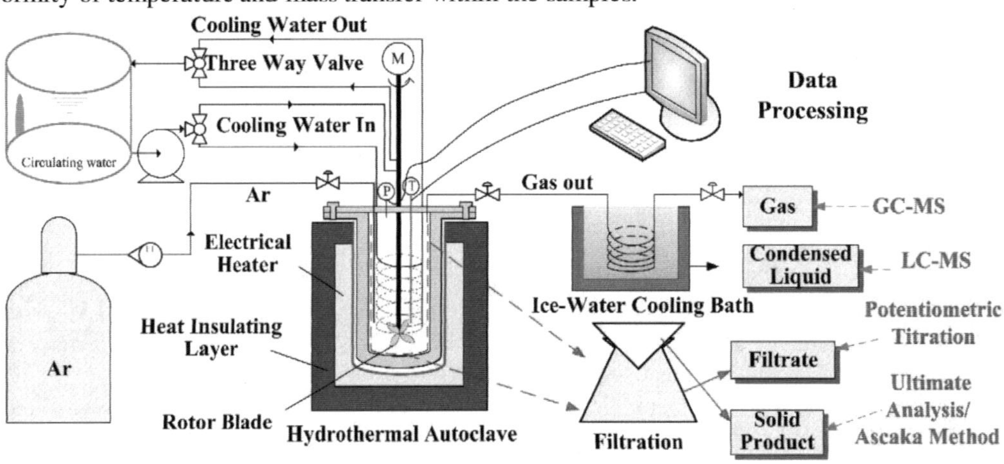

Figure 1. The illustration of experimental procedures and schematic diagram of HT dechlorination

2.3. Determining Cl removal efficiency

The HT dechlorination efficiency (DE) can be expressed by the amount of chlorine removed, which was equal to that in the separated liquid and the gaseous product. As rare chlorine was observed in the gaseous products, the chlorine contained in the separated liquid was assumed to be the total amount of chlorine removed from PVC during the HT dechlorination process. The amount of Cl in the liquid sample was determined by potentiometric titration. In this method, the double saltbridge saturated calomel electrodes was used as the reference electrode, and the silver electrode was used as the indicator electrode. The end of titration was judged by the extreme point of potential changes, and the corresponding consumption of $AgNO_3$ standard solution was noted as V_1. Then the DE can be calculated by the following equation.

$$\text{HT dechloination efficiency (DE, \%)} = \frac{\text{Cl content in liquid}}{\text{Cl content in PVC+Cl content in coal}} \times 100\% \quad (1)$$

The chlorine content of solid particles was measured by using Eschka mixture method referring to GB/T 3558-2014 (eqv ISO 587:1997), which was introduced here briefly.

1 g sample of residue and 3g of Eschka mixture are well mixed with each other in a crucible, covered with an additional 2g of Eschka mixture to ensure that no chlorine is lost during combustion. After that, the mixtures are ignited gradually in a muffle furnace at 680±20 °C for 3 hours. After

cooling, the incinerated mixture is transferred to a beaker with hot water, acidified with 20 ml of 1:1 nitric acid to dissolve the mixture, and filtered. The chloride is then determined by potentiometric titration. Accordingly, the chlorine in the solid product could be derived from equation (2).

$$X = \frac{0.035453 \times c(V_1 - V_2)}{m} \times 100 \qquad (2)$$

V_1: consumption of $AgNO_3$ standard solution (ml)
V_2: consumption of $AgNO_3$ standard solution for blank test (ml)
c: concentration of $AgNO_3$ standard solution, 1mol/L
m: quality of samples (g)

3. Results and discussion

3.1. HT dechlorination of PVC with High-alkali coal

3.1.1. Effect of coal particle sizes on the DE.

Figure 2 shows the effect of three coal particle size-0.054~0.075, 0.075~0.1 and 0.1~0.22mm on DE of PVC. As indicated, at a HT holding time of 60 mins, the DE of samples treated at a HT temperature of 260 and 300 °C both experience an upward trend, while it remains stable at around 44% for the sample treated at a HT temperature of 240 °C for the samples with a diameter of 0.054~0.075 and 0.075~0.1 mm. However, the DE of HT 240 and 300 °C noticeably keep going up between PS 0.075~0.1 and 0.1~0.22mm, while HT 260 °C, with 89.15% of DE, falls slightly back to 87.94%. It can be concluded that the DE increase with the rise of particle size, which could be due to the effect of coating by ferric hydroxide (Fe $(OH)_3$) colloid. For example, the PS 0.054~0.075 of PCV/HAC composites was more incline to act as the colloidal nucleus than any other group of PS, resulting in the locking of chlorine by shell of ferric hydroxide ($Fe(OH)_3$) colloid. Furthermore, Patel et al. (1992) had studied the thermal degradation of virgin and HCl-treated PVC in powder form, as well as of PVC films of different thicknesses as a function of time and temperature[19]. They had found that the rate of dehydrochlorination of even the thinnest film significantly exceeds that of the powder samples. In addition, the rate of HCl evolution grow with film thickness, which could be explained by noting that in thicker films the HCl generated takes longer to diffuse out of the sample. Hence, the autocatalysis was more effective in the thicker films, when considering the autocatalytic effect of evolved hydrochloric acid (HCl).

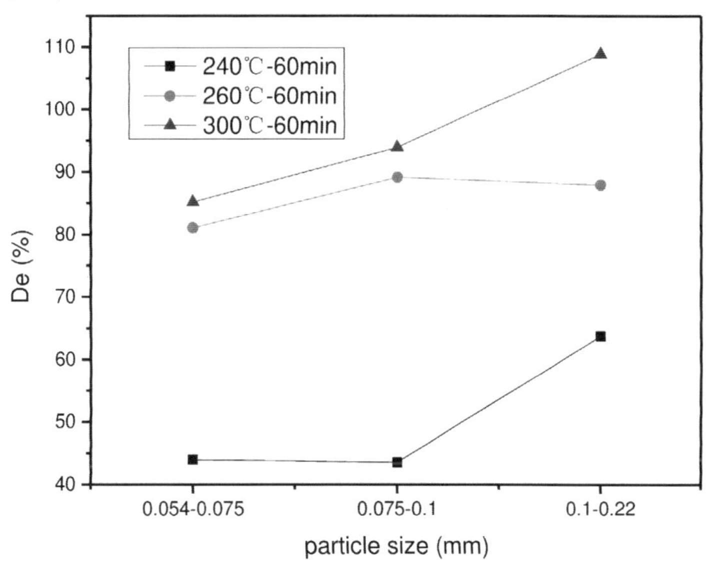

Figure 2. PS dependence of DE

3.1.2. Effect of HT temperature on the DE.

The effect of temperature on HT dechlorination of PVC with high-alkali coal is illustrated in Figure 3. As has been anticipated, the DE of three kind of particle size-0.054~0.075, 0.075~0.1 and 0.1~0.22mm, all increase with HT temperature (from 240, 260 to 300°C). Moreover, from almost the same value 44%, the DE of particle size-0.054~0.075 and 0.075~0.1mm both soar to 81% and 89%, respectively. Because of the relatively high value of 64% under 240 °C, DE of PS 0.1~0.22mm doesn`t go up so much compared with the other two PS, yet up to 89%. Besides, DE of PS 0.1~0.22mm keeps going considerably up to 100%, while DE of the other two PS grow slightly. Poerschmann et al. (2015) have studied the hydrothermal carbonization of poly (vinyl chloride) [20]. They found that dehydrochlorination increased with increasing reaction temperature and quantitative hydrodechlorination was observed beyond temperatures of ~235 °C. Through the analysis of all DTA/TGA results, Castro et al. (2012) supposed that the release of chlorine starts at 252±2 °C [21]. Considering the competitive reaction model of dechlorination proposed by Wu et al. (1994), the sequences of reaction was [22]: (1) PVC converted into intermediate species and HCl; (2) the intermediates converted into polyene and some volatiles.

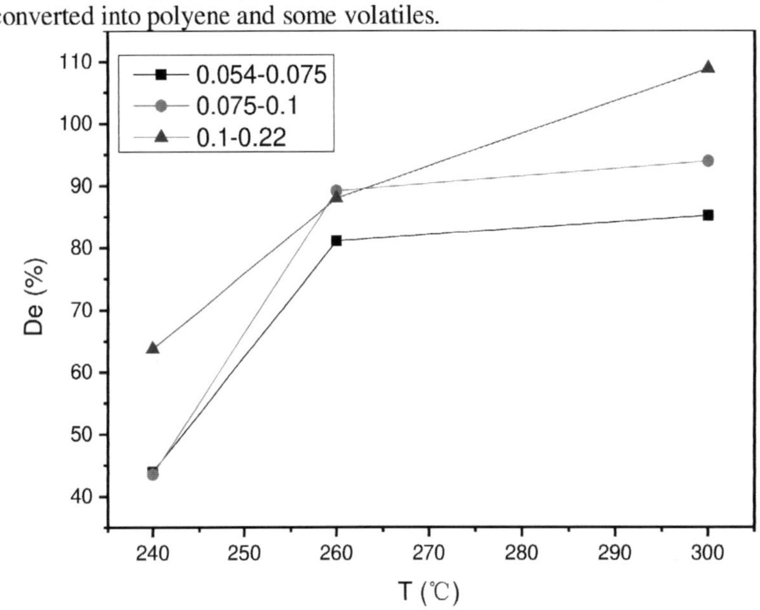

Figure 3. Temperature dependence of DE

3.1.3. Effect of HT holding time on DE.

Figure 4 displays the effect of HT holding time on DE. 30 minutes of holding time here means the sample was firstly heated to pre-determined temperature (240, 260 and 300 °C) and then removed from the heater to cool down after holding the temperature for 30 minutes. As indicated in Figure 3, the DE of the three PS all keep rising with the extension of holding time. However, between the holding time of 60 and 90 minutes, the DE of PS 0.1~0.22 experience a marginal downward trend, from 88% (60 mins) to 84% (90 mins), while the other two rise consistently, from 44% and 89% with 60minutes to 59% and 99% with 90 minutes respectively. It came to conclusion that the DE went up with the increase of holding time, which could be explained by the autocatalytic effect of evolved hydrochloric acid (HCl). It was accordant with the effect of coal particle size on HT. For example, the autocatalysis of PVC/HAC composites over 90 min took more thoroughly than that over 60 min and 30min. In addition, the downward trend experienced by the 0.1~0.22 of PS could be due to that the larger size of composites locked the chlorine in the polymer backbone because of its steady structures such as aromatic rings. Another reason could be that the effect of holding time could be negligible when compared with that of particle size.

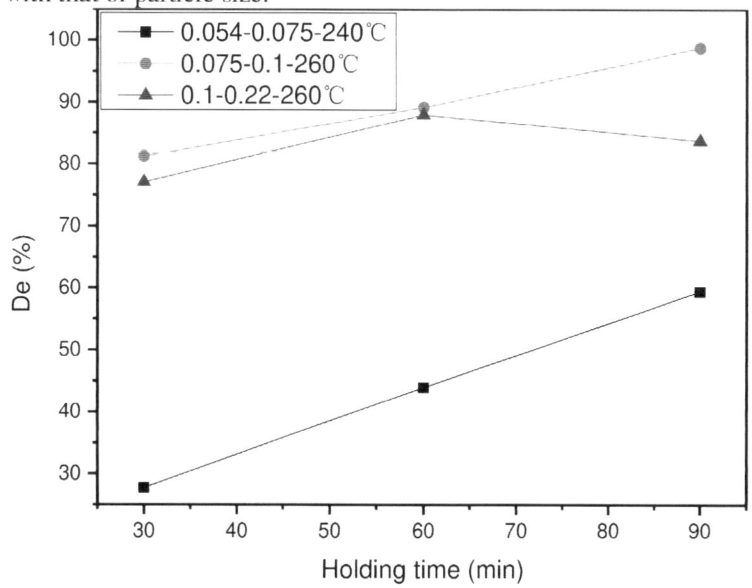

Figure 4. Holding time dependence of DE

3.2. Chlorine balance during HT dechlorination of PVC

To validate whether the organic chlorine has been totally transferred into chloridion over the co-hydrothermal treatment of PVC and high-alkali coal, it is crucial to evaluate the chlorine mass balance by measuring the chlorine both in solid and liquid. The data shown in Figure 5, Figure 6 and Figure 7 represented the chlorine mass balance of particle size-0.1~0.22, 0.075~0.1 and 0.054~0.075mm during co-hydrothermal treatment. In the figures, the chlorine in liquid and solid represented the DE and chlorine in residues respectively. As illustrated, the gross of chlorine in liquid and solid was approximately 100% in all operating conditions, suggesting that the chlorine in PVC was mainly removed in the form of Cl^- during the HT dechlorination process. Castro *et al.* (2012) studied the kinetics of thermal de-chlorination of PVC under pyrolytic conditions [21] and found that the chloride could be fixed in the form of an aqueous solution of HCl or calcium chloride. Takeshita *et al.* (2004) found that the chlorine of PVC dissolved in water as hydrochloric acid, and there were no detrimental chlorinated compounds in the liquid and gas fractions after HT at 300°C [14]. During dehydrochlorination, chlorine can be released from PVC, via elimination of HCl or nucleophilic substitution with water molecules as nucleophiles, respectively [20] and the HT of PVC is effected on the releasing of organic chlorine. Additionally, the chlorine loss of PS 0.054~0.075 was slightly higher than the two others, suggesting that there was a little chlorine in the gaseous product apart from the losing component in operating process and measuring error. Tsuji *et al.* (1998) had reported the observation of the hydrogen chloride(HCl) in fractionated gas [23], but Takeshita *et al.* (2004) thought that the hydrogen chloride dissolved in the water soon after its elimination from the PVC, due to its high solubility in water [14]. Anyway, the chlorine loss of PS 0.054~0.075 was also accordant with the effect of coal particle size on HT, that is, in smaller PS, gas products was more easily to escape from HT systems, which could be explained by noting that in smaller PS the HCl generated takes shorter to diffuse out of the PVC/HAC composites sample.

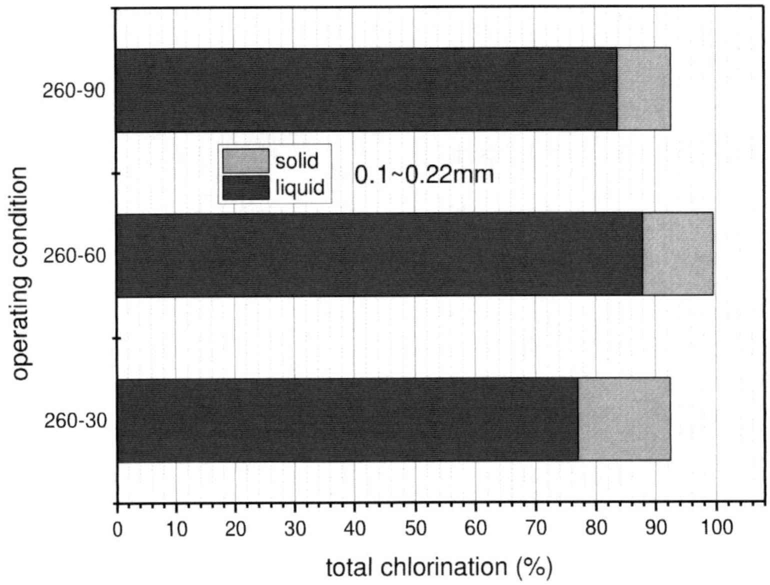

Figure 5. The change of chlorine in solid and liquid CPVC products with PS 0.1~0.22mm

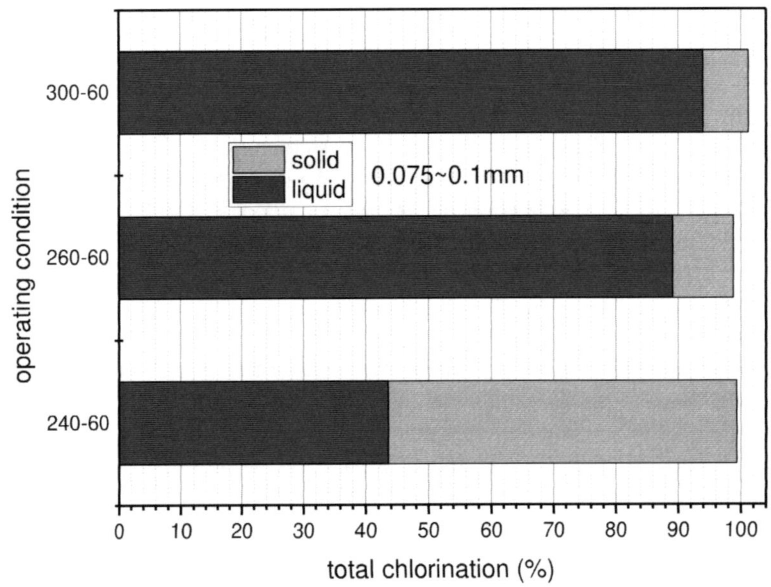

Figure 6. The change of chlorine in solid and liquid CPVC products with PS 0.075~0.1mm

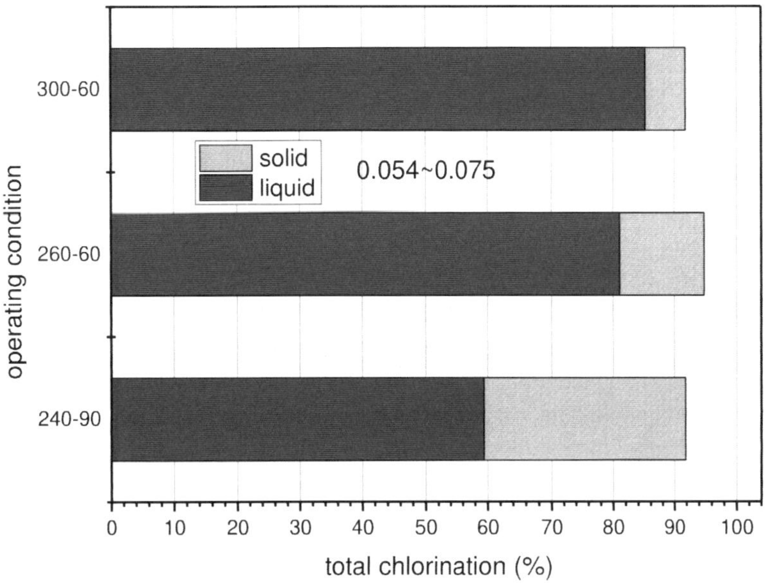

Figure 7. The change of chlorine in solid and liquid CPVC products with PS 0.054~0.075mm

4. Conclusions

HT dechlorination of PVC and coal has been proved to be an effective method of dechlorination of PVC according to previous analyses. The results are summarized as follows:

- High-alkali coal can accelerate the dechlorination process of PVC during HT. The DE increased with the rise of: 1) particle size (0.054~0.22mm), 2) holding time (30~90 min) and 3) temperature(240~300 °C). the DE was 100% when 300 °C and 60min of holding time, while the DE was 98.77% when 260 °C and 90min of holding time. When 300°C and 60min of holding time, the 85.18% of DE with first PS was lower than the 93.93% of DE with second PS and 100% of DE with third PS.
- The organic chlorine mainly transferred into chloridion in aqueous solution in HT process. Little chlorine existed in gaseous products and it increased with the decrease of particle size.

Acknowledgement

The authors gratefully acknowledge the financial support of the Natural Science Foundation of Jiangsu Province NO.BK20150174 and the Fundamental Research Funds for the Central Universities under grant NO. 2015QNA15 and No. 2015XKMS057.

References

[1] 2014 Global Demand for PVC to Rise by About 3.2%/Year to 2021 *Additives for Polymers* **2014** 10-1
[2] Goodman 2013 *The Microwave Induced Pyrolysis of Problematic Plastics Enabling Recovery and Component Reuse* (London: Imperial College London)
[3] Sadat-Shojai M and Bakhshandeh G R 2011 Recycling of PVC Wastes *Polymer Degradation and Stability* **96** 404-15
[4] López A, Marco I D, Caballero B M, Laresgoiti M F and Adrados 2011 Dechlorination of fuels in pyrolysis of PVC containing plastic wastes *Fuel Processing Technology* **92** 253-60
[5] Keane M A 2007 Catalytic conversion of waste plastics: focus on waste PVC *J. Chemical Technology and Biotechnology* **82** 787-95
[6] Sánchez-Jiménez P E, Perejón A, Criado J M, Diánez M J and Pérez-Maqueda L A 2010 Kinetic model for thermal dehydrochlorination of poly(vinyl chloride) *Polymer* 51 3998-

4007
[7] Saeed L, Tohka A, Haapala M and Zevenhoven R 2004 Pyrolysis and combustion of PVC, PVC-wood and PVC-coal mixtures in a two-stage fluidized bed process *Fuel Processing Technology* **85** 1565-83
[8] Starnes W H 2002 Structural and mechanistic aspects of the thermal degradation of poly(vinyl chloride) *Progress in Polymer Science* **27** 2133-70
[9] Troitskii B B, Troitskaya L S, Yakhnov A S, Novikova M A, Denisova V N, Cherkasov V K and Bubnov M P 1997 Investigation of Autocatalytic thermal degradation of poly(vinyl chloride) by esr spectroscopy *Polymer Degradation and Stability* **58** 83-9
[10] Akimoto M, Murayama Y, Ochiai Y, Inomata S, Sakai T, Sangu T and Yanagi K 2004 Fuel oil production from municipal waste plastics pretreated with aqueous alkaline solutions *J. Chemical Engineering of Japan* **37** 1478-87
[11] Lu J, Ma S B and Gao J S 2002 Study on the pressurized hydrolysis dechlorination of PVC *Energy & Fuels* **16** 1251-5
[12] Inoue T, Miyazaki M, Kamitani M, Kano J and Saito F 2005 Dechlorination of polyvinyl chloride by its grinding with Koh and Naoh *Advanced Powder Technology* **16** 27-34
[13] Yoshioka T, Furukawa K and Okuwaki A 2000Chemical recycling of rigid-pvc by oxygen oxidation in naoh solutions at elevated temperatures *Polymer Degradation and Stability* **67** 285-90
[14] Takeshita Y, Kato K, Takahashi K, Sato Y and Nishi S 2004 Basic study on treatment of waste polyvinyl chloride plastics by hydrothermal decomposition in subcritical and supercritical regions *J. Supercritical Fluids* **31** 185-93
[15] Zhou J B, Zhuang X G, Alastuey A, Querol X and Li J H 2010 Geochemistry and mineralogy of coal in the recently explored Zhundong Large coal field in the Junggar basin, Xinjiang Province, China *Int. J. Coal Geology* **82** 51-67
[16] Yang Y M, Wu Y X, Zhang H, Zhang M, Liu Q, Yang H R and Lu J F 2016 Improved sequential extraction method for determination of Alkali and Alkaline earth metals in Zhundong Coals *Fuel* **181** 951-957
[17] Li W Q, Wang L Y, Qiao Y, Lin J Y, Wang M J and Chang L P 2015 Effect of atmosphere on the release behavior of alkali and alkaline earth metals during coal oxy-fuel combustion *Fuel* **139** 164-70
[18] Naruse I, Kamihashira D, Khairil, Miyauchi Y, Kato Y, Yamashita T and Tominaga H 2005 Fundamental ash deposition characteristics in pulverized coal reaction under high temperature conditions *Fuel* **84** 405-10
[19] Patel K, Velazquez A, Calderon H S and Brown G R 1992 Studies of the solid - state thermal degradation of PVC. I. autocatalysis by hydrogen chloride *J. Applied Polymer Science* **46** 179-87
[20] Poerschmann J, Weiner B, Woszidlo S, Koehler R and Kopinke F D 2015 Hydrothermal carbonization of poly(vinyl chloride) *Chemosphere* **119** 682-9
[21] Castro A, Soares D, Vilarinho C and Castro F 2012 Kinetics of thermal de-chlorination of PVC under pyrolytic conditions *Waste Management* **32** 847-51
[22] Wu C H, Chang C Y, Hor J L, Shih S M and Chen L W 1994 Pyrolysis kinetics of waste PVC at high conversion *The Chemical Engineering Journal and the Biochemical Engineering Journal* **55** 87-92
[23] Tsuji T, Ikemoto H, Irita S, Sakai N, Shibata T, Uemaki O and Itoh A 1998 Thermal and catalytic degradation of poly(vinyl chloride) *Nippon Kagaku Kaishi* **8** 546-50

Dynamic response characteristics analysis of the doubly-fed wind power system under grid voltage drop

Y Chen[1, 3], J Wang[1], H H Wang[2], L Yang[1], W Chen[1] and Y T Xu[1]

[1]School of Electrical Engineering, Wuhan University, Wuhan 430072, China
[2]Dispatching Control Center, China Southern Power Grid, Guangzhou, P. R. China

Email: chenyiwhu@whu.edu.cn

Abstract. Double-fed induction generator (DFIG) is sensitive to the disturbances of grid, so the security and stability of the grid and the DFIG itself are under threat with the rapid increase of DFIG. Therefore, it is important to study dynamic response of the DFIG when voltage drop failure is happened in power system. In this paper, firstly, mathematical models and the control strategy about mechanical and electrical response processes is respectively introduced. Then through the analysis of response process, it is concluded that the dynamic response characteristics are related to voltage drop level、 operating status of DFIG and control strategy adapted to rotor side. Last, the correctness of conclusion is validated by the simulation about mechanical and electrical response processes in different voltage levels drop and different DFIG output levels under DIgSILENT/PowerFactory software platform.
Keywords: Time doubly fed induction generator; mechanical response characteristics; electrical response characteristics; voltage drop; operating status; control strategy

1. Introduction

With the reduction in resources reserves and decline in environmental quality, it is popular to develop clean and renewable energy. Wind energy is the most promising green energy in all renewable energy and in recent years wind power technology is improved [1]. Doubly fed induction generators (DFIG) are preferred for wind turbine application [2-4], accounting for 70% to 80% ratio of total installed capacity of wind power, because they offer variable speed, keep the size of the controllers small so as to reduce the costs and decouple control of active and reactive power [5-6].However, the security and stability of the grid and the DFIG itself are under threat with the rapid increase of DFIG because of the diffidence between DFIG and conventional synchronous generators[7-10], including the diffidence of electrical dynamic response characteristics in the case of grid failure, weaker disturbance rejection with the small capacity converter and so on[11-12]. Voltage drop is the most common and serious failure in a variety of grid failure [13-14]. So it is necessary for grid stability to study on mechanical and electrical response processes of DFIG under grid voltage drop.

Nowadays, there are lots of researches on response characteristics of wind turbine during the grid failure at home and abroad. According to [15-16], the conclusion is drawn that different control strategies will lead to different response processes by the analysis of a multi-coupling model for wind turbines. From [17], it can be concluded that electrical response process of DFIG is more severe with

[3] Address for correspondence: Y Chen, School of Electrical Engineering, Wuhan University, Wuhan 430072, China. Email: chenyiwhu@whu.edu.cn.

deeper voltage falling and greater slip value. In [18], the differences of dynamic response of the DFIG are discussed with different drive system model and different rotor-side controller parameters; According to [19], Electrical parameters of DFIG including its Stator flux, rotor flux, voltage and current are related to not only parameters and Control Strategy of Wind Turbines but also the Grid operation status.

On the base of previous studies, the dynamic response characteristics are related to voltage drop level， operating status of DFIG and control strategy adopted to rotor side. However, while electrical response characteristics are adequately studied, mechanical response characteristics are ignored. In this paper, firstly, mathematical models and the control strategy about both mechanical and electrical response processes is respectively introduced. Then through the analysis of response processes, it is concluded that the dynamic response characteristics are related to voltage drop level， operating status of DFIG and control strategy adopted to rotor side. Last, the correctness of conclusion is validated by the simulation about mechanical and electrical response processes in different voltage levels drop and different DFIG output levels.

2. Analysis of mechanical part

Mechanical part is mainly composed of wind turbine, shaft and pitch angle control. Wind turbine catches wind energy and turns it into rotational kinetic energy. Shaft transmits Mechanical energy from wind turbine to generator. Pitch angle control Tracks power change quickly and maintains stability. The parameters of mechanical part which should be paid attention to are rotor speed, wind turbine speed, pitch angle and mechanical power. Mechanical response processes is relatively slow.

2.1. Wind turbine

For a horizontal axis wind turbine, the amount of mechanical power P_w that a turbine produces in steady state is given by:

$$\begin{cases} P_w = \frac{1}{2}\rho\pi R^2 C_P(\beta,\lambda) V_{eq}{}^3 \\ C_P(\beta,\lambda) = 0.22(\frac{116}{\lambda_1} - 0.4\beta - 5.0)e^{\frac{-12.5}{\lambda_1}} \\ \lambda_1 = \cfrac{1}{\cfrac{1}{\lambda+0.08\beta} - \cfrac{0.035}{\beta^3+1}} \\ \lambda = R\omega_{tur}/V_{eq} \end{cases} \tag{1}$$

Where, ρ is the air density. R is the turbine radius. V_{eq} is the equivalent wind speed and C_P is the power coefficient, which for pitch controlled wind turbines depends on both the pitch angle β and the tip speed ratio λ. ω_{tur} is the rotor turbine speed.

According to equation (1), mechanical power is related to β 、 V_{eq} and ω_{tur} . In steady-state operation, when V_{eq} is a constant, the fan will be running at the maximum power point, where the ω_{tur} and P_w are in the most appropriate state. However, in power system fault, the ω_{tur} will be accelerated and P_w will be decreased because of unbalance torque to maintain stability.

2.2. Shaft

In steady-state analysis, the shaft is always a one mass model ignoring torsional vibration because of electro-mechanical decoupling and the effect of inverters. However, in serious power system fault, a detailed model, mostly a two mass model is used to describe the shaft for better analysis of its transient characteristics , as follows:

$$\begin{cases} 2H_{tur}\dfrac{d\omega_{tur}}{dt} = T_{tur} - K_s\theta_s - D_{tur}\omega_{tur} \\ 2H_{gen}\dfrac{d\omega_{gen}}{dt} = K_s\theta_s - T_E - D_{gen}\omega_{gen} \\ \dfrac{d\theta_S}{dt} = \omega_0(\omega_{tur} - \omega_{gen}) \end{cases} \tag{2}$$

Where, subscripts of tur and gen respectively denote the parameters of wind turbine and generator. H is inertia time constant. ω is rotor speed. ω_0 is synchronous speed. K_s is stiffness coefficient D is damping factor . θ_S is the angular difference between the two ends of the flexible shaft. T_{tur} and T_E are mechanical torque and electromagnetic torque.

According to the model above, ω_{gen} has an oscillating increase and ω_{tur} increases slowly without oscillation under power system short-circuit fault.

2.3. Pitch control

The aims of pitch control are to track maximum power point and to limit mechanical power in high wind speed preventing the fan running too fast. The principle is that if β is bigger, and then P_w will be lower according to equation (1). Under grid voltage drop, the rotor will increase speed because electromagnetic power of fan is limited causing torque to become unbalanced. To make the fan and power system safe, it is need to increase β to reduce mechanical power of fan on the basis of principle above.

There are two common methods used for pitch control. One is that β is adjusted according to the change of active power of fan, another is that β is adjusted according to the change of rotor speed. However, in both of the two methods the reference is fixed value set in advance which makes the adjustment valid only in the high wind speed. So in this paper, the reference will be changed on the base of the wind speed. Given fast degeneration of active power which is not good for the stability of pitch angle, the second method is chose, as it shown in figure 1.

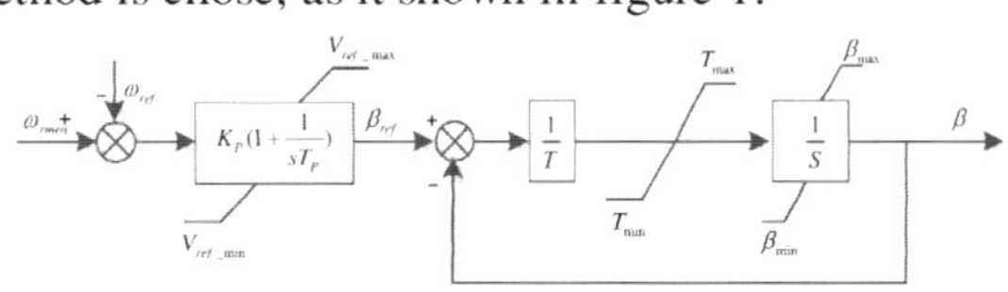

Figure 1. Pitch angle control

When measurement of rotor speed is bigger than the reference of it according to current wind speed, the output of β will be larger than 0° making the pitch angle to change.

2.4. Mechanical response process

The deeper the voltage drops, the greater mechanical power differs from electromagnetic power and the greater parameters in mechanical part change.

Once grid occurs short circuit fault, the voltage will drop and active power transmit will be limited leading to the unbalance between mechanical torque and electromagnetic torque and acceleration of the rotor speed ω_{gen} . Then according to equation (2), ω_{gen} has an oscillating increase and ω_{tur} increases slowing. The difference between ω_{tur} and the reference speed ω_{ref} will make the pitch control work to increase β according to Figure 1.The increase of β will reduce the mechanical power P_w, which will affect operation of shaft and change θ_s、ω_{tur}、ω_{gen}. It has been cycling on each other until the fault is cleared.

3. Analysis of electrical part

Electrical part is mainly composed of generator model, rotor-side converter and its control model, and special control strategies in low voltage. Generator model is the key component to achieve electromechanical energy conversion. Rotor-side converter and its control model is the key part to realize active and reactive power decoupling. Special control strategies in low voltage are the key method to prevent secondary attack caused by the disconnection of the fan. The parameters of this part which should be paid attention to are active power, reactive power, voltage and current. Electrical response processes is relatively fast. To simplify the analysis, the grid-side power converter and the power outputted by the rotor are not considered here.

Doubly fed induction generator model, rotor-side converter and its control model are referred to [5], which are typical. In this paper, special control strategies in low voltage are mainly introduced.

3.1. Special control strategies in low voltage

Grid codes will require wind turbines to have Low Voltage Ride Through (LVRT) capability to prevent suffered secondary attack caused by the disconnect of the fan in voltage drop fault [20]. It is means that DFIG can continue to run and provide reactive power for the grid to help the voltage rise. However, an abrupt voltage drop can result in overvoltage or overcurrent in the rotor windings which may destroy the rotor-side converter, so it needs special control strategies to protect it. Currently, there are two ways: additional hardware and to change inverter control strategy. In this paper, combination of LVRT converter control and crowbar resistance is used to protect DFIG and make it play the best effect in the low voltage.

3.1.1. Crowbar control.

An abrupt voltage drop can result in over-currents which may destroy the rotor-side converter. To avoid the situation happening, increasing crowbar resistance can make rotor windings short and let the current flow to bypass. However, when the crowbar plays a role, DFIG will be in asynchronous state and need lots of reactive power which is harmful to the recovery of voltage. So the appropriate time of crowbar in operation is necessary and many papers have done research on this [21]. The control strategy in this paper is that when the rotor current is up to double nominal value, make the crowbar valid and cut off it automatically after 60ms.

The control strategy is shown in figure 2. The value of current decides whether crowbar is valid.

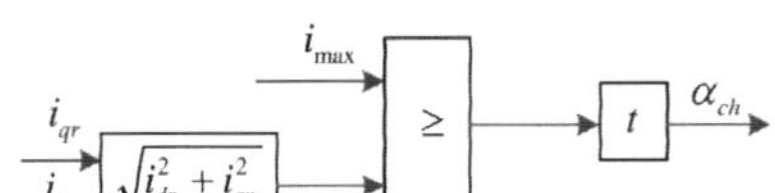

Figure 2. Crowbar control

In figure 2, $i_{\max}$ is the maximum current which can flow the rotor-side converter. t is the continue time that crowbar is valid. α_{cb} is the state of crowbar, $\alpha_{cb}=1$ means crowbar is valid, and $\alpha_{cb}=0$ means crowbar is invalid.

3.1.2. Low voltage ride through, LVRT.

The crowbar working time is short than the fault, so the rotor speed will still accelerate. According to the original control strategy, the active power will increase to decrease the rotor speed which will make the current lager and make crowbar valid again. And it needs crowbar produce reactive power to help the voltage recover. So the new strategy is needed, and low voltage ride through control strategy is designed in this paper. Once the grid has short-circuit fault, the reactive power compensation is first to be considered in the base of the degree of voltage drop. Then under the limitation of maximum current, the active current is relatively set to provide margin for the reactive current. The aims of this strategy are supporting the grid voltage recovery as well as protecting DFIG safe.

In the voltage drop, the references of reactive and active power are provided according to the following equation:

$$\begin{cases} Q_{ref1} = U_{WT} I_{dr}, I_{dr} \le I_N \\ I_{dr} = 1.5 I_N (0.9 - U_{WT}) \end{cases} \tag{3}$$

$$P_{ref1} = \min[P_{ref}, U_{WT}(1 - I_{dr})] \tag{4}$$

Where U_{WT} is the terminal voltage of DFIG, and $0.2p.u \le U_{WT} \le 0.9p.u$. I_{dr} is dynamic reactive current provided by DFIG. I_N is nominal current. P_{ref} is the active power reference in the maximum power point.

The low voltage ride through control design is shown in figure 3. The essence is to, change active and reactive power reference value input in the original closed-loop control by increasing the voltage detection aspect.

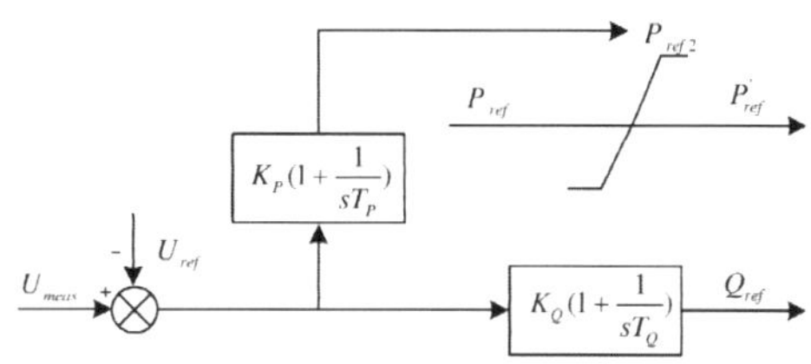

Figure 3. Low voltage ride through control

3.2. Electrical response process

Dynamic electrical characteristic of DFIG is affected by excitation control, which is related to the voltage drop level and operating status, and has a fast degeneration. When the voltage drop is shallow, excitation current is controlled to maintain the stability of active and reactive power because of the constant power control strategy. When voltage drop is heavy, special strategy -low voltage ride through control or crowbar control is valid to protect the rotor converter and help the grid recovery.

If the crowbar is valid or not, it is decided to the voltage drop level and operating status of DFIG before short fault. And DFIG will be in asynchronous machine state in which the reactive power is 0 and active power is lower. Under low voltage ride through control, DFIG mainly provides reactive power and its active power is limited which is involved in voltage drop level

The voltage drop can be divided into four levels according to the situation of converter control operation. When Voltage drop is shallow, generally u> 0.9, LVRT control and crowbar will be both invalid. Voltage drop is medium, 0.2<u<0.9, only LVRT control will be valid. When Voltage drop is deep, 0.2<u<0.9, LVRT control and crowbar will be both valid. When the voltage drop is over safety value, the DFIG will be removed.

4. Simulation results

To verify the analysis about the DFIG dynamic response characteristics in voltage drop and effectiveness of the control scheme, the following model is built in DIgSILENT/ PowerFactory. Wind farm is consisted of 30 units of 2MW DFIG and is access to single machine infinite bus system by 20/0.69kV double-winding step-up transformer. And DFIG run in rated power and low power (20%) two states. The different voltage drop levels are completed by adjusting grounding impedance in three-phases short-circuit.

4.1. Rated power state

When the wind speed is greater than the rated wind speed 11.5m/s rate, the DFIG is in rated power state. Three-phase short circuit occurs in the 0.5s, lasted 0.625s and grid voltage drop, respectively, to 92%, 82%, and 45% of the rated voltage. The mechanical response process and electrical response process are separately shown in figure 4 and figure 5.

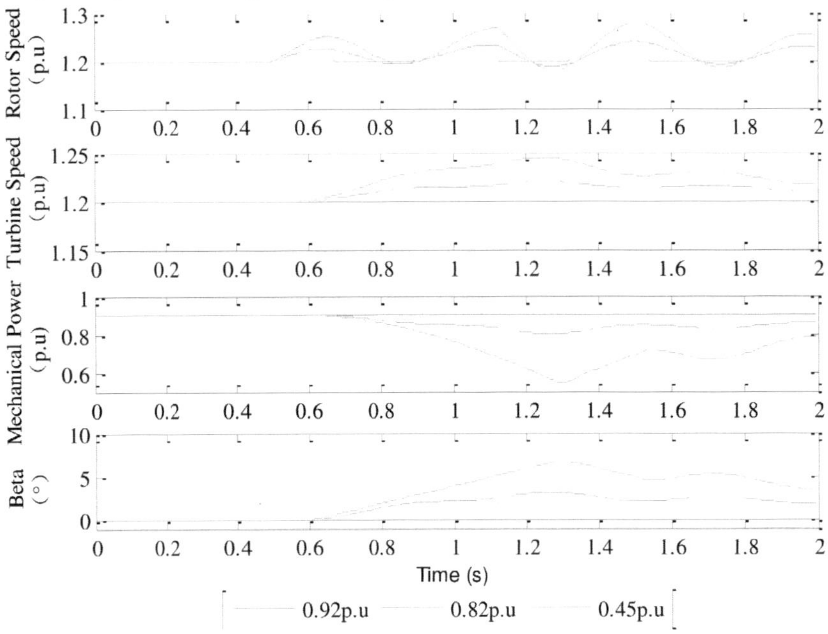

Figure 4. Mechanical response processes of DFIG, when it operated with rated power

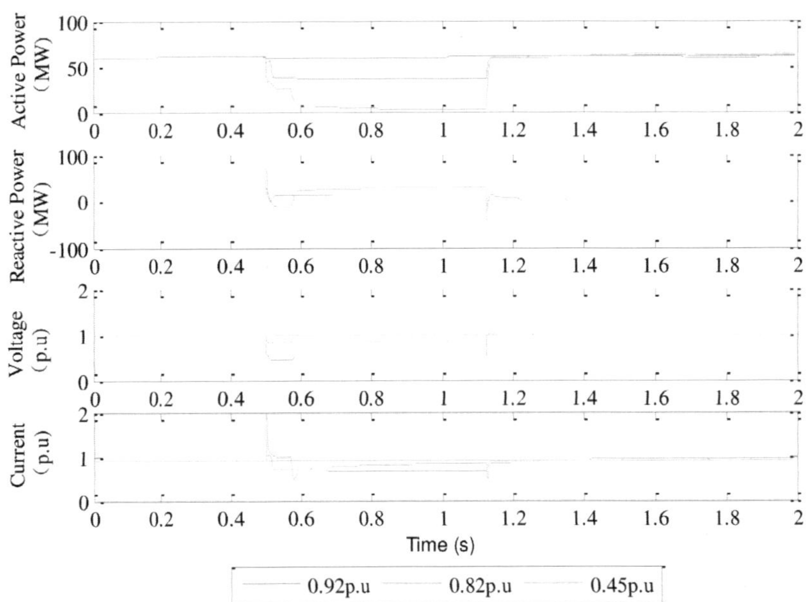

Figure 5. Electrical response processes of DFIG, when it operated with rated power

As seen from figure 4 and figure 5, when the voltage drop is shallow (0.92p.u), all amounts have been adjusted in accordance with the existing control instructions, and special control strategies are not valid.

As indicated by figure 4, mechanical responses are similar in medium and deep large voltage drop

(0.2<u<0.9). Once the fault happens, ω_{gen} has an oscillating increase, ω_{tur} starts to increase after a period of time, β increases and active power decreases. Once the fault is cleared, ω_{gen} immediately begins to decrease, the remaining parameters come to recover after a while. Voltage drops deeper, the change of all parameters are larger.

As indicated by figure 5, the electrical response process will be different with different voltage drop level.

In medium voltage drop, the crowbar is invalid as the current instant surge is too small to meet 1.5p.u. Low voltage ride through control is taken resulting that the active power is limited and decreases quickly, reactive power increases fast to support voltage recovery, the voltage dips in the duration of fault and has a recovery in the process, and the current increases. In deep voltage drop, the crowbar is effective as the current instant surge is big enough to meet 1.5p.u which makes the converter invalid. The reactive power is zero, active power decreases and the current increases in that process. After 60ms, the crowbar is removed and low voltage ride through control is taken. Then reactive power is produced, and active power decreases more and the current decrease to the maximum value allowed. After clearing the fault, although mechanical parameters still oscillate, the electrical parameters recover quickly due to the excitation control.

It can be verified that different level of voltage drop will cause different mechanical and electrical response characteristics which are decided by the strategy taken in rotor side. What's more, Mechanical response is lagging, and electrical response is particularly rapid.

4.2. Low power state

When the wind speed is lower than the rated wind speed 11.5m / s rate, the DFIG is in at maximum power tracking state where its output size is determined by wind speed. In this paper, wind speed 9m/s is selected, and the DFIGs are running at 20% of rated power. Three-phase short circuit occurs in the 0.5s , lasted 0.625s and grid voltage drop, respectively, to 92%、82%、45%、and 32% of the rated voltage. The mechanical response process and electrical response process are separately shown in figure 6 and figure 7.

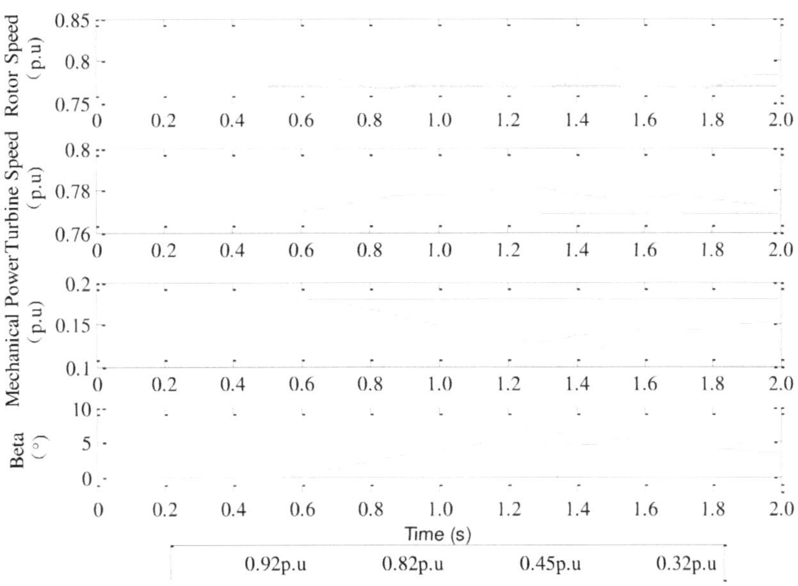

Figure 6. Mechanical response processes of DFIG, when it operated with low power

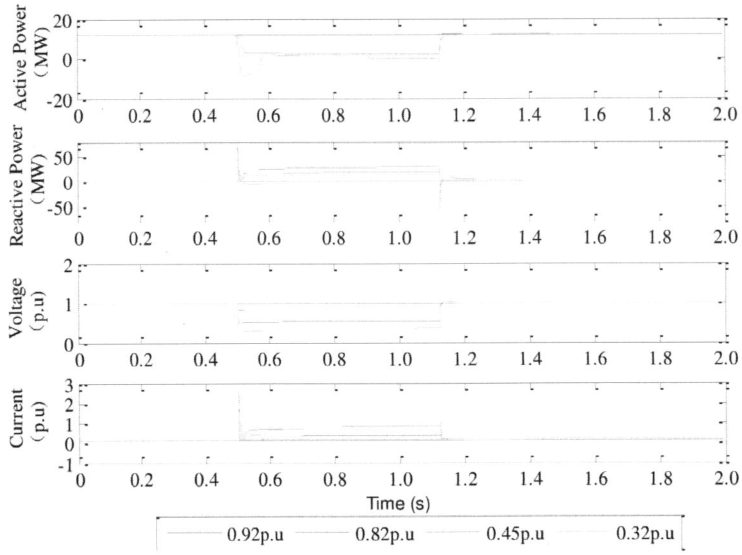

Figure 7. Electrical response processes of DFIG， when it operated with low power

As seen from figure 6 and figure 7, when the voltage drop is shallow (0.92p.u), all amounts are almost unchanged.

Compared with figure 3 and figure 6, dynamic response is different in the same voltage drop level but different speed wind. All amounts in mechanical part don't response to the medium voltage drop because the output is maintained in 20% of nominal power which isn't over the lamination and the torques are equal. The mechanical responses under deep voltage drop(0.45p.u、0.32p.u) are similar to that happen in rated power, but the degree is smaller.

Compared with figure 4 and figure 7, in medium power voltage (0.82p.u), the reactive power is only related to the voltage level and same to that provided in rated power state. There is no change in active power. In deep voltage drop, whether crowbar is valid, it is related to the current at the moment when the short circuit happens, that is to say it is to related to operating status of DFIG. While the crowbar is valid in high wind speed, it is not valid in low speed when other conditions are same. When the voltage drop is deeper (0.32p.u), the crowbar is put in and its response process is similar to that in rated power state.

It can be concluded in this case that reactive power compensation is only related to the voltage drop level, while active power and crowbar are limited to the wind speed level.

5. Conclusions

In this paper, the model and control system about mechanical and electrical parts of double-fed induction generator are studied and it is conclude the mechanical and electrical response characteristic under grid fault which is verified in DIgSILENT/ PowerFactory. Conclusions are as below

- Under fault condition, mechanical response characteristic is affected by electrical characteristics. The changes of beta and mechanical power will be different because of original active power value and it after the fault. Mechanical response characteristic is hysteresis;
- Under fault condition, electrical response characteristic is decided by rotor-side control strategy including vector control for rotor voltage, low voltage ride through control and control for crowbar. What's more, using which strategy above is decided by voltage drop level、operating status of DFIG;
- After the fault is cleared, since the excitation control leads to electromechanical decoupling, Oscillation in the mechanical part does not affect the changes in the electrical part.

In conclusion, the dynamic response characteristics are related to voltage drop level、operating status of DFIG and control strategy adopted to rotor side. The conclusion is significant for transient stability of power system with large-scale wind farm large and safe operation of the wind turbine.

Acknowledgments
The authors would like to appreciate the financial support of National Natural Science Foundation of China under Grant 51477121.

References

[1] Zhao J J, Lv X, Fu Y and Hu X G. 2015 Frequency regulation of the wind/photovoltaic/diesel microgrid based on DFIG cooperative strategy with variable coefficients between virtual inertia and over-speed Control *Transactions of China Electrotechnical Society* **30** 59-68
[2] Zhang Y X, Tong R, Zhao J and Xuan W B. 2013 Transient characteristics analysis and low voltage ride-through scheme of doubly-fed wind turbine generators *Automation of Electric Power Systems* **37** 7-11
[3] Ji X Q, Liu G B, Li D and Li K J. 2014 An methodology for on-line power quality monitoring of grid-connected doubly fed wind turbine *Power System Protection and Control* **42** 145-50
[4] SLOOTWEG J G 2003 Wind power: modeling and impact on power system dynamics (Delft, the Netherlands: De1lt University of Technology) 24-5
[5] Zi P, Zhou X X, Tian F, Hou J X, Zhang S and Tao X Y. 2015 Electromechanical transient modeling of wind turbines based on doubly-fed induction generator *Proc. the CSEE* **35** 1106-14
[6] Tsili M and Papathanassiou S 2009 A review of grid code technical requirements for wind farms *IET Renewable Power Generation* **3** 308-32
[7] Zhang M L, Xu J Y and Li J J 2013 Research on transient stability of sending power grid containing high proportion of wind power *Power System Technology* **37** 740-5
[8] Nian H, Cheng P and He Y K 2015 Review on operation techniques for DFIG-based wind energy conversion systems under network faults *Proc. the CSEE* **35** 4184-97
[9] Wang Y P, Zheng T, Wang Z P, Cao Y R, Zhu S Y and Yang G S. Impact analysis of different slips on the short circuit current of DFIG after crowbar operation *Power System Protection and Control* **43** 8-14
[10] Bu Q S, Gao Lei, Song L L, Song S, Chen N and Gao B T. 2015 Short circuit current and voltage frequency inconsistency in wind farm and its effect on distance protection and relay protection test *Power System Protection and Control* **43** 21-7
[11] Yu L, Chen Y C, Chen G C and Wu G X. 2011 A low voltage ride-through control strategy of doubly fed induction generator *Transactions of China Electrotechnical Society* **26** 30-6
[12] Flannery P and Venkataramanan G 2007 Evaluation of voltage sag ride-through of a doubly fed induction generator wind turbine with series grid side converter *Proc. of the 38th Annual Power Electronics Specialists Conference* 1839-45
[13] Hu J B, Sun D, He Y K and Zhao R D. 2006 Modeling and control of dfig wind energy generation system under grid voltage dip *Automation of Electric Power Systems* **30** 21-6
[14] Kong X Y, Xu Y H and Tao S 2015 Sensitive equipment immunity assessment based on a new voltage sag description *Transactions of China Electrotechnical Society* **30** 165-71
[15] Wang C Q, Song G B, Chi Y N, Liu K, Shen Q Y and Tang H Y. 2015 Fault characteristics analysis of wind power system *Automation of Electric Power Systems* **39** 52-8
[16] Miao F L, Shi H S and Zhang X Q 2015 Modellfng of wfnd trbines coupled in multi-domafn and dynamic rtesponse analysis *Proc. the CSEE* **35** 1704-12
[17] Meng Y F, Hu S J, Wang L L and Xu H H. 2013 Characteristics analysis and coordinated control of the doubly-fed wind power system under grid transient fault *Power System Protection and Control* **41** 106-13
[18] Pan X P, Ju P, Wu F and Jin Y P. 2015 Discussion on model structure of dfic-based wind

turbines *Automation of Electric Power Systems* **39** 7-14

[19] López J, Sanchis P, Roboam X and Marroyo L. 2007 Dynamic behavior of the doubly fed induction generator during three-phase voltage dips *IEEE Transactions on Energy Conversion* 22 709-17

[20] Zhang J H, Wang J, Chen X Y, Mo Y P and Xin F L. 2012 Analysis of DFIG-based wind generation LVRT control strategy *Power System Protection and Control* **39** 28-33

[21] Zhang M and Jiang H L 2014 Adaptive low voltage ride-through of doubly-fed induction generators based on crowbar with a parallel dynamic resistor *Transactions of China Electrotechnical Society* **29** 271-8

IOP Conf. Series: Earth and Environmental Science **40** (2016) 012072 doi:10.1088/1755-1315/40/1/012072

Catalytic pyrolysis of model compounds and waste cooking oil for production of light olefins over La/ZSM-5 catalysts

F W Li, S L Ding, L Li, C Gao, Z Zhong, S X Wang and Z X Li[1]

School of Chemistry and Chemical Engineering, Guangxi University, Nanning, China

E-mail:zhixiali@hotmail.com

Abstract. Waste cooking oil (WCO) and its model compounds (oleic acid and methyl laurate) are catalytically pyrolyzed in a fixed-bed reactor over La modified ZSM-5 catalysts (La/ZSM-5) aiming for production of C2-C4 light olefins. The LaO content in catalysts was set at 0, 2, 6, 10 and 14 wt%. The gas and liquid products are analyzed. The La/ZSM-5 catalyst with 6% LaO showed higher selectivity to light olefins when WCO and methyl laurate were pyrolyzed, and olefin content was 26% for WCO and 21% for methyl laurate. The catalyst with 10% LaO showed high selectivity to light olefins (28.5%) when oleic acid was pyrolyzed. The liquid products from WCO and model compounds mainly contain esters and aromatic hydrocarbons. More esters were observed in liquid products from methyl laurate and WCO pyrolysis, indicating that it is more difficult to pyrolyze esters and WCO than oleic acid. The coked catalysts were analyzed by temperature-programmed oxidation. The result shows that graphite is the main component of coke. The conversion of WCO to light olefins potentially provides an alternative and sustainable route for production of the key petrochemicals.

1. Introduction

Light olefins such as ethylene and propylene are important basic industrial materials. Currently, light olefins are mainly produced from petroleum hydrocarbons [1]. It is well known that petroleum hydrocarbons are running low and that limits bloom in light olefin industry. Bio-oil, an alternative energy source, has been considered as a potential substitute for fossil fuels [2, 3]. Waste cooking oil (WCO) is a typical bio-oil, and millions of tons of WCO are produced in China each year. Development of new technique for effective use of WCO is one of the most pressing problems facing this country [4, 5], and helps to improve the environmental problems in China.

WCO consists mainly of triglycerides and a small amount of free fatty acids [6, 7]. Catalytic pyrolysis is a simple and efficient way to convert WCO to valuable materials. During catalytic pyrolysis process, oxygen atoms are removed by releasing H_2O, CO_2 and CO via dehydration, decarboxylation and decarbonylation mechanisms. The liquid product supplies a new material for preparation of biofuel which has similar components to that of fossil based fuels. The gas product with considerable light olefin selectivity has drawn widespread attention [8].

[1] Address for correspondence: Z X Li, School of Chemistry and Chemical Engineering, Guangxi University, Nanning, China. E-mail:zhixiali@hotmail.com.

2. Materials and methods

2.1. Synthesis of the La/ZSM-5 catalysts

H-type ZSM-5 zeolite (Si/Al=25) was obtained from Nankai University Catalyst Plant. (Tianjin, China). Other reagents were of analytical grade and used as received. WCO was collected from University Restaurant and used as received. La/ZSM-5 catalyst was prepared by impregnation method. $La(NO_3)\cdot 6H_2O$ in its desired amounts was dissolved in deionized water to form a metallic salt solution. ZSM-5 zeolite was firstly dried at 105 °C for 90 min and dispersed in the metallic salt solution. The mixed suspension was stirred at 40 °C for 30 min and 80 °C for 3 h to remove the water. The obtained solid samples were dried, tableted, and calcined at 550 °C for 5 h. The LaO content in the resulting catalysts was 2, 6, 10 and 14%, respectively. The corresponding catalyst was denoted LaO2%, LaO6%, LaO10% and LaO14%, respectively. The obtained catalysts were sieved to collect the particles passing through a 20-40 mesh screen, which were kept for the subsequent experiments.

2.2. Catalytic cracking of three kinds of oil samples

The catalytic pyrolysis experiments were carried out in a flowing fixed bed reaction system. The system was mainly composed of an inner quartz tube reactor (inner diameter: 20 mm; length 30 mm), an electric heating sleeve, a syringe located on the top of reactor (was used to inject oil samples into reactor) and a condenser. For a typical case, one gram of the sieved catalyst was used. Before the reactions, the reactor was flushed with nitrogen at a flow rate of 40 ml/min, and heated to 600 °C. 2 ml of oil samples was then pushed into the reactor by the syringe at constant speed in 2 min, and then reacted for another 2 min, thus, the total reaction time was 4 min. The formed gas and liquid products during reaction were condensed, collected and analyzed.

2.3. Product analysis

The liquid product was firstly methyl esterified according to the standard methods of GB/T 17376-2008. The components in products were analyzed by GC-MS (GC7890A; MS5975C, Agilent) equipped with HP-5MS capillary column. The inlet and detector temperature was maintained at 290°C. The oven temperature program was set as follows: hold at 60 °C for 5 min, heated to 300 °C at 10 °C/min, and hold at this temperature for 16 min.

The gaseous hydrocarbon products were analyzed using a gas chromatography (GC, FULI-9790II) equipped with a flame ionization detector (FID). A column of HP-PLOT/Q was used. A GC equipped with a packed column (TDX-01) connected with a Residual Gas Analyzer (RGA) was used to separate and analyze CO, CO_2 N_2 and He (carrier gas). The following temperature program was used: hold at 20 °C for 4 min, heated to 240 °C at 15 °C/min, and hold at 240 °C for 15 min. The qualitative and quantitative analysis was performed using standard gases (CH_4, C_2H_4, C_2H_2, C_2H_6, C_3H_8, C_3H_4, C_3H_6, C_4H_8, C_4H_{10}, CO, CO_2, and He).

The coke on catalyst formed during pyrolysis process was analyzed using temperature-programed oxidation (TPO) connected with RGA system. The used catalysts were temperature programmed at a rate of 10 °C/min from 100 °C to 900 °C by electric furnaces in He and O_2 mixed atmosphere (He: 40 ml/min; O_2: 2 ml/min). The released CO, CO_2 and H_2O were constantly monitored.

3. Results and discussion

3.1. Catalytic conversion of oleic acid

Figure 1 shows the product distribution for the catalytic pyrolysis of oleic acid over different La/ZSM-5 catalysts at 600 °C. GC-RGA data in Figure 1 (a) shows that the catalyst with 10% LaO had better olefins selectivity (28.5%), which is 5.45% higher than that over LaO-free catalyst. The LaO-free catalyst showed a high alkane selectivity (17.3%). A small amount of alkynes, CO and CO_2 were detected in gas products. Some other gas products are formed but cannot be identified with the present standard gases. The gaseous product selectivity obtained over different La/ZSM-5 catalyst was summarized in Table 1.

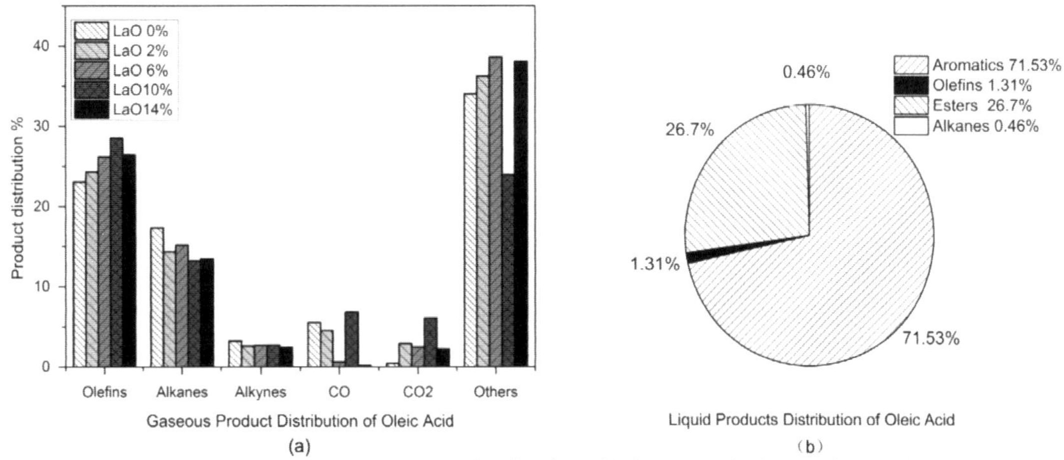

Figure 1. Gaseous (a) and liquid products distribution (b) from catalytic pyrolysis of oleic acid.

Figure 1 (b) shows the components in liquid products obtained by pyrolysis of oleic acid over LaO 10% catalyst. Liquid products contained 26.7% esters (e.g. methyl decanoate, methyl laurate, methyl palmitate, methyl stearate, etc.), 71.5% aromatic hydrocarbons (e.g. toluene, ethylbenzene, paraxylene and its derivatives, etc.) and a small amount of olefins (1.31%) and alkanes (0.46%). The results indicate that the presence of catalyst facilitates the production of light olefins. A lot of aromatic hydrocarbons as by-product were formed during catalytic conversion of oleic acid to light olefin. There is still considerable amount of oleic oil remained in liquid product, indicating the conversion reaction is incomplete. Deoxygenation of carboxyl group of oleic acid was achieved by releasing CO, CO_2 and H_2O via decarbonylation, decarboxylation and dehydration mechanisms. C_3H_8 is suggested to be an intermediate product which will form various liquid hydrocarbons through alkylation and isomerization reaction during the cracking process [9].

3.2. Catalytic conversion of methyl laurate

Figure 2 shows product distribution for catalytic pyrolysis of methyl laurate over LaO 6% catalyst. GC analysis (Figure 2 (a)) shows that LaO 6% catalyst had higher olefin selectivity (31%) and the

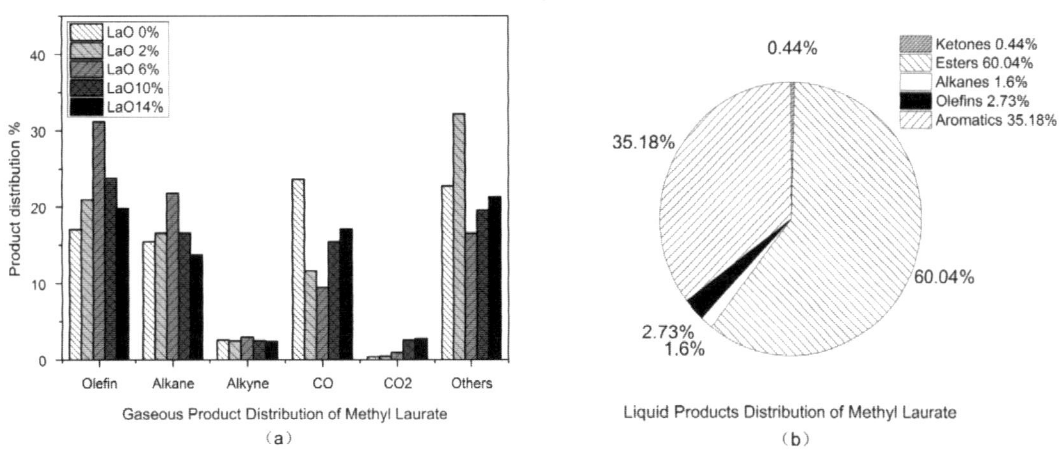

Figure 2. Gaseous (a) and liquid products distribution (b) from catalytic pyrolysis of methyl laurate.

Table 1. The components of gas products from catalytic pyrolysis of oleic acid over various La/ZSM-5 catalysts.

Gases		The components of gas product over different catalysts (%)				
		LaO 0%	LaO 2%	LaO 6%	LaO 10%	LaO 14%
Olefin	C_2H_4	9.52	9.14	10.09	11.16	10.26
	C_3H_6	10.03	12.1	12.74	14.82	13.71
	n-C_4H_8	3.52	3.08	3.36	2.53	2.46
Alkane	CH_4	3.8	2.89	2.9	3.16	3.05
	C_3H_8	10.24	7.08	7.94	5.4	5.89
	n-C_4H_{10}	3.29	4.37	4.34	4.65	4.52
Alkyne	C_2H_2	3.26	2.63	2.72	2.72	2.46
Inorganic Gas	CO	5.55	4.55	0.6	6.85	0.18
	CO_2	0.44	2.94	2.52	6.11	2.27
	Others	33.99	36.24	38.01	23.95	38.08
C_2H_4+C_3H_6+C_4H_8 Selectivity (%)		23.07	24.32	26.19	28.52	26.43
Yield (ml)						
Total gas		677	752	766	797	556
C_2H_4+C_3H_6+C_4H_8		156.20	182.92	200.59	227.29	146.95

selectivity of olefins decreased as the LaO content increased. Compared with the LaO-free ZSM-5 catalyst, the olefin and alkane selectivity of LaO 6% catalyst increased by 14.11% and 6.36%, respectively. The LaO-free catalyst showed the highest CO selectivity (23.63%). The gaseous products and the product selectivity obtained at different La/ZSM-5catalysts are summarized in Table 2. The all gas products contained a higher CO content and lower CO_2 content, indicating that deoxygenation of ester group in methyl laurate mainly followed the decarbonylation mechanism.

Figure 2 (b) shows GC-MS analysis of liquid product obtained from catalytic pyrolysis of methyl laurate over LaO 6% catalyst. As observed in Figure 2(b), the liquid product mainly contained 60.04% esters (including methyl decanoate, methyl laurate, methyl palmitate, methyl oleate and its derivatives), 35.18% aromatic hydrocarbons (including indenes, naphthalenes, toluene, ethylbenzene, paraxylene and its derivatives, etc.), and a small amount of olefins (2.73%), alkanes (1.6%) and ketones (0.44%). There are a high content of esters remaining in liquid product, indicating an incomplete conversion reaction. Compared with oleic acid, liquid products from cracking methyl laurate contained more esters and less aromatic compounds, suggesting that esters are more difficult to be cleaved to form aromatics than aliphatic acids.

3.3. Catalytic conversion of WCO

Figure 3 shows gasous and liquid product distribution from catalytic pyrolysis of WCO over various La/ZSM-5 catalysts. GC analysis (Figure 3(a)) shows that there was the highest olefin selectivity over LaO 6% catalyst, the content of olefins in gas product was 26%, which was 4.24% higher than that of LaO-free ZSM-5 catalyst. The selectivity to olefins decreased as LaO content increased over 6%. The selectivity to alkanes decreased as LaO content in catalysts increased. The gaseous products and the selectivity obtained over different La/ZSM-5 catalysts are summarized in Table 3. As shown in Table 3, a high selectivity to propylene (10.39%) and butylene (9.53%) was obtained, suggesting the high conversion activity of LaO 6% catalyst.

Table 2. The composition of gas products from catalytic pyrolysis of methyl laurate over various La/ZSM-5 catalysts

Gas		The composition of gas products obtained with various catalysts (%)				
		LaO 0%	LaO 2%	LaO 6%	LaO 10%	LaO 14%
Olefin	C_2H_4	5.91	7.28	10.97	8.24	6.9
	C_3H_6	7.89	9.82	15.18	12.18	10.19
	n-C_4H_8	3.25	3.82	5.01	3.37	2.72
Alkane	CH_4	3.35	3.01	3.94	3.21	3.14
	C_3H_8	8.81	9.66	12.07	8.81	6.7
	n-C_4H_{10}	3.26	3.88	5.78	4.54	3.89
Alkyne	C_2H_2	2.54	2.45	2.9	2.46	2.36
Inorganic Gas	CO	23.63	11.63	9.45	15.43	17.09
	CO_2	0.36	0.43	0.92	2.54	2.71
	Others	22.75	32.14	16.53	19.57	21.3
C_2H_4+C_3H_6+C_4H_8 Selectivity (%)		17.05	20.91	31.16	23.8	19.8
Yield (ml)						
Total gas		507	582	625	592	530
C_2H_4+C_3H_6+C_4H_8		86.44	121.73	194.74	140.90	104.96

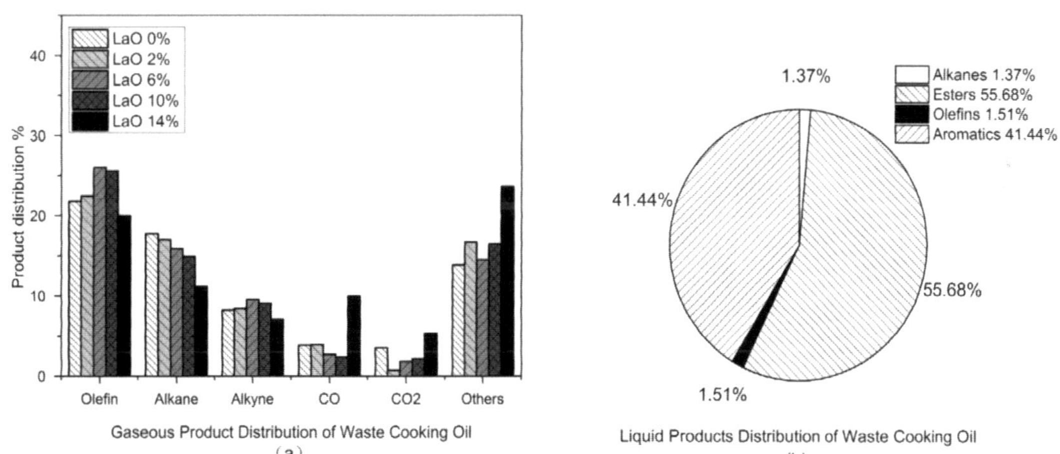

Figure 3. Gaseous (a) and liquid product (b) distributions from catalytic pyrolysis of WCO.

Figure 3 (b) shows the result of GC-MS analysis of liquid product from catalytic pyrolysis of WCO over LaO 6% catalyst. It was found from Figure 3(b) that the liquid products mainly contained 55.68% esters (including methyl caprate, methyl enanthate, dodecanoic acid methyl ester, methyl tetradecanoate, etc.), 41.44% aromatic hydrocarbons (including indenes, naphthalenes, toluene, ethylbenzene, paraxylene and its derivatives, etc.), and a small amount of olefins (1.51%) and alkanes (1.37%). The composition of liquid product was similar to that from catalytic pyrolysis of methyl laurate, and there were a lot of esters remaining in products, indicating a lower conversion of WCO. By comparison, it is more difficult to pyrolyze WCO and methyl laurate than aliphatic acids by using present catalysts.

Table 3. The composition of gas products from catalytic pyrolysis of WCO at various LaO.

Gases		The composition of gas products over various catalysts (%)				
		LaO 0%	LaO 2%	LaO 6%	LaO 10%	LaO 14%
Olefin	C_2H_4	3.84	3.38	3.37	2.86	2.5
	C_3H_6	8.92	9.13	10.39	9.84	7.62
	n-C_4H_8	8.24	8.42	9.53	9.04	7.1
Alkane	CH_4	9.35	9.8	12.57	12.47	10.4
	C_3H_8	10.68	10.38	8.38	7.77	5.09
	n-C_4H_{10}	3.52	3.55	3.06	3.26	1.98
Alkyne	C_2H_2	3.24	3.27	4.16	4.31	3.61
Inorganic Gas	CO	3.84	3.9	2.69	2.32	9.97
	CO_2	3.49	0.74	1.81	2.12	5.34
	Others	13.86	16.71	14.51	16.48	23.63
C_2H_4+C_3H_6+C_4H_8 Selectivity (%)		21.78	22.48	26.02	25.57	20
Yield (ml)						
Total gas		767	792	822	792	547
C_2H_4+C_3H_6+C_4H_8		167.06	178.06	213.87	202.49	109.39

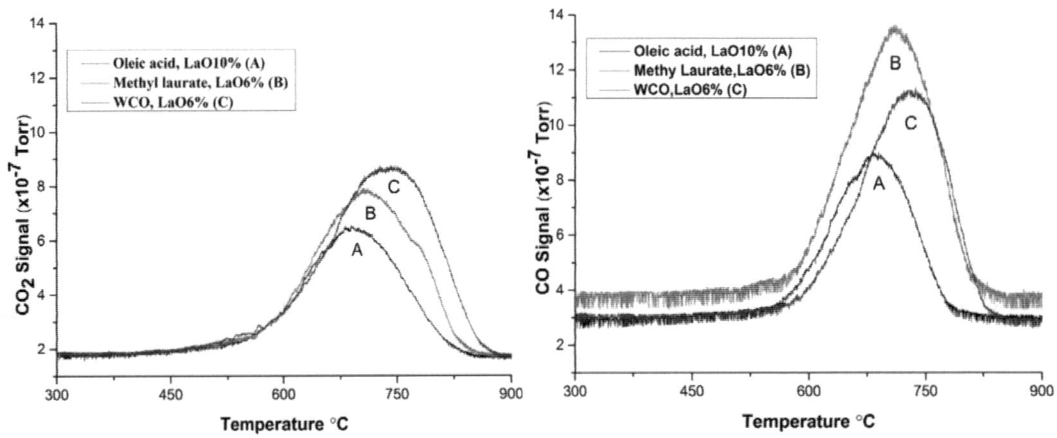

Figure 4. TPO profiles of the three catalysts after cracking of the three oils.

3.4. Coke analysis
The coke deposits and carbon content on catalysts formed during catalytic pyrolysis reactions were analyzed via TPO-RGA system. The result is shown in Figure 4. The order of the temperature corresponding to peak height (Tp) in TPO curves of three oil samples (Figures 4(a) and (b)) are: WCO > methyl laurate > oleic acid. The order of the total amount of CO and CO_2 are: methyl laurate > WCO > oleic acid. These results suggest that more graphite-type carbon with high thermal-stable structure was formed in catalytic pyrolysis process of WCO and methyl laurate. The Tp of coke deposited on LaO6% catalyst (740 °C) shifted towards lower temperatures (670 °C) as the LaO content increased to 10%, indicating that La addition likely causes the decrease in the acidic properties of catalysts. CO and CO_2 are formed due to incomplete and complete oxidation of the deposited carbon, respectively. H_2O and organic gas had not been detected in TPO curves, indicating few H

content of coke. These results suggest that graphite is likely the main component in the coke deposited on present catalysts [10, 11].

4. Conclusions

Light olefins were produced from catalytic pyrolysis of WCO and its model compounds. A high selectivity of light olefins up to 26% was obtained by catalytic pyrolysis of WCO over La/ZSM-5 catalyst with 6% LaO content. The obtained light olefins mainly contained propylene (10.39%) and butylene (9.53%). The liquid composition of WCO pyrolysis product remained a lot of esters (55.68%) which was similar to that from catalytic pyrolysis of methyl laurate, indicating a lower conversion of WCO and methyl laurate by using present catalysts. Graphite is the main component in the coke deposited on present catalysts. In the future research, the effects of several parameters such as reaction temperature and liquid hourly space velocity will be investigated to further improve the light olefins yield. The mass balance analysis of reactants will be conducted and the characteristics of the catalysts will be examined in the future work.

Acknowledgements

The authors are grateful for supports from the National Natural Science Foundation of China (Nos.21266002 and 21566004), the Guangxi Natural Science Foundation (2015GXNSFAA139036) and the Scientific Research Foundation of Guangxi University (Nos. XGZ120081 and XTZ140787).

References

[1] Sadrameli S M 2016 *Fuel* **173** 285
[2] Ramkumar S and Kirubakaran V 2016 *Energy Convers.Manage.* **118** 155
[3] Liu Y C, Savas A J and Avedisian C T 2013 *Fuel* **108** 824
[4] Keske C M H, Hoag D L, Brandess A and Johnson J J 2013 *Biomass Bioenergy* **54** 89
[5] Zhao X, Wei L, Julson J, Qiao Q and Dubey A 2015 *New Biotechnol.* **32** 300
[6] Wu X and Leung D Y 2011 *Energy* **88** 3615
[7] Li H, Yu P and Shen B 2009 *Fuel* **90** 1087
[8] Shena Z, Kea M, Yua P and Hua H 2015 *J. Mol. Catal. A: Chem.* **396** 120
[9] Zhao X, Wei L, Cheng S, Huang Y, Yu Y and Julson J 2015 *Fuel Process. Technol.* **139** 11
[10] Zhang D Z, Li X J, Liu S L, Zhu X X, Xin W J, Xie S J, Zheng P and Xu L Y 2011 *Chin. J. Catal.* **32** 1747
[11] Kim S, Sasmaz E and Lauterbach J 2015 *Appl. Catal. B* **168-169** 212

Optimization of thermal performance of Ranque Hilsch Vortex Tube: MADM techniques

K D Devade[1, 3] **and A T Pise**[2]

[1]India College of Engineering and Management, Pune, India-410506
[2]Dy. Director, DTE, Mumbai, Maharashtra, India -400001

E-mail: kiran.devade@gmail.com

Abstract. Thermal performance of vortex tube is noticeably influenced by its geometrical and operational parameters. In this study effect of various geometrical (L/D ratio: 15, 16, 17, 18; exit valve angle; 30^0, 45^0, 60^0, 75^0, 90^0; cold end orifice diameter: 5, 6 and 7mm, tube divergence angle: 0^0, 2^0, 3^0, 4^0) and operational parameters (inlet pressure: 2 to 6 bars) on the performance of vortex tube have been investigated experimentally. Multiple Attribute Decision Making (MADM) techniques are applied to determine the optimum combination of the vortex tube. Performance of vortex tube was analysed with optimum temperature difference on cold end, COP for cooling. The MADM (Multiple Attribute Decision Making) methods, namely WSM (Weighted Sum Method), WPM (Weighted Power Method), TOPSIS (Technique for Order Preference by Similarity to Ideal Solution) and AHP (Analytical Hierarchy Process) are applied. Experimental best performing combinations are obtained for Length to Diameter ratios 15, 16, 17 with exit valve angle as 45^0,75^0 and 90^0 at orifice diameter 5mm for inlet pressure of 5 and 6 bar pressure. Best COP, efficiency and cold end temperature difference are 0.245, 40.6% and 38.3K respectively for the combination of 15 L/D, 45^0 valve angle, 5mm orifice diameter and 2 bar pressure by MADM techniques.

1. Introduction

The vortex tube is a device that splits compressed air into two different temperature air streams viz. cold and hot. Vortex tube consists of hollow vortex cavity, exit valve, cold end orifice, hot end and entry nozzles. Vortex cavity can be cylindrical, divergent or convergent. Cold end consists of orifice and nozzles for supplying compressed air. On the hot end, exit valves are placed to vary the temperature and cold mass fraction. The advantages of vortex tube are constructional simplicity, less cost, easy repairs, smaller size, light weight, quick response; capability to reach a mark temperature immediately. Vortex tubes are majorly used for plastic blow molding, spot and panel cooling, vacuum forming, cleaning, drying, separating gas mixtures, DNA application and liquefying natural gas. [1]

Vortex tube invented by Ranque [2] and Hilsch [3], works on compressed air and provides two different temperature streams as outlet. When compressed air enters the vortex cavity through tangential

[3] Address for correspondence: K D Devade, India College of Engineering and Management, Pune, India-410506. Email: kiran.devade@gmail.com.

nozzles, air expands on its entry and attains high velocity. This high velocity air starts moving towards the hot end inside the tube. This movement of air is called as swirl flow. As air reaches hot end the valve on hot end restricts the air flow. The pressure on valve end increases slightly and the reversal of air flow takes place. The air now flows to cold end of the tube through tube center forming two streams of air. The air stream at periphery is compressed by this central layer and there is energy transfer from central layer to peripheral layer. This way the central air gets cooled and the peripheral layer gets heated and two streams are obtained at different temperatures.

Saidi and Valipour [4] conducted experiments with vortex tubes with different L/d ratios and tube geometries to investigate the effect of geometry on the operational characteristics of vortex tube, the major investigation was; optimum value of L/d is in the range $20 \leq L/d \leq 55.5$. C. Gao [5] obtained cold air temperature difference of 28°C at L/d ratio equal to 64.7, when the tested tubes had L/d ratio equal to 8, 32.7 and 64.7.

Pourmahmoud and Bramo [6] deduced that the best cold air temperature difference of 43.96 K is obtained when the length to diameter ratio was 9.3 among the experiments on six different tubes of 8, 9.3, 10.5, 20.2, 30.7, and 35 L/d ratio. Aydin et al. [7] Experimented on four different tubes with L/d 10, 20, 30 and 40. It was reported that tube with L/d 30 gives maximum cold air temperature difference of 45.9K. The major intention of study by Kirmaci [8] was to investigate the effect of the nozzle number. It was investigated that using 2 nozzles produce best result. Similarly Polat and Kirmaci [9] conducted experiments with 5 different nozzle no. (2, 3, 4, 5 and 6) and working fluid used was Air, O2, N2 and Ar to conclude that maximum temperature difference is obtained for 2 nozzles with air. Promvonge and Eiamsa-ard [10] experimented to investigate effect of orifice diameter. The used orifice diameter was in the range of 0.6d to 0.9d and it was concluded that 0.5d orifice yields the highest temperature reduction. Prabhakaran and Vaidyanathan [11] experimented for orifice diameter and concluded that minimum cold air temperature is obtained for 0.5d. Nimbalkar and Muller [12] investigated experimentally and numerically that cold orifice diameter of 0.5d is responsible for maximum energy separation. For investigation of nozzle diameters Prabhakaran and Vaidyanathan [13] performed experiments, but they couldn't establish relationship between nozzle diameter and tube diameter. Markal et al. [14] Tested the effect of the exit valve angles and concluded that effect of valve angle is generally negligible. Experiments performed by Devade and Pise [15] state that 450 and 600 valve angle produce best cooling and heating respectively. While on the other hand Dincer et al. [16] Reported best performance at angle of 300 and 600. Experiments by Chang et al. [17] were based on the influence of divergence angle on the performance of vortex tube. The experimental result show that performance of vortex tube is enhanced by using a divergent tube and 4° divergent angle yields the highest temperature reduction. CFD analysis was done to consider L/D ratio as design criteria by Pour Mahmoud et.al [18] with focus on stagnation point and length of tube

Literature review states that performance of vortex tube varies considerably with changes in L/d, d_o N, φ, d_n and working fluid. The range of selected geometry as used in experiments is far wide. For ex. L/d ratio is taken then the ranges used are from 0.6 to hundreds. Close end study has not been conducted; this prompted to make use of L/d ratio in close range of 15 to 20. The literature also says that valve angle has very limited effect on performance to validate this valve angles are chosen in steps of 150 from 300 to 900. As mentioned in Nimbalkar et al. [12] Optimum performance of tube is at d/d_o equal to 0.5 to validate this $d_o$is selected as 5, 6, and 7 mm where 6 mm represents d_o/d equal to 0.5. The divergence angle has also been set to 00and 40. The purpose is to combine most of the parameters from the literature for getting true optimum performance. Vortex tube has the potential to replace conventional refrigeration system and get commercialized to be implemented in number of applications. Optimization studies have also been conducted by many researchers, Ersoyogulc et.al [19] used Rule-Based Mamdani-Type Fuzzy modelling for optimization. Suresh Kumar et.al [20] used taguchi approach for optimizing the performance of vortex tube. ANN is used by Uluer et.al [21] for modelling performance of counter flow vortex tube. Graphical and experimental optimization by Devade and Pise [22] on vortex tube for geometry and operational parameters. Pinar et.al [23] applied taguchi method for assessment of performance of vortex tube. In the present study MADM methods are used for optimization of geometrical combinations.

2. Experimental method

The experimental study of selected parameters is done using the setup shown in Figure 1 The experimental setup consist of a compressor, an air reservoir, pressure regulator , Rota-meters for measuring the flow rates of inlet air and cold air, pneumatic pipes, connectors, vortex tube, digital temperature indicator and thermocouples. The details of measuring instruments are as given in table 1. [1]

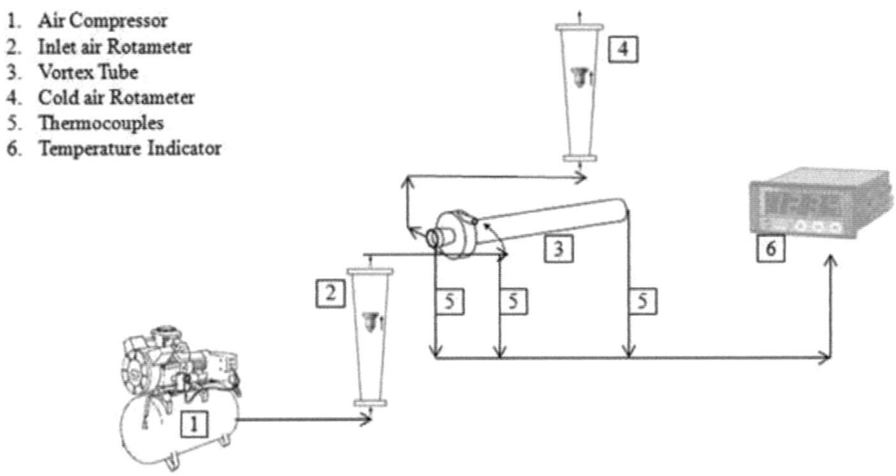

Figure 1. Experimental setup.

Table 1 Instrument details **[1]**.

Instrument	Range	Accuracy
Rota meters	0 to 500LPM	± 1 LPM
Pressure Regulator	0 to 10 bar	±0.1bar
RTD	-50 to 150°C	±0.1^{O}c

2.1. Data reduction

Cold Mass Fraction (CMF) is the ratio of mass of cold to total mass of air entering the vortex tube.

$$CMF = \frac{m_c}{m_i} \quad (1)$$

Coefficient of Performance (COP) is the ratio of cooling effect produced to the energy input required by the compressor.

$$COP_{act} = \frac{R_E}{W_{comp}} \quad (2)$$

Refrigeration effect/ cooling effect of the vortex tube is the total enthalpy drop in air emerging from cold end.

$$R_E = m_c c_p(t_i - t_c) \quad (3)$$

Work required by compressor is the isothermal work being minimum assuming the compression as isothermal is given by

$$W_{comp} = m_i R t_i ln\frac{p_d}{p_i} \quad (4)$$

The cold end temperature drop is the difference in temperature at inlet and temperature of air at cold end.

$$\Delta t_c = t_i - t_c \quad (5)$$

The temperature drop without vortex effect, because of pure expansion is static temperature which is given by

$$\Delta t_c' = t_i\left[1 - \left(\frac{p_d}{p_i}\right)^{\frac{\gamma-1}{\gamma}}\right] \quad (6)$$

The relation between the static temperature drop and actual temperature drop is given by

$$\Delta t_{rel} = \frac{\Delta t_c}{\Delta t_c'} \quad (7)$$

The adiabatic efficiency is proportional to product of CMF and relative temperature drop as

$$\eta_{ad} = CMF(\Delta t_{rel}) \quad (8)$$

Efficiency of compression is the only input parameter and is calculated as

$$\eta_{comp} = \frac{ln\frac{p_d}{p_i}}{\frac{\gamma}{\gamma-1}\left[\left(\frac{p_d}{p_i}\right)^{\frac{\gamma-1}{\gamma}} - 1\right]} \quad (9)$$

Theoretical COP can be calculated by

$$COP_{th} = \eta_{ad}\eta_{comp}\left(\frac{p_d}{p_i}\right)^{\frac{(\gamma-1)}{\gamma}} \quad (10)$$

The tube is experimented to record the temperatures and mass flow rates for various combinations as listed in Table 2. Experimental parameters.

Table 2. Experimental parameters.

Parameters	
Tube L/D ratio	15,16,17,18
Tube Divergence Angle	0 and 4°
Number of Nozzles	2 No's
Orifice Diameters	5,6,7 mm
Valve Angles	30^0, 45^0, 60^0, 75^0 and 90^0
Pressure at Inlet	2 to 6 bars in step of 1 bar

Experiments are performed systematically with the parameters used is given in table 3. The sets are formed for illustration following the sequence of $L/d - \emptyset - d_o - \theta - p_i$. i.e. 15-0-5-30-2 to 6

Table 3. Experiment sets of vortex tube.

L/D-Div. Angle-Orifice Diameter-Valve Angle-Inlet Pressure
15-0-5-30-2 to 6; 15-0-5-45-2 to 6; 15-0-5-60-2 to 6; 15-0-5-75-2 to 6; 15-0-5-90-2 to 6.
15-0-6-30-2 to 6; 15-0-6-45-2 to 6; 15-0-6-60-2 to 6; 15-0-6-75-2 to 6; 15-0-6-90-2 to 6.
15-0-7-30-2 to 6; 15-0-7-45-2 to 6; 15-0-7-60-2 to 6; 15-0-7-75-2 to 6; 15-0-7-90-2 to 6.
15-4-5-30-2 to 6; 15-4-5-45-2 to 6; 15-4-5-60-2 to 6; 15-4-5-75-2 to 6; 15-4-5-90-2 to 6….. For all combinations.

2.2. Observations

Using the coded system given earlier the observations for temperature, pressure flow meter are recorded for all tubes with various L/d ratios and divergence angle $\emptyset$ for various $d_o$and θ observations are made for all pressures p_i. The data is analyzed using the relations provided earlier forCMF, COP, R_E, W_{comp}. The analytical results are discussed below.

3. Results and discussion

The experiment is performed with the empirical relations provided in literature with some ranges of L/d to decide the optimum geometrical combinations under provided conditions. The results are presented in terms of effect of L/d, d_o and effect of θ on COP andCMF.

3.1. Effect of L/D at specific Θ on COP and Δt_c.

The figure 2-6 show the effect of L/d ratio at a specific valve angle θ on COP andΔt_c. It can be seen that the combination of L/d and θ has mixed effects, it is observed that COP increases for some combinations of L/d and θ and decreases for other combinations.

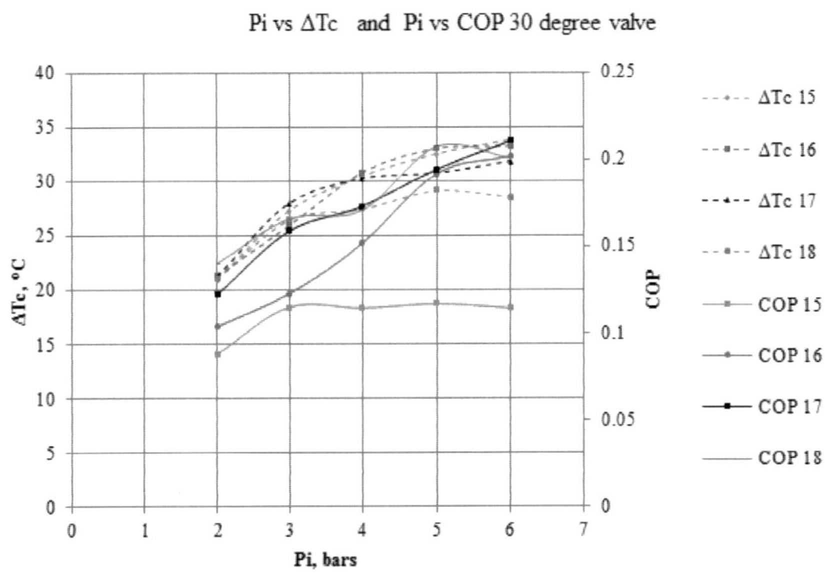

Figure 2. Effect of L/D on COP and Δtc for 30^0 valves.

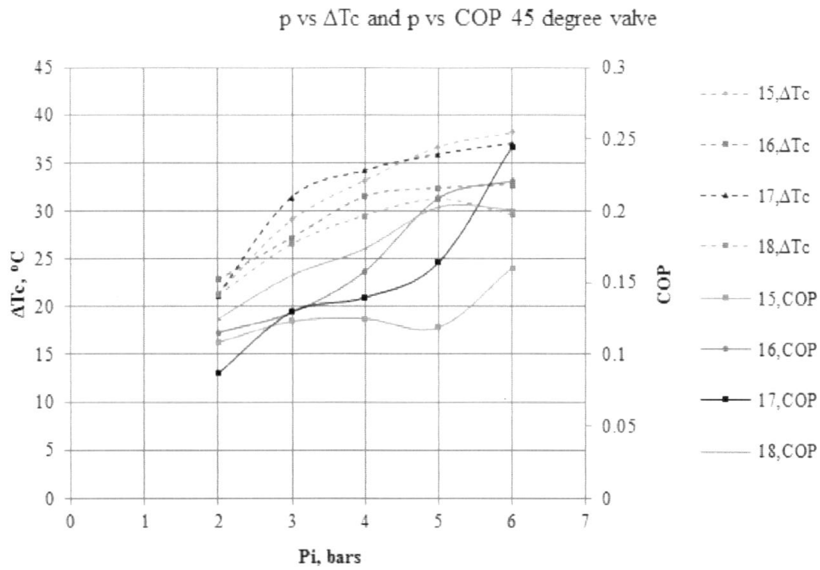

Figure 3. Effect of L/D on COP and Δtc for 45^0 valves.

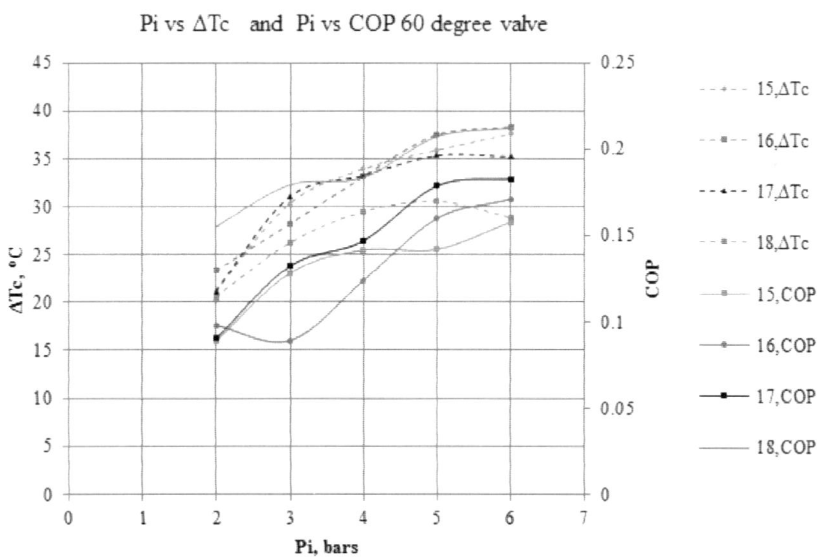

Figure 4. Effect of L/D on COP and Δtc for 60^0 valves.

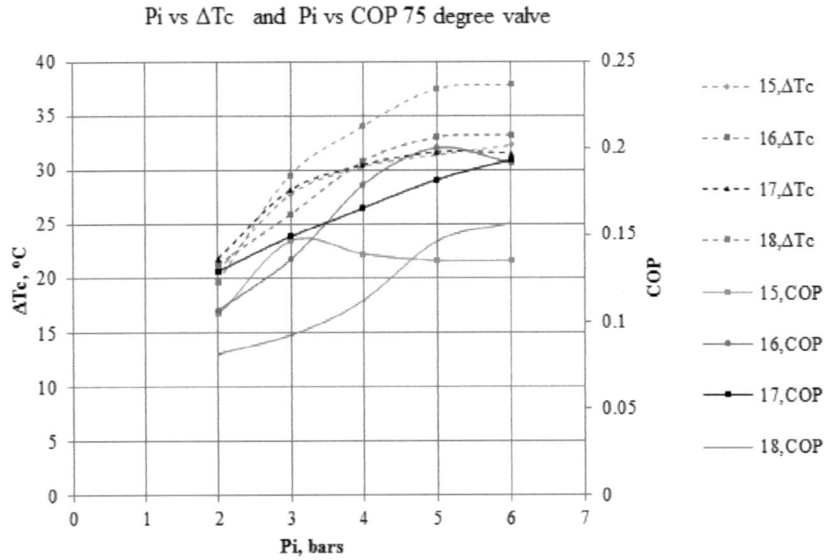

Figure 5. Effect of L/D on COP and Δtc for 75^0 valves.

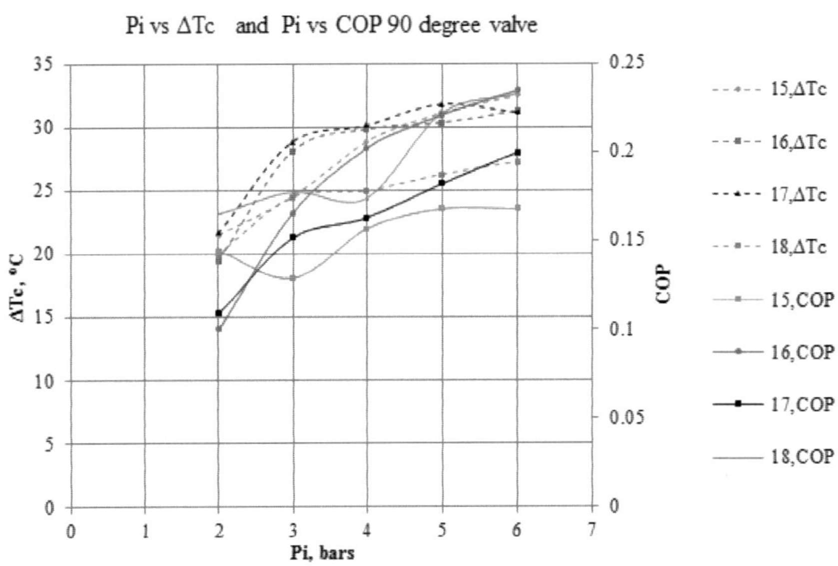

Figure 6. Effect of L/D on COP and Δtc for 90^0 valves.

The mechanism of energy separation acts between peripheral vortex flow of air and the reversal of air flow at hot end. On hot end as the flow is restricted by valves based on the angle of valve the flow reversal and mass of air getting reversed changes. For more flat surface of valve more mass of air is reversed but since the peripheral air mass is reduced the energy transfer from central layer to peripheral layer is reduced. As length of tube increases the transaction length for heat exchange also increase between two layers.

The length of tube also has effect on energy separation in this way. But on hot end flow reversal starts after stagnation point. Stagnation point usually gets located inside the tube and it is away from the flat surface of valve. It is logical to say that as length of tube increases the energy transfer should increase, but stagnation point puts limitation on this. With increase in length the velocity of peripheral decreases and this may be the reason that length of stagnation region also increases. Because of this the mass of air getting reversed gets hampered. This might be the reason for mixed results getting produced. Thus it cannot be said that a definite combination of L/d and θ produces better result. Experimental results show that as p_i increase all tubes produce considerably good COP but still the COP obtained is less than unity. The reason for this is changes in cold mass of air.

3.2. Effect of orifice diameter d_o:

To analyze the effect of orifice diameter d_o as well as valve angles on COP and CMF, a ratio called as filling ratio is defined. This ratio considers the volume blocked by the valve and volume available inside the cone. The blocked volume is calculated as ratio of valve volume V_{valve} to cone volume V_{cone}. This represents blocked volume or filled volume in percentage.

Figures 7-10 show the Effect of orifice diameter d_o on COP and CMF for various L/d ratios it can be observed that as d_o increases the COP and CMF also increases. This confirms the results of Nimbalkar *et.al.* [12]

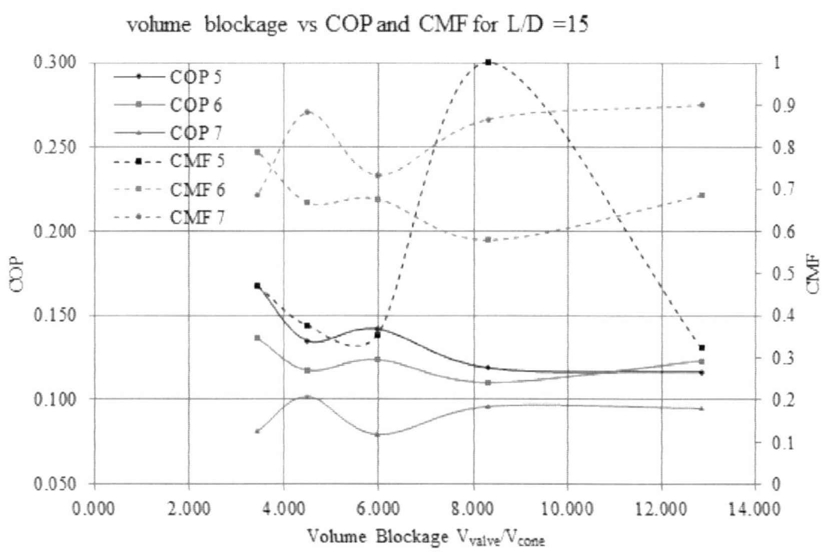

Figure 7. Effect of Orifice diameter on COP and CMF for L/D =15

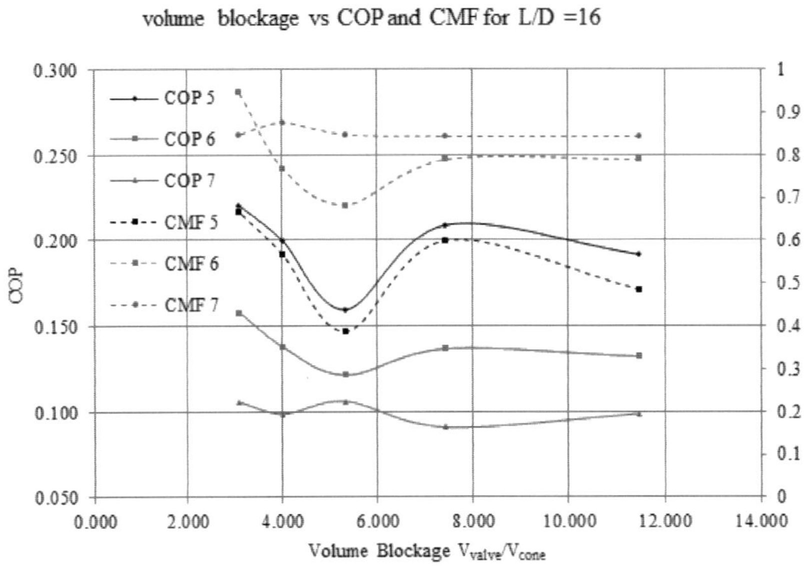

Figure 8. Effect of Orifice diameter on COP and CMF for L/D =16

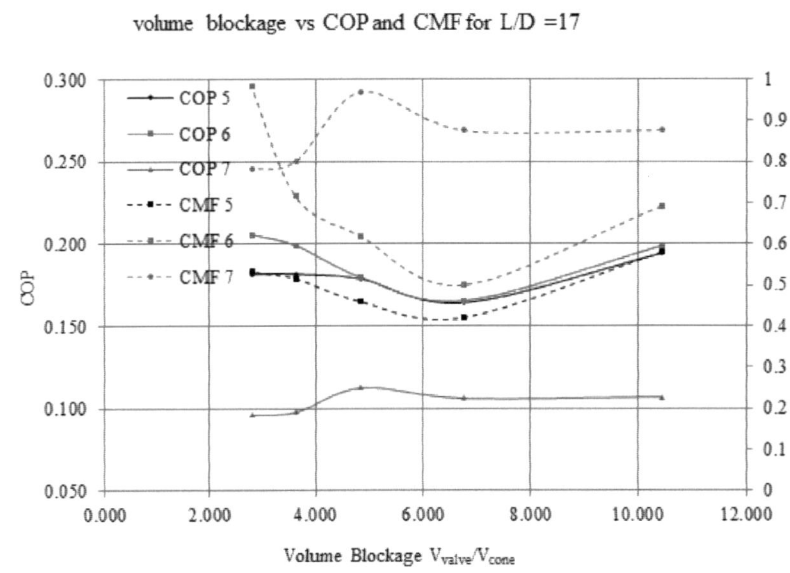

Figure 9. Effect of Orifice diameter on COP and CMF for L/D =17

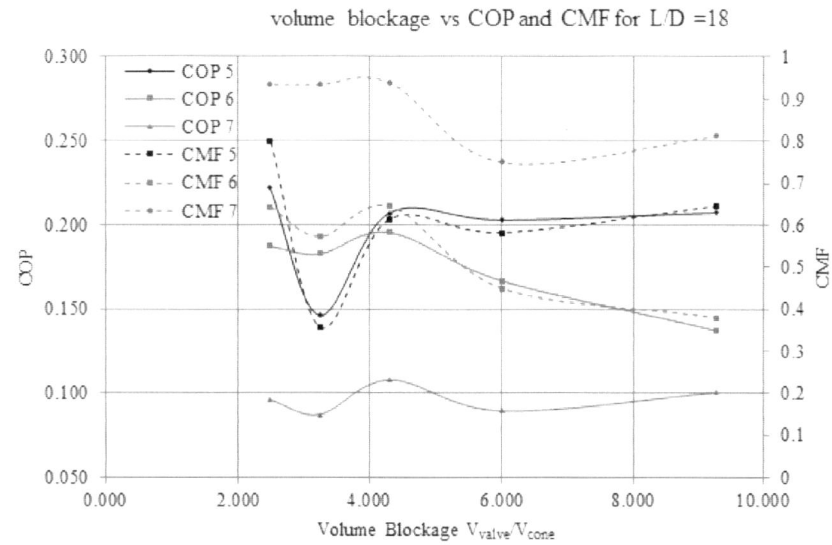

Figure 10. Effect of Orifice diameter on COP and CMF for L/D =18

The major reason is when the ratio of d_o/d is less than 0.5 the cold mass of air at periphery is restricted at the walls of orifice. Because of this restriction the cold air gets mixed with the fresh air. This also produces secondary circulation on orifice end. As d_o/d is 0.5 or more back flow is reduced and cold air comes out of the orifice without reversal. This helps to increase mass of cold air at exit and thus improves COPandCMF. This is the probable reason that increase in orifice diameter is addressed and that to in relation of $d_o/d \geq 0.5$. The results also show the change in $d_o$with L/d gives mixed results i.e. for some L/d ratios $d_o/d \leq 0.5$ produces good result for COP and CMF while with change in L/d for $d_o/d \geq 0.5$ produces good for COP and CMF. This mixed nature of the results prompt to make use of MADM methods to select the best combination for getting better COPandCMF.

4. MADM (Multiple attribute decision making)

MADM is a decision making method when there are multiple parameters affecting the end output. MADM methods are discrete with a limited number of pre specified alternatives. These methods require both intra and inter- parametric comparisons for unbiased judgment or decision for the considered problem. [24]

The decision making system in MADM methods is based on four main components, namely: alternatives (Set of experiments), attributes (resulting parameters), weight (index of influence for variation), or strength of each resulting parameter (attribute) and measure of performance of each set of experiment (alternative) with respect to the others. Some of the commonly used methods under MADM are Weighted Sum Method (WSM), Weighted Power Method (WPM), Analytic Hierarchy Process (AHP) and Technique for Order Preference by Similarity to Ideal Solution (TOPSIS). [24]

4.1. Weighted sum method (WSM)

WSM was introduced by Fishburn in 1967 [25]. This is the simplest method. Here each resulting parameter (attribute) is given a weight and sum of all weights must be equal to 1. Each set of experiment (alternative) is assessed with regard to every resulting parameter (attribute). The data presented for each set of experiment is normalized based on beneficial or non-beneficial parameter. The composite score of a set of experiment is given by [26]

$$C_i = \sum_{j=1}^{N} w_j (n_{ij})\, norm \tag{11}$$

4.2. Weighted Power Method (WPM)

This method is similar to WSM. The main difference is that instead of addition, there is multiplication. This method is introduced by Miller and Star [27]. The composite score is given by

$$C_i = \prod_{j=1}^{N} [(n_{i,j}) norm]^{w_j} \tag{12}$$

4.3. Analytic hierarchy process (AHP)

It is one of the popular techniques introduced by Satty [28]. It decomposes a decision making problem into an organized orders of objectives, resulting parameter (attributes). AHP can proficiently deal with objective as well as subjective features. The main procedure of using AHP using the radical root method (Geometric Mean Method, GMM) is as follows.

- Step 1: Determine the objectives and evaluation attributes
- Step 2: Determine relative importance of different attributes. This step comprises of relative comparison of attributes against attributes this becomes matrix A1. The relative importance weights are given as designed by Satty [28] for the attributes comparison.

$$\begin{array}{c} \\ b_1 \\ b_2 \\ b_3 \\ b_4 \end{array} \begin{array}{c} \begin{array}{cccc} b_1 & b_2 & b_3 & b_4 \end{array} \\ \begin{bmatrix} 1 & b_{12} & b_{13} & b_{14} \\ b_{21} & 1 & b_{23} & b_{24} \\ b_{31} & b_{32} & 1 & b_{34} \\ b_{41} & b_{42} & b_{43} & 1 \end{bmatrix} \end{array} = B_{N\,x\,N} \tag{13}$$

- Step 2.1 Find relative normalized weight of each attribute by calculating mean of i-th row and as

$$GM_j = [\prod_{j=1}^{N} b_{ij}]^{1/N} \tag{14}$$

- Step 2.2 Normalise the geometric mean of rows in the comparison matrix

$$w_j = GM_j / \sum_{j=1}^{N} GM_j \quad \text{..This becomes matrix A2.} \tag{15}$$

- Step 2.3 Calculate matrices A3 and A4 such that

A3 = A1*A2 and A4 = A3/A2

- Step 2.4 Determine the maximum Eigen value λ_{max} that is the average of the matrix A4.
- Step 2.5 Calculate consistency Index CI as

$CI = (\lambda_{max} - N)/(N-1)$The smaller the value of CI the smaller is the deviation.

- Step 2.6 Calculate consistency ratio CR as

$CR = CI/RI$, RI is random index, given by Satty.

- Step 3: compare the alternatives pair wise with respect to each attribute.
- Step 4: Obtain overall or composite scores for the alternatives by using WSM, WPM method. [24]

4.4. Technique for order preference by similarity to ideal solution (TOPSIS)

TOPSIS was developed Hwang and Yoon [29]. This method is based on the concept that the chosen alternative should have the shortest Euclidean distance from the ideal solution and the farthest negative ideal solution. The main procedure of the TOPSIS for selection of best alternative from the available is as follows [30, 31]:

- Step 1: determine objectives and evaluation attributes.
- Step 2: Establish decision matrix comprised of attributes and alternatives.
- Step 3: Establish normalized matrix as

$$R_{ij} = n_{ij}/(\sum_{j=1}^{N} n_{ij}^2)^{1/2} \tag{16}$$

- Step 4: Decide relative importance weights using either AHP or any other suitable method
- Step 5: obtain normalized weighted matrix as

$$B_{ij} = w_j R_{ij} \tag{17}$$

- Step 6: obtain ideal best and worst solutions as B^+ and B^-
- Step 7: obtain separation measures as Euclidean distance (the distance of the solution from ideal or worst solution) from the ideal solution as

$$S_i^+ = \left[\sum_{j=1}^{N} \left(B_{ij} - B^+\right)^2\right]^{1/2} \tag{18}$$

$$S_i^- = \left[\sum_{j=1}^{N} \left(B_{ij} - B^-\right)^2\right]^{1/2} \tag{19}$$

- Step 8: The relative closeness of a particular solution from ideal solution is given by the Euclidean distance. (Here as S_i^- appears in numerator the higher the distance from the worst solution the closer is the solution to ideal one.)

$$p_i = (S_i^- / S_i^+ + S_i^-) \tag{20}$$

- Step 9: A set of alternatives is generated in the descending order in this step to indicate most preferred and least preferred solutions. [32]

5. Application of MADM

The decision matrix in the given problem consists of total 375 combinations presented as $L/d - \emptyset - d_o - \theta - P_i$. Where L/d takes values as 15,16,17,18 $\emptyset$ varies as 0 and $4^O d_o$varies from 5, 6, 7mm θ varies in steps of 15^0 in the range 30 to 90 and inlet pressure varies in step of 1 bar from 2 to 6. Thus in total there were many alternatives to choose the best from. The attributes are cold end temperature t_c, COPandCMF. MADM methods are used since the experimental results do not lead to predict the combination of parameters producing best result. The results are conflicting.

The decision matrix for illustration is given in table 4. The table provides Final set of values selected from the best performance of each L/d ratio. The selection is based on MADM methods. Here Out of the best performing combinations MADM will be used to order the tubes in hierarchy. The range of all parameters is as shown in table 2.

Table 4. Decision Matrix for Illustration.

	TC	CMF	COP
15-0-6-45-2	66.4	0.923076923	0.103644
15-0-7-90-4	52.2	0.896551724	0.090148
15-0-7-45-3	62.7	0.846153846	0.093995
15-0-7-90-6	49	1	0.083057
15-0-7-90-5	50.6	0.941176471	0.083068
15-4-5-45-2	48.1	1	0.305182
15-4-5-45-3	55.5	1	0.230179
15-4-5-45-4	59.9	1	0.190488
15-4-5-45-5	63.4	1	0.169185
15-4-5-45-6	64.8	1	0.149352
16-4-6-90-5	46.9	0.947368421	0.106654
16-4-6-90-4	45.1	0.8	0.104375

16-4-6-30-6	54.5	0.925	0.09264
16-4-7-75-4	41.8	0.928571429	0.095466
16-4-6-45-6	54.5	0.857142857	0.088922
17-4-6-90-2	40.7	0.756756757	0.14284
17-4-6-90-3	44.5	0.904761905	0.122115
17-4-6-90-4	45.5	0.979591837	0.102501
17-4-6-90-5	47.2	0.98245614	0.092424
17-4-7-60-5	40.4	0.967741935	0.093052
18-4-5-90-2	41	0.5	0.139263
18-4-7-90-3	31	0.916666667	0.100402
18-4-5-45-2	50.4	0.380952381	0.116153
18-4-6-60-6	44.8	0.806451613	0.087671
18-4-5-45-5	59	0.580645161	0.083825

5.1. Using AHP method

AHP method is used to find out relative importance weight of each attribute. This is done by comparing the attributes against attributes and giving scale of importance. The A1 matrix is formed by comparing attributes t_c, COP and CMF against each other and allocating relative importance like when t_c is compared with COP then COP is more important that t_c, and when t_c is compared with CMF then $CMF$is more important than t_c but less important than COP.

$$\begin{array}{cc} & \begin{array}{ccc} t_c & CMF & COP \end{array} \\ A1 = \begin{array}{c} t_c \\ CMF \\ COP \end{array} & \begin{bmatrix} 1 & 1/3 & 1/5 \\ 3 & 1 & 1/3 \\ 5 & 3 & 1 \end{bmatrix} \end{array}$$

The Geometric mean matrix of the above relative importance is given by Equation. 14

$$GM_j = \begin{bmatrix} 0.4055 \\ 1 \\ 2.4662 \end{bmatrix}$$

The weights of the attributes which makes the Matrix A2 is given by Equation. 15

$$A2 = w_j = \begin{bmatrix} 0.1047 \\ 0.2583 \\ 0.6370 \end{bmatrix}$$

The matrix A3= A2*A1 is given by,

$$A_3 = \begin{bmatrix} 0.3182 \\ 0.7848 \\ 1.9355 \end{bmatrix}$$

The matrix A4=A3/A2 is given by

$$A_4 = \begin{bmatrix} 3.0385 \\ 3.0385 \\ 3.0385 \end{bmatrix}$$

The maximum Eigen value λ_{max} that is the average of the matrix A4 is 3.0385. Consistency index is $= 0.0192$, for R=0.52 as given by Satty [28] for number of attributes =3the consistency ratio is $CR = 0.03703$, the consistency ratio is much less than 0.1hence the weights are acceptable for the analysis and the decided weights are, $w_{tc} = 0.10147$ $w_{cmf} = 0.2583$ $w_{cop} = 0.6370$

Now we apply the MADM methods to sort among the decision matrix for the best alternative. For doing this the normalized matrix is formed on the basis of beneficial and non-beneficial attributes. Here all T_c, COP and CMF are beneficial attributes i.e. we have to have the maximum values for all attributes. The normalized matrix is calculated and is given as,

$$\begin{array}{c} 15-0-6-45-2 \\ 15-0-7-90-4 \\ 15-0-7-45-3 \\ . \\ . \\ . \\ 18-4-6-60-6 \\ 18-4-5-45-5 \end{array} \begin{bmatrix} 1 & 0.9231 & 0.3396 \\ 0.7861 & 0.8966 & 0.2954 \\ 0.9443 & 0.8462 & 0.3080 \\ . & . & . \\ . & . & . \\ . & . & . \\ 0.6747 & 0.8065 & 0.2873 \\ 0.8886 & 0.5806 & 0.2747 \end{bmatrix}$$

The normalized data is used to calculate composite scores of each alternative. For this WSM, WPM and TOPSIS is used the TOPSIS normalized matrix is calculated using Equation. 16.

$$\begin{array}{c} 15-0-6-45-2 \\ 15-0-7-90-4 \\ 15-0-7-45-3 \\ . \\ . \\ . \\ 18-4-6-60-6 \\ 18-4-5-45-5 \end{array} \begin{bmatrix} 0.2597 & 0.2078 & 0.1556 \\ 0.2041 & 0.2019 & 0.1354 \\ 0.2452 & 0.1905 & 0.1412 \\ . & . & . \\ . & . & . \\ . & . & . \\ 0.1752 & 0.1816 & 0.1317 \\ 0.2307 & 0.1307 & 0.1259 \end{bmatrix}$$

TOPSIS normalized weighted matrix is as follows for TOPSIS using Equation. 17.

$$\begin{array}{c} 15-0-6-45-2 \\ 15-0-7-90-4 \\ 15-0-7-45-3 \\ . \\ . \\ . \\ 18-4-6-60-6 \\ 18-4-5-45-5 \end{array} \begin{bmatrix} 0.02719 & 0.05368 & 0.09914 \\ 0.02137 & 0.05214 & 0.08623 \\ 0.02568 & 0.04921 & 0.08991 \\ . & . & . \\ . & . & . \\ . & . & . \\ 0.01834 & 0.0469 & 0.08386 \\ 0.02416 & 0.03376 & 0.08018 \end{bmatrix}$$

5.2. WSM method

The overall or composite score of an alternative for WSM method is given by equation 11 and the alternatives are ordered in descending order.

5.3. WPM method

The overall or composite score of an alternative for WPM method is given by equation 12 and the alternatives are arranged in descending order.

5.4. TOPSIS method

Using the TOPSIS normalized weighted matrix presented above and calculating the separators as to distinguish the alternatives on the basis of Euclidean distances, equation 18 and 19 are used to check how far the solution is from the worst and the ideal solution. Equation 20 gives the Euclidean distance from the worst solution. Hence the maximum the value the close the solution is to ideal solution. Hence the values are arranged in descending order. Composite scores for WSM, WPM and TOPSIS methods for the nonlinear data are calculated and the results are arranged descending order. Table 5 shows the results and composite scores in descending order. The results give the best possible combination for getting optimized solution from the experimental data.

Table 5. MADM Comparison of results.

Preference	L/d-ϕ-do-Θ-pi	TOPSIS		SAW		WPM
1	15-4-5-45-2	0.966408	15-4-5-45-2	0.971136001	15-4-5-45-2	0.966798
2	15-4-5-45-3	0.669491	15-4-5-45-3	0.826259727	15-4-5-45-3	0.820008
3	15-4-5-45-4	0.499512	15-4-5-45-4	0.750355214	15-4-5-45-4	0.732705
4	15-4-5-45-5	0.411276	15-4-5-45-5	0.711410737	15-4-5-45-5	0.683445
5	15-4-5-45-6	0.332415	15-4-5-45-6	0.672222132	15-4-5-45-6	0.632708
6	17-4-6-90-2	0.281886	15-0-6-45-2	0.559476316	17-4-6-90-2	0.545072
7	18-4-5-90-2	0.251621	17-4-6-90-3	0.55875684	17-4-6-90-3	0.521417
8	17-4-6-90-3	0.216663	17-4-6-90-2	0.55779273	15-0-6-45-2	0.492343
9	16-4-6-90-5	0.175486	16-4-6-90-5	0.541274842	16-4-6-90-5	0.486726
10	15-0-6-45-2	0.171441	17-4-6-90-4	0.53872136	18-4-5-90-2	0.482264
11	17-4-6-90-4	0.17059	17-4-6-90-5	0.521110425	17-4-6-90-4	0.477168
12	17-4-6-90-5	0.152734	16-4-6-30-6	0.518235673	16-4-6-90-4	0.457684
13	18-4-7-90-3	0.152292	15-0-7-45-3	0.513631846	15-0-7-45-3	0.44964
14	18-4-5-45-2	0.150331	15-0-7-90-6	0.508929398	16-4-6-30-6	0.449229
15	17-4-7-60-5	0.149248	17-4-7-60-5	0.507895722	17-4-6-90-5	0.448787
16	15-0-7-90-6	0.147381	16-4-7-75-4	0.505024717	16-4-7-75-4	0.445805
17	16-4-7-75-4	0.145738	15-0-7-90-4	0.502058386	18-4-7-90-3	0.444682
18	16-4-6-30-6	0.144357	15-0-7-90-5	0.496282828	17-4-7-60-5	0.441718
19	16-4-6-90-4	0.143586	16-4-6-90-4	0.495616985	15-0-7-90-4	0.435972
20	15-0-7-90-5	0.136321	18-4-7-90-3	0.49521934	16-4-6-45-6	0.429129
21	15-0-7-45-3	0.135814	16-4-6-45-6	0.492947036	15-0-7-90-6	0.42283
22	15-0-7-90-4	0.134333	18-4-5-90-2	0.48448434	15-0-7-90-5	0.417699
23	16-4-6-45-6	0.125981	18-4-6-60-6	0.461945265	18-4-6-60-6	0.410126
24	18-4-6-60-6	0.109951	18-4-5-45-2	0.420326241	18-4-5-45-2	0.409235
25	18-4-5-45-5	0.071188	18-4-5-45-5	0.417992794	18-4-5-45-5	0.376862

Thus the best combination revealed by all these methods is 15-4-5-45-2, 15-4-5-45-3, 15-4-5-45-4, 15-4-5-45-5, and 15-4-5-45-6. Thus it can be concluded that the L/d 15 at 4 degree divergence angle and 5 mm orifice diameter at 45 degree conical valve angle at all inlet pressures performs best.

6. Conclusion

Vortex tube is tested with reference to the literature data experimentally for optimum performance under the tested condition. It has produced lowest cold end temperature of -14.8^{0}C and COP 0.305 with maximum CMF as 1. The tube has shown best results for 4^0 divergence angle as compared to plain tube tested at 0^0. The Nimbalkar relation of $d_o/d \geq 0.5$ is also verified for obtaining maximum CMF. Application of MADM methods like WSM, WPM and TOPSIS have come up with the optimum tube combinations as, $L/d\,15, \emptyset - 4, d_o - 5, \theta - 45$, at all inlet pressures$p_i$. The limitations of the study so far are the tube has been tested with only 2 entry nozzles, hence in future study can also be conducted with increasing number of nozzles, N_n and the close range of divergence angle, $\emptyset$. If the results can be optimized with these vortex tube may find wider range of applications in the commercial field.

Nomenclature

Capital Letters

A	matrix	
B	Ideal and worst values,	
b	row column element	
C	composite score	
c	specific heat	[$KJKg^{-1}K^{-1}$]
CI	Consistency Index	[Non Dimensional]
CMF	Cold Mass Fraction	[Non Dimensional]
COP	Coefficient of Performance	[Non Dimensional]
CR	Consistency Ratio	[Non Dimensional]
d	Diameter of Tube	[mm]
GM	Geometric Mean	
L	Length of Tube	[mm]
m	Mass of air	[Kg]
N	Matrix Size	
P	Pressure	[Kpa]
R	Row element	
RE	Refrigeration effect	[$KJsec^{-1}$]
RI	Random Index	
S	Euclidean Separation distance	
T	Temperature	[K]
ΔT	Temperature Difference [K]	
V	volume	[m^3]
W	Compressor Work	[$KJsec^{-1}$]
w	weightage	

Greek Letters

$\emptyset$	Divergence angle
Θ	Conical valve angle
λ	maximum Eigen value
η	efficiency

Subscripts

ad	adiabatic
c	cold end
cone	divergent cone
comp	compressor
d	discharge condition
i	inlet condition, i-th column
j	j-th row
max	maximum
n	number

norm	normalized value
o	orifice
rel	relative
valve	hot end conical valve
1	relative importance value matrix
2	weight matrix
3	geometric mean matrix
4	operation matrix
Superscripts	
+	Highest value in range of data
-	Lowest value in range of data
'	Stagnation value

Reference

[1] S. M. Murthy, K. D. Devade and A. T. Pise 2015 An Experimental Study on Effect of Area Ratio and Mach Number on Thermal Performance of Diverging Vortex Tube *nternational Journal of Innovative Research in Advanced Engineering* **2** 114-20

[2] G. Ranque 1993 Experiments on expansion in a vortex with simultaneous exhaust of hot air and cold air *Journal of Physics* 112-14

[3] M. Hilsch 1947 The use of expansion of gases in a centrifugal field as the cooling process *Revised Scientific Instrumentation* **18** 108-13

[4] M. Valipour and M. H. Saidi 2003 Experimental modeling of vortex tube refrigerator *Applied Thermal Engineering* **23** 1971-80

[5] C. M. Gao 2005 Experimental Study on the Ranque-Hilsch Vortex Tube CIP-DATA LIBRARY TECHNISCHE UNIVERSITEIT EINDHOVEN geboren te HuBei China

[6] A. R. Bramo and N. Pourmahmoud 2010 CFD simulation of length to diameter ratios effects on energy separation in a vortex tube *International Journal of Research and Reviews in Applied Science* **6** 1-16

[7] O. Aydin B. Markal and M. Avci 2010 New vortex generator geometry for a counter flow Ranque-Hilsch vortex tube *Applied Thermal Engineering* **30** 2505-11

[8] V. Kirmaci 2009 Exergy analysis and performance of counter flow vortex tube *International Journal of Refrigeration* **32** 1626-33

[9] K. Polat and V. Kirmaci 2011 Application of the output dependent feature scaling in modeling and prediction of performance of counter flow vortex tube having various nozzles numbers at different inlet pressures of air, oxygen, nitrogen and argon *International Journal of Refrigeration* **34** 1387-97

[10] P. Promvonge and S. Eiamsa-ard 2005 Investigation on the Vortex Thermal Separation in a vortex tube refrigerator *Science Asia* **31** 215-23

[11] J. Prabhakaran and S. Vaidyanathan 2010 Effect of orifice and pressure of counter flow vortex tube *Indian Journal of Science and Technology* **3** 374-76

[12] S. U. Nimbalkar and M. R. Muller 2009 An experimental investigation of the optimum geometry for the cold end orifice of a vortex tube *Applied Thermal Engineering* **29** 509-14

[13] J. Prabhakaran and S. Vaidyanathan 2010 Effect of diameter of orifice and nozzle on the performance of counter flow vortex tube *International Journal of Engineering Science and Technology* **2** 704-07

[14] B. Markal, O. Aydin and M. Avci 2010 An experimental study on the effect of the valve angle of counter-flow Ranque–Hilsch vortex tubes on thermal energy separation *Experimental Thermal and Fluid Science* **34** 966-71

[15] K. D. Devade and A. T. Pise 2012 Investigation of Rerigeration Effect Using Short Divergent Vortex Tube *International Journal of Earth Sciences and Engineering* 378-84

[16] K. Dincer, S. Baskaya, B. Z. Uysal and I. Ucgul 2009 Experimental investigation of the performance of a Ranque–Hilsch vortex tube with regard to a plug located at the hot outlet *International Journal of Refrigeration* **32** 87-94

[17] K. Chang, Q. Li, G. Zhou and Q. Li 2011 Experimental investigation of vortex tube refrigerator with a divergent hot tube *International Journal of Refrigeration* **34** 322-27

[18] N. Pourmahmoud, R. Esmaily and A. Hassanzadeh 2015 CFD investigation of vortex tube length effect as a designing criterion *International Journal of Heat and Technology* **33** 129-36

[19] A. S. Ersoyogulu, K. Dincer, A. Berber, Y. Yilmaz and G. Onal 2015 Rule-Based Mamdani-Type Fuzzy Modeling of Exergy Efficiency Performances of CounterFlow Ranque-Hilsch Vortex Tubes for Square Cross sectional Area *International Journal of Mining, Metallurgy & Mechanical Engineering* **3** 135-41

[20] S. G. Kumar, G. Padmanabhan and D. B. Sarma 2014 Optimizing the Temperature of Hot outlet Air of Vortex Tube using Taguchi Method *Procedia Engineering* **97** 828-36

[21] O. Uluer, V. Kirmaci and S. Atas 2009 Using the artificial neural network model for modeling the performance of the counter flow vortex tube *Expert Systems with Applications* **36** 12256-63

[22] K. D. Devade and A. T. Pise 2016 Effect of Mach , valve angle and length to diameter ratio on thermal performance in flow of air through Ranque Hilsch vortex tube *Heat and Mass Transfer* 1-8

[23] A. M. Pinar, O. Uluer and V. Kirmaci 2009 Statistical Assessment of Counter-Flow Vortex Tube Performance for Different Nozzle Numbers, Cold Mass Fractions, and Inlet Pressures Via Taguchi Method," *Experimental Heat Transfer* **22** 271-82

[24] R. V. Rao 2007 Decision Making in the manufacturing Environment, London: Springer Verlog London Ltd.

[25] P. C. Fishburn 1967 Additive Utilities with Incomplete Product Sets: Application to Priorities and Assignments *Operations Research* 537-42

[26] J. J. Wang, Y. Y. Jing, C. F. Zhang, G. H. Shi and X. T. Zhang 2008 A fuzzy multi-criteria decision-making model for trigeneration System *Energy Policy* **36** 3823-32

[27] D. W. Miller and M. K. Starr 1969Executive decisions and operations research. Englewood Cliffs N.J.: Prentice Hall

[28] T. L. Satty 1980 The Analytic Hierarchy Process. New York: McGraw Hill

[29] C. L. Hwang and K. P. Yoon 1981 Multiple attribute decision making - methods and applications. Berlin Heidelberg: Springer-Verlag

[30] J. Kittur 2015 Using the PROMETHEE and TOPSIS multi-criteria decision making methods to evaluate optimal generation in *Power and Advanced Control Engineering (ICPACE), IEEE*, Banglore.

[31] R. V. Rao Introduction to Multiple Attribute Decision-making (MADM) Methods 2013 in *Decision Making in the Manufacturing Environment*. London, Springer 27-41

[32] R. V. Rao 2008 Evaluation of environmentally conscious manufacturing programs using multiple attribute decision-making methods in *Proceedings of the Institution of Mechanical Engineers, Part B: Journal of Engineering Manufacture*

Study of methane hydrate as a future energy resource: low emission extraction and power generation

L Chen[1, 2, *], H Yamada[3], Y Kanda[3], H Sasaki[3], J Okajima[1], Y Iga[1], A Komiya[1] and S Maruyama[1]

[1] Institute of Fluid Science, Tohoku University, Katahira 2-1-1, Aoba-ku, Sendai 980-8577, Japan
[2] Overseas Research Fellow, Japan Society for the Promotion of Science (JSPS), Japan
[3] Graduate School of Engineering, Tohoku University, Aoba 6-6, Aramaki-aza, Aoba-ku, Sendai 980-8579, Japan

E-mail: chenlintohoku@pixy.ifs.tohoku.ac.jp; chenlinpkucoe@gmail.com

Abstract. With the fast increase of world energy consumption in recent years, new and sustainable energy sources are becoming more and more important. Methane Hydrate is one promising candidate for the future energy supply of humankind, due to its vast existence in permafrost regions and near-coast seabed. This study is focused on the effective low emission utilization of methane hydrate from deep seabed. The Nankai Trough of Japan is taken as the target region in this study for methane hydrate extraction and utilization system design. Low emission system and power generation system with CCS (Carbon Capture and Sequestration) processes are proposed and analyzed for production rate and electricity generation efficiency problem study. It is found that the gas production price can reach the current domestic natural gas supply price level if the production rate can be improved. The optimized system is estimated to have power efficiency about 35%. In addition, current development and analysis from micro-to-macro scale methane hydrate production and dissociation dynamics are also discussed into detail in this study.

1. Introduction

Methane hydrate (MH), a substantial energy resource with vast existence in permafrost region and under seabed, has been proved to be of high energy density and extractable in the near future [1-3]. Indeed, it has been reported that the overall gas hydrate reservoirs on earth contain around 2.0×10^{14} m^3 to 1.2×10^{17} m^3 of methane gas (STP). This 'fire from underground' is believed to harbor more than twice the amount of known fossil fuels on earth according to reports from geologists [4]. The exploring and production of methane gas from methane hydrate reservoirs have attracted major concerns of both the public and scientific communities in recent years.

Many countries have started methane hydrate Research and Development programs to test the feasibility of extraction technologies. The world first tests were in Russia in Messoyakha, which has become the only commercialized methane hydration production site since 1969 [4]. Canada put its

* Address for correspondence: L Chen, Institute of Fluid Science, Tohoku University, Katahira 2-1-1, Aoba-ku, Sendai 980-8577, Japan ; Overseas Research Fellow, Japan Society for the Promotion of Science (JSPS), Japan. E-mail: chenlintohoku@pixy.ifs.tohoku.ac.jp; chenlinpkucoe@gmail.com.

research production site at Mallik of Mackenzie Delta region since 1970 and tried thermal method production (Mallik 2002 test) and depressurization method (Mallik 2007 test) [5]. For oceanic methane hydrate exploration, Japan has tried the assessment and production tests in Nankai Trough and obtained successful gas production around 6 days and obtained 1.3×10^4 m^3 methane gas (see Figure 1; [6]). In addition, China, Korea, India and other countries have also made small scale explorations and production tests in recent years.

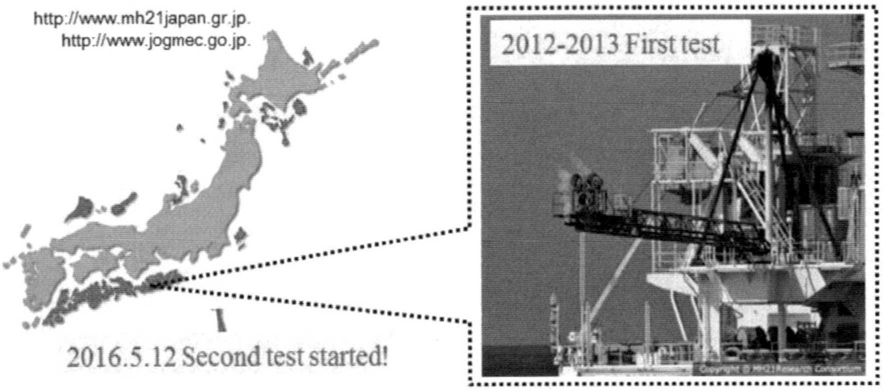

Figure 1. Methane hydrate existence and production test in Japan [6].

For the situation of Japan, the first series research projects were started around 1992-1995, when the fundamental investigations of methane hydrate were proposed by Industrial Technology Research Institute (now called AIST: National Institute of Advanced Industrial Science and Technology). The first production experiment using hot water circulation method was made in 2002. Then a lot of real explorations and tests are executed in Nankai Trough of Japan, such as the geological survey in 2004, the proving test of depressurization method experiment in 2008, Nankai Trough real production test in 2012 and new round of test since May of 2016.

Those above projects and trials have made major progress in the production analysis and environmental assessment. However, there still exist several bottlenecks of real large scale commercialized and safe supply with methane gas. It is estimated by Sloan [2] that only the production rate is higher than 5×10^5 m^3 (STP) per day can the method be considered as economically reasonable. Also, due to the various conditions of widely spread methane hydrate reservoirs, there is no generally applicable method for stable production. According to major reviewers [1-4], the knowledge of methane hydrate dissociation and production from underground or below the sea bed are still not sufficient in the current stage.

In this study, a new low emission extraction and utilization system is proposed and analyzed for the feasibility and system power generation efficiency. The basic system construction, generation system and efficiency dependence on input parameters are discussed. Also the detailed analysis and investigations into the multi-scale methane hydrate dissociation dynamics are also introduced in this paper.

2. Low emission utilization system development

The basic dissociation and extraction process of methane gas from methane hydrate reservoirs is shown in Figure 2. According to the phase equilibrium curve in Figure 2(a), there are several ways of extracting methane gas from methane hydrate: increase the temperature (thermal stimulation), reduce the pressure (depressurization), adding inhibits or combined method of those. The dissociation reaction is expressed in Equation (1) under reservoir conditions.

$$CH_4 \cdot 5.75H_2O_{(s)} \rightarrow CH_{4(g)} + 5.75H_2O_{(l)} \quad \Delta H_{\mathrm{MH}} > 0 \tag{1}$$

Due to the high pressure and low temperature existence condition of methane hydrate existence in reservoirs, especially for undersea reservoirs taking the form of alternating sand and mud layers, it is very difficult to systematically extract and store the methane gas for possible commercial use.

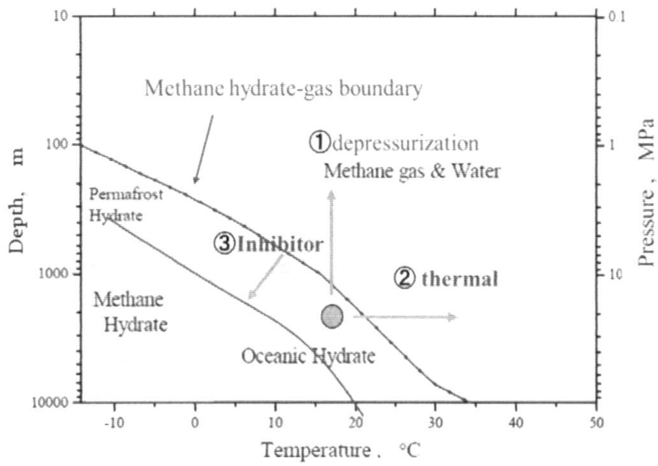

Figure 2. Basic Methane Hydrate dissociation process, phase diagram and methods [7].

In the current stage, systematic analysis and novel designs of utilization systems are yet to be proposed [2, 4]. Kawata *et al.* [8], Maruyama *et al.* [9], Chong *et al.* [3] proposed several representative systems. That is not only for an energy-balanced model (thermal consideration), but also for a more practical way of CO_2 utilization/processing, by CCS or by a consideration of CO_2-hydrate replacement of CH_4-hydrate [3, 9]. It is usually concluded difficult for single numerical modeling to cover the complex thermal, fluid dynamic and mechanical changes of reservoir conditions [3].

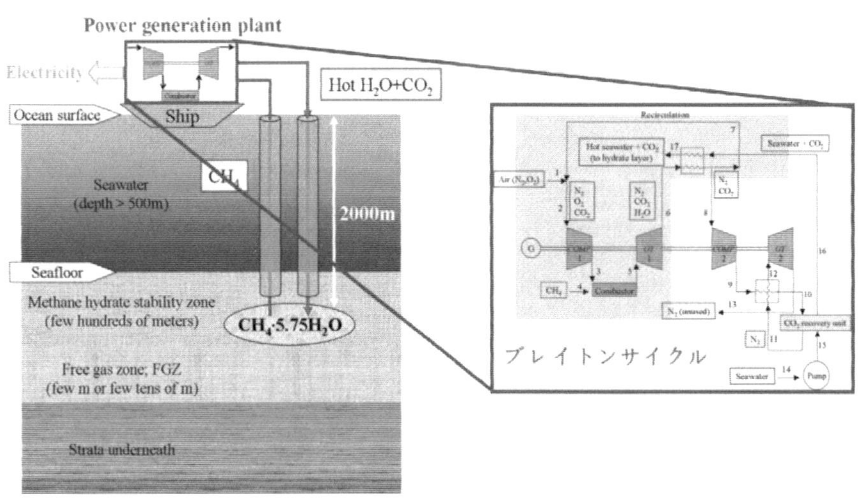

Figure 3. Representative methane hydrate utilization system (with simultanesous CCS process) [9].

In this study, the detailed development and analysis of utilization and power generation model is discussed, which consists of production process, on spot gas turbine power generation, hot water and

CO_2 re-injection, and electric transportation systems, as shown in Figure 3. Such system combines production estimation and CCS process, which may be one future trend of efficient utilization.

3. Low emission utilization system development

3.1. Macro-scale utilization system and challenges

For the proposed macro-scale utilization system, the production rate tests and relationship with the gas production price is conducted. Using the current systems and generation design, the electricity production will have different price levels, as shown in Figure 4. It can be seen from Figure 4 that in the current production rate (according to the test of Nankai Trough test in 2001), the price is much higher than the market price. And only when the production rate is improved to be around 8.0×10^4 m^3/day the production would be compatible with the market price.

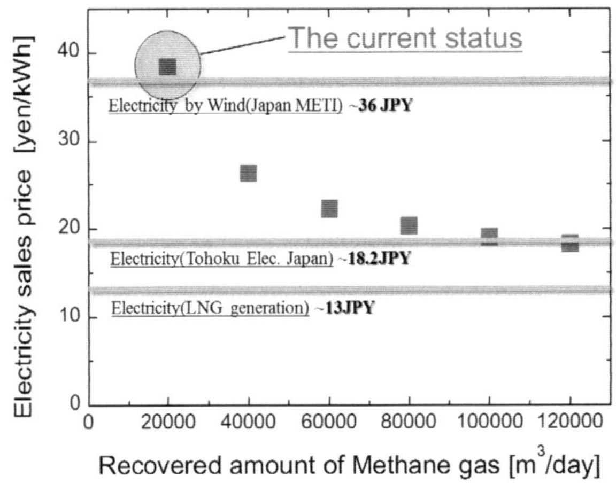

Figure 4. Methane gas production price estimation and comparisons.

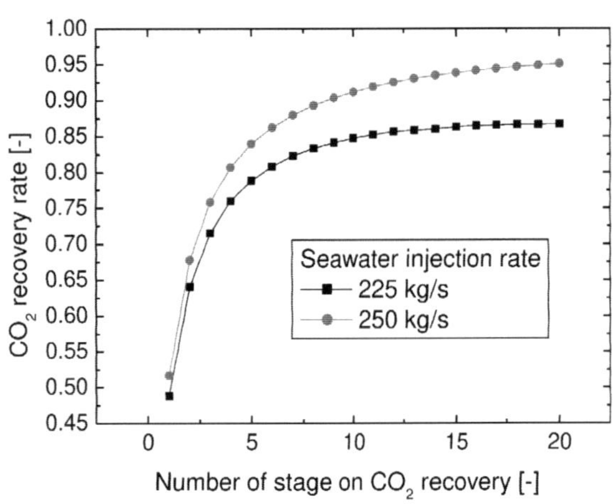

Figure 5. (a) CO_2 recovery rate

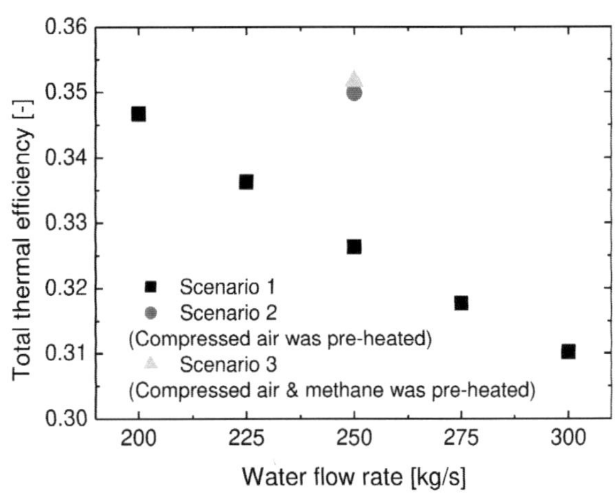

Figure 5. (b) system thermal efficiency

Figure 5. Low emission system efficiency analysis. (a) CO_2 recovery rate; (b) system thermal efficiency.

Also in the current analysis, the CO_2 recovery rate and the efficiency of power generation are analyzed, as shown in Figure 5. It can be seen from Figure 5 (a) that when the recovery stage increase, the recovery rate will be increase. Also the increase of seawater injection would also lead to an increase of recovery rate. Indeed, the consideration of CO_2 recovery and re-injection would help increase the storage rate and substitute of the underground methane hydrate cages. The optimized system is estimated to be able to achieve power efficiency about 35% for the proposed system. The optimized system is estimated to be able to achieve power efficiency about 40% for the proposed system. However, as shown in Figure 5 (b), when the seawater injection is increased, the thermal efficiency would decrease, due to the increased pump power needed to do this. And an effective use of exhaust heat and water compressing heat would also increase the thermal efficiency.

3.2. Lab-scale experiments and tests

For the system efficiency improvement, the understanding of underground gas extraction and flow condition analysis is needed. In order to do that in lab environment, a core-scale experimental prototype is set up in this study, as shown in Figure 6. As it is very difficult to use the real core from real dee-sea methane hydrate reservoirs, in this analysis, the mimicking strategy of using water-ethanol inside porous sand is utilized to form the low permeability condition. The basic design is shown in Figure 6 (a) and by controlling of volume fraction of ethanol inside the porous sand, the formation of ice would be achieved and thus to generate the low permeability conditions needed. The permeability measurement results are shown in Figure 6 (b). This design is critical for the test of basic core characteristics and would be replaced by mixing of methane hydrate and sand easily to conduct real dissociation test in laboratory experiments [10-11].

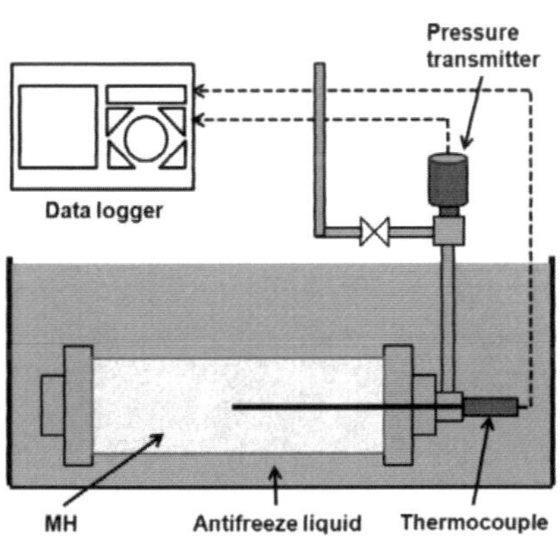

Figure 6. (a) basic core system design

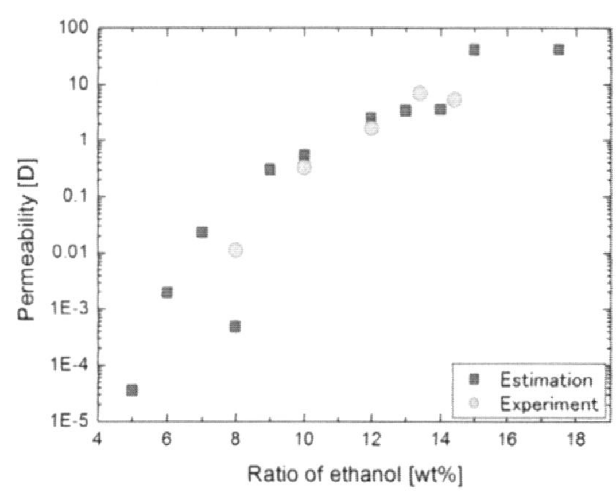

Figure 6. (b) permeability measurement results

Figure 6. Permeability control experiment.

3.3. Micro-scale mechanisms and visualizations

In this study, the concentration field near the reaction surface was visualized by using the phase-shifting interferometer system [12]. In this experiment, the MH dissociation phenomenon was observed in the reaction chamber shown in Figure 7 (a). Experiments were carried out in the temperature controlled room under the temperature condition of 2.0 °C. Temperature and pressure were set to about 5°C and 9.5 MPa after injection of methane gas. To observe the MH dissociation phenomenon pressure in the chamber was reduced by 0.1 MPa to below equilibrium pressure at 5 °C in the chamber. In the phase-shifting interferometer system, an arbaa prism was used as shown in Figure 7 (b). The arbaa prism splits an incident beam into four output beams with different information. At the output of this prism, all three of the output beams are filtered by the polarizer to obtain three types of interference fringes.

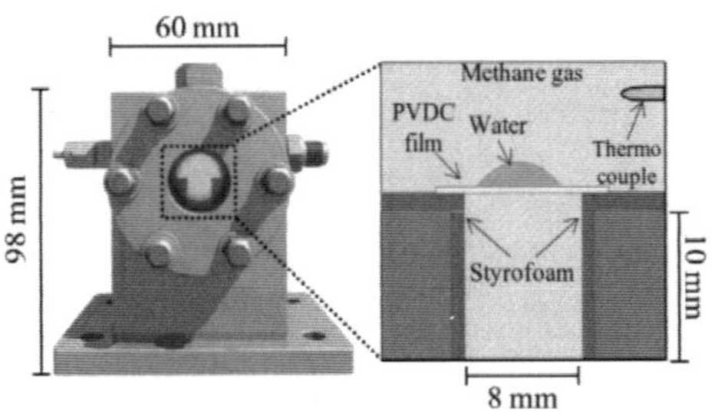

Figure 7. (a) chamber design

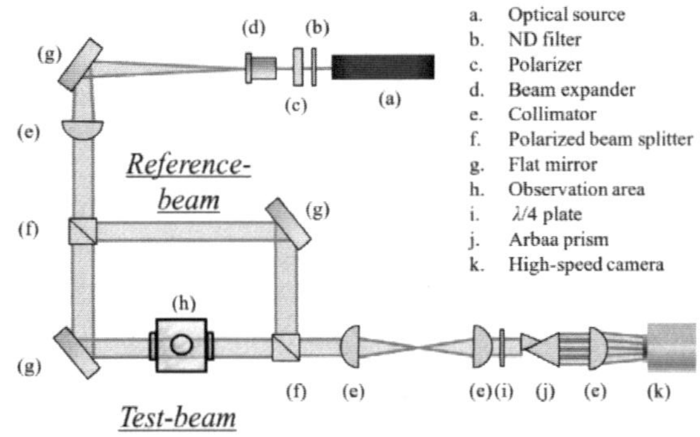

Figure 7. (b) phase-shifting interferometer

Figure 7. Optical experimental system design.

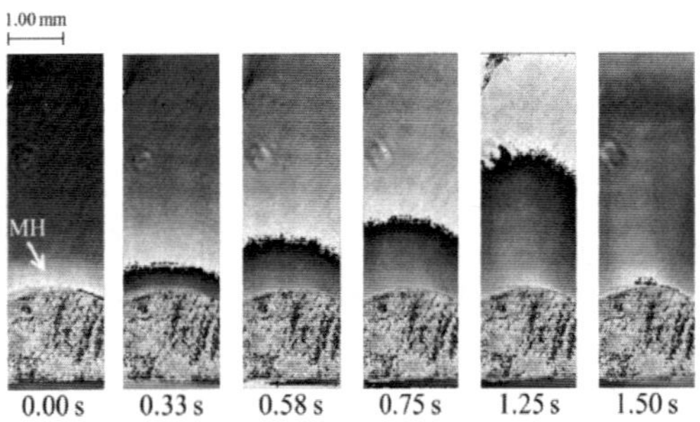

Figure 8. (a) Sequential phase-shifted data

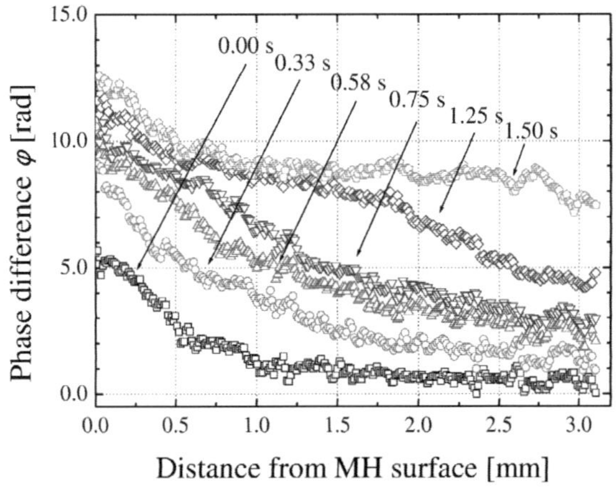

Figure 8. (b) Phase difference distribution of methane

Figure 8. Phase-shifting experimental results during MH dissociation process.

To visualize the dissociation process of MH, the pressure in the chamber was depressurized below equilibrium pressure (about 4.1 MPa) [12]. The sequential phase-shifted data during MH dissociation process is shown in Figure 8(a). The phase difference distribution of methane in the gas phase during MH dissociation process is plotted in Figure 8(b). Changes of the phase difference in those figures indicate density variation due to heat and mass transfer near the solid-gas interface. In this experiment, it was found that unsteady density distribution of methane in the gas phase existed during MH dissociation processes and it is possible to test quantitatively of the interface dynamics of methane hydrate dissociation process.

3.4. Lab-scale production behaviors

Using the basic dissociation apparatus as discussed in Figure 6 (a), dissociation experiment is conducted. The basic dissociation process can be divided into three stages in core scale, as shown in Figure 9. The first stage is the quick temperature drop process due to the fast dissociation when the right end of the chamber is exposed to low pressure. As the dissociation reaction is endothermic kind, the temperature drops quickly. Then the heat use and boundary conduction from ambient reach a balanced states for the long duration of core temperature region. After the dissociation is finished, the temperature will recovery to the ambient value. This result is important as to support the basic thermal analysis in large scale system extraction.

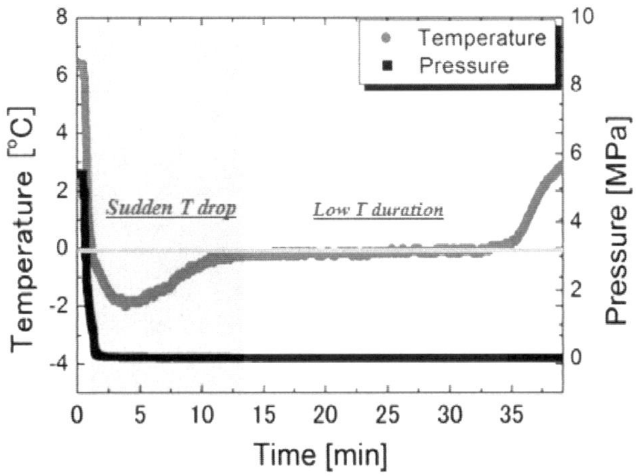

Figure 9. Basic core-scale dissociation parameter behaviors.

3.5. Core-scale numerical simulation and comparison

The basic high-pressure operating system of core-scale dissociation process in this study is simplified to be a 3D numerical model and with also an extension to three-dimensional model analysis in later discussions [13-14]. As shown in Figure 10, the physical model is a horizontally operated cylinder core, where an axis is applied for simplicity of model. The internal region of the model can be applied with porous media with artificial bearing layer porosity, permeability and other heat transfer/flow parameters [12, 14, 15-18].

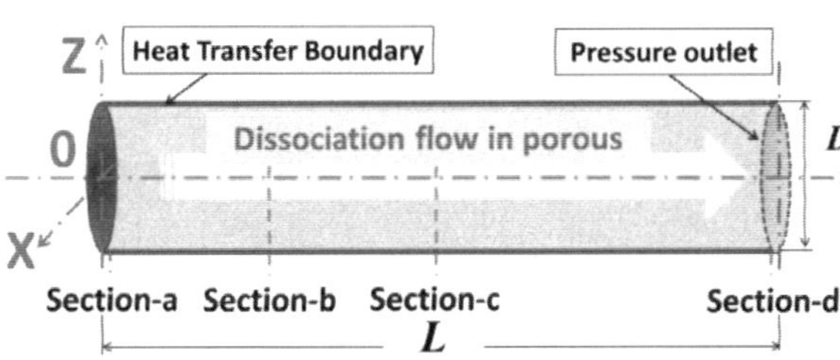

Figure 10. (a) simplified 3-D core model

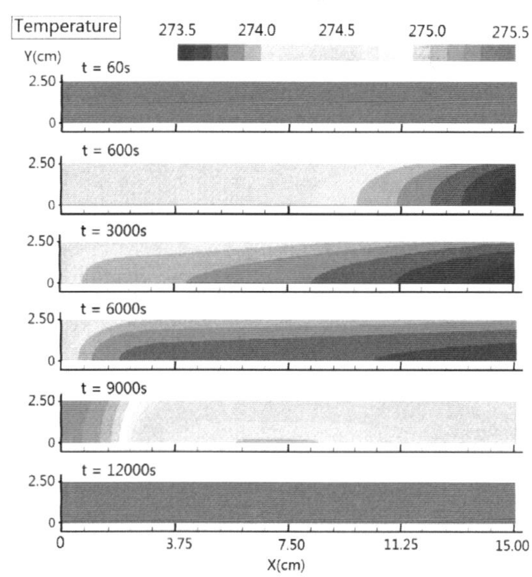

Figure 10. (b) temperature evolution of the core

Figure 10. Basic modeling and temperature results for low permeability dissociation case (porous case; adiabatic boundary; half view)

In the simulation, the model scale is set to be within a lab scale experimental core from L = 10~30 cm length and D = 2.0~5.0 cm diameter core. The numerical results are shown in Figure 10 (b) and Figure 11 for basic dissociation process and the ice formation. From Figure 10 (b) it can be seen that the basic dissociation process is delayed due to the low permeability in the core. Also the core-scale dissociation is governed by the boundary heat transfer conditions, as the dissociation curves are shaped by the boundary heat transfer conditions. In Figure 11, it can be seen that the ice forming dominant the dissociation core and resides in the bottom part due to gravity effect, leaving the methane gas in the upper part. This result agrees with experimental result, where the ice is found for the non-porous dissociation chamber, as shown in Figure 11 (b). However, the detailed effects of ice on the methane gas production, water flow and hydrate dissociation kinetics is still not well understood. The question still remains that how the local or gross thermal field conditions are will affect the global gas production, and how is the effect on core-scale? More studies/proposals are recommended for better understanding of those related phenomena and methods.

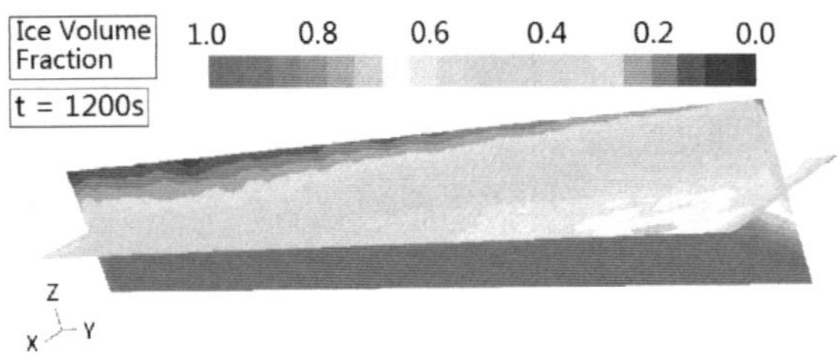

Figure 11. (a) ice volume fraction

Figure 11. (b) ice formation inside experimental core

Figure 11. Three dissociation (3D) model and real experimental core with ice formation.

4. Conclusion

In this study, the systematic analysis of a proposed methane hydrate extraction and power generation system is conducted. The Nankai Trough of Japan is taken as the target for methane hydrate extraction and utilization system design. Low emission system and power generation system with CCS (Carbon Capture and Sequestration) processes are proposed and analyzed. The optimized system is estimated to be able to achieve power efficiency about 35% for the proposed system. Detailed system analysis and dissociation dynamics studies are also included in this paper. It is hoped that the current study can be useful for related new energy and sustainability development of the industry.

Acknowledgement

The support from JST-CREST (Program: Breakthrough on Multi-scale Interfacial Transport Phenomena in Oceanic Methane Hydrate Reservoir and Application to Large-Scale Methane Production) and JSPS Overseas Researcher Grant (No. 16F16068) are gratefully acknowledged by the authors. The authors are grateful for the assistance of permeability experiments from Mr. Guillaume Lacaille (former graduate student of Tohoku University, Japan).

References

[1] Boswell R 2009 Is gas hydrate energy within reach? *Science* **325** 957-8
[2] Sloan E D 2003 Fundamental principles and applications of natural gas hydrates *Nature* **426** 353-9
[3] Chong Z R, Yang S H B, Babu P, Linga P and Li X S 2016 Review of natural gas hydrates as an energy resource: Prospects and challenges *Appl. Energ.* **162** 1633-52
[4] Makogon Y F, Holditch S A and Makogon T Y 2007 Natural gas-hydrate-a potential energy source for the 21st century *J Petro Sci Eng* **56**14-31
[5] Yamamoto K and Dallimore S 2008 Aurora-JOGMECNR Can Mallik 2006-2008 gas hydrate

research project progress *Fire in the Ice* **7** 1-5

[6] JOGMEC 2013 Gas produced from methane hydrate. Japan Oil, Gas and Metals National Corporation. http://www.jogmec.go.jp/english/news/release/release0110.html.

[7] Kvenvolden K A 1993 Gas hydrate: geological perspective and global change *Rev. Geophys.* **31** 173-87

[8] Kawata Y, Matsuhashi R, Masuda Y and Fujita K 2005 Investigation of the methane hydrate development by hybrid-type power generation with LNG *J. Japan Inst. Energy* **84** 68-75

[9] Maruyama S, Deguchi K, Chisaki M, Okajima J, Komiya A and Shirakashi R 2012 Proposal for a low CO_2 emission power generation system utilizing oceanic methane hydrate *Energy* **47** 340-7

[10] Yamada H, Lacaille G, Chen L, Gonome H, Shoji E, Okajima J, Komiya A and Maruyama S 2016 Experimental evaluation of methane hydrate core mimicking sample design and permeability control *In Proc. of: The 53th Heat Transfer Symposium of Japan*

[11] Lacaille G, Yamada H, Shoji E, Chen L, Okajima J, Komiya A, and Maruyama S 2016 Estimation and measurement of permeability inside methane hydrate mimicking porous media *J. Fluid Sci. Tech.* (In Press)

[12] Kanda Y, Gonome H, Shoji E, Okajima J, Komiya A and Maruyama S 2015 Visualization of concentration boundary layer in the vicinity of methane hydrate dissociation *Proc. of China-Japan Joint Workshop on Flow Dynamics and Transport Phenomena*

[13] Chen L, Yamada H, Kanda K, Lacaille G, Shoji E, Okajima J, Komiya A and Maruyama S 2016 a Numerical analysis of core-scale methane hydrate dissociation dynamics and multiphase flow in porous media *Chem. Eng. Sci.*(Doi: doi:10.1016/j.ces.2016.07.035)

[14] Chen L, Lacaille G, Yamada H, Kanda K, Shoji E, Okajima J, Komiya A and Maruyama S 2016b Numerical simulation of core-scale methane hydrate dissociation flow and heat transfer in porous media (ID: 1583) *Proc. of: The 53th Heat Transfer Symposium of Japan* Osaka, Japan.

[15] Moridis G J and Collett T S 2004 *Recent Advances in the Study of Gas Hydrate* (New York: American Institute of Chemical Engineers and Kluwer Academic 75-88

[16] Nazridoust K and Ahmadi G 2007 Computational modeling of methane hydrate dissociation in a sandstone core *Chem. Eng. Sci.* **62** 6155-77

[17] NETL report 2011 Energy resource potential of methane hydrate: an introduction to the science and energy potential of a unique resource *National Energy Technology Labotaory Report*, U.S. Department of Energy

[18] Ruan XK, Song Y C, Zhao J F, Liang H F, Yang M J and Li Y H 2012 Numerical simulation of methane production from hydrates induced by different depressurizing approaches *Energies* **5** 438-58

Indoor weather related to the energy consumption of air conditioned classroom: Monitoring system for energy efficient building plan

W Rattanongphisat[1,2 4], A Suwannakom[1] and A Harfield[3]

[1] Department of Physics, Faculty of Science, Naresuan University, Phitsanulok, Thailand
[2] Center of Excellence for Energy Technology and Environment, Faculty of Engineering, Naresuan University, Phitsanulok, Thailand
[3] Department of Computer Science and Information Technology, Faculty of Science, Naresuan University, Phitsanulok, Thailand

E-mail: warapornr@nu.ac.th

Abstract. The current research aims to investigate the relation of indoor weather to energy consumption of air conditioned classroom by design and construct the indoor weather and energy monitoring systems. In this research, a combined temperature and humidity sensor in conjunction with a microcontroller was constructed for the indoor weather monitoring system. The wire sensor network for the temperature-humidity sensor nodes is the Controller Area Network (CAN). Another part is using a nonintrusive method where a wireless current transformer sending the signal to the data collection box then transmitted by the radio frequency to the computer where the Ethernet application software was installed for the energy monitoring system. The results show that the setting air temperature, outdoor ambient temperature and operating time impact to the energy consumption of the air conditioned classroom.

1. Introduction

The record shows the increase of the global surface temperature of nearly 1°C during 1880-2010 [1]. Evidently, the increase of CO_2 level has shown that it is relevant to the climate change. The highest temperature in Thailand sets a new record this summer, April 2016, for 44.5 °C and other countries in Southeast Asia set similar new records. A significant impact of the indoor weather is the human comfort and because more than half of human activities take place in building thus affects to industrial, public sector and residential buildings. Air conditioning systems become necessary as can be seen by high demand in the market. Consequently, energy consumption rapidly increases. The measuring of energy consumption for educational buildings in Osaka City University found that the air conditioning system consumed nearly 60 per cent of the total building energy consumption [2]. In the view of building energy management, more data recorded would greatly contribute to the better energy analysis and control thus results in the energy efficient building [3]. Although, the two factors,

[4] Address for correspondence: Department of Physics, Faculty of Science, Naresuan University, Phitsanulok, Thailand. Center of Excellence for Energy Technology and Environment, Faculty of Engineering, Naresuan University, Phitsanulok, Thailand. Email: warapornr@nu.ac.th.

education and economic, are considered relevant to the pattern of energy use, it was found that those factors failed to change the customer behaviour for energy conservation. This reveals that social, personal and psychological are more relevant to customer habit pattern. The role of energy monitoring is significant to building energy planning and efficiency. A research reports on the possibility of reducing energy consumption in residential housing with the use of smart meters or advance metering infrastructure and community based social marketing [4]. It was reported that there was 20% peak demand reduction, 10% peak demand shifting and 10% average total energy use reduction from the household participating in the program. Another benefit of the monitoring system is that customers could see how rapid the energy consumption increase during different electrical appliances operation thus awareness has been raised. Other study showed an enhancement of energy efficiency in the Heating Ventilation and Air Conditioning system (HVAC) by the wireless technology [5]. The employment of wireless sensor network included temperature sensors, CO_2 sensors and webcam for controlling the HVAC suggested that the operation of HVAC during the occupation of the room show the wide range of energy reduction. An overview of data collection for building monitoring is reported [6]. Various approaches for energy efficient in air conditioning are summarized and reported [7]. Building energy management system is widely accepted that it could help deduct the energy consumption of the building. Particularly, the building energy management system involved with the heating, ventilation and air conditioning system (HVAC). Air conditioning system control temperature, part of air humidity and could integrate with air quality and circulation for a conditioned space. The level of awareness on building energy consumption and efficiency varies from one country to another country. Different building policies and measures were adopted as to promote building energy efficiency. Many countries achieved the energy reduction according to the analysis of official recorded data from the past to recent date. Building code and building labels and certifications are the two key parameters for the successful of the building energy efficiency [8]. The HVAC system consumes high energy thus requires first attention among the electrical appliances in building. Regarding the old and oversize system, high energy and cost were spent unnecessary. A proposal of zone operation was considered to solve or improve energy efficiency of central heating and cooling systems. The predicted mean vote index is an important key to indicate thermal comfort and use to control HVAC by developing Artificial Neural Network (ANN) model to indicate the suitable time for turning on and off [9].

In this paper, the indoor weather and energy monitoring system of an air conditioned classroom, at the Department of Physics, Naresuan University was designed and fabricated. A combined temperature and humidity sensor in conjunction with a microcontroller which connected through the Controller Area Network (CAN) was installed for the indoor weather monitoring system. A nonintrusive method using a wireless current transformer and Ethernet application software connection was employed for the energy monitoring system. The visualisation of the energy consumption as well as the indoor temperature and humidity and how they related to the occupant use of energy. That could raise the occupancy awareness. Furthermore, the data analysis could be further used for the energy efficient building planning.

2. Methodology

The two monitoring systems were designed and constructed for monitoring the indoor weather and energy consumption of air conditioned classroom. The dimension of the auditorium classroom is $15.3\times19\times5.5$ m^3. The function of this classroom includes lecturing, seminar, student camp meeting and scientific movie show. In the current test of the monitoring system, the lecture and seminar are the main activity considered.

2.1. Indoor weather monitoring system

A 172-seat air conditioned classroom was selected to investigate the relation of indoor weather to energy consumption. The HIH6131 temperature-humidity sensors in conjunction with STM32F072CBT6 microcontroller connected through the Controller Area Network (CAN) for the indoor weather monitoring system. The specification of a combined temperature-humidity sensor is shown in table 1 as well as the equation for this sensor written in equation (1) and (2) [10]. The integrated temperature and humidity sensors are installed at various positions in the classroom as can be seen in figure 1. The data from the sensors communicate with microcontroller and the Controller Area Network then transfer via wireless router thru the internet network for collecting and showing on the computer screen in real time.

$$\text{Humidity (\%RH)} = ((\text{Humidity 14 bit ADC output}) / (2^{14} - 2)) \times 100 \quad (1)$$

$$\text{Temperature (C°)} = ((\text{Temperature 14 bit ADC output}) / (2^{14} - 2)) \times 165 - 40 \quad (2)$$

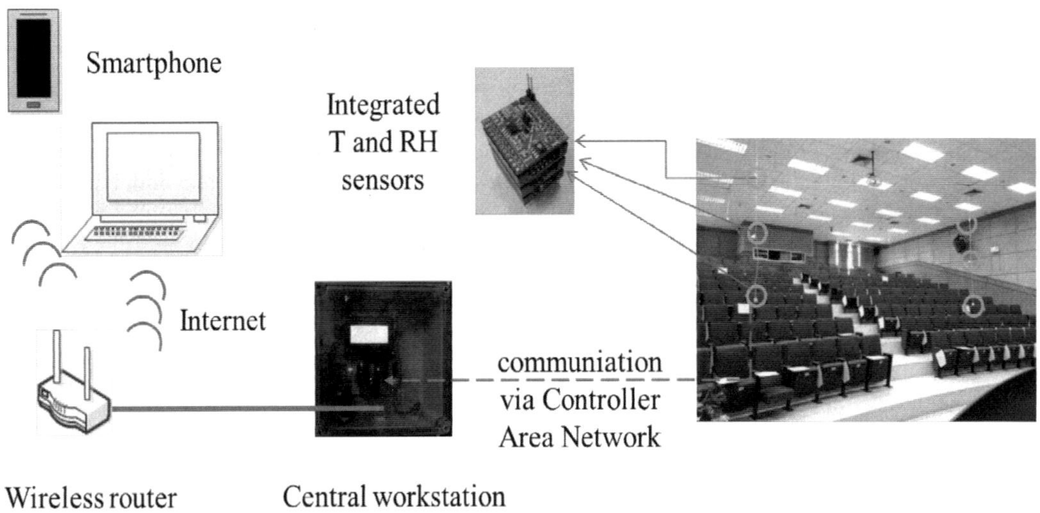

Figure 1. The indoor weather monitoring system for an air conditioned classroom.

Table 1. Specification of combined humidity and temperature sensors [10].

Criteria		Specification
Voltage		2.3 – 5.5 Vdc
Current	Sleep mode Full operation mode	1 μA 650 μA
Measurement	Time	50 ms
Temperature	Range	5 °C – 50 °C
Temperature	Accuracy	±0.5 °C
Humidity	Range	10 - 90 %RH
Humidity	Accuracy	±4% RH

2.2. Energy monitoring system

An auditorium classroom equipped with an individual air conditioner was selected to demonstrate the energy monitoring system. The air conditioner is TRANE model no. TTV 250 AD11A0A00A, fan motor 3.7kW using R-22 refrigerant. The energy monitoring system incorporates with the wireless current (CT) sensors, CT1 56/3000 from Midori Co.Ltd., and operating software for power and energy display were installed at the demonstrating room and monitoring room respectively as shown in figure 2. The data are stored in a cloud system. The smart CT sensors clamp around the electric cable in the electric control box as to measure the power consumed by the air conditioner. Three CT sensors were used to measure the current of the 3-phase 4 wires power line. Data then send to the collection terminal box wirelessly and through the laptop computer for real-time monitoring. The SMV Ethernet program developed by Midori company was installed and used for displaying the real-time power and energy consumption daily.

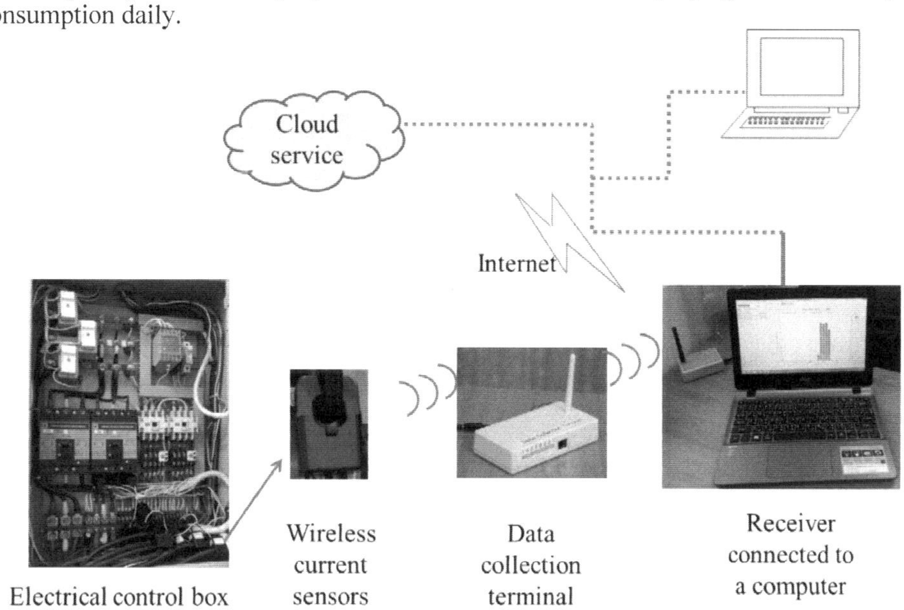

Figure 2. Energy monitoring system.

3. Results and Discussion

Figure 3 shows the period where the teaching class and seminar were took place in the classroom during the weekday. From Monday to Friday, it can be observed that heavy use was on Monday, Thursday, Tuesday and Wednesday respectively while less activities is on Friday. Although, the time table is fixed for this classroom during the semester, the room may be booked for other activities such as student science camp, student meeting and special lecture from an invited speaker. In that case, the daily load profile of the air conditioned classroom can be different from what is shown in figure 3. The energy monitoring data collection can be used to analyse the pattern of energy utilization and also suggested the efficient use of the demonstrated classroom.

The alternating current measurement using wireless smart current sensors and calculation procedure in the software developed by Midori company for power and energy consumption analysis were calibrated with the measurement from a Yogokawa power meter. The test were carried out during term time from September to December 2015. The data analysis reveals that the ambient temperature does relate to the temperature setting of the air conditioner and results in the amount of consumed electricity. The different between the outside and inside temperature show the altered requirement of energy which can be seen in figure 4. The analysis on the past year recorded data, the ambient temperature and humidity taken from the meteorological station in Phitsanulok province, can be seen from figure 5 and figure 6. The pattern of environmental ambient temperature and relative humidity shows that the temperature is quite high throughout the year as well as relative humidity. Thus, it can be seen from the results that the energy consumption tends to be increased when the gap between the average ambient and room temperature is higher. Another point is that the compressor running time is definitely correlated to the energy consumption when the average room temperature reached the setting point then the fan is only operated thus during this period the lower energy consumption can be observed. Prior study by the authors suggested the improvement of energy efficient use for this air conditioned classroom can be found when the temperature set point of an air

IOP Conf. Series: Earth and Environmental Science **40** (2016) 012075 doi:10.1088/1755-1315/40/1/012075

conditioning system increase by 1-2°C as to reach the thermal comfort of people who lives in hot and humid climate.

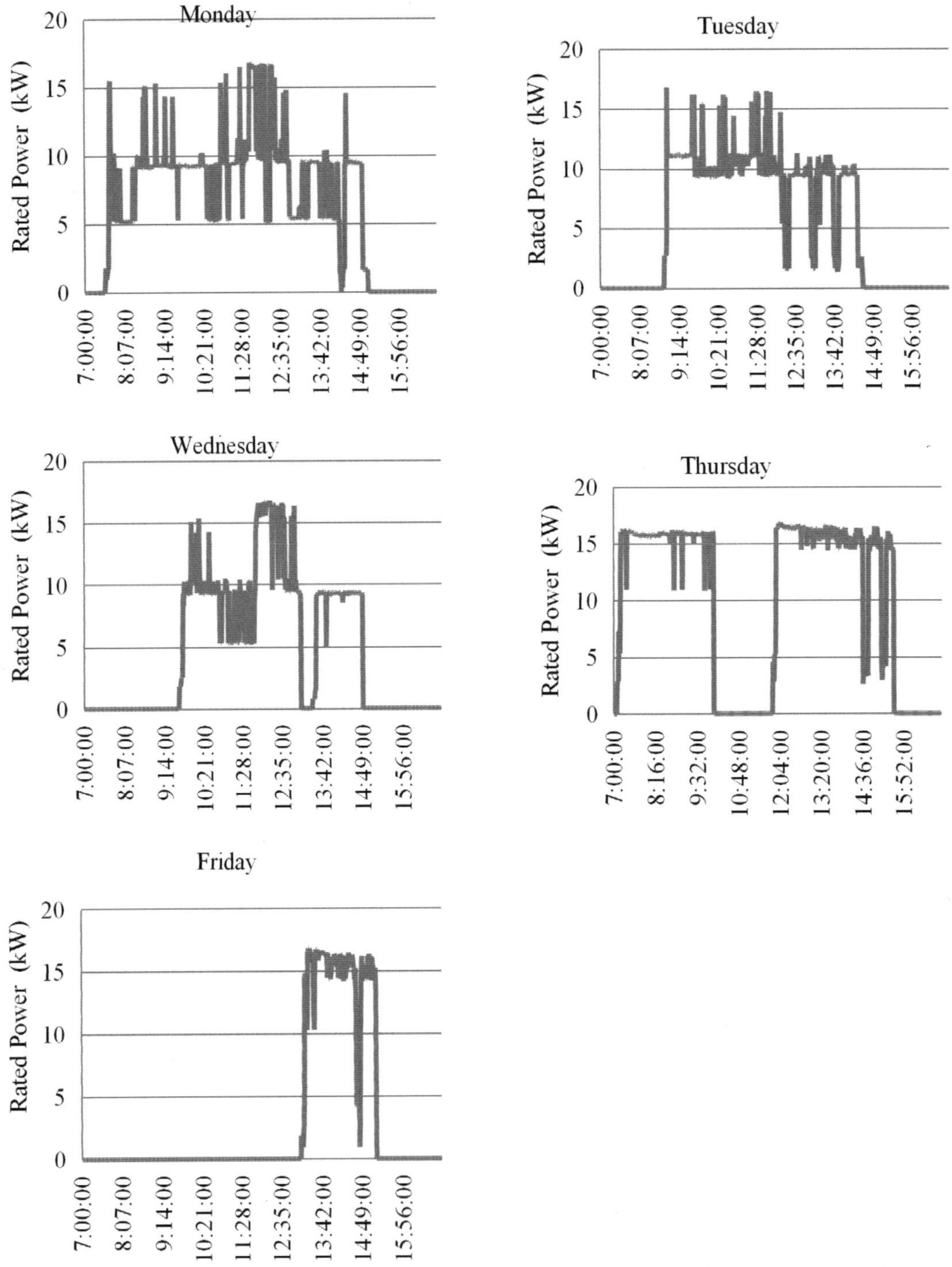

Figure 3. Rated power during weekday

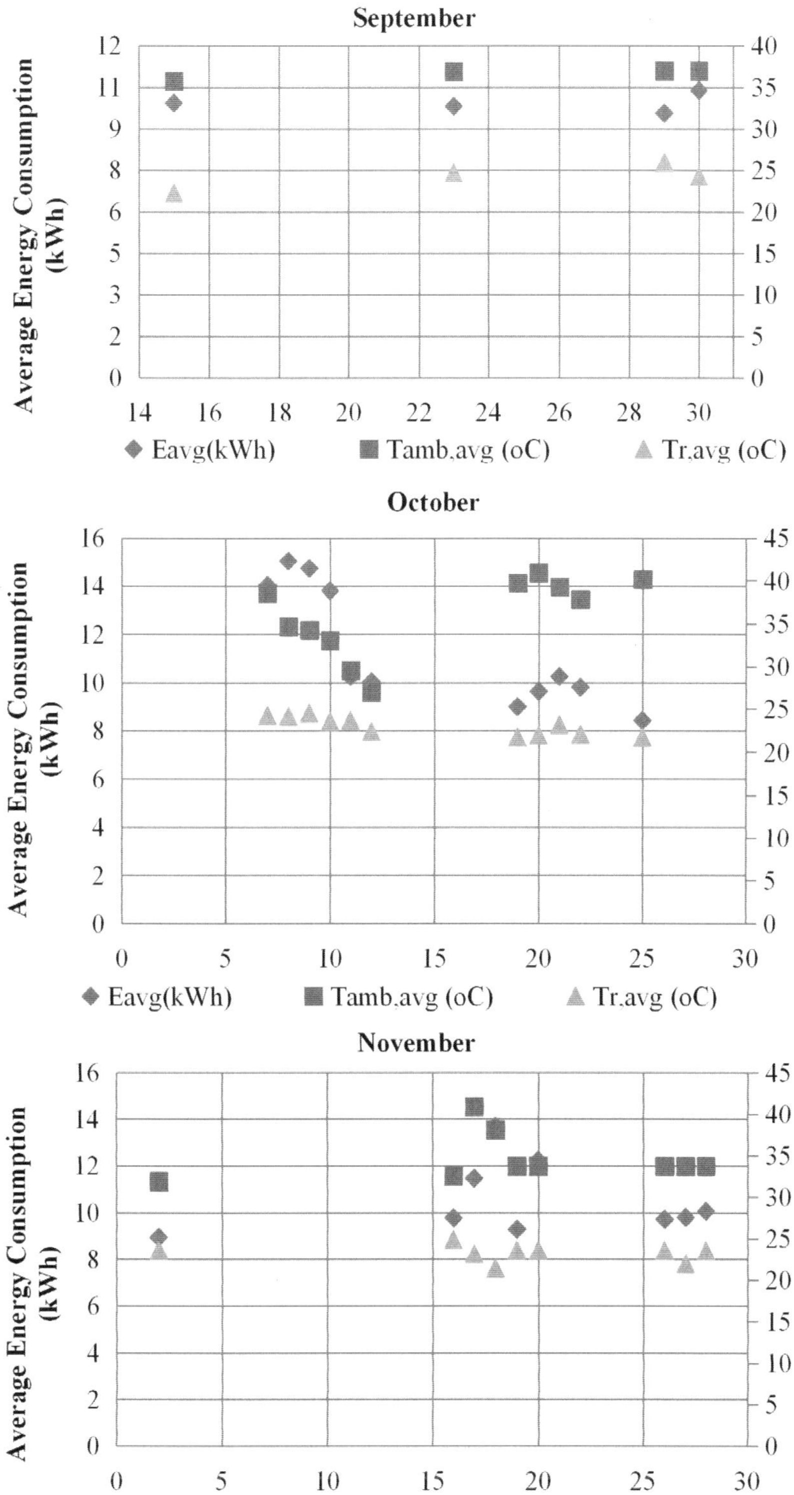

Figure 4. Energy consumption versus temperature during term time.

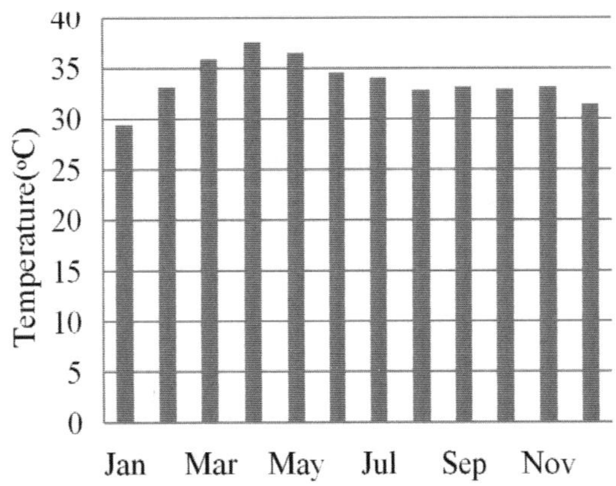

Figure 5. The average maximum temperature at Phitsanulok province in the past year.

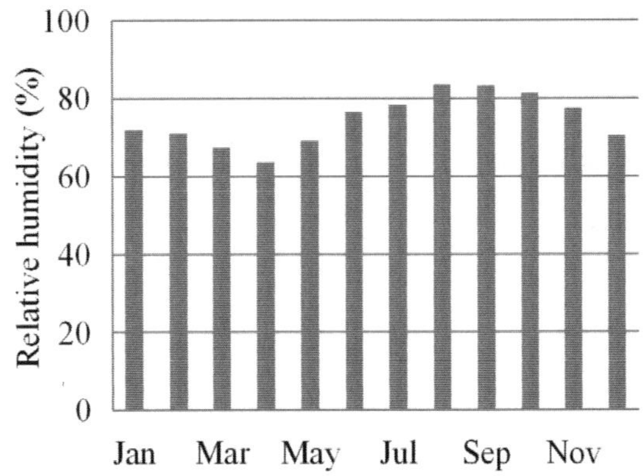

Figure 6. The monthly average relative humidity at Phitsanulok province in the past year.

The measuring power results in the average hourly energy consumption during the first semester last for 4 months from September to December 2015 is shown in figure 7. The rated energy consumption within an hour use of air conditioner varies from the minimum of 8kWh to the maximum of 15 kWh depended on the outdoor weather.

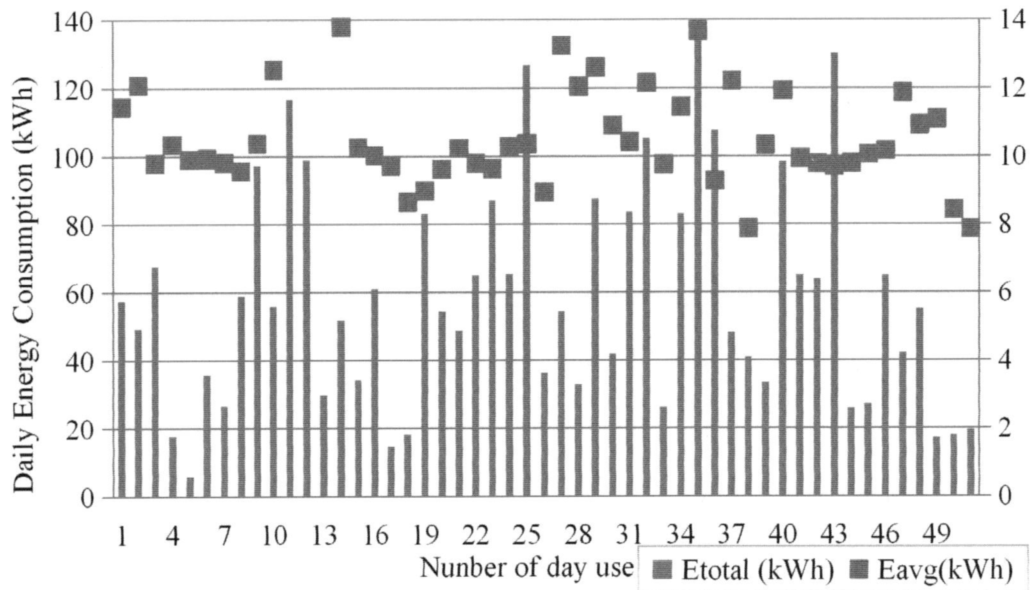

Figure 7. Energy consumption during the term time from September to December 2015.

Considering the applications of the monitoring system, an air conditioned space for various activities from low activities such a residence to high activities such an indoor gym can be beneficial by closely monitoring the real time rated power. It can be used to adjust the appropriate temperature setting on the HVAC system to reduce the energy consumption. Furthermore, the data obtained from monitoring can assist in efficient energy utilization planning (by setting usage policies). Although, the temperature - humidity sensors are cheap, the controller network is relatively expensive. However, the sharing of the controller network for several air conditioned room or for the large and high energy consumption air conditioned room could be considered to receive high profit.

4. Conclusions

The average energy consumption of the air conditioning system is 10.30 kWh which ranges from minimum to maximum of 8 and 15 kWh. The results showed the relation of indoor weather and energy consumption as well as understanding the characteristic of energy consumption by the air conditioner in the educational building. The data analysis can be used to plan for the energy efficient building by cutting down the energy use. The energy can be monthly saved for 206 kWh by switching off during noon for only one hour when the room is unoccupied. Further study will need to be carried out on the reduction of energy consumption by increase the temperature set point combined with some ventilation strategies.

Acknowledgments
This research project was financial supported by Naresuan University under the project code R2558B010. Highly thanks to Midori Co.Ltd. for supporting the smart sensors and operating software. Special thanks to Prof. K Tanaka for technical suggestion and Mr. O Hanaoka for the CT sensor technical support. Further thanks to our students in the Energy Technologies Building Laboratory, ARM Laboratory and Mobile Computing Lab for their assistant in the lab.

References

[1] IPCC, 2014: Climate Change 2014: Synthesis Report. Contribution of Working Groups I, II and III to the Fifth Assessment Report of the Intergovernmental Panel on Climate Change [Core Writing Team, R.K. Pachauri and L.A. Meyer (eds.)]. IPCC, Geneva, Switzerland, 151
[2] Yuan J, Farnham C, Emura K 2015 Development and application of a simple BEMS to measure energy consumption of buildings *Energ Buildings* **109** 1–11
[3] Levermore G J 1992 Building energy management systems: an application to heating and control London : E & FN SPON, 275-315
[4] Anda M, Temmen J 2014 Smart metering for residential energy efficiency: The use of community based social marketing for behavioural change and smart grid introduction. *Renew Energ* **67** 119-27
[5] Leavey A Fu Y Sha M Kutta A Lu C Wang W Drake B Chen Y Biswas P 2015 Air quality metrics and wireless technology to maximize the energy efficiency of HVAC in a working auditorium *Build Environ* **85** 287 –97
[6] Guerra-Santin O Tweed CA 2015 In-use monitoring of buildings: An overview of data collection methods *Energ Buildings* **93** 189–207
[7] Chua K J Chou S K Yang WM Yan J 2013 Achieving better energy-efficient air conditioning – A review of technologies and strategies *Appl Energ* **104** 87–104
[8] Allouhi A, Fouih Y El, Kousksou T, Jamil A, Zeraouli Y, Mourad Y 2015 Energy consumption and efficiency in buildings: current status and future trends *J Clean Prod* **109** 118-130
[9] Garniera A, Eynardb J, Caussanel M, Grieu S 2015 Predictive control of multizone heating, ventilation and air-conditioning systems in non-residential buildings *Appl Soft Comput* **37** 847–62
[10] Honeywell Humidlcon Datasheet http://sensing.honeywell.com/honeywell-sensing-humidicon-hih6100-series-product-sheet-009059-6-en.pdf

IOP Conf. Series: Earth and Environmental Science **40** (2016) 012076 doi:10.1088/1755-1315/40/1/012076

Overall review of feed-in tariff and renewable portfolio standard policy: A perspective of China

Q Y Yan, Q Zhang[1], L Yang and X Wang

School of Economics and Management, North China Electric Power University, Beijing, 102206, PR China

E-mail: zhangqian881111@163.com

Abstract. A major share of China's total carbon dioxide (CO_2) emissions is from the electric power sector. To solve this problem, Chinese government has implemented many renewable energy policies in the electric power sector. In China, the most popular renewable energy policies are Feed-in tariff (FIT) and renewable portfolio standard (RPS). This paper first introduces the current development of renewable electricity generation. Second the design plan and implement of FIT and RPS in China's thermal electricity generation sector are summarized in this paper. Third this paper establishes a complementary mode of FIT and RPS which can provide a stable environment to make the FIT and RPS work together. Finally, based on the above analysis, this paper proposes relative suggestions for the implementation of FIT and RPS in China making recommendation for the development of electricity generation from renewable energy.

1. Introduction

Electricity not only plays a major role in the sustainability of social but also is the important component for a country's economic development. In China, electricity consumption rose from 77.63 Mtce in 1991 to 427.37 Mtce in 2009, with an average annual growth rate of about 9.93% [1]. After rapid growth since the early 1990s, China overtook the US in 2011 as the largest electricity consumer. In 2014, thermal power account for an estimated 73% of China's electricity production, and about 90% of that thermal power was coal-fired, resulting in massive quantities of CO_2 emissions and great pressure to reduce it. According to the International Energy Agency, China is the world's largest producer of CO_2 emissions, passing the US in 2007. During the 2015 Paris United Nations Climate Change Conference, Chinese government announced that the country would decrease its CO_2 emissions per unit of gross domestic product (GDP) by 60% - 65% between 2005 and 2030. China also pledged to peak CO_2 emissions by around 2030 and try to achieve it as soon as possible. For China, this pledge meant that the Chinese government promised a maximum level of CO_2 emissions, and China would endeavor to control its CO_2 emissions and develop the renewable energy industry's scale to build a competitive renewable energy industry system. Compared with other sectors, the electric power sector is one of main sectors for the Chinese CO_2 emissions. In 2010, almost 40% of all CO_2 emissions in China are from this sector. To meet China's targets for the reduction of CO_2 emissions, Chinese government has begun to implement the renewable energy power generation.

It is widely known that the Chinese government has made great efforts to develop renewable energy electricity generation. Recently, China is striving to increase its investment in renewable. In 2014,

[1] Address for correspondence: School of Economics and Management, North China Electric Power University, Beijing, 102206, PR China. E-mail: zhangqian881111@163.com.

China's investment in renewable was up to US$89.5 billion increasing almost 32% compared with the previous year. China is now the largest investor in renewable and its investment for renewable is nearly 73% more than the second largest investor, the US. Chinese government also implemented some renewable energy policies to reduce CO_2 emissions from its electric power sector. The development of renewable energy policies in China can be tracked back to 2004, when Chinese government announced to formulate a renewable energy policy to develop renewable energy. The renewable energy polices used in the electricity power sector can be summarized here as follows:

- Promote the implementation of FIT policy in the electricity power sector, and the Chinese government provides a fixed price based on the learning curve associated with a particular technology.
- Make clear a specific fraction of renewable generation in power.
- Government financial support for renewable energy projects.
- Encourage various demand side management (DSM) programs.

Although these renewable policies have help increase the energy efficiency of the electric generation sector, China still faces with many increasing problems in the electricity generation, such as energy security, environmental pollution. The reasons for these problems are the high generation cost, the uneven distribution resource situation and immature technologies.

The balance of this paper can be organized as follows. Section 2 presents a literature review on the renewable energy policies. Section 3 presents the current situation of renewable energy development in China's power sector. Section 4 discusses the implementation of FIT and RPS in China and presents arrangements when FIT and RPS work together. Section 5 concludes the paper.

2. Literature review

2.1. FIT

FIT is a policy mechanism designed to encourage the generator to increase investment in renewable energy technologies. Currently, FIT is the most popular renewable energy policy in the world. FIT is account for a greater share of renewable energy development than RPS policy. In the FIT scheme, the power grid operators require to purchase electricity produced by renewable energy sources at a price determined by the regional public authorities and guaranteed for a specified period of time [2]. FIT is represented by Germany, Denmark, Spain, and China. These countries have made a great progress in renewable energy industry by implementing FIT scheme. In the European Union between 2000 and the end of 2009, more than 15,000MW of solar photovoltaic (PV) power and more than 55,000 MW of wind power has been developed under the FIT scheme [3]. In total, FIT promote approximately 75% of global PV and 45% of global wind deployment. That means that FIT can help to solve the energy security and environmental pollution at the same time.

Currently, many scholars propose that FIT is the most widely used renewable energy policy and efficient at rapidly developing the renewable energy. In the literature, some researches focus on how to design more efficient FIT mechanism. Lesser and Su propose an innovative two-part FIT which is easy to implement considering the strengths of traditional FIT and based on market mechanism [4]. del Río González P provides an overview of Spain's FIT system in the last ten years and uses a political economy approach to interpret the practical consequences of serial FIT reforms and their main design principles [5]. Couture and Gagnon present an overview of seven different FIT payment models for electricity produced by renewable energy sources and suggest that different FIT payments methods have important impacts on investor risks, and total rates of renewable energy development [6]. Sijm proposes that FIT is an effective implement to promote the development of electricity produced by the renewable energy sources in the short term but may become hard to sustain as it may suffer from some major drawbacks in the long term [7]. Sun makes a comparison between three regulation types of FIT by using a three-stage game model and assesses their differences of optimal regulated price and social welfare [8].

In fact, different countries and regions execute different types of FIT policy. In general, they can be summarized as market dependent and market-independent FIT models [6]. The market-independent

FIT which is called as fixed-price policies, because it provides a fixed or minimum price for electricity produced by renewable energy sources delivered to the grid [9,10]. Market-dependent FIT is known as premium price policy, or feed-in premiums, since additional premium payment is added on the market price [9,10]. Most countries with FIT policy select the market-independent FIT, such as Germany, Denmark and Spain. However, market-dependent FIT policy contributes to an increased integration of renewable energy source into the electricity market, resulting in a more efficient combination of electricity supply with demand [11]. Therefore, market-dependent FIT policy has been introduced in several European Union countries in the past few years, such as Czech Republic, Denmark, Germany (since 2012) and Italy. In China, power grid operators are required to purchase electricity produced by registered renewable energy sources at a fixed price set by the National Development and Reform Commission (NDRC).

2.2. RPS

RPS is a regulation which requires electricity supply companies to provide electricity produced by renewable energy sources, which as a percentage of total energy use, to their customers. RPS is implemented by several countries, including UK, Italy, Poland and 29 of 50 US states. Although RPS is relatively new among the renewable energy policies, it has developed rapidly in recent years. The reason is that RPS maintains an incentive for electricity supply companies to improve their renewable energy technologies and reduce generation cost, reduces the government involvement, and can help to combine the regulated market outcome with an environmental target [12-16]. The researches about the RPS policy mainly focus on renewable energy target, tradable green certificate, renewable energy technologies and the impact of RPS on the carbon emissions.

In the literature, Bird et al. examine the impacts of RPS on the US electricity sector with three different renewable energy targets [17]. UK has been a pioneer in the use of RPS with the introduction of Renewable Obligation Certificates (ROCs). In UK, the electricity supply companies require to buy ROCs with a percentage of total energy sales [18]. Berry and Jaccard think that it is a main challenge to determine the renewable energy target in RPS scheme. They suggest that during the process of setting target the government should take into consideration the impacts of renewable energy target on generation cost, electricity price and the level of competition among fuels and technologies [19].

Tradable green certificate (TGC) is a supporting policy for the RPS scheme. Renewable energy target or quota support scheme based on TGC involves a renewable energy electricity quota set by the national, regional or local government. The electricity supply companies must show that they have obtained the renewable energy quotas. The electricity supply companies either producing renewable energy electricity by themselves or buying green certificates from the other electricity providers to obtain the quotas. In different countries, TGC scheme pursue different targets. Generally, the objectives of different countries are concentrated on increasing the electricity produced by renewable energy sources, driving investment in renewable energy electricity by a cost-efficient manner, and inducing innovation and technical changes [20-23]. At present, several scholars highlight that TGC is an efficient scheme to increase the electricity produced by renewable energy sources and reduce CO_2 emission. Bergek and Jacobssonanalyze the outcome of the Swedish TGC system and point out that TGC is an effective scheme for increasing the supply of "green electricity" and decreasing social cost [20]. Morthorst introduces the international TGCs to an open power market and analyzes the cost effectiveness and the contribution on greenhouse gas reduction [24]. MátyáTamás et al. make a comparison between FIT and TGC about their implementation effects by using the data from the UK market, the results show that TGC is more efficient in social welfare than FIT [25].

However, implementation of TGC always faces with many problems and market risks. In addition, performance of TGC scheme suggests that not all the goals, which are made by the government and electricity supply companies, are satisfied [26-28]. Amundsen and Mortensen assess the implementation of Denmark TGC scheme and point out that an increase in the minimum requirement will result in the increase of electricity price and the decrease of total supply [29]. Lemming analyzes financial risk of TGC market from existing renewable energy sources and potential investors and points out that TGC price fluctuations are efficient in decreasing the financial risk [30]. And Nielsena et al. point out that it is difficult to build a unified TGC market because of the different types of TGC in Europe [31,32].

As can be seen from the literature review, a lot of work has been done towards the FIT and RPS, which concentrate on the impact of FIT and RPS on economic, technology and environment. However, few studies exist for the relationship between FIT and RPS. Richkerson et al. consider that FIT and RPS can be formed to work together synergistically [33, 34]. There is no research made in-depth study on the arrangement of the implementation of FIT and RPS in developing countries. At present, China is in a critical period to reform electric system and drive investment in electricity generated by renewable energy sources. China began to use FIT in 2005 and RPS in 2007, respectively. What's effects of the implementation of FIT and RPS? Does FIT and RPS can work together in China? What's the arrangement if FIT and RPS work together in China? These problems will be discussed in the rest of this paper.

3. Renewable energy development in China's power sector

As China's economic development continuously growing, China's power sector develops rapidly. China has become one of the largest energy production and consumption countries [35]. Coal is the dominant fuel in China's power sector which leads to huge pressure to solve environmental problems. On March, 2016, China's thirteenth "Five-Year Plan" was approved. It is the first plan to include specific guidance on energy consumption control. In this plan, China announces that it would limit its overall energy use to 5 billion-tonne coal equivalent by 2020. It indicated that China should endeavor to develop renewable energy electricity in the next five years. At present, China already owns the largest installed capacity of wind and hydroelectric power, as well as the vast majority of solar heating and biogas installations in the worldwide. In 2013, China overtook the whole of Europe and become the largest country in installed capacity. China is also the world's leading investor in renewable energy industry.

3.1. Hydropower

China is one of the world's leading investors in hydropower. At the end of 2013, China's installed hydroelectricity generating capacity is 280 gigawatt (GW) [36]. In China, almost three-quarters of the total installed renewable energy electricity are hydro. China increased 19,370 megawatt (MW) of new hydropower generating capacity which helps China's total installed capacity achieve to 320 GW in 2015.

Figure 1 shows the hydropower generating capacity in China between 2000 and 2015. Due to the implementation of effective policies, China's installed capacity of hydropower increased from 79 GW in 2000 to 320 GW in 2015. The percentage of hydropower in China's total power generation increased from 16.4% in 2000 to 17.4% in 2012, as shown in Figure 2. China has almost met the objectives about its hydropower development which is set in the twelfth "Five-Year Plan". However, the pumped storage capacity has not been developed at the same rapid rate as the traditional hydropower. In 2015, China's installed pumped storage capacity is 23 GW which is lower than the planned goal of 40 GW in the twelfth "Five-Year Plan". This means that there exists considerable developing space for China's hydropower generation. Chinese government promised that the installed capacity of hydropower would reach to 420 GW by 2020 which include 70 GW of pumped storage capacity [37].

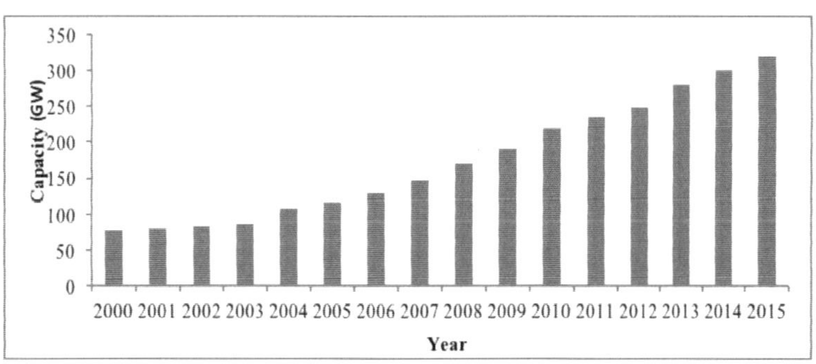

Figure 1. China's hydropower generating capacity from 2000 to 2015

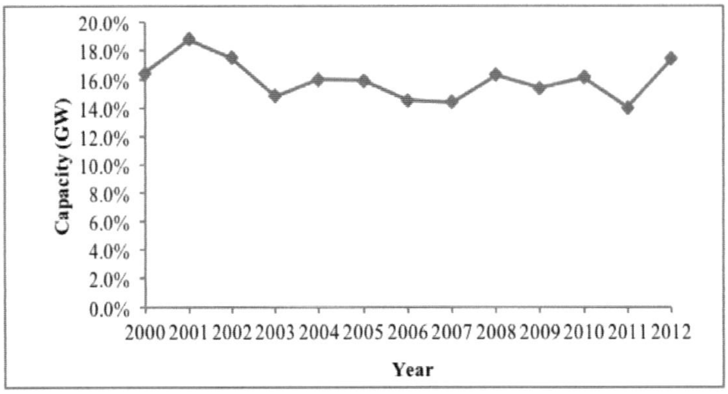

Figure 2. The proportion of China's hydropower from 2000 to 2012

3.2. Wind power

At present, China's wind energy resources can produce about 1 billion kW of wind power. However, the current exploitation of wind resource only generates 0.26 million kW [38]. This suggests there is considerable space for the wind power developing. Figure 3 shows China's on-grid wind power installed capacity between 2006 and 2012. The cumulative wind power installed capacity increased from 2.67 GW in 2006 to 62.66 GW in 2012, which is an annual growth rate of 69.2%. With the high development rate, China overtook the US and become the main user of wind energy in the world. Although the percentage of wind power in China's total electricity generation was only 2.02% in 2012, wind power had become the third largest source of Chinese electricity generation after thermal and hydro.

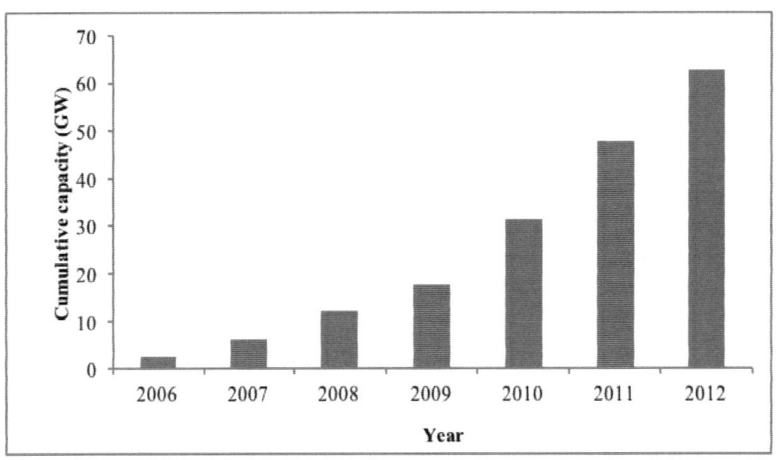

Figure 3. China's on-grid wind power installed capacity from 2006 to 2012

About 80% of the total wind capacity in the wind power sector is state-owned, and the state-owned wind power installed capacity reached to 50.77 GW by 2012. Sino-foreign joint venture invested 8.83 GW in 2012, accounting for 14% of total. However, only 1% (5%) of the total wind capacity results from the foreign (private) ownership. In China, the main producers of wind energy resources consist of the north, northeast, southeast coastal areas and local island areas [39]. Among these areas, the north China generates almost 70% of the total installed wind capacity. In addition, Hebei and Gansu provinces keep their installed wind capacities at a high level.

3.3. Solar power

China is also having a great boom in solar photovoltaic (PV) projects. In the end of 2008, China began to develop the solar PV. Recently, installed solar PV has increased quickly. The installed solar PV increased from 0.8 GW in 2010 to 7 GM in 2012, and about 4 GW were added in 2012. In 2013, China become the world's leading user of solar PV, and the total installed solar PV capacity reached to 35.78 GW by end-June 2015. At present, more than 66% of solar energy resources in China are concentrated in Qinghai, Xinjiang, Tibet, Inner Mongolia, Sichuan and Gansu provinces.

The country's National Energy Administration recently said that China aims to triple its solar PV generation capacity to 143 GW by the year 2020. This means that annual increments of solar PV generation capacity must maintain over 15 GW.

3.4. Biomass power

China is rich in biomass energy resources that can be categorized into forestry, energy crops, animal and sewage, and waste of agricultural. Table 1 lists the provinces with top ten biomass energy installed capacity in 2012. From the table, we conclude that China's biomass energy resources are unevenly distributed.

Table 1. The provinces with top ten biomass installed capacity in 2012

ID	Province	MW
1	Shandong	1,089
2	Henan	640
3	Jiangsu	554
4	Heilongjiang	524
5	Hubei	475
6	Guangdong	442
7	Anhui	388
8	Hebei	367
9	Inner Mongolia	320
10	Zhejiang	290

Figure 4 is China's biomass power installed capacity and generation from 2000 to 2012. Figure 4 shows that installed biomass power capacity increased from 1.0 GW in 2000 to 7.7 GW in 2012, with an annual growth rate of 18.54%. Figure 4 also shows that the installed biomass power generation increased from 2.2 Terawatt Hour (TWh) in 2000 to 33.7 TWh in 2012, which is an annual growth rate of 25.53%. Before the wind power caught up, biomass power is the second largest source of Chinese renewable energy power capacity.

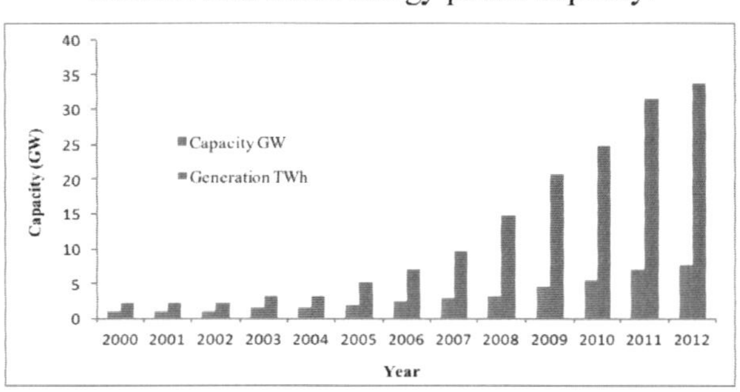

Figure 4. China's biomass power installed capacity and generation from 2000 to 2012

4. The problems of the implementation of FIT and RPS and complementary mode

The above analysis indicates that China has great success in the development and utilization of renewable energy, and developing renewable energy electricity in the future is very prospective. The development of renewable energy contributes to the supporting of renewable energy policies. However, in the process of implementing FIT and RPS in China, there exist some crucial barriers.

4.1. The problems of the implementation of FIT

In early 2003, FIT was introduced in China to support the development of wind power. Under the FIT, power grid operators have to purchase energy generated from registered renewable energy producers where the buying price is set by the NDRC.

The tariff amount at first was determined by bidding or negotiation based on a case-by-case basis. However, the prices with and without bidding may be far from each other. In particular, the former is about 0.38-0.5 RMB/kWh, while the latter is about 0.5-0.8RMB/$_k$Wh. This may result in intense competition among large state-owned renewable power generators, and speculative bids that are usually not sufficient to implement the project [41]. In July 2009, the NDRC proposed the "Notice on Improving the Pricing Policy for On-Grid Wind Power Prices" to solve this problem. In the notice, the NDRC divided the country into four wind energy resources areas and provided a wind power price ranging from 0.51 RMB/kWh to 0.61 RMB/kWh in the different areas.

In 2014, NDRC raised the hydropower price in the "Notice on improving the Pricing Mechanism for Hydropower Prices". According to the notice, the hydropower tariffs will be determined by the average electricity price of power demand regions. This plan can help improve the investment to the hydropower sector and achieve the goals of twelfth "Five-Year plan" and thirteenth "Five-Year plan".

In 2011, NDRC proposed the first solar FIT policy in the "Notice of Improving the Pricing Policy for On-Grid Solar Photovoltaic Power Prices". This notice provided a solar PV tariff ranging from 1.0 RMB/kWh to 1.15 RMB/kWh based on the data in the PV project. However, the 2011 notice did not consider the intensity of solar radiation in different areas of China. Solar projects always focused on the western part of China, such as Gansu and Qinghai, where the energy demand is not as high due to lower population density and level of economic development [42]. To improve this situation, the NDRC proposed the "Notice on Promoting the Healthy Development of the Solar PV Industry through the Price Leverage Effect" in 2013.

Governmental designated price and governmental guided price are the two stages of China's biomass power tariff. At first, Chinese government applied governmental designated price as its biomass power tariff. In this stage, the government set the biomass power tariff of the area where the facility was located for fifteen years. And the value of this kind price was the sum of 0.25 RMB/kWh and the 2005 benchmark desulfurized coal price in the area [38]. In July 2010, NDRC proposed the "Notice on Improving the Pricing Policy for Biomass Power Prices". This notice provided a unified price of 0.75 RMB/kWh for biomass power projects.

Although Chinese government endeavored to improve renewable energy policies to develop its FIT, it still faces with a few problems. First, China's FIT scheme did not take into consideration the China's uneven resources distribution. In China, the resources are concentrated in the regions of Northwest, North, and South, and the developed administrative regions are short of resources. Therefore, some electricity supply companies in the developed administrative regions have little interest to produce the renewable power, which hinders the implementation of FIT scheme. Second, China's tariff for renewable power was insufficient to provide incentives for generators compared with the international average level. The private generators are reluctant to get involved in other energies to produce power. Third, the permanent benchmark tariff can not provide guidance for the generators to obtain certain return on investment. It's hard for the generators to make a response to market price signals which result in increasing peak-valley deviation and obvious anti-peaking effect [43].

4.2. The problems of the implementation of RPS

In China, RPS is a relatively new policy to promote the development of renewable energy. To meet the mid-term (2007-2010) and long-term (until 2020) development objectives for renewable energy, NDRC introduced the RPS scheme in 2007. Under RPS scheme, China's power generators require to produce a certain fraction of renewable energy electricity in their total electricity generation, especially wind, solar and biomass. Table 2 shows the renewable energy goals of RPS in China. Table 2 indicates that China's biomass power would achieve to 30 GW by 2020, with an annual growth rate of 18.2%. The hydropower, pumped hydro, solar PV and on-grid wind power would increase to 350 GW, 70 GW, 100 GW and 230 GW by 2020, respectively.

In the process of developing the RPS scheme, there are four factors that influence the renewable power. They are targets, eligible technologies, penalty and monitor mechanism, and TGC. At present, many aspects of these factors are left undeveloped in China.

Table 2. The overview of China's renewable energy goals

Items	Current	Goals
	2012	2020
Biomass power	8 GW	30 GW
Hydropower	249 GW	350 GW
Pumped hydro	20 GW	70 GW
Solar PV	5.4 GW	100 GW (70 GW by 2017)
Concentrated solar power	0.014 GW	3 GW
On-grid wind power		200GW onshore 30 GW offshore

First, with regard to the renewable energy target, the thirteenth "Five-Year Plan" sets ambitious targets about non-fossil fuel which increase from 12% in twelfth "Five-Year" Plan to 15%. In order to make the RPS scheme working, the targets need to be widely accepted. Although thirteenth "Five-Year Plan" has set the explicit target, it did not divide the long-term target into smaller short-term targets. For the electricity supply companies, the local government cannot verify the percentage of renewable energy electricity in their total electricity generation.

Second, the high cost of the renewable power technologies would hinder implementation of RPS. For instance, the generation costs for wind is 500-600 RMB/MWh. However, the wind farms that are far away from the local power grids must pay higher cost for their electricity transportation and construction. Therefore, low-cost renewable energy sources would like to control the market and prevent the development of emerging and more expensive renewable energy sources.

Third, there is a lack of thorough policy system in the renewable energy industry to encourage the investor's confidence in China. In particular, the lack of incentive mechanisms and supervision mechanisms in China indicate that the RPS scheme is still weak or voluntary. In fact, a lack of incentive mechanisms will influence the renewable power capacity, the motivation of market participants and the formation of the market. In addition, the lack of supervision mechanisms would result in unfair competition in the RPS scheme.

The last but not the least, China has not established the TGC trading market net. In March 2016, the National Energy Administration (NEA) announced that China is planning to build its TGC trading market in the future. The new plan aims to support the RPS scheme and trade the renewable power among the registered renewable energy sources. NEA estimated that the solar PV and wind power will obtain great benefit from the TGC mechanism. The lack of TGC will hinder the fulfillment of renewable energy target in RPS scheme. In TGC trading market, generators falling to fulfill the target can choose to purchase green certificate to meet targets. Hence, the TGC trading market is an important support mechanism in the process of implementing the RPS successfully.

4.3. Complementary mode of FIT and RPS

From the above analysis, China has made a great progress in increasing renewable power by implementing of FIT and RPS. However, there exist a lot of problems in the process of implementing the renewable energy policies. Both the FIT scheme and RPS scheme have the advantages and disadvantages. In fact, FIT scheme plays an important role in meeting targets of RPS scheme. This paper establishes the complementary mode of FIT scheme and RPS scheme based on China's power market structure. The complementary mode takes into account three interactions, as described below.

4.3.1. Renewable power price under FIT and RPS.
In the complementary mode, the renewable power price is determined by the FIT scheme. Price is the most effective instrument to adjust the mechanism in a market economy. In practice, reasonable price can help to ensure the profit of renewable energy sources. The government should improve the attractive price policy and adjust the price when the cost of power generation changes. FIT scheme provides enough profit to the renewable energy sources and make sure that the sufficient supply can come online. The complementary mode of FIT and RPS would use the power price which is set by the State Electricity Regulatory Commission based on the FIT.
4.3.2. The quota allocation under FIT scheme and RPS scheme.
There are abundant renewable energies, huge potential of renewable energy development and market demand in China. However, due to the uneven distribution resource situation and no penalty in FIT scheme, the electricity supply companies have no motivation to generate renewable power. The complementary mode of FIT and RPS require the electricity supply companies to satisfy the minimum percentage of renewable energy electricity in the total power generation under FIT scheme. That means that the electricity supply companies have to obey the RPS scheme. At the same time, the power grid operators have to purchase renewable energy electricity generated from registered renewable energy sources at a buying price set by FIT scheme.
4.3.3. Support for renewable energy technologies under RPS and FIT.
Although there exist long-term utility contracts, new or emerging technologies still cannot secure finance [11]. In complementary mode of FIT and RPS, a renewable energy subsidy fund would be built. This fund is mainly supplied by a fixed surcharge for total electricity purchases. This fund is established to compensate the electricity supply companies with new renewable energy technologies. The subsidy should be enough high to support the additional investment risk faced by the electricity supply companies. Appropriately structured FIT scheme will take into consideration the risk premium for new renewable energy technologies and provide the long-term assurance that the electricity supply companies require.

5. Conclusions

It is necessary to promote sustainable economic growth by using renewable energy. Currently, Chinese government has proposed a series of renewable energy policies to solve its environmental problems and support the renewable energy development. Among these policies, the FIT and RPS are the two most popular renewable energy policies. This paper elaborates and presents three issues about the FIT and RPS. They are the current situation of renewable energy development in China's power sector, the problems of the implementation of FIT and RPS, and the complementary mode of FIT and RPS. The main conclusions drawn from the analysis are summarized as follows.

First, China already owns the largest installed capacity of wind and hydroelectric power, as well as the vast majority of solar heating and biogas installations in the worldwide. In the twelfth "Five-Year Plan", renewable energy resources were identified as an emerging strategic. This plan also set overall targets for renewable energy sources. In the plan, China's biomass power would increase to 30 GW by 2020, with an annual growth rate of 18.2%. The hydropower, pumped hydro, solar PV and on-grid wind power would increase to 350 GW, 70 GW, 100 GW and 230 GW by 2020, respectively.

Second, although Chinese government endeavored to improve renewable energy policies to develop its FIT, it still faces with a few problems. With regard to the FIT scheme, there exist three problems faced by the Chinese government: (1) China's FIT scheme did not take into consideration the China's uneven resources distribution; (2) China's tariff for renewable power was insufficient to provide incentives for generators compared with the international average level; (3) the permanent benchmark tariff cannot provide guidance for the investment.

Third, for the RPS scheme, the problems can be summarized as: (1) the thirteen "Five-Year Plan" did not divide the long-term target into smaller short-term targets; (2) the high cost of the renewable power technologies can have an important effect on implementation of RPS; (3) there is a lack of thorough policy system in the renewable energy industry to encourage the investor's confidence in China.

From the future perspective, there exists considerable space for the development of renewable energy. Due to the technology development and improvement in renewable energy policies, the renewable energy will be gradually moving into high gear in the next thirteenth "Five-Year Plan". We suggest that the policy-markers establish the complementary mode of FIT and RPS that can provide a stable environment to make the FIT and RPS work together. In the complementary mode, the electricity supply companies have to obey the RPS scheme. At the same time, the power grid operators are required to purchase energy generated from registered renewable energy sources at a buying price set by FIT scheme. This complementary mode plays an important role in promoting a low-cost, diversified and large-scale development for renewable energy. This mode also can help China built the TGC trading market and meet emission reduction targets.

Acknowledgments
The China Scholarship Council, Project of the Production-Study-Building for Joint Training of Graduated Students in Beijing and Fundamental Research Funds for the Central Universities (Grant No. 2016XS77), supports this research.

References
[1] China Energy Statistical Yearbook 2000-2013 *China Energy Statistical Yearbook* China Statistical Press
[2] Menanteau P, Finon D and Lamy M L 2003 *Energy Policy* **31** 8
[3] Couture T D, Cory K, Kreycik C and Williams E 2010 Policymaker's guide to feed-in tariff policy design *National Renewable Energy Laboratory*
[4] Lesser J A and Su X J 2008 *Energy Policy* **36** 3
[5] Del Río González P 2008 *Energy Policy* **36** 8
[6] Couture T and Gagnon Y 2010 *Energy Policy* **38** 2
[7] Sijm J P M 2002 The performance of feed-in tariff to promote renewable electricity in European countries *Energy research Centre of the Netherlands*
[8] Sun P A 2013 A comparative study of renewable energy industry regulation on feed-in tariff based on pricing strategy of value standard method *J. Appl. Math.* 368296
[9] Mendonça M 2007 *Feed-in Tariffs: Accelerating the Deployment of Renewable Energy* (Trowbridge: Cromwell Press)
[10] International Energy Agency (IEA) 2008 *Deploying Renewables: Principles for Effective Policies* 978-92-64-04220-9
[11] Cory K, Couture T and Kreycik C 2009 *Feed-in tariff policy: design, implementation, and RPS policy interactions" Oak Ridge* DE-AC36-08-GO28308
[12] Wiser R, Namovicz C, Gielecki M and Smith R 2007 *Electricity J.* **20** 4
[13] Fershee J P 2008 *Energy L. J.* **29** 1
[14] Yin H and Powers N 2010 *Energy Policy* **38** 2
[15] Michaels R J 2008 *Energy L. J.* **29**
[16] Rossi J 2010 *Conn. Law Rev.* **42** 5
[17] Bird L, Chapman C, Logan J, Sumner J and Short W 2011 *Energy Policy* **39** 5
[18] Butler L and Neuhoff K 2008 *Renew. Energy* **33** 8
[19] Berry T and Jaccard M 2001 *Energy Policy* **29** 4
[20] Bergek A and Jacobsson S 2010 *Energy Policy* **38** 3
[21] Kildegaard A 2008 *Energy Policy* **36** 9
[22] Haas R, Panzer C, Resch G, Ragwitz M, Reece G and Held A 2011 *Renew. Sustain. Energy Rev.* **15** 2
[23] Jacobasson S, Bergek A, Finon D, Lauber V, Mitchell C, Toke D and Verbruggen A 2009 *Energy Policy* **37** 6
[24] Morthorst P E 2003 *Energy Policy* **31** 13
[25] Mátyá Tamás M, Bade Shrestha S O and Zhou H Z 2010 *Energy Policy* **38** 8
[26] Agnolucci P 2007 *Energy Policy* **35** 6
[27] Menanteau P, Finon D and Lamy M L 2003 *Energy Policy* **31** 8

[28] Toke D and Lauber V 2007 *Geoforum* **38** 4
[29] Amundsen E S and Mortensen J B 2001 *Energy Econ.* **23** 5
[30] Lemming J 2003 *Energy Policy* **31** 1
[31] Nielsen L and Jeppesen T 2003 *Energy Policy* **31** 1
[32] Verhaegen K, Meeus L and Belmans R 2009 *Renew. Sustain. Energy Rev.* **13** 1
[33] Richerson W and Grace RC 2007 *"The debate over fixed price incentives for renewable electricity in Europe and the United States: Fallout and future directions. Washington DC: Heinrich Boll Foundation"* http://www.boell.org/docs/Rickerson_Grace_FINAL.pdf
[34] Grace RC, Rickerson W and Corfee K 2008 *California feed-in tariff design and policy options. California Energy Commission* CEC-300-2008-009D
[35] Zhemin J 2008 *Energy in China* **30**
[36] China National Renewable Energy Centre (CNREC) 2014 *Activities within the China National Renewable Centre* China.
[37] World Resources Institute (WRI) 2014 *What are China's national climate and energy targets?* http://www.chinafaqs.org/files/chinainfo/ChinaFAQs_table_China_climate_energy_targets_0.pdf.
[38] Zeng M, Liu XM, Li N and Xue S 2013 *Renew. Sustain. Energy Rev.* **25**
[39] The State Electricity Regulatory Commission 2009 *Research report on the wind power development in China*
[40] Zhang N, Lior N and Jin H 2009 *Energy* **36** 6
[41] Zhao ZY, Zuo J, Fan LL and Zillante G 2011 *Renew. Energy* **36** 1
[42] Winston and Strawn 2014 *Feed-in tariff handbook for Asian renewable energy systems*
[43] Zhang P D, Yang Y L, Tian Y S, Yang X T, Zhang Y K, Zheng Y H and Wang L S 2009 *Renew. Sustain. Energy Rev.* **13** 9

Fine study on single sand body and measures for tapping the potential of residual oil during polymer flooding in Pubei reservoir of Daqing

Y J Meng[1]

Oil Recovery Plant No.7, Daqing Oilfield Corp. Ltd., Daqing, Heilongjiang, 163517, China

E-mail: genius510lc@163.com

Abstract. In order to effectively guide the narrow channel sand body oil fields to exploit, according to the sand body distribution characteristics and geological genesis of narrow channel sand body oil fields, the type of single sand body is clarified. By means of identification of logging curves and correlation of well-tie profile, the internal structure of single sand body is recognized. and then the remaining oil genesis, distribution characteristics and the potential areas for polymer flooding are clarified by combining numerical simulation technology and dynamic analysis technology, and the remaining oil potential tapping method is designed by taking into consideration various factors including the characteristics of the remaining oil, reservoir property and product dynamic character. The result shows that the single sand body is divided into five types including multiphase channel superposition, distributary channel, single channel, sheet sand and lenticular sand. Potential remaining oil mainly are distributed in thick oil layers of multiphase channel superposition type and distributary channel type in which channel sands were developed and sedimentary environment are stable inner front facies and lake regressive inner front facies. The remaining oil is developed by optimizing the parameters of polymer flooding and combining many different measures. The study provides technical support for the efficient exploration for polymer flooding.

1. Introduction

The earlier study on the single sand body is only based on the sedimentary microfacies and sand body spatial distribution features, while the study of the internal structure feature of single sand body is neglected. In recent years, along with the continuous development of fine reservoir description technique, the sand body description continually becomes fine; the depicting of internal structure feature of single sand body gradually becomes perfect. In 1985 Miall proposed a analytical method of architectural-elements used for fluvial deposition[1], and in 1988 he published the research contents of architectural-elements and interface about fluvial deposition[2].

Based on Miall's studies and sedimentary cycle characteristics of the single sand body and interlayer distribution, domestic scholars did a large number of researches about fluvial depositional interface for different sedimentary settings. In 2000 Ma Shizhong et al. established horizontal and

[1] Address for correspondence: Y J Meng, Oil Recovery Plant No.7, Daqing Oilfield Corp. Ltd., Daqing, Heilongjiang, 163517, China. E-mail: genius510lc@163.com.

vertical sedimentary patterns of single lateral accretion body of meandering point bar and studied its three-dimensional configuration and interface on the basis of datum such as vertical sequences of point bar or lateral accretion body type of Daqing placanticline[3]. Then aiming at Daqing Oilfield, they used the sealed coring datum and the logging response characteristics, studied the internal structure of underwater distributary channel that belongs to Songliao basin of large fluvial-delta sedimentary System, and established internal structure of channel sand body and vertical sequence interface order[4]. With the deepening study of internal structure model of single sand body, much attention has been paid to the effect of the interlayer distribution on remaining oil distribution [5-7]. In recent years, the relevant studies are getting more and more perfect, and have been well applied in many blocks.

In this paper the research area is located in Putaohua oilfield, which is in the southern of Daqing Placanticline, and belongs to delta front subfacies depositional system which is controlled by the northern provenance. The channel sedimentary sand orientation is from north to south, with sheet, zebraic, interrupted zebraic or lens-shaped distributions, and it is characterized by thin, narrow and relatively large lateral variation. Many adjustments of well pattern thickening and injection-production system have done since the area being developed. But now the area has entered extra high water cut period and the remaining oil becomes so scattered that there is much more difficulty for tapping the potential of remaining oil. In order to seek out a new way to tapping the potential of remaining oil and enhance recovery by well pattern thickening and polymer flooding, a test area of polymer flooding has been carried out. Because the area is a reservoir with medium-low permeability and thin-narrow sand body, it's necessary that the fine study on the internal structural feature of single sand body and the precise description of remaining oil distribution should be done, and then make a advisable method of remaining oil potential tapping to ensure the development effect of polymer flooding.

2. Identification and classification of single sand body

Single sand body is a sand unit, relative independent on development but genetically related on geological formation. From deposition perspective, single sand body has the same cause that is vertical sequential deposition or superimposed erosion [8-9]. According to sedimentary characteristics of the area and definition of single sand body, the sand body assemblages that are continuous in vertical and in plane itself but discontinuous with upper and lower sand body will be recognized as a single sand body[10-12].

Combined with block sand body sedimentary background and characteristics, the single sand body is divided into five types: multiphase channel superposition, distributary channel, single channel, sheet sand and lenticular sand.

Multiphase channel superposition single sand body consists of distributary main channel, underwater distributary channel and shallow shoal underwater distributary, among them, mainly is the underwater distributary main channel; distributary channel single sand consists of distributary main channel, underwater distributary channel and shallow shoal underwater distributary, among them, mainly is distributary main channel and distributary shallow river; Single channel single sand body consists of inside the leading edge of the river, transitional channel and non- body sheet sand body; sheeted sand monosandbody substantially consists of body and non-body sheet sand; lenticular mould single sand is of isolated lenticular distribution.

3. Internal structure characteristics of single sand bodies

Reference to Maill's analysis method of structural element and the internal structure of channel and interface classes vertical sequence Proposed by Ma Shizhong. And according to the internal structure interface division of the core data from the two sealed coring wells near experimental zone to demarcate experimental zone's logging curve interface, to carry out the analysis of the internal structure of the single sand body. Pu 68-Jian 862 well's core internal structure analysis diagram is shown in figure 1. The wash surface A can be seen at the bottom of the figure. The microelectrode curve increases abruptly at the interface, which is level 4. The calcareous siltstone developed at the top. And the peaked microelectrode curve shows characteristics of high amplitude and huge difference.

The physical interlayer B developed at the middle of sand layer. Microelectrode curve and spontaneous potential curve amplitude decrease, which shows the interface is level 3. According to the structure interface of core data demarcate logging curve (Figure 2) and the interfaces of level 3 to level 5 can be identified on the logging curve. And the interface of level 5 is equal to boundary of sandstone series or thin layers. A stable thick mudstone generally develops at the interface; the interface of level 4 is equal to boundary of sedimentary units, interlayers mostly developed at interface, which are mainly composed of argillaceous interlayer and there are also a few physical interlayer and calcareous interlayer; the interface of level 3 is equal to boundary of uniphase channel. What develop at the interface mostly are physical property interlayers, and argillaceous interlayer and calcium sandwich are less; and interface of level 1 and 2 represent the level interface of stratification, which cannot be identified and contrasted laterally on the well logging curve.

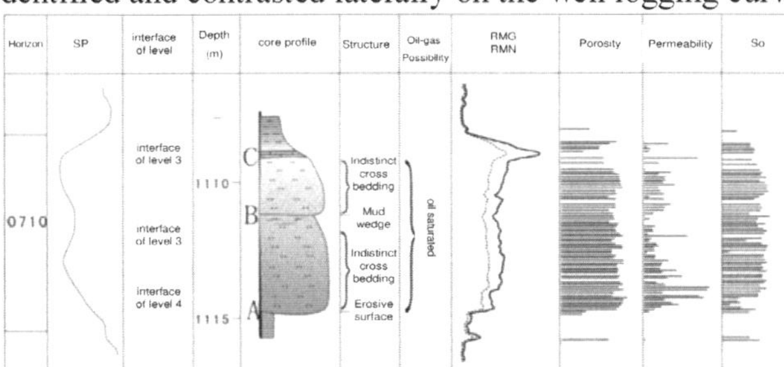

Figure 1. Core internal structure analysis of 7P68-J862 coring Well

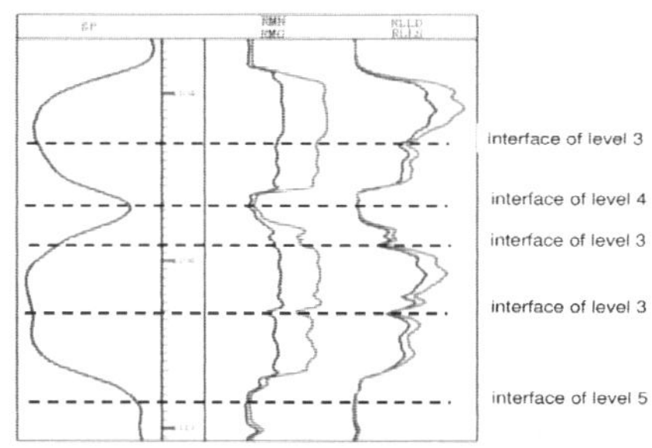

Figure 2. Single sand body level interface

On the basis of sedimentary cycle and interlayer distribution, the internal structure of reservoir is recognized subtly by means of the logging curves and correlation of well-tie profile. The result shows that channel's width of channel type single sand body's single genetic structure is narrow and most of width is less than a well spacing. The degree of channel's curvature is little, dominated by low sinuosity and straight channels. Then, internal channel of the multiphase channel superposition type single sand body is given priority to the multiphase channel superposition in the vertical. And the channel positions developed in different stages are unstable. They are superimposed on a plane and present cross mesh, schistose, dendritic locally. Physical property interlayer and argillaceous interlayer developed at vertical dismembered superimposed location, the average frequency of interlayers is 0.19/m; the internal sedimentation time of distributary channel type single sand body is relatively few and the channel positions developed in different stages are relatively stable. The channel branches or intersections are dendritic on the plane. The scope of interlayers is smaller and average frequency of interlayers is 0.18/m; the channel positions of single channel type single sand body that developed in different stages are stable. Migration swing is small. Channels are superposed and banded mostly. The scope of interlayers is smaller and average frequency of interlayers is 0.11/m; the channel positions of sheet sand type single sand body that developed in different stages are smaller, which are zebraic on the plane, one-period sheet sand body distribution of irregular thin bedded sheet structure, interlayer is given priority to argillaceous interlayer, followed by physical interlayer and average frequency of interlayer is 0.14/m (Figure 3).

IOP Conf. Series: Earth and Environmental Science **40** (2016) 012077 doi:10.1088/1755-1315/40/1/012077

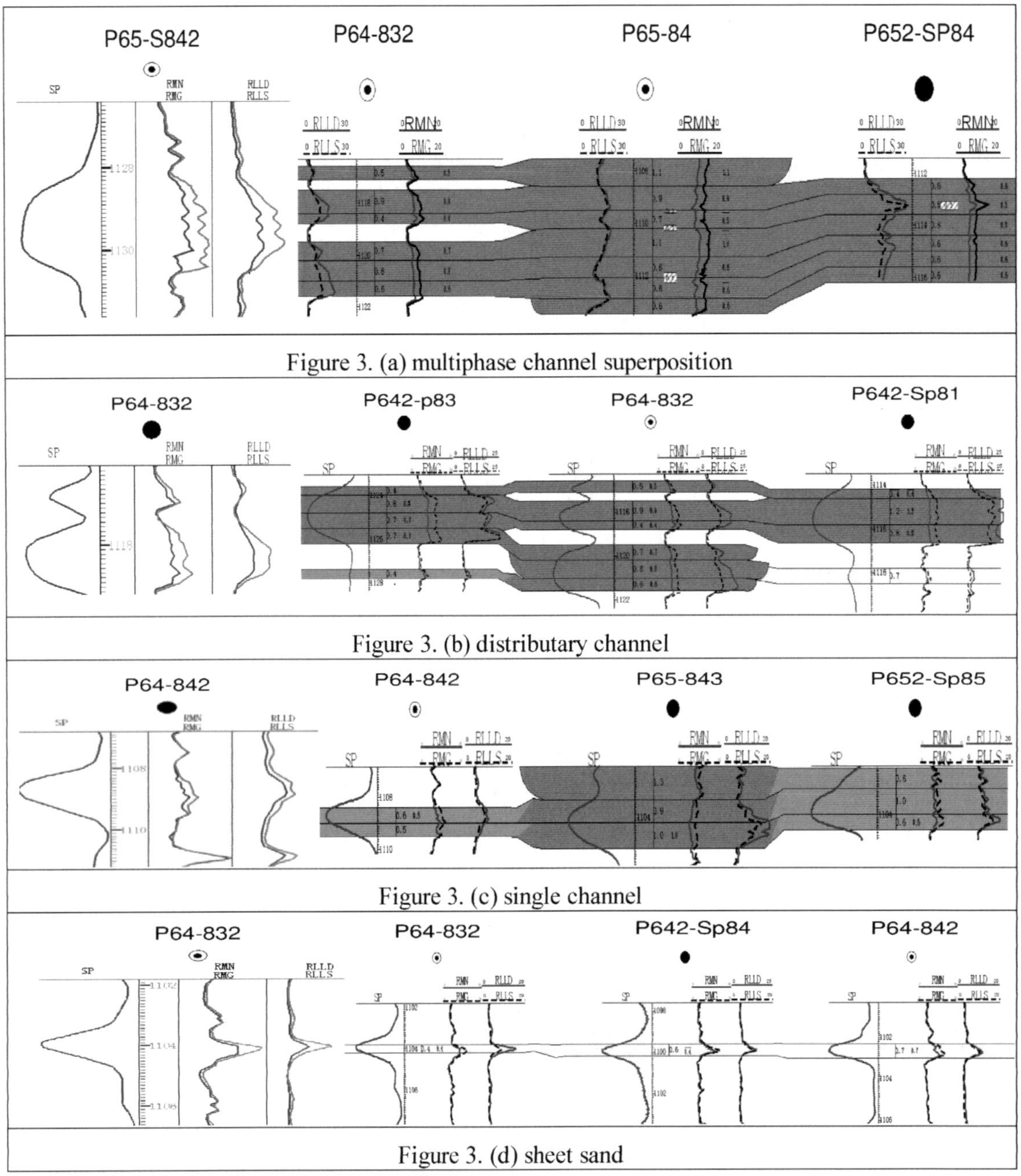

Figure 3. (a) multiphase channel superposition

Figure 3. (b) distributary channel

Figure 3. (c) single channel

Figure 3. (d) sheet sand

Figure 3. The reservoir section of single sand body formation

4. Remaining oil distribution characteristics of single sand body

4.1. Types of remaining oil

By using numerical simulation technique to accurately describe the remaining oil distribution, and according to the genesis of remaining oil [14], we classified remaining oil into 3 categories: on the top of thick oil layer type, reservoir injection-production faultiness type, poor reservoir type (Figure 4). Among them, the type of remaining oil on the top of thick oil layer's sand body are main channel sand body that effective thickness mostly over 1.0m. Entirety development is stable, connected situation is good, injection-production relation is perfect, in-layer heterogeneity are mainly positive rhythm and compound positive rhythm, this type of remaining oil is major target of polymer flooding potential tapping; the type of reservoir injection-production faultiness mainly because of the small sand body's development scale, narrow band, and connected direction is single or formed without connection, this type of remaining oil can be target of polymer flooding potential tapping selectively; remaining oil of the poor reservoir type mainly because of the thin reservoir and poor physical property, it has two types: one is the thin and poor reservoir which has large distributions area, the other is on edge sand body's poor physical property part, this kind of remaining oil mainly distributed in sand sheet's sand body, can be the major target of water flooding after polymer flooding potential tapping.

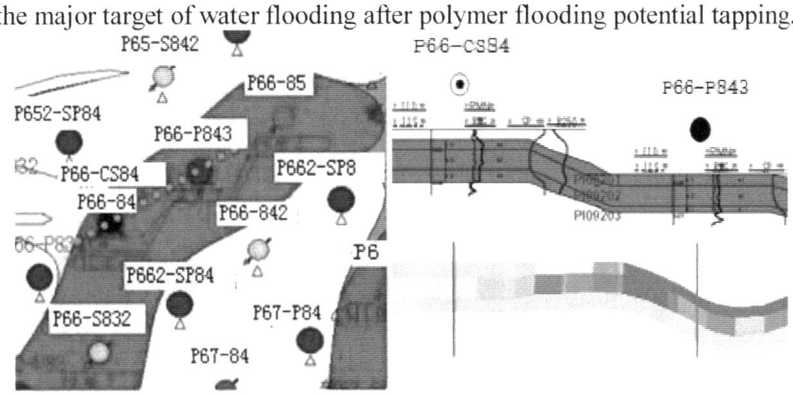

Figure 4. (a) top of thick oil layer type

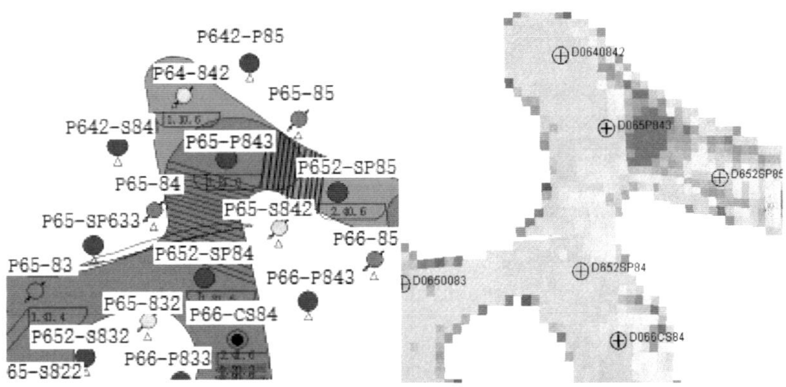

Figure 4. (b) reservoir injection-production aultiness type

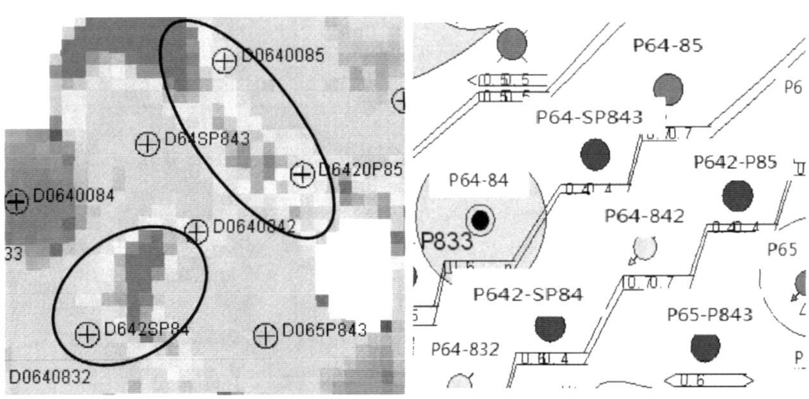

Figure 4. (c) poor reservoir type

Figure 4. The remaining oil type of single sand bodies.

4.2. Types of remaining oil

By using numerical simulation technique to analyze remaining oil distribution of different kinds of single sand body [13-14], the results show (Table 1) that remaining oil mainly distributed in multiphase channel superposition, distributary channel and sheet sand. From the type of single sand body, remaining oil mainly concentrate on multiphase channel superposition and distributary channel's single sand body that developed channel sand. Sedimentary environment are stable inner front facies and lake regressive inner front facies, is the major target of polymer flooding potential tapping. From different types of remaining oil, thick oil layer take the most percentage of remaining oil reserve, most is thick-injection and thick-production, cover 71.7% of remaining oil reserve, so the thick-injection and thick-production layer is the major layer; and the kind of reservoir injection-production faultiness's remaining oil reserve mainly is unidirectional response, cover 8.3% of remaining oil reserve; the kind of poor layer's remaining oil reserve narrow distributed, only cover 6.7% of total remaining oil reserve.

On the basis of numerical simulation's result, combine with the information of two sealing core drilling well in Pubei oilfield, analyze single sand body internal structure's control action to remaining oil (Figure 5), the results show that level 4 interface control channel sand's whole remaining oil distribution, bottom water flushed is serious, remaining oil concentrate in the middle-upper part; if level 3 interface develop argillaceous interlayer (or calcareous interlayer)[15], remaining oil of interlayer upper-lower part structure mainly concentrate in the upper part of each structure, show that remaining oil in the same sand body concentrated piecewise, if develop physical property interlayer or nondeveloped interlayer, the control degree of interface to remaining oil is poor or has no effect, remaining oil in interface upper structure is more concentrated.

Table 1. The remaining oil distribution of different single sand bodies

Sand body type	permeability ($10^{-3}\mu m^2$)	degree of reserve recovery (%)	Percentage of remaining oil reserve (%)						
			Thick layer		injection-production faultiness			Poor layer	total
			Thick injection and thick production	Thin injection and thick production	unidirectional response	Production without injection	Injection without production		
multiphase channel superposition	310.1	38.85	19.8	1.1	1.6	0.4	0.5	1.1	24.4
distributary channel	335.3	37.04	26.2	2.1	2.5	0.7	0.5	1.2	33.2
sheet sand	223.5	30.64	18.9	3.5	1.3	0.5	0.5	3.1	28.0
single channel	260.9	31.39	6.1	0.0	1.9	0.8	0.4	0.8	10.0
lenticular sand	132.6	14.41	0.8	0.0	1.0	1.2	1.1	0.4	4.4
total	284.3	34.52	71.7	6.7	8.3	3.6	3.0	6.7	100.0

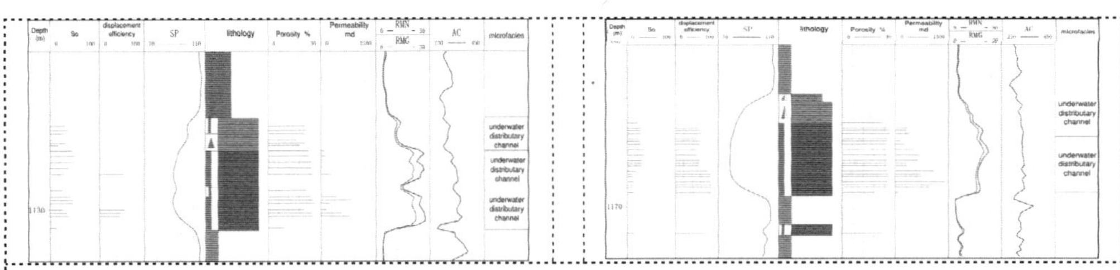

Figure 5. Core integrated histogram of 7P73-J912 coring Well

5. The method of remaining oil potential tapping for single sand body

Single sand body seen as a unit, advisable method of remaining oil potential tapping is made for distinct type of remaining oil by considering the inner structural feature of single sand body, the connecting relationship of sand body between injection well and production wells, the distribution feature of remaining oil, and production situation of wells.

5.1. Remaining oil of the top type of thick reservoir

This type remaining oil falls into two subclasses, thick-layer injection with thick-layer production and thick-layer injection with thin-layer production. In the concentrated area of remaining oil of hick-layer injection with thick-layer production, based on injection-production balance rule, production wells of polymer flooding parameters are optimized. On the premise of feasible injection pressure, the injection rate should be promoted in well field of low injection rate. Injection concentration should be regulated firstly in well field of high injection rate, in order to guarantee equilibrium driving of injection and unhurried-smooth ascension of injection pressure. At the same time, in view of response producer in different step, flowing pressure should be reasonably controlled. Then measure of augmented injection and raise liquid could be developed for production well. Remaining oil can be sufficiently tapped and polymer flooding effect could be broadened. As to remaining oil of thick-layer injection with thin-layer production, the thin and poor reservoir of injection side causes poor producing degree and the concentrated remaining oil staying in the reservoir of production side. Combining with production performance situation of reservoir of full well and sidetrack drilling of horizontal hole on the injection side reservoir, imbibing situation of thin-poor reservoir could be improved. Fracturing measure achieves synchronous tapping of injection-production wells. Based on above redeveloping measures, nine well groups of injection-production wells have been regulated after polymer injection test. The regulation wells of parameter optimization of injection wells are 21 and the regulation wells of fracture of injection wells are 2. The regulation wells of parameter optimization of production wells are 29. The regulation effect is visible. The average daily oil production of 9 well groups is 1.6 tons more than water flooding, and the water cut descents 7.2%. Moreover, by means of comprehensive analysis, it obtains that 16 wells are fit for fracturing in this type of remaining oil. And 4 wells present latent capacity of sidetrack drilling. The control reserve of latent capacity wells is 11.67×10^4t. There is a huge potential for further development.

5.2. Remaining oil of imperfect injection-production relationship type

In order to carry out polymer flooding test, the well pattern is encrypted and injection/production relationship is regulated. But due to the narrow channel development, the partly zebraic is more obvious after regulation well pattern and the development breadth of channel is less than 150m. That causes imperfect injection/production relationship of partly well pattern. During polymer flooding, the main potential target is unidirectional response type of remaining oil, by means of regulation of injection/production parameter optimization, augmented injection and liquid extraction. Since the test of polymer flooding carried out, the regulation wells of parameter optimization regulation for edge injection/production wells are 32. The regulation wells of acidulation for injection well are 2. Daily augmented injection is 30m^3 and submergence depth is about 450m. The average production fluid of single production well is 6.3 tons less than water flooding. The average production oil of single production well is 0.9 tons more than water flooding and water cut is 3.8% less. By means of analysis, there are 6 production wells with fracture potential. There are 2 injection wells with acidulation potential and 1 well with fracture potential. That supplies foundation for efficient tapping during polymer flooding.

5.3. Remaining oil of poor reservoir type

Seen from injection/production relationship, this type of remaining oil falls into two subclasses, thin-layer injection with thin-layer production and thick-layer injection with thin-layer production. The regulation potential of remaining oil of poor reservoir type is low. And this type can not be the main tapped target during polymer flooding. But some individually potential layers could be synchronously reformed and tapped with the main layers, in combination with production situation, enrichment situation of remaining oil and connection situation.

6. Conclusions

- According to the characteristics of single sand body distribution, the single sand body is divided into five types including multiphase channel superposition, distributary channel, single channel, sheet sand and lenticular sand. Channel single sand body which genesis structure body channel is small, bending is little, mainly is low sinuosity and straight channel. The interlayer at the interface mainly is the material and the clay. Single sheet sand body's single causes structure distribution of irregular thin sheet shape, interlayer mainly is argillaceous interlayer.
- Remaining oil mainly is distributed in thick oil layers of multiphase channel superposition type, distributary channel type and sheet sand type. The main potential of polymer flooding is distributed in. multiphase channel superposition type and distributary channel type which developed channel sand. The type of remaining oil is dominated by the top of thick oil layer, followed by the imperfect injection production relationship. From the perspective of the internal structure, the interlayer of the interface has a certain effect on the distribution of the remaining oil. The control degree of the remaining oil in the internal structure is higher than that of the material layer, which is more than the thickness of the calcareous interlayer.
- The main potential layer for polymer flooding are distributed in completeness injection production relationship of thick oil layers and better reservoir properties of unidirectional effect sand body. According to the types of sand body can take note the optimization and adjustment of production parameters, injection production well acidizing and fracturing and increasing injection measures, In view of this type of sand body can adopt injection production parameter optimization adjustment, injection production well acidification and fracturing and other injection measures to achieve the injection production synchronization potential, to ensure that the injection agent balance. For the poor thin layer of remaining oil in the layer can specifically tapping the potential with the main synchronization measures.

References

[1] MIALL A D 1985 Architecture element analysis: A new method of facies analysis applied to fluvial deposit *J. Earth—Science Review* **22** 261-308

[2] MIALL A D 1988 Architectural Elements and Bounding Surfaces in Fluvial Deposits:Anatomy of the Kayenta Formation(Lower Jurassic) *J. Southwest Colorado. Sedimentary Geology* **55** 233-262

[3] Ma S Z and Yang Q Y 2000 Sedimentary model of meandering point bar,three-dimensional configuration and its heterogeneous model *J. Journal of Sedimentary* **18** 241-6

[4] Ma S Z, Wang Y B, Cui Y, Ge Z J and Hao L Y 2006 Establishment of the model for inter-architecture of underwater distributary channel in oil-gas area *Journal of Daqing Petroleum Institute* **30** 1-3

[5] Ma S Z, Xu F Z and Tian X L 2014 Heterogeneity of river channel sand reservoir and its effect on remaining oil distribution *J. Heilongjiang University of Science & Technology* **24** 440-5

[6] Pan Z C and Ma S Z 2011 A study on internal architecture of single body in low permeability oil layer and remaining oil tapping *J. Science Technology and Engineering* **11** 5403-6

[7] Li Z P, Lin C Y, Peng X H, Bu L X and Wang Y 2011 The control of interbed on the remaining oil of monosandboby on minghuazhen formation of sourthern area of shallow strata of gaoyou depression *Journal of Oil and Gas Technology (J. Jianghan Petroleum Institute)* **33** 23-7

[8] Dou H W 2011 Delta front subfacies reservoir recognition method of single sand body *J. Oil and Gas Technology (J. Jianghan Petroleum Institute)* **33** 165-7

[9] Cheng K, Zhao L and Dan L H 2014 Study on single sand body battery of wells fine injection-production adjustment of Pucheng oil field eastern block Saerxia reservoir *J. Yangtze University(Nat Sci Edit)* **11** 78-81

[10] Du W, Ma S Z and Fan G J 2011 Study on disjoining of complex sand-body and interlayers—

with P I 2 layer in the east of the fault block in Beiyi district of Daqing oilfield as an example *J. Science Technology and Engineering* **11** 3637-40

[11] Zhang J, Zhang L H, Hu S Y and Nan L Y 2003 The genesis and characteristics and identification of intercalations in non-marine reservoir with clastic rock *J. Well Logging Technology* **27** 221-4

[12] Lv X G, Ma F S and Tian D H 1994 Study on interlayer lithology, physical property and its distribution *J. Petroleum Exploration and Development* **21** 80-7

[13] Luo W 2014 Study on remaining oil types and distribution features of Daqing placanticline peripheral oilfields *J. Yangtze University(Nat Sci Edit)* **11** 82-4

[14] Lu Y, Zhang E P and Jia D J 2011 Division of single sand bodies and influence to remaining oil distribution of Xinmin oilfield *J. Inner Mongolia Petrochemical Industry* **8** 134-5

[15] Ding S M, Ji M, Shi J, Li J F, Cheng T J, Liu Y M and He Z G 2004 Type of mudstone barriers and its residual oil control *J. Jianghan Petroleum Institute* **26** 130-4

Study on model current predictive control method of PV grid-connected inverters systems with voltage sag

N Jin[1], F Yang[1 5], S Y Shang[2], T Tao[3], J S Liu[4]

[1]College of Electric and Information Engineering, Zhengzhou University of Light Industry, Zhengzhou 450000, China
[2]Pingdingshan Industrial College of Technology, Pingdingshan 476001, China
[3]State Grid Xuchang power supply company, Xuchang 461000, China
[4]Suoling Electric Co., Ltd, Zhengzhou 450000, China

E-mail: 1050048189@qq.com

Abstract. According to the limitations of the LVRT technology of traditional photovoltaic inverter existed, this paper proposes a low voltage ride through (LVRT) control method based on model current predictive control (MCPC). This method can effectively improve the photovoltaic inverter output characteristics and response speed. The MCPC method of photovoltaic grid-connected inverter designed, the sum of the absolute value of the predictive current and the given current error is adopted as the cost function with the model predictive control method. According to the MCPC, the optimal space voltage vector is selected. Photovoltaic inverter has achieved automatically switches of priority active or reactive power control of two control modes according to the different operating states, which effectively improve the inverter capability of LVRT. The simulation and experimental results proves that the proposed method is correct and effective.

1. Introduction

With the capacity and scale of PV generator continues to expand, grid-connected technology has become the research the focal point [1]. Under the condition of large scale new energy power station grid-connected, all the countries in the world provide the terms of the LVRT, which can make PV power station connected with network quickly or resume supply rapidly during grid faults [2, 3]. When the traditional double closed-loop vector control strategy is used [4], the control circuit contains multiple PI controllers, therefore, it is difficult to adjust the control parameters of PI in practical engineering. the PR current controller[5] is only used to control the fundamental frequency, but it is not clear how it will control to reactive power. Reactive power is injected into the grid through Static Var Compensator when a fault occurs in power system in literature [6], but SVC can't provide enough reactive power and bad dynamic performance. Solution of the LVRT through based on energy storage devices is proposed in reference [7], this approach can sufficiently solve the problem of power fluctuation, but the cost is too expensive.

[5] Address for correspondence: F Yang, College of Electric and Information Engineering, Zhengzhou University of Light Industry, Zhengzhou 450000, China. Email: 1050048189@qq.com.

This paper comprehensively and deeply studies the single stage PV grid-connected inverter control system, and the method of MCPC with voltage sag is advanced. According to the grid voltage drop, the reactive current should be injected for reasonable compensation, and to ensure that the inverter can continue to maintain the grid. According to the current predictive model to predictive direct the active current id and reactive current iq , the optimal space voltage vector is selected by minimizing the cost function. In addition, the control method avoids the traditional control method of the external reactive power compensator and current controller with different frequencies, so it reduces the system cost and apply the inverter capability of LVRT which is easy to implement.

2. Another section of your paper

LVRT refers to the power grid malfunction or disturbance may lead to voltage sag of power access point. PV power system keep connected with the grid even in the situation of voltage sag within a certain rang and the prescribed time[8].

In 2011,State Grid Corp promulgated the technical regulation for PV power station access to power grid[9], It is pointed out that the large and medium-sized PV power stations should have a certain LVRT capability and provide dynamic reactive power support during the process of LVRT. LVRT requirement is shown in Figure 1. If line voltage is above the limit curve, The PV power station should be connected with network; or it will be separated from the grid.

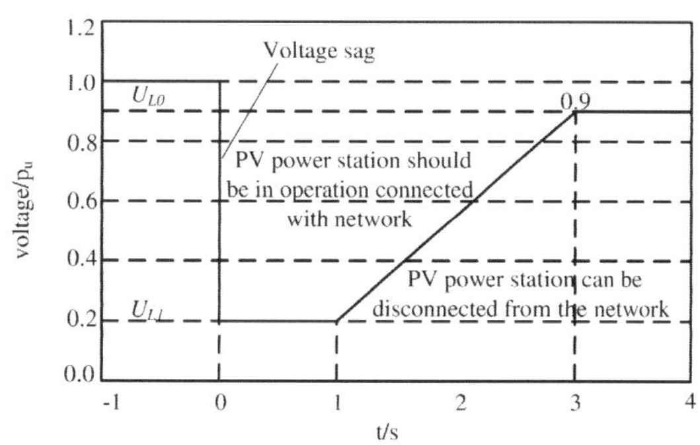

Figure 1. LVRT requirements of the large and medium-sized PV power stations

The German Grid Code has also provided for the relationship between the reactive current and the voltage drop depth[10], as shown in Figure 2. The reactive current to support the grid is not required when the voltage fluctuations in the range of -10% ~ +10%; The corresponding reactive current should be supplied 20% of the rated current in 10% the voltage sag; A 100% reactive current must be supplied if the voltage drop is up to 50%; Even after the voltage drop returns to the over 50%, the voltage support must be maintained.

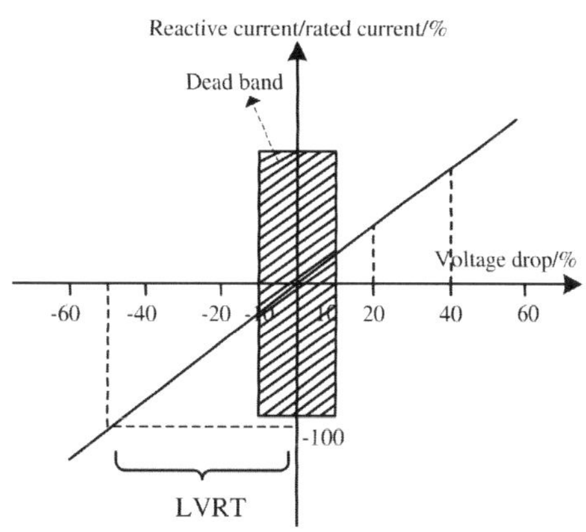

Figure 2. The required reactive current to support the voltage(Germany)

3. Low voltage ride through control technology

3.1. The current predictive mathematical model

Figure 3 displays a Three-phase Single Stage Grid-connected inverter, which is connected to main power system by using a line resistance R and a filtering inductance L. As reported by Kirchhoff law, the equations (1) of state under three-phase stationary frame is obtained as follows.

$$L\frac{\mathrm{d}}{\mathrm{d}t}\begin{bmatrix} i_a \\ i_b \\ i_c \end{bmatrix} + R\begin{bmatrix} i_a \\ i_b \\ i_c \end{bmatrix} + \begin{bmatrix} e_a \\ e_b \\ e_c \end{bmatrix} = \begin{bmatrix} u_{aN} \\ u_{bN} \\ u_{cN} \end{bmatrix} \tag{1}$$

where e_a, e_b, e_c are three-phase grid voltage, i_a, i_b, i_c are grid-connected inverter output current, u_{aN}, u_{bN}, u_{cN} are output voltage of the inverter.

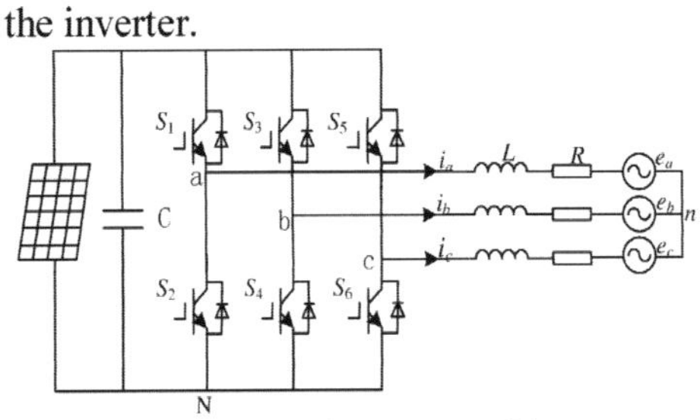

Figure 3. Three-phase Single Stage Grid-connected inverter

As reported by Park transformation, equations (2) can be rewritten in dq-frame

$$L\frac{\mathrm{d}}{\mathrm{d}t}\begin{bmatrix} i_d \\ i_q \end{bmatrix} + \begin{bmatrix} R & -\omega L \\ \omega L & R \end{bmatrix}\begin{bmatrix} i_d \\ i_q \end{bmatrix} + \begin{bmatrix} e_d \\ e_q \end{bmatrix} = \begin{bmatrix} u_d \\ u_q \end{bmatrix} \tag{2}$$

where e_d, e_q, i_d, i_q are d, q components of the grid-connected voltage and current under dq-reference frame, u_d ,u_q are d, q components of the inverter output voltage under dq-reference frame.

The three-phase single stage grid-connected PV inverter that switches status $S_i\,(i{=}a,b,c)$ is defined as follows:

$$S_i = \begin{cases} 1 & \text{Turn-on of the up bridge arm, turn-off of the down bridge arm} \\ 0 & \text{Turn-off of the up bridge arm, turn-on of the down bridge arm} \end{cases} \tag{3}$$

In the model of three-phase grid-connected PV inverter shown in Figure 3, which existence seven various voltage vectors. The generated voltage space vectors is displayed in Figure 4.

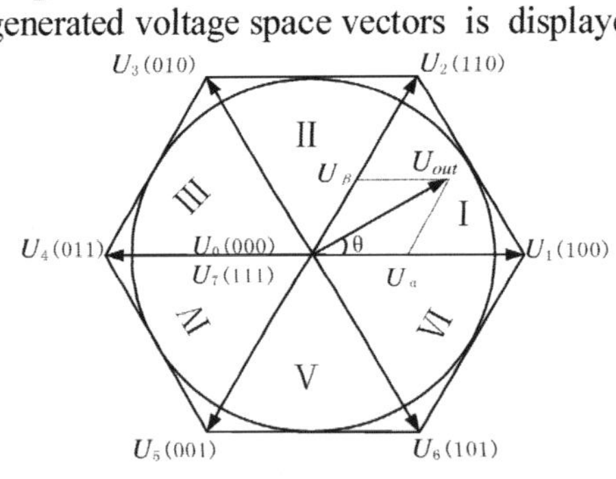

Figure 4. Diagram of grid connected inverter voltage space vector

The output voltage of the grid connected inverter is $U_i(i=d,q)$

$$\begin{bmatrix} u_d \\ u_q \end{bmatrix} = \frac{2}{3}\begin{bmatrix} \cos\theta & \sin\theta \\ -\sin\theta & \cos\theta \end{bmatrix}\begin{bmatrix} 1 & -1/2 & -1/2 \\ 0 & \sqrt{3}/2 & -\sqrt{3}/2 \end{bmatrix}\begin{bmatrix} S_a U_{dc} \\ S_b U_{dc} \\ S_c U_{dc} \end{bmatrix} \quad (4)$$

where U_{dc} is the voltage of DC bus, θ is the space angle of power grid.

In the light of the resistance R is very small, the effect of R could be overlooked. After discretization, equations (2) can be written as

$$\begin{bmatrix} i_d(k+1) \\ i_q(k+1) \end{bmatrix} = \frac{T}{L}\begin{bmatrix} u_d(k)+\omega L i_d(k) - e_d \\ u_q(k) - \omega L i_q(k) - e_q \end{bmatrix} + \begin{bmatrix} i_d(k) \\ i_q(k) \end{bmatrix} \quad (5)$$

The formula (5) is the predictive function based on model current prediction control.

3.2. Model current predictive control principle

PV inverter control structure[11-13] is shown in Figure 5.The optimum space voltage vector minimizing the cost function will be applied to the following sampling period. Thus the voltage vector corresponding to the switching state S_a, S_b and S_c are acquired by switching state optimization, in order to control switching of switch tube.

So as to select the optimal voltage vector, a cost function g has been set up. The predicted all current values will be compared through the cost function. To choose the minimum value function of voltage vector will be applied at the next moment. The value of error absolute sum between the predictive current and the reference current is used as the cost function:

$$g(i) = \left| i_d^* - i_d(k+1) \right| + \left| i_q^* - i_q(k+1) \right| \quad (6)$$

where i_d^* and i_q^* are reference values of the active and reactive current.

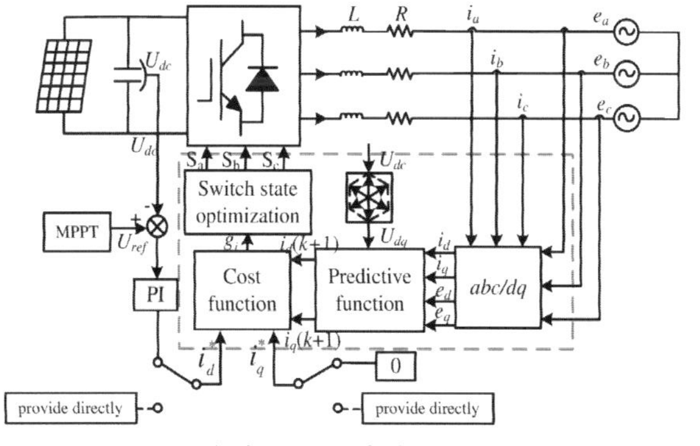

Figure 5. Control diagram of photovoltaic inverter

4. Low voltage ride through control strategy

The LVRT technology based on grid connected current control should supply reactive current to the grid according to the depth of the voltage drop, in order to support the grid voltage and recover power network[14].

When no fault occurs in the power grid, Priority must be given to meeting the active current. The output reference voltage U_{dc_ref} by maximum power point tracking (MPPT) technology for PV power generation. The error signal between the DC side voltage Udc and the reference voltage U_{dc_ref} is the input of proportional-integral (PI) regulator adjustment to i_d^*, then a reactive current reference value is given by the user directly. when the fault occurs, the inverter uses reactive priority control.

In the reactive priority control, the reference value of reactive current can be got with the voltage drop range between 10% and 50%:

$$i_q^* = 2\frac{U - U_0}{U_N} I_N = 2\frac{\Delta U}{U_N} I_N \tag{7}$$

When the voltage drop is more than 50%, the reference value of the reactive current can be expressed as

$$i_q^* = -I_N \tag{8}$$

In fact, considering the 1.1 times overload capability of the inverter itself[15], which allows the maximum continuous output current of the inverter to be $1.1i_N$. The active current of the inverter output is shown in the formula(9),where iN is the rated current of the power grid.

$$i_d^* \le \sqrt{(1.1i_N)^2 - (i_q^*)^2} \tag{9}$$

The three-phase output voltage and current of the inverter do the *abc*/*dq* rotation transformation by the synchronous rotation angle. The instruction current is responsible for the voltage sag detection、reference active and reactive current calculation. In normal conditions, The value of error absolute sum between the predictive current and the reference current is used as the cost function, thus the link of switch state optimization control switch tube conduction and shut off. When the power grid voltage sags, the reactive power priority control limits the active current to ensure the voltage recovery of generate reactive power by the inverter.

LVRT control flow chart of PV grid connected power generation system is shown in Figure 6, where S(n) is the target switch state. *m*, *i* are the variable Parameters. *i* correspond to respective eight voltage vectors of inverter output from 0 to 7. g_i is the value of evaluation function in the ith switch state.

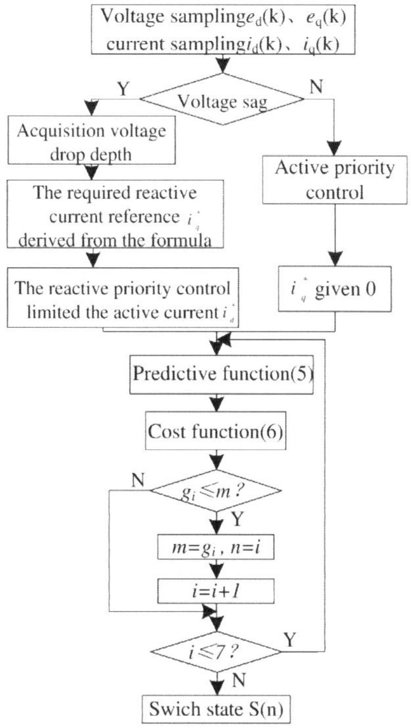

Figure 6. Flow chat of LVRT control

5. Simulation results

The simulation model of PV system is designed as illustrated in Figure 3 by MATLAB/simulink, the purpose is researching characteristics of three-phase symmetrical faults. The simulation waveforms of the different situation are compared and analyzed on the following.

5.1. Comparative analysis before & after LVRT control input

Three phase symmetrical faults occurs in photovoltaic grid connected generation system at 0.08 s, the grid voltage drops to 0.6 p.u, then the fault is cleared at 0.14 s. The specific parameters are shown in Table 1.

Table1. The simulation of system parameters.

Parameters	Value
DC source voltage Udc	350V
Grid line voltage e	110 V
Filter inductance L	20mH
Grid frequency f	50Hz
Power grid sag mode	Symmetrical sag
Duration of voltage sag	0.06s

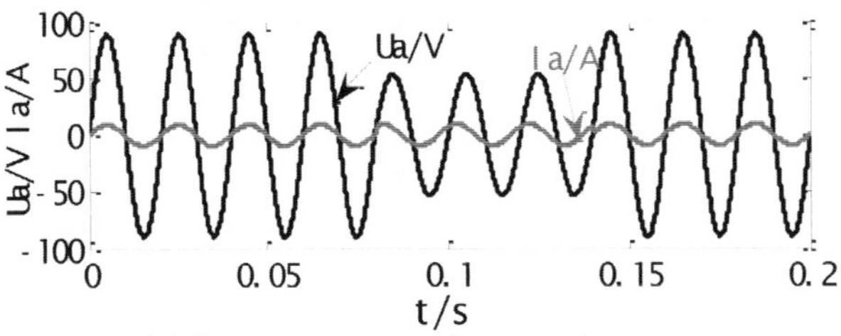

(a) Instantaneous voltage and current

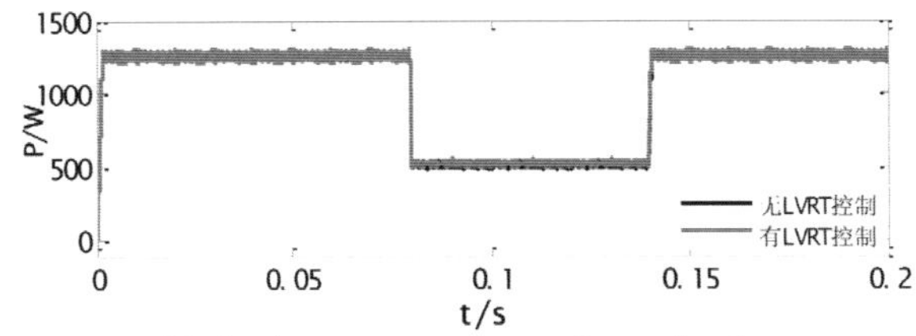

(b) Active power comparison chart

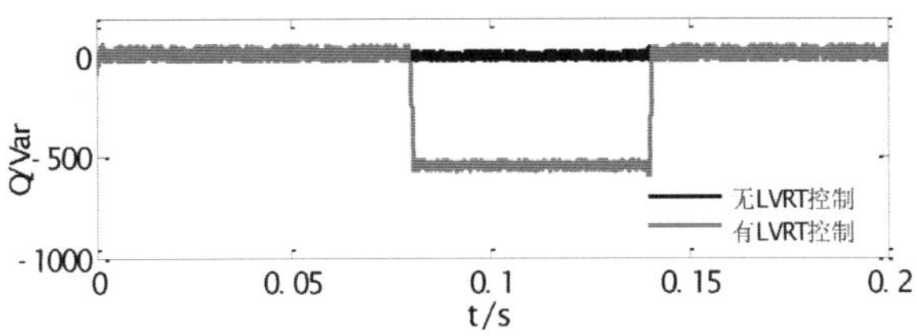

(c) Reactive power comparison chart

Figure 7. Comparison of power before & after LVRT control

As is evident from (a) of Figure7, the output current of inverter remains stable under LVRT control during the fault period. The grid connected generation system should be injected into the power system after the voltage drop occurred by taking the reactive priority control, then the phase difference between the voltage and current is changed significantly. Figure 7 (b) and7 (c) showed that the active power of the inverter output decreases during the voltage sag, the output reactive power increases during the voltage sag after the LVRT control strategy adopted. At this point PV power plant can provide priority a certain reactive power and become the reactive-load compensation equipment. After the voltage is restored to normal , active and reactive power can be quickly recovered to the steady state value before fault.

5.2. Comparative analysis with the traditional controller

In this section, the model predictive current control is compared with the traditional PI controller in system responsiveness. At 0.08s the active reference value is changed, which of the simulation results as shown in Figure 8. Where I_{d1} is the output current of the inverter active component under the PI control . I_{d2} is the output current of the inverter active component under the current predictive control. I_{dref} is the reference current value.

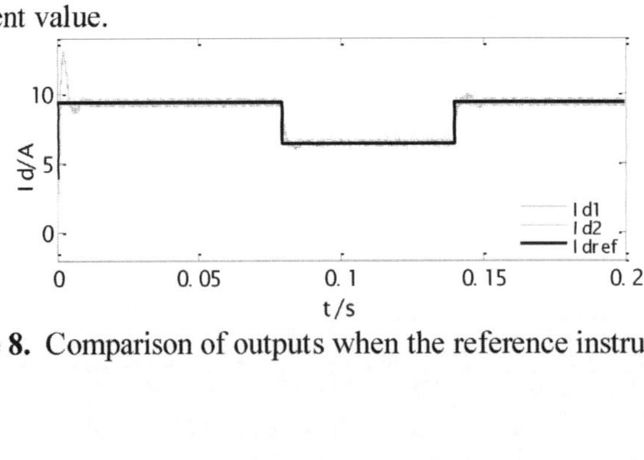

Figure 8. Comparison of outputs when the reference instructions changed

The simulation results show that the current predictive control and PI control was changed in accordance with the current dynamic directives in the dynamic changes of reference current value. However, the response speed of the current predictive control is faster than the PI controller. The responses of model predictive current control is affected by sampling frequency and common mode voltage; with overshoot by PI control needs to set PI parameters, which will increase its control complexity and compute sharply.

5.3. Comparative analysis of different drop depth

Power grid voltage drop 0.02 P.U at 0.04s; drop 0.4 P.U at 0.08s; drop 0.7 P.U at 0.012s, the fault is cleared at 0.16 s.

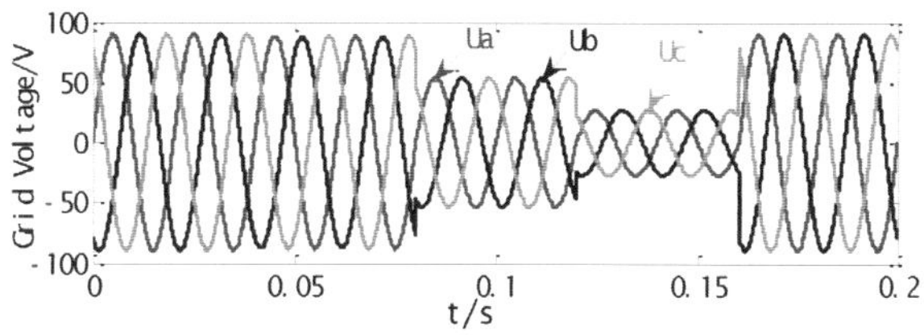

(a) Three-phase grid connected voltage

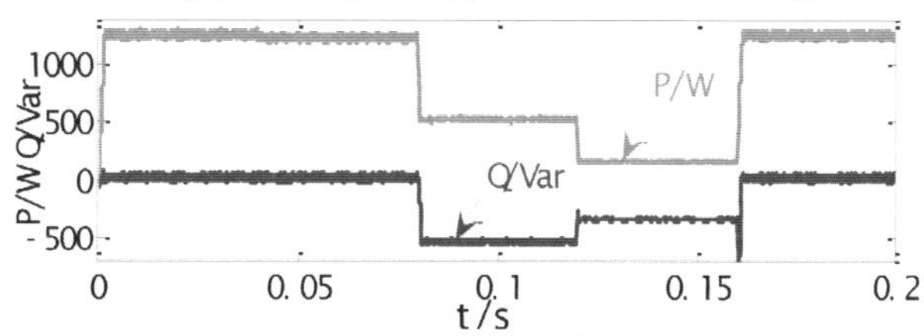

(b) Active power P, reactive power Q

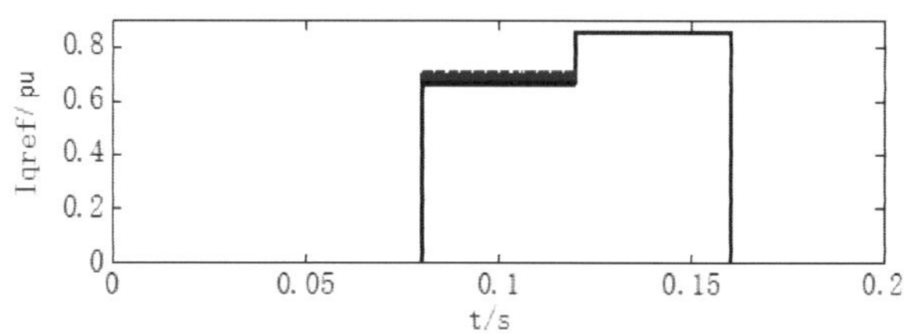

(c) Reactive current provided during voltage sag

Figure 9. Comparison of outputs when depth of grid voltage sags changed

It can be seen from Figure 9 that the PV power system must provide reactive current during voltage sag, the reactive current of the linear output when the voltage drop depth is 0.4p.u; When the voltage sag depth is 0.7p.u, and thus offer 100% reactive current. It shows that the reactive current output of the inverter can be changed according to the curve law shown in Figure 2.

6. Experiment

6.1. Experiment platform

To verify what be proposed control algorithm feasibility and effectiveness further, a three phase grid connected experimental platform based on PE-PRO is built as shown in Figure10. The control system is realized by the TMS320F28335 control chip of TI company, power device IGBT is 7MBP50RJ120, test instrument is DLM4000 YOKOGAWA series Mixed-signal oscilloscopes, APL-II DC power supply and so on.

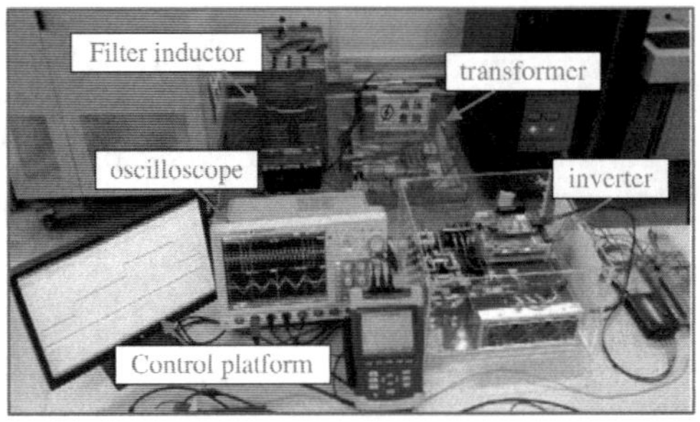

Figure 10. Experimental setup

6.2. Experiment result

Three phase AC voltage balance drops 60%, the experimental by using MCDPC plus compensation as displayed in Figure 11. The results is that the power $p = 1.1$kW jump into $p = 0.5$kW, $q = 0$ jump into $q = 0.58$kW. A phase grid voltage and current would be phase-shifted, which can achieve the reactive power compensation of the inverter during the voltage sag, and maintain the voltage stability of the network.

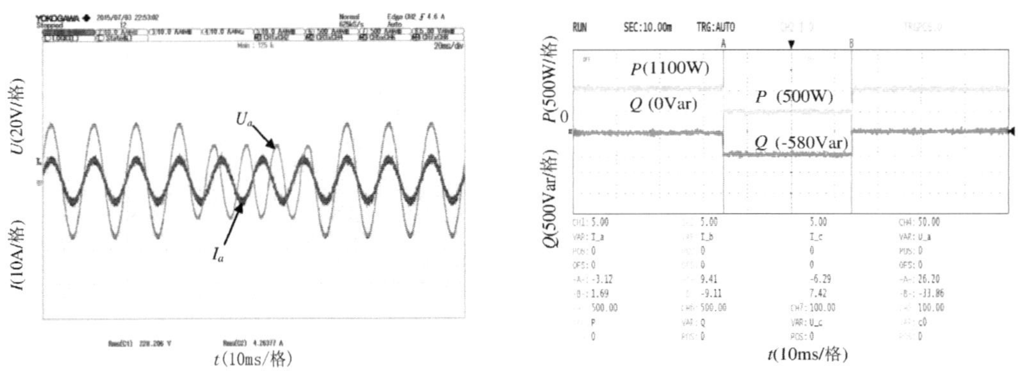

(a) Voltage and current waveform (b) power waveform

Figure 11. experimental waveforms

7. Conclusion

By using the model current predictive control method, the output voltage can be controlled under balanced conditions. When the control strategy is improved, using good dynamic response of MCPC strategy may quickly realize the control of grid connected inverter. Three phase symmetrical fault simulation examples are carried out on MATLAB/simulink, and the precision of the model is analyzed. This control strategy is utilized to regulate reactive power output in real-time, according to the grid voltage drop depth ; and to improve the voltage of PCC; to support the continuous operation of PV power generation system; to help the rapid recovery of power grid.

References

[1] X Y Zhen, Z D Yin, Y F Wang, Z Sun 2011 Overview of Low Voltage Ride Through in Photovoltaic Power Generation *Power System and Clean Energy* **27** 65-8

[2] S Alepuz, S Busquets-Monge, J Bordonau, P Cortes, J Rodríguez, R Vargas 2008 Predictive current control of grid-connected neutral-point-clamped converters to meet low voltage ride-

through requirements *Power Electronics Specialists Conference*, 2008. *PESC* 2008. *IEEE*. *IEEE* 2423-8

[3] S Alepuz, S Busquets, J Bordonau, J Pontt, C Silva, J Rodríguez 2007 Comparison of control strategies to meet low voltage ride-through requirements in distributed power generation systems *Industrial Electronics*, 2007. *ISIE* 2007. *IEEE International Symposium on*. *IEEE* 2619-24

[4] X L Yuan, P F Song, F J Fan, Y Chen 2013 Low Voltage Ride Through Control Strategy for PV Inverter *Dian li Ji shu/ Power Electronics* **47** 67-9

[5] C H Benz, W Franke, F W Fuchs 2010 Low voltage ride through capability of a 5kW grid-tied solar inverter *Power Electronics and Motion Control Conference (EPE/PEMC), 14th International*. *IEEE* T12-13-T12-20

[6] M Castilla, J Miret, J L Sosa, J Matas, L G De Vicuna 2010 Grid-fault control scheme for three-phase photovoltaic inverters with adjustable power quality characteristics *Power Electronics, IEEE Transactions on* **25** 2930-40

[7] B Wang, Z Shi, T Zhu, Z Y Meng 2007 Design of controller with ultracapacitor-battery hybrid energy on a stand-alone PV system *Energy Engineering* **5** 37-41.

[8] Y F Meng, S J Hu, L Wang, H H Xu 2013 Characteristics analysis and coordinated control of the doubly-fed wind power system under grid transient fault *Power System Protection and Control* **41** 106-13.

[9] Y S Wu, C H Chang, Y M Chen, C S Cheng, C W Liu, Y R Chang 2012 The current control of PV inverter for low voltage ride through *15th International Power Electronics and Motion Control Conference* LS1d.4-1 - LS1d.4-6

[10] E.ON Netz 2006 Grid code—High and extra high voltage. E.ON Netz GmbH, Tech. Rep. [Online].Available:http://www.eonnetz.com/Ressources/downloads/ENENARHS2006eng.

[11] A M Almaktoof, A K Raji, M T E Kahn 2008 Modelling and simulation of three-phase voltage Source Inverter using a model predictive current control *International Journal of Innovation, Management and technology* **5** 9-13

[12] P Cortés, M P Kazmierkowski, R M Kennel, D E Quevedo, J Rodríguez 2008 Predictive control in power electronics and drives *Industrial Electronics, IEEE Transactions on* **55** 4312-24

[13] S Effler, A Kelly, M Halton, K Rinne 2008 Automated optimization of generalized model predictive control for DC-DC converters *Power Electronics Specialists Conference*, *IEEE* 134-9

[14] Q J Mao 2009 Compensating Reactive Power Loss of Wind Farm With Reactive Power Generated by Wind Turbine Generators *Power System Technology* **19** 175-80

[15] B Chen, X D Zhu, L Z Zhu, J P Lai 2012 Strategy for reactive control in low voltage ride through of photovoltaic power station *Power System Protection and Control* **40** 6-12

Energy-efficient vertical transportation with sensor information in smart green buildings

H Bahn[1]

Department of Computer Science & Engineering, Ewha Womans University, Seoul, 120-750, South Korea

E-mail:bahn@ewha.ac.kr

Abstract. In modern smart green buildings, sensors can detect various physical status of a building such as temperature, humidity, motion, and light, which can be used for smart living services. This paper presents an energy-efficient vertical transportation by making use of indoor sensor technologies. Specifically, sensors detect elevator users before they push the call button, and then inform to the elevator control system through building networks. By using this information, our system generates a reservation call and controls the moving time and direction of each elevator efficiently. Simulation experiments with a variety of traffic situations show that our elevator control system exhibits significantly better performance than the conventional system that does not use sensor information with respect to passengers' waiting time and energy consumption.

1. Introduction

Due to the recent advances in indoor sensor technologies, various smart living services in a modern green building are being realized. For example, indoor sensors can collect temperature, humidity, motion, light, and sound, which might be used for energy saving, comfort, healthcare, and security [1]. HVAC (Heating, Ventilating, and Air Conditioning) is a representative system that can utilize indoor sensor technologies. HVAC controls the air condition and the temperature of a building according to the weather and the existence of human beings, thereby leading to an energy-efficient building infrastructure.

Vertical transportation is another important service that can be involved in indoor sensor technologies in a smart green building. There is a report that HVAC and vertical transportation are the two major complaints of building tenants [2]. To alleviate this situation, this paper presents a novel elevator control system that utilizes indoor sensor technologies. As human behaviors and movements can be detected precisely with current sensor technologies such as video, audio, optical, and floor sensors [1], [3], [4], [5], an elevator control system can recognize the arrival of users before they push the actual call buttons. Thus, our system exploits this information for enhancing the efficiency of elevator control, leading to reduced waiting time and energy consumption. Although a lot of studies on vertical transportation have been performed, utilizing indoor sensor technologies for elevator control is in its infancy.

[1] Address for correspondence: H Bahn, Department of Computer Science & Engineering, Ewha Womans University, Seoul, 120-750, South Korea. E-mail:bahn@ewha.ac.kr.

Figure 1. (a) Video sensors

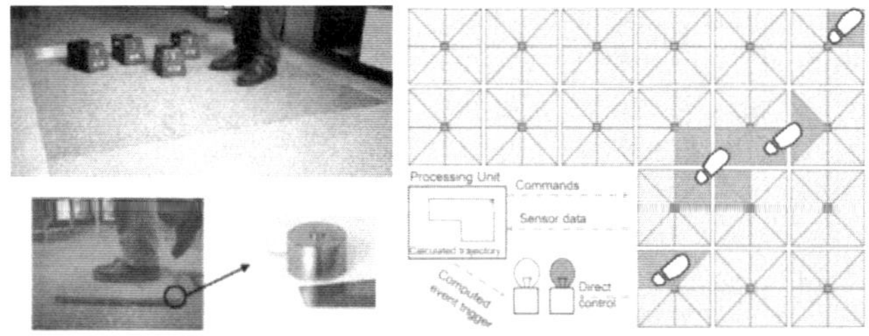

Figure 1. (b) Floor sensors

Figure 1. User tracking systems with various sensor technologies [3, 4, 5].

The goal of elevator control systems is usually mentioned as the minimization of waiting time, riding time, and the energy consumption of the systems. Previous works have mostly focused on the minimization of the average waiting time as passenger's dissatisfaction grows significantly as the waiting time becomes long [7]. This paper focuses on the minimization of both energy consumption and passenger's waiting time.

Elevator control is a complex optimization problem due to the complicated elevator dynamics, uncertainty of traffic patterns, and multiple goals to be considered concurrently. Due to this reason, previous works utilize well-known optimization techniques like fuzzy logics and genetic algorithms [8], [9], [10], [11], [21]. Also, some facilities such as cameras and additional buttons are adopted to collect more information on waiting passengers [10]. Some research predicts traffic patterns using peak time distribution analysis [9]. However, prediction and adaptation used in existing research is limited as elevator control systems recognize passengers' arrival only after they push the elevator call buttons. Unlike previous works, the proposed elevator control system collects passenger information by sensor technologies before they actually arrive, which will be utilized in efficient elevator control. Specifically, our system generates a reservation call for candidate passengers detected by sensors and controls the moving direction and the moving time of elevators efficiently. To validate the efficiency of the proposed elevator control system, we perform simulations with a variety of traffic conditions. Our experimental results depict that the proposed elevator control system performs significantly better than the legacy system that does not use sensor information with respect to the average waiting time, the worst case waiting time, and the energy consumption. In particular, the improvement of the average waiting time and the energy consumption is in the range of 15-30% and 28-31%, respectively.

The remainder of this paper is organized as follows. Section 2 discusses previous studies on sensor technologies for smart green building environments. Section 3 describes the proposed elevator control system in detail. Section 4 validates the proposed system through simulation experiments. Finally, Section 5 concludes this paper.

2. Sensor technologies for smart green buildings

Recently, smart green building prototypes are developed that contain various indoor sensors such as video sensors, floor sensors, and battery-backed wireless sensors. We can collect a lot of context information from these sensors, which might be exploited for smart green building services. Kidd *et al.* present a context-aware home prototype [3]. It consists of several living spaces such as bedroom, bathroom, living room, dining room, and office, and video sensors are installed in each space for tracing human movements. With this information, human behaviors can be analyzed and predicted. Figure 1.(a) shows a person tracked by the video sensor in this prototype building.

Orr and Abowd present a smart floor for tracking human movements and identities [4]. They created a system for identifying users by making use of their footstep force profile. This also allows the tracking and the prediction of tenants in a smart building. Addlesee *et al.* present an active floor, which is similar to the smart floor [5]. The active floor is the square grid of conventional tiles that is supported at the corners by cylindrical load cells, which send weight changes of about 50 grams to the location system. With this information, the system can predict the future locations of people. Similarly, Steinhage and Lauterbach present capacitive sensor arrays which are embedded in floor carpets in order to track the movement across a large area [14]. Figure 1. (b) shows these floor sensors. Want *et al.* present the active badge system for tracking people in an office building [13]. In their system, people wear an active badge, and it periodically delivers signals, which are used to trace the location of people.

Sensor networks have also been widely studied for smart green building infrastructure. Gao *et al.* propose an intelligent light control system based on wireless sensor networks [19]. Their system reduces the energy consumption of a smart green building significantly. Eliades *et al.* present the sensor placement problem in smart buildings in order to monitor and protect indoor air quality against contaminations [20]. They formulate the problem as a multi-objective optimization problem to save the sensor cost as well as the impact of contamination events. Sreedharan *et al.* design a real-time sensor system that combines information from multiple sensors to accurately detect a sudden release of toxic contaminants in a smart building [21].

Smart parking providers also make use of sensor technologies to monitor the real-time availability of parking spaces [15]. Finding the best path to an exit with sensor networks is also attempted [16]. Brunette *et al.* develop wireless sensor nodes that can provide contextual information including human activity, environment, and RFID tags, which can be used in ubiquitous computing applications to automatically adjust their behavior according to the situation [18].

Sensor information in a smart green building can also be used efficiently in vertical transportation systems. Strang and Bauer use RFID tags to inform the destination floor before passengers enter the elevator car [17]. Kwon *et al.* also use context information by using floor sensors, but their scheduling is not applicable to group elevator systems [6]. Mitsubishi Electric Corporation developed an RFID-enabled elevator system [12]. The RFID tag informs the arrival of passengers and their destination, which are used for the elevator control and security. By combining RFID and cameras, their system discerns if a person wants to use an elevator or is just walking near an elevator [12].

3. Energy-efficient elevator control with sensors

Legacy elevator control systems cannot recognize passengers before they actually push the elevator request buttons. Thus, waiting time and energy consumption are increased significantly. Take, for example, a typical situation where a passenger pushes a request button at the 1st floor right after the elevator goes from the 1st up to the highest floor. Then, the passenger should wait during a full round trip time of the elevator, and the energy consumption of the elevator is also increased because of the long moving distance.

To address this issue, we present a novel elevator control system called ESG (Elevator control for Smart Green buildings). ESG obtains passengers' arrival information in advance from multiple sensors and exploits the information to control elevator systems. As mentioned in Section 2, with the rapid advances in sensor and wireless network technologies, there is no difficulty in detecting and transferring the passenger information in advance. Given information on subsequent passengers beforehand, ESG controls elevators more efficiently with respect to the waiting time and energy consumption.

Now, let us see the details of ESG. Figure 2 shows the overall architecture of ESG; it is composed of three subsystems, namely the Control Subsystem (CS), the Reservation Subsystem (RS), and the Assignment Subsystem (AS). CS controls the moving or stopping of the elevator. RS collects and processes passengers' arrival information detected by sensors. When a passenger approaches the elevator, multiple sensors detect the passenger's location in advance, and then inform this information to the elevator control system through building networks. RS calculates the time when the passenger will arrive at the elevator and then generates a reservation call accordingly. AS receives the

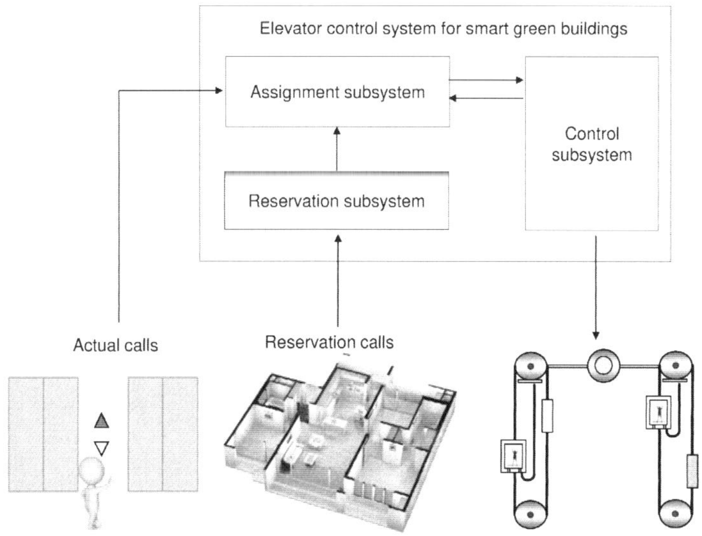

Figure 2. Basic architecture of the proposed elevator control system.

reservation call from RS, and decides two parameters; DT (delay time) and MD (moving direction). DT is the time interval that should be delayed before the elevator begins its movement for the reservation call, and MD is the moving direction, i.e., upward or downward, of the elevator. If a reservation call arrives during the idle time of the elevator car, the car does not move immediately at right, but delays its movement as long as it can reach the requested location before the passenger arrives there. Determining DT is important to reduce the moving distance of the car, potentially related to the energy consumption, since other requests towards the same direction during the delayed time can arrive. We define DT as the time difference between the passenger's moving time to the elevator door (denoted by *PT*) and the elevator car's moving time to the reserved floor. In particular, the elevator car's moving time depends on the distance of the movement. If the distance is longer than a certain threshold (denoted by *MIN*), a uniform motion interval exists in the middle of the moving. Otherwise, the car moves only as acceleration and deceleration motions. Equation (1) represents how *DT* is calculated.

$$DT = \begin{cases} PT - 2\sqrt{\dfrac{DIST}{ASC}} & \text{if } DIST < MIN \\ PT - \left(2\sqrt{\dfrac{MIN}{ASC}} + \dfrac{DIST - MIN}{USC}\right) & \text{otherwise} \end{cases} \quad (1)$$

$$PT = AHT - RCT$$
$$DIST = |CE - PF_i| \cdot FH$$

where CE is the current floor of the elevator, PF_i is the current floor of the passenger i, FH is the height of a floor, $DIST$ is the distance between the elevator's and the passenger i's floors, MIN is the minimum distance for uniform motion, ASC is the uniform acceleration of the car, USC is the uniform velocity of the car, AHT is the actual call time, and RCT is the reservation call time. By expression (1), delay time of an elevator car before starting its movement can be calculated.

Now, we will explain our scheduling system by describing the workings of the algorithm with appropriate examples. Figure 3 depicts a simple situation to show the effectiveness of delayed movement. Suppose that two passengers P1 and P2 want to ride the elevator. P1 is at the 18th floor and aims at going down to the 1st floor. P2 is at the 6th floor and aims at going up to the 18th floor. In this simple example, we suppose the logical time that is increased by one whenever an elevator goes up or down for a floor and ignore all other time components like boarding time. We suppose that sensors detect passengers 30 time units before they arrive at the elevator door. Thus, P1 and P2 make actual calls at t_{31} and t_{37}, but their reservation calls are generated at t_1 and t_7, respectively. When we do not apply the delayed movement, the car starts moving at t_1 towards the 18th floor to pick up P1. Although P2 makes a reservation at t_7, the car already passed by the 6th floor at that time. As a result, the wait time of P2 becomes longer, and also the elevator consumes additional energy because of the increased moving distance. In contrast, if we adopt delayed movement, the elevator delays its movement for 13 time units, and then starts moving at t_{14}; thus, P2 can take the elevator at this turn. Accordingly, the total execution time of the elevator system is reduced, which also leads to significant energy-savings.

MD (moving direction) is another critical parameter for the waiting time and the energy consumption. Figure 4 shows an example situation of MD's effects. Suppose that reservation calls for passengers P1 and P2 are issued at t_1 and t_6, from the 12th and 6th floors, respectively, and the passengers will actually arrive in front of the elevator door 30 time units later; the elevator car initially stops at the 8th floor and the destination of P1 and P2 are the 15th and 18th floors, respectively. Then, in the legacy elevator control system, the car first goes up to the 12th floor at t_{31} to pick up P1. P1 rides the car at t_{35}, and arrives at the destination floor at t_{38}. Then, the car goes down to the 6th floor to pick up P2. In this way, the waiting time of P2 becomes 11 time units. In contrast, in our system, the car first goes down to the 6th floor to pick up P2, and then goes up to the 12th floor to pick up P1. Accordingly, the waiting time of the passengers and the moving distance of the elevator are significantly reduced.

Now, we will describe how the aforementioned algorithm can be applied in group elevator control systems. The operation of ESG for group elevator systems is composed of two phases: the allocation phase and the processing phase. When a reservation call is issued, the elevator control system first decides which elevator should be allocated to that request. For efficient allocation, ESG calculates the passengers' expected waiting time for each car. Then ESG allocates the car incurring the minimum expected waiting time to that request. The expected waiting time (P_{WT}) of elevator car x from the i-th to the j-th floor is calculated as

$$P_{WT}(x, i, j) = T_{MOVE} + T_{ADDITIONAL}$$

$$T_{MOVE} = \begin{cases} 2\sqrt{\dfrac{DIST}{ASC}} & \text{if } DIST < MIN \\[2ex] 2\sqrt{\dfrac{MIN}{ASC}} + \dfrac{DIST - MIN}{USC} & \text{otherwise} \end{cases} \tag{2}$$

$$T_{ADDITIONAL} = T_{OPEN} + T_{BOARDING} + T_{CLOSE}$$

where $DIST$ is the distance between the passenger and the car, MIN is the minimum distance for uniform motion, ASC is the uniform acceleration of the car, and USC is the uniform velocity of the car. T_{OPEN}, $T_{BOARDING}$, and T_{CLOSE} are the time to open the elevator door, the time to board the elevator, and the time to close the elevator door, respectively. If the car needs to visit some additional floor during the movement from i to j, multiple T_{MOVE} and $T_{ADDITIONAL}$ are added to P_{WT}.

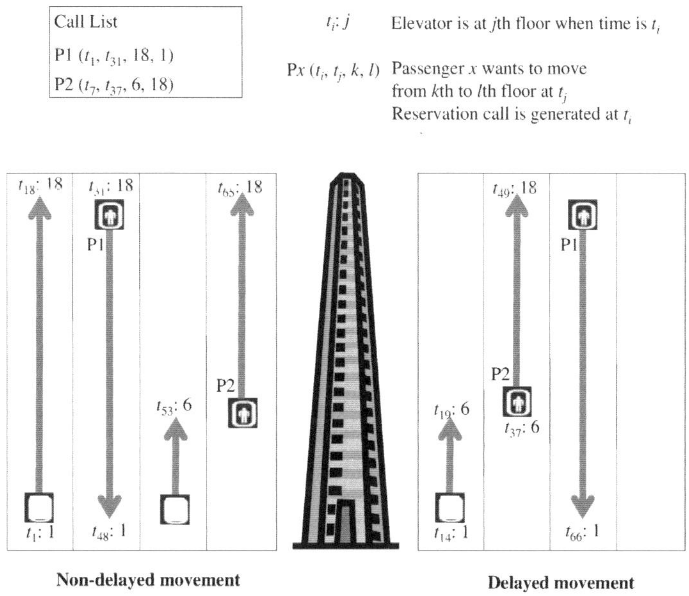

Figure 3. A simple situation that contrasts non-delayed and delayed movements.

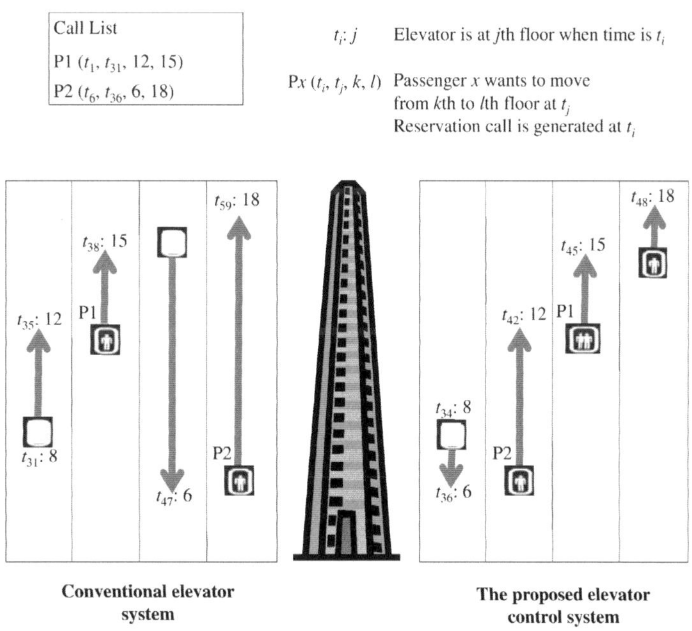

Figure 4. A simple situation of adopting Moving Direction (MD) in the proposed system.

After the allocation phase is completed, ESG inserts the passenger's reservation call into the request queue of the selected car. ESG, then, processes the passenger's reservation call based on the original ESG algorithm.

IOP Conf. Series: Earth and Environmental Science **40** (2016) 012079 doi:10.1088/1755-1315/40/1/012079

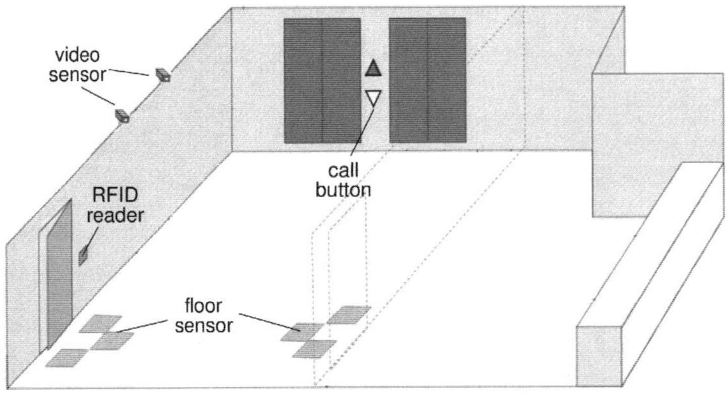

Figure 5. Prototype configurations to collect sensor data.

4. Experimental results

To validate the effectiveness of multiple sensor devices in detecting passenger information, we equipped RFID, video, and floor sensors in a twenty-story building for residence, and collected sensed data for enrolled users. The building authenticates the enrolled users at the gate of the underground parking area or the main entrance of the ground floor using an RFID tag or a password. We first collect sensor data at this time to make a reservation call. Less than 120 seconds are needed for a candidate passenger to move from this location to the elevator door. Along the hallway towards the elevator, floor sensors are located, and they can detect the moving direction of people through the path of the step. This information is collected about 60 seconds prior to the actual call at the elevator. Video sensors are also located at the ceiling of the hallway, and they can also recognize the movement of people. The distance from this location to the elevator is about 30 seconds or less. We set the default moving direction of reservation calls to "up" at the parking area and the ground floor. Note that a reservation call is dropped from the request queue when it is not continued to appear in the subsequent sensors. For example, when a reservation call is generated 120 seconds prior to the actual call through RFID tags but there is no corresponding reservation call through video or floor sensor, the request is dropped from the scheduling queue.

In the residence area ranging from the 2nd to 20th floors, each home has its front door and there are floor sensors along the hallway. 60 seconds or so are needed to move from the front door to the elevator including the locking time of the door. Video sensors are also located at the ceiling of the hallway, and the distance to the elevator is about 30 seconds or less. We set the default moving direction of reservation calls to "down" at residence floors. Figure 5 depicts our experimental environments of sensing systems.

We perform simulation experiments to evaluate the efficiency of ESG. We use similar conditions with previous works [8], [10]. The number of floors and elevators is set to 20 and 6, respectively, and each elevator car can accommodate 20 people. The passengers' traffic is generated by Poisson process with the average arrival rate ranging from 5 to 35 passengers/min as usual [8], [10]. To reflect the peak-time traffic in real situations, we also generate passengers' traffic by a non-homogeneous Poisson process where the arrival rate itself changes according to another Poisson process. We compare ESG with CS (current system) that does not use reservation calls. We use three criteria: waiting time (average case), waiting time (worst case), and energy consumption.

According to the aforementioned prototype sensor settings, we perform three ESG configurations: ESG-30, ESG-60, and ESG-120. In ESG-30, the reservation call is generated 30 seconds prior to the passenger's actual call. Similarly, ESG-60 and ESG-120 generate reservation calls 60 and 120 seconds

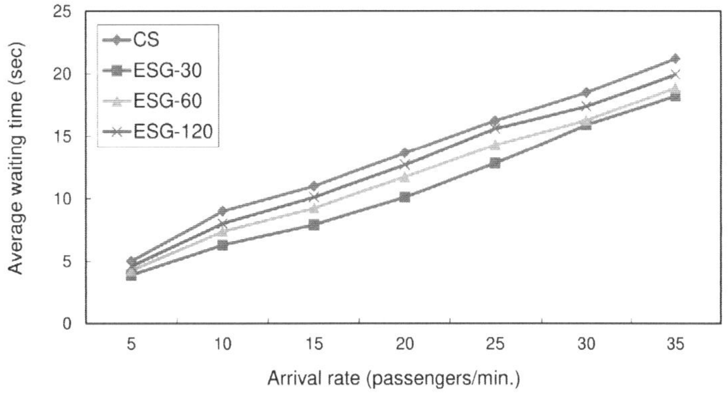

(a) waiting time – average case

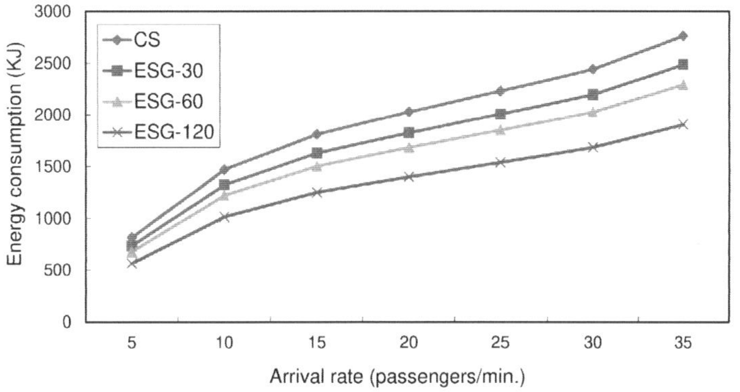

(b) Energy consumption

Figure 6. Comparison of the current system (CS) and the elevator control for smart green buildings (ESG).

prior to actual calls, respectively. In reality, the optimum time period for the reservation call should be defined empirically depending on passengers' and elevators' moving time and the location of sensors. An elevator's moving time can be estimated precisely by the law of motion as in Equation (1), but a passenger's moving time varies depending on the location of sensors and each passenger's step. If we overestimate the passenger's moving time, the passenger should wait long for the elevator car. If we underestimate the passenger's moving time, the elevator car may arrive before the passenger arrives. As a closer sensor estimates the moving time of the candidate passenger more precisely, in our empirical study, a reservation call from a distant sensor is dropped from the request queue when it appears from a closer sensor.

Location of sensors is also important to determine the time period for reservation calls. A distant sensor provides the control system with passengers' information earlier and thus using this information allows the elevator car to pick up many passengers at a time. This eventually leads to the reduction of energy consumption. In contrast, a close sensor provides more accurate information, and thus the waiting time of passengers can be minimized. For now, we show the results for a spectrum of policies by utilizing multiple sensors and do not provide the tuning issue of the optimal time period.

Figure 6. (a) shows the average waiting time of CS and ESG as the arrival rate is varied. As depicted in the figure, ESG exhibits consistently better performance than CS irrespective of the arrival rate. Specifically, the average waiting time of ESG-30 is better than CS by 22% on average and up to

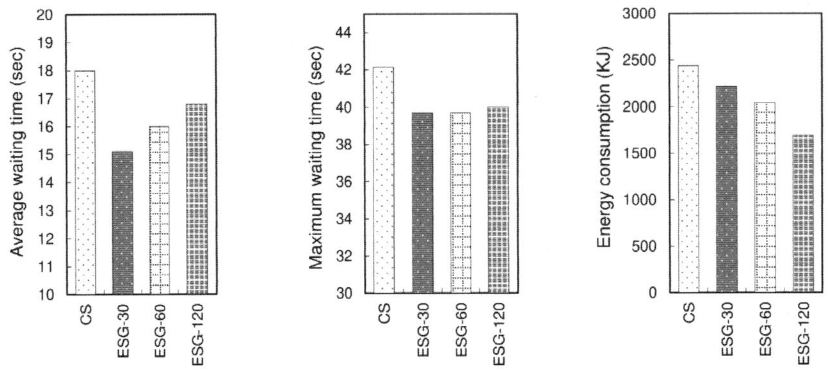

(a) waiting time – average case (b) waiting time – worst case (c) Energy consumption

Figure 7. Comparison of the current system (CS) and the elevator control for smart green buildings (ESG) as the traffic is non-homogeneous.

30%. The performance enhancement of ESG-120 is relatively small as many passengers ride the elevator at a time. Figure 6(b) depicts the amount of energy consumed while operating the elevator system. ESG again exhibits better performance than CS. Among the three configurations, ESG-120 performs the best as more people can ride the car at a single round trip time compared to other elevator control systems. The gain of ESG-120 over CS is 28.6% on average and up to 31.2% with respect to the energy consumption.

Figure 7 depicts the results for non-homogeneous traffic patterns. Again, ESG performs better than CS for all performance measures. As shown in Figure 7(a), ESG-30 exhibits 16.1% better performance than CS with respect to the waiting time of average case. Similar to the results in Figure 6, the performance enhancement of ESG-120 is not large with respect to the waiting time of average case. However, this is not the case for the waiting time of the worst case as shown in Figure 7(b). Irrespective of ESG configurations, the improvement of the worst case waiting time is in the range of 6.5-6.8%. Figure 7(c) depicts the energy consumption of the elevator system. Similar to the homogeneous traffic cases, ESG-120 performs the best; the improvement is 30.7% compared to CS in terms of the energy consumption.

5. Conclusion

This paper presented a novel elevator control system, called ESG, for smart green buildings. ESG uses multiple sensor devices to collect passengers' arrival information before they arrive at the elevator. Specifically, three types of sensor devices, RFID, video, and floor sensors are used to estimate passenger information precisely. By utilizing this prior information, ESG controls the elevator system more efficiently than legacy elevator control systems. We generate a reservation call for candidate passengers detected by sensors and control the moving direction and the moving time of elevator cars for the reservation calls. Experimental results with various traffic conditions showed that ESG reduces the average waiting time and the energy consumption by 15-30% and 28-31%, respectively, compared to legacy elevator systems. As a future work, we plan to extend ESG to collect the destination floor of passengers in advance by using a variety of sensors including smartphone sensors. We expect that more complete information can be used for efficient scheduling.

Acknowledgment

This research has been supported by the National Research Foundation of Korea (NRF) grant funded by the Korea government (MSIP) (No.2011-0028825).

References

[1] Essa I 2000 Ubiquitous sensing for smart and aware environments *IEEE Personal Communications* **7** 47–9
[2] So AT and Liu SK 1996 An overall review of advanced elevator technologies *Elevator World* **44** 96–103
[3] Kidd C D, Orr R J, Abowd G D, Atkeson C G, Essa I A, MacIntyre B, Mynatt E, Starner T E and Newstetter W 1999 The aware home: a living laboratory for ubiquitous computing research *Lecture Notes in Computer Science* **1670** 191–8
[4] Orr R J and Abowd G D 2000 The smart floor: a mechanism for natural user identification and tracking *Pro. Int. Conf. Human Factors in Computing Systems* (The Hague, Netherlands) 275–6
[5] Addlesee M, Jones A, Livesey F and Samaria F 1997 The ORL active floor *IEEE Personal Communications* **4** 35–41
[6] Kwon O, Bahn H and Koh K 2006 A context-aware elevator scheduling system for smart apartment buildings *Proc. Int. Conf. Advances in Hybrid Information Technology* (Jeju Island, Korea) 362–72
[7] Brand M E and Nikovski D N 2004 Risk-averse group elevator scheduling *Proc. Int. Congress Vertical Transportation Technologies* (Istanbul, Turkey) 1–8
[8] Lee S and Bahn H 2005 An energy-aware elevator group control system *Proc. IEEE Int. Conf. Industrial Informatics* (Perth, Australia) 639–43
[9] Brand M E and Nikovski D N 2004 Optimal parking in group elevator control *Proc. IEEE Int. Conf. Robotics & Automation* (New Orleans, LA) 1002–8
[10] Kim J H and Moon B R 2001 Adaptive elevator group control with camera *IEEE Trans. Industrial Electronics* **48** 377–82
[11] Eguchi T, Hirasawa K, Hu J and Markon S 2004 Elevator group supervisory control systems using genetic network programming *Congress on Evolutionary Computation* **2** 1661–7
[12] Shimbun N S 2004 Elevators get smart *RFID in Japan*
[13] Want R, Hopper A, Falcão V and Gibbons J 1992 The active badge location system *ACM Trans. Information Systems* **10** 91–102
[14] Steinhage A and Lauterbach C 2008 Monitoring movement behavior by means of a large area proximity sensor array in the floor *Proc Workshop Behavior Monitoring and Interpretation* 15–27
[15] Doherty L, Simon J and Watteyne T 2012 Wireless sensor network challenges and solutions *Microwave J.* **55** 22–34
[16] Christakos C 2006 Sensor networks applied to the problem of building evacuation: an evaluation in simulation *Proc. IST Mobile and Wireless Summit* (Mykonos, Greece) 134–8
[17] Strang T and Bauer C 2007 Context-aware elevator scheduling *Proc. Int. Conf. Advanced Information Networking and Applications Workshops* (Niagara Falls, Ont.) 276–81
[18] Brunette W, Lester J, Rea A and Borriello G 2005 Some sensor network elements for ubiquitous computing *Proc. IEEE/ACM Int. Conf. Information Processing in Sensor Networks* (Los Angeles, CA) 388–92
[19] Gao Y, Lin Y and Sun Y 2013 A wireless sensor network based on the novel concept of an I-matrix to achieve high-precision lighting control *Building and Environment* **70** 223–31
[20] Eliades D, Michaelides M, Panayiotou C and Polycarpou M 2013 Security-oriented sensor placement in intelligent buildings *Building and Environment* **63** 114–21
[21] Sreedharan P, Sohn M, Nazaroff W and Gadgil A 2011 Towards improved characterization of high-risk releases using heterogeneous indoor sensor systems *Building and Environment* **46** 438–47

IOP Conf. Series: Earth and Environmental Science **40** (2016) 012080 doi:10.1088/1755-1315/40/1/012080

Real options analysis for photovoltaic project under climate uncertainty

Kyeongseok Kim[1], Sejong Kim [2] and Hyoungkwan Kim[1,3]

[1]School of Civil & Environmental Engineering, Yonsei University, Seoul 03277, Korea
[2]ABIK Inc., Yongin-si, Gyeonggi-do 16825, Korea

Email: hyoungkwan@yonsei.ac.kr

Abstract. The decision on photovoltaic project depends on the level of climate environments. Changes in temperature and insolation affect photovoltaic output. It is important for investors to consider future climate conditions for determining investments on photovoltaic projects. We propose a real options-based framework to assess economic feasibility of photovoltaic project under climate change. The framework supports investors to evaluate climate change impact on photovoltaic projects under future climate uncertainty.

1. Introduction

Solar power plays an important role as a renewable energy, accounting for 139GW in the 2013 world energy market [1]. Photovoltaic (PV) cells transform solar energy into electricity through a semiconductor junction [2]. PV systems has the advantage of being able to be installed in any place with any size, unlike hydropower and wind power. However, climate-related risks affect the production of solar energy. Two major risk factors in solar power are temperature and insolation changes. Thus, these climate conditions should be considered for optimum investment in photovoltaic projects.

The 2014 fifth assessment report (AR5) of the Intergovernmental Panel on Climate Change (IPCC) predicted that the climate change will alter temperature and insolation in many areas [3]. In PV projects, the energy production is determined by such climate factors as insolation and temperature, and efficiency of the system. Traditional method of economic analysis could not capture the uncertainty latent in PV projects. Therefore, this study proposes a framework to assess economic feasibility of PV projects under future climate uncertainty. Real options analysis is known to have the ability to consider uncertainty of future scenarios. Real options analysis has originated from financial options by Black and Scholes [4]. Trigeorgis [5], Mun [6], and Copeland and Antikarov [7] improved upon this methodology. Real option analysis can consider management flexibility and volatility of project returns, contrary to traditional discounted cash flow [5]. The proposed framework can provide more accurate analysis of the PV project because flexible response to variability of climate factors is systematically factored in.

---

[3] Address for correspondence: Hyoungkwan Kim, School of Civil & Environmental Engineering, Yonsei University, Seoul 03277, Korea. E-mail: hyoungkwan@yonsei.ac.kr.

2. Framework for investment decisions

2.1 Steps of the framework

Figure 1 shows the three steps of the proposed framework: identification of project conditions (step 1) selection of key variables (step 2), real options analysis (step 3), and decision-making (step 4).

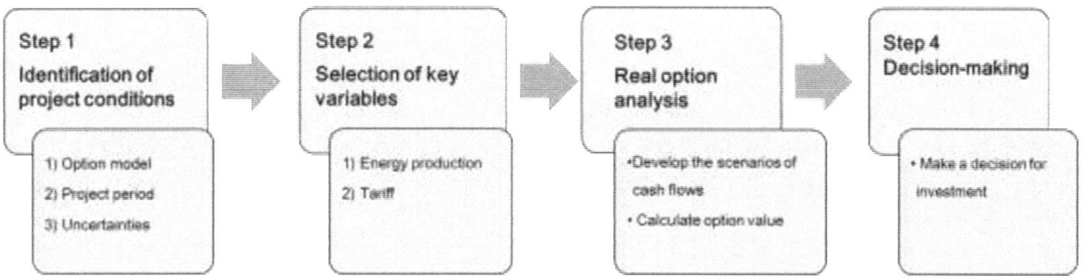

Figure 1. Real options-based framework for photovoltaic projects under climate uncertainty

2.2 Evaluation procedures in the proposed framework

2.2.1 Step 1. Identification of project conditions

Step 1 is to identify the project conditions and the project goal. The investors want to reduce the impact of uncertainty and obtain economic feasibility. If market declines severely, management could execute the options to abandon or wait or contract. Investors need to identify project period, uncertainties of the project, and economic conditions such as investment costs, project period, discount rate, risk-free rate, operation and maintenance cost (O&M cost), and inflation,

2.2.2 Step 2. Selection of key variables

Step 2 is the selection of key variables for the real options analysis. Although PV projects can be affected by a range of factors, this study considers two major factors: energy production and tariff (electricity selling price). These two factors directly affect the cash flows of PV projects. The potential energy production (E) is given by

$$E = \eta A Q \tag{1}$$

where photovoltaic energy production (E in kWh) is a function of the overall efficiency (η), the area of array (A in m^2), and the insolation (Q in kWh/m^2) [8]. According to the Representative Concentration Pathway (RCP) 8.5, 6.0, 4.5, and 2.6 scenarios, insolation is expected to increase in 2100 by 3.6%, 2.5%, 1.9%, 1.1%, respectively [9]. RCPs are separated by four pathways. RCP2.6 is a severe relief scenario, RCP8.5 is very high green house gas emissions scenario, and RCP 4.5 and RCP 6.0 are two intermediate scenarios [3]. Tariff is another significant factor for an investment decision in energy markets. Tariff is determined by the market condition which is consisted with supply and demand. Future tariffs are estimated using historical data. In this study, the future tariff is assumed to follow a leaner regression equation.

2.2.3 Step 3. Real options analysis

Step 3 is the real options analysis for the project. Using the electricity production and tariff information generated in the first step, three cases are developed: best, worst, and moderate (most-likely) cases. The present value of the most-likely cash flow case is the underlying asset value (S_0) used for the options analysis. The volatility (σ) of the project value is estimated using the following equation [10]:

$$\sigma = \frac{ln(\frac{S_{best}}{S_{worst}})}{4\sqrt{t}} \tag{2}$$

where S_{best} is the underlying asset value under the best scenario, S_{worst} is the underlying asset value under the worst scenario, and t is the production period of the project.

The binomial lattice model is used with up movement (u), down movement (d), risk-neutral probability (q), and option value (C) represented by Equation (3) – (6) [10].

$$u = e^{\sigma\sqrt{\Delta t}} \tag{3}$$

$$d = \frac{1}{u} \tag{4}$$

$$q = \frac{(1+r-d)}{u-d} \tag{5}$$

$$C = e^{-rt}[qC_u + (1-q)C_d] \tag{6}$$

where Δt is an incremental time step, r is the risk-free interest rate, and C_u and C_d are the option values associated with up and down movements, respectively.

A real options analysis utilizes management flexibility to try lessen uncertainties such as deferral, abandonment, growth, and multistage. Investors should resolve investment decision. The investment cost is the strike price in real options analysis, which could be implemented when the return is larger than the strike price. Otherwise, investors must wait until the project produces expected return. Real option analysis investigates project value by binomial lattice which solve option model using risk-neutral probabilities [6] (Figure 2).

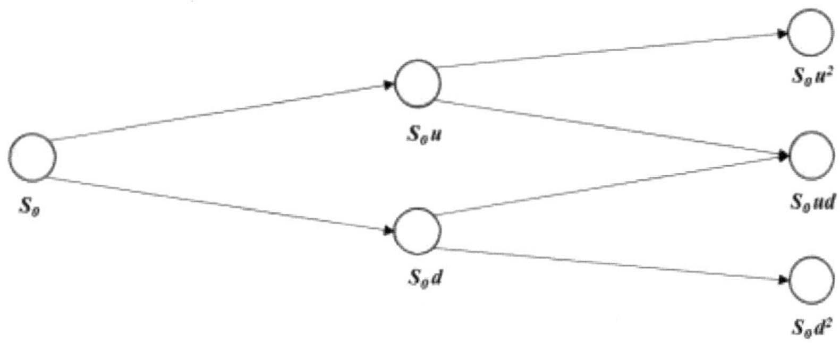

Figure 2. Binomial lattice approach for real options analysis

2.2.4 Step 4. Decision-making

Step 4 is the step for decision making. If the option value of each node of the binomial lattice is larger than 0, the project is worth investment; otherwise, the project better be abandoned. The result is a good indicator as to the project profitability and a guidance to reasonable selection and execution of the project.

3. A case study of a PV project in South Korea

A PV project is developed inside a thermal power plant in South Korea with the capacity of 1MW in 2016 and start to operate from 2017. As previously mentioned, step 1 is to identify the project condition.

Table 1. Identification of project conditions.

Factor	Value
Investment cost	US$ 2.73 Millions
Capacity	1 MW
Efficiency	15%
Power loss	5%
O&M cost	Yearly 0.7% % of the investment cost
Risk-adjusted discount rate	10.0%
Risk-free interest rate	5.0%
Tax rate	27.5%
Climate change scenarios	RCPs
Development period	1 years (2016)
Production period	20 years (2017 - 2036)
Option model	Option to abandon

Table 2. Information of technical specifications and climatic conditions in case study.

Category	factor	Specification
Technical data	Module type	Single–crystalline silicon
	Module size	1,580L*802W*46T (mm)
	Number of module	6,240 pieces
	Mounting type	Ground
	Angle of the panel position	30%
Climatic conditions	Average outdoor temperature	Spring 12.5 ℃
		Summer 24.3 ℃
		Autumn 14.7 ℃
		Winter 1.7 ℃
	Daylight hour per year	2,105 hours
	Daylight hour per day	5.76 hours
	Average humidity	69%
	Average days of yellow dust	6.5 days

Table 1 shows the project conditions of the case study. The case study of PV project is designed for an average annual generation of 1,246.3 MWh in 2017. However climate uncertainty may alter the energy production and tariff may sharply fluctuate. Thus, we investigate the economic feasibility and establish the strategy for investment of the PV project under climate change.

In this study, the type of option is abandonment. Option to abandon is used to select the go/no-go investment decision for obtaining economic feasibility under climate uncertainty using real options analysis. The PV project will be operated for 20 years. However the investor has an option to abandon and resale the asset to the land owner within five year from start of operation at the 50% of investment cost. Table 2 shows the information of technical specifications and climatic conditions in case study [11].

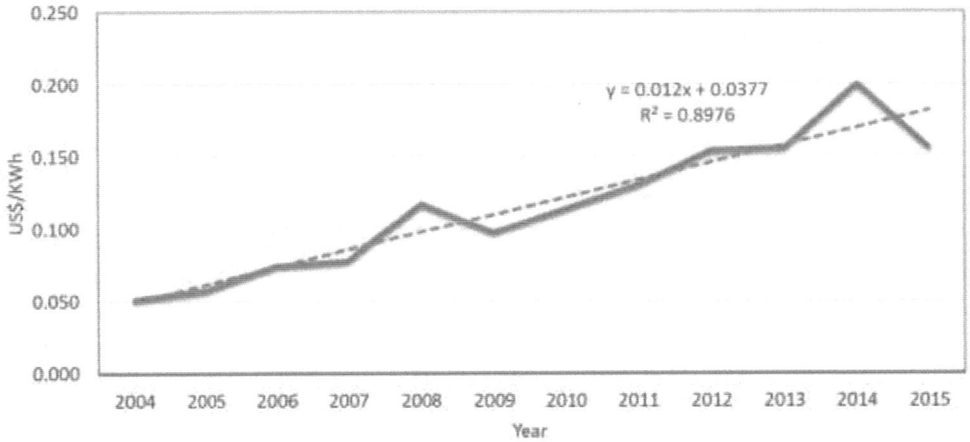

Figure 3. Historical annual tariff between 2004 and 2015

In step 2, the key variables were selected to develop the project cash flows. Table 3 shows the values of the two key variables for each case: moderate, best, and worst. These values were utilized to create three scenarios of cash flows. The underlying asset was generated through the moderate scenario of the future cash flow. The insolation increase of 3.6% by 2100 on the basis 2010 under RCP8.5 is converted to the best case of annual increase rate of 0.040%. The worst case is 0.012% in RCP2.6. The moderate case is the average value of 0.026%. In case of tariff, we could simulate future tariff during the project period using historical data between 2004 and 2015. However, Schröder et al. [12] found that investment and O&M costs of PV power will be less than a third of the current cost by 2050. Although the tariff of the renewable energy such as wind and PV energy declines due to rapidly changing technologies, the historical data of KPX [13] shows the increase during 12 years as shown in Figure3. Linear regression was used for simulation of future tariff during the project period. The best case of tariff is assumed to be 200% and worst is 50% of the moderate value.

Table 3. Key variables of the case study project under the best, moderate, worst cases.

Variable	Best	Moderate	Worst
Tariff	T × 200%	T × 100%	T × 50%
Energy production (annual increase)	0.040%	0.026%	0.012%

Figure 4 shows the revenues in the three scenarios: best, moderate, worst scenario. The volatility of the project was calculated by present value of underlying asset under the best and worst scenarios.

IOP Conf. Series: Earth and Environmental Science **40** (2016) 012080 doi:10.1088/1755-1315/40/1/012080

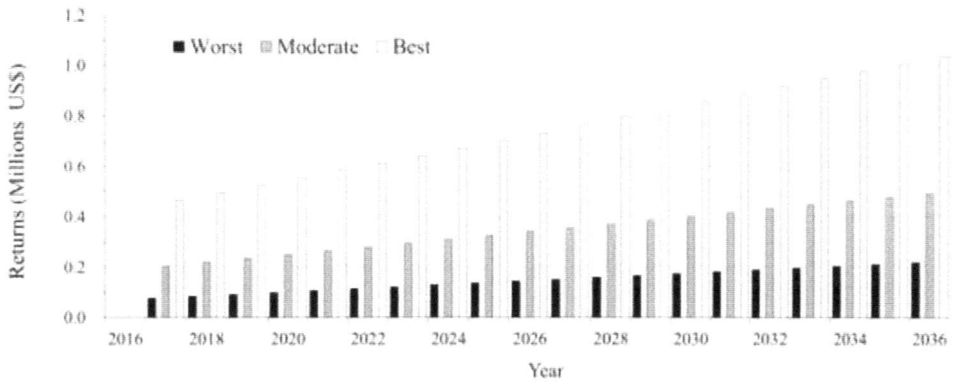

Figure 4. Project returns of three scenarios

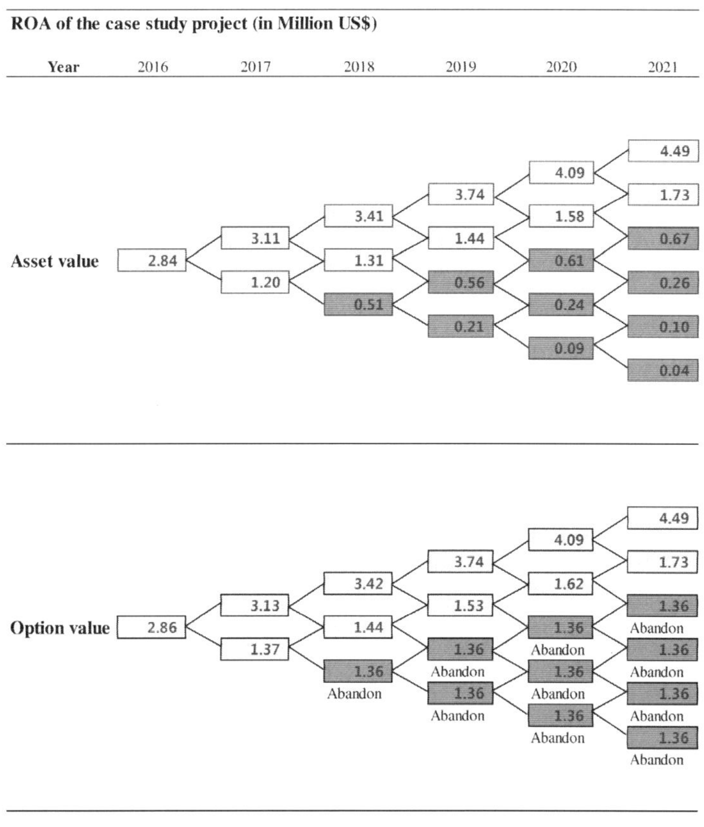

Figure 5. Binomial lattice for real options analysis

Using step 3, we conducted the real options analysis. Figure 5 shows the binomial lattice calculation which yielded the underlying asset values and option values between 2016 and 2021. Investors have the right to abandon and resale the PV power asset and will obtain 50% of the investment cost. US$1.36 million is the fixed salvage value for the next five years.

In step 4, investors may decide to use the American put option for maximizing project value. The abandonment option can be exercised at any time up to the expiration date. Depending on the outcome of the binomial lattice, the option value of US$ 2.86 million would make the project economically feasible in the year 2016. If the project uncertainty is clear, the project will continue. If the project becomes unprofitable, investors could exercise the abandonment option from 2018 to 2021, as shown in red boxes in Figure 5. The traditional net present value (NPV) of the project was US$ 0.28 million; but the option value was US$ 2.86 million. This result means that the project was quite feasible in case that the abandonment option is included in the project value.

4. Conclusions

This study proposed a framework that copes with the uncertainty of climate change for PV projects. The contributions of this study to the body of knowledge are three-fold. First, the proposed framework can help investors find proper investment strategy in PV projects. Second, the proposed framework can be extended to predict the energy production under different climate scenarios. Third, real options analysis has an advantage to reflect management flexibility to exercise options until project value become positive in PV projects.

Out study is confined to only major factors such as tariff and energy production. Other technical and socio-economic factors that could determine the economic feasibility of PV projects were not considered. Notwithstanding, the outcomes of this study gives an idea for a role of government. Governments can provide abandon and resale options to investors for improved economic feasibility of PV projects.

Acknowledgments

This work was supported by grants from the National Research Foundation of Korea (NRF), funded by the Korean government (MSIP) (No. 2011-0030040 and NRF-2014R1A2A1A11052499).

References

[1] REN21 2014 *Renewables 2014 Global Status Report: REN21 Secretariat Paris, France*
[2] Crook J A, Jones L A, Forster P M and Crook R 2011 Climate change impacts on future photovoltaic and concentrated solar power energy output *Energy & Environmental Science* **4** 3101-9
[3] IPCC 2014 *Climate Change 2014: Synthesis Report: Intergovernmental Panel on Climate Change*
[4] Black F and Scholes M 1973 The Pricing of Options and Corporate Liabilities *J. Political Economy* **81** 637-54
[5] Trigeorgis L 1996 *Real options: Managerial flexibility and strategy in resource allocation* (Massachusetts: MIT press)
[6] Mun J 2002 *Real options analysis: Tools and techniques for valuing strategic investments and decisions* John Wiley & Sons
[7] Copeland T E and Antikarov V 2003 *Real options: a practitioner's guide* (New York: Texere)
[8] Kim D S and Wang Y D 2013 Economic analysis of rural green-village planning with solar energy considering climate change *J. Korean Society of Rural Planning* **19** 25-36
[9] KMA 2012 *Understanding and Application of Climate Change: Korea Meteorological Administration*
[10] Kodukula P and Papudesu C 2006 *Project valuation using real options: a practitioner's guide* J. Ross Publishing
[11] KMA 2008 Meteorological resources analysis for optimal use of solar energy
[12] Schröder A, Kunz F, Meiss J, Mendelevitch R and Von Hirschhausen C 2013 *Current and*

IOP Conf. Series: Earth and Environmental Science **40** (2016) 012080 doi:10.1088/1755-1315/40/1/012080

prospective costs of electricity generation until 2050 DIW Data Documentation 68
[13] KPX 2016 *Electric power statistics information system* http://epsis.kpx.or.kr/epsis/ekmaStaticMain.do?cmd=004069&flag=&locale=EN (accessed Jan, 2016)

The influence of thermal insulation position in building exterior walls on indoor thermal comfort and energy consumption of residential buildings in Chongqing

D Wang[1], W Yu, X Zhao, W Dai and Y Ruan

Faculty of Urban Construction and Environmental Engineering, Chongqing University, Chongqing 400045, China

E-mail: wang18239978874@163.com

Abstract. This paper focused on the influence of using position of thermal insulation materials in exterior walls on the indoor thermal comfort and building energy consumption of residential building in Chongqing. In this study, four (4) typical residential building models in Chongqing were established, which have different usage of thermal insulation layer position in exterior walls. Indoor thermal comfort hours, cooling and heating energy consumption of each model were obtained by using a simulation tool, Energyplus. Based on the simulation data, the influence of thermal insulation position on indoor thermal comfort and building energy consumption in each season was analyzed. The results showed that building with internal insulation had the highest indoor thermal comfort hours and least cooling and heating energy consumption in summer and winter. In transitional season, the highest indoor thermal comfort hours are obtained when thermal insulation is located on the exterior side.

1. Introduction

In 2014, China and the United States (US) signed a joint declaration on climate changing and cooperation of clean energy, and China formally indicated that its carbon emissions are expected to reach a peak in 2030 [1]. [2] pointed out that in 2013 China's total building energy consumption accounted for 19.5% of the total energy consumption, which showed an increasing trend every year. Thus, reducing building energy consumption is become an important part of the national goal of achieving energy conservation and emission reduction. To reduce energy consumption of buildings, the most critical way is to reduce the energy consumption resulted by improve indoor environment especially the thermal environment [3].

Chongqing is a typical city in hot summer and cold winter climate, located in the transitional zone between Qinghai Tibet Plateau and the Yangtze River Plain in China. Typical meteorological yearly data of Chongqing shows that the annual average temperature is 18.4℃, the average temperature of the hottest month, July is 30.95℃, the average temperature of the coldest month, January is 8.1℃, and the annual average relative humidity is 80.6% [4]. It can be seen that Chongqing is hot in summer and cold in winter, as a result, residents have higher demands for cooling and heating. With the rapid increase of residents' income, the residents' demand to improve indoor environment is becoming an

[1] Address for correspondence: D Wang, Faculty of Urban Construction and Environmental Engineering, Chongqing University, Chongqing 400045, China. E-mail: wang18239978874@163.com.

urgent need [5]. However, the improvement of indoor thermal environment in Chongqing may result in a greater amount of energy consumption. Therefore, designers should simultaneously consider building energy consumption and thermal comfort when design buildings.

Adopting proper thermal insulation form is one of the effective ways to reduce building cooling and heating energy consumption [6]. [7] also pointed out that the design of building thermal insulation has to be considered a priority in the energy saving optimization. With the publication of energy saving design standards [8] and [9], building energy saving requirements will be more stringent and thermal insulation materials have been increasingly used in engineering.

Changing the position of the thermal insulation layer in the envelope will not only change thermal resistance, but it will have different effects on the heat transfer process between indoor and outdoor [10]. Many scholars have analyzed using position of thermal insulation materials in envelope:

In 2000, Bojic [11] studied the influence of different location of thermal insulation layer on the annual cooling load and the daily maximum cooling load in two high-rise residential buildings in Hong Kong by using HTB2 simulation software. The results showed that cooling load can be different when the insulation layer are placed in different locations in envelope, and the effect of increasing thermal insulation thickness on reducing cooling load becomes weak when the thickness of insulation material reaches a certain value. In 2001, Kossecka [12] designed six (6) typical external thermal insulation walls, with different thermal insulation layer placement. Six cooling and heating loads of the single residential with different walls were calculated using DOE-2.1E. It was concluded that the location of the insulation layer in the exterior wall significantly affects the energy demand of the building. In 2006, Ozel [13] set up twelve (12) different kinds of exterior wall through changing the position of thermal insulation layer in exterior wall, and analyzed the influence of each exterior wall on the time lag and decrement factor during typical summer and winter days. Analyses of the results showed that the best distribution of insulation layer was in three equal parts at both sides and the middle section of the wall. In 2013, Tummu [14] studied the effects of different thermal insulation materials and thermal insulation layer location in the exterior wall on the heat transfer in the bedroom, living room and studio room in Thailand by using measurement and simulation methods. The data showed that insulation layer located in internal plane had better performance than the external one in all the rooms. In 2015, Ruan [15] studied the effects of external insulation and internal insulation in exterior wall, and pointed out that under the intermittent energy consuming mode, external insulation would increase cooling load in summer and the internal insulation could reduce the cooling load and heating load.

From the above research, we can see that the existing research mainly focused on the influence of thermal insulation material's location on the heat transfer and cooling and heating energy consumption of buildings. However, there is limited research on the influence of thermal insulation material's location on the indoor thermal comfort. In these previous energy saving optimization research, the indoor thermal comfort is somewhat not studied. Therefore, it necessary to explore whether the use of thermal insulation materials would affect the indoor thermal comfort when outdoor meteorological parameters are appropriate, and heating and cooling is not needed especially in the spring and autumn. In this study, an in-depth analysis of the influence of thermal insulation material's location in the exterior wall on the indoor thermal comfort is carried out. This study also investigated the influence of thermal insulation material's location in the exterior wall on residential building energy consumption in Chongqing area.

From the indoor thermal comfort and energy consumption perspective, this study established four (4) typical residential building models with different insulation layer location in the exterior wall in Chongqing. Heating and cooling energy consumption under air conditioning and indoor thermal comfort hours under natural ventilation were obtained by using the dynamic simulation software, Energyplus. Thorough analysis of the simulation results in different seasons, the influence of thermal insulation material's location in the exterior wall on indoor thermal comfort and energy consumption in Chongqing were obtained. This study provided a basis for designers to design the using position of thermal insulation materials and fills up the gaps in related research.

2. Methodology

This study established four typical residential building models with different insulation layer location in the exterior wall in Chongqing. Annual heating and cooling energy consumption under air conditioning and hourly indoor operating temperature and relative humidity under natural ventilation were obtained by using the dynamic simulation software, Energyplus. Base on the acceptable thermal comfort zone model and hourly indoor operating temperature and relative humidity, the annual indoor thermal comfort hours are obtained. Indoor thermal comfort hours and cooling and heating energy consumption of each building model are obtained and compared in every season, so the influence of thermal insulation material's location in the exterior wall on indoor thermal comfort and energy consumption is obtained. Then, through a comparison of indoor and outdoor meteorological parameters and statistical the standard deviation, maximum and minimum of operating temperature in each model, the influence of thermal insulation material's location in the exterior wall on indoor thermal comfort and energy consumption are explained.

2.1. Thermal insulation placement

The difference between the 4 models established in this study is that the exterior wall has different locations of thermal insulation layer, as shown in figure 1:

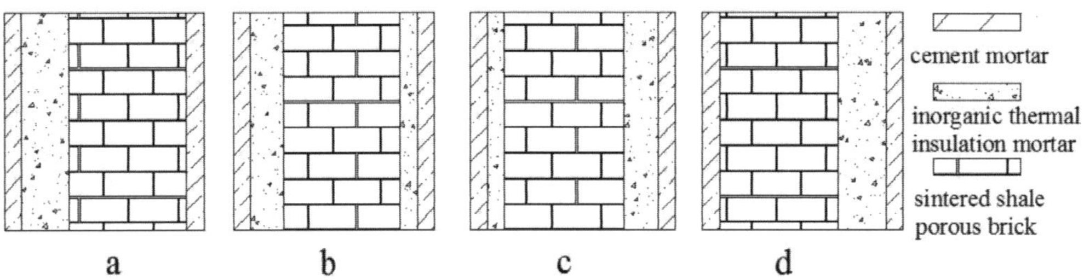

Figure 1. Thermal insulation layers in the four exterior walls.

The four walls are composed of outside to inside:

A. 10mm cement mortar + 30mm inorganic thermal insulation mortar + 200mm sintered shale porous brick + 10mm cement mortar

B. 10mm cement mortar + 20mm inorganic thermal insulation mortar + 200mm sintered shale porous brick + 10mm inorganic thermal insulation mortar + 10mm cement mortar

C. 10mm cement mortar + 10mm inorganic thermal insulation mortar + 200mm sintered shale porous brick + 20mm inorganic thermal insulation mortar + 10mm cement mortar

D. 10mm cement mortar + 200mm sintered shale porous brick + 30mm inorganic thermal insulation mortar + 10mm cement mortar

Model A is external insulation (thermal insulation layer is located on the external side). Model D is internal insulation (thermal insulation layer is located on the internal side). Model B and C are composed of different thickness of external and internal thermal insulation. The physical properties of the materials used for the four external walls are given in table 1.

Table 1. Physical properties of external wall materials.

Material name	Thermal conductivity coefficient W/(m • K)	Density kg/m3	Specific heat J/kg • k
Inorganic thermal insulation mortar	0.07	280	1050
Sintered shale porous brick	0.58	1400	1050
Cement mortar	0.93	1800	1050

Inorganic insulation mortar is a common thermal insulation material which has the following advantages; low cost, convenient construction and so on. It has been widely used in the hot summer and cold winter area [16]. Based on the material thickness of the external wall and the physical properties of the material, the thermal performance parameters of the external wall are calculated as shown in table 2.

Table 2. Thermal performance of exterior wall.

Number	Total heat transfer coefficient W/(m • K)	Thermal inertia index
1	1.05	3.48
2	1.05	3.48
3	1.05	3.48
4	1.05	3.48

Owing to the same material and thickness, the exterior walls have the same thermal performance parameters. Meanwhile, the influence of the different thermal insulation location on thermal comfort and energy consumption are easier to analyse, when the thermal performance is same.

2.2. Simulation tool

This study used Energyplus V8.4.0 as the dynamic simulation tool for calculating the hourly indoor operating temperature, relative humidity and annual energy consumption. Energyplus is a building energy analysis tool developed in 2001 by the US Department of Energy, based on the most popular features and capabilities of BLAST and DOE-2 and has been widely used in throughout the world [17]. According to building model information from user's setting, Energyplus can calculate the annual energy consumption of the building as well as the hourly indoor thermal and humidity condition. Firstly, the building model is built without the air-conditioning system, and the operating temperature and relative humidity of the indoor in natural ventilation are calculated, which is the basic data for the evaluation of indoor thermal comfort. Thereafter, adding the air conditioning system, building cooling and heating energy consumption is obtained.

2.3. Indoor thermal comfort hours and acceptable humidity hours

Taking into account the people's psychological and behavioral adjustment, Yao [18] established the Adaptive Predicted Mean Vote (aPMV) based on automatic control theory, 'black box'. The aPMV is a thermal comfort evaluation model for evaluating the thermal comfort of the human body in non-air-conditioned environment. According to the survey data, Wei [19] puts forward the acceptable thermal comfort zone (when $-1 \leqslant aPMV \leqslant 1$) of the residents in non-air-conditioned environment in Chongqing. As shown in figure 2, the vertical axis is the operating temperature, the horizontal axis is the moisture and the curve is relative humidity. The area surrounded by the dotted line is the acceptable thermal comfort zone. Based on the indoor air quality standard [20], the acceptable thermal comfort zone sets acceptable relative humidity range between 30%~80%.

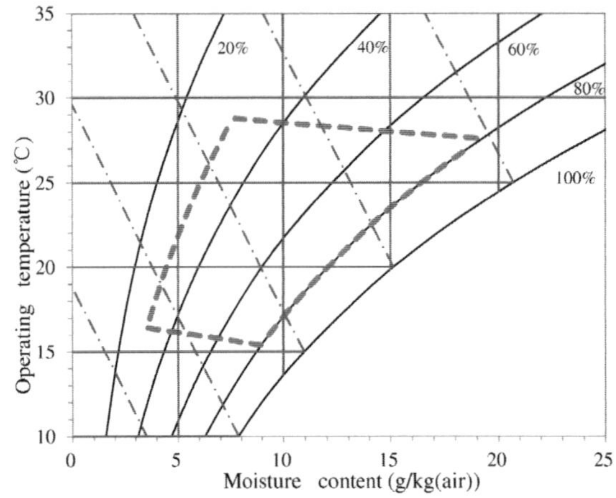

Figure 2. Acceptable thermal comfort zone in Chongqing.

This study made the following definitions:

Thermal comfort hours: indoor operating temperature and humidity points are in the range of acceptable thermal comfort zone in non-air-conditioned environment.

Acceptable humidity hours: indoor humidity points are in the range of acceptable relative humidity in non-air-conditioned environment.

2.4. Model specification

Figure 3 shows the layout of the building. The building has three (3) floors, with floor height of 2.8m, and there are two (2) residence s in each floor. The house construction area is about 80m^2, which is the mainstream size of residence having three (3) people in Chongqing [21]. Figure 4 shows the drawing of the building.

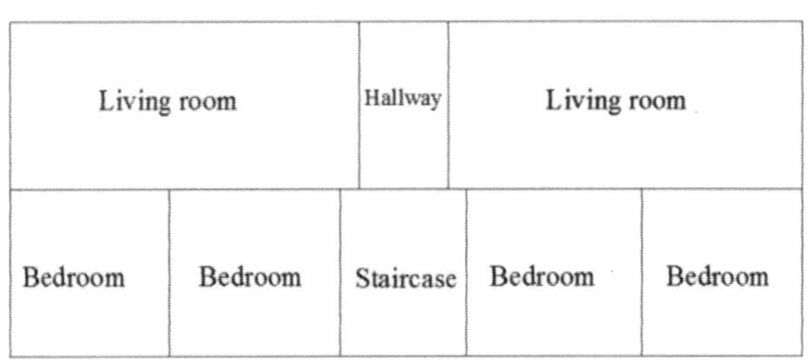

Figure 3. Typical residential building plane.

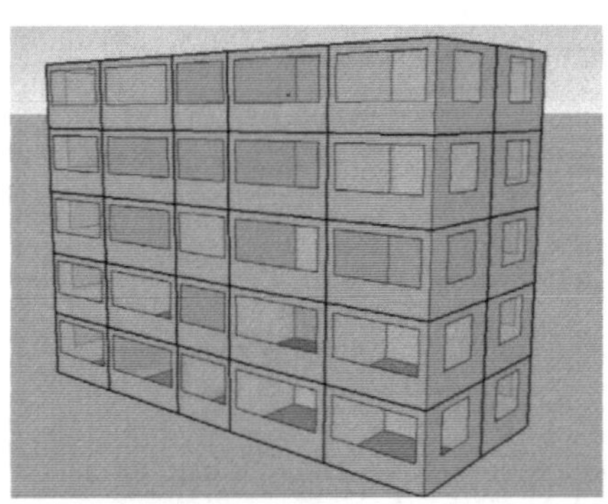

Figure 4. A drawing of typical residential building.

The building envelope met the requirements of design standard for energy efficiency, 65% of residential building [8], as shown in table 3. The external wall was set in section 2.1 part of this paper.

Table 3. Setting parameters of envelope.

Variable	Value
Shape coefficient	0.39
Roof heat transfer coefficient W/(m·K)	0.69
Roof thermal inertia index	2.55
Exterior window heat transfer Coefficient W/(m·K)	2.01
North window wall ratio %	45
South window wall ratio %	45
East window wall ratio %	25
West window wall ratio %	25

Based on the design standard for energy efficiency, 65% of residential building [8], cooling and heating periods were set. Cooling period (summer): June 1st - September 30th. Heating period (winter): December 1st to 30th and January 1st to February 28th. The heating setpoint was 18℃ and cooling setpoint was 26℃. In the rest of the time (transitional season), the air-conditioning system was not operating. Due to the fact that the research object is the residential building, air conditioning system set to a split type air conditioner and the coefficient of performance was 3.5. If the outdoor temperature was in the range of 18-26℃, the indoor ventilation was 5 times/h, otherwise was 1 time/h. Occupancy was 25 m2/person and occupancy design load was 60 W/person (Seated, quiet). The lighting design load was 5 W/m2, and the equipment design load was 4.3 W/m2. Schedules for air conditioners, equipment and lighting used the template from EnergyPlus.

3. Results

Table 4. The annual results.

Number	Annual energy consumption kWh/m^2	Annual thermal comfort hours	Annual acceptable humidity hours
1	22.671	3423	8242
2	22.643	3425	8233
3	22.637	3428	8228
4	22.631	3435	8225

Table 4 shows the values of annual results of the four different building models. With the reduction of external insulation thickness and increase of internal insulation thickness, building energy consumption decreased from external insulation building 22.671 kWh/m^2 to internal insulation building 22.631 kWh/m^2. Meanwhile, thermal comfort hours increased from external insulation building 3423 to internal insulation building 3435. Annual thermal comfort hours and energy consumption showed an opposite trend. The internal insulation had the least acceptable humidity hours, which was 8225, accounting for 93.89% of the total hours of the year. All the four building models have good wet environment, so this study did not analyze the humidity. The following was a further analysis of the seasonal energy consumption, thermal comfort hours and operating temperature.

3.1. Summer

Table 5. Cooling energy consumption and thermal comfort hours in summer.

Model	Cooling energy consumption kWh/m^2	Thermal comfort hours

1	14.654	241
2	14.634	242
3	14.634	246
4	14.629	260

Table 5 shows the cooling energy consumption and thermal comfort hours of different building models in summer. It can be seen that the external insulation building has the most cooling energy consumption which was 14.654 kWh/m^2. Model 2 and 3 had the same cooling energy consumption which was 14.634 kWh/m^2. The internal insulation building has the least cooling energy consumption which is 14.629 kWh/m^2. The external insulation building had the least thermal comfort hours, which were 241. With the reduction of external insulation thickness and increase of internal insulation thickness, thermal comfort hours gradually increased. The internal insulation building had the most thermal comfort hours, which were 260. Thermal comfort hours and cooling energy consumption showed opposite trend.

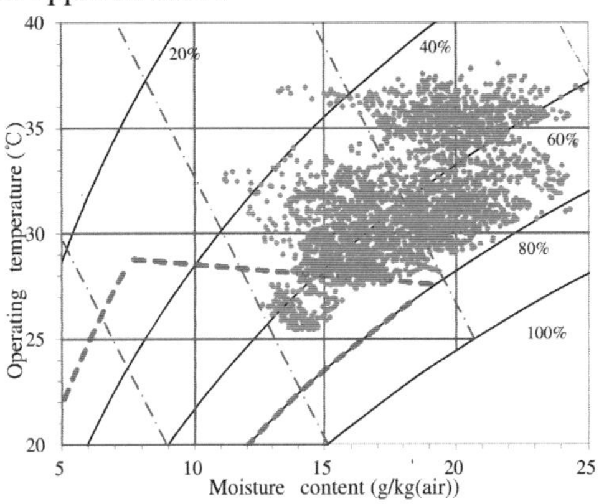

Figure 5. Indoor operating temperature and humidity points in summer.

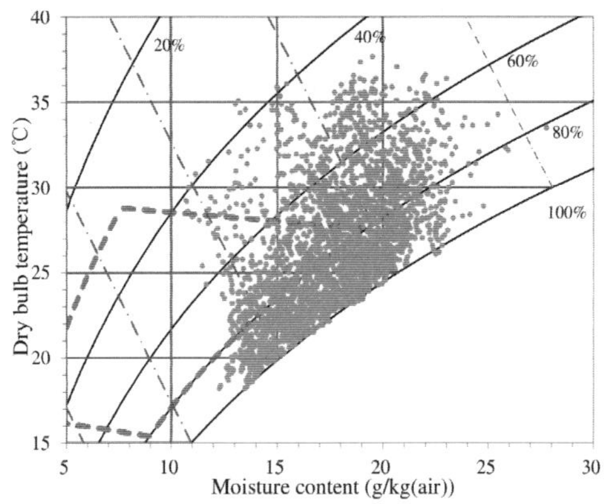

Figure 6. Outdoor dry bulb temperature and humidity points in summer.

By taking internal insulation building as an example, the distribution of indoor operating temperature and humidity points in summer is shown in figure 5. It can be seen that the points were mainly distributed above the thermal comfort zone. Only some points below fell within the acceptable area. The operating temperature was basically distributed between 25℃ and 38℃. Therefore, in order to improve the indoor thermal comfort, it is necessary to reduce the indoor operating temperature. Figure 6 shows the comparison of outdoor meteorological parameters [4], with acceptable thermal comfort zone in summer. Since there is no annual hourly outdoor operating temperature data in Chongqing, using the dry bulb temperature displayed the outdoor thermal environment in Chongqing. It can be seen that outdoor temperature was mainly distributed between 18℃ and 38℃.

Table 6. Operating temperature distribution in summer.

Model	Average value of standard deviation of daily operating temperature℃	Maximum operating temperature℃	Minimum operating temperature℃
1	0.525	38.012	25.586
2	0.532	38.052	25.544
3	0.536	38.072	25.505
4	0.539	38.079	25.457

Table 6 shows the operating temperature distribution in summer. It can be seen that the internal insulation building has the most average value of standard deviation of daily operating temperature which was 0.539℃, the most maximum operating temperature was 38.079℃ and the least minimum operating temperature was 25.457℃. This means that indoor temperature of internal insulation building had the largest fluctuation, mostly being affected by the outdoor weather and had the worst thermal stability. With the reduction of internal insulation thickness and increase of external insulation thickness, the average value of standard deviation of daily operating temperature and maximum operating temperature gradually decreased, and the minimum operating temperature gradually increased, which means the influence degree on indoor thermal environment of the outdoor gradually weakened. The external insulation building had the least average value of standard deviation of daily operating temperature was 0.525℃, the least maximum operating temperature was 38.012℃ and the most minimum operating temperature was 25.586℃. The thermal stability of the external insulation building was excellent, and it was not easily affected by the change of the outdoor environment.

Based on the analysis of indoor and outdoor meteorological parameters and thermal stability, it can be concluded that:

Thermal comfort: Due to poor thermal stability, the internal insulation building could change the indoor temperature more quickly than the other building models. When the outdoor temperature rose to a lower range, some points of the internal insulation building may be reduce to the acceptable thermal comfort zone. Due to the better thermal stability, external insulation was least affected by the outdoor environment. Although indoor temperature decreased, the indoor temperature couldn't change rapidly. When the outdoor temperature increased to a lower range, the indoor temperature in external insulation building was higher than in the internal insulation building, and the temperature in external insulation building may not be reduced to the range of acceptable thermal comfort zone. So in the summer, the thermal comfort hours increases with the reduction in external insulation thickness and increase in internal insulation thickness.

Energy consumption: Due to the poor thermal stability, internal insulation was mostly affected by the outdoor environment. When the outdoor temperature increased to a higher range during daytime, the internal insulation building had the highest hourly operating temperature and the highest hourly cooling load. When the outdoor temperature decreased to a lower range during the night, the internal insulation building had the lowest hourly operating temperature and the lowest hourly cooling load. However, the internal insulation building had the lowest total cooling energy consumption in summer.

3.2. Winter

Table 7. Heating energy consumption and thermal comfort hours in winter.

Model	Heating energy consumption kWh/m^2	Thermal comfort hours
1	8.017	47
2	8.009	50
3	8.003	54
4	8.003	56

Table 7 shows the heating energy consumption and thermal comfort hours of different building models in winter. It can be seen that the external insulation building has the most heating energy consumption was 8.017 kWh/m^2. Model 3 and internal insulation building had the least heating energy consumption, which was 8.003 $kWh/m^2/m^2$. The external insulation building had the least thermal comfort hours, which was 47.With the reduction of external insulation thickness and increase of internal insulation thickness, thermal comfort hours gradually increased. The internal insulation building had the most thermal comfort hours, which was 56. Thermal comfort hours and heating energy consumption showed opposite trend.

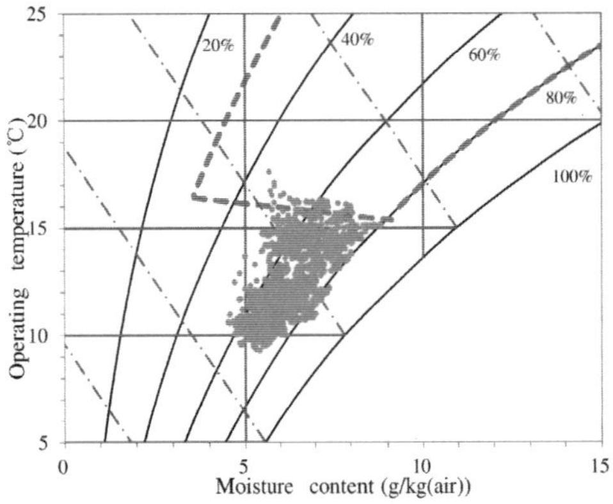

Figure 7. Indoor operating temperature and humidity points in winter.

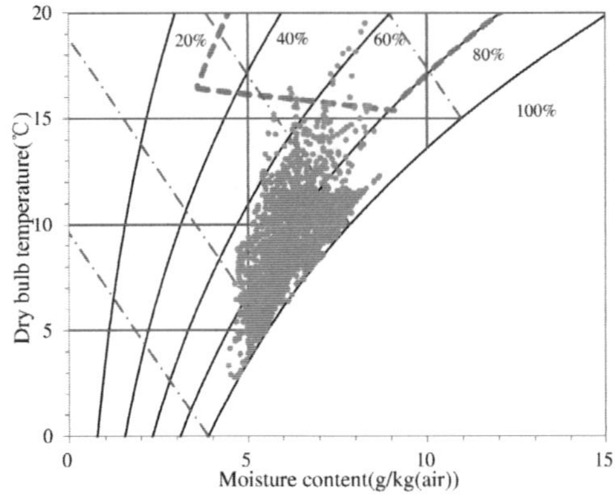

Figure 8. Outdoor dry bulb temperature and humidity points in winter.

By taking internal insulation building as an example, the distribution of indoor operating temperature and humidity points in winter is shown in figure 7. It can be seen that the points were mainly distributed below the thermal comfort zone. Only some points above fell within the acceptable area. The operating temperature was basically distributed between 9℃ and 17℃. Therefore, in order to improve the indoor thermal comfort, it is necessary to improve the indoor operating temperature. Figure 8 shows the comparison of outdoor meteorological parameters [4], with acceptable thermal comfort zone in winter. It can be seen that the outdoor temperature was mainly distributed between 3℃ and 15℃.

Table 8. Operating temperature distribution in winter.

Model	Average value of standard deviation of daily operating temperature℃	Maximum operating temperature℃	Minimum operating temperature℃
1	0.297	17.395	9.354
2	0.300	17.491	9.334
3	0.302	17.560	9.316
4	0.305	17.618	9.294

Table 8 also shows operating temperature distribution in winter which was similar to summer. With the increase of external insulation thickness and reduction of internal insulation thickness, the thermal stability of the building gradually improved. The thermal stability of the external insulation building was the best and internal insulation building was the worst.

From the analysis of indoor and outdoor meteorological parameters and thermal stability, it can be concluded that:

Thermal comfort: Due to poor thermal stability, the internal insulation building could rapidly change indoor temperature than the other building models. When the outdoor temperature increased to a higher range, some points of the internal insulation building may increase to the acceptable thermal comfort zone. Due to the better thermal stability, external insulation was least affected by the outdoor environment. Although indoor temperature increased, the indoor temperature could change rapidly. When the outdoor temperature increased to a higher range, the indoor temperature in external insulation building was lower than in the internal insulation building, and the temperature in external insulation building may not increase to the acceptable range of thermal comfort zone. Consequently, in the winter, the thermal comfort hours increased with the reduction of external insulation thickness and increase of internal insulation thickness.

Energy consumption: Due to the poor thermal stability, internal insulation was mostly affected by the outdoor environment. When the outdoor temperature increased to a higher range during daytime, the internal insulation of building had the highest hourly operating temperature and the lowest hourly heating load. When the outdoor temperature decreased to a lower range during the night, the internal insulation building had the lowest hourly operating temperature and the highest hourly heating load. However, the internal insulation building had the lowest total heating energy consumption in winter.

3.3. Transitional season

Table 9. Thermal comfort hours in transitional season.

Model	Thermal comfort hours
1	3135
2	3133
3	3128
4	3119

The transitional season does not belong to the cooling and heating period, as a result, this part only analyze thermal comfort hours. As it can be seen from table 9, thermal comfort hours of the four building models were all longer than 3000 hours. The lowest was internal insulation building, and thermal comfort hours was 3119, reaching 84.94% of the total hours of the transitional season. With the reduction of internal insulation thickness and increase of external insulation thickness, thermal comfort hours gradually increased. The external insulation building had the most thermal comfort hours, which were 3135, reaching 85.38% of the total hours of the transitional season.

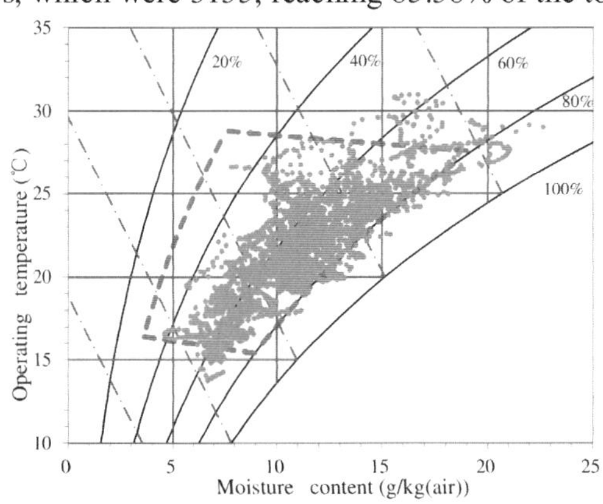

Figure 9. Indoor operating temperature and humidity points in transition season.

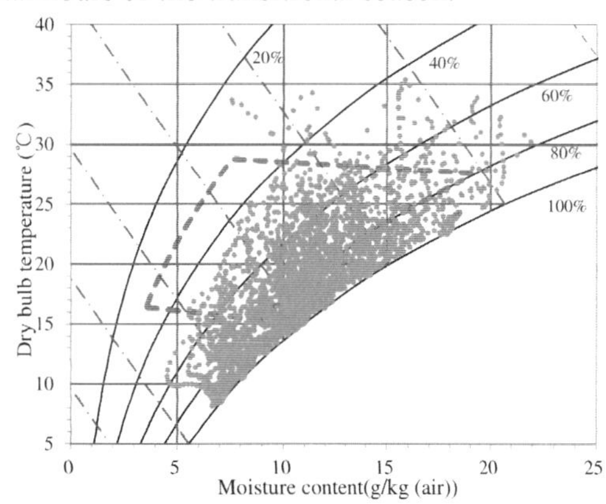

Figure 10. Outdoor dry bulb temperature and humidity points in transition season.

When considering internal insulation building as an example, the distribution of indoor operating temperature and humidity points in transitional season is shown in figure 9. It can be seen that the points were mainly distributed in the acceptable thermal comfort zone, which means that there was good indoor thermal environment in transitional season. Figure 10 shows the comparison of outdoor meteorological parameters [4], with acceptable thermal comfort zone in transition season. It can be seen that the outdoor thermal environment of the transition season was clearly better than other seasons, and 61.37% hours met the requirements of the 5 ventilation times/h, which could help improve the indoor temperature.

Table 10. Operating temperature distribution in transition season.

Model	Average value of standard deviation of daily operating	Maximum operating temperature℃	Minimum operating temperature℃

	temperature℃		
1	0.686	30.704	13.776
2	0.693	30.804	13.769
3	0.699	30.887	13.764
4	0.705	30.957	13.759

Table 10 also shows operating temperature distribution in transition season which has the same law as summer. With the increase of external insulation thickness and reduction of internal insulation thickness, the thermal stability of the building was gradually improved. The thermal stability of the external insulation building was the best and internal insulation building is the worst.

Based on the analysis of indoor and outdoor meteorological parameters and thermal stability, it can be concluded that:

Thermal comfort: Due to poor thermal stability, the internal insulation building could quickly change the indoor temperature than the other building models. When the outdoor temperature increased to a higher range, some points of the internal insulation building may increase beyond the acceptable thermal comfort zone. Due to the better thermal stability, external insulation was least affected by the outdoor environment. Although indoor temperature increased, could not quickly change the indoor temperature. When the outdoor temperature increased to a higher range, the temperature was lower than in the internal insulation building, and the temperature in external insulation building may still be in the range of acceptable thermal comfort zone. At the same time, the internal insulation building could reduce the indoor temperature more quickly than the other building model. When the outdoor temperature reduced to a lower range, some points of the internal insulation building may be lower than the low limit of acceptable thermal comfort zone. Therefore, in the transitional season, the thermal comfort hours increased with the reduction of internal insulation thickness and increase of external insulation thickness.

Thus, the external insulation was the most appropriate thermal insulation form in the transitional season, which could increase the thermal comfort hours. However, all the four building models had a comfortable indoor environment in the transition seasons.

4. Conclusions

Based on the above analysis, the following conclusions were drawn:

- Under the condition of the same thermal insulation material thickness, the residential buildings in Chongqing with internal insulation had the most thermal comfort hours in summer and winter. On the contrary, in the transitional season, it was external insulation. From the perspective of the whole year, the thermal comfort hours of internal insulation building was the largest.
- In Chongqing, the residential buildings with internal insulation had the lowest cooling and heating energy consumption.
- The residential building with internal insulation had poor thermal stability and was more likely to be influenced by the outdoor temperature. In summer, building with internal insulation had the highest hourly operating temperature and hourly cooling load during daytime, and the lowest hourly operating temperature and hourly cooling load during the night. In winter, building with internal insulation had the highest hourly operating temperature and the lowest hourly heating load during daytime, and the lowest hourly operating temperature and the highest hourly heating load during the night. The internal insulation was suitable for the building which had personnel long stay in summer night and winter day. Contrary to internal insulation, the external insulation was suitable for the building which had personnel long stay in winter night and summer day.

Acknowledgments

The work was supported by National Natural Science Foundation of China (51408079), by Construction Plan of Chongqing Science and Technology Development Base (International Science and technology cooperation project) (cstc2013gihz90002) and by Chongqing graduate research innovation project (CYB15040).

References

[1] Zheng B, Zhang Q, Borken-Kleefeld J, Huo H, Guan D, Klimont Z, Peters G P and He K 2015 How will greenhouse gas emissions from motor vehicles be constrained in China around 2030? *Appl. Energ.* **156** 230-40Another reference
[2] Building Energy Research Center 2015 *China Building Energy Efficiency Annual Development Report* (Beijing: China Architecture & Building Press)
[3] Nicol J F and Humphreys M A 2002 Adaptive thermal comfort and sustainable thermal standards for buildings *Energ BuildingS* **34** 563-72
[4] Meteorological information center of China Meteorological Administration 2005 *Meteorological Data for Thermal Environment Analysis of Chinese Architecture* (Beijing: China Architecture & Building Press)
[5] Zhang Y 2012 *Report on The Research Progress of Indoor Environment and Health in China* (Beijing: China Architecture & Building Press)
[6] Daouas N 2011 A study on optimum insulation thickness in walls and energy savings in Tunisian buildings based on analytical calculation of cooling and heating transmission loads *Appl Energ* **88** 156-64
[7] Omakl K and Yüksel B 2003 Optimum insulation thickness of external walls for energy saving *Appl Therm Eng* **23** 473-9
[8] Urban and Rural Construction Committee of Chongqing 2010 *Design Standard For Energy Efficiency 65% of Residential Building DBJ50-071-2010* (Chongqing)
[9] 2010 Ministry of Housing and Urban-Rural Construction of the People's Republic of China 2010 *Energy Saving Design Standard of Residential Buildings in Hot Summer and Cold Winter Regions JGJ 134-2010* (Beijing: China Architecture & Building Press)
[10] Al-Homoud D M S 2005 Performance characteristics and practical applications of common building thermal insulation materials *Build Environ* **40** 353-66
[11] Bojic M, Yik F and Sat P 2001 Influence of thermal insulation position in building envelope on the space cooling of high-rise residential buildings in Hong Kong *Energ Buildings* **33** 569-81
[12] Kossecka E and Kosny J 2002 Influence of insulation configuration on heating and cooling loads in a continuously used building *Energ Buildings* **34** 321-31
[13] Ozel M and Pihtili K 2007 Optimum location and distribution of insulation layers on building walls with various orientations *Build Environ* **42** 3051-9
[14] Tummu P, Chirarattananon S, Hien V D, Chaiwiwatworakul P and Rakkwamsuk P 2013 Thermal performance of insulated walls enclosing residential spaces in Thailand *Energ Buildings* **61** 323-32
[15] Ruan F, Qian X Q, Zhu Y T, Wu M L, Qian K L and Feng M P 2015 Wall Insulation Effect on Building Energy Efficiency with the Intermittent and Compartmental Energy Consuming Method *Applied Mechanics & Materials* **744-746** 2340-7
[16] Wang W 2008 *Researching the Mix Design of Polymer Modified Thermal Insulation Mortar and Its Testing Method* (Zhejiang: Zhejiang University)
[17] Stadler M, Firestone R, Curtil D and Marnay C 2006 On-site generation simulation with EnergyPlus for commercial buildings *Lawrence Berkeley National Laboratory*
[18] Yao R, Li B and Liu J 2009 A theoretical adaptive model of thermal comfort - Adaptive Predicted Mean Vote (aPMV) *Build Environ* **44** 2089-96
[19] Yu W 2011 *Building Multi-objective Predicting Model Based on Artificial Neural Network* (Chongqing: Chongqing University)
[20] General Administration of Quality Supervision, Inspection and Quarantine of China 2002

Indoor Air Quality Standard GB/T 18883-2002 (Beijing)
[21] Construction technology development center of Chongqing 2005 *Building Energy Saving Design Standard Teaching Tutorial* (Chongqing: Chongqing Building Energy Conservation Association)

3Microcanonical entropy as a function of energies for the Triangular-Lattice Blume-Capel magnetic system

W Kwak[1], S Park[1] and S Y Kim[2, 3]

[1]Department of Physics, Chosun University, Gwangju 61452, Republic of Korea
[2]School of Liberal Arts and Sciences, Korea National University of Transportation, Chungju 27469, South Korea

E-mail: sykimm@ut.ac.kr

Abstract. The triangular-lattice Blume-Capel magnetic system is investigated using efficient Wang-Landau Monte Carlo algorithm. From extensive Wang-Landau Monte Carlo simulations with massive large-scale computing, the microcanonical entropy as a function of energy variables, the most fundamental quantity in statistical thermodynamics, for the triangular-lattice Blume-Capel ferromagnet and antiferromagnet at the same time is evaluated for the first time. The important properties of the microcanonical entropy as a function of energy variables are discussed for the triangular-lattice Blume-Capel magnetic system.

1. Introduction

Magnetic materials have been widely known since ancient times. The English word Magnet originated from Magnesia, ancient civilization in western Turkey. East Asian have used a loadstone compass more than two thousand years ago. Magnetic materials are usually ferromagnetic ones. In ferromagnet, microscopic magnetic spins interact strongly each other, and most spins align to the same direction even at a room temperature, resulting in macroscopic strong magnetic field [1]. Above Curie temperature (named after Pierre Curie, Nobel prize winner in 1903), because microscopic magnetic spins align randomly due to large thermal fluctuations, ferromagnet becomes paramagnet.

In addition to paramagnet and ferromagnet, antiferromagnet is also an important magnetic material, first introduced by Neel (Nobel prize winner in 1970) [1]. In antiferromagnet, neighboring magnetic spins align antiparallel each other, not resulting in macroscopic magnetic field. Therefore, it is very difficult to detect an antiferromagnetic material and to distinguish it from a paramagnetic material. Above Neel temperature, antiferromagnet changes its phase into paramagnet. Antiferromagnet plays an important role in giant magnetoresistance [2] and high-temperature superconductivity [3]. Antiferromagnetic materials are important in various industrial applications including hard disk drives.

Modern theory of magnetic materials, phase transitions, and critical phenomena began with the simplest magnetic system, the spin-1/2 Ising system [4], to understand paramagnetism-ferromagnetism phase transition. In the spin-1/2 Ising magnetic system, a microscopic magnetic spin can take 1 (upward) or −1 (downward). Simple and easy theory for understanding the spin-1/2 Ising system is the Weiss molecular field theory [1], well explained in textbooks on magnetic materials or phase transitions. The results of the Weiss molecular field theory are reasonable but not accurate. The

[3] Address for correspondence: S Y Kim, School of Liberal Arts and Sciences, Korea National University of Transportation, Chungju 27469, South Korea. E-mail: sykimm@ut.ac.kr.

exact solution for the square-lattice spin-1/2 Ising system [5] was first obtained by Onsager (Nobel prize winner in 1968) with transfer-matrix method. It is now called the Onsager revolution [6] because modern theory of phase transitions and critical phenomena is based on the Onsager solution for the square-lattice spin-1/2 Ising system.

2. Blume-Capel magnetic system

More complex magnetic system is the spin-1 Ising system whose exact solution even on simple square lattice has not been known. In the spin-1 Ising magnetic system, a microscopic magnetic spin can take 1 (upward), 0 (vacancy), or −1 (downward). The Blume-Capel magnetic system [7, 8] is an extension of the spin-1 Ising system with a vacancy-favoring term. The Blume-Capel system has been extensively studied to understand the behaviors of He^3-He^4 mixtures. Also, it has been studied to understand various systems such as multicomponent fluids, metamagnets, molecule-based chain magnets, and ternary alloys. For example, the square-lattice Blume-Capel system [9] shows second-order phase transitions, first-order phase transitions, and a tricritical point separating these different transitions.

The physical origin of the vacancy-favoring term (the so-called crystal field D) for the Blume-Capel magnetic system arises from non-central potentials for metal atoms coordinated with various ligands in the crystal. The Blume-Capel system with $D>0$ prefers the spin vacancy and shows second-order phase transitions, first-order phase transitions, and a tricritical point. The Blume-Capel system with $D<0$ is similar to the spin-1/2 Ising system. In this work, we investigate the triangular-lattice Blume-Capel magnetic system whose properties are not known, using efficient Wang-Landau Monte Carlo algorithm [10]. From extensive Wang-Landau Monte Carlo simulations with massive large-scale computing, we obtain the microcanonical entropy as a function of energy variables of the triangular-lattice Blume-Capel ferromagnet and antiferromagnet at the same time for the first time.

The triangular-lattice Blume-Capel magnetic system is defined by the Hamiltonian

$$H=-J\sum_{<i,j>}\sigma_i\sigma_j+D\sum_i\sigma_i^2, \tag{1}$$

where J is the coupling constant, D is the crystal field (also called the single-spin anisotropy parameter or the spin impurity chemical potential), $<i,j>$ represents a sum over all nearest-neighbor pairs, and $\sigma_i(=-1,0,\text{or }1)$ is the microscopic magnetic spin at the lattice site i. Positive coupling constant $(J>0)$ defines the Blume-Capel ferromagnet, whereas negative coupling constant $(J<0)$ indicates the Blume-Capel antiferromagnet.

3. Microcanonical entropy of the triangular-lattice Blume-Capel magnetic system

If we define the density of states, $g(E,S)$, with two energy variables $E=\sum_{<i,j>}\sigma_i\sigma_j$ and $S=\sum_i\sigma_i^2$, the partition function of the Blume-Capel system, $Z=\sum_{\{\sigma_n\}}e^{-\beta H}$, can be written as

$$Z(T;J,D)=\sum_E\sum_S g(E,S)e^{\beta J(E-\frac{D}{J}S)}, \tag{2}$$

where $\beta=1/k_BT$ (k_B is the Boltzmann constant and T is temperature). Then, we obtain the microcanonical entropy, the most fundamental quantity in statistical thermodynamics, according to the Boltzmann formula $M(E,S)=k_B\ln g(E,S)$. In this work, we evaluate the microcanonical entropy as a function of E and S for the Blume-Capel magnetic system on a 12 × 12 triangular lattice with

Wang-Landau Monte Carlo simulations, as shown in Figure 1. It should be noted that this kind of figure for the triangular-lattice Blume-Capel magnetic system is obtained for the first time.

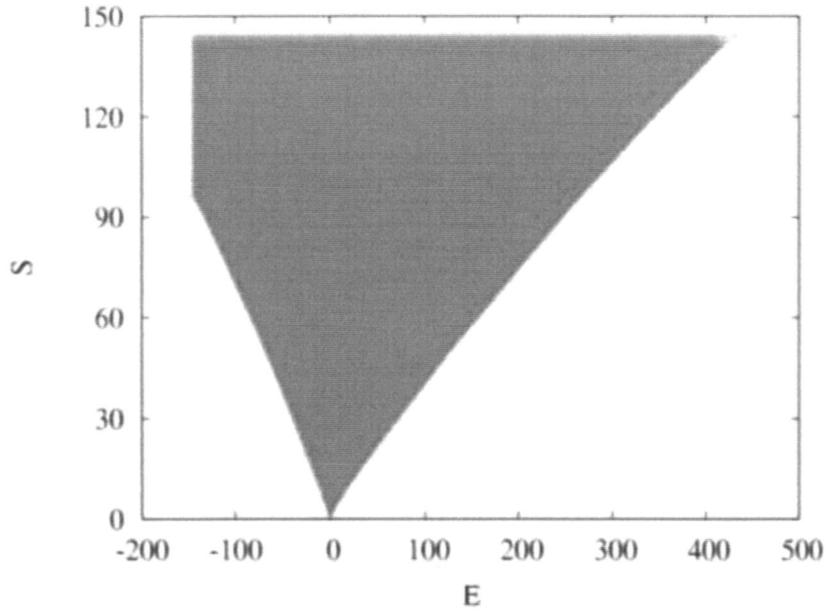

Figure 1. Top view of the microcanonical entropy $M(E,S)$ (in units of k_B) for the Blume-Capel magnetic system on a 12 × 12 triangular lattice, where two energy variables E and S are defined as $E=\sum_{<i,j>}\sigma_i\sigma_j$ and $S=\sum_i\sigma_i^2$ with $\sigma_i=-1,0,$ or 1.

As shown in the figure, the microcanonical entropy has four vertices in the $E-S$ plane. The microcanonical entropy is not allowed outside the polygon defined by four vertices. The vertex $(E,S)=(0,0)$ is the spin vacancy state $\sigma_i=0$ for all i. The density of states is $g(0,0)=1$ for the spin vacancy state. The density of states on the straight line from the vertex $(E,S)=(-144,144)$ to $(-144,96)$ corresponds to the antiferromagnetic ground states, whereas the vertex $(E,S)=(432,144)$ to the ferromagnetic ground states. The density of states is $g(N_b,N_s)=2$ for the ferromagnetic ground states, but the density of states is highly degenerate for the antiferromagnetic ground states.

Furthermore, the partition function can be written as

$$Z(T;J,D)=\sum_E\sum_S\exp\left[\frac{M(E,S)}{k_B}+\frac{J}{k_BT}\left(E-\frac{D}{J}S\right)\right]. \tag{3}$$

If we define the combined energy $\tilde{E}=E-\frac{D}{J}S$, the partition function can be simply written as

$$Z(T)=\sum_{\tilde{E}}\exp\left[\frac{M(\tilde{E})}{k_B}+\frac{J\tilde{E}}{k_BT}\right] \tag{4}$$

for a fixed value of D/J. Now the microcanonical entropy $M(\tilde{E})$ is dependent on the only one energy variable $\tilde{E}$.

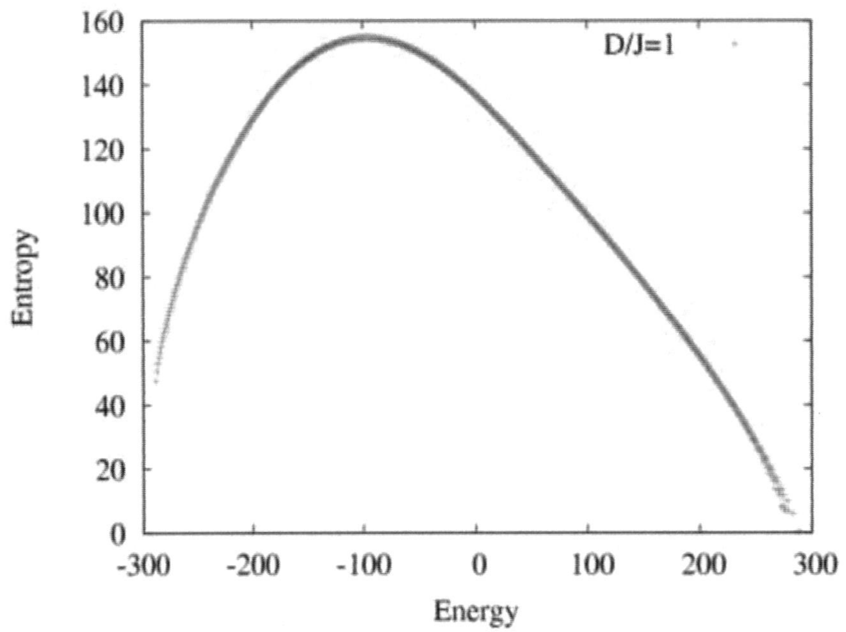

Figure 2. Microcanonical *entropy* $M(\tilde{E})$ (in units of k_B) of the Blume-Capel magnetic system on a 12 × 12 triangular lattice for $D/J = 1$, where the combined *energy* $\tilde{E} = E - \frac{D}{J} S$.

Figure 2 shows the microcanonical entropy $M(\tilde{E})$ of the Blume-Capel magnetic system on a 12 × 12 triangular lattice for $D/J = 1$. As shown in the figure, the maximum entropy is located at $\tilde{E} = -96$. The antiferromagnetic ground states is located at $\tilde{E} = -288$, and the ferromagnetic ground states at $\tilde{E} = 288$. The energy interval $\tilde{E} = [-288, -96]$ corresponds to the Blume-Capel antiferromagnet. Similarly, the energy interval $\tilde{E} = [-96, 288]$ corresponds to the Blume-Capel ferromagnet.

4. Conclusion

We have studied the triangular-lattice Blume-Capel magnetic system using Wang-Landau Monte Carlo method to calculate its microcanonical entropy. From extensive Wang-Landau Monte Carlo simulations with massive large-scale computing, we have evaluated the microcanonical entropy as a function of energy variables (exchange energy and vacancy-favoring energy), the most fundamental quantity in statistical thermodynamics, for the triangular-lattice Blume-Capel ferromagnet and antiferromagnet at the same time for the first time.

Given the microcanonical entropy, we can investigate the important properties of the triangular-lattice Blume-Capel magnetic system through various methods such as canonical analysis [11, 12, 13, 14], microcanonical analysis [15, 16, 17], and partition function zeros [18, 19, 20]. In canonical analysis, we can obtain the canonical thermodynamic functions (such as specific heat and other response functions) as a continuous function of temperature. In microcanonical analysis, we can use the whole information on the microcanonical entropy, directly analyzing its topological and geometric properties. If we have the microcanonical entropy of a given system, we can calculate its partition function zeros from which we can understand its phase transitions and critical phenomena accurately.

References

[1] Aharoni A 2007 *Introduction to the Theory of Ferromagnetism*, 2nd edition (Oxford: Clarendon Press)
[2] Fert A 2008 *Rev. Mod. Phys.* **80** 1517
[3] Sheahen T 1994 *Introduction to High-Temperature Superconductivity* (New York: Plenum Press)
[4] Ising E 1925 *Z. Phys.* **31** 253
[5] Onsager L 1944 *Phys. Rev.* **65** 117
[6] Domb C 1996 *The Critical Point* (London: Taylor and Francis)
[7] Blume M 1966 *Phys. Rev.* **141** 517
[8] Capel H W 1966 *Physica* **32** 966
[9] Kim S Y and Kwak W 2014 *J. Korean Phys. Soc.* **65** 436
[10] Wang F and Landau D P 2001 *Phys. Rev. Lett.* **86** 2050
[11] Kim S Y 2009 *New Phys.: Sae Mulli* **58** 660
[12] Kim S Y 2009 *J. Phys. Studies* **13** 4006
[13] Lee J H, Kim S Y and Lee J 2011 *Comput. Phys. Comm.* **182** 1027
[14] Lee J H, Kim S Y and Lee J 2015 *AIP Advances* **5** 127211
[15] Hwang C O, Kim S Y, Kang D and Kim J M 2007 *J. Stat. Mech.* **7** L05001
[16] Lee J H, Kim S Y and Lee J 2012 *Phys. Rev. E* **86** 011802
[17] Kim S Y and Kwak W 2015 *New Phys.: Sae Mulli* **65** 1017
[18] Kim S Y and Creswick R J 1998 *Phys. Rev. Lett.* **81** 2000
[19] Kim S Y 2004 *Phys. Rev. Lett.* **93** 130604
[20] Lee J 2013 *Phys. Rev. Lett.* **110** 248101

Investigation of parameter limit of selecting polymer flooding potential block at the Pubei Oilfield

Y N Wang [1], J H Zhang[1,3], D Guan [2], C Liu [1]

[1]Key Laboratory for Enhancing Oil/Gas Recovery of Ministry of Education in Northeast Petroleum University, Daqing, Heilongjiang, 163318,China
[2]Oil Recovery Plant No.3, Daqing Oilfield Corp. Ltd., Daqing, Heilongjiang, 163318, China

E-mail: dqzhjh@126.com

Abstract. The Pubei oilfield belongs to the reservoir with low permeability and poor physical property. It has entered the late period of high water cut stage, so it needs polymer flooding to produce remaining oil. In order to pursue benefits and avoid risk, it is necessary to select potential block of polymer flooding by optimum parameters. In the paper, the limit of permeability for selecting potential block of polymer flooding is calculated by using both reservoir engineering method and economic analysis theory.

1. Introduction

The Pubei oilfield, a low permeability reservoir, has entered the late period of high water cut stage, with the production declining significantly[1-2]. The polymer flooding pilot test carried out from 2000 to 2006 achieved better results that cumulative oil increment is 6.17×10^4 t and enhanced oil recovery is 5.84%. The polymer flooding field test with dense well pattern has been carried out since 2014, which enhanced oil recovery is forecasted 7.65 %. Various types of field test showed that polymer flooding can further enrich the means of producing remaining oil potential and have the feasibility of practice in the narrow and thin sandstone oilfield [3-4].

Because it need capital investment more than water flooding, it is necessary to select potential block of polymer flooding by optimum parameters [5]. The permeability is a vital parameter that can assess whether polymer flooding is technically feasible. So it is crucial to calculate the limits of permeability for selecting potential block of polymer flooding. The author calculates the limits by using both reservoir engineering method and economic analysis theory, and confirms the rationality by taking advantage of the existing experimental results and the analysis of injection profile resulting from measured datum of polymer flooding well.

2. Limits of economic output of single well

According to the principle of profit and loss balance, the limits of economic output of single well is the output when the total costs of the oil well and the total incomes are equal [6]. The basic formula is:

[3] Address for correspondence: J H Zhang, Key Laboratory for Enhancing Oil/Gas Recovery of Ministry of Education in Northeast Petroleum University, Daqing, Heilongjiang, 163318, China. E-mail: dqzhjh@126.com.

$$P_o \cdot Q_{\min} \cdot R = V \cdot Q_{\min} + F \tag{1}$$

Where:

$Q_{\min}$—Limits of economic annual output of single well, t;

R —Commodity rate of crude oil, %;

P_o—Price of crude oil, Yuan/t;

V—Operation cost of crude oil, Yuan/t;

F —Fixed annual costs, including the cost of new well, pharmacy and ground construction costs, yuan.

So the limits of economic output of single well can be calculated as follows:

$$Q_{\min} = \frac{F}{(P_o \cdot R - V)} \tag{2}$$

Among them, the definition of the commodity rate of crude oil refers to the proportion of the commodity volume of crude oil production accounted for the total output of crude oil. The commodity of crude oil is equal to the total output of crude oil minus the amount of own oil consumption and the evaporation loss during production. But the commodity rate of crude oil in this paper is 100%. Fixed cost and operating cost of crude oil are used to learn from the cost of polymer flooding test block of Pubei oilfield, crude oil prices in accordance with $70 / barrel to calculate (although the price is lower than $70 at present, this study can be seen as a technology applied in the future). At last the limits of economic annual output of single well of polymer flooding is calculated by using the second formula, and the value is 993t.

3. Technical and economic limits of polymer flooding reservoir

The research that theoretical formula deduction and calculation of permeability limit for polymer flooding reservoir needs to use the limits of economic annual output of single well model and the basic theory of the fluid mechanics in porous medium.

The rheological properties of the polymer flooding reservoir are basically in accordance with the power law model, and the inflow performance relationship of the polymer flooding oil well can be obtained according to the flow law of the power law fluid. Due to the influence of shear degradation, the rheological parameters (consistency coefficient and flow index) of the formation fluid are changing continuously along the flow direction, which makes it difficult to solve and the accuracy is poor. Therefore, some scholars give a semi- empirical equation on the basis of theoretical analysis [7].

$$p_e - p_{wf} = Aq^{nb} \tag{3}$$

Thereof:

$$A = a\left[\frac{\mu_\varepsilon \left(r_\varepsilon^{1-n} - r_w^{1-n}\right)}{(2\pi h)^n (1-n)K}\right]^b \tag{4}$$

Where:

q —Daily output of single well, m^3/d;

p_e—Supply edge pressure, MPa;

p_{wf}—Bottom flowing pressure, MPa;

μ_ε —Effective viscosity of polymer solution, Pa.s;

n —Polymer solution flow index, $0<n<1$;

K —Permeability of formation, μm^2;

h —Effective thickness, m;

r_ε —Drainage radius, m;

r_w —Bottom hole radius, m;

a,b—regression coefficient, determined by the measured data.

Table 1. Parameter list of calculating regression coefficients

No.	Well number	p_e	p_{wf}	q	μ_e	n	K	h	r_e	r_w
1	Pu642- SP84	17.5	4.68	4.0	0.02419	0.8	0.255	14	150	0.031
2	Pu64-85	17.5	4.65	1.1	0.02419	0.8	0.212	5.8	212	0.031
3	Pu64- SP833	17.5	6.81	1.2	0.02419	0.8	0.244	10.8	150	0.031
4	Pu652- SP84	17.5	4.45	5.1	0.02419	0.8	0.288	11.2	150	0.031
5	Pu652- SP85	17.5	5.2	1.8	0.02419	0.8	0.216	10.7	150	0.031
6	Pu65-P843	17.5	6.23	1.8	0.02419	0.8	0.011	12.4	150	0.031
7	Pu67-82	17.5	3.87	1.5	0.02419	0.8	0.279	3.5	212	0.031
8	Pu67-842	17.5	8.74	1.1	0.02419	0.8	0.307	13.2	212	0.031
9	Pu67- SP832	17.5	3.8	0.2	0.02419	0.8	0.085	2	150	0.031
10	Pu67- SP853	17.5	7.73	1.1	0.02419	0.8	0.259	12.4	150	0.031

According to the datum of ten wells listed in Table 1, making regression according to the third formula, the values of a is 24.66 and the values of b is 0.3.

Assuming $U = h^n \cdot K$, then a , b , $Q_{\min}$ and other parameters in the Table 2 are substituted into the third formula, and the minimum value of U can be calculated. When effective thickness is taken as the average (in the paper, the average effective thickness is 9m in the Pubei oilfield), the minimum value of the permeability K can be calculated as $170\times10^{-3}\mu m^2$.

Table 2. Parameters of calculating the limits of polymer flooding reservoir

p_e	p_{wf}	μ_e	n	r_e	r_w	a	b
17.5	5.24	0.02419	0.8	150	0.031	24.66	3

Based on the above calculation, we determine that the limit of the permeability for polymer flooding in Pubei oilfield is $170\times10^{-3}\mu m^2$.

4. Proof the permeability limits e of polymer flooding reservoir is reasonable

4.1. Existing experiment on the adaptability of polymer molecular weight to core permeability

The injection capacity of the two kinds of polymers are determined under the condition of 800mg/L of salt resistant polymer concentration, 1000mg/L of the common polymer concentration[8-9]. Research results (Figure 1) show that when the air permeability of reservoir is greater than $150\times10^{-3}\mu m^2$, there is no blocking when injecting salt resisting polymer (8 million molecular weight) or common polymer (12 million molecular weight). Therefore, the reservoir whose air permeability is greater than $150\times10^{-3}\mu m^2$ is adaptive with the low and middle salt resisting polymer and the common middle polymer. So the limit of the permeability for polymer flooding is consistent with the experimental results.

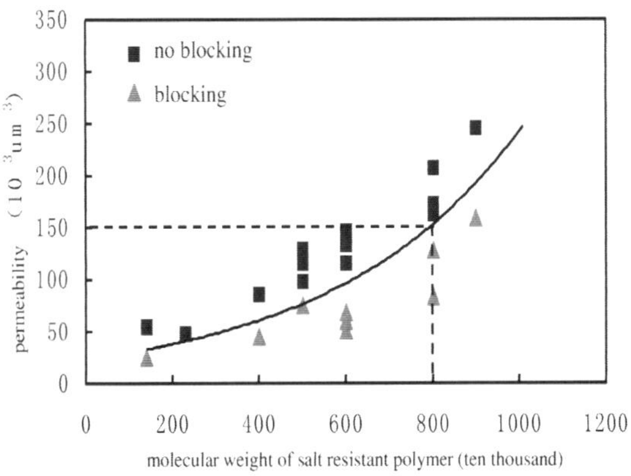

Figure 1. (a)

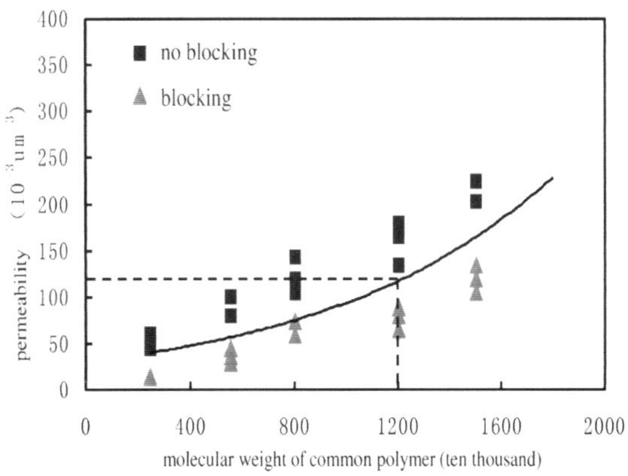

Figure 1. (b) 5d

Figure 1. The adaptability of polymer molecular weight to core permeability.

4.2. Comparative analysis of injection profile of polymer flooding well

According to the pilot test of 8 wells from continuous injection profile data of polymer flooding (molecular weight of salt resisting polymer is 8 million) in Pubei Oilfield, the result shows that from the beginning of blank water flooding to 2006, the decline range of relative fluid amount of accepting fluid zones, which permeability is less than $100\times10^{-3}\mu m^2$, is large after injection of polymer solution. But the amount in accepting fluid zones whose permeability is more than $150\times10^{-3}\mu m^2$ increases during polymer flooding, increased from 76.41% during water flooding to 90.98% in 2006. It demonstrates that the main accepting fluid zones are channel sand whose permeability is more than $150\times10^{-3}\mu m^2$. So the limit of the permeability for polymer flooding is consistent with the analysis result of injection profile.

Table 3. The entry profile statistics of Pubei oilfield

Permeability range ($10^{-3}\mu m^2$)	Number of plies	Thickness (m)				The relative amount of accepting liquid (%)						
		Sandstone	Effective	Thickness ratio (%)	Absorbent thickness ratio (%)	Water flooding	2001	2002	2003	2004	2005	2006
0-50	12	7.6	0.7	0.7	0	4.30	0.2	0.4	0.5	0.2	0.4	0.4
50-100	19	13.1	7.9	7.3	31.6	9.46	0.9	2.1	3.1	3.8	3.9	3.1
100-150	21	16.6	7.7	7.1	42	9.84	3.1	5.6	6.4	4.2	5.2	5.6
150-200	7	5.3	4.2	3.9	76.2	2.10	6.5	4.8	5.6	3.4	4.1	4.27
200-250	6	9.2	7.8	7.2	84.6	5.44	7.6	6.7	7.0	8.1	10.3	9.82
250-300	11	20.7	17	15.7	84.7	6.59	11.3	11.3	10.2	14.9	14.3	14.14
300-350	18	43.4	37.6	34.8	100	39.54	36.8	30.0	26.2	28.3	29.0	31.31
350-400	3	10.6	10.2	9.4	100	12.23	11.6	12.3	13.8	12.0	12.7	11.62
400-500	1	5.8	5.5	5.1	100	4.49	5.0	11.2	12.0	13.0	12.0	9.81
More than500	2	10.4	9.6	8.9	100	6.02	17.1	15.7	15.2	12.3	8.2	10.01

5. Conclusions

- According to the principle of profit and loss balance, the limit of economic annual output of single well by polymer flooding is 993t by calculation.
- Using the limits of economic annual output of single well model and the basic theory of the fluid mechanics in porous medium, the limit of the permeability is $170\times10^{-3}\mu m^2$ for polymer flooding in Pubei oilfield.
- This method has been confirmed by Existing experimentation and analysis of injection profile resulting from measured datum of polymer flooding well, so it can be applied to select potential block of polymer flooding at Pubei oilfield. The method maybe applied to other oilfields, but it needs confirmation prior to the application.

References

[1] Liu W *et al.* 2010 Integrated heterogeneity evaluation of thin sandstone reservoirs and remaining oil distribution- a case study of Pu1 oil formation, No.3 fault block, Pubei Oilfield *J. Petroleum Geology and Recovery Efficiency* **17** 37-40

[2] Zhang Z H, Yang Z Y and Guo Z 2005 Feasibility of the ASP flooding in northern Putaohua oilfield in Daqing *J. Petroleum Geology & Oilfield Development in Daqing* **24** 82-4

[3] Wang D M, Cheng J C, Wu J Z 2005 Application of polymer flooding technology in Daqing Oilfield *J. Acta Petrolei Sinica* **26** 74-8

[4] Guo Z,Wu L J and Mei Q T 2001 Feasibility of the polymer flooding in northern Putaohua oilfield *J. Petroleum Geology & Oilfield Development in Daqing* **20** 80-2

[5] Wen J Q 2015 The effect evaluation of polymer flooding in No.1 zone of northern eastern block *J. Chemical Engineering & Equipment* **7** 79-81

[6] Han L C, Han D J and Zhang X L 2008 Determinationg of technical and economic limits for horizontal well in waterflooding reservoir *J. Xinjiang Petroleum Geology* **29** 79-81

[7] Chen T P and Hu J B 2000 *The petroleum engineering* (China: Petroleum Industry Press) p 231

[8] Li D S, Xu Z S and Zheng F P 2000 A laboratory study on ASP flooding in the low permeable oil field of North Putaohua *J. Petroleum Geology & Oilfield Development in Daqing* **19** 37-9

[9] Shatat R, Ariffin A, Norulaini N, Rahma A, Ab Kadir M O and Kadir A 2008 The effect of molecular weight and charge density on floc sizedistribution of palm oil mill effluent flocculated with cationic polyelectrolytes *J. Basic and Applied Sciences* **4** 95-103

IOP Conf. Series: Earth and Environmental Science **40** (2016) 012084 doi:10.1088/1755-1315/40/1/012084

Temperature distribution of air source heat pump barn with different air flow

X He[1], J C Li[1,3] and G Q Zhao[2]

[1]College of Mechanical Engineering, Gui Zhou University 550025, China
[2]College of Mechanical Engineering, Guizhou University of Engineering Science 551700, China

E-mail: ljcgd211@163.com

Abstract. There are two type of airflow form in tobacco barn, one is air rising, the other is air falling. They are different in the structure layout and working principle, which affect the tobacco barn in the distribution of temperature field and velocity distribution. In order to compare the temperature and air distribution of the two, thereby obtain a tobacco barn whose temperature field and velocity distribution are more uniform. Taking the air source heat pump tobacco barn as the investigated subject and establishing relevant mathematical model, the thermodynamics of the two type of curing barn was analysed and compared based on Fluent. Provide a reasonable evidence for chamber arrangement and selection of outlet for air source heat pump tobacco barn.

1. Introduction

Tobacco cured is one of the most important stages in the tobacco production, and the equipment of tobacco barn is the key factor to determine the baking performance. At present, the most widely used is bulk curing barn [1], it can be divided into air rising and drop according to the indoor airflow direction of tobacco filled room [2], the two types curing barns were used, however, it needs to be proved by theoretical and practical baking to distinguish what kind of airflow pattern is more conducive to tobacco curing.

In the actual curing process, there are many factors affecting the curing performance of two air flow forms, including the filling of tobacco filled room, physical properties of tobacco leaf, baking process (stage length, temperature, wind speed etc.) At present, people can only decide the layout and structure of tobacco barn according to curing experience, lack of theoretical guidance. Therefore, it's necessary to seek a better method to simulate and analyse the temperature and airflow distribution of the two types of tobacco barn, and obtain better tobacco barn structure. At present, numerical simulation of the temperature and air distribution in the curing barn has been studied by scholars through the computational fluid dynamics software, analysed the uniformity of temperature and air distribution in the tobacco filled room, they obtain reasonable air mass for tobacco curing[3]. However, it did not put forward a reasonable structure and layout of the curing barn. In order to solve this problem, based on the above research, thermodynamic analysis and comparison of the two types of

---

[3] Address for correspondence: J C Li, College of Mechanical Engineering, Gui Zhou University 550025, China. E-mail: ljcgd211@163.com.

curing barn are completed using Fluent to obtain a better tobacco barn form which has better curing performance.

2. Analysis of the structure and curing process of the two types of tobacco barn

2.1. The comparison of two types of curing barn

The air rising and drop curing barn are almost identical in the shape and structure, the main difference is that the installation and location of each component in the heating room is different, the position of the air inlet and outlet on the heat insulation wall are also different. The air inlet of air rising curing barn is located in the bottom of heat insulation wall, the outlet in the top, and the condenser in the heating room is installed on the upper part of the axial flow fan, as shown in Figure 1 (a). The air inlet position and the layout of the condenser and the axial flow fan of the air drop curing barn are opposite to the air rising. As shown in Figure 1. (b).

2.2. Analysis of curing process of The Two types curing barn

The process of tobacco curing to dry the water in tobacco leaves is heat exchange between hot air and tobacco leaves, take air source heat pump as the research object. The indoor air at first is heated by the condenser in the heating room, and then send into tobacco filled room from the inlet by the axil flow fan. For air rising curing barn, hot air gradually rising from the bottom of the tobacco filled room to the to, after fully contacting with the tobacco leaves, the air return to the heating room from the upper outlet, while the air flow form is opposite in air drop curing barn. The air flow of the two in curing process is as shown in Figure 1.

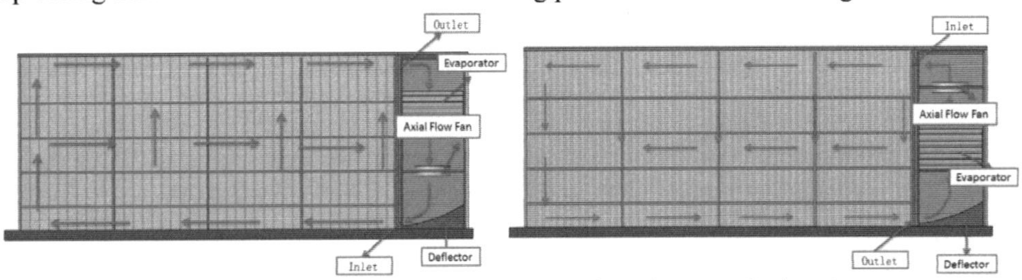

(a) Air Rising Curing Barn (b) Air Drop Curing Barn

Figure 1.Comparison of Air Flow Form of Two Types of Curing Barn

3. Computational Models

3.1. Computational model of the curing barn

In order to compare the temperature and air distribution in the tobacco filled room of the two types of tobacco barn, the other conditions are consistent of the two types of calculation model, except that the inlet and outlet are swapped. The structural parameters of the tobacco filled room are shown in Table 1.

Table 1. Parameters of the calculation model of curing barn

Types	Length (mm)	Width (mm)	Height (mm)
Tobacco Filled Room Size	8000	2700	3550
Inlet Size	×	2600	400
Outlet Size	×	1400	440

3.2. Research hypothesis of computational domain model

This study uses $k-\varepsilon$ two equation three-dimensional turbulence models. In order to simplify the problem, and to facilitate the mathematical description and solving, under the same curing condition, without considering tobacco leaves in the tobacco filled room. And make the following approximation:

(1)There are room door, observation window, humidity window and so on in the tobacco barn, from the point of view of simplifying mathematical model, it is considered that the physical parameters are same to the wall;

(2)The air in the curing barn is considered as ideal gas that can't be compressed;

(3)The airflow is low speed incompressible flow, the heat consumption can be ignored caused by the viscous force of fluid;

(4) Because the temperature of the hot air is not high, so only the convection is considered, the radiation is not;

(5) Without considering the influence of air leakage (the tightness of air in curing barn is good); and the ventilation and drainage of the barn is smooth.

3.3. *Flow control equation*

For the study of the curing barn, the flow of air in tobacco filled room is a steady three-dimensional incompressible turbulent flow, so the flow and heat transfer in the curing barn should satisfy the following control equations:

(1) Continuity equation:

$$\frac{\partial \rho}{\partial \tau}+\frac{\partial}{\partial x_i}(\rho \mu_i)=0 \tag{1}$$

In equation (1):

ρ—Fluid Density; μ_i—Fluid velocity along i direction.

(2) $N-S$ Equation:

$$\frac{\partial}{\partial \tau}(\rho \mu_i)+\frac{\partial}{\partial x_j}(\rho \mu_i \mu_j)=-\frac{\partial \rho}{\partial x_i}+\frac{\partial \tau_{ij}}{\partial x_j}+\rho g_i+F_i \tag{2}$$

In equation (2):

ρ—Static pressure; τ_{ij}—Stress vector; ρg_i—Gravity component along i direction;

F_i—Other energy sources due to resistance and energy.

(3) Energy conservation equation:

$$\frac{\partial}{\partial \tau}(\rho h)+\frac{\partial}{\partial x_i}(\rho \mu_i h)=\frac{\partial}{\partial x_i}(k+k_\tau)\frac{\partial_T}{\partial x_i}+S_h \tag{3}$$

In equation (3):

h—Entropy; k—Molecular conductivity; k_τ—Transmission rate due to turbulent flow; S_h—Defined volume source.

(4) k-ε equation

k Equation:

$$\rho\frac{Dk}{D\tau}=\frac{\partial}{\partial x_i}\left[\left(\mu_l+\frac{\mu_\tau}{\sigma_k}\right)\frac{\partial k}{\partial x_i}\right]+G_k+G_b-\rho\varepsilon \tag{4}$$

ε Equation:

$$\rho\frac{D_\varepsilon}{D_\tau}=\frac{\partial}{\partial x_i}\left[\left(\mu+\frac{\mu_t}{\sigma_\varepsilon}\right)\frac{\partial \varepsilon}{\partial x_i}\right]+C_{1\varepsilon}\frac{\varepsilon}{k}\left(G_k+C_{3\varepsilon}+G_b\right)-C_{2\varepsilon}\rho\frac{\varepsilon^2}{k}$$

(5)

In equation (4), (5):

μ_l—Laminar viscosity coefficient; μ_t—Turbulent viscosity coefficient;

$$\mu_\tau=\rho C_\mu\frac{k^2}{\varepsilon} \tag{6}$$

G_k—Turbulent kinetic energy generated by a laminar velocity gradient;

G_b—Turbulent kinetic energy generated by buoyancy;

$C_{1\varepsilon}, C_{2\varepsilon}, C_{3\varepsilon}, \sigma_k$ and σ_ε —Empirical constant; C_μ —Turbulence constant.

3.4. *Physical properties and related parameters*

The whole peripheral structure of the barn use the same high thermal insulation material: polyurethane, the medium indoor used to cure tobacco is the heated air. The physical parameters of the selection of insulation material and air: density, thermal conductivity, pressure specific heat capacity and viscosity are shown in Table 2.

Table 2. Physical properties and related parameters

Material	Density (kg/m^3)	conductivity (W/m·K)	Specific heat capacity at constant pressure (J/kg·K)	viscosity (kg/m·S)
Polyurethane	35~40	≤0.025	2380~370	×
Air	1.225	0.0242	1006.43	1.79e-05

3.5. *Boundary condition*

The air inlet of the curing barn is used as the inlet boundary, and the type selection is set as inlet velocity; the air outlet of the barn is set as the outlet boundary, and the pressure of the airflow is the standard atmosphere, the temperature is same to the external environment. The wall of the curing barn and the roof are treated as convection heat transfer boundary condition, and the floor is approximately treated as an adiabatic boundary condition.

4. Calculation result analysis and processing

4.1. *Analysis of temperature and velocity field*

The air supply condition in actual curing and the experimental result were considered, so set the inlet air velocity as 5m/s, the temperature as 320K, and take the section y=0 as observation object.

It could be seen from the speed vector diagram in Figure2. (a), (b), the air speed in the air rising curing barn is a little higher at the inlet and outlet, and the rest of the speed distribution is basically uniform. However, the speed in air drop barn at the inlet and outlet is observably higher than other domain and the distribution is not uniform; it could also be seen from Figure 2. (c), (d) that: the distribution of temperature in the curing barn is more uniform than that of the drop type. Therefore, it is concluded that the air rising type is more conducive to the tobacco curing.

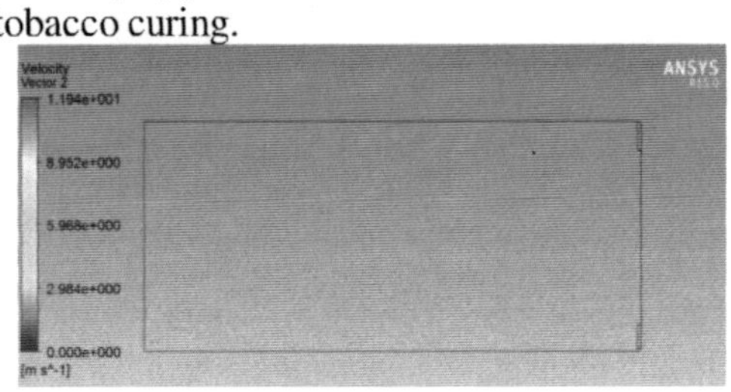

(a) velocity distribution in the barn of air rising type

(b) velocity distribution in the barn of air drop type

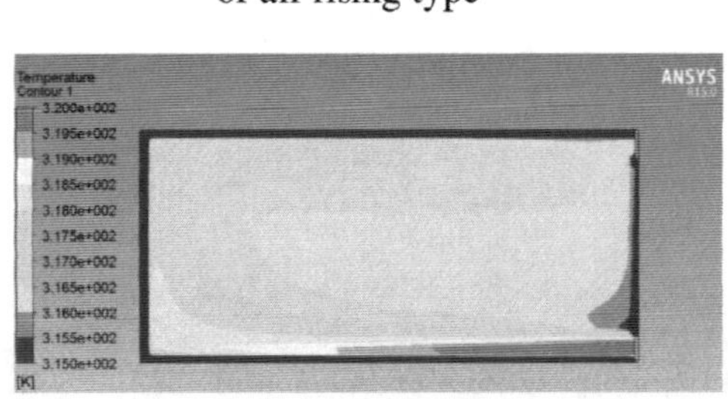

(c) temperature distribution in the barn of air rising type

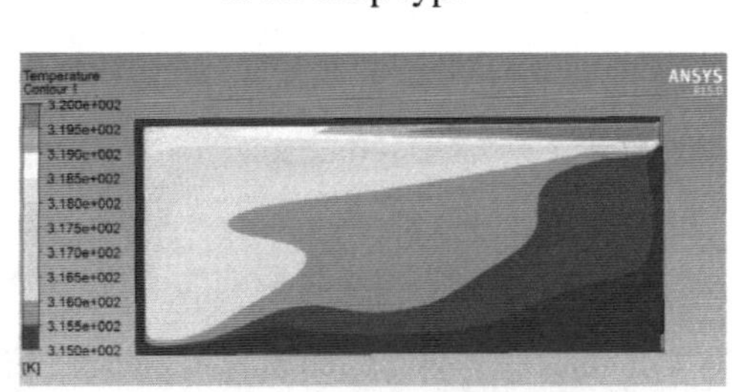

(d) temperature distribution in the barn of air drop type

Figure 2. Temperature and velocity vector diagram of two kinds of curing barn

4.2. Data node selection

Temperature is an important factor for tobacco curing, in order to fully prove that air rising type curing barn is better for curing, thermodynamic analysis and comparison were completed after the change of initial inlet temperature condition. This paper compared the results through selection of temperature data node of the curing barn, and then calculated the mean and standard deviation of these data [4-6].

Air source heat pump barn belong to bulk curing barn, the National Tobacco Bureau issued a corresponding technical specification for construction of bulk curing barn, and the specification provides the basic structure, main equipment and technical parameters of the barn [7]. The number of the tobacco-shed in the barn is generally designed as three. In order to make the analysis more reasonable, the two kinds of curing barn are both considered the temperature distribution in three directions: the length, width and height the. The data collection points are selected as follows:

Figure 3. (a) Cross section selection in width direction collection

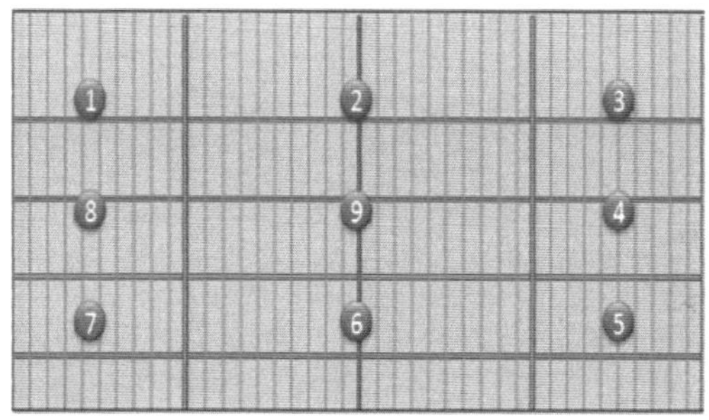

Figure 3. (b) Points selection in length direction

Figure 3. Data node selection

(1) Three sections were selected in the width direction of the tobacco filled room, as shown in Figure 3. (a):

Section 1: $y = 0.7$, Medium section: $y = 0$, Section 2: $y = -0.7$;

(2) 9 data collection points are selected on each cross section, as shown in Figure 3. (b)

Table 3. Temperature value of data node

	Air rising			Air drop		
	Section 1	Medium section	Section 2	Section 1	Medium section	Section 2
1	316.8	316.9	316.8	315.8	316.5	315.5
2	316.5	316.8	316.6	316	316.2	315.7
3	316.4	316.6	316.1	315.5	315.7	315.3
4	316.1	316.2	316	315.1	315.5	315.2
5	316.1	316.4	315.8	315	315.4	315.1
6	316.2	316.5	316.2	315.3	315.6	315.7
7	316.6	316.8	316.6	316.2	316	315.8
8	315.9	316.5	316.4	315.6	315.9	315.5
9	316	316.1	316	315.4	315.4	315.4
Mean		316.4			315.6	
standard deviation		0.317			0.364	

Inlet velocity: $V = 5m/s$ Inlet temperature: 320K

4.3. Data analysis and processing

Data points are selected according to the above rules, under different initial conditions, each barn can get 27 data. Tobacco curing is carried out in stages and the temperature of each stage is gradually increased, the temperature range from 35℃ to 70℃, selected initial temperature from 35℃, then selected by the intervals: 5℃, and calculated the standard deviation according to the equation: $\sigma = \sqrt{\frac{(x_1-\bar{x})^2+(x_2-\bar{x})^2+\cdots+(x_n-\bar{x})^2}{N}}$ The standard deviation of temperature node of the two barns under different initial inlet temperature is obtained as follows:

Table 4. Temperature standard deviation of curing barn under different temperature

Temperature		35℃	40℃	45℃	50℃	55℃	60℃	65℃	70℃
Air rising	Mean	307.25	312.23	317.26	322.26	327.26	332.26	337.26	342.26
	standard deviation	0.083	0.081	0.079	0.082	0.083	0.084	0.084	0.084
Air drop	Mean	306.84	311.86	316.85	321.86	326.85	331.85	336.85	341.93
	standard deviation	0.106	0.169	0.111	0.115	0.105	0.109	0.108	0.286

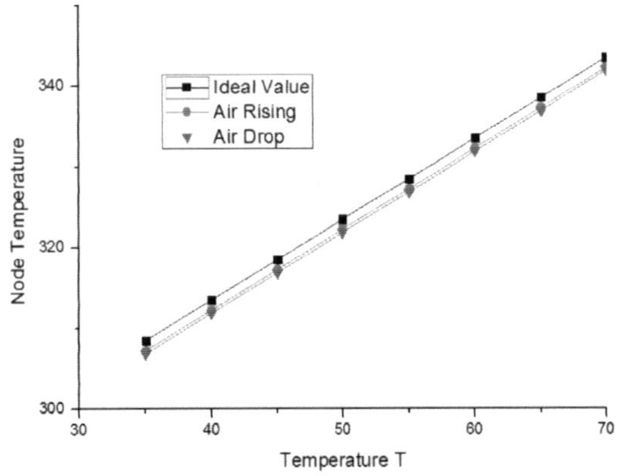

Figure 4. (a) mean of node temperature of the two under different temperatures

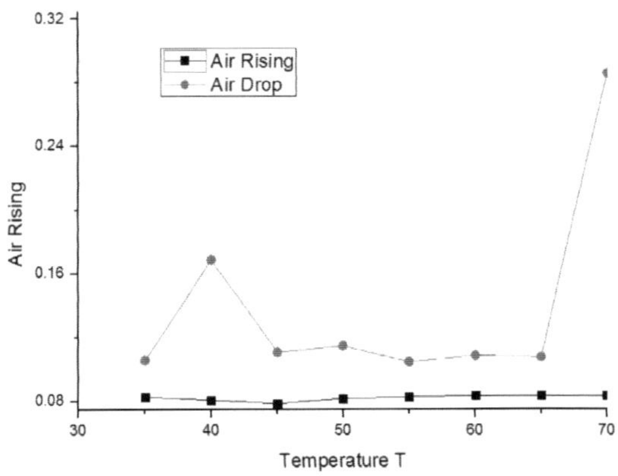

Figure 4. (b) standard deviation of node temperature of the two under different temperatures

Figure 4. Node temperature data solving

As Table 2.shows, when inlet temperature is 320 K, standard deviation of air rising type is less than air drop, and the node temperature is closer to current inlet temperature. Standard deviation reflects the degree of dispersion between node temperature and mean in current condition. The value is lower, the temperature are more uniform. Therefore, when the inlet temperature is 320K, temperature distribution is more uniform in air rising type curing barn.

It could be concluded from these standard deviations in Table 3: under the same initial conditions, temperature distribution is more uniform in air rising type curing barn, therefore, air rising type curing barn is more conducive for tobacco curing.

5. Conclusions

It is the first time to use Fluent to analyze the distribution of temperature and airflow of the two types of curing barns, and calculate the standard deviation through uniformly selecting the Collection Points. The main conclusions are as follows:

- The temperature of each node in air rising type curing barn is closer to the inlet temperature under the same inlet temperature;
- The standard deviation of each node temperature in the air rising curing barn is smaller than that of the air drop under the same inlet temperature.
- Due to the more uniform distribution of temperature and airflow, the air rising type is more conducive to the tobacco curing.

It can be used as a basis for selecting the most suitable tobacco curing structure in the actual construction.

Acknowledgments
This work is supported by the science and Technology Department of Guizhou province of industrial research projects which name is Key technology research and application of air source heat pump curing barn (Grant No. [2014]3026) and Digital precision molding technology and equipment innovation base for graduate students of Guizhou University (GZU CXJD [2014]001).

References

[1] Xu X H, Sun F S, Wang Y, Liu J Y, Zhao B, Deng Q Q and Ding Z Y 2008 Current situation and further direction on research and application of bulk curing barn in China *Chinese Tobacco Science* **4** 54-6, 61
[2] Wang N R, Xu Z H, He M X, Yin Y Q and Wei Z Y 2011 The quality difference of tobacco leaves cured in bulk curing barns of airflow in different directions *Chinese Tobacco Science* **2** 81-5
[3] Ou Y J, Wang Y Q, Luo H L, Cui G M, Zhao G K and Huang Wei 2015 Numerical simulation of temperature and velocity distribution in bulk curing barn *Chinese Agricultural Science Bulletin* **31** 279-84
[4] Gu S Y, Liu D and Xiang L L 2015 Research on the key factors influencing the temperature uniformity of cooking oven *Building Energy & Environment* **34** 76-82
[5] Xiang L L, Liu D, Reng Y Guo Z B and Wu M F 2016 Optimization study of the uniformity of the temperature field of an oven cavity *Building Energy & Environment* **35** 45-50
[6] Ma J J, Sun X S and Li X M 2014 Numerical analysis of temperature uniformity for climatic test chamber *Equipment Environmental Engineering* **11** 48-54
[7] Cai X C Hu W Liang Y G and Wang H G 2008 Study on curing properties of different types of intensive curing barn *Agriculture of Yun Nan Provence* 27-8

IOP Conf. Series: Earth and Environmental Science **40** (2016) 012085 doi:10.1088/1755-1315/40/1/012085

Design for the turbine of solar chimney power plant system with vertical collector

Y Zhou, B Gao[1], H R Dong and K Hao

College of Electromechanical Engineering, Qingdao University of Science & Technology, Shandong, Qingdao, China

E-mail: 790375901@qq.com

Abstract. Based on the design theory of the wind turbine, a turbine applied to solar chimney power plant with a vertical collector is designed, and the performance of the wind turbine is conducted by wind tunnel test. The results point out that the experimental performance is consistent with the designed performance. Then the number of blades and tip speed ratio are conducted, as well as we obtained the optimum blade number and tip speed ratio of turbine for the system.

1. Introduction

The solar chimney power plant system, which was first put forward by Professor Schlaich in 1978, can convert solar radiation into electrical energy without generating pollution[1~3]. Many scholars have carried out a lot of research on the operation characteristics, flow field, energy conversion process, and the structure of power plant system. which proved that the solar chimney power generation system has the advantages of low power generation technology, operation and maintenance[3].However, due to its large area and high chimney, the application of it is limited. Qing-Ling Li [4] combined the solar power generation technology with the building and designed a kind of vertical solar chimney power plant system applied in the urban area, as shown in figure 1. In this new kind of solar chimney power plant system, the vertical solar chimney is made from transparent materials supplemented with the enclosed air absorbing solar energy and flowing upward[4]. At the same time, the external cold air enters into system constantly, thereby forming an air circulation. The turbine driving a generator is installed at the bottom of the chimney, where the differential pressure is greatest, so that the energy of the air pressure difference can be used best.

For the solar chimney power plant system with vertical heat collector, Yan Zhou [4~8] introduced its structure and the principle of power generation. A mathematical model considering the influence of size, solar radiation, and environmental conditions have been made; A new type of heat collector and accumulation device has been designed, and the shape of the flow channel has been optimized; And the relationship between system power and the height, width, and thickness of the air layer of the heat collector has been analyzed. However in the power station system, the main equipment that converts the of other forms of energy into mechanical energy is the turbine. The performance of the which is also directly affect the improvement of the system energy conversion process.

[1] Address for correspondence: B Gao, College of Electromechanical Engineering, Qingdao University of Science & Technology, Shandong, Qingdao, China. E-mail: 790375901@qq.com.

The design of turbine in solar chimney power plant system with vertical heat collector began in a paper devoted to the study of the characteristics of a solar chimney power plant published by Backstrm and Gannon. In this paper, the theoretical analysis method was used to explain the influence of turbine fluid flow and load coefficient on the efficiency of turbine[9]. F. Denantes and E. Bilgen developed an efficient model of counter-rotating turbines for a solar chimney[10]. The air flow condition of 200MW solar chimney power plant was calculated and the turbine was designed by Puprecht. Ting-Zhen MING et al. designed and analysed the heat collector, wind turbine, and conducted the influence of each parameter on the system [11].

Figure 1. The chart of solar chimney power plant system with vertical heat collector.

Although the temperature field of the solar chimney is similar to that of the traditional solar chimney power plant, the speed gradient of the system is obviously larger, and the pressure gradient is less[12]. Therefore, as far as the solar chimney power plant system with vertical heat collector, the main energy of hot air flow in the system is kinetic energy, the static pressure energy and heat energy is relatively small. According to the characteristics of energy in the system, this paper intends to design a turbine applied to solar chimney power plant system with vertical heat collector based on the design principle of the wind turbine, and make a detailed study of the performance.

Blade is the base and key components for the wind turbine, it directly affects the performance and cost of the whole machine. Wind turbine blade design is divided into aerodynamic shape design and structural design, the main task of aerodynamic shape design is to choose appropriate airfoil and calculate the chord length of the blade and installation angle to determine the external surface of the blade. At present, the design of the turbine blade design theory is a lot, the more famous of which have Schimitz theory, Clauert theory and Wilson theory; Because Wilson theory makes the improvement of Glauert theory, analyzes the tip loss and lift drag ratio on the optimum performance of the blade. Therefore, the theory is widely used in blade aerodynamic performance calculation. Based on the design theory of the wind turbine, this paper uses Matlab software to solve the geometric parameters of wind turbine blade profile and the physical model of the turbine is established by using SolidWorks software, then and the performance of the wind turbine is investigated by wind tunnel test.

2. Calculation of turbine design

2.1. General parameters of turbine design

In the solar chimney power plant system with vertical heat collector, the main parameters of turbine blade design include: impeller diameter, blade number, blade airfoil, the airfoil chord length and installation angle. According to the data provided by the literature [13~14], the average wind speed of the system is V = 5m/s. In the design calculation, the wind energy utilization coefficient is set as Cp = 0.35, wind turbine electrical and mechanical efficiency $\eta_1 g \eta_2$=0.9 . As the flow channel diameter of the experimental power station is 0.45m, it can be assumed that the diameter of the wind wheel is D = 0.43m.

2.2. Determination of turbine power

The power of the turbine equipment of the experimental power station can be calculated according to the calculation formula provided by the literature [14]:

$$P = 0.5\rho V^3 A C_p \eta_1 \eta_2 \tag{1}$$

ρ is air density, its value varies with the change of air temperature. In the the solar chimney power plant system with vertical heat collector, turbine is placed in the top of the collector. The simulation results point out that the air temperature increases with the increase of the height of the heat collecting plate. That is, the temperature at the top of the collector is lowest, so assuming that the temperature is 30℃, and the air density is 1.169 kg/m3. A is swept area of the wind wheel. The power of the turbine can be calculated: $P = 2.45W$.

2.3. Determine the number of blades, tip speed ratio and airfoil

There is a certain relationship between the number of blades and the power of turbine. In general, under the conditions of low flow rate, multiple-blade has high efficiency and start-up characteristics. 6 blade turbine with wonderful balance refuse to produce disturbance to the turbine, Therefore, the number of blades are determined to be 6; At the same time according to the relationship of tip speed ratio and the number of blades, tip speed ratio is assumed to be 3.

The performance of the airfoil directly affects the efficiency of the overall generator. The aerodynamic performance of the blade, the quality of the blade and the difficulty of manufacturing are comprehensively considered when the airfoil is selected. Excellent airfoil should possess a larger lift coefficient and a smaller drag coefficient, which makes the lift drag ratio tend to reach the maximum. NACA63-215 airfoil is commonly used in the traditional wind turbine airfoil. At different angles of attack, it has a different lift drag ratio. When angle of attack is 7 degrees with the maximum lift drag ratio of 14, the airfoil lift coefficient is $C_l = 0.7$, drag coefficient is C_d=0.05 [15].

2.4. Blade design model

Based on the Wilson approach to establish the design procedure of the blade, the design method of high precision will obtain chord length and installation angle of each leaf element.

2.4.1. Objective function

When the partial loss of the blade is considered, the local optimum power coefficient of the blade dC_p can be determined:

$$\frac{dC_p}{d\lambda} = -\frac{8}{\lambda_0^2} b(1-a)F\lambda^3 \tag{2}$$

Where: b is circumferential induction factor; a is axial induction factor; F is loss coefficient; λ is tip speed ratio at r of wind wheel radius. According to equation (2) the power factor of the whole blade reach the maximum value, it is necessary to make the maximum value of $\frac{dC_p}{d\lambda}$ from each leaf blade surface.

2.4.2. *Constraint condition*

In order to determine the maximum value of the objective function in the formula (2), the constraint conditions must be satisfied:

$$b(1+b)\lambda^2 = a(1-aF) \tag{3}$$

Iterative computation of each leaf element a, b, F, and F by the formula (4), (5), (6) to determine:

$$F_{tip} = \frac{2}{\pi}\cos^{-1} e^{\frac{-B}{2}\frac{R-r}{R\sin I}} \tag{4}$$

$$F_{hub} = \frac{2}{\pi}\cos^{-1} e^{\frac{-B}{2}\frac{r-R_{hub}}{r\sin I}} \tag{5}$$

$$F = F_{tip} \cdot F_{hub} \tag{6}$$

Where: F_{tip} is tip correction factor; F_{hub} is leaf root correction factor; R is impeller radius, m; r is blade element section to the center of the wind wheel, m; I is flow angle of leaf element (°).

The calculation formula for blade chord:

$$\frac{BcC_l}{r} = \frac{8\pi aF(1-aF)}{(1-a)^2} \cdot \frac{\sin^2 I}{\cos I} \tag{7}$$

The calculation formula for the installation angle of the blade is:

$$\tan I = \frac{(1-a)V}{(1+b)wr} = \frac{(1-a)}{(1+b)\lambda} \tag{8}$$

$$\beta = I - \alpha \tag{9}$$

Where: β is Blade mounting angle, (°). c is blade chord length, m.

2.5. *Wilson optimization design steps*

Table 1. Parameters of turbine blade.

Radius r(m)	Axial induction factor a	Circumferential induction factor b	Inflow angleI(°)	Chord L(m)	Installation angleβ(°)	Loss coefficient F
0.05	0.3303	0.3202	34.0902	0.0559	27.0902	0.8481
0.065	0.3229	0.1969	30.1383	0.054	23.1383	0.9479
0.08	0.3253	0.1362	26.3419	0.0506	19.3419	0.9675
0.095	0.3273	0.0998	23.2424	0.0468	16.2424	0.9757
0.11	0.3293	0.0762	20.7022	0.0431	13.7022	0.9768
0.125	0.3315	0.0603	18.595	0.0397	11.595	0.9727
0.14	0.3345	0.049	16.8182	0.0368	9.8182	0.9619
0.155	0.3391	0.0411	15.2793	0.0342	8.2793	0.9381
0.17	0.3488	0.0358	13.8569	0.0321	6.8569	0.884
0.185	0.3734	0.0339	12.3261	0.0307	5.3261	0.7445
0.2	0.3869	0.0309	11.793	0.0306	4.793	0.6703

(1) Blades will be divided into equal parts along the radial direction of blade, the first airfoil with radius r = 0.05m, r = 0.015m for every part.

(2) For a per leaf element, with formula (2) as the objective function, type (3) and (6) as the constraint condition and objective function for solving the optimization function, we can acquire inducing factor a, b and the correction factor F.

(3) According to the formula (8), the installation angle of each section is solved.

(4) Equation (7) can be used to solve the chord length C.

In the design of blade process, Matlab programming tool plays a very important role. In the choice of optimization function, we select the fmincon function, and nonlinear multi constraint problem is usually solved with this function,thus the accuracy is higher.

The aerodynamic shape of the blade is obtained as shown in table 1.

3. 3D solid modeling of blade

According to the coordinate geometry transformation principle, the 3D coordinates of each section in the space are solved. The 3D space coordinate is saved as the notepad file, and the SolidWorks software is directly introduced into the model to extract the diagram of each blade airfoil section and The 3D solid model of blade, that are shown in figure 2 and figure 3.

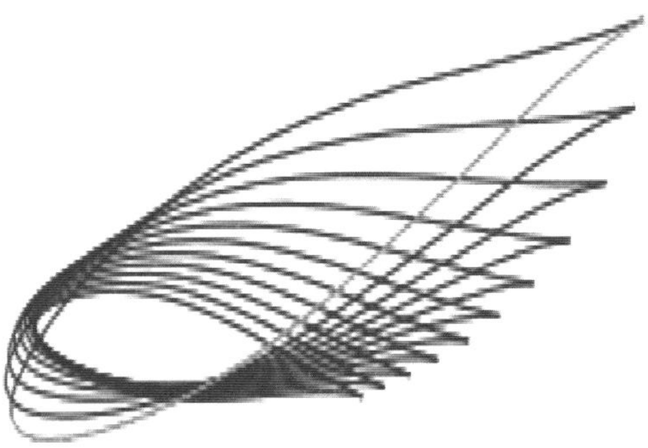

Figure 2. The diagram of each blade airfoil section.

Figure 3. The 3D solid model of blade.

4. Experimental study on turbine performance

4.1. Experiment principle

The low speed wind tunnel used in this experiment is shown in figure 4. The experimental section is a cylindrical pipe, the ratio of which to the actual channel is 1:1, and the turbine is made according to the actual design size.

Figure 4. Schematic diagram of experimental set-up.

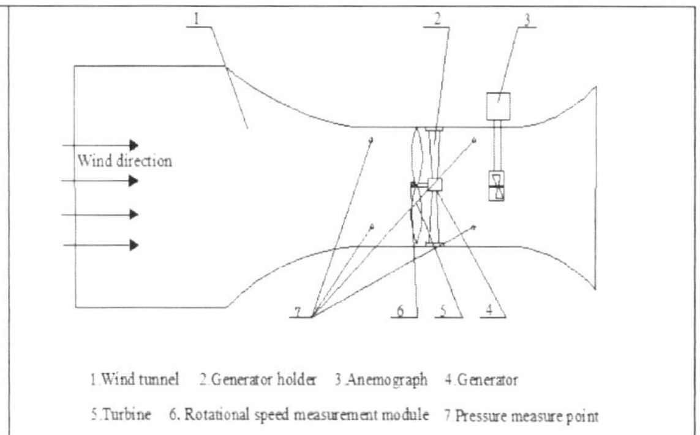

Figure 5. Schematic diagram of experimental principle.

Figure 5 is the experimental principle diagram, the turbine connects with the generator directly using coupling. The generator carries on the load, so as to measure the current and voltage of the generator exporting, and deduce the turbine power. Therefore turbine's performance can be comprehensively analysed. In the system of wind tunnel, the air enters the passage from the left side of the nozzle as shown in the diagram, and after contraction section, its speed greatly improves, then it enters the experiment section of the system.

Turbine blades are made by 3D printing technology. The blade model is shown in figure 6. Rotating speed is measured by speed measure module that includes a STM32 development board and photoelectric rotary encoder. We use the standard GM8901 handheld wind speed meter to measure the air speed of the experiment section. The position of the pressure measure point is shown as figure 6, and we use the differential pressure gauge to measure the differential pressure.

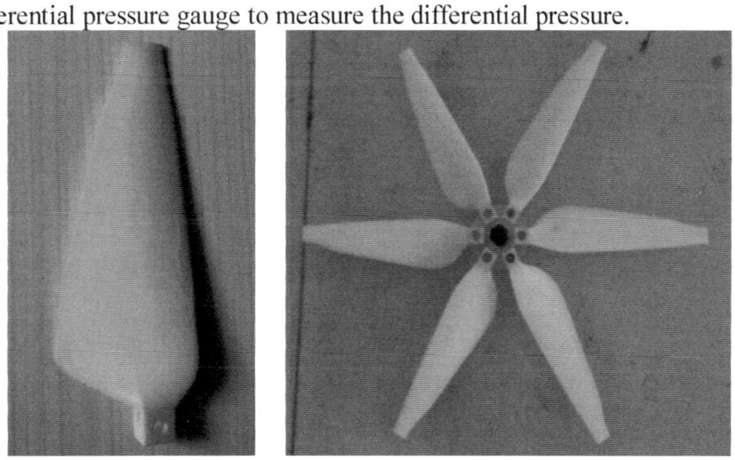

Figure 6. (a) Single blade Figure 6. (b) A six-blade turbine

Figure 6. Blades of turbine.

4.2. Experimental results

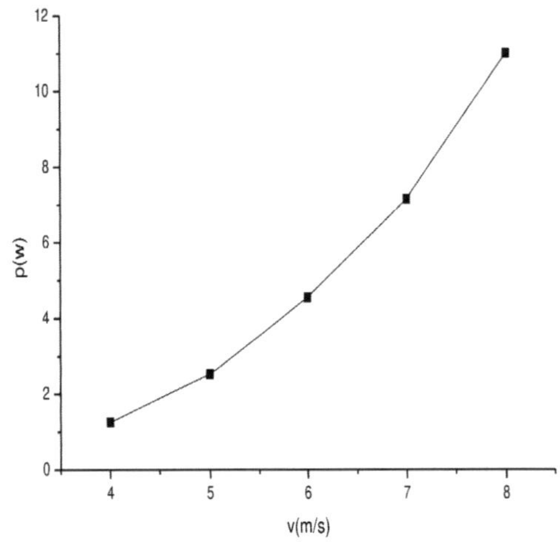

Figure 7. Relationship between power and wind speed.

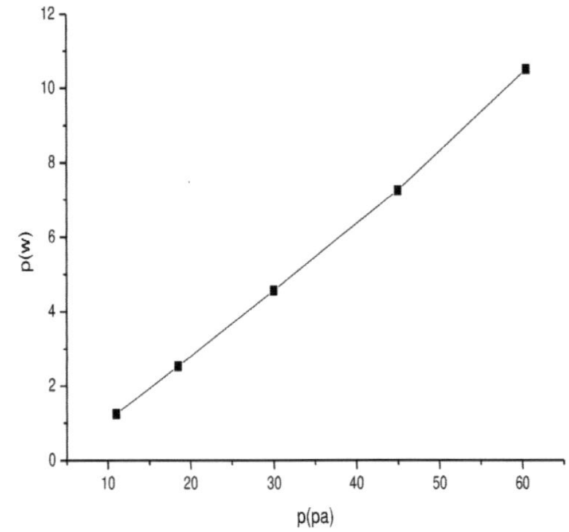

Figure 8. Relationship between power and differential pressure.

Figure 7 and figure 8 are relationship between power and wind speed and differential pressure. We can observe that the power of turbine increase with turbine inlet speed and pressure difference increases. The greater the inlet speed and pressure difference, the greater the output power of the turbine. And it is observed that the turbine power changes in accordance with the exponential relationship with the variation of the wind speed at the turbine inlet, and changes in accordance with the linear relationship with the variation of between and after the turbine, which indicates that the turbine can utilize the air energy better than the pressure energy.

We can also find that the starting wind speed of the 6 blade turbine is about 4m/s. But in the experiment we notice that after the turbine starts rotating, even when the wind speed is less than 4m, turbines will not immediately stop rotating. Through testing, when the wind speed reduces to 3m/s, turbine will stops rotating. The difference in speed between of them is about 1m/s. The experimental data obtained are referred to in figure 9.

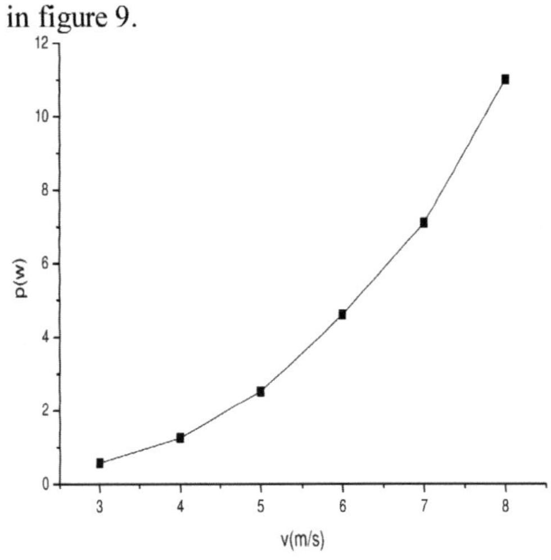

Figure 9. Relationship between power and wind speed.

4.3. Comparative analysis of designed and experimental values of blade performance

Figures 10 and 11, respectively point out that relationship between designed and experimental power and inlet speed and differential pressure. As can be observed from the figure, the designed values and experimental values are similar, and there is the same trend. There are some differences between experimental and design values, and the greater the wind speed are, the greater the difference is. There may be some reasons for the error, on the one hand, in the design of the blades, the wind energy utilization coefficient, tip speed ratio and other parameters are assumed to be an approximation according to experience. On the other hand, measuring and reading test dates have unavoidable errors in the experiment.

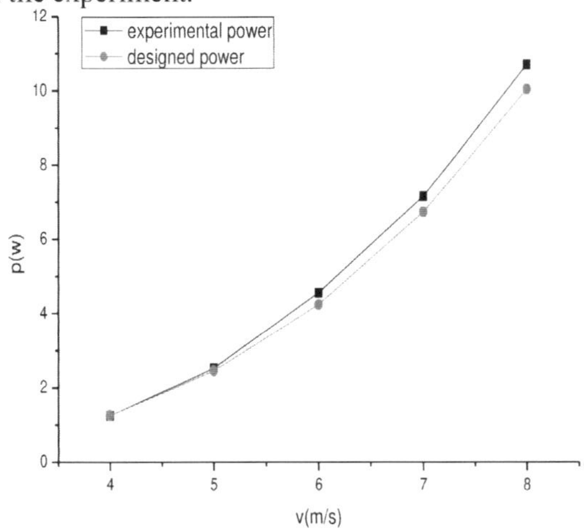

Figure 10. Relationship between experimental and designed power and wind speed.

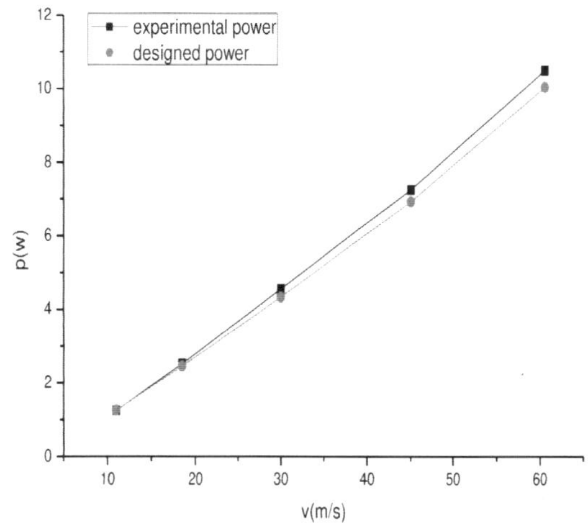

Figure 11. Relationship between experimental and designed power and differential pressure.

5. Optimization of turbine

5.1. Optimization of blade number

In order to obtain the optimum blade number of the turbine for vertical solar chimney power plant, the turbine of the other number of blades must be tested. Since wind speed of this system is relatively small, in order to extract a small starting wind speed, so we try to use multi-blade turbine. Due to the limited size of the hub, when the number of blades reaches 9, the hub can not meet the size requirements of the number. Therefore, the redesigned turbines with blade number of 5, 7, 8 are conducted. Understood by the design of the blades, we can know the greater the number of blades, the blade smaller the width of the blade. The experimental dates are shown in figure 12.

As can be observed from the figure 12, the power of the turbine with different number of blades is slightly different, especially at low wind speed. When the wind speed is increased, we can observe that the number of 6 blades of turbine power is the highest, followed by 5 blades, 8 blades of turbine power is minimum. As far as starting wind speed, it can be concluded that the number of minimum starting wind speed is 8, whose value is 3.5m/s. The starting wind speed of 5 blade turbine is the highest, whose value is 4.2m/s. Therefore, taking into account the cost, power, starting wind speed and other factors, we selected the optimum number of the turbine blades is 6.

5.2. Optimization of tip speed ratio

In the above design, the tip speed ratio is assumed to be 3. When the wind speed is constant, change the tip speed ratio amounts to alter the rotating speed. Therefore, when the wind speed is 5m/s, the rotating speed of the turbine in the off-design condition is changed to acquire the best tip speed ratio.

Figure 13 is the relationship between turbine power and tip speed ratio. We can conclude that the power of the turbine increases firstly and then decreases with the increase of the tip speed ratio. The turbine has the maximum power when the tip speed ratio is 1.9.

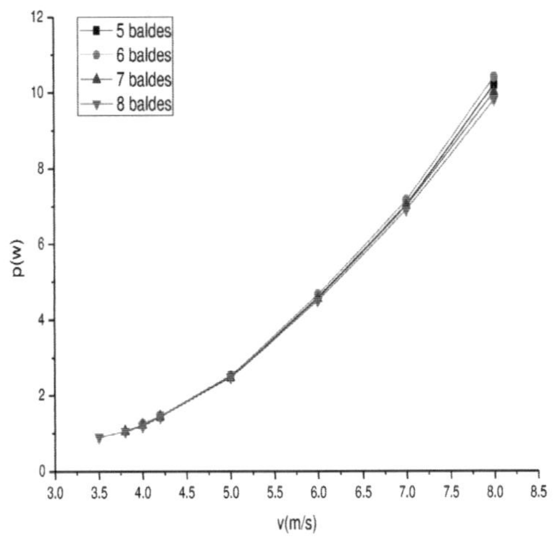

Figure 12. Relationship between turbine power with different blade numbers and wind speed.

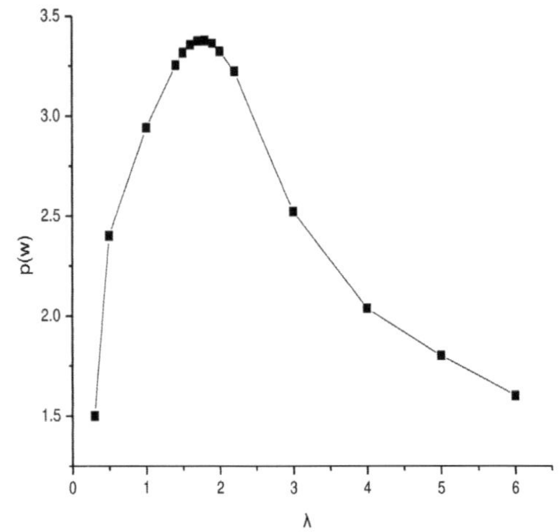

Figure 13. Relationship between turbine power and tip speed ratio.

Therefore, the turbine has the optimum blade number and tip speed ratio. This is because when the number of blades and tip speed ratio is large, it will block the wind passing through the turbine, thereby reducing the power of the turbine.

6. Conclusion

Based on the design theory of the wind turbine, a turbine applied to solar chimney power plant with a vertical collector is designed, as well as the performance of the turbine is analyzed by experiments. The following conclusions are obtained:

- Based on the Matlab platform, we adapt Wilson design theory to solve parameters of each blade airfoil section. At the same time, the NACA63-215 airfoil is chosen as the airfoil of the blade design. Finally, the physical model was built with Solidworks software.
- The performance of the wind turbine is conducted by wind tunnel test, which is compared with the design power. The results refer to the experimental power is consistent with the design power.
- Through the experimental method, we come to conclusion the optimum blade number is 6, the optimum tip speed ratio is 1.9.

Acknowledgement
This research was financially supported by the National Natural Science Foundation.

References

[1] Schliach J 1995 The solar chimney: Electricity from the sun (Stuttgart: Axel Menges)
[2] Robert R 1981 Spanish solar chimney nears completion *MPS Review* **6** 21-3
[3] Haaf W, Friedrich K, Mayer G and Schlaich J 1983 Solar chimney. Parta: Principle and construction of the pilot plant in Manzanares *Int Jof Solar Energy* **2** 3-20
[4] Li Q L 2008 *A solar chimney power plant with vertical collector* China 200820232887 **7**
[5] Zhou Y, Lu H B, Gong Y Y, Wang L, Li Q L 2015 Design for micro turbines in solar chimney power plants with vertical collectors *J. Engineering Thermophysics* **36** 1542-6
[6] Jie X Q 2014 *Study of heat transfer characteristics with temperature pressure coupling natural convection of solar chimney power plant system* (Qingdao, Qingdao University of Science and Technology)
[7] Gong Y Y, Lu H b, Zhou Y, Li Q L 2015 Effect of turbines on performance of solar chimney power plant system with vertical collectors *Thermal power generation* **44** 12-6

[8] Zhou Y, Li J H and Li Q L 2010 CFD simulation of a solar energy hot air stream power plant system installed in a high building *J. Engineering for Thermal Energy and Power* **25** 226--9
[9] Von BackstrÊm T W and Gannon A J 2004 Solar chimney turbine characteristics *Solar Energy* **76** 235-41
[10] Denantes F and Bilgen E 2005 Counter-rotating turbines for solar chimney power plants *Renewable Energy* **31** 1873-91
[11] Ming T Z, Liu W, Xu G L, Xiong Y B, Guan X H and Pan Y 2008 Numerical simulation of the solar chimney power plant systems coupled with turbine *Renewable Energy* **33** 897-905
[12] Lu H B 2014 *Design and anslysis for blades of turbines in solar chimney power plants with vertical collectors* (Qingdao, Qingdao University of Science and Technology)
[13] Bao F 2008 *Geometric design and aerodynamic simulation for the wind turbine blade* (Dalian: Dalian University of Technology)
[14] Zhang L D, Ren L C, Chen R S and Mao J D 2008 Study on shape design and 3D solid modeling of wind turbine blade *Acta Energiae Solaris Sinica* **29** 1177-9
[15] Su M J 2006 *Pnuematic characterstics property studies of blade airfoil profile for horizontal axis wind turbine* (Liaoning, Liaoning Technical University)

Application of new geological modeling technology in secondary development in Daqing oil field

C L Li, J H Zhang[1], H K Li and C Liu

Northeast Petroleum University, Institute of Petroleum Engineering, Daqing, Heilongjiang, China

E-mail: dqzhjh@126.com

Abstract. The reconfiguration on underground geology system is one of the key techniques for secondary development in Daqing oil field. The geological modeling is the unique method to characterize new knowledge system of reservoir. The development history of maturing field is long. The structure of maturing field is complex and the distribution of oil remaining is highly scattered. The difficulty of adjustment and potential tapping is great. In viewing of demand for secondary development, the strategies and methods of geological modeling are proposed. According to the characteristics that many faults crosscut each other, the clue of fractional simulation—key horizon controlling—overall structural modeling is carried out to accurately and effectively build fine structural models. In order to approximate the real microfacies simulation effect, microfacies modeling technology of multiple iterations and geology tendency under vertical and lateral geology tendency constraint is used. And the attribute models could approximate the real parametric distribution. Moreover, in viewing of the key and potential reservoir sand, the countermeasure on configuration modeling by different stochastic simulation methods and step simulation is proposed to rapidly build geologic models. The geologic models are scientific and feasible. The above-mentioned countermeasures and methods have been used in secondary development of Daqing oilfield and the effect was well. This new technology presents directive sense.

1. Introduction

Because of the increasingly severe development situation, most of maturing land fields have been at later stake of high water cut and high degree of reserve recovery [1-3]. Yet, contribution of production and reserves of these oil fields is still greater. The potential tapping of remaining and recoverable reserves is still the essential measure for maturing field. Because of this, the secondary development idea is proposed—the new development system of maturing field is rebuilt using new technologies and methods and idea, in order to greatly raise the final recovery rate [4].

Rebuilding underground knowledge system is one of the key technologies for secondary development. In order to deepen oil remaining distribution regularities, three-dimensional geological modeling is the unique method to research oil remaining by using the three-dimensional seismic structure describe and distributing feature of reservoir. For geologic feature of maturing field with complex blocks and demands of secondary development, the corresponding countermeasures of

[1] Address for correspondence: J H Zhang, Northeast Petroleum University, Institute of Petroleum Engineering, Daqing, Heilongjiang. E-mail: dqzhjh@126.com.

triaxial geological modeling technology are proposed to meet demands of secondary development [5-6].

2. Countermeasures on structure modeling technology in complex blocks

Structural modeling that is the carrier of attribute models, could ascertain the spatial position of geologic model and contact relation of interlamination. Therefore, building feasible geologic models is the base of triaxial geological modeling [7-8] and it is the triaxial visualization of achievements on previous structural knowledge. Multitude faults and complex collocation connection are the classic geology features of maturing field. It is difficult to match the research report of geology with structure. In order to meet demands of secondary development, the study scale of geology should be more refined [9-10]. In addition, the study scale of geology should be research from reservoir to layer, until to several meters single sand body. It is well known that the completely presentation of feature of thin-numerou layers is difficult for modeling.

The Daqing oilfield, with a long development history, has been at later stake of high water cut and high degree of reserve recovery. Take Pubei area for example, the fault system is complex and the blocks are fracture (Figure 1).

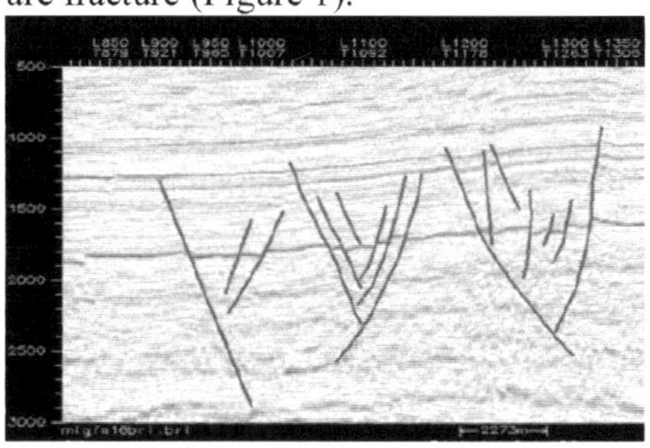

Figure 1.(a) Crossing type

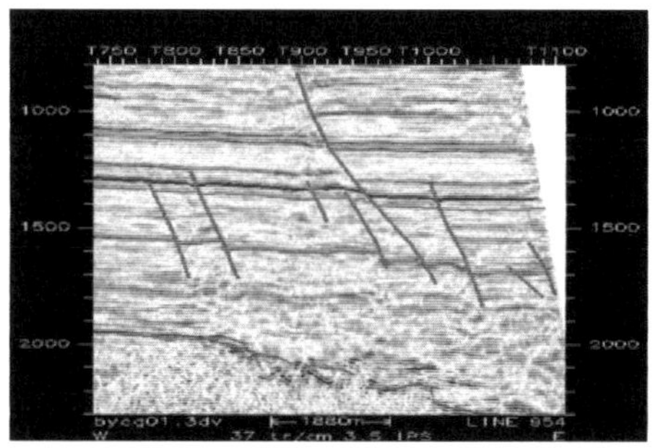

Figure 1.(b) Listric type

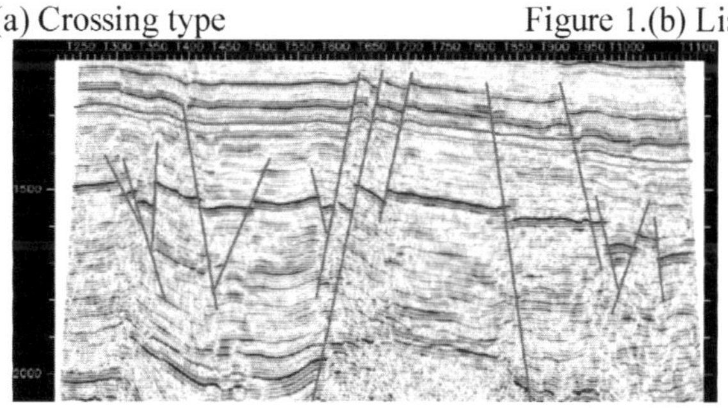

Figure 1. (c) "Y" type

Figure 1. Diagram of some fault types.

In viewing of difficulty and complexity of faults and layers, the technology countermeasures on structural modeling are proposed: fractional simulation, key horizon controlling, overall structural modeling. Structural model is composed with fault model and bedding model. As an important part structural model, effectiveness and accuracy of fault model affect particularity of subsequent modeling. Bedding model is the three-dimensional distribution of depositional interface. For the structural model of maturing field subdivided to single sand body, because of the multitude beds and thin sand body and different type fault, the fault model is established firstly, then bedding model is established.

Fault system is divide into3 levels as the scale. The development time of second level fault is long, the fault throw is large, propagation range is long. The third level fault performs control functions for local structure and oil/gas distribution. The other is the forth level fault with short development time and small fault throw. The research is constructed form different hierarchies. First of all, the second and third level faults with stability development are simulated, in order to obtain the fault model that controls structure format. In addition, the obtained fault model could carry out quality supervision of the fault parameters and reciprocally crosscutting relationship between faults. Based on these, combining with analysis on local microtectonics, the fault model is established by mutual analysis and quality supervision. Bedded plane model reflects the superimposed form of sedimentary association. Because of the maturing field with multitude single sand bed and complex faults, it is hard to match layers and fault system. In consequence, based on fault model, the key single sand beds are built firstly to control layer stratigraphic framework and are use as restraint to build bedded plane of others single sand bed. The references on optimizing single sand bed are as follows: (1) Overall structure could be essentially controlled and bedding separation is not too large. (2) Fault framework of key bedding should present representativeness and could essentially express fault alteration. (3) The single sand bed essentially stand for oil horizon system. The optimized key horizons are initially simulated and mutually edited in order to accord with the knowledge on actual geology and be well fitted with fault model. Base on horizon model of key single sand horizon, building entire structural model of single sand bed could minimize the amount of work on beds system modeling and is easy to carry out quality control. The model is made to correspond in structure.

3. Countermeasures on modeling technique of sedimentary microfacies

Using multi scientific information, sedimentary microfacies modeling characterizes distribution of sedimentary microfacies [11]. As research and development of modeling technique, it is an inevitable tendency to truly reflect the subsurface reservoir feature. Structural models are the base of modeling and the attribute models are the attribute models manifestation. At first, the facies model that is built by using interpolation as algorithm tendency using well logging data is used for restraint on facies control. However, as the development of exploratory development technique, the knowledge of reservoir is deeper. Now, various methods are used to restrict predication on reservoir attribute [12-13]. The knowledge of sand and mudstone could not meet the necessaries of fine fine structural models or the secondary development of maturing field. Hence, precipitation facies are used. Because different precipitation facies have different distribution features of physical property data and. Under the restraint by precipitation facies, the attribute models come into real parametric distribution of geology (Figure 2-3). Thus precipitation facies models affect accuracy of attribute models and accuracy of subsequent numerical reservoir simulation.

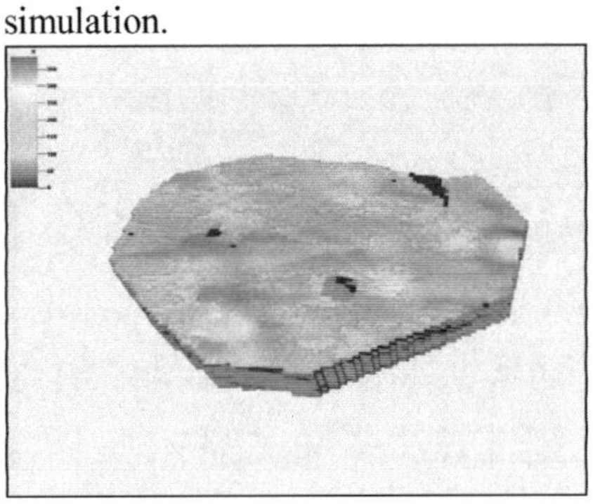

Figure 2. Attribute model without restraint

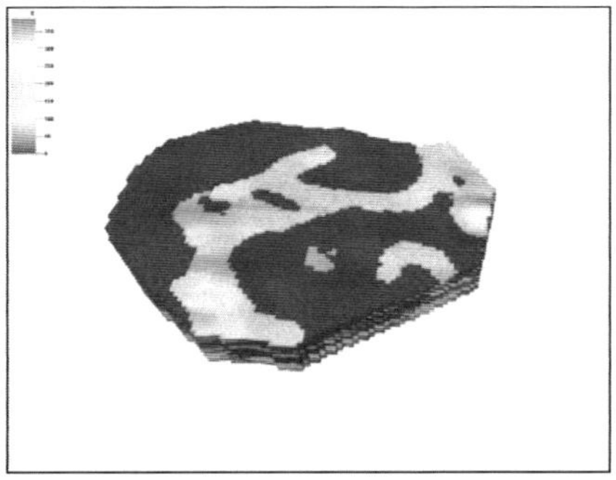

Figure 3. Attribute model restraied by precipitation facie

Based on the method of sequential indicating simulation, microfacies models are built by using the restraint method of vertical and lateral geology tendency. According to subdivision result of microfacies in wells, traditional microfacies modeling using variation function analysis to interpolate on plane is not exact to reflect the cognitive achievements of geology. In addition, the antecedent fine geology achievement could not be used in microfacies modelling [14].

Except for restraint method of vertical and lateral geology tendency, the method of multiple iteration of geologic body are used in simulating process. In order to adjust simulation result to geology tendency, this method is used to achieve automatic contraction and effects on denoising. The simulation effect of denoising and convergence of geologic body attribute after iteration is better than the simulation effect with iteration (Figure 4). However, it is not the more iterate frequency, the better simulation effect. The simulation effect is also affected by coincidence degree of trend surface and thickness of microfacies classified by well point.

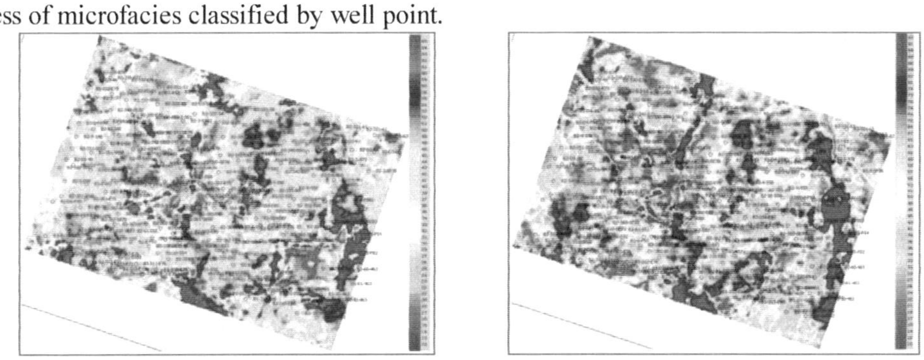

Figure 4. (a) No-iteration result of microfacies

Figure 4. (b) Iteration result of microfacies

Figure 4. Iteration of microfacies modeling

4. Fine modeling of internal structure of key single sand body

The redevelopment of the maturing fields is on the basis of rebuilding geologic model and reservoir model to form the new underground cognizing system, and according to the accurate geological research to find the potential oil sand body. Especially to the "double high" field, that reservoir mainly is fluvial sedimentary facies, remaining oil distribution in the "border, finely laminated, poor" parts of sand body or structure, need us take focus on the research of some "small and fat" sand body, depict to its internal structure and translate into geologic model, supply numerical simulation to subtly anticipate remaining oil distribution and find important breakthrough area of remaining oil "dessert". Geologic model become the important vinculum to link geological research achievements and numerical reservoir simulation.

The key to single sand body internal structure detailed modeling is description of point bar internal lateral accretion layer, and the difficulty is how to use suitable means to translate the tectonics understanding into model. Using different stochastic simulation method nested technology in the modeling process, stepwise simulate spatial distribution of meandering river point bar and lateral accretion layer. Means on the basis of structural model combine with microfacies research achievement, first using seismic sedimentary facies simulation method to build river channel and point bar model, then using multipoint geostatistics method to simulate lateral accretion layer, put lateral accretion layer model embed into facies model and get configuration model.

Seismic sedimentary facies simulation is on the basis of seismic sedimentology, using high resolution 3D seismic data and manifold technology to delicate depict sedimentation system. Commonly used strata slice, well-log and seismic integration technique to predict sand body boundary and thickness trend, geological cognition restrictive revision to build channel and point bar model. Point bar internal configuration simulation is the key point of refined modelling, but the precondition is that make fine dissection of point bar internal configuration, which is the only way of old oil field redevelopment geologic research. After deeply analyze point bar internal configuration, have comprehensive comprehension about point bar internal lateral accretion layer's parameters (curvature, trend, dip angle), reserve quantitative knowledge base to configuration simulation. The configuration is simulated by marked point process simulation algorithm and using multi-point geological statistics random simulation. Select the multipoint geostatistics for two main reasons: ① the multiple-point statistics colligates the advantage of method based on target and pixel; ②fine geological research of secondary development provides operability for training images. Due to the side product layer is thin, the simulation results' non-convergence is easy to produce with grid precision simulation model of facies model. So a suitable grid size must be explored by continuous cross-validation before the simulation. That is to ensure the precision of configuration model by the way of secondary encryption.

After determining the grid size of configuration simulation, establish the training image of the side product layer. The establishment of the training image is based on the configuration study quantitative knowledge base, then build up training images(Fig.5) of the side product layer by using software, according to some determined parameters including side product layer's curvature, trend. The training images reflect quantitative distribution pattern of lateral accretion layers. And simulation results(Fig.6) are approximatethe to the research achievement of geology of microfacies sand body.

Figure 5. Training image of lateral accretion layers

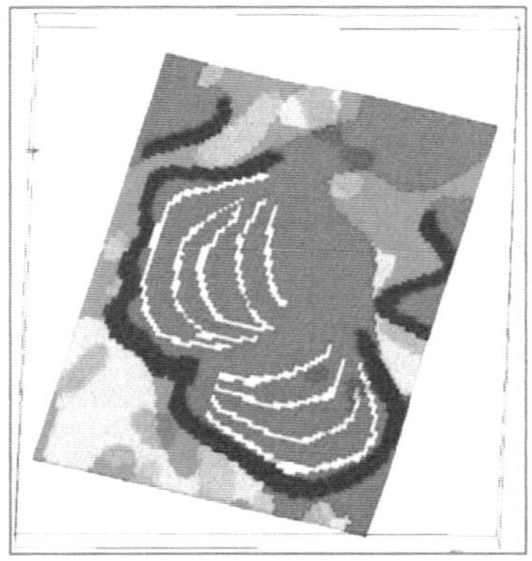

Figure 6. Simulation results of lateral accretion layers

5. Conclusion

- In view of the structure modeling of the complex fault block old oilfields' secondary development, puts forward the technical countermeasure of the key layer controlling-distributed simulation-the overall modeling. Effectively solves the difficult problem of many single sand layers, complex fault and well-seismic matching and forms a quick and accurate modeling thinking.
- Sedimentary microfacies simulation puts forward a double geological trend constraint method including plane trend and vertical trend. At the same time, taking multiple iterations in the simulation process achieves a result of denoising, smooth, making full use of geological results and having high goodness of fit with geological trend and provides a good reference for microfacies simulation.
- In view of the "small-fat" oil-rich single sand body of the secondary development, focusing on dissecting, establishing internal configuration model, putting forward using different random simulation method nested using technologies to simulate microfacies -configurations gradually. Multipoint geostatistics method is applied to simulate configurations and can make the simulation results fine and reliable

References

[1] Chen L F and Ebughu 2003 Secondary development of maturing field *Foreign Oil Field Engineering* **19** 13-4
[2] Wang P, Shao X J and Ma P H 2010 Technical Measures for Secondary Recovery in Maturing Fields and Itspractice *Drilling & Production Technology* **33** 48-52
[3] Liu Y M, Hou J G and Cai M J 2009 Method of detailed reservoir description in secondary development of old oilfield *Science & Technology Review* **27** 46-9
[4] Wu Z M 2013 Research on secondary development *China New Technologies and Products* 2 131-2
[5] Wang W 2015 Secondary developing methods for the waterflooding of lamadian oilfield *Petroleum Geology & Oilfield Development in Daqing* **34** 69-73
[6] Hu W R 2007 Overview of oilfield secondary development in China *Special Oil and Gas Reservoirs* **14** 1-4
[7] Wu W X and Li C L 2014 The research on anatomical techniques of internal configuration of II -type reservoir *Int. J. Petroleum Science and Technology* **8** 69-78
[8] Chong R J and Yu X H 2011 The keys to quality control of 3D reservoir modeling *Marine Geology Frontiers* **27** 64-9
[9] Guan S W and He D F 2011 Theories and technical frmaeworks of complex structural modeling *Acta Petrolei Sinica* **32** 991-1000
[10] Guang S W, Chen Z X, Li B L and Zhang C 2010 Discussions on the character and

interpretation model of Kelasu deep structures in the Kuqa area *Petroleum exploration and Development* **37** 531-6

[11] Wu S H, Wu J C, Li S J and Wu Z Y 2003 3D microfacies mode ling of pingqiao horizontal well region of Ansai oilfield *Acta Sedmentologica* **21** 266-71

[12] Li S H, Guo Y K, Li W F, Wang Z F and Shi X C 2006 3D reservoir model for SN31 oilfield *Natural Gas Geocience* **17** 811-4

[13] Liao X W 2004 Stochastic reservoir modeling restricted by geological conditions *Petroleum Exploration and Development* **31** 92-4

[14] Wu S H 2001 Geological constraint principles in reservoir stochastic modeling *J. the Universtiy of Petroleum, China* **25** 55-8

Energy conservation behavioural intention: attitudes, subjective norm and self-efficacy

J W C Lee[1] and A Tanusia

Faculty of Business, Nilai University, Persiaran Universiti, Putra Nilai, 71800 Nilai, Negeri Sembilan, Malaysia

E-mail: wclee@nilai.edu.my

Abstract. This paper examines the influence of attitude towards energy conservation at home, the attitude in the campus, subjective norm, and self-efficacy on energy conservation behavioural intention among students in a private university using the Theory of Planned Behaviour (TPB). Data was collected from about 194 students using a questionnaire developed from current literature on TPB. Data analysis using Smart PLS version 3.2.4 found that attitude towards energy conservation at home has an indirect significant influence on attitude towards energy conservation behavioural intention via the mediating effect of attitude towards energy conservation in the campus. Self-efficacy and subjective norm are also positively related to energy conservation intention. The study also indicates the suitability of the TPB in predicting behavioural intention through attitudes, subjective norms, and self-efficacy. Results suggest that energy education is vital in creating a positive attitude towards energy conservation while facilities managers in institutions need to formulate appropriate policies and regulations to inculcate the right attitude and behaviour towards energy saving.

1. Introduction

The private university in which this study was conducted has been facing escalating electricity bills for a number of years even when student intakes have remained constant over the same period. The university management in its initial response to address the escalating bills was to introduce LED lightings in classrooms and offices as well as in the hostels. While the results of these changes did see some savings, the management felt that more could be done to reduce the electricity bills. While innovative energy saving technology could be introduced into buildings, the responsibility still rests on the users of this technology to optimize its usage such that costs could be relatively lower than before the adoption of the new technology. On this account, it was decided to conduct a research to examine student attitude towards electricity conservation in the campus and how attitudes, subjective norm and self-efficacy could determine the level of intention to save energy on the campus.

Intentional studies in the literature have their underpinnings in the Theory of Reasoned Action (TRA) [1], and its extension the Theory of Planned Behaviour (TPB) [2] with the latter including perceived behavioural control or self-efficacy, as an addition to the attitude and subjective dimensions of the former. The TPB has been used to investigate determinants of intention to purchase a wide array of products/services such as green products, organic products, and hybrid cars as well as the intention

[1] Address for correspondence: J W C Lee Faculty of Business, Nilai University, Persiaran Universiti, Putra Nilai, 71800 Nilai, Negeri Sembilan, Malaysia. E-mail: wclee@nilai.edu.my.

to recycle, or to exercise [3]. Past studies have used TPB theory in energy-related behavioral studies [2], energy use and energy conservation behavior [4, 5]. This paper attempts to investigate intention to conserve electricity consumption among students in a university campus using a modified TPB. By doing so, this study aims to provide the university, researchers and policy makers an understanding of how attitudes, expectations of others and self efficacy could influence energy conservation intention.

2. Literature review - The Theory of Planned Behaviour (TPB)

The TPB [2] postulated that behaviour can be determined by an individual's beliefs and evaluations of the likely outcomes of the behavior, the normative expectations of significant others and motivation to comply with these expectations, and the individual's perceived ability to control the behavior. In combination, these factors could influence attitude toward the behaviour. When aggregated, attitudes, subjective norm, and self-efficacy could manifest into a behavioural intention. In short, the more positive the attitude and the stronger the subjective norm as well as the greater the self-efficacy in controlling the behaviour, the more inclined is the person's towards performing the said behavior.

2.1. Attitudes

The literature on intentional studies has shown that attitudes are often used to measure pro-environmental behaviour [6]. Past scholars working within the context of pro-environmental behaviour (attitude, values and belief) have reported that people with positive attitude and mindset are more prone towards pro-environment intention and behaviour [7]. Previous studies have also revealed pro-environmental behaviour being associated with energy conservation and consumption [8, 9 & 10]. To solve energy-related problem does not only require the use of technology but the behavior and attitude of each individual to solve the problems. Nair *et al.* [11] showed positive pro-environmental behaviour does not necessarily lead to an actual reduction in energy usage. It appears that even with new evaluation on renewable energy sources and awareness on climate changes, pro-environmental behaviour fails to limit the consumption of energy. Past researches explain that with constant change in attitudes, it has a very small impact on pro-environmental behaviour [7, 8]. The reason in changing attitudes on energy saving is more on minimizing cost (time, effort, money) rather than doing it for the welfare of the environment [9]. Wells *et al.* [6] however noted that attitudes associated significantly with a specific environmental behaviour will have an influence on that behaviour both at home and in the workplace.

2.2. Subjective Norms

Subjective norms refer to how influential the opinions of significant others in the live of an individual on the individual performing a certain behaviour [2, 12]. As one of the predictors of intention, subjective norms are influenced by beliefs and the thoughts of performing an act [13, 14]. This lead to individual act as per expected by their family, friends and the society. A study by Goldstein *et al.* [15] and Thaler & Sunstein [16] found social norm to be visible and salient in making a powerful impact on behavioural intention. These will prompt users to exhibit energy saving behavior and reduce energy consumption by 40 percent. For students living in campus dormitories, with utilities being part of accommodation provided, there is a lack of awareness of conserving energy and it poses a problem to the university [17, 18]. Another study conducted among students living in campus dormitories found subjective norms and peer influence can be used as an effective method of encouraging students to conserve energy consumption [18]. Costanzo *et al's.* [19] energy conservation model based on social-psychological constructs show that social influence, diffusion and reference groups including friends, family, and other social networks play significant roles in promoting and maintaining energy conservation. On a similar note, Stern ([20], p. 1229) opined that "the personal opinions and actions of one's friends may have a more powerful influence over household energy choices than expert advice, even if the latter is better informed".

2.3. Self-efficacy

According to Bandura [21] an individual's capabilities to act on certain behaviour defines his self-efficacy over that behaviour. Abrahamse and Steg [5] found that users are more inclined towards conserving energy if they realise that there is a negative outcome on the environment. Strong self-efficacy and internal locus of control suggest an ability to determine one's decisions, circumstances and behaviours strongly influenced by internal factors, such as one's own motivation and actions. On the other hand a low self-efficacy suggests a person's decisions and actions are being largely controlled by external factors beyond the person's control [5, 22].

2.4. Energy Conservation Intention

Intention refers to a person's subjective probability dimension that connects that person to a particular behaviour [23]. Energy saving intention is associated with TPB constructs which measure an individual's inclination towards saving energy and evaluating positively or negatively on energy saving [24]. Previous research has found that energy saving intention in households were significantly influenced by perceived behavioral control and positive attitudes towards energy conservation [5]. Furthermore, intentions to save energy are associated with cost/benefit considerations [5]. The TPB assumes socio-demographics variables influence intentions and behavior indirectly [25]. Extant literatures have recorded the use of TPB to examine energy-related behavioural intention [2], such as energy conservation [26] and energy use [27, 28].

3. Methods

A quantitative research approach in the form of a predictive study was used in this research. A total of 194 students constituted the sample. Data collection was by means of a survey questionnaire derived from the extant literature. The questionnaire for this research comprises of six parts. Part 1 consists of a profile of respondents; part 2 consists of student's attitude at home; part 3 consists of student's attitude on campus; part 4 consists of subjective norms; part 5 consists of self-efficacy and part 6 consists of intention to conserve energy consumption. The indicators for these constructs were adapted from Wells *et al.* [6] (attitude at home, attitude in campus) and Ajzen [2] (subjective norms, energy conserving intention), and Chen *et al.* [29] (self-efficacy) anchored on a 5-point Likert scale (1 = strongly disagree to 5 = strongly agree). (Please see Appendix). The Smart PLS ver. 3.2.4 structural equation modeling technique was used to run a confirmatory factor analysis (CFA). This technique was used following the rationale given by Ringle *et al.* [30] who suggests that Smart PLS-SEM techniques should be used for testing an incremental theory. This rationale was also adopted by Yeap *et al.* [31]. While acknowledging that the TPB is a well-established model, the authors have decided to include two variations of attitudes, namely, attitude towards energy saving at home and attitude on the campus. In this sense, we are testing an incremental model hence justifying the use of Smart PLS. The study also conducted tests on convergent and discriminant validity of the proposed model.

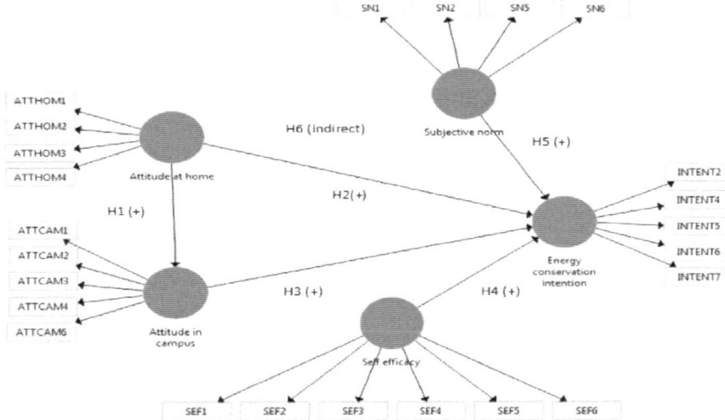

IOP Conf. Series: Earth and Environmental Science **40** (2016) 012087 doi:10.1088/1755-1315/40/1/012087

Figure 1. Research model of the study.

Based on the above literature review a research model was conceptualized for this paper (See Figure 1 which was adapted from Ajzen [2]). Subsequently, the following hypotheses were developed:

H1: The more positive the attitude towards conserving energy at home, the more positive is the attitudes toward conserving energy in campus.

H2: A more positive attitudes towards conserving energy at home will bring about a greater energy conservation intention in campus.

H3: The more positive the attitude towards conserving energy in campus, the greater is the energy conservation intention in campus.

H4: The higher the self-efficacy of students, the higher is their energy conservation intention in campus

H5: The higher the subjective norm among students, the higher is their energy conservation intention

H6: Attitude towards energy conservation in campus mediates the effect of attitude towards energy conservation at home on intention to conserve energy in the campus.

4. Results

4.1. Profile of the respondents

Table 1 indicates the respondents' profiles. Female students (58.2 %) slightly outnumbered male students (41.8 %). The majority of the respondents were between 18 to 23 years old (91.8 %). Most of the students staying in the hostel were Malaysians (91.2%), with Chinese ethnicity (58.8 %) forming the majority. About 43.3% of the students are pursuing diploma courses while 39.7% are doing degree courses. Overall about 91.2% of students are pursuing business courses.

Table 1. Profile of the respondents.

Profile	Frequency	Percent
Gender		
Male	81	41.8
Female	113	58.2
Age		
18-23 years	178	91.8
24-30 years	16	8.2
Nationality		
Malaysian	177	91.2
Non-Malaysian	17	8.8
Ethnicity		
Malay	19	9.8
Chinese	114	58.8
Indian	40	20.6
Others	21	10.8
Program		
Foundation	28	14.4
Diploma	84	43.3
Degree	77	39.7
Masters	5	2.6
Faculty		
Business	177	91.2

Science	17	8.8

4.2. Measurement model evaluation

The research model was tested using the SMART PLS ver 3.2.4 software [30]. Tests were conducted to determine the validity and reliability of the measures and to test the hypotheses [32]. The bootstrapping method was also used to test the significance of the path coefficients and the loadings [32]. Table 2 explained the convergent validity of measurement model. As suggested by Hair [32], both loadings and composite reliability (CR) were more than 0.7; and the average variance extracted (AVE) were more than 0.5. Items SN3, SN4, INTENT 3, ATTHOM5, ATTHOM6 and ATTCAM5 were omitted from the structural model because of loadings < 0.70. INTENT 1 was also omitted because of cross-loading with Attitude at the campus. Adopting the procedures of Fornell and Larcker [33] the discriminant validity of the measures was examined in Table 3. The diagonal values in bold represent the square roots of the AVEs (average variance extracted) and they are all greater than the correlations in the off diagonal section of the table [33]. Hence, Table 2 and Table 3 have established the convergent and discriminant validity of this study.

Table 2. Convergent validity.

Construct	Item	Loadings	[a] CR	[b] AVE
Attitude in campus	ATTCAM1	0.903	0.931	0.732
	ATTCAM2	0.887		
	ATTCAM3	0.866		
	ATTCAM4	0.864		
	ATTCAM6	0.749		
Attitude at home	ATTHOM1	0.790	0.903	0.699
	ATTHOM2	0.857		
	ATTHOM3	0.834		
	ATTHOM4	0.860		
Energy conservation intention	INTENT2	0.814	0.916	0.687
	INTENT4	0.840		
	INTENT5	0.727		
	INTENT6	0.890		
	INTENT7	0.864		
Self-efficacy	SEF1	0.798	0.924	0.670
	SEF2	0.834		
	SEF3	0.837		
	SEF4	0.851		
	SEF5	0.832		
	SEF6	0.753		
Subjective norm	SN1	0.807	0.872	0.631
	SN2	0.816		
	SN5	0.739		
	SN6	0.815		

[a]CR = Composite reliability = (square of the summation of the factor loadings)/ [(square of the summation of the factor loadings) + (square of the summation of the error variances)]

[b]AVE=Average Variance Extracted = (summation of squared factor loadings)/(summation of squared factor loadings) (summation of error variances)

Table 3. Discriminant validity.

	Attitude at home	**Attitude in campus**	**Energy conservation intention**	**Self-efficacy**	**Subjective norm**
Attitude at home	**0.836**				
Attitude in campus	0.632	**0.855**			
Energy conservation intention	0.441	0.673	**0.829**		
Self-efficacy	0.483	0.654	0.687	**0.818**	
Subjective norm	0.499	0.636	0.646	0.724	**0.795**

Diagonal values in bold are the square root of the average variance extracted (AVE). Values below the diagonals are correlations among the constructs.

Diagonal values (bolded) should be larger than the correlations in order to establish discriminant validity.

4.3. Evaluation of the structural model

This study evaluated the structural model by examining the R^2, Beta (β) and the corresponding t-values [31]. Bootstrapping procedure (refer Table 4) was applied to obtain t-values, report predictive relevance (Q^2) and effect sizes (f^2) [31]. There were four predictors for energy conservation intention, namely, attitude at home, attitude in campus, subjective norm and self-efficacy. Attitude at home ($\beta = -0.045$, $p > 0.05$) has no direct effect on energy conservation intention in campus. However attitude at campus ($\beta = 0.359$, $p < 0.01$), self-efficacy ($\beta = 0.328$, $p < 0.01$) and subjective norms ($\beta = 0.202$, $p < 0.01$) were positively related to energy conservation intention and together with attitude at home could explain about 58% of the variance in intention to conserve energy ($R^2 = 0.578$).

Table 4. Results of hypotheses testing.

Hypothesis	Relationship	Std. beta	Std. error	t-value	Decision	R^2	f^2	Q^2
H1	Attitude at home -> Attitude in campus	0.632	0.068	9.359***	Supported	0.400	0.666	0.283
H2	Attitude at home -> Energy conservation intention	-0.045	0.071	0.636	Not Supported	0.578	0.003	0.385
H3	Attitude in campus -> Energy conservation intention	0.359	0.104	3.460***	Supported		0.127	
H4	Self- efficacy -> Energy conservation intention	0.328	0.082	3.992***	Supported		0.105	
H5	Subjective norm -> Energy conservation	0.202	0.096	2.112*	Supported		0.041	

	intention				
H6	Attitude at home -> Attitude in campus-> Energy conservation intention	0.227	0.073	3.104**	Supported

*** p < 0.05 ** p < 0.01 *** p < 0.001**

The Preacher and Hayes' [34, 35] method of bootstrapping to identify indirect effect was applied to test H6. An indirect effect of attitude towards energy conservation at home on energy conservation behavioural intention via attitude in campus was identified (β = 0.227 was significant with a t-value of 3.104). Following the procedures of Preacher and Hayes [34, 35], mediation was present as the indirect effect, 0.227, at 95% confidence level with Boot CI: [LL=0.089, UL=0.379] shows that the lower limit (LL) and upper limit (UL) does not straddle a 0 in between. Hence, H6 was supported as the mediating effect is statistically significant.

Concomitantly, f^2 is used to measure the effect size for predictive regression equations, i.e. f^2 = 0.02 (small effect); 0.15 (medium); 0.35 (large) [32]. Table 4 indicates that effect size of relationships range from no effect (1 relationship), small (1 relationship), medium (2 relationships) to large effect size (1 relationship). Q^2 measures a model's predictive relevance (PR); if $Q^2 > 0$, then there is PR; if < 0, then no PR [32]. From Table 4, the two Q^2 values are more than 0 with 0.283 and 0.385 respectively, indicating that there is predictive relevance in the structural model.

5. Discussion and Conclusion

In this study, the predictors of energy conservation intention among students staying in a campus hostel were examined. The research instrument was adapted and developed from Ajzen [2], Wells et al. [6] and Chen et al. [29]. All three constructs of attitudes in campus, subjective norms and self-efficacy were significantly and positively related to energy conservation intention. However, attitude at home was found to have an insignificant influence on energy conservation intention in campus but has an indirect significant impact on intention to conserve energy via the mediating effect of attitude in campus. The findings in this study imply that attitude at campus was strongly influenced by attitude at home which unexpectedly was not translated into intention to conserve energy in campus. Energy conservation intentions are directly driven by attitude at campus, subject norms and self-efficacy.

The findings for attitudes at campus with energy conservation intention are consistent with past research findings by Kollmuss & Agyeman [8], Gaspar & Antunes [9] and Guerra Santin [10]. These researchers found that by having a positive attitude and mindset, the more prone their intention towards energy saving. For subjective norms, the findings are consistent with Goldstein *et al.* [15], Taler & Sunstein [16] and Hassan *et al.* [18] where family, friends and peers were found to influence and encourage students in conserving energy. The findings for self-efficacy towards energy conservation intention, is consistent with studies by Abrahamse & Steg [5], Costanzo *et al.* [21] and Stern [22]. Students believe on their own decision as well as their family and peers influence on energy conservation intention.

The finding of attitude on energy conservation in campus mediates the relationship between attitude at home and behavioural intention to conserve energy in campus is also supported by Wells *et al.* [6]. This result seems to suggest a spread effect of attitudes at home into the campus environment. However, a direct effect was not discernible on behavioural intention to save energy. A plausible reason could be that students do not see themselves as responsible to carry the financial burden of investment measures as well as non investment measures to save energy in the campus. In this context, the inculcation of positive attitudes on energy conservation in schools and campuses should start as early as possible. Facilities managers including university accommodation managers should take heed of this need to educate and inculcate good attitudes towards energy saving.

The above views lead us to recommend the following:

5.1. Environmental education

Using pedagogical approach, the education programmes should be concerned with raising awareness on the needs to conserve energy consumption in the campus. The programmes should focus on improving energy-related attitudes, behavioral intention and strategies for action. Moreover, the educational awareness on energy conservation can be incorporated into curriculum activities for students. Through these programmes, students do not only acquire knowledge and cognitive skills but also affective attitudes which could influence other students and their peers on reducing energy use. In this respect, Ntona *et al.* [36] suggested that the education programme could include two levels, namely (1) imparting knowledge on the usefulness and importance of energy resources and (2) scientific and technical knowledge on the optimum utilization of energy resources.

Specifically, environmental education should inculcate in students sensitivity towards caring of the environment and its resources and understanding of what and why energy crises occur. At the same time opportunities should be given to them to deliberate on courses of action and strategies to address energy crises in the future. More critical is the issue of developing the right attitudes, values nd acceptable practices toward energy conservation as the findings in this study have indicated that attitudes towards energy conservation in the campus have the biggest effect size on behavioural intention in energy saving.

Peers, friends and role models play an important role in influencing behavioral intention; to create energy awareness, students should be affiliated to environmental awareness clubs and non-governmental organizations (NGOs) outside the campus. The student representative councils (SRCs) need to promote awareness on energy saving. Hence, both clubs and SRCs can promote energy saving awareness by event sponsoring, fund raising as well as related activities.

5.2. Policies and regulations

A new approach on policies and hostel regulations related to conserving energy should be developed to motivate and promote energy awareness to on-campus students. This is to encourage the university community to participate in raising awareness as well as change their attitudes and behaviour in conserving energy. Specifically, student handbooks should clearly specify behaviour such as switching off lights, fans and air-conditioning in hostels and classrooms when not in use. Incentives in the form of rental discounts can be offered to hostel blocks which register the lowest monthly consumption consistently for 3 consecutive months could be introduced.

In addition, the university management dealing with students and accommodation must be student friendly at all times hence creating a conducive environment for students to appreciate and value the university's initiatives to conserve energy in all its facilities. It is believed that students, who were "angry" with the university administration because they felt that they have been shortchanged or mistreated by the accommodation office, have shown a negative attitude towards such energy saving initiatives.

The overall objective is to inculcate through these educational approaches and policies in institutional framework of lifelong learning that subsequently can bring about a comprehensive and integrated strategy towards promoting more positive energy conservation practices in campuses and work places.

6. Limitations and suggestions for future research

It is suggested that future research examine the perceptions of both private university and public university students as the latter's operating costs are funded by state funds and grants while the former has to source for funds on its own largely from students' fees. This disparity may have implications in influencing student attitudes towards energy saving in the campuses. Secondly, the sample of this study constitute only one private university hence the findings may be limited and cannot be generalized to all private and public universities in the country. Thirdly, it would be interesting to examine the mediating effect of the affective attitude of students on energy conservation intention since both the attitudes examined in the study are cognitive or instrumental attitudes towards energy saving. Lastly, it is suggested that the study be extended to more campuses in the country so that a more representative sample could be selected. The results from such a study could give greater accuracy on information with regards to attitudes, subjective norm, and self-efficacy of students as predictors of energy conservation intention.

Appendix.

Constructs	Items	Source
Attitude at home	ATTHOM1. I believe reducing energy consumption in the home has an effect in tackling climate change. ATTHOM2. Reducing my home's energy consumption would help protect the environment. ATTHOM3. It is my responsibility to reduce energy consumption I use at home. ATTHOM4. It is important to conserve energy at home. ATTHOM5. Doing things like switching off air conditioning and lights when I am not in my room is important in reducing energy wastage. ATTHOM6. I try as far as possible to use natural lights from outside my windows instead of switching on my lights.	Adapted from Wells et al. [6]
Attitude in campus	ATTCAM1. Doing things like switching off lights, fans and air conditioning when not in use is important in reducing our campus' energy consumption. ATTCAM2. I believe that reducing energy consumption in the campus has an effect on climate change. ATTCAM3. It is every student's responsibility to reduce energy consumption in the campus. ATTCAM4. It is important to conserve energy consumption while staying and studying in the campus. ATTCAM5. I am still concerned about turning lights and air-conditioning on even though the university pays for the electricity bills. ATTCAM6. I am concerned about conserving energy even though new technologies will be developed to address the energy problems for future generation.	Adapted from Wells et al. [6]
Subjective norm	SN1. In general, people who are important to me would support my efforts to conserve energy consumption. SN2. People who are significant in my life think that I should conserve energy wherever possible. SN3. My close friends think I should not waste energy consumption. SN4. My close friends think I should join energy	Adapted from Ajzen [2]

	saving awareness campaign. SN5. The campus management expects of me to switch off all lights, fan and air-conditioning if not in use. SN6. I value the opinion of people who are significant in my life when it comes to making a decision on energy conservation.	
Self-efficacy	SEF1. I will be able to achieve most of the goals that I have set for myself concerning energy conservation. SEF2. When facing difficult decisions on energy conservation, I am certain that I will accomplish them. SEF3. In general, I think I can obtain energy conservation outcomes that are important to me. SEF4. I believe I can succeed at most any energy conservation endeavour to which I set my mind. SEF5. I am confident that I can perform effectively on many different tasks relating to energy conservation. SEF6. Compared to other people, I can do most energy conservation task very well.	Adapted from Chen et al. [29]
Energy conservation intention	INTENT1. I will try to conserve electricity consumption when I am in the campus. INTENT2. I intend to switch off lights and air-conditioning when not in use. INTENT3. I plan to open windows rather than turning on air-conditioning when it's hot. INTENT4. I will make sure to switch off the electrical-electronic equipment after each class session. INTENT5. I intend to join energy awareness campaign in campus. INTENT6. It is my responsibility to reduce energy resources I use in campus to save cost. INTENT7. I will encourage my classmates to conserve energy all the time.	Adapted from Ajzen [2]

Acknowledgements
The authors wish to acknowledge assistance from Nilai University for financial support for travel, accommodation and presentation of the research paper in NEFES 2016.

References

[1] Fishbein M and Ajzen A 1975 Belief, attitude, intention, and behaviour: An introduction to theory and research. Reading, Massachussetts: Addison-Wesley Publishing House

[2] Ajzen I 1991 The theory of planned behavior *Organizational Behavior and Human Decision Processes* **50** 179-211

[3] Kraft P, Rise J, Sutton and Roysamb E 2005 Perceived difficulty in the theory of planned behaviour: Perceived behavioural control or affective attitude *British Journal of Social Psychology* **44** 479-96

[4] Harland P, Staats H and Wilke H A M 1999 Explaining pro-environmental intention and behavior by personal norms and the theory of planned behaviour *J. Appl. Soc. Psychol.* **29**

2505–28
[5] Lindenberg S and Steg L 2007 Normative, gain and hedonic goal frames guiding environmental behavior *J. Soc.* 117–37
[6] Wells V K, Taheri B, Gregory-Smith D and Manika D 2016 The role of generativity and attitudes on employees home and workplace water and energy saving behaviours *Tourism Management* **56** 63-74
[7] Abrahamse W and Steg L 2011 Factors related to household energy use and intention to reduce it: the role of psychological and socio-demographic variables *Hum.Ecol.Rev.* **18** 30-40
[8] Kollmuss A and Agyeman J 2002 Mind the gap: Why do people act environmentally and what are the barriers to pro-environmental behavior? *Environ. Educ. Res.* **8** 239–60
[9] Gaspar R and Antunes D 2011 Energy efficiency and appliance purchases in Europe: Consumer profiles and choice determinants *Energy Policy* **39** 7335–46
[10] Guerra Santin O 2011 Behavioural patterns and user profiles related to energy consumption for heating *Energy Build* **43** 2662–72
[11] Nair G, Gustavsson L and Mahapatra K 2010 Factors influencing energy efficiency investments in existing Swedish residential buildings *Energy Policy* **38** 2956–63
[12] Rivis A and Sheeran P 2003 Descriptive norms as an additional predictor in the theory of planned behaviour: a meta- analysis. *Curr. Psychol.Dev.Learn. Pers.Soc.* **22** 218–33
[13] Schultz P W, Khazian A and Zaleski A 2008 Using normative social influence to promote conservation among hotel guests *Soc. Influ.* **3** 4–23
[14] Nolan J M, Schultz P W, Cialdini R B, Goldstein N J and Griskevicius V 2008 Normative social influence is under detected *Pers.Soc.Psychol.Bull.* **34** 913–23
[15] Goldstein N J, Cialdini R B and Griskevicius V 2008 A room with a viewpoint: using social norms to motivate environmental conservation in hotels *J. Consum. Res.* **35** 472–82
[16] Thaler R H and Sunstein C R 2008 *Nudge: improving decisions about health, wealth, and happiness* (New Haven: Yale University Press)
[17] Poortinga W, Steg L and Vlek C 2004 Values, environmental concern, and environmental behavior: a study into household energy use *Environ Behav* **36** 70–93
[18] Hassan G M, Hirst R, Siemieniuch C and Zoba A F 2009 The impact of energy awareness on energy efficiency *Int J Sustain Eng* **2** 284–97
[19] Costanzo M, Archer D, Aronson E and Pettigrew T 1986 Energy conservation behavior: The difficult path from information to action *Am. Psychol.* **41** 521–8
[20] Stern P 1992 What psychology knows about energy conservation *Am. Psychol.* **47** 1224–32
[21] Bandura A 1986 *Social foundations of thought and action: A social cognitive theory* (NJ: Prentice-Hall)
[22] Sheeran P and Abraham C 2003 Mediator of moderators: Temporal stability of intention and the intention-behavior relation *Pers. Soc. Psychol. Bull.* **29** 205–15
[23] Vining J and Ebreo A 2002 Emerging theoretical and methodological perspectives on conservation behaviour In Handbook of Environmental Psychology; Bechtel, R.B., Churchman *A., Eds.* (New York, NY, USA: Wiley) 541–58
[24] Pedersen M 2008 Segmenting residential customers: energy and conservation behaviours *Proceedings of the 2008 ACEEE Summer Study on Energy Efficiency in Buildings,* Pacific Grove, CA, USA **7** 229–41
[25] Ajzen I and Fishbein M 1980 Understanding Attitudes and Predicting Social Behavior Englewood Cliffs (NJ: Prentice Hall)
[26] Harland P, Staats H and Wilke H A M 1999 Explaining pro environmental intention and behavior by personal norms and the theory of planned behavior *J. Applied Social Psychology* **29** 505-28
[27] Steg L and Vlek C 2009 Encouraging pro-environmental behaviour: An integrative review and research agenda *J. Environ. Psychol.* **29** 309–17
[28] Greaves M, Zibarras L D and Stride C 2013 Using theory of planned behaviour to explore

environmental behavioral intention in the workplace J. *Environmental Psychology* **34** 109-20

[29] Chen G, Gully S M and Eden D 2001 Validation of a New General Self-Efficacy Scale *Organizational Research Methods* **4** 62-83

[30] Ringle C M, Wende S and Becker J M 2015 Smart PLS 3. SmartPLS GmbH: Boenningsted Retrieved from http://www.smartpls.com.

[31] Yeap J A L, Ramayah T and Soto-Acosta P 2016 Factors propelling the adoption of m-learning among students in higher education *Electronic Markets*

[32] Hair J F, Hult G T M, Ringle C M and Sarstedt M 2014 *A primer on partial least squares structural equation modeling (PLS-SEM)* (Los Angeles: Sage Publication)

[33] Fornell C and Larcker D F 1981 Evaluating structural equation models with unobservable variables and measurement error *J. Marketing Research* **18** 39–50

[34] Preacher K J and Hayes A F 2004 SPSS and SAS procedures for estimating indirect effects in simple mediation models *Behavior Research Methods, Instruments, & Computers* **36** 717–31

[35] Preacher K J and Hayes A F 2008 Asymptotic and resampling strategies for assessing and comparing indirect effects in multiple mediator models *Behavior Research Methods* **40** 879–91

[36] Ntona E, Arabatzis G and Kyriakopoulos G L 2015 Energy saving: views and attitudes of students in secondary education *Renewable and Sustainable Energy Review* **46** 1-15

Institute of Physics
Dirac House, Temple Back
Bristol BS1 6BE UK

ISSN: 1755-1307
ISBN 978-1-5108-3050-9

CURRAN ASSOCIATES INC.
proceedings.com